Sundance Square, Fort Worth, Texas
(Photo courtesy of Fort Worth Convention and Visitors Bureau)

DETECTOR RESEARCH AND DEVELOPMENT

FOR THE

SUPERCONDUCTING SUPER COLLIDER

1990

Barry Greer,

The best to you in your future endeavors . . .

Valerie Kelly

PROCEEDINGS OF THE SYMPOSIUM ON

DETECTOR RESEARCH AND DEVELOPMENT FOR THE SUPERCONDUCTING SUPER COLLIDER

OCTOBER 15 – 18, 1990
FORT WORTH, TEXAS

EDITORS

Tom Dombeck, Valerie Kelly, and George P. Yost

Sponsored by

Superconducting Super Collider Laboratory
United States Department of Energy, and
Universities Research Association, Inc.

Published by

World Scientific Publishing Co. Pte. Ltd.
P O Box 128, Farrer Road, Singapore 9128
USA office: 687 Hartwell Street, Teaneck, NJ 07666
UK office: 73 Lynton Mead, Totteridge, London N20 8DH

Library of Congress Cataloging-in-Publication Data

Symposium on Detector Research and Development for the Superconducting Super Collider (1990 : Fort Worth, Tex.)
Proceedings of the Symposium on Detector Research and Development for the Superconducting Super Collider, October 15–18, 1990, Fort Worth, Texas / editors, Valerie Kelly, George P. Yost, and Tom Dombeck; sponsored by Superconducting Super Collider Laboratory, United States Department of Energy, and Universities Research Association, Inc.
p. cm.
Includes bibliographical references and index.
ISBN 9810204450
1. Superconducting Super Collider--Congresses. 2. Superconducting Super Collider--Instruments--Congresses. 3. Nuclear detectors--Congresses. 4. Calorimetry--Congresses. I. Kelly, Valerie. II. Yost, George P. III. Dombeck, Tom. IV. Superconducting Super Collider Laboratory. V. United States. Dept. of Energy. VI. Universities Research Association (U.S.) VII. Title.
QC787.P7S96 1991
539.7'3--dc20 91-11834
CIP

Printed in Singapore by Chong Moh Offset Printing Pte. Ltd.

Preface

The Symposium on Detector Research and Development for the Superconducting Super Collider was held at the Worthington Hotel in Fort Worth, Texas, October 15–18, 1990. About 500 participants discusses this extremely active field. It was generally agreed that an astonishing amount of progress has been made in spite of a constricting funding situation. These proceedings, in one volume, contain the invited and contributed papers documenting that progress.

A highlight of the conference was the announcement during the conference that Professor Jerome Friedman of MIT, who had chaired the first session of the Symposium, had been awarded the 1990 Nobel Prize in Physics.

A treat for all the participants was a concert at the Meyerson Symphony Hall in Dallas. The performance was a gift to the Symposium from the outgoing president of the Texas National Research Laboratory Commission, Mort Meyerson.

On behalf of the entire Symposium, we would like to thank Phyllis Hale, Head of the SSC User's Office, and her hard-working staff, for a smoothly-organized conference. We wish also to thank Larry Cormell, who organized the industrial exhibits. Terry Johnson and those who assisted him in coordinating our computing and communications. Many people worked behind the scenes and we want to express our appreciation to those, too numerous to mention, who were nevertheless key elements.

On behalf of the organizing committee,

V. Kelly, G. P. Yost and T. Dombeck

Table of Contents

PLENARY SESSIONS

PARALLEL SESSIONS

1. Silicon Tracking Devices

Chairman: H. Sadrozinski
University of California, Santa Cruz

2. Wire Chambers

Chairman: W. Ford
University of Colorado

3. Scintillating Fiber Tracking

Chairman: J.E. Elias
Fermi National Accelerator Laboratory

4. Liquid Ionization and Silicon Calorimetry

Chairman: P. Mockett
University of Washington

5. Scintillation Calorimetry

Chairman: H. Kasha
Superconducting Super Collider Laboratory

6. Front-End, Data Acquisition and Triggering Electronics

Chairman: J. Dorenbosch
Superconducting Super Collider Laboratory

7. Superconducting Magnets

Chairman: T. Fields
Argonne National Laboratory

8. Computing and Simulations

Chairman: T. A. Gabriel
Oak Ridge National Laboratory

9. Particle Identification and Muon Systems

Chairman: A. Skuja
University of Maryland

10. Radiation Damage

Chairman: D. Groom
Lawrence Berkeley Laboratory

11. Detector Engineering

Chairman: R. Stefanski
Superconducting Super Collider Laboratory

Plenary Sessions

FRONT-END ELECTRONICS DEVELOPMENT FOR THE SSC

Michael Levi
Lawrence Berkeley Laboratory*
1 Cyclotron Road
Berkeley, CA 94720

Abstract

This is a status report on electronics development undertaken by the Front-End Electronics Collaboration. The overall goal of the collaboration remains the development by 1992 of complete, architecturally compatible, front end electronic systems for calorimeter, wire drift chamber, and silicon strip readout. We report here a few highlights to give a brief overview of the work underway. Performance requirements and capabilities, selected architectures, circuit designs and test results are presented.

Introduction

The purpose of this project is to identify architectural and technical issues in the electronics designs for SSC detector front-ends. We have focused on issues that will significantly effect development and construction of the detectors. There remain significant issues where the success or failure of the research and development will alter the implementation of some of the detector concepts.

The electronics for signal detection and triggering for an SSC detector will be one of the most sophisticated electronics systems ever built by the high energy physics community. Signal processing must be performed on over one trillion analog samples per second. In comparison to previous experiments, at the SSC much larger sampling speeds, wider dynamic range, and larger channel counts are encountered and in the presence of significant constraints on power consumption, radiation hardness, and cost. Fortunately, high density CMOS and bipolar processes are available that can meet these challenges and provide the necessary signal processing power, signal shaping, analog to digital conversion and sophisticated addressing schemes including sparse or selective data readout and trigger buffering and selection.

In this report we will review the development efforts centered on: 1) architectural considerations and Level 1/Level 2 storage, 2) wire chamber drift time measurement, 3) liquid ionization calorimeter electronics, 4) signal shaping, 5) analog storage of liquid ionization or scintillator calorimeter signals, and 6) radiation hardness studies,. This work is a summary of the combined efforts of the Collaboration members.[1] Further discussion can be found in the Front-end Electronics Collaboration proposals.[1,2]

Architecture

One of the most important goals in the development of a front end electronics system for experiments at the SSC is the choice of an overall architecture which meets the requirements of all detector systems. While the characteristics of a particular detector may force minor variations in this overall architecture, the cost is likely to be minimized, and the reliability and ease of debugging optimized, if one can find an architecture which is nearly the same for all systems. It is in fact primarily for this reason that we included wire chambers, liquid ionization and scintillator calorimeters, cathode pad chambers, and silicon strip chambers within the scope of the overall front end electronics subsystem development.

The term "Front-end Architecture", is broadly used to describe the topology and functionality of the system of electronics that is physically located on the detector. In the case of an SSC detector, this includes a major portion of the total electronics system. The design of this portion includes the implementation, in integrated circuit form, of sophisticated components, including a complex system of data and control signals to "glue" these components into a cohesive system.

It is impractical (if not impossible) to move the data from every channel off the detector for every interaction. Usually, a trigger decision of a 1-2 µs latency is made to judge the value of the stored signals. These signals are then either discarded or digitized, processed, and transferred to computer systems. It is therefore desirable to filter the data through multiple levels of triggering. Thus, these systems usually require time delay via temporary storage of the signal to match the latency of the trigger and the bandwidth of the data processing system.

In the most likely scenario, data is stored during two successive levels of triggering prior to event readout. The architecture discussed could be utilized as well with a single trigger level if higher rejection factors are achieved with the Level 1 trigger or higher bandwidth of the readout is implemented. It is our judgement that for some systems and at the highest luminosities (e.g. > 10^{33}) it will be necessary to include Level 2 storage. In any event we have not yet identified any particular cost or penalty of including it other than a modest increase in circuit complexity.

Fig. 1 shows the basic organization of the data flow paths of the front end electronics. The portion above the dashed line - located on the detector - receives signals from the detectors, filters them with Level 1 and Level 2 triggers (trigger not shown), and delivers packets of digitized data off the detector for event processing.

---

* This work has been supported by the Director, Office of Energy Research, Office of High Energy and Nuclear Physics, Division of High Energy Physics, of the U.S. Department of Energy under Contract No. DE-AC03-76SF00098.

It is expected that two basic building blocks will be required - Front-end IC's (FE's) and data collection circuits (DCC's). Front-end integrated circuits accept signals directly from the detector elements, process and store the signals during the time necessary to form Level 1 and Level 2 triggers, and output digitized data selected by the Level 2 triggers. DCC's interface to a small data bus structure that interconnects a small number (e.g. 16) of FE's. They control the collection of digitized data from the FE's, and organize it for transmission off the detector.

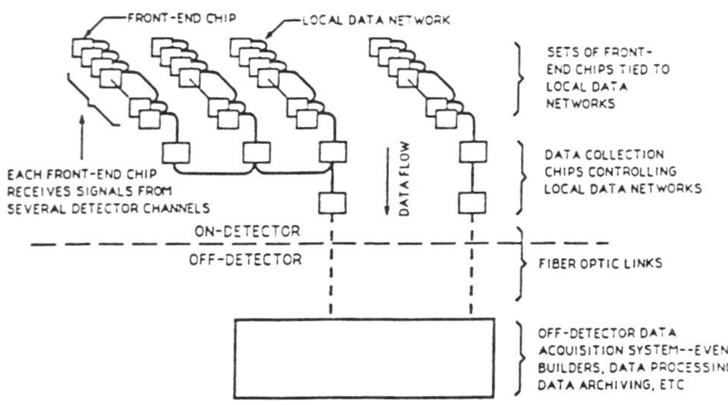

Fig. 1. The basic organization for the data collection scheme. The front end IC's (FE's) accept the signals from the detector and the Data Collection circuits collect the data from a set of FE's and prepare it for transmission.

Given the desire for Level 1 and Level 2 storage, there is still a broad range of possibilities. The various solutions may be distinguished by:

1. The Level 1 and Level 2 storage buffers are physically distinct.
2. The Level 1 and Level 2 storage buffers are not physically distinct, but only distinguished by address pointers (virtual pipelines).
3. Both the Level 1 and Level 2 storage are digital.
4. Both the Level 1 and Level 2 storage are analog.
5. The Level 1 storage is analog, but the Level 2 storage is digital.
6. The transfer from Level 1 to Level 2 occurs within a bunch crossing (16 nsec).
7. The transfer from Level 1 to Level 2 is buffered.

All these possibilities have been considered in our study of architectures. However, we are not pursuing a system in which Level 1 is digital with high priority since the utilization of 62 MHz ADC's implies rather large power dissipation and is not generally applicable throughout the detector subsystems. In addition for very high density systems (such as silicon microstrip detectors) there is insufficient space.

One of the simplest approaches is to transfer the information stored in Level 1 to Level 2 as soon as a Level 1 Accept for that particular crossing has been received. Since one may need to transfer information on subsequent crossings, either because another Level 1 trigger was received or because it is desirable to read out the data from several crossings for a given Level 1 trigger, the transfer must be accomplished in 16 nsec. This then requires a rather fast amplifier. To understand how much power is required for such a transfer, an operational amplifier was designed in a 1.6 micron CMOS process. A single stage folded cascode was used to enable maximum speed at minimum power. The conclusion is that even with a relatively advanced process, it requires approximately 4 to 5 mW to obtain bandwidths of 230 - 250 MHz at restricted dynamic range. These bandwidths are still not sufficient. Therefore this approach was abandoned (at least temporarily) in favor of a system incorporating either virtual pipelines or buffered Level 1 to Level 2 transfers.

In summarizing each of the investigations, we considered separately "data driven" systems which naturally sparcify the data and are appropriate for low occupancy detectors, and systems in which sampling is performed regularly every 16 nsec as is likely to be used for the calorimeter readout.

Besides these specifications, we have followed several other considerations in developing the architecture. We have tried to design a system with inherently low power consumption. This means that fast analog-to-digital (ADC) data conversion is avoided. We want to minimize the number of digital control lines connecting the front end circuit with the trigger and DAQ system. We therefore have avoided centralized bookkeeping that forces us to move information between the trigger system and the front end.

We are also constrained by the practical considerations of implementing sensitive low-noise analog circuitry on the same chip where a significant amount of fast digital logic is occurring. We note that because of the asynchronous nature of the system, where one can have data being digitized at the same time that analog data is being recorded or transferred, the potential for noise pick-up is much more severe than earlier collider environments. We have therefore tried to reduce the total number of fast signal lines in proximity to the front end IC's. Other ways of reducing noise pick-up would be the use of low-voltage differential signals with slow risetimes. This drives us toward a system with as few components (and functionality) as possible in the front end.

Clock Driven Architecture

In the case of a clock-driven system, the simplest conceptual system is shown in Fig. 2. The data enters a FIFO in which the data is moved in synchronization with the machine RF. The FIFO is then a clocked analog delay element for the Level 1 trigger. At the end of the Level 1 delay, the data enters a Level 2 FIFO, now clocked by the Level 2 trigger decisions. Data which corresponds to events successfully satisfying a Level 2 trigger are then digitized and passed to the Level 3 trigger along read-out fibers.

As an alternative to this arrangement, with only a slight overhead in terms of complexity, is to leave the data in the storage element it is deposited in and use a pointer system to label data according to its Level 1, Level 2, or read-out status. The data does not move but remains in the same storage element until being digitized or discarded (overwritten). Since the order of Level 2 triggers in the storage array is not necessarily in any sequence, a scheme for keeping track of which trigger belongs to which storage element must be implemented. This "virtual pipeline" architecture, shown in Fig. 3, makes maximal use of the fact that Level 1 and Level 2 decisions are taken in order (monotonic). The primary objective is to minimize power requirements and produce a read-out architecture with minimal dead-time.

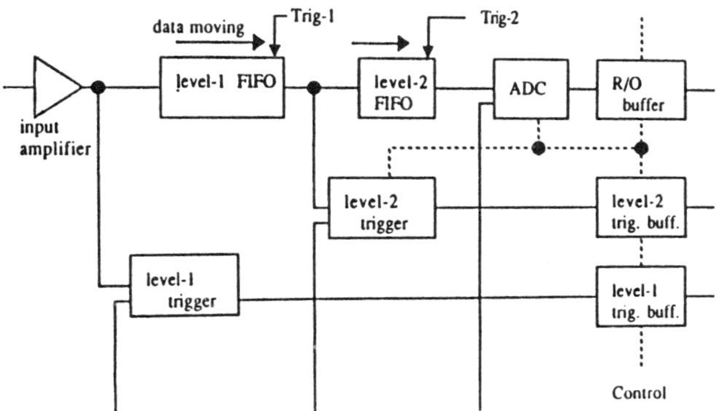

Fig. 2. Simplest Analog Front End

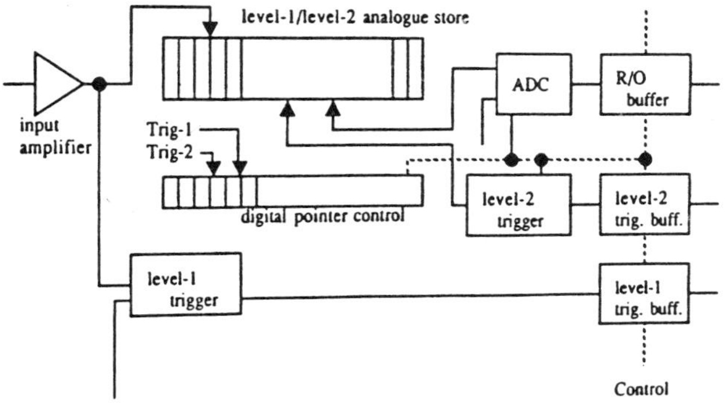

Fig. 3. Single Analog Store, Pointer-Address Architecture

It is also inherent in this scheme that there is not a one to one correspondence between a particular Level 1 or Level 2 address and a particular storage location. A significant consideration is the fact that, unlike the scheme of Fig. 2, there is a potential need for calibration constants for each cell in the storage array. Hence it is very desirable that the variations in gain and offset between the various storage locations be sufficiently small that they may be ignored. This is a very desirable goal for any system utilizing analog storage.

Data Driven Architecture

The case of data driven storage folds in another layer of complexity to each of the stages presented in the clock-driven schemes. The potential advantage is that considerably less data need be stored.

The technique is to record only when there is data present. An input pulse which exceeds an input threshold causes the data to be stored, along with a number recording the RF bucket to which the data belongs. Since both data and address must be stored this architecture is applicable only to low occupancy channels.

In spite of the greater complexity, calculations suggest that data-driven systems become clearly advantageous when the channel occupancy is less than ~ 10^{-3}. At higher occupancies a more careful consideration of the specific trade offs becomes necessary.

The number of buffers in the Level 1 storage, Level 2 storage, and the readout FIFO depend on the occupancy of the detector the front end IC is designed to read out, the Level 1 and Level 2 trigger latencies, and the Level 1 and Level 2 trigger rejection rates. Our model assumes that the occupancy is relatively low (< 10^{-3}), although a similar architecture would probably work for a device that is not data-driven or that has high occupancy.

Throttling the Data Flow

In the data-driven scheme, the prevention of data corruption due to the overflow of buffers or bandwidth must be performed on-chip, since the trigger system itself has no way of knowing how many buffers are available to process data pending Level 1 and Level 2 accepts. When the data-processing capability of the front end is insufficient, some data loss is unavoidable.

It is possible to have situations where the amount of Level 1 storage is inadequate to handle all of the signals from the channel. Although the number of buffers should normally be sufficient to avoid this problem, one can imagine scenarios where this could occur. One obvious case is that of a relatively "warm" channel that has an occupancy of 30% or more. In this case, the amount of data that could be generated is vast, far exceeding the internal bandwidth of the chip. To handle such cases, it is probably necessary to provide some mechanism by which the entire channel can be disabled, by for example providing a bit that would disable the signal from the comparator. However, there are several places in the chip where some additional control must be provided.

Several overflow conditions must be detected:

1. If a flood of data comes in, the first problem will be that all of the available Level 1 storage buffers will become full.

2. The Level 2 FIFO may become filled.

3. A similar problem exists in the readout FIFO, though we believe that it can be made fast and deep enough to take whatever data rate the Level 2 digitization can produce.

Loss of data is automatic in only the first two situations. It would be preferable to signal those cases where loss of data might have occurred. This can be accomplished by signaling these different error conditions to the data acquisition by inserting the necessary error information into the data stream.

Drift Chamber Electronics

For a high precision drift tube or straw tracking system, it is necessary to accurately measure the time of arrival of the first electron at the anode. While many possible schemes exist for making this measurement, our initial judgement was that an inherently analog measurement scheme would offer both lower power and greater resolution than an equally complex digital system. In addition, we realized that it would be necessary to incorporate all of the desired system features (connection to trigger and DAQ systems, for instance) in any usable design in order to keep power and mass of the final system under control.

The minimum set of specifications necessary for a successful time measuring device would seem to be:

- < 0.5 ns time accuracy, ~ 0.2 ns time resolution
- Deadtimeless operation
- High rate capability
- Local storage of data during the trigger decision

In the straw tube tracking systems being proposed for SSC, each straw tube sensor will require a preamplifier, shaper, discriminator, time converter and sparsification unit. Analog bipolar technology offers optimal performance tradeoffs for the preamplifier, shaper and discriminator. It's high gain-bandwidth at low power is unmatched by any other commercially available technology as is the noise performance for shaping times of less than 20 ns and the matching of the control voltage between transistors. CMOS technology offers significant power advantages for functions such as analog memory and digital logic when operating below 100 MHz. For the time conversion and sparsification blocks, it is clearly the technology of choice.

Measurements with Straw Tubes

Under the *Generic Research* program we have developed a fast, low power and low noise monolithic preamp and shaper based on design goals suitable for a proportional drift tube tracking detector at SSC.[3] A prototype version was fabricated in the AT&T ALA200 Semi-Custom bipolar process. The design is based on a cascoded common emitter preamplifier configuration requiring only a modest supply voltage of 3 to 4 volts. The power dissipation is therefore relatively low.

The shaper has one zero matched to cancel the dominant pole of the preamplifier and three equivalent integrations that limit the bandwidth and maintain pulse symmetry. The pulse symmetry realized by use of multiple pole shaping is demonstrated in Fig. 5.

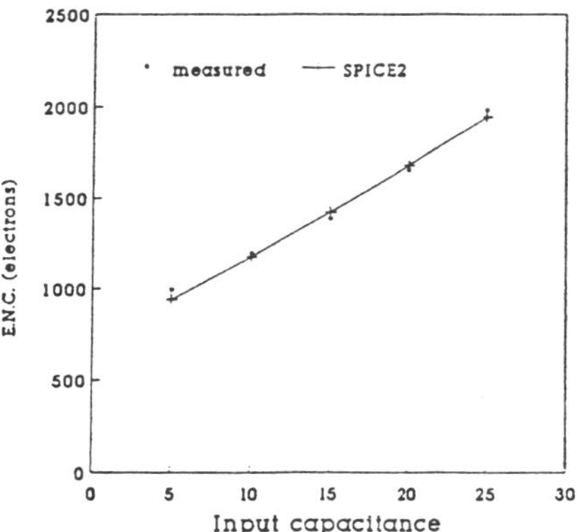

Fig. 4. Comparison of simulated and measured noise levels for a 5 nsec peaking time.

The intrinsic amplifier noise, shown in Fig. 4, is primarily determined by the size of the input transistor and it's quiescent current. In the design stage, the noise performance was optimized for an amplifier with a pure detector capacitance of 5 - 10 pF. The quiescent current was set to about 0.5 mA and an input transistor of 75μm emitter length was used to achieve a base resistance of about 15Ω.

Fig. 5 shows a signal in a 2 meter prototype straw tube at Indiana University due to a cosmic ray. The waveform in this figure shows the output of the prototype amplifier when attached to a passive detector tail cancellation network. The signal that extends beyond the 5 ns measurement time creates a *detector tail* that must be compensated or cancelled in high rate systems. The detector tail is eliminated by adding a set of poles and zeroes to the shaping function.[4]

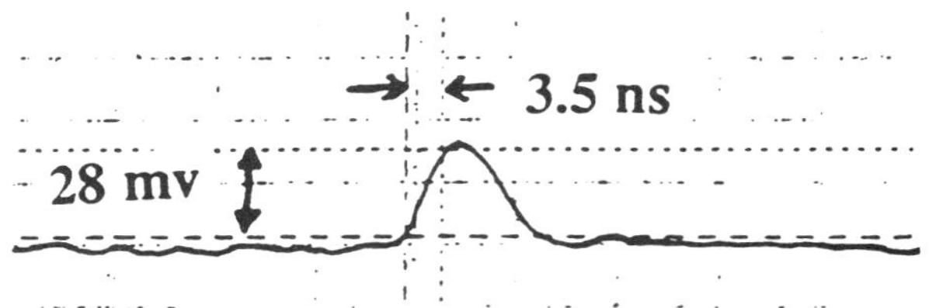

Fig. 5. Signal from a 2 meter straw tube filled with CF^4. The waveform shows the output of the prototype amplifier attached to a detector tail cancellation circuit. The scale from this digitizing oscilloscope is 10 nsec/div on the abscissa, and 20 mV/div on the ordinate.

Time to Voltage Converter/Analog Memory

We have implemented in CMOS a data driven architecture which is suitable for wire chamber readout (and could be extended to silicon microstrip detectors). The circuit includes a time-to-voltage converter (TVC) needed to convert the output of the preamp/shaper to a drift time measurement. This required the development of threshold discriminators, and an analog comparator.[5,6] The block diagram of the TVC is shown in Fig. 6. The accuracy of the TVC is shown in Fig. 7.[7] In the design of the prototype time-to-voltage converter/analog memory unit (TVC/AMU), we have limited the number of Level 1 analog memory locations to eight, which should suffice given the necessarily modest occupancies of any viable tracking system. A block diagram of the TVC/AMU is shown in Fig. 8.

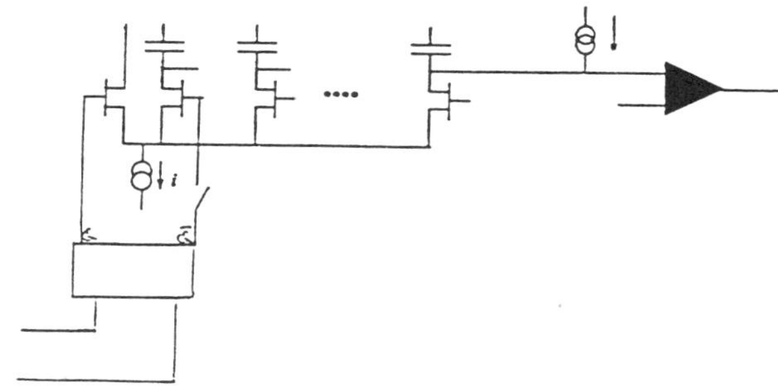

Fig. 6. TVC block diagram.

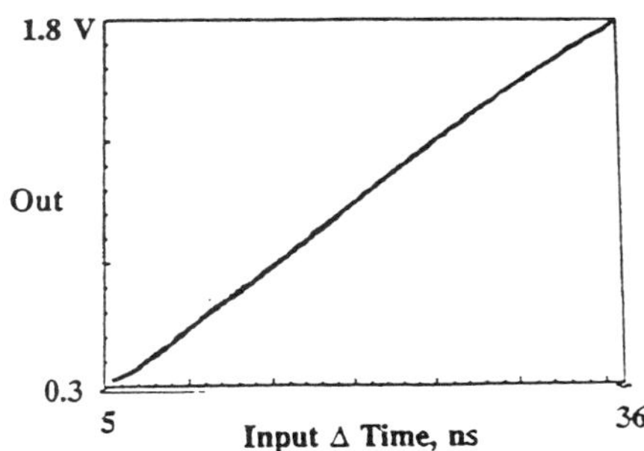

Fig. 7. TVC output voltage vs. input time difference with 8 independent channels superimposed. In scale shown 1 ns = 50 mV.

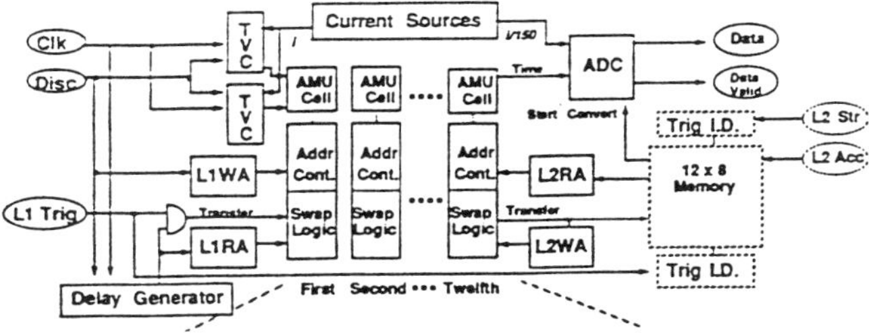

Fig. 8. TVC/AMU block diagram.

Additional control logic is included in the circuit to implement the data driven architecture, including logic to manipulate Level 1/Level 2 buffers, and time stamp information required to retrieve the data.

The system consists of

- A circular Level 1 storage buffer, containing both an analog memory and storage for a relatively small (~ 4-bit) local bunch counter (LBC). This buffer stores all data pending a Ll trigger decision.
- Two Level l storage write and read address counters (LlWA and LlRA).
- A Level 1 Read Address FIFO (LlRF) with a depth of about 4 elements and a width of order 15-20 bits. This feature effectively buffers the Ll-L2 transfers and allows for a relatively low bandwidth analog transfer without increasing the deadtime of the system.
- A circular Level 2 storage buffer, providing storage for analog data and ~ 10 bits of digital data and consisting of order 5-10 locations.
- Two Level 2 write and read address counters (L2WA and LlRA).
- An ADC that will digitize the data associated with events passing the Level 2 trigger.
- A digital readout FIFO of up to 5-10 elements.

Each storage location consists of an analog memory and a 4 bit digital memory, which will store the value of a local bunch counter (LBC). The LBC counter is incremented at each beam crossing and wraps around itself (the number of bits for this counter is determined by the maximum number of crossings that need to be read out for each Ll trigger).

Given the non-deterministic nature of a data driven architecture, it will be necessary to include in each data packet a crossing identifier. The simplest time stamp, would be a crossing counter - advancing every 16 nanoseconds. However, for Level 2 storage times of 50 μs or more plus realistic DAQ pipeline delays of many tens of microseconds, a 13 or 14 bit crossing counter would be necessary in order to avoid ambiguity. Instead, a more robust scheme should be implemented, with much longer unambiguous periods. Here, only the Level 1 trigger Accepts are counted. Even at 10^5 Hz an eight bit Trigger I.D. Counter would remain unambiguous for 2.5 ms.

In the case of multiple samples per trigger, it is necessary to have a small two to four bit counter, enabled by Level 1 Trigger, and clocked by the crossing clock, to completely specify the time. This counter is enabled by Level 1 Trigger true and counts until Level 1 Trigger goes false.

Simulation Studies of Front End Architectures

The front-end and data acquisition architectures we have been considering are sufficiently complex that their overall system behavior is difficult to predict without detailed simulations of the various components. This is particularly true in our case where the front end architecture alone is quite complex. A component of our R&D effort has therefore involved the development of behavioral models of the various front end architectures. This effort is particularly relevant to the task of defining the interfaces between the front end, trigger and data acquisition (DAQ) systems. Our initial task was a detailed gate-level simulation of the prototype TVC/AMU chip.

We have started the development of a comprehensive behavioral model of the front end system using the Verilog software package as the platform for this simulation. The first task was the development of a model for the signals generated by the trigger system, as those form one of the primary inputs to the front end chip. We have also developed a model for the output of a wire chamber discriminator, which forms the other primary input to the front end chip, as the initial thrust of this work has been to model a wire chamber readout system. We have implemented a model of a wire chamber readout chip employing the buffered Ll-L2 transfer scheme, using as input the simulated trigger and wire chamber signals.

We have tried to define a "minimal" interface between the front end system and the trigger and DAQ systems. One proposal for the trigger interface consists of the set of signals as described:

- **L1 Synch** that signals to the front end when a Level 1 decision is available (can be generated from the system clock); and
- **Ll Accept** that communicates the Level 1 trigger decision for that crossing.
- **L2 Synch** that signals to the front end when a Level 2 decision is available for the next Level l trigger still waiting for a Level 2 decision; and
- **L2 Accept** that communicates the Level 2 decision for the next unprocessed Level l trigger.

These signals appear to be the minimum set of signals necessary between the trigger and the front end chip. They assume that the Level 1 trigger is synchronous, with a decision for a crossing becoming available a fixed number of crossings later. They also assume that the Level 2 trigger is monotonic (the next Level 2 trigger decision is always made for the earliest Level 1 trigger still to be processed), and is asynchronous, i.e. the Level 2 trigger decision is available a variable number of crossings after the Level 1 trigger.

This scheme has several advantageous features: for example, it is not necessary to propagate an event ID flag at each Level 1 or Level 2 trigger as the front end chip is responsible for generating this information.

We have assumed that the digitized information after a Level 2 trigger is "pushed" to a data collection circuit (DCC) where the data from several chips are collected and formatted into a packet that is then transferred via the DAQ to the Level 3 system.

An example of the information provided by this simulation is shown in Fig. 9, where we show the occupancy distributions of the Level 1 buffer, Level 2 buffer and the Level 1 read FIFO for this system with the following system parameters:

• A Level 1 acceptance rate of 0.1%, with a delay of 60 crossings.

• A Level 2 acceptance rate of 2.0%, with a gaussian Level 2 decision time distribution with mean of 500 crossings and a width of 500 crossings. The minimum separation between Level 2 decisions was 0.1 μs.

• A channel occupancy of 0.05 hits/crossing with a pulse width of 20 ns.

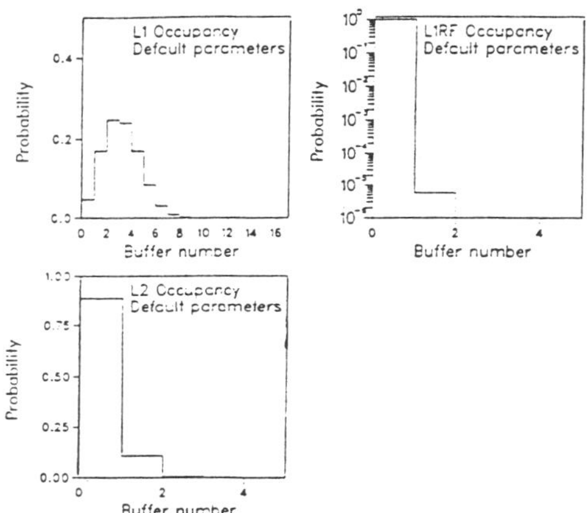

Fig. 9. The occupancies of the Level 1 buffer, the Level 2 buffer, and the Level read address FIFO in the Verilog simulation. Parameters of the simulation are given in the text.

The simulation employed 16 Level 1 buffers, 4 Level 2 buffers, and 4 buffers in the Level 1 read FIFO. On each Level 1 accept, any hits occurring within 3 crossings of the Level 1 accept were transferred to the Level 1 buffer. A Level 1 to Level 2 transfer time of 100 ns and a Level 2 to DAQ transfer time of 100 ns was assumed (neither of these are necessarily realistic). The Level 1 occupancy shows a mean of ~ 2.7 hits, which is in good agreement with the naive calculation when one takes into account the 20 ns pulse width included in the simulation. The mean occupancy of the Level 2 buffer is ~ 0.1, which is also in good agreement with the naive calculation. There was essentially no loss of data due to Level 1/Level 2 buffer overflow conditions. However, preliminary studies with this simulation have shown that the actual loss of data due to buffer full conditions grows more rapidly than the loss calculated if one assumes that the buffer occupancies obey Poisson statistics. This is a result of the delays involved in the Level 1/Level 2 and the Level 2/DAQ data transfer.

Although the parameters of this simulation are not representative of all of the expected features of the circuit or the trigger system, they illustrate the information that is available from such a behavioral model.

Analog Signal Processing for Calorimetry

The basic components of the signal processing chain for scintillation and liquid ionization calorimetry are depicted in Fig 10. The components comprise:

1. Preamplifier and charge calibration system.
2. Data recording chain, with additional gain (x 1 and x 64), first and second level buffer (analog "pipeline") and multiplexed readout.
3. Trigger forming chain, with additional gain, summing amplifiers, fast pulse shaping.
4. Control circuits for calibration, sampling, memory and multiplexing.

Analog sums of signals from large sections of the calorimeter will be formed for trigger purposes. These will result in a much smaller number of fast outputs, of the order of 1-3 x 10^3, for which more sophisticated shaping circuits may be employed. Several integration times (e.g., 50, 100, 150 nsec for ionization calorimeters and 16-48 nsec for scintillator calorimeters) may be used, so that pileup and electronic noise can be minimized for different experimental conditions.

The data recording chain contains memory units required to store information until Level 1 and Level 2 trigger decisions. The data are initially stored for each sample taken at 16 nsec intervals. None of the known techniques for analog storage and for fast analog to digital conversion can cover the large dynamic range required (15 to 17 bits), due to thermal noise limitations. Thus at least two data storage channels per signal channel with overlapping scales are needed, each covering 11-12 bits.

Preamplifiers are followed by two fast amplifiers with different gains, indicated as x 1 and x 64 in Fig. 10, which drive the memory units. The impulse response from the detector electrodes to the preamplifier output may have a rise time of no less than about 40-50 nsec, for ionization calorimeters, due to the charge transfer time from the electrodes to the preamplifier. For scintillator calorimeters the peaking time might be as short as 16 nsec though 32-48 nsec is more likely to allow adequate integration of the current from the calorimeter. In order not to increase further the response time of the system, the fast amplifiers should have a rise time of less than about 15 nsec.

The fast amplifier response, together with the charging time constant of the memory unit, provides sufficient band limiting at high frequencies to satisfy the Nyquist criterion.

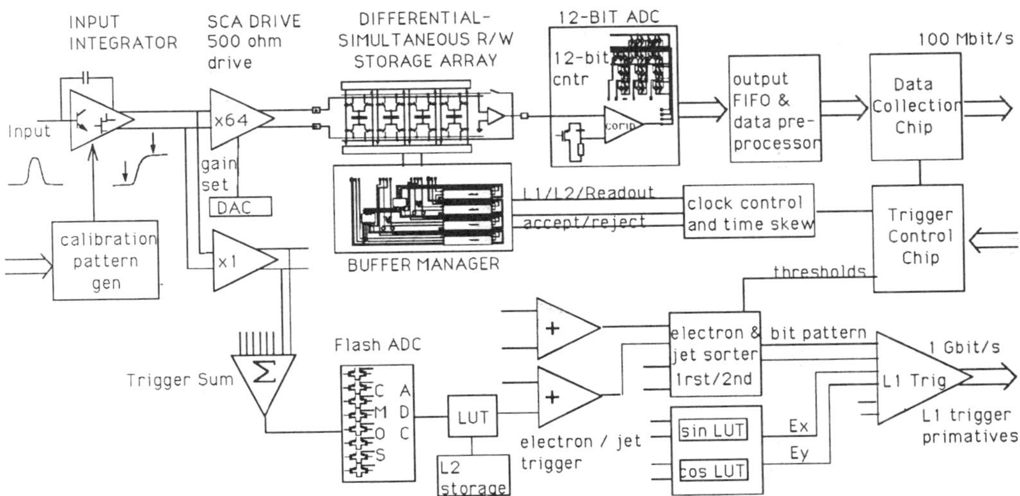

Fig. 10. Calorimeter front-end block diagram

Low frequency cutoff may be realized in the preamplifier and additionally in the fast amplifier, to provide a short enough decay time in the response that the pileup in any single channel does not restrict the dynamic range. Thus no elaborate pulse shaping would be performed prior to the memory. The actual optimal shaping for a given counting rate will be realized as a weighted sum of a number of samples. This shaping should be programmable so that the pileup and the electronic noise could be minimized for a particular experiment. This type of signal processing is equivalent to pulse shaping by conventional time invariant circuits. A wide range of weighting functions can be realized.

Preamplifiers for Ionization Calorimeters

Charge preamplifiers for ionization chamber calorimeters must have a low noise, a short response time and a low power dissipation. In addition, their apparent input resistance created by the feedback should be within a certain range to provide an aperiodic transfer of charge in the presence of inductance in the connections between the detector and the preamplifier. One of the most difficult requirements is a very large dynamic range. In the electromagnetic part of the calorimeter the dynamic range will be at least 2×10^4 and may be as high as 10^5. A linear response over a large dynamic range requires certain currents and voltages resulting in a high power dissipation. It appears that a power dissipation in the range between 75 and 200 mW is necessary depending on the type of capacitance matching. For liquid argon, the preamplifiers should be able to operate immersed in the liquid for optimal signal to noise and minimum charge transfer time.

The input amplifying device should have the device time constant C_{fet}/g_m as short as possible (<0.5 nsec), the equivalent series noise resistance should be close to the theoretical value ~ $2/3\ g_m$, and the parallel shot noise should be negligible. Junction field-effect transistors satisfy all these requirements. However, their optimum operating temperature range with respect to noise is 120 K<T<300 K. Development of devices which could operate at ~ 90 K (directly in liquid argon) is being pursued in collaboration with industry (INTERFET Corp., Garland, TX.).

Monolithic JFET Preamplifier Processing

A monolithic preamplifier using exclusively n-channel diffused JFET's has been designed and is now being manufactured by INTERFET Co. (Garland, TX) by means of a dielectrically isolated process which allows preserving as much as possible the technology upon which discrete JFET's are based.[8]

The new technique will still allow the use of an epitaxially formed channel and a diffused gate, as in standard JFET processing. The epitaxially grown lattice is free of defects created by ion implantation which only in part can be annealed.

Dielectric Isolation Method

Fig. 11 depicts the basic steps needed to achieve dielectric isolation. A standard slice of the needed resistivity is first oxidized and then patterned through a photoresist process. The slice is then etched by means of an orientation dependent etching process to form "V" grooves which will achieve electrical insulation. Another layer of oxide is then applied followed by a polysilicon grow to fill the grooves and to build up a thickness comparable to the original wafer. The slice is then turned over and most of the original silicon is lapped away until the tips of the V-grooves are exposed. This results in single crystal silicon islands ("tubs") isolated by silicon oxide and supported by a thick polysilicon substrate. A single layer of n-type epitaxial silicon is grown over the dielectric insulation substrate and provides a uniform channel of controlled thickness for the JFET.

Planar Process:

If a standard planar process is used with multiple tubs, adjacent JFET's will be connected by back-to-back diodes and a high value resistor due to the polysilicon bridge as in Fig. 12. This should not cause isolation problems except perhaps increased noise in the input device.

"Mesa" Isolation:

If true isolation is needed, V-grooves can be etched to leave a free-standing "mesa" of silicon on a silicon dioxide base. Metallization across this groove to connect devices on adjacent "mesas" may degrade the yield, so it may be better to use the simpler planar process.

Double Epitaxy and V-Groove Isolation Method:

This method utilizes an n-type silicon substrate on top of which a p-type layer for the back gate and an n-type layer for the channel are epitaxially grown. Next, V-shaped grooves are etched deep enough to reach into the n-substrate thereby separating the back side gates. A true isolation is not achieved with this method since adjacent devices are connected by back-to-back diodes through the common substrate.

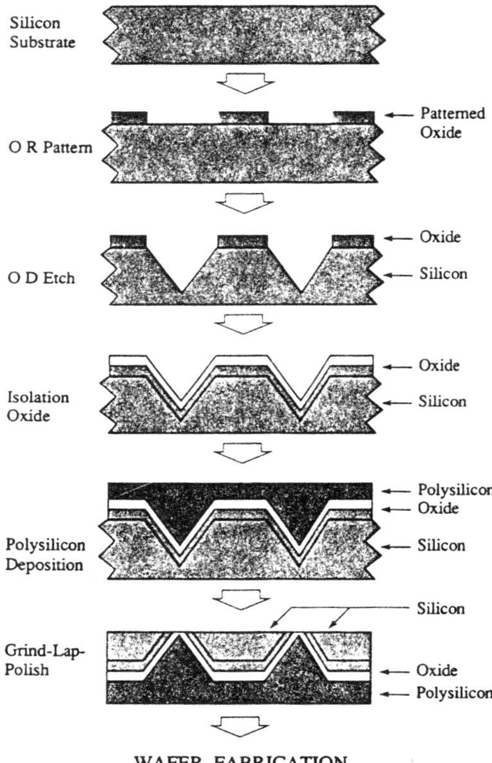

Fig. 11. Dielectric Insulation Wafer Fabrication. A silicon slice is first oxidized. After patterning, anisotropic etching is used to produce the "V" grooves which will become the barriers between circuit elements. Oxide and polysilicon are successively grown over the entire surface. The slice is then flipped over and most of the original silicon is lapped away until the tips of the V-grooves are exposed. This results in isolated single crystal islands ("tubs").

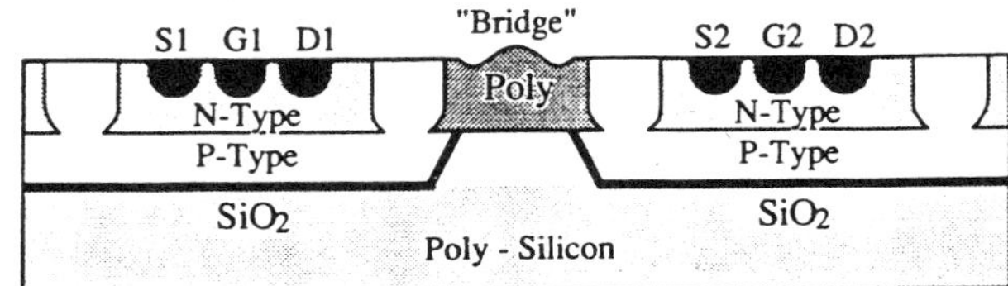

Fig. 12. Dielectric Isolation Process. Devices built in adjacent "tubs" are connected by back-to-back diodes in series with a high value resistor due to the polysilicon bridge shown above.

First Experimental Results

Up to the present only a preliminary silicon lot using the double epitaxial method has been manufactured, packaged and delivered. This first lot was manufactured on available material, which has a higher *p*-epi resistivity (about 1 Ωcm, instead of the design value of about 0.01 Ωcm). This increases the gate spreading resistance in series with the back gate, thus degrading the noise performance.

Fig. 13 compares the g_m and g_m/I_d characteristics for a monolithic and discrete JFET of W = 11400 μm and drain-to-source spacing of 16 μm (corresponding to an effective gate length of about 5 μm). At higher currents (I_d >1.5mA) they are almost identical.

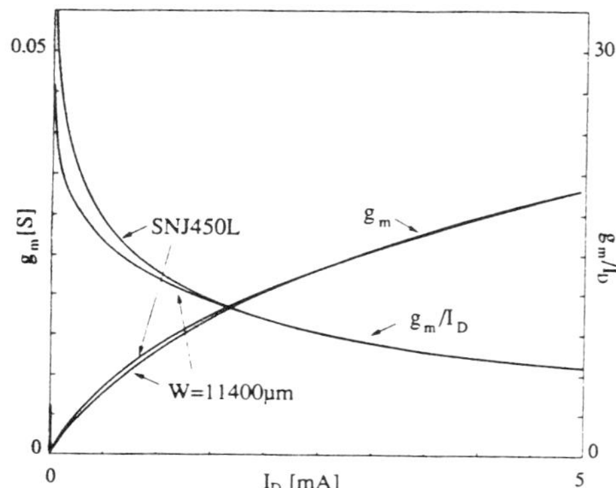

Fig. 13. Comparison of Discrete and Monolithic JFET's with L = 5 μm, W = 11400 μm.

Preamplifier Circuit and Experimental Results

An all JFET charge sensitive preamplifier has been manufactured in the dielectrically isolated process (see Fig. 14 for schematic). In applications with large capacitance detectors the input device J1 should be made as large as the production technology in order to take advantage of the better transconductance to current ratio attainable at low current density and reduce the noise at fixed current. For the prototype, a channel width W = 11400 μm, corresponding to an input capacitance C_{iss}~50 pF has been chosen.

The noise performance has been evaluated and is consistent with the additional gate spreading resistance in series with the back gate due to the low *p*-epi doping.

As soon as an improved version of the preamplifier is ready, it will undergo low temperature testing and evaluation of its resistance to neutron and ionizing radiation as well to electrostatic discharge.

Both circuit performance and neutron resistance can be improved by reducing the channel length (and thus the active volume) of JFET's. A systematic study of the optimum doping levels in the channel region is necessary for the development of a JFET with optimum noise performance near liquid argon temperature (90 K).

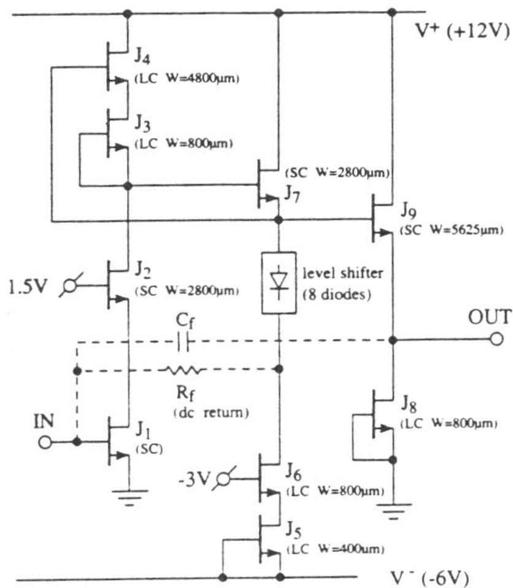

Fig. 14: Circuit Schematic of the JFET Charge Sensitive Preamplifier. SC: devices with short channel, L = 5 μm; LC: devices with long channel, L = 7 μm. The feedback resistor and capacitor (Rf and Cf.) are shown connected with dotted lines since they are outside the monolithic chip.

Voltage Programmable Shapers

As discussed above, calorimeter signals, as well as signals from other detectors, will require some form of shaping prior to analog storage. In principle, digital signal processing can be employed off-line to implement any desired weighting function, but even in this case, an anti-aliasing filter must be employed prior to analog storage. A more complete shaper implemented in hardware is however essential if one is to include the signals in a first or second level trigger. Off-line signal processing can still be used for enhancement of signal to noise ratio.

The challenge in design of integrated circuit shapers is to compensate for the process variations which inevitably occur in fabrication while maintaining good linearity over a wide dynamic range. The signal shape should be reproducible at the few percent level, while resistor and capacitor variations for even high quality bipolar processes are typically 25%. Various techniques for the design of externally programmable filters exist in the literature and are being studied for this application. MOSFET-C filters have been extensively described but it appears that they do not provide adequate linearity over the several volts of dynamic range required.

Translinear circuit techniques have been used for several decades in accurate analog multipliers, square and square root circuits and other analog function modules. These circuits utilize the accurate exponential current-voltage relationship of bipolar transistors explicitly to generate desired functions. We are investigating a new class of voltage programmable filters which utilize translinear circuit methods.

Our goal is to implement a bipolar pulse shape with a peaking time somewhere in the range of 75 to 150 ns for ionization calorimeters. For scintillation calorimeters, shapers of peaking time 32 - 48 nsec will be considered. For the bipolar pulse shape the required transfer function is composed of two differentiator sections and two integrator sections. The basic cell used to implement these functions consists of a resistor, a capacitor, and a current programmable current gain cell; a "Gilbert gain cell". The gain cell is used to provide Miller multiplication of the capacitor and therefore provide tuneability of the filter. In addition, the gain cell configuration is known to provide high linearity over a wide dynamic range. A bipolar implementation of the basic cell is shown in Fig. 15.

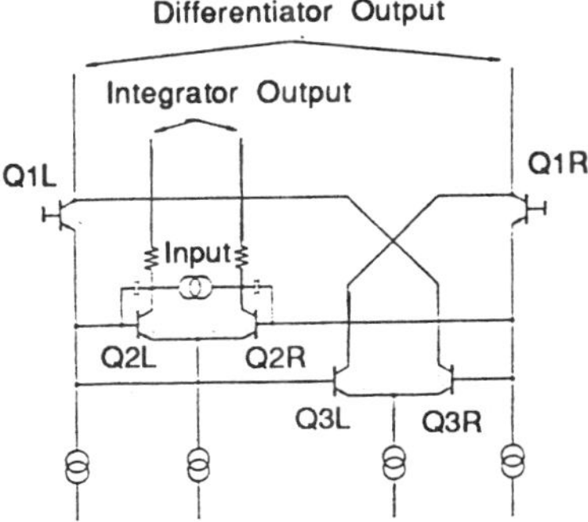

Fig. 15. Schematic of bipolar implementation of basic "Gilbert cell".

The gain of the cell is the ratio of the Q2 (and Q3) collector current to the Ql collector current and is externally programmable. Furthermore, the gain is constant for signals almost up to the limits of the bias current of the transistors. Cells formed in this way are fully differential and may be cascaded to form a variety of filter transfer functions. A bipolar pulse shape formed by two differentiators and two integrators is shown in Fig. 16 for several values of gain cell gain. We are doing detailed design work on this approach.

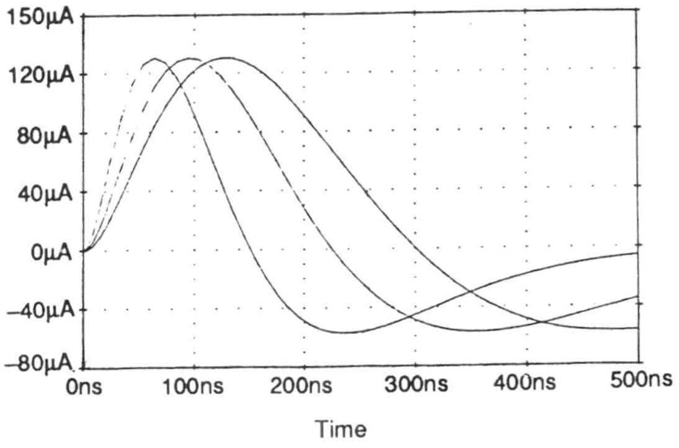

Fig. 16. Output pulse shape simulation of two differentiators and two integrators for several values of the "Gilbert cell" gain (i.e. different collector currents).

Level 1/2 Analog Storage for Calorimetry

The development of readout electronics for calorimeters is focusing on further refinements of the switched capacitor array (SCA) analog memory under development at LBL.[9] The analog memory will allow for the retention of data during the trigger decision.

The SCA circuit contains 2048 sample and hold cells subdivided into 16 parallel channels of 128 cells per channel. Each of the 16 channels has a dedicated analog input which is connected to a bus distributing the input signal to 128 sample and hold cells. Each sample and hold cell consists of a complementary CMOS transmission gate (complementary switch) and a 1 pF double polysilicon capacitor. An externally supplied reference voltage is applied to the bottom plate of all capacitors. The voltage stored on each sample and hold capacitor then corresponds to the difference between the input signal from a given channel at the time its sample and hold switches are opened and the applied reference voltage. The reference voltage can be adjusted to shift the baseline for better level matching.

An address decoder turns on and off sample and hold switches using a "break before make" action. During the readout sequence each sample capacitor is placed in the feedback path of the reconstruction op-amp. There is one reconstruction op-amp per channel. Before readout of each sample, stray charge on the signal distribution busses are cleared by shorting each amplifiers inverting input nodes and output nodes through a reset switch. This prevents charge remaining on parasitic nodes from the previous samples from influencing the next read out sample.

SCA Circuit Performance

Non-uniformities in the sample and hold elements limit dynamic range unless corrections are applied. The SCA readout scheme is not sensitive to capacitor values to first order, but non-uniformity of the sample and hold switches can be important, as the charge injection from these switches (due to gate to drain capacitance) will then vary. These switches are complimentary, and therefore the charge injection is canceled to first order, but variations still exist.

In order to achieve 12 bit dynamic range the pedestal variation of each of the storage elements in a channel must be less than one part in 4096 of the full scale voltage range. A channel operating with a full scale range of 4 volts would then require less than 1 mV of pedestal variation per element.

To achieve 12 bit dynamic range 8.3 charging time constants must elapse in each 16 nsec write operation to charge the sample and hold capacitors, this fast input bandwidth requires a large input charging switch. For increasing switch sizes the amount of parasitic charge injection from the switch will also increase. In the version tested, with a 2 nsec charging time constant, the cell to cell pedestal variation at a 50 MHz acquisition rate was 0.4 mV rms.

The measured incoherent noise on any given single cell when operated with a 62.5 MHz acquisition rate is less than 0.6 mV rms. Thus, the total dynamic range (single cell) could exceed 8000:1, or 13 bits. In previous versions of the IC the clocking noise feedthrough was well below one millivolt; however, in the current version inadequate separation of the on chip digital and analog power lines have increased the clocking noise feedthrough to several millivolts. The clocking noise is absent in a differential manifestation of the IC and is expected to be absent in the next corrected version of the IC. The measured parameters of the SCA are summarized in Table 1.

Table 1. Measured performance parameters of a simultaneous read-write switched capacitor array integrated circuit. Measurements were performed at a 50 MHz acquisition rate, but not utilizing the simultaneous read write capability.

Analog Store Parameters	Achieved Performance
No. channels/chip	16
No. of elements/channel	128
Power consumption/channel	23 mW
Non-linearity	+/- 0.9%
Cell to cell pedestal variation	0.3 mV rms
Charging time constant	2 ns
Channel dynamic range	4000
Maximum sample rate	90 MHz
Maximum readout rate	200 KHz
Noise floor	0.7 mV rms
Input voltage range	0.5-5.5 Volts
Capacitor droop rate	0.1 mV/msec
Output settling time (0.02%)	1.2 μsec
Gain variation from unity	+/- 1.5%

Operation of the integrated circuit has been verified on a sample of test pulses simulating the SSC calorimeter detector signals. Shown in Fig. 17 are a sequence of successively faster single cycle sinewaves, the pulses simulate the signals seen from the slowest to the fastest of the considered calorimeter detectors.

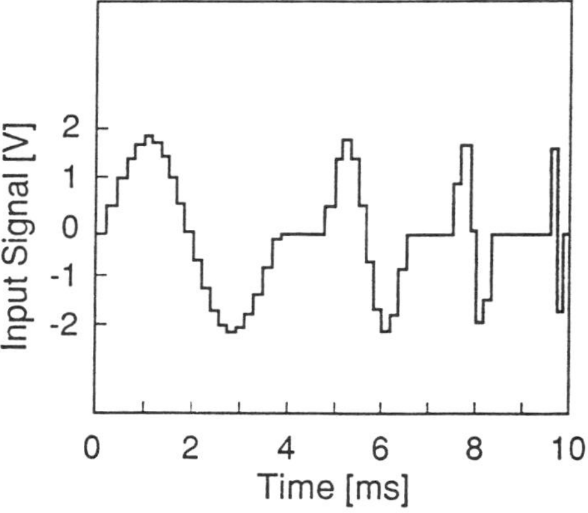

Fig. 17. Shown is the recorded input signal, 400, 200, 100, and 40 nsec period sinewaves of duration equal to the period. The SCA acquired this waveform while running at an acquisition frequency of 50 MHz.

Virtual Multiple Pipelines and the ALP

In the classical implementation of trigger delay pipelines, multiple physically distinct pipelines buffer data during multiple levels of trigger decisions. One principal problem is that the transfer of analog data with high dynamic range within 16 ns is extremely difficult and necessarily power consuming. The virtual multiple pipeline approach simulates the existence of essentially any number of sequential pipelines within one analog memory (SCA).[9]

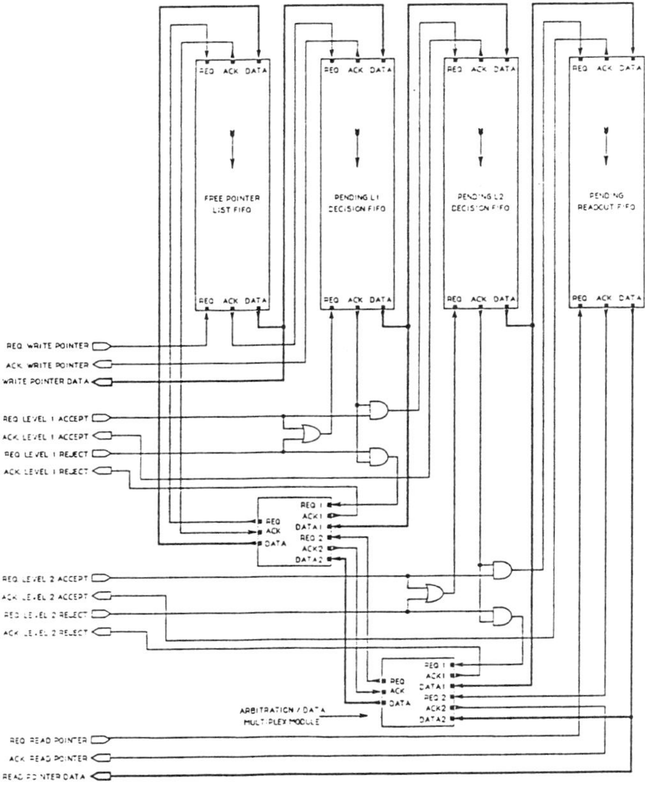

Fig. 18. Block diagram of the address list processor, showing the four FIFO memories for the address pointer lists for: unused cells, pending Level 1, pending Level 2, and pending read-out.

An address list processor (ALP) substitutes the movement of analog data with digital (the address of the analog data). The address list processor, presently in fabrication and shown in Fig. 18, contains four digital first-in-first-out (FIFO) memories. A prototype FIFO has already been fabricated. The four FIFO's contain addresses of SCA storage locations. The presence of a given address in a given FIFO indicates the state of the corresponding storage location. In this diagram, the four possible states are free (empty), pending Level 1 trigger decision, pending Level 2 trigger decision, or pending readout. With each beam crossing, a data word moves from the free list to the pending Level 1 trigger list. From there, data may move back on to the free list (on a Level 1 reject signal) or onto the pending Level 2 trigger list (on a Level 1 accept). From that list, the data may also move back onto the free list (on a Level 2 reject) or onto the pending readout list (on a Level 2 accept). The pending readout list buffers readout requests during the readout of prior events.

Radiation Effects on JFET's

The SSC poses the most severe test for semiconductor devices because of the anticipated high luminosity and confined geometry of detectors within the collider. We expect fast neutrons of about 1 MeV up to a few times $10^{12}/cm^2$ per year and up to 10 Mrad per year of ionizing radiation can be expected both within the central tracking region and in the calorimeter at the luminosity of $10^{33}cm^{-2}sec^{-1}$.[10] Fast neutrons produce displacements in the active detector volume which act as generation centers for excess leakage current. Photon interactions and most of the energy loss by very energetic, light, charged particles create ionization which leads primarily to surface effects at the interface of silicon and its passivating oxide. Radiation testing has been pursued on silicon detectors, on JFET's, on the AT&T CBC-U bipolar process, and on the radiation hard 1.2 micron CMOS process of UTMC.

Hybrid preamplifiers developed in the BNL Instrumentation Division are representative of analog components that would be at risk to ionizing radiation and can be studied conveniently. A set of BNL I0-535 single unit preamplifiers were exposed up to 1.45 Mrad of ^{60}Co at room temperature in air and the noise power spectrum was checked at intervals during the exposure.

Measurements were made over a range of amplifier shaping time constants; 100 pF was added to the preamplifier inputs to enhance high frequency series noise. Half of the preamplifiers were radiated with power on and half were not powered. Fig. 19 shows the increase of the noise in rms electrons (σ) after the irradiation, mainly for preamplifiers which were powered and at higher shaping time constants, suggesting a parallel noise source. The input JFET is an INTERFET 2N6451, which is guard-ringed,[11] and is expected to be reasonably tolerant to ionizing radiation. However, an unguarded 2N4416 is placed across the input as a protection diode and is a clear suspect. Fig. 19 shows that on removal of this diode, the noise in the powered devices returned to the pre-irradiation value. The effect power on is explained by the fact that under this condition some bias appears across oxides adjacent the channel, directing holes to the Si-SiO_2 interface forming interface states therefore charging the interface positively.[12] The fixed positive charge inverts the light *p*-diffusion, isolating the device (in the unguarded geometry), creating a parasitic leakage path and increasing the gate leakage current, the observed parallel noise source. In the unpowered preamplifiers, most of the ionization produced in the oxide recombined and little charge was directed to the interface.

Several preamplifiers from this experiment were annealed at 65°C and later at 120°C for periods of 1000 hours. In agreement with others, little improvement was noted at 65°C, but apparent full recovery was observed after the 120°C anneal. However, a small further increment of dose, 10^5 rad, returned the parallel noise to nearly its post radiation value. The 120°C anneal permits electrons from the silicon to tunnel to the interface charge and neutralize it but does not substantially decrease the interface trap concentration.

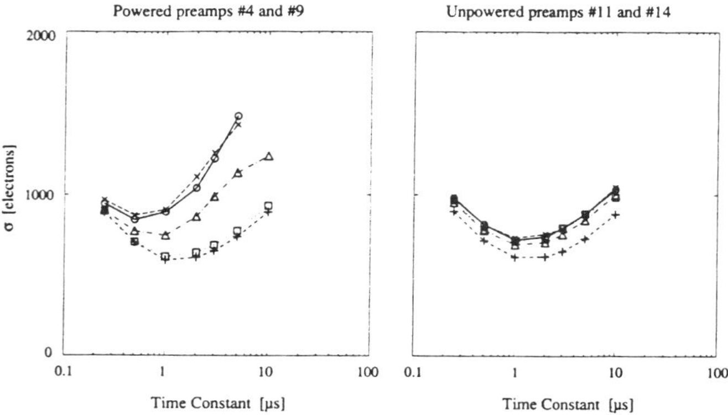

Fig. 19. The effect on observed noise σ (in rms electrons) of removing the protection diode from preamplifiers irradiated at room temperature with (at left) and without (at right) power applied. Legend: + = pre-irradiation, Δ = 0.5 Mrad, o = 1.1 Mrad, × = 1.45 Mrad with diode protection, ȯ = 1.45 Mrad without diode protection.

Powered and unpowered type I0-475 preamplifiers designed for operation in liquid argon were irradiated to 12 Mrad in a cryostat held at 120 K, and were also exposed to 10^{13} neutrons/cm^2 while under power and at ~120 K. A substantial increase in parallel, low frequency noise was observed only in the powered devices and a relaxation of the noise was observed when power was removed. Some increase in noise was observed at low frequencies. At pulse shaping times of 100 nsec or less, the noise increase was negligible.

Radiation Effects on Bipolar's

Measurements were performed on the degradation of bipolar transistors, particularly as characterized by decreases in the value of the current gain h_{fe}, after exposure of bipolar transistors from the AT&T CBC-U process to both neutrons and photons. The results are summarized in Fig. 20 and 21. Given a reasonably conservative design, circuits implemented in this process should operate after doses of 10^{14} n/cm^2 and up to doses of > 2 Mrads of ionizing radiation.

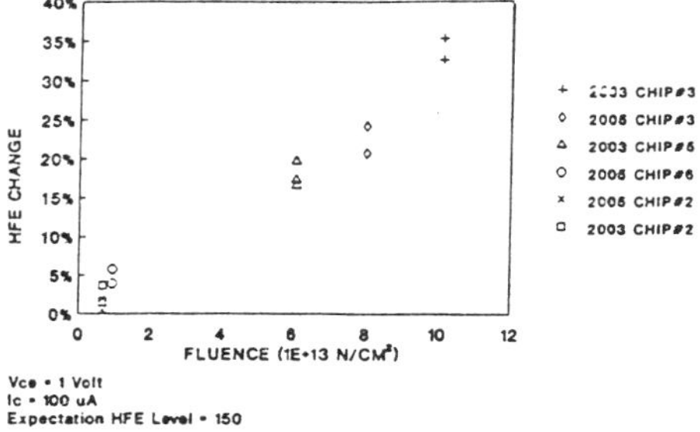

Fig. 20. Fractional decrease in current gain, h_{fe}, for npn transistors as a function of neutron fleuence.

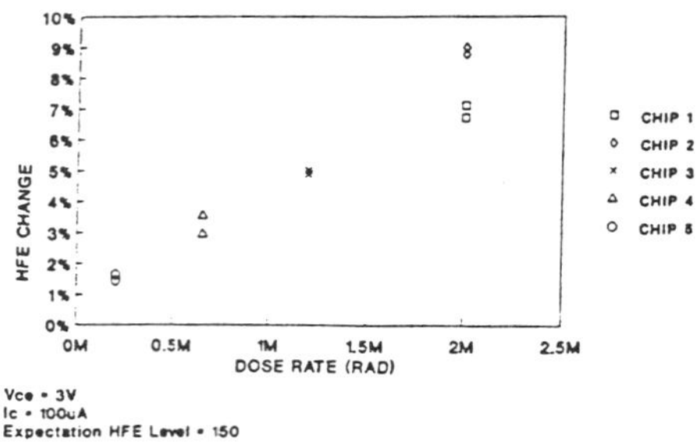

Fig. 21. Fractional decrease in current gain, h_{fe}, for npn transistors as a function of total dose exposure.

References

[1] A. Bhattacharya, C. Blocker, M. Botlo, G. Brandenburg, R. Cashmore, J. Dorenbosch, K. Einsweiler, R. Ely, M. French, J. Harder, R. Jared, F. Kirsten, S. Kleinfelder, A. Lankford, M. Levi, O. Milgrome, F.M. Newcomer, R. Nickerson, J. Oliver, V. Radeka, K. Ragan, S. Rescia, L. Rogers, P. Seller, P. Sharp, J. Siegrist, P. Sinervo, W. Smith, S. Tedja, R. Van Berg, J. Van der Spiegel, H.H. Williams, T. Yau, *Front End Electronics Development for SSC Detectors,* submitted to the SSC Laboratory, October 1, 1990.

[2] A. Bhattacharya, et al., *Front End Electronics Development for SSC Detectors,* submitted to the SSC Laboratory, October 2, 1989.

[3] F.M. Newcomer, et al., *High Speed Integrated Circuits for SSC Applications*, Nucl. Instr. Meth **A283** (1989) 806.

[4] R.A. Boie, A.T. Hrisho, P. Rehak, *Signal Shaping and Tail Cancellation for Gas Proportional Detectors,* Nucl. Instr. Meth., **A192** (1982) 365.

[5] F.M. Newcomer, S. Tedja, *A Low Power Timing Discriminator for SSC Applications*, in these proceedings.

[6] D. Christian et. al., *The Development of Two ASIC's for a Fast Silicon Strip Detector Readout System*, IEEE Trans. Nucl. Sci. **NS-36** (1989) 507.

[7] A.E. Stevens, et al., *A Time to Voltage Converter for Colliding Beam Detectors*, IEEE Trans. Nucl. Sci. **36-1** (1989); and *A Time to Voltage Converter and Analog Memory Unit for Colliding Beam Detectors*, Journal of Solid State Circuits, **24** (1989) 1748.

[8] V. Radeka, et al., *Monolithic JFET Preamplifier for Ionization Chamber Calorimeter*, pub. in these proceedings.

[9] S. Kleinfelder, IEEE Trans. Nucl. Sci **NS-37** (1990). S. Kleinfelder, M. Levi, O. Milgrome, *Test Results of a 90 MHz Integrated Circuit Sixteen Channel Analog Pipeline For SSC Detector Calorimetry,* pub. in these proceedings.

[10] S. Kleinfelder, M. Levi, O. Milgrome, *Toward a 62.5 MHz Analog Virtual Pipeline Integrated Data Acquisition System,* LBL-29608, submitted to Nucl. Instr. Meth.

[11] Radiation Levels in the SSC Interaction Regions, D.E. Groom ed., SSC-SR-1033 (1988).

[12] D.J. Allen, et al., *Gamma-Induced Leakage in JFET's,* IEEE Trans. Nucl. Sci. **NS-31** (1984) 1487.

[13] P.S. Winokur, et al., *Dependence of Interface State Buildup on Hole Generation and Transport in Irradiated MOS Capacitors*, IEEE Trans. Nucl. Sci. **NS-23** (1976) 1580.

PRECISION ELECTROMAGNETIC CALORIMETERS FOR THE SUPERCONDUCTING SUPERCOLLIDER

Harvey B. Newman*
Lauritsen Laboratory
California Institute of Technology
Pasadena, CA 91125

Abstract

High resolution electromagnetic calorimeters will have unique physics discovery potential at the SSC: in the search for Higgs particles in the mass range between 80 and 180 GeV, and in the search for new physics signatures involving electrons and photons beyond the Standard Model. The high resolution, speed, and radiation resistance requirements have pointed the way towards two leading technologies: BaF_2 crystals, and liquid Xenon (LXe), which have been proposed as options for the L* Detector at the SSC. The conceptual designs, R&D status and key development issues for the proposed BaF_2 and LXe calorimeters are summarized in this report.

INTRODUCTION: PHYSICS OF PHOTONS AND ELECTRONS

Over the last 30 years, precision measurements of electrons and photons, as well as muons, have been a key factor in most of the important discoveries in our field. Examples include the discovery of two neutrinos, the J/Ψ, τ, W, Z^0, and radiative quarkonium transitions for charm and bottom. Apart from W and Z, none of these discoveries were predicted at the time the accelerators which found them were built.

The most important discoveries at the SSC thus could result from the detection of entirely new phenomena at the 1 TeV scale, outside of the Standard Model or its simplest extensions. These discoveries could appear as new multilepton or multiphoton final-states, or as new configurations of leptons, photons and jets. In order to isolate the new physics signals from the backgrounds of the *"Standard"* physics, such as the inclusive production of Z's and W's together with jets, the detector must be able to cleanly separate leptons and photons from jets and single hadrons, and to reconstruct multilepton, multiphoton, and lepton-photon invariant masses precisely.

In past year, a tremendous effort has been devoted to search for the Higgs at LEP. Using precision measurements of electrons, muons and photons, together with jets, the LEP experiments will be able to explore a Higgs mass range up to $M_H \leq 50$ GeV at LEP Phase I by 1992, and up to $M_H \leq 80$ GeV at LEP Phase II by 1996.

There exists a gap between the upper limit for Higgs detection at LEP Phase II (80 GeV) [1] and the lower limit for Higgs detection at the SSC using the four lepton final state (150 GeV) [2]. Recent studies in the U.S. ([3], [4]) and in Europe ([5]) have shown that a precision electromagnetic calorimeter can close this gap by detecting the distinctive decay mode $H^0 \rightarrow \gamma\gamma$, and by reconstructing the Higgs with a mass resolution $\Delta M/M \approx 0.5\%$. A precision electromagnetic calorimeter also will complement the physics discovery potential of a precision muon detector, by increasing the sensitivity for Higgs detection in the mass range from 150 GeV upwards, through measurements of $e^+e^-e^+e^-$ and $e^+e^-\mu^+\mu^-$ final states.

Requirements for a Precision Electromagnetic Calorimeter at the SSC

In order to detect the intermediate-mass Higgs, and to search for new physics with sufficient sensitivity, the calorimeter should have the following features:

- **High Speed:** gating the signal in less than 15 nsec (in a single beam crossing);

*Work supported in part by U.S. Department of Energy Contract No.DE-AC03-81-ER40050.

- **Excellent Energy Resolution:** $\Delta E/E = (1.3/\sqrt{E} + 0.5)\%$, as has been achieved in L3.
- **Good Position Resolution:** Δx and $\Delta y \approx 1$ mm for electromagnetic showers. ($\Delta\theta \approx \Delta\phi \approx 0.1°$);
- **Excellent e/π, γ/jet, and e/jet Separation:** $\sim 10^{-4}$;
- **Good Radiation Resistance:** $\geq 10^7$ Rads.

Of the proposed electromagnetic detectors which are currently under development, only continuous sampling calorimeters with high light output can meet the resolution requirement. The two leading technologies, for which substantial R&D is underway, are:

- A BaF_2 crystal calorimeter, closely following the experience developed in L3 for the BGO calorimeter over the last ten years.
- A Liquid Xenon calorimeter.

The physics goals, conceptual design, R&D progress, and the key issues for being able to reach the performance goals at reasonable cost for these two technologies, are summarized in the remainder of this report†.

$H^0 \rightarrow \gamma\gamma$ Search for $M_{H^0} \in (80, 160)$ GeV

As summarized in Table 1 the importance of covering this Higgs mass range [6] stems from the fact that:

- If $M_H > M_{Z^0}$ for the lightest Higgs particle, then the minimal supersymmetric extension to the Standard Model is ruled out.
- Nonminimal supersymmetric theories require $M_H \leq 200$ GeV. If $M_H > 200$ GeV, then SUSY is ruled out.
- Most importantly, if no Higgs exists with $M_H \leq 200$ GeV, then there **MUST** be new physics below the Planck scale.

Recent Monte Carlo studies at Caltech and elsewhere have shown that the BaF_2 calorimeter (summarized below and discussed further in reference [7]) should be able to find the Higgs in the 80 to 160 GeV mass range. The LXe calorimeter is expected to have similar physics capability.

The distinctive feature of isolated photons in the $H^0 \rightarrow \gamma\gamma$ decay mode can be used to reject the QCD background. The main background in the $H^0 \rightarrow \gamma\gamma$ search is direct photon production: $q\bar{q} \rightarrow \gamma\gamma$ (59 pb) and $gg \rightarrow \gamma\gamma$ via a box diagram (493 pb) [8]. This background is reduced by event selection cuts:

- $|\eta_\gamma| < 2.8$;
- $E_T^\gamma > 20$ GeV;
- $|\eta_{\gamma\gamma}| < 3$ ($\Rightarrow$ to reduce $qq \rightarrow \gamma\gamma$).
- $|\cos\theta^*_\gamma| < 0.8$ ($\Rightarrow$ to reduce $gg \rightarrow \gamma\gamma$).

The background cross sections after event selection cuts are reduced to 91 pb for $gg \rightarrow \gamma\gamma$ and 14 pb for $q\bar{q} \rightarrow \gamma\gamma$. Since the Higgs width in this mass region is very narrow ($\sim$10 MeV) a very high resolution electromagnetic calorimeter, such as the BaF_2, is needed to detect the signal from $H^0 \rightarrow \gamma\gamma$, which has a typical cross section of 0.2 pb [9].

There are copious π^0's, and thus photons, produced at the SSC. The QCD two jet cross section is 1.2 mb for $\hat{P}_T > 20$ GeV and $|\eta| < 2.8$. There are also processes producing one real photon and a QCD jet, such as $qg \rightarrow q\gamma$ (119 nb) and $qq \rightarrow g\gamma$ (6 nb), which will fake a two-isolated-photon final state with a higher probability. As discussed in reference [7] an isolation cut is used to reject isolated photons from a jet by a factor of 10^{-4}. The rate of two isolated photons from two jets is thus reduced by a factor of 10^{-8}, assuming that the two jets are statistically independent.

Figure 1a) shows the signal, superimposed on all backgrounds, as measured by the BaF_2 calorimeter. The corresponding signals, with signal/background ratios of 5, 9, 15, 14 standard deviations for Higgs masses of 80, 100, 120 and 150 GeV, respectively, are shown in Figure 1c) with the BaF_2 calorimeter. The corresponding signals before and after background subtraction for a calorimeter with conventional resolution $(15\%/\sqrt{E} + 1\%)$ are shown in Figures 1 b) and d). It is clear from these graphs that a very high resolution calorimeter, such as BaF_2 , is required to do this physics.

$H^0 \rightarrow ZZ/ZZ^* \rightarrow e^+e^-e^+e^-$ Search for $M_{H^0} \in (140, 800)$ GeV

The cleanest signatures for the heavy Higgs search in this mass region are four isolated electrons or muons in its final state: $H^0 \rightarrow ZZ/ZZ^* \rightarrow e^+e^-e^+e^-$, where Z^* represents a virtual Z with mass $< (M_{H^0} - M_{Z^0})$ for M_H less than $2M_Z$. The proposed BaF_2 calorimeter has excellent capability to detect 4 isolated final-state electrons. All of the QCD background, including contributions from charm, bottom and top decays (computed with top mass = 150 GeV), is rejected with an isolation cut (with a reduced radius R=0.3). This is discussed further in reference [7].

†Other techniques, including liquid Argon and scintillating fiber fine-sampling calorimetry which have been discussed at this conference, do not appear to be capable of reaching the necessary energy resolution. See reference [5].

Table 1: Detection of The Higgs and Physics Implications.

Experiment	M_H Range	Implications
LEP Phase I (1992)	$\leq$ 50 GeV	Minimal SUSY
LEP Phase II (1996)	$\leq$ 80 GeV	Required by Minimal SUSY
BaF_2 or LXe at SSC	$\leq$ 200 GeV	Required by nonminimal SUSY. If not, new physics below M_{Planck}.
Lepton Detector at the SSC	200 GeV — 1 TeV	Standard Model

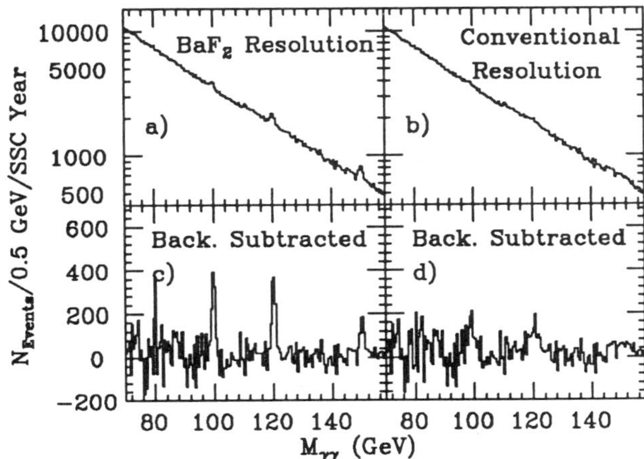

Figure 1: $H^0 \rightarrow \gamma\gamma$ Detection for electromagnetic calorimeters with BaF_2 energy resolution (a and c), and with conventional resolution (b and d).

Search for New Physics

The high resolution electromagnetic calorimeter also will be used to search for new massive gauge bosons through their decays into e^+e^-. This will allow us to explore the energy regime where high energy symmetries, such as superstring-inspired E6, break down to form the gauge structure observed at low energies.

BARIUM FLUORIDE CRYSTAL CALORIMETER

Experience With L3 BGO Performance and Detector R&D

The high resolution BaF_2 crystal calorimeter closely follows the experience with the L3 BGO calorimeter. Barium fluoride has been chosen because of its fast scintillation in the UV (0.9 nsec decay time), and the high radiation resistance of high purity crystals.

Since 1981, many physicists involved in the development of the BaF_2 calorimeter have participated in the design and construction of L3 at LEP: a 4π detector which is optimized for the precision measurement of electrons, photons and muons, combined with good resolution and granularity for the measurement of hadron jets. L3 measures electrons and photons with an energy resolution of $\sigma_E/E = 1.3\%/\sqrt{E} + 0.5\%$, and a position resolution of ~1 mm. L3 also measures a 100 GeV particle decaying into a pair of muons with a mass resolution of 1.7%. The hadron calorimeter in L3 measures hadron energy with an energy resolution of $55\%/\sqrt{E} + 5\%$, and an angular resolution for jets of $\Delta\Theta = 3.0°, \Delta\Phi = 2.5°$. The excellent electron/pion separation has been achieved by the L3 detector through:

- Precision measurement of the transverse shower shape, using the energy sharing among neighboring crystals;
- Matching the momentum measured in the inner time expansion chamber (TEC), with the energy deposited in the BGO electromagnetic calorimeter; and
- Measurement of the energy deposited in the hadron calorimeter, behind an energy cluster in the electromagnetic calorimeter.

The performance of the L3 BGO calorimeter is illustrated in Fig. 2, where the BGO energy resolution and position resolution (~1 mm) is used to reconstruct clear π^0 and η^0 signals in the inclusive two-photon spectrum in hadronic events at the Z^0.

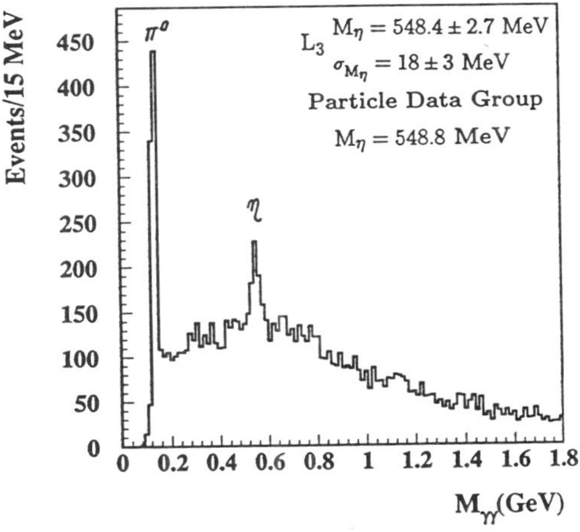

Figure 2: The Inclusive $\gamma\gamma$ Spectrum Measured with the L3 BGO Calorimeter.

The L3 BGO has also been used in a search for multiphoton resonance production at LEP [10]. Because of its high resolution, L3 has been able to set upper limits on the relative branching ratio of 10^{-4} over the mass range up to 40 GeV. With more data at LEP, it is expected to reach the 10^{-5} level of sensitivity in 1991. The example of a search for a 40 GeV scalar particle decaying into a 4γ final state, with a branching ratio $\sim 10^{-3}$, is shown in Figure 3

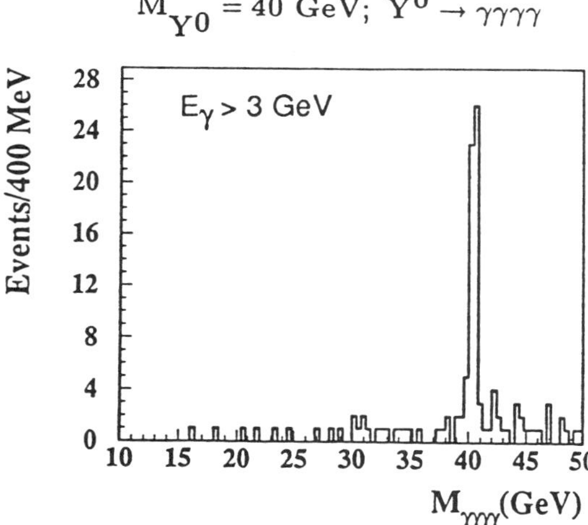

Figure 3: The invariant mass spectrum expected for the decay of a new 40 GeV scalar $Y^0 \to \gamma\gamma\gamma\gamma$, as measured by the L3 BGO calorimeter.

Over the last decade, the Shanghai Institute of Ceramics (SIC), in collaboration with Caltech, Carnegie Mellon, Princeton, LAPP Annecy, Lausanne, Rome, Aachen, Lyon, and Geneva, have systematically performed large scale R&D for L3 on:

1. Mass production at low cost of large, radiation resistant, high optical quality crystals;
2. Cutting crystal surfaces to 200 μm accuracy;
3. Methods of coating crystal surfaces to obtain uniform pulse height response;
4. Electronics and data acquisition systems to collect data with a linearity of 10^{-3} and a dynamic range of 10^6 (20 bits);
5. Cooling system to ensure that the temperature is constant to 0.1° C over large volumes;
6. Production of a thin carbon fiber structure to support the crystals with a minimum amount of inactive material in the detector;
7. Calibration systems to ensure that the response of all crystals is known at all times;
8. Most importantly, continued R&D efforts to produce fast crystals with high radiation resistance for high intensity pp colliders.

The success of these R&D efforts enabled the L3 group to build the precise BGO electromagnetic calorimeters consisting of 12,000 BGO crystals, to measure electrons and photons with an energy resolution of $1.3\%/\sqrt{E} + 0.5\%$.

Research on the development and calibration of a high precision BaF_2 crystal calorimeter for the SSC has continued at SIC (Shanghai), in collaboration with the Beijing Glass Research Institute (BGRI), Caltech, BNL, KEK, LAPP (Annecy), and AccSys Technology, Inc. Work in the U.S. for this program has been supported, in part, by the DoE Generic R&D program for the SSC [11]. Los Alamos, Oak Ridge, Livermore, Princeton, CMU and many institutes in Europe, China, the USSR and India have joined this effort in the last three months [3].

BaF_2 Detector Design

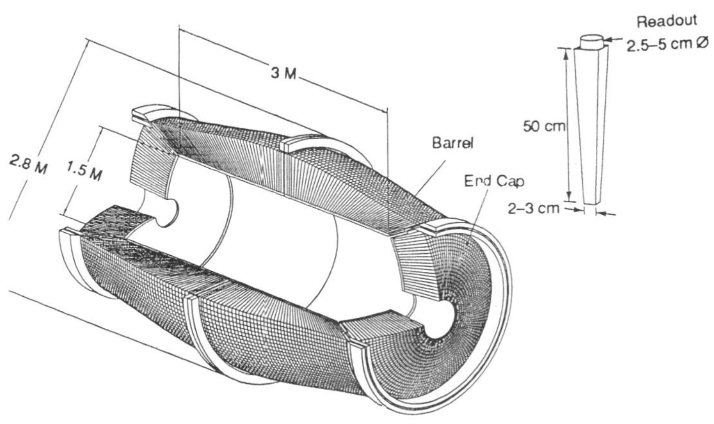

Figure 4: Layout of the L* BaF_2 electromagnetic calorimeter.

Figure 4 shows the conceptual design of the proposed L* BaF_2 calorimeter, including the central barrel and endcaps.

In order to keep the BaF_2 calorimeter as compact as possible, we have adopted a cylinder and endcap geometry which closely surrounds the inner tracker. This geometry is also important in keeping the total volume of the crystals as small as possible. The crystals are supported by an ultra-light carbon fiber structure, engineered according to the design developed for L3.

Each BaF_2 crystal calorimeter element is 50 cm (24.5 radiation lengths) long, and is a truncated pyramid pointing to the center of the interaction region, with a small angular offset ($\sim$10 mrad) to suppress photon leakage. The cross section of the crystals is approximately square, with the shape changing along θ because of the varying distance to the interaction point. The crystals range in size from 2×2 cm^2 in the front for the smallest crystal, to 5.2×5.2 cm^2 in the back for the largest crystal. The typical crystal covers a $\Delta\eta \times \Delta\phi$ interval of 0.04 × 0.04.

The calorimeter has two main parts:

- A central barrel calorimeter with an inner radius of 75 cm and an outer radius of 140 cm, covering the rapidity range $|\eta| \leq 1.45$ ($26° \leq \theta \leq 154°$).
- Two endcaps, located at $z = \pm$ 150 cm, covering the rapidity range $1.45 \leq |\eta| \leq 2.87$.

The detector is segmented into two end caps and two half length cylinders, which will be constructed and assembled separately.

The total crystal volume of the BaF_2 calorimeter is 11.7 m^3, with a total crystal weight (without mechanical support) of 57.1 tons. Table 2 shows the features of the proposed BaF_2 calorimeter.

BaF_2 Scintillation Light

BaF_2 is a unique scintillator which has fast decay components at 195 and 220 nm with decay time constants of 0.87 and 0.88 nsec [12]. Because of the high speed of the fast component and the evidence of high radiation resistance [13], BaF_2 has attracted a great deal of interest in recent years [14,15,16]. There is, however, a stronger scintillation component at 310 nm which has a decay time of about 600 nsec. Because of the 16 nsec beam crossing time, the slow component could lead to pile-up and thus could degrade the high resolution at the SSC. The suppression of the slow component thus is a key issue in designing a high energy-resolution BaF_2 calorimeter for the SSC.

The intensity ratio of the fast BaF_2 scintillation components (peaked at 195 and 220 nm) to the slow component (peaked at 310 nm) has been measured to be about 1:5 [12]. Studies by a group at Delft University [17] and C. Woody [18] have shown that the fast-to-slow ratio (F/S) can be improved to 1:1 or better by the addition of lanthanum doping.

Figure 5 shows the emission spectra for a sample of pure BaF_2 , and a BaF_2 sample doped with 1% of lanthanum. The peak intensity of the slow component is reduced by a factor of about five, with little change to the fast components, in the lanthanum-doped sample.

Fast UV-Selective Readout

Since the fast components and the slow component of BaF_2 are at different wavelengths, they are spectroscopically separable. Figure 5 also shows the quantum efficiencies of two photomultipliers (PMT): a special UV-sensitive, solar-blind PMT with a Cesium Telluride (Cs-Te) photocathode (Hamamatsu R3197) and a PMT with a bialkali photocathode (Hamamatsu R2059). Both PMT's have a quartz windows. The Cs-Te The Cs-Te photocathode is mainly sensitive to the fast scintillation light, and also yields a F/S ratio of 1:1 to 1.5:1 for typical undoped BaF_2 samples.

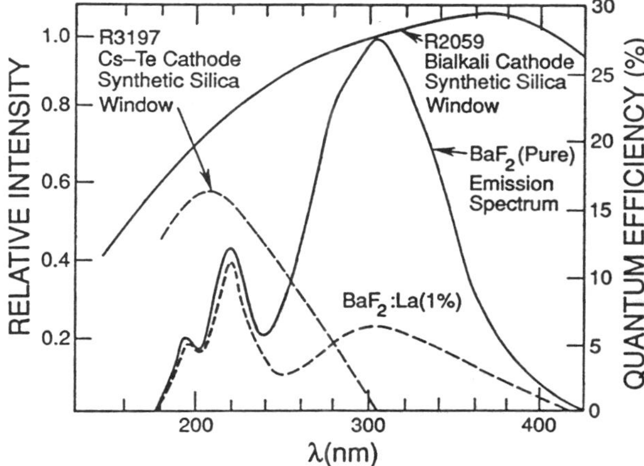

Figure 5: BaF_2 scintillation spectra (pure and doped with 1% lanthanum), and the spectral response of bialkali and Cs-Te photocathodes.

Figure 6 shows pictures of the scintillation light pulses recorded on a digital sampling scope at Caltech, using a bialkali photocathode (a and d), and a Cs-Te photocathode (b and e). Figures 6c and 6f show the scintillation pulse obtained with a Cs-Te photocathode, together with the results from a BaF_2 crystal doped with 1% lanthanum. The rise time of the scintillation light pulse in these pictures was completely dominated by the 2.3 nsec rise time of the PMT's. An F/S ratio of 5:1 is obtained with the lanthanum doped crystal.

Higher F/S ratios (more complete suppression of the low component) can be obtained with pure BaF_2 samples by using a new photocathode K-Cs-Te recently developed by Hamamatsu [19]. Figure 7 shows a comparison of the spectral responses of the new solar-blind K-Cs-Te photocathode and a the widely used Cs-Te photocathode. A F/S ratio of approximately 10:1 can thus be obtained by using this K-Cs-Te photocathode for pure BaF_2.

As detailed in reference [3], the fast UV light will be read out, and the slow component will be fully suppressed by:

- A vacuum photodiode with a K-Cs-Te photocathode and a quartz window as the photosensitive device. This suppresses the slow 310 nm light from BaF_2 by a factor of 50.
- A preamplifier mounted on the base of the photodiode, with a 45 nsec decay time, and a typical rms noise of 1000 photoelectrons. Once most of the

Table 2: Features of the BaF_2 Electromagnetic Calorimeters.

Barrel	$(\mid \eta \mid \leq 1.45)$
No. of Crystals	10,944
Crystal Volume	9.2 m^3
Crystal Weight	44.9 Tons
Crystal Size (Typical)	$(3 \times 3) \times (5 \times 5) \times 50 cm^3$
Endcaps	$(1.45 \leq \mid \eta \mid \leq 2.87)$
No. of Crystals	7,100
Crystal Volume	2.5 m^3
Crystal Weight	12.2 Tons
Crystal Size (Typical)	$(2.3 \times 2.3) \times (3.1 \times 3.1) \times 50\ cm^3$
Photodetector	UV Vacuum Photodiode
Electronics	Low Noise Preamp; Fast Shaper
Gating Time	<15 nsec

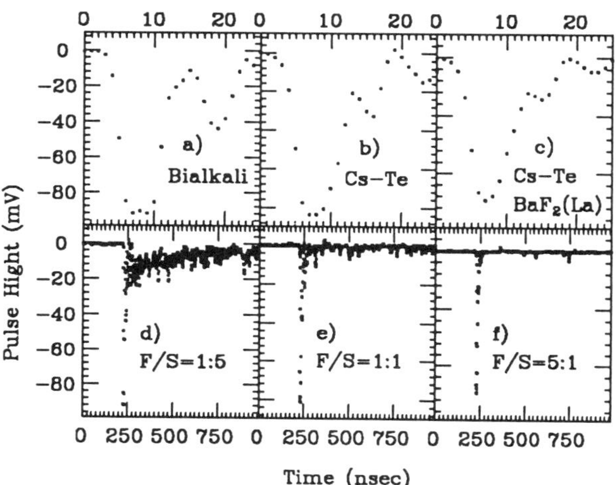

Figure 6: BaF_2 scintillation light pulse measured with a bialkali PMT with quartz window (a and d), a Hamamatsu R3197 UV selective PMT (b and e), and a Hamamatsu R3197, on a BaF_2 crystal doped with 1% lanthanum (c and f).

slow component has been suppressed by the UV-selective photocathode, the preamp should be able to operate with a very large dynamic range. The small capacitance of the vacuum photodiode allows a very low noise preamplifier design.

- A fast shaper with 3-10 nsec peaking time, and with two pole-zero cancellations: one to fully suppress the residual slow component ($\sim$ 600 nsec) and the other for the preamplifier decay (45 nsec).

The first three UV-selective vacuum phototriodes (R4406) were delivered by Hammamatsu to Caltech in early October. These are to be followed by the first

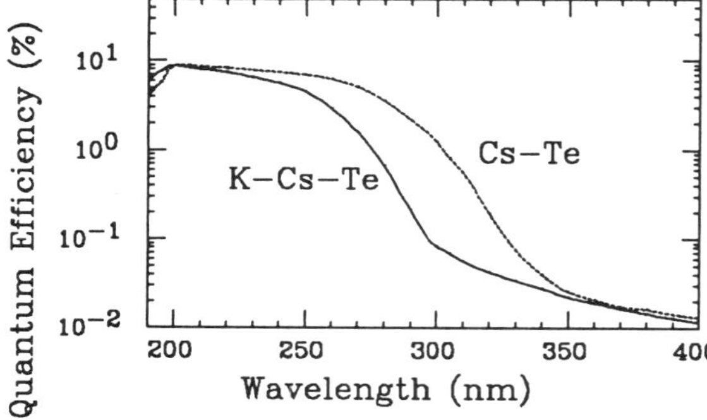

Figure 7: Quantum efficiency of a Hamamatsu K-Cs-Te photocathode compared with a conventional Cs-Te photocathode.

production series of 50 tubes immediately following our tests at Caltech, which are now underway.

Hammamatsu is also planning to develop a thin proximity-focused UV-selective vacuum photodiode, with high photoelectron collection efficiency and very high magnetic field immunity by the middle of 1991, in collaboration with Caltech.

Based on developments by V. Radeka's group at BNL, a preliminary version of the preamp and shaper circuits has been developed, as shown in Fig. 8. The first prototype series of circuits will be built and tested at ORNL and LASL as part of the Subdetector R&D program by March, 1991. The circuits will then be optimized in terms of shaping speed versus noise, and will be used during the first BaF_2 beam test in mid-1991.

The results of a PSPICE simulation for these circuits for a BaF_2 pulse with a F:S ratio of 1:1 show that the signal returns to baseline, down to a level of $\sim 10^{-4}$ within 35 nsec (see reference [7]). This will enable the

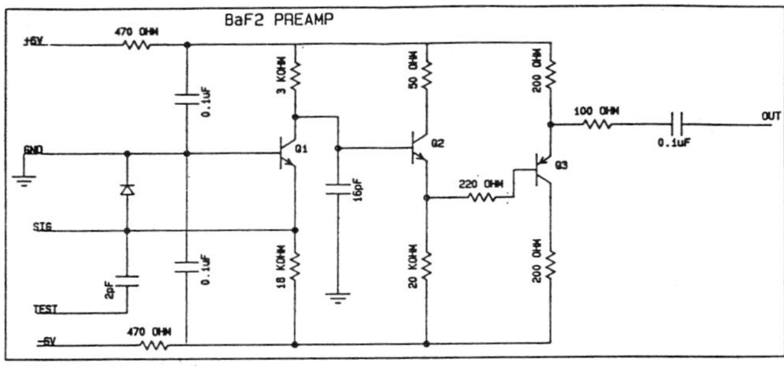

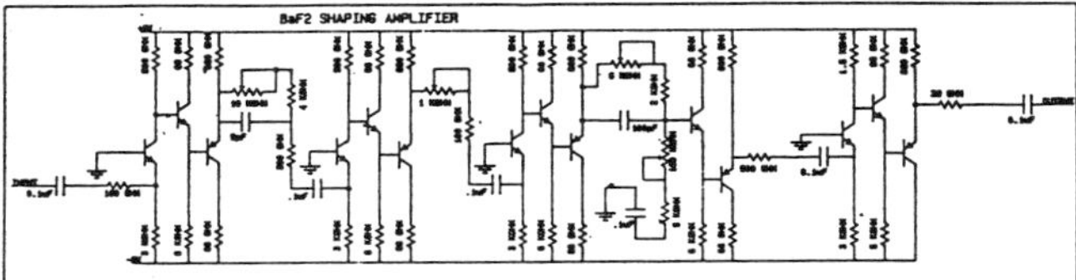

Figure 8: Circuit diagrams of the fast preamp and shaper.

BaF_2 calorimeter to be read out a single beam crossing, and to reduce pile-up effects to a minimum.

UV Transmittance

Fig. 9 shows that the transmittance of a typical 25 cm long BaF_2 crystal from the first batch made by SIC/BGRI is very close to meeting our specifications. A crystal manufactured by Merck is shown for comparison. The factors limiting the UV transmittance of BaF_2 crystals have been fully analyzed by SIC, as presented in Reference [3]. It is expected that the next batch of crystals, to be delivered by the end of 1990, will meet or exceed the specifications.

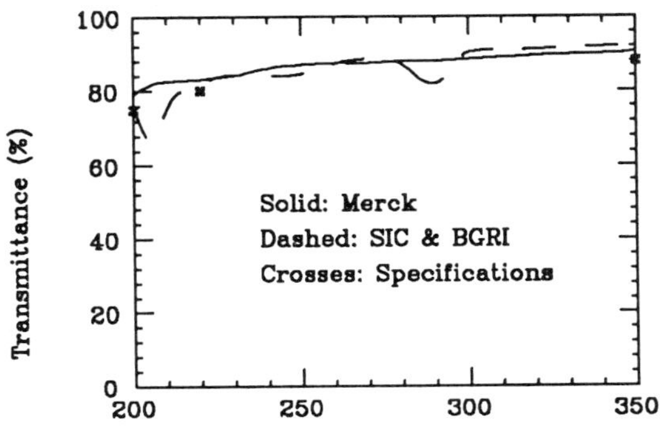

Figure 9: Optical transmissions of 25 cm long crystals delivered by SIC/Beijing and by Merck, as a function of the wavelength. The crosses represent the required specifications.

UV Light Collection Uniformity

The light uniformity of the crystals delivered by SIC/BGRI has been measured at Caltech with a collimated ^{137}Cs source. Figure 10 shows the result of the measurements using a) a PMT with a full-spectrum response; and b) a solar-blind (Cs-Te) PMT. With simple aluminum wrapping, the fast component (F) measured with a Cs-Te phototube (R3197) shows almost uniform response, while the response of the slow component (S) shows a slight increase of the response with increasing distance from the PMT. This is due to a compensation between two effects in light collection: bulk attenuation and optical focusing caused by the tapered shape of the crystal.

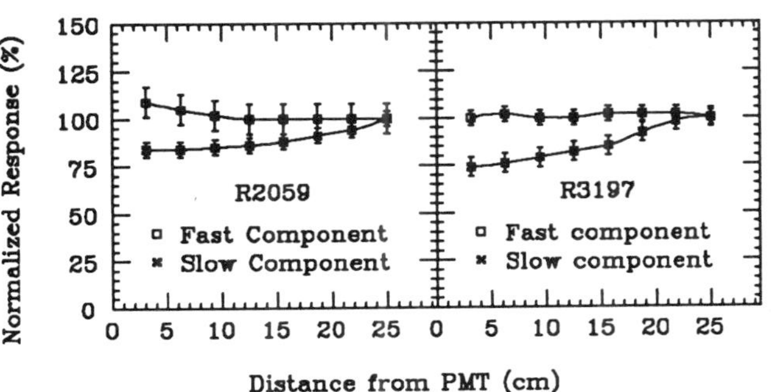

Figure 10: Normalized light collection response measured with a collimated ^{137}Cs source running along the axis of a 25 cm long BaF_2 crystal.

It has thus been shown, based on the first Caltech trials, that the fast UV light can be collected with a deviation from uniformity of $\sim 2\%$, using simple techniques in laboratory tests. Note that the uniformity achieved with the aluminum wrapping is well within the specifications, and is better than a typical BGO crystal in L3.

Part of the BaF_2 Subdetector R&D program will study how to achieve light collection at this level of uniformity, in mass production.

BaF_2 Radiation Resistance

The radiation resistance of BaF_2 crystals provided by several suppliers have been tested at Caltech, SIC, BNL, IHEP (Beijing) and CERN. These tests have shown that BaF_2 is among the most radiation resistant crystals known.

Early work done by S. Majewski and D. Anderson [13] showed that no color centers were formed in BaF_2 up to a dose of 1.3×10^5 Gy in an 800 GeV proton beam. Other work, in which BaF_2 was irradiated with charged particles or photons, [13] confirmed this early observation. A BaF_2 crystal from Optivac (Merck) recently measured by Woody *et al.* showed a light transmission reduction of $\sim 1\%$ after a dose of 2 MRads of ^{60}Co gamma rays. As illustrated in Figure 11, a BaF_2 crystal measured by Charpak *et al.* showed that after 50 MRads (corresponding to decades of operation of the

SSC at design luminosity, for a typical crystal) the light transmission was reduced by $\sim 5\%$.

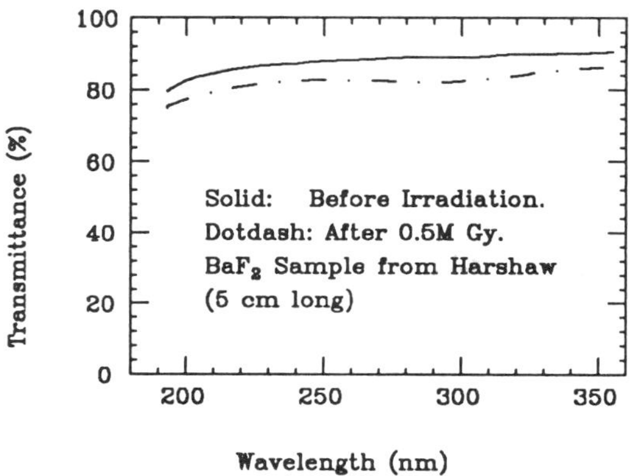

Figure 11: The transmittance as a function of wave length before and after 0.5 MGy dose measured by Charpak *et al.* for a 5 cm long BaF_2 crystal.

The results on BaF_2 radiation resistance can be summarized as follows:

- Some BaF_2 samples are extremely radiation resistant.
- The radiation damage caused by γ-ray or neutron irradiation is recoverable by annealing the crystal at 500°C for 3 hours [20]. This indicates that neutrons, as well as photons, do not cause permanent damage in BaF_2 crystals. The results also indicate that the BaF_2 damage mechanism from neutrons is the same as that from photons. Figure 12a) shows the light transmittances before and after 10^{14} neutrons/cm^2 irradiation. Also shown in the figure is the full recovery after annealing and repolishing.
- Most samples suffer a small initial damage, during the first small dose (up to 1 kGy), followed by no further change for up to extremely high doses (0.1 MGy). Figure 12b) shows the typical saturation effect after 1 kGy up to 0.2 MGy and a 10^{12} neutron/cm^2 dose. The damage is also fully recovered after annealing and repolishing.

All of the above observations, along with the fact that the yield of fast UV light is independent of the temperature, confirm that the fast component in BaF_2 is produced by the well-known 'cross scintillation' mechanism [21]). This mechanism is known to occur in crystals with very high radiation resistance, and has been previously studied in other pure fluoride crystals.

The saturation effect which we have observed confirms that radiation damage in BaF_2 is not caused by an intrinsic color center in the bulk material of crystal, such as O^{--} vacancies in BGO [22], but externally-introduced impurities. By controlling the level of impurities in BaF_2, one is thus able to obtain very radiation

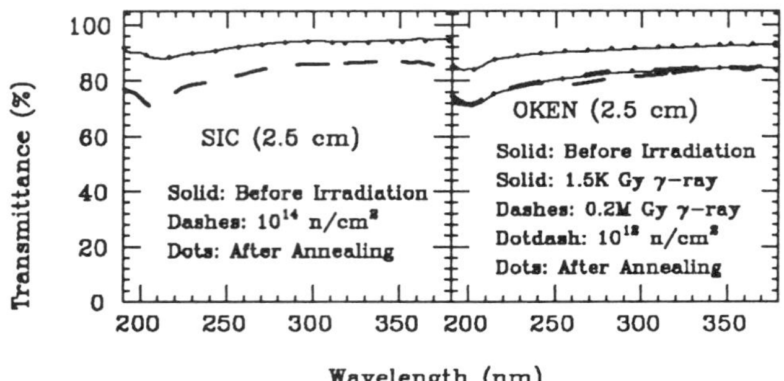

Figure 12: Transmittances before and after irradiation showing a) recovery from neutron damage; and b) typical saturation effect.

hard crystals. We thus conclude that one can readily obtain radiation hard crystals in small quantity.

A further investigation of the correlations between the quantity and type of impurity in the crystal, and the small initial damage will be carried out at SIC and BGRI, Caltech and CMU. It is expected that large size, highly radiation resistant BaF_2 crystals will be produced following this investigation. The two main approaches for ensuring that the mass-produced BaF_2 crystals are radiation-hard are ([23],[3]):

- Purification of the raw materials, down to the sub-ppm level for critical impurities.
- Doping the crystal with rare-earth compounds. This method has been successfully used in L3 to increase the radiation resistance of the BGO crystals in the small angle luminosity monitors by an order of magnitude.

BaF_2 Calibration

Precise, frequent calibration *in situ* is vital to maintain the high resolution of the precision detector. As shown in our GEANT study on BaF_2 resolution [4], the dominant contribution to the resolution of BaF_2 calorimeter is the uncertainty in the calibration.

RFQ Calibration

Our primary calibration will use a novel technique based on a Radio Frequency Quadrupole accelerator (RFQ) [24]. This technique has been developed by Caltech and AccSys over the past six years. As illustrated in Figure 13, the first calibration RFQ will be installed in L3 in February, 1991, to calibrate the L3 BGO electromagnetic calorimeter. A detailed description of the RFQ calibration system can be found in Section A.6 in reference [3]. By bombarding a fluoride target with a 3.85 MeV proton beam, bursts of 6 MeV photons totalling up to 30 GeV equivalent per pulse, for a typical crystal, are produced. The feasibility of this calibration mode has been proven in experimental tests with BaF_2 counters and a 7 × 7 L3 BGO crystal array at an RFQ facility at AccSys Technology. A calibration precision of 0.4% was achieved in the test, with a calibration time of two minutes (also see [7]).

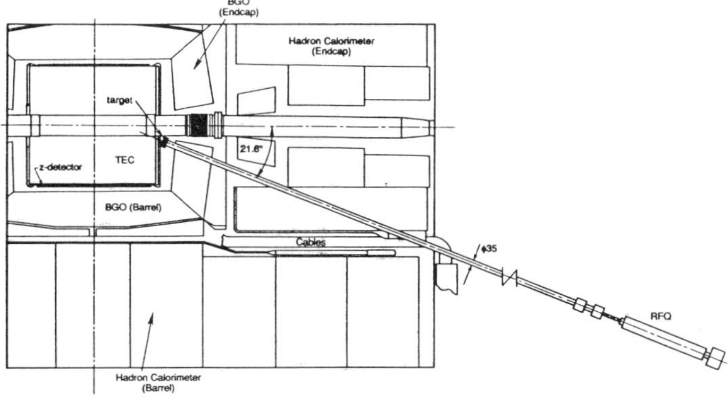

Figure 13: Schematic view of the RFQ system to be installed in L3 in February, 1991.

MIP Calibration

Complementary to the RFQ calibration, we will also use the high flux of minimum ionizing particles (MIPs) produced by the collider as a calibration souce. A MIP passing through a BaF_2 crystal longitudinally would deposit 0.33 GeV energy in the crystal, and would be readout by the crystal with a few percent resolution. With a statistics of a few hundreds tracks, the peak position or the calibration point can be determined to 0.4%. The multiplicity of charged hadrons at the SSC is large enough to provide a calibration within 12 hours during SSC running.

The daily calibrations with these techniques will be confirmed by the measurement and invariant mass reconstruction of inclusive Z^0, Υ and J production, where the vector bosons decay into e^+e^-. With a p_T cut of 10 GeV on the vector boson, a typical crystal in the central calorimeter will see 20 e^+ or e^- from Z^0decay, and 200 from Υ decay [25] for each week of running.

Summary of Resolution

A GEANT simulation was carried out to calculate the intrinsic resolution of the BaF_2 calorimeter. The simulation was calculated for an array composed of 121 (11 × 11) BaF_2 crystals with the proposed size: (3 cm×3 cm) in front, by (5 cm×5 cm). Effects included in the simulation are:

- 250 μm carbon fiber wall between crystals;
- shower leakage because of summing a limited number of crystals;
- 0.30 radiation lengths of dead material, representing the beam pipe, tracker, and carbon fiber mechanical support, at the front of the BaF_2 array.

Table 3 lists the resulting BaF_2 resolution, including the intrinsic contribution, and the other contributions listed above. Assuming a 0.4% precision calibration, achieved, for example, in our RFQ test in 1988 [24], it is clear that the BaF_2 calorimeter will be able to reach the design goal of $1.3\%/\sqrt{E} + 0.5\%$ resolution for electrons and photons.

Table 3: Energy Resolution (%)

E (GeV)	5	10	100	500
Photoelectrons	0.2	0.1	0.04	0.02
Noise	0.8	0.40	0.04	0.008
Calibration	0.40	0.40	0.40	0.40
GEANT	0.67	0.60	0.42	0.36
Total	1.1	0.83	0.58	0.54

π/e Separation

Because of the high uniformity and fine granularity of the proposed BaF_2 calorimeter, one can use the detailed transverse shower information to distinguish an electron from a charged pion. The ability to match the momentum measured in the central track detector, or the energy measured in the hadron calorimeter, with the energy measured in the electromagnetic calorimeter provides an additional powerful rejection.

Table 4: Acceptance of pions & electrons.

E (GeV)	10	100	1000
After	Lateral	Shower	Cut:
π (10^{-2})	32.6 ± 0.4	34.7 ± 0.5	39.4 ± 0.7
e (%)	99.3	99.5	99.5
After	E_{BaF_2}	and P_{trk}	match:
π (10^{-3})	< 0.1	12.3 ± 0.9	388 ± 7
e (%)	97.3	99.2	99.5
After	HCAL	Veto:	
π (10^{-4})	< 1	2.0 ± 1.4	< 2.2
e (%)	97.3	96.4	95.0

The results of a detailed GEANT simulation [3] are summarized in Table 4. As seen in the table, the complete detector will be able to separate isolated electrons and pions with a rejection ratio of $\sim 10^{-4}$.

BaF_2 R&D PROGRAM

The goal of the BaF_2 R&D program is to build a high precision electromagnetic calorimeter prototype, composed of an array of 144 large BaF_2 crystals, and to test and calibrate the array using high energy electron and pion beams, together with a hadron calorimeter prototype, at either BNL, CERN or Serpukhov. The array will also be tested at an RFQ calibration test stand, which will be setup at Caltech or AccSys. The research and development program will extend over two years. It will focus on the following six aspects:

- **BaF_2 crystal mass production:** to produce large size BaF_2 crystals at low cost, which meet the mechanical and transparency specifications and are radiation hard up to 10^5 Gy. We also want to ensure that the cost of the crystals is at or below the goal of \$ 2.5/cc;
- **BaF_2 crystal quality control and radiation hardness tests:** to establish the means of verifying that each crystal meets the specifications; to study UV-reflective crystal coatings, and other surface treatments, to provide uniform light output and high light yield; to study the correlation between radiation resistance of BaF_2 crystals and impurities in the crystal;
- **BaF_2 readout development:** to develop a selective readout which is sensitive only to the fast scintillation component of BaF_2 (0.9 nsec decay time), while fully suppressing the slow scintillation component; to construct prototype preamplifiers and fast shaper cricuits. We will also study the design of the higher-level digital readout.
- **A BaF_2 prototype crystal array:** to construct a BaF_2 array consisting of 144 BaF_2 crystals. The following related engineering problems will be studied:
  - Mechanical design and construction of a prototype section of the carbon fiber mechanical support.
  - Thermal control for stable operation of the readout.
  - Shielding to reduce electronic noise.
- **An accurate RFQ calibration facility:** to build a RadioFrequency Quadrupole (RFQ) accelerator facility which will produce a pulsed, low energy proton beam to bombard a fluoride target. The photons produced by the target will be used to produce an accurate calibration [24].
- **Test beam and cosmic ray (MIP) bench:** to test the prototype BaF_2 crystal matrix with high energy electron and pion beams, to verify the resolution and radiation hardness; a cosmic ray test bench also will be used, to test the feasibility of calibration with minimum ionizing particles.

Production of BaF_2 Crystals

One of the most important aspects of the BaF_2 R&D program is to develop the mass production techniques for large BaF_2 crystals with high quality and low cost. This program has been initiated by the Shanghai Institute of Ceramics (SIC) and the Beijing Glass Research Institute (BGRI) in China. Recently, the Leningrad State Optical Research Institute in U.S.S.R. has also proposed to develop BaF_2 crystals for the SSC. The prices quoted for mass production by these institutes, following the R&D program, are shown in Figure 14. For comparison, the price available from a major commercial vendor – Merck (BDH) [26], is also shown in the figure.

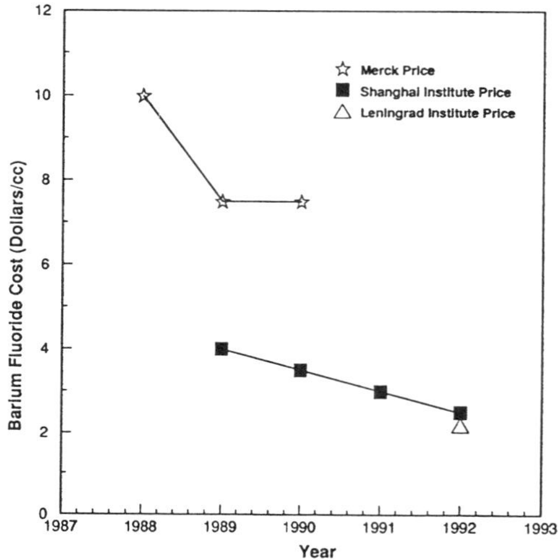

Figure 14: The BaF_2 price (\$/cm^3) for different vendors plotted as a function of time.

Starting from 1989, SIC began developing mass production techniques for BaF_2 crystals, including a new technique involving platinum crucibles (from material on hand from the L3 BGO production) in an inert atmosphere. SIC also invested in a large vacuum furnace facility, and began a collaboration with the Beijing Glass Research Institute (BGRI), which has 6 years of experience in BaF_2 crystal growth with their large vacuum ovens.

BGRI has been working on crystal growth technology for many years. BaF_2 crystals 30 cm in length and 10 cm in diameter has been grown at BGRI by using a modified Bridgeman method in vacuum. This technique may produce a crystal ingot approximating the final shape. The ingots produced by the Bridgeman method are only slightly larger than the final size, to allow for machining. The final growing parameters, using this method, will be chosen following completion of the first (pre-production) series of 49 crystals, which is now underway, and will be completed in the next few months.

The research and development program for BaF_2 crystal mass production at SIC and BGRI will focus on:

- A micro-compositional chemical analyses of impurities in raw materials and crystals;
- The effect of impurities on UV transmittance and radiation resistance;
- The effect of doping on properties of crystals, including radiation resistance;
- Research on pre-processing of the raw materials;
- Optimization of crystal growth technology on large crystals;
- Systematic studies of mechanical processing of BaF_2 crystals;
- Methods of eliminating internal stresses in large BaF_2 crystals;
- Crystal structure and defects studies;

Following the R&D period, SIC and BGRI have planned to mass produce the crystals at the rate of 1000 pieces per month. Production would start at the beginning of 1993 and would be completed in 1995.

Purity of Materials and Chemical Analysis

Raw materials of sufficient purity, and special preparation methods for the oxide powders, were used by SIC to successfully produce the BGO crystals for L3. Similarly, SIC and BGRI have studied the requirements for the BaF_2 powder which is used to grow the BaF_2 crystals. They have determined the following requirements for the production of transparent, radiation resistant crystals:

- The level of harmful impurities, including Fe, Mn, Co, Cr, Cu and other transition metals, must be $\leq 10^{-6}$.
- The total content of less harmful impurities, such as alkali-earth metals, must be $\leq 10^{-4}$.
- O_2 and OH^- must be carefully eliminated from the crystal, in order to avoid radiation damage. Pretreatment of the BaF_2 powder will be used prior to growing, to eliminate these impurities.
- Pb is particularly harmful, as it is associated with a 205 nm absorption band, in the region of the spectrum where the fast scintillation light is emitted. It will be eliminated, down to a level well below 1 ppm.

BaF_2 R&D Schedule: Milestones

Table 5 lists the major milestones of the BaF_2 R&D program.

Table 5: BaF_2 R&D Milestones

Date	Milestone
12/90	Finalize crystal specifications
12/90	Complete 3 prototype readout channels (Radeka design)
1/91	Finish measurements of BaF_2 mechanical properties
2/91	Complete crystal transparency test facility
3/91	Complete prototype readout (BNL design) test
3/91	Completion of crystal light yield test facility
4/91	Complete BaF_2 array support structure design
5/91	Complete crystal dimension measurement facility
6/91	Complete delivery of the first 50 BaF_2 crystal pairs
6/91	Complete radiation resistance test facility
6/91	Complete test of various photosensitive devices
6/91	Finish design of preamp/amplifier
7/91	Complete construction of cosmic ray test bench
8/91	Finish design of buffered readout
9/91	Complete test readout design
10/91	Complete first batch crystal test
11/91	Complete delivery of second 50 BaF_2 crystal pairs
12/91	Complete the first 50 readout channels
1/91	Complete BaF_2 array support structure production
2/92	Complete test of the first 50 channels
3/92	Complete delivery of last 60 pairs of BaF_2 crystals
6/92	Complete test of 110 readout channels
6/92	Complete 144 BaF_2 crystal array production
6/92	Begin production array beam tests
9/92	Determine crystal calibrations with cosmic ray test bench
12/92	Finish beam tests with production array

LIQUID XENON CALORIMETER

Liquid Xenon (LXe) has the advantages of very high intrinsic radiation resistance of the noble liquid, and higher light output than BaF_2 crystals. It is relatively fast (decay time $\approx$ 20 nsec), and it has a radiation length of 2.8 cm. It has therefore been proposed as an alternative technology for the L* electromagnetic calorimeter.

A collaboration of physicists from MIT, Waseda, Aachen, IHEP (Beijing) and several other universities and laboratories in the U.S., U.S.S.R., Japan, France, Germany and China [27] has begun an R&D program to develop an LXe calorimeter for the SSC. The R&D goals, progress and principal technical issues are summarized below.

LXe Detector Concept

The detector conceptual design is shown in Figure 15: 5 m in length and 2.8 m in diameter, divided into $\approx 3 \times 10^4$ cells. The depth of the active volume is 24.5 radiation lengths (68.5 cm of liquid) over most of its surface, but 22 X_0 for $-0.5 < \eta < 0.5$.

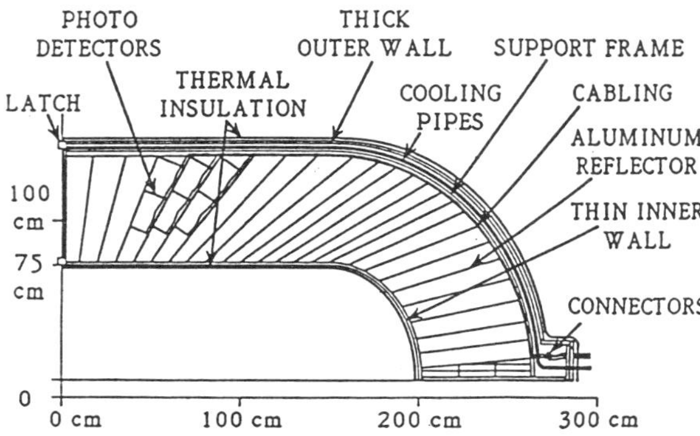

Figure 15: Schematic view of one quadrant of the proposed Liquid Xenon calorimeter.

The LXe calorimeter uses multiple layers of UV-photodiodes (without window) and fast amplifiers, developed and manufactured by MIT and Waseda, submerged directly in the liquid. The goals include detecting photons and electrons with an EM energy resolution of < 0.7%, and with a spatial resolution of their impact points to $\sim$ 1.5 mm.

LXe R&D Progress

Progress in the development of the LXe calorimeter has been made in several areas:

- MIT and Waseda have successfully developed 5 cm diameter, low capacitance photodiodes on a 400 μm Si wafer covered with a 20 nm gold mesh. The photodiodes have a quantum efficiency of 50% for the 170 nm light emitted by the liquid Xenon. They are now ready to produce the first 100 diodes, which are planned to be used in a beam test with a 5x5 cell LXe calorimeter prototype in 1991.
- Fast amplifiers also have been developed, with a 10 nsec peaking time. The aim is to have 100 amplifier channels (200 FADC channels) ready in 1991 for a 5×5 cell LXe prototype, and 400 channels (800 FADC's) ready for an 11×11 cell LXe prototype in 1992.
- Calibration with 5.5 MeV α particles gives a 17% resolution (shown in Figure 16) when used with the fast amplifiers. The measured pulse height of corresponds to 4×10^4 photoelectrons, with a resolution dominated by the electronic noise of the amplifier. The temperature dependence of the scintillation yield was measured to be approximately $-0.4\%/^\circ$C in the temperature region between -110°C and -75°C.

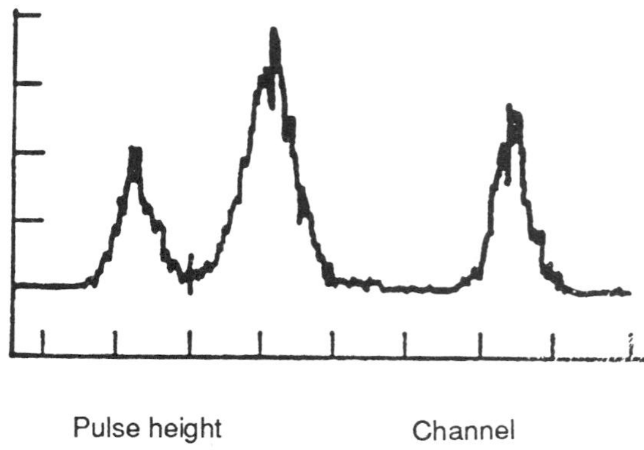

Figure 16: Pulse height distribution of 5.5 MeV α's in LXe, measured with fast amplifiers. The signals are (from the left): pedestal, α signal, and a test pulse used to measure the number of photoelectrons in the signals.

- An energy resolution of $\delta E/E < 0.7\%$ for heavy ions at $E > 1.6$ GeV with fast amplifiers has been achieved in beam tests at KEK and Riken. A typical result, with 2.5 GeV ^{40}Ar ions, is shown in Figure 17. Heavy ion tests with a full length LXe prototype cell are scheduled to start in November, 1990.
- Monte Carlo studies have shown that thin UV aluminum reflecting foils may be used, together with a diode at either end of a cell, to achieve uniform light collection.
- Monte Carlo results have also shown that scintillating liquid xenon detectors might be able to achieve compensation by using fast signals from low velocity nuclear spallation fragments, such as protons, resulting in an electron-to-pion response ratio $e/\pi = 1 \pm 0.1$. It is planned to experimentally confirm this by beam-testing a prototype detector together with a compensating hadron calorimeter.

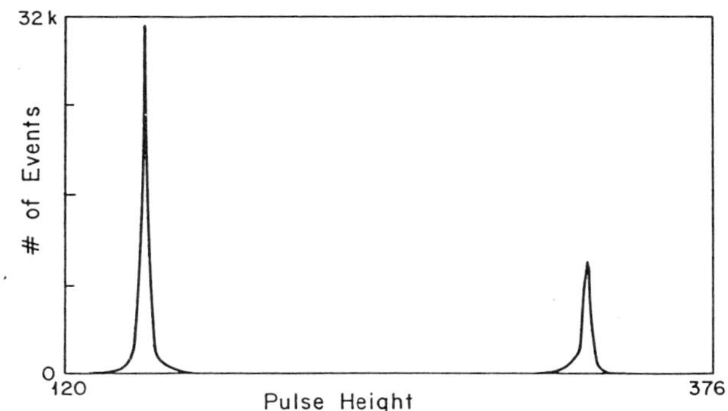

Figure 17: Pulse height distribution obtained with 2.5 GeV ^{40}Ar ions in LXe. The peak on the left is the pedestal. The signal is the peak on the right.

- It has been shown that the gains of the electronics (ADC and amplifiers) can be calibrated using direct α hits on the diodes, which produce the equivalent of 150 MeV energy in liquid xenon. The data shows the peak position for an individual hit can be measured to an accuracy of 1.5% with a fast amplifier.
- The attenuation length in LXe of the 170 nm scintillation light from α's has been measured, by varying the distance between the α source and the photosensor. The light attenuation was found to be small in a test setup, over a distance of 1 to 9.5 cm. The data were consistent with the variation of light collection expected from light acceptance and 82.5% reflectance at the metal walls. It is planned to determine the attenuation length to higher accuracy with a full size LXe cell in the November, 1990 heavy ion beam test.

Prototype Cell at MIT for Cosmic μ's

At MIT, a full length (65 cm) LXe prototype chamber with foam insulation has been completed, which will be tested with cosmic ray muons. It will also provide information on the practicality of foam versus vacuum insulation, and on the requirements for purity of the liquid.

Prototype Cell for Heavy Ion Tests

A full size 65 cm long prototype cell with vacuum insulation has also been constructed and tested at MIT. This chamber has 7 beam windows, and a diode at each end. The two diodes will be read out in coincidence, to measure the uniformity over a distance of 65 cm under realistic conditions, using heavy ion beams at Riken entering each of the 7 thin windows.

Calibration

Because of the high light output per unit energy deposited in the liquid Xenon, one can use α's, cosmic muons, or the RFQ system (see above) to track the calibration of the detectors after they have been calibrated in test beams. It is planned to mount one source for each cell in the liquid, and one source directly on the diode to monitor the diode and electronics chain. As in the case of BaF_2 , the calibration process can be monitored by reconstructing inclusively-produced vector bosons.

LXe R&D Milestones

The major milestones in the planned LXe R&D program are as follows:

- Fourth Quarter 1990 :
  1. Start the construction of a 5×5 cell prototype
  2. Radiation hardness tests of photodiodes at liquid xenon temperatures
  3. UV reflector development
- Third Quarter 1991 : Beam test of the 5×5 prototype
- Fourth Quarter 1991 :
  1. Start the construction of the 11×11 cells as one real sector of the EM calorimeter
  2. Start engineering design of the full "Olive" detector
- Third Quarter 1992 : Beam test of the 11×11 prototype together with a compensated hadron calorimeter to measure the e/π response ratio.

Liquid Xenon Mass Production

Four commercial companies: Air Liquide, Matheson, Spectra Gases and Union Carbide, have submitted letters stating that each of them can produce up to 15 m^3 of new LXe by 1999 at a price of about 2.5 M\$/m^3 of LXe, if there is sufficient guarantee/encouragement from SSC/DOE to induce any one of them to start new production of xenon in 1991. The current procurement plan for xenon is shown in Figure 18.

TECHNICAL CHALLENGES for PRECISION ELECTROMAGNETIC CALORIMETERS

The leading technical challenges for the BaF_2 and LXe precision electromagnetic calorimeters are:

- **BaF_2 :**
  Mass production of high quality (highly transparent and radiation hard) crystals at low cost.

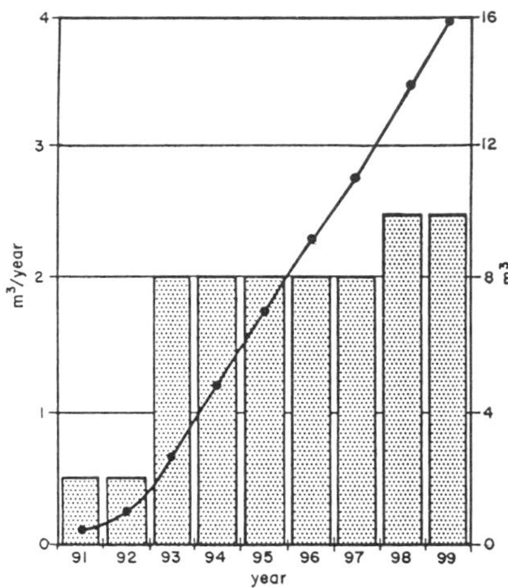

Figure 18: Xenon procurement plan 1991–1999. Xenon procurement in m^3 per year (left scale) and total accumulated to date (right scale).

- **Liquid Xenon:**
    - o Mass production at low cost, in time for SSC startup.
    - o Silicon photodiode radiation damage, which may be helped by operation in the cryogenic liquid.
    - o UV (170 nm) transmittance and light collection uniformity.
    - o Large system design, with a cryostat consistent with the parameters of a large SSC experiment.

Other technologies, which are also being considered for precision electromagnetic calorimetry, are at a more generic stage of development than either BaF_2 or LXe. In order to be considered as serious contenders, they must overcome the following general problems:

- **Fine-sampling calorimeters:** Energy resolution.
- **Other crystals or scintillating glasses:** Radiation damage and speed.
- **Other cryogenic liquids:** Speed.
- **Cherenkov radiators:** Light yield and radiation resistance.

All of the above technologies must meet the challenge of mass production at low cost, and precision calibration.

CONCLUSIONS

- Over the past two years, many studies for the SSC (and for the LHC) have established that high resolution measurements of electrons and photons will be an extremely important part of the SSC physics program.
- Many promising technologies for high resolution electromagnetic calorimetry are therefore under study.
- Because of their superior intrinsic properties, and substantial progress in R&D, there are two leading technologies: BaF_2 crystals and LXe.
- Prototype BaF_2 and LXe calorimeters are under construction, with first beam tests planned for 1991.

ACKNOWLEDGEMENTS

I wish to thank Profs. Z.Y. Wei, Z.W. Yin, and Drs. R.Y. Zhu, A. Gurtu, H. Ma, C. Woody, T. Matsuda and F. Takasaki, for their collaboration. Many useful discussions with Prof. M. Chen, and Drs. P. LeCoq, M. Schneegans, R. Hamm, S. Majewski, G. Charpak, D. Schigocki, J. Virdee, W. Wallraff, and S. Ting are gratefully acknowledged.

References

[1] CERN Green Book, *ECFA Workshop on LEP 200*, Aachen, September, 1986.

[2] J. Gunion et al., *The Higgs Hunter's Guide*, UCD Preprint, 89-4 (1989).

[3] Caltech *et al.*, *A Precision BaF_2 Crystal Calorimeter for the Superconducting Supercollider*, Subdetector R&D Proposal to the SSC Laboratory, September, 1990.

[4] L* Collaboration, *Expression of Interest to the Superconducting Super Collider Laboratory*, May 1990.

[5] J. Virdee, Talk at the ECFA/LHC Workshop, Aachen, October, 1990.

[6] J. Ellis. T. Sjöstrand and F. Zwirner, discussion at CERN, May, 1990.

[7] R. Y. Zhu, *A Precision Barium Fluoride Crystal Calorimeter for the Superconducting Supercollider*, talk presented at this conference.

[8] D. Dicus and S. Willenbrock, Phys. Rev. **D37** (1988) 1801.

[9] H. Benson and T. Sjostrand, *A Manual to the Lund Monte Carlo for Hadronic Processes*, PYTHIA version 5.3, November, 1989.

[10] H. Newman, *Search for New Particles in L3 at LEP: Beyond the Standard Model*, talk presented at the XXVth ICHEP, Singapore, August, 1990.

[11] H. Ma *et al.*, *Development and Precise Calibration of a High Energy Barium Fluoride Crystal Array*, **CALT-68-1517**, 1988.

[12] M. Lavel *et al.*, Nucl. Instr. and Meth. **A206** (1983) 169;
P. Schotanus *et al.*, Nucl. Instr. and Meth. **A259** (1987) 586;
P. Schotanus *et al.*, IEEE–NS **34** (1987) 76.

[13] S. Majewski and D. Anderson *et al.*, Nucl. Instr. and Meth. **A241** (1985) 76;
A. J. Caffrey *et al.*, IEEE Trans. Nucl. Sci. **NS-33** (1986) 230;
S. Majewski *et al.*, Nucl. Instr. and Meth. **A260** (1987) 373;
G. Charpak *et al.*, CERN Preprint, CERN-EP/90-41 (1990).

[14] D. F. Anderson *et al.*, Nucl. Instr. and Meth. **A228** (1984) 33.

[15] G. Charpak *et al.*, CERN-EP/90-41.

[16] R.Y. Zhu, *A High Resolution Barium Fluoride Crystal Calorimeter*, **CALT-68-1656**, 1990, in *"Super Collider II"* and *"Super Collider I"*, Plenum Press, edited by McAshan, (1989) 573.

[17] P. Schotanus *et al.*, IEEE–NS **34** (1987) 272, and Nucl. Instr. and Meth. **A281** (1989) 162.

[18] C.L. Woody *et al.*, IEEE–NS **36** (1989) 536.

[19] S. Suzuki, Hammamatsu Corp., private communication.

[20] Z. Wei *et al.*, *A Study on Light Yield and the Surface Treatment of Barium Fluoride Crystals*, **CALT-68-1659**, 1990, submitted to Nucl. Instr. and Meth.

[21] See for example, P. LeCoq, talk presented at the ECFA/LHC Meeting on Detector R&D, Aachen, October 1990.

[22] Z. Wei *et al.*, *Radiation Resistance and Fluorescence of Europium Doped BGO Crystals*, **CALT-68-1628**, 1990, accepted by Nucl. Instr. and Meth.;
R.Y. Zhu *et al.*, *A Study of Radiation Damage in Doped BGO Crystals*, **CALT-68-1392**, accepted by Nucl. Instr. and Meth.;
T.Q. Zhou *et al.*, Nucl. Instr. and Meth. **A258** (1987) 464.

[23] SIC and BGRI, *Letter of Intent for Production of the Full Quantity of BaF_2 Crystals for the L^* Detector*, September, 1990.

[24] H. Ma *et al.*, Nucl. Instr. and Meth. **A274** (1989) 113;
H. Ma *et al.*Nucl. Instr. and Meth. **A281** (1989) 469
R.Y. Zhu *et al.*, in *Super Collider I*, Plenum Press, edited by McAshan, (1989) 587.

[25] H. Bergstroem *et al.*, Penn State Preprint PSU/TH/63, March, 1990.

[26] Merck is a subsidiary company of BDH Limited, England, in Germany.

[27] M. Chen *et al.*, *Liquid Xenon Prototype Detector Subsystem Proposal*, 1989 and 1990;
M. Chen, talk presented at this conference.

COMPENSATING SILICON CALORIMETRY FOR THE SSC/LHC COLLIDERS

SICAPO Collaboration

E. Borchi [1], C. Leroy [3], P.G. Rancoita [2], A. Seidman [2,a]

[1] INFN-Florence and at University, I-50125 Florence, Italy

[2] INFN-Milan, I-20133 Milan, Italy

[3] Université de Montréal, Montréal (Québec) H3C 3J7, Canada

[a] and Tel-Aviv University, 69 978 Tel-Aviv, Israel

Abstract

The compensation condition ($e/\pi = 1$) which mainly determines the performances of the silicon calorimeters to be used at the SSC/LHC is discussed. This condition is achieved via the local hardening effect. Moreover it can be also realized through a filtering effect.

Introduction

Future experiments at the Superconducting Super Collider (SSC) and the Large Hadron Collider (LHC) will provide the opportunity, to search for new physics: for instance, to search for heavy leptons, supersymmetric particles, the Higgs particle(s), and the additional vector bosons predicted in many possible extensions of the Standard Model.

The possibility to achieve these research programs will rely on highly performant detection systems. High luminosities and high multiplicities in a high-level radiation environment will characterize the operation of these colliders and will challenge the performances of the hadron calorimeter which will constitute the core element of the central detection system. For instance, the positive identification of neutrinos will request the condition that the amount of missing energy is larger than the uncertainty on the visible energy; this means that the multiparticle energy resolution must be better than $K/\sqrt{E}$, where $K = E_\nu/\sqrt{E_{tot}}$ [1]. Since an appreciable fraction of the energy of the jet can be carried by a 'leading particle', the instrumental resolution for jets, however, approaches that for single particles. This is also critical for the detection of narrow large-mass resonances decaying into jets (like W and Z). If a energy resolution of $50\%/\sqrt{E}$ can be maintained for jets, W and Z particles decaying into jets could be distinguished by their mass event by event (whereas with a resolution of $100\%/\sqrt{E}$ for instance, the mass peaks of centrally produced W and Z would overlap).

In summary, the physics goals, and the processes that limit the effective energy resolution, fix the required calorimeter energy resolution at $\sim 50\%/\sqrt{E}$ [1]. That energy resolution can be obtained by using an hadron calorimeter with silicon as active medium. Silicon has the follwing characteristics which make it a good candidate for facing the exceptional experimental conditions encountered at the future colliders. Silicon indeed allows compact and flexible construction, fine granularity, fast charge collection, easy calibration, and has good radiation hardness.

In addition such a calorimeter with silicon readout allows the achievement of the compensation condition ($e/\pi = 1$) requested to obtain the energy resolution of $\sim 50\%/\sqrt{E}$. This last point is at the core of the discussion presented here in the framework of the SICAPO program.

e/mip, h/mip and e/π ratios

The electromagnetic processes at work during cascading have large cross sections, and nuclear interactions play no role in the energy deposition. Therefore the electromagnetic cascading depends essentially on the density of electrons in the absorber medium and is precisely described [2] by Quantum Electro-Dynamics (QED). As a consequence the behaviour of the electromagnetic showers in a calorimeter is well understood and so is the response of calorimeters to electromagnetic cascading. In the process of energy deposition, the incident energy is shared among the large number of secondary particles created by bremsstrahlung and electron pair production.

The hadronic shower is propagated through a succession of various inelastic interactions leading to multiparticle production characterized by a multiplicity increasing logarithmically with the available energy ($\approx$ $ln\ s$). From the primary energy, half is carried away by leading-particles and the second half absorbed in the production of secondaries, mainly pions and nucleons.

The fact that only a small fraction of the energy is deposited in the active layers and contributes to the calorimeter signal is the origin of sampling fluctuations (σ_{samp}). Since it is mainly generated by the fluctuation in the total number of charged particles crossing the active layers, the contribution of this source of fluctuations is proportional to the square root of the thickness of the absorber plates (t_{abs}).

A large amount of the deposited energy in a calorimeter goes in breaking nuclei (binding energy) or in low energy neutrons and therefore is partially not visible. In addition pion and muon decays produce secondary particles like ν which are undetected. The amount of energy converted into excitation or break-up of the nuclei, fluctuates from event to event. These intrinsic fluctuations (σ_{intr}) contribute to the energy resolution. Therefore, the energy resolution of the calorimeter can be generally written as:

$$\sigma/E = C(\sqrt{t_{abs}})/\sqrt{E}\ +\ \phi((e/\pi)-1)$$

where $C = (\sigma^2_{intr} + \sigma^2_{samp})^{1/2}$.

The e/π signal ratio is defined as:

$$e/\pi = \frac{(e/mip)}{(e/mip).f_{em} + (h/mip).(1-f_{em})} \quad (1)$$

The ratio e/mip, appearing in (1) measures the energy deposited in the sen sitive part of the calorimeter by electron and photon showers relative to minimum ionizing particles mip. The energy shared (E_s) by the minimum ionizing particle in the silicon detectors is given by

$$E_s = \frac{(dE/dx)_{Si}.X_{Si}.E}{(dE/dx)_{Si}.L_{Si} + \Sigma(dE/dx)_i.L_i} \quad (2)$$

where, X_{Si}, L_{Si}, L_i, $(dE/dx)_i$, and E are the depletion depth, the thickness of the Si detector, the thickness of the absorber i, the average energy loss per unit of length in the absorber i, and the incoming particle energy, respectively.

The e/mip ratio is therefore defined as

$$e/mip = \epsilon_{vis}(e)/E_s$$

where $\epsilon_{vis}(e)$ is the visible electromagnetic energy measured in the calorimeter, and E_s is given by (2).

A ratio π/mip can be defined and measures the energy deposited by hadrons relative to mip.

The ratio π/mip is defined as:

$$\pi/mip\ =\ \epsilon_{vis}(\pi)/E_s$$

where $\epsilon_{vis}(\pi)$ is the visible hadronic energy measured in the calorimeter.

Normally, after the primary interaction, decays of hadronic resonances created during the degradation of the energy of the incident hadrons and charge exchange reactions, produce π^0 (mainly) and η which will propagate electromagnetically without any further nuclear interactions and consequently deposit their energy in the form of electromagnetic showers. As a result any hadronic shower has a purely hadronic and a purely electromagnetic component. The size of this electromagnetic component is largely determined by the production of π^0 (and η) in the first interaction.

In the extreme case where there is no generation of an electromagnetic component in the hadronic shower, similarly to π/mip, one can define, a h/mip ratio:

$$h/mip\ =\ \epsilon_{vis}(h)/E_s$$

where $\epsilon_{vis}(h)$ is the visible pure hadronic energy measured in the calorimeter in absence of electromagnetic component in the shower.

Therefore the total visible energy can be expressed now as:

$$\epsilon_{vis}(\pi) = E_s(f_{em}(\frac{e}{mip})+(1-f_{em})(\frac{h}{mip})) \quad (3)$$

where f_{em} is the average fraction of the converted electromagnetic energy resulting from the decays of π^0's, η's and vector mesons produced during the hadron cascade. The average fraction converted into electromagnetic shower increases with energy. For instance, the total number of π^0 in a shower initiated by π^+ (in Fe) slowly increases with energy as [3]:

$$n_{\pi^0} \approx 5\ ln\ E(GeV) - 4.6$$

Since the electromagnetic fluctuations are partially compensated by the energy dissipated by relativistic charged particles and by fast protons, f_{em} is expected to increase also smoothly with the incident energy [4]. Eq. (1) can be rewritten as:

$$e/\pi = (e/h)/[1 - f_{em}.(1 - e/h)] \qquad (4)$$

Eq. (4) shows that e/h = 1 obviously implies $e/\pi = 1$ and vice versa.

The nuclear effects, which dominate pure hadronic showers, have no counterpart in electromagnetic showers. Therefore, the calorimeter response to the electromagnetic and non-electromagnetic part of the hadronic shower is different, with the consequence that the electromagnetic (e) to hadronic (π) signal ratio $e/\pi \neq 1$. The performance of hadron calorimeter is then determined by their relative response to the electromagnetic (e) and hadronic (π) shower, measured by the e/π signal ratio.

The local hardening effect

The equalization of the response of an hadron calorimeter with silicon readout to electromagnetic and hadronic showers can be achieved by suppressing the electromagnetic response of the calorimeter when high-Z (Pb, W, U) absorber plates are used. This suppression mainly occurs via multiple scattering when the silicon detectors are inserted in between thin low-Z plates (G10, polyethylene). The thickness of these low-Z plates must be negligible in radiation length compared to that of the high-Z absorber plates.

A large fraction of the initial electron energy goes into the production of low-energy photons during the shower development. These soft photons interact almost exclusively in the absorber and most of them will transfer their energy partly (Compton effect $\sim$ Z of the absorber material) or totally (photoelectric effect $\sim Z^5$ of the absorber material) to the electrons of the absorber. As a consequence, low-energy photons from the electromagnetic shower development may convert into electrons sufficiently close to the surface of the absorber plate for them to escape and normally to contribute to the measured signal. These low-energy electrons may be multiply-scattered. In a high-Z material, the low-energy electrons form larger angles α with respect to the shower axis than in low-Z material:

$$< cos\ \alpha > \approx cos(21 MeV/\epsilon\pi)$$

where ϵ is the critical energy as given by [5]:

$$\epsilon = B(ZX_0/A)^a\ [MeV] \qquad (5)$$

A is the atomic mass, B = 2.66 and a = 1.11. The radiation length X_0 (in g/cm^2) i.e. the mean-path-length of an electron in a material (A,Z), is given approximately by [3]:

$$X_0 \approx 180\ (A/Z^2)\ [g/cm^2] \qquad (6)$$

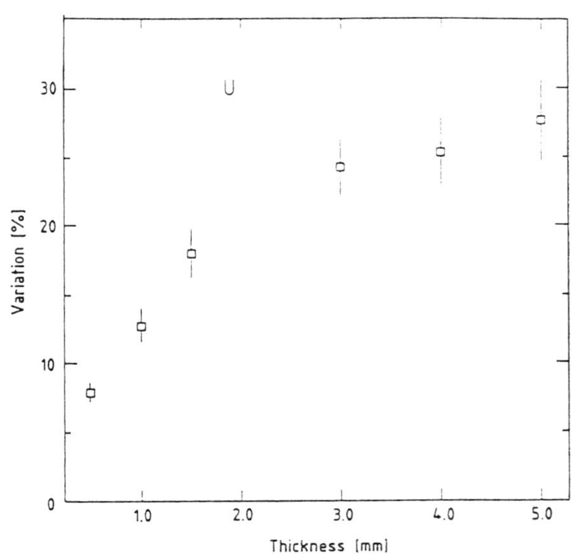

Figure 1. The variation (in %) of the response of a Si/U calorimeter (see ref. 6) as a function of the thickness of G10 absorbers located at the rear side (only) of the detectors.

The number of large-angle electrons (decreasing as ϵ^2) is large in high-Z material and small in low-Z material. Multiple scattering increases the path-length of the electrons: their path-length, d, becomes $d' \equiv d/ < cos\alpha >$. The net effect of multiple scattering is to create a flux of backward-going electrons in the absorber. These soft electrons can be absorbed by the G10 plates inserted between the active and absorber planes and, therefore, are responsible for the decrease of the visible energy: the local hardening effect [6]. It was observed, for instance, that the response of a Si/U calorimeter can be reduced by 29% in the case of the 5.0mm G10 absorber located at the rear side (only) of the detectors (see fig. 1). This suppression leads to a reduction by about 20% of the ratio of the electromagnetic to the non-electromagnetic shower component. The degree of hardening depends on the thickness of the low-Z absorber (small so far in radiation length compared to the high-Z material) allowing the fine tuning of the calorimeter response to electromagnetic showers. In a first approximation, the e/mip ratio for an absorber i behaves as:

$$e/mip \;\approx\; (d'_{Si}/d_{Si})/(d'_i/d_i)$$

Any multiple scattering for a mip and the energy loss in the active medium of the calorimeter have been neglected. It shows that e/mip clearly depends on the ratio d'/d from the multiple scattering. Since $d'/d > 1$ in a high-Z absorber and $d'_{Si}/d_{Si} \approx 1$, one has $e/mip < 1$.

The filtering effect

For a given absorber σ_{samp} can be easily modified and eventually decreased by changing the frequency of the sampling (ie changing the number of active plane per interaction lengths (λ) and/or changing the ratio between the relative thicknesses of the readout and the absorber [7]. σ_{intr} is sensible to the number of neutrons released by the absorber per GeV of incident energy [7] and therefore can be minimized by an adequate choice of the absorber. For instance $\sigma_{intr}/E = 45\%/\sqrt{E}$ for an uranium absorber and $18\%/\sqrt{E}$ for a Fe absorber.

Having an energy resolution of $\sigma/E = 50\%/\sqrt{E}$, as a goal, leads to the concept of a calorimeter with a Fe dominated absorber and a fraction of high-Z metal. This create a situation where the compensation condition has to be achieved in a silicon calorimeter with a combination of low-Z (Fe) and high-Z (the choice is Pb) as the passive medium. In that case, the thickness of the low-Z material is no longer negligible compared to the thickness of the high-Z material when expressed in radiation length. Interspersing low-Z layers in a high-Z material lead to sizeable effects. The combination of Pb and Fe leads to an electromagnetic shower energy transformation effect, generated by a sharp transition from small to large values of the critical energy (ϵ = 7.4 MeV and 21.0 MeV for Pb and Fe, respectively). Because the value of ϵ coincides with that value of the electron energy below which ionization energy loss starts to dominate the energy loss by bremsstrahlung, the increase of the critical energy of the absorber favors the ionization energy loss with respect to the energy loss by radiation. The energy spectra of incident electrons become softer when moving from a high-Z absorber to a low-Z (Fe) absorber. The low-Z (Fe) part of the absorber produces a filtering effect [8] on the electrons. As a result the energy of the shower is transformed and the response of the calorimeter to the incoming shower is dramatically modified.

The degree of lowering of the critical energy obviously depends on the fraction of Fe and Pb in the absorber [8]. fig. 2 shows the e/mip ratio as a function of the Pb fraction, f, in the p assive absorber (f $\equiv$ $L_{Pb}/(L_{Pb}+L_{Fe})$) for various calorimeter configurations (L_{Fe} = $1.99X_0$). For the configurations, FePb-Si-FePb and PbFePb-Si-PbFePb, where the forward-generated electrons in Pb can enter the silicon detectors directly, there is an increase of the value of the e/mip ratio up to f $\approx$ (5 – 8)%. For the PbFe-Si-PbFe configuration, where the forward-generated electrons in Pb enter now the Fe absorber, e/mip is steadily decreasing and equal to 0.60±0.02 at f = 0.36.

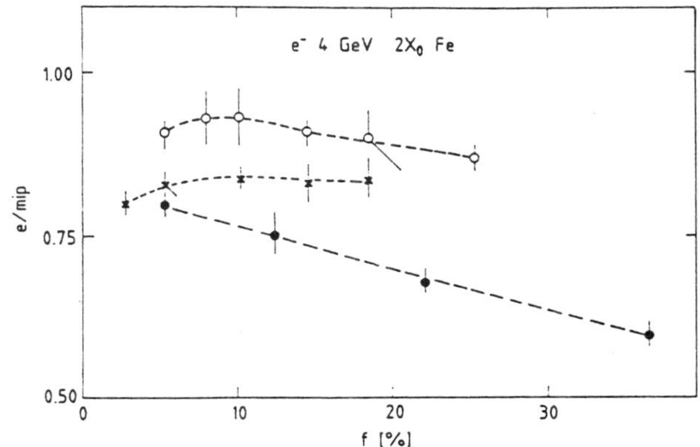

Figure 2. e/mip as a function of Pb fraction, f, in the passive absorber (f $\equiv$ $L_{Pb}/(L_{Pb}+L_{Fe})$; L_i is the thickness of the absorber i) for various calorimeter configurations (Fe is 1.99 X_0 thick) and for 4 GeV incoming electron energy: (•) PbFe-Si-PbFe, (×) FePb-Si-FePb, and (o) PbFePb-Si-PbFePb.

In summary, the filtering effect is obtained when combinations of low-Z and high-Z materials are used as absorbers. The action of Fe on Pb, by modifying the critical energy during the electromagnetic shower development, has a net effect to decrease the number of electrons (the probability of energy losses by collisions is increased) for the forward going electrons. The effect depends both on the Pb fraction (f) (since the fraction of low energy electrons is proportional to f) and on the Fe sampling thickness (because this thickness defines the maximum absorbable electron energy).

Compensating calorimetry

An experimental program has been conducted by the SICAPO Collaboration at the CERN-PS with a calorimeter having silicon as active medium and using uranium as absorber in view to demonstrate the achievement of the compensation condition. The results [6,9] are summarized in section 5.1. Another program is in progress by the same collaboration in order to demonstrate the possibility to achieve the same goal with a full size Si/Fe + Pb hadron calorimeter [8].

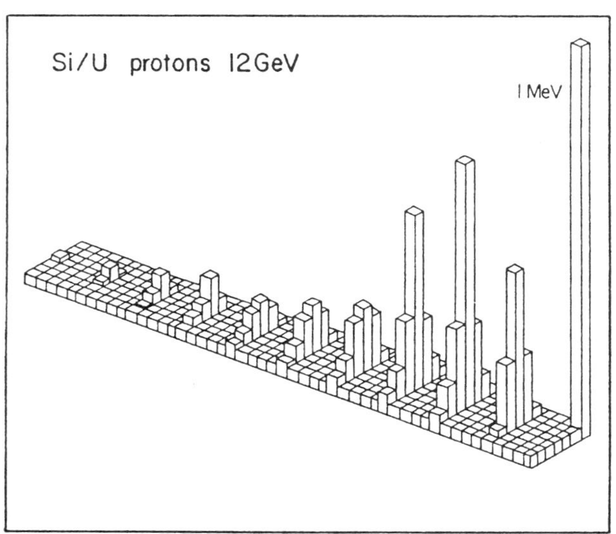

Figure 3. The measured visible energy for 12 GeV protons as a function of the calorimeter depth and the strip location. Each strip is, on average, 36 cm long and 4 cm wide. The total depth is 50 cm and the lateral coverage is 28 cm.

The calorimeter used to perform these studies contains aborber plates up to a 60×60 cm^2 size. The length available is 4 m along the beam direction, and therefore easily accomodates 10 λ (see (9) below) of absorber (U, Fe, Pb). The silicon mosaics can be located at any position. The readout electronics is inside the calorimeter and organized in readout units of 24 preamplifiers each. Each preamplifier is associated to 5 (or 5 groups of detectors) located in 5 subsequent mosaics along the beam direction.

Absorber made of Uranium

The silicon mosaics have a fan-out supporting structure of G10 0.2 and 1.0mm thick on the rear and front side respectively. The detectors are trapezoids with an active area of 28 cm^2 and are operated at a depleted layer width of 200 μm. The active area of a mosaic is about 524 cm^2 which provides a transversal area equivalent to a circle of radius 1.23 λ_U. Two mosaics with a larger active area have been build and used to measure the longitudinal and lateral shower development of the hadron shower (see fig. 3). As already mentioned, for standard mosaic area the lateral losses are about 5%.

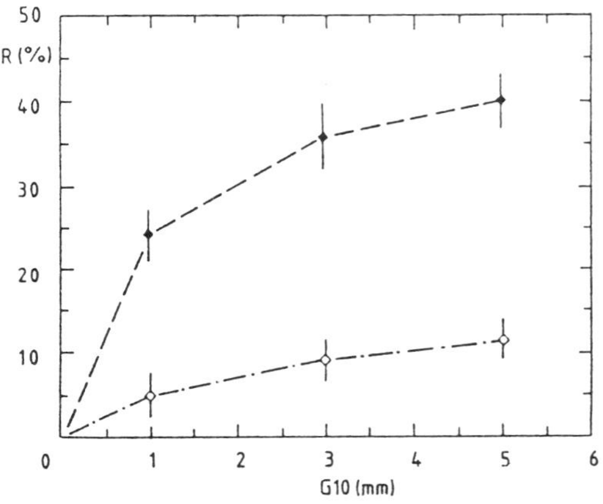

Figure 4. Reduction of the visible energy sensed by the calorimeter for incoming electrons (♦) and protons (◇) versus the thickness of the G10 plates (in addition to the two mosaic supports 0.2 and 1.0 mm thick, respectively) inserted in the front and at the rear of the silicon mosaics. The lines are to guide the eye.

Evidence for the achievement of the compensation condition has been found with a Si/U hadron calorimeter by using the local hardening effect [9]. For instance, as shown in fig. 4, the effect of 5 mm thick G10 plates, inserted in front and at the rear side of the Si readout mosaics, is to reduce the electromagnetic visible energy by about 40% while the hadronic visible energy is reduced by about 12% only. This different behaviour for electrons and protons is the essence of the realization of the compensation condition in Si/U hadronic calorimetry. At these incoming proton energies, the average fraction f_{em} of the converted electromagnetic energy in a hadronic shower is about 25-30%. The part of the visible energy that is due to the electromagnetic component of the shower is expected to be reduced by the local hardening effect when low-Z (G10) plates are inserted. However, the reinteractions of the hadronic particles in these plates increase the ionizing hadronic component of the shower, since the A-value of G10 is lower than that of uranium. Thus, when the sampling fraction (defined as the absorber material thickness in units of interaction lengths in between two active planes) is increased by the additional G10 plates, the expected decrease (about 5.7% with 5 mm thick G10 plates at the front and at the rear side of the Si mosaics) of the part of the visible energy part that is due to the pure hadronic component of the shower is largely attenuated. Consequently, the addition of low-Z G10 plates allows the tuning of the e/π ratio, mainly by attenuating the calorimeter response to the electromagnetic component of the hadronic shower.

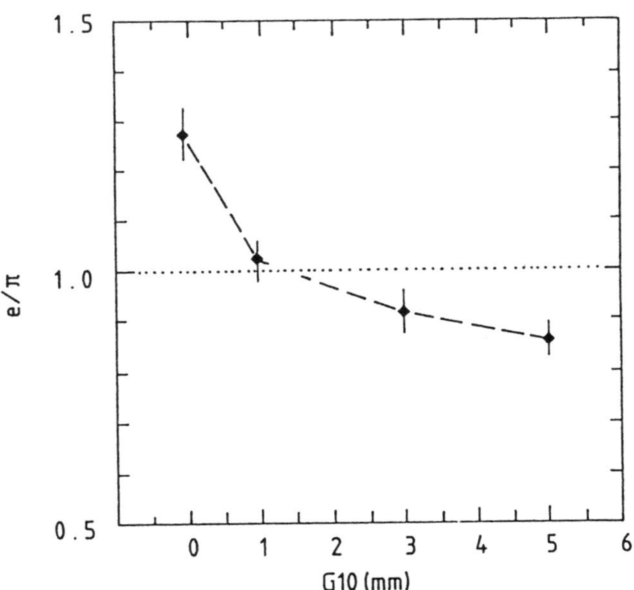

Figure 5. Measured values of the e/π ratio (see text) versus the thickness of additional G10 plates. The line is to guide the eye.

In fig. 5, the e/π values (from an average of measurements at incoming particle energies of 8, 10, and 12 GeV [9]) are shown versus the thickness of the G10 plates. The thickness of the low-Z G10 plates to be located in front and at the rear side of the Si mosaics in order to achieve the compensation condition (e/π = 1), is estimated [9] to be $t_g = (1.2 \pm 0.2)$ mm. The local hardening effect has allowed the value of the e/π ratio to be modified by about 34%, when changing to a calorimeter configuration where 5 mm thick G10 plates are inserted both at the front and at the rear of the silicon detectors.

Absorber made of a combination of Fe & Pb

The 51 silicon mosaics used in the Si/Fe + Pb calorimeter are made of 400 Si detectors, with a 2×2 cm^2 area each and are fully depleted at 400 μm. The leakage current is typically $\approx$ 120 nA. The detector are in groups by 10 (in parallel on the mosaics). The Fe + Pb absorber thickness between the mosaics in 2.1 cm. The final calorimeter depth will be 6.25 λ (see (9) below) which will allow to fully contain 50 GeV protons.

The ability to modify $(e/mip)_{Fe+Pb}$ implies the possibility to tune the e/π ratio with the fraction of Pb in the absorber. Using (3) and (4), the visible energy can be written as:

$$\epsilon_{vis}(\pi) = (e/\pi).[\frac{(h/mip).E_s.(1-f_{em})}{(1-(e/\pi)f_{em})}] \quad (7)$$

and in the case of an absorber composed of Fe and Pb, h/mip is given by:

$$(1/L).[(\frac{L_{Pb}}{\lambda_{Pb}})(\frac{h}{mip})_{Pb} + (\frac{L_{Fe}}{\lambda_{Fe}})(\frac{h}{mip})_{Fe}] \quad (8)$$

with $L \equiv L_{Pb}/\lambda_{Pb} + L_{Fe}/\lambda_{Fe}$.

λ is the nuclear interaction length and is given by:

$$\lambda = \frac{A}{(N_{av}\rho\sigma_i)} \; [cm] \approx 35 \; A^{1/3} \; [g/cm^2] \quad (9)$$

where A, N_{av}, ρ, and σ_i, are the mass number, the Avogadro number, the density of the material and the proton inelastic cross section, respectively.

The estimates of h/mip from Monte Carlo predictions [10] allow to compare ϵ_{vis} obtained from $(h/mip).E_s$ (eq. (7) when the compensation condition e/π = 1 is achieved) with ϵ_{vis} measured for the PbFe-Si-PbFe configuration.

The comparison provides the exact proportions of Pb and Fe that are necessary to achieve the compensation condition [1,8].

In fig. 6 we show the reduction of the electromagnetic shower energy, $\Delta\epsilon_{vis}/\epsilon_{vis}$, which is due to the local hardening effect), for the PbFe-Si-PbFe configuration when low-Z G10 absorbers, 5 or 10 mm thick are inserted in the back of the silicon detectors in front of the Pb absorbers.

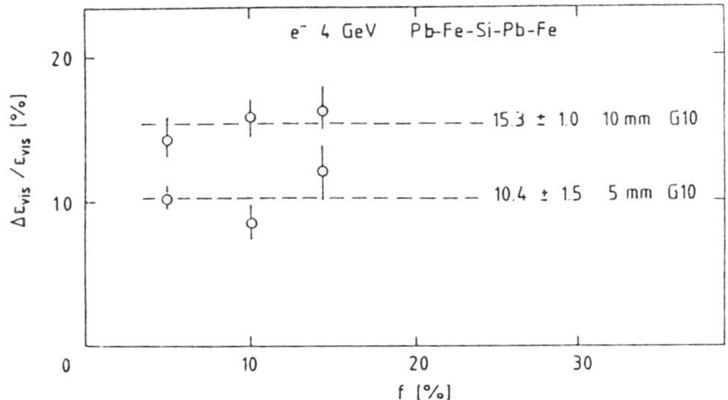

Figure 6. Reduction of the electromagnetic shower due to the local hardening effect, $\Delta\epsilon_{vis}/\epsilon_{vis}$, as a function of the Pb fraction in the absorber at 4 GeV for the PbFe-Si-PbFe configuration when G10 absorbers 5 or 10 mm thick are inserted at the rear side of the Si detectors (in front of the Pb absorbers).

Radiation damages

In view of the very high neutron fluences (The maximum annual fluence foreseen in the SSC calorimeter at very forward angles is about 10^{13} n/cm^2) radiation hardness of the calorimeter is an important constraint. The energy spectrum of the neutrons, produced by the interaction of multi-TeV protons with the environment, is peaking around 1 MeV [11]. The interactions of these fast neutrons by elastic scattering with the silicon atoms, induce deep defect levels in the silicon band gap which take the form of vacancies, multivacancies, vacancy-impurity pairs, and cluster of defects. These defects alter the electrical properties of the silicon detectors and consequently the use of Silicon detectors in the SSC environment requires detailed studies in order to understand the defect formation processes and of the electrical behavior of irradiated silicon detectors. It is therefore important to perform measurements on silicon samples in order to correlate the damages caused by neutrons with the degradation of detectors.

It has been recognized [12] from TSC [13] and DLTS [14] analysis that the main effect of neutron irradiation on detectors lattice is the formation of three point defects at $E_1 = (0.16\pm 0.2)$, $E_2 = (0.25\pm 0.02)$ and $E_3 = (0.40\pm 0.01)$ [eV] below the conduction band edge, to be associated to A center, double-minus charge state center and to the sum of single-minus charge state center and E center respectively. By using the TSC techniques, the E_3 component is resolved [12,15]: (0.39±0.01)[eV] for the single charge state of divacancy and (0.42±0.02)[eV] for a vacancy-phosphorous complex. It has to be noted that neutron irradiation also causes formation of defect clusters that are invisible with these techniques and which participate in defining the electrical characteristics of irradiated samples. A correct study of clusters could be done with Hall effect measurements. Studies on electron-capture cross sections showed that the shallow A center,together with the deeper level at 0.40 [eV], could give an important contribution to the leakage current.

By far the most important effect of the radiation is the increase of the leakage current since very high values of the current and the related electronic noise impose a limit on the operability of the detectors and hence the lifetime of the calorimeter.

If ΔI is the difference of the detector current before and after irradiation, V is the depleted volume, and ϕ is the irradiation fluence, the leakage current constant α is defined by:

$$\Delta I/V = \alpha\phi$$

where $\Delta I = I_i - I_0$; I_i is the leakage current after irradiation and I_0 the leakage current before irradiation.

Experimental studies of the electrical behaviour of n-type silicon detectors irradiated with fluences up to $\sim 10^{12}$ [neutrons/cm^2] showed a linear increase of leakage current vs the fluence of irradiation with a proportionality constant [15]:

$$\alpha = (4.8 \pm 0.9) \times 10^{-17} [A/cm]$$

for 1 MeV neutrons from a Cf^{252} source at room temperature (20°C) and no bias applied. This value is consistent with other results reported elsewhere [16], [17], [18] and obtained with shorter irradiation time.

Detectors have been also irradiated up to a measured neutron fluence of 0.7 10^{13} n/cm^2 by the SICAPO Collaboration in a Fe calorimeter at a 24 GeV primary proton beam at the CERN-PS. The measured value of α (=(5.4±1.4)10^{-17} A/cm) and the point defect found are in agreement with the finding obtained with the 1 MeV neutrons from the Cf^{252} source. This program of study by the SICAPO Collaboration is continuing presently.

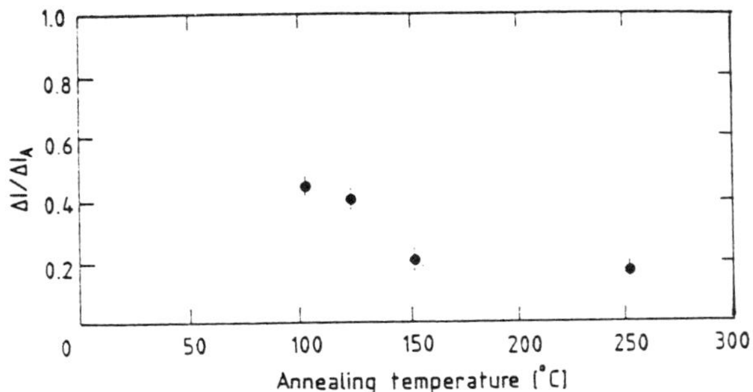

Figure 7. The percentage leakage current recovery $\Delta I/\Delta I_A$ is shown as a function of the annealing temperature (°C). An ~ 80% recovery appears with a ~ 200 °C annealing during 50 hours.

The question of the damage annealing becomes of importance in order to allow the maintenance of the good operation of the detector and of the calorimeter at the end. Room temperature self annealing and short term heat treatments seem to be likewise important and have been investigated carefully. Electrical and lattice characteristics recovery of ~ 80% occur with a 200° C annealing during 50 hours (see fig. 7) [12,15]. Re-irradiation and re-annealing of the previously annealed samples give approximately the same result concerning leakage current measurements, so that it seems that detectors behavior after irradiation and annealing is not affected by previous irradiation-annealing cycles, if these cycles have brought to a good electrical and lattice recovery of the samples. Neutron damage results reported previously by Kraner and collaborators [18] are consistent with our findings.

The dependence on the measuring frequency of the capacitance-voltage (C-V) characterisitics of the detectors with the high resistivities (in practice $\rho > 1k\Omega - cm$) has been also investigated [19]. The C-V characteristics are frequency independent up to neutron fluence of $\phi = 7\times10^{10}/cm^2$ [19] in agreement with data reported earlier [12]. At a neutron fluence of $\phi = 8\times10^{10} n/cm^2$, the detector high frequency ($f > 50kHz$) C-V characteristics becomes essentially bias-independent equaling the detector geometrical capacitance [19]. The frequency dependence of the C-V characterisitcs, being affected by the neutron fluence, is depending on the defect concentration over the dopant concentration.

Electronics

The silicon detectors have a linear response to the deposited energy and therefore are free of saturation effects. Their fast charge collection (with a transit time ¡ 10 ns) allows a fast associated electronics [20]. A preamplifier employs a mixed bipolar CMOS (BiCMOS) technology developped by the SGS-Thompson Microelectronics. The rise time of the preamplifier is ≈ 7.5 ns for a maximum output voltage swing of 5 V with 150 pF input capacitance for a RC-CR shaping time (τ_m) of 20 ns.

The Equivalent Noise Charge was ENC = 17 $ke_{r.m.s.}$ in a first version and has been reduced to ENC = 4 $ke_{r.m.s.}$ in a new version. The linearity of the device has been tested up to 1 GeV of energy deposited in the detector (1 mip = 120 KeV).

The ENC equation shows two noise components, the series and parallel noises. The series noise is proportional to the input capacitance and the inverse of τ_m while the parallel noise is proportional to τ_m. So at low τ_m value and high input capacitance, the parallel noise contribution to ENC may become negligible. It has been shown [15] that the noise σ_n (for the first version of the preamplifier and expressed in dimension of energy) increases by 9% for an increase of the detector leakage current from a few nA to 200 μA. This results that the fast VLSI preamplifier has a noise performance which is almost independent of the leakage current value.

In general, the use of this fast electronics has several advantages. At first, it naturally matches the interbunch crossing time, and secondly minimize the pile-up events. For the Silicon detectors, in particular, it minimizes the effect of the leakage current on the calorimeter performances.

Discussion and conclusions

The SICAPO Collaboration results demonstrate the effectiveness of the local hardening effect in modifying the response to the incoming showers of a hadronic calorimeter, with silicon readout and a combination of low-Z and high-Z materials as absorbers. This local hardening effect increases the reduction of the visible energy beyond the reduction normally expected from energy sharing. In the case of the Si/U used in the measurements [9], a ratio e/π = 1 is obtained resulting from the insertion of G10 plates (1.2 ± 0.2)mm thick in front and at the rear side of the silicon detectors.

The use of a combination of Fe (low-Z) and Pb (high-Z) materials as absorbers allows the transformation of the electron energy distribution of the incident showers ("the filtering effect"). The action of the low-Z on the high-Z material, by modifying the critical energy during the electromagnetic shower development, leads to a yield of soft electrons (and very few photons).

The flux of backward-going electrons (resulting from multiple scattering in the high-Z absorber) are absorbed by the thin (in radiation length) G10 plates inserted in the calorimeter configuration between the rear of the active and the passive media ("the local hardening effect").

The local hardening effect has been also measured for polyethylene layers inserted in front and at the rear side of the silicon mosaics. The level of reduction of the electromagnetic visible energy $\Delta\epsilon_{vis}/\epsilon_{vis}$ is similar to that obtained for G10 [1]. Therefore polyethylene can be used instead of G10 to achieve the compensation condition (e/π = 1). The possible use of polyethylene presents the advantage that it can play the role of a neutron moderator as indicated by recent activation studies [21] and may then contribute to reinforce the calorimeter radiation hardness.

All these facts support the choice, for the next generation of colliders, of an hadron calorimeter with silicon as active medium and a combination of Fe and Pb as absorbers. The choice of an absorber made of a combination of Fe and Pb permits the achievement of the compensation condition through also the combined effect of hardening and filtering resulting from the insertion of G10 planes between the read-out and the absorber consisting of a mixture of low-Z and high-Z metals [8]. As a consequence a linear response of the calorimeter to hadronic showers and a good energy resolution, improving as the incident energy increases, can be obtained. In addition an increase of the radiation hardness is possible with the insertion of polyethylene foils which have basically the same properties than G10, and increase the pure hadronic response (i.e. h/mip value).

References

[1] E. Borchi et al. (SICAPO Collab.), Silicon sampling hadronic calorimetry: a tool for experiments at the next generation of colliders, presented by C. Leroy at the Int. Conf. on Advanced Technology and Particle Physics, Como, 1988 [Nucl. Instrum. Methods A279 (1989)], 57; C. Bertrand et al. (SICAPO Collab.), Silicon Calorimetry for the SSC, Summary report of the silicon calorimetry working group, Proceedings of the Workshop on Calorimetry for the Superconducting Super Collider, The University of Alabama, Tuscaloosa, Alabama, USA, March 13-17th 1989, Editors R. Donaldson and M.G.D. Gilchriese (World Scientific, Singapore, 1990) 489.

[2] Y.S. Tsai, Rev. Mod. Phys. 46 (1974) 815.

[3] U. Amaldi, Physica Scripta Vol. 23 (1981) 409.

[4] $f_{em}(E) \approx 0.1 ln\ E(GeV) for 10 < E < 100 GeV$.

[5] O.I. Dovzhenko and A.A. Pomanskii, Sov. Phys.-JETP 18 (1964) 187.

[6] F. Lemeilleur et al. (SICAPO Collab.), Phys. Lett. B222 (1989) 518; F. Lemeilleur et al. (SICAPO Collaboration), IEEE Trans. Nucl. Sci. Vol. NS-36 (1989) 331.

[7] C. Leroy, Y. Sirois and R. Wigmans, Nucl. Instrum. Methods A252 (1986) 4; R. Wigmans, Nucl. Instrum. Methods A259 (1987) 389.

[8] E. Borchi et al. (SICAPO Collab.), Phys. Lett. B222 (1989) 526; E. Borchi et al. (SICAPO Collaboration), IEEE, Trans. Nucl. Sci. Vol. NS-37 (1990) 1186.

[9] A.L.S. Angelis et al. (SICAPO Collab.), Phys. Lett. B242 (1990) 293.

[10] J.E. Brau and T.A. Gabriel: Theoretical studies of hadronic calorimeters for high-luminosity, high-energy colliders, presented by J.E. Brau at the Int. Conf. on Advanced Technology and Particle Physics, Como, 1988 [Nucl. Instrum. Methods A279 (1989)], 40.

[11] D.E. Groom, paper presented at the Int. Conf. on Advanced Technology and Particle Physics, Como, 1988 [Nucl. Instrum. Methods A279 (1989)], 4.

[12] E. Borchi et al. (SICAPO Collab.), Radiation Damages Effects on Silicon Detectors, presented by L. Vismara at the Int.

Conf. on Advanced Technology and Particle Physics, Como, 1988 [Nucl. Instrum. Methods A279 (1989)], 277.

[13] M.G. Buehler,W.E. Phillips, Solid State Electr. 19, 777 (1976).

[14] D.V. Lang, J. Appl. Phys. 45, 3023 (1974).

[15] E. Borchi et al. (SICAPO Collab.), Study of operating condition of semiconductors for calorimetry in LHC/SSC radiation environment, paper presented at the Int. Conf. on Advanced Technology and Particle Physics, Como, 1990.

[16] M. Hasegawa et al., "Radiation Damages on Silicon Detectors by Neutrons, Radiation effects at the SSC, SSC-SR-1035, 93.

[17] G. Lindstroem et al., Radiation Damage Effects in Silicon Detectors, Proceedings of the Workshop on Calorimetry for the Superconducting Super Collider, The University of Alabama, Tuscaloosa, Alabama, USA, March 13-17th 1989, Editors R. Donaldson and M.G.D. Gilchriese (World Scientific, Singapore, 1990) 489.

[18] H.W. Kraner, Z. Li, and K.U. Posnecker, Fast Neutron Damage in Silicon Detectors, Proceedings of the Int. Conf. on Advanced Technology and Particle Physics, Como, 1988 [Nucl. Instrum. Methods A279 (1989)], 266.

[19] Z. Li, W. Chen, and H.W. Kraner, Neutron Radiation Damage Studies on Silicon Detectors, paper presented at this conference.

[20] A. Gola, G. Pessina, and P.G. Rancoita, Nucl. Instrum. Methods A292 (1990) 648.

[21] E. Borchi et al. (SICAPO Collab.), presented by R. Steni at the ECFA Study Week on Instrumentation Technology for High-Luminosity Hadron Colliders, Barcelona, Spain, 14-21 September 1989.

RECENT DEVELOPMENTS IN WIRE CHAMBER TRACKING AT SSC

Harold Ogren
Department of Physics, Indiana University
Bloomington, Indiana 47405

Abstract

All of the major SSC proposed detectors use wire chambers in their tracking systems. The feasibility of wire chambers in an SSC detector has now been established by a number of groups planning detectors at SSC.The major advances during the past year in understanding straw tube drift chambers are presented and several innovations in gaseous wire chambers are discussed.

Introduction

I will try to review the work that has gone on during the past year on wire chambers at SSC. The major part of this will have to do with straw drift chambers, since they are present in all three major detectors proposed at SSC. This talk is divided into two major parts. I will quickly review the wire chambers being planned for SSC detectors and then outline the major areas of research and development for the SSC environment. The R&D section will concentrate on progress in drift cell design, electronics and signal processing, and engineering aspects of the tracking designs.

Wire Chambers at SSC

SDC

The wire chamber tracking system that is being planned for SDC is shown in Fig. 1. It consists of a central tracking system built from straw drift tubes contained in modules as shown in Fig. 2. This design has been worked on primarily by the Wire Tracking Subsystem Group. [1] The central tracker uses 4 mm diameter straws with lengths varying from 4 meters to 2 meters from the outside radius of 1.7 meters to the inside 0.70 meter layers. The total number of modules in the two z halves is 1088 and the total number of straws is about 188,000. As we will see, a good deal of the R&D concerns itself with producing a simple construction and assembly scheme for this large number of straws. The modular approach for the construction will allow each module to be tested with final electronics before insertion into the tracker. The modules also group the straws in precise position and support the wire load of approximately 12 Kg per module. The tracking system in the intermediate region is a radial wire chamber as shown in Fig. 3. The tracking in the intermediate region is being worked on primarily by the groups from the United Kingdom. The present intermediate tracker consists of a radial wire chamber module with a wire "bunch tagger chamber" immediately in front. This tagger has sufficient timing resolution to select the proper crossing time for each event and tag the event for the somewhat slower radial wire chambers. There would be five radial modules required for a precise measurement of the track momentum in the forward direction.

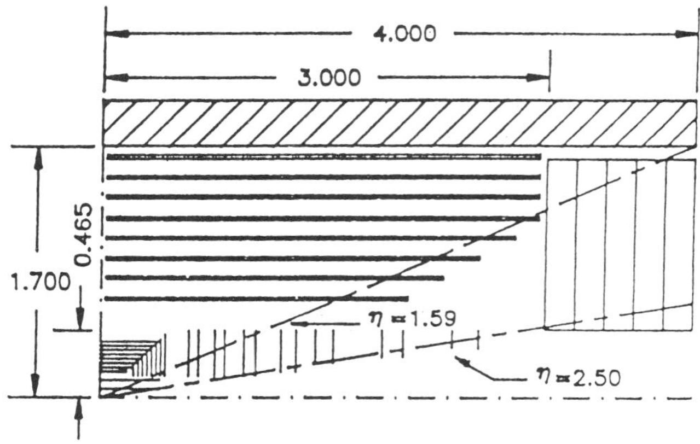

Fig.1. Tracking system for the SDC detector. The straw chamber superlayers surround the silicon system and radial wire chambers are used in the intermediate angle region.

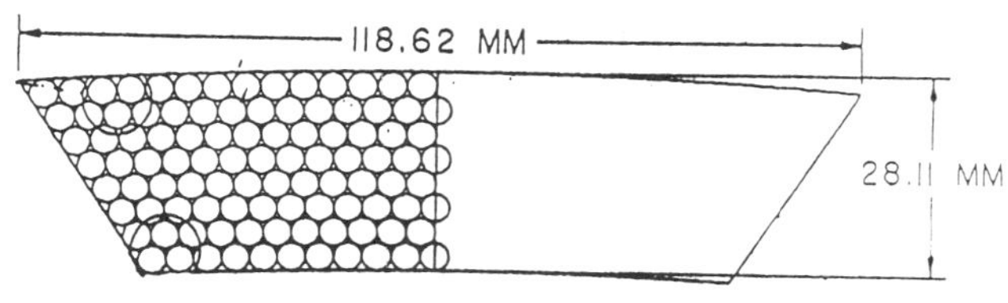

Fig.2. Straw module holding 196 straws. The module walls are constructed from a 250 μ carbon fiber composite.

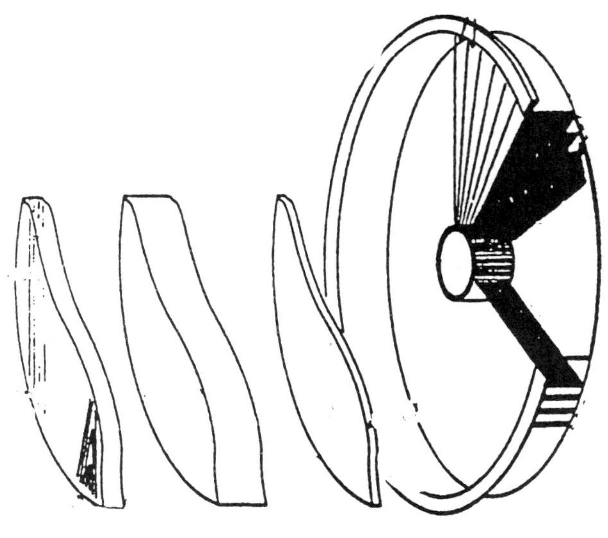

Fig. 3. An exploded view of the radial wire chamber with a bunch tagger and a transition radiation section in front.

Work is also being done on a "Hybrid" detector, consisting of axial straw superlayers and scintillating fiber stereo layers. This has been pursued in some detail by the Hybrid Tracking Subsystem Group.[2] The Hybrid tracker is shown in Fig. 4. The straw layers are glued to a precision cylinder, supported from each end by a conical support. Stereo scintillating fibers are attached to a separate support cylinder on the outer 2 or 3 superlayers. The tracker would use 4 mm straws and 750 micron scintillating fibers.

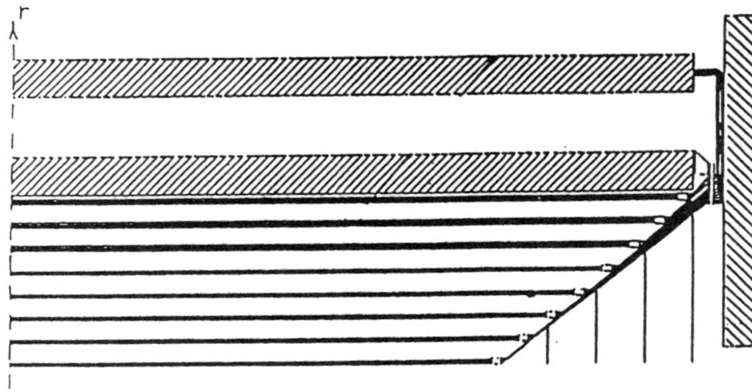

Fig. 4. Superlayer structure for the Hybrid Chamber.

EMPACT

The EMPACT tracker is shown in Fig. 5. As shown in the detailed close-up, straw chambers are used for tracking and transition radiation detectors. These drift tubes are arranged in modules that span a small delta phi, delta z region. There are a total of 385K readout channels for this straw tracker. Although TDCs will not be used for each channel, the pulse height will be digitized for electron id. This wire tracking and electron id work is being pursued by Boston University and Brookhaven National Laboratory.

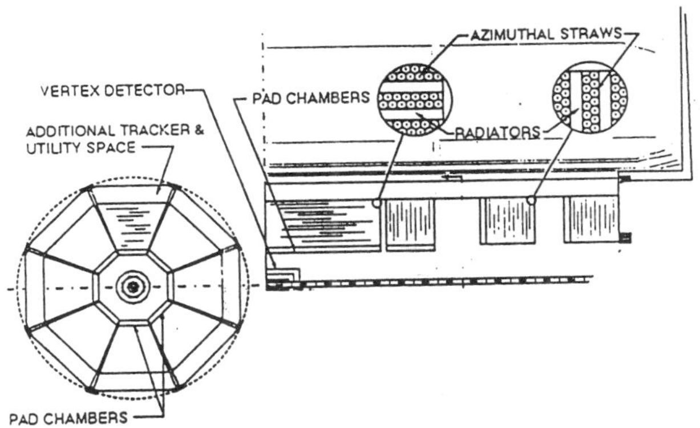

Fig. 5. End and side views of the EMPACT detector. The TRD chambers use straw tubes.

LSTAR

The central wire and scintillating fiber tracker for the LSTAR detector can be seen in Fig. 6. The straw tube chamber fills the region from a radius of 40 cm out to 75 cm and is about 2.8 meters long. The straw tracking system has about 75K channels with TDCs on all wires. As in the Hybrid system, the stereo tracking is done with scintillating fibers. The groups working on the straw chambers are from Boston University, New Mexico University, Indiana University and Los Alamos National Laboratory.

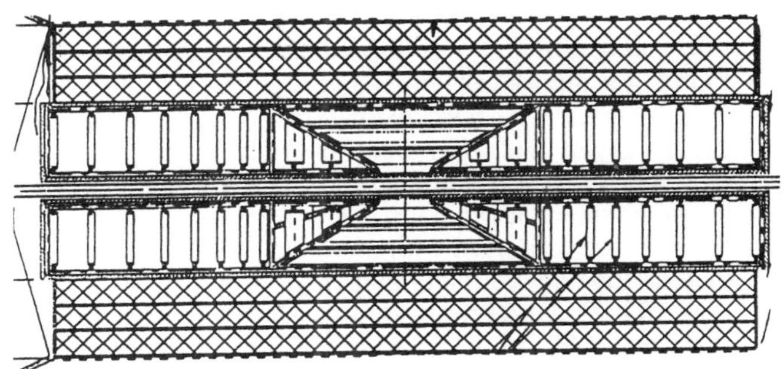

Fig. 6. The tracker for the LSTAR detector. The outer straw and scintillating fiber modules surround a silicon system.

Major Areas of R&D for the SSC Environment

Drift Cell Research

One of the central concerns in tracking at the SSC with straw chambers is shown in Fig. 7. This plot shows the occupancy, hit rate, and current for a straw drift cell in the SDC detector as a function of the radius. The cells were assumed to be half the length of the detector. The calculation was was carried out at KEK[3] and used a PYTHIA particle generator, and GEANT for detector simulation. All known materials were included in the model, and looping tracks in the magnetic field were used. We see that we can expect occupancies near 16% at the inner radius, and currents of up to .6 μ amps. Similar calculations have been carried out by other groups[4] and are in agreement with these basic numbers.

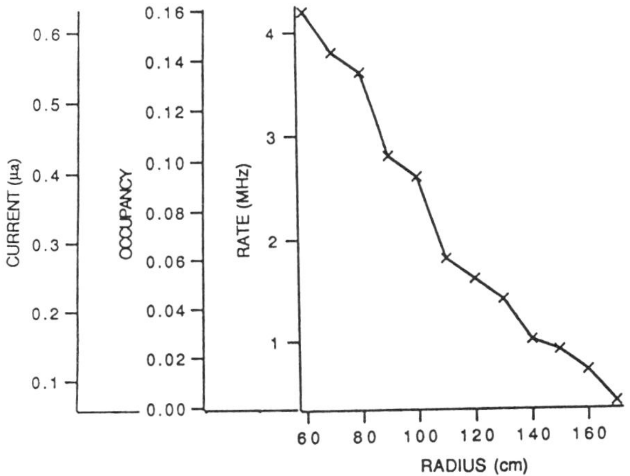

Fig. 7. The radial dependence of the hit rate, occupancy, and current draw versus radius of superlayers.

Although the occupancy is high, a Monte Carlo study at Colorado[5] has shown that the number of hits found in a superlayer remains very high even in the inner super- layers. The number of good 7 hit segments in a super- layer with 8 layers is about 90% at design luminosity at 75 cm radius and climbs to about 98% at the outer super- layer. This is is a very encouraging study and is being continued with more realistic tube efficiencies to determine the optimal number of straw layers.

These simulation studies show us what we can expect at SSC and what sort of detectors will be required. However, what must now be addressed is whether such detectors exist or can be built. This past year there has been a considerable amount of work on straw drift chamber R&D.

Gas selection:

CF_4-Isobutane(20%) appears we can be the most promising gas for use at SSC. In Fig. 8 the drift time distribution for a 4 mm straw drift tube filled with CF_4-Isobutane(20%) is shown. It has a drift time of about 19 ns. Additional measurements with 20% Isobutane show that straw chambers are fully efficient above 1900 volts.

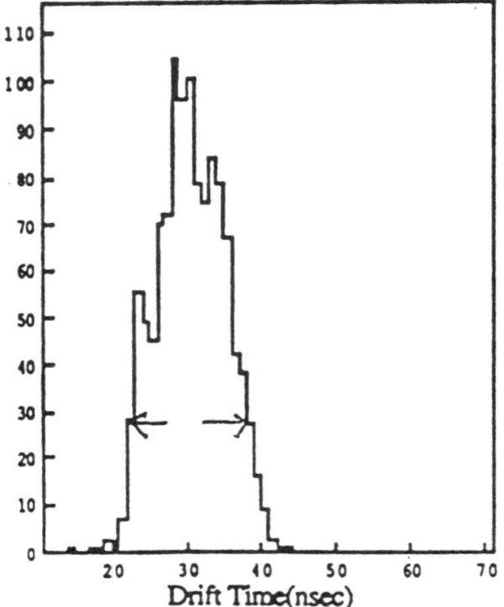

Fig. 8. Drift time distributions in a 4 mm straw for CF_4-Isobutane (20%).

Detailed measurements of the gain have been made by several groups[6]. As shown in Fig. 9, the gain of the CF_4-Isobutane chambers is about 30,000 at 1900 volts for 4 mm straws and 25 μ wire diameter. The measurements shown in the figure were made on a 3.5 meter straw system. At Princeton University the single cluster timing distributions have been measured.[7] These were used to determine the diffusion limit of spatial resolution for 3.5 mm drifts. As can be seen from Fig. 10, the diffusion limit for CF_4-Isobutane is about 40 μ. All of these measurements indicate that a fast efficient gas with good resolution in already in hand for the design of straw chambers.

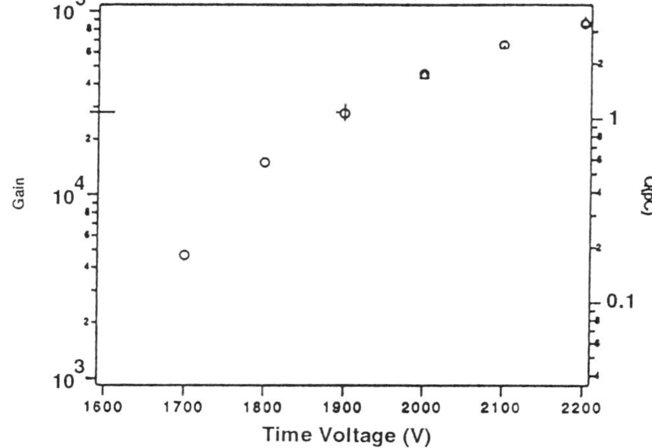

Fig. 9. Gain measurement on 3.5 meter straw chamber for CF_4-Isobutane (20%) 4 mm straw 25 m wire.

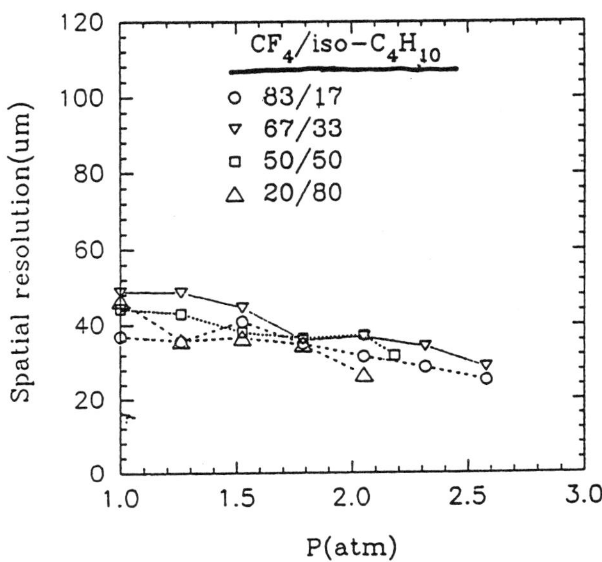

Fig. 10. Diffusion limit for CF_4-Isobutane for 3.5 mm drift.

Wire lifetimes

There are a number of questions to be addressed concerning the lifetime of straw chambers at SSC. It is expected that wire gain and long term contamination will be a function of the gas used in the drift cell. However, the other area of concern is the mechanical or structural damage due to the high radiation dose of components. During this past year there has been considerable progress in this areas.

In Fig. 11 we show some of the work that has been done at LBL.[8] The wire gain of CF4 - Isobutane(20%) is compared to Ar/ethane with a known contaminant injected in the chamber. The CF_4 gas show little sign of damage , but the Ar/ethane gain is seriously degraded. It appears that the CF_4/Isobutane mixture has the ability to clean the signal wires. This has been also reported by a group at TRIUMF.[9] In Fig. 12 a previously contaminated chamber running on Argon/Ethane is shown to return to its original state when running in CF4-Isobutane. The damaged area was isolated to a small region in the middle of the wire. After an accumulated charge of about 1 C/cm it had completely recovered.

These measurements give us confidence that wire gain should remain constant at the SSC over the lifetime of the experiment for luminosities far above the design.

Neutron Damage

There were two reports at this conference on neutron damage to straw chambers. The North Carolina State reactor PULSTAR [10] was used to irradiate mylar straws and glued straws in chamber arrays with thermal neutron fluences up to 10^{16} cm^{-2} and fast neutrons fluences up to 10^{15} cm^{-2}. The basics straw components are confirmed to survive at much higher fluences than those expected at SSC.

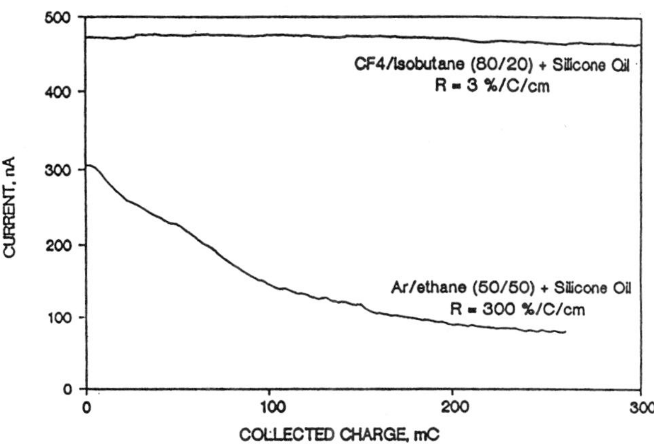

Fig. 11. The gain of a contaminated straw chamber system with CF4- Isobutane(20%) and Ar-ethane (50%) as a function of total charge.

A dramatic test of an operating straw chamber has been reported by a group at Boston University.[11] A 25 straw array was subjected to a 0.5 MeV neutron flux of 7 x 10^7 / cm^2/ sec . The straw chambers used CF4 gas and was running typically at 5 MHz. and drawing 15 μa of current. The total fluence of neutrons was 6.5 x 10^{14}/ cm^2 and no damage or gain loss was reported. Both of these tests give us confidence that the radiation levels predicted for SSC will present no obstacle to the operation of straw chambers.

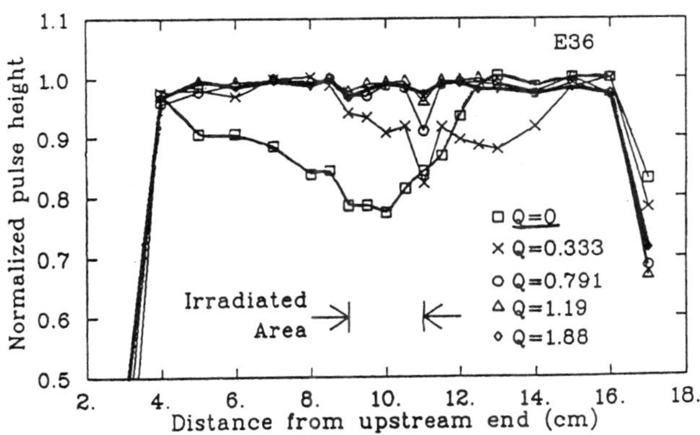

Fig. 12. The pulse height of a straw drift chamber with a contaminated central region (10 cm) for different amounts of total charge deposited.

High Rate tests

The principle long range activity of the groups working on straw chambers is to demonstrate that tracking in a high rate beam is possible. Ideally this should be with a full length (3 meter) straw system. Preparations are underway for such tests.

At Indiana University construction has started on a number of short 64 straw modules. These can be used for trigger and electronics tests by members of the SDC collaboration. A full length module is now under design at Oak Ridge and Indiana University. By next summer a full scale module should be ready for testing. At Duke University a full length cylinder of radius 0.8 meters is being constructed. This prototype superlayer can be used for high rate tests, and for verifying construction techniques.

Electronic related R&D

Signal size

There are a large number of electronics related issues under study. These include measurements of the signal size, effects of straw capacitance and attenuation in the straw chambers. There is also design of the preamps, shapers and fast TDC underway with an emphasis on high density, low power consumption, and radiation hardness. Finally, a design for a fast level one trigger has been worked out and will soon be tested on a straw module.

Detailed calculations of the signal size expected with 3 meter straw chambers have been carried out at KEK[12] for a 7 mm straws with a slow gas (Ar/ Ethane). It is expected that a 4 mm straw diameter and a fast gas like CF4- Isobutane will give more that twice the signal size. From these calculations we can estimate the time jitter and time walk expected in a 3 meter straw at SSC. There does not seem to any difficulty in principle to attaining 100-150 micron position resolution.

In Fig. 13 a pulse in a 3.6 meter straw chamber is shown along with the reflection pulse from the unterminated end.

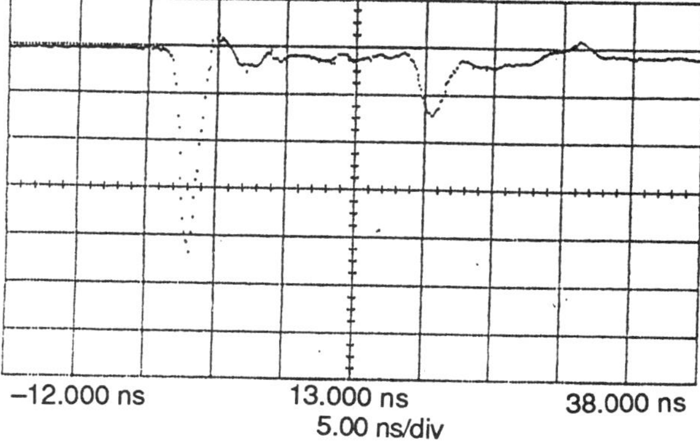

Fig. 13. The signal from an Fe 55 source for a 3.6 meter straw system. The reflection pulse can be seen to be attenuated.

Termination has also been applied to eliminate the reflected signal.The attenuation of reflected pulse in the figure indicates one of the areas of interest in R&D. This attenuation comes about due to the resistivity of the cathode material and the signal wire. The resistance of both wire and cathode were about 100 ohm/meter in the test shown in the figure, whereas the coaxial impedance of the straw is about 300 ohms.

The signal attenuation in straw chambers has been measured by several groups[13] and is shown in Fig. 14. The average attenuation length is about 5 meters. There are several ways to increase the attenuation length. Work is in progress on with lower resistance cathodes and larger diameter wires. The effect of appreciable attenuation would require an increase of gain in the straws and might require a more elaborate analysis to attain the desired time resolution.

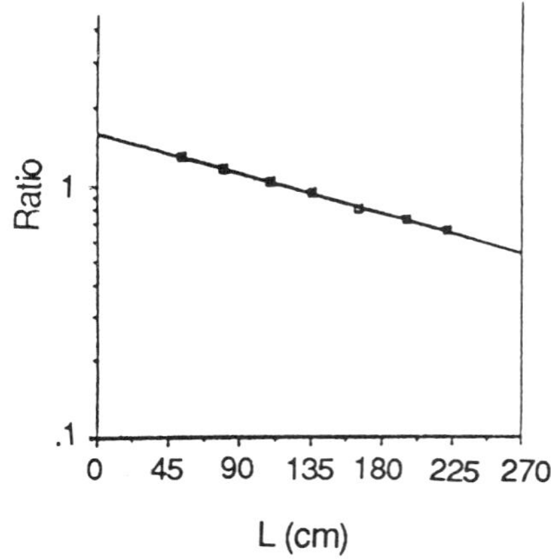

Fig. 14. Measurements of the signal attenuation for a 2.3 meter straws.

Signal electronics

The drift chamber electronics has been worked on by groups from University of Pennsylvania[14] and the Japanese laboratory, KEK[15]. The system that is being designed by the group from the US is shown in Fig. 15. It consists of a preamplifier, shaper, discriminator, and time to voltage convertor, and an analog memory for the Level 1 and Level 2 triggers. Their design has progressed significantly in the past year. They now have prototype chips with low threshold, low power and radiation hardness sufficient for SSC. The prototype preamplifiers have been used on several of the straw chambers, in fact they were crucial in the measurements of the single cluster characteristic presented by Princeton University. We anticipate that in the next year we will have a sufficient quantity of the amplifiers and shapers to test several straw modules. The time to voltage converters (TVC) and the analog storage are shown in somewhat more detail in Fig. 16. We anticipate that the first prototypes of this will also be available next year. KEK is working on an alternative method (the TMC) of measuring the time and storing it for a level 1 trigger.

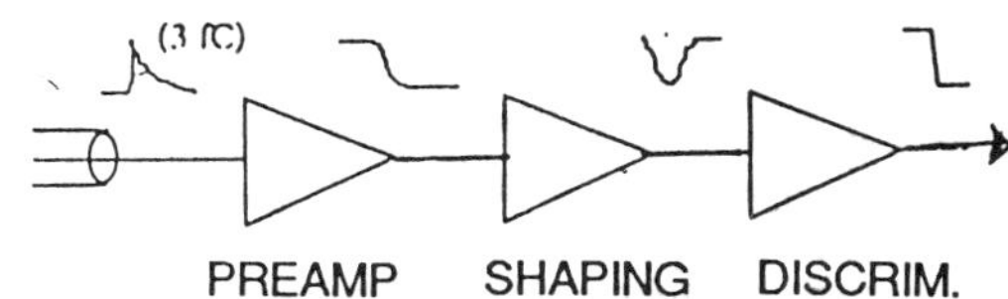

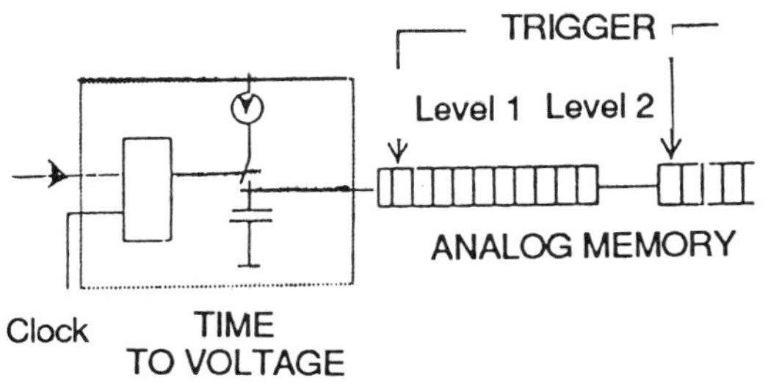

Fig. 15. A diagram of the straw chamber electronics.

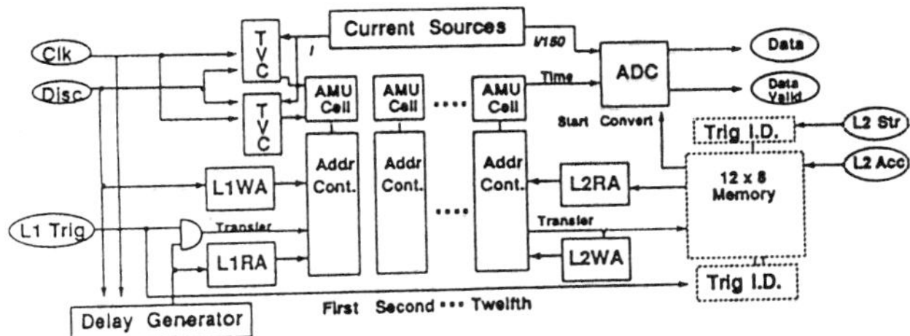

Fig. 16. Details of the TVC and level 1 and 2 triggers.

This Time Memory Converter or TMC is shown in Fig. 17. The chip has progressed to the fabrication stage, and should be available to test in a chamber by the end of the year. At University of Colorado there has been an attempt to design the layout for all of the front end electronics for a straw module. The resulting board layout is shown in Fig. 18. This is an existence proof that a compact layout of preamplifiers, shapers TVC along with all data and read out lines will fit in the modular scheme. The trigger electronics has also made real progress this year. At the University of Michigan[16] a real time trigger has been designed to find stiff tracks in a straw superlayer. The technique and the circuit is shown in Figure 19. The scheme relies on the layout of straw wires along radial rays from the intersection region. In this case a stiff track will produce a fixed relationship between the drift times in three successive straw layers. A synchronizer circuit will trigger on a stiff track within 30 ns of the arrival of the last pulse. The circuit has been designed and constructed. Testing will begin this fall.

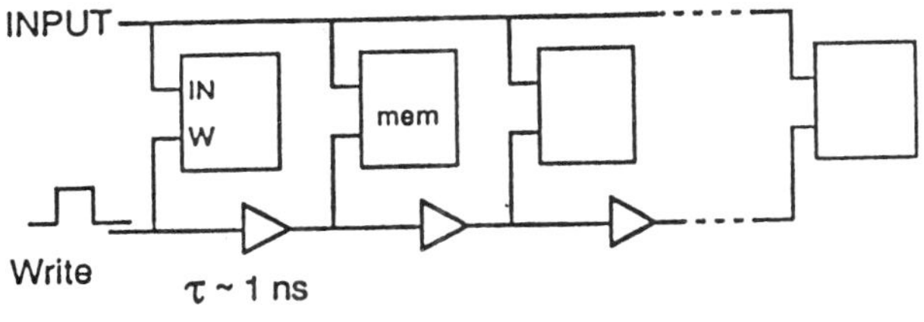

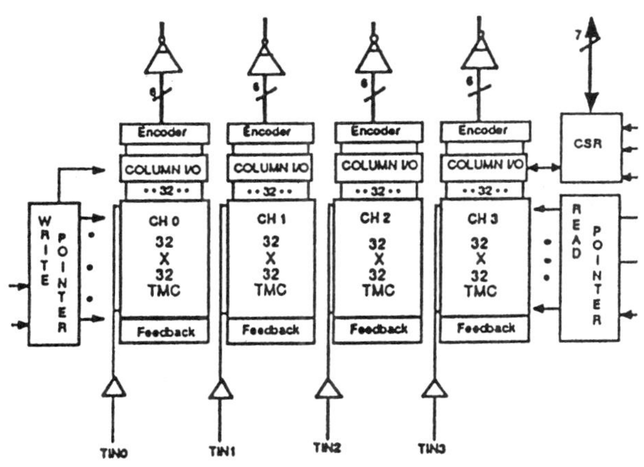

Fig. 17. The Time Memory Chip for recording and storing timing information from the straw chambers.

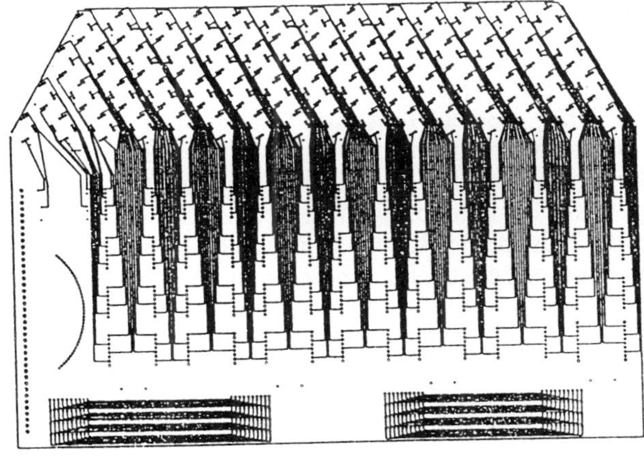

Fig. 18. A prototype layout of the front end electronics for a straw module.

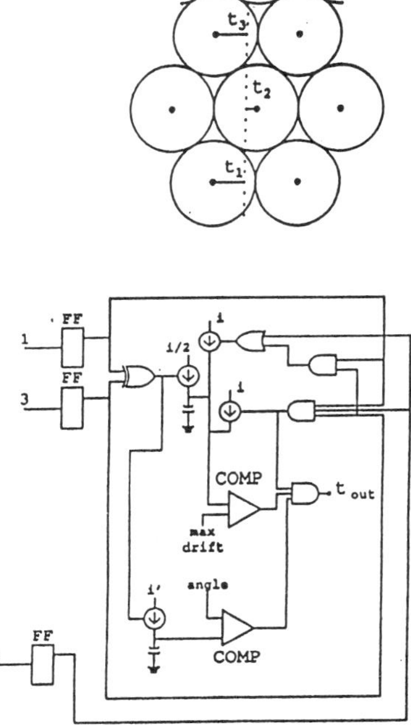

Fig. 19. The synchronizer circuit for triggering on triads of straw in the straw superlayer.

This has been a very productive year for the electronics R&D. Many circuits are now in the prototype stage and are ready for testing. The next step is to measure the high rate capabilities of the electronics and to confront the important details of chamber design such as proper straw resistance, proper wire attenuation, problems of capacitance, shielding and grounds.

Engineering R&D

Prototype Construction

Straw chamber prototypes have been built at Indiana University, Duke University, Boston University, and Princeton University. The problems of wire support and straw support have been solved in a number of different ways. At Duke a 25 straw system has been constructed. The straws were laid out and glued together on a flat grooved substrate. The straw length was 2.7 meters and the diameter was 4 mm. Measurements of the horizontal position in various rows show that the straws position for this configuration could be held to about 2 mils. This same system was instrumented and run at high voltage successfully. There is an ongoing program to study ways to accurately layout and glue the straws to the support cylinders in this hybrid scheme. A large prototype full length cylinder of 0.8 meter radius is being worked on. At Princeton University there is also work on building large structures of 7 mm diameter straws for a BCD experiment. Several straw clusters have been constructed and a novel method of pressurizing the straws to handle them has been worked out.[17] At Indiana University a 6 straw 3.6 meter long assemble was tested with electronics from University of Pennsylvania and attenuation in the system measured.[18] The major work at Indiana now, however, is the construction of modular arrays to hold the straws in position. A 64 straw module shown in Fig. 20 has been constructed and is now under testing. The support shell is 300 μ thick carbon fiber composite. The end plates have been designed to have very little material. The positioning of the straws and wires is accomplished by self centering of the wire supports within the shell. This wire support is shown in Fig. 21. It consisted of two identical molded pieces that snap together to form a cylinder with opposing V shaped passages at either end. It has the added advantage that the straw chamber can be strung after being assembled, or can be rewired if there is wire breakage.

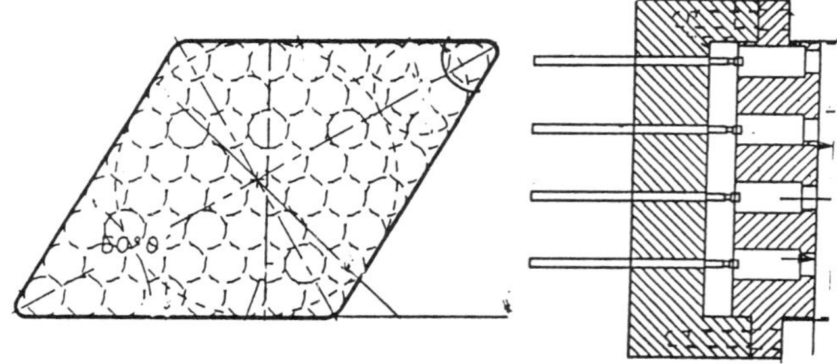

Fig. 20. Detail of the end and side view of a 64 straw module.

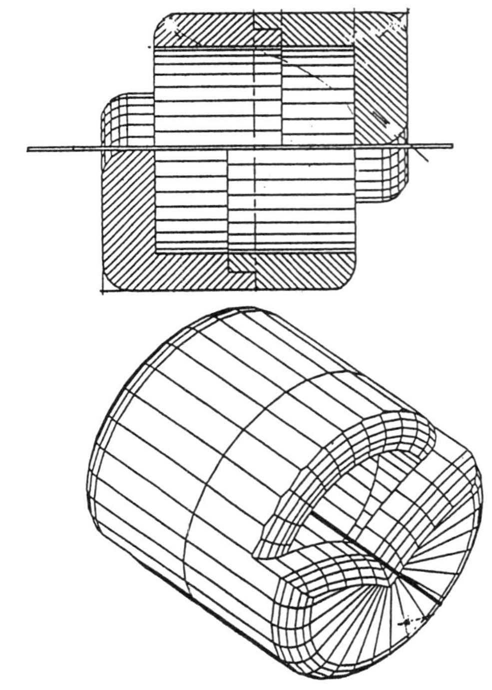

Fig. 21. Two view of a "double V" wire support.

Design Studies

Modular Tracker

All of the tracker engineering construction techniques are driven by the two somewhat incompatible considerations: First the amount of material must be kept small and second the straws and wires must be aligned with about 50 micron precision and be stable in that configuration.

The calculation of the radiation length versus eta for the SDC tracking system shows that in the region of the endplate of the central tracker there is more that 20% of a radiation length. It is difficult to reduce this significantly, since each superlayer has less than 1% of a radiation length for perpendicular traversals, and about 1/2 of this is from the straws themselves. The engineering challenge is to design a support system with 60-75 m tolerance with light weight materials. One support scheme that has been designed by Westinghouse[19] is shown in Fig. 22. The modules are held in position by a rigid support frame at three points. These frames have to only support the modules and not the wire tension, so they can be very light. They would be made from light weight carbon composites. Each module has a set of alignment pins that locate and hold it in position in the support disks. Four of the eight superlayers would be stereo layers. These are constructed by using the same modules as in the axial layers, but rotating them by 3 degrees. Modules could be assembled into the frame as the final operation after all electronics testing was complete and modules could be replaced in the case of accidental damage.

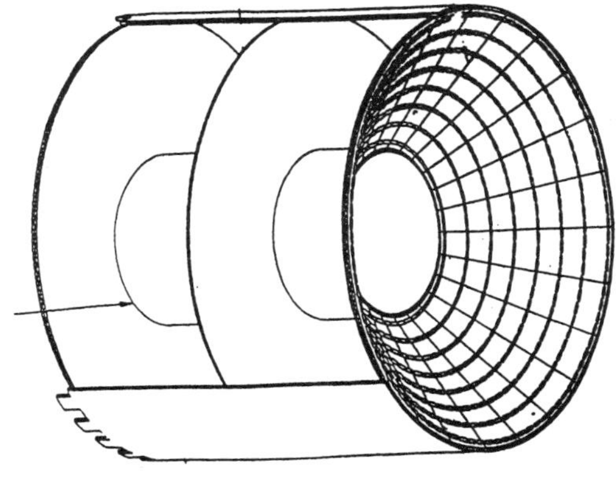

Fig. 22. An assembly drawing of one half of the modular drift chamber tracker showing the support system.

Hybrid Tracker

The hybrid central tracker is shown in Fig. 4. The design work on this tracker has been done at Oak Ridge National Laboratory.[20] In this concept the straws are supported on full length cylinders that are held on the ends by a cone and ring support. The details of the endplate with utilities are shown in Fig.23. This endplate must support wire tension for the entire cylinder, and also serve as the precision wire positioner. The amount of material in the support cylinder is comparable to the modular approach, however, there would need to be extra cylinders for the scintillating fiber layers. During this coming year both techniques will be designed in much more detail.The major R&D for reduction of material is in the endplates and utilities. This is being worked on at ORNL.

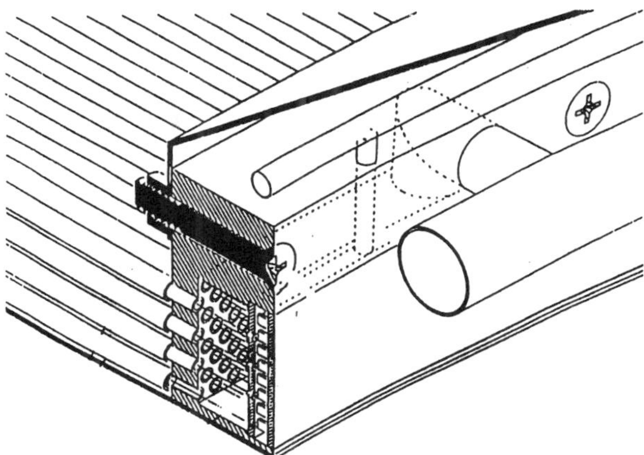

Fig. 23. Details of the endplate for the Hybrid design, showing utilities.

TRD Tracker

At CERN [21] there have been several very impressive tests of a straw tracking system coupled with TRD possibilities. A large array of straw tubes was tested in a CERN beam line to verify that the TDR tracking cell would give the resolution required by EMPACT. They found that using only the cell hits (and not drift time) they were able to achieve an position resolution of 0.4 mm. At Boston University a strawless detector is being developed.[22] The idea is to combine two foam molds with cylindrical depressions into a straw chamber. This has the advantage of low mass, ease of construction, and economy of handing. It may well be excellent solution to the short straw arrays required for EMPACT.

Intermediate Tracker

Radial Chamber

The intermediate tracker design is being worked on by several groups in the United Kingdom.[23] The design is based on the radial chambers being built for the H1 experiment at HERA. The details of the design are shown in Fig. 3. and Fig. 24. The present design involves 5 radial chambers position along the z axis. They each span a 50 cm to 150 cm radial space. Each has 300 azimuthal cells with a maximum drift distance of 15.7 mm. Within each module there are 8 layers of drift cells. The long drift times could be a problem at SSC where the bunch crossing is 16 ns., so a plane of tagging counters is placed in front of each radial chamber. These " bunch tagging counters" are shown in Fig. 25. The tagging counter time resolution is about three crossing times. This has been shown to be sufficient to result in good pattern recognition in the radial chamber. The R&D effort at present is concentrating on current draw in the voltage grading resistors. Because of the higher currents in the chambers at SSC this could be a difficult problem. Also the Lorentz angle for the drift gas (CF4) must be measured[24]. This work is already underway and first results were presented at this conference. One of the primary advantages of the radial wire chamber is that the geometry matches the physics in the forward region and an excellent tracking system can be built with very few electronics channels. It is anticipated that the total channel count from both ends will be less than 50K. There is also an active design program on electronics for radial chambers at Rutherford Laboratory. Alternative devices such as microstrip chambers are also under study.

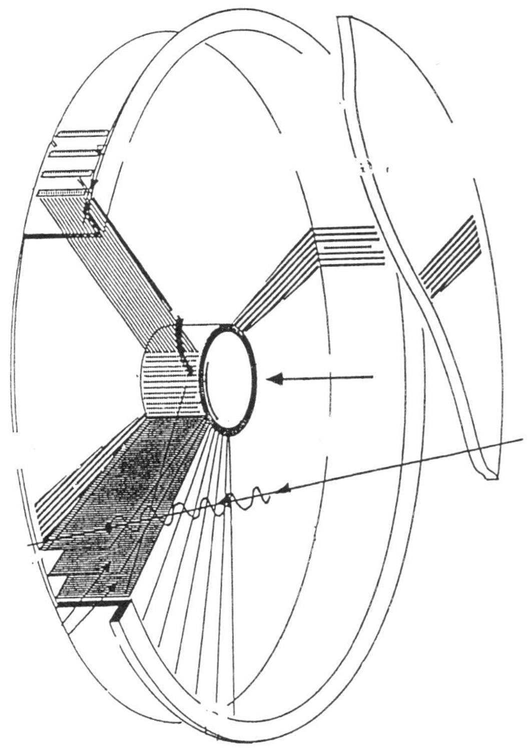

Fig. 24. Details of a radial wire chamber. The SDC version would have larger inner and outer radii and 300 sectors.

New Gaseous Chambers

In the intermediate region there have been a number of proposed detectors in addition to the radial wire chambers. Groups from Japan [25]and Texas[26] have been developing a new type of gaseous discharge chamber called the knife edge chamber with a signal and cathode structure printed on silicon. Reports from both groups were presented at this conference. In the case of the Texas design, an intermediate tracker has been proposed and the design started. Another technique being looked into at University of Rochester[27] is a resistive cathode drift cell. This scheme effectively uses a drift pad with hexagonal pads to fill the forward area. Each pad has a graded voltage for electron to drift toward the center pin where they avalanche. Scanning Tunneling Microscope points could be use for the discharge pins. Work is under way to design in intermediate tracker with such a system.

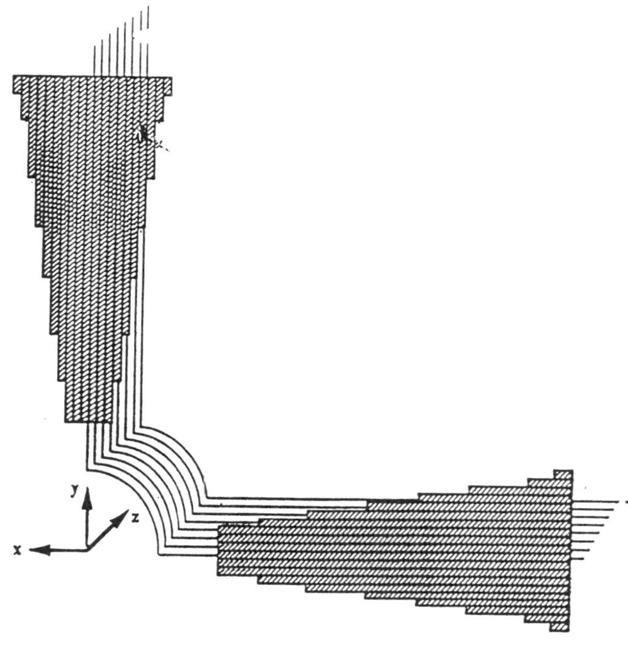

Fig. 25. Details of the bunch tagger counter which covers each sector and tags the proper bunch crossing time.

Summary

Wire chambers can play an important role at the SSC. Although straw drift chamber are a mature technique, there have been a number of innovations in straw drift chambers during the past year. All of the principle SSC detectors have been developing straw chambers and have concluded that such systems are feasible. The counting rates in the straws are high, but Monte Carlo studies indicate that tracking looks promising at and above design luminosity. The chamber lifetimes with CF_4 gases look very good.and radiation effects appear to be no problem at SSC for CF_4 gas mixtures.

In the intermediate region there is a radial wire chamber design under study and there are newer gaseous chambers that look promising.

Full scale tests are being planned to verify tracking at high rates and to confirm the triggering aspect of straw chambers.

[1] The members of the Wire tracking Subsystem group are Indiana University, University of Colorado, University of Pennsylvania, University of Liverpool, University of Glasgow, Rutherford Laboratory, KEK, Westinghouse Science and Technology Center, LBL and Oak Ridge National Laboratory.

[2] The members of the Hybrid Tracking subsystem are CBAF, Duke University, Florida State University, General Electric Canada, KEK, North Carolina State University, Northeastern University, Oak Ridge National Laboratory, Quantum Research Services, Supercomputer Calculations Research Institute, TRIUMF, and University of Pennsylvania.

[3] Asai- International Workshop on detectors for SSC, April, 1990, p260. The original plot was only for hit rate, which was scaled for the other parameters based on calculations at Indiana University.

[4] M. Corden, Proceeding of this conference. He predicts 17% at the first superlayer, and 3% at the final superlayer.

[5] W. Ford.and M Lohner "Track Reconstruction in Straw Superlayrs", Proceedings of this conference.

[6] C. Neyman, "Gain of a 3.65 m Straw Drift Tube", IUHEE-90-12.

[7] C. Lu "Investigations of a Single electrons' behavior in a proportional drift tube." Proceedings of this conference..

[8] J. Kadyk, J. Vavra. and J. Wise " Use of Straw tubes in High-Radiation Environments" Slac-pub 5306, and J. Kadyk,Proceedings of this conference.

[9] R.Henderson " Etching of Anode Wire Deposits with CF_4/Isobutane" Proceeding of this conference.

[10] W. Dunn " Radiation Damage Studies of Straw Tube and Scintillating Fiber components" Proceedings of this conference,

[11] B. Zhou " New Results on SSC Tracking Studies" WorkShop on Major SSC Detectors , Tuscon, Ar. ,Feb. 18,1990 ; S.Ahlen Private communication.

[12] H. Iwosaki " A time resolution study of Straw Chambers.". Kek, 89-158

[13] C. Neyman Attenuation Studies of 3.5 meter Straw Tubes" IUHEE-90-7 ; S. Oh, "Performance of a Prototype 3 meter Straw chamber", Proceedings of this conference.

[14] F. Newcomer et. al "A fast Monolithic Preamp and Shaper for High Rate Gas Tracking Detectors", Penn electronics Publication to appear in IEEE Trans. Nuclear Physics

[15] Y. Arai, "Development of TMC Chip and On-Chip Processing" " Proceeding of International Workshop on Solenoidal Detectors for the SSC, 1990 p 453.

[16] J. Chapman, " Synchronizer Development Update" , Proceedings of this conference.

[17] J. Armitage, "A Straw Tracking System", Proceedings of this conference.

[18] Ogren "Progress Report on 4 mm Straw chambers" International Workshop on Solenoidal Detectors for the SSC, 1990 p 302.

[19] R. Swensrud, " Design of a modular Straw tracking chamber", Proceedings of this conference.

[20] T. Ryan, " Design of the Hybrid Tracker and Utilities" Proceedings of this conference.

[21] V. Polychronakos "Particle Identification and Tracking with Transition Radiation Detectors", Proceedings of this conference.

[22] S. Whitaker, Private communication

[23] University of Liverpool, University of Glasgow, and Rutherford Laboratory are working on this design. D. Saxon " Forward Tracking with Enhanced Electron Indentification", Proceedings of this conference.

[24] J. Baily, " A measurement of the Electron Drift in Fast Gases" Proceeding of this conference.

[25] A.Maki, " The knife edge chamber R&D at KEK" , Proceedings of this conference.

[26] P. McIntyre, "Knife Edge chambers", High resolution tracking for he SSC detectors." Proceedings of this conference.

[27] Sills, "Advanced Field Shaping Drift Chambers for SSC Muon Tracking", Proceedings of this conference.

SILICON TRACKING AT THE SSC

A. Seiden
Santa Cruz Institute for Particle Physics
University of California, Santa Cruz, CA 95064

Abstract

The interaction rates and particle densities in events make silicon detectors an attractive choice for charged-particle tracking at the SSC. Several R&D efforts are underway to solve the technical problems unique to SSC application of such detectors. These R&D areas include optimization of detector hardness, development of fast low-power radiation-hard readout, development of highly stable low-mass supports, and cooling of many dispersed electronics heat sources. Additionally, progress is being made in the area of simulations and optimization of the detector geometrical arrangement.

Introduction

R&D on silicon detectors is being pursued by several groups with interests covering most of the major proposed detectors aimed at SSC physics. The major efforts include work on

1) Pixel detectors [1].
2) Silicon drift detectors [2].
3) Strip detectors of two types [3]:
   a) short strips, where the detector is segmented and readout every 1 cm.
   b) double-sided, long-strip detectors with a readout on average every 12 cm.

This interest is motivated by the excellent precision and two-track separation achievable in silicon (about an order of magnitude better than with typical gas drift detectors) as well as a speed of response which is well matched to the 16 nsec repetition rate of the SSC machine.

The silicon tracking devices being proposed are much larger than presently operating detectors in both area and channel count, necessitating significant R&D. Developments in detectors, fast radiation-hard VLSI electronics, and precision mechanical structures are required.

The work on detector development has typically involved a manufacturer and a group of physicists working together on the design of the detailed detector electrode structure, biasing network, as well as measurements of radiation performance. Some of the vendors and groups involved are listed below:

- Several Japanese Groups (Nagoya, Hiroshima and collaborators), working with Hamamatsu, Japan.
- Groups from U.K., Italy, Kracow and CERN, working with S.I., Oslo.
- Dubna group, working with the Institute for Space Device Engineering, Moscow.
- Oklahoma and Yale groups, working with EGG Solid State Products Group.
- Monolithic Integration of Detectors and Electronics on the same wafer is under development by an LBL Group, and by Hawaii and the Stanford Center for Integrated Systems.

The silicon detectors in use today show a moderate degree of radiation hardness as discussed below. Work to understand the limits to functionality with radiation are in their infancy within the particle physics community. Work in this direction will require further cooperative work between physics groups and potential vendors.

Proposed Detector Types

Figure 1 shows a picture of a proposed pixel detector [1]. The readout electronics and detection element cover the same area and are electrically connected by indium bump bonds. The small area of each pixel results in very small leakage currents per channel as well as very small capacitance, resulting in superb signal-to-noise. The number of electronics channels needed is very large putting a very high premium on low-power per channel. The pixel detectors are being proposed for the inner 10 cm of radial distance at the SSC. Thus the radiation hardness requirements on detector and electronics are greatest for this detector type.

Figure 2 shows a pattern of hits in a beam test for a pixel detector with bonded electronics taken from astrophysical applications. The signals stand out nicely, with the charge shared by a small number of pixels. Such detectors should be able to distinguish particles in very dense jets.

An alternative candidate for a vertex detector is the silicon drift detector. A suggested arrangement is shown in Fig. 3 [2]. In this arrangement the array provides full tracking over a large rapidity range. The detectors are organized in a tower configuration which would be particularly attractive in a nonmagnetic detector or in tracking very high momentum tracks in a magnet.

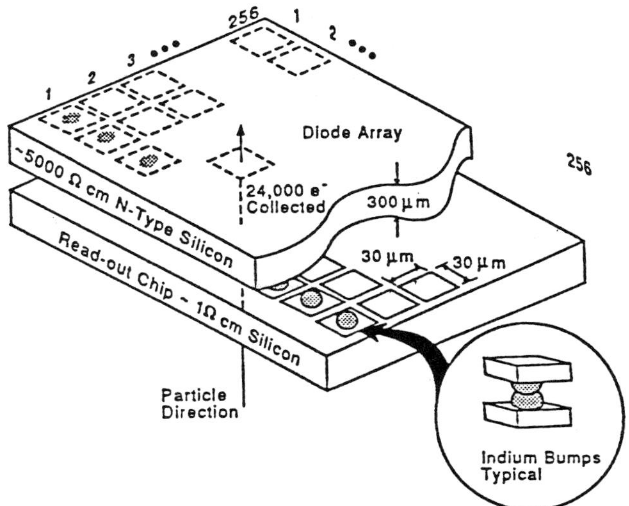

Fig. 1 Schematic representation of a silicon PIN diode detector.

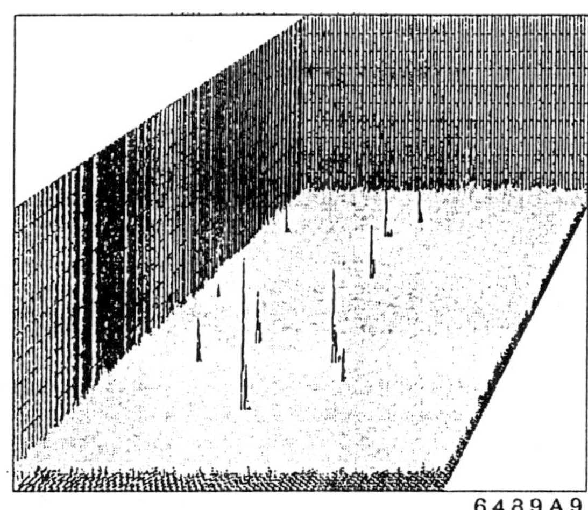

Fig. 2 Pulse heights in pixel detector placed in a particle beam. Peaks come from passage of individual particles.

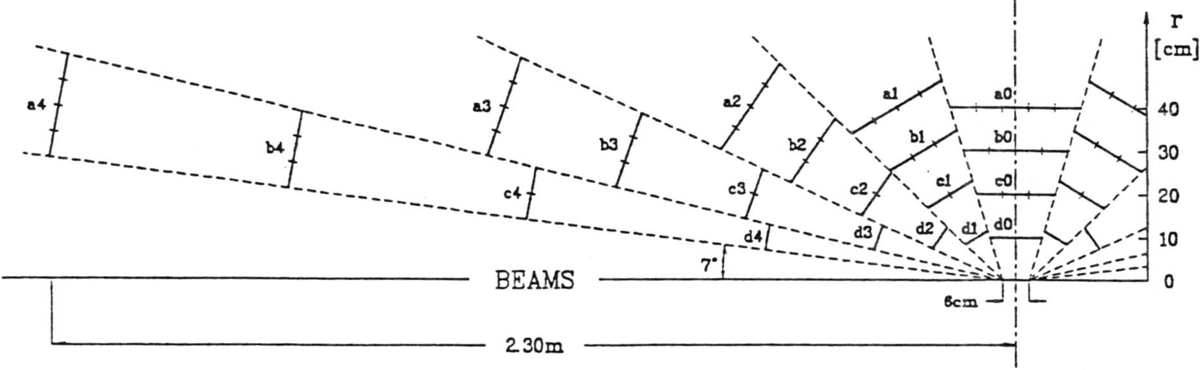

Fig. 3 Cross section along the beam axis of the inner tracking detector based on silicon drift detectors. Detector has a full azimuthal coverage and rapidity coverage $|\eta| \leq 3$. Lines indicate the cross section of rings of detectors centered around the beam axis. Each line represents close layers of silicon drift detectors.

The silicon drift detector has electronics at the edge of the wafer. Using the channel number in conjunction with the drift time allows two dimensional information with a modest number of channels. To unfold the beam crossing time from the drift time two detectors, arranged as in Fig. 4, have to be used.

The principal disadvantage of the silicon drift detector is the dependence of the drift time on the electric field within the detector. This requires a careful control of the detector temperature and more importantly may make it more sensitive to radiation damage than other types of silicon detectors. More quantitative information on the radiation sensitivity should be available next year.

Figure 5 shows a quarter section of an alternative detector configuration composed of silicon strip detectors. It contains a cylindrical barrel section and forward planar disk shaped detectors. The transition from barrel to endcap planar geometry is made at 45° to minimize material. The primary (high resolution) coordinate readout

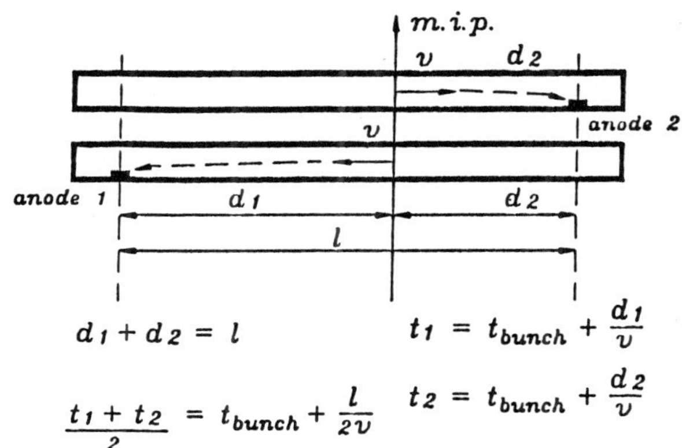

$$d_1 + d_2 = l \qquad t_1 = t_{bunch} + \frac{d_1}{v}$$

$$\frac{t_1 + t_2}{2} = t_{bunch} + \frac{l}{2v} \qquad t_2 = t_{bunch} + \frac{d_2}{v}$$

Fig. 4 View (not to scale) of two silicon drift detectors placed in such a way that a particle crosses both detectors. The produced electrons drift in opposite directions with the same speed v. The sum of drift distances for any position of the incident particle is constant and equal to the distance between anode1 and anode2. The average of two arrival times gives the time of the bunch crossing.

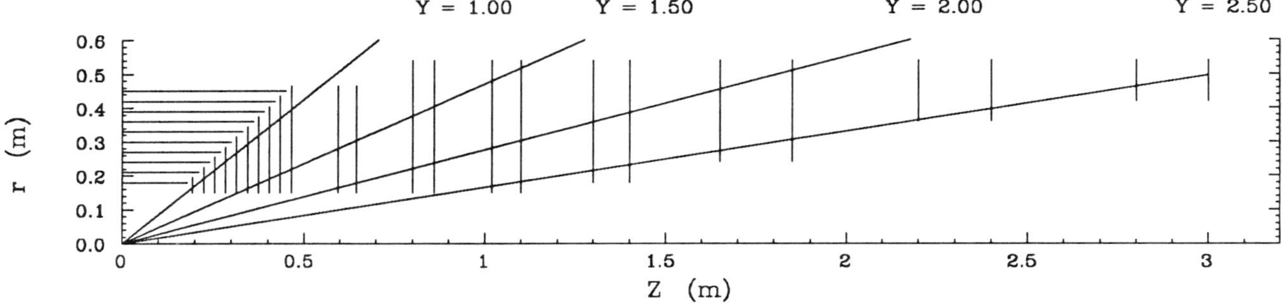

Fig. 5 SSC silicon microstrip tracker.

is the ϕ coordinate for both barrel and planar detectors since this is the coordinate exhibiting bending in a solenoidal magnetic field. The other coordinate (z in the barrel, r in the forward planes) is gotten using small angle stereo (5 mrad stereo angle) for the second coordinate of a double-sided detector, or by segmenting the single-sided readout into short segments of length about 1 cm.

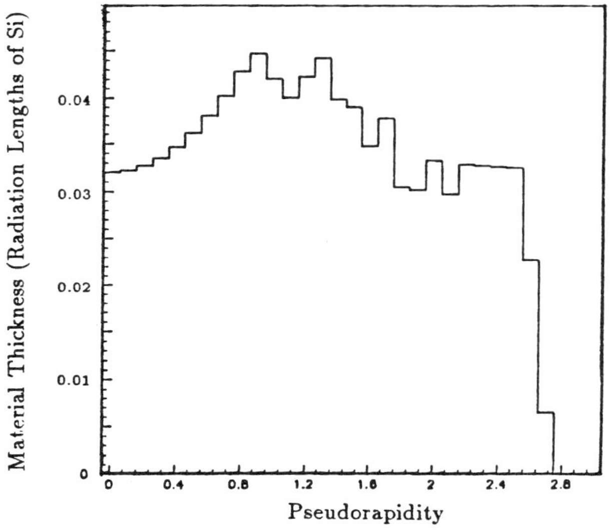

Fig. 6 Material for silicon alone.

Figure 6 shows the material in radiation lengths crossed by a track coming from the origin for the silicon alone. The average is about 3.5% and illustrates the importance of minimizing the material in support structures since the 3.5% is already a significant number. For the detector in Fig. 5 the total silicon area is about 38 m^2 and the number of electronics channels would be about 11 million for the double-sided detector option assuming 50 μm readout pitch throughout the detector. In the discussion below we present further results on physics simulations and radiation damage for this large strip array.

Physics Considerations

The trackers in Figs. 3 and 5 are very long in order to cover a large range in rapidity. This complicates mechanical support problems, but is necessary at a hadron collider where particles of interest often emerge at small angles to the beam. Figure 7 shows the expected efficiency to contain all four leptons from the "Standard Model" Higgs boson decay. A detector covering rapidities up to about 2.5 (10° to beam) is a practical compromise giving reasonably good acceptance without having to cover impractically small production angles.

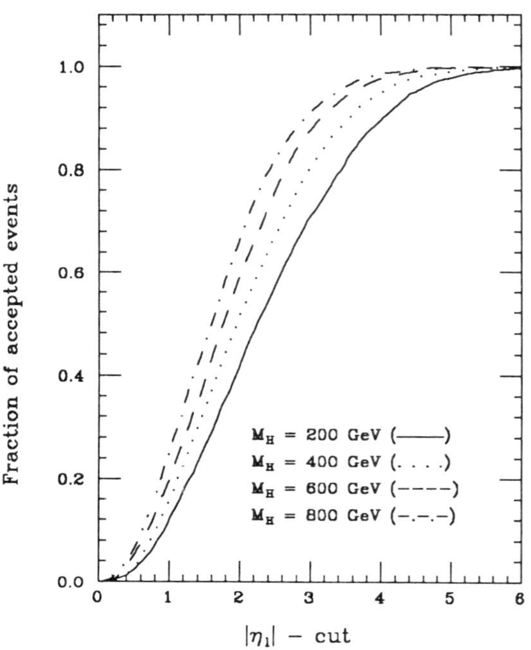

Fig. 7 Acceptance for all four leptons from $H^0 \rightarrow Z^0Z^0$ decay vs. rapidity coverage.

The presence of silicon tracking in a detector should contribute significantly to Higgs boson searches, top quark studies (top quarks yield B's and thus can be tagged by vertexing) and very high transverse momentum jets. Table 1 below lists some of the special capabilities brought to a magnetic detector by a silicon tracker.

We can look at some of the capabilities for a silicon tracker within the context of a large magnetic detector

Table 1. Special Capabilities of Silicon Tracking System

(1) Measurement of dE/dx in a few inner layers: to search for oddly ionizing particles and to reject overlapping electron-positron pairs coming from photon conversions.

(2) Vertexing: look for long-lived particles and special events yielding B's or τ's in the final state. Assignment of leptons among vertices. For multiple events in a time bucket, allows assignment of tracks to each event.

(3) For magnetic detector contributes to momentum resolution. It makes a particularly important contribution to momentum resolution at large rapidities, where tracks are measured with a small lever arm.

(4) Very good efficiency over $|y| \leq 2.5$ to reconstruct high $P_\perp$ isolated tracks, e.g., for the four charged leptons in H^0 decay.

(5) Very good lepton identification for the overall detector: for e by matching with the calorimeter, for μ by matching with the outer muon identifier.

(6) Track reconstruction in high $P_\perp$ jets: measurement of multiplicity and most tracks in events with $P_\perp$ up to 1 TeV. This could be important in understanding jet parentage and allows reconstruction of leptons in jet environments.

[4]. Figure 8 shows the impact parameter resolution for the large strip array of Fig. 5 as well as for such an array supplemented by pixel detectors. Impact parameter resolutions significantly smaller than the average B decay length can be achieved down to 1 GeV/c transverse momentum for the full silicon tracker.

Perhaps the most important goal for an SSC detector is the search for the Higgs boson. This has been looked at in some detail for Higgs masses of 300 GeV and 800 GeV for the large strip array using a Monte Carlo program [5]. The analysis is outlined in Table 2 below.

Figure 9 shows the reconstructed Z^0 mass using the tracking detectors alone (both Si and Si plus outer tracking devices). The low mass tail comes from radiation by electrons in the material of the detector. For this simulation the detector support structure has been assumed to double the material provided by the silicon. The good mass resolution can be recovered by using the calorimeter measurement of the electron with its nearly overlapping Bremsstrahlung photons, provided the radiation is not too great. This illustrates the importance of minimizing the overall material.

Figure 10 shows the efficiency for reconstructing all four leptons from the H^0 as a function of the number of simultaneous uncorrelated overlapping events. This shows the potential for high luminosity tracking using the large strip array, since the efficiency remains very high even for 12 overlapping events (the largest number in the study).

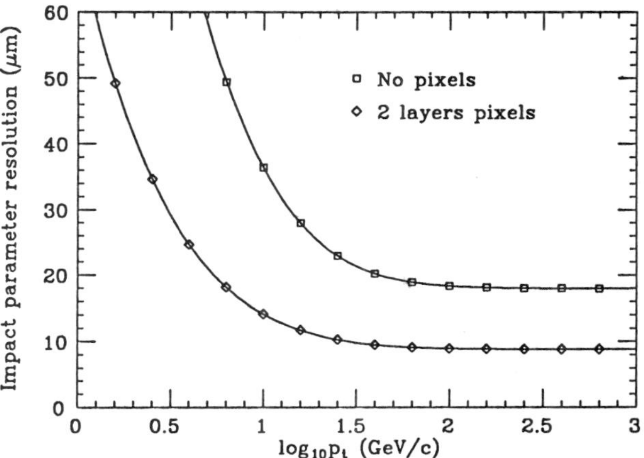

Fig. 8 Impact parameter resolution at the interaction point as a function of momentum for the pixel/silicon vertex detector system for a 90° track.

Table 2. Analysis: $H^0 \to Z^0 Z^0 \to \mu^+\mu^-e^+e^-$

(1) Generate Monte Carlo hits in detector array.

(2) Use pattern recognition program to associate points into charged tracks.

(3) Fit tracks to get momentum vectors.

(4) Calculate invariant mass of pairs of $e^+e^-, \mu^+\mu^-$ to find Z^0's.

(5) Calculate efficiencies to find H^0 within rapidity acceptance of the detector array, based on mass cuts for the Z^0's.

The limitations in luminosity for the silicon at present come mainly from radiation damage. Fixed charge damage sites change the electric field within the silicon requiring changes in voltage which eventually can lead to breakdown. This appears to be a more serious limitation than leakage current generation in the detectors which has received wide attention in the past.

In the coming year, we can expect more careful measurements to clarify the issues. Several SSC R&D groups with interests both in tracking and calorimetry will be looking at damage mechanisms. A practical question is whether we can improve hardness by the reduction of specific impurities, e.g., phosphorous or oxygen, in the detectors.

Figure 11 gives the fluence of particles for the large strip array within the large SDC solenoid per year of running at the SSC design luminosity. Present indications are that the limit to detector lifetime occurs around 10^{14} particles/cm^2. Thus even the innermost silicon detectors should survive about 10 to 20 years at the design luminosity. The fluence of charged particles, rather than

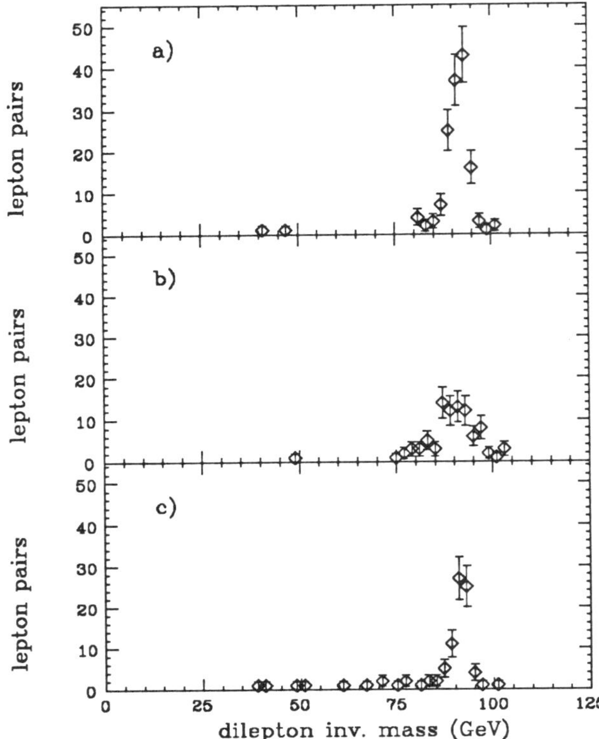

Fig. 9 Reconstructed invariant mass of lepton pairs from Higgs to Z^0 decays. a) Using silicon strip tracker only, for events with Higgs mass = 300 GeV. b) Events with 800 GeV Higgs mass. c) Using combined tracking system for 800 GeV Higgs.

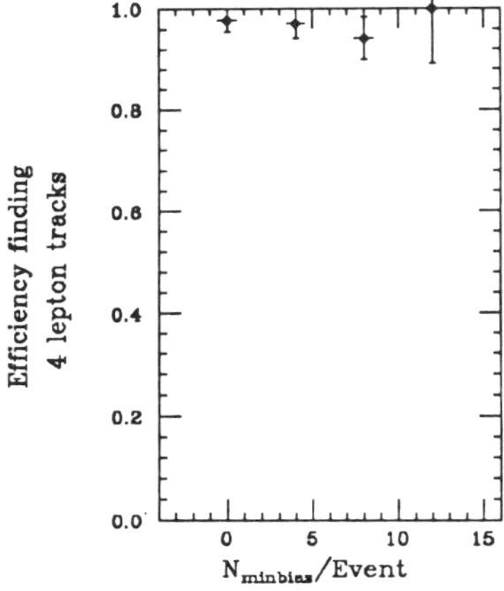

Fig. 10 Efficiency for finding all four leptons from $H^0 \rightarrow Z^0Z^0 \rightarrow e^+e^-\mu^+\mu^-$ vs. number of overlapping minimum bias events.

albedo neutrons, dominates the damage within a 30 cm radius. For radii of about 5 cm, the lifetime becomes only 1 to 2 years. Further work on improving hardness is quite important for detectors this close to the beam.

Developments in Electronics

The electronics for silicon readout has to satisfy a number of very stringent requirements. These are:

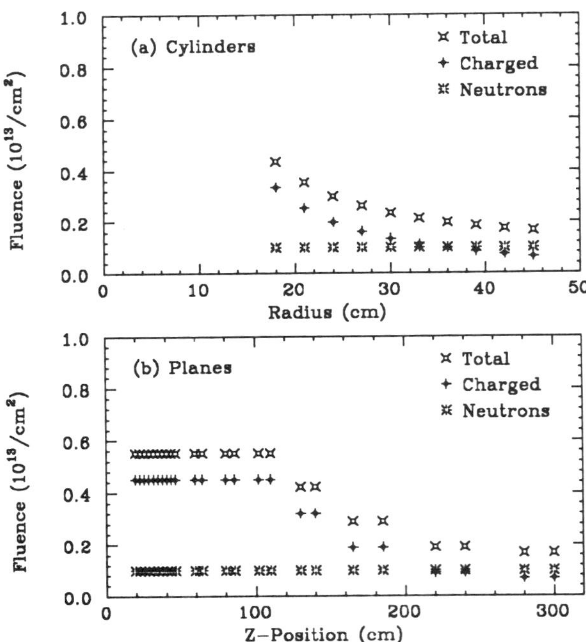

Fig. 11 Maximum particle fluence in one year running at $\mathcal{L} = 10^{33}\text{cm}^{-2}\text{sec}^{-1}$, normalized to 1 MeV neutrons, for cylindrical detectors (a) and planar detectors (b). The fluence of neutrons and charged particles (normalized to neutron damage coefficient) separately indicated. Only the worst case subdetector at each radius or Z position is plotted.

(1) It has to function after dosages of 5 to 10 Mrad of charged particles and $10^{14}/\text{cm}^2$ of 1 MeV neutrons.

(2) To do timing at the single beam crossing level requires fast front ends with risetimes in the 20 nsec range.

(3) Analog or digital circuitry has to store the large amount of information and output the information for the interesting crossings chosen by external triggers.

(4) In the case of pixels the electronics must fit into the pixel area, in the other cases into a linear dimension of about 50 μm.

(5) Because of the large number of channels, the minimization of power per channel is a major goal. As an example, for the large strip detector 1 mW per channel is a design goal.

(6) The signal-to-noise has to be good enough so that random counts are well below the count rate from the many uninteresting low transverse momentum tracks in SSC events.

Some of the major efforts in electronics are listed in Table 3.

Significant progress is being made on all the critical electronics issues. Radiation hardness of high speed, small feature size bipolar has been demonstrated as well

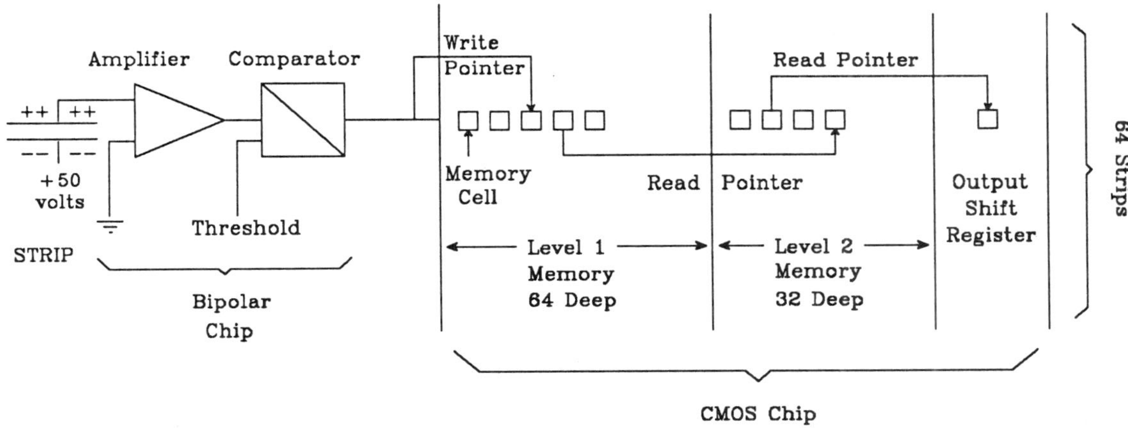

Fig. 12 Readout – VLSI Chips.

as the digital use of hardened CMOS [6]. Careful tests of analog CMOS will be made this coming year. CMOS designs which meet the power requirements are being tested [7,3]. Circuit designs for adequately low power bipolar now also have been completed [3].

An example of a readout scheme for the large strip array is shown in Fig. 12. In this case, the front end is shown as a bipolar chip and the storage, which is digital and clock driven, is in a separate radiation-hardened CMOS chip. An attractive alternative to two chips would be the use of BiCMOS.

Table 3. Front-end Electronics Development Work

The KEK group is working with NTT using 0.8 μm Bipolar process.

UCSC and Silicon Dynamics are working on a full custom Dielectric Isolated Bipolar chip.

Oak Ridge National Laboratory and Oklahoma are testing several Bipolar processes (from VTC and Harris).

The LBL group is working on low-power CMOS design (radiation-hard version to follow using UTMC process).

The U. of Penn. group is studying CMOS analog structures made by UTMC.

Hughes Aircraft plus LBL are developing a pixel readout.

BNL, Milan and Munich are working on a readout for silicon drift detectors.

Mechanical Considerations

The major work on mechanical issues has been done for the large strip array. This detector will allow position determination with an accuracy of about 15 μm per measurement. To avoid degrading the detector performance it is important that the mechanical system which holds the detectors provide position stability of about 5 μm. In addition the amount of material that can be tolerated is at most a few per cent of a radiation length.

Figure 13 shows a picture of the suggested support framework planned for the detectors. The detailed design has been done by an engineering group at the Los Alamos National Laboratory. The central detector wafers are glued together into pairs to form silicon boxes. These are supported on three cylinders which are in turn held on a large frame which also supports the forward planes. Figure 14 shows the units from which the frame is made. Figure 15 shows the frame in more detail. It is made of cylindrical tubes which connect together. To provide adequate positional stability, the material for the detector framework has to be chosen to avoid creep as well as dimensional change under radiation. It must be built in such a way as to avoid putting stresses on the brittle silicon detectors as temperatures change.

A highly desirable property for the frame would be a negligible coefficient of thermal expansion. This can be achieved by the use of graphite–metal matrix composite materials, which also have adequate strength and reasonably good thermal conductivity. Figure 16 shows the coefficient of thermal expansion versus the composition of the material. A value of zero with mostly graphite, which has a large radiation length, is possible.

Figure 17 shows the material intersected by a particle for the space frame support as a function of azimuth at a 30° production angle, which is chosen as an example. Shown in the figure is an all beryllium support (best material from a radiation-length point of view, but not best thermal properties) as well as the metal matrix material. The flat baseline is the material from the beampipe and outer detector enclosure. The average number of radiation lengths is less than 1.5%.

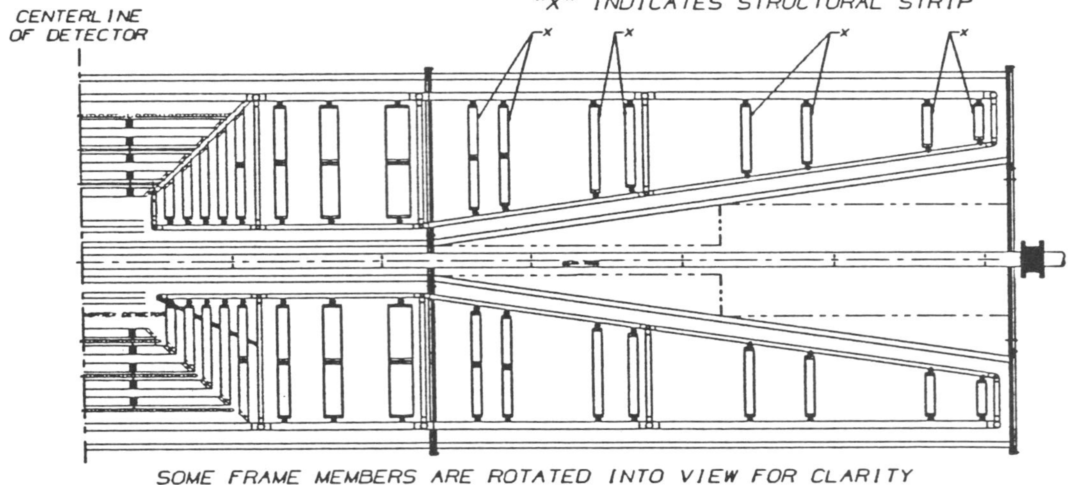

Fig. 13 Schematic drawing of silicon detectors and support members.

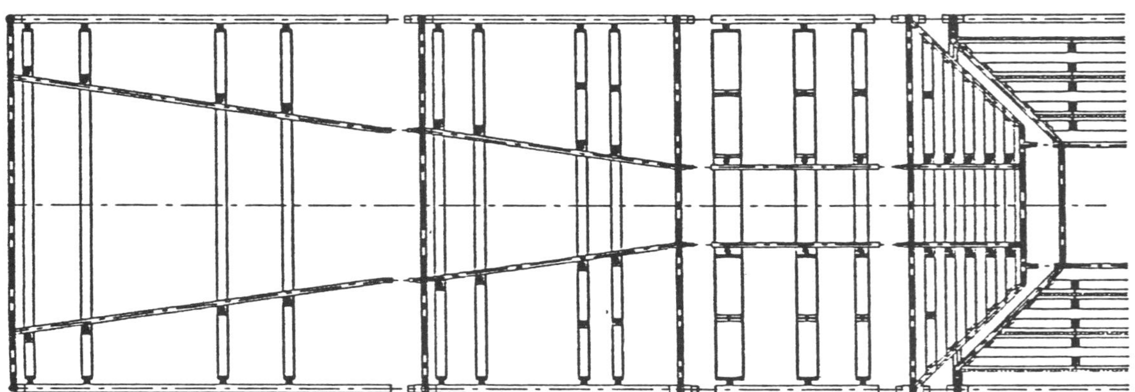

Fig. 14 Schematic drawing of pieces from which the detector is assembled.

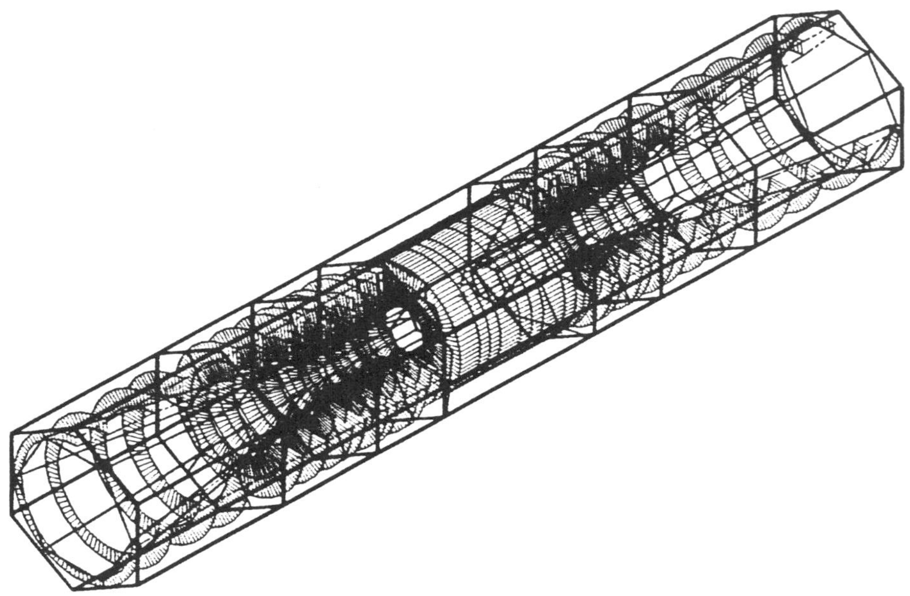

Fig. 15 Support framework for detectors in the forward region.

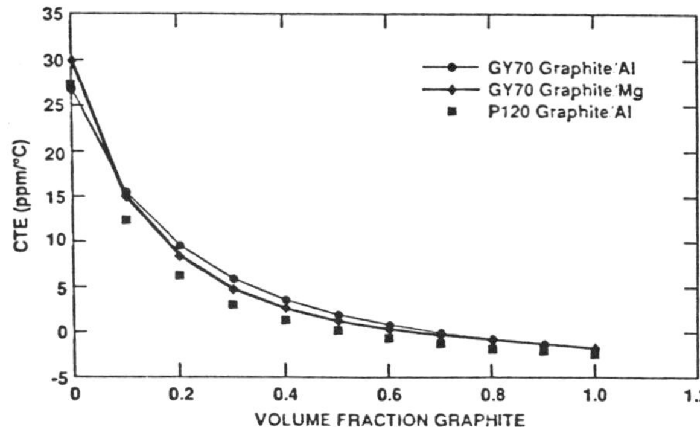

Fig. 16 Isotropic thermal expansion.

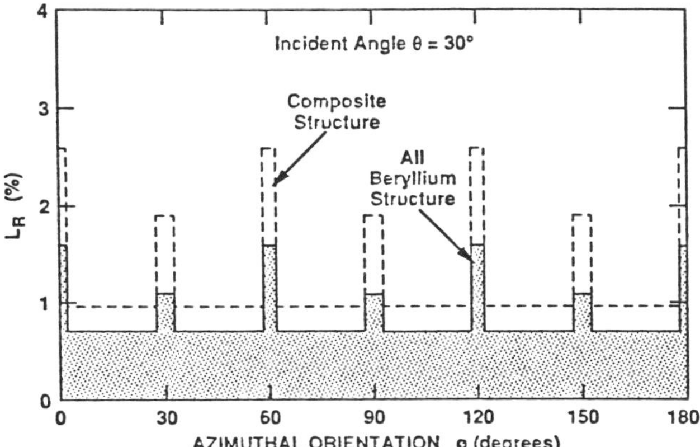

Fig. 17 Per cent radiation lengths crossed by a particle produced emanating from the origin at 30° to the beam as a function of azimuthal angle.

An end view of the barrel detector is shown in Fig. 18. The paired detectors, tilted to compensate for the magnetic fields influence on charge collection in the silicon, and the three support cylinders are shown. The most likely material for the cylinders are thin carbon fiber skins sandwiching a corrugated ribbon filler. The silicon pairs and servicing rings which carry coolant and cables are shown in Fig. 19.

The plan for the cooling of the detector electronics involves evaporative cooling using heat pipes. Such a system is extremely efficient, involves small fluid flows, and will equilibrate the temperature globally. The elements involved in such a system are:

(*a*) Reservoir to provide fluid, which requires a few cm diameter input line.

(*b*) Arteries to feed fluid to circuits.

(*c*) Wick structure to take heat from the electronics. This is a set of capillaries which soak up the liquid. This is self-regulating—the amount of heat drives the absorption rate of liquid by the wick. The temperature is set by the liquid evaporation temperature.

(*d*) The vapor can fill the detector cavity (if it is non-reactive) or can be returned along sealed paths.

(*e*) Outside the detector volume a condenser turns the vapor back to liquid.

(*f*) Our present choice of cooling fluid is a hydrocarbon. This allows a range of temperature choices, however has the disadvantage of involving a flammable substance.

Conclusions

Silicon tracking detector R&D for the SSC is a very active field involving scientists from the U.S., Japan, Western Europe and the USSR. Work has focused on:

(1) Tracking simulations, which point to excellent performance possibilities.

(2) Detectors. Here work is just beginning to optimize detectors for the SSC. Present-day detectors already offer the possibility of some high luminosity capability.

(3) Work on electronics has been a very active area and we should have several circuits for testing (and further optimization) this year.

(4) Mechanical support issues are under active investigation with construction of prototype support and cooling structures a very high priority for next year.

We can expect the upcoming two-year period to be used to optimize all of the areas above.

References

1. SSC Detector R&D Proposal: Development of Technology for Pixel Vertex Detector, U.C. Berkeley Space Science Laboratory, J. Ahrens, Spokesperson, October 2, 1989; *ibid*, 1990.
2. W. Chen *et al.*, SSC Detector Subsystem R&D Proposal to Develop Track and Vertex Detector Based on Silicon Drift Devices, BNL, 1989; W. Chen *et al.*, SSC Detector Subsystem R&D Interim Report on Silicon Drift Devices for Tracking and Vertex Detection, BNL, 1990.
3. A.J. Weinstein *et al.*, Subsystem R&D Proposal to Develop a Silicon Tracking System, U.C. Santa Cruz, 1989; *ibid*, 1990.
4. Expression of Interest, Solenoidal Detector Collaboration, Spokesperson, G. Trilling (May, 1990).
5. J. Hylen *et al.*, Silicon Tracker Conceptual Design Report, SCIPP 90/24 (September, 1990).
6. H.F.-W. Sadrozinski *et al.*, Radiation-hard Front-end Electronics and Silicon Microstrip Detectors, these proceedings.
7. Front End Electronics Development for SSC Detectors, H.H. Williams, Spokesperson [1990], University of Pennsylvania.

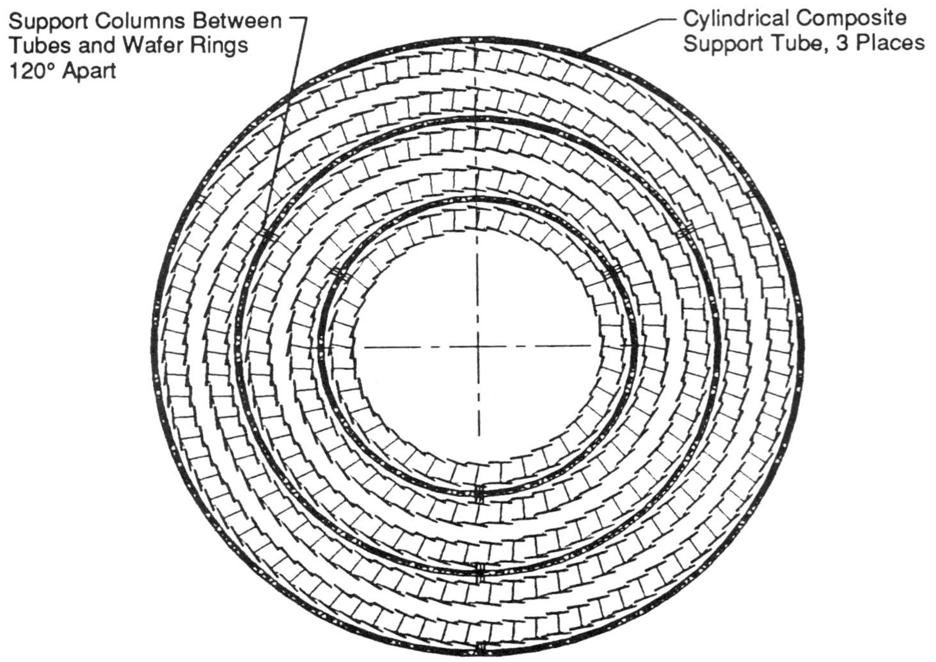

Fig. 18 End view of barrel detector.

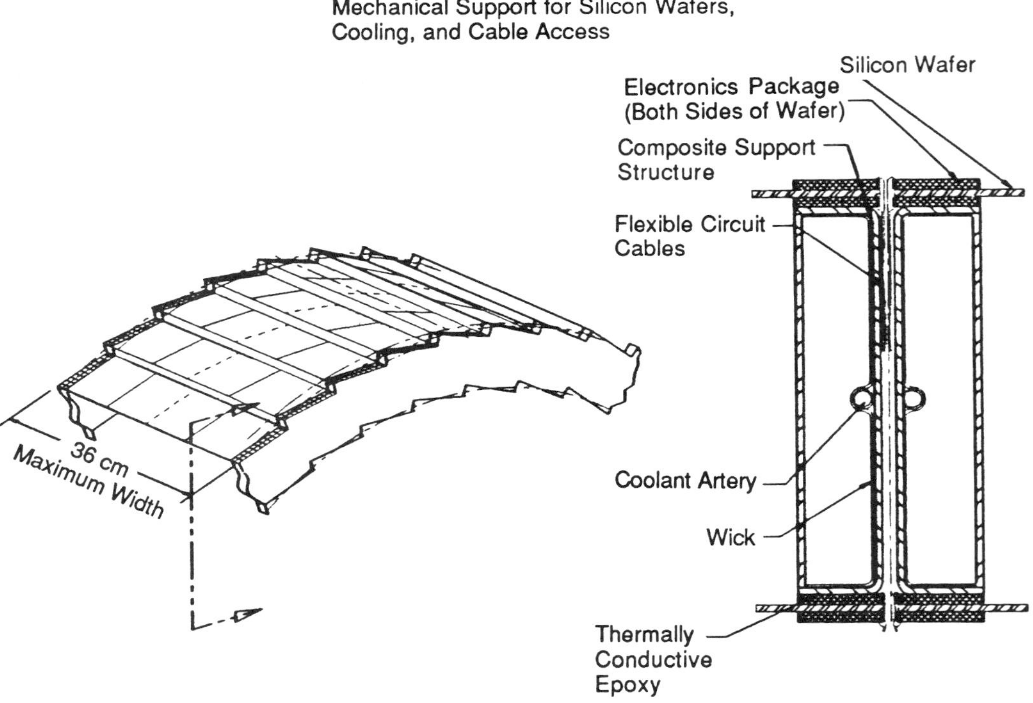

Fig. 19 Silicon boxes and end cooling rings. Design assumes double-sided detectors with end readout. Maximum detector length readout is assumed to be 18cm.

PRECISION MUON TRACKING

Ulrich BECKER
Massachusetts Institute of Technology
Cambridge, Massachusetts 02139, USA

A detector for precision measurement of Muons at the SSC is described which uses high resolution magnetic analysis in a solenoidal field with multisampling drift chambers and minimal material in the path. L3 expertise and specific R+D results to facilitate the construction of such a detector are described.

Introduction

To get the importance of precision tracking in proper perspective let's review the evolution of Physics based on precision detection of e and μ. The discovery [1,2] of $J \rightarrow e^+ e^-$ (Fig. 1) and $\Upsilon \rightarrow \mu^+ \mu^-$ were enabled by double arm spectrometers with resolutions of 0.2 and 2.0% $Z \rightarrow e^+ e^-$ was discovered [3] by UA1 at CERN having good $e^+ e^-$ separation and resolution. Detecting $\mu^+ \mu^-$ in collider detectors is possible (Fig.2) in low background environments. However, the resolution does not permit to see the 3 states separated, which the fixed target spectrometer of E605 at FNAL clearly resolves (Fig.3).

HIGH RESOLUTION MUON DETECTORS FOR COLLIDERS

A new generation of precise collider detectors [4] started with the design of the L3 experiment at LEP, CERN. Muons were directly detected by their bend in a 0.5 T solenoidal field by measuring their deflection of 3.7 mm at 45 GeV accurate to 2.4% (Fig.4). Multiple scattering and alignment were limited to <30 μm over the measuring length 1 of

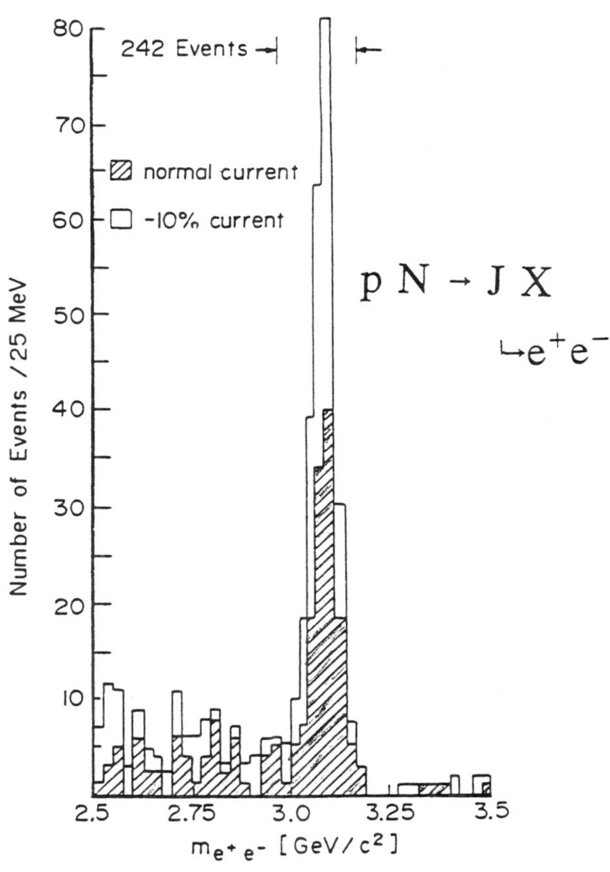

Fig. 1 Discovery of J at BNL [1]

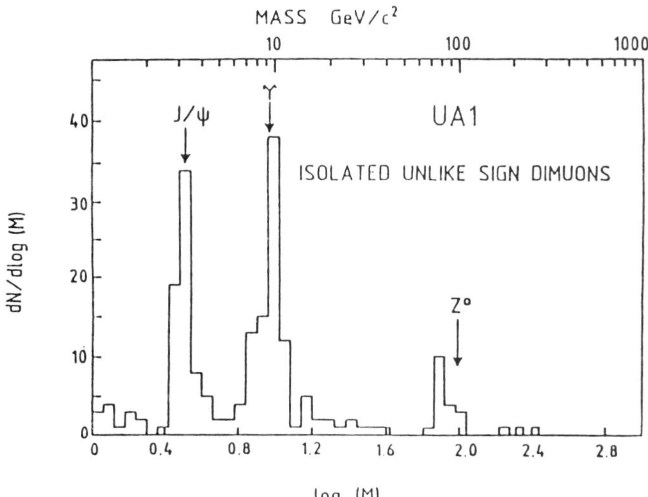

Fig. 2 Dimuon resonances observed by UA1.

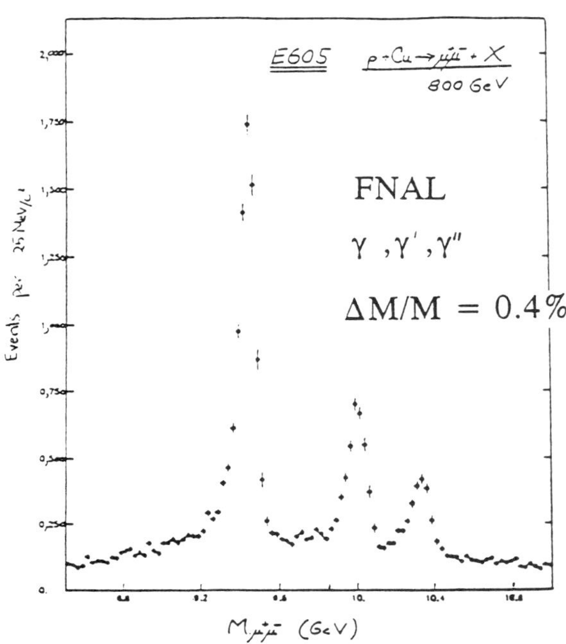

Fig. 3 ϒ-resonances resolved by 0.4% mass resolution of E605.

the spectrometer having an analysis power of $Bl^2 = 4.2\ Tm^2$. The layers of multisampling (16-24-16 wires) drift chambers define the track to about $250\ \mu m/\sqrt{16} = 50\mu m$. This corresponds to :

$$\frac{\Delta S}{S} = \frac{\Delta P}{P} = 2.4\% \text{ at } 50 \text{ GeV}$$

The critical alignment of the three chamber layers of each module is ensured by optical straightness

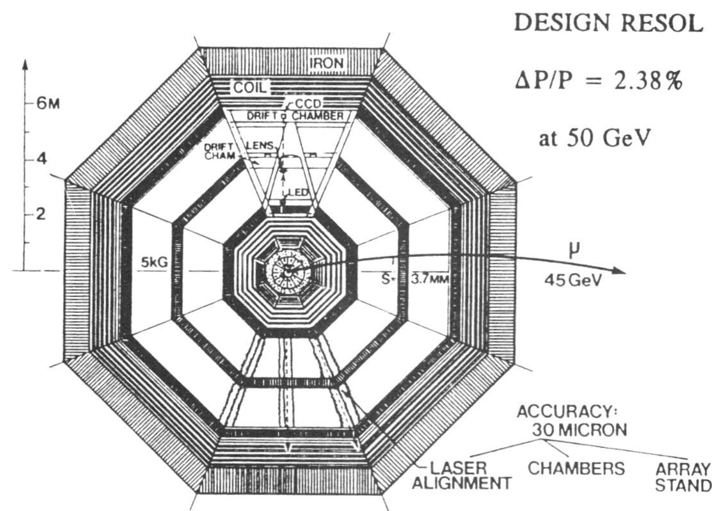

Fig. 4 L3 concept of muon detection.

monitors [5]. The middle chambers are adjusted initially and tracked by these monitors there-after. The resulting corrections are small. The alignment was verified by UV lasers and cosmics. However, the most stringent and realistic test come from the $Z \to \mu\mu$ reconstruction of L3 events from LEP. Indeed, the predicted resolution of

$$\Delta p/p = 2.4\% \text{ at } 46 \text{ GeV}$$

is obtained , see Fig. 5.

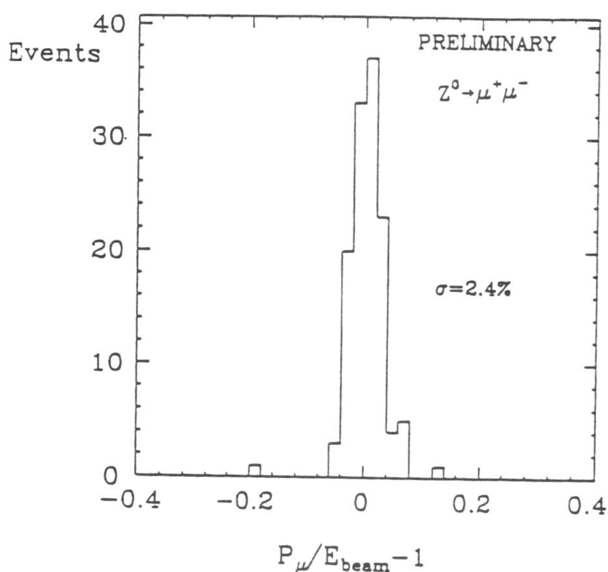

Fig. 5 Measured resolution of the L3 spectrometer using $Z \to \mu^+ \mu^-$

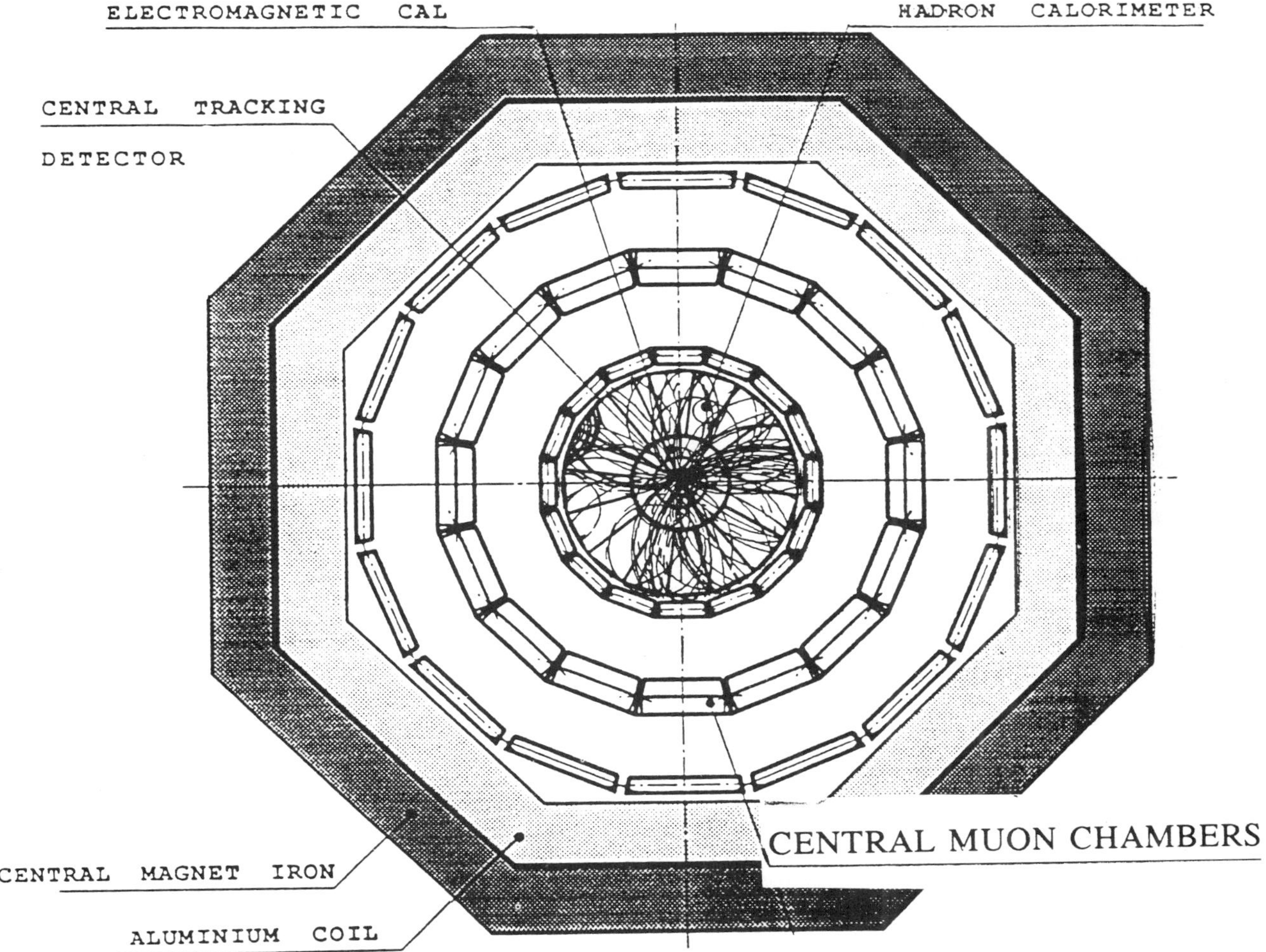

Fig. 6 L* detector, end view with computer generated $H \rightarrow 4\mu$ event. Also shown: the hadronic tracks which will be absorbed in the calorimeter.

A DETECTOR FOR SSC L* [6]

Owing to the high energy of SSC the events will be very complex. As Fig. 6 shows, the early absorption of hadrons in the calorimeter projects the clear signature of $H \rightarrow 4\mu$ into the outer muon detector of L*, as shown in Fig. 6. Comparing the measured muon momenta :

$$p(\text{muon}) = p(\text{vertex}),$$

with the vertex determination ensures good muon identification. Muons exiting materials are known to be associated by low energy tracks confusing the position in the next chamber. Here, due to the field of the solenoid, those low energy tracks are swept away, not impairing the muon momentum determination.

For multimuon event reconstruction full coverage of the detector is extremely important. Fig. 7 gives the side view of the detector covering all angles down to 2^0 to the beam.

This is where the particle rates will be highest. However, having blocked out the hadrons by calorimeters, the muon chambers are exposed to moderate rates only. 150kHz as highest rate on a long wire demand attention for ageing, but poses no real problem.

The aim for resolution is given by the physics of identifying Z's up to 0.5 TeV in the best possible manner, since :

$$\Gamma_z = 2.5 \text{ GeV}$$

250 GeV

21^0

250 GeV

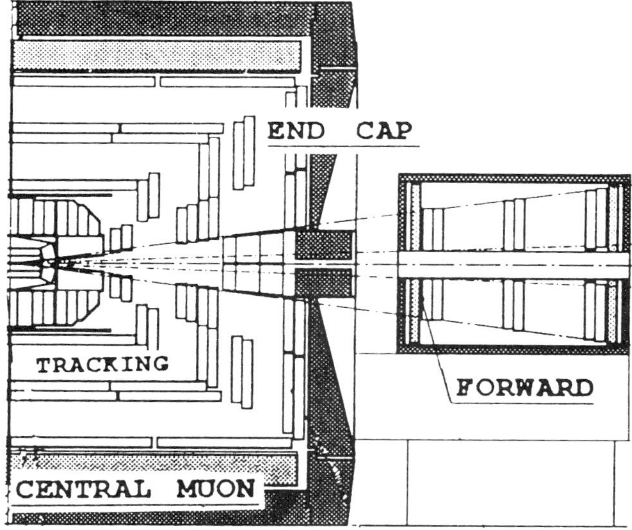

Fig. 7 Side view of L* detector.

$$\frac{\Delta m}{m} = \frac{1}{2}\left[\left(\frac{\Delta p_1}{p_1}\right)^2 + \left(\frac{\Delta p_2}{p_2}\right)^2 + \left(\frac{\Delta\Theta}{\tan\Theta/2}\right)\right]^{1/2}$$

$$< \frac{\Gamma_z/2.3}{m_z} = 1.22\%$$

Hence $\Delta p/p$ = 1.2% at 250 GeV
or = 2.4% at 500 GeV is required.

The resolving power of the spectrometer

$$\frac{\Delta p}{p} \;\alpha\; \frac{1}{Bl^2}$$

is cho sen to be large by B=0.75T, and l = 5.25m for the effective measuring length. Multisampling chambers with resolutions of 150μm/wire determine the track with 32,(64) fold averages to

$$\varepsilon_1 = \frac{150\mu}{\sqrt{32}} = 27\ \mu m \qquad (\text{and } \varepsilon_2 = 19\mu m).$$

As shown in Fig. 8 this translates into a sagitta error of

$$\Delta S = [\varepsilon_1^2/2+\varepsilon_2^{\,2}+\Delta^2_{sys}+\Delta^2_{ms}]^{1/2}$$

$$= [27^2/2+19^2+20^2+15^2] = 37\mu m.$$

In 0.75T the sagitta for a 0.5 TeV muons is s= 1610μm, hence

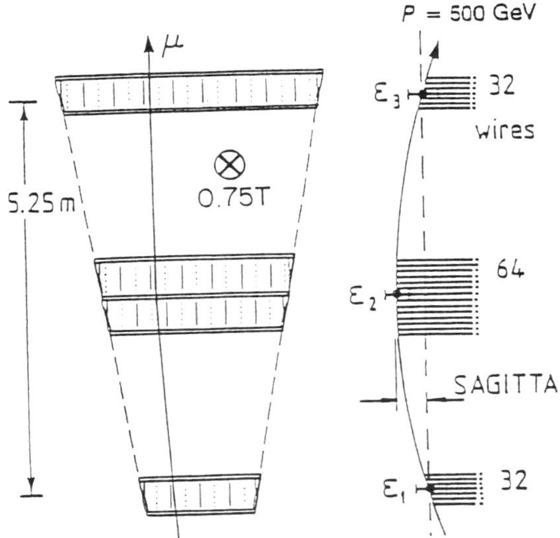

Fig. 8 Sagitta measurement by multisampling chambers.

$$\frac{\Delta s}{s} = 2.4\% \text{ as demanded.}$$

As seen in Figs. 6,8 the middle chamber layer consists of 32 and 32 wires offset by 1/2 cell width as schematically depicted in Fig.9.
Besides the technical simplification to construct units of 32 wires only, this has the following advantages :

a) Every normal track has a self calibration by T_1+T_2 = constant.

b) The left-right ambiguity of the drift chambers is resolved.

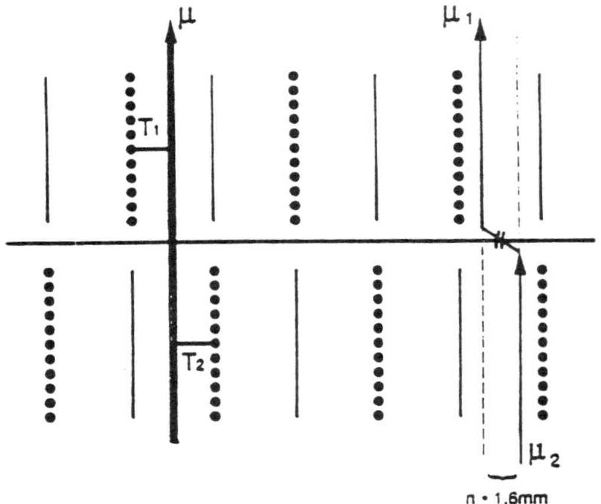

Fig. 9 Middle chamber

c) Tracks from wrong (by n) beam crossings drift n.1.6mm apart and can be rejected.

d) Similarly due to ToF, cosmics will show ~ 4 mm discontinuities enabling their elimination.

Summarizing the accuracy requirements we have :

average tracks in chambers: 20 μm (30μm), array stands known to < 20 μm, calibration by UV lasers and cosmics to < 20 μm.

Based on L3 experience these demands can be met. In particular, the alignment knowledge to 20μm of the relative chamber positions will be provided by the optical straightness monitors of a prealigned matched triplet of LED, lens, and quadrant receiver, as seen in Fig. 10. These systems are derived from L3 experience and record accurate to 1μm.

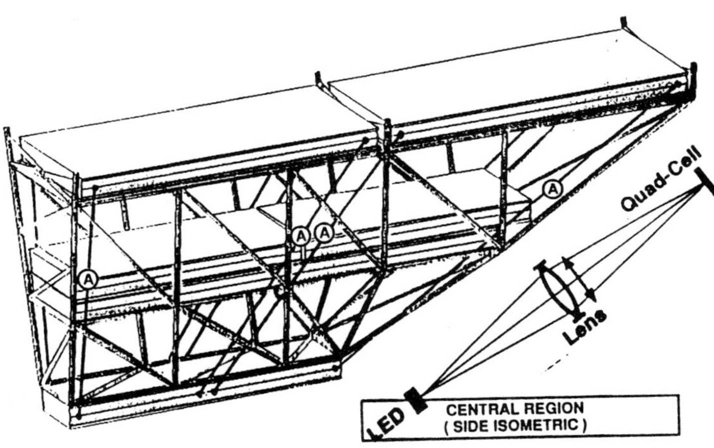

Fig. 10 Central detector module aligned by optomechanical system.

However the larger size of L* and the SSC energies necessitate specific research and development.

R&D Tasks

The PRECISION MUON CHAMBER COLLABORATION addresses the following areas [7]:

Chambers : MIT, LLNL, John Hopkins, Purdue, Princeton.

Gas Research : MIT, Harvard, ORNL

Structural Design : Draper Lab, MIT

Alignment : ETH Zurich, Univ. of Geneva.

Data transfer : ETH Zurich, Univ. of Geneva,

engineering coordination by Prof. Samuel C.C.Ting, MIT.

MUON CHAMBER R&D

The necessity for R&D can be seen from the comparison of L* to L3.

	L*	L3
Wires, total	1200 k	(260k)
Wires/plane	32/64	(16/24)
Chambers	304	(80)
Accuracy	20μm	(30)

In L3 the accuracy of wire positioning was given by optically flat (<2μm) glass pieces arranged in "bridge" like structures to <5μm accuracy as the interferometer measurements in Fig. 11

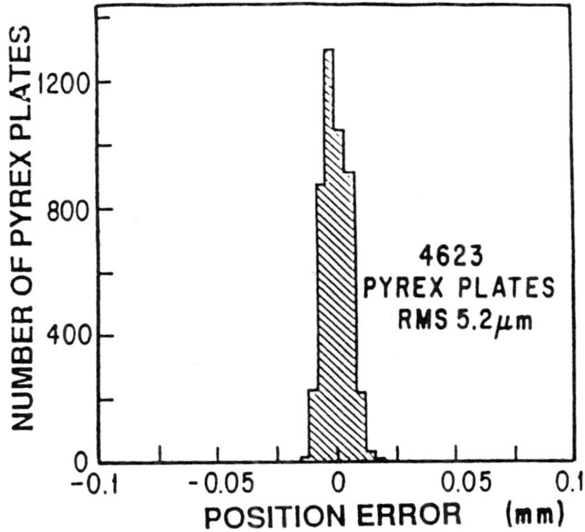

Fig. 11 Accuracy of locating glass plates.

demonstrate. Having achieved < 10μm of final location accuracy of wires in L3 ; the < 8μm requirement for L* is reasonable considereing the similar wire location procedures.

The principal components of the L3 chambers are given in Fig. 12.

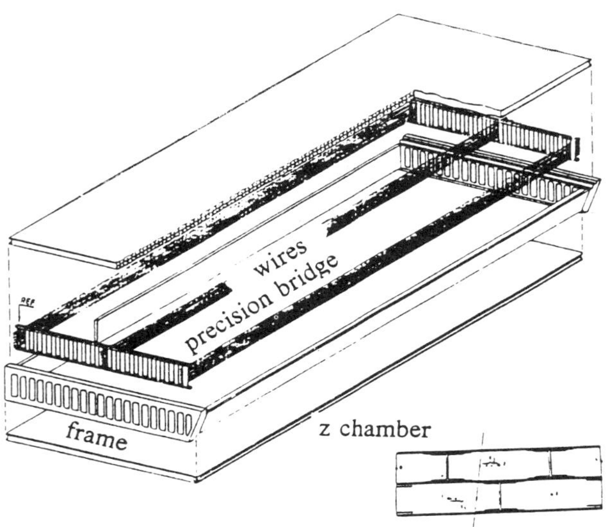

Fig. 12 L3 precision chamber

The chambers were accurate yet mass produced (80) with well documented quality control and test procedures. From this knowledge experts [8] recommended to:

> Eliminate 60% of all wires by using foils wire meshes instead of wire cathode planes, see fig. 13.

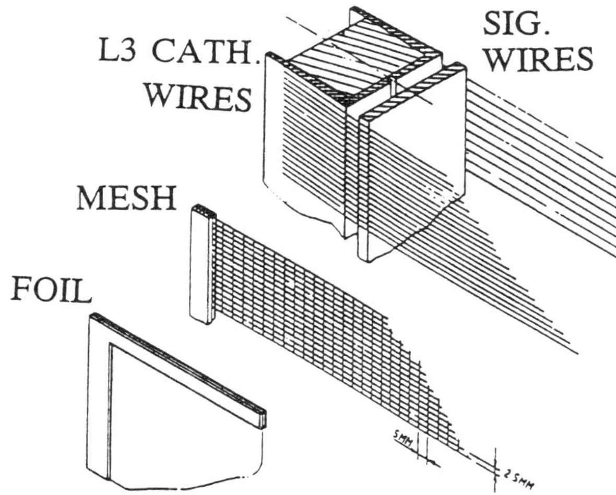

Fig. 13 Cell structure of precision chamber, showing the cathode wires to be replaced.

Foils glued on prestressed frames showed considerable wrinkling. Calculations confirmed that flat foils will be extremely difficult to achieve. Exposing aluminuminized capton foils to corona discharges soon revealed damage in the aluminum coating.

Meshes behaved extremely well under the same conditions and could easily be made flat on transfer frames. Using large winding machines, signal wire planes with 32 signal wires were produced. Those are affixed to the opposite side of the transfer frame.

Stacking such mass produced frames will result in a chamber, see fig.14a.

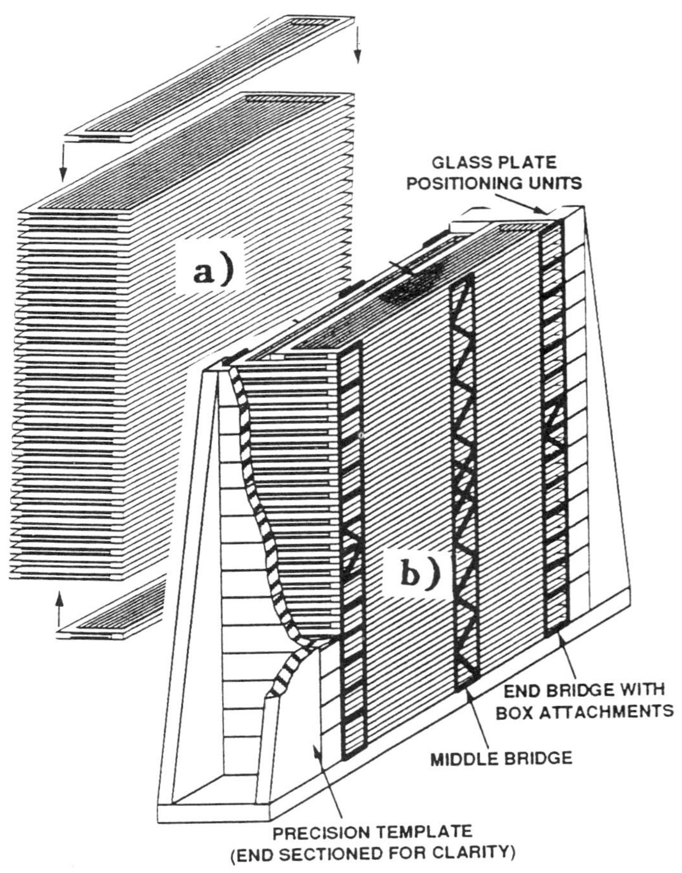

Fig. 14 Chamber mass production
a) stacking frames
b) joining by C fibre and template

Attaching thermally stable carbon fiber bars to the sides synthesizes the stock into a precise chamber (Fig. 14b), which will be surrounded by a gas enclosure. Electrostatic calculations for 32 wire planes have been carried out, so that the predicted accuracy for a particle is 27µm and the slope is determined to 0.6 mrad.

In summary [8], cathode planes saving 60% of the wires were computed, built, and tested. Foils are not suitable. Meshes behave well and a manufacturer has been located. A mass production scheme keeping the good features of the L3 construction has been devised.

CHAMBER GAS R&D

is needed [7] due to the following changes

	L*	L3
Total volume	2250 m^3	(250)
Flammable	no	(Ar:Eth)
Rates, max.	150k	(3k)
Mag. field	0.25 T	(0.5)

The larger L* volume forbids the use of the flammable L3 gas. The increased accuracy requirement cannot be met by increased pressure because this will automatically result in gas safety [9] classification III. The aim is to find and check out a mixture of class I, thus easing chamber construction and avoiding costs from safety precautions in construction, assembly, test procedures, and running of the system.

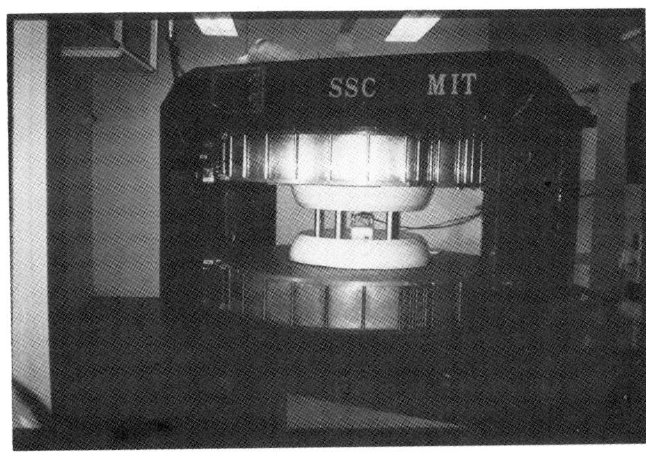

Fig. 15 MIT Cyclotron with test chamber during a measurement.

R&D carried out by our group in 1982 [10] indicated Argon : CO_2 as an non flammable accurate gas. However, its sparking behavior prevents use in a large system.

The MIT cyclotron was reactivated after 20 years to provide a homogeneous magnetic field for a new test chamber,

Fig. 15. An N_2-laser beam injected via a mov able mirror provides a track at various distances. The wire signal is used to measure the drift velocity. Pick up pads mounted behind the wire locate also the orthogonal coordinate to 60µm accuracy, and thus determine the drift deflection angle in the magnetic field, see Fig. 16.

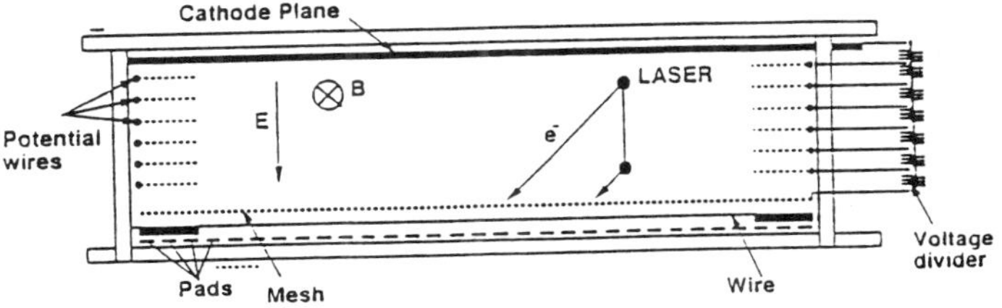

Fig. 16 Top view of the test chamber

Calibration measurements with Argon Methane = 90:10% confirmed our and previous results [10]. The gas will not be used because it has big diffusion and $\alpha = 40^0$ at 0.75T.
Fig. 17 shows new measurements using Argon : CO_2 with 2-propanol suppressing the sparking behavior.

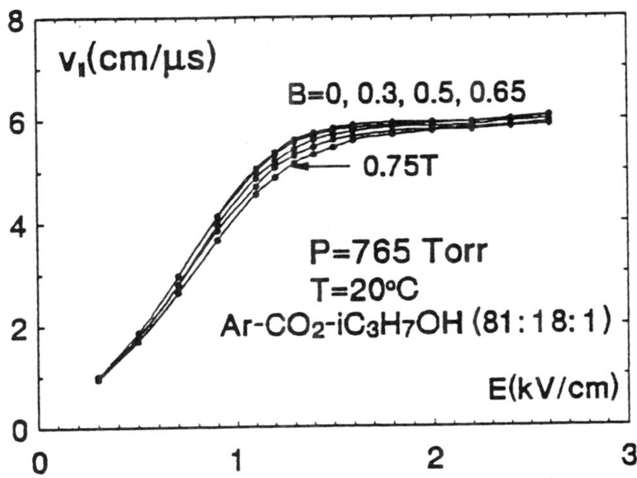

Fig. 17 Drift velocity vs. electrical field for Ar:CO_2 mixture with additive.

As seen the drift velocity shows a remarkable insensitivity to the magnetic field which also is expressed in the very small deflection angles in figure 18. This is a very desirable feature.

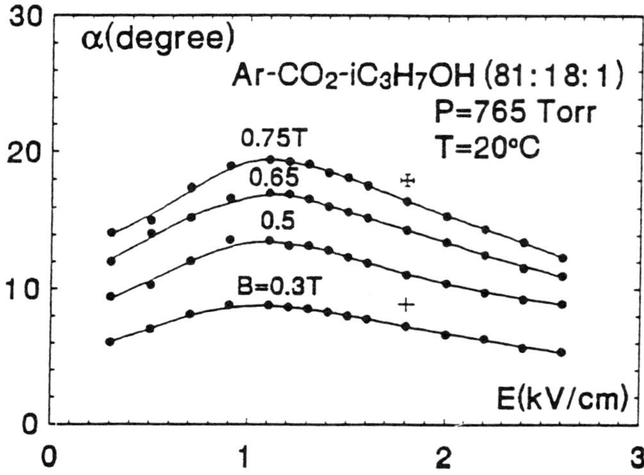

Fig. 18 Deflection angles vs.E for different B values

Fig. 19 gives the accuracy measured for Ar:Ethane in L3 in comparison to the design goal. CO_2 based mixtures carry the promise of smaller diffusion and will be studied at Oak Ridge.

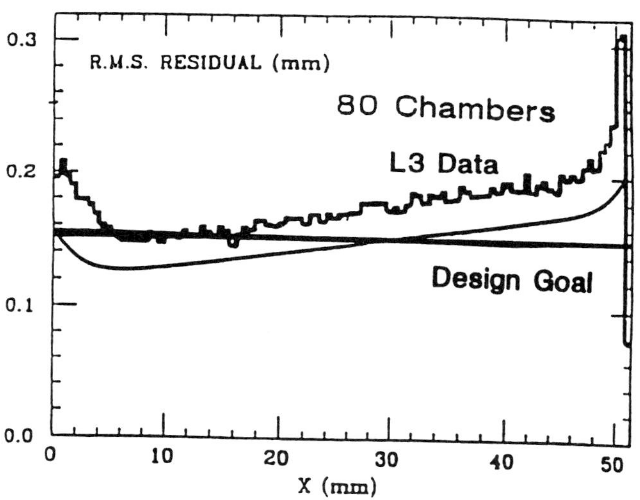

Fig. 19 Single wire resolution as measured in L3 compared to L* design goal.

Systematic studies of ageing with sources, pulsed X-ray tubes, and UV irradiations are prepared. Here, the more important question of what happens before as addressed.

Accelerated ageing was observed in a r = 12mm drift tube filled with Ar : CH4 = 90:10 %. UV feedback from wire amplification causes photoeffect at the wall creating the multiple pulses seen in Fig. 20a).
Studying Ar : CO_2 = 82:18 %, random self-ionization of the gas is observed, Fig. 20b, which explains the unreliability of this mixture. However, a 1% admixture of 2-propanol eliminates this problem completely, see fig.20c.

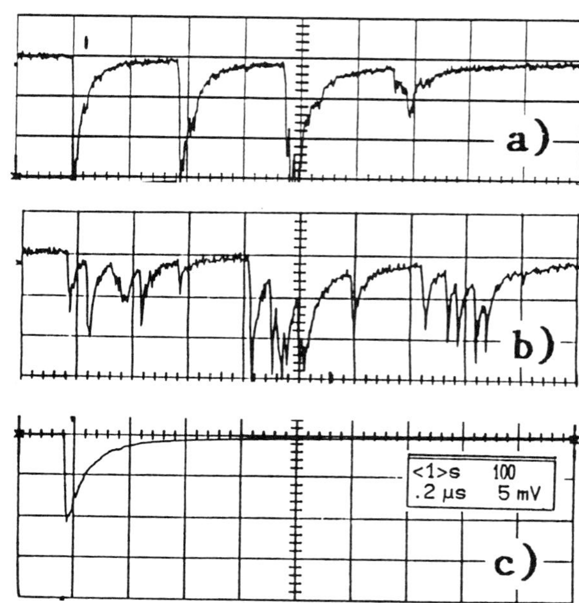

Fig. 20 Signals recorded by digital scope
a) Ar:CH4 at $2\text{-}10^5$ gain
b) Ar:CO2 = 82 : 18 with selfionization
c) Ar:CO2 + 2-propanol gives a clean signal

To summarize the progress in this part of the R&D :

A measurement set up has been completed. Measurement of drift velocities and deflection angles have been carried out for known and new gas mixtures.

A guided approach to evaluate admixtures was developed. Next in the program 10 gas mixtures will be systematically studied regarding diffusion (ORNL), υ,α, reliability, and ageing.

ARRAY STRUCTURE R&D

Here the necessity of R&D is driven by the larger sizes in L* :

	L*	L3
Structural length	12m	(6)
Stability over	5m	(2.7)
Forward detector	yes	no
Alignment knowledge	20μm	(30)

Fig. 21 gives an impression of the complexity, with design carried out by C.S. Draper Lab. as in L3 before. The

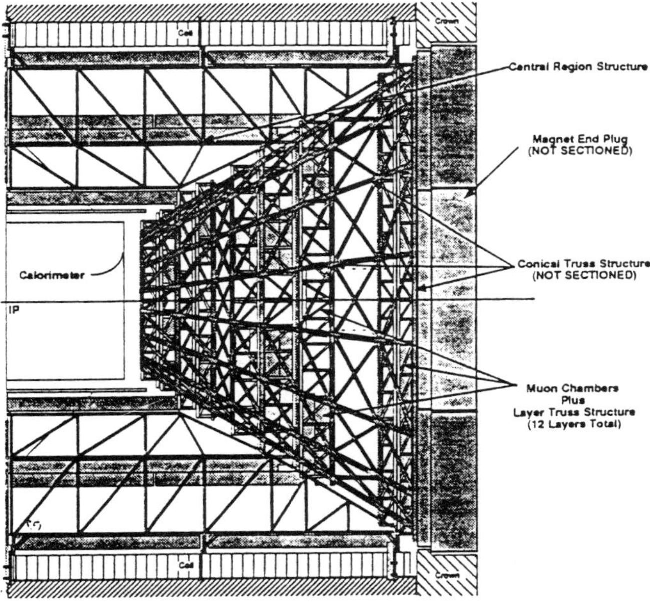

Fig. 21 Structures of Central and End cap detector

individual modules of the central part, (see Fig. 10) must be known to < 20 μm for accurate momentum reconstruction. The device to achieve this very developed for L3. Fig. 22 gives the optical straightness monitor tracking the movements of a L3 detector module during 15 days of running with the magnet powered. The deflections are small and stable to 4μm. These modules had been subject to rotation tests before, showing sinusoidal deflections of 700 μm amplitude. They proved reproducible to the measuring accuracy of 6μm as predicted by the design and analysis program NASTRAN [12].

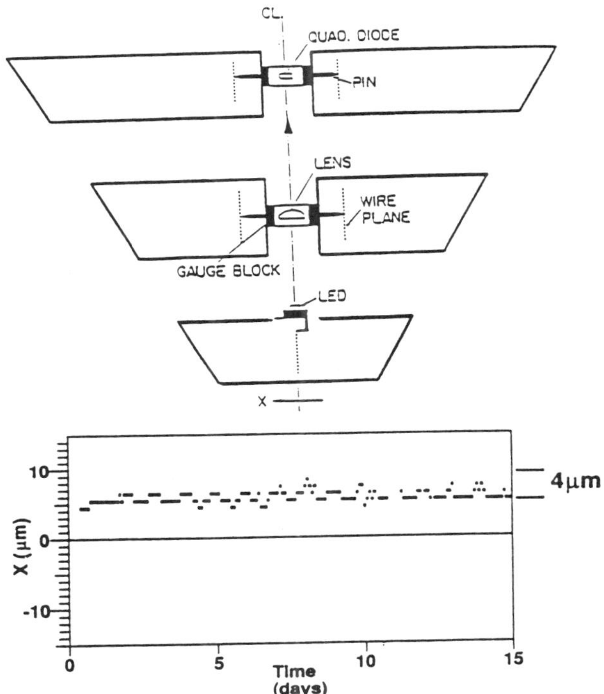

Fig. 22 Alignment recording in L3 during 15 day run

Using this very analysis, the original EOI structure [6] showed too large deflections and was replaced by the new full truss structure of Fig. 10, having much smaller deflections (1mm).

The end caps region initially foreseen as an array of simple x-y chambers could not be made modular.

Detailed studies of a radial concept, Fig.23 where the middle chamber wires

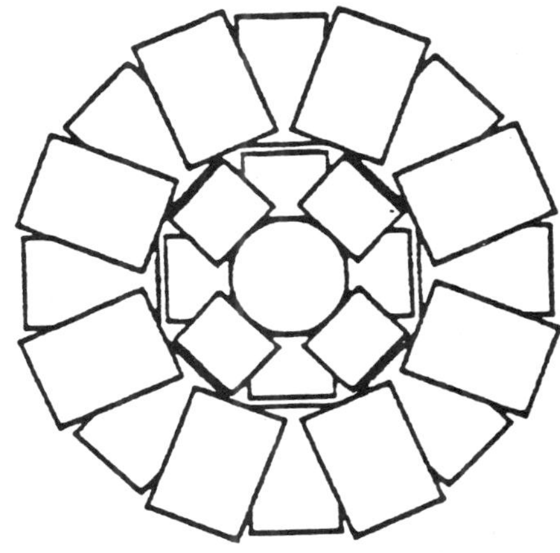

Fig. 23 End cap detector with chambers in radial arrangement

point toward the beam, gave the following results :

i) About 50% of the wires can be saved. This is because all wires now are measuring the critical coordinate.

ii) Despite seemingly higher complexity the chambers can be arranged in momentum measuring modules.

This critical feature is indicated in Fig.24 by shading and allows alignment according to the needs of momentum measurement. Fig. 24 also shows how two modules are put together without interlocking. Note the alternating square and triangular module support structures.

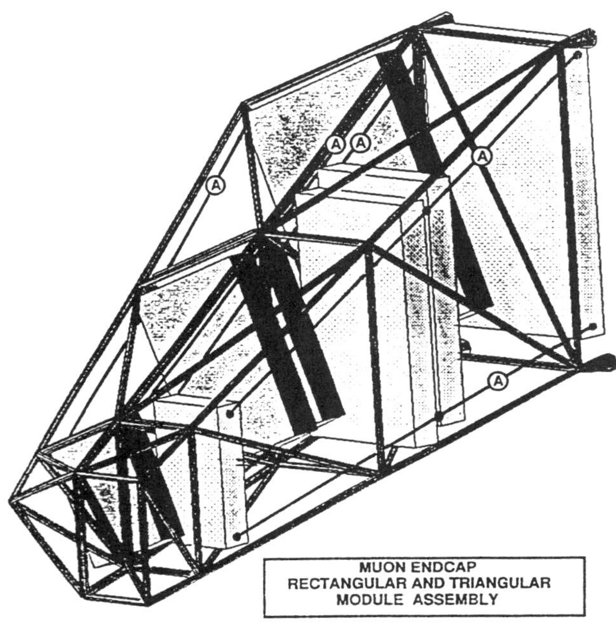

Fig. 24. Two modules of the end caps

Much attention is paid to the joints of the truss assembly (see Fig. 25) to make them transmit only linear forces in low stress configuration. This feature is essential for the computability of the entire structure.

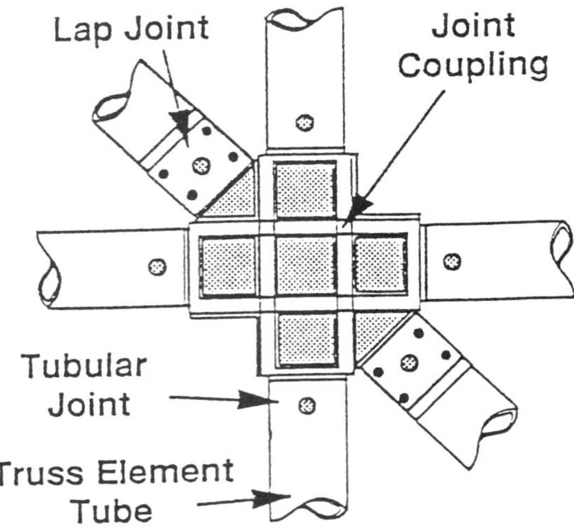

Fig. 25 Truss joint

The R&D Progress comprises :

a) An improved structure for the Central Detector with tolerable deflections (1mm).

b) A new End Cap design enables modularity and saves ~ 50% of the wires.

c) A straight forward structure for the Small Angle Region.

Thereby the concepts for all regions are fixed and detailed design can proceed.

Compared to the experimental challenges of the past all this appears well in reach.

1974 proportional chambers with σ = 1mm operating in 10^9 Hz. Highest ageing resistance known.

1976 drift chambers with fieldshaping profiles [14] cover 800 m^2 at the ISR and Mark J experiments

1979 thin drift tubes developed for vertex detector near the beam pipe [15]. Now known as straw tubes. Design for a large chamber [16] for high rates at Isabelle reaches σ = 130 μm by gas R&D.

1982 80 L3 multisampling chambers [4] of (2.4x6m size) reach σ = 50μm with alignment to < 30 μm.

1990 R&D for SSC demands continues as described above.

CONCLUSION

SSC is unique in its energy of 40 TeV and intensity of 10^{33} cm^{-2} s^{-1}. The muon physics will provide clean signal. Muon detectors ought to be unique in resolution and coverage to provide the necessary match of capabilities. This is well within reach.

Acknowledgements

I am indebted to Professor Samuel C.C. Ting and my colleagues of L3 whose efforts demonstrated this new concept. In particular, I like to aknowledge Drs. M. White, Y.H. Chang and C. Grinnell for their R&D results.

References

[1] J.J. Aubert et al Phys. Rev. Lett 33 (1974), 1404.

[2] S.W. Herb et al Phys. Rev. Lett. 39 (1977), 252.

[3] G. Arnison et al Phys. Lett. 126 B, (1983), 398.

[4] L3 Collaboration, Technical Proposal to LEP (5/1983) and B. Adeva et al Nucl. Instr. and Meth. A285 (1990) 35.

[5] P. Duinker et al. Nucl. Instr. + Meth. A273 (1988) 814.

[6] Expression of Interest to the SSC Laboratory (EOI). The L* Collaboration, May 1990. U. Becker et al. Nucl. Instr. A253 (1986) 15.

[7] A Research and Development Program to Continue the Development of Precision Instrumentation for the Study of Muons in the TeV Region. (Original Proposal) The Precision Muon Chamber Collaboration SSC PC-001, n.d. (rec'd Nov. 1989) 296pp. SSC Library Only.

[8] Progress on the R&D Program to Continue the Development of Precision Instrumentation for the Study of Muons in the TeV Region with Second Year Request. Submitted to the SSC Laboratory by the Precision Muon Chamber Collaboration, September 1990.

[9] Fermilab for Mandatory Engineering Standard Storage + Use of Flammable Gases in Physics Experiments SD-45B. Rev. Oct. 1988.

[10] M.C. Ma U. Becker, A.H. Walenta MIT-LNS Technical Report 129 (1982)

[11] C. Grinnell, W. Toth CSDL-R-1885 C.S. Draper Laboratory Cambridge, MA 02139.

[12] N.K. Chabra Draper Pub. CDSL-R-2011 (1987) unpublished.

[13] Ar : Methylal (-4 ^{0}C) gas mixture

[14] U. Becker et al Nucl Instr.+ Meth 128 (1975) 593.C. Willmott, et al Nucl Instr.+ Meth A263 (1988) 10.

[15] Mark J collaboration Phys. Rep. 63 (1980) 7. U. Becker et al. Nucl. Instr. + Meth. 180 (1991) 61.

[16] U. Becker et al. Nucl. Instr + Meth. 196 (1982), 381.

SCINTILLATING FIBER CALORIMETRY

Richard Wigmans
CERN, Geneva, Switzerland

Abstract

In the past $2\frac{1}{2}$ years, the SPACAL Collaboration have developed the compensating lead/scintillating fibre calorimeter technique from its conceptual stage towards a mature technology allowing the construction of high-precision particle detectors. Recently, we tested a 20-ton prototype calorimeter containing 176,855 scintillating fibres. Preliminary results from these tests are described.

1. INTRODUCTION

Although large calorimeters have been applied since about two decades in particle physics experiments, it is only since a few years that the factors which determine (the limitations in) their performance, and in particular the compensation mechanism, are sufficiently well understood to allow for dedicated improvement projects [1,2] . The project that is the subject of this talk was initiated after it was theoretically predicted [1] , and subsequently proven experimentally [3] that a lead/plastic-scintillator sampling calorimeter can be made compensating, a feature which until that moment was believed to be a unique property of uranium devices. The project aimed for optimizing the technique of compensating lead/plastic-scintillator calorimetry by using large numbers of scintillating plastic fibres as active material, rather than the traditionally used scintillator plates.

In order to make a lead/plastic-scintillator calorimeter compensating, the detector should have a small sampling *fraction*, considerably smaller than the typical value for current devices employing scintillator or LAr readout. Sampling fluctuations will, therefore, tend to dominate the energy resolution. This can be counterbalanced by using a high sampling *frequency*, hence the choice of fibres as active material. In addition, scintillating fibres produce extremely fast signals and allow constructing a very hermetic device, thus satisfying some of the conditions considered most essential for SSC calorimetry.

In the first phase of this project, known as the Spaghetti Calorimeter (or SPACAL) project, the feasibility and the purely calorimetric properties of such a detector were investigated. The project was carried out at CERN in the LAA framework, by a steadily growing group of people, comprising to date ~ 50 physicists, engineers, technicians and students from 12 different institutions.

2. EXPERIMENTAL RESULTS FROM SPACAL

In the past $2\frac{1}{2}$ years, most of the efforts of the SPACAL group went into optimizing the lead/scintillating fibre calorimeter technique. The aspects that were studied included techniques for incorporating large numbers of scintillating fibres in a block of lead, the optical properties of the fibres (e.g. maximizing the attenuation length) and the light collection uniformity. The results of the prototype studies were described in a number of articles [4,5,6]

Various generations of prototypes were constructed, starting with small e.m. modules of $\sim$ 10 kg, followed by hadronic modules of $\sim$ 100 kg, and by a 2-ton e.m./hadronic calorimeter. Recently our final detector, a 20-ton integrated e.m./hadronic calorimeter with 155 readout channels, was exposed for the first time to particle beams. After the technical problems involved were solved, the actual construction of this detector took about 4 months. Being obtained recently, some of the hadron data that will be shown in the following are to be considered highly preliminary. They were obtained on-line, concern raw data, without the proper calibration factors, etc. Analysis is going on and final results will be available in the near future.

2.1 Electromagnetic shower detection

2.1.1 Energy resolution

In spite of its small sampling *fraction* (2.3%), the SPACAL detector provides good energy resolution for e.m. shower detection, because of the high sampling *frequency*. This is illustrated in fig. 1 , where the signal distribution for 80 GeV electrons is shown. The resolution σ/E amounts to 1.4%, which corresponds to $12.5\%/\sqrt{E}$. This result was obtained for $\theta_z = 90^o$, i.e. by shooting the particles perpendicular to the fibre direction, which is the most favourable situation. More relevant for operation at the SSC is the performance for very small angles θ_z. We measured that the energy resolution scales as $\sigma/E = 13\%/\sqrt{E} + c(\theta_z)$, where c amounts to $\sim$ 1% for $\theta_z = 3^o$. For smaller angles, c increases further, and for $\theta_z < 1^o$ deviations from a Gaussian line shape were observed. We demonstrated experimentally that these deviations are due to particles entering the detector at a fibre position and depositing an anomalously large amount of energy (δ-rays) in this one fibre [6] .

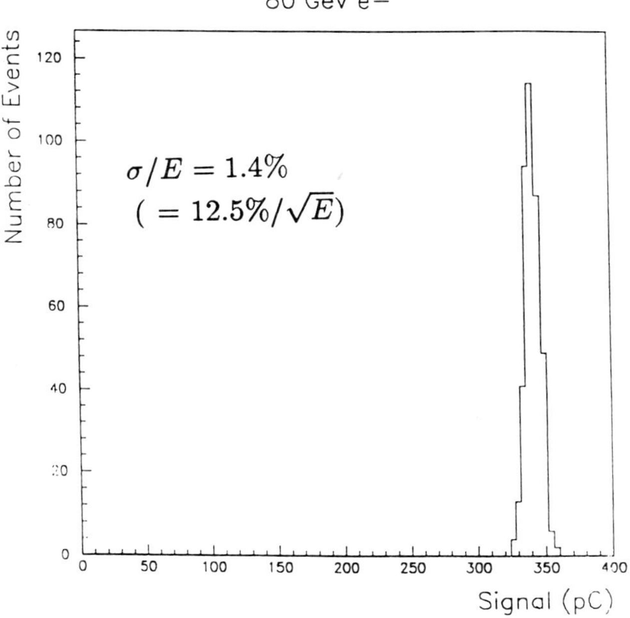

Figure 1: *The signal distribution for 80 GeV electrons measured with the SPACAL detector at 90° incidence*

The origin of the constant term is also understood. Figure 2 shows the electron signal as a function of the

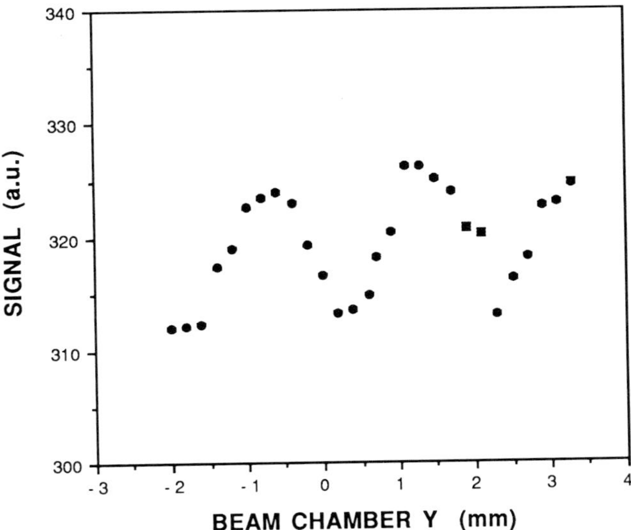

Figure 2: *The calorimeter signal as a function of the y-coordinate of the impact point. Data for 80 GeV e^-.*

y-coordinate of the impact point. One sees an oscillating pattern of which the minima correspond to the situation where electrons enter the detector in the lead, and the maxima to the situation where the particles enter the detector in a plane composed of fibres and lead. The 1.9 mm period corresponds exactly to the distance between two subsequent layers of fibres.

The lateral shower dimensions are such that only a small number of fibres effectively contribute to the signal and the sampling fraction depends, therefore, on whether the electrons enter the detector in a fibre or in the absorber material. If one selects only electrons entering the detec-

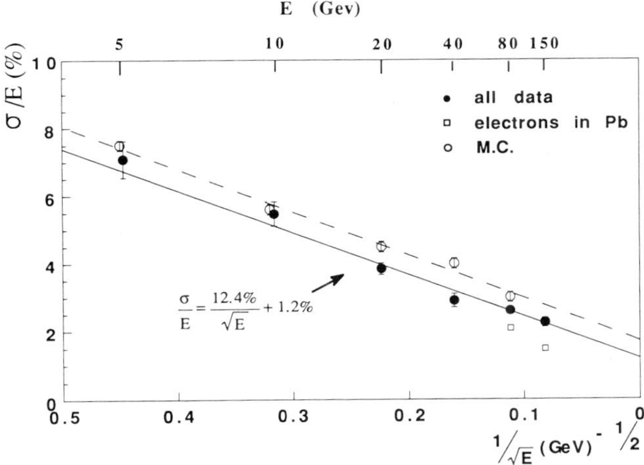

Figure 3: *The energy resolution for electrons as a function of energy, at $\theta_z = 3^0$. The open circles concern all events, the closed circles are for events that enter the detector in the lead. Monte Carlo predictions are shown as squares.*

tor in the lead, the energy resolution clearly improves, to the extent that the constant term in the energy resolution practically disappears. This is illustrated in fig. 3 , which also shows the results of a Monte Carlo simulation (GEANT).

Submillimetric knowledge of the impact point, perhaps achievable with a preshower counter, would therefore significantly improve the energy resolution for electrons. To avoid channeling, an SSC Spaghetti Calorimeter should be designed in such a way that electrons cannot enter it under angles less than e.g. 3°. A possible way of doing this is shown in fig. 4 .

2.1.2 Signal uniformity, hermeticity and linearity

Good energy resolution is only meaningful if the signal uniformity is good as well, i.e. if the variations of the mean signal with the impact point across a detector cell are small compared to the energy resolution. Several iterations of improvements were needed to achieve this goal [6] . Figure 5 shows results of the most sensitive measurement, a scan with an electron beam that was moved in small steps from the centre of a detector cell to the point were 3 modules join, and further into the crack between two modules. All individual modules were calibrated with an electron beam sent into their centre.

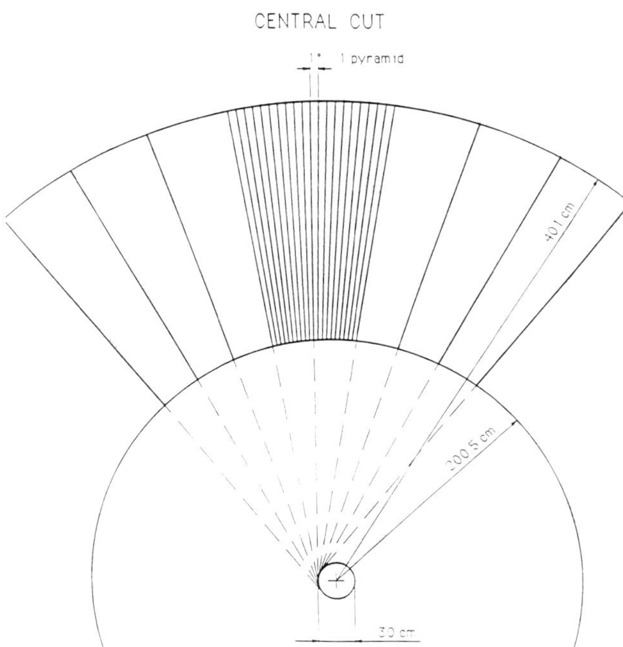

Figure 4: *A possible solution for the problems with electrons entering the detector at very small angles with the fibre direction. The modules are not pointing towards the interaction point but to a cylinder around the beam pipe. Lateral cross section of a possible SSC detector.*

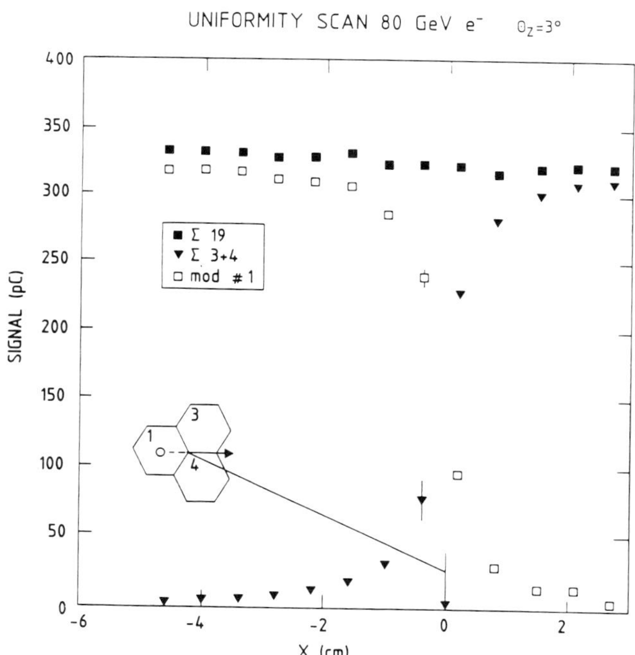

Figure 5: *The signal uniformity for e.m. shower detection. See text for details.*

The total calorimeter signal measured in this *crack scan* is practically independent of the impact point of the particles. This is even more true for a scan with pions and for the measured energy resolution. We conclude that the particles practically do not notice that they are detected by a modular device and, therefore, that the calorimeter is excellent in terms of signal uniformity and hermeticity.

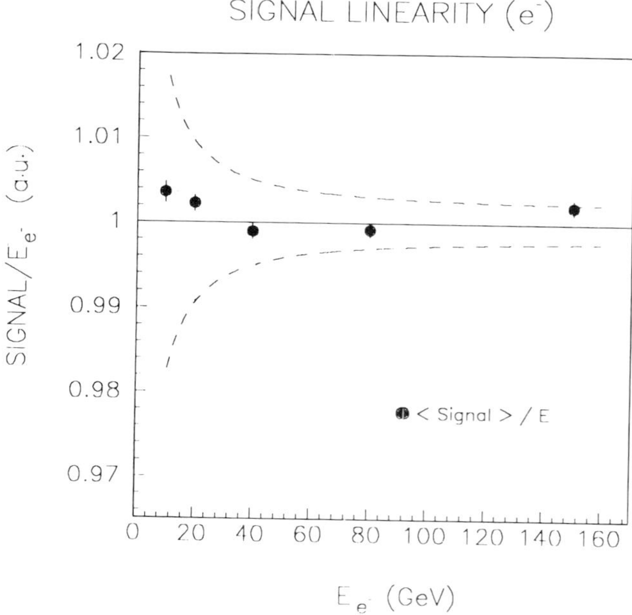

Figure 6: *The signal per unit energy measured for e.m. showers*

As can be seen from figure 6 , the calorimeter is also very *linear* for e.m. shower detection. The signal per unit energy changes by less than 1% over the energy range 10 - 150 GeV.

2.2 Hadrons and jets

Being very recently obtained, the results in this subsection have to be considered highly preliminary, and further improvements from the off-line data analysis may be expected.

2.2.1 Compensation, shower containment

The original motivation for starting this project was the expectation that a lead/fibre calorimeter could be made compensating, provided that it had the proper sampling fraction. In fig. 7 the e/π signal ratio is shown as a function of energy. Measurements were done at 5, 10, 20, 40, 80 and 150 GeV. The ratio is indeed sufficiently close to 1.0 (typically 1.05) for making the undesirable consequences of non-compensation negligibly small [1] .

Figure 8 displays the relation between the e/π signal ratio and the lateral size of the detector, a measurement made possible thanks to the fine lateral granularity. The figure shows that the showers are very well contained in

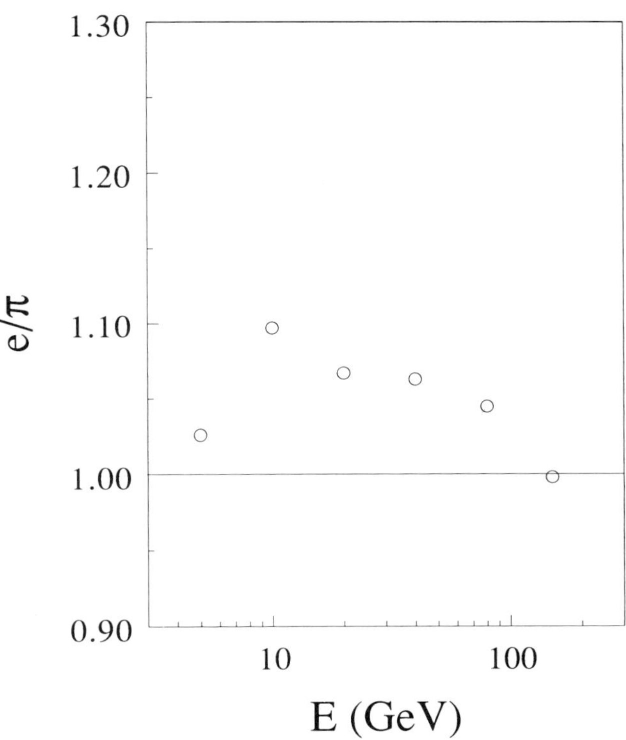

Figure 7: *The e/π signal ratio as a function of energy. Preliminary results.*

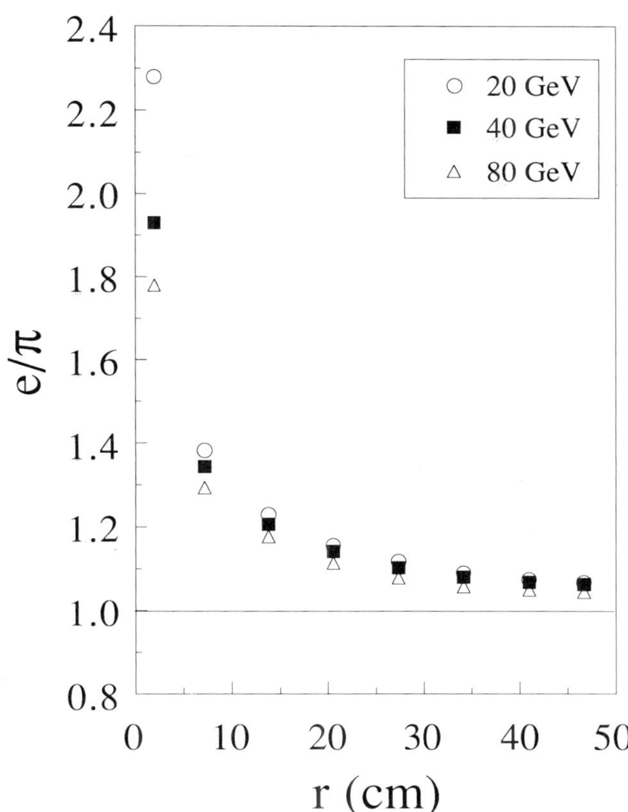

Figure 8: *The e/π signal ratio as a function of the radial detector size. Preliminary results.*

our detector and that in practice a region with a radius of $\sim$ 30 cm around the impact point is probably sufficient for an accurate evaluation of the particle energy. The longitudinal leakage from this 10 λ_{int} deep detector was measured to be negligibly small ($\ll$ 1%) at all the mentioned energies.

2.2.2 Energy resolution for hadrons and jets

The energy resolution was measured for negatively charged pions at energies ranging from 5 - 150 GeV. The results are shown in fig. 9 . A least squares fit to the data yields: $\sigma/E = 31.0\%/\sqrt{E} + 2.6\%$. Much work has been and still is going into understanding the relatively large constant term. The data shown in fig. 9 are on-line data, obtained with calibration factors that were not more accurate than $\sim$ 5%. Using the correct values off-line gave already a significant improvement. Light attenuation in the fibres was identified as the main contributor to the constant term. Thanks to the good position resolution this effect could be completely eliminated event-by-event for single pions at $\theta_z \neq 0$. We expect the final value of the constant term to be at the 1% level.

Contrary to e.m. showers, the energy resolution for pions turns out to be *independent* of the angle of incidence θ_z (fig. 10), which can be understood since the process responsible for the deviations from a Gaussian lineshape in the case of e.m. showers has no equivalent for pions.

Figure 11 shows the hadronic energy resolution as a function of the lateral detector size. It illustrates that the improvement is indeed only marginal when a radius beyond $r \sim 30$ cm ($1.5\lambda_{int}$) is considered.

The detection of jets was simulated by placing a paraffine target upstream of the calorimeter. The pions were required to interact in this target, while the interaction products were detected with the calorimeter. Measurements were performed at 2 target distances, at 44 and 147 cm from the calorimeter front face, respectively. At the close target position, the energy resolution for detecting the reaction products was even better than for single pions (fig. 12), especially at the highest energies. For the far target position, the results were worse, which can be understood from the fact that secondaries may miss the detector in this case. Off-line analysis shows that the constant term in the energy resolution for multiparticle detection decreases by $1-2\%$ compared to the single-pion case, depending on the multiplicity of the interaction. These results demonstrate the decreased effect of light attenuation on the energy resolution[4] .

SPACAL Energy Resolution for Pions at 0°

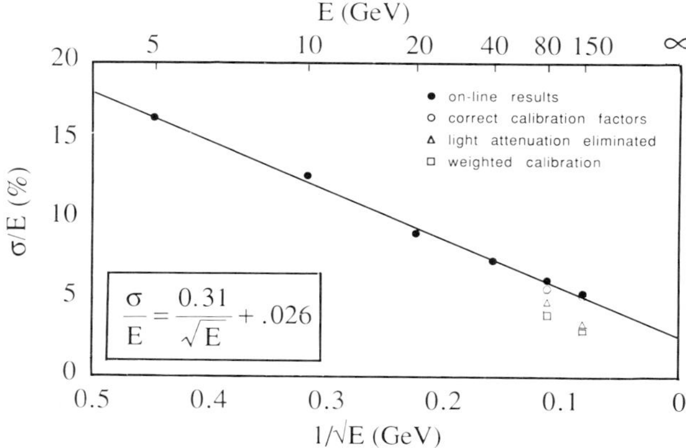

Figure 9: *The energy resolution for 5 - 150 GeV π^- particles. The improvements in the resolution at 80 and 150 GeV resulting from various steps in the off-line analysis are shown as well. Preliminary results.*

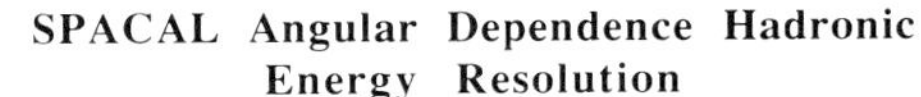

Energy Resolution

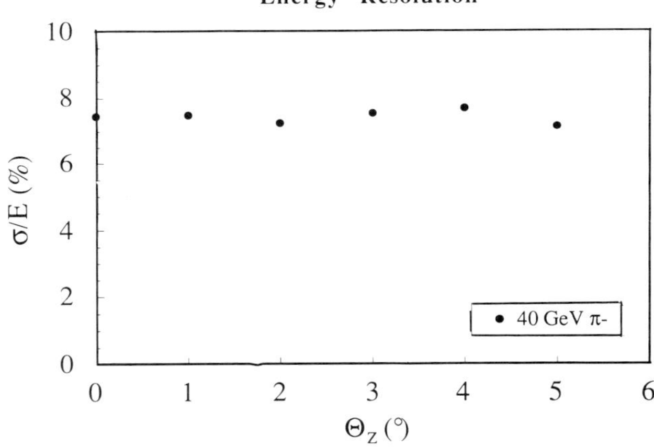

Figure 10: *The energy resolution for 40 GeV π^- as a function of the angle of incidence θ_z. Preliminary results.*

2.2.3 Position resolution

Thanks to the fine lateral granularity, the calorimeter also yields a good position resolution. The impact point of the particle was determined from the centre of gravity of the energy deposited in many calorimeter cells. The coordinates obtained in this way were compared with the ones derived from the wire chambers installed upstream of the detector.

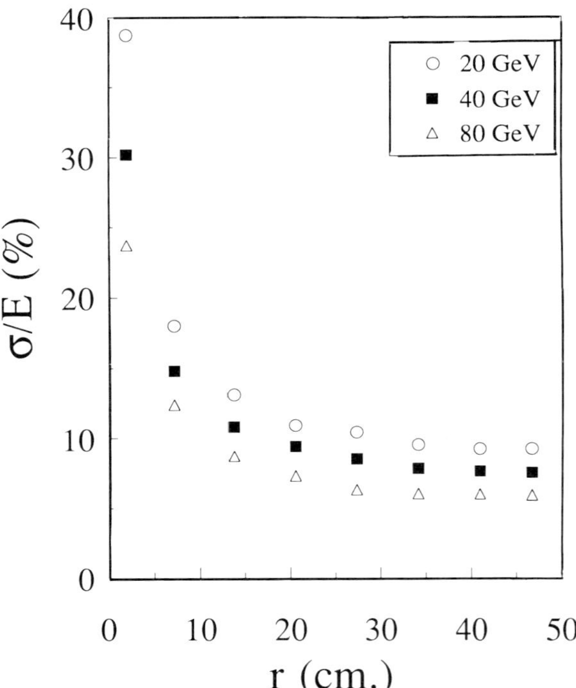

Figure 11: *The energy resolution for pions as a function of the lateral detector size. Preliminary results.*

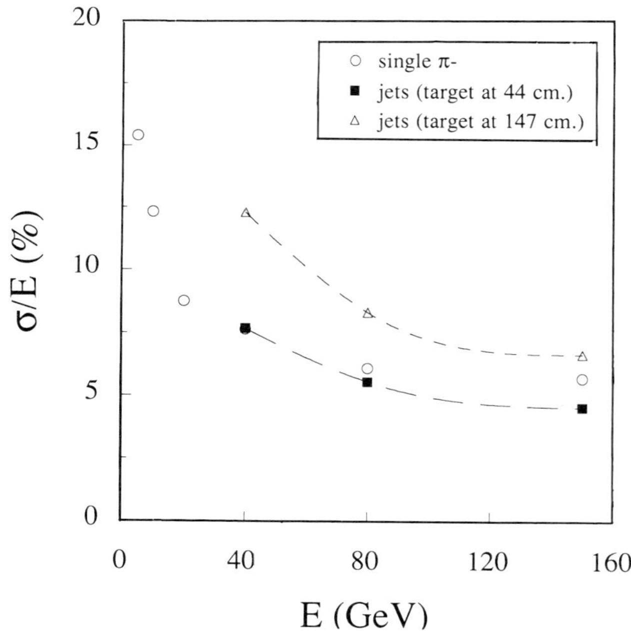

Figure 12: *The energy resolution for the reaction products of a pion interacting in a target upstream of the detector, compared to the single-π resolution, as a function of the pion energy. Preliminary results.*

The position resolution was found to depend on the impact point, especially for e.m. showers, where up to 95% of the shower energy may be contained in one detector cell. The position resolution is worst when the particles hit a cell in the centre, and best in the region where 3 calorimeter cells join. This is illustrated in fig. 13 , where

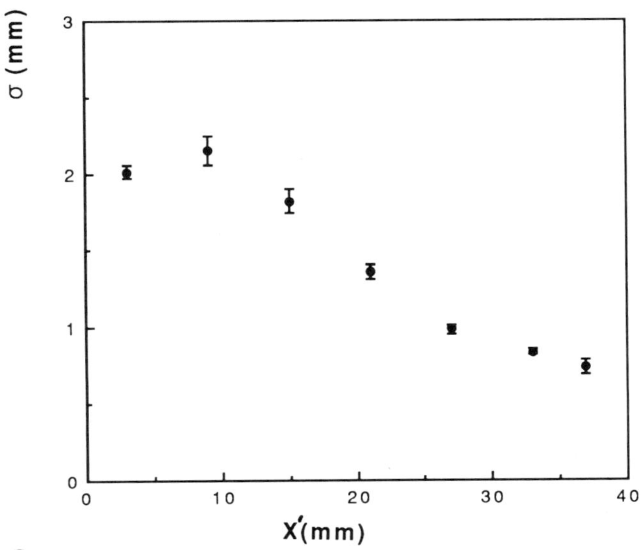

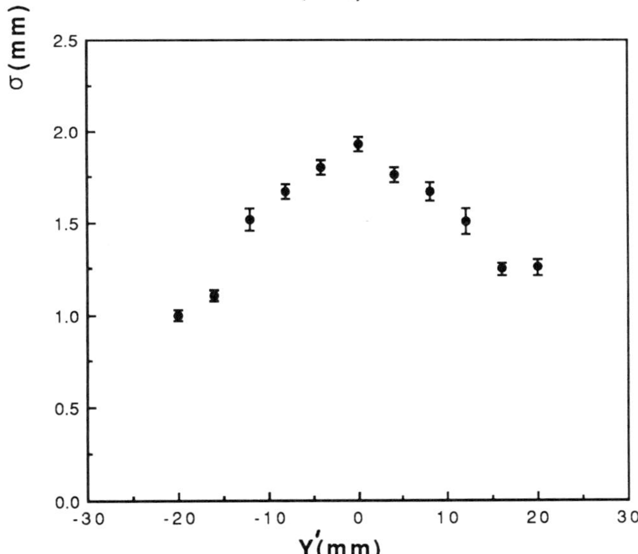

Figure 13: *The position resolution for e.m. showers as a function of the coordinates $x\prime$ (running between the centres of two adjacent hexagonal cells) and $y\prime$ (perpendicular to $x\prime$ along one of the sides of the hexagon).*

the position resolution is given for 80 GeV e^- as a function of the coordinates of the impact point. Averaged over the whole hexagonal tower, the resolution amounted to 1.6 mm at this energy. Figure 14 shows the position resolution as a function of energy for electrons (fig. 14 a) and pions (fig. 14 b), measured in the centre of a detector cell, which represents the worst case. The results scale as $1/\sqrt{E}$ with an offset of 2.4 mm in the case of pions.

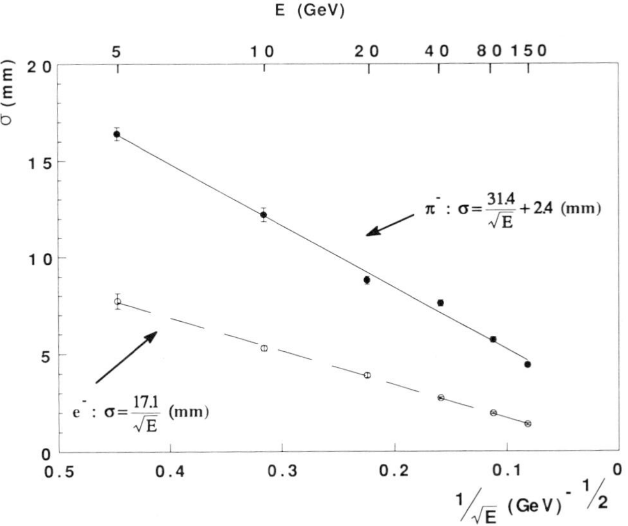

Figure 14: *The position resolution as a function of the energy for electrons and pions, together with the results of least squares fits.*

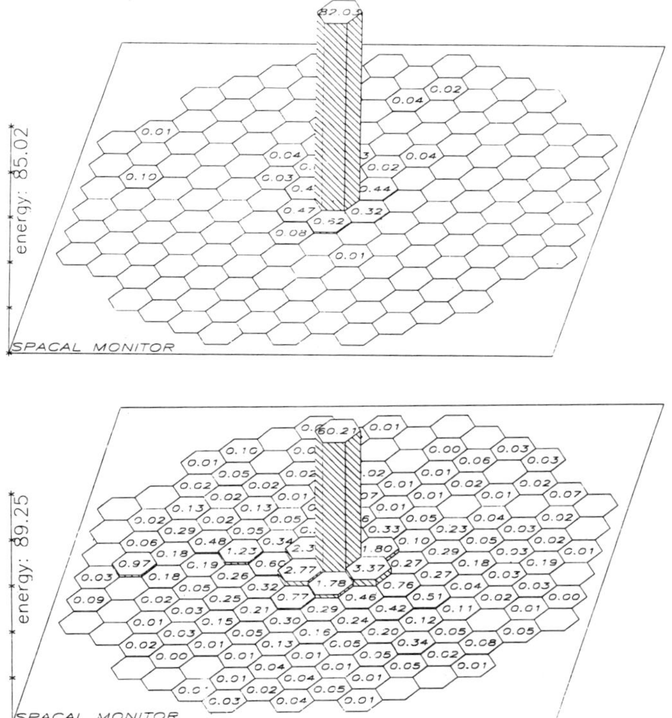

Figure 15: *Event displays of typical e.m (top) and hadronic (bottom) showers developing in the SPACAL calorimeter.*

Event displays of typical e.m. and hadronic showers are shown in fig. 15 . This figure illustrates that although the energy of a pion shower is distributed over a large number of detector cells, a large fraction of it is deposited in the cell hit by the entering particle ($\sim 50\%$). Because of this feature, it is quite possible to *distinguish separate hadrons* at distances much smaller than the 30 cm needed for an accurate measurement of their energy. An example of a multiparticle event, obtained with the interaction trigger,

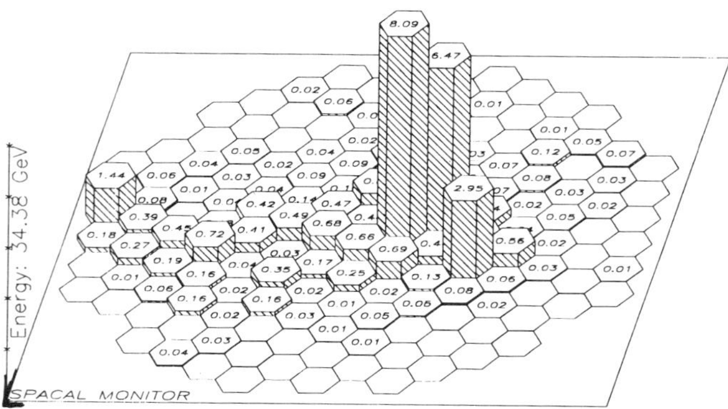

Figure 16: *Event display of a pion interacting in an upstream target, the reaction products being detected by the calorimeter.*

is shown in fig. 16 . Pion-pion separation through analysis of the energy deposit pattern was achieved at the 95 + % level for distances down to 8 cm at 80 GeV (fig. 17). One might exploit this feature for measuring the detailed structure of jets, including the recognition of electrons from heavy-flavor decay (see sect. 2.4).

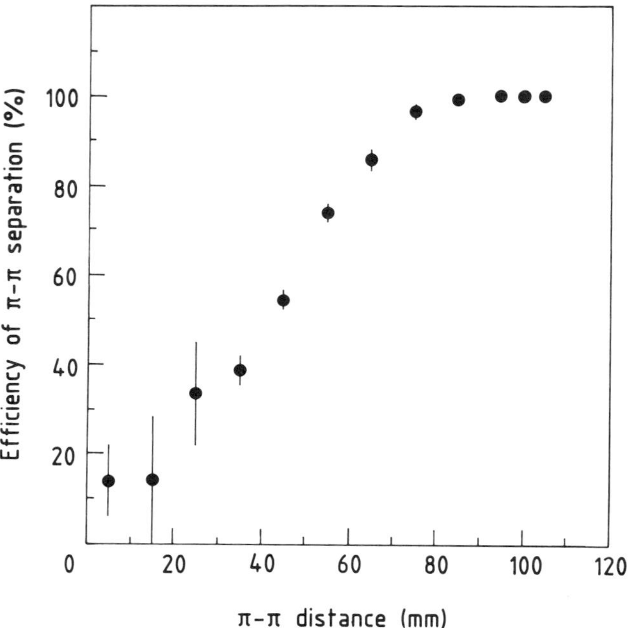

Figure 17: *The probability of recognizing a 80+80 GeV π^- pair as a function of their relative distance d, using a criterion that recognized single pions in 95% of the cases as such.*

Also for this purpose, a finer granularity than the one used sofar, is likely to pay off in even better performance.

2.3 Muons

The Spaghetti Calorimeter can also contribute relevant information for the detection of muons, in particular in the case of a catastrophic energy loss in the calorimeter, which will frequently happen at high energies. The signals for minimum-ionizing particles are completely separated from the pedestal and also allow an accurate determination of the energy loss by individual muons traversing the full depth of the calorimeter (fig. 18).

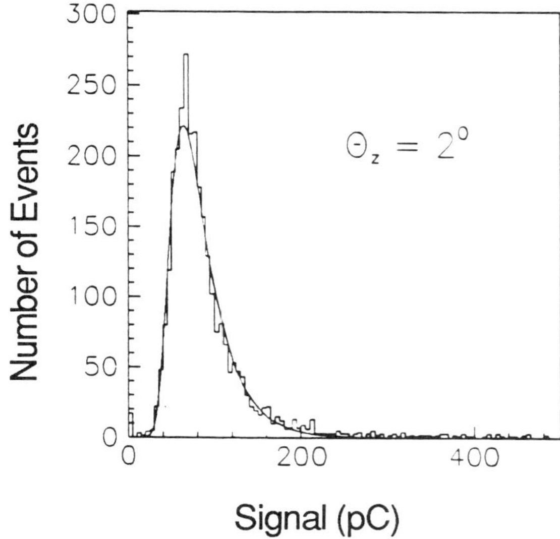

Figure 18: *Signal distribution for 150 GeV muons traversing the Spaghetti Calorimeter at $\theta_z = 2^0$.*

2.4 Electron/pion separation

For the identification of *isolated* electrons, a variety of methods were developed. The classical methods exploit the differences in the shower profiles between e.m. and hadronic showers, which are most pronounced for high-Z materials. We studied e/π separation with a number of algorithms acting on the lateral energy deposit pattern in our fine-grained calorimeter. Some results of these studies, in particular the ones using simple algorithms suitable for trigger purposes, were recently published [7] . H. Paar presented these results in his contribution to this Symposium [8] .

An example of a more sophisticated algorithm, useful in the off-line data analysis, is shown in the following. A quantity R_p was defined as

$$R_p = \frac{\sum_i r_i \, E_i^{0.4}}{\sum_i E_i^{0.4}}$$

where E_i are the energies deposited in individual towers i and r_i is the distance between the centre of tower i and the centre of gravity of the energy deposit profile. Figure 19 shows that electron and pion showers are very cleanly separated using this parameter, which is a measure of the effective shower width.

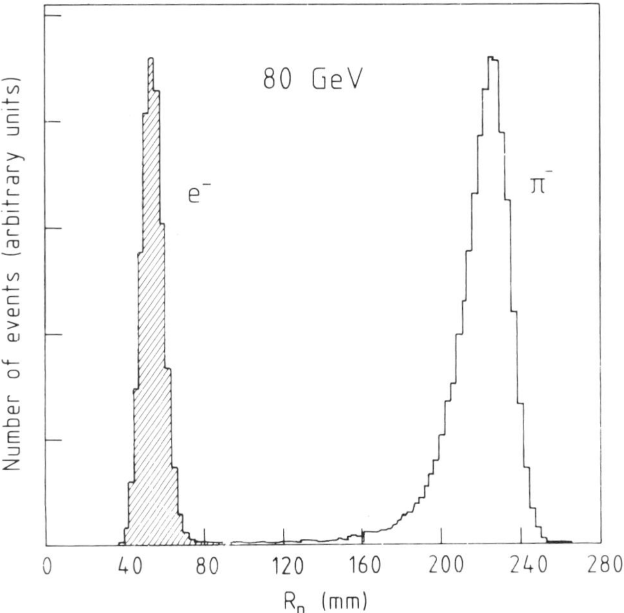

Figure 19: *Distribution of the effective width R_p (see text) for electrons and pions at 80 GeV.*

However, a small fraction (order 10^{-3}) of the pion events cannot be separated from electrons with such a criterion. These are presumably events where (almost) all the pion energy is transferred to π^0's in the first interaction, and therefore the showers look laterally very much like e.m. ones.

The starting point of such showers occurs at a certain depth z inside the calorimeter (with a probability distributed as $\exp(-z/\lambda_{int})$, so that one may distinguish them from genuine electron showers using the shift between the lateral centre of gravity and the impact point. This works very well.

We selected the pion events from fig. 19 that could not be distinguished on the basis of their lateral shower profile, i.e. at $R_p < 100$ mm. The distribution of their centre of gravity displacement with respect to the pion impact point is given in fig. 20 . For comparison, the distribution for electrons is given as well. Thanks to the good position resolution, all the pions interacting beyond a few cm in depth can be distinguished from electrons in this way. The final e/π separation achieved by combining both the lateral and longitudinal shower characteristics is shown in fig. 21

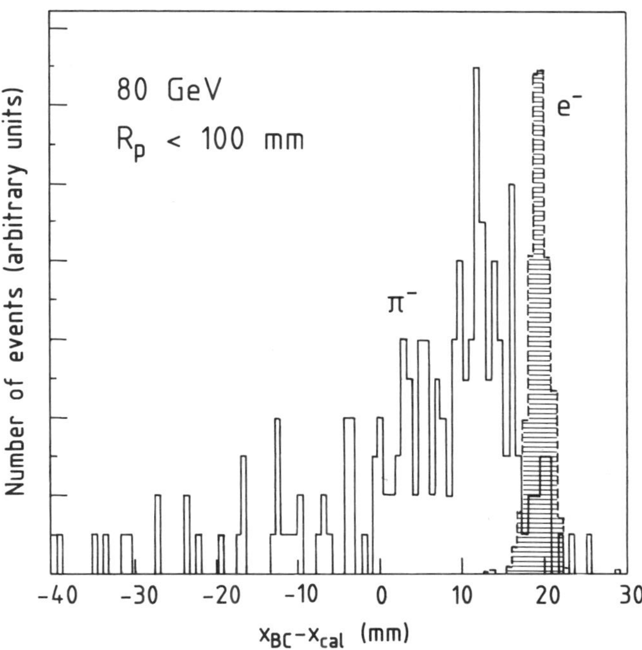

Figure 20: *Distribution of the displacement of the shower centre of gravity with respect to the particle impact point for electrons and for pions that produce showers that are laterally indistinguishable from electrons. Data for 80 GeV particles.*

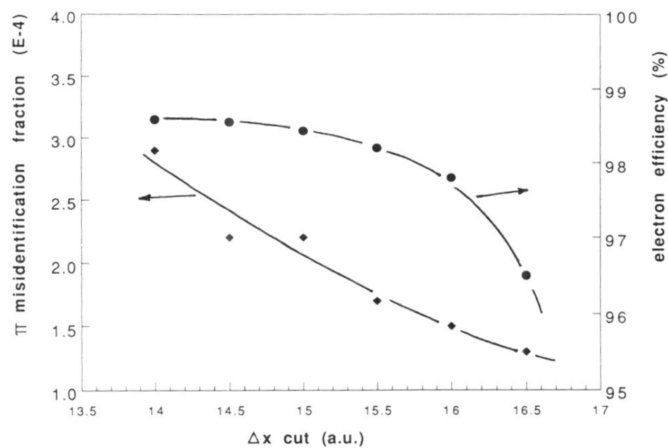

Figure 21: *The electron identification efficiency and the fraction of misidentified pions for a combination of longitudinal and lateral cuts. See text for details.*

This figure gives the electron efficiency and the fraction of pions passing the cuts for a combination of longitudinal and lateral cuts. For example, at 80 GeV the fraction of misidentified pions amounts to 1.5 10^{-4}, for 98% electron efficiency.

A further improved e/π separation and increased redundancy was obtained when preshower counter information was used in combination with the calorimeter data itself

[7] . Also, the explicit longitudinal calorimeter segmentation that is possible in projective structures may work in this sense. A. Para showed experimental evidence for this in his contribution to this Symposium [9] . The (large differences in) lateral shower characteristics also make it possible recognizing electrons *inside jets*. By superposition of electron and pion showers, we found that electrons impacting at only 7.4 cm from a pion of equal energy could be identified in 93% of the cases using a criterion that predicted a fake electron in about 0.1% of the cases.

In the course of this project, a new way to identify electrons was exploited [5] The method is based on small, but significant differences between the *time structures* of e.m. and hadronic shower signals. These differences, which are among others due to the different longitudinal shower development, can be exploited thanks to the high signal speed of this detector. The method is especially interesting since it has the potential of offering an extremely fast electron trigger (no pattern recognition needed).

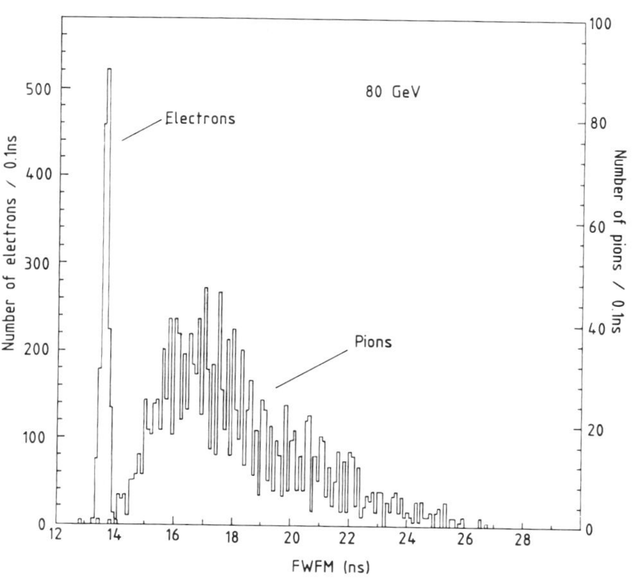

Figure 22: *Distribution of the widths, measured at 20% of the amplitude, of e.m. and hadronic shower signals at 80 GeV.*

How powerful this method can be, is illustrated in fig. 22 which shows the distribution of the widths (measured at 20% of the amplitude) of e.m. and hadronic shower signals. The e.m. signals are all nearly identical ($\sigma \sim 130$ ps!), while practically all the hadronic signals are significantly broader. We achieved e/π separation at the 10^{-3} level (80 GeV) just using this method.

These results show that in spite of the absence of longitudinal segmentation, this detector performs very well in terms of e/π separation.

2.5 Signal speed

The lead/fibre calorimeter is definitely the fastest hadron calorimeter available today. The width of the signal is mainly determined (and limited) by the difference in the velocity at which the scintillation light travels inside the fibres ($c/1.6$, the index of refraction of the fibre core) and the velocity of the shower particles that generate the light. Due to the negligible cross section for thermal-neutron capture slow shower tails, characteristic for uranium calorimeters, are absent.

Of course, neutrons *do* contribute, quite substantially, to the hadronic signals from the calorimeter. It is the main reason why the detector is compensating. Elastic $n-p$ scattering in the plastic is the relevant process. The time constant for this component was estimated, from the production energy spectrum of the neutrons and from the mean free path in between collisions, to be about 10 ns [6] . The calorimeter signals are so fast that the neutron component can be observed and measured event-by-event. This is illustrated in fig. 23 . Figure 23 a shows a typical electron signal after it has passed through a compensating filter that removes the contributions of cables and other RC components. The filtered signal is well described by a Gaussian with a FWHM of 4 ns. In fig. 23 b a typical hadronic event is shown. The Gaussian component is clearly broader than for the e.m. case, reflecting the effect of the larger longitudinal area in which the light is produced. In addition, however, there is an exponential tail with a time constant of 10 ns, that is very likely caused by neutrons (fig. 23 c).

One example, illustrating the advantage of the fast detector response was already given in the previous subsection. Another advantage, crucial for SSC experiments, concerns the fact that one can use a very short gate width without deteriorating the calorimeter performance. Also, the effect of pile-up will be as small as can be. We measured the e/π signal ratio and the energy resolution as a function of the gate width. The results are given in figs. 24 and 25 . It turns out that the calorimeter performance remains acceptable down to gate widths of the order of the SSC bunch crossing time.

3. *SUMMARY AND CONCLUSIONS*

The Spaghetti Calorimeter performs up to and in some respects even better than the expectations on the basis of which the project was started 3 years ago. It is an extremely compact, hermetic and fast device, with equal response to electrons and pions. Due to the frequent shower sampling and to the fact that it is compensating, the energy resolution is very good. The compactness (effective radiation length 7.5 mm, effective nuclear interaction length 21 cm) pays off in good position resolution and particle-particle separation possibilities, since the showers are very narrow.

Electron identification, thought to be a weak point at the start of the project because of the lack of longitudinal segmentation, turned out to be no problem.

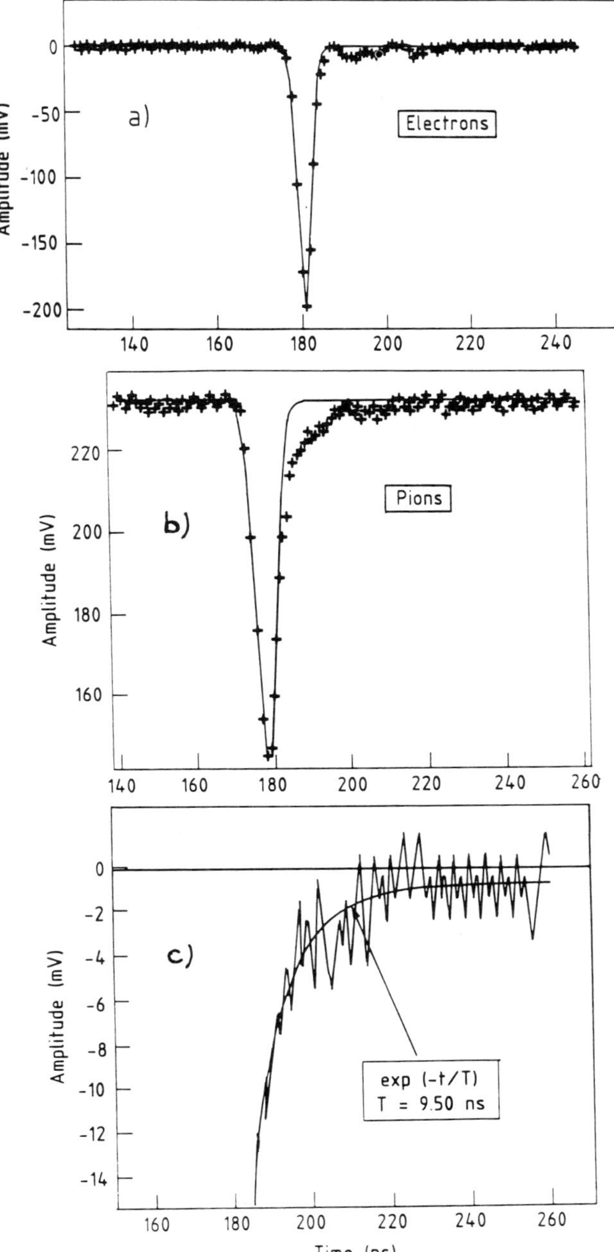

Figure 23: *An example of a digized filtered 150 GeV electron signal (a) and pion signal (b), together with the results of Gaussian fits to the data points. After subtraction of the Gaussian component for the pio. signal, an exponential tail remains (c). The bin size is 1 ns and the $t = 0$ point is arbitrary.*

Thanks to the powerful possibilities offered by the fine lateral granularity, by the fast response and by the good position resolution, e/π separation turned out to be possible at the 10^{-4} level (isolated particles), without even using a preshower counter.

Further improvements/redundancy of these results can be obtained from such a preshower counter or from explicit longitudinal segmentation in projective structures. The pencil-like nature of the showers combined with the fine lateral granularity make it possible to do a good job on electron identification in the vicinity of hadronic activity.

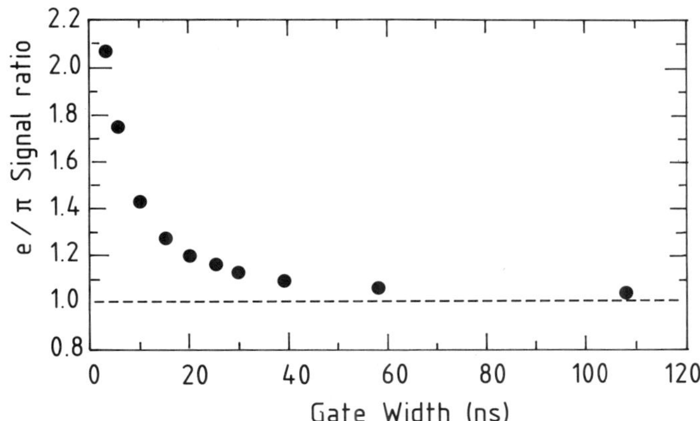

Figure 24: *The e/π signal ratio as a function of the gate width. Preliminary results.*

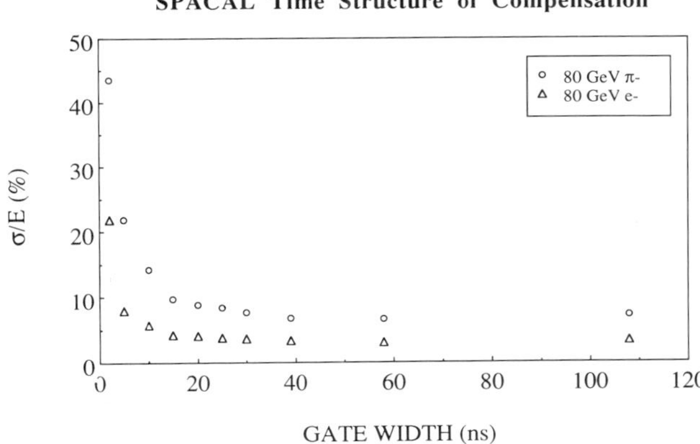

Figure 25: *The e.m. and hadronic energy resolution as a function of the gate width. Preliminary results.*

Encouraged by these and other results, the SPACAL Collaboration has submitted a proposal [10] for a follow-up of the project, that is specifically aimed at application of this type of detector in a Supercollider environment. In this new phase of the project the emphasis will therefore be on

- Readout, triggering and data acquisition.
- High-rate and radiation hardness studies.
- Calibration and stability monitoring.

- Engineering of a hermetic device with a mass of $\sim 10^4$ tonnes, integrated in a system with several other detectors.
- Construction of a prototype with a (quasi-)projective geometry.

This project has in the meantime been approved by the CERN Research Board.

References

[1] R. Wigmans, Nucl. Instr. and Meth. **A259** (1987) 389.

[2] H. Brückmann *et al.*, Nucl. Instr. and Meth. **A263** (1988) 136.

[3] E. Bernardi *et al.*, Nucl. Instr. and Meth. **A262** (1987) 229.

[4] F.G. Hartjes and R. Wigmans, Nucl. Instr. and Meth. **A277** (1989) 379.

[5] R. DeSalvo *et al.*, Nucl. Instr. and Meth. **A279** (1989) 467.

[6] D. Acosta *et al.*, Nucl. Instr. and Meth. **A294** (1990) 193.

[7] D. Acosta *et al.*, Electron-pion discrimination with a scintillating fibre calorimeter, preprint CERN-PPE/90-172 (1990), submitted to Nucl. Instr. and Meth.

[8] H.P. Paar, contribution to these Proceedings.

[9] A. Para, contribution to these Proceedings.

[10] SPACAL Collaboration, Scintillating fibre calorimetry at the LHC, Proposal CERN/DRDC/90-23 DRDC/P1 (1990).

SUPERCONDUCTING AIR CORE TOROIDS FOR HADRON COLLIDERS

Michael Marx
State University of New York at Stony Brook

Abstract

Muon spectrometers for hadron colliders based on superconducting air core toroids hold the promise of improved resolution over iron toroids and improved acceptance over solenoids. Significant design efforts for complete systems and for forward spectrometers for both SSC and LHC are described.

Muon systems for collider experiments determine to a large extent the scale - in size, weight, and often cost - of the entire experiment. Therefore, the specifications for such a system must be rigorously justified. The rationale for the measurement of muons rests with the increased statistics for exploring decays of heavy particles into both muons and electrons (e.g., a factor of four for Higgs to Z^0Z^0 compared to electron measurement alone); the difference in systematics and backgrounds for electrons and muons that can provide confirmation of marginal signals; the requirement to measure muons in the forward direction, permitting determination of decay asymmetries for new heavy objects; and the ability to identify and measure muons closely accompanied by other particles.

The function of the muon system is to trigger on and then precisely measure the momenta of muons in single and multiple muon events. One goal set for such a muon system is to measure muons over the full acceptance with resolution similar to that for measuring electrons with calorimetry. In the SSC/LHC regime one benchmark is to reconstruct Z^0's decaying from heavy objects, close to the level of the intrinsic Z^0 width. The muon system should also have sufficient acceptance and resolution to measure the asymmetry of a new Z'. The trigger system must recognize muons above a minimum p_T, rejecting calorimeter punchthrough particles and muons from the decay of light quarks, and be flexible enough to increase the threshold as the luminosity increases. The limitations on the measurement system, for this as for all SSC muon systems, is determined by the capability of the first chambers after the calorimeter to withstand the rates and find muon tracks in the presence of real and fake muon backgrounds, exacerbated by the burden of accompanying electromagnetic showers.

Until recently the designs for collider muon systems relied on utilizing either iron toroids outside the calorimetry volume, or measurement of the muon trajectory in air with a solenoidal magnet. Thus measurements of muons were compromised either by the limited acceptance of a solenoid, or by the limited resolution achievable with iron.

A muon spectrometer based on a system of superconducting air core toroids provides a unique opportunity for improving a muon system at the SSC [1]. The toroidal field is efficient because it is always perpendicular to the particle trajectories, and the closed field obviates the need for massive quantities of iron for flux return. In addition, the toroidal field allows the possibility of tuning the resolution attainable with the system by separate adjustments of the currents and hence the field in the central and end regions, as well as independent adjustment of the field lengths. Work is in progress on spectrometers using toroids to cover the full solid angle, and also on systems of forward toroids used to augment solenoidal detector geometries.

Two different magnet configurations are feasible. The first provides access to the interior magnetic field volume so that muon detectors can be positioned within the magnetic field to measure the trajectory's sagitta with minimal multiple scattering. The second has no access to the interior and all muon trajectory measurements are made outside the toroid in a nonmagnetic environment. Preliminary design studies indicate that toroids of sufficient bending power can be built with relatively little material to scatter muons.

Although large superconducting air core toroidal magnets have not previously been used in high energy physics experiments, they have been successfully developed and applied in the plasma confinement fusion research program. The approach here is to adapt this proven technology for an SSC experiment. The feasibility of using large superconducting air core toroids has been verified in an earlier study commissioned by the SSCL. [2] The conclusion of that report was that although the sizes of the toroidal magnets under consideration are larger than any previously built, the technology required to construct such a system is well within state-of-the-art, and therefore it can be constructed with a conservative design.

Forward Toroids for a Solenoid Detector

The proposed design for a solenoidal detector layout is shown in Figure 1. In the forward direction designs for two types of air core toroids has been considered to match the resolution obtainable with the solenoid in the central region.

Figure 1 SDC detector with a superconducting air toroid at 14 meters, covering $\eta = 1.5$ to 2.5.

Figure 2 shows a discrete coil toroid design accomplished by Hitachi Ltd. of Japan. [3] The parameters of the magnet are given in Table 1. It utilizes 32 discrete coils, with indirect cooling of an aluminum stabilized conductor. The individual dewars are also aluminum resulting in a total thickness of 3 or less radiation lengths for a trajectory traversing the entire coil.

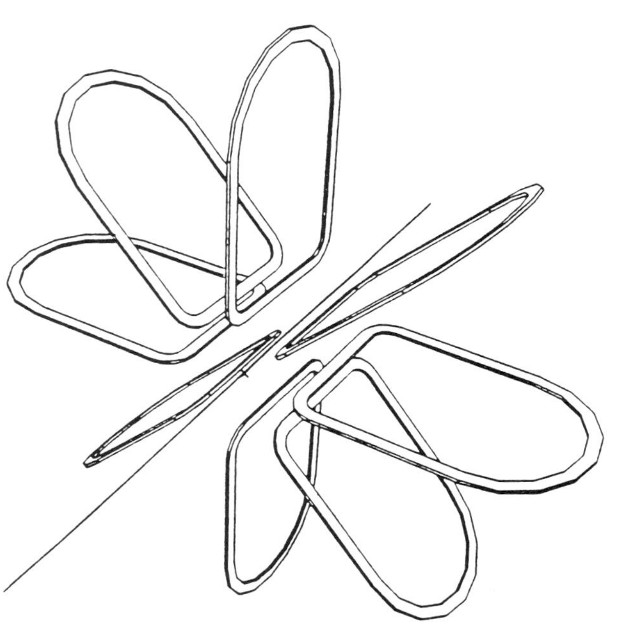

Figure 2 Discrete-coil air-core toroid.

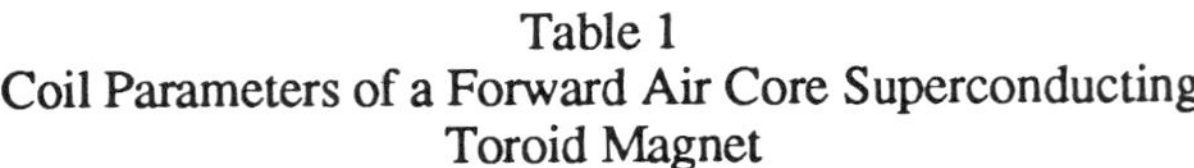

Table 1
Coil Parameters of a Forward Air Core Superconducting Toroid Magnet

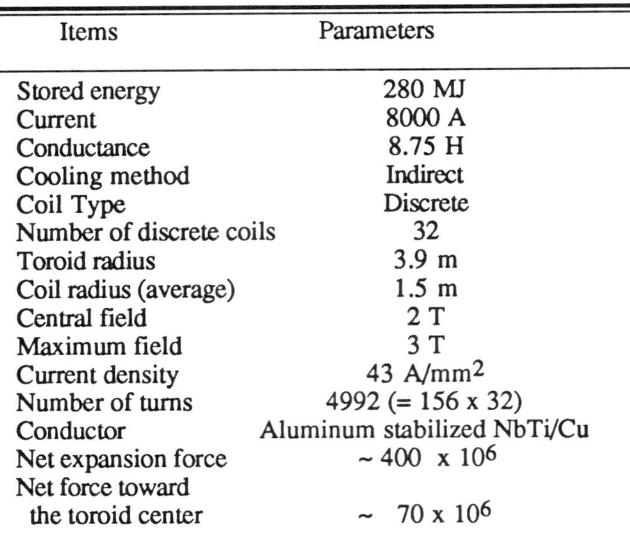

Items	Parameters
Stored energy	280 MJ
Current	8000 A
Conductance	8.75 H
Cooling method	Indirect
Coil Type	Discrete
Number of discrete coils	32
Toroid radius	3.9 m
Coil radius (average)	1.5 m
Central field	2 T
Maximum field	3 T
Current density	43 A/mm^2
Number of turns	4992 (= 156 x 32)
Conductor	Aluminum stabilized NbTi/Cu
Net expansion force	~ 400 x 10^6
Net force toward the toroid center	~ 70 x 10^6

Figure 3 shows a schematic of a continuous coil design accomplished by Advanced Cryo Magnetics (CA). [4]

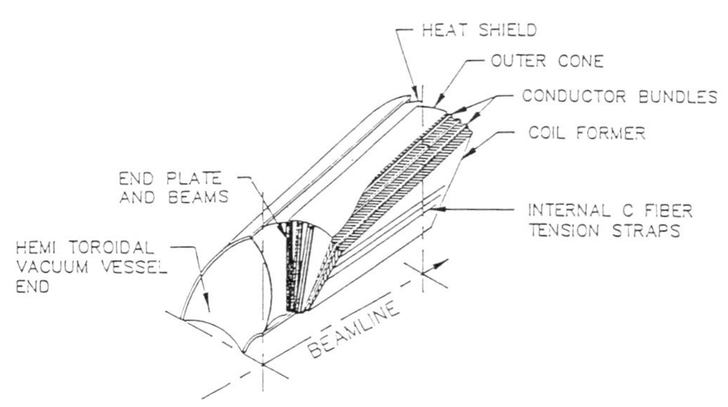

Figure 3 Air core toroid designed by Advanced Cryo Magnetics, Inc.

The objectives of the design were to fulfill the physics requirements, minimize the cost, and keep the fabrication as simple as possible. In addition new technology requiring extensive development was avoided.

The magnets are designed to be wound with aluminum stabilized NbTi conductor that is bath cooled. By operating at a modest current density, and with the conductor fully exposed to the liquid helium, bath cooling provides complete stability and excellent reliability. A quench can occur in this type of magnet only if the liquid helium level is allowed to drastically decrease, and even for this unlikely event, the venting arrangement is such as to make the quench harmless. This is the design criterion used on the large bubble chamber magnets that have been so successful. Both the vacuum vessel and the helium vessel are fabricated from aluminum to minimize the interference with the high energy particles. The magnetic pressure on the 2 flat ends is supported by carbon fiber looks that tie the ends together. All the stressed members, including the carbon fiber loops, operate at low stress relative to the material strength, further increasing the system reliability. The magnet has a complicated topography and considerable effort was devoted to developing an assembly procedure that did not require elaborate winding machines, fixtures, or precise machining, all of which can have a large impact on the cost.

This conceptual design meets all the physics requirements and is economical to build. Utilizing fully stabilized conductor, the operation of the coil to full field is assured. The overall reliability is further enhanced by the low stress levels of the support systems of the coil.Table 2 shows the design parameters for this continuous toroid.

Table 2
A Superconducting Toroidal Magnet for the Argonne National Laboratory
Superconducting Super Collider Detector

Items	Parameters
Minimum radius	2.3 m
Maximum radius	5.95 m
Length	3 m
Peak field	2.7 T
Front end plate thickness at I.D., total	9.78 cm
Rear end plate thickness at O.D., total	5.26 cm
Torus internal volume, carbon	0.02%
Stored energy	~ 300 MJ
Operating current	7000 A
Conductor	Al stabilized NbTi
Conductor size	0.39 cm x 2 cm
Current density	9000 A/cm^2
Type of stabilization	Bath cooled
Ampere turns	34 x 10^6
Conductor length (@ 7000)	51,000 m
Conductor weight	26,300 lbs.
Cold weight including conductor	135,400 lbs.
Warm weight	37,500 lbs.
Total weight	172,900 lbs.
Liquid helium volume	15,000 liters

Toroidal Spectrometer for the LHC [5]

A study has been initiated at CERN in the LAA Project for the design of a toroidal muon spectrometer with discrete air coils for a multi-TeV hadron collider. [6] The configuration is shown in Figure 4 in a side view and in Figure 5 transverse to the beam. It consists of a long central toroid with small toroids inserted at the ends. The design allows access to the field regions to permit trajectory measurements within the field. The parameters of the system are given in Table 3

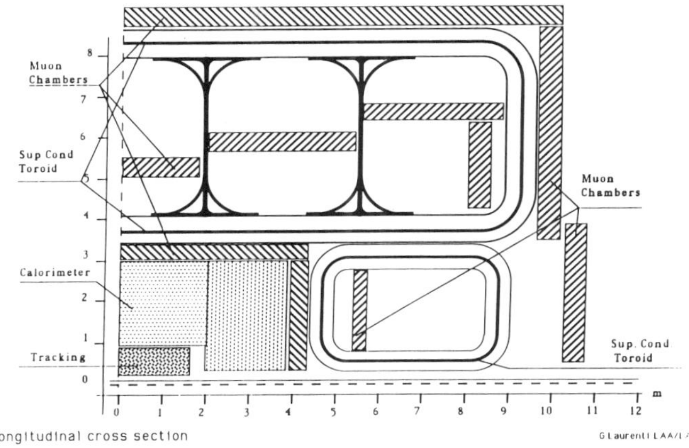

Figure 4 Side view of quadrant of the LHC detector.

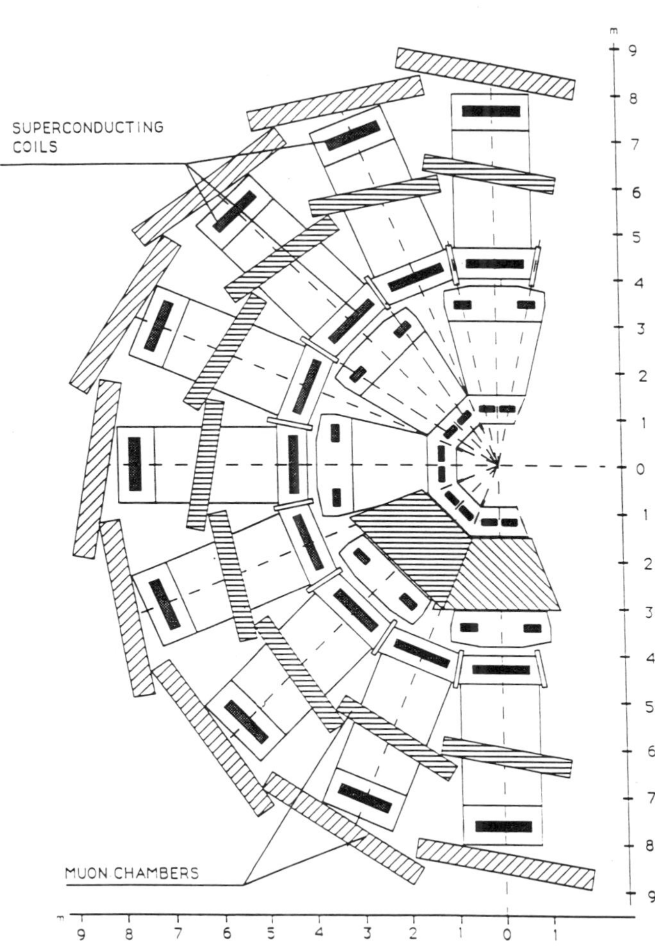

Figure 5 Transversal cross section of the LHC detector

Table 3
Coil Dimensions for a Toroidal Muon Spectrometer with Discrete Air Coils for a multi-TeV hadron-hadron collider

	Central Detector	Forward Detector
# of coils	16	16 (rotated by 11.25 deg. relative to central coils)
width	1 m	0.3 m
largest straight section	8.5 m	2 m
smallest straight section	2.6m	
radius of arcs	1 m	1 m
distance centers of 2 opposite coils	12 m	4 m
current per coil	2700 kA	1200 kA
central field	2 T	2.5 T
maximum field	2.7 T	3.7 T
outer diameter	16 m	6 m
length	19 m	4 m

The toroidal magnetic field was generated by a set of separated super conducting coils, all to be contained in their own cryostat (Fig. 6). A set of 16 large coils was covering the central region, two sets of 16 smaller coils were used as end caps to cover the forward regions. This layout was chosen for ease of construction and accessibility. A big advantage of this construction is that no iron return yoke is necessary, giving a light, compact detector. In order to achieve high momentum resolutions no iron is used at all, since due to multiple scattering the momentum resolution in an iron spectrometer is restricted to about 10%.

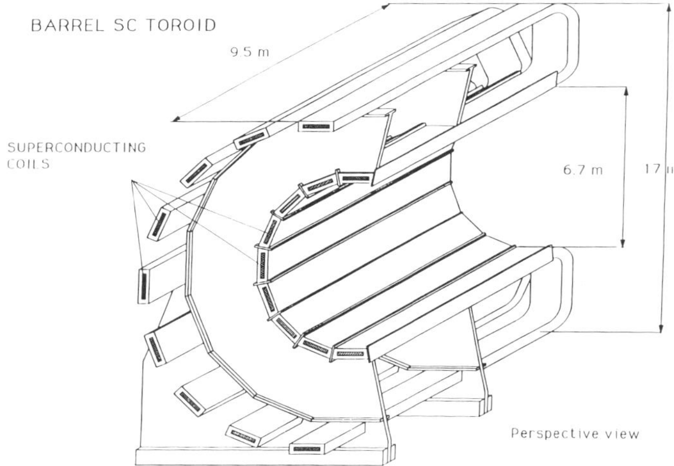

Figure 6 Perspective view of the LHC detector showing superconducting coils

A three dimensional field calculation shows that a good field uniformity is achievable in the proposed modular design. A more homogeneous coil would require more complicated cryostats and reduce the accessibility to the inner muon chambers and other inner detector parts, without introducing benefits on the momentum measurement accuracy.

One can conclude that an integral toroidal muon spectrometer with discrete air coils can give an excellent momentum resolution without "dead" angles. However, it was also found that the forces on the conductors are large and that the amount of stored magnetic energy is considerable. To absorb these forces a strong support between the superconducting coils and cryostats may be necessary, which will make the thermal insulation difficult. The stored magnetic energy must be released in a controlled way in case of a magnet quench and thus a complicated safety system has to be built. The superconducting wires have to be constructed in such a way that they are able to stand the high magnetic field. Since at least one layer of detection elements has to be within the high field region, special detectors are necessary. In the following sections some ideas to solve these problems will be given. A lot of research still has to be done to reach a final design.

Since it is non-trivial to operate a detector, based on drifting charges in a gas, in such a high magnetic field, it is necessary to test prototype chambers in fields of 1 to 3 T. Especially the operation of one of our new prototypes for a round chamber in the forward region [7], based on gas amplification around the sharp edge of a blade, could be sensitive to a high magnetic field. In this chamber, contrary to a wire chamber, not all drifting electrons will end up in the highest field region. Applying a high magnetic field could give rise to an asymmetry in the signal of electrons approaching the blade from different sides: from one side the electrons would be pushed towards the high field region, from the other side they would be pushed away from it. This effect has to be measured and, if large, it should be decreased e.g. by the use of a different gas an/or introducing an asymmetry in the chamber. For the drift chambers of the central region, consisting of normal wire chambers, the influence of the magnetic field on the drift time has to be determined.

Superconducting toroidal spectrometer for the SSC

A system of superconducting toroids is a central feature in an experiment proposed for the SSC - EMPACT. [8] For this proposal a specific system of toroidal magnets has been considered as the basis both for a proof-of-principle and for an investigation of detail design alternatives. The magnetic field volume will serve as a refractive medium alone, as there are no position measurements in the field region. Comparisons of measurement of sagitta with external measurements (i.e. point-line) indicate one needs typically 4 to 5 times as many points to measure tracks with the same momentum resolution, assuming the same point precision and field integral. The higher number of points then requires more stringent alignment to achieve this resolution. [9] In order to optimize momentum resolution, the design maximizes bending power while minimizing material thickness (radiation lengths) during the refractive process. Although considerable mechanical analysis effort is required to optimize parameters, initial results indicate that this technique is both promising and effective [10].

The configuration chosen for study is shown in Fig. 7. Rectangular cross section toroids with peak surface magnetic fields of 1.5 T and 4.5 T have been considered for the central and end toroids, respectively. The preferred superconductor is NbTi in a stabilized, forced-cooling system with an aluminum alloy conduit. Most of the stabilizer is aluminum.

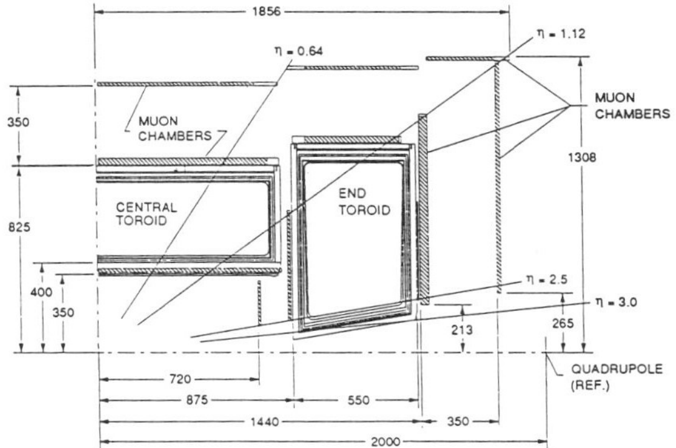

Figure 7 Layout of EMPACT Muon System Showing Key Dimensions (cm) of Toroids and Muon Detectors

The magnetic field strength within the coil varies inversely with radius and is well confined. It is purely azimuthal and homogeneous since the superconducting coils are closely spaced and uniformly wound. A bending length (∫B dl) of approximately 3.6 T m can be obtained at $\theta = 90°$. This will increase with η due at first to increased path length in the central toroid, and then to increased magnetic field in the end toroid, reaching a value of 18.6 T m at $\theta = 9.4°$. The stored energy in each of the three magnets (one central and two end toroids) will be approximately 1 GJ.

A conceptual design of air core toroidal magnets that are radiation thin and permit high resolution measurement of muon momenta has been carried out [10]. A primary design issue has been the structural stability under both operating and fault conditions rather than stress analysis alone. The structure must support the magnetic pressure $B^2/2m_0$ directed everywhere outward from the magnetic field volume. Since the field-radius product $B \times r$ is constant for a toroid, there exists a net centering pressure on the cylindrical portions of the coil, i.e., a force directed radially inward. The annular end faces of each coil are subjected to equal and opposite axial forces $\pm F_z$. The axial force is independent of the coil shape between the inner and outer radii. The centering force is shape-dependent. Forces scale as B^2. Stresses are constant for constant aspect ratio. By detailed mechanical analysis, it was found that keeping the structural spans at 5° azimuthally reduced the average amount of material and therefore the radiation lengths and multiple scattering effects compared to larger spans. The structural material is aluminum alloy 2219-T87, a well documented material for use at low temperatures, with a radiation length of 7.95 cm.

The approach that is suggested is to use bi-axially tensioned membranes (i.e., plates) placed in azimuthal planes that tie together the inner and outer cylinders and also connect the end faces to each other. The result is a tensioned vane that directly reacts the axial forces and supports, in part, the inner cylinder of the coil by using the radially outward pressure on the outer cylinder of the coil. This arrangement also improves the stability of the structure by controlling cylinder distortions and raising the buckling mode number. The chosen subdivision of 72 sectors, each with a 5° included angle, is selected to permit thin annular disks with reasonable deflections. Figure 8 shows the details of the 5° magnet segment.

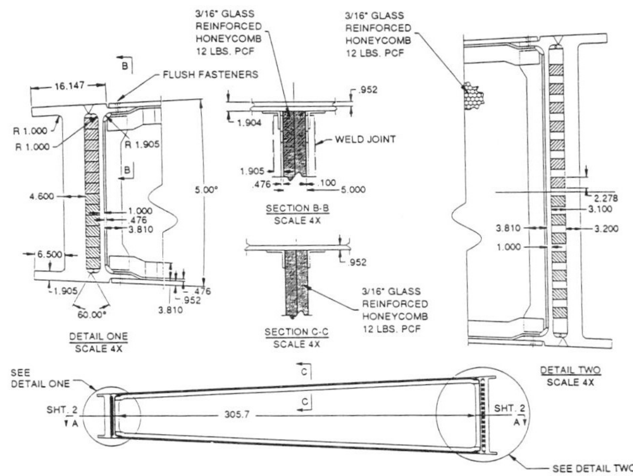

Figure 8 Degree Cross Section (All Dimensions in cm)

This geometry has been evaluated over the angular range from the minimum detection angle of 9.02° (h = 2.54) to 90°. The ratio of rms multiple scattering angle to magnetic bend angle, a figure of merit for the design, is shown in Fig. 9. This does not include the scattering effects of supports, cables, electronics or the utilities required for the magnet itself that will reduce transparency over localized areas. The region of poorer resolution, $30° < \theta < 45°$, corresponds to the overlap between magnets where a ray must traverse the material of both coils.

Muon trajectories through the vanes will suffer increased multiple scattering (about a factor of 4 or 5) compared to neighboring trajectories and hence will have worse momentum resolution. Careful design of assembly joints may limit the degraded fraction of azimuth to less than approximately 5%.

A considerable effort has been expended on assembly and systems considerations in order to confront and evaluate the problems inherent in realizing a toroidal magnet based detector as the design evolves. An ongoing engineering group has been set up to consider the size, weight, and transportability of the major toroid elements, as well as the integration of muon detectors, calorimeter, supports, and systems.

A key feature in the implementation of the design is the high multiplicity of identical coil elements. The basic subdivision of 5° sectors is shown in cross section and

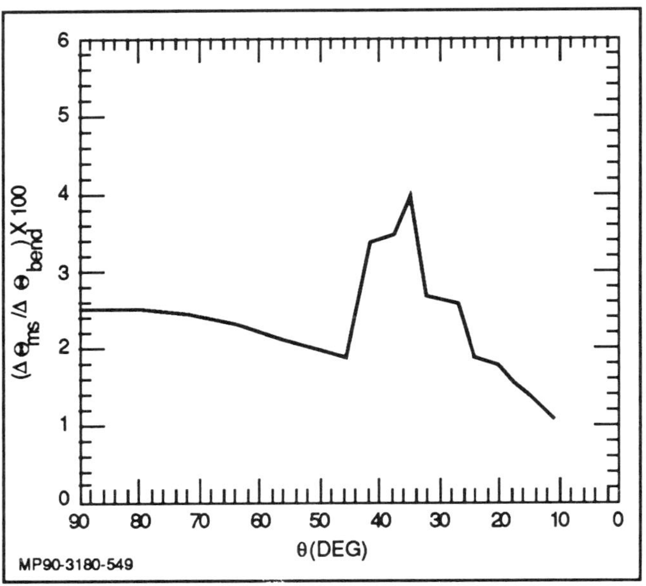

Figure 9 Ratio of RMS Multiple Scattering Angle to Magnetic Bend Angle for Toroids versus Polar Angle

perspective in Figs. 8 and 10. These coils can be fabricated in industry, and the requirement of 72 identical central coils and 144 end coils provides savings, reliability, and spares, and a high degree of optimization in fabrication techniques. The coils could be fabricated and completely tested before shipment to the site where the central coils will be assembled into 45° units for underground assembly. The end toroids can be completely assembled above ground and installed as a unit.

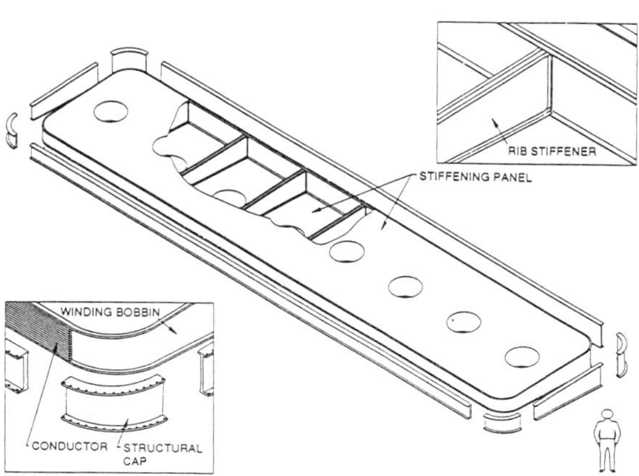

Figure 10 Central Toroid 5 Degree Coll Assembly

The central toroid dewar is designed to provide support for the calorimetry system and for support and alignment of the inner muon chambers (Fig. 11). It is designed to permit access and replacement of the muon chambers located between the dewar and calorimeter walls without removing the calorimeter. The toroid dewar inner wall is a double wall cylinder assembly; the outer wall is the vacuum boundary for the toroid, and the inner wall is the support tube for the calorimeter modules. The walls are joined together by eight longitudinal ribs that also support the muon chambers. A preliminary finite element analysis of the dewar/calorimeter support structure has been done (Fig. 12). This analysis prefers 12 ribs instead of 8, permitting thinner walls, but increasing the muon channel count. The end dewar walls are not only vacuum enclosures but also provide support for calorimeter cabling and vacuum plenums for calorimeter pumping.

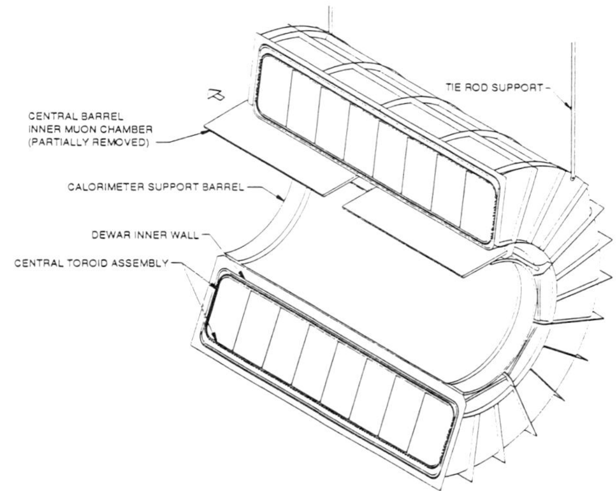

Figure 11 Central Toroid Assembly

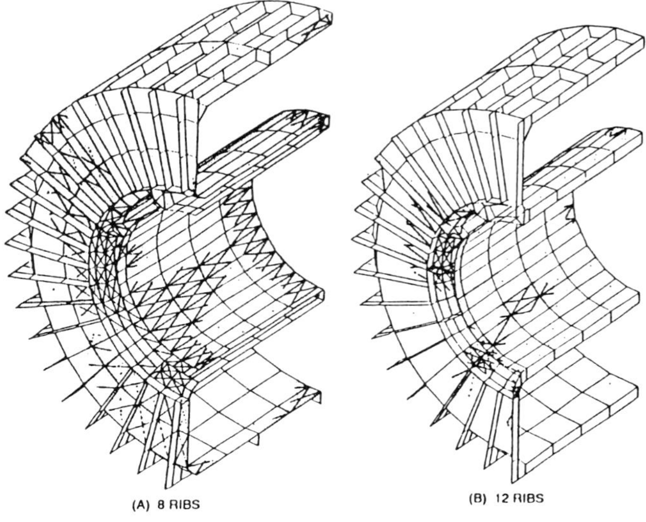

Figure 12 Stress Plots for Central Toroid Dewer for (A) 8 Ribs and (B) 12 Ribs

The central components are independently supported on fixed overhead frames giving minimal interference with the coverage of the muon chambers, and the central toroid structure/conductor assembly is supported by room temperature tie rods (Fig. 13). The outer central muon detectors are supported from intermediate fixed frames. The end toroid structure, dewar, and muon chambers are similarly supported, except the support frames which are mounted on rollers to permit lateral movement of the total assembly for installation, maintenance, and repair of otherwise inaccessible systems (Fig. 14).

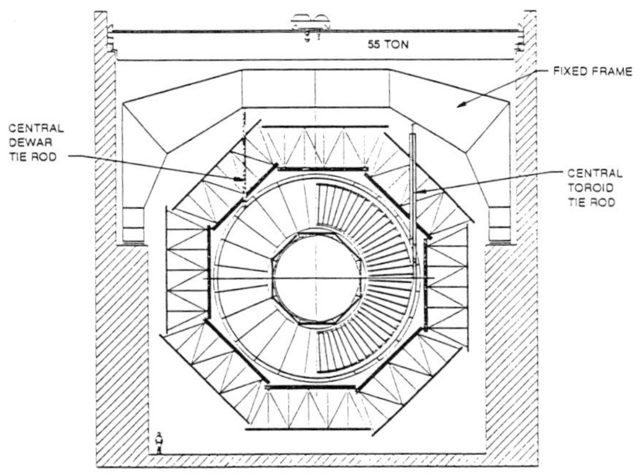

Figure 13 Section through Central Toroid Showing Overhead Support Frame and Muon Detector Packages

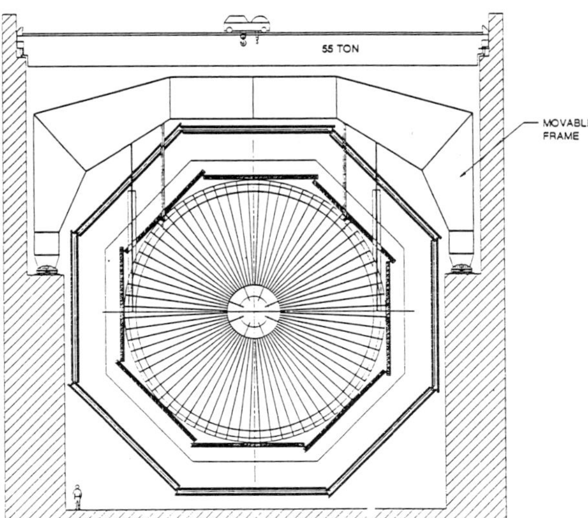

Figure 14 View of End Toroid Showing Movable Overhead Support Frame

A promising design for a cable-in-conduit conductor (CICC) [11] for the air core superconducting toroids was carried out with the aim of reducing the radiation length compared to earlier designs. The cross section is shown in Fig. 15. Aluminum alloy was substituted for the usual stainless sheath to enclose the conductor, and pure aluminum was used for the major portion of the stabilizer copper that is found on conventional CICC conductors. The sheath would be continuously formed about the wrapped cable, obviating the need for a full length seam weld. The individual NbTi/Cu multifilamentary strands would be clad with stabilizing aluminum before cabling, sheathing, and compaction to approximately 30% void. Although the conductor concept appears reasonable, it is an extension of the CICC art that remains to be demonstrated. Among other things considered were low temperature resistance of the stabilizer, stability margin, maximum quench pressure, sheath strength and loads, stored energy, dump voltage, conductor current density, helium path length, and conductor mass. All were found to be acceptable. If conductor development by producing and testing samples is successfully completed, the conductor and torus into which it is wound should be economical and perform satisfactorily.

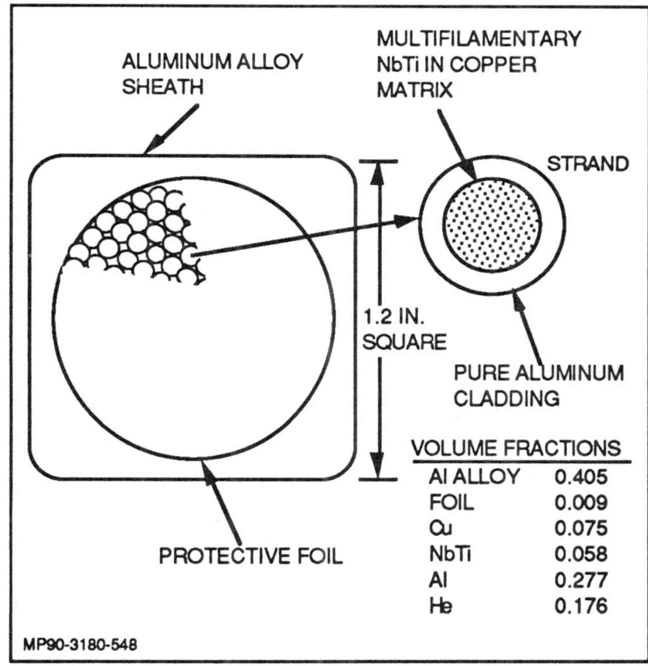

Figure 15 Cross Section of Cable-in-Conduit Conductor

The muon track reconstruction quality has been studied for the conceptual magnet model. The study was based on the assumption of groupings of multilayer devices, such as drift or streamer tubes. The groupings are represented as providing two "superlayers" of detectors to determine the trajectory before and after the magnet. The number of layers in these superlayers was varied to find the optimal distribution to minimize the momentum resolution. A crude representation of the material at smaller radii than the muon system was made by assuming the presence of a spherical "calorimeter" with a thickness equal to 450 radiation lengths. This material has very little influence on the momentum resolution but strongly affects the precision of the other track parameters.

A program was developed that incorporated the mechanical details of the proposed superconducting magnet and took into account multiple Coulomb scattering and measurement errors. The program is flexible with regard to input parameters and is well suited to optimization studies. The output of the program is the correlated error matrix for the χ^2 estimates of the five parameters that describe a particle track (P, θ, ϕ, X_0 ,Y_0) giving momentum, directions and intercepts. For simplicity, particles were assumed to lose no energy by ionization or radiative processes, and multiple scattering was taken to be Gaussian. The full details of this program are found in [12].

The results from this program were found to be in excellent agreement with the output of a well tested fitting and tracking program that had been developed earlier for the SLD detector [13]. Figures 16, 17, and 18 show the resolution of track parameters as a function of muon momentum and polar angle for toroidal magnets with B_0 = 4.5 T and B_1 = 1.5 T. B_0 (B_1) is the maximum field at the minimum radius of the end (central) toroid.

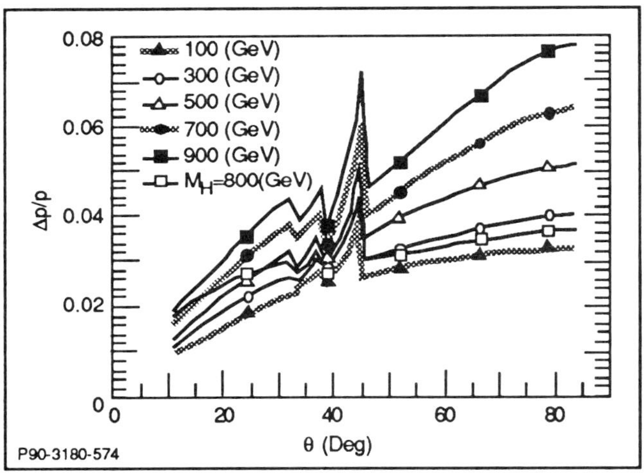

Figure 16 Momentum Resolution as a Function θ for the Baseline Design

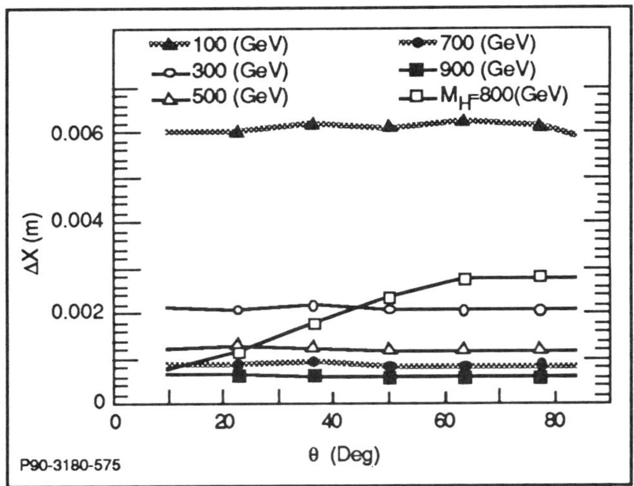

Figure 17 Resolution on Vertex Position for Baseline Design

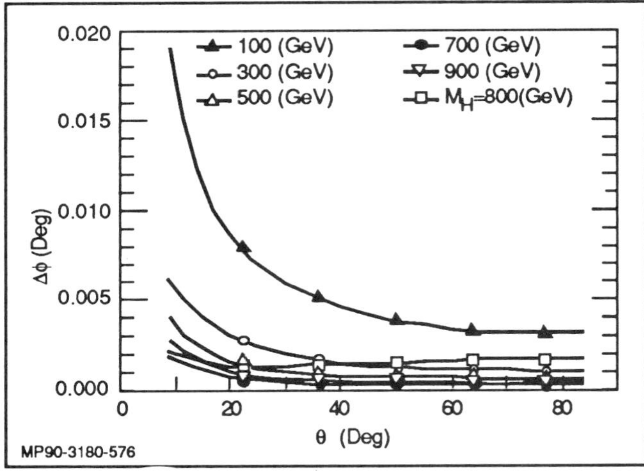

Figure 18 ϕ Resolution for Baseline Design

A series of studies were made of the performance characteristics of magnet/detector systems to give an indication of the range of magnet performance. The momentum resolution for a change in B_1 to 2.25 T is shown in Fig. 19 (compare to Fig.16). The changes in the complicated structure that appears in the region 31° – 45°, as one varies B_1, arise from the interplay of the change in bending power due to variation of the magnetic field and the associated change in the multiple scattering in the superconductor in the coils and support structure. Another important factor in this region is the gap between the end and central toroid support structures that is required for access and mechanical support of other parts of the detector system such as the calorimeter and tracking system. We have also studied the sensitivities to detector misalignment as described in [9] All these aspects are under study and will be optimized.

To study the effects of these changes, a sample of 800 GeV Higgs decays into two Z^0's decaying into muons was produced using ISAJET. The track parameters were modified by the previously described program, then analyzed. The Higgs particles were given the expected width, but the Z^0's were generated with zero natural width to isolate the effect of the track fitting on the reconstruction process. The samples

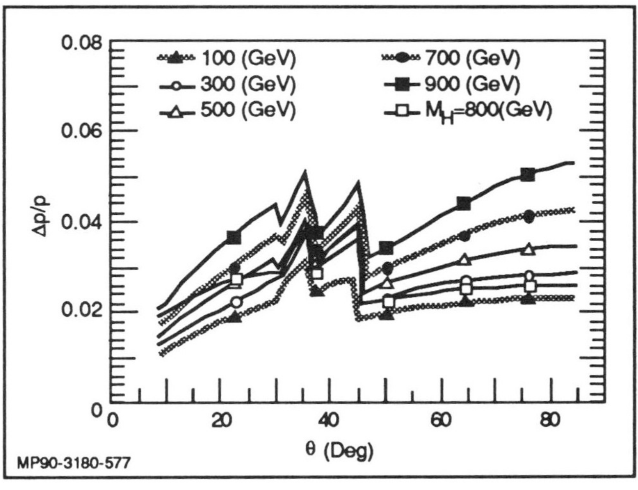

Figure 19 Momentum Resolution for Muons with Central Toroid Field at 2.25 T

were selected to include only events for which all four muons fell within the acceptance of the model detector. This resulted in an efficiency of 77%. In Fig. 20, the effect on the mass resolution for variations of both the end and central toroid field strengths is shown. The effect of a vertex constraint is displayed for comparison. Such studies will be used in conjunction with the detailed designs for the magnets to obtain parametric information on cost and schedule for various performance characteristics. These studies of physics and fabrication data, in conjunction with efforts on detector integration, will give an excellent opportunity to design for cost and performance.

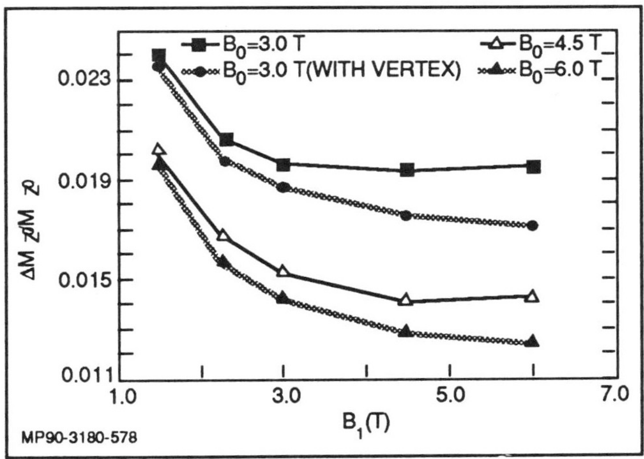

Figure 20 Z^0 Mass Resolution
$H \rightarrow Z^0 + Z^0 \rightarrow \mu^+\mu^-\mu^+\mu^-$

In summary, it seems possible to design a detection system, based on a superconducting toroidal magnet, that has excellent acceptance over a large range of rapidity and with resolution properties for muons equivalent to the calorimetric resolution available for electrons.

REFERENCES

[1] M. Marx, "An Alternative Approach to a High P_T SSC Experiment," SSC-219, 1989.

[2] P. Bonanos and J. Luton, "Toroidal Detector Scoping Study," SSCL-N-697 (1990).

[3] S. Mori, in *Proceedings of the Symposium on Detector Research and Development for the Superconducting Super Collider,* T. Dombeck, V. Kelly and G. Yost (eds.), World Scientific Publishing, Singapore, 1990.

[4] J. Purcell, in *Proceedings of the Symposium on Detector Research and Development for the Superconducting Super Collider,* T. Dombeck, V. Kelly and G. Yost (eds.), World Scientific Publishing, Singapore, 1990.

[5] F.. Bergsma, *et al.,* "A study of a toroidal muon spectrometer with discrete air coils for a multi-TeV hadron-hadron collider," presented at the European Committee for Future Accelerators Large Hadron Collider Workshop, Aachen, 4-9 October, 1990.

[6] Q. Linn, in *Proceedings of the Symposium on Detector Research and Development for the Superconducting Super Collider,* T. Dombeck, V. Kelly and G. Yost (eds.), World Scientific Publishing, Singapore, 1990.

[7] G. Ambrosi, et al., "Detector tests in a high magnetic field and muon spectrometer triggering studies," CERN/DRDC/9042, DRDC/p.10, Geneva, 1990.

[8] 'The EMPACT Experiment at the Superconducting Super Collider," SSC–EoI–0006 (1990).

[9] L. Rosenson, "The Effects of Misalignment on the Track Reconstruction in EMPACT," EMPACT Note 239 (1990).

[10] P. Bonanos, "Toroid Conceptual Design," EMPACT Note 183 (1990).

[11] J. Luton, "The Feasibility of Low Mass Conductors for Toroidal Superconducting Magnets for the Superconducting Super Collider (SSC)," EMPACT Note 202 (1990).

[12] L. Rosenson, "Momentum Resolution and Track Reconstruction Quality for the EMPACT Detector," EMPACT Note 224 (1990).

[13] R. Verdier, SLD report, in preparation

SCINTILLATING FIBER DETECTORS*

Fiber Tracking Group (FTG)

B.Abbot[h], D.Adams[i], M.Atac[a,b], C.Anway[a], A.Baumbaugh[b], B.Baumbaugh[d], P.Berge[b], M.Binkley[b], J.Bishop[d], N.Biswas[d], A.D.Bross[b], C.Buchanan[a], N.Cason[d], R.Chaney[j], D.Chrisman[a], A.Clark[b], D.Cline[a], H.Cohn[e], M.Corcoran[i], R.Davies[h], J.Elias[b], E.Fenyves[j], D.Finley[b], G.W.Foster[b], J-M.Gaillard[b], J.Godfrey[d], H.Goldberg[c], H.Hammack[j], X.Huang[d], J.Jaques[d], V.Kenney[d], R.Kephart[b], D.Koltick[h], J.Kolonko[a], K.Kondo[k], R.A.Lewis[g], J.Marchant[d], R.McIlwain[h], S.Margulies[c], H.Mietteneni[i], R.Moore[g], R.J.Mountain[d], T.Okusawa[f], J.Piekarz[d], A.Pla-Dalmau[b], R.Ruchti[d], R.Scalise[g], W.Shephard[d], E.Shibata[h], G.A.Smith[g], J. Solomon[c], K.Takikawa[k], S.Tkaczyk[b], and R.Wagner[b]

University of California at Los Angeles[a]
Fermi National Accelerator Laboratory[b]
University of Illinois at Chicago[c]
University of Notre Dame[d]
Oak Ridge National Laboratory[e]
Osaka City University[f]
Pennsylvania State University[g]
Purdue University[h]
Rice University[i]
University of Texas at Dallas[j]
Tsukuba University[k]

Presented by R. Ruchti

INTRODUCTION

The decade of the 1990s will see the construction and commissioning of the Superconducting Super Collider as the one of the World's premier scientific instruments. To meet the challenges of experiment at this new machine, it is essential to have successful development of technologies capable of providing charged-particle tracking and fast triggering on high p_T tracks to the highest luminosities, $L = 10^{34}$ cm^{-2} sec^{-1}. Tracking detectors based on scintillating fibers represent one of these important initiatives, particularly for large volume, general-purpose detectors.

The Fiber Tracking Group (FTG) is developing a high-rate tracking subsystem appropriate for SSC applications in general and for the SDC experiment in particular, based on scintillating fiber technology. Typical SSC detector designs contain 10^6 fibers and associated readout channels. Additionally, the group is actively pursuing the development of a 10^5 fiber central tracker for the upgrade of the D0 detector for operation at the Fermilab Main Injector. The use of scintillating fibers as a detection medium affords the prospect for tracking charged particles with high resolution, low occupancy, and very short resolving time, and for providing first-level triggering based on the transverse momentum of charged particles.

PHYSICS CHALLENGE: THE DETECTION OF RARE PROCESSES

An important challenge for any SSC experiment is the provision of an effective tracking detector. Because of the high luminosity environment and the short bunch separation at the SSC, any tracking detector must have fast response, low occupancy, good spatial resolution, and good resistance to radiation damage.

A central tracking system should be capable of reconstructing low-mass states such as the Z^o with a precision of 1-2%, and should enable cross-checks of the electromagnetic and hadronic calorimeter calibration over a wide momentum range. Further, expected topologies such as the Higgs boson, detected via the decay $H \rightarrow Z^oZ^o \rightarrow qqll$ and the decay

*Work Supported by the US Department of Energy, the SSC Laboratory, the Texas National Research Laboratory Commission, and participating Universities and Labs.

$H \rightarrow Z^oZ^o \rightarrow$ 4 leptons, should be identifiable, with at least a 3σ charge sign determination for tracks of $p_T \leq 1$ TeV/c.

At an SSC detector operating at machine luminosities $L = 10^{34}\ cm^{-2}\ sec^{-1}$, any viable tracking device must operate reliably in a high radiation environment (typically $< 10^6$ rads in 10 years at a radius r = 60 cm from the interaction region due to beam interactions and neutron albedo). The tracking system, which must have low material thickness, should be capable of reconstructing all tracks, have good rejection against spurious tracks, and should be capable of separating tracks resulting from multiple interactions in a bunch crossing. A number of tracking options are being considered, but so far no technique has been proven. At small radii (typically r < 50 cm) silicon detectors (strips and/or pixels) are being considered for track reconstruction. At larger radii, the possibility of a stand-alone scintillating fiber tracker is potentially very attractive. It should be noted that the advantages of these two detector types are complementary.

Since the first successful use of fiber tracking in a collider environment [1], significant progress has been made [2]. Fiber trackers have the following attractive features.

1. Good tracking precision: multilayer structures of 750μm diameter fibers can result in a measurement resolution of 60μm per layer [3,4] and provide good two track resolution.

2. Fine granularity: for fibers of 750μm diameter and 4m length placed at radii r = 60cm, the worst case occupancy from all sources is expected to be < 3.2% (< 21%) per fiber, per bunch crossing, at a luminosity $L = 10^{33}\ (10^{34})\ cm^{-2}\ sec^{-1}$. [5] For fibers placed at larger radii, the occupancies are substantially lower.

3. Good time resolution: an intrinsically prompt response;

4. Insensitivity to magnetic fields and to RF noise;

5. Simple electronics: binary (yes/no) readout is used;

6. No power dissipation in the tracking volume; and

7. The possibility to develop a level 1 trigger based on transverse momentum of individual charged particle tracks [4,6,7].

However, several technological developments are required before a realistic scintillating fiber central tracker for SSC detectors can be claimed. Figure 1 shows a hypothetical central tracking system [8] which covers the kinematic range $|\eta| \leq 2.5$. For $|\eta| \leq 1.6$ the geometry is cylindrical. Tracking in this region is accomplished with four superlayers, each comprised of four sublayers of fibers of 750μm diameter and lengths ≤ 4m. Two of these sublayers contain axial fibers (x) which measure $r\phi$ coordinate, and two provide narrow angle stereo (u,v) measurements (refer to Figure 2). As shown in Figure 3, the placement of component fiber layers within a sublayer can be adjusted to provide optimal spatial resolution. Four layers of fibers of 750μm diameter each offset by 1/4 of a diameter, lead to a position resolution of 60μm per sublayer[3]. Because of the flexibility of the fibers, stereo layers can be wrapped in a helical fashion, and different stereo angles can be chosen for each superlayer to avoid ghosts in pattern recognition. Typical choices are angles in the 4°-10° range, with larger stereo angles at smaller radii to improve the resolution in longitudinal coordinate. For the intermediate region $1.6 \leq |\eta| \leq 2.5$, the flexibility and field immunity of scintillating fibers permits novel arrangements of detection elements. Figure 4 shows one possibility in which sublayers of spiral right and left coordinates are combined with a sublayer of fibers positioned azimuthally [8]. With this particular configuration, development of a fast trigger on stiff

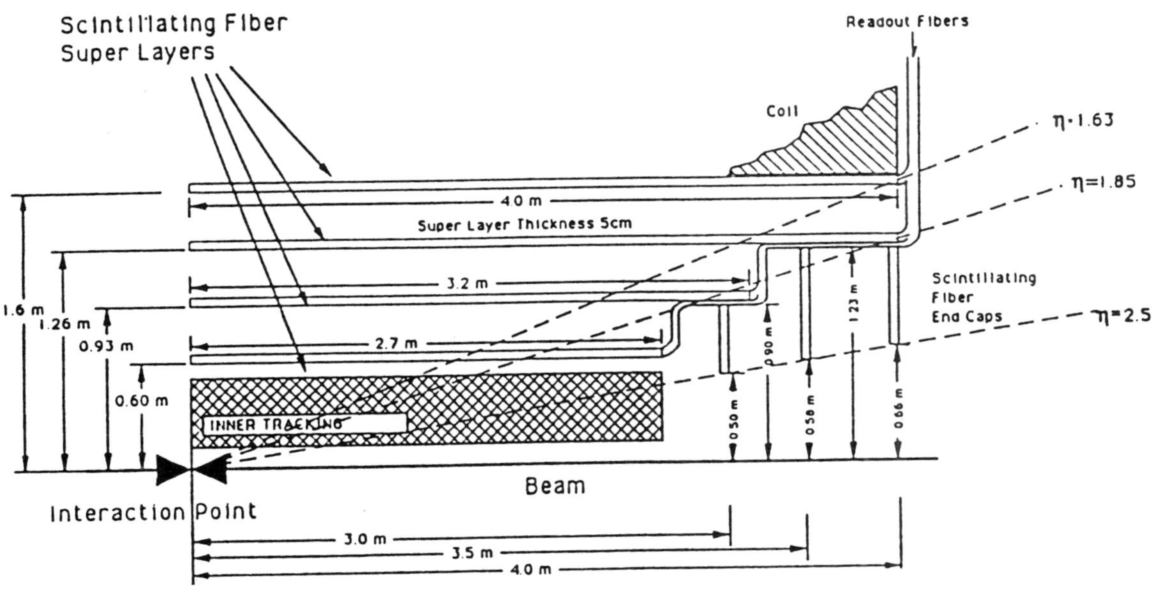

Figure 1. Schematic of a central tracker based on scintillating fibers.

tracks is straightforward, through the use of look up tables.

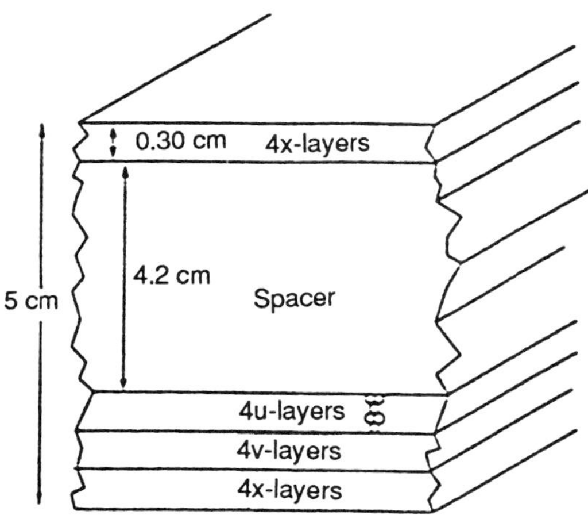

Figure 2. Schematic of a Scintillating Fiber Superlayer

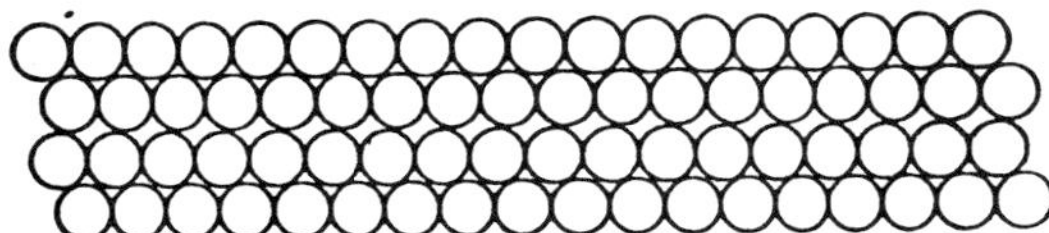

Figure 3. Placement of component fiber layers within a sublayer for optimal spatial resolution. The effective spacing of the layers is 1/4 of a diameter, by shifting the lower doublet relative to the upper doublet.

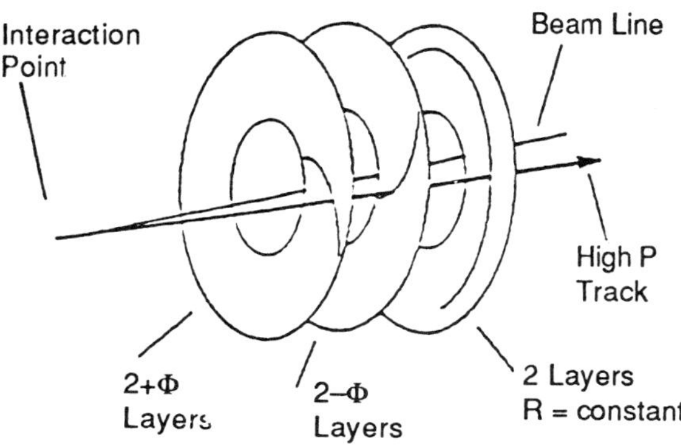

Figure 4. Possible fiber superlayer structures for intermediate tracking, $1.6 < |\eta| < 2.5$.

For the fiber tracker, the basic detection element is an active scintillator fiber of ≤ 4m length, coupled via a 4m-5m long, clear fiber waveguide to a high-rate photodetector. The performance goals are: the detection of minimum ionizing particles resulting from a beam-beam collision with an efficiency exceeding 95% per layer; the determination of the longitudinal coordinate and impact parameter to 1mm for each track; unambiguous assignment of each track to a beam crossing (16 nsec time interval); momentum resolution sufficient to reconstruct the mass of the Z^0 to < 2% in a leptonic decay; momentum resolution sufficient to make a 3σ determination of the charge of a 1 TeV/c particle; and radiation resistance adequate for all components to survive for several years of beam without significant degradation of capabilities.

Experimental Challenge: The Detection of Scintillation Light

The challenge of scintillating fiber tracking is the efficient and effective detection of scintillation photons created by minimum ionizing particles in a large volume detector. A schematic of a fiber tracking element is shown in Figure 5. It includes: a scintillating fiber; a "clear" fiber-optic waveguide which is optically coupled (spliced) to the scintillating fiber; and optical coupling of the waveguide fiber to a photodetector. Of these, the scintillating fiber is the active detection element. The clear fiber waveguide acts as a convenient means of propagating the scintillation information to the location of a photodetector whose position may be remote (typically several meters) from the actual detector region.

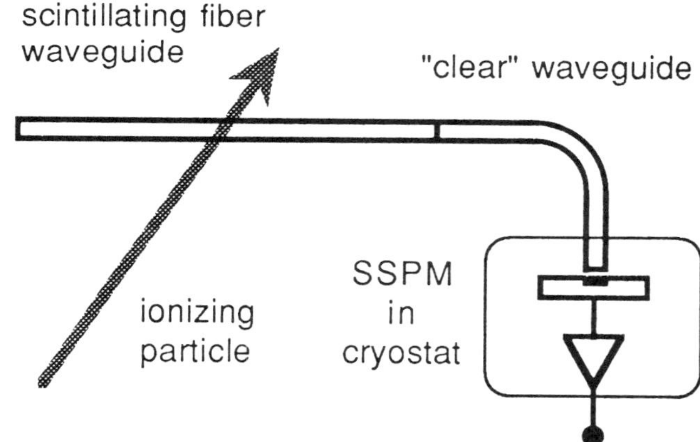

Figure 5. Schematic of a single fiber tracker channel.

The nature of the photodetection problem is indicated numerically in the Equations 1-3 below [9,10,11]:

$$I(z_s+z_w) = I_o\, F\, A(z_s,z_w)\, G_{pm}\, Q_{pm} \quad (1)$$

with:

$$I_o = (dE/dx)\, \Delta x_{eff}\, V_{eff}\, (Q_{sc}/\Delta E_{eff})\, C_{uv} \quad (2)$$

and:

$$A(z_s,z_w) = \exp(-z_s/\Lambda_s)\, T_1 \exp(-z_w/\Lambda_w)\, T_2 \quad (3)$$

where:

$I(z_s+z_w)$ is the detected number of photoelectrons detected at the photodetector located at a total distance z_s+z_w from the scintillation site, where z_s is the effective path length in scintillation fiber material and z_w is the effective path length in the waveguide fiber material.

I_o is the average yield of scintillation photons produced by a minimum ionizing particle in the scintillating fiber, and averaged over the effective cross sectional area of the fiber.

The various quantities which contribute to I_o include:

dE/dx the mean energy loss per unit length.

Δx_{eff} is the effective path length of scintillator through which the particle passes.

V_{eff} is the fraction of the fiber volume occupied by the core, hence active scintillation material.

Q_{sc} is the quantum efficiency of the scintillator.

ΔE_{eff} is the effective deposited energy to generate a scintillation photon, ($\Delta E_{eff} \sim 1239.8/\lambda$(nm) in eV).

Λ_{uv} is the mean free path for waveshifting between primary and secondary dyes.

C_{uv} is the efficiency of UV light conversion between primary and secondary dyes.

F is the fraction of light emitted from the secondary dye (i.e. the resultant scintillation light) which is trapped by total internal reflection in the fiber waveguide material [9].

$A(z_s,z_w)$ is the overall attenuation factor for light propagating in the waveguide from the scintillation site to the photodetector. It includes separate exponential terms for attenuation in the scintillation and waveguide fibers which depend in part on the reflection coefficients between core and cladding and in part on optical absorption. The attenuation lengths Λ_s and the glass capillary and plastic fiber cool, the plastic, having greater thermal contraction, pulls away from the glass wall, allowing it to be withdrawn. may themselves be functions of path length, in which case the attenuation is not a single exponential. Additionally, there are two transmission coefficients: T_1 is the optical transmission at the splice between scintillator and waveguide fibers; and T_2 is the optical transmission through the cryostat boundary.

G_{pm} is the the geometric efficiency for light collection at the photodetector.

Q_{pm} is the quantum efficiency of the photodetector.

A summary of numerical values for these paramaters are given in Table I for a 1mm diameter scintillating fiber composed of polystyrene and the dyes PTP and 3HF. Where possible, measured values are listed. Otherwise the parameter values have been calculated or estimated from Monte Carlo simulation.

Photoelectron Yield

Inserting these values into Equation 1, and assuming fiber scintillators and waveguides each of 4 meter length, $z_s = z_w =$ 4m, we expect to detect an average of $I(z_s+z_w) = I(4+4) = 4.5$ photoelectrons from such a fiber tracking element of 1mm diameter using currently available fiber technology. For 750μm diameter fibers, we expect ≥ 3 photoelectrons. While these yields represent a major achievement already, they are marginal for a large scale system. For implementation in an SSC central tracker composed of 10^6 fiber elements as discussed above, it is desireable to have a safety margin of a factor of ~ 2 in light yield, recognizing that additional light losses will be incurred when an actual detector is constructed. We now discuss the key issues involved, to achieve this goal.

Table I. Values of parameters used in Equations 1 - 3.

Parameter	Value	Status	Ref.
dE/dx	2 MeV/cm	Meas	
Δx_{eff} (norm. inc)	0.78mm	Calc	
V_{eff}	0.94	Calc	
Q_{sc}	0.04	Meas	[nuc]
ΔE_{eff}	4.77eV	Meas	[16]
Λ_{uv}	1.3mm	Calc	
C_{uv}	0.7	Calc	[Fig.7]
F (min,max)	0.031,0.048	Calc	[9]
Λ_s	6m	Meas	[10]
Λ_w	10m	Meas	[14,15]
T_1	0.95	Meas	[18]
T_2	0.9	Est	
G_{pm}	0.96	Calc	[Fig.11]
Q_{pm}(min)	0.6	Meas	[19]

Values are based on a 1mm diameter scintillating fiber. The fiber is assumed to be composed of a polystyrene core of 970μm diameter containing PTP and 3HF primary and secondary dyes. The cladding wall is assumed to be 15μm thick PMMA. The status column indicates whether the value is measured, calculated, or estimated.

Discussion of Fiber Detector Characteristics

For a scintillating fiber tracking element to provide high detection efficiency, hence $I(z_s+z_w)$ large, the following requirements must be satisfied:

1. The effective quantum efficiency of the active fiber element must be high, with local light emission within the fiber. As shown in Table II taken from references [12,13], a number of new promising candidate scintillation materials have been developed. Many of these have fluorescence efficiency which exceeds that of standard scintillators such as polystyrene/PTP/DMPOPOP. Even more impressive is the combination of high efficiency with long wavelength fluorescence, a feature which is essential for good optical

transmission in polystyrene as indicated in Figure 6 [1,14,15]. Attenuation lengths in excess of $\Lambda_s \geq 5$m are desireable.

Table II. Light output before and after 10 Mrad radiation dose from a ^{60}Co source in arbitrary units.

Scintillator[a]	λ^b (nm)	Light output Before	After	After 13 d[c]
MOPOM	420	0.79	.04	.42
MOPOM+.01%BBQ	470	0.95	.13	.63
MOPOM+.02%BBQ	470	0.96	.16	.69
MOPOM+.01%K27	500	0.95	.27	.75
MOPOM+.02%K27	500	0.95	.27	.75
O415A	415	0.92	.49	.79
O415A+.01%BBQ	470	1.01	.42	.86
O415A+.02%BBQ	470	1.02	.34	.89
O415A+.01%K27	500	0.97	.57	.87
O415A+.02%K27	500	0.99	.53	.88
O408	410	0.88	.48	.76
O408+.02%BBQ	470	1.02	.34	.87
O408+.02%K27	500	0.97	.57	.89
DAT	375	0.73	.16	.46
DAT+DMPOPOP	430	0.86	.31	.70
DAT+.01%3HF	530	0.85	.40	.63
pT	360	0.84	.21	.41
pT+DMPOPOP	430	0.88		

[a]Fluor concentrations are by weight, the concentrations of primaries are 1% (1.25% for DAT). [b]Wavelength of peak emission. [c]Annealed 10 days in oxygen and 3 days in air.

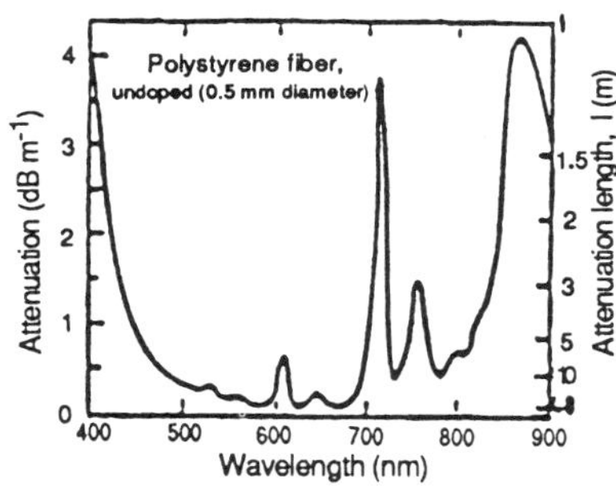

Figure 6. Optical attenuation in 500μm clear polystyrene waveguide as a function of wavelength [1,14,15].

There are several important contributions to this high efficiency. First is the energy level overlap between primary dye and the solvent polystyrene. As indicated in the references [12,13], the efficiency reaches a maximum when the primary concentration is 1% by weight in the material as expected from a non-radiative energy transfer mechanism (Forster Transfer) between the solvent and primary [16,17]. Second is energy overlap between the primary and secondary dyes, for efficient waveshifting (following Beer's Law)[16]. The efficiency of this waveshift is a function of the fiber waveguide material (here assumed to be a polystyrene core with acrylic cladding) and the mean-free-path Λ_{uv} of photons emitted by the primary and absorbed by the secondary. Numerically, Λ_{uv} is inversely proportional to the product of the molar concentration and wavelength-dependent, molar extinction coefficient of the secondary dye. The fiber efficiency is displayed as a function of Λ_{uv} in Figure 7 based on a Monte Carlo simulation program discussed in reference [9], assuming a uniform distribution of dyes in the fiber. For a standard scintillator in which DMPOPOP is the secondary, Λ_{uv}~200μm. Hence ~92% of the light would be effectively waveshifted in a 1mm diameter scintillating fiber of this material, hence C_{uv}=0.92 [18]. By contrast, the mean-free-path for a fiber incorporating 3HF as the secondary is Λ_{uv}~1300μm, which leads to C_{uv}=0.70. Note that there is a minimum of efficiency C_{uv}~0.55. which is determined by total internal reflection of light at the cladding/air interface, i.e. the exterior wall of the fiber. (This minimum efficiency level may well be influenced (lowered) through uv absorption by the polystyrene particularly in the presence of radiation damage.)

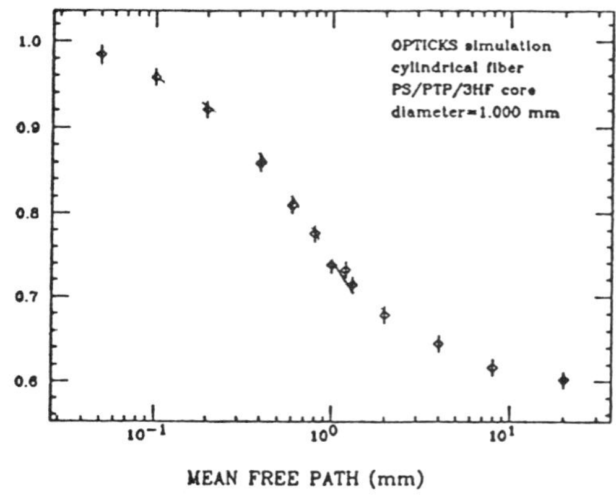

Figure 7. Fiber Efficiency (C_{uv}) as a function of Λ_{uv} (See text)

2. In general the fluorescence decay time of a scintillator is usually determined by the secondary dye and should be as fast as possible. Under SSC conditions, this would require return to baseline in a time of order 16ns. Scintillators containing secondaries such as K27, BBQ, and their variants have fluorescence decay times in the range $10 \leq \tau \leq 12$ ns, whicle others such as G1 and 3HF are in the $3 \leq \tau \leq 6$ ns range.

3. The chemical properties of the scintillator should be selected so that the material is radiation-resistant and has good optical transparency to the produced scintillation light (i.e. low self-absorption). Radiation resistance in polystyrene based scintillators depends, in part, upon emission wavelength. As

shown in Figure 8, the optical transparency of bulk polystrene has a marked cutoff at a wavelength $\lambda \sim 500$ nm for an applied dose of 10 Mrad of ^{60}Co gamma radiation. The expected integrated radiation dose from all sources for an inner fiber superlayer at r = 60cm is expected to be ~ 1 Mrad in 10 years of operation at luminosities $L = 10^{34}$ cm^{-2} sec^{-1}, based on calculations using information from the LBL Report on Radiation Damage [19]. As indicated in Table I, many of the newly developed scintillation materials have the potential to survive in such a hostile environment, but additional radiation damage measurements will be performed to confirm this.

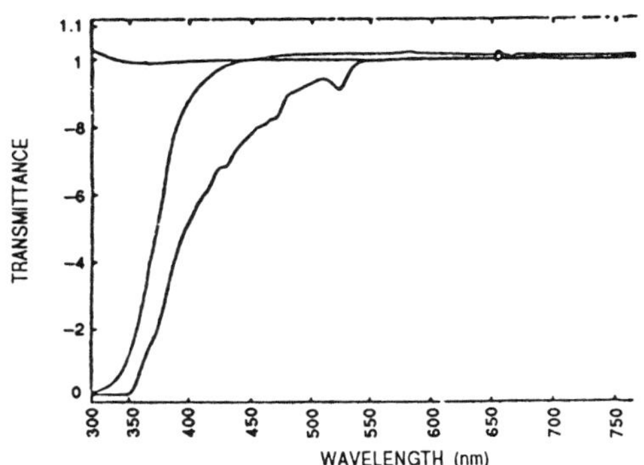

Figure 8. Optical transmission in undoped polystyrene as a function of wavelength, prior to irradiation (upper), immediately after irradiation (lower), and after annealing (intermediate).

4. The mechanical properties of the scintillating fiber materials and the effects of the fiber drawing process must be optimized so that a high reflection coefficient (R ~ 1) can be maintained at the interface between the core and cladding of the fiber waveguide. The characteristics of this interface become yet more important as the ratio of refractive indices of the core and cladding is increased. As discussed in reference [9], the fraction of scintillation light trapped by total internal reflection (F in Equation 1) depends sensitively on this ratio, since a larger ratio of refractive indices implies a larger critical angle. Increasing the critical angle means more boundary encounters (internal reflections) per unit distance along the fiber. Hence higher reflection coefficients R are required to realize a net increase in overall detected light yield.

To date, our studies have been focussed principally on fibers with polystyrene cores and acrylic (PMMA) claddings, because of the considerable experience in drawing such materials by private industry. Refractive indices of the core and clad are n=1.59, n=1.49 repectively at λ=580nm, and the cladding wall thickness is typically 15μm for a 1mm diameter fiber. It should be noted that there are a number of important new initiatives in the area of core and cladding for fibers, particularly low index claddings, and these are being pursued.

5. Optical splicing between scintillation and waveguide fibers should be physically robust and provide a very high transmission coefficient ($T_1 > 0.95$, where T_1 is the coupling coefficient utilized in Equation (3) above). Splicing of scintillation fibers to waveguide fibers has been accomplished by a thermal fusion method [20]. Highly reproducible, robust splices have been produced by this method, with typical optical transmission in excess of 95% integrated over all wavelengths of interest (see Figure 9).

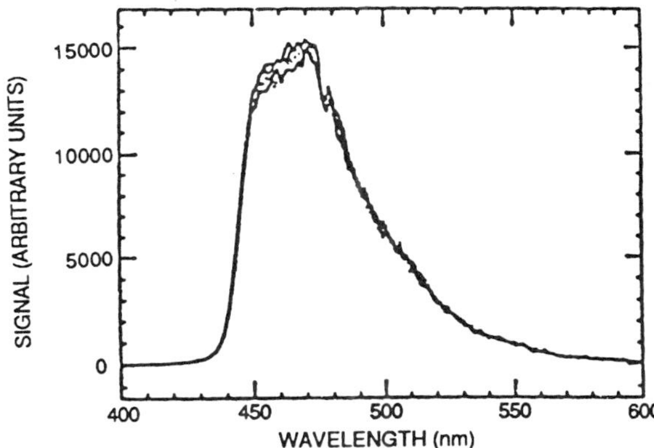

Figure 9. Optical transmission as a function of wavelength for spliced fibers. Solid Lines indicate measurements on the near (detector) side of the splice, dotted lines indicate measurements on the far side of the splice.

6. Clear (non-scintillating) optical fiber waveguides should have long attenuation length for scintillation light (Λ_w~10m). As discussed in Sections 1 and 3 above and illustrated in Figure 6, such transmission is possible in clear polystrene at wavelengths in the visible region of the optical spectrum, $480 \leq \lambda \leq 700$nm, and depends critically on the purity of the materials used to make the core and cladding, as well as the mechanical preparation of preforms and the quality of the fiber drawing.

7. Coupling of the waveguide fiber to the photodetector must be straightforward. For devices which operate at room temperature such as phototubes, this poses no problem. For cryogenically operated solid-state photodetectors such as Solid State Photomultipliers (SSPMs) [21,22] and Visible Light Photon Counters (VLPCs) [23], this requires extra care in the design and maintenance of cryogenics, the fiber-optic link between ambient temperature and the cryogenic environment in which the photodetectors reside, and mechanical infrastructure. Polystyrene fibers, both scintillating and non-scintillating, have been found to provide reliable optical interconnection between room temperature and 7K where the SSPMs and VLPCs operate efficiently. Such fibers have been cycled many times between these temperature extremes without mechanical failure in laboratory tests.

An assessment of the coupling efficiency between fiber waveguide and photodetector, T_2 in Equation (3) above, has been provided by waveguide simulation [9]. Figure 10 shows light rays emergent from the end of a scintillating fiber of 500μm diameter, having polystyrene core and acrylic cladding, assuming uniform illumination by scintillation light over the cross sectional area of the fiber, generated several meters from the end of the fiber. Figure 11 shows the fraction of emergent rays from a 1mm diameter fiber which strike a photodetector facing the end of the fiber, as a function of separation distance between fiber and detector, and for various detector sizes. As long as the photodetector is positioned within 1/4 diameter from the end of the fiber, geometric efficiency is excellent (G_{pm}~96%) and crosstalk to adjacent photodetector elements is minimal.

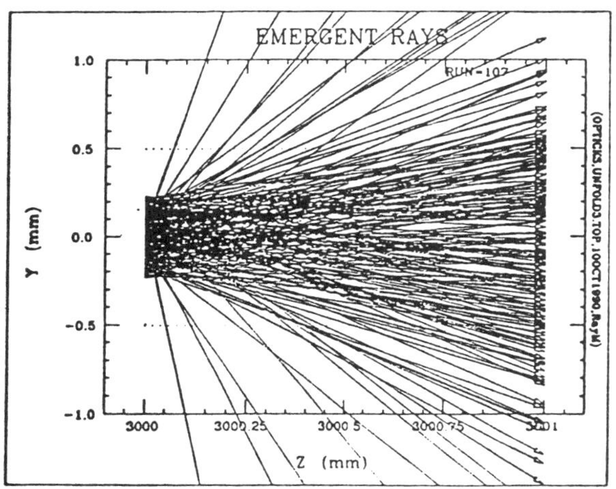

Figure 10. Monte Carlo simulation of light rays emerger from the end of a fiber scintillator of polystyrene core and PMMA cladding.

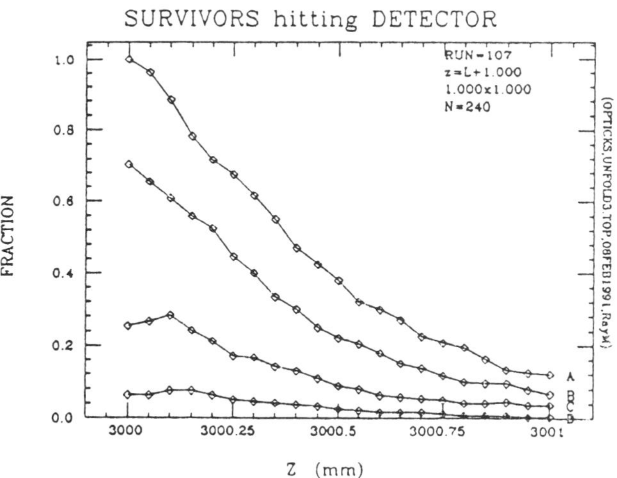

Figure 11. Fraction of light emergent from a fiber waveguide, detected by an VLPC/SSPM as a function of geometry.

8. It is essential that the quantum efficiency of the photodetector be as high as possible at the wavelengths of interest (Q_{pm}> 50%) and, additionally, that the photodetector have fast response. The Solid State Photomultilplier (SSPM) developed by Rockwell [21] incorporates both of these features. Figure 12 shows a schematic of the SSPM. A variant of this structure, VLPC, which is sensitive to visible light only, is under contractual development with Rockwell International Science Center as part of the FTG Subsystem program for SSC applications. As shown in Figure 13, the quantum efficiency of the SSPM structure is Q_{pm}~60% at all visible wavelengths λ>400nm. For the VLPC structure, Q_{pm} is expected to reach 80%. These values are to be contrasted with typical quantum efficiencies of photomultiplier tubes utilizing bialkali or multialkali transmission photocathodes which are at least a order of magnitude lower at wavelengths λ>530nm.

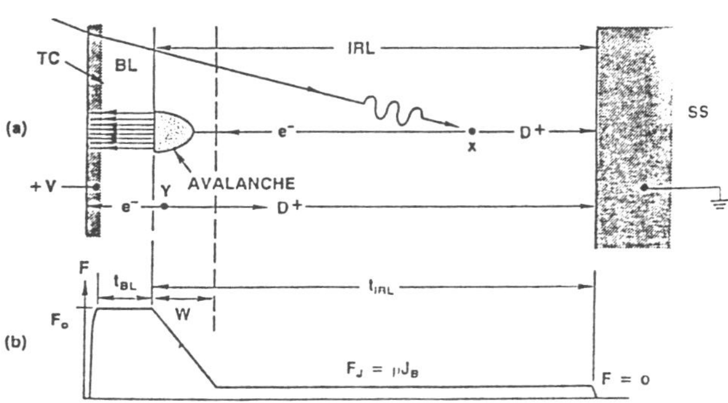

(a) SSPM LAYER CONFIGURATION WITH SCHEMATIC REPRESENTATION OF THE GENERATION OF BIAS CURRENT AND THE EFFECTS FOLLOWING ABSORPTION OF A PHOTON.
(b) ELECTRIC-FIELD PROFILE

Figure 12. Schematic of the SSPM device.

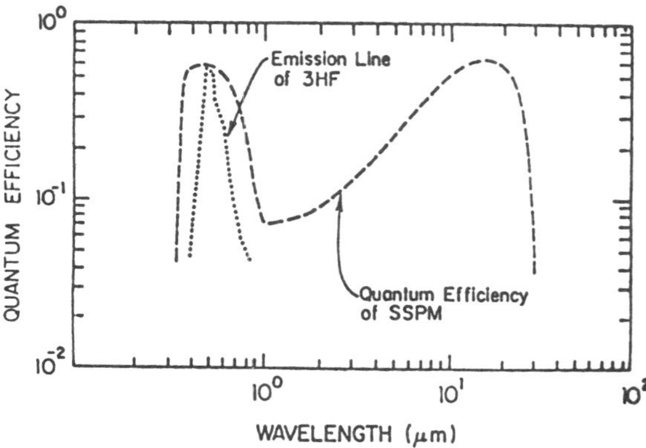

Figure 13. Quantum Efficiency of the SSPM as a function of wavelength.

For operation with high gain, good gain dispersion, and good signal to noise characteristics, the SSPM must be cryogenically cooled to ~7K. Typical gain is 5×10^4, and as shown in Figure 14, the device is a marvelous single photon counter. Peaks corresponding to 1-10 simultaneously detected photoelectrons are easily recognizable in these distributions [23]. Such spectra are obtained routinely when light yields are measured for scintillating fibers which fluoresce at wavelengths in the visible region, and illustrate graphically the capability of the SSPM, and its significance for fiber tracking.

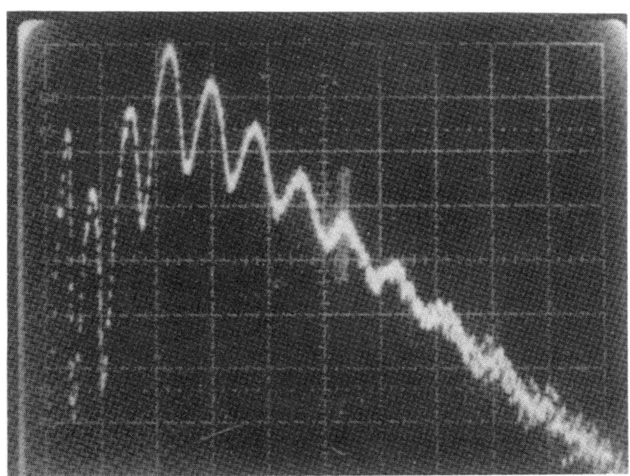

Figure 14. Spectrum of simultaneously detected photoelectrons observed using an SSPM to detect light from a 500μm diameter scintillating fiber with emission at 530nm. Up to 10 distinct photo- electron peaks are discernable. Vertical scale is logarithmic.

An additional important feature of the SSPM/VLPC structure is that devices can be prepared as discrete detectors or in linear arrays and mosaics of detection elements of arbitrary size and complexity. For SSC applications, a linear array of detector elements has been selected as the architecture for initial development [23]. A number of preamplifier options are under consideration for use with the VLPC/SSPM devices, some of which operate cryogenically [24].

Cryogenic heat loads for SSPM/VLPC operation have been estimated [25]. A tracking system of 10^6 scintillating fibers, each coupled to its own VLPC channel, cryogenically cooled preamp, driver and signal cable, constitutes a 150W load for a helium refrigeration system, under high luminosity conditions ($L = 10^{34}$ cm^{-2} sec^{-1}). This is to be compared to a Tevatron ring magnet which requires ~10W of refrigeration.

9. Lastly, the key components of a fiber tracking elements: fiber scintillator, fiber wavguide, photodetector, and ancillary readout electronics must be studied for radiation sensitivity. Important sources of damage are gamma, charged particle and neutron fluences. Such measurements are planned during the next year. However it should be noted that the photodetection elements will be situated outside of the calorimetry where radiation levels are expected to be low.

Improvements to Detected Photoelectron Yield

Possible improvements in overall photoelectron detection efficiency are summarized in Table 3. These include improved quantum efficiency of the VLPC devices, improved scintillation materials and fiber manufacture, and careful study and implementation of new cladding materials. While it is unlikely that each of the listed bènefits will be realized in full measure, nevertheless it is expected that a combination of these will lead to sufficient improvement that an overall factor of two in detected photoelectrons can be realized..

Table III. Potential Improvements to Detected Photoelectron Yield

Item	Factor
Improved QE of VLPC over SSPM	1.4
Improvement in dye C_{uv}	1.3
Improvement in fiber manufacture	1.5
Improvement in light trapping using low refractive index claddings. (Must be accompanied by simultaneous improvement in reflection coefficient in order to be realized.)	1.5 - 3

TRACKING SYSTEM PERFORMANCE

Online Implementation

As discussed in above, the strengths of a fiber tracker are its intrinsic speed and high granularity. These features are being exploited in the development of a Level 1 trigger capability. The principal requirement at this level is to identify charged particle trajectories with a minimum transverse momentum $p_T \geq 10$ GeV/c. Such a trigger must correctly identify high p_T tracks from interesting physics processes, while having excellent rejection power against fake high p_T tracks created by chance correlations (pileup) of minimum bias events.

One possible trigger scheme is: first to select a hit in the tracking superlayer at largest radial distance from the interaction point, then pick a second hit at the intermediate tracking layer affording the maximum sagitta, and finally search for hits in other layers as a redundancy check. This procedure has been studied in simulation [7]. Figure 15 shows the probability per beam crossing for detecting a fake high p_T track based on fiber tracking information only. At a luminosity of 10^{33}, requiring the participation of 3/4 of the tracking elements in the identification of a trajectory leads to a probability of triggering on a fake high p_T track at the 10^{-4} level; at higher luminosity, 10^{34}, one must require all the available tracking elements to achieve a probability of 10^{-1}. [26] Such identified trajectories can be simultaneously associated with deposited energy in calorimetry towers, again at the timing of Level 1, leading to powerful discrimination for electrons at the earliest stages of the trigger chain.

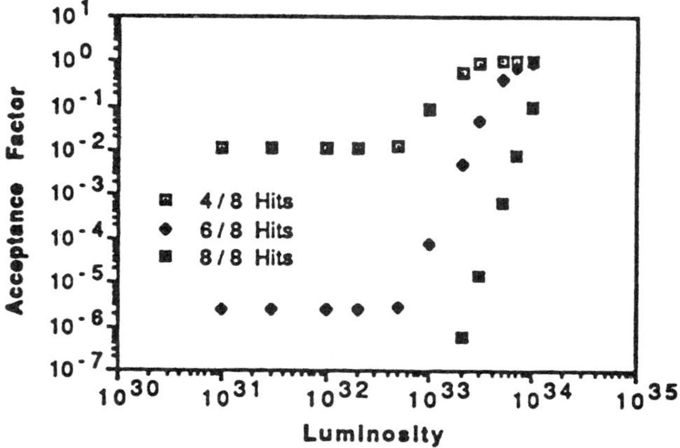

Figure 15. Simulated acceptance for fake tracks in a 650μm scintillating fiber tracker. Vertical scale is probability per beam crossing.

As discussed in references [4,6], a level 1 hardware trigger architecture is being developed which provides rapid determination of the sagitta of a particle track. Initial prototyping is based on discrete technology, whereas ultimate implementation will incorporate ASIC technology. The technique involves identifying clusters of hits in fiber sublayers, then correlating pairs of sublayer clusters within a superlayer, and finally connecting track segments between superlayers. Figure 16 taken from reference [4] indicates schematically the appearance of a low momentum track and a high Pt track, as they would appear in the two sublayers which form a superlayer. The illustrated condition is 3% occupancy.

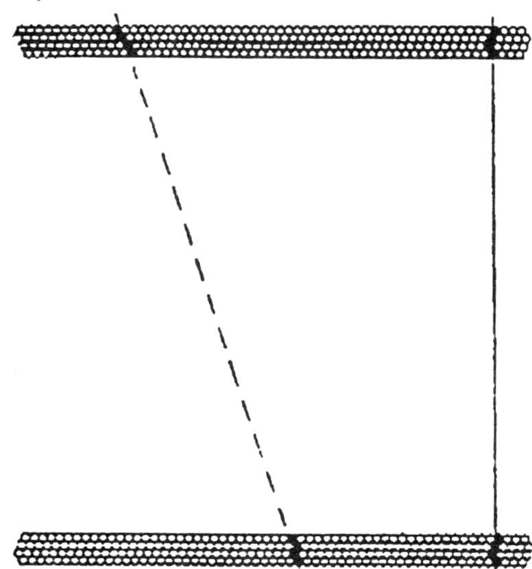

Figure 16. Scintillating fiber superlayer with a radial track (solid line) and a low momentum track (dashed line). Occupancy is 3%. Clusters are shown in the sublayers.

Functionally, the trigger starts from the outer superlayer, and linking proceeds toward the origin, assuming a beam constraint. The search road for a 10 GeV/c track is highly restrictive, since the sagitta for such a trajectory is 2cm over a radial distance of 1.6m. Reconstructed tracks and associated quality factors (whether found in three or four superlayers) are then sent to trigger decision logic. For a track, the azimuthal coordinate is partly encoded by the physical cable and partly by data. The momentum is encoded as the sagitta in units of fiber diameter. The resolution in momentum is better than $0.004p_T$ at the trigger level, and will be implemented with pipelined AND/OR logic. Figure 17 shows a schematic of the trigger architecture.

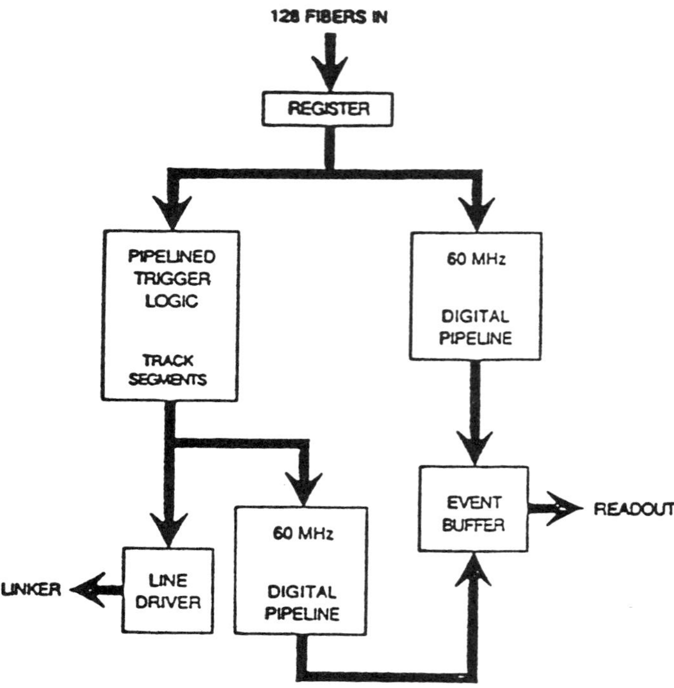

Figure 17. Functional block diagram for digital ASIC combining pipeline delay with track segment finding for the trigger.

Offline Implementation

A substantial effort is underway to develop a detailed simulation and analysis code for fiber tracking based on GEANT [5,27]. Momentum resolution for central tracking is found to be excellent - particularly when incorporating a vertex constraint or tracking information from a vertex or microvertex tracking system. The expected performance for the rapidity interval $|\eta|<1.6$ is: $dp_T/p_T \sim 0.2p_T$ (TeV/c) with no beam constraint, and $dp_T/p_T \sim 0.1p_T$ (TeV/c) with beam constraint [28].

CONCLUSIONS

Scintillating Fibers can provide a powerful means for tracking charged particles in SSC experiments. Substantial Research and Development work still remains, particularly in the areas of scintillation materials and waveguides, and the incorporation and interfacing of these structures with solid state photodetectors. A subsystem program is being carried out by the Fiber Tracking Group directed at these important issues, and substantive beam tests are planned as part of this program, beginning early in 1991 at Fermilab.

REFERENCES

[1] J. Allitti, et al, IEEE Transactions on Nuclear Science, Vol. 36, No. 1(1989) 29-34.

[2] For a comprehensive review of the field and recent progress see: The Proceedings of the Workshop on Scintillating Fiber Detectors for the SSC, Fermilab, November 14-16, 1988, S. Reucroft and R. Ruchti, eds., and various contributions by FTG to these proceedings.

[3] This intrinsic positional accuracy is determined from $750\mu m/(4\sqrt{12})$ which is $55\mu m$. When combined in quadrature with a fiber placement/alignment accuracy of $25\mu m$, the overall measurement resolution is expected to be of order $60\mu m$.

[4] J. Elias, et al, these proceedings.

[5] D. Adams, et al, these proceedings.

[6] G.W. Foster, et al, these proceedings.

[7] R. A. Lewis, et al, these proceedings.

[8] The hypothetical tracker shown is one of the prototypical designs under consideration for the Solenoid Detector Collaboration (SDC). See also D. Koltick, et al, these proceedings.

[9] R.J. Mountain, et al, these proceedings.

[10] B. W. Baumbaugh, et al, paper submitted to the Proceedings of the 1990 IEEE Nuclear Science Symposium, Crystal City, VA, October 1990.

[11] R. Ruchti, lectures on Scintillating Fiber Detectors at the ICFA Summer School on Instrumentation in Particle Physics, Rio de Janeiro, July, 1990.

[12] J. Piekarz, et al, these proceedings.

[13] A. Pla--Dalmau, et al, paper submitted to the Proceedings of the 1990 IEEE Nuclear Science Symposium, Crystal City, VA, October, 1990.

[14] C. Hurlbut, Bicron Corporation, private communication.

[15] T. Kaino, M. Fujiki and K. Jinguji, Review of the Electrical Communication Laboratories, Vol. 32, No. 3, (1984) 478-488.

[16] I. Berlman, *Handbook of Fluorescence Spectra of Aromatic Molecules*, Academic Press (1971).

[17] J. B. Birks, *Theory and Practice of Scintillation Counting*, Pergamon Press (1964).

[18] In the UA2 experience, little improvement was found when the scintillating fibers used for this experiemt were aluminized, in an effort to trap more UV light within the fiber. Figure 7 illustrates that the improvement could be at most 8-10%, consistent with their observations. However, it should be noted that significant improvement (>30%) could be expected for fibers containing secondary dyes such as 3HF, which are in low concentration and/or have low molar extinction coefficients.

[19] D. Groom, et al, Radiation Levels in SSC Interaction Regions, SSC-SR-1033 (1988).

[20] M. Atac, G. W. Foster, and M. Lundun, "A Simple Method for Fusing Plastic Fibers", Fermilab Technical Note FN-537, March, 1990.

[21] M. D. Petroff and M. G. Stapelbroek, IEEE Transactions on Nuclear Science, Vol. 36, No.1, (1989) 158-162.

[22] M. D. Petroff and M. Atac , IEEE Transactions on Nuclear Science, Vol. 36, No.1, (1989) 163-164.

[23] M. Atac, et al, these proceedings.

[24] R. Chaney, et al, these proceedings.

[25] S. Margulies, et al, these proceedings.

[26] As indicated in reference 7, this performance is superior to other large-volume tracking alternatives such as straw tube detectors.

[27] C. Anway, et al, these proceedings.

[28] A summary of system resolution is provided in the Expression of Interest to the SSC Laboratory by the Solenoidal Detector Collaboration, 24 May, 1990, and in the subsequent Letter of Intent.

LIQUID ARGON CALORIMETRY FOR THE SSC

H. A. Gordon

Brookhaven National Laboratory, Dept. of Physics, Upton, NY 11973

Abstract

Liquid argon calorimetry is a mature technique. However, adapting it to the challenging environment of the SSC requires a large amount of R & D. The advantages of the liquid argon approach are summarized and the issues being addressed by the R & D program are described.

Introduction

The liquid argon(LAr) approach to calorimetry has a long history beginning with the pioneering work of Willis and Radeka[1]. A number of large devices have been built and done physics (Mark II at SLAC, ISR806, E706 at Fermilab, and NA31 and HELIOS at CERN). Some very large collider based systems are presently being installed and tested (SLD at SLAC, DØ at Fermilab, and H1 at HERA). These systems constitute an invaluable base of experience upon which to design a detector for the SSC. However the SSC will push the designs further in many directions.

This paper will summarize the recent progress in LAr R&D much of which was presented at this symposium. This will be done in the context of a list of the advantages of the LAr approach and the issues (sometimes called problems) being actively worked on. Much of this work has been organized in the U.S. via the Major Subsystem Proposal[2].

I. Advantages

A. Gain = Unity - Calibration straight forward

Calibration in ionization chamber calorimeters (argon, warm liquid, or silicon) is predicated on controlling the tolerances in the sampling fraction with careful mechanical design, monitoring the purity of the liquid (or effect of radiation damage in the case of silicon), and finally monitoring the gain of the electronics chain with precision pulsers. This pulser calibration needs careful attention in the design, but once implemented, the calibration becomes a straight forward exercise. Typically channels show the same response to within a few percent even without calibration. This can be contrasted to the system requiring the measurement of light which typically has large variations in the response of each channel, and from channel to channel.

B. Linearity over full dynamic range ($> 10^5$)

Linear response over the full dynamic range is an important requirement in doing physics at the SSC. On the high end of the energy scale, the electromagnetic (EM) calorimeter must search for $Z' \rightarrow ee$ up to masses of 8 TeV for a luminosity of $10^{34} cm^{-2} s^{-1}$. Depending on the transverse size of the EM towers and the number of longitudinal subdivisions, most of this energy may at times fall into a single channel - perhaps 5 TeV. The low end of the energy scale is specified by one of two conditions. First if optimizing for $H \rightarrow \gamma\gamma$, the design would probably aim for energy resolution of $\frac{\approx 7\%}{\sqrt{E}}$. In that case, the least count should not contribute to the energy resolution so the low end of the energy scale should be below 0.3 GeV (for a 20 GeV γ). The second condition is that the least count not compromise the position resolution for EM showers. When the shower is nearly centered in a tower, the small energy in adjacent towers must be measured with sufficient accuracy ($\approx$0.025 GeV). This leads to a dynamic range $> 10^5$. This dynamic range is not possible to achieve with state of the art photomultiplier tubes. It is not easy to achieve in ionization calorimeters, but preamplifiers do have that dynamic range.

C. Uniformity

All ionization chamber calorimeters whether using liquids or silicon can have a uniform response over the region of a single tower or the entire calorimeter. However, in order to achieve the type of performance demanded for the discovery of a low mass $H \rightarrow \gamma\gamma (< 0.5\%)$ all mechanical tolerances of the fabrication need to be controlled to this level. This has been achieved, for example, by the NA31 collaboration at CERN[3]. In a scan with electrons across the 1.8 m face of their calorimeter with 80 GeV/c electrons, the variation in the mean measured energy was $< 0.2\%$.

D. Good Energy Resolution

The energy resolution for EM showers depends mainly on the thickness of the absorber plates. In order to achieve better resolution, the thickness must be reduced. For example NA31[3] has achieved $\frac{\Delta E}{E} = \frac{7.5\%}{\sqrt{E}} \oplus 0.5\%$.

A new concept in calorimetry dubbed the accordion has been developed in Europe [4]. The signal boards serve as the conduit for connecting the calorimeter longitudinally allowing the fastest signal transfer to the preamps and maximizing the hermeticity. This allows a low inductance connection to be easily achieved. The EM prototype tested at CERN this past summer had extraordinary performance. The energy resolution for 1.8 mm lead plates clad with 0.1 mm stainless steel for electrons was measured to be $10.2\% \pm 0.5\% / \sqrt{E} \oplus 0.03\% \pm 0.05\%$. The position resolution at 125 GeV for electrons was 0.700 ±0.015 mm. The signal/noise for minimum ionizing particles was 30/1. All of this was achieved with 100 ns shaping. There are plans for further testing next year and a large group is active in Europe developing various aspects of this design.

Another aspect of energy resolution in the hadronic response is the famous e/π ratio. Measurements by the HELIOS collaboration for 3.4 mm uranium plates and G10 signal boards separated by 2×2 mm argon gaps have shown e/π to be in the neighborhood of 1.1 decreasing with energy with 100 ns shaping time [5]. This is in a similar range to the recent SPACAL results [6].

Part of the work of the Subsystem group is to be able to predict the e/π response of a calorimeter with simulation programs. At this Symposium, results were shown by the Oak Ridge/Tennessee/Mississippi collaboration which uses CALOR89 and HETC [7]. Data taken by the DØ collaboration were compared to the Monte Carlo and show good agreement.

E. Minimal systematic variations

Once a large LAr system is cold and operating, there are minimal changes in the response. For example, the NA31 collaboration has shown variations of less than 0.2% in the average pulse height over months of operation[3]. The variations would have been even less if the precise temperature changes would have been unfolded from the data. The monitoring of the liquid is typically done with a few radioactive sources. The inherent stability of the argon atom is what comes into play. There are always suggestions for improving LAr performance by adding various materials. However one must always insure that such additions do not harm the stability. For example, photosensitive dopants may improve the e/π response [8]. In the case that this does help, one must make sure that the dopants are uniformly mixed as a function of time and not susceptible to radiation damage.

F. Small Fluctuations in the waveform

Since there are $\approx 10^6$ electrons produced per GeV (depending on the sampling ratio), the waveform from LAr does not suffer from statistical fluctuations. This means that the trigger can sample a very small fraction of the waveform and still accurately predict the energy. Also since neutrons have little impact on LAr, the prompt signal is representative of the final e/π. This case can be contrasted with the scintillator case where there are much fewer photons/GeV and the e/π has a time dependence which implies that a fast trigger will have a different value for the energy than the final measurement.

G. Easy to segment

Towers are usually defined with arbitrary shapes in either projective or slightly non-projective arrays. The ganging together of the signal in depth can present a problem if it interferes with the mechanical details or introduces dead area or cracks.

H. Hermeticity

The physics performance of a liquid argon calorimeter is dependent on the impact of the dead material on the hermeticity. Although many large devices have been engineered in the last few years (D0, SLD, and H1), none of them addressed the scope of a 5000 ton LAC as suggested in the LBL study of 1987 [9]. A recent study of this problem has found that with the proper design, no significant deterioration in the physics performance of such a large calorimeter is necessary [10].

The study was guided by input from the community in two workshops - Boulder, Co. in Dec. 1988 [11] and Tuscaloosa, Alabama in March, 1989[12]. An integrated design, modeling and analysis package was used to allow rapid iteration of the 3D model as well as to provide the ability to trace rays through the design to keep track of the live and dead material in great detail. Several requirements guided the design: maximum hermeticity for $|\eta| < 3$; access to the central tracking; provision for 1 m^2 of space to allow tracking cables to exit; meeting ASME Division II code; making sure the assembly, installation, cable routing are sensible; and self-supporting modules less than 10 tons.

After several iterations, a baseline design, while not unique, evolved which met these requirements. This was inspired by the SLD structural design. It features independent barrel and endcap modules, shell and washer assemblies with self-supporting modules, flat heads on the barrel vessels, elliptical heads on the endcap vessels, and aluminum throughout. In particular the use of aluminum in this design allows there to be ≤ 1 radiation length in front of most of the electromagnetic calorimetry and less than a few radiation lengths of dead material in the path of any shower. This guarantees excellent hermeticity and physics performance. The energy resolution is calculated taking into account the dead material encountered for each ray in η. Although this energy resolution distribution is not a constant, the deviation from a smooth curve does not change the response of the calorimeter enough to affect the missing energy distribution.

The results of this study show that a liquid argon system for the SSC can be built from a mechanical engineering standpoint and that such a device would achieve excellent physics performance.

I. Radiation Hard

All the normal components in a LAr system are radiation hard. Even the preamps have been tested with ^{60}Co and neutron exposures. For example for doses of up to 12 Mrad or up to 10^{13} neutrons/cm^2, silicon JFETs at 77^oK suffered no deterioration in their performance [13]. The accordion [4] used such JFETs and also some GaAs FETs which have intrinsically even better performance. The extreme forward calorimeter is a special concern due to the Grad exposures for small radii. There extensive testing of all components will be required.

J. Experience with large systems

There is a wealth of experience from the design, fabrication and installation of NA31 and NA34 at CERN, D0, SLD, E706 at Fermilab, H1 at HERA. Each of these experiments has focused on slightly different specifications and construction techniques.

II. Issues (Problems)

A. Speed

New results were presented in this Symposium which showed the excellent performance that can be achieved with liquid argon if certain procedures are followed. The important steps are: using low inductance leads to gang the signals together; coupling the electrodes to the preamps with a transformer, particularly the hadronic sections; locating the preamps close to the electrodes; using appropriate bipolar shaping and sampling the shaped signal every 16 ns. The shaping time would be normally fixed for the highest luminosity e.g. $10^{34} cm^{-2}s^{-1}$. As the shaping time gets longer, the signal to noise improves. However, the pileup noise also increases with increasing shaping time. Therefore, the optimum is to balance the thermal noise and pileup noise for each luminosity as described by Cleland et al [14]. One would probably choose a short fixed shaping time, 50 ns. However when running at lower luminosity one can achieve the performance of a longer shaping time by adaptive signal processing of the 16 ns samples in the neighborhood of the bucket of the event. This adaptive signal processing would be accomplished by the dedicated low level computing before the data reaches level 2. A figure of merit is the total noise within a cone of $\Delta R = 0.2$ which is important for determining if leptons are isolated. The recent calculations show this number to be ≈ 1.3 GeV at 10^{33} and 2.6 GeV at 10^{34} including both thermal and pileup noise. This is sufficient for finding isolated electromagnetic showers (e or γ).

B. Detailed electrode structure

The challenge is to design an electrode structure that uses low inductance longitudinal ganging and does not introduce cracks and dead space. The accordion concept described above accomplishes these constraints. Also, traditional sandwich designs as used in D0 or SLD could be modified so that the cracks between modules become totally non-projective. This might lead to the adding together of signals from adjacent modules into towers which are projective electrically but not physically.

C. Cooling of preamps, and feedthroughs design

The preamps designed with a dynamic range of 10^5 have a power dissipation of 75 mW. If no transformer is used, the heat dissipation will be higher by at least a factor of 2. If all this heat is concentrated in a small volume, tests have shown that with no argon head pressure, boiling will occur [15]. However, it is possible that by mounting the preamp to an appropriate heat conductor so that the heat is dissipated over a larger area, no active cooling will be required. This will be studied as part of the Subsystem proposal [2].

The large detectors being designed for the SSC have hundreds of thousands of channels. Since the preamp is located in the liquid argon, the signal lines from the preamp to the shaping circuits can be 100 Ω twisted pair line. Still, the design of a high density cold feedthrough may need some attention. SLD has achieved 1.6 connections/cm^2 while H1 has 6.5/cm^2.

D. Absorber material

The Liquid Argon Subsystem proposal is exploring various choices of absorbers [2,16]. This work is building on the data and experience of the NA34(HELIOS) calorimeter which measured $e/\pi \sim 1.1$ with uranium absorber and 100 nsec shaping time. For an SSC detector, the use of uranium may prove to be too costly, so it would be useful to make an accurate measurement of the performance of a lead liquid argon calorimeter. Some EGS studies have suggested that e/π will more nearly equal one if a 12 mm lead plate is used instead of a 6 mm plate [17]. It is also suggested [17] that thin iron cladding of the lead may reduce e/π even further. A stack 100 cm × 100 cm is being built with a depth of 7.5 λ. It will be segmented transversely into 40 X alternated with 40 Y strips and divided longitudinally into 3 sections. Each unit cell will consist of 1 lead plate followed by a 2 mm argon gap followed by a 1.6 mm G10 printed circuit signal board and then another 2 mm argon gap. The interleaving of the X and Y strips in depth provide information about the transverse profile of the shower and a direct measurement of the sampling fluctuations. The connection scheme will follow closely the HELIOS method of achieving low-inductance charge transfer to the preamplifiers. The preamplifers will be the same type that were used in HELIOS which operated in the liquid argon.

The electronics external to the cryostat also follows the HELIOS chain [16]. First there is an intermediate amplifier attached to the outside wall of the cryostat, very close to the signal feedthroughs. A good electrical and mechanical connection between the vessel and the intermediate amplifier housing defines a Faraday cage. The output of the intermediate amplifier drives a balanced line using a transformer which connects to a variable gain amplifier. These amplifiers are adjusted to a fraction of a percent by using the results of the calibration system. Each channel can be adjusted separately so that the contribution to an analog energy sum is equalized. Each channel can also be turned off to measure the cross-talk. Fast sums of all the channels are made with separate shaping amplifiers. These will be built for shaping (peaking) times of 50, 100, and 200 ns to be able to measure the variation of e/π with the shaping time. The stack will be exposed to a beam of electrons and π's at the AGS with momentum in the range from 0.5 GeV/c to $\approx$ 20GeV/c. The intensity of the beam can be as high as 10^7 particles/sec. This will allow the study of pileup effects directly. These beam tests will be carried out in 1991.

E. $\eta \geq 5$

In the forward calorimeter, radiation levels exceed 3 Grad for a small region [18]. This region may stress any of the materials that would be used - for example the resistive epoxy, the signal boards, or any cables. Tests of all materials used obviously need to be done.

F. Redundancy in electronics increases cost

Since the preamps will be located in the liquid, their reliability needs to be assured. They would be treated like electronics going on a space flight - lots of testing plus adding redundancy. The NA34 hadronic calorimeter used 572 preamps which survived in the liquid for several cooldowns, without any preamplifier failures. This demonstrates that there is experience with having preamps survive in the liquid. However, one would likely also insist on redundancy. The simplest method of redundancy involves the separate readout of alternate signal boards. This interleaving doubles the number of separate channels that otherwise would be required. A related consideration is that in order to minimize the time dispersion of the signal for a longitudinal section, it is necessary to keep the physical longitudinal depth of a module from exceeding 50-60 cm. If the section becomes longer than that, the accuracy of the determination of which bucket a large signal is in would be compromised. Therefore, several longitudinal (interleaved) channels are required at least in the hadronic sections which increases the number of channels. These additional channels probably do not dominate the cost of the calorimeter, but need to be kept in mind.

Conclusion

The prospect of using liquid argon calorimetry for an SSC experiment is extremely viable. The vast experience already available gives confidence that large systems can be built mechanically and electrically. New results from beam tests and calculations show that the signal/noise and physics performance of properly designed systems is excellent for the SSC. Some issues require further R&D.

This Research has been supported by the U.S. Dept. of Energy under Contract #DE-AC02-76CH00016. We gratefully acknowledge the contributions of Veljko Radeka to much of the work described here.

References

1. W.J. Willis and V. Radeka, Nucl. Instr. Methods **120**, #221 (1974).
2. G.S. Abrams *et al.*,Progress Report on Work Done in FY90 and Request for FY91 Funding for Research and Development of a Liquid Argon Calorimeter for the SSC, Sept. 1990.
3. H. Burkhardt *et al.*, Nucl. Instr. and Meth. **A268**, 116(1988).
4. B. Aubert *et al.*, "Liquid Argon Calorimetry with LHC-Performance Specifications",CERN/DRDC/90-31
Aug., 1990.
5. HELIOS (NA34) Data, SSC Detector R&D at BNL, BNL Report 52244 (1990), p. 14.
6. R. Wigmans, SPACAL Results, theseproceedings.
7. T. Handler *et al.*,these proceedings.
8. D. F. Anderson, Nucl. Inst. and Meth. **A245**, 361 (1985).
9. G. G. Hansen *et al.*, "Report of the Large Solenoid Detector Group, Proceedings of the Workshop on Experiments, Detectors and Experimental Areas for the Supercollider, Berkeley, CA, Eds. Rene

Donaldson and M.G.D. Gilchriese, World Scientific, Singapore 340 (1988).

10. T. Adams *et al.*, Summary Report on an Engineering Design of a Hermetic Liquid Argon Calorimeter for the SSC, submitted to Nucl. Inst. and Methods., Jan. 1990. Further details are also contained in T. Adams et al., "Evolution of a Reference Design for a Hermetic Liquid Argon Calorimeter for the SSC", BNL Internal Report, Jan. 1990.
11. Proceedings of the Workshop on Cryostat Systems for Generic Hermetic Liquid Argon Calorimetry for the SSC, Boulder, CO, BNL Internal Report, Dec. 1988.
12. T. Adams *et al.*, "Status Report on an Engineering Design of Hermetic Liquid Argon Calorimetry for the SSC" in the Proceedings of the Workshop on Calorimetry for the SSC, Tuscaloosa, AL, 13-17 March, 1989.
13. SSC Detector R&D at BNL,BNL Report 52244 (1990), p. 80.
14. W. E. Cleland, E.A. Scholle and E.G. Stern, "Pileup and Thermal Noise in a Liquid Ionization Calorimeter". these proceedings.
15. T. Adams *et al.*, "Test and Modeling of heat Transport in Liquid Argon", these proceedings.
16. D. Makowiecki *et al.*, "A Measurement of e/π for a Fast Lead Liquid Argon Calorimeter," these proceedings.
17. P. M. Mockett and M. Boulware, "Electromagnetic Shower Sampling Efficiency in Liquid Argon Sampling Calorimeters and the Sampling Efficiency Dependence on Low-Z Cladding of High-Z Absorber Plates," these proceedings.
18. I.L. Azhgirey and N.V. Mokhov, " Radiation Doses and Background Particle Fluxes in EMPACT Experiment at the SSC", EMPACT note 329 (Oct., 1990).

DEVELOPMENTS IN WARM LIQUID CALORIMETRY*

Morris Pripstein
Lawrence Berkeley Laboratory
1 Cyclotron Rd.
Berkeley, CA 94720

for the WALIC Collaboration

Abstract

We provide an overview of some of the recent progress in warm liquid calorimetry and an assessment of the critical issues and present status of the R&D. Recent results are encouraging for the development of fast, hermetic, compensating warm-liquid calorimeters for use at the future (SSC/LHC) hadron colliders.

1. Introduction

The advantages of liquid ionization calorimetry are well known. Direct collection of charge leads to a stable, well calibrated and uniform response. There is flexibility and ease of segmentation in both depth and surface area, relatively high resistance to radiation, and, with the development of the electrostatic transformer, insensitivity to magnetic fields. A vigorous R&D program pursued world wide in the past few years has shown that some organic liquids at ambient room temperature (so-called "warm liquids") can make excellent calorimeters [1]—[3]. The yield of electrons, taking into account dE/dx, electron lifetime and drift velocity is comparable with that for liquid argon; and the relatively short drift times are an advantage in suppressing pile-up. Warm liquids can provide superior detectors, because they require neither cryogenic equipment nor thermal insulation; this enhances simplicity, flexibility, and hermeticity. Furthermore, as hydrogenous materials, organic liquids can provide a compensated response - equal sensitivity to hadronic and electromagnetic particles, leading to better resolution and linearity over a large energy range, without having to resort to expensive and exotic materials or techniques.

These features strongly suggest that warm-liquid calorimetry may result in a more effective and less expensive detector. Nonetheless, although very promising, this technology is still new and is as yet unproven for use in a large calorimeter system. One of the main immediate goals of the R&D program needs to be a "proof-of-principle" with the construction and testing of a large prototype test beam module, containing all the features necessary to satisfy the performance requirements at the SSC and the LHC. This is being pursued by a portion of the WALIC collaboration [3] as part of its comprehensive program to develop the warm-liquid technology.

* This work was supported in part by the Director, Office of Energy Research, Office of High Energy and Nuclear Physics, Division of High Energy Physics and the Office of the Superconducting Supercollider of the US Department of Energy under Contract No. DE-A C03-76SF00098.

2. Critical Issues

2.1 Liquid Purity

Much progress has been made in the routine production of high-purity warm liquids, particularly 2,2,4,4- TMP (C_9H_{20}) and TMS ($(CH_3)_4Si$). The ionization electron lifetime in the liquid is very sensitive to impurities. For signal collection to be insensitive to modest changes in the liquid purity, one would like purities corresponding to electron drift lifetimes of more than ten times the signal shaping time of 50-100 ns. To achieve this, the impurity concentration must be less than 100 ppb oxygen equivalent.

For TMP, the UA-1 collaboration [4] has achieved lifetimes of several hundred microseconds and the Saclay and LBL groups of the WALIC collaboration [3] have reached lifetimes of about 100 µs. For TMS, College de France [3] has obtained lifetimes of 100µs and the Penn and Japanese groups [3] have achieved lifetimes of the order of 1µs, with intentionally modest cleaning procedures for R&D purposes. The latter result can be easily improved upon.

With regard to choice of liquid, more of the WALIC R&D effort has focussed for the time being on TMP rather than TMS, simply because of the safety issue, since TMP has a much higher boiling point. However, TMS has better signal/noise properties because of its higher mobility. An even better and slightly safer liquid than TMS, but unfortunately much more expensive, is TMG. In a contribution at this conference, Yuta et. al. [5], report a lifetime in TMG of 16 ±7 µsec. These results indicate that liquid purification is a tractable issue.

2.2 Materials Compatibility and Long-Term Stability

This is one of the most important issues, since it has a profound impact on the design and costs of large warm-liquid calorimeters, especially for the "swimming-pool" type configuration, with the absorber immersed in the liquid.

UA-1 [6] has shown that their earlier TMP samples, with free electron lifetimes of about 15 µs, did not suffer any reduction in lifetime after three years in their sealed calorimeter boxes consisting of stainless steel and ceramic.

This stability is much greater than is necessary for designs such as the swimming-pool which involve the recirculation and repurification of the liquid. The choice of this design concept, over that of the sealed UA-1 container, is dictated by the desire for significant reduction in the number of high voltage and signal feed-throughs, ease of construction, lower costs, and better hermeticity. Thus, it is essential to determine the compatibility of various materials (and their surface treatment) in contact with warm-liquid, that is, to determine if the needed liquid purity can be maintained.

Recent WALIC results on a variety of metals and plastics [3], [7], [8] are shown in Figure 1, where the ionization electron lifetimes in the liquid are plotted as a function of the contact time of the material with the liquid.

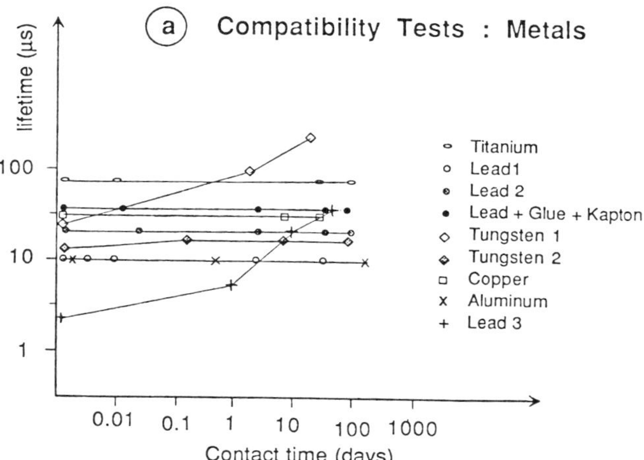

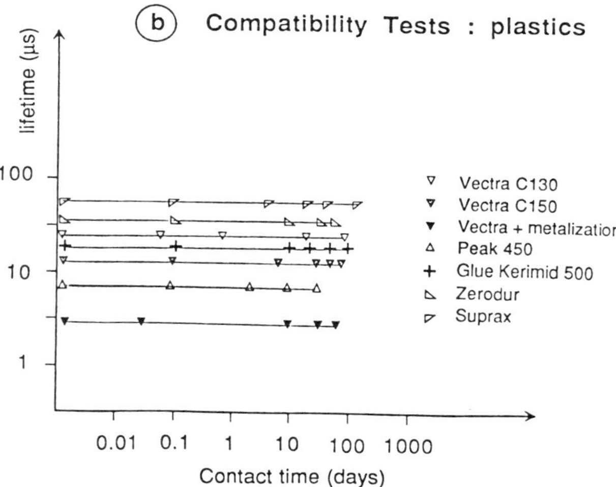

Figure 1. Material-Warm liquid compatibility tests on (a) metals and (b) plastics. Plots are of ionization electron lifetime in the liquid (μs) versus material-liquid contact time in days. All measurements are in TMS except for Aluminum and lead sample #3 which were done in TMP.

In all cases, there is no degradation of liquid purity with time. Moreover, in some cases, tungsten sample #1 and lead sample #3, the lifetimes rise dramatically, suggesting that these samples, subjected to their particular cleaning procedures, act as getters! This needs to be studied more systematically. In any case, the results indicate that there is a wide variety of materials compatible with warm-liquids, including all those specific materials being considered for a swimming-pool calorimeter, or for the so-called prism plastic calorimeter (ppc) design being developed by the College de France group [8]. Further independent tests by the Penn group [9] on Pb-TMS and by the Japanese groups [10] on Pb-TMG also demonstrate compatibility.

2.3 Radiation Resistance

The deterioration of warm-liquids due to radiation may not be as important for a swimming-pool calorimeter design, since the liquid can be recirculated and repurified whenever necessary. Nonetheless, radiation damage of the other material in the calorimeter may release contaminants into the liquid which may perhaps not easily get flushed out. This requires much more extensive investigation in the near future and is being pursued by the WALIC collaboration.

As far as the liquids per se are concerned, the most comprehensive study in TMP and TMS to date has been performed by R.A. Holroyd of BNL [11] in which he exposed the liquids to radiation from an intense Cobalt-60 source, up to doses of 10^5 grays (i.e. 10^7 rads). Even for this maximum dose, only about 1% of the liquid suffers from radiolysis decomposition. The conversion products are predominantly other saturated and unsaturated hydrocarbons or silanes, which do not attach electrons, and , therefore, do not significantly decrease the free-electron lifetime. The drift velocity of ionization electrons is very nearly unchanged and the electron lifetime in his setup drops from about 60 μs to about one μs at 10^7 rads, which is still quite acceptable. However, the gas pressure from radiolysis builds up linearly with dose. Since there have to be expansion tanks to accommodate possible temperature changes, the gas build up should not be a problem. In any case, significant radiolysis occurs only at 10^7 Rads, or above a radiation level that involves only a very small portion of the calorimeter in the very forward direction.

A new result presented at the conference by Yuta et al [10] is for radiation damage to TMG exposed also to a Cobalt-60 source. They report that the free ion yield, G_{fi}, extrapolated to a 2mm gap chamber, changes by only 2% at 1.2×10^3 grays· This is consistent with Holroyd's results.

To conclude, although prospects look very promising, let me re-emphasize the need for further tests of radiation resistance of entire calorimeter prototypes and not just of the liquid alone.

2.4 Fast Signal Response and Signal/Noise

Although liquid ionization calorimeters have been traditionally used with slow readout, their signals have a fast risetime, which can be exploited to obtain fast readout.

2.4.1 Ionization Current and High Voltage

Figure 2 shows the current signal versus time for a liquid ionization calorimeter, assuming uniform ionization in the gap. The current rises rapidly to a peak current, I_{max}, and then drops linearly (assuming very pure liquid) to 0 at a time, t_d, corresponding to the drift time for an electron to traverse the entire gap.

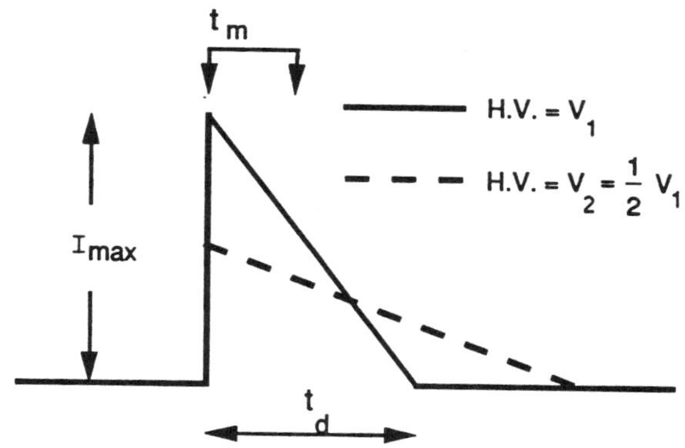

Figure 2. Induced current waveform as a function of time in liquid ionization chambers for a uniform energy deposition in the interelectrode gap, for two different voltages. t_d is the electron drift time across the gap, t_m indicates the signal measurement time interval and I_{max} is the peak ionization current.

The peak current, I_{max}, is a function of the density of ionization (dE/dx), the free electron yield (G_{fi}) and the drift velocity of the free electrons (V_d):

$$I_{max}=G_{fi} \times dE/dx \times V_d$$

In warm liquids, unlike liquid argon, V_d and G_{fi} do not saturate as a function of the applied electric field. Figures 3 and 4 show recent measurements of the dependence of these quantities on electric field [12], [13] for several warm liquids.

Thus, for higher electric field (see solid line in Figure 2), I_{max} increases and the drift time, t_d (= gap width/drift velocity) decreases. For the short shaping times ($t_m<t_d$) required at the SSC/LHC, it is clear from Figure 2 that signal/noise depends on the peak current, I_{max}, rather than the total charge. Moreover, to reduce pileup, it is necessary to reduce the drift time t_d. Thus for both effects, it is advantageous to operate warm-liquid calorimeters at higher fields than for liquid argon.

An important part of the WALIC R&D program has been the successful development of calorimeter prototypes which operate at high electric fields. To this end, a small, highly segmented, prototype TMP calorimeter (16 towers of 64 one-mm gaps) was constructed and tested at 60 kV/cm. Recently, a full-size tower for the large "proof-of-principle" swimming-pool TMP calorimeter module under construction (described below) has been tested extensively at 50 kV/cm! Although designed for electric fields greater than 80 kV/cm, the plan is to operate initially at 30-50 kV/cm.

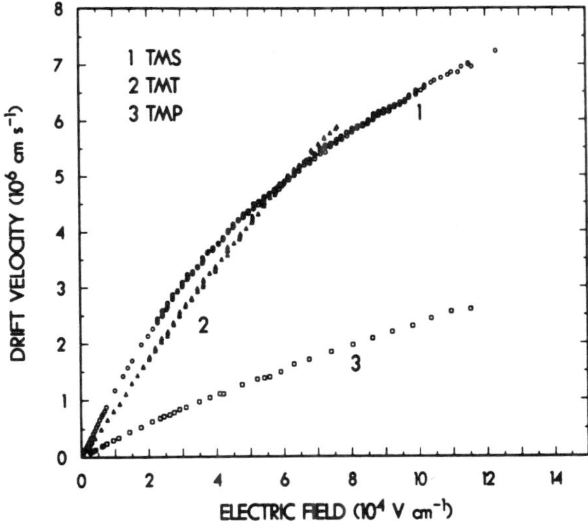

Figure 3. Electron drift velocity as a function of applied electric field in TMS, TMT and TMP. From reference [12].

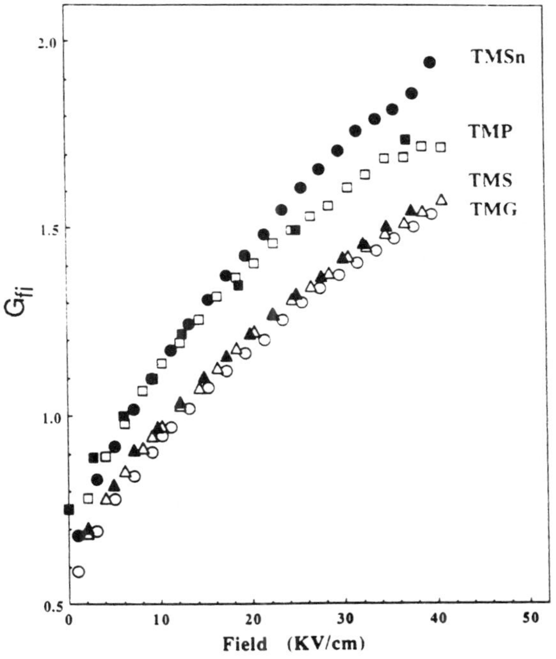

Figure 4. Free ion yield, G_{fi}, as a function of applied electric field for various warm liquids. From reference [13].

Figure 5 shows the results of a simple model calculation by Wenzel [14] that compares the multigap signal-to-noise (S/N) of liquid argon, TMP and TMS for various shaping times and voltages, under the particular assumption of fixed liquid depth of 10 cm and area of 0.01 m^2 per tower plate. For liquid argon the electric field is taken to be 10 kV/cm. The performance is not sensitive to this value. For TMP and TMS, there are two sets of curves, corresponding to high voltages of 5 kV and 10 kV across the gaps. It is desirable to maximize both S/N and, for pileup reasons, the so-called "ballistic factor", $B(\approx t_m/t_d)$, which measures the effective fractional charge utilization. Warm-liquid calorimeters (TMP or preferably TMS) operating at the achieved high voltages and indicated short signal shaping times would meet performance specifications at the SSC/LHC.

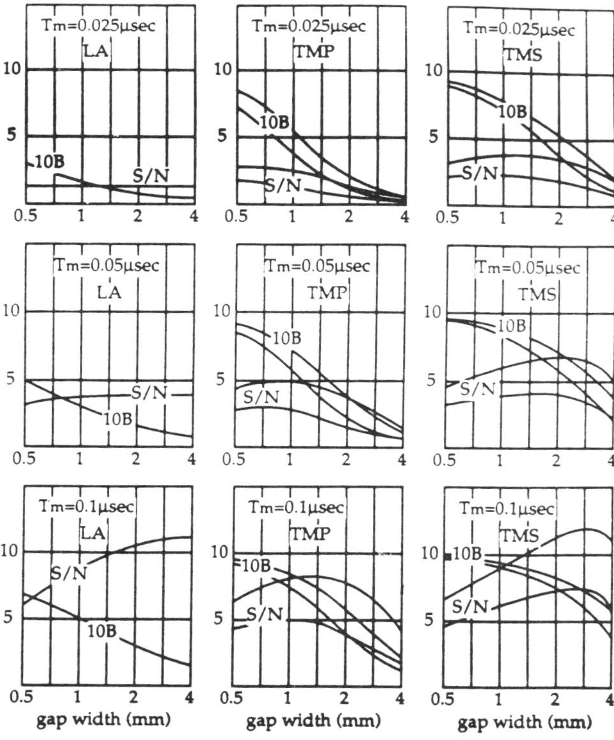

Figure 5. Signal-to-noise (S/N) and pile-up sensitivity ("Ballistic factor", B) for liquid argon, TMP and TMS, for various pulse shaping times, T_m. From reference [14].

2.4.2 Charge Transfer Time and the Electrostatic Transformer

As described in the fine article by Radeka and Rescia [15], one of the factors affecting the signal response speed from liquid ionization calorimeters is the charge transfer time from tower plates to the preamps. This transfer time depends on the tower impedance, as well as the length and impedance of the connecting cable. To minimize this, it is essential to reduce considerably the tower capacitance, C_d.

In typical liquid ionization calorimeters, C_d is of the order of 5nf, resulting in transfer times of about 500 ns, which is large compared to beam crossing times of about 16 ns at SSC/LHC. If the capacitance could be reduced to about 300 pf, then the transfer time would be correspondingly reduced to about 30 ns. In addition, the reduced capacitance would be a better match to the preamp capacitance, thereby improving the signal-to-noise as well.

One way to reduce the tower capacitance is to gang the drifting gaps in series rather than in parallel. Recall that for a tower with *n* gaps connected in parallel, the total capacitance, $C_d=nC$ where C is the single gap capacitance, while for *n* gaps in series the resulting capacitance is $C_d = C/n$. The WALIC collaboration has been developing and testing a scheme of ganging tower electrodes in combination of series and parallel connections to substantially reduce the overall tower capacitance. The arrangement would then act as an electrostatic transformer (EST) [16], so-called because it behaves from the preamplifier point of view very much the same as a ferrite-core transformer. This is discussed more fully in reference 16. Unlike the ferrite-core transformer, the operation of the EST is unaffected in the presence of a strong magnetic field. Considerable study has been made with computer simulations and testing with mockups and small calorimeter prototypes, [16], [17], yielding promising results. The aforementioned "proof-of-principle" hadronic test beam module is being constructed with the electrostatic transformer concept and will provide the first large scale beam test.

3. Compensation Studies

Recent Monte-Carlo calorimetry models [18] suggest that compensation — equal response to electrons and hadrons, e/h=1 — can be achieved for sampling calorimeters with hydrogenous active media (e.g., warm-liquids) and passive absorbers of material other than uranium, such as lead or iron, which are cheaper and easier to work with. To this end, the WALIC collaboration is presently running an experiment at Fermilab (E-795) to make a systematic study of calorimetry properties by measuring e/h as a function of absorber material (lead and iron) and as a function of absorber-to-TMP thickness. The TMP calorimeter is highly modularized, to allow for changes of configuration to be made simply. The TMP is isolated from the absorber by being contained in thin sealed boxes of stainless steel of the UA-1 type [1], [6]. Each box is 30 x 60 cm^2 with a 4-electrode plane in the middle of 2.5 mm of TMP, resulting in two 1.25 mm liquid gaps per box. The operating electric field was 6.5 kV/cm.

A key feature of the calorimeter is that each component (absorber plates, TMP boxes) is independently suspended by hanging frames from the support structure (Figure 6). After each change of configuration, the stack was recompressed. The TMP gap thickness was then monitored by measuring the capacitance of each electrode and found to be extremely stable. This was also confirmed by the signals from muon beams. The entire calorimeter is enclosed in a perforated aluminum Faraday Cage, which reduces the external electrical noise by a factor of 50,000. More details can be found in reference 19.

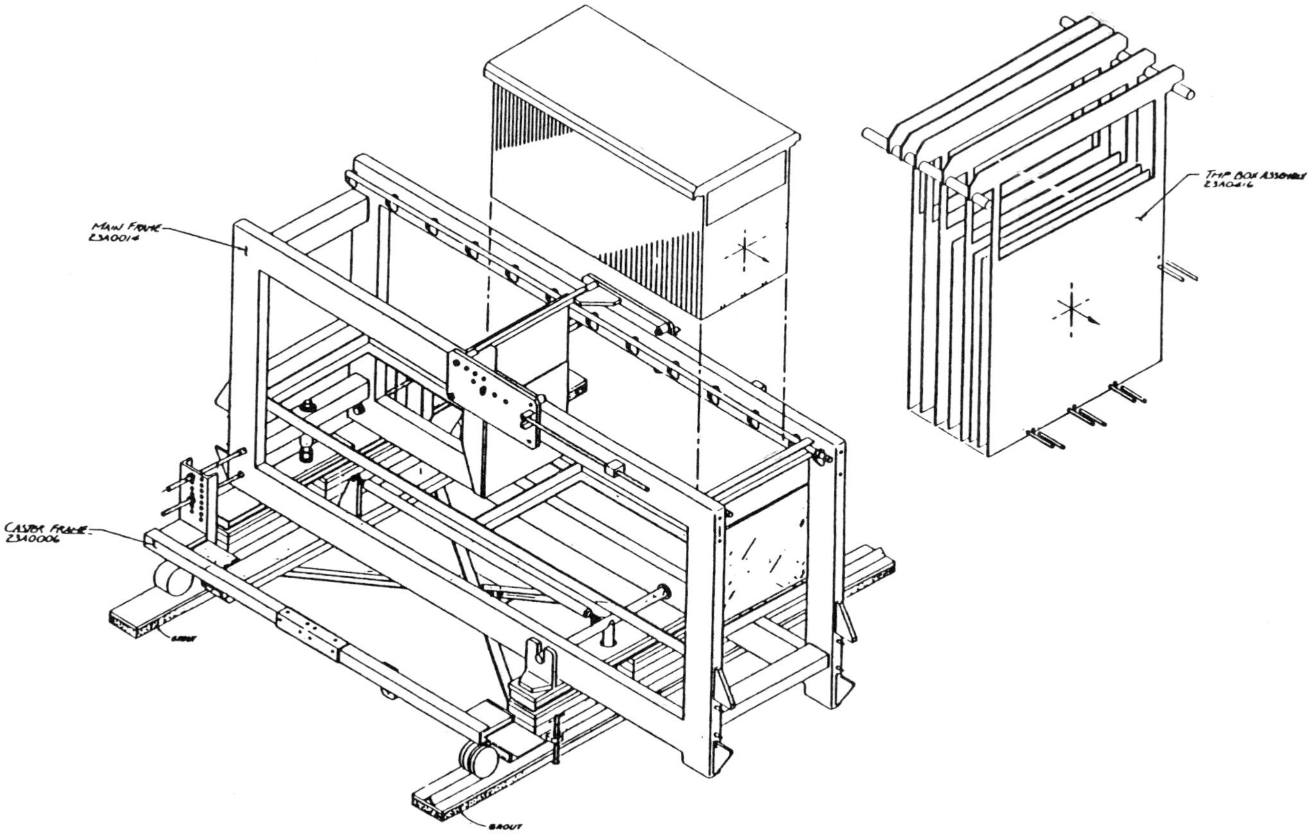

Figure 6. Isometric layout of WALIC hanging frame TMP calorimeter.

The first measurements were done with an electromagnetic configuration (e.m.) to measure the response to e.m. showers only. This consisted of a 6.35 mm lead plate followed by a single TMP box (4-electrode plane, 2.5 mm of liquid), repeated 26 times, for a total of 30 radiation lengths. In this configuration, the calorimeter was exposed to positron beams between 2.5 and 175 GeV/c and to a 175 GeV/c muon beam.

Figure 7 shows the response of the e.m. configuration (the sum of the signals from the 26 electrodes along the beam) to the 175 GeV/c muons and the pedestal noise. The signal/noise is about 3.8, as expected. This can be extrapolated to a full depth warm-liquid calorimeter for the SSC with a short 100 ns shaping time to give a signal/noise for muons of about 6.5!

Figure 8 shows preliminary results of linearity and resolution for this e.m. configuration, but with the data still uncorrected for the large beam momentum spread which explains the sizeable value of the constant term in the fitted resolution. These data have been compared to Monte Carlo data (Geant 3.14 which predicts a resolution of 18%/√E) and the agreement on the shape of the shower profiles is excellent.

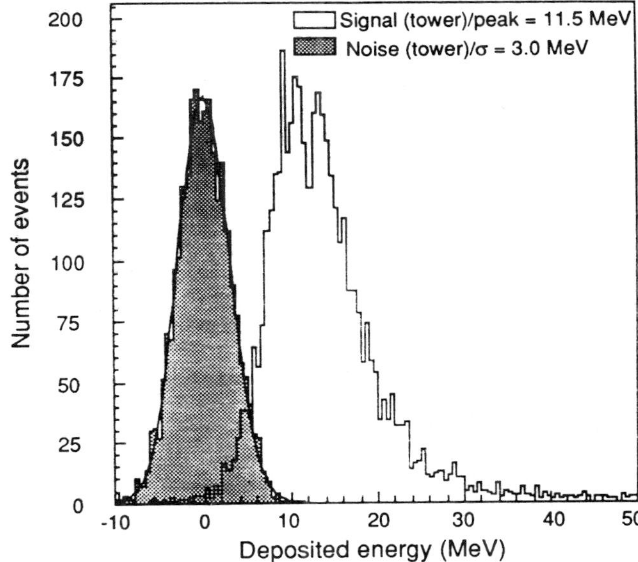

Figure 7. Muon signal and pedestal signal in a 26-plane tower.

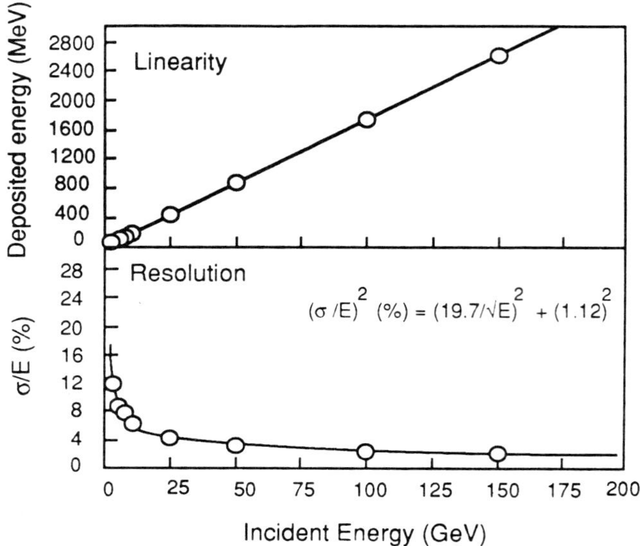

Figure 8. Signal response linearity and energy resolution as a function of incident positron beam energy, for the 26-plane electromagnetic (e.m.) configuration. Preliminary results.

To measure hadron showers, it was necessary to mount two TMP boxes in a detector plane hanger, one above the other, to cover a cross-section of 60 x 60 cm^2. With only one-half of the TMP boxes, comprising 34 TMP planes out of the eventual 68 planes (each 60 x 60 cm^2), several ("hadronic") configurations of lead and TMP, arranged to a depth of about 6.5 λ, were exposed to both positron and pion beams. Details are given in reference 19. Because of the limited number of TMP boxes, a number of the gaps contained "dummy TMP boxes" made up of CH_2 with the same nuclear content as the TMP planes in order to achieve the same shower development as if TMP planes were used. The beam energies were low for these first runs to avoid excessive leakage out the back because the calorimeter was only instrumented to half its depth. The full hadronic runs to study e/h will be done in the spring of 1991.

The analysis is still ongoing, but we show some preliminary results in Figure 9 for e/h for two extreme configurations to indicate some general features. First, it is important to remark that the TMP boxes were operated at an electric field of only 6.5 kV/cm, where the signal saturation, or Birk's constant (K_B), is very large according to measurements by the WALIC collaboration [20] and by Ochsenbein [21]. For these conditions, the response to hadron showers is expected to be substantially lower than at higher electric fields [18]. From Wigman's predictions [18] of e/h versus K_B, and the WALIC measurements of K_B [20], we obtain the curves in Figure 9, which agree quite well with the data. Wigman's predictions, together with the measured K_B dependence on electric field [20], [21], would suggest that one can achieve compensation, e/h $\approx$ 1, for a lead-TMP calorimeter if operated at high electric fields, 30 - 50 kV/cm, and with a lead-to-TMP ratio of about 4-5 to 1. These, in fact, are the design conditions for the "swimming-pool" test beam module now under construction, and to be tested in the coming year.

Another important feature of the next WALIC E-795 run at Fermilab, is the installation and testing of a small tungsten-TMP electromagnetic calorimeter in the "swimming-pool" configuration, built by the Harvard group. This is described in the paper by Brandenburg et al. [22] contributed to this conference.

Thus, the goal is to have by fall 1991 a complete data set on compensation measurements for lead-TMP and iron-TMP, as well as a comprehensive test of the tungsten-TMP e.m. calorimeter.

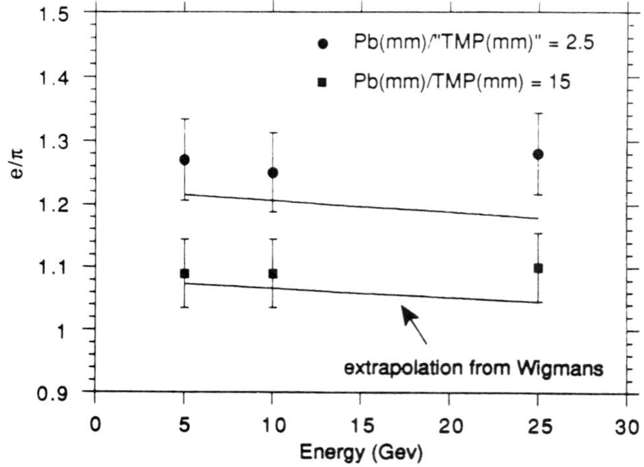

Figure 9. Preliminary results of e/π as a function of the incident beam energy for two extreme hadronic lead-TMP ratios of the calorimeter. The error bars represent 5% systematic errors.

4. "Proof-of-Principle" Test Beam Module

Even as each of the critical issues confronting this new technology gets successfully resolved, ultimately one needs to build and test a large "proof-of-principle" prototype. Part of the WALIC collaboration [3] is now building a large test beam module (TBM), designed in the "swimming - pool" configuration (the absorber inside the liquid volume), which is to satisfy the safety, hermeticity, hadronic compensation, resolution and time response requirements for an SSC/LHC detector. Furthermore, it is possible to extrapolate this design to a full-size SSC calorimeter module without major conceptual changes.

The design of the TBM itilizes many of the features that are envisioned being used in an actual calorimeter for an SSC detector, such as the "electrostatic transformer" readout, fine transverse segmentation in a tower configuration, and materials that lend themselves to efficient mass production techniques. The most time consuming part of the design phase of the TBM has been creating a module that uses only materials known to be compatible with TMP. It has also been a challenge to eliminate any volume that would trap air, since oxygen is very electro-negative and its presence would decrease the free-electron lifetime in TMP. Another design feature is that it can operate at electric fields up to at least 50 kV/cm.

The TBM is envisioned to consist of two modules, each one being about five interaction lengths deep, and 60 cm on each transverse side, comprising a five-by-five array of 12x12 cm^2 towers. The towers are read out in the electrostatic transformer configuration, summed so that there are four depths for the total of ten interaction lengths. An interesting feature of the design is that each layer of ten gaps is assembled from prefabricated, self-contained one-by-five arrays. A conceptual assembly drawing is shown in figure 10, and a detail of the tie-rod and insulator interface is in figure 11.

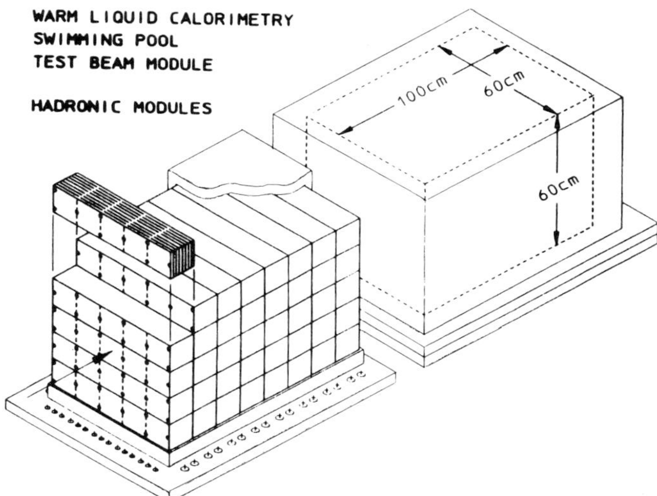

Figure 10. A conceptual drawing of a swimming-pool module (TBM) showing how a one-by-five preassembled tower module is assembled into the stack.

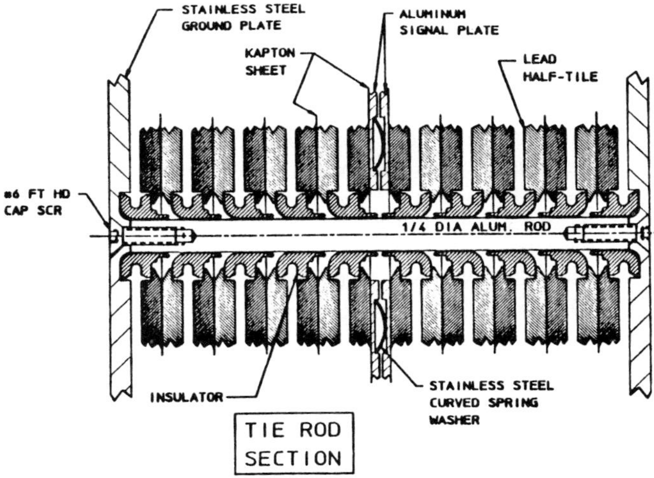

Figure 11. Detail of the ceramic insulator-lead tile interface of the test beam module (TBM).

The insulators were designed so as to maximize the surface leakage path in order to achieve the desired higher electric fields. The tiles are lead and ground plates are steel. High voltage tests performed so far in TMP have sustained 50 kV/cm across the gap. Each unit of this ten-gap building block has one signal plate in the center that collects the signal from two "electrostatic transformers" with a "turns ratio" of five. The advantage of this construction technique is that a one-by-five is small enough and sufficiently self contained that it can be built, cleaned, and tested independently of the final assembly of the large modules.

Figure 12 shows the voltage distribution and signal flow from the electrostatic transformer cell. The arrows in the TMP gaps show the direction of the ionization drift electrons across the gap.

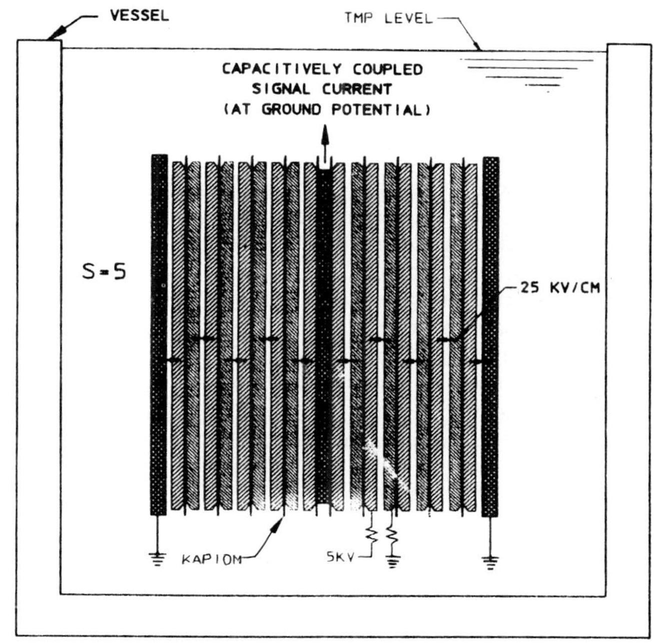

Figure 12. Electrostatic transformer cell configuration corresponding to the tower stack shown in Figure 11. The arrows in the gaps represent the ionization electron drift across the gap.

As mentioned earlier, all the components to be used in the test beam module have been tested for compatibility, with TMP and TMS. In addition, a ten-gap tower unit was recently tested by the Penn group [9] in TMS, using cosmic rays and they found no decrease of the free-electron lifetime.

If funds become too limited, only one of the two modules will be constructed and tested. In any case, several prototypes of TBM subsections are also being built in order to expose them to meaningful system radiation damage studies. These are intended to be completed in the coming year as well.

Finally, with the ideas incorporated in the test beam module, a detailed design of a full-size 4π warm liquid calorimeter for an SSC detector (SDC) has been made [23]. Among the outstanding features are the excellent hermeticity and the modularity which considerably eases the problems and schedule of assembly.

5. Conclusions

Considerable progress has been made recently in the development of warm-liquid calorimetry technology, which is very encouraging for application to SSC/LHC detectors. Nonetheless, more R&D needs to be done culminating in construction and testing of large "proof-of-principle" prototypes which satisfy performance requirements for the SSC/LHC colliders. These tests are planned to be done in the coming year, with at least two different design concepts: the swimming pool approach described above and the prism plastic calorimeter (PPC) [8] by the College de France group. In particular, there is still a great need for more extensive radiation-resistance studies of entire warm liquid calorimeter systems and not just of the liquid alone. This is especially crucial for consideration of this technology in a forward calorimeter whose the radiation levels are very high. However, radiation studies thus far, coupled with the fact that the liquid can be recirculated and repurified when necessary, hold out much promise for this technology. Nonetheless, this additional testing must be done. (Of course, the latter need applies even to all the other, more conventional, technologies).

Engineering design studies of full-size 4π warm liquid calorimeter have been made which exploit the hermeticity and modularity allowed by this technology.

Finally, there is an especially active R&D program being pursued which we expect will provide most of the remaining needed information soon.

6. Acknowledgements

I would like to thank all my colleagues in the WArm LIquid Calorimetry (WALIC) collaboration from France, Japan and the U.S. who have contributed significantly to the progress in this R&D program while making the whole research effort a lot of fun and a stimulating experience.

References

[1] M.G. Albrow et al., "Performance of a Uranium/ Tetramethylpentane Electromagnetic Calorimeter", Nuclear Instruments and Methods **A265**, 303-318 (1988);
K. Ankoviak et al., "Construction and Performance of a Position Detector for the UA1 Uranium-TMP Calorimeter", Nuclear Instruments and Methods **A279**, 83-90 (1989).

[2] J. Engler, H. Keim and B. Wild, "Performance Test of a TMS Calorimeter", Nuclear Instruments and Methods **A252**, 29-34 (1986).

[3] WArm LIquid Calorimetry (WALIC) Collaboration: B. Aubert et al., "Fast Hermetic Calorimetry Using Warm Liquids: Proposal for Generic Detector R&D for the SSC/LHC", LBL-Group A Physics note 971 (1987);
B. Aubert et al., "Warm Liquid Calorimetry Large Subsystem Proposal", LBL-Group A Physics note 972 (1989);
J. Busenitz et al., "Warm Liquid Calorimetry Large Subsystem R&D Progress Report", LBL-Group A Physics note 973 (1990).

[4] E. Radermacher, "Calorimetry with Warm Liquids: Summary and Prospects for LHC Experiments", Proceedings of the LHC Aachen Workshop, October 4-9, 1990, Vol. 3, pp. 341-347.

[5] H. Yuta et al., "Radiation Damage to a Warm Liquid Calorimeter", contribution to this conference.

[6] A. Givernaud, "Performance of a UA1 Uranium-TMP Calorimeter Module", Proceedings of the Workshop on Calorimetry for the Supercollider, Tuscaloosa, Alabama, March 13-17, 1989, pp. 287-312.

[7] B. Aubert et al., "Results on Material Compatibility Studies in TMS, "Walic Collaboration, to be submitted to Nuclear Instruments and Methods.

[8] L. Dobrzynski et al., "The Prism Plastic Calorimeter (PPC): A Warm Liquid Calorimeter for the LHC", College de France Report LPC 90 53 (November 27, 1990).

[9] R. Hollebeck, private communication.

[10] H. Yuta et al., "Material Compatibility Test for a Warm Liquid Calorimeter", submitted to this conference.

[11] R.A. Holroyd, "Effects of Radiation Damage to TMP, TMS and Liquid Argon Solutions", Proceedings of the Workshop on Radiation Effects in the SSC, M.G.D. Gilchriese ed., SSC-SR-1035 , June, 1988, pp. 335-347.

[12] H. Faidas, L.G. Christophorou, D.L. McCorkle, and J.G. Carter, "Electron Drift Velocities and Electron Mobilities in Fast Room-Temperature Dielectric Liquids and Their Corresponding Vapors", Nuclear Instruments and Methods **A294**, 575-582 (1990).

[13] S. Geer, R.A. Holroyd and F. Ptohos,"Field Dependent Free Ion Yields of Room Temperature Tetramethyl Liquids", Nuclear Instruments and Methods **A287**, 447-451 (1990).

[14] W.A. Wenzel, "Electronic Signal to Noise Ratio for LA, TMP and TMS", Solenoidal Detector note # SDC-90-00071 (August 9, 1990).

[15] V. Radeka and S. Rescia, "Speed and Noise Limits in Ionization Chamber Calorimeters", Nuclear Instruments and Methods **A265**, 228-242 (1988).

[16] J. Colas, M. Pripstein and W.A. Wenzel, "The Electrostatic Transformer", Nuclear Instruments and Methods **A294**, 583-590 (1990);
J. Colas, "Speed of Response, Pile-up and Signal to Noise Ratio in Liquid Ionization Calorimeters",

Proceedings of the Workshop on Calorimetry for the Supercollider, March 13-17, 1989, pp. 553-568;
J. Colas and W.A. Wenzel, "Electrostatic Transformer for Large Towers", ibid, pp. 643-648;
J. Colas et al., "Electrostatic Transformer for Large Liquid Ionization Calorimetrs", presented by W.A. Wenzel at IEEE Conference in San Francisco, Ca, January 19, 1990.

[17] G. Abrams et al., private communication and contribution to this conference.

[18] R. Wigmans, "Calorimetry at the SSC", Proceedings of the workshop on Experiments, Detectors and Experimental Areas for the Supercollider, Berkeley, Ca, July 7-17, 1987, pp. 608-635.

[19] P. Ghez, "Warm Liquid Calorimetry: Preliminary Results for E-795", to appear in the Proceedings of the Fermilab Conference on Calorimetry, October 29-November 1, 1990.

[20] B. Aubert et al., "Saturation of Ionization Signals in TMP and TMS at Different Angles and Electric Fields", Nuclear Instruments and Methods **A286**, 147-154 (1990).

[21] S. Ochsenbein, "Room Temperature Liquid Ionization Chambers", SIN Report, presented at the NATO ASI Conference, Portugal, July, 1987.

[22] G.W. Brandenburg, A. Daw, S. Geer. J. Oliver, C. Sadowski, "A Room Temperature Liquid Calorimeter Prototype for the SSC", paper contributed to this conference.

[23] A. Ciocio et al., "A Warm Liquid Calorimeter Concept for the SSC", Solenoidal Detector note # SDC-90-00126 (November 2, 1990).

Parallel Sessions

Silicon Tracking Devices

1

SILICON DRIFT DEVICES FOR TRACK AND VERTEX DETECTION AT THE SSC

W. Chen, H. Kraner, Z. Li, C. Ng, V. Radeka, P. Rehak, S. Rescia
Brookhaven National Laboratory, Upton NY 11973

J. Clark, S. Henderson, L. Hsu, J. Oliver, R. Wilson
Harvard University, Cambridge, MA 02138

M. Clemen, T. Humanic, D. Kraus, G. Vilkelis, B. Yu
University of Pittsburgh, Pittsburgh, PA 15260

K. McDonald, C. Lu, M. Wall
Princeton University, Princeton, NJ 08544

A. Vacchi
Rockefeller University, New York, NY 10021

G. Bertuccio, E. Gatti, A. Longoni, M. Sampietro
Politecnico di Milano, Piazza Leonardo da Vinci 32, Milano

P. Holl, L. Strüder
Max-Planck-Institut for Extraterrestrial Physics, D–8046 Garching

E. Lorenz
Max-Planck-Institut for Physics, D–8000 Munich 40

U. Faschingbauer, A. Wörner, P. Wurm
Max-Planck-Institut for Nuclear Physics, D–6900 Heidelberg 1

J. Kemmer
Technische Universität München, D–8046 Garching

Abstract

We report on the recent progress in the study of Semiconductor Drift (Memory) Detectors intended for an inner tracking and vertexing system for the SSC. The systematic studies and the calibration of the existing detectors and the simulated performance in the actual SSC environment are highlighted.

Introduction

This collaboration is committed to develop a vertex and an inner tracking detector for the SSC based on Semiconductor Drift (Memory) Detectors. We believe that silicon drift detectors are ideal detector elements to be used for the charged particle tracking close to the interaction region of the SSC for the following reasons:

1. the unambiguous position resolution of only several μm in two perpendicular directions
2. attribution of hits belonging to different bunch crossings with a timing accuracy better than $1 ns$ and
3. the intrinsic signal pipeline within the very volume of the silicon drift detector.

This contribution is organized as follows. It starts with a brief description of silicon drift detectors, followed by an analysis of rate capabilities of silicon drift detectors in the SSC environment. The main part of the contribution describes the systematic studies done on a multi–anode detector. The next section reports on the Monte Carlo simulation of the silicon drift detectors performance in the SSC environment. A short list of conclusions closes this contribution.

Semiconductor Drift Detectors - Description

Silicon drift detectors are a relatively new kind of semiconductor detector able to provide very precise position and ionization measurements with a very modest

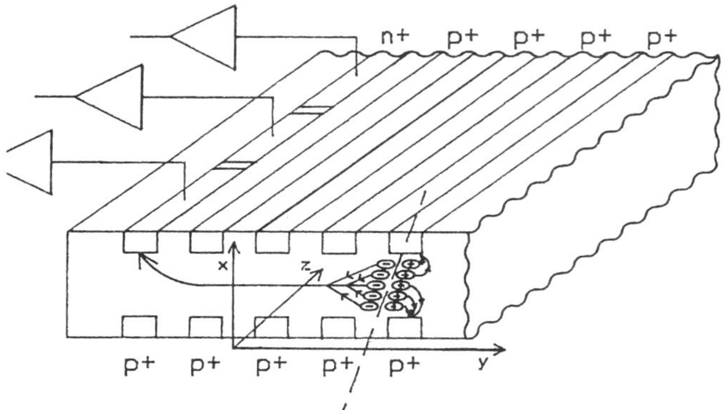

Figure 1: Perspective view (not to scale) of a semiconductor drift detector. Electrons created by an ionizing particle are transported long distances parallel to the detector surface. The anode is divided into short segments to measure the coordinate perpendicular to the drift direction.

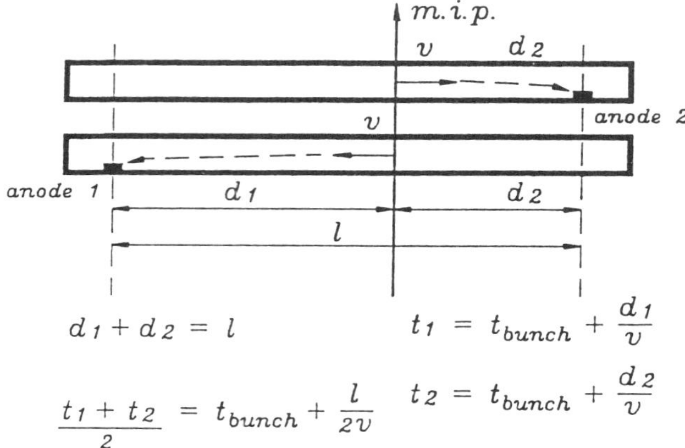

Figure 2: View (not to scale) of two silicon drift detectors placed in such a way that a particle crosses both detectors and the produced electrons drift in opposite directions with the same speed v. The sum of drift distances for any position of the incident particle is constant and equals the distance between anode 1 and anode 2. The average of two arrival times gives the time of the bunch crossing.

amount of electronics. A perspective view of the drift detector[1] is shown in Fig. 1. In principle, the electric field of thedrift detector forces electrons liberated by an ionizing particle to drift parallel to the large semiconductor surface to the anode. The transit time of electrons inside the detector measures the distance of an incident particle from the anode.

As a direct consequence of this electron transport method, the anode capacitance is much lower than that of a classical semiconductor detector of the same dimensions. The amplifier noise can be made much smaller, which is the main reason behind the excellent position resolution obtained from silicon drift detectors.

The measured position resolution in a test beam was $10\mu m$ over a drift distance of $4000\mu m$. These results[2] were achieved in fact with external amplifiers which were not matched to the very low anode capacitance.

We would like to stress that the maximum drift time is not the dead time of the silicon drift detectors. Due to the fact that all electrons are continuously drained at the anode, silicon drift detectors have no dead time. Electrons drifting toward the anode have their charges screened by the electrode structure of the silicon drift detector and do not interfere with the other signals or with the signal processing at the anode.

Fig. 1 shows the collection anode divided into individual pad–anodes. The position of the responding anode(s) measures the coordinate perpendicular to the drift direction. Due to the transverse diffusion, the electrons produced by the ionization arrive as a gaussian shaped charge cloud in the anode region. Ordinarily, charge will be collected on more than one anode. The charge division method yields the position of the crossing particle in the second coordinate with a precision of about 4% of the anode pitch. Thus, a multi–anode silicon drift detector provides unambiguous two dimensional position information. The position information is similar to that which is expected from proposed silicon pixel devices. The multi–anode silicon drift detector provides better resolution with four orders of magnitude fewer read–out channels.

Rate Capabilities of Silicon Drift Detectors

In the high rate environment of the SSC a silicon drift detector stores at any given time hits of particles originated at different bunch crossings. Fig. 2 shows a simple method to time hits in a silicon drift detector. Two silicon drift detectors are placed parallel one to another at a small distance.The drift fields in the two detectors are arranged in such a way that electrons drift in opposite directions in these two detectors. Following a simple calculation described in Fig. 2, we see that there is a constraint for the sum of the drift times measured in both detectors with respect to the time of the particle crossing. The time resolution of each drift detector is better than $1ns$. The mean value of the drift times of the pair of drift detectors and hence the bunch crossing timing can be determined with a resolution better than $1ns$. We see that we have no problem with the bunch crossing association in the case of a single particle within

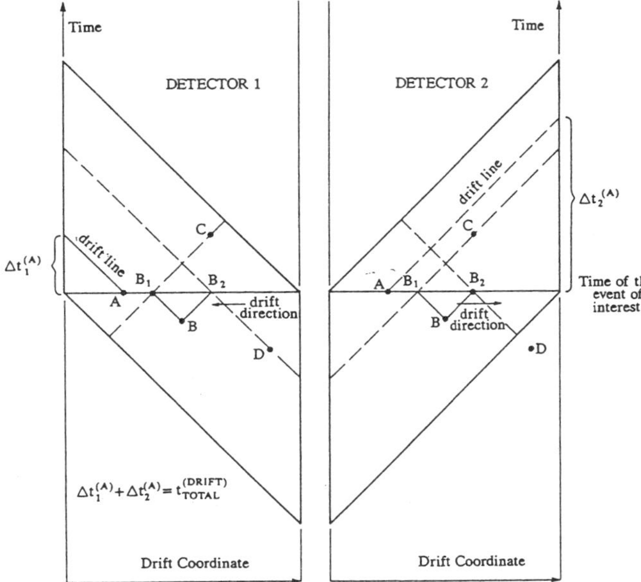

Figure 3: The drift of electrons in a silicon drift detector can be visualized by drift lines on a space–time plot. The drift coordinate is the horizontal axis, and the time is the vertical axis. The two diagrams represent the two parallel detectors drifting electrons in opposite directions.

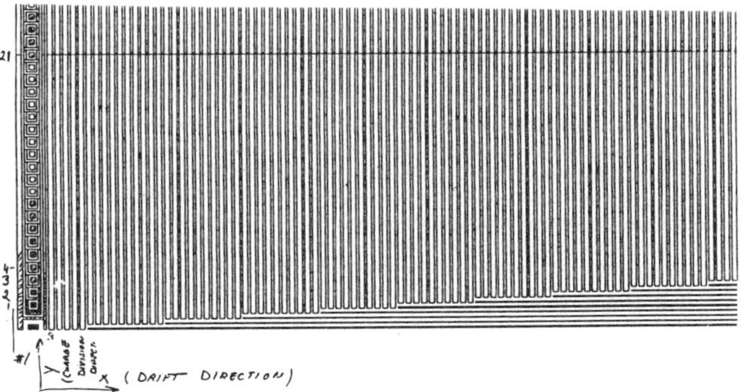

Figure 4: P^+ mask of an existing 41 anode drift detector

an anode and memory time of the detector. In the case of several particles produced at different bunch crossings and detected at the anode during the memory time of the silicon drift detector the analysis is more complicated.

In Fig. 3 we see a diagram visualizing the drift of electrons in a space–time plot. The vertical axis represent the time, the horizontal axis the drift distance. Hits on the detector originating within the same beam crossing are distributed along a horizontal line. As they drift, they move upwards along diagonal lines called drift lines until intersecting the vertical axis. The drift line of each detector points in the electron drift direction.

One can see from this diagram that there is a minimum time over which digitization must be recorded to acquire all information for a given beam crossing time. Within this minimum time, there is still a large area in the space–time from which hits are being collected. Hits early enough in time, far enough from the anodes, would be near the bottom of this diagram. The space–time area enclosed within the thick lines in Fig. 3 is mapped onto the one–dimensional vertical time axis.

The method of identifying hits produced at the time of the event of interest is the method displayed in Fig. 2. Hit "A", for instance, will drift in detector 1 diagonally up and to the left until intersecting the vertical axis approximately at 1/4 of the total drift time. In detector 2, this intersection is approximately at 3/4 of the total drift time. The sum is simply the total drift time. Hits occurring out of time will add to some other value.

In the case of several hits, with the projection onto the time of interest there is the possibility that an incorrect pairing of two hits out of time in two detectors will mimic a hit in time of interest. Let us consider the hit "B" in Fig. 3. The intersection of the drift line of hit B in detector 1 (2) with the time of the event line is called B_1 (B_2). Hits B and B_1 are projected onto the same point on the time axis of detector 1. Hit "C" in Fig. 3 was placed on the drift line of detector 2 passing through the point B_1. Again, hits C and B_1 are projected onto the same point on the time axis of detector 2. In the described case hit B in detector 1 can be paired with the hit C in detector 2 to produce an in time phantom hit called B_1. Similarly, in Fig. 3 there is a construction of the symmetrical phantom hit. The hit B in detector 2 can be paired with the hit D in detector 1 to produce a phantom hit called B_2.

The probability for an out of time hit B to produce an in time phantom hit after being incorrectly paired with the other hit C from a uniformly populated space–time area is $\simeq 2 \times \delta t / t^{drift}_{total}$ where δt is a combined time accuracy of the drift detector (2ns) and t^{drift}_{total} is the maximum drift time (2 μs). This probability in the case of a single pair is about 0.2%. In the case of n out of time hits the probability of a phantom in time hit would be n^2 times larger.

We can see that silicon drift detectors can operate in rates producing up to 2–4 out of time hits per anode, within the memory time of the detector without compromising the performance. We may take advantage of the multiple hit capability to decrease the amount of electronics associated with the silicon drift detector tracking system.

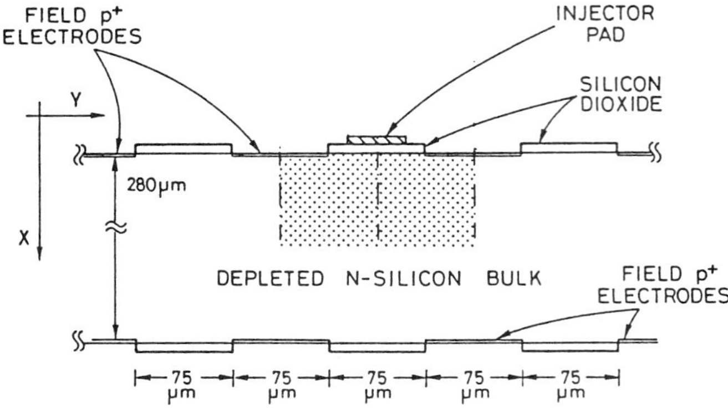

Figure 5: Cross section of the Silicon drift detector in the injector region. The z-direction is perpendicular to the plane.

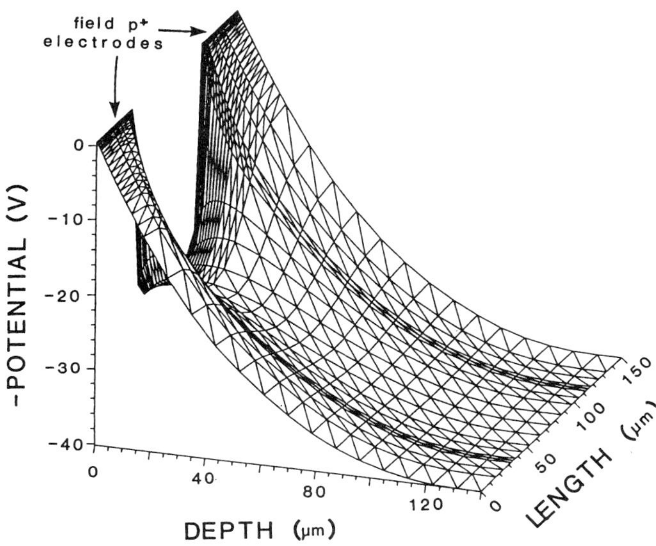

Figure 6: Potential energy for electrons calculated within the dotted area on the previous Figure.

Experimental apparatus

Most of the measurements presented in this section were performed with a multi-anode (41) linear drift detector, shown in Fig. 4. This is a rectangular detector made of $300\mu m$ thick 10 $k\Omega cm$ NTD n-type silicon. The detector is 1.5 cm long in the drift direction and 1 cm wide. The lateral resolution is obtained by charge sharing between 41 anodes, spaced 250 μm apart. There is an implanted voltage divider directly on the detector, located right at the center, above anode No. 21. The detector was manufactured in 1984 at TU Munich and tested at the SPS beam at CERN.[2]

To simulate the ionization produced by a minimum ionizing particle, infrared light pulses were used. A GaAs laser diode was the source of light pulses, having duration of a few ns at a wavelength of about 900nm. Light pulses were carried in a $100\mu m$ diameter fiber and focused onto the detector surface by microscope optics. The spot size was demagnified down to a $5\mu m$ diameter spot. The detector assembly was mounted on an x–y stage movable by stepping motors and controlled by a PC. The mechanical precision was 2–$3\mu m$ with about the same degree of reproducibility. To measure the temperature of the silicon drift detector a platinum thin film temperature sensitive resistor was glued onto a corner of the detector. Although the temperature could be read within 0.1K, the temperature of the sensor might not correspond precisely to the detector temperature. All the equipment was bought with funds from the SSC R&D program.

Temperature Dependence of the Drift Velocity and its Calibration

The silicon drift detectors are quite sensitive to temperature changes. The drift velocity in silicon depends on the absolute temperature as $T^{-2.4}$. Due to the precision of the drift coordinate measurement, silicon drift detectors are sensitive to temperature variations of about 0.1K. Stabilizing the overall temperature to better than 0.1K may not be practical. Rather, it seems easier to calibrate the drift velocity.

A practical calibration of the drift velocity must be accomplished by simple and reproducible means, ideally by pulsing an electrode at a given distance from the anode.[3] Such an injecting electrode was implemented on a silicon drift detector to be used in the UA6 experiment at the CERN SPPS.[4] The injector is a small metallic gate deposited on the silicon dioxide present between the field p^+ strips, at the desired position. A cross section of the detector around an injector pad is shown in Fig. 5. The injector is a metal on silicon (MOS) capacitor. The injection mechanism is related to the actual shape of the potential energy for electrons in a silicon drift detector with working biases. The shape of the potential energy for electrons is shown in Fig. 6 for the dotted region of Fig. 5.

The positive charge at the $Si - SiO_2$ interface causes a downward bending of the potential at the interface. There are many electrons confined in the surface region extending up to the saddle point. The number of accumulated electrons in this surface region is the maximum that the region can hold. An additional electron falling into this region would deform the potential energy upward and one electron would escape into the main valley of the detector. (This dynamic equilibrium was used by the program to calculate the potential within the detector shown in Fig. 6.)

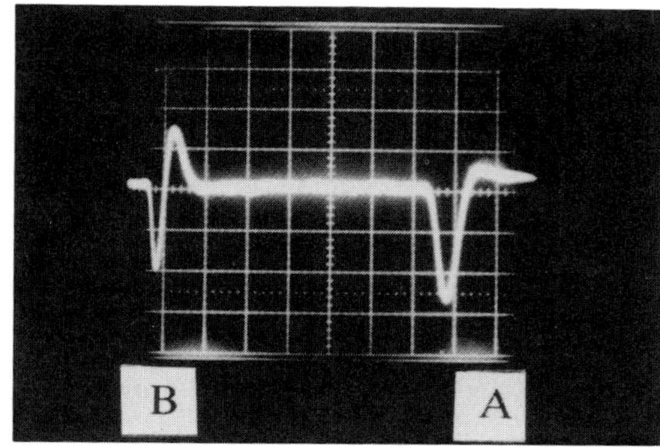

Figure 7: A–pulse corresponding to the collection of the injected charge after an electron drift time of $3.8\mu s$. B–injection start pulse.

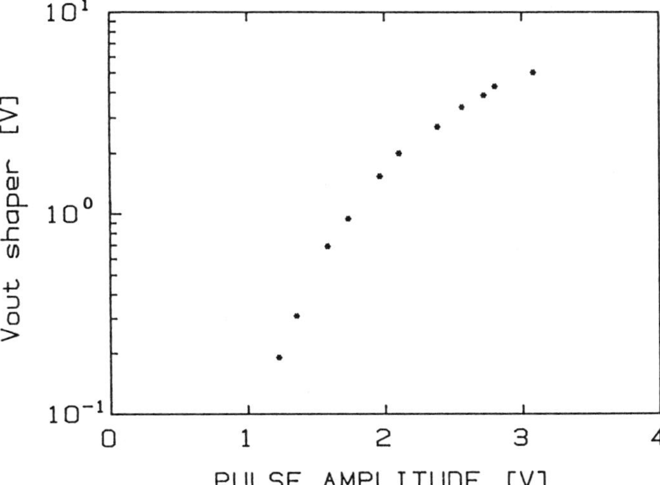

Figure 8: Injected charge as a function of the amplitude of the voltage pulse applied to the injector electrode. Note the log scale for the injected charge.

Accumulated electrons are candidates for being injected into the main valley of the detector. Let us assume some negative charge is brought on the metallic gate covering the SiO_2, by applying a negative voltage pulse to the gate. A portion of the positive charges is neutralized and electrons from the surface region are pushed above the saddle point into the main valley. After the injection and the return of the gate potential to the original value, the electrons are replenished from the path along the z–direction. Thus, the pulse repetition rate may approach 0.5Mhz in the detector, without appreciable rate dependent effects.

We have implemented and tested injectors at a distance of about 2cm from the anode. The size of the metal gate was $45 \times 125\mu m^2$. The drift field was 492

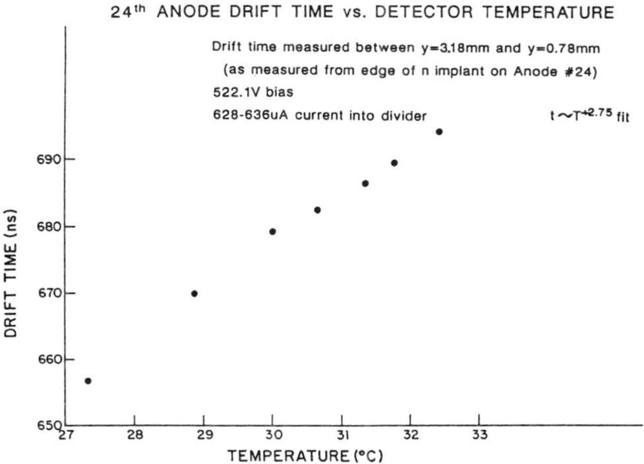

Figure 9: Drift time for 2.4mm drift distance as a function of the detector temperature.

V/cm corresponding to an electron drift time of $3.8\mu s$. Fig. 7 shows the waveform of the pulse–shaped anode signal. Pulse A corresponds to about 25000 electrons, equivalent to the charge liberated by the passage of a minimum ionizing particle through the detector. Pulse B is the prompt pulser signal applied to the p^+ electrode surrounding the collecting anode.

Fig. 8 shows the dependence of the injected charge on the amplitude of the pulse applied to the electrode. For small pulses, the linear part (on a semilogarithmic scale) of the curve has a slope smaller by a factor of 10 than that of an ideal junction diode. We believe we understand the behavior. For the detector calibration Fig. 8 shows that charges up to several times the charge liberated by a minimum ionizing particle can be injected into the detector.

The tested injection scheme is not the only one considered by the collaboration. We have a different injection scheme for the drift velocity calibration in the existing, but not yet tested cylindrical silicon drift detector. We are studying a third calibration scheme for the Mark–1 SSC silicon drift detector.

The dependence of the drift time on the temperature of the detector was measured in the 41 anode detector described at the beginning of this section. Fig. 9 shows the dependence of the drift time on the temperature. The drift field was 278V/cm and the net drift distance 2.4 mm. The best fit gives the drift time to be proportional to the 2.75 power of the absolute temperature. Removing one extreme point from the sets of measurements the power decreases to 2.5, which is in reasonable agreement with the published value of 2.42, knowing that the measured temperature is not necessarily the detector temperature.

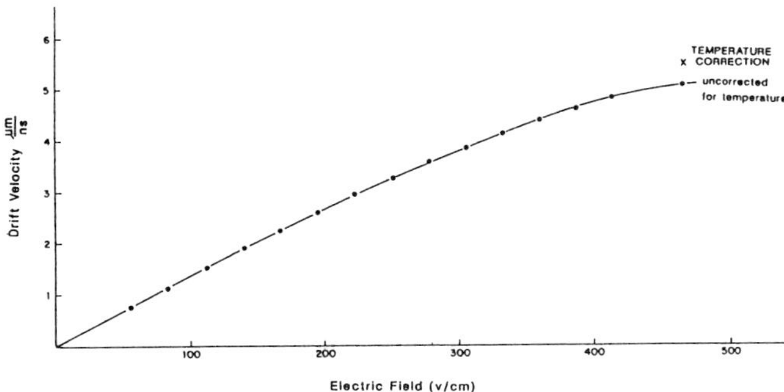

Figure 10: Drift velocity in a hyperpure silicon detector as a function of the drift field at the detector temperature. Correction corresponds to 300K.

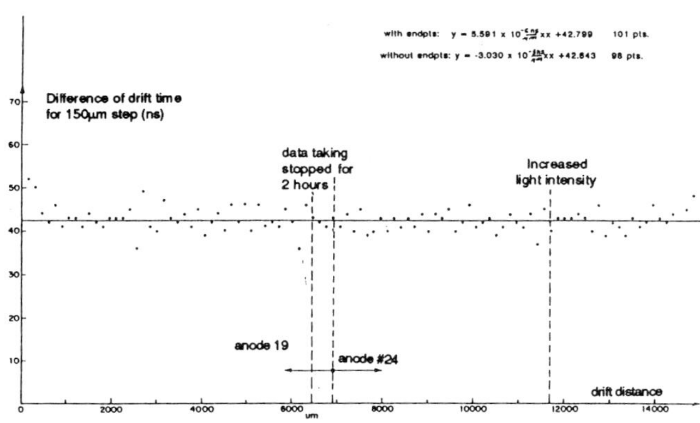

Figure 11: Differences of drift time for $150\mu m$ steps along the drift direction.

At this point it is important to comment briefly on the dependence of the drift velocity on the drift field. There should be a simple linear dependence with the constant of proportionality known as the electron mobility. Fig. 10 shows the measured dependence. The measurement was performed for most of the points at 300K. At higher drift fields heating of the detector due to the power dissipated in the voltage divider was observed. The cross indicates the amount of the temperature correction. There are two important consequences of the curve shown on Fig. 10:

1. The mobility is lower than the published value already at the lower portion of the curve ($1330 cm^2/Vs$ as compared to $1500 cm^2/Vs$).
2. We see a small saturation effect for the drift fields as low as 200V/cm.

The saturation of the drift velocity is a mixed blessing for the performance of silicon drift detectors. Although the ideal isochronism is compromised, the stability is improved and the effects of the external magnetic field are reduced.

Differential non–linearities in the position measurement along the drift coordinate

A very careful measurement of the dependence of the drift time on the drift distance was performed. The position of the detector was repeatedly moved one period of the drift electrode structure, that is, $150\mu m$. The drift field was 278V/cm and the temperature 300K. The drift time was read on the scope as the edge of a zero crossing discriminator output which followed a differentiated pseudo–gaussian shaping amplifier. We did not have a TAC with suitable dynamic range for the measurement.

The integral linearity line is too straight to indicate any deviation and is not shown. Fig. 11 shows the differential non–linearity, that is, the plot of differences of the drift time measurements for each $150\mu m$ step. All points are scattered about the fitted line which shows a slope very near zero. The constant value of the time difference indicates that the drift velocity is constant along the drift direction for the detector under study. There are a few higher points at the short drift distances and one much higher point at the end of the drift region. The non-linearity at the small values of the drift distances is due to a non-ideal drift field in the anode region which penetrates into the periodic region of the detector. The voltage in the anode region can be adjusted to remove this non-uniformity, or a correction can be applied later in the off–line analyses. The non-linearity at the end of the detector cannot be corrected in hardware and a software correction is required.

The measurement of differential non–linearities plotted in Fig. 12 shows an interesting structure. The measurement is similar to the previous one, however, the step was decreased to $50\mu m$ and only the first 4mm of the detector was scanned. At the left side of the plot we can see again the penetration of the electric field from the anode region. The most striking feature is the appearance of 3 bands into which all measured points fall. Moreover, there is a periodicity of 3 measurements corresponding to the $150\mu m$ pitch of the detector drift structure. The drift time measurement corresponding to the second position difference is longer than the third one and the first time interval is the shortest of all. An explanation of the structure is shown on the upper right corner of Fig. 12, where the periodic structure of the detector together with the light injection points are shown.

The electric field close to the detector surface does not have a constant component along the drift direction as is the case in an ideal detector or the central part of the detector under study. The drift field is zero right at the

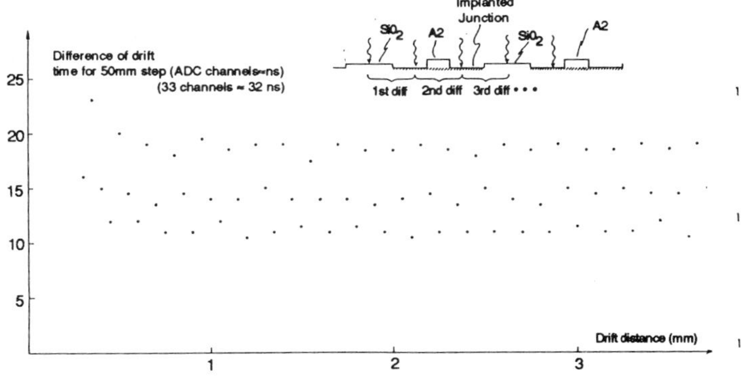

Figure 12: Drift time differences for $50\mu m$ steps along the drift direction.

rectifying junction and 3 times stronger than the average field at the $Si - SiO_2$ interface. There is, however, a large difference in the location across the thickness of the detector of the charge liberated by our laser light and by an ionizing particle. The absorption length of 900nm light in silicon is about $20\mu m$, whereas a fast particle produces electron–hole pairs uniformly across the detector thickness. The drift time differences are therefore amplified with the light test. The analysis of the effect for minimum ionizing particles is published in Ref. 2. The error is about $5\mu m$ when no correction for the effect as a function of the drift coordinate is applied, and can be reduced to $2\mu m$ with a fine correction equivalent to a non–constant drift velocity. The effect cannot be completely corrected due to the fluctuations of the density of the ionization of a fast particle crossing the detector.

Independence of the drift time on the coordinate transverse to the drift direction

In an ideal drift detector the drift time is independent of the transverse coordinate. In a multi–anode detector there is one constant per anode which takes into account slightly different delays of the signal cable, differences in the signal processing electronics, etc. There may be other imperfections within the detector which may require individual calibration of anodes and in an extreme case, a complicated correction may be required. The most obvious imperfection is the non–uniformity of the silicon doping which deforms the drift trajectories and changes the local drift velocity.

Fig. 13 shows the drift time measured on several anodes for drift distances ranging from 0.3 to 5.7mm. The drift field was 335V/cm. A more detailed analysis of the data shown in Fig. 13 indicates that the drift velocities for anodes in the center of the detector are slightly lower than the drift velocity closer to the detector sides. This difference can be attributed to the higher temperature of the detector in the center due to the presence of the

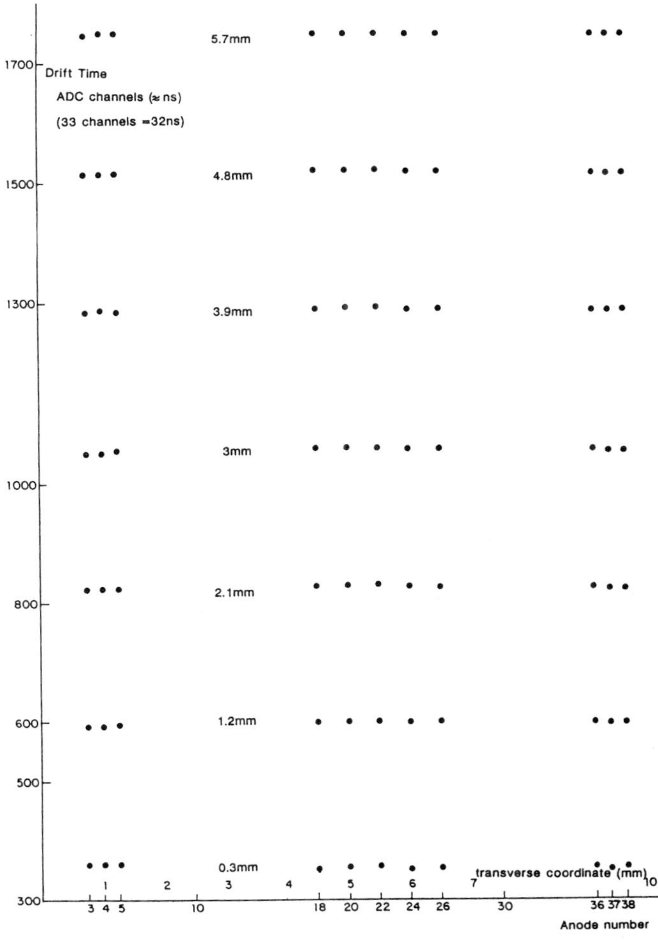

Figure 13: Drift time as a function of the transverse coordinate for several drift distances.

voltage divider implanted directly on the center of the detector.

The uniformity of the drift coordinate measurement after the adjustment of the first measured delay (at 0.3mm) to the same value, and removing the temperature dependence, was about $6\mu m$.

It is obvious that a correctly designed silicon drift detector should have a uniform temperature when properly biased. The voltage divider must not cause temperature gradients within the detector.

Independence of the transverse coordinate measurement on the drift coordinate

This is a complementary study relative to the study of the previous subsection. Multi–anode silicon drift detectors measure the transverse coordinate by the charge division among anodes where the diffused charge is collected. Interpolation among anodes leads to an increase

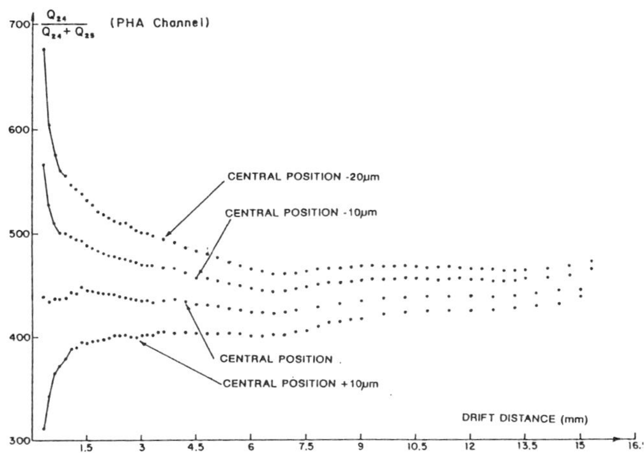

Figure 14: Measurement of the transverse coordinate as a function of the drift distance.

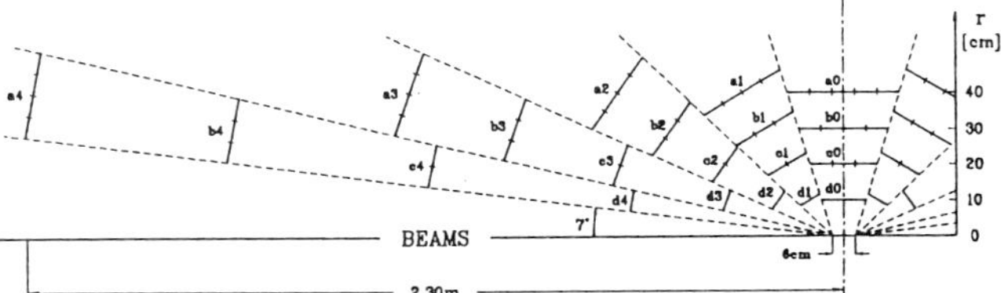

Figure 15: Cross section along the beam axis of the inner tracking detector based on silicon drift detectors. Detector has a full azimuthal coverage and rapidity coverage $|\eta| \leq 3$. Lines indicate the cross section of rings of detectors centered around the beam axis. Each line represents close layers of silicon drift detectors.

in the precision of the measurement of the transverse coordinate to as low as several percent of the inter–anode spacing. This measurement assumes that the signal electrons drift on straight lines perpendicular to the field defining electrodes of the detector. Doping non–uniformities of the silicon may distort drift trajectories and interfere with the measurement of the transverse coordinate.

Fig. 14 shows the ratio of the charge collected at anode 24 divided by the sum of charges collected at anodes 24 and 25 versus the drift distance. The light spot was positioned close to the middle position between the two anodes. For shorter drift distances where the charge diffuses only to these two anodes the measured ratio is a very sensitive function of the position. For longer distances the charge spreads among more anodes and the sensitivity of the ratio to the transverse position is decreased. The plotted ratio is not the best method of measuring the transverse coordinate for longer drift distances. At the time of measurement we did not have a read–out system capable of reading more than one ADC channel and calculating the centroid of the charge.

Fig. 14 shows the four series of measurements, each series taken with the position of the spot moved by $10\mu m$ in the transverse direction. If we attribute all deviations from the perfect lines to the imperfection of the detector (there may be a systematic error due to the DC offset of one of the linear amplifiers used during the measurement) than the systematic error in the measurement of the transverse coordinate is about $10\mu m$. The error can be decreased to $4\mu m$ by calibrating out the linear part of the systematic deviations.

Silicon Drift Detector Performance in the SSC Environment

The Monte Carlo of the Silicon Drift Chamber was done within the GEANT framework. Two generators were used, ISAJET and a simple few-particle generator used for cross-checking. The detector modelled is the 8 anode silicon drift detector sized to fit a 4 inch wafer. The area per anode is 1.6cm in the drift direction and 3.2cm in the transverse direction. This detector measures precisely only one coordinate and is equivalent to a microstrip detector but needs much fewer channels. It is a good candidate for layers of the inner tracking detector farther away from the intersection region. (Candidate layers are called a,b and maybe c in Fig. 15.) In this simulation, however, we have placed this large size silicon drift detector into the d–layer, which is only 10cm away from the SSC beams. The reason for placing such a relatively coarse and large detector so close to the beams was to study the possible problems of the detector for a smaller number of generated events.

The Monte Carlo consisted of two parts. The first was a generator and tracking program that found the hits in the detectors, made the digitization and wrote the results to tape. The second was a reading and reconstruction program that read this "data" back. The original hit banks were also included on tape for comparison with the reconstructed results. Two detector layers were modelled simultaneously in all event generations. One detector in a required position, and another parallel detector immediately behind the first. The anode placement was identical in the parallel detector, but with the opposite drift direction to reconstruct the beam crossing time.

The detector performance was modelled using a parameterization of the drift of the charge left by the original ionizing particle. The width of the charge cloud as it drifts has been calculated and compared to measurements, as a function of total charge and total

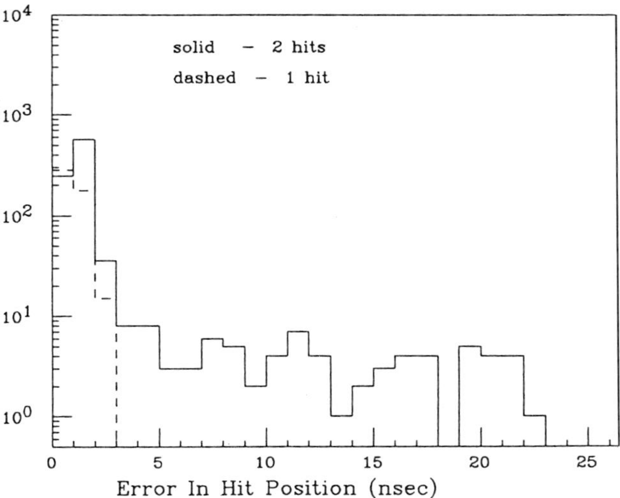

Figure 16: Reconstruction Accuracy for 1 and 2 hits randomly placed within an Anode

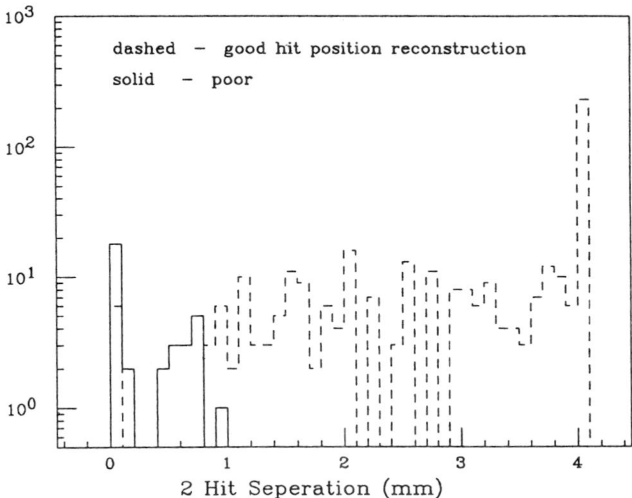

Figure 17: Seperation between the 2 randomly placed hits when the reconstruction error was small, and when it was large.

drift time. This time evolution of the charge cloud accounted for self-repulsion and diffusion. The position of a hit tracked to the detector was then drifted in software towards the anode region. The drift electric field was chosen to be 600 V/cm (for a drift velocity of the charge carriers of $8\mu m/ns$). The software simulated a gaussian shaping amplifier with a shaping sigma of 20 nsec. Due to the diffusion and repulsion of the charge cloud, the typical width of the gaussian signals after the digitization is nearer to 30 nsec.The output of the shaping amplifier was digitized in a fake Flash ADC in 16 nanosecond steps.

The values of the drift field, shaping amplifier time constants and the sampling frequency were taken from the drift detector to be used in the NA–45 experiment at CERN. The parameters of the design were frozen more than a year ago at rather conservative values. We hope to run silicon drift detectors faster at the SSC, which requires a higher value of the drift voltage and a higher bandwidth of the read–out of used electronics. We used the conservative values again to see the problems with the detector.

The reconstruction program would sequentially scan each digitization stored on tape and search for the expected gaussian-shaped response of a hit. The center of the gaussian response was fit using both a simple center of gravity calculation and also a chi-squared minimization with the fitting program MINUIT. The result was compared to the original hit banks. This method was repeatedly applied to a given event, until all signals over a certain threshold were found and subtracted. Some improvement was found over the center of gravity method, when using the MINUIT fitting. It is thought that even this could be improved upon. The method presented here is the simplest, using the center of gravity calculation.

The strategy of the Monte Carlo was to start with the generation of a single particle to compare the digitization for this single hit with that expected. The reconstruction efficiency and accuracy were determined, to judge the method. Two particles were then simulated from the same bunch crossing to judge the interference in reconstructing each position and the bunch crossing time. Two particles, however, from different bunch crossings were then generated.

Fig. 16 shows the error in the reconstruction of a hit. Note carefully that the vertical scale is logarithmic. Each hit has an expected drift time; the difference between this time and the reconstructed drift time is shown here in nanoseconds. A bin of 1 nsec here corresponds to 8 microns. The 2-hit case shows considerable interference in the drift time reconstruction. Fig. 17 shows the spatial separation between the two hits, for two cases. The first case (shown dashed,) is where the drift time reconstruction error in Fig. 16 was less than 5 nsec for both hits. The solid line histogram shows the case when the drift time reconstruction error is greater than 5 nsec.

The drift time reconstruction of the two hits becomes poor when their separation from each other is less than about 1 mm. With the width of the output of the pseudo–gaussian shaper having an average sigma of 30 nsec, 1 mm corresponds to about 4 sigma separation between the two gaussian signals. This is the separation level at which one may expect problems to begin with this simple method. We have published[5] a detailed study of the double particle resolution showing deterioration of the resolution only by a factor of two at 2 sigma separation. It should be stressed

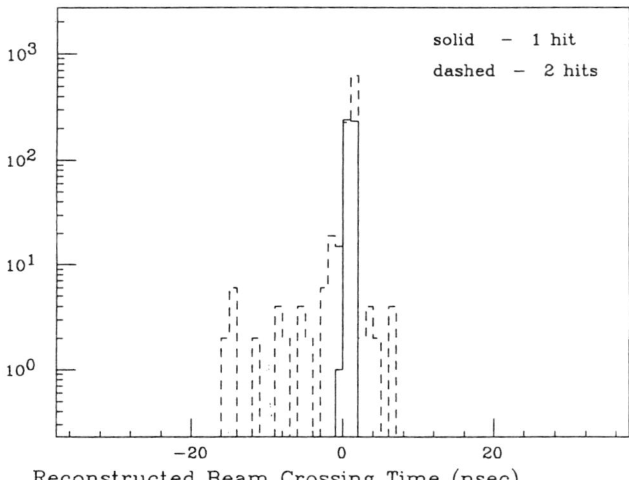

Figure 18: Beam Crossing Time reconstructed from 1 and 2 hits in an anode

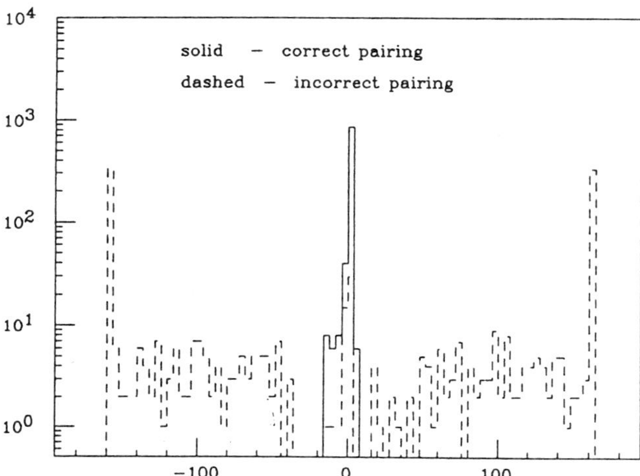

Figure 19: Reconstructed beam crossing time for two hits in two parallel detectors, for correct pairing between the two detectors, and for incorrect pairing.

that a more sophisticated method would be expected to give a smaller separation between neighboring signals before appreciable interference results.

We see that the reconstruction method used up to now in the software simulation is not the optimum one and provides results inferior to the real measurements.

This "interference separation" also affects the beam crossing time reconstruction. Fig. 18 shows the reconstructed beam time crossing, for the case of 1 hit in an anode in the detector and for 2 hits in the same anode. It is expected here that perfect performance would show two sharply peaked curves at 0 nsec. It was found

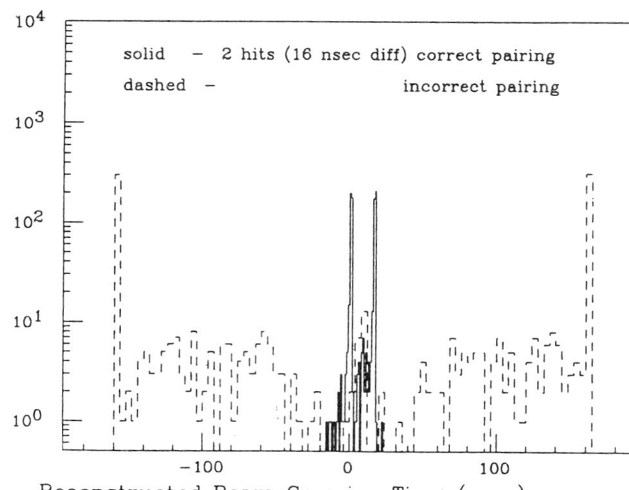

Figure 20: Reconstructed beam crossing time for two hits in two parallel detectors from two consecutive beam crossings, for the correct pairing between the two detectors, and for the incorrect pairing.

that the tails in the 2-hit case occurred when the 2 hits were again within 1mm of each other. Again, it is anticipated that a more sophisticated reconstruction method of separating two nearby hits could be more usefully applied.

Fig. 19 shows the contamination due to incorrect pairing of two hits in two parallel detectors. The solid histogram shows the correct pairing, the random dashed background the incorrect. The peaks near $\pm 160ns$ are underflow and overflow. The central peak is contaminated at a level of several percent. The background due to incorrect pairing shows very weak structure. This will be studied further, to guarantee that incorrect pairing in the reconstruction between the two detectors does not lead to systematic errors in the beam reconstruction time.

Fig. 20 shows the same reconstructed crossing times for two hits, where the hits are separated by 16 nsec. The correct pairings displays the expected time difference.

What has been demonstrated then is that with a simple method, a reasonably accurate reconstruction can be made of 1 and 2 particles in an anode. The failures even with this method are at the level of a few percent.

The Monte Carlo tracking was applied using ISAJET minimum bias events. Table 1 shows the population of each of the detectors nearest the interaction region for one such test. We can see:

- The placement of the detectors corresponds approximately to the same rate in each detector. This is close to the ideal situation.
- Every detector receives about 40 hits over the detector, or an average of 5 hits per anode. This number

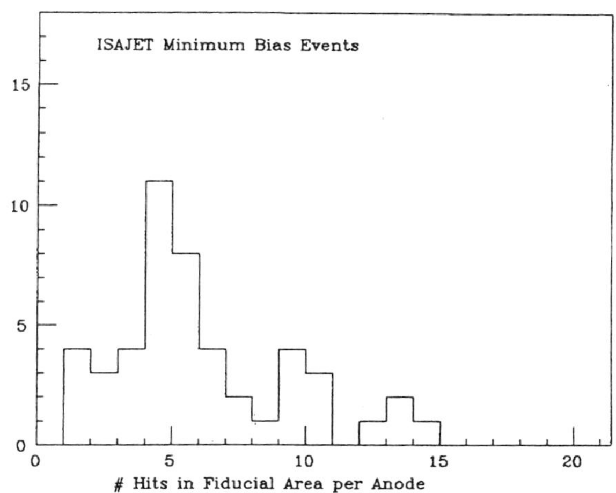

Figure 21: The distribution of the number of hits per anode is shown, for ISAJET minimum bias, in the fiducial area of the detector.

Table 1: Hit counts vs. detector position in a $6.4 \times 6.4 cm^2$ detector in $2 \mu s$ for minimum bias events at luminosity of $10^{33} cm^2 s^{-1}$

Position	Number of hits
D4	47
D3	45
D2	44
D1	37
D0	40

of hits is only about a factor of two higher than an acceptable number of hits. That means that rather large detector elements of 1.6 cm × 1.6 cm are sufficient for a luminosity of $10^{33} cm^{-2} s^{-1}$ at 10 cm distance from the SSC beams. Fig. 21 shows the hit distribution per anode in this fiducial area accumulated over several Monte Carlo runs. We know from the analysis of the previous subsection that 5 hits accumulated during the memory time of the silicon drift detector are too many. The remaining plots are for now only of academic interest, to show that the performance of the detector deteriorates only slightly with higher occupation numbers. This feature, however, may become important with higher luminosities at the SSC in the future. The ability to reconstruct these hits was examined, and the results are shown in Fig. 22. In the 1-hit case, this efficiency is understandably 100%, dropping off by approximately 5% over each of the next few hits, and reaching a plateau at about 75-80%. To be able to

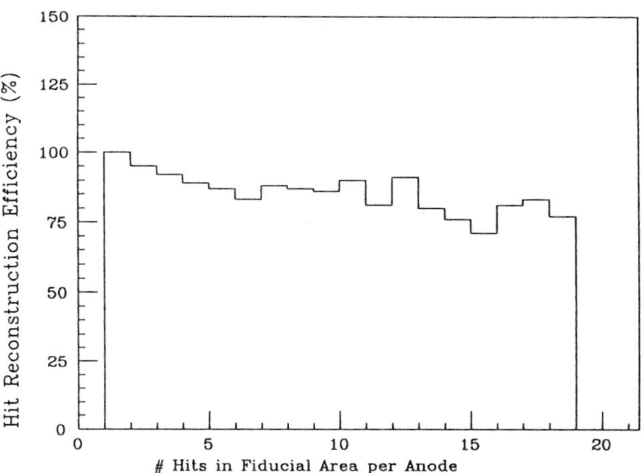

Figure 22: The efficiency for identifying and fitting a hit signal in the presence of other pileup hits is shown. Here the method of identifying a hit uses a center of gravity calculation about a peak, and subtraction according to the known amplitude for a minimum ionizing hit.

obtain such high multiplicities a special run of two jet event was performed.

Conclusions

- It is possible to monitor and to calibrate out the changes in the drift time in silicon drift detectors with temperature. The calibration is achieved by a simple pulsing of an electrode on the detector.
- The systematics of the silicon drift detector provides a resolution about $10 \mu m$ in both directions without the use of a special set of calibration constants. To improve the resolution to $5 \mu m$, a special set of detector calibration constants is required.
- There is no problem with the memory time of silicon drift detectors used at the SSC at the luminosity of 10^{33} at a distance of 10 cm from the beams. A pair of drift detectors solves completely the "confusion" problem.

References

1. E. Gatti and P. Rehak, Nucl. Instr. and Meth. 225, 608 (1984).
2. P. Rehak et al., Nucl. Instr. and Meth. A248, 367 (1986).
3. E. Gatti at al.,Nucl. Instr. and Meth. A295, 489-491, (1990).
4. E. Gatti et al.,Nucl. Instr. and Meth. A273, 865-868, (1988).
5. E. Gatti at al., Nucl. Instr. and Meth. A274, 469 (1989).

VERTEX DETECTOR TECHNOLOGY FOR THE SSC

P. Skubic[f], J. Arens[c], H. Attias[g], K. Barger[e], T. Collins[a], G. Jernigan[c], G. Kalbfleisch[f], D. Kaplan[f], P. Karchin[g], G. Kramer[a], J. Kuehler[f], M. Lambrecht[f], E. McCliment[e], W. Ross[g], S. Shapiro[d], J. Sinnott[g], S. Utku[g], C. Wilburn[b], S. Worley[a]

[a]*Hughes Aircraft Company,* [b]*Micron Semiconductor Inc.,* [c]*Space Sciences Laboratory, University of California,* [d]*SLAC,* [e]*University of Iowa,* [f]*University of Oklahoma,* [g]*Yale University*

Abstract

An overview of a SSC R&D program for silicon vertex detector development is presented. The current test program with silicon microstrip and pixel detectors is discussed and selected results of beam tests are presented including measurements of position resolution as a function of angle of incidence. Plans for future tests are also discussed.

Introduction

We have a R&D program[1] in progress to develop detectors capable of reconstructing particle decay vertices in experiments at the SSC. This program includes development of VLSI readout electronics, radiation damage measurements[2], mechanical design, Monte Carlo simulations, and development of solid state tracking detectors in collaboration with industry. This work[3] is aimed at developing high resolution vertex detectors to be used at hadron colliders to measure secondary vertices from decays of particles with intermediate P_t such as charm and bottom particles.

A vertex detector must measure the impact parameter of tracks of decay secondaries relative to the primary vertex. The *r.m.s.* impact parameter resolution due to multiple scattering is approximately given by $\sigma = (0.015R/P_t)\sqrt{X/X_o}$ where X/X_o = fraction of radiation length of material up to and including the first detector layer, R = radius of the first position measurement, P_t = transverse momentum (GeV/c) of the secondary particle with respect to the beam. Since σ increases with R, the radius of the detector should be kept as small as possible. Such a detector in the SSC environment could be exposed to ionizing radiation doses of 10 Mrad per year and must provide good spatial resolution over a wide range of incident angles. Therefore, the vertex detector should have good intrinsic position resolution (< 10 μm), good signal-to-noise, and material must be minimized to reduce multiple scattering. Because of these requirements, our R&D work has concentrated on silicon tracking detectors including double-sided microstrip and pixel detectors, and associated VLSI readout circuits.

Silicon Microstrip Detectors

AC-coupled silicon microstrip detectors were used in a 227 GeV/c pion test beam at Fermilab during the summer of 1990 to study the position resolution as a function of incident angle. Single-sided detectors (made by the Center for Industrial Research in Oslo, Norway) which had a strip pitch of 25 microns with every alternate strip read out[4], and double-sided detectors (made by Messerschmitt Bolkow Blohm in Munich, Germany[5]) of the design used for the Aleph experiment at LEP were used. SVX-D chips[6] were used to read out the silicon microstrip detectors which were used to measure positions in only one view. LBL designed SRS and SDA CAMAC modules[7] were built by the University of Oklahoma to control the SVX chips and read data out to the VAX on-line computer. The layout of the test beam and silicon microstrip results are described in more detail in a talk at this conference by P. Karchin.[8]

The SVX chips were operated in double-sampling mode with a sample time of 2.4 μs. All 128 channels per chip were read out on every event. A noise of 2.9 ADC counts ($\sim$ 1800 electrons) was obtained during the beam test with the Oslo detectors. The average signal for minimum ionizing particles at normal incidence was 43 ADC counts when summed over two adjacent readout strips. (Since alternate strips were read out, charge sharing occurs if a particle passes through an intermediate non-readout strip.) Therefore the maximum (when all charge is collected by a single strip) signal-to-noise ratio was 14.8.

Bench tests at Oklahoma were carried out with 34 SVX-D chips with no detector connected. Thirty of these gave an average noise per channel of 1.5 ADC counts ($\sim$ 900 electrons) with the 2.4 μs sample time. The other 4 chips were rejected because of significantly larger noise. The chips were operated successfully with a clock rate up to 20 MHz. Fig. 1 shows the negative gain linear response in ADC channels for two amplifier channels of the same chip to a calibration pulse which varied in 100 steps from 0 to 200 mV (gain 8 setting). The responses of the remaining 126 channels of the chip lie between the two curves shown.

During the beam test, a hodoscope of from three to five detectors was placed in a precision stand which allowed selected planes to be rotated about a horizontal axis. In one configuration, data where taken with the two outside detectors at normal incidence with respect to the beam and the center detector was rotated to angles of 0, 25, 35, and 45 degrees with respect to normal incidence with the rotation axis parallel to the diode strips. Fig. 2 shows the difference between measured position in the center plane and the predicted position from the outside planes for 35° which gave an *r.m.s.* position resolution of 15.6 ± 0.7 μm.

In order to verify that the detector was working as expected, a Monte Carlo program was run which simulated operation of the microstrip detectors and SVX readout circuit. This program tracks particles through the silicon microstrip detector and simulates deposited ionization charge according to the expected dE/dx distribution for silicon. The collected charge at each amplifier was smeared according to a Gaussian distribution with an *r.m.s.* of 1800 electrons to simulate the effect of amplifier noise. Fig. 3 shows the position resolution versus angle of incidence for the Monte Carlo compared to the test beam measurements and shows good agreement between the data and the simulation.

Pixel Detectors

Beam tests were also performed with hybrid pixel detectors which consisted of a 256 x 256 diode array sensor with 30 μm x 30 μm elements (made by Micron Semiconductor Inc.) bump-bonded with Indium to a readout chip manufactured by Hughes Aircraft Company. This hybrid detector has been used previously in infrared imaging applications for space science and astronomy.[9] Two independent stand-alone data acquisition systems, one built by Hughes Aircraft Co. and the other built by UC-Berkeley Space Sciences Laboratory, were used in the beam tests and both systems performed well. Detectors of this type have many advantages compared to long microstrip detectors including good multi-track resolution, simultaneous measurement of positions in two dimensions, good signal-to-noise, and resistance to radiation damage effects.

The readout was designed to function as a camera and therefore all 65,536 elements were read out on every event in a minimum time of 16 ms. The detectors (and readout chips) were operated at room temperature. Fig. 4 shows a 128 pixel square region of a typical event with pedestals subtracted for an array with incident 227 GeV/c pions using the UC-SSL data acquisition system. We expect these data to be useful in measuring properties such as charge sharing which will determine the optimized design of future readout circuits. Further results from the pixel beam tests are presented in a talk by S. Shapiro at this conference.[10]

Future Plans

During the second half of the Fermilab fixed target run, we expect to continue our program of beam tests of silicon tracking detectors. A hodoscope of up to five pixel detectors will be used to measure their position resolutions as a function of incident angle. Similar measurements will be made with double-sided silicon microstrip detectors which will be supplied by Micron Semiconductor.

In parallel with the test beam work, design will proceed on the next generation pixel detector readout circuit which will subtract pedestals and have sparse readout and time slicing capability. The schedule for this R&D program is described in a talk by J. Arens at this conference and a preliminary design for the readout architecture is presented by O. Barkan.

VLSI readout design will be carried out by a collaboration of Fermilab and Oak Ridge National Laboratory, and radiation damage tests will be done with microstrip and pixel detectors, and with electronic readout circuits as part of the continuing R&D program. This future work will be coordinated with the Silicon Tracking Subsystem Collaboration, of which Oklahoma and Yale are members, and future development of detectors will continue in collaboration with Hughes Aircraft Company and Micron Semiconductor Inc.

References

[1] This work was supported by U.S. Department of Energy contracts DE-AC03-76SF00515, DE-AS05-80ER10629, and DE-AC02-76ER03075/A048.

[2] E. J. Kennedy, et. al., "Radiation Effects on JFETS, MOSFETS, and Bipolar Transistors, as Related to SSC Circuit Design", these proceedings.

[3] H. Castro et al., R&D proposal to Fermilab (E784/BCD) Jan. 2, 1989.

[4] M. Caccia et al., Nucl. Inst. Meth. **A260** (1987) 124.

[5] P. Holl et al., IEEE Trans. Nucl. Sci. **36** (1989) 251.

[6] Kleinfelder, S. A., et al., IEEE Trans. Nucl. Sci. **35** (1989) 171.

[7] F. Kirsten and C. Haber, IEEE Trans. Nucl. Sci. **37** (1990) 288.

[8] H. Attias et al.,"Beam Tests of Silicon Microstrip Detectors with VLSI Readout", these proceedings.

[9] S. Shapiro, et al., Nucl. Instr. Methods **A275**, 580 (1989).

[10] Stephen Shapiro, et al., "Performance Measurements of Hybrid PIN Diode Arrays", these proceedings.

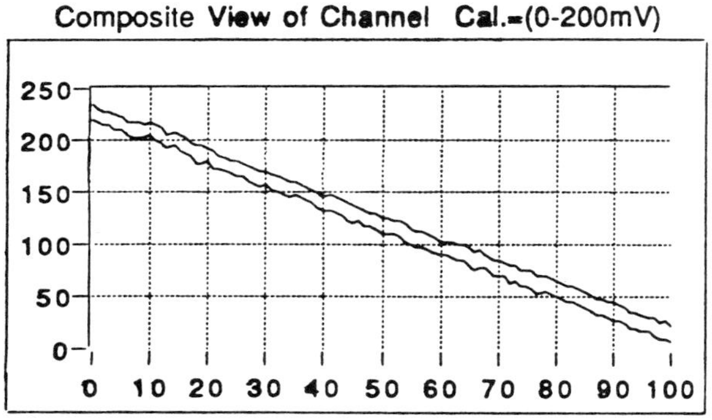

Composite View of Channel Cal.=(0-200mV)

Figure 1: Negative gain response in ADC counts (ordinate) of two channels of a typical SVX-D VLSI readout chip to a calibration pulse which varied from 0 to 200 mV in 100 steps. The responses of the other 126 channels lie between the two curves shown.

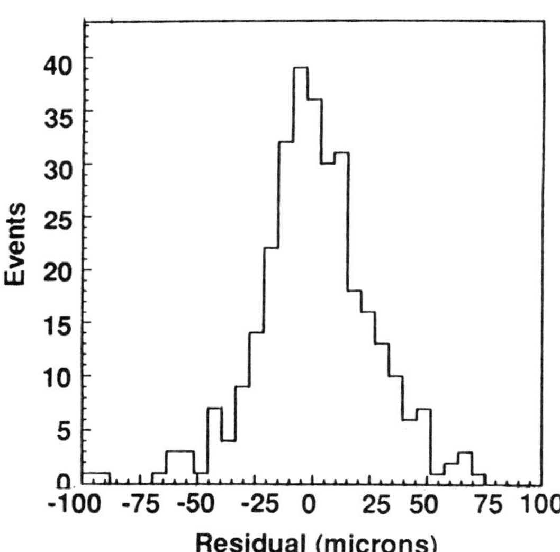

Figure 2: Residual in μm between the measured and predicted cluster position for tracks with an incident angle of 35°.

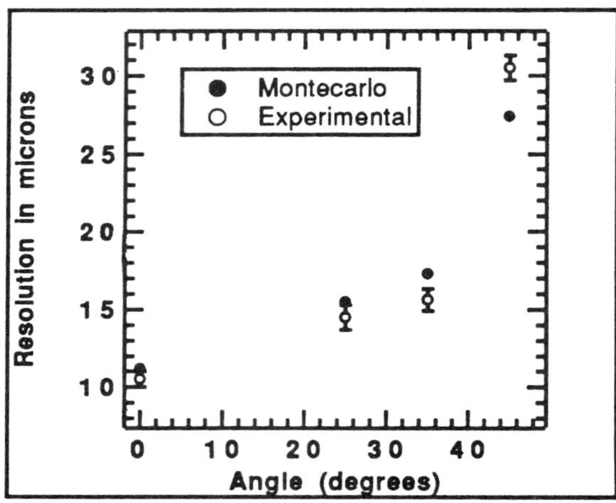

Figure 3: Position resolution versus track angle of incidence for Monte Carlo simulation and for test beam data.

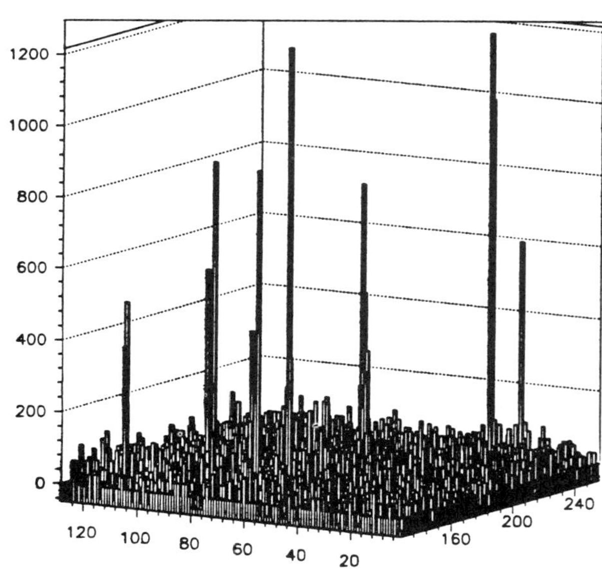

Figure 4: Event display showing pulse heights in ADC counts of a 128 pixel square region of the hybrid pixel detector operated at room temperature and illuminated with 227 GeV/c pions.

BEAM TESTS OF SILICON MICROSTRIP DETECTORS WITH VLSI READOUT

H. Attias[c], K. Barger[a], G. Kalbfleisch[b], D. Kaplan[b], P. Karchin[c], J. Kuehler[b], M. Lambrecht[b], E. McCliment[a], W. Ross[c], J. Sinnott[c], P. Skubic[b], S. Utku[c]

[a] *University of Iowa,* [b] *University of Oklahoma,* [c] *Yale University*

Abstract

AC-coupled silicon microstrip detectors with SVXD readout were operated in a high energy charged beam at Fermilab. Results are reported on signal-to-noise ratio, spatial resolution and detector response as a function of angle of incidence.

Introduction

We have recently tested single and double sided AC coupled silicon microstrip detectors read out with Berkeley SVX version D VLSI chips[1]. The detectors were operated in a 227 GeV/c negatively charged pion beam at Fermilab. This work is part of an ongoing R&D program[2, 3] aimed at developing high resolution vertex detectors to be used at hadron colliders to measure secondary vertices from charm and bottom decay. Detector planes with spatial resolution of $\sim 5\mu m$ are required to provide measurement of the vertex separation to an accuracy of $\sim 20\mu m$. Since the decay products emerge over the full 4π solid angle, detectors must provide good spatial resolution over a large range of angle of incidence.

Silicon Microstrip Detectors

The single-sided AC-coupled microstrip detectors used in this test were manufactured by the Center for Industrial Research in Oslo, Norway[4]. The strip pitch is 25 μm with every other strip having direct readout. The alternate strips are readout by capacitive coupling to their neighbors. The strip length is 40 mm and there are 257 strips covering a length of 6.4 mm. The detector thickness is approximately 300 microns. The detectors fully deplete at approximately 40 volts. The leakage current at full depletion is typically 50 μA and is dominated by the current through the guard ring. Bias is provided through polysilicon resistors integrated onto the wafer.

The double-sided AC-coupled detectors were made by Messerschmitt Bolkow Blohm in Munich, Germany[5] and are of the design used for the Aleph experiment at LEP. The wafer size is 5.1 cm by 5.1 cm and 300 microns thick. The strip pitch is 50 μm on the ohmic (n-doped) side and 25 μm on the junction (p-doped) side. The readout pitch is 50 μm on both sides and the strips on both sides are capacitively coupled. Bias is provided through a 10 μm channel gap. The isolation on the ohmic side is provided by p-implants between the strips. Leakage currents are typically 50 μA at full depletion of about 150V.

The detectors were initially tested using a Model MP-1000 Micro-circuit Probing System from Wentworth Labs. For each detector an I-V curve was taken to establish that the detector exhibited the normal diode characteristic. We found that for the SI detectors, the leakage current varied for the first few hours after bias was applied and then stabilized. Sometimes the initial current was much smaller ($< 1\mu A$) than the final current ($\sim 50\mu A$). This behavior is consistent with measurements made by the manufacturer.

The single-sided detectors were glued to a G-10 supporting board using conductive epoxy. We found that special care had to be taken to prevent stray silver grains in the epoxy from making a short circuit between the detector substrate and the printed-circuit board on which the SVX chip is mounted, which are separated by a gap of only about 500 μm. The gap is small to minimize the length of wire bonds between the detector and readout chip. The double-sided detectors were mounted in Delrin (plastic) holders using non-conductive epoxy. Again, special care had to taken to prevent contact between the PC boards and the detector substrate.

The wire bonding was done with a Model TU-909 Ultrasonic Aluminum Wedge Bonder from Mech-El Industries. The G-10 or Delrin boards holding the detectors were fixed into place under a microscope so that the strips on the detector were parallel to the edge of the aluminum mounting plate to the accuracy of less than a strip width over the length of a strip.

Arrangement in the Test Beam

In the test beam, the detector assemblies were mounted in a holder that allowed the detector assemblies to be rotated about a single axis at any desired angle to the beam. The detectors could be removed and replaced while maintaining the same mechanical alignment to better than a strip width. Scintillation counters just upstream and just downstream of the silicon holder detected beam particles over a 1 inch by 1 inch cross-section, while a veto counter protected against interactions and halo. The beam rate was typically 8 KHz through these counters. A third, smaller beam counter with a cross-sectional area of 2mm by 2.5 cm. provided a "tight" trigger.

Readout

The SVX chips were read out with the Berkeley programmable readout sequencer ("SRS") and flash-ADC ("SDA") CAMAC modules[6]. These modules were located a few feet from the silicon test stand. A Camac module containing a bus transceiver chip handled communication to and from the SVX chips via a conventional flat multi-conductor cable with alternating ground and signal lines.

The SRS-SDA readout system was used in an asynchronous "fixed target" mode, similar to that used by CDF in their test beam studies[7] using an earlier version of the SVX chip. We used double correlated sampling (which is appropriate for AC- coupled detectors) with a sample time of 2.4 μ sec. The pulse height was read out from every channel at a rate of 4.8 μsec per channel. The SDA memories were read out with the Fermilab "Vax Online" software package into a VAX 3100 workstation.

Results

The data analysis reported here is based on only part of the data set taken during the 1990 fixed-target running period at Fermilab. In this data, only the single-sided detectors and the junction side of the double sided detectors are used. We found for the double-sided detectors from MBB that the integrated capacitors on the ohmic side could not withstand the 150 volt bias without punch through. Thus, these detectors were operated as single-sided devices. Up to four detectors have been operated measuring the same view.

The pulse height distributions from individual strips demonstrate the basic operation of the detector system. A typical pulse height distribution from a single strip is shown in Figure 1a. The prominent structure is the Gaussian pedestal. This is due to triggers where the beam particle did not go through the strip in question. Since the area of each strip is only a small fraction (.002 to .02 depending on the trigger used) of the area of the trigger counter, most of the time the pulse height read out from a given strip is due to the intrinsic noise in the system. The noise sigma is 3.1 counts. We also observed the phenomenon of "common mode" noise where the pulse heights from all the channels in a single chip move up or down in concert from event to event. By subtracting out this noise on an event- by- event basis, the noise sigma is reduced to 2.9 counts.

The logarithmic plot of Figure 1b reveals both the noise pedestal, which is Gaussian over 4 decades, and the signal due to track hits which appears as a tail to the left. Because the SVX has negative gain larger pulse heights give lower ADC counts. Since the microstrip detector has capacitive charge division readout of every other strip, the single channel pulse height distributions do not have the characteristic Landau shape. This shape is expected for the summed pulse height distribution from a cluster of adjacent strips and is discussed later. The overflows on the right are due to non-Gaussian noise which occurs much less frequently than the Gaussian noise.

The event display in Figure 2 shows data with 4 detector planes measuring the same view. The pulse heights from the channels in different planes are offset vertically in the display. The readout sense for each plane and the approximate alignment have also been taken into account in these displays. A 4-hit track is clearly seen. The variation in the number of strips per cluster (due to every other strip readout) and the pulse height per cluster (due to the Landau distribution) can be seen by comparing hit clusters in different planes.

The signal pulse height distribution and the spatial resolution can be studied by pointing to one of the planes with tracks based on the other planes. In the pointed-to plane, the pulse heights in adjacent strips are summed to insure that all the charge from a single particle traversal is taken into account. The resulting spatial residual for normal incidence is shown in Figure 3. A Gaussian fit to this distribution gives a sigma of 9.7 μm. This implies that the resolution of the target plane is 8.4 μm *r.m.s.* The cluster pulse height distribution is shown in Figure 4a for normal incidence and in Figure 4b for 45°. At normal incidence, the mean signal is 43 ADC counts and the mean cluster size is 1.75 strips while at 45° the mean signal is 58 counts and the mean cluster size is 4.9 strips. This implies a signal-to-noise ratio of 11.2 at normal incidence and 9.0 at 45°. Analysis for the spatial resolution as a function of angle of incidence is in progress.

References

[1] Kleinfelder, S. A., et al., IEEE Trans. Nucl. Sci. **35** (1989) 171.

[2] H. Castro et al., R&D proposal to Fermilab (E784/BCD) Jan. 2, 1989.

[3] S. Dhawan et. al., "Performance Limits of Silicon Strip Front-End Electronics for the SSC Bottom Detector", DOE contract DE-AC02-76ER03075/A048.

[4] M. Caccia et al., Nucl. Inst. Meth. **A260** (1987) 124.

[5] P. Holl et al., IEEE Trans. Nucl. Sci. **36** (1989) 251.

[6] F. Kirsten and C. Haber, IEEE Trans. Nucl. Sci. **37** (1990) 288.

[7] F. Bedeschi et al., IEEE Trans. Nucl. Sci. **36** (1989) 35.

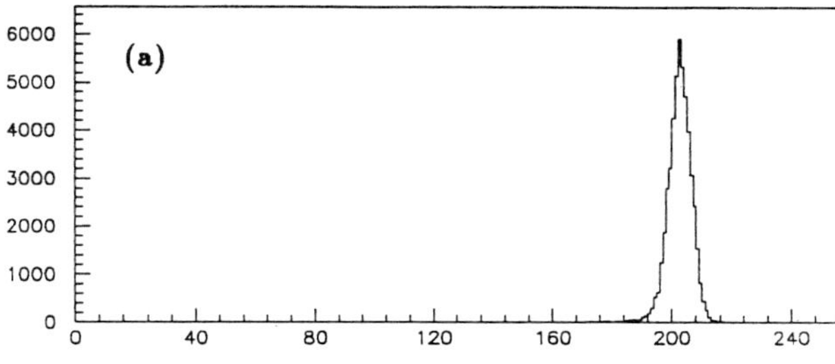

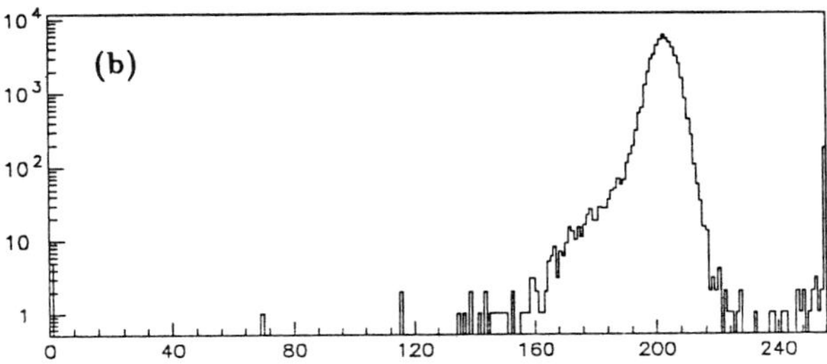

Figure 1: Number of events versus pulse height in ADC counts from a single strip, with linear event scale in (a) and logarithmic scale in (b).

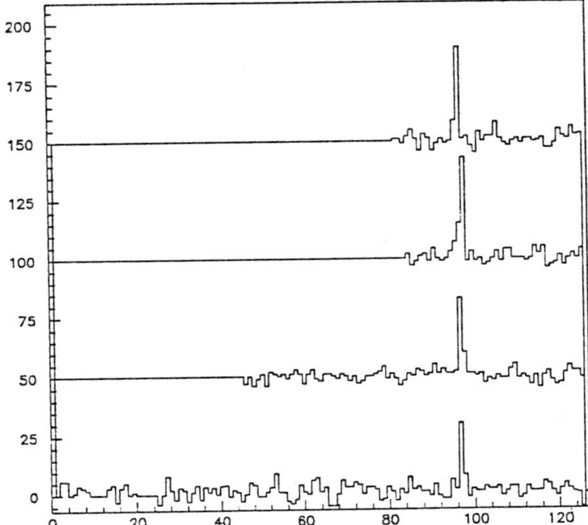

Figure 2: Event display showing pulse height in ADC counts versus strip number for 4 planes.

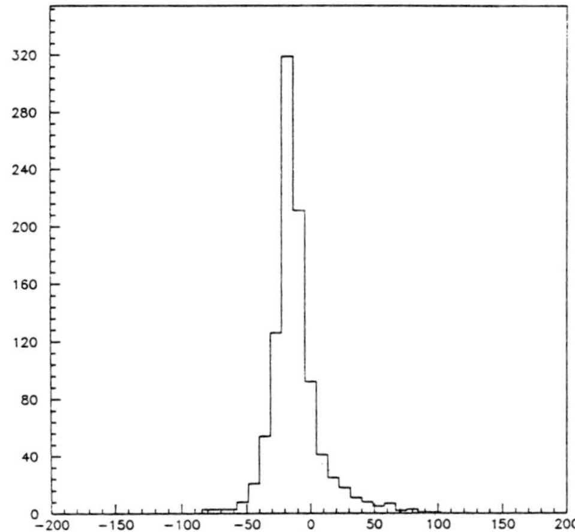

Figure 3: Number of events versus residual in μm between the measured and predicted cluster position.

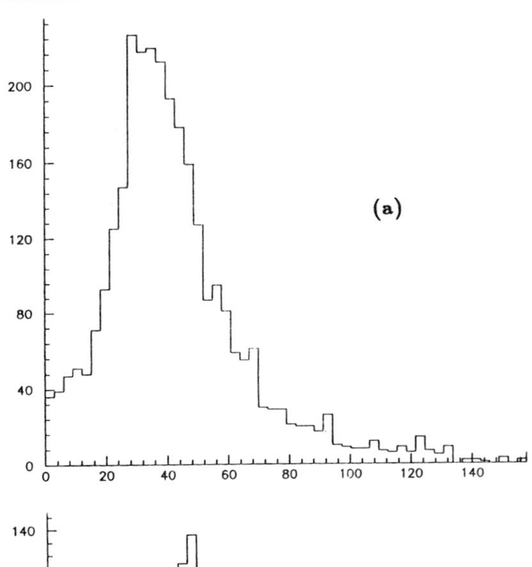

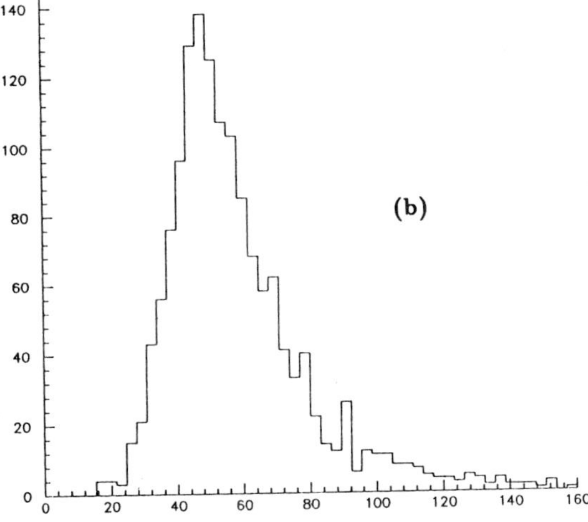

Figure 4: Number of events versus pulse height in ADC counts for clusters in planes at normal incidence (a) and 45° (b).

A PRECISION VERTEX CHAMBER FOR EMPACT/TEXAS*

V. Cook and P.M. Mockett
University of Washington Seattle, Washington 98195

Abstract

Vertex chambers are used in many detectors to improve secondary vertex identification. In a non-magnetic detector, such as EMPACT/TEXAS, there is another function for a precision vertex chamber; identification of low momentum particles that may scatter before detection in a tracking chamber and give rise to false secondary vertices. We discuss a vertex chamber that uses multiple scattering to tag low momentum tracks so they can be ignored in vertex construction.

Introduction

The physics goals for research at the SSC require among other things precise measurement of tracks near the primary interaction vertex. Such measurements are required to enable identification of short lived particles that travel a very short distance (100 μm , more or less) from the primary interaction vertex before decaying. Detection of a secondary vertex signals the presence of heavy quark jets. Tagging b-jets in this way is an important tool for top quark and Higgs boson searches, and after they are discovered, for collecting enough events to measure the properties of these particles.

The EMPACT/TEXAS detector is designed to have excellent electromagnetic and hadronic calorimetry as well as excellent capability for detecting and measuring the energy of muons. To achieve these goals it is proposed to build the detector without a central magnetic field. While a detector of this sort has significant advantages in design and operating capabilities, it also is subject to confusion created by low momentum charged particles which scatter in the beam pipe or other matter before they can be measured in a tracking chamber. Such particle tracks could appear to come from a "fake" secondary vertex.

We propose a vertex chamber using silicon pixels (or possibly silicon drift devices) that will achieve the following goals:

- Location of secondary vertices with a resolution of better than 10 μm .
- Exclusion of at least 95% of the low momentum ($<$ 1 Gev/c) charged tracks while keeping at least 90% of high momentum ($>$ 10 Gev/c) charged tracks for use in vertex fitting.
- Energy resolution sufficient to distinguish minimum ionizing particles (mips) from twice-mips.
- Total material at 90° less than 0.1 radiation length.
- Better than 95% solid angle coverage for $\eta < 2.5$.
- Radiation hardness sufficient to enable useful operation up to 10 Mrad integrated dose.

We discuss a conceptual design that we believe will satisfy these goals but acknowledge that much work remains to be done to verify that such a device can be constructed within reasonable budget constraints.

Our conceptual design assumes silicon pixels since at this stage we believe such devices are likely to have the required characteristics. Silicon drift detectors are being studied for this purpose by Pavel Rehak and his colleagues at BNL[1] and their progress will be reported at this meeting. Silicon pixels are under development at a number of laboratories around the world[2]. We are currently working with the Pixel Detector Development Collaboration (an SSC Sub-System group) lead by LBL and the Berkeley Space Sciences Laboratory (SSL) with strong support and technical expertise provided by Hughes Aircraft Company[3]. We expect to evaluate these two technologies during the coming year and select the most promising for EMPACT/TEXAS requirements by the end of 1991.

Conceptual Design

Three layers of chips are the minimum required to accomplish the tasks listed above. We consider combinations of cylinders coaxial with the beam and planes perpendicular to the beam to cover the desired solid angle, with no detector element closer than 5 cm from the beam line. With silicon chip thickness of 600 μm , each detector layer comprises 0.64% radiation length.

*Work Supported by NSF Grant PHY-8822028

We add material to the central chip layer to make a 3% radiation length. There are several strategies for determining track straightness using the three layer positions. We choose to use the inner two layers to determine the incident track direction and its extrapolated intersection at the outer layer. The deviation between this point and the actual hit in the outer layer is used as a measure of scattering. The optimum position for the central layer is midway between the outer two layers. The rejection probability for a track of momentum p and velocity βc is given by

$$P_{rej} = exp\{-\alpha/[1 + (p_O/p\beta)^2]\}$$

where α depends on the deviation "cut" (d) selected for rejecting low momentum tracks and measuring resolution in one dimension (σ);

$$\alpha = d^2/12\sigma^2$$

and p_O is a function of x, the number of radiation lengths in the central layer, and L, the separation between layers;

$$p_O = (0.0141 \frac{Gev}{c} \sqrt{x}L)/(\sqrt{6}\sigma)$$

Using the values

$$\sigma = 5\mu m$$

$$x = 0.03$$

$$L = 5cm$$

one finds a p_O of 10 Gev/c. Figure 1 shows the rejection probability as a function of $p_O/p\beta$ for two values of the minimum probability at high momentum. The change in performance resulting from changing the parameters of the detector can be seen with this universal curve by calculating the appropriate p_O, and adjusting the particle momentum scale.

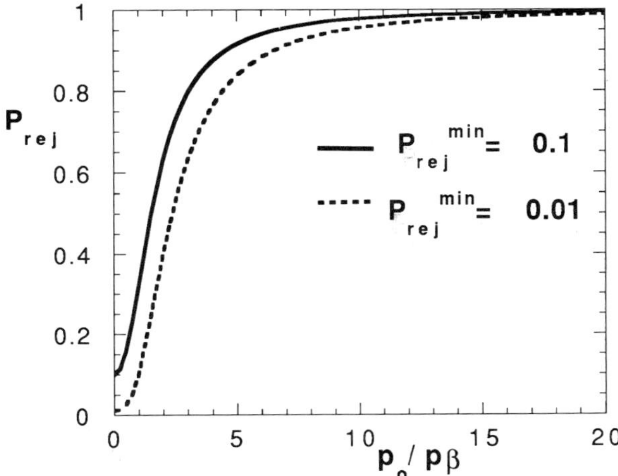

Figure 1: Rejection probability as a function of $p_O/p\beta$.

For d= 40 μm and the assumed properties given above, 96% of the tracks with $p\beta$ below 1 Gev/c are rejected and 92% of those above 10 Gev/c are accepted. These numbers can be "tuned" by varying the detector parameters and by selecting the deviation cutoff parameter d.

The Pixels.

LBL and Hughes Aircraft are developing "smart pixel arrays" utilizing micro- silicon diodes (as small as 30 μm x 30 μm) fabricated on a high resistance silicon chip for the particle detector array[4], and a separate readout electronics chip which is then indium bump-bonded to the detector diode chip[5] (See Figure 2a). The detector diodes are typically 300 μm thick and operate at room temperature with a signal/noise of better than 50. Spatial resolution of 2 to 3 μm has been achieved. This is possible because of charge-sharing between pixels.

Chips that have been made at LBL and Hughes are 1 cm square, but the experts believe that chips as large as 1 inch square should be possible. Since there is a cost saving (per unit area) in using larger area chips we are pursuing this option.

The parameters we assume for the detector pixel chips for our conceptual design are:

- Pixel Size → 50 μm x 50 μm by 300 μm thick
- Chip Size → 1 inch square (250,000 pixels/chip)
- Heat/Chip →<2 Watts
- Resolution → 5 μm

A drawing of the basic chip showing the major structural elements is given in Figure 2b.

Readout Electronics

Hybrid sparse-readout circuitry is being developed at Hughes Aircraft and at LBL. The philosophy of these two parallel efforts is complementary and after both have produced results, the most successful elements of the two designs will be combined to produce the final device. The basic readout schemes are quite similar and we can consider the Hughes prototype, which will be ready for testing this fall, for purposes of this discussion.

Chip logic and control circuitry consists of the following:

- COLUMN and ROW RAM to store hit addresses for each "event". An event occurs when any pixel signal exceeds its comparator threshold.
- A timing COUNTER and logic to "time stamp" each event within one beam crossing (16 ns). This time and COL/ROW RAM address are stored in CAM.

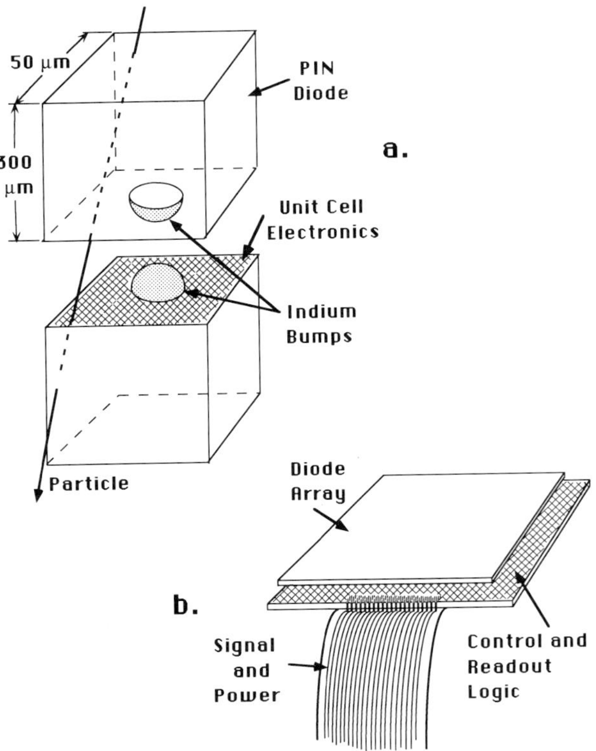

Figure 2: a. Silicon PIN Diode unit cell. b. Vertex detector chip.

- On receipt of a Level 1 trigger "time" a search for a match in the CAM is made. Sample and Hold processing is initiated sequentially for all rows in each hit column found in a time-matched event. An ADC on each chip provides digital output.

Readout times will be typically a few μs. Note that WRITE, READ and RESET operations can proceed in parallel so operation is essentially without deadtime. Of course if more than one hit on a given pixel (or neighbor) occurs before readout, that data will be contaminated.

Mechanical Design.

A simplified sketch of the detector is shown in Figure 3. Three layers of detector chips (the minimum required) are shown. The center layer contains a material layer (probably either 350 μm of copper or 2200 μm of silicon) to provide the multiple scattering needed. The silicon chips are tilted so the plane of each chip is perpendicular to a line from the center of the interaction region. The mechanical structure must be light (few radiation lengths) and rigid. We are considering beryllium as well as various carbide foams. Silicon Carbide and Boron Carbide foam are particularly attractive. Both can be machined and cast, have good rigidity and long radiation lengths. Density of the foam is typically 10% of the solid material. Boron Carbide foam has a radiation length about 2 m so a 5mm thick support structure would contribute only 0.25% radiation length.

Cooling the detector is a major concern. With 2500 1-inch square chips 2 to 3 KW of heat will be generated and must be removed from the detector. We are considering several possible schemes including cold gas flow, liquid nitrogen, and evaporative cooling using liquids with vapor points between 0 ° C and 100 ° C.

References

[1] Pavel Rehak et al., Nucl. Instr. and Meth. **A248**, 367 (1986).

[2] *Proceedings of the International Workshop on Silicon Pixel Detectors for Particles and X-rays*, Leuven, Belgium, (1988), Nucl. Instr. and Meth. **A275**, 1989.

[3] *The Pixel Detector Development Collaboration*, Hughes Aircraft Company, 1990 (unpublished).

[4] S.L. Shapiro et al., Nucl. Instr. and Meth. **A275**, 580 (1989)

[5] S. Gaalema, IEEE *Trans. on Nucl. Sci.* **NS-32**, No. 1:417 (1985).

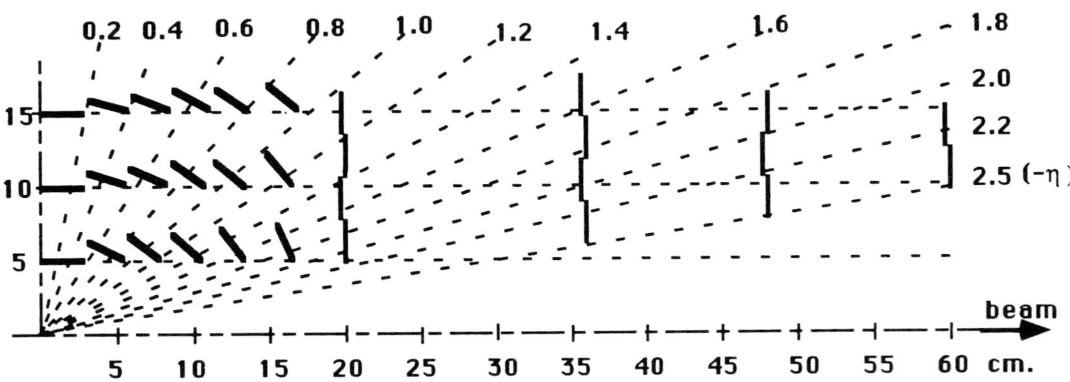

Figure 3: Sketch of quarter section of the detector.

PERFORMANCE MEASUREMENTS OF HYBRID PIN DIODE ARRAYS*

Stephen L. Shapiro
Stanford Linear Accelerator Center, Stanford University, Stanford, CA 94309
John F. Arens and J. Garrett Jernigan
Space Sciences Laboratory, University of California, Berkeley, CA 94720
Gordon Kramer
Hughes Aircraft Company, El Segundo, CA 90245
Tim Collins and Stuart Worley
Hughes Aircraft Company, Carlsbad, CA 92009
Colin D. Wilburn
Micron Semiconductor Inc., Longwood, FL 32250
and
Pat Skubic
University of Oklahoma, Norman, OK 73019

Abstract

We report the successful development of hybrid PIN diode arrays and a series of room-temperature measurements in a high-energy pion beam at FNAL. A PMOS VLSI 256 × 256 readout array having 30 μm square pixels was indium-bump bonded to a mating PIN diode detector array. Preliminary measurements on the resulting hybrid show excellent signal-to-noise at room temperature.

Introduction

In this paper, we present a progress report on the program to develop two-dimensional silicon PIN diode hybrid arrays which can be used as building blocks for vertex detectors [1,2]. Funded by the SSC Laboratory as part of the Generic Detector Research Program, this development effort has seen the fabrication and testing of three different hybrid arrays [3]. The latest array, a PMOS VLSI 256 × 256 pixel array having 30 μm square pixels, has been tested in a 250 GeV/c pion beam at Fermilab. Preliminary results of these tests will be discussed.

The PMOS Array

The PIN diode array hybrids were implemented by indium-bump bonding PIN diode arrays made by Micron Semiconductor Inc. to mating VLSI readout arrays made by the Hughes Aircraft Co. The first two arrays have been discussed in earlier publications [1,2] and are only briefly described here.

The first array (a 10 × 64 array having pixels 120 μm on a side) has ten readout channels per array, one for each column. The readout structure provides random access to any pixel. There is no power necessary during the time charge is being detected and stored by the array; however, during the read cycle, 10 mW (1 mW per channel) is required. The readout electronics is an NMOS VLSI circuit which has been made radiation hard to 1 MRAD of ^{60}Co gamma rays.

The second array, a 256 × 256 array, having pixels 30 μm on a side, has two readout channels per array. Random access readout is provided via row and column shift registers. Data stored in the pixel represented by the intersection of row and column address is presented to the readout node. No power is necessary for storing data, and only 2 mW is necessary for readout. The readout electronics is an NMOS VLSI circuit, which is not particularly radiation hard.

The third array (fabricated earlier this year) is similar to the second, except that the readout electronics, though identical to the second array, is implemented in a PMOS VLSI circuit. This variation is necessary to preserve the option of using these arrays at room temperature. Traditionally, high-energy physicists employ diode detectors, which collect holes rather than electrons, and the original diode arrays from Micron Semiconductor were of that variety. The PMOS array is more appropriate to collect the holes, rather than the electrons generated by the passage of charged particles through the 300 μm-thick silicon PIN diodes. The PMOS circuit has a much larger dynamic range of operation, and dark current integrated over the time necessary to read the 32,768 pixels on each readout node no longer develops voltages that drive the circuit to its supply rails. Tests were made in the laboratory at the Hughes Aircraft Co.,

*Work supported by Department of Energy contract DE-AC03-76SF00515.

Figure 1. Photograph of two independent test systems in the beam line at Fermilab.

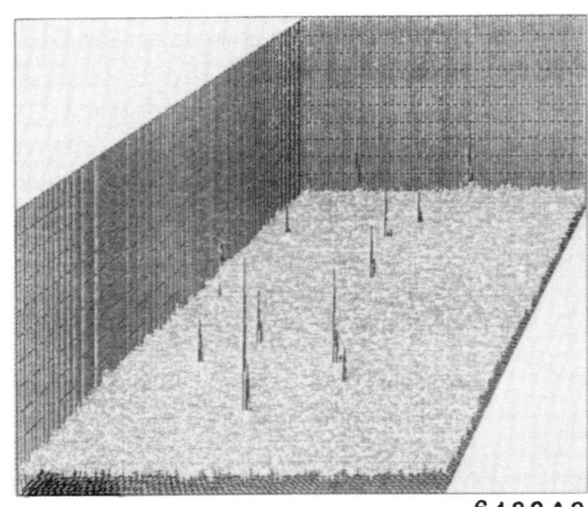

Figure 2. A three-dimensional plot of minimum ionizing particles incident on a detector array, demonstrating excellent signal-to-noise and the power of two-dimensional arrays to eliminate confusion in complex events.

Carlsbad, and at the Space Sciences Laboratory of the University of California, Berkeley, confirming their operation at room temperature, prior to our beam tests at FNAL.

Beam Tests at FNAL

The data being presented were obtained in a test performed at FNAL in a high-energy pion beam in August 1990. Two separate and independent readout systems were employed, one fabricated by the Hughes Aircraft Co., and one fabricated by the SSL/SLAC collaboration. Several detector arrays were tested in each system, one at a time. All of the arrays worked well. Data were taken using 250 GeV/c pions, impinging perpendicular to the arrays and at 15 degrees to the array normal. Data were also taken with material placed in front of the arrays to generate showers and jets, and—as a final calibration—alphas and 59.5 KeV photons from $^{241}A_m$ were collected.

Figure 1 is a photograph of the two Dewars, each containing an array, one behind the other, in the beam line. The beam is incident from the right. Although the arrays were operated at room temperature, they were contained and supported in Dewars capable of operation at cryogenic temperatures. During these runs, power line surges and electromagnetic pickup were detected which caused hardware failures and some excess noise. The runs were successful in demonstrating the detection of high-energy particles, using a room temperature detector with a good signal-to-noise ratio.

The two readout systems which were employed reflect the differing uses to which they are normally put. The Hughes system is a copy of a system used for testing arrays. This system takes data slowly, at frame rates of up to 120 per second, but has the flexibility to measure array parameters should that be necessary. The SSL/SLAC system is more typical of a high-energy physics data acquisition system.

Figure 2 is a three-dimensional display of a number of high-energy pions incident perpendicular to the array. Evident in this presentation of the data is the good signal-to-noise ratio of each detected pion. Also evident is the power inherent in two-dimensional arrays to eliminate confusion normally present when a large number of particles are incident on a single detector. The arrays used in this detector have 256 × 256 pixels each 30 μm square; thus the arrays are about 7.5 mm on a side or 0.5 cm^2 in area.

Figure 3 is a two-dimensional representation of a track (probably a delta ray) angling its way through the array. From the uniform staircase-like track, one can get a feel for the ultimate spatial resolution which will be available with these detectors, once they are properly calibrated.

Figure 4 is another two-dimensional representation of what is believed to be a nuclear fragmentation event. Events like this are rare; only one like it was detected in the thousands of events recorded. The gray scale shows the pulse height information that is retained by each pixel. The black area to the side of the cluster is an artifact of the readout electronics external to the hybrid array. This artifact can be removed in the data analysis. The electronics will be modified later to remove this effect.

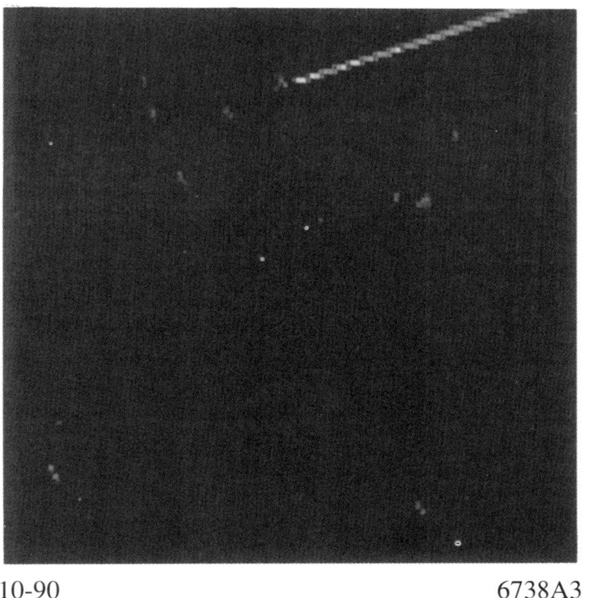

10-90 6738A3

Figure 3. A two-dimensional plot of a minimum ionizing track traversing the detector array at an angle, demonstrating potentially excellent spatial resolution.

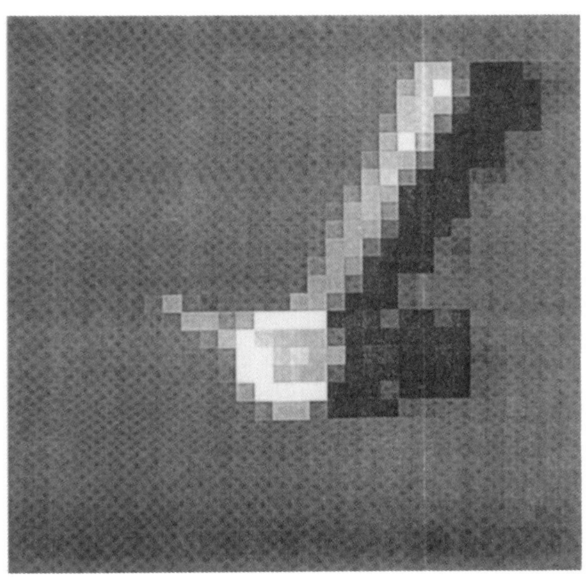

10-90 6738A4

Figure 4. Data containing a nuclear fragmentation event, demonstrating the pattern recognition power of retaining the pulse height information available in each pixel.

Figure 5 represents preliminary data. There has not been time in the preparation of this paper to analyze the data in detail. However, the main features of the data are quite clear. Figure 5 is a histogram of the pulse heights of the minimum ionizing particles, as they traverse the array.

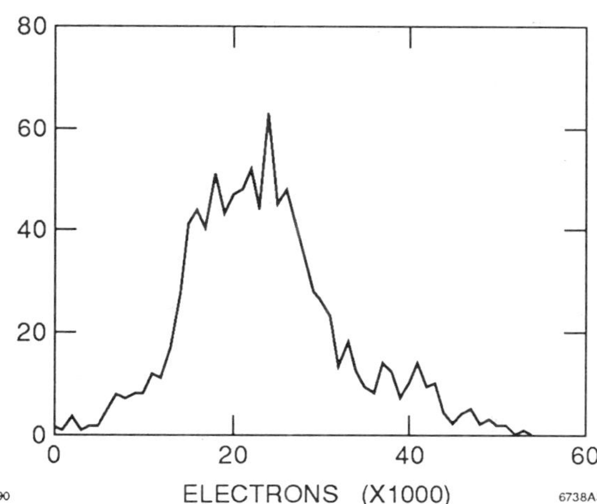

Figure 5. A histogram of pulse height for a data set containing minimum ionizing particles traversing the array normally.

It must be emphasized that no gain calibration has been attempted, as of this time. No correction has been applied to equalize the gains from each of the two chip outputs, nor has the variation of gain across the face of the array been studied. Noise attributable to the integration of dark current during the 50 ms readout time has not been minimized in this analysis. All this notwithstanding, the peak of the pulse-height spectrum for the minimum ionizing particles is in the proper ratio to the observed peak in the 59.5 KeV photon spectrum from Americium data.

In conclusion, the test run at FNAL demonstrated room-temperature operation of a two-dimensional array of pixels. The signal-to-noise ratio was good, approximately 50:1. The data has not been fully analyzed, and the arrays are not yet calibrated; but there is consistency between the minimum ionizing data and the 59.5 KeV photon data.

References

[1] S. Shapiro, W. Dunwoodie, J. Arens, J. Gernigan, and S. Gaalema, "Silicon PIN Diode Array Hybrids for Charged Particle Detection," Nucl. Instrum. Methods A**275**, 580 (1989).

[2] S. Gaalema, G. Kramer, S. L. Shapiro, W. Dunwoodie, J. Arens, and J. G. Jernigan, "Silicon PIN Diode Hybrid Arrays for Charged Particle Detection: Building Blocks for Vertex Detectors at the SSC," Proc. Int. Industrial Symposium on the Supercollider, New Orleans, LA, 1989, p. 173.

[3] J. G. Jernigan, J. Arens, G. Kramer, T. Collins, J. Herring, S. L. Shapiro, and C. Wilburn, "Performance Measurements of Hybrid PIN Diode Arrays," presented at the 1990 Int. Industrial Symposium on the Supercollider, Miami Beach, Fl, 1990, SLAC–PUB–5211.

DEVELOPMENT OF A CUSTOMIZED SSC PIXEL DETECTOR READOUT FOR VERTEX TRACKING[1]

Ozdal Barkan, Eugene L. Atlas, Wayne L. Marking, Stuart Worley, Ghassan Y. Yacoub
Hughes Aircraft Company, Carlsbad, CA 92009

Gordon Kramer
Hughes Aircraft Company, El Segundo, CA 90245

John F. Arens, J. Garrett Jernigan
Space Sciences Laboratory, University of California,
Berkeley, CA 94720

Stephen L. Shapiro
Stanford Linear Accelerator Center,
Stanford University, Stanford, CA 94309

David Nygren, Helmuth Spieler, Michael Wright
Lawrence Berkeley Laboratory, Berkeley, CA 94720

Abstract

We describe the readout architecture and progress to date in the development of hybrid PIN diode arrays for use as vertex detectors in the SSC environment. The architecture supports a self-timed mechanism for time stamping hit pixels, storing their xy coordinates and later selectively reading out only those pixels containing interesting data along with their coordinates. The peripheral logic resolves ambiguous pixel ghost locations and controls pixel neighbor readout to achieve high spatial resolution.

A test lot containing 64 X 32 pixel arrays has been processed and is currently being tested. Each pixel contains 23 transistors and six capacitors consuming an area of 50 μm by 150 μm and dissipating about 20 μW of power.

Introduction

Pixel arrays (2-dimensional array of small semiconductor detector elements) provide significant technical advantages for use as vertex detectors [1,2]. These include good track resolution, small capacitance, high signal-to-noise ratio, increased radiation resistance [3,4]. High interaction rates and the increased quantity of data (10^8 data channels for a typical vertex detector [3]) necessitate sparse readout circuitry within the detector chips. We have developed a pixel array chip architecture that supports a self-activated mechanism for time stamping hit pixels, storing their xy coordinates and later selectively reading out only those pixels containing interesting data along with their coordinates. Test chips that contain a 64 x 32 array of pixels (50 μm x 150 μm each) have been processed.

Requirements/Assumptions Concerning SSC Detectors

Our chip architecture was developed with the goal of being able to handle the worst case assumptions about SSC interaction rates and background radiation. Pixels and the periphery architecture can be simplified if less demanding assumptions are made.

It is assumed that every 16 ns there is a beam crossing. On the average there are 1.5 events/beam crossing and only 1 percent of them will be interesting. A 1 cm^2 chip at a radius of 3 cm from the beam axis will be hit during 30 different beam crossings while waiting for the Trigger level 1 signal. (A Trigger level 1 signal is externally generated, signals the occurrence of an interesting event and has a delay of 1-2 μs from the beam crossing. This delay will be fixed once the calorimeter design is completed.) For each beam crossing when the array is hit, up to 25 pixels will be hit or will be a neighbor of such a hit pixel. (It is desirable to read neighbors of hit pixels to determine the centroid of a particle track). It is also assumed that these 25 pixel locations will not be hit again until the data is read or discarded.

Requirements/specification for the chip include [5,6] quiescent power dissipation of 0.1 - 1.0 W/cm^2, dynamic range of 500 (determined by 100,000 holes maximum input charge and 200 electron rms noise), time stamping with 16 ns time resolution, readout of the charge stored and xy addresses in about 1 μs, radiation hardness to 10 MRad, and testability.

Prototype Architecture and Block Diagrams

A first prototype array for the final SSC vertex detector has been designed and will be going into fabrication in December 1990. Figure 1 shows the periphery architecture and Figure 2 shows the pixel architecture. A number of prototyping steps are envisioned to reach the final radiation hard version of the SSC array.

[1]This work was supported by the SSC Laboratory under contract numbers SLAC P.O. # U.S. 839760.M, LBL P.O. # 4566510 and DOE contract # DE-AC03-76SF00515..

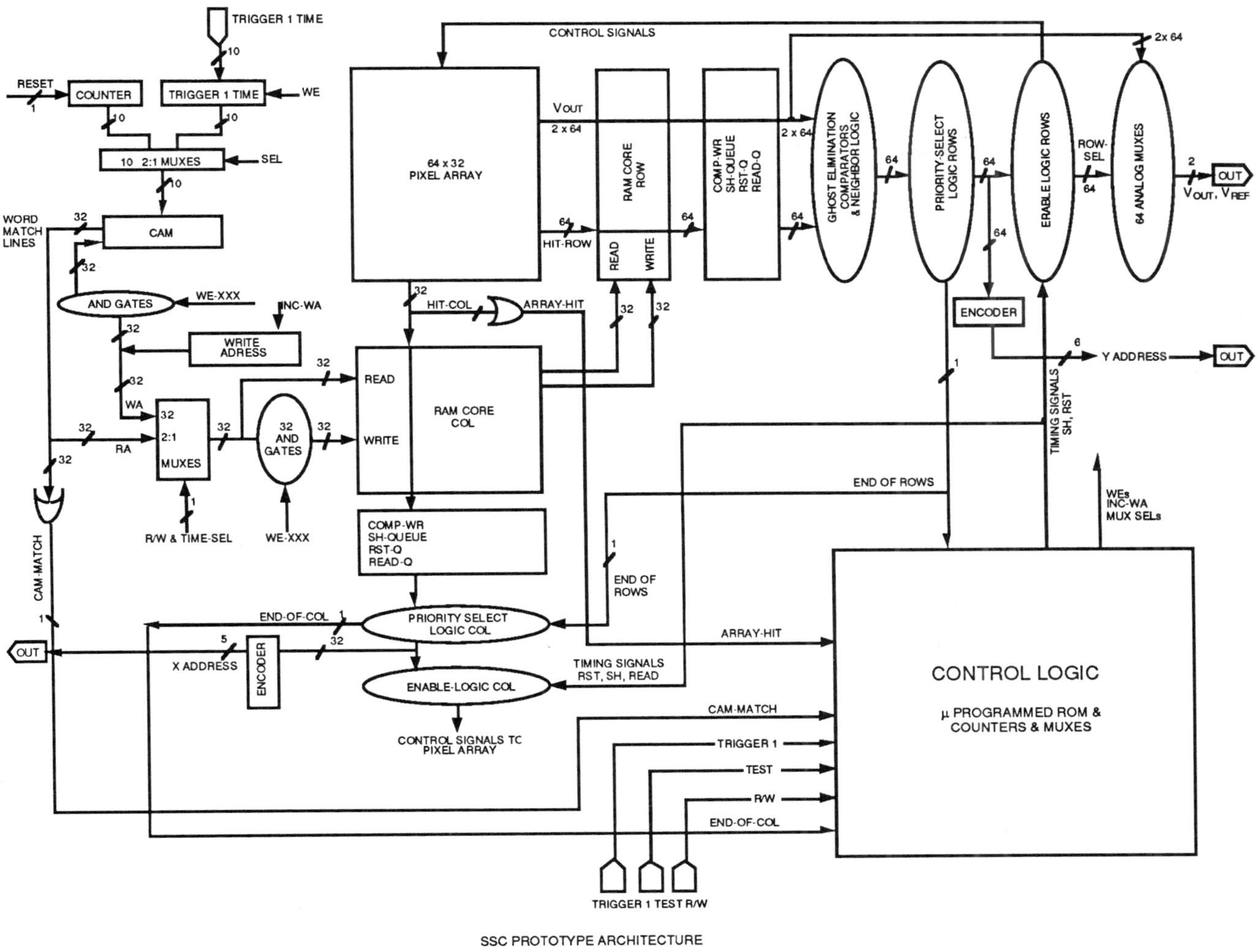

SSC PROTOTYPE ARCHITECTURE

Figure 1. Pixel Detector SSC Prototype Architecture.

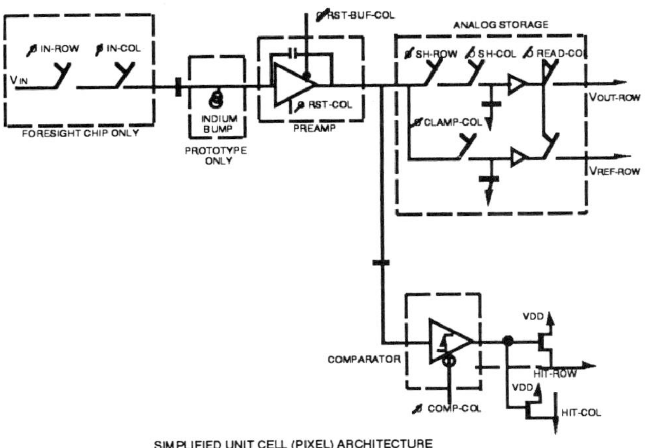

Figure 2. Simplified Unit Cell (Pixel) Architecture.

A brief overview of chip operation is as follows:
If the array is hit, the time of the hit is stored in the content addressable memory (CAM), along with the hit pixel locations that are stored in the random access memory (RAM). The array keeps storing new hit information until the arrival of a Trigger level 1 signal. At that time the chip searches the CAM for a hit at the time of interest indicated by the Trigger level 1 signal. If a hit exists, it reads those hit locations one at a time and sends the stored charge information and address to be processed. Hit data that does not correspond to the time of interest is not read out and pixels are reset to receive future hits.

This architecture resolves ambiguous hit locations (due to multiple pixels hit at one time) and does not require continuous clocking of pixels or periphery logic, eliminating switching noise and power dissipation. All control signals are generated internal to the chip using μ-programmed control logic. The CAM and the RAM do not require encoding and decoding. Dynamic storage with no refresh circuitry could also be used since, typically, we are no longer interested in data after a few microseconds. These characteristics result in a smaller periphery and an improved fill factor (> 90%). The chip has a test mode of operation for production testing, as well as an inside-SSC test mode to disconnect bad pixels from the logic.

The unit cell has within it a transimpedance amplifier, charge storage and analog buffering, and a fast comparator to signal hits to the row and column circuitry.

The pixel has both digital and analog parts in close proximity due to the density required. Great care has been taken to minimize the coupling of the digital signals to the sensitive analog portion of the circuit. Separate analog and digital supply rails are provided in the unit cell to isolate the analog circuitry from any spikes resulting from digital switching. DC bias lines are used at the cell edges to prevent cell-to-cell crosstalk.

Several capacitor types are used; PS1 to Metal 1 and Metal 1 to Metal 2 for low value capacitors (2.5 - 10 fF) and PS1–Pwell for higher values (50 - 300 fF).

In the near future we expect advances in both scaling and silicon-on-insulator (SOI) multilevel technologies to allow the unit cell size to shrink to 50 µm X 50 µm or smaller with current circuit complexity. We also expect the advances to result in a faster, better performing unit cell.

Preprototyping Arrays

Before committing to full prototype production we have decided to process test chips to verify the pixel functionality and the speed of time stamping and analog readout. We have used an L_{eff} = 1.2 µm single poly, double metal (pitch ~ 3.5 µm) CMOS process which guarantees 100 MHz operation at room temperature for worst case process conditions. We are also going to relayout and process a single pixel at Hughes Aircraft using a Hughes radiation hard SOS technology to evaluate the radiation hardness of the design. None of these test chips contains detectors. Instead, electrical voltage inputs to pixels are used for test and verification of operation. The prototype will contain PIN detectors to be built by LBL.

Test Data

The test arrays have arrived and are under test.

FY91 Plans

We plan to process our current prototype design in early 1991. A 6
design with AC coupled detectors (to eliminate dark current buildup due to radiation damage induced leakage) will be processed during the second half of FY91. A rad hard SOS version of this second design will also be processed after verification of the non-rad hard version.

References:

[1] S. Shapiro, W. Dunwoodie, J. Arens, and S. Gaalema, "Silicon PIN Diode Array Hybrids for Charged Particle Detection," Nucl. Instrum. Methods **A275**: 580 (1989).

[2] J. G. Jernigan, J.F. Arens, G. Kramer, T. Collins, J. Herring, S.L. Shapiro, and C. Wilburn, "Performance Measurements of Hybrid PIN Diode Arrays," Presented to the 1990 Int. Industrial Symposium on the Supercollider, Miami Beach, FL (1990) SLAC-PUB 5211.

[3] H. Spieler, "Integrated Microsystems as a Driving Force in Modern Detector Designs," Proceedings of the International Conference on the Impact of Digital Microelectronics and Microprocessors on Particle Physics, Trieste, Italy, March 28-30, 1988.

[4] D. R. Nygren, "Silicon Tracking Devices for the SSC," Proceedings of the Summer Study on High Energy Physics in the 1990s, pp 754-761, Snowmass, Colorado, June 27 - July 15, 1988.

[5] G. Kramer, et al, Memorandum of Understanding SSCPC-025 "Proposed Plan for Pixel Array Development for the SSC," (1989) as modified by O. Barkan, Preliminary Design Review HTC, Carlsbad, CA, May 2, 1990.

[6] G. Kramer, et al, "The Pixel Detector Development Collaboration," Summary Report for FY'90 and Proposed Effort for FY'91, Presented to the SSC Laboratory, September 1, 1990.

POSITION SENSITIVE DETECTION OF CHARGED PARTICLES BY AMORPHOUS SILICON DIODE LAYERS AND ASSOCIATED THIN FILM ELECTRONICS

V. Perez-Mendez, G. Cho, J. Drewery, I. Fujieda, T. Jing, S. Kaplan,
D. Wildermuth; Lawrence Berkeley Laboratory, UCB, Berkeley, CA 94720
R. Street, Xerox Parc, Palo Alto, CA 94306

Abstract

Hydrogenated amorphous silicon (a-Si:H) p-i-n diodes can detect minimum ionizing particles (MIPs) either by direct electron-hole pair production, or when used as photodiodes coupled to evaporated layers of $C_sI(T\ell)$ scintillator. Pixel readout is achieved by depositing the a-Si:H and $C_sI(T\ell)$ layers onto an array of a-Si:H thin film transistors.

Introduction

a-Si:H p-i-n diode layers can be readily used for the detection of minimum ionizing particles by direct e, h pair production when the i layer is sufficiently thick. From our measured value of W = 4.8 ± 0.3 eV, we estimate and have measured that ~80 e, h pairs are produced by MIPs/μm of i layer [1]. p-i-n layers > 50/μm have been produced [2]. These layers are produced in large areas by plasma enhanced chemical vapor deposition at a rate of 1-3 μm/hour. A PECVD machine is shown in Fig. 1. The R.F. power decomposes the gases to form the a-Si:H layers. p layers are made by decomposing a mixture of B_2H_6+Si:H_4; n layers from PH_3+Si:H_4; i layers from pure Si:H_4. The substrate which is held at ~250°C can be glass, stainless steel or polyimide (Kapton) plastic sheet. The same type machine can be used for depositing the thin film transistor (TFT) electronics. However, because of the size limitations of presently available precision lithography machines, electronic arrays are limited to 30 x 30 cm.

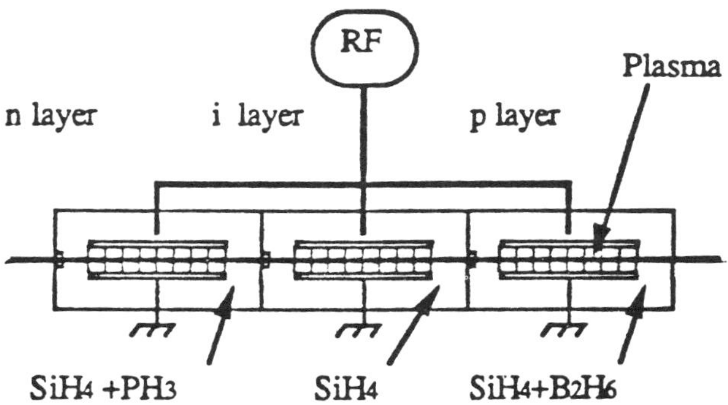

Fig. 1: Multi-chamber system for p-i-n a-Si:H deposition.

We have measured the properties of these thick diodes (3). They can be fully depleted at bias potentials of 20 - 30 V/μm. Electron charge collection is faster than 50 nsec whereas full hole collection requires > 5 μsec. The a-Si:H is also very radiation resistant. We have shown that fast neutron fluxes up to 10^{14}n/cm^2 produce negligible damage; this is also readily annealed away by heating to ~150°C for 2 hours (3). Because of the low signal levels when electrons are collected only, pixel arrays require individual amplification at each pixel point. We have designed and will be testing such TFT amplifier arrays (4).

An alternative approach is to make the a-Si:H layers relatively thin and to use them as photodiode arrays to detect the scintillation light of a suitable coupled scintillator such as C_sI. Because the scintillator can be made as thick as necessary, the signal produced by MIPs is larger and hence the required amplification/pixel is minimal, thus simplifying the TFT electronic arrays. Furthermore, because of the high biases that we can put on thin p-i-n diodes, the internal E fields are quite high. Thus electron collection <10 ns, hole collection < 200 ns.

In this paper we describe construction and measurements of 1-5 μm a-Si:H photodiode arrays coupled to evaporated layers of $C_sI(T\ell)$ 300 - 1000 μm thick (5). With the large signal produced it is then possible to simplify the readout and make it an a-Si:H TFT routing matrix only.

a-Si:H Photodiode - C_sI Layer Detectors

A typical p-i-n photodiode coupled to a $C_sI(T\ell)$ layer is shown in Fig. 2. The structure has a transparent, conductive indium tin oxide layer (ITO) as a top contact

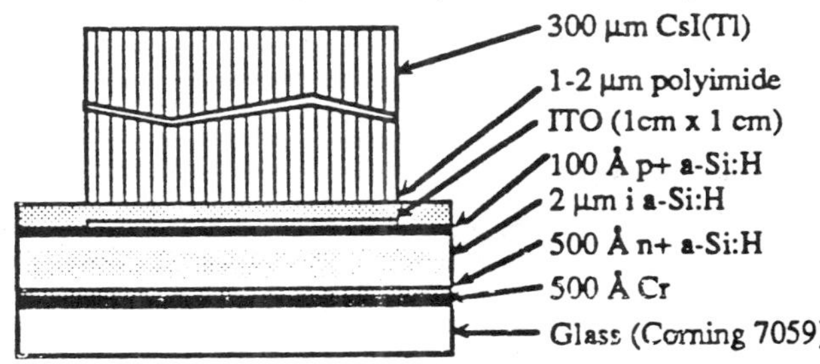

Fig. 2: Monolithic X-ray sensor consisting of $C_sI(T\ell)$ and a-Si:H photodiode.

followed by a p layer ~50 nm, an i layer, an n layer, all deposited on a patterned bottom chromium layer to form the pixel area. The $C_sI(T\ell)$ layer is made by vacuum evaporation of $C_sI(T\ell)$ crystal chips [5] at a deposition rate of 300 - 400 μm/hour onto the heated substrate (200°C). A thin polyimide layer is spun on to make the C_sI bind well. When the $C_sI(T\ell)$ cools on the substrate it forms vertical columnar structures which partially collimate the scintillation light. This is similar to the process used in x-ray image intensifiers where the x-ray sensitive layer is C_sI(Na) (6). Fig. 3 shows the light emission spectra of $C_sI(T\ell)$ and the response of a-Si:H and photodiode layers. The 15-20% light losses in the photodiode response are due to absorption in the polyimide (~5%), in the ITO (~5%), and in the p layer of the top contact.

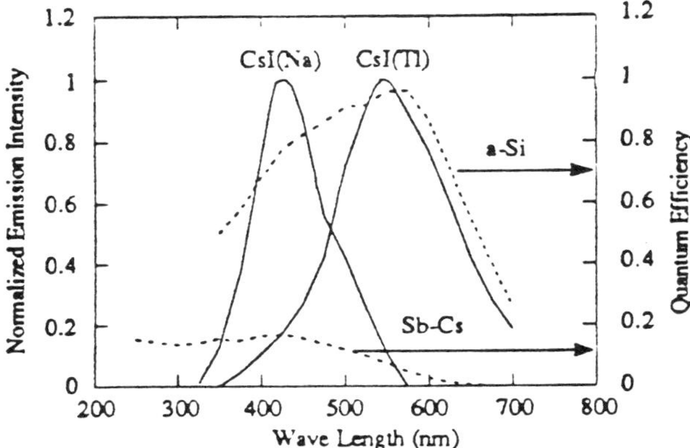

Fig. 3: Normalized emission spectra of CsI(Na) and CsI(Tl) and quantum efficiency of an a-Si:H photodiode and a photocathode.

The efficiency and position resolution of the CsI-a-Si:H combination were measured with a calibrated pulsed x-ray machine (5). The point response spread function which was measured using a facsimile machine linear readout a-Si:H head and TFT electronics with elements 70x70 μm (7) show the collimating effect of the CsI(Tl) structure, as shown in Fig. 4. Incidentally, this fax readout head has a-Si:H photodiode sensors and a-Si:H readout TFT electronics.

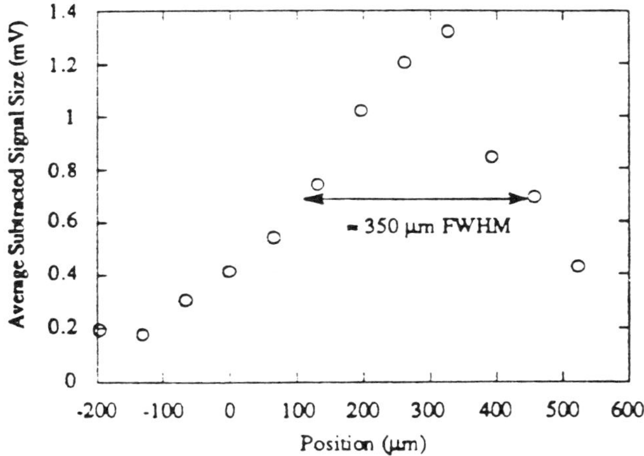

Fig. 4: Line spread functions for the CsI(Tl) obtained with X-ray.

The response to electrons from a Bi-207 beta source is shown in Fig. 5. The charge collection > 10,000 e, h pairs is well above noise. An important factor in scintillator detector devices is their potential damage due to radiation. In the case of CsI(Tl) and others it has been shown (8) that the light loss in large crystals is due mostly to light transmission attenuation. The relatively high resistance to radiation damage expected from this is confirmed - as shown in Fig. 6, by measurements on our thin layers compared to 1 cm crystals.

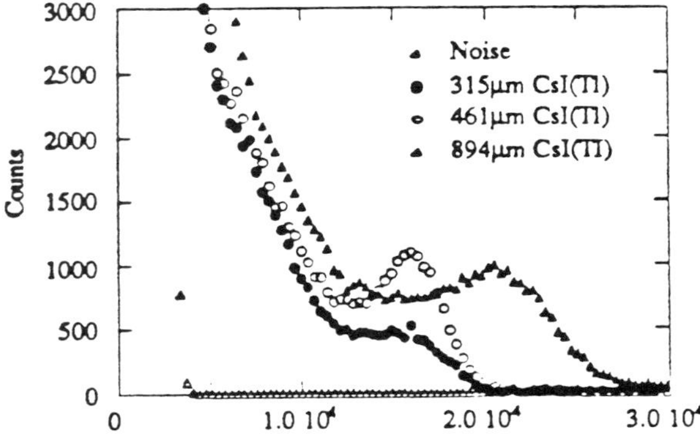

Fig. 5: Detection of β particles from a Bi-207 source by an evaporated CsI(Tl) layer coupled to a photodiode.

A matrix readout for diode arrays is shown schematically in Fig. 7. 64 x 40 arrays have been made (9). 128 x 128 and larger arrays are under construction. We propose to use the SVX low noise linear chip (10) to read out the lines of the matrix sequentially.

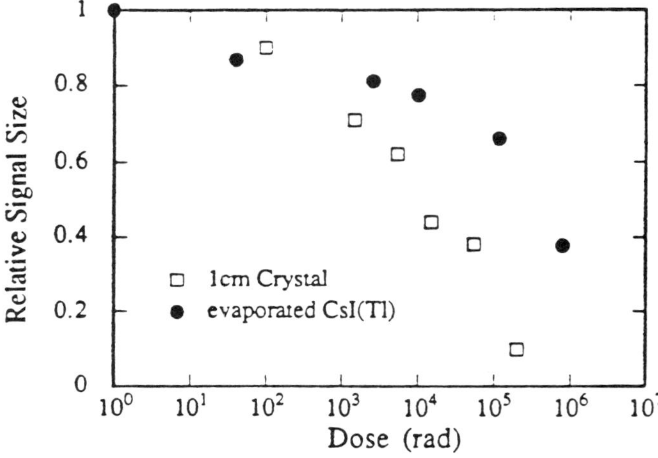

Fig. 6: Light yield decrease due to radiation damage for a thin CsI(Tl) layer.

Summary and Conclusions

We have described the properties of thin a-Si:H photodiode layers coupled to CsI(Tl) evaporated layers 300-1000 μm thick. Pixel characteristics are obtained by making the metal (chromium) contacts discrete and of the required size. In our present application a single a-Si:H TFT is connected to each pixel to form a matrix array which is read out onto a linear C-Si chip (SVX). The readout speed can be increased by replacing the a-Si:H switching TFT by polycrystalline TFT arrays and using a faster version of the SVX.

Faster scintillator layers than our $C_sI(T\ell)$ are available. For example, ultra pure C_sI with some very short (20 nsec) time constants could be used. Because the emission spectrum of this material peaks in the blue and extends to shorter wave lengths the ITO and p layers would have to be thinner or have a different structure such as p-polysilicon [11]. Other scintillators such as GdSO could also be considered.

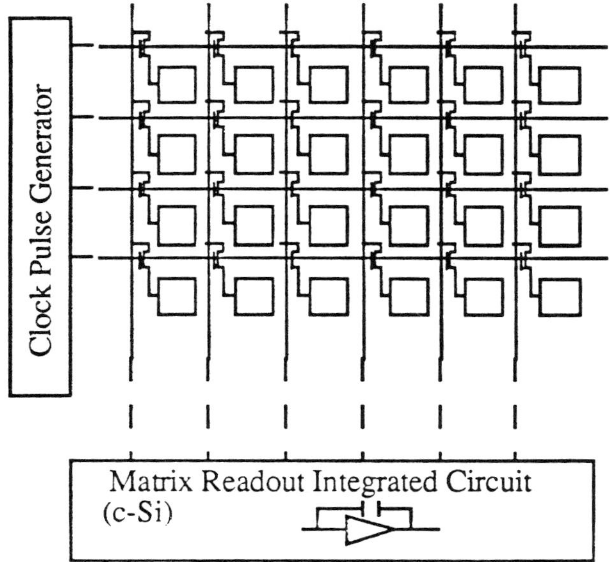

Fig. 7: Readout electronics for the 2D array.

We would like to thank .Mr. C. Borlinghaus (Harshaw, Solon, Ohio) for supplying us with raw $C_sI(T\ell)$ crystal chips for the evaporated layers and Dr. Gene Weckler from EG&G (Sunnyvale, CA) for the use of the fax head readout.

This work was supported by the Director, Office of Energy Research, Office of High Energy and Nuclear Physics, Division of High Energy Physics, of the US Department of Energy under Contract No. DE-AC03-76F00098.

References

[1] V. Perez-Mendez, G. Cho, J. Drewery, I. Fujieda, S. Kaplan, S. Qureshi and R. A. Street. LBL 28339 (Feb. 1990) To be published as a chapter in "Physics and Applications of Amorphous and Microcrystalline Semiconductor Devices", J. Kanecki Ed. , Artech House, Boston.

[2] R. E. Hollingsworth, J. Xi, R. Buitrago [Glasstech Solar], Presented at MRS meeting, San Francisco, April 1990. To be published in MRS Proceeedings Vol 192 (Oct 1990).

[3] V. Perez-Mendez, G. Cho, I. Fujieda, S. Kaplan, S. Qureshi, R. Street, MRS Proceedings, 149, (1989), pp. 621-630.

[4] G. Cho, J. Drewery, J. Fujieda, S. Kaplan, V. Perez-Mendez, S. Qureshi, R. Street, IEEE Trans. Nuc. Sci. NS 37 (1990), pp.1142-1148.

[5] I. Fujieda, G. Cho, J. Drewery, S. Kaplan, V. Perez-Mendez, D. Wildermuth, R. Street, LBL 29177 (Oct. 1990). To be published IEEE Trans. Nuc. Sci. 37 (Feb 1991).

[6] A.L.N. Stevels, A. D. Schrama de Pauw, Philips Res. Reports 29 (1974), pp. 340-352.

[7] Seiko-Epson fax head with readout electronics made by EG&G (Sunnyvale, CA).

[8] M. Kobayashi, S. Sakuragi, Nuc. Inst. & Methods 254 (1987) pp. 275-280.

[9] R. Street, S. Nelson, L. Antonuk, V. Perez-Mendez, presented at MRS meeting San Francisco, April 1990. To be published in MRS Proceedings Vol 192 (Oct 1990).

[10] S. A. Kleinfelder, W. Carithers, R. Ely, C. Haber, F. Kirsten, H. Spieler, IEEE Trans. Nuc. Sci, NS 35 (1988) pp. 171-175.

[11] R. L. Welsfield, C. C. Tsai. To be published in MRS Proceedings Vol. 192 (Oct 1990).

NEUTRON RADIATION DAMAGE STUDIES ON SILICON DETECTORS*

Z. Li, W. Chen, and H. W. Kraner

Brookhaven National Laboratory, Upton, NY 11973

Abstract

Effects of neutron radiation on electrical properties of Si detectors have been studied. At high neutron fluence ($\phi_n \geq 10^{12}$ n/cm^2), C-V characteristics of detectors with high resistivities ($\rho \geq 1$ kΩ-cm) become frequency dependent. A two-trap level model describing this frequency dependent effect is proposed. Room temperature anneal of neutron damaged (at LN_2 temperature) detectors shows three anneal stages, while only two anneal stages were observed in elevated temperature anneal.

Introduction

It is well known that, upon neutron exposure, silicon junction detectors experience changes in their electrical properties. These changes include detector leakage current degradation [1-4] and carrier removal or dopant compensation [3]. For an annual fluence of fast neutrons of about 10^{12} n/cm^2, (the SSC value) the leakage current density in a typical 300 μm thick detector is expected to go up more than 2 μA/cm^2 [1]. At this fluence, capacitance-voltage (C-V) characteristics of a detector using high resistivity Si (typical resistivity: $\rho \simeq 2-6$ kΩ-cm) may also deviate from the normal $V^{-1/2}$ dependence. This change in C-V characteristics, which will be described in this work, results from changes of detector material parameters such as effective dopant density, levels of defects and their concentrations, which are in turn closely related to the detector electrical properties. In this work, a systematic study of neutron irradiated Si detectors has been conducted, which includes the increase of detector leakage current as a function of neutron fluence, dependence of damage constant on detector resistivity, frequency dependent C-V effect as a function of detector resistivity and neutron dose, room temperature (RT) and elevated temperature anneals of irradiated detectors, and deep level transient spectroscopy (DLTS). A model describing both the frequency dependent C-V effect and dopant compensation is also proposed.

Experimental Method

The Si p^+-n junction detectors used in this study were made on n-type <111> wafers, with resistivities ranging from 10 Ω-cm to 30 kΩ-cm. The neutron source used for low fluence (<10^{11} n/cm^2) is a 16Ci PuBe that is certified to yield $3.68\times10^7(\pm3\%)$ n/sec of ^{9}Be(a,n) neutrons which have an average energy of about 4.5 MeV. At 5 cm this source produces a flux of 1.06×10^5 n/cm^2-sec. For high fluence (up to 10^{13} n/cm^2), fast neutrons from 10 keV to 2.2 MeV were obtained from the ^{7}Li(p,n) reaction using 4 MeV protons from a van de Graaff accelerator, with the flux of about 4.5×10^8 n/cm^2-sec. Various groups of detectors were exposed to neutron radiation up to the fluence of 8×10^{12} n/cm^2. Detectors used for RT anneal study were irradiated at liquid nitrogen (LN_2) temperature and were kept at this temperature immediately before the measurements. A rapid thermal process (RTP) was used for elevated temperature anneal. Anneal ambients were N_2 and forming gas (4% H_2/N_2), and anneal times are 20 sec and 1 min. All I-V and C-V measurements were done at RT.

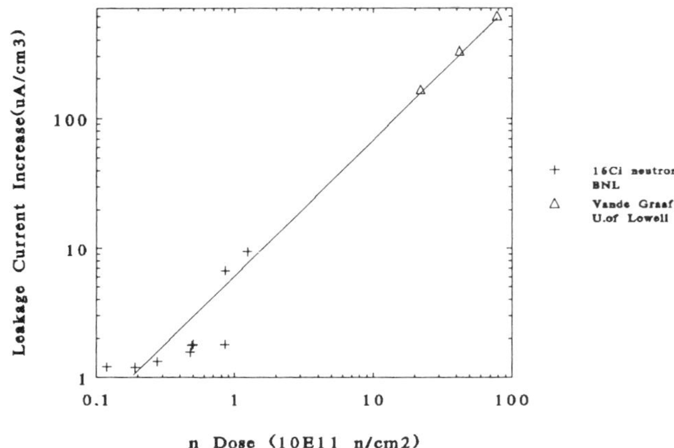

Fig. 1. Increase of detector volume leakage current with fast neutron fluence up to 8×10^{12} n/cm^2. A linear fit is also shown.

Results and Discussion

Leakage Current

Data presented in Fig. 1 is the average volume leakage current increase (averaged over detectors with resistivities from 10 Ω-cm to 4 kΩ-cm) as a function of neutron fluence ϕ_n. As it is shown in the figure, detector leakage current goes up nearly linearly with neutron fluence over the range of 10^{10} n/cm^2 to 10^{13} n/cm^2. Figure 2 shows the dependence of damage constant K (defined as $1/\tau_h = 1/\tau_{hi} + \phi_n/K$, where τ_h is the minority carrier life time) on Si resistivity with a two-Shockley-Read-Hall-level fit [1]. Little resistivity dependence, particularly when $\rho \leq 5$ kΩ-cm, has been found.

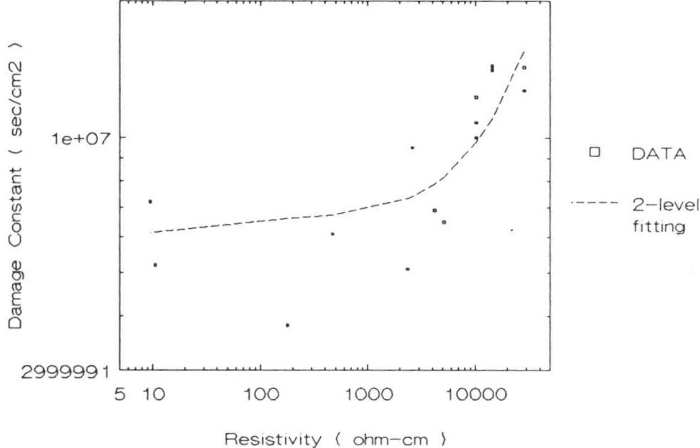

Fig. 2. Two-level fit of the data of damage constant as a function of detector resistivity.

Frequency-Dependent Capacitance

As it is seen in Fig. 3a, the detector C-V characteristics are frequency independent up to the neutron fluence of 7×10^{10}/cm^2, which is in good agreement with the data presented by Vismara [5]. Figure 3b shows the frequency-dependent C-V curves after the detector was exposed to a neutron fluence of 8×10^{12} n/cm^2. At this fluence, the detector high frequency (f > 50 kHz) C-V characteristics become essentially bias-independent, equaling the detector geometrical capacitance, $\epsilon_{Si}\epsilon_o/d_{Si}\cdot$Area. It is interesting to note that the full depletion voltage, V_d, over the entire wafer thickness, d_{Si} = 300 μm in this case, has been reduced from about $V_d \simeq 60$ volts to about 20 volts, indicating a nearly 60% dopant compensation.

Figure 4 shows the values of the detector capacitance at a given voltage as a function of measuring frequency, C(f), with neutron fluence ϕ_n being the parameter. It is clear that the degree of the dependence of capacitance on frequency is affected strongly by neutron fluence ϕ_n. Since the ratio of defect concentration and dopant concentration N_t/N_d is proportional to ϕ_n, one can conclude that this degree of frequency dependence is strongly dependent on N_t/N_d. This argument is also supported by the data shown in Fig. 5, where the C(f) measurements for silicon detectors with various resistivities have been plotted. For low resistivity silicon detectors, where $N_t/N_d \ll 1$, little or no frequency dependence has been observed.

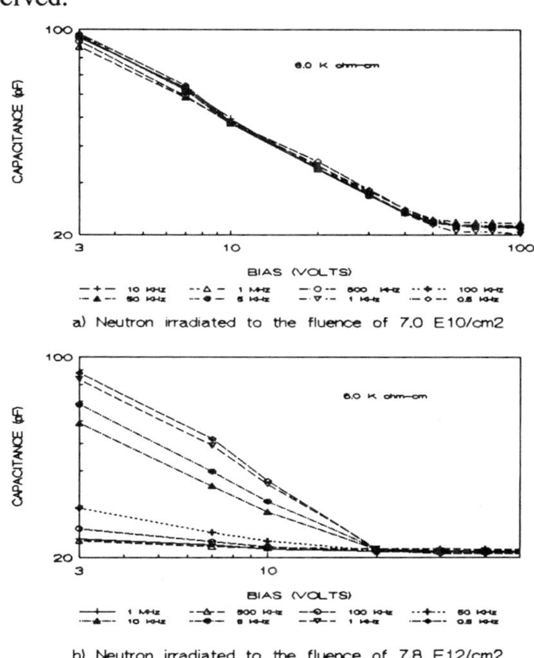

Fig. 3. Detector C-V characteristics at various frequencies of detectors irradiated to the neutron fluence of a) 7.0×10^{10}/cm^2 ; and b) 7.8×10^{12}/cm^2.

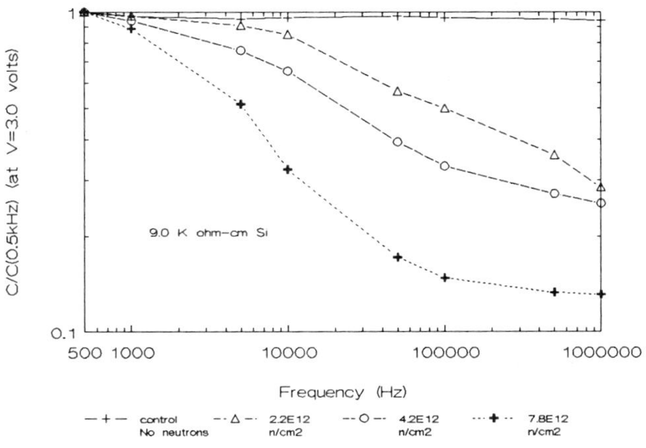

Fig. 4. Data of detector capacitance (at 3.0 V) vs frequency as a function of neutron fluence.

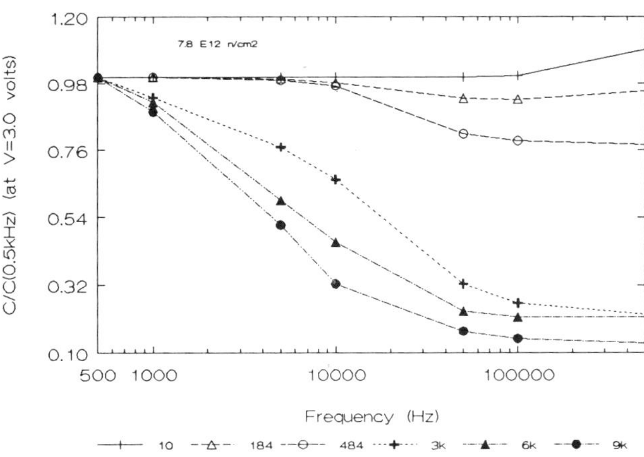

Fig. 5. Data of detector capacitance (at 3.0 V) vs frequency as a function of Si resistivity.

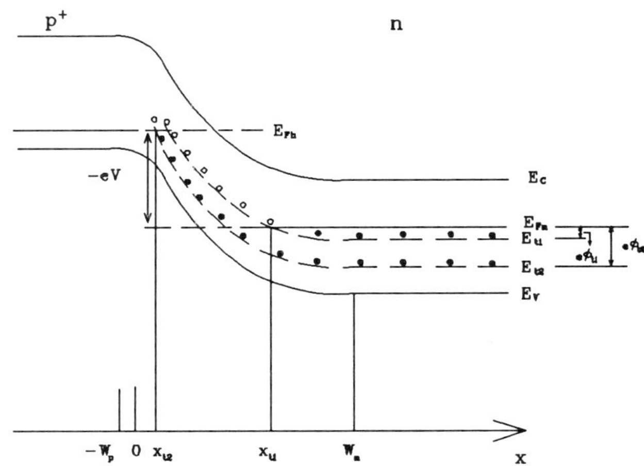

Fig. 6. Energy band diagram for the proposed two-level model describing the observed frequency-dependent C-V effects.

$C(\omega, V)$ Model

Frequency-dependent C-V effect was usually modeled in the literature by using one defect level that is responsible only for split of C-V curves with frequency, but not the dopant compensation. A new model describing both the effects of frequency-dependent C-V characteristics and dopant compensation is proposed in this study. As it is shown in Figure 6, two acceptor-like trap levels are used in this model. Trap level number 1, denoted as E_{t1}, N_{t1}, ϕ_{t1}, etc., is assumed to obey the Fermi-Dirac statistics with the Fermi level being the quasi-fermi level for electrons E_{Fn}, and accounts for the frequency dependence in the model. In the formulism, E is the level energy, N is the level concentration, and ϕ is the difference between E and fermi level $(E_F - E_t)$. The second trap, denoted as E_{t2}, N_{t2}, ϕ_{t2}, etc., is assumed to obey the Fermi-Dirac statistics with the Fermi level being the quasi-Fermi level for holes, and accounts for the dopant compensation [6]. This model is more extensive than the early one proposed by Sah, et al. [7] because it takes into account both the frequency dependence of trap density at $x = x_{t1}$ to the small ac signal and the effect of dopant compensation. Since all solutions are in their analytical forms, it gives physical insights more clearly than those derived from numerical models [9-10]. There are other models in the literature that simply do not explain the effect of voltage-independence of capacitance measured at high frequencies [10-14].

The junction capacitance as a function of angular frequency $\omega = 2\pi f$ and voltage V can be derived as (details of the model will be published elsewhere [15]):

$$C(\omega, V) = \epsilon_{Si}\epsilon_o \frac{N_{t1}(\omega) + N_{t2}(\omega)\dfrac{\Delta x_{t2}}{\Delta x_{t1}} + (N_d' - N_{t1})}{N_{t1}(\omega)x_{t1} + N_{t2}(\omega)x_{t2}\dfrac{\Delta x_{t2}}{\Delta x_{t1}} + (N_d' - N_{t1})W_n} \tag{1}$$

Where $N_d' = N_d - N_{t2}$ is the compensated dopant density, W_n is the depletion width (into the n-side of the junction), and $N_{t1}(\omega)$, $N_{t2}(\omega)$ are frequency-dependent defect densities related to equilibrium defect densities N_{t1}, N_{t2} and time constant τ_{t1} and τ_{t2} as follows:

$$N_{t1}(\omega) = \frac{N_{t1}}{1+\omega^2\tau_{t1}^2} \qquad N_{t2}(\omega) = \frac{N_{t2}}{1+\omega^2\tau_{t2}^2} \tag{2}$$

Quantities x_{t1}, x_{t2}, $\Delta x_{t1}/\Delta x_{t2}$ and W_n are obtained analytically by solving the Poisson equation:

$$\epsilon_{Si}\epsilon_0 \frac{d^2\phi}{dx^2} = \rho(x) = \begin{cases} 0 & x \le -W_p \\ -eN_A & -W_p < x \le 0 \\ eN_d & 0 < x \le x_{t2} \\ eN_d' & x_{t2} < x \le x_{t1} \\ e(N_d' - N_{t1}) & x_{t1} < x \le W_n \\ 0 & x > W_n \end{cases} \tag{3}$$

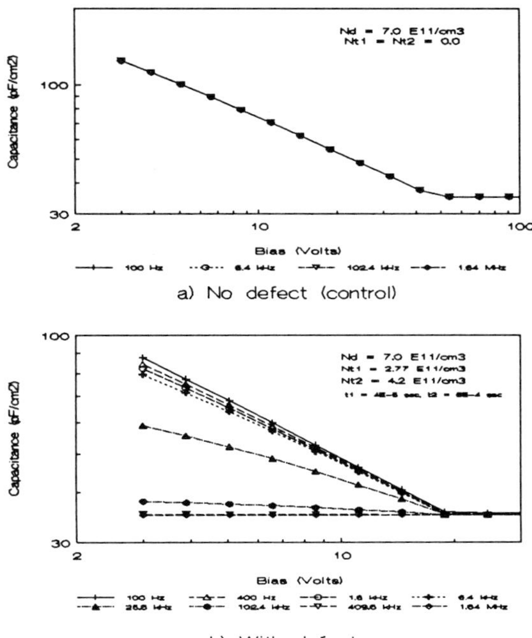

Fig. 7. Calculated frequency-dependent C-V characteristics with a) No defects (control); and b) two defect levels.

Figure 7 shows $C(\omega,V)$ curves calculated by the model using Eq. (1), Eq. (2) and Eq. (3). Values of ϕ_{t1} and ϕ_{t2} used in the calculations are 0.2 V and 0.4, respectively. Other parameters used in the calculations are listed in the figure. As it is shown in the figure, $C(\omega,V)$ curves predicted by the model are in good agreement with data shown in Fig. 3. Figure 8 shows the degree of the C-V frequency dependence as a function of N_{t1}/N_d', while it is demonstrated in Fig. 9 that the transition frequency f_t is approximately proportional inversely to the time constant of trap 1, i.e., $f_t \sim 1/\tau_{t1}$.

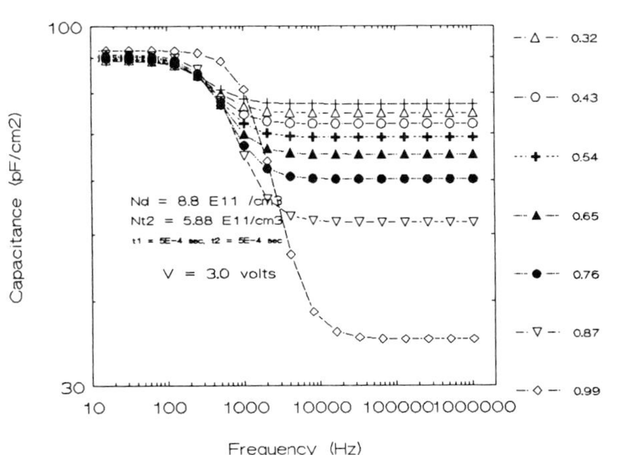

Fig. 8. Calculated C-f plots as a function of N_{t1}/N_d'.

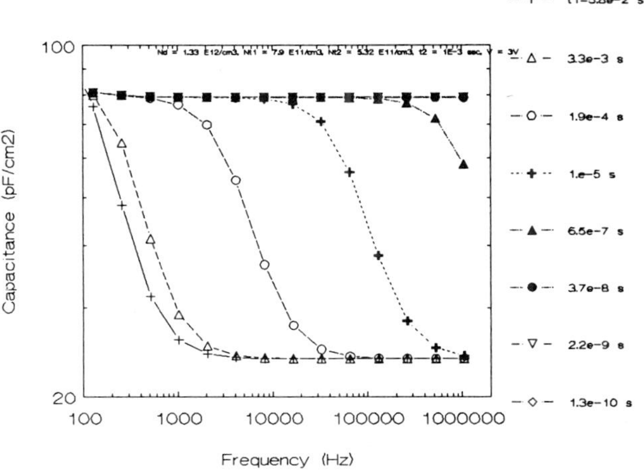

Fig. 9. Calculated C-f plots as a function of τ_{t1}. The transition frequency f_t is $1/\tau_{t1}$.

A good agreement between the experiment data and calculation is illustrated in Figure 10, where calculated C vs. f and data of a p^+n detector with 3 kΩ-cm resistivity have been plotted. The model predicts that $\tau_{t1} = 4\times10^{-5}$ sec in this case.

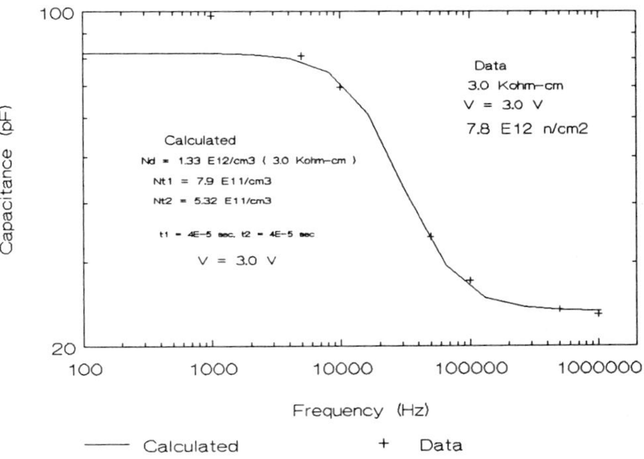

Fig. 10. Comparison between calculated and measured C-f data.

Figure 11 shows data of the deep level transient spectroscopy (DLTS) of a neutron irradiated Si detector. Two electron traps (E_c – 0.4 eV and E_c – 0.55 eV) and two-hole traps (E_v+0.3 eV and E_v + 0.54 eV) were found. The first electron trap, E_c – 0.4 eV, may be identified as a double vacancy center (V-V) and/or phosphorus-vacancy center (P-V). The deeper hole trap, E_v + 0.3 eV, may be responsible for dopant compensation as it is the second level proposed in the model (Fig. 6). The energy of another electron trap, E_0, was not determined. It may be a V-V center in a different charge state.

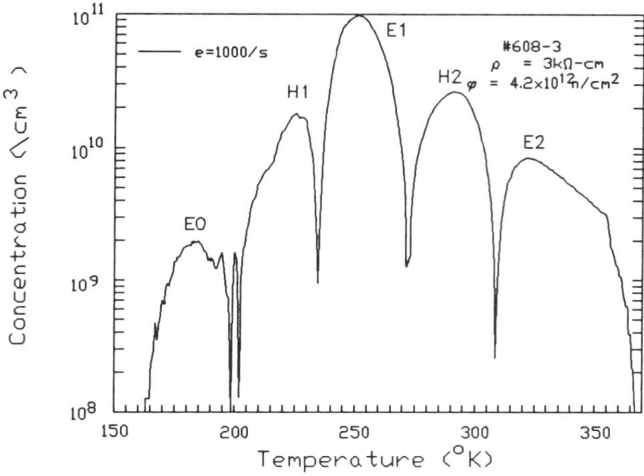

Fig. 11. A typical DLTS plot of a detector damaged with a neutron fluence of 4.0×10^{12} n/cm^2.

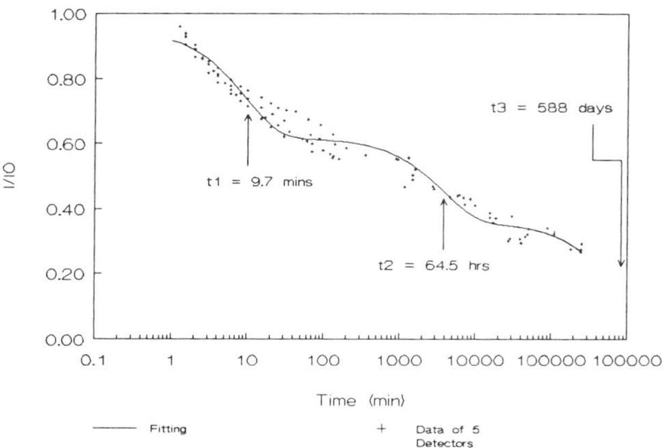

Fig. 12. Room temperature anneal data of detectors neutron damaged at LN_2 temperature. A three-term fit is also shown.

RT Anneal

Three time constants have been observed during RT anneal of neutron irradiated detectors. Figure 12 shows the RT anneal data of a group of five detectors irradiated at LN_2 temperature. A three-term fitting $(I/I_0 = \sum_{n=1}^{3} a_n e^{-t/\tau_n})$ curve is also plotted in the figure. It is found that adding or taking away a term in the fitting would only make the fitting much worse. It is clear that in the first anneal stage (the first 10 min) about 30% of the total damage (in terms of leakage current) is annealed out. In the next 2-3 days, another 25% of the total damage no longer exists. In one and one-half years, only 20% of the total damage will be left. The longest time constant observed here, $\tau=588$ days, agrees very well with that observed by Lindström, et al. [16], where a value of $\tau=563$ days was reported.

However, since our radiation was done in liquid nitrogen and detectors were kept there immediately before measurements, we are able to observe true short term annealing effects which cannot be observed if RT self-annealing is taking place during radiation. In the first anneal stage, the net current decrease of a detector damaged with neutron fluence of 8×10^{12} n/cm^2 has been found to be: $\Delta J = 600\ \mu A/cm^3 \times 30\% = 180\ \mu A/cm^3$. If this leakage current recovery is associated with the annealing of one single level of the traps, one can obtain the minority carrier generation lifetime (τ_h) of this trap level using,

$$\Delta J = \frac{en_i}{2\tau_h} \tag{4}$$

where $n_i = 1.0\times10^{10}/cm^3$ is the Si intrinsic carrier density at RT (293 K). The total trap concentration of this level, N_t, can then be obtained using

$$N_t = \frac{1}{\sigma v_T \tau_h} = \frac{1}{\sigma v_T \cdot (\frac{en_i}{2\Delta J})} \tag{5}$$

Where σ is the trap cross section and v_T the electron thermal velocity. Using a typical value of 5×10^{-15} cm^2 for σ and 9×10^7 cm/sec for v_T at RT, and Eq. (5), one obtains that in the first 10 min of RT anneal, as much as $N_t = 5\times10^{11}/cm^3$ of traps of a single level have been annealed out. This level may be related to vacancy-boron (V-B) complex since: a) There are enough borons in Wacker high resistivity n-type silicon (typical values: $N_p \simeq 8\times10^{12}/cm^3$, $N_B \simeq 2.5\times10^{12}/cm^3$, [17]), and b) V-B defect complex has been observed to anneal out at about 270 K in 15 min [18].

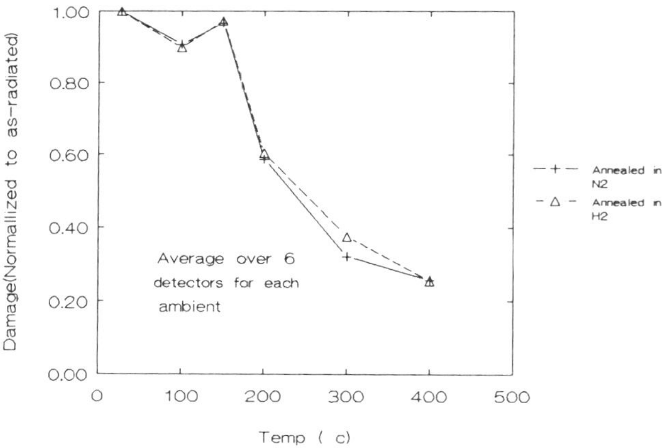

Fig. 13. Elevated temperature anneal data of neutron damaged detectors to the fluence of 1×10^{12} n/cm^2. Anneal ambients were N_2 and forming gas (4% H_2/N_2) and anneal time was one minute.

Elevated Temperature Anneal

Figure 13 and Fig. 14 are the data of the effect of elevated temperature anneal on detector leakage current. The anneals were performed in two ambients: N_2 and 4% H_2/N_2, and no ambient effect was observed. Two anneal stages around T = 150°C to 300°C were observed throughout the experiment. As is shown in Fig. 13, the first anneal stage is between 100°C to 200°C and the second stage is between 200°C to 300°C. Since these detectors were kept in RT atmosphere for about a day before being annealed at elevated temperatures, the anneal stages we have seen here are probably the annealing out of the level that have RT annealing time constants of about 3 days and 1.6 years, respectively. The level in the first anneal stage may be identified as V-P complex and the level in the second anneal stage may be identified as V-V complex, since V-P has been observed to anneal out at 150°C and V-V to anneal out at 300°C [19].

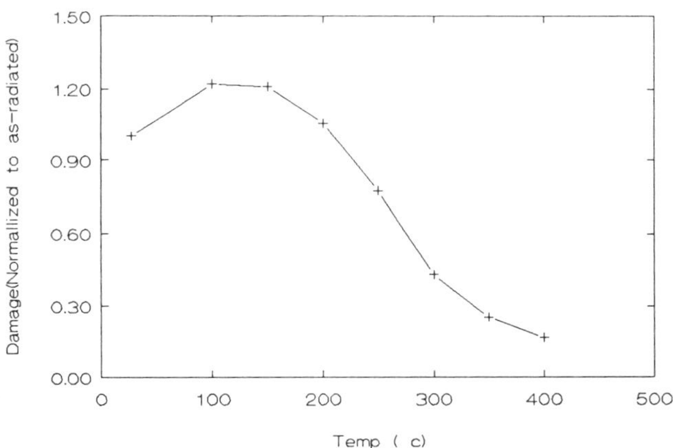

Fig. 14. Elevated temperature anneal data of neutron damaged detectors to the fluence of 4.0×10^{12} n/cm^2. The anneal conditions are one minute in forming gas ambient.

Conclusion

In the neutron fluence range of 10^{10} n/cm^2 to 10^{13} n/cm^2, silicon detector leakage current goes up nearly linearly with neutron fluence. A small dependence of damage constant on silicon resistivity has been observed. This dependence may be related to the V-P complex after radiation. Considerable information about trapping nature of radiation induced defects can be obtained from the frequency dependence of the $C(V)$ characteristics of junction diodes (silicon detectors), specifically, defect densities can be determined by the degree of this frequency dependence and the defect time constant (or electron emission rate $e_n = 1/\tau_t$) can be obtained from transition frequency $\tau_t = \tau_{t1} = 1/f_t$.

Three RT anneal stages have been observed. The first or the fastest (~10 min) anneal stage may be related to the annealing out of V-B complex. The annealing stages observed at elevated temperature is probably the annealing out of the levels that are to anneal out in about 3 days and 1.6 years at RT, respectively.

References

[1] H. W. Kraner, Z. Li, and K. V. Pösnecker, "Fast Neutron Damage in Silicon Detectors," Nucl. Instrum. Methods **A279**, 266-271, (1989).

[2] G. Lindström, E. Fretwurst, H. Herdan, M. Rollwagen, P. Thomsen, and R. Wunstorf, "Radiation Damage Effects in Silicon Detectors," DESY 89-105, August (1989).

[3] H. Dietl, T. Gooch, D. Kelsey, R. Klanner, A. Loffler, M. Pepe, and F. Wickens, "Radiation Damage in Silicon Strip Detectors," Nucl. Instrum. Methods **A253**, 460-466, (1987).

[4] H. W. Kraner, "Radiation Damage in Silicon Detectors," Nucl. Instrum. Methods **225**, 615-618, (1984).

[5] L. Vismara, "SICAPO Collaboration: Radiation Damage Effects on Silicon Detectors," paper presented at the Intl. Conf. on Advanced Technology and Particle Physics, Como, Italy, June 13-16, 1988.

[6] I. M. Lannoo and J. Bourgoin, *Point Defects in Semiconductors*. (Solid-State Sciences Series, **22**). New York: Springer Verlag, 1981, pp. 154-162.

[7] C. T. Sah and V. C. K. Redd, "Frequency Dependence of Reverse Biased Capacitance of Gold-Doped Silicon P^+N Step Junctions," IEEE Trans. Electron Devices **ED-16**, 345-349, (1969).

[8] C. W. Gwyn, "Calculated Small Signal Characteristics for Irradiated PN Junctions," IEEE Trans. Nucl. Sci., **NS-19**, 355-361, (1972).

[9] E.P. Eernisse, "Accurate Capacitance Calculations for PN Junction Containing Traps," Appl. Phys. Lett. **18 (5)**, 183-186, (1971).

[10] Y. Zohta, "Frequency Response of Gold Impurity Centers in the Depletion Layer of Reverse-Biased Silicon P^+-N Junctions," J. Appl. Phys. **43 (4)**, 1713-1716, (1972).

[11] Y. Zohta and Y. Ohmura, "Determination of the Sparial Distribution of Deep Centers from Capacitance Measurements of PN Junctions," Appl. Phys. Lett. **21 (3)**, 117-119, (1972).

[12] G. Vincent, D. Bois, and P. Pinard, "Conductance and Capacitance Studies in Gap Schottky," J. Appl. Phys. **46 (12)**, 5173-5178, (1975).

[13] Y. Tokuda and A. Usami, "Admittance Studies of Neutron-Irradiated Silicon P^+-N Diodes, J. Appl. Phys. **48 (4)**, 1668-1672, (1977).

[14] Y. Tokuda and A. Usami, "Studies of Defects in Neutron-Irradiated P-Type Silicon by Admittance Measurements of N^+-P Diodes," J. Appl. Phys. **49 (2)**, 603-607, (1978).

[15] Z. Li and H. W. Kraner, "Studies of Frequency Dependent C-V Characteristics of Neutron Irradiated P^+-n Silicon Detectors," presented at the 1990 IEEE Nuclear Science Symposium, Arlington, VA, October 23-27, 1990.

[16] E. Fretwurts, H. Herdan, G. Lindström, U. Pein, M. Rollwagen, H. Schatz, P. Thomsen and R. Wunstorf, "Silicon Detector Developments for Calorimetery: Technology and Radiation Damage," Nucl. Instrum. Methods **A288**, 1-12 (1990).

[17] Private communication with Wacker.

[18] G. D. Watkins, "The Lattice Vacancy in Silicon," Sokrates T. Pantelides, ed., *Deep Centers in Semiconductors*. New York: Gordon and Breach Science, 1986, pp. 167-183.

[19] L. C. Kimberling in *Radiation Effects in Semiconductors 1976*, Institute of Physics Conference Series, no. 31. Bristol, U.K.: The Institute of Physics, 1977.

STUDY OF RADIATION DAMAGE TO *P-I-N* PHOTODIODES USING 63 MeV PROTONS

D.E. Pellett and J. Bacigalupi
Physics Department
University of California, Davis
Davis, CA 95616

Abstract

Previous studies indicate that displacement damage in silicon has an approximate linear dependence on nonionizing energy loss for a variety of incident particle types and energies [1]. This may allow extrapolation of lower energy bulk damage measurements to the SSC environment. We report on irradiations of *p-i-n* photodiodes with 63 MeV and 27 MeV protons for fluences up to 3.1×10^{13} p/cm^2. Preliminary results on reverse current, room temperature annealing and changes in dopant concentration are compared with data on similar devices at other incident proton energies.

Introduction

Pixel vertex detectors are being considered for use in SSC experiments within 10 cm of the beamline [2]. Here, the dominant radiation damage is expected to be due to charged hadrons (chiefly pions) from primary beam-beam collisions [3]. The fluence is estimated to be 1.2×10^{13} per cm^2 at 10 cm from the beamline for one year at 10^{33} luminosity and behaves like $1/r_\perp{}^2$. The irradiation will engender bulk damage and long-term ionization in the detector and its readout circuitry [4]. Bulk damage effects such as increase in generation current, charge trapping and change in dopant concentration will limit the useful life of *p-i-n* diode detector arrays. Prototypes must be tested in fluences equivalent to $\approx 10^{14}$ high energy charged hadrons per cm^2.

Previous studies [1] show that in many cases, bulk damage in semiconductor devices is approximately linearly dependent on the non-ionizing energy loss (NIEL) in the material (essentially, the energy which goes into dislodging ions rather than atomic electrons). This suggests that bulk damage resulting from different incident particles and energies could be correlated on the basis of NIEL, allowing the use of low energy facilities to simulate the bulk damage to an SSC detector. Even if significant deviations exist, one might still be able to perform useful tests of prototype detectors if appropriate bulk damage factors are measured at the energies of interest.

Recent calculations of NIEL have been made by Burke [5] and Van Ginneken [6]. For charged pions or protons with kinetic energy in the range 1 MeV–10 GeV, NIEL decreases with energy, reaching an approximately constant plateau. The plateau value of NIEL in Si for pions (above 200 MeV kinetic energy) is calculated to be 1.3 KeV cm^2/g. For protons, NIEL is 2.8 times higher at 63 MeV and 1.5 times higher on the plateau. Accurate determination of NIEL becomes difficult at high energies due to the number of possible processes and the uncertainties in the cross sections. For example, the calculations of Burke and Van Ginneken agree at the energies of the present experiment but show a 27% disagreement for 800 MeV protons. Calculations agree for 10 GeV pions, however [7].

Experiment

Radiation damage studies were performed using Hamamatsu S1723-06 *p-i-n* photodiodes [8]. Similar devices have been used elsewhere for such applications [9,10]. These have an area of 1 cm^2, wafer thickness of 200 μm, and are mounted in a ceramic package with metal leads.

The conditions for irradiation of each diode are shown in Table 1.

Table 1. Diode Irradiation Conditions.

Diode	E (MeV)	Fluence (cm^{-2})	Bias (V)
D0	63	8.5×10^{11}	40
D1	63	3.1×10^{13}	30
D2	63	6.5×10^{12}	30
D3	63	1.3×10^{12}	30
D4	27	6.9×10^{11}	40
D5	63	7.3×10^{12}	0

The devices were irradiated individually in 30–60 min exposures using the proton radiation test beam at the UC Davis Crocker Nuclear Laboratory (CNL) cyclotron [11]. Fluence was measured with a secondary emission monitor calibrated with a Faraday cup. The beam profile has been determined using activation of aluminum blocks, thermoluminescent dosimetry and radiachromic film [12]. The beam spot is approximately uniform over a 40 cm^2 area although only the central 1 cm^2 was used. The accuracy of the fluence measurements for these tests was ±10%.

Preliminary Results

We present the results of a preliminary analysis of the irradiated diodes. These are compared with other proton data and NIEL calculations.

Leakage Current, Annealing

Bulk damage causes the diode reverse current, I_{leak}, to increase proportional to fluence, Φ, through production of generation centers throughout the depleted volume, $\mathcal{V}$, of the detector. The constant of proportionality is $\alpha = \Delta I_{leak}/\Phi\mathcal{V}$. Measurements of this quantity at 63 and 27 MeV were made using diodes D0, D3 and D4. At regular intervals (typically 5 min) during irradiation, the cyclotron beam was turned off and I_{leak} measured as a function of fluence. For D3 and D4, curves of I_{leak} *vs.* bias voltage, V were also periodically measured and recorded using a Hewlett Packard 4145B semiconductor parameter analyzer.

In the early part of the run, I_{leak} was found to be only slowly varying with V above 20 V (typically a 3% rise between 20 V and 40 V, 1.3% between 30 V and 40 V), as expected for full depletion. As the run progressed, however, I_{leak} began to deviate toward higher values at large V. For diode D3, a sharp rise eventually developed at the end of the run with a knee at $\approx$ 30 V and a current in excess by a factor of $\approx$ 2.2 at 40 V. This current excess decayed away after the run. After 15 hours, the deviation, while still visible, was less than 1% at 40 V. The leakage at 20 V seemed relatively unaffected, tracking well with fluence throughout the run. Similar effects were observed for the other two diodes, although detailed $I-V$ measurements were not made for D0 during the run. In order to eliminate this transient effect, only the the first several I_{leak} measurements in the runs were used to determine α.

As irradiation progressed, the devices became radioactive. The radioactivity decayed rapidly, however. No activity was measured for D0 (< 1 mR/hr near the surface) $\approx$ 16 hours after the exposure. No correction has been made for this effect in the α measurements.

The diode temperatures were monitored and the currents corrected to 20°C. The depletion depth was taken to be 170 ± 20 μm.

The individual α measurements for D0 and D3 agree within 3% when the D3 measurement is corrected to 40 V bias. We find $\alpha = 1.3 \pm .2 \times 10^{-16}$ A/cm at 63 MeV and $2.1 \pm .3 \times 10^{-16}$A/cm at 27 MeV. The latter agrees well with the measurement of Lindström *et al.* for 25 MeV protons, 2.3×10^{-16} A/cm [13]. The ratio, α(27 MeV)/α(63 MeV) is found to be 1.55 ± 0.22, in reasonable agreement with the NIEL ratio, 1.4.

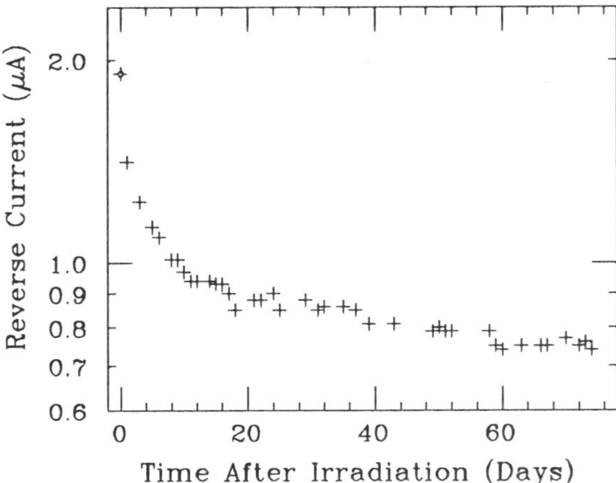

Figure 1: Time dependence of reverse current (normalized to 20° C) for diode D0 following irradiation. The point at t = 0 was extrapolated using α (see text).

Periodic measurements of diode leakage currents were made following irradiation to evaluate the effect of room temperature annealing. A semilogarithmic plot of leakage current *vs.* time for diode D0 is shown in Fig. 1. The point at t = 0 was extrapolated using our measured value of α and the total fluence.

As was observed by Lindström *et al.*, the annealing is not well represented by a single exponential. After 70 days, however, the leakage current is only slowly varying and has fallen by a factor of 2.5 from the initial extrapolated value (also in agreement with the 25 MeV data of Lindström *et al.*).

Ziock *et al.* [10] have presented α measurements for 800 MeV protons irradiating a number of the same type of Hamamatsu photodiode used in the present study. There were two runs, one with fluence $\approx 10^{12}$ cm^{-2} in 128 hours ("low flux") and the other with $\approx 10^{14}$ cm^{-2} in 174 hours ("high flux"). The leakage was measured at 40 V bias during and immediately after irradiation as well as at two later times: 29 and 58 days for low flux and 21 and 50 days for high flux. Average values of α were presented for the two runs separately and were found to agree within the errors. Because of the complicated time dependence of the leakage current which we observe at short times, it is most convenient to make a comparison at a time in the neighborhood of their longer measurement interval, where the current is slowly varying. We compare with our data on diode D0 after 58 days and extrapolate their results to the corresponding time using their exponential parametrization of the time dependence. The resulting ratios of 63 MeV to

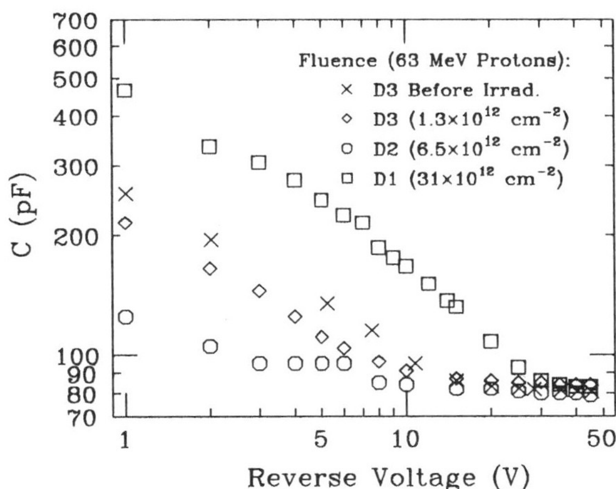

Figure 2: Diode C *vs.* V as a function of fluence.

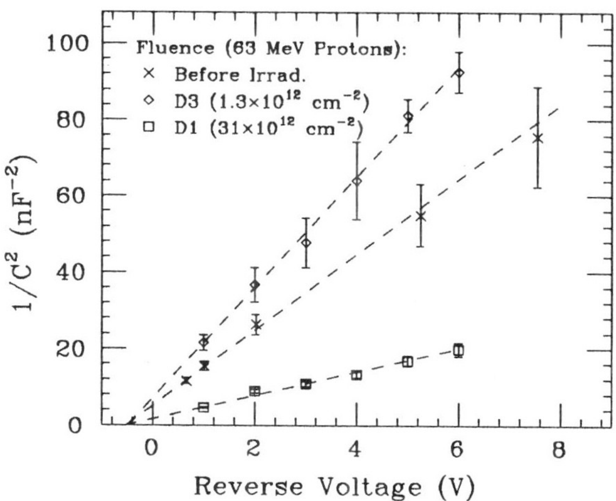

Figure 3: Plot of $1/C^2$ *vs.* V as a function of fluence.

800 MeV leakage currents are $2.0 \pm .3$ and $1.8 \pm .2$ using their high and low flux data, respectively. As mentioned above, NIEL calculations show moderate inconsistancies in this energy regime. Burke calculates the corresponding NIEL ratio to be 2.1 whereas Van Ginneken finds 1.6. The experiment favors an NIEL value between the two.

Changes in Dopant Concentration

After irradiation, the low frequency (1 kHz) diode capacitance, C, was measured as a function of bias voltage, V. The results for D1, D2 and D3 as well as for D3 before irradiation are shown in Fig. 2. At sufficiently high bias, all reach a plateau indicating full depletion. At a given low bias voltage, D3 and D2 show progressively lower capacitance with increased fluence, whereas D1, subjected to the greatest fluence, shows a marked increase. D2 is close to full depletion at 3 V.

For a diode far from full depletion and with an abrupt junction,

$$1/C^2 = 2(V_{bi} + V)/eK_S\epsilon_0 N, \qquad (1)$$

where e is the magnitude of the electron charge, K_S is the dielectric constant of silicon, V_{bi} is the junction built-in voltage and N is the dopant concentration in the lightly-doped region [14].

In Fig. 3, $1/C^2$ is plotted against V for D3 (before and after irradiation) and D1. The data are well-fitted by straight lines with reasonable V_{bi}. We use (1) to extract N as a function of fluence. Since D2 depletes at low voltage, N must be small, but is difficult to estimate accurately in this manner. As shown in Fig. 4, these results are consistent with a linear change of dopant concentration with fluence, starting with n-type material of 4 kΩ cm resistivity, as expected for the unirradiated diode. Diode D1 appears to undergo "type reversal" (change from n-type to p-type). Defining $\beta = \Delta N/\Phi$, we find $\beta = 0.16 \pm .02$ cm^{-1}. Note that this preliminary result depends heavily on the D1 measurement.

Our present data at 27 MeV are insufficient to yield a significant β measurement.

Lindström *et al.* measure $\beta = 0.53$ cm^{-1} for 25 MeV protons. This a factor 3.3 higher than our measurement at 63 MeV and substantially larger than one would expect on the basis of the NIEL ratio, 1.4. At higher energies, Dietl *et al.* find $\beta = 0.05 \pm .015$ cm^{-1} for $\approx$ 200 GeV hadrons [15] whereas Nakamura *et al.* report $\beta = 0.019 \pm 0.003$ cm^{-1} and evidence of type reversal for 800 GeV protons [16]. These comparisons are among different types of devices under different conditions, so other factors may be playing an important role. Nonetheless, the decrease of β with energy appears to be significantly more rapid than that of NIEL.

Effects of Dose Rate, Bias During Irradiation

No significant long-term effects were observed due to dose rate or bias voltage during irradiation in these tests. In particular, we compared the leakage current approximately forty days after irradiation for the five diodes irradiated with 63 MeV protons. The proton flux varied by almost two orders of magnitude in the various exposures. The bias voltage for D5 was 0 V during irradiation, whereas the others were biased to 30–40 V. When normalized to the same fluence, the maximum deviation of the leakage current of any diode from the average was 8%, well within experimental errors. The C–V curve for diode D5 was not significantly different from that of D2.

Conclusions

We have investigated the correlation of p-i-n diode reverse current due to bulk damage with non-ionizing energy loss (NIEL). Proportionality was found between

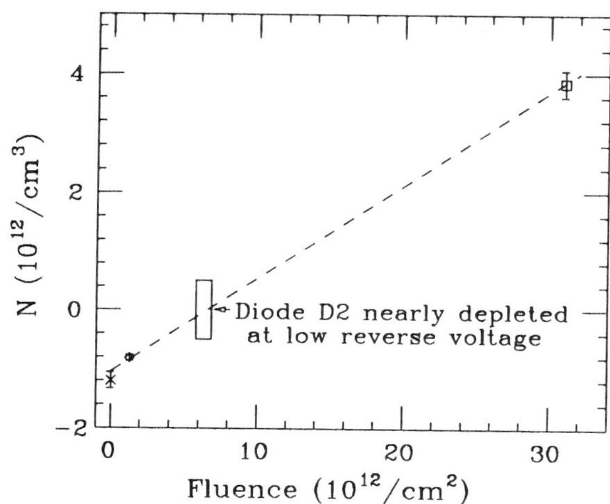

Figure 4: Dopant concentration (N) *vs.* fluence (preliminary). Positive N denotes p-type material. The open rectangle indicates the estimated range of possible values for diode D2.

the two quantities within the uncertainties in the experiments and calculations for 27, 63 and 800 MeV protons (the latter using results of Ziock *et al.*). A preliminary measurement of the rate of change of dopant concentration with fluence, β, was made at 63 MeV. This quantity appears to decrease more rapidly than NIEL with increasing proton energy. Evidence was found for semiconductor type reversal at high fluence. Since the observed effects appear to be linear with fluence and scalable in energy (although with different dependences for α and β), we conclude that short, moderate-energy proton exposures are useful for radiation tests of prototype silicon detectors subject to high fluences of charged particles at the SSC.

Acknowledgements

We thank Dr. Carlos Casteneda of CNL for performing the dosimetry for the experiment. It is a pleasure to acknowledge useful comments and suggestions from E. Burke, S. Holland, H. Kraner, D. Nygren, T. Ohsugi, S. Parker, H. Sadrozinsky, H. Spieler and A. Van Ginneken. We are grateful for the help of W. Greig, P. Kiefer and the CNL cyclotron crew in performing the experiment.

References

[1] G. P. Summers *et al.*, "Correlation of Particle-Induced Displacement Damage in Silicon," IEEE Trans. Nucl. Sci., **NS-34**, 6, 1134–1139 (1987).

[2] See, for example, G.H. Trilling *et al.*, *Solenoidal Detector Collaboration Expression of Interest*, May 1990, p. 23.

[3] D. E. Groom, "Radiation Levels in SSC Detectors," in *Proceedings of the Summer Study on High Energy Physics in the 1990s*, Snowmass, Colorado, 1988, pp. 711–716.

[4] See, for example, Victor A. J. Van Lint, "The Physics of Radiation Damage in Particle Detectors," Nucl. Instr. and Meth. **A253**, 453-459 (1987).

[5] Edward A. Burke, "Energy Dependence of Proton-Induced Displacement Damage in Silicon," IEEE Trans. Nucl. Sci. **NS-33**, 6, 1276-1281 (1986).

[6] A. Van Ginneken, "Non Ionizing Energy Deposition in Silicon for Radiation Damage Studies," FNAL Report FN 522 (October, 1989).

[7] Edward A. Burke, private communication.

[8] K. Yamamoto *et al.*, "Highly Stable Silicon *p-i-n* Photodiode," Nucl. Instr. and Meth. **A253**, 542-547 (1987).

[9] M. Hasegawa *et al.*, "Radiation Damage of Silicon Junction Detectors by Neutron Irradiation," Nucl. Instr. and Meth. **A277**, 395-400 (1989).

[10] H. J. Ziock *et al.*, "Measurement of Proton Induced Radiation Damage to CMOS Transistors and PIN Diodes," IEEE Trans. Nucl. Sci. **NS-37**, 3, 1238-1241 (1990).

[11] The proton radiation test beam at CNL was set up and instrumented through a joint effort of CNL and the U. S. Naval Research Laboratory.

[12] Ken Murray, KEM Science, private communication.

[13] G. Lindström *et al.*, "Radiation Damage Effects in Silicon Detectors," DESY Report 89-105 (August, 1989).

[14] S. M. Sze, *Physics of Semiconductor Devices*. New York: Wiley-Interscience, 1969, ch. 3, p. 90.

[15] H. Dietl *et al.*, "Radiation Damage in Silicon Strip Detectors," Nucl. Inst. and Meth. **A253**, 460-466 (1987).

[16] M. Nakamura *et al.*, "Radiation Damage Test of Silicon Multistrip Detectors," Nucl. Inst. and Meth. **A270**, 42-55 (1988).

OPERATION OF SILICON MICROSTRIP DETECTORS IN A HIGH RADIATION ENVIRONMENT

J.S. Kapustinsky, D.M. Alde, J.G. Boissevain, R.G. Jeppesen, D.W. Lane, M.J. Leitch, J.W. Lillberg, T.A. Lopez, P.L. McGaughey, J.M. Moss, and J.C. Peng
Los Alamos National Laboratory
P.O. Box 1663
Los Alamos, New Mexico 87545

B.M. Brooks, L.D. Isenhower, and M.E. Sadler
Abilene Christian University
Abilene, Texas

L.M. Lederman, and M.H. Schub
University of Chicago
Chicago, Illinois

C.N. Brown, W.E. Cooper, K. Gounder, Y.B. Hsiung, and C.S. Mishra
Fermi National Accelerator Laboratory
Batavia, Illinois

G. Gidal, M.S. Kowitt, and K.B. Luk
Lawrence Berkeley Laboratory
Berkeley, California

D.M. Kaplan, V.M. Martin, R.S. Preston, and J. Sa
Northern Illinois Umiversity
DeKalb, Illinois

R.L. Childers, and C.W. Darden
University of South Carolina
Columbia, South Carolina

G. C. Kiang, and P. K. Teng
Academia Sinica
Taiwan, Republic of China

Abstract

A Silicon Microstrip Spectrometer was recently installed and operated in an 800 GeV proton beamline at Fermilab as a major new component of experiment E789. The detectors received an estimated radiation exposure of up to 7.8 x 10^{12} minimum ionizing particles per cm^2 over a period of two months. We report on the changes in detector performance that we have observed following preliminary data analysis.

Introduction

Silicon microstrip detectors are a major new component of Fermilab fixed target experiment E789, designed to study the rare decays of the B and D mesons in a high radiation environment. During a recent test run, ten detectors were installed in the 800 GeV proton beamline for E789. The detectors are positioned five above, and five below the beam axis. The innermost strips of the up and down arm of the Spectrometer are located at an angle of 20 mrad from the target relative to the beam axis. The detector plane that is farthest upstream, YS1, is 40 cm downstream of the target, and 8 mm above the beam. Strip rates were monitored at several locations on the YS1 detector. The target related beam rate at YS1 varied by a factor of three across the first 400 strips. Beam monitoring was precise for the data runs, but not for additional beam exposure during setup and calibration. The total integrated beam intensity is known to about 15%. We estimate the total local integrated flux from the measured rates at YS1 to be 7.8 x 10^{12} minimum ionizing particles per cm^2, (195 KRad), at the innermost strips.

Detectors

The E789 silicon detectors were manufactured by Micron Semiconductor. They are single-sided N-type detectors with ion implanted P-type strips. The silicon wafers are 300 microns thick, and 5 cm x 5 cm in area. The strips are oriented horizontally. The strip pitch is 50 microns. The strips are fanned out on a printed circuit board and dc coupled to custom designed fast preamplifiers[1]. The amplified signals are discriminated and latched. No pulse height information is recorded.

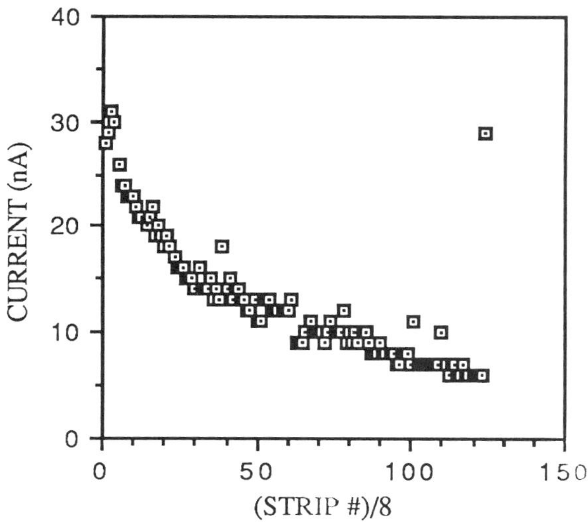

Figure 1. Strip current across YS1T.

Leakage Current

Prior to installation, the leakage current was measured for each detector at it's full depeletion voltage. The total current for a single detector before beam exposure was typically 1 μA. Individual strip currents were mostly uniform across the detector. The typical voltage required to fully deplete the detectors was 35V.

Individual strip currents for the YS1 detector are shown in figure 1. The data were recorded 18 days after the last beam exposure. In a previous radiation damage study performed at the Los Alamos Meson Physics Facility[2], (LAMPF), in a 0.8 GeV proton beam, we measured a damage constant, (α), for bulk generation current of 1.8×10^{-17} A/cm. Using this value for α, a strip current of 30 nA, and the relationship, $I_{leak} = \alpha F$, the local fluence, (F), we derive for the innermost strips of YS1T is $5.6 \times 10^{12}/cm^2$. The discrepancy between this estimate and the estimate based on scaler rates might be explained by short term annealing occuring during beam down periods.

The reverse bias current vs bias voltage characteristic, (I/V), from four of the detectors, as measured 11 days after the run, is shown in figure 2. The total current has increased by more than a factor of 20 on each of the detectors. With the exception of the detector that is labeled YS4B, the detectors exhibit good diode characteristics. The detector YS4B is a prototype that required much rework to the printed circuit fanout board to repair shorts and open circuits. The strip current profile that we measure does not match the beam intensity across this detector. We see clusters of high leakage current that are distributed across the wafer. These hot spots exhibit a linear dependence on bias voltage and dominate the I/V characteristic. Either the wafer was initially defective or surface damage was incurred during the repair.

Figure 3 shows the same I/V characteristics measured 25 days after the run. There is clear evidence that the detectors are annealing. The average decrease in leakage current for all 10 detectors is 32% in 25 days.

The I/V characteristics also suggest that the voltage at which the detectors are fully depleted is lower than the initial 35V. A decrease in depletion voltage is consistent with acceptor site creation in the N-type bulk of the silicon, as has been described by Dietl et al.[3]. A recent study by Ziock[4] indicates a damage rate, (β), for acceptor site creation of β =.066/cm. The E789 detectors have an initial dopant concentration of approximately 5×10^{11} donors per cm^3. If we assume that the average fluence across an entire detector was $4.9 \times 10^{12}/cm^2$, and use the value for β of .066/cm, then more than 50% of the initial donor sites have been effectively replaced by acceptors in that detector. Depletion voltage is proportional to the donor concentration[5], and can

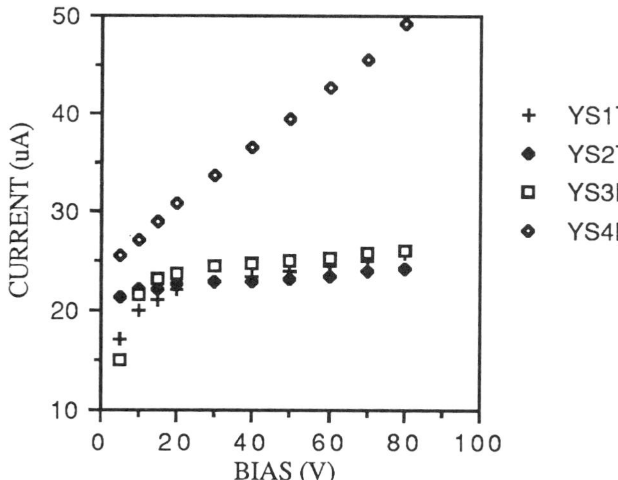

Figure 2. I/V charateristic, 11 days after irradiation.

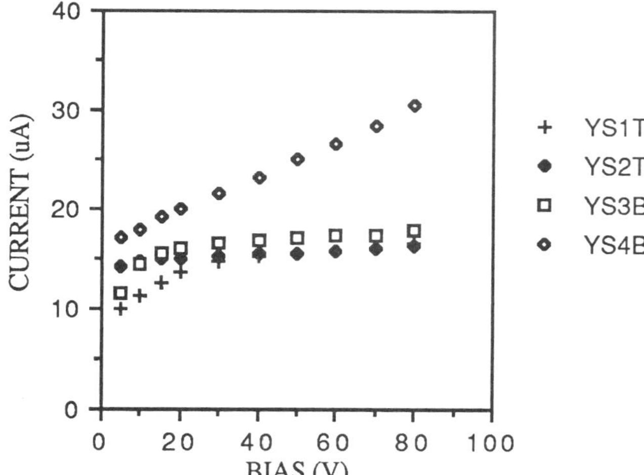

Figure 3. I/V characteristic, 25 days after irradiation.

be approximated by :

$$V_d = (q \cdot N_d \cdot d^2) / 2\varepsilon, \quad (1)$$

where V_d is the depletion voltage, q is the unit electron charge, N_d is the donor concentration, d is the wafer thickness, and ε is the dielectric constant for silicon. The damage to the E789 detectors, however, is not uniform across the detector. The areas of the detector that have received the most damage should be depleting at lower voltages than other less damaged regions. Figure 4 is the I/V characteristic for three strips from detector YS1B. Strip 16 is closest to the beam, and therefore has received the most damage of the three. The current plateau for strip 16 occurs at about 10V, compared to about 20V for the other two, indicating that it is depleted at a lower voltage.

During the run, with beam on, detector currents increased well above 100 μA. This very high value of current would decrease rapidly after turning the detectors off for several hours. In the LAMPF study we observed similar detector behavior, and attribute this component of the total current to surface current.

Detector Performance

Preliminary results indicate that the Silicon Spectrometer performance is good. The silicon enables us to construct a target profile from a 200 micron wire target with a sigma of 100 microns. The sigma of the residual distribution for the location of tracks in the silicon is 14 microns.

Approximately 10% of all tracks in the silicon fire adjacent strips. There is no evidence thus far that the level of charge sharing has changed with detector damage.

The detector bias voltages were all initially set at 40V. We measured the detector efficiencies on-line to have an average value of 90%. After about 40 days of beam time detector efficiciencies fell by about 10%. In order to restore good efficiency it was necessary to raise the bias voltage. Since the voltage required to deplete the detectors had decreased, we conclude that the increased electric field in the detector reduced the effects of charge trapping. For the remainder of the run, we operated the detectors at 80V. A sharp reduction in strip pulse height was observed during the previous LAMPF study, and occured between 5.5×10^{12} and 1.4×10^{13} protons/cm^2. This loss in amplitude was also recovered by raisng the bias voltage.

Conclusion

The damage effects of ionizing radiation on silicon microstrip detectors include increased leakage current, and the creation of acceptor sites in the bulk of the silicon. Acceptor sites initially lower the voltage required to fully deplete the detector. In addition, acceptor sites cause a loss of charge collection efficiency due to trapping. Increasing the electric field strength within the detector can restore efficiency.

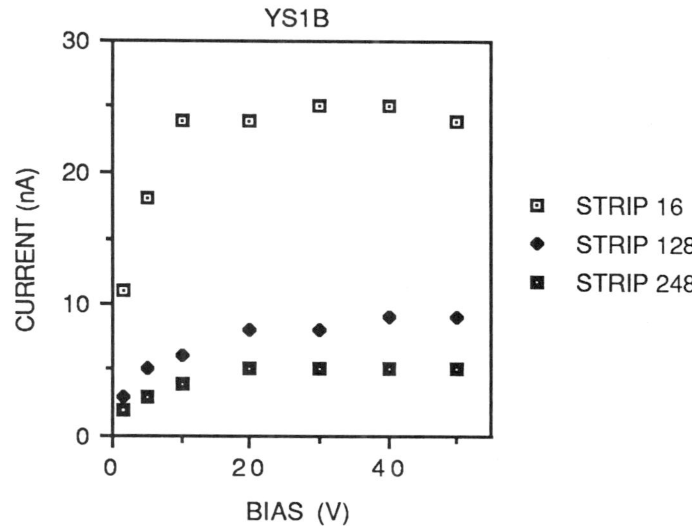

Figure 4. The I/V characteristic for three strips from detector YS1B. Strip 16 is closest to the beam.

References

[1] D. Christian et al., " The Development of Two ASIC's For A Fast Silicon Strip Detector Readout System," IEEE Trans. on Nucl. Sci., NS-36, 507-11, (1989).

[2] C.S. Mishra et al., 15th APS Division of Particles and Fields General Meeting (DPF-90), Fermilab Conf. 90/107, Houston, Texas, January 3-6, 1990.

[3] H. Dietl et al., " Radiation Damage in Silicon Detectors," Nucl. Instr. and Meth., A253, 460-66, (1987).

[4] H. Ziock, private communication.

[5] E. Gatti, " Processing the Signals From Solid-State Detectors in Elementary-Particle Physics," Revista Nuovo Cimento, Vol. 9, No. 1, 1-146, (1986).

RADIATION DAMAGE MEASUREMENTS ON DOUBLE-SIDED Si STRIP DETECTORS

Mikko Laakso, Petteri Nieminen, Tom Schulman, Peter Weilhammer+, Szymon Gadomski++

Research Institute for High Energy Physics
University of Helsinki
Tekniikantie 17,SF-02150 Espoo, FINLAND

+CERN
Geneva, SWITZERLAND

++Institute of Nuclear Physics
Cracow, POLAND

Abstract

First studies on the radiation effects on double sided AC-coupled Si strip detectors have been done with doses up to 10 krad. Even at such low doses radiation induced oxide charge creates an electron accumulation layer at the Si-SiO2 interface, that connects the strips to each other through the guard ring of the detector. Qualitative simulations indicate that this can be avoided using a modified gate electrode design at the end of the strip.

Introduction

Ionizing radiation effects in DC coupled Si detectors have been thoroughly studied, and the changes in detector parameters such as leakage current increase are fairly well known[1-3]. First radiation damage studies on AC coupled detectors have shown that an electric field across the oxide influences strongly the leakage current behaviour as a function of dose[4]. In double-sided detectors this oxide charge induces also an increase in the electron accumulation layer density at the n-side Si-SiO_2 interface, which results in a decrease in the interstrip resistance between the n^+ -strips and eventual signal loss and resolution degradation.

In this work double-sided strip detectors were irradiated with γ - and β -radiation to find out how radiation affects the n^+ -strip separation behavior.

Experimental setup

The detectors used in this work were manufactured by SI in Oslo, Norway. A view of the layout of the n-side of the detector is shown in figure 1.

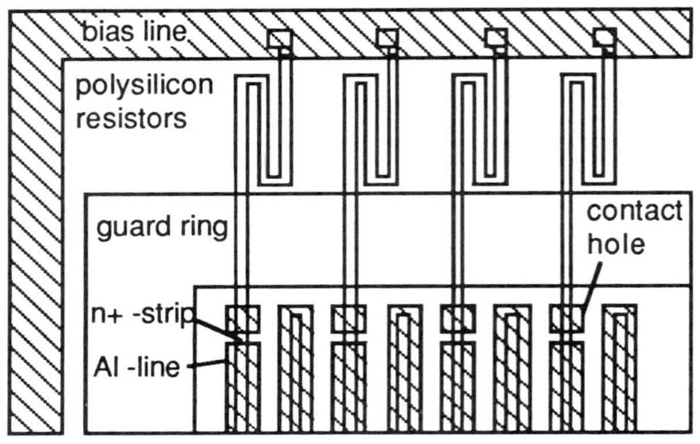

Figure 1. The detector n-side layout

The strips are biased via a common bias line through polysilicon resistors. The strip pitch is 50 μm. The aluminium readout lines are extended on both sides of the n^+ -strips to create a field induced depletion region between the strips. The tested detectors are divided into three areas with different widths of the Al-lines: 15, 25 and 35 μm. A negative gate voltage of a few volts on the Al-lines is enough to separate the strips from each other[5]. At the end of the Al-line is a contact hole to provide an ohmic contact to the n^+ -strip. The active detector area is surrounded by a guard ring.

The leakage current was measured as a function of bias voltage and as a function of time (at constant bias voltage) at the p-side separately for the active area and the guard ring. The depletion layer C-V -curve was measured at high frequency (1 MHz). The polysilicon resistance R_p was measured from the p-side resistors.

The connection of the strips via the common bias line through the resistors R_p results in a resistor network at the n-side of the detector and thus the measurement of the interstrip resistance R_i between the n^+ -strips is not straightforward. The interstrip resistance R_i was measured using three different methods.

Method 1

In this method the resistance value R_m between one strip and the bias line was measured. With a very large interstrip resistance value this resistance should be the polysilicon resistance value Rp, and its value should decrease with decreasing Ri as the other polysilicon resistors are coupled parallel to the measured resistor. The value of the interstrip resistance R_i can then be solved from equation (1), which is the solution of a multi resistor network.

$$R_i = \frac{4R_pR_m^2}{R_p^2 - R_m^2} \qquad (1)$$

Equation (1) is accurate enough for networks with at least 20 strips. A principal source of error in this method is the rapid increase of the derivative of equation (1) when R_m approaches R_p: a small error in the measured R_m results in a large error in R_i.

Method 2

In the second method the bias line connecting the strips to each other through polysilicon resistors is cut in order to isolate a number of strips from the network. This makes it possible to measure the resistance between two strips directly. Because of the layout of the detector the cutting of the bias line, however, leaves at least one strip unbiased, which may affect the measurement.

Method 3

The interstrip resistance can also be measured directly by applying a voltage between a contact hole and the bias line and shorting the next contact hole to the bias line via a current meter. This method, however, is strongly affected by the leakage current, which also flows through the current meter and the resolution decreases rapidly when R_i reaches high values.

The γ - radiation effects were studied using a ^{60}Co source with a dose rate 2.2 krad/h. During the irradiation the detectors were biased at 60 V and the negative gate voltage on the n-side Al-lines was 10 V. The Al-lines and the diodes at the p-side were grounded.

For the electron irradiation a ^{90}Sr source with a maximum dose rate of about 4.8 krad/h was used.

Results

After 10 krad γ -radiation an increase of 0.15 nA/cm^{3}rad of the active area leakage current was observed. No effect on the polysilicon resistors and the depletion layer C-V characteristics was observed.

A change in the n^+ -strip separation behavior is observed after 1 krad γ -radiation. In figure 2 the R_m value of method 1 is plotted as a function of the negative gate voltage on narrow Al-lines. R_p of this strip was 555 kΩ.

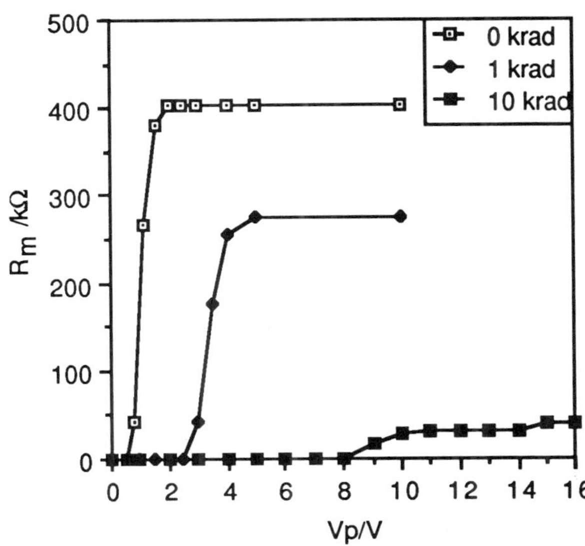

Figure 2. Resistance between one strip and bias line (R_m) as a function of negative gate voltage.

At zero dose R_m reaches a value of about 400 kΩ, which would give 2.4 MΩ for R_i (according to equation 1). After 1 and 10 krad R_i reduces to 900 kΩ and 18 kΩ, respectively. It turns out, however, that the reason for the decreasing R_m value is not the decreasing R_i but an oxide charge induced electron accumulation channel under the polyresistor, which connects the strips through the guard ring. Because the bias line and the guard ring are connected for biasing purposes, the accumulation channel resistances R_g between individual strips and the guard ring can be considered resistors parallel to the polysilicon resistors. Thus, if we assume that R_i remains high, which can be verified using the method 3, the value that R_m approaches is not R_p, but R_p and R_g parallel. Because R_p remains constant we can see that even at zero dose R_g has a finite value of about 1.5 MΩ and after 10 krad R_g has decreased to about 50 kΩ. An increase in the gate voltage needed to separate the strips can also be seen from figure 2. This change was smaller for the wider Al-lines.

A similar behavior of the strip separation was observed when detectors were exposed to electrons from the ^{90}Sr source and the interstrip resistance was measured using method 2. These measurements also revealed a large variation between the channel resistances R_g of different strips even before irradiation. The R_g values were also time dependent. We also made a crude observation that the channel resistance could be increased by applying a negative voltage with a probe needle on the region between the end of the strip (the contact hole) and the guard ring. This also indicated that the accumulation channel is formed under the polysilicon line and not at the other end of the strip.

Computer simulations

For a better understanding of the behavior of the detector structures under different conditions a two-dimensional simulation program was used. Figure 3 shows a model of the cross-section of the area between the end of the n^+ -strips and the guard ring. The dotted lines show artificial contacts to the n^+ -strips. These contacts are needed for proper biasing of the device. The real biasing conditions can not be simulated with a two dimensional model.

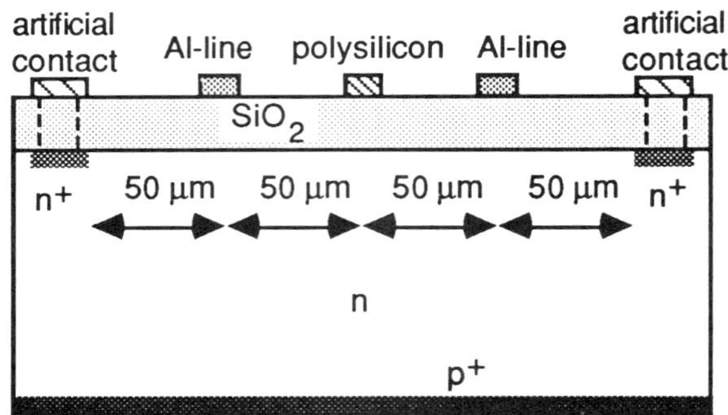

Figure 3. Simulation geometry no1. Wafer thickness 300 µm, oxide thickness 1 µm.

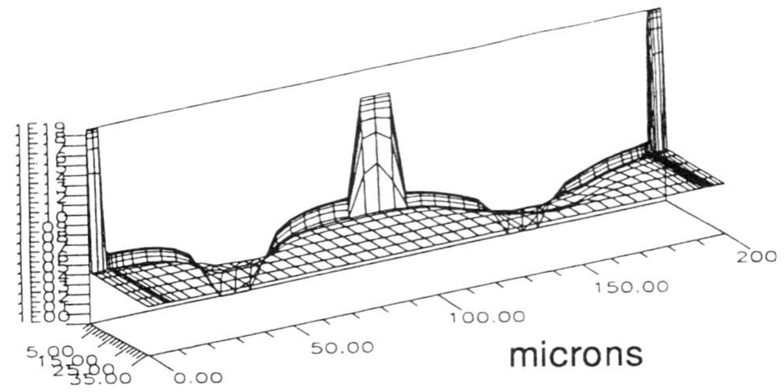

Figure 4. Electron density at the Si-SiO_2 interface. Simulation geometry 1.

A situation with a gate voltage of -50 V on the Al-lines and zero voltage on the polysilicon line was simulated to see if the gate voltage is able to cut the accumulation channel under the polysilicon line. Figure 4 shows the electron density distribution at the Si-SiO_2 interface for a silicon oxide charge of 10^{10} $1/cm^2$. It is clearly seen that the accumulation layer is removed under the Al-lines but that the electron density is still high under the polysilicon line. The gate voltage is unable to cut the accumulation channel, which connects the strip to the guard ring. The observed high channel resistances for some strips may be explained by the different biasing of a real detector structure.

In figure 5 is a modelled view of a cross section along an n^+ -strip.

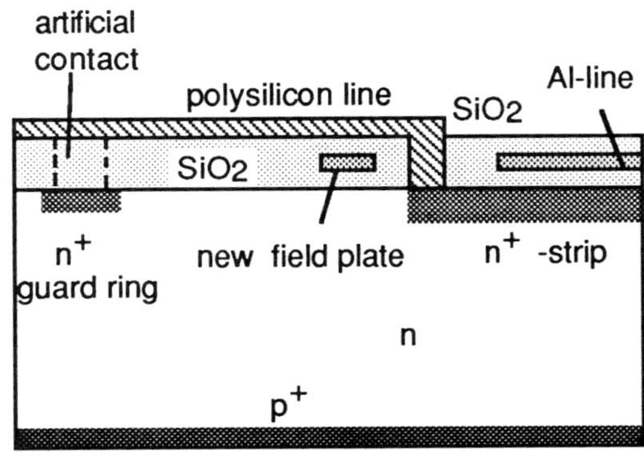

Figure 5. Simulation geometry 2.
Wafer thickness 300 μm, oxide thickness 1 μm.

In this new design an additional aluminium field plate is placed under the polysilicon line in order to cut the accumulation channel. This design could be achieved by extending the Al-lines to surround the contact holes. Figure 6 shows the simulated electron density distribution for this cross section when the oxide charge is 10^{11} $1/cm^2$ and the field plate voltage is -10 V. The electron accumulation channel under the polysilicon line is clearly cut.

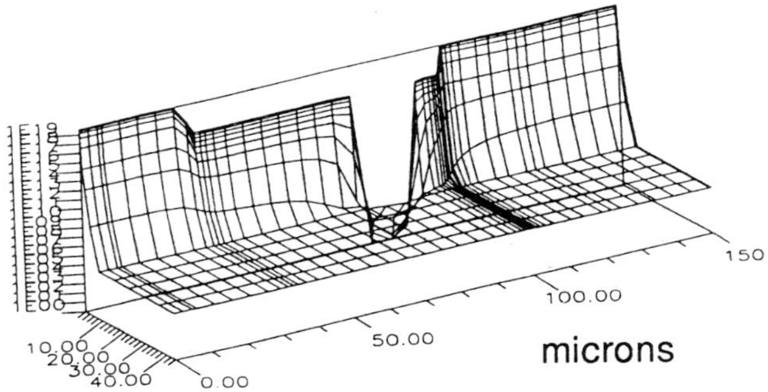

Figure 6. Electron density at the Si-SiO_2 interface.
Simulation geometry 2.

Conclusions

After a radiation dose of 10 krad oxide charge induced electron accumulation channels, which connect the strips to the guard ring, are observed on the n-side of an AC coupled double sided strip detector. Using the AC coupled readout strips as separation gates the strips can be separated from each others, although an increase in the required gate voltage is observed with increasing dose. Wider Al-lines needed less gate voltage to separate the n^+ -strips. Qualitative simulation results indicate that the observed connection to the guard ring can be interrupted with an appropriate gate electrode design at the end of the strip.

We expect that in general biasing structures based on accumulation layer resistance might show wide variation of resistance for geometrically identical structures and as a function of time. This is likely due to varying surface charges on top of the oxide. We also anticipate that such structures change their characteristics already after small irradiation doses.

References

[1] H. W. Kraner, "Radiation Damage in Silicon Detectors", Nucl. Instr. Meth. **225, 615-618, 1984.**

[2] T. Ohsugi et. al., "Radiation Damage in Silicon Microstrip Detectors", Nucl. Instr. Meth. **A265**,105-111, 1988.

[3] M. Campanella et. al., "The Effect of Radiation on Ion-implanted Silicon Detectors", Nucl. Instr. Meth. **A243**, 93-97, 1986.

[4] H. Dijkstra et. al. ,"Developments and Prospects for Capacitively Coupled Silicon Detectors", IEEE Trans. Nucl. Sci. Vol.**36**, No.1, 591, 1989.

[5] H. Dijkstra et. al.," A New Microstrip Detector with Double-sided Readout", IEEE Trans. Nucl. Sci., Vol. **37**, No. 3, 11-53, 1990.

RADIATION-HARD FRONT-END ELECTRONICS AND SILICON MICROSTRIP DETECTORS

H.F.-W. Sadrozinski, N. Cartiglia, J. Dewitt, D.E. Dorfan, B. Hubbard,
K. O'Shaughnessy, D. Pitzl, W.A. Rowe, A. Seiden, E. Spencer
Santa Cruz Institute for Particle Physics
University of California, Santa Cruz, CA 95064

P. Ferguson, C.M. Hoffman, W.W. Kinnison, E.C. Milner, W.F. Sommer, H.J. Ziock
Los Alamos National Laboratory, P.O. Box 1663, Los Alamos, NM 87545

P. Giubellino
University of Turin, Corso M. D'Azeglio 46, I-10125 Turin, Italy

J. Ellison
University of California, Riverside, CA 92521

Abstract

As part of generic and subsystem R&D for the SSC, we have investigated the radiation resistance of Silicon microstrip detectors and associated electronics. We present data on the radiation hardness of custom-designed dielectric isolated bipolar transistors, of logic gates from a recent radiation-hard CMOS test run and of elements of ac-coupled Silicon strip detectors.

Introduction

A proposed high precision tracking device for the SSC consists of about 11 million channels of silicon strip detectors arranged within a radial distance of less than 50 cm around the SSC beam line [1]. One of the most important requirements of this system is radiation hardness to both ionizing dose and neutrons. Using the assumptions of Ref. 2, we predict a maximum exposure of a few MRad of minimum ionizing particles and an equivalent neutron dose of close to 10^{14} neutrons/cm^2 due to neutrons and minimum ionizing particles over the lifetime of the detector [3]. In a systematic study we have identified radiation-hard electronic technologies, built prototype front-end electronics, and tested the degradation of key transistor parameters in exposures corresponding to over 10 years of SSC running. In a parallel effort we have exposed silicon diodes, microstrip detectors and test structures to ionizing radiation and neutrons to understand the radiation resistance of all parts of ac-coupled double-sided microstrip detectors.

Radiation Testing of Front-end Electronics

We have reported on our progress in identifying radiation-hard technologies before [4-6], and we will give here only an update. A large effort in Japan addresses the radiation hardness of bipolar front-ends [7]. Because of the high-collision frequency of the SSC, we designed a "digital" read-out system, recording for every beam pulse only the addresses of strips with pulse height above a certain threshold. Consideration of signal-to-noise and power consumption led to a system of mixed technologies: the amplifier chip (AACC) is being built in bipolar technology for low noise, while the digital buffer chip (DTSC) is built in CMOS to minimize the power consumption. The present system is slower than that required for the SSC but will allow meaningful extrapolation. We now have gone through the first test runs with the two chips and report on the radiation hardness of the components.

Analog Amplifier Comparator Chip (AACC)

The design of the AACC has been described in detail in Ref. 8. In brief, it contains a low-power, low-noise amplifier with a peaking time of 20 nsec on pulses from silicon strip detectors [9] and a comparator firing at about 30% of the anticipated mean pulse height from minimum ionizing particles. The SPICE simulations give a signal-to-noise ratio of 20. The chip has been layed out for 64 channels in 85 μm pitch in dielectric isolated bipolar (D.I.) technology in a full custom process developed by Silicon Dynamics, Inc. [10], and is in fabrication. In order to examine the radiation hardness of the D.I. process we have placed transistors on backside wafers produced by AT&T and exposed them to the 1 MeV neutron beam at the Los Alamos National Laboratory. We were interested in a possible degradation of the current gain β with large neutron fluences. Fig. 1 shows the result of the test: after fluences of about $5 \cdot 10^{13}$ n/cm^2, β had changed by less than 10%, even for very small current densities. This

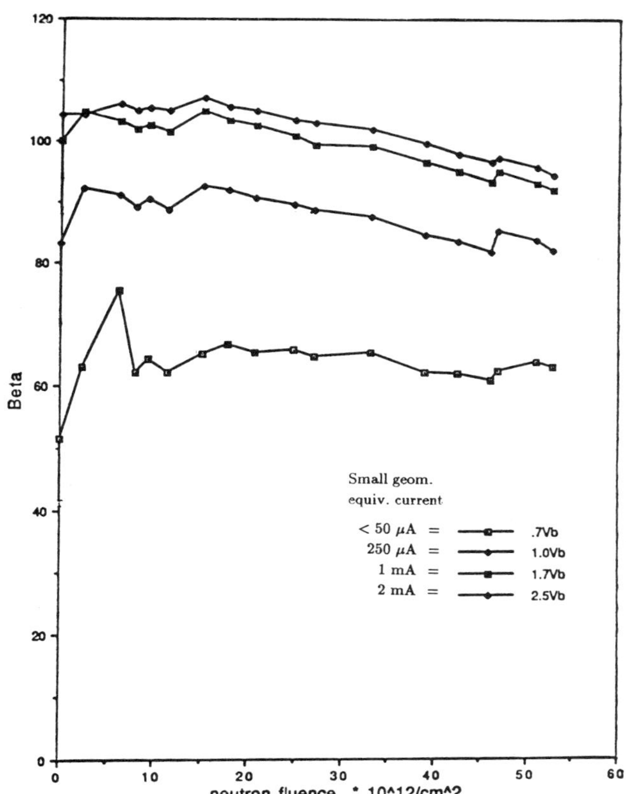

Fig. 1 Variation of the current gain β of the dielectric isolated bipolar transistors with the neutron fluence.

indicates that the D.I. bipolar process exhibits sufficient radiation hardness for experimentation at the SSC.

Digital Time Slice Chip (DTSC)

The design of the DTSC has been described in detail in Ref. 11. It is a two-level data buffer with first- and second-level buffers, designed for 10 MHz operation. The architectural goal is data pipelining until trigger decisions are made, automatic data reduction, and ease of external interface. For low-power consumption, the DTSC is designed as SRAM's and implemented in CMOS. A test version has been manufactured in radiation-hard 1.2 μm CMOS by UTMC [12]. The test die (Fig. 2) has 3 parts and they are undergoing different tests: the 64 channel DTSC and the test frame (with RAM array, mini buffer and sense amp test circuits) are undergoing functional tests while the test structures are being irradiated to test the radiation hardness of the design. Fig. 3a shows the threshold voltage shift ΔV_{th} of p and n transistors and Fig. 3b the change in transition voltage ΔV_G of the NAND gate and INVERTER after doses of up to 3.2 MRad of Co^{60} exposures. The transistors show the expected small shifts in V_{th} [5] while the logic gates exhibit a much smaller change because the effects of n and p transistors compensate each other. As observed before [5] the damage saturates at larger doses. Another test device for radiation damage is a ring oscillator: we have measured the frequency of a 51 stage ring oscillator and found a frequency of 34.0 MHz before and 32.6 MHz after irradiation with 3.2 MRad Co^{60}. Radiation tests of larger circuits and of the DTSC are underway.

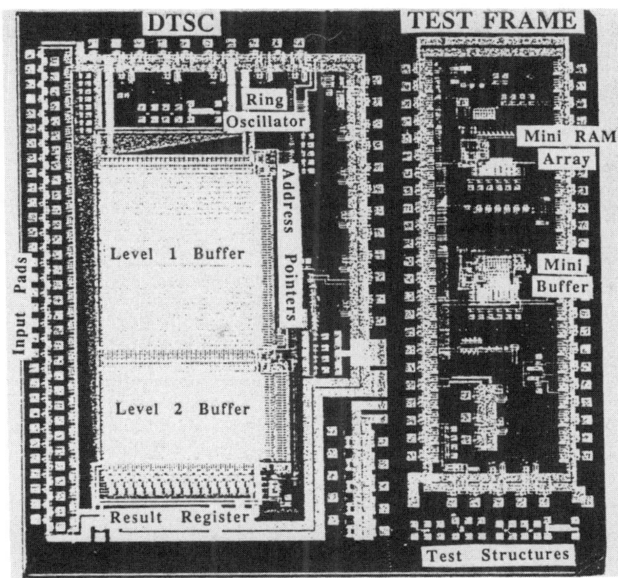

Fig. 2 Layout of the DTSC die produced by UTMC in 1.2 μm radiation-hard CMOS.

Radiation Tests of Silicon Diodes and Strip Detectors

Existing ac-coupled double-sided detectors have been developed for experimentation at LEP [13] where the radiation exposure is factors of 100 smaller than anticipated at the SSC. The aim of our program was to develop detectors which can function even after exposure to the SSC beams for 10 years. A parallel effort is underway by groups in Japan [14] in collaboration with Hamamatsu Photonics. Our main purpose was to test elements of the detectors built in different technologies to identify the most radiation-resistant designs. This effort is being pursued in close collaboration with groups in the United Kingdom and Italy [15]. Possible anticipated radiation effects are the increase in leakage current, type inversion, and failures of capacitors and resistors. Additionally, surface effects due to accumulated oxide charges in the oxide-silicon interface could influence the interstrip isolation and capacitance. Below a few of our results are given. More details can be found in Refs. 5,15, and 16.

- We have measured the leakage current constant for 800 MeV protons and found it in good agreement with the one measured for 12 GeV protons [17], independent of bias and flux.
- We have measured the change in the values of biasing resistors: polysilicon resistors increase their value by about 15% up to 200 kRad of Co^{60} dose, but are constant afterwards up to an accumulated dose of 2.8 MRad (see Fig. 4). Irradiation with 1 MeV neutrons up to a fluence of $3 \cdot 10^{13}$ neutrons changed the value of the polysilicon resistors by less than 2%. Diffusion resistors showed no change in ionizing radiation but in-

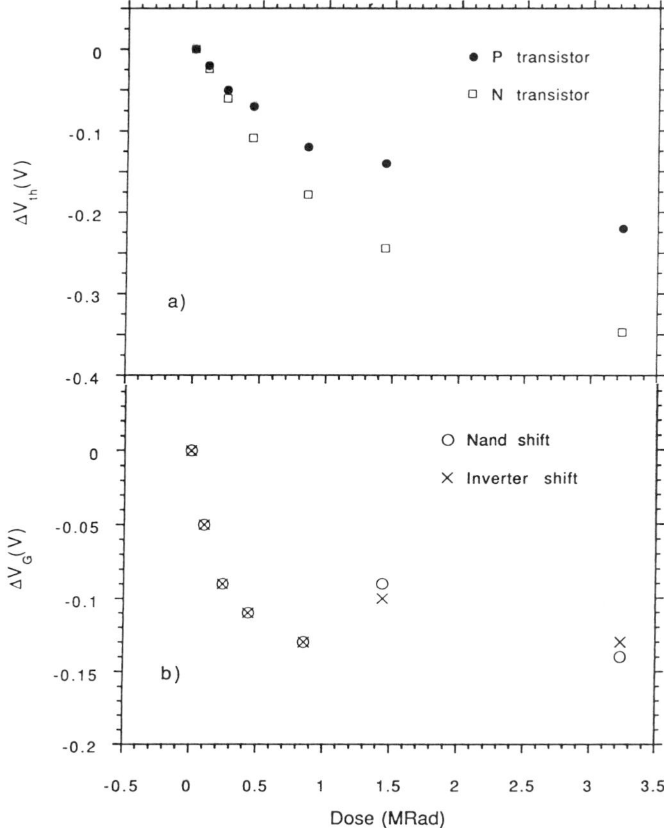

Fig. 3 a) Shift of the threshold voltage as function of Co^{60} dose for representative n and p transistors in the chip of Fig. 2. b) Shift of the transition voltage of the NAND gate and the INVERTER built with the transistors of Fig. 3a.

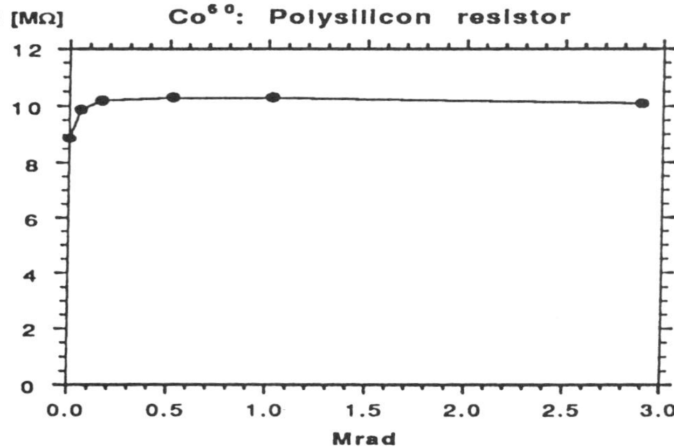

Fig. 4 Polysilicon resistor of strip 33 on detector 11 under photon irradiation.

creased their value by less than 10% during a neutron fluence of $3 \cdot 10^{13}$ neutrons/cm^2 exhibiting saturation after about $3 \cdot 10^{12}$ neutrons/cm^2.

- We have verified that the isolation between p strips is maintained and increases during irradiation with ionizing radiation. We have observed and measured the build-up of oxide charges in the interface and can deduce that they will pose no problem for the operation of the p-type strips.
- Coupling capacitors made of p-implant and overlayed silicon dioxide and metal traces did not change their values by more than 0.5% during irradiation with 2.8 MRad of Co^{60}.

Conclusions

We have irradiated dielectric isolated bipolar transistors of the design being fabricated for an fast, low power amplifier chip and found very good stability of β beyond neutron fluences of $5 \cdot 10^{13}$ neutrons/cm^2. We have built a prototype digital storage chip in 1.2 micron CMOS and find radiation resistance of logic gates beyond a dose of 3.2 MRad. Measurements on critical elements of ac-coupled detectors show excellent radiation resistance.

REFERENCES

1. SSC Subsystem Proposal for a Silicon Tracker, SCIPP 89/39 and SCIPP 90/28; SDC Expression of Interest, May 1990; A. Seiden, Invited Talk at this Conference.
2. D. Groom *et al.*, SSC-SR-1033.
3. H.F.-W. Sadrozinski, SCIPP 90/02; G. Hall, Imperial College Note IC/HEP/90/1.
4. H.F.-W. Sadrozinski, *Nucl. Instrum. & Methods* **283**, 810 (1989).
5. H. Ziock *et al.*, *IEEE Trans. Nucl. Sci.* **NS-37,** 1238 (1990).
6. H.F.-W. Sadrozinski *et al.*, 2nd London Conference on Position Sensitive Devices, London, U.K., September 1990.
7. N. Ujiie and H. Ikeda, KEK 90-48.
8. D.E. Dorfan *Nucl. Instrum. & Methods* **A279**, 186 (1989).
9. R. Sonnenblick *et al.*, 2nd London Conference on Position Sensitive Devices, London, U.K., September 1990.
10. Silicon Dynamics Inc., Ames, IA, 50010, USA.
11. J. DeWitt, *Nucl. Instrum. & Methods* **A288**, 209 (1990).
12. United Technologies Microelectronic Center (UTMC), Colorado Springs, CO, USA.
13. B.S. Avset *et al.*, *IEEE Trans. Nucl. Sci.* **NS-37-3,** 1153 (1990); E. Focardi *et al.*, *IEEE Trans. Nucl. Sci.* **NS-36,** 40 (1989); P. Holl *et al.*, *IEEE Trans. Nucl. Sci.* **NS-36,** 251 (1989).
14. T. Ohsugi *et al.*, HUPD-9014 (1990).
15. G. Hall *et al.*, 2nd London Conference on Position Sensitive Devices, London, U.K., September 1990.
16. D. Pitzl *et al.*, International Conference on Advanced Technology and Particle Physics, Como, Italy (1990).
17. T. Ohsugi *et al.*, *Nucl. Instrum. & Methods* **A265**, 105 (1988).

TESTS OF GaAs DETECTORS FOR MINIMUM IONIZING PARTICLES IN A HIGH RADIATION ENVIRONMENT

S.P. Beaumont
Department of Electrical and Electronic Engineering
University of Glasgow, U.K.

R. Bertin, A. Zichichi
CERN
Geneva, Switzerland

S. D'Auria, C. del Papa, F. Fiori, B. Lisowski, V. O'Shea
Dipartimento di Fisica dell'Universita' and INFN
Bologna, Italy

M. Edwards
Rutherford Appleton Laboratory
Chilton, Didcot
U.K.

J.G. Lynch, C. Raine, I.O. Skillicorn, K.M. Smith, R.M. Turnbull
Department of Physics and Astronomy
University of Glasgow, U.K.

M. Nuti, P.G. Pelfer
Dipartimento di Fisica dell'Universita' and INFN
Florence, Italy

Presented by D.H. Saxon
Department of Physics and Astronomy
University of Glasgow, U.K.

Abstract

This report describes results obtained on GaAs Schottky diode charged particle detectors by the Glasgow and CERN-based LAA groups. Their radiation hardness and potential speed, more than competitive with silicon detectors, make GaAs detectors attractive for applications in high radiation environments.

Introduction

We first discuss the choice of semi-insulating GaAs as a detector material, the manufacture of our Schottky diodes [1], and the response of the diodes to alpha, beta and gamma radiation and to test beam pions. Results are presented of gamma—and neutron—irradiation effects on GaAs diodes, and of measurements of the speed of response of the diodes using laser pulse excitation. The report concludes with a summary of our current understanding of GaAs diode detectors.

The Advantages of GaAs

GaAs is a direct band-gap semiconductor, with a band-gap of 1.43 eV. The radiation length is 2.3 cm, (four times shorter than silicon), but this is partly compensated by the higher specific ionization loss of 5.6 MeV/cm. GaAs wafers are generally prepared by the LEC (Liquid-encapsulated Czochralski) method. Higher purity may be achieved by Molecular Beam Epitaxy, (MBE), but this is limited to layers of at most a few tens of microns in thickness by the slow rate of deposition. Vapor Phase Epitaxy (VPE) may overcome the latter limitation in the future [2]. Liquid Phase Epitaxial growth, (LPE), formerly employed for pure and doped GaAs wafers, has now been largely superseded by MBE and VPE for most applications.

The best results obtained in the early 1970's [3] with LPE GaAs diode detectors were difficult to reproduce but established that good energy resolution for X- and gamma-rays was possible. While development in LPE detectors has been inhibited by the variability of the wafer material, improvements in LEC commercial crystal growth techniques [4] have resulted in more uniform wafers of higher purity and mechanical strength, and with a greatly improved surface quality.

Finally, the high electron mobility in GaAs devices, (almost six times that of silicon at best), offers the prospect of high speed particle detection and signal processing.

Diode Manufacture and Electrical Tests

Diodes of 500, 300 and 125 microns thickness and 1 and 3 mm contact diameter were manufactured in the University of Glasgow Department of Electrical and Electronic Engineering by evaporating a Ti-Au Schottky barrier onto one side of a semi-insulating GaAs chip and an ohmic contact of Ni-Ge-Au onto the other. No attempt was made to deposit a passivating layer on the surface of the diodes.

A typical diode displays a reverse bias leakage current which is significantly higher than that for typical silicon detectors. Measurements on diodes with a guard ring electrode surrounding the Schottky contact suggest that the leakage is not predominantly due to surface effects, but probably more to bulk generation.

Tests with Radioactive Sources and Beam

Each diode was tested with alpha-, beta- and gamma-ray radioactive sources. The charge released by each was known from the energy required to generate an electron-hole pair in GaAs, namely 4.2 eV. The measured charge collection efficiency increased with bias voltage as shown in Figure 1.

The inefficiency is interpreted as due to very rapid trapping of the holes released by the ionizing particles, the observed signal being due to the electrons migrating to the collecting electrode. This interpretation is supported by comparing the pulse height spectrum with an alpha source next to the ohmic contact with that obtained with the source at the Schottky electrode. Simulation of the charge trapping by a simple model gave the predicted charge collection efficiency variation, also shown in Figure 1, (cf. [1], [5]). The variation in charge collection efficiency among the diodes of an array on one chip is given in Figure 2. Figure 3 shows the pulse height spectrum obtained in a 6 GeV/c pion test beam [1].

Speed of Response

The picosecond laser facility at the University of Florence was used to excite a GaAs diode with a 1.5 picosecond pulse. The output signal from the diode, observed using a high speed oscilloscope, is shown if Figure 4. The very fast initial pulse is followed by a tail of around 4 nanoseconds in length, thought to be due to trapping effects in the charge transport mechanism. More detailed studies of the trapping and de-trapping process may lead to better understanding of the charge collection inefficiency discussed above.

Radiation Hardness

Several diodes, subjected to gamma irradiation up to a total dose of 17 MRad, showed only a very small change in leakage current, and the response to radioactive sources was almost completely unaffected. A set of diodes was also subjected to neutron irradiation by a fluence of 7.10^{14} n/cm^2 at the R.A.L. test facility in the ISIS accelerator [6]. A comparison is given in Figure 5 of the pulse height spectrum obtained with a collimated Ru-106 source, (equivalent to minimum ionizing particles), before and after the neutron irradiation. The GaAs diode has clearly deteriorated, but the signals are still easily resolved, and the detector continues to be usable.

Alternatives to Semi-insulating Material

We have concentrated on diodes made from semi-insulating GaAs, the cheapest available material. Previously, L.P.E. GaAs diodes have been used successfully for gamma ray spectroscopy. More recently, high purity V.P.E. GaAs wafers have been produced at growth rates exceeding 100 microns per hour [2]. The density of trapping centers in L.P.E. and V.P.E. materials is reported to be less than in semi-insulating wafers, so that the charge collection efficiency may be higher in epitaxial diodes. We hope to test samples of both these alternative materials within the next few months.

GaAs Read-out Electronics

GaAs FET pre-amplifiers operating at low temperatures have recently been used in a liquid argon calorimeter read-out system, [7]. In addition, a SPICE analysis of a GaAs bipolar pre-amplifier design predicts noise and power consumption figures which are significantly lower than corresponding values for silicon, [8]. The expected higher radiation resistance of the GaAs circuits make this a worthwhile study. Integration of optical read-out onto the semi-insulating wafer could offer useful advantages in particle physics, which we intend to investigate further.

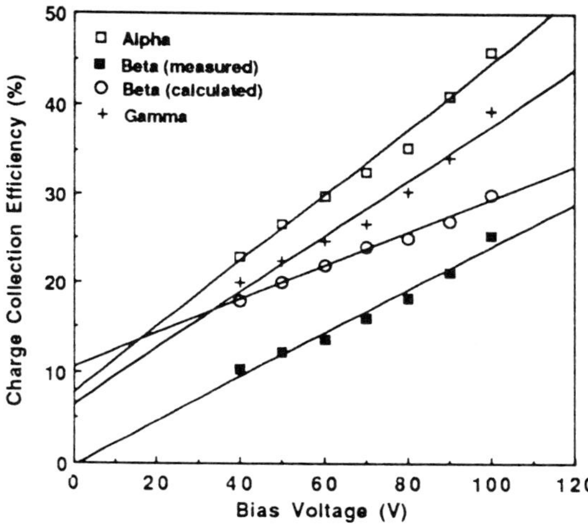

Figure 1: Charge collection efficiency variation with bias voltage, for alpha-, beta- and gamma-rays.

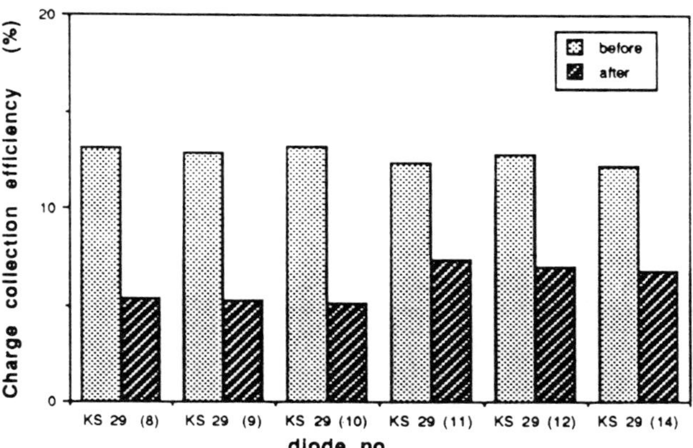

Figure 2: Charge collection efficiency variation for the diode array, before and after neutron irradiation by $7.10^{14}\,n/cm^2$.

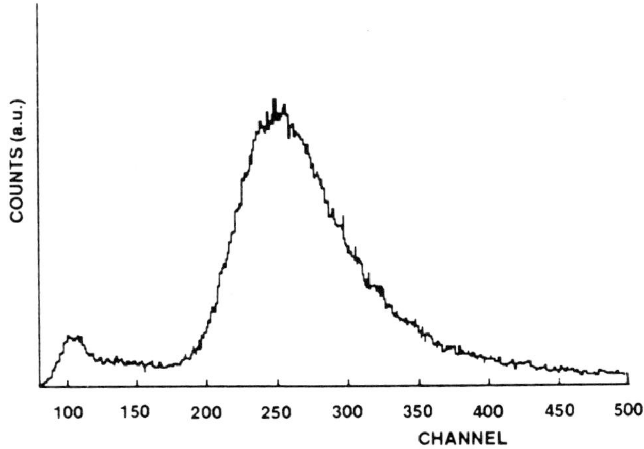

Figure 3: Pulse height spectrum obtained in a GaAs detector with 6 GeV/c pions.

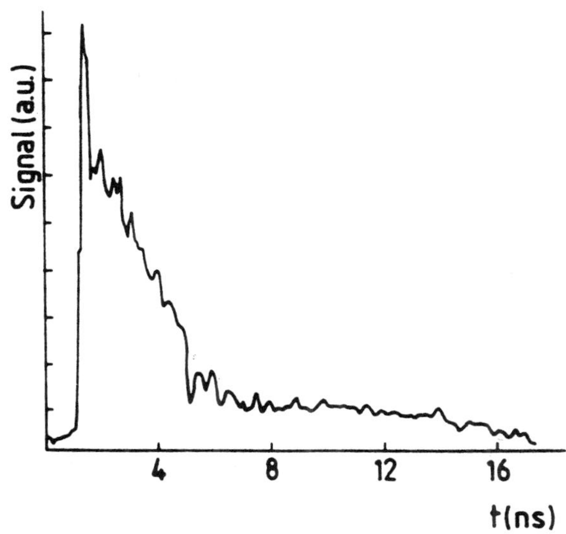

Figure 4: Oscilloscope trace of GaAs detector response to a 1.5 picosecond laser excitation pulse.

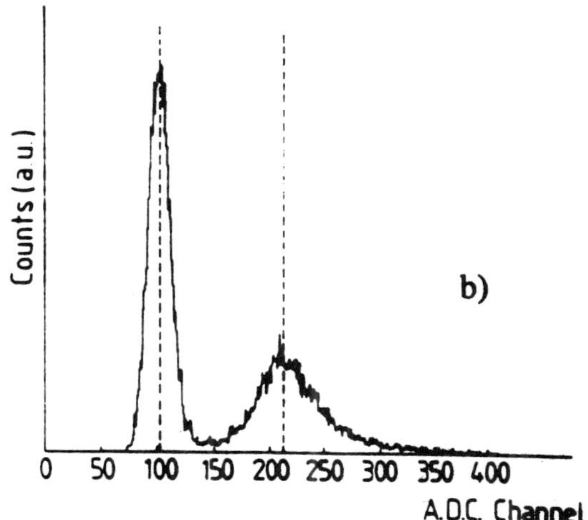

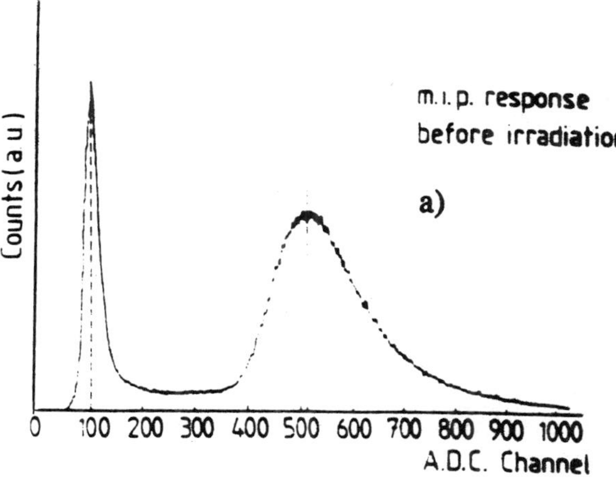

Figure 5: Pulse height spectrum from Ru-106 source, corresponding to minimum ionizing particles, (a) before, and (b) after irradiation of the GaAs detector by $7.10^{14}\,n/cm^2$.

Summary of Results Obtained to Date

In summary, GaAs Schottky diodes work satisfactorily as charged particle detectors, with essentially 100% detection efficiency for minimum ionizing particles. They tolerate radiation loads at the level expected in several years of running at a radial distance of only a few centimeters from the intersection point of the Superconducting Super Collider at nominal luminosity. The output signal of only a few nanoseconds is very satisfactory for the new colliders. While these results are already encouraging, further study of the charge trapping which we have observed in semi-insulating GaAs may enable further improvements in the performance of diodes manufactured in this material.

References

[1] R. Bertin et al., *A Preliminary Study of GaAs Solid-State Detectors for High Energy Physics*, Nucl. Instr. and Meth. **A294**, 211 (1990).

[2] K. Gruter et al., *Deposition of High Quality GaAs Films at Fast Rates in the LP-CVD System*, Jour. of Crystal Growth **94**, 607 (1989).

[3] J.E. Eberhardt, R.D. Ryan and A.J. Tavendale, *High-resolution Nuclear Radiation Detectors from Epitaxial n-GaAs*, Appl. Phys. Lett. **17**, 427 (1970).

[4] R.N. Thomas et al., *Status of Device-qualified GaAs Substrate Technology for GaAs Integrated Circuits*, Proc. I.E.E.E. **76**, 778 (1988).

[5] C. del Papa, C. Raine, *New Results on GaAs Solid-State Detectors for High Energy Physics*, Contributions to the 4th San Miniato Topical Seminar on Experimental Apparatus for High Energy Physics and Astrophysics, June 1990. (To be published).

[6] M. Edwards and D.R. Perry, *The Radiation Hardness Test Facility*, R.A.L. Report RAL-90-065, August 1990.

[7] A. Alessandrello et al., *Gallium-arsenide Charge-sensitive Preamplifier for Operation in a Wide Low-temperature Range*, Nucl. Instr. and Meth. **A289**, 426 (1990).

[8] P.H. Sharp, *Developments in VLSI*, Proceedings of the Eloisatron Workshop on New Technologies for 200 TeV Physics, Erice, September 16–20, 1990.

CABLING FOR AN SSC SILICON TRACKING SYSTEM

Hans Ziock, Jan Boissevain, Bradly Cooke, and William Miller
Los Alamos National Laboratory, Los Alamos, NM 87545

ABSTRACT

As part of the Superconducting Super Collider Laboratory (SSCL) funded silicon tracking subsystem R&D program, we examine the problems associated with cabling such a system. Different options for the cabling plant are discussed.

Introduction

A silicon microstrip tracking detector for an SSC experiment is an extremely complex system. The system consists of approximately 10^7 detector channels, each of which requires a communication link with the outside world and connections to the detector bias voltage supply, to a DC power supply for the onboard electronics, and to an adjustable discrimination level. The large number of channels and the short time between beam interactions (16 nanoseconds) dictates the need for high speed and large bandwidth communication channels, and a power distribution system that can handle the high current draw of the electronics including the large AC component due to their switching. At the same time the constraints imposed by the physics measurements require that the cable plant have absolutely minimal mass and radiation length.

Considerations for DC Power Cabling

The cross section of a quadrant of such a detector system is shown in Fig. 1. The system is cylindrically symmetric about the z-axis. Therefore in the central region the system consists of a series of cylindrical shells, while in the forward region it consists of sets of annular rings. The system being investigated for the subsystem work is approximately 1 meter in diameter and 6 meters in length. It is estimated to use 20,000 watts of power, although one hopes to reduce this value. If the electronics operate at 5 volts, the main power bus must carry 4,000 amps. The power distribution is assumed to be independent for each half (±z) of the system so that each bus carries 2000 amps. It is further assumed that each power bus, including the associated secondary lines and the ground lines (a factor of 2), is 2x2.5m long on average and has a total voltage drop of $\Delta V = 2\text{x}0.1$ volts.

Using Ohm's law its resistance is $R = \Delta V/I = 0.2\text{V} / 2000\text{A}\ 1.0\times10^{-4}\Omega$. Therefore the cross-sectional area of the power bus is $500\text{cm}\times\rho/R = 13.25\text{cm}^2$, where $\rho \equiv$ resistivity $= 2.65\times10^{-6}\Omega$-cm for aluminum. Accounting for both the supply and ground lines, the total cross section of the cable is 26.5 cm^2. The power dissipated in the cable is $\Delta V\times I = 400\text{W}$. To minimize the path length of any particle through the power bus, it is clearly advantageous for the power bus to have the same cylindrical symmetry as the system and to be at as large a radius as possible. For a power bus in the form of a cylindrical shell, area $\equiv A = 26.5\ \text{cm}^2 = 2\pi rt$, or the thickness $\equiv t = A/2\pi r$. (Note the 1/r dependence.) Assuming that the bus is at the inner face of the vessel that encloses the silicon system at say r = 55cm, one finds t = 0.077cm = .0086X_o (radiation lengths) for aluminum. Using the values given in Table 1, one finds t=0.0337X_o for copper and t=0.0033X_o for beryllium.

TABLE 1

Metal	resistivity @20°C [ρ] (Ω-cm)	radiation length [X_o] (cm)	figure of merit $\rho/X_o/\{\rho/X_o\}_{Al}$
Cu	1.67×10^{-6}	1.43	3.92
Al	2.65×10^{-6}	8.90	1.00
Be	$4.0\ \times10^{-6}$	35.30	0.38

In addition to the metal, the cables will require insulating layers. A likely candidate is Kapton, which has a radiation length of 28.6 cm, a dielectric constant of 3.5, and a dielectric strength of ~ 200V/micron. It will therefore be a negligible contributor to the total number of radiation lengths. However, the total X_o values for the cable are not insignificant. The copper option would seem to be immediately ruled out. Furthermore, due to constraints on the cable routing imposed by the geometry, some particles

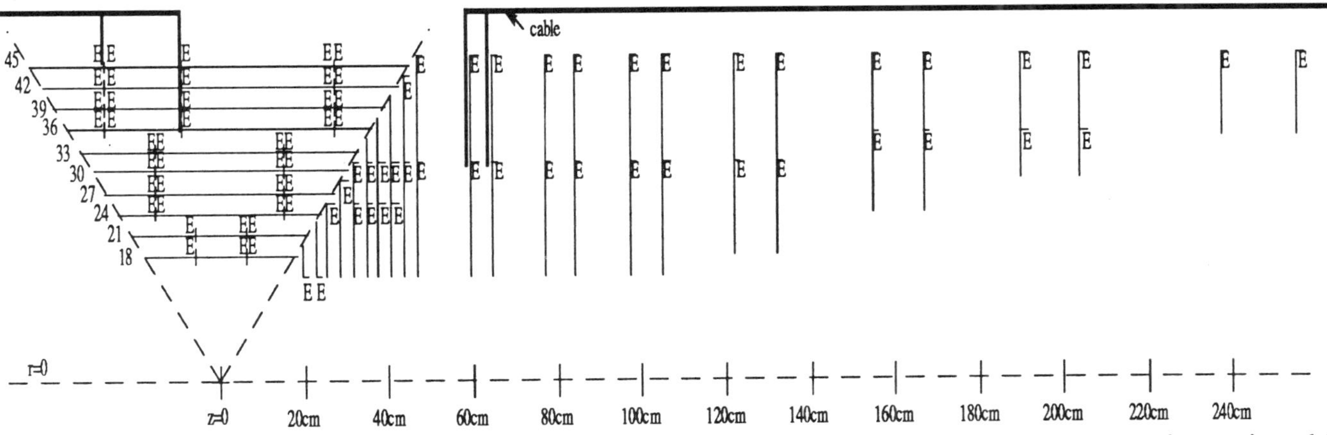

Figure 1. Layout of the silicon microstrip tracker system. The system is a cylinder of revolution about the z-axis and is symmetric about z=0. The exception is the position of the electronic packages (E) in the central region, of which both halves are shown. The positions of the electronic packages are not yet finalized. The routes of a few sample cables are shown.

will have multiple crossings of parts of the power bus. The only options available to further reduce the total material in a particle's path are to use thinner conductors, and to combine the electrical distribution job with that of mechanical support, using some existing structure in the system. The first option increases the already large power dissipation in cable itself, possibly necessitating active cooling for the conductor. The second option is equally untenable. The only structures available in the present design that could double in the role of power distribution are the enclosure vessel and the space frame[1]. As the secondary power lines must continually branch off of the main power bus to feed the individual layers of detector electronics, the transition region from main to secondary lines would probably be very complex, and negate any potential material savings. The requirement of spatially closely coupled ground and supply lines makes the integrated approach even more difficult if not impossible. The close coupling minimizes any magnetic forces and torques on conductors that run perpendicular to the magnetic field, and minimizes induced voltages and forces caused by a quench (see the next section).

The most reasonable power distribution scheme (see Fig. 1 and 2) would have a cylindrical power bus shell, which consists of a series of supply and ground conductors (S&G lines) running along its length. Three layers of conductor (ground, supply, ground), would form a single S&G line, and would be separated by insulating layers. In essence one has a stripline with segmented ground planes. At the z position of a layer of detectors, an appropriate number of S&G lines would run radially inward to supply and ground planes on the cooling ring[1] for that layer. The electronics would be mounted directly on the cooling ring. The bias supply voltage could be supplied in an analogous manner. As several orders of magnitude less current is required for that job, the amount of metal in that bus is likewise reduced.

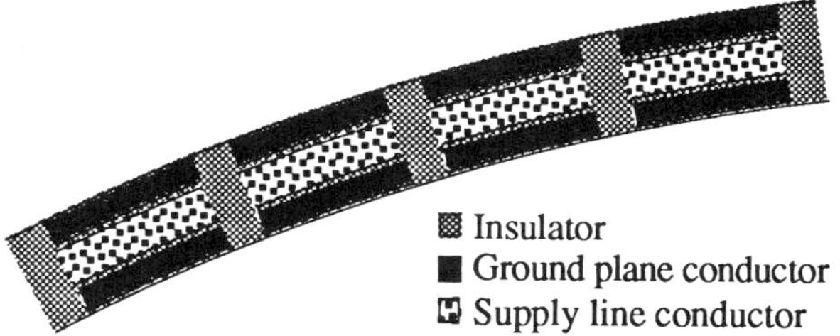

Figure 2. A cross section of a section of the power bus.

Magnetic Forces and Quench Effects

In the worst case, the cable layout is such that a single S&G line supplies a ring of electronics at r=50cm. Assuming the system properties given below in Table 2, one has a current of 50 amps in such a cable. The force on a wire of length l, which carries a current I orthogonal to a uniform magnetic field is F = lIB. A 0.1m long radial supply wire, which is carrying the current of 50A, feels a force of 50A×.1m×2T = 10 newtons. That is clearly not negligible. By running the supply and ground lines back to back, and mechanically coupling them, the net force is 0, although a torque can be induced. By keeping the distance between the supply and ground lines small, or orienting the cables such that the vector between them is parallel to the magnetic field lines, the torque goes to zero.

TABLE 2

Total power consumption	20,000 W	
Channels	10,000,000	
Power/channel	2.0 mW	
Pitch	50µm ≡	200 channels/cm
Strips on both faces	→	400 channels/cm
Power/cm width	0.8 W/cm	
Power for r=50cm ring	250 W	
DC supply voltage	5 V	
Supply current (ring)	250 W / 5V ≡50A	
Magnetic field (B_0)	2.0 Tesla	
B field decay (quench)	$B_0 \exp(-t/\tau)$, $\tau > 20$ seconds	

The emf induced during a quench is $d\phi/dt = (dB/dt)_{max}A$, where ϕ (= B_0A) is the flux and A is the area of the conducting loop threaded by the flux. During a quench, dB/dt is largest at t=0 and has a value of $-B_0/\tau$. By keeping the ground and supply lines back to back $A \approx 0$. If the supply and ground lines are separated by say 1 mm over a 50 cm length and are oriented so that the maximum number of field lines pass through this area (worst case), then the maximum induced emf is 5×10^{-5} volts, which is negligible. Eddy currents will be further reduced by the circumferential subdivision we propose for the power bus. Therefore, any induced current and force resulting from the quench will also be negligible.

Considerations for AC Power Cabling

An estimate of the spectral components of the power fluctuation seen at the supply line interconnect to a dynamically switching logic or amplifier circuit can be made as follows. An outflowing digital pulse will produce an approximately equivalent inflowing pulse from the power supply interconnect. For a pulse with a 50% duty cycle and 10% rise and fall time (this is a typical digital pulse), the DC and first 6 AC components of its normalized, power spectral-distribution are:

DC → 0.4363, f_o → 0.4837, $2f_o$ → 0.0418, $3f_o$ → 0.0156, $4f_o$→ 0.0179, $5f_o$ → 0.0000, $6f_o$ → 0.0035, where $f_o = 1/T$ and T = clock frequency.

What the above results illustrate, is that greater than 50% of the power required to generate a pulse is AC power. Somehow the main power bus to wafer level interconnects must be able to neutralize and contain these AC components. Failure to do so could lead to wafer to wafer coupling through the power bus and through electromagnetic radiation, and pulse-shape deterioration due to a band-limited power supply interconnect or inadequate localized power supply filtering. (Note: detector strips with high gain amplifiers make good E&M receivers and with sufficient coupling, excellent oscillators.)

Cylinder to Wafer Interconnects

Based on preliminary work we have conducted on metallic transmission lines for SSC high speed communication channels[2], the stripline configuration shown in Fig. 2 provides a good compromise between bandwidth, radiation length and shielding. The structure allows for multiple power leads and, if applicable, communication channels in a flat tape package. One of its main advantages is that the width, w, of the metal conductor can be made wider to decrease both DC and AC resistance. In addition, the characteristic impedance of the line can be decreased, providing a lower dynamic AC impedance.

The utility of providing multiple, uncoupled power-supply leads to a single wafer should not be underestimated, especially for wafers that consume large amounts of power and contain a large number of digital circuits that switch simultaneously. In order to filter out most of the spectral components f_o and higher, capacitors will need to be placed at the wafer/power cable interface. Tantalum capacitors are relatively small and quite radiation resistant. The exact amount of metal ribbon cable and capacitance needed per wafer will depend on such factors as power dissipation, clock frequency, wafer geometry and layout rules.

Communication Links

The design of the communication link from the silicon system to the external world will be driven by the required bandwidth. That bandwidth will be determined in part by how many communication channels there are (namely how finely the system is segmented). For a given channel, the data rate has two components: 1) the average rate caused by isolated particles randomly striking a detector, and 2) the rarely occurring peak rate which is caused by a large number of correlated particles passing through the small part of the system read out by a single communication channel. The data transfer rate for component 1) can be estimated as follows: The η coverage for a 1 cm wide section of the central wafer which extends from z=-6cm to z=6cm at r=18cm is given by

$\tan\theta_{start}$	= 18cm/-6cm	$\rightarrow$ 108.435 deg,
$\tan\theta_{stop}$	= 18cm/6cm	$\rightarrow$ 71.565 deg,
η_{start}	$= -\ln(\tan(.5\theta_{start}))$	= - 0.327,
η_{stop}	$= -\ln(\tan(.5\theta_{stop}))$	= 0.327,
$d\eta$	$= \eta_{stop} - \eta_{start}$	= 0.655.
circum. of ring	$= 2\pi\times 18$cm	= 113.097 cm

ASSUME:

1) $dN/d\eta$/event = 7.5, where N is the # of charged particles,
2) An added factor of 2.2 for interactions and loopers for N,
3) At $L=10^{33}/cm^2$, 1.6x60MHz = 9.6×10^7 events/sec }

Then the rate of particles through a 1 cm wide section of the central wafer which extends from z=-6cm to z=6cm at r=18cm: $= 2.2\,(dN/d\eta)\times d\eta\times 9.6\times10^7 / 113.097$
$= 9.17\times10^6$/sec.

If on average a typical particle causes two adjacent strips to be above threshold, and the detector has strips on both its sides, then the number of hit strips is four times the rate or 3.67×10^7/sec/cm width. Assuming that one communication channel services a 5cm wide section, then the number of hit strips is a factor of 5 larger. Eleven bits are needed to uniquely identify any one of the 2000 strips in the 5cm segment. The corresponding data transfer rate is 2 Gbit/sec to get all the information off. Assuming an event reduction factor of 1000 given by the level 1 trigger, the rate drops to 2 Mbit/sec.

The data transfer rate for component 2) follows: A 4TeV p_t jet can contain as many as 100 charged particle tracks that pass through the detector region covered by one readout channel[3]. Assume that 5 μsec are available to transmit the hit information for use in the level 2 trigger. Once again we multiply by a factor of 2 for the average number of hit adjacent strips per track, a factor of 2 for strips on both sides of the detector, and a factor of 11 for the number of bits. The required bandwidth is then $100\times2\times2\times11 / 5\times10^{-6}$/sec = 880 Mbit/sec. Clearly if one uses the level one trigger rate reduction, it is component 2) that determines the needed bandwidth. Note also that while component 1) is directly proportional to the luminosity (L), component 2) is independent of L. The required bandwidth and considerations, such as attenuation, power, crosstalk, and material-mass[2],[4], would argue for fiber optic communication links. A typical fiber optic cable is 100 microns in diameter, is made of SiO_2, and is sufficiently radiation resistant. Even for 1 fiber per wafer (~10,000 overall), the total cross-sectional area of all the fibers is only $1cm^2$. Two way communication will at most double that area.

Short Strips

Although the preceding arguments have been made in the context of long strip double-sided detectors, the results apply equally well to the short single-sided strip detector option. The power consumption is approximately equivalent for both. Overall the communication channel bandwidth is also almost the same for the two options as the number of tracks is independent of the detector type. The number of hits is therefore the same for the two options within a factor of 2. (Single-sided: only 1 hit/track, double-sided: 2 hits/track). There are approximately 8 times as many short strip detectors overall. That will require an additional 3 bits for the strip specification.

Conclusion

The constraints imposed by the requirements of minimal mass and large bandwidth drive one in the direction of optical communication channels, and power buses that are in the form of stripline like ribbon cables at large radii. The conductors will have to be low Z metals such as aluminum or beryllium, and the cable plant design and routing must be carefully considered in the mechanical design of the system. Any reductions in the total power consumption will be very helpful. Furthermore, issues such as crosstalk between individual detector channels and power supply isolation will have to be carefully addressed. With sufficient effort, a functional design can be achieved.

References

[1] W. Miller *et al.*, "Mechanical Engineering and Design of Silicon-Based Particle Tracking Systems", these proceedings

[2] B. Cooke *et al.*, "SSC High Speed Communication Channel and Interconnects", these proceedings.

[3] A. Palounek, LBL, private communication.

[4] J . Ogle *et al.*, "Applications of Electro-optic Modulators for Wide Bandwidth Recording of Transient Signals", these proceedings.

DESIGN STUDY OF REAL TIME PROCESSING OF SILICON TRACKER HITS FOR PARTICLE MOMENTUM AND VERTEX DETERMINATION*

Carl Pfeiffer
Hughes Aircraft Company, El Segundo, CA 90245

Abstract

A strawman design of a hardwired, special-purpose Momentum and Vertex Processor was constructed, which operates upon hits received from a three-barrel silicon tracker. Estimates of primary and secondary vertices and momentum measures are calculated after each beam crossing (every 16 ns) for up to 128 tracks. The throughput is $0.5*10^{12}$ operations per second and data latency due to processing is 768 ns. It is concluded that this processor can be built with state-of-the-art components.

Introduction

A silicon tracking system for inner tracking at the Superconducting Super Collider (SSC) might contain three barrels shingled with chips containing pixels or strips, along with a series of vertical end plates also supporting such chips. One approach to dealing with the large number of pixels needed for good particle position resolution is to build a "smart readout", which uses an external trigger to only send "interesting events" [1]. Another approach is to read out all hit data, and later dump the uninteresting data from the Processor Farm. This scheme has the potential advantages of simplifying the readout chip and allowing the silicon hit data to be used to augment a first-level trigger decision.

A hybrid scheme might send hits only from the outermost barrels of a many-barrel system, where the hit rate per chip is low (proportional to the inverse square of the barrel radius), to augment the first-level trigger decision. Hits from events which satisfy a hypothesis test based upon both silicon and calorimeter data would later be read out. In any case, data from more than 50 particle tracks crossing each detector barrel at every beam crossing (16 ns) could be delivered to the Processor Farm, which poses a formidable computational task.

Assuming that the readout problem can somehow be solved, a signal processor strawman design will be described which can treat 128 tracks per beam crossing time, each penetrating three silicon barrels, to produce primary and secondary vertex

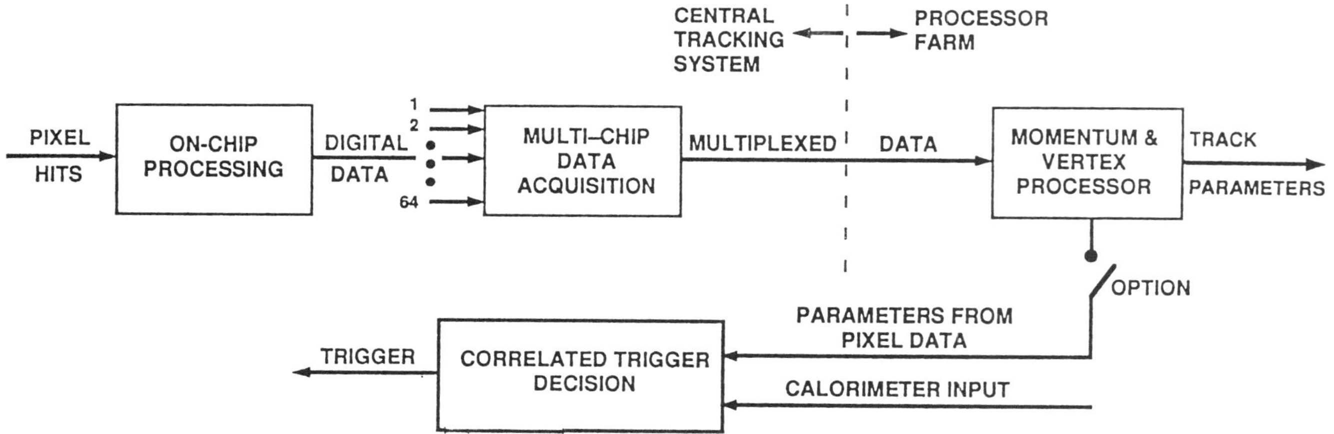

Figure 1. Strawman Signal Processing Architecture.

* This work was funded by a Hughes Research and Development project, in support of the SSC program.

estimates, as well as momentum and goodness-of-fit measures. End-plate detectors have not been treated, but it is reasonable to assume that they could be treated separately from, in parallel with, and similarly to the barrel detections (which doubles the processing hardware).

The Strawman Design

Referring to Figure 1, the signal processing architecture postulated for this paper includes: (1) silicon pixel or strip detectors, with on-chip hit detection and position estimation; (2) on-chip analog-to-digital conversion, with the capability to send estimates of all hit positions to the Processor Farm; (3) load-leveling and data sorting in a Data Acquisition Chip serving multiple detector chips; (4) high speed data transmission to the Processor Farm via laser-diode-driven fiber optics or air path; and (5) high speed calculation of particle trajectory parameters in a hardwired, special-purpose Momentum and Vertex Processor (MVP).

A typical inner-tracker array of silicon detector chips for an SSC experiment might consist of a three-barrel configuration. Small resolution elements can be constructed by on-chip interpolation of neighboring hit positions, but two, concatenated 16-bit words should adequately describe the axial and circumferential coordinates of a particle hit on a barrel ($4.3*10^9$ two-dimensional elements). Assuming that the MVP is designed to handle 128 tracks per beam crossing (each 16 ns), the output from the three-barrel data acquisition chips to the Processor Farm will be $768*10^9$ bits/sec. This data rate can be handled with fewer than 100 laser diodes, each driving a fiber optic or air path at $10*10^9$ bits/sec, which is the state-of-the-art of this technology.

Each concatenated, 32-bit word describing a particle hit on a barrel can be treated as both the address and the content of the word, i.e., these words are inherently "content-addressable." It is reasonable to assume that they can be sorted according to magnitude while being formed by the Data Acquisition Chip. A rapid fetch by binary search can later be done in the MVP with seven, hardwired instructions, as will be described below.

The architecture of the MVP is shown in Figure 2. It consists of 48 Beam Crossing Processors (BCP), each containing 128 Track Processor Elements (TPE) to deal with the 128 tracks per beam crossing (Figure 3). Each BCP is supported by a high-speed memory containing space for 3*128 32-bit words for the three detector barrels.

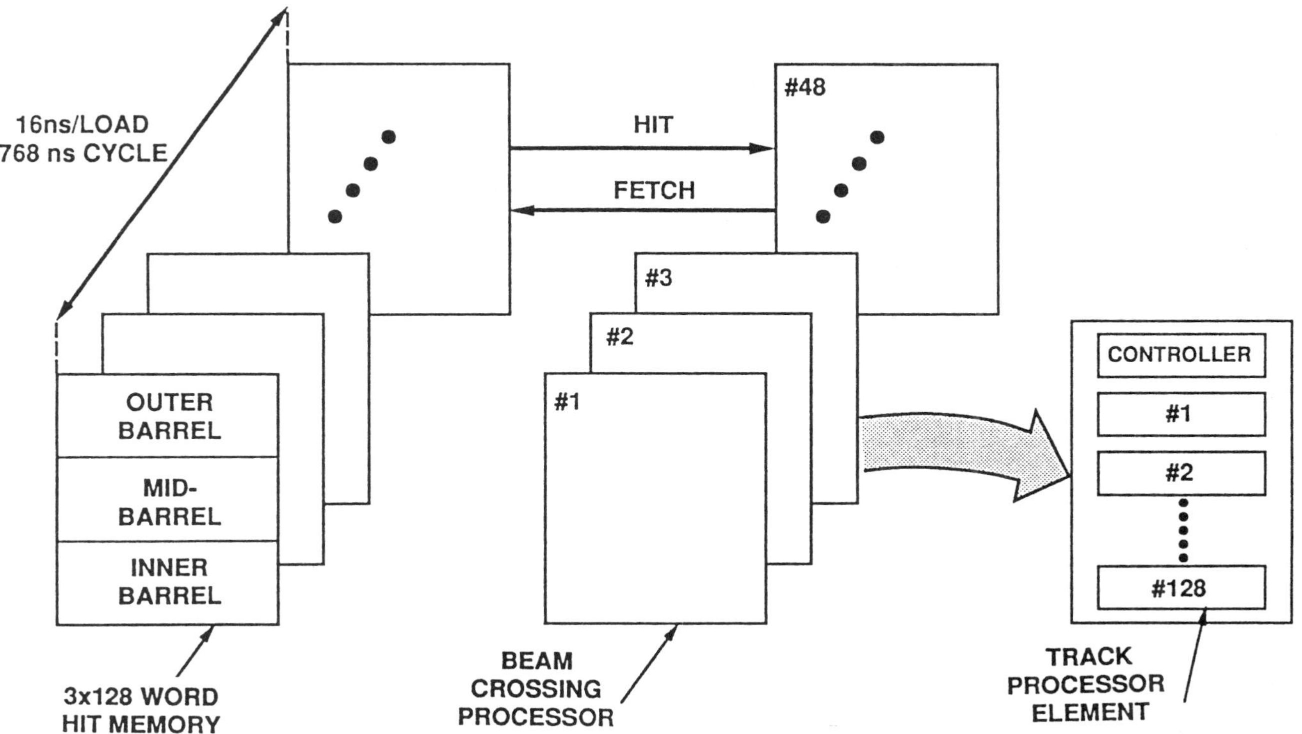

Figure 2. Momentum and Vertex Processor.

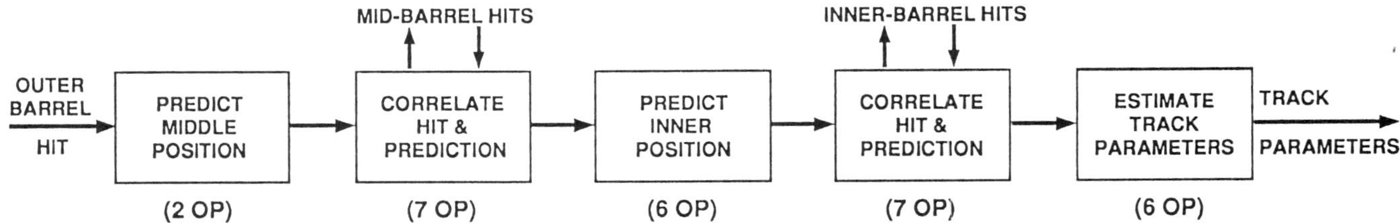

Figure 3. Track Processor Element.

A single particle trajectory can be processed in a TPE in 768 ns at a 40 MHz instruction rate (see following section), which implies that (768/16)=48 beam BCP's will be needed for "real-time processing", i.e., to keep up a beam crossing every 16 ns. Each BCP memory would therefore be refreshed every 768 ns, with a 16 ns load time. These rates are achievable with state-of-the-art components (custom VLSI).

The total throughput of the system is $0.52*10^{12}$ operations per second, assuming 65 operations per track and 128 tracks processed every 16 ns. This throughput would be doubled if a similar system independently operated on the end-plate silicon detectors.

Momentum and Vertex Processor Algorithms

Following Hylen [2], approximate equations for a particle in a magnetic field (e.g. for the SDC experiment) are

$$\varphi_i = K\, r_i + \varphi_o - b/r_i$$

$$z_i = r_i \tan \lambda + z_o$$

where i denotes an observation or barrel number (i = 1,2,3); φ = helix angle; z = axial coordinate; r_i = radius of ith barrel; K = $(2*\text{radius of curvature})^{-1}$; b = impact parameter and λ = helix pitch angle. Momentum components can be calculated from

$$\tan \lambda = [p_z/p_t]$$

$$p_t = [\text{constant}/K]$$

where p_z and p_t are the axial and transverse components of the particle momentum, and the constant is proportional to the magnitude of the magnetic field. Similar equations might apply to the EMPACT experiment, but with the apparent curvature of a particle trajectory caused by scattering in the silicon layers. Thus, calculation of the trajectory parameters K, φ_o, b, $\tan \lambda$ and z_o in the MVP provides estimates of primary and secondary vertices and particle momentum.

The first task of the MVP is track assembly, i.e., finding those triplets of observations which most nearly form straight lines (Figure 3). Starting with the outer barrel, the mid-barrel position is predicted by dropping a perpendicular to the beam axis (2 operations = 2 OP), and the inner barrel position is predicted by extrapolating the outer and mid-barrel positions (6 OP). The predicted values are used to find the nearest hits in the sorted, 128-word MVP memory, using a binary search algorithm. The predicted position is subtracted from the middle word in the memory and, depending on the sign of the difference, the 64th or 192nd word is tested next. This process is continued, with the number of candidate words being reduced by one-half after each test. The hit nearest the prediction is therefore found after seven instruction cycles. (Remembering words "close" to the prediction would also allow hit clusters to be identified).

Given the three observations of φ and the three observations of z at the three barrel crossings, the five trajectory parameters can be calculated as linear functions of the data, i.e.,

$$K = C_{11}\, \varphi_1 + C_{12}\, \varphi_2 + C_{13}\, \varphi_3$$

$$b, \varphi_o = \text{similar forms}$$

$$\tan \lambda = C_{41}\, z_1 + C_{42}\, z_2 + C_{43}\, z_3$$

$$z_o = C_{50}\, (z_1 + z_2 + z_3)$$

where the C_{ij} are input values (programmable). Setting b = 0, the sum-of-squares (SOS) of least squares data residuals for both the φ and z coordinates can be calculated as measures of "goodness-of-fit," i.e.,

$$\text{SOS}\ (\Delta\varphi\, |b = o) = (E_1\, \varphi_1 + E_2\, \varphi_2 + E_3\, \varphi_3)^2$$

$$\text{SOS}\ (\Delta z) = (F_1\, z_1 + F_2\, z_2 + F_3\, z_3)^2$$

where $\bar{E}$ and $\bar{F}$ are eigenvectors of the rank-one projection operators inherent in the least-squares fitting algorithm. These equations were used to estimate the instruction counts shown on Figure 4.

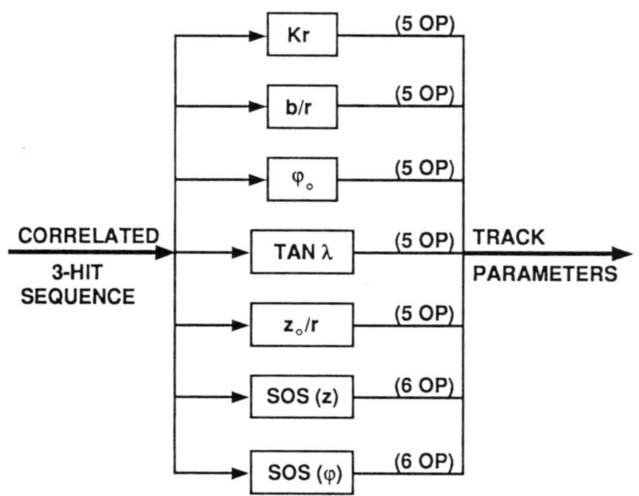

Figure 4. Estimate Track Parameters.

A total of 59 instructions per track was obtained for a track processor element, but, with the track parameters being calculated in parallel (6 instruction cycles serves all), only 28 instruction cycles are needed. Adding 10% for overhead, and assuming a 25 ns instruction time (40 MHz), the 768 ns processing time per track was obtained.

Conclusion

This paper has developed a strawman processor design to show that real-time processing of silicon tracker hits for momentum and vertex determination is feasible. Other hardwired MVP architectures are possible, such as 128 pipelines of N processors, each doing one or more of the 59 track instructions every 16 ns. An average (instead of maximum) of 128 tracks per beam crossing might then be processed, which would significantly increase system capability. Programmable processors should also be considered [3,4]. For example, the Hughes 3-D Computer [3] has been shown to be an attractive candidate for signal processing applications which are similar to this one. Custom VLSI always wins over programmable in the weight and power race, but one is often willing to pay a price for flexibility.

Future work will address these issues, leading to more detailed timing and sizing analyses and comparative assessments of candidate system architectures.

Acknowledgements

The author is indebted to the members of the SDC, BCD and EMPACT collaborations who provided the physical insight needed to write this paper, and for their stimulating discussion and critique of the ideas presented here. Edward Gussin and Richard Burney of Hughes Aircraft Company provided invaluable assistance in the design of the strawman processor.

References:

[1] O. Barkan, et al, "Development of a Customized SSC Pixel Detector Readout for Vertex Tracking," Presented to the Symposium on Detector R&D for the SSC, Forth Worth, TX, October 15-18, 1990.

[2] J. Hylen, "Tracking Leptons in Higgs Decay," Presented at July 1990 Snowmass Conference.

[3] M.J. Little, et al, "The 3-D Computer," Proceedings of IEEE Int. Conf. on Wafer Scale Integration, San Francisco, CA, January 1989.

[4] R.M. Lea, "ASP Modules: Cost-Effective Building-Blocks for Real-Time DSP Systems," J. of VLSI Signal Processing, 1, 69-84, 1989.

MONTE CARLO SIMULATIONS OF A CENTRAL TRACKING SUB-SYSTEM

W. W. Kinnison, D.M. Lee, H.-J. Ziock
Los Alamos National Laboratory
Los Alamos, New Mexico 87545

Abstract

A GEANT-based Monte Carlo Code has been developed to model the performance of the central tracker for an SSC experiment. A summary of the current capabilities of the Monte Carlo will be presented followed by a more complete description of the results of a study of the details of detector performance.

Introduction

Various studies of the tracking criteria of a *generic* SSC detector have been carried out at previous workshops. All of those studies have used event-level Monte Carlos. By that we mean that they have used as input the resolution functions of various detector components, the central tracker for instance, and determined what the physics capabilities of a detector with those resolution functions would be. Our group at Los Alamos has recently put together a fairly detailed Monte Carlo of a detector which begins at the component level. The initial purpose of the new Monte Carlo was to derive (and therefore validate) the resolution functions which had previously been used in event level studies. More recently, it has been used to study issues of data acquisition and triggering techniques which are inaccessible to an event level code.

This report will describe the new Monte Carlo by first describing the current capabilities and some of its general features. Next, results of resolution functions and detector performance which will impact the design of the data acquisition system will be summarized.

Scope of the Program

The Monte Carlo program that we have developed is based upon GEANT 3 [1] and uses ISAJET 6.25 [2] for event generation. Both the geometry specifications as well as which parts of the code are active during a run are determined by input data cards. These two features are quite useful since the exact specifications of detector subcomponents can be quickly modified in order to try different geometries. Also, entire sub-detectors of the SSC detector can be left out in order to focus computer time on the parts which are relevant for a particular study.

Currently, the code can model the performance of a silicon tracker, a straw tube tracker and a scintillating fiber sub-system. In addition, some information has been included to take into account an electromagnetic and hadronic calorimeter, as well as, muon chambers and a magnet. On the other hand, the elements other than the central tracker are far less developed and we will therefore only describe the central tracking simulation in this report.

The first step of the program is the simulation section. The simulation includes all active elements of each sub-system of the detector as well as a very close approximation of the amount and location of support structures. Via GEANT, all physics processes of charged and neutral particle interactions with material are taken into account. The energy loss in active detector elements is recorded and, after all tracks have been followed through the entire detector, it is digitized. That latter process results in an output which is very much like a data tape. For the silicon we record the strip number and the width of the cluster of contiguous struck strips and time of the strip firing, in units of 16 ns for each digitization. In that process, only strips which have a total of 20 KeV of energy deposited are assumed to have fired their discriminators and are thus recorded. For the straws, the equivalent TDC output is recorded (with respect to the beam crossing) and for the scintillating fibers we record which fiber was struck, i.e. we assume that only discriminators are connected to the fibers. In all cases, an appropriate time evolution, including drift times, particle flight times, and propagation times down long sense wires are taken into account.

The results of the simulation can be recorded on disk as a file tape or immediately input into the second section, the event reconstruction section. The event reconstruction follows the same scheme as would be used in most high-energy collider experiments. That is, for each detector sub-system, coordinates are recreated from the digitized data, pattern recognition is performed to get track candidates from sub-systems of similar resolutions which are then fit, and finally, tracks from the various sub-systems are joined for particle identification and kinematic event analysis. In our current scheme, the results of the track fit in the silicon is used to do primary and secondary vertex finding, but the particle identification phase is incomplete due to not having a full model of the calorimeters and muon chambers.

At each stage of the process, enough information is carried forward to allow a comparison of the results with the generated Monte Carlo data. In particular, we can use the code to derive the coordinate, vertex, and track fitting resolutions which can be input into event level simulators to help determine the physics capabilities of the detector. (Note that that could also be done with this Monte Carlo and not resort to event-level simulation by generating enough events of a specific type, but at the current level of SSC detector designs, event level simulation is a more practical approach.) We can also use the digitization results themselves to provide information which is relevant to the suitability of various data acquisition schemes.

Simulation Results

The results which are summarized here use the detector geometry from the Silicon Tracker Subsystem R&D collaboration. (See Reference 3 for a description of that geometry.) Two types of data were used in the simulations. The first type is data which have a Higgs event (M_H = 300 GeV/c^2) along with a number of overlapping minimum bias events corresponding to a luminosity of 10^{33}. (Actually, the overlapping events were 4 GeV P_T two-jet events.) In the discussion which follows we call these normal data. The second kind of data, which we call jet data, were high P_T two-jet events, again overlapping minimum-bias events were included. The normal data were to study the performance of the detector for the majority of the triggers (the Higgs tracks do not perturb those results), and the second set were to examine how the detector performs when there are a large number of very nearby tracks. We will first present the component level performance which will have an impact upon data acquisition followed by the higher level resolution function data.

Figure 1 shows the distribution of the number of strips which are above threshold for each contiguous cluster of strips of the silicon detector. Though the distribution has an average of 3.4 one can see there are long tails primarily arising from hadronic interactions in the silicon wafers. This quantity is independent of whether it is normal data or jet data. The second quantity which will be of importance in the read-out design is the number of clusters per 18-cm long by 3.2-cm wide wafer which is shown in Figure 2. These data, combined with the information that there are approximately 10 charged particles per unit of rapidity per interaction provide information on the amount of data which are required to be read out or buffered for each SSC trigger in a given geometry.

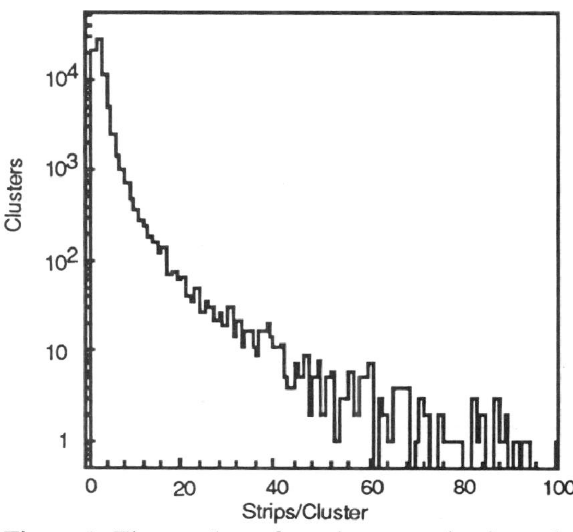

Figure 1: The number of contiguous strips in each cluster read out of the silicon detector.

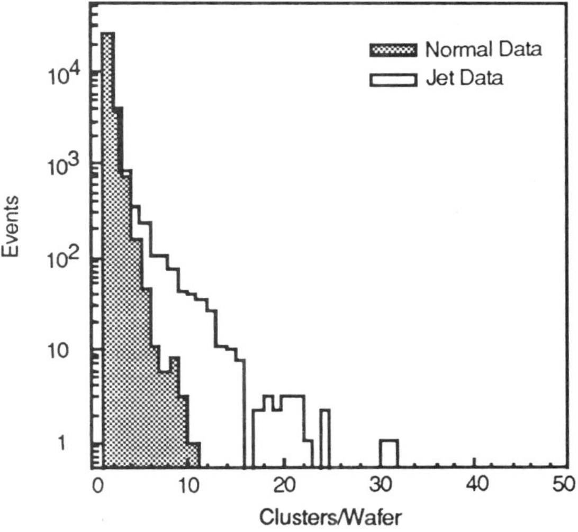

Figure 2: The number of clusters per wafer for both normal and jet data. Even in jet data only 3.3% of the events have more than three clusters per wafer.

Next one must answer the question of how the resolution is affected by the overlap of the cluster from one track with that of a nearby track. Figure 3 shows the number of actual tracks which are found to contribute to a cluster used for coordinate reconstruction. The actual number of times that two or more tracks contributed to the same cluster is 1.3% for normal data and 2.7% for data in jets. Consequently, one would expect that the normal estimate of the individual coordinate resolution would be a close approximation to the actual prediction of the Monte Carlo. That is indeed the case. The Monte Carlo predicts that in the barrel of the detector the r-φ resolution is a square distribution with a width equal to the pitch (50 μm in our simulation) and the z resolution in the barrel is a triangular distribution with a base of 1 cm (which comes from the 5 mrad stereo angle and 50 μm pitch). Similarly, in the forward detector, the φ-angle is associated with the 50 μm pitch and the radial resolution of the coordinate has a triangular distribution with a base of 1 cm. The silicon detector has an inefficiency due to a 0.5 mm dead region around the edges of every wafer and a triangular dead region due to the stereo read-out. In our Monte Carlo, those factors contribute to a 97% efficiency per coordinate.

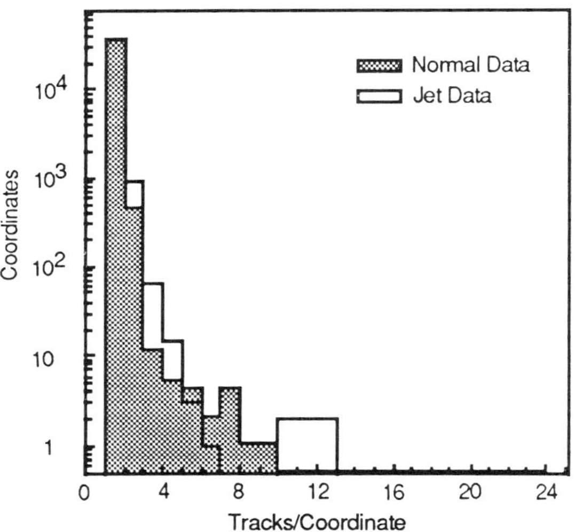

Figure 3: The number of tracks which contributed to each reconstructed coordinate. In an ideal detector only one track would contribute to each reconstructed coordinate.

As a last item, we will report on the pattern recognition capabilities of the tracker that we have investigated. In our approach, we construct vectors made up of coordinates from successive layers of the detector. From those vectors, a local estimate of φ_0 (the initial direction of the track in the r-φ plane), κ (the curvature) and $\cos\theta$ of the track can be estimated [4]. One then searches for clusters of points in that 3-space to find associations of coordinates to tracks. Figure 4 shows the efficiency of finding at least 8 of the ten possible coordinates per track correctly. The algorithm that we use for the individual vector finding attempts to have a low P_T cut-off of 300 MeV and that is the source of the low efficiency below about 500 MeV P_T. The efficiency levels off at about 94% due to the individual coordinate efficiency.

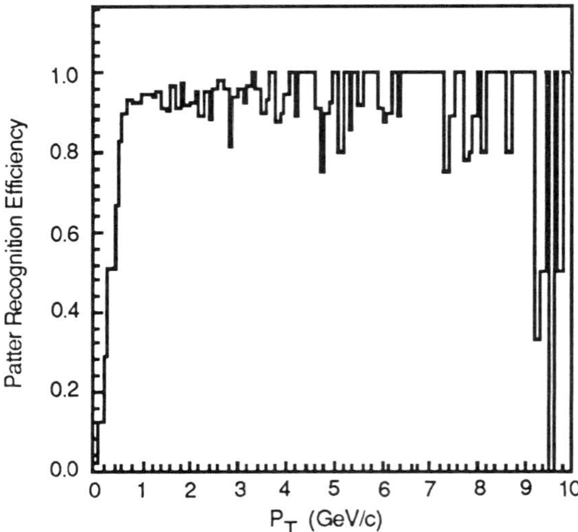

Figure 4: The pattern recognition efficiency as a function of P_T. The efficiency shown is for finding 8 good coordinates out of the 10 possible coordinates for the silicon tracker subsystem.

Conclusions

In conclusion, we have developed a Monte Carlo code which has been used to begin a study of component level simulations of a silicon tracker sub-system. Its primary utilization would be to provide input into event level simulators and study issues of importance to data acquisition and pattern recognition.

References

[1] R. Brun, *et al.*, "Geant 3", CERN Report CERN DD/EE/84-1.

[2] F. E. Page, *et al.*, "ISAJET 6.25", CERN Pool W999.

[3] W. Miller *et al.*, "Mechanical Engineering and Design of Silicon-Based Particle Tracking Systems", these proceedings.

[4] Alan Weinstein, private communication.

Wire Chambers

2

PROGRESS IN THE DEVELOPMENT OF A TRACKING TRANSITION RADIATION DETECTOR

J.S. Whitaker, J. Beatty, J.T. Shank, R. J. Wilson
Boston University, Boston, MA

V.A. Polychronakos, V. Radeka, D. Stephani
Brookhaven National Laboratory, Upton, NY

H. Beker, R.K. Bock, M. Botlo, C.W. Fabjan, J. Pfennig, M.J. Price
CERN, Geneva, Switzerland

W.J. Willis
Columbia University, New York, NY

T. Akesson
Lund University, Lund, Sweden

V. Chernyatin, B. Dolgoshein, P. Nevsky, M. Potekhin, A. Romanjuk, V. Sosnovtsev
Moscow Physical Engineering Institute, Moscow, USSR

I. Gavrilenko, S. Maiburov, S. Muravjev, A. Shmeleva
P.N. Lebedev Institute of Physics, Moscow, USSR

Abstract

We report on progress towards the development of an integrated transition radiation detector and charged particle tracker. Mechanical design and simulation of a detector has been pursued; a prototype device with 240 channels has been constructed and tested. Innovative construction techniques have been developed.

Introduction

The purpose of the TRD/Tracker is to provide charged particle tracking in the r-z plane and to provide particle identification capabilities that are independent of and complementary to calorimetric methods. The tracking goals include observation of the charged particle multiplicity and topology, reconstruction of the primary vertex or vertices, and assignment of charged particles to the correct vertex. Particle identification goals include the independent validation of electron candidates selected by calorimetric signatures, the rejection of false electron candidates that rise from accidental overlaps of low momentum charged particles with photon-induced electromagnetic showers in the calorimeter, and the identification of electrons arising from Dalitz decays or from photon conversions. Performance requirements depend on the specific physics channels under discussion and on the desired level of background rejection. Broadly speaking, rejection of hadrons and of hadron-photon overlaps by a factor of 100 to 1000 is desirable.

In a non-magnetic tracking volume, the rejection of false electron candidates can be accomplished in several ways: by a TRD as we are studying, or by a fine-grained preshower device [1], or possibly by a multilayered high-resolution silicon tracking device in which low momentum particles can be identified by their excessive multiple scattering [2]. While there are alternative technologies for achieving the necessary electron identification, the TRD approach still stands out as a leading candidate. Transition radiation detectors were discussed in the Expressions of Interest submitted to the SSCL by the High Luminosity collaboration and the by the EMPACT collaboration.

Significant progress has been made by the Transition Radiation Detector Collaboration in several areas. Monte Carlo studies of the performance of a detector system in the SSC environment have been pursued. A prototype detector of 240 channels was constructed, and its particle identification and tracking performance was tested in a beam of pions and electrons. Design calculations and mechanical prototyping of detectors have been started. Studies of drift times, signal shapes, and radiation damage have been done. Study of optical modulators for readout of the TRD has been begun. Progress in each of these areas will be summarized below.

Monte Carlo Studies

We have pursued Monte Carlo studies of the performance of a tracking TRD detector for the SSC. Figure 1 shows the design under study. The detector consists of modules that are built up with double layers of 4 millimeter diameter

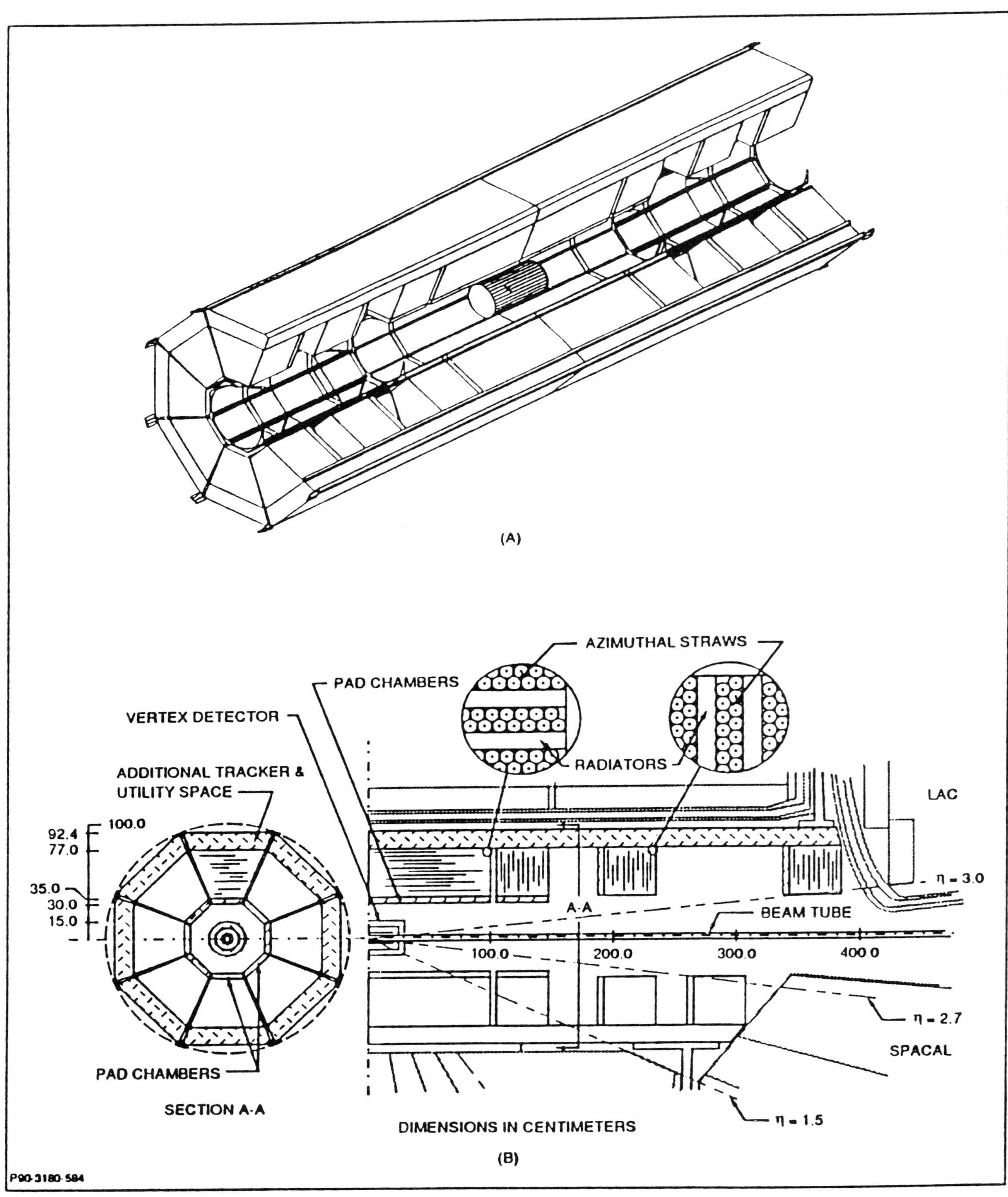

Fig. 1 **Baseline Design for TRD/Tracker with Pad Chambers and Vertex Detector Options. (A) 3D model (B) Section views showing straw and radiator orientations and octant layout. Above the Beam is the Liquid Argon (LAr) Option; Below the Beam is SPACAL Option.**

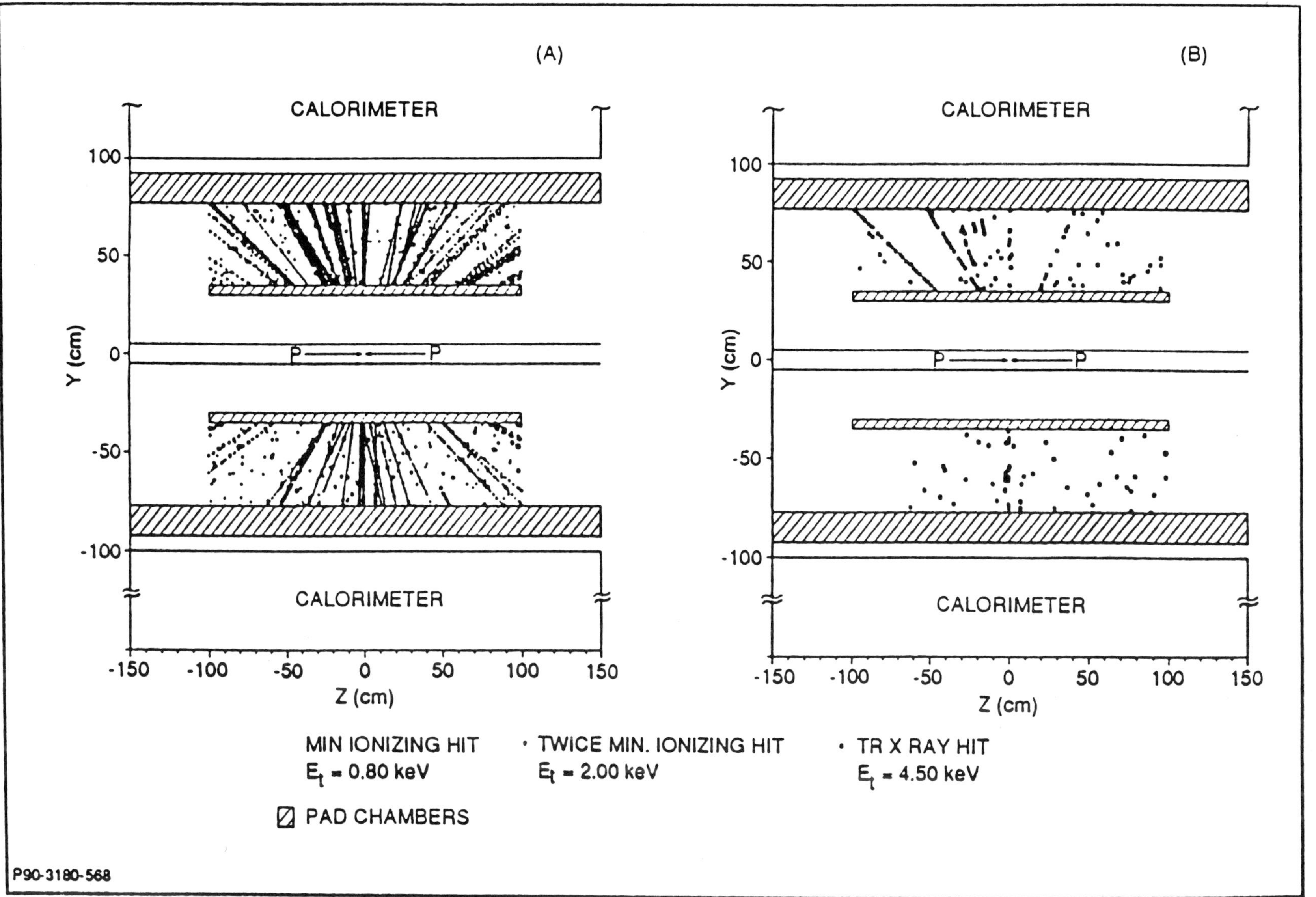

Fig. 2 TRD/Tracker Simulation for 500 GeV Higgs Decay to Electrons and 15 Minimum Bias Events; (A) All hits, (B) Hits over TR Threshold

straws oriented azimuthally, interleaved with radiator material that is taken to be polyethylene foam. The straws, filled with a mixture of xenon and a quenching gas, are operated as proportional chambers with a gain of approximately 10^4. The straws detect both the ionization by the charged particles that pass through the gas and the transition radiation x-rays emitted by particles as they pass through the radiator material. The x-rays are emitted colinearly with the particle. Pulse height discrimination allows the separation of charged particle ionization with and without an accompanying x-ray interaction on a statistical basis. Electrons and other particles with high Lorentz gamma are identified by having a larger number of high pulse height hits along their trajectory than would be expected from the dE/dx of low momentum hadrons.

Twenty-four repetitions of the radiator-plus-straws module provide the fine-grained sampling and many independent measurements of energy deposition that are required for good tracking and efficient electron identification. There are 300,000 to 400,000 channels in the full system. The thickness at normal incidence is approximately 10% of a radiation length: roughly 1/2 in the polyethylene foam radiators, 1/4 in the xenon mixture, and 1/4 in the straws and wire. Maximum drift time in a straw is approximately 40 nanoseconds. Occupancy at design luminosity is calculated to be several percent per straw.

Figure 2 shows a simulation of the production and decay of a 500 GeV Higgs into four electrons. The ISAJET code was used to create the event. The simulation included dE/dx fluctuations, random noise, and photon conversions. Pileup of low-P_T background events was included; the event in figure 2 includes 15 background events, characteristic of three times design luminosity for the SSC. The simulation assumes an electronic integrating time that is longer than the maximum drift plus shaping time.

The tracking performance of the system in the r-z plane is evident from the event display. Calculations show that the track position projected back to the beam line or out to a radius of one meter is determined with a resolution that is a fraction of a millimeter. These Monte Carlo data have been used as input to a study of the application of massively parallel computation to the pattern recognition and tracking problem in such a device; the status of that work is discussed in the paper by R. J. Wilson *et. al.* in these proceedings. The paper by R. Bock *et. al.* in these proceedings presents work done on the study of trigger processing of TRD data.

The figure of merit for the TRD's performance in particle identification is the rejection power, defined as $R=\varepsilon_\pi/\varepsilon_e$, where ε_π or ε_e is the probability for a pion or an electron to satisfy electron selection criteria. Figure 3 shows the calculated pion rejection power of this device versus pion momentum, for ε_e = 90%. The calculation was done by throwing tracks on top of high-P_T events such as that shown in figure 2, with an average of 5 overlap events as is expected for operation at the SSC design luminosity. The TRD meets the goal of a factor of 100 in pion rejection. Comparable rejection power is found when high-P_T two-jet events are used for the underlying events. The performance is nearly unchanged even when the random background is increased from the 0.5% used in events such as figure 2 to 5%.

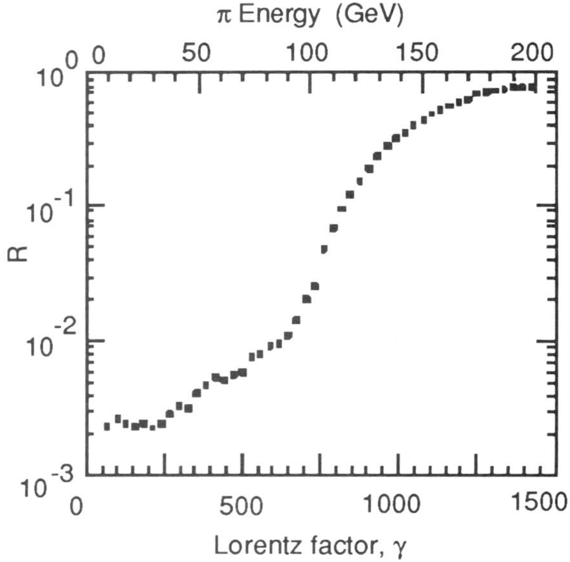

Figure 3. Pion rejection R of the TRD plotted versus the Lorentz factor and pion energy.

Prototype Beam Test Results

A straw tube transition radiation detector was tested this past June and July in the X5 test beam of the SPS at CERN. The experimental setup is depicted in figure 4. The detector consists of 4 TR modules each containing 60 straw tubes. The straw tubes are made of 64 μm thick polycarbonate and are placed in holes drilled into a polyethylene foam block which acts as the radiator. The foam has a density of 59 Kg/m^3 and an average wall thickness of 5 μm. Previous tests of this foam [3] show that it has an xray yield 85% of a regular stack of foils. The gas used in the straw tubes for most of the running was 50% Xe 50% CO_2.

A small lead glass calorimeter was placed behind the TR modules to allow us to distinguish between pions and electrons. Additional particle identification was obtained from two Cerenkov detectors which were part of the beam line. The electronics for this test consisted of preamps mounted near the ends of the straws connected via approximately 45 meters of coaxial cable to fast(12 ns) shaping amplifiers, whose outputs were digitized by LeCroy 2282 ADCs. The data were acquired by a Macintosh II and the MacUA1 system software [4]. Later runs were made with a LeCroy 9424 digital scope connected to one channel just after a special very fast preamp. This allowed us to record the waveform so we could understand the effect of the shaping amplifier.

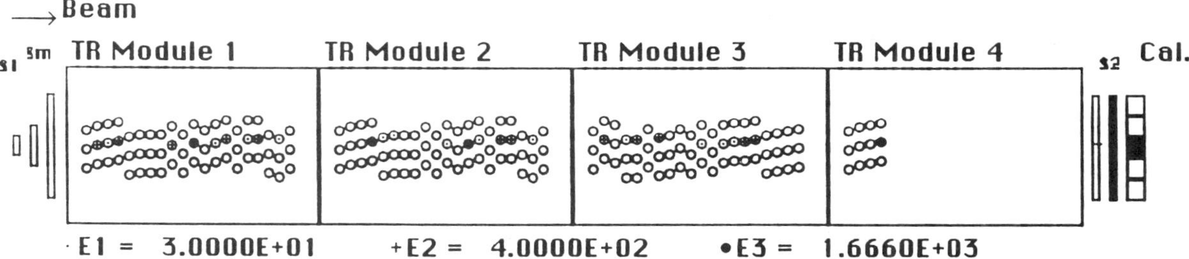

Figure 4. Schematic of the prototype beam test setup.

The straw tubes were calibrated by monitoring 4 tubes which had Fe55 sources permanently attached. Channel to channel pedestal and gain variations were corrected by periodically recording data from a single pulse injected into every preamp. The 5.9 keV xray peak was typically resolved with a σ of 10%. The X5 test beam can produce single electrons, pions or muons in the energy range 10-110 GeV.

Analysis of the data is in progress. Results are discussed in detail in the paper by V. Polychronakos *et. al.* in these proceedings. Both the tracking performance and the TRD performance can be studied with these data. At normal incidence, the electron spectrum has 18% of the entries above 5 KeV compared to less than 5% in the pion spectrum. The measured pulse height spectra for 30 GeV pions have been histogrammed for different ranges of the impact parameter of the beam particle with respect to the straw center. These data have been input to our Monte Carlo program which simulates the full central TRD/tracker for the SSC. Work on comparison of the x-ray yields observed in the test with our MC predictions is continuing. The preliminary results from this work show good agreement with earlier MC predictions for the single particle performance of the TRD/Tracker.

Component Development and Tests

Radiator Materials

Studies have been begun on a polyethylene foam produced by BXL Industries Ltd. Their HD foam has properties that make it an attractive material for use in the TRD. It is blown with nitrogen from bulk polyethylene that has been cross-linked by exposure to approximately 4 Megarads of Co-60 radiation.[5] This gives a dense foam with no blowing agent residues and may be more stable under ageing and irradiation than chemically cross-linked foams. Due to its higher cross-linking, the thermal conductivity of this foam is 20% higher than the chemically cross-linked foams of the same density. This higher thermal conductivity is advantageous for thermal management as discussed below. The compressive strength of the HD foams is three times larger than that of the LD foams, making it structurally superior material as well.

Radiation damage studies of the BXL foam have been begun. Samples were exposed to a neutron flux at a reactor at the University of Arizona[6] and to Co-60 radiation at the Boston University exposure facility. The neutron exposure was $5*10^{15}$ neutrons per square centimeter. This is several orders of magnitude higher than the flux to be expected at the SSC at a distance of 50 cm from the beam line in one year's running at design luminosity [7, p. 26]. Shrinkage of the linear dimensions by 0.3 to 0.5 % was observed at this very high dose. The Co-60 exposure was to 15 Megarads, also several orders of magnitude above the annual expected SSC dose [7, p. 181]. Shrinkage of linear dimensions was less than or near the 0.3% sensitivity of the measurements. The material had a distinct odor after the measurement, whilst it was odorless before the exposure, indicating some effect of the radiation. Material properties however were not discernably different for the exposed and unexposed samples. Effects at smaller doses will be studied in the coming year.

Thermal Management

One concern with the TRD design is the management of the heat load in the system. The temperature variation within the straw array should be less than about 5 degrees Celsius for gas gain variations to be less than about 10%. The heat load per channel is expected to be of order 15 mw if preamps are located at the straw ends. This heat load, amounting to roughly 400 watts/m^2, will require fluid cooling. There is an additional heat deposition in the gas of the straw detectors as the positive ions due to the gas amplification of the ionization by charged particles are transported through the applied potential. This heat load is uniformly distributed along the straw. Since the foam in which the straws are imbedded is a good thermal insulator there is the potential for problematic temperature variations. We have used NASTRAN codes to model the two dimensional temperature profile to be expected in the TRD module, taking into account the anisotropic heat conduction through the foam, the straw material, and the gas, assuming heat sinks at the straw ends and at the front and back surfaces. The heat load was injected at ten points in this calculation; heat carried by the flowing gas was neglected.

The results are shown as isothermal contours in figure 5. The temperature variation is below 5 degrees for operation at design luminosity. Heat transport by the flowing gas supplements conduction and would keep the temperature variation lower still. However, at luminosities above about 5 times the SSC design luminosity the gain variations associated with the temperature variation would exceed the 10% goal. Rough calculations indicate that gas flows of order 1 cm/sec would be required to keep temperature variations acceptably low when running at ten times design luminosity.

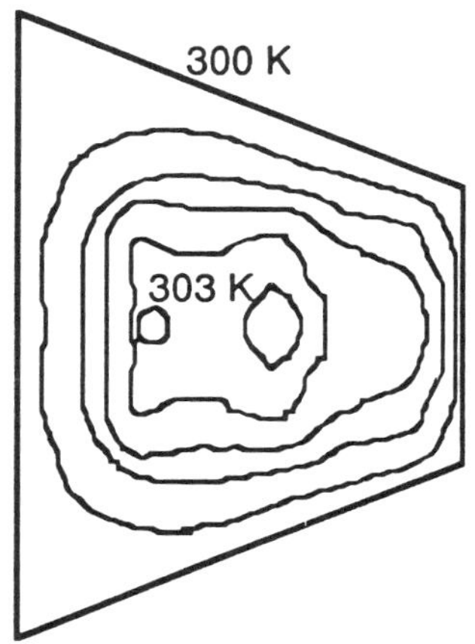

Figure 5. Isothermal contours in the TRD module due to heat deposition in the gas by drifting positive ions.

A set of chambers is being fabricated to test details of straw plugs, wire supports, and gas manifolds. These same chambers will be used in beam tests at the Bates Linear Accelerator to evaluate radiator materials. The plugs that provide mechanical support and localize the wire have been injection molded. The goal in this design is to provide precise wire support with a maximum aperture for gas supply while keeping the length of the end region to a minimum.

Study of Gas Mixtures

The gas mixture used in the TRD is selected to meet a number of criteria: sufficient xenon concentration to give good x-ray absorption in a few millimeters for x-rays in the energy range of 5 to 20 KeV; good pulse height resolution to facilitate pulse height discrimination; high drift speed for low occupancy; and good performance in a high radiation environment. Extensive tests have been done at the Moscow Physical Engineering Institute to explore the properties of various gas mixtures. Xenon-carbon dioxide mixtures have been found to be attractive due to their relatively high drift velocities, good pulse height performance, and excellent ageing behavior. Results on drift times in 4 mm diameter proportional straws are are shown in figure 6. The variation of drift time with anode wire diameter is due to the dependence of drift velocity on electric field. Further results are presented in reference [8]. Work is continuing on ageing and on the performance of other mixtures; trinary mixtures of xenon, CO_2, and CF_4 are particularly interesting for their high drift velocity and good resolution.

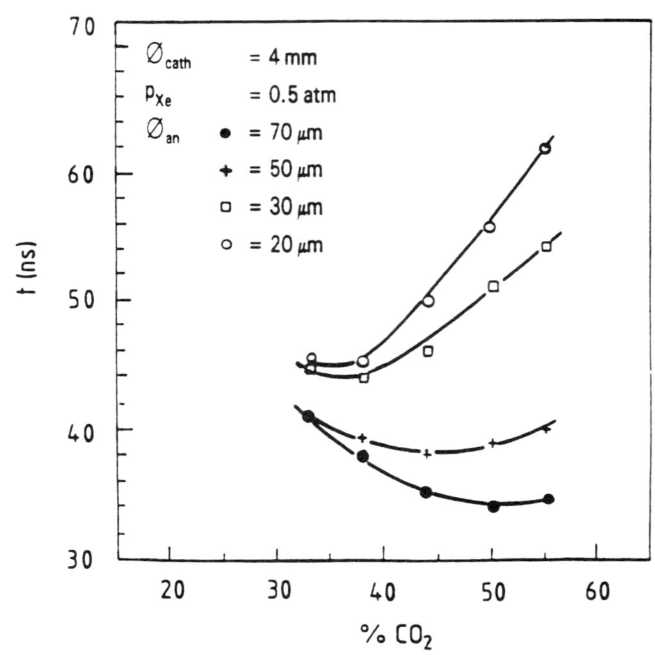

Figure 6. Drift times in various Xe/CO_2 mixtures.

Optical Modulator Development

Readout of the TRD proportional straws is a daunting project: the number of channels is large and there is a high premium on minimizing the space lost between modules to the electronics, cabling, gas supply, and cooling. One very attractive option is to read out the straws with optical modulators. In this approach, the proportional straw is attached directly to an electro-optic structure, with no preamplifier. The proportional signal alters the optical properties of the device and modulates a laser signal that carries the information to the outside. This technology has several advantages: it has a high dynamic range, a high bandwidth, power dissipation at the detector is very low, and the optical fibers that carry data to the outside offer a low-mass compact solution to the cabling problem. Similar systems have been used for special applications on a small scale.[9] A prototype system is under development; first results are described in these proceedings in the paper of V. Radeka et. al. Optical modulator readout of several hundred channels of TRD is part of the beam test planned by the SSC Subsystem TRD Collaboration for the coming summer.

Strawless Detector Development

Estimates of the cost of a tracking TRD based on straw proportional chambers and experience with straw tracking systems indicate that a large fraction of the cost is in the manufacture of the straws and in the assembly of straws and end pieces. We have begun to explore the possibility of making the proportional chambers without the use of the straws. Our goal in this work is to mold layers of half-straws into the surface of the radiator material; the foam will be metalized to provide a continuous cathode and to serve as a barrier to permeation of xenon into the foam. Chambers would be constructed of pairs of such molded components; stringing of this detector would be done ``open-faced" in a labor-efficient process.

Some studies of mechanical properties and fabrication techniques for the BXL foam have been made at Boston University. The material does not machine well to .005 inch tolerance except by grinding. Precision fabrication is possible, however, by foaming into appropriate molds. This technique is being pursued in conjunction with United Foam Plastics, a company that does precision molding and metalization of various foams including the BXL products. The prototype device is based on the foam molding shown in figure 7.

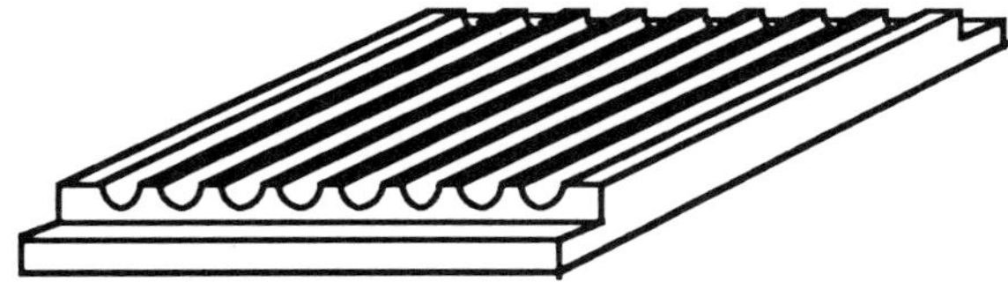

Figure 7. Polyethylene foam molding for a TRD on the half-shell.

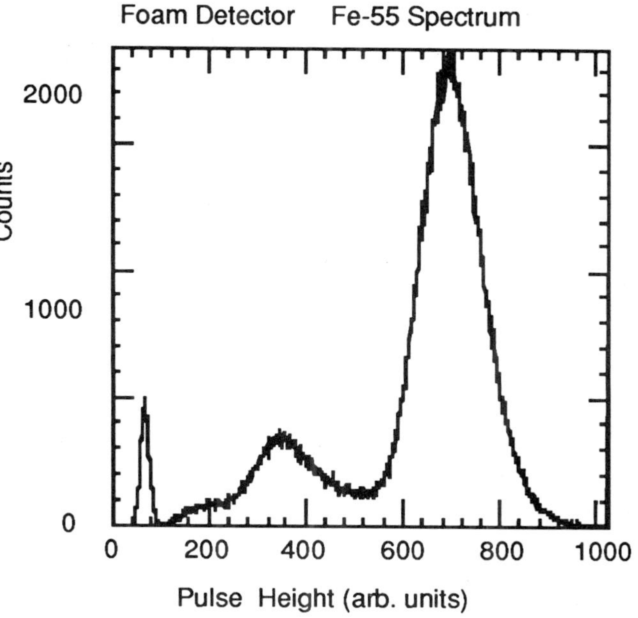

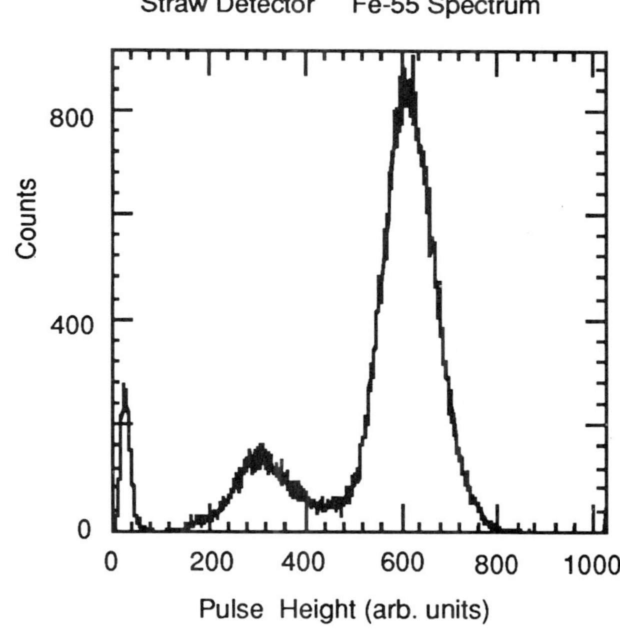

Figure 8. Pulse height spectra for Fe-55 from both a 4 mm straw proportional chamber and from the strawless molded foam detector.

The molding was made with BXL LD70 foam. There are eight 4 mm diameter half cylinders with 10 cm active length. The molded pieces were then metallized with a vacuum sputtering of aluminum. Measurement of the surface resistivity indicates that the aluminum thickness is of order 500 Angstroms.The Prototype detector was constructed with printed circuit boards glued to the ends of the molding. wires of 25 μm diameter were glued and soldered to the pc boards, and gas manifolds were formed by gluing additional g-10 pieces in. The detector was completed by attaching a similar molding above.

This detector operates as a proportional chamber in the usual way. A pulse height spectrum from exposure to Fe-55 x-rays is shown in figure 8, along with a comparable spectrum from a proportional straw. The measurements were made with 50% argon 50% CO_2. In both spectra the primary peak at 5.95 KeV and the escape peak at 2.8 Kev are clearly evident. The primary peak in the straw detector is resolved with a σ of 10%, and the resolution in the foam detector is quite comparable. Work is underway to study the mass production of proportional chambers using this technique.

Summary

Significant progress has been made in a number of areas in the development of a tracking transition radiation detector: the design has been refined through simulations and calculations, a 240 channel prototype device has been tested, and components are being developed and tested. In the coming year we expect to construct a larger system with 1200 channels, to be equipped with faster electronics, and to test this device in a particle beam.

References

[1] See for example R. E. Ansorge et. al., Nucl. Instr. and Meth. 265 (1988), and J. Alitti et. al., CERN-EP/88-126, September 1988.

[2] P. Mockett, to be submitted to Nucl. Instr. and Meth., and P. Mockett and S. Shapiro, EMPACT EOI, section 2.3.2.

[3] B. Dolgoshein et. al., to be submitted to Nucl. Instr. Meth.

[4] Cittolin, S. et. al., MACUA1, CERN-UA1 TN 90-01.

[5] Dr. David Eaves, BXL Plastics Limited, private communication.

[6] The neutron study was done by Geoff Forden of the University of Arizona.

[7] SSC-SR-1035, Radiation Effects at the SSC, M. D. Gilchriese, Editor. SSC Central Design Group, June 1988.

[8] B. Dolgoshein, M. Potekhin, A. Romanyuk, and V. Sosnovtsev, CERN-EP/89-161, December 1989.

[9] M. Lowry, paper in these proceedings.

TEST BEAM PERFORMANCE OF A TRACKING TRD PROTOTYPE

J.T. Shank, J.S. Whitaker,
Boston University, Boston, MA

V.A. Polychronakos, V. Radeka, D. Stephani
Brookhaven National Laboratory, Upton, NY

H. Beker, R.K. Bock, M. Botlo, C.W. Fabjan, J. Pfennig, M.J. Price
CERN, Geneva, Switzerland

W.J. Willis
Columbia University, New York, NY

T. Åkesson
Lund University, Lund, Sweden

V. Chernyatin, B. Dolgoshein, P. Nevsky, M. Potekhin, A. Romanjuk, V. Sosnovtsev
Moscow Physical Engineering Institute, Moscow, USSR

I. Gavrilenko, S. Muravjev, A. Shmeleva
P.N. Lebedev Institute of Physics, Moscow, USSR

Abstract

A Tracking Transition Radiation Detector prototype has been constructed and tested. It consists of 240 straw tubes, 4 mm in diameter, imbedded in a polyethylene block acting as the radiator. Its performance as an electron identifier as well as a tracking device for minimum ionizing particles has been determined.

Introduction

Lepton identification is expected to play a crucial role in addressing the TeV scale physics which will become accessible with the commissioning of the Superconducting Supercollider (SSC). Transition radiation is potentially a powerful tool for particle identification as its measurement is complementary to calorimetry since the radiation process does not significantly affect the particle energy and direction. Our collaboration has been engaged in a R&D program to develop a transition radiation detector and integrated tracking device for possible use with an experiment at the SSC or the European Large Hadron Collider (LHC).

Prototype Construction and Testing

The detector consists of 4 TR modules each containing 60 straw tubes. The straw tubes are made of 64 μm thick polycarbonate and are placed in holes drilled into a polyethylene foam block which acts as the radiator. The foam has a density of 67 Kg/m^3 and an average wall thickness of 5 μm . Previous tests of this foam [1] show that it has an X-ray yield 85% of that of a regular stack of polyethylene sheets. The gas used in the straw tubes for most of the running was 50% Xe 50% CO_2. A small lead glass calorimeter was placed behind the TR modules to allow distinction between pions and electrons. Additional information about particle identification was obtained from two Cerenkov detectors in the beam line.

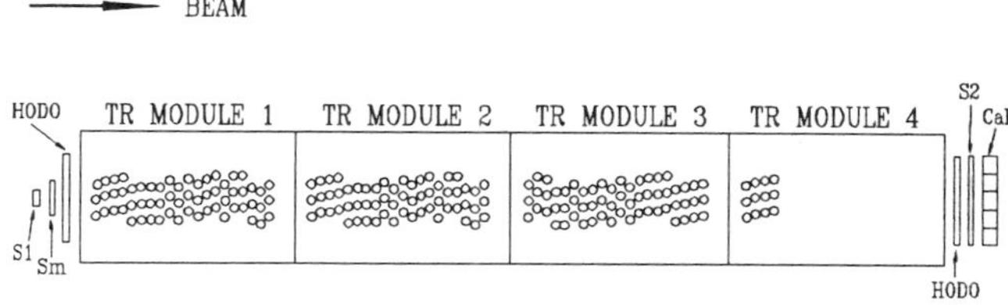

Figure 1: Experimental setup.

The overall setup is depicted in Fig. 1. The electronics for this test consisted of preamplifiers mounted near the ends of the straws connected via approximately 45 meters of coaxial cable to fast(12 ns FWHM) shaping amplifiers, whose outputs were digitized by LeCroy 2282 charge integrating ADCs. The data were acquired by a Macintosh II and the MacUA1 system software [2]. Later runs were made with a LeCroy 9424 digital scope connected to one channel just after a special very fast (6 ns FWHM) shaper. This allowed us to record the waveform in order to study the effect of different shaping times. The straw tubes were calibrated by monitoring 4 tubes which had ^{55}Fe sources permanently attached. Channel to channel pedestal and gain variations were corrected by periodically recording data from a signal pulse injected into each preamplifier. The 5.9 keV X-ray peak of a typical ^{55}Fe spectrum was measured with a full width at half maximum (FWHM) of 22%. During construction of the modules the straw-to-straw uniformity of the gas gain was measured for a batch of more than 200 straws. The rms deviation of the gains was found to be $\approx$ 1%. The setup was installed in the X5 test beam of the CERN Super Proton Synchrotron (SPS).

Results and Conclusions

Typical runs consisted of single particles passing through the center of the detector at right angles to the straws in order to determine the isolated pion rejection of the detector.

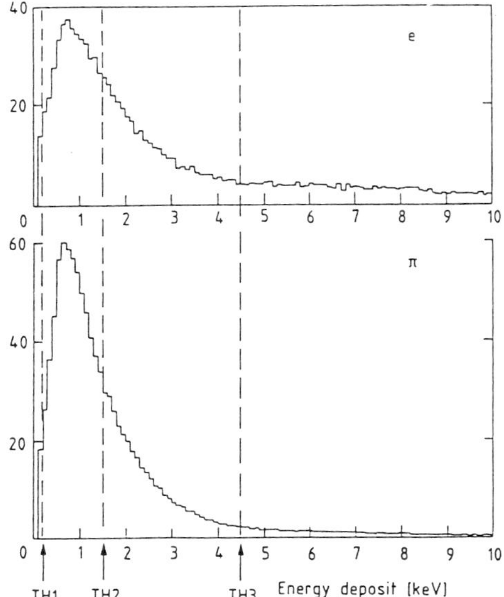

Figure 2: Pulse height distributions.

In Fig. 2 are shown pulse height spectra of all straws crossed by the beam for 10 Gev/c electrons and pions. There is a large excess of energy deposition above 4 keV for the electrons , as expected from TR X-rays. At 10 GeV/c the average number of straws along a track with greater than 5 keV energy deposition is 1.4 for pions, compared to 6.9 for electrons. The average number of straws traversed per track is 32. We have determined the TR photon yield as a function of γ in the interval $70 < \gamma < 6 \times 10^4$ using the following beams:

- pions at 10,20,50 and 75 GeV/c

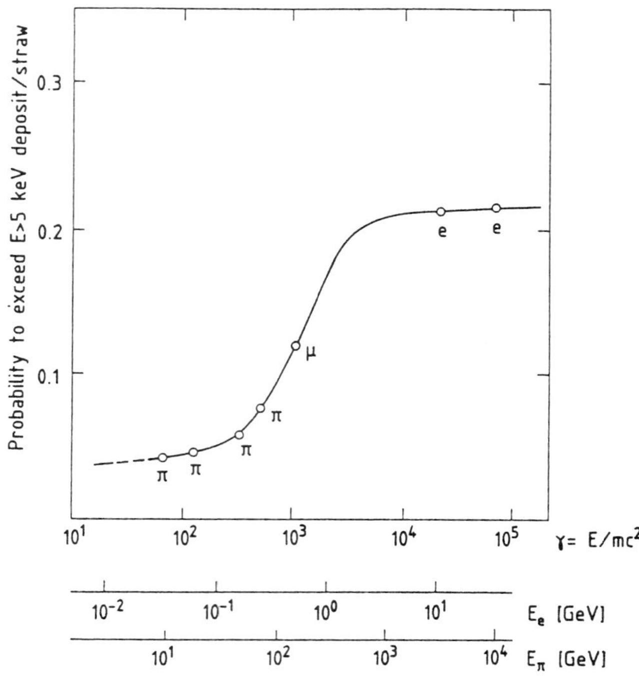

Figure 3: TR photon yield as a function of γ.

- muons at 110 GeV/c
- electrons at 10 and 30 GeV/c

The results are presented in Fig. 3 and indicate both the contribution of the relativistic rise in the energy loss, dE/dx, ($E_\pi < 50$ GeV, $\gamma < 350$), and of the TR photons ($\gamma \geq 350$).

In order to determine the electron/pion discrimination capability, two parameters must be fixed: the energy threshold, E_3, above which a hit is considered a TR X-ray, and the required number of such hits along the track, N_3. The track is called an electron if it has greater than N_3 straws with energy deposition greater than E_3. Further discrimination of false electrons from various sources can be achieved by utilizing two additional thresholds. The first of these thresholds, E_1, is set at $\approx$0.2 KeV and provides 90% detection efficiency for relativistic particles and is also used for tracking. The second threshold, E_2 = 1.5 KeV, (located between the one- and two-particle dE/dx distributions) has 50% efficiency for minimum ionizing particles and 80% efficiency for conversions. In Fig. 4 are shown scatter plots of the frequency of high energy clusters ($E > E_3$), N_3, versus energy deposits satisfying the condition $E_1 < E < E_2$, N_{12}, for measured 30 GeV/c electrons (a), 10 GeV/c pions (b), and convolutions of the data to simulate two pions (c), three pions (d), and two electrons (e). The solid lines represent two-dimensional maximum likelihood cuts for 90% electron efficiency. The importance of the two-threshold algorithm in rejecting overlapping pions or electrons (e.g. from conversions) is evident.

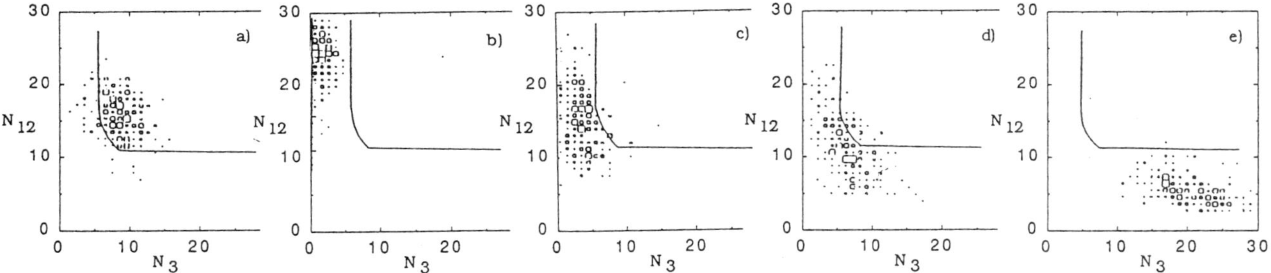

Figure 4: Pulse heights with $E_1 < E < E_2$, N_{12} vs $E > E_3$ (see text)

The tracking performance of the detector was measured by reconstructing two tracks from one: the first track uses the odd number straws on the original track, the second uses the even straws. Each of these "tracks" is an independent measurement of the same original track.

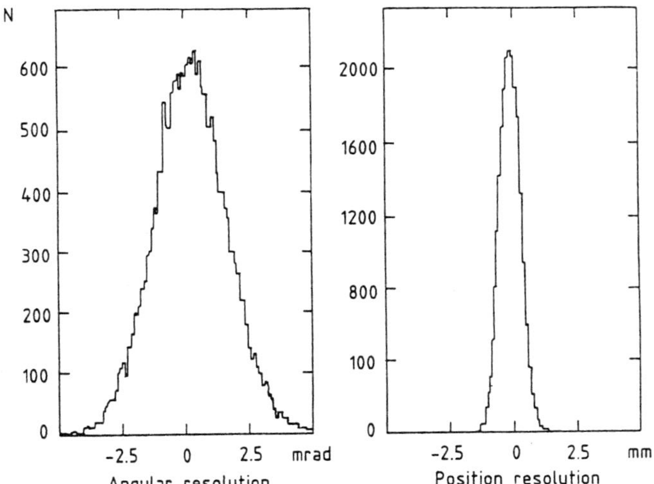

Figure 5: $\delta\alpha$ and $\delta\beta$ for fitted tracks.

Straight line fits are made to each track

$$y = a_1 x + b_1 \qquad y = a_2 x + b_2$$

and the differences in the slopes and intercepts

$$\delta_a = (a_1 - a_2)/2 \qquad \delta_b = (b_1 - b_2)/2$$

are plotted in Fig. 5. Fig. 5(b) shows the error on track reconstruction to be 0.4 mm.

Runs were made with isolated pions and electrons at various energies and with the TR modules rotated so the beam crossed the straw tubes at angles of 42 and 63 degrees. Noticeably better performance was observed for these runs reflecting the longer radiator traversed as well as the diminished space charge effects because the anode charge is not collected at a single point.

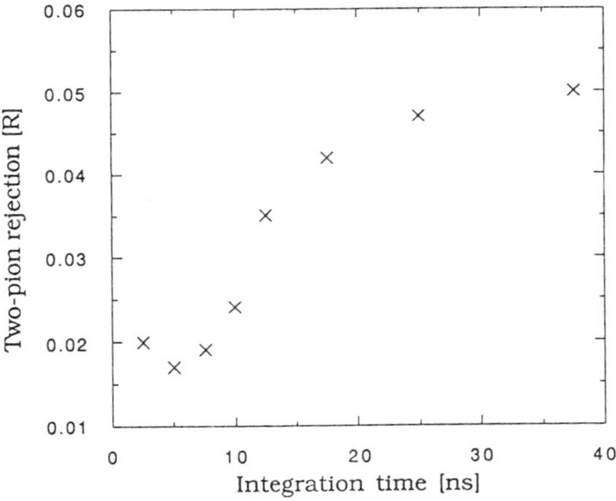

Figure 6: Two-pion rejection as a function of effective integration time.

The absorption by a Xe atom of a TR photon produces several primary electrons due to the Auger effect. Their range is much smaller than that of a delta-ray of energy equal to the energy of the TR photon. The produced ionization cluster is therefore much more localized. One could then improve the discrimination between TR photons and spatially extended ionization clusters by very fast shaping. We used the data from the recorded waveforms of the straw instrumented with the 6ns (FWHM) shaper to test this hypothesis. Electron and pion "tracks" were obtained by convolution and the single and double pion rejection was determined as a function of the effective integration time. No significant improvement was found for single pion rejection. The discrimination against two pions, however, improved by a factor of 2.5 as shown in Fig. 6. The saturation below 10ns reflects the finite extend of the ionization cluster due to diffusion.

We wish to thank L. Gatignon for his help with the X5 beam.

References

[1] B. Dolgoshein, *et al.*, to be published.

[2] S. Cittolin, *et al.*, MACUA1, CERN-UA1 TN 90-01.

A TRANSITION RADIATION DETECTOR AS A TRIGGERING DEVICE

R.K.Bock and J.Pfennig (CERN, Geneva)

(presented by R.K.Bock)

1. The problem definition

For potential use in future collider detectors, a transition radiation detector has been proposed [1] whose capability in e/hadron discrimination can be put to use in real time. In this contribution, we discuss the data handling aspects of this device, characteristic for a class of triggering devices at supercolliders.

We use as a TRD model a straw configuration called 'halo model', schematically shown in fig.1. Straws are arranged between 70 and 120 cm of radius around the beam pipe, in multiple planes (a total of about 900), and about 580 straws to a plane. In order to get optimal tracking capabilities from this detector, every alternate plane carries the straws in a reverse arrangement, thus allowing full stereo views. Note that some multitrack ambiguities can arise in such a tracking arrangement, but due to the non-parallelism of straws they are minimized.

Readout of the roughly 500 000 straws (on outer end only) is proposed to be largely centralized off the detector, to avoid the various problems encountered by on-detector intelligence: radiation, obstruction, and power dissipation. For the intended trigger, this siting issue is secondary, except that full access to all channels must be possible. We assume that the front-end readout takes care of signal extraction for individual bunch crossings, and remains passive until a (calorimeter-based) signal for a given time slice is received, indicating both that the corresponding bunch crossing establishes one or more electron candidate(s), and where these candidates are to be found ('pointer'). The assumption is made that such bunch crossing candidates do not occur more frequently, on average, than every 10 μsec., i.e. the first-level trigger reduces from the bunch crossing rate by a factor of 1000. It has become customary to term this kind of reduced-rate/fine-grain/full-precision/simple-algorithm/local trigger a 'second-level' trigger, for obvious reasons.

We have chosen the halo model for straws because it shows the typical second-level trigger problems in an extreme form: bandwidth of transmission, locality of algorithms, component performance for both low- and high-level aspects of the algorithm.

2. The triggering algorithm

The basis of discriminating electrons from hadrons in a TRD is a statistical analysis of pulse heights of all digitizings belonging to a track, so that a measure for the probability of X-ray emission (and detection) can be determined. This probability is a function sensitive to the Lorentz factor γ of the track. Hence the finding of a (stiff) track pointing towards the calorimeter cell signalling an electron candidate is a first local step, its (statistical) pulse height analysis a second local step, and a decision taking into account correlations of (multiple) signals, physics cuts, etc., is a concluding global step of the algorithm. All computations have to be preceded, for practical reasons of bandwidth and performance, by a data selection step under control of the first-level pointers.

3. Data flow through the TRD trigger

Due to the fact that tracks can pass through roughly one quarter of the longitudinal extension of the detector and due to the statistical nature of track detection, a single algorithm running in a single hardware unit is conceivable only in a serial von Neumann machine. We are interested in parallelizing and/or pipelining the process, hence we propose to subdivide the 2nd level trigger into three phases, each of them implemented on a different type of hardware and running a dedicated algorithm. The following sections will expand on the proposed practical solutions, as we envisage them at this early state of design.

3.1 Phase I algorithm: data distribution

All data belonging to bunch crossings that passed the 1st level trigger selection, will first be synchronized and then transported from the experiment site to the main trigger electronics in a counter house. The synchronization has been introduced to allow simple point to point communication between the readout logic and the trigger, as well as between the phases of the trigger. The TRD trigger, internally, will operate synchronously.

At the input side the full data rate is 100 GBit/s. All incoming data will be kept during a period of some msec in public addressable memory, to allow other components of the experiment to monitor the TRD and the 2nd level trigger performance. The data transport at this level could be based on the HPPI standard [2].

Only a fraction of the data is relevant for tracking in the trigger. Using the 1st level trigger information and limiting the number of electron candidates to four (independent calorimeter candidates), one sixth of the data will be mapped into 4 data windows that will undergo further processing. Next the data is mapped from the TRD specific straw representation into plane rectangles of pixels. The plane contains the beam axis and the firing calorimeter cell, and for each potentially interesting track such a plane containing all space points belonging to it can be constructed.

Data mapping and selection are based on loadable Look-Up tables, where the data packing is handled by DSPs or similar hardware (no buses, but of the order of 512 local point-to-point connections).

3.2 Phase II algorithm: searching for information

Every 10 μsec (one time slot) the data of 4 to 8 planes (depending on the calorimeter resolution) generated from each of the 4 data windows is sent to the phase II hardware. This is acomplished using the some type of HPPI links mentioned above.

The phase II processors are divided into 10 groups, called banks, to work in parallel on different time slots, and that are used in a pipelined fashion. This is required in order to extend the interval that is availlable for processiing from 10 to 100 μsec. Each time slot is assigned to a processor bank on a round-robbin schedule. On the input side, the data can be further reduced by requiring digitizings to be inside of roads defined by the calorimeter information from the first-level trigger.

For processing of phase II, a massively parallel architecture can be put to use. A solution that has been studied so far is the SIMD 'associative string processor' (ASP) [3]. An ASP implementation of phase II would use 10 Fastbus crates with a total of abouty 640.000 processing elements, on 80 boards, as well as 40 controller/road-builder boards. The processing power of this solution allows to check for about 128 plane areas, each one covering a trapezoid of 16 TRD slices out- and 32 TRD slices inside (per time slot).

In our present thinking, the track finding algorithm is based on shifting (distorting) the trapezoid and histogramming in parallel. The shift (angle) for the highest value is saved for each bin, and should, in an ideal case, coincide with a track. Note that a balance has to be found between calorimeter tower size, vertex area, and histogram bin size.

3.3 Phase III algorithm: decision taking

As a result of Phase II the n (about 64) best peaks would be read out (still under control of the ASP Hardware) and transferred to n corresponding processors in the phase III hardware. Processors in phase III are assumed to be conventional RISCs, programmed under C in a real time operating system environment. It will be at this level that sensitive physics- or luminosity-dependent cuts and limitations are introduced.

The interface of the second-level trigger to other components of the detector is via public addressable memory. The 3rd level trigger, for example, polls for the second-level status and data, and can therefore be treated as completely asynchronous. Assuming this interface to be implemented in buses like SCI [4], even the complete TRD raw data could be read out occasionally without affecting the trigger operation. The total latency is expected to lie below 4 msec, and raw data can safely be kept available for a time span of 20 msec.

4. Highlighting some of the innovative ideas

The discussion above contains a number of ideas not currently found in detector readout systems. We want to highlight them briefly in this section.

The internally *synchronous operation* of the entire second-level trigger facilitates hardware implementations enormously. At least phases I and II can readily seen to be operating in fixed time frames (i.e. time does not depend on the data). Abandoning thresholding of data for compression ('zero-suppression', not very effective at the TRD's high occupancy), and data-independent algorithm execution are a prerequisite for this.

The *separation of data and control* flows is made possible by the synchronous operation and the use of simple point-to-point connections. As fixed-size data blocks are transmitted, block transfers can be initiated. Control information, on the other hand, is low in volume and needs broadcasting, hence can be carried by buses.

The data is kept in *distributed memories*: no internal data collection is required. The availability of buses like SCI will, nevertheless, allow access to the full data set for monitoring and testing, albeit at a reduced rate.

Due to the complexity of the architecture and the vital impact of trigger functioning on physics, a *self-testing mechanism* must be built into the system. We foresee to use the idle time slots, invariably occurring as a consequence of synchronous operation, when a no-trigger situation occurs, to inject sets of test patterns instead of real data. A clever way of testing all parts of the trigger will have to be devised.

5. Conclusion

We have discussed above a possible architecture for executing the TRD trigger algorithm at a frequency of 100 KHz, using a number of innovative ideas not all restricted to this particular detector, and based on architectural components available in 1991. The evolution of technology will make solutions of 1993 or 1995 simpler and cheaper, and may also favor alternative architectures. We are thus confident that the TRD, even in its most complicated form, can perform its triggering task in real time, if the device performance and the physics environment turn out to be as predicted today.

We consider it our task in the coming two or three years to demonstrate the various parts of our solution in hardware, emulating as closely as possible the expected situation at future supercolliders.

6. References

[1] EMPACT Expression of Interest, M.D.Marx, coordinator, May 1990.

Integrated High-Rate Radiation Detector and Tracking Chamber for the LHC, Proposal, B.Dolgoshein, spokesman, CERN - DRDC/90-38

[2] HPPI, High-Performance Parallel Interface, Specifications, ANSI X3T9.3/88-023 (February 1990)

[3] M.Lea, ASP: A Cost-effective Parallel Microcomputer , IEEE Micro Oct.1988.

[4] SCI Scalable Coherent Interface, IEEE P1596.

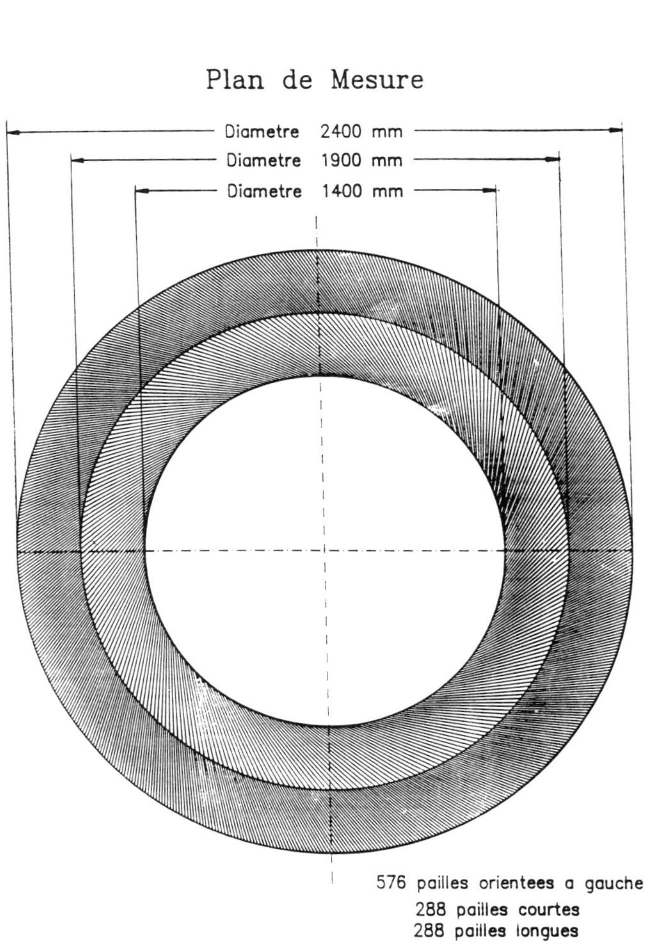

Fig 1: TRD Halo Model

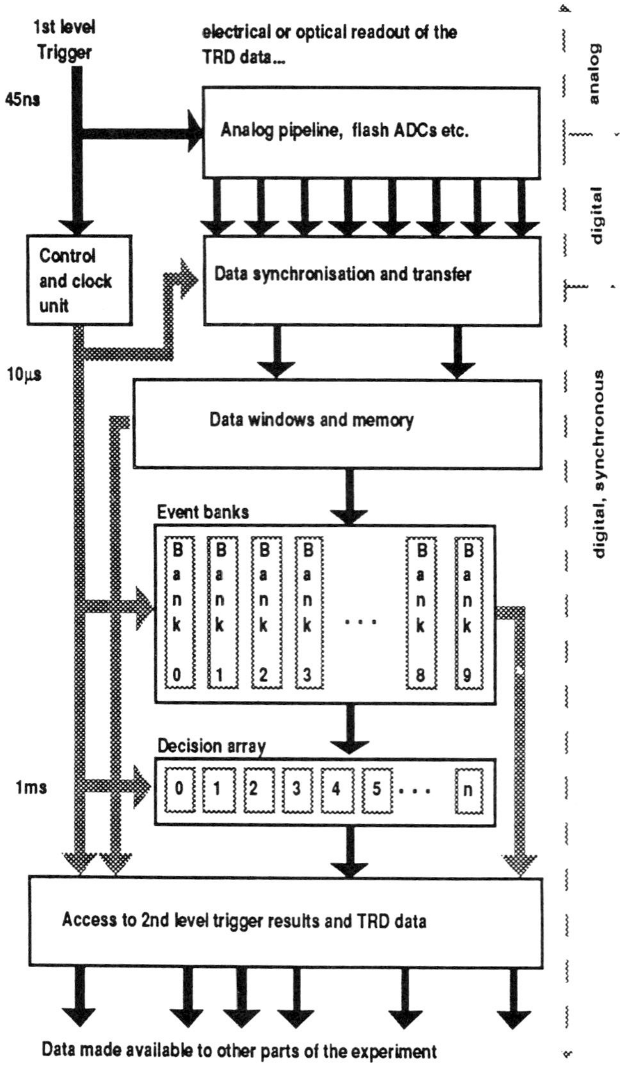

1st level Trigger
electrical or optical readout of the TRD data...
45ns
Analog pipeline, flash ADCs etc.
Control and clock unit
Data synchronisation and transfer
10µs
Data windows and memory
Event banks
Bank 0 Bank 1 Bank 2 Bank 3 ... Bank 8 Bank 9
Decision array
1ms
0 1 2 3 4 5 ... n
Access to 2nd level trigger results and TRD data
Data made available to other parts of the experiment
analog
digital
digital, synchronous

Fig 2: TRD Data Flow

OPTIMISATION OF THE TRANSITION RADIATION DETECTOR

J. C. Palathingal
Department of Physics
University of Puerto Rico, Mayagüez, P. R. 00708

The first results of a continuing investigation for an optimal design of the transition radiation detector and an efficient method of analysis of the detector signals are presented. The studies show that an improvement of about an order of magnitude over current models can be achieved through a proper choice of parameters and an improved method of analysis of the detector signals. The optimal features of a 30-cm TRD and the estimated performance are described. The basic aspects of analyzing the detector output are outlined.

Introduction

Theoretical and experimental studies of the past have shown [1-3] that the transition radiation produced by GeV-range electrons are primarily soft x rays which are sparcely produced, and also easily absorbed in the radiator itself. As a result, in the earlier designs, the typical rate of detection of photons was around 0.1 per cm of TRD length [4]. The need for reduced absorption hence necessitates the use of thin radiators in a row interposed with photon detectors. Such a modular design was pursued in the current studies. The results could be summarized thus:

1. Designs with photon detection rates up to 0.3 per cm length of TRD are possible.
2. Improvements of nearly an order of magnitude in pion-electron rejection ratio can be brought about.
3. The method of analysis of the signals is most important.
4. The multiwire proportional counters normally employed as photon detectors should have a minimum thickness that could be consistent with absorption of an adequate proportion of photons.
5. Very thin foils (for example, 13 microns) are less preferable to foils of medium thickness (example, 25 microns) since the number of modules in the TRD can be reduced with the latter, and the rejection ratio could be realized by the overall length of the TRD.

Parameters investigated and results

The studies hitherto were conducted over CH_2 foils, of two thicknesses, 12.7 and 25.4 microns, stacked with gaps 8 times larger. The TRD has a length very close to 30 cm, in all cases. Photon detectors of Xe gas of two thicknesses, 0.6 cm and 1.0 cm were considered. A space of 0.55 cm was allowed between modules for thin dry-nitrogen chambers on each side of the Xe counters and for electrical wiring. The most appropriate number of radiator foils was found to be close to 90. Much beyond this, self-absorption could be too serious. Too-thin radiators result in an avoidable multiplication of the number of modules of which a good too many yield no photon signal at all (Table 1).

Since the ionization produced by an electron or pion in the MWPC is a major contributor to the signal output, optimizing the detector thickness is quite significant. A 0.6-cm thick Xe detector (in combination with a 2-cm radiator of 25.4 microns foils) could yield superior performance as compared to one of thickness 1 cm (Table 2). The TR+ionization spectrum (normalized to unity) produced by electrons of energy in the GeV range in the 0.6-cm thick photon detector is given in Fig. 1, along with

Table 1. Photon Detection Rates

Foil Thickness	12.7 Microns			25.4 Microns			
# Foils	40	88	132	44	66	88	132
Rad. length (cm)	0.46	1.0	1.5	1.0	1.5	2.0	3.0
Module length (cm)	1.6	2.15	2.65	2.15	2.65	3.15	4.15
Photons/module	0.35	0.53	0.65	0.58	0.74	0.86	1.04
Photons/cm	0.22	0.25	0.25	0.27	0.28	0.27	0.25
Average signal (keV)	7.2	9.5	10.0	10.0	11.0	12.4	14.4

the normalized ionization spectrum produced by pions.

Analysis of Signals

In the approach followed, whichever signal from the MWPC having height above a preset threshold will be considered to be a trigger of the module. In order to identify a particle as an electron, a minimum predetermined number of modules in the TRD array should register trigger signals simultaneously, and the probability that an electron could do so determines the detection efficiency. The selection of the threshold level (Table 2) is critical. At a preset threshold, the probability of a pion causing a number of triggers that could misidentify it as an electron must be minimal. The minimum number of triggers that defines an electron event is decided on the basis of the experiment-required efficiency for electron detection. The ratio of the probabilities that an electron or a pion could produce this many or more triggers simultaneously determines the pion rejection factor. A range of efficiencies, from 60 to 98% were looked into, and, as would be expected, the rejection factor dropped drastically with increasing electron detection efficiency.

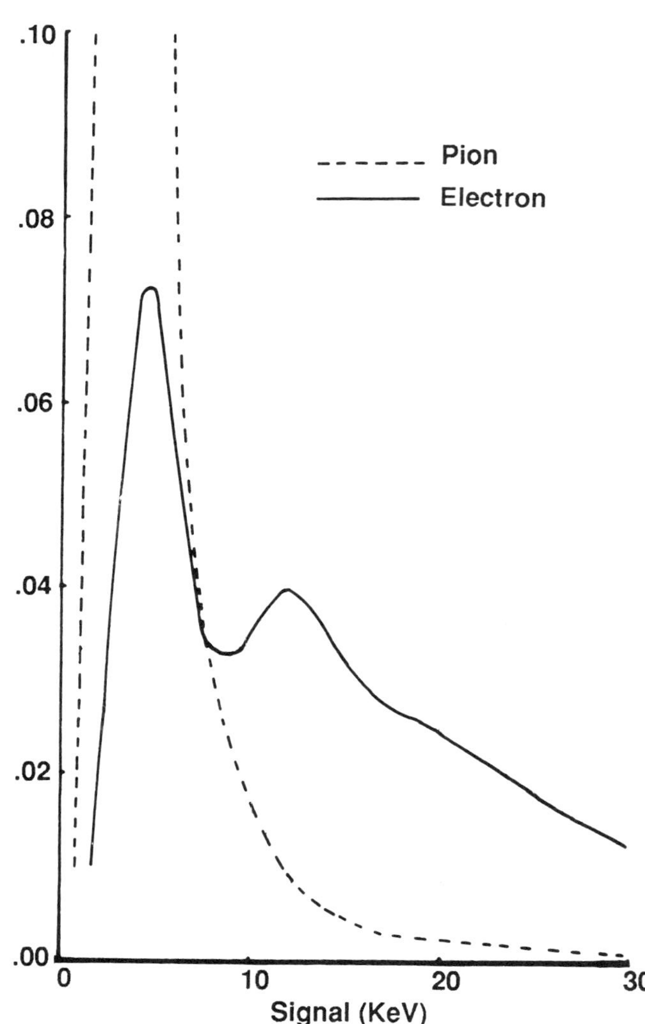

Figure 1. Normalized spectrum from a single module (25.4 micron foils, 2.0 cm radiator, 0.6 cm Xe detector)

Table 2. Pion Rejection Factor for Two Detectors

Discr. Level (keV)	Rejection Factor 0.6 cm	1.0 cm
5	85	10
7	170	75
10	475	290
12	270	310
15	120	225

Table 3. Best Pion Rejection for a 30-cm TRD

Foil Thickness (microns)	Radiator Length (cm)	Optimum Discr. (keV)	Pion Reject Factor
12.7	0.46	4.0	220
	1.0	10.2	650
	1.5	8.0	185
25.4	1.0	10.0	620
	1.5	10.5	620
	2.0	10.5	630
	3.0	11.2	210

Conclusion

The studies conducted so far have yielded promising results (Table 3). Investigations are being pursued in greater detail on the different parameters already looked into, and also over other features such as optimum separation between foils and a more efficient gas mixture to be used for photon detection.

References

(1) X. Artru, G. B. Yodh and G. Menessier, "Practical Theory of the multi- layered transition radiation detector", Phys. Rev. D12, 1289-1306 (1975).

(2) J. Fischer, S. Iwata, V. Radeka, C. L. Wang and W. J. Willis, "Lithium Transition Radiator and Xenon Detector Systems for Particle Identification at High Energies", Nucl. Instr. Meth., 127, 525-537 (1975).

(3) C. W. Fabjan and H. G. Fischer, "Particle Detectors", Reports Progr. Phys., 43, 1003-1063 (1980).

(4) B. Dolgoshein, "Transition Radiation Detectors and Particle Identification", Nucl. Instr. Meth., A252, 137-144 (1986).

FORWARD TRACKING WITH ENHANCED ELECTRON IDENTIFICATION†

EHIT Collaboration

J.M.Bailey[(a)], G.A.Beck[(a)], P.J.Bussey[(b)], P.Colrain[(b)], J.B.Dainton[(a)], E.Gabathuler[(a)], J.C.Hart[(c)], J.W.Hiddleston[(c)], S.J.Maxfield[(a)], N.A.McCubbin[(c)], J.M.Morton[(a)], G.D.Patel[(a)], B.T.Payne[(c)], C.Raine[(b)], E.J.Romans[(c)], D.P.C.Sankey[(a)], D.H.Saxon[(b)], I.O.Skillicorn[(b)], D.J.White[(c)]

†Supported by the UK SERC and the US DoE through SSClab R & D project SSC-PC-024.

[(a)] Department of Physics, University of Liverpool, P O Box 147, Liverpool L69 3BX, England
[(b)] Department of Physics and Astronomy, University of Glasgow, Glasgow G12 8QQ, Scotland
[(c)] Rutherford Appleton Laboratory, Chilton, Didcot, Oxon OX11 0QX, England

Abstract

Design concepts and developments are described for the construction and use of charged particle tracking detectors in the intermediate pseudorapidity range ($1.4 < \eta < 2.4$). Electron identification is enhanced by simultaneous detection of transition radiation X-rays and ionisation energy loss.

Physics Agenda and Detector Concept

The physics agenda set the detector requirements. A representative range of processes to consider includes Higgs searches, t-quark physics (Wb, Ws, H^+b, $\tilde{t}\tilde{\gamma}$, $\tilde{W}\tilde{b}$, Z^oc decays), Z' search and triple gauge boson vertex studies [1].

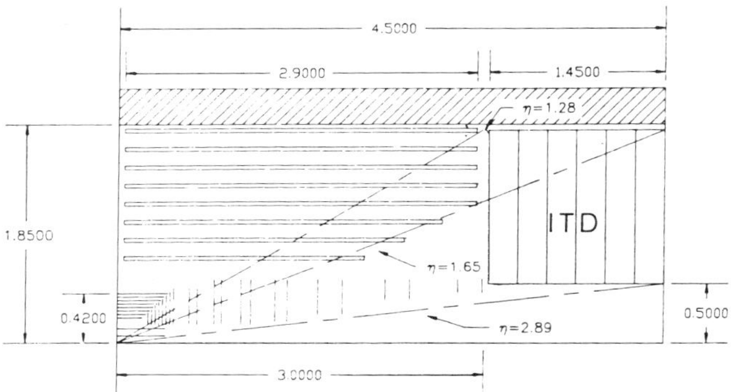

Figure 1 a) Charged track detector layout for use in 2T solenoidal magnetic field; 6 ITD modules are shown [2,3]

The (s)fermion channels have "large" cross-sections and detailed signatures, and the boson channels have low rates (at high masses) and simple signatures. Both types of process identify tracking, $e^{\pm}$ identification and charge sign determination over an angular range exceeding ±2 units of rapidity as prime requirements. In congested events it is a great benefit to have $e^{\pm}$ identification ($^{dE}/_{dx}$ + TRD) located on the electron track, as well as calorimeter signatures.

Figure 1 shows a detector concept for an Intermediate angle Track Detector (ITD) [2,3] using detectors that measure $\phi.r$ (150 μm accuracy) and r (2 cm accuracy) at fixed z (beam direction) in a 2T solenoid. The momentum resolution is compared to that of a dipole detector in figure 2. The accuracies are comparable but a solenoid is preferred because of azimuthal uniformity, advantages in systematic errors and track matching to an inner track/vertex detector (e g silicon in SDC, see figure 1), and in suppression of background caused by (very) low momentum secondaries produced in the beam pipe [4]. The measurement is obtained using radial-wire drift chambers that measure ϕ-differences directly. To obtain given a momentum error the number of sense planes needed is only half that of a Cartesian grid of the same cell-size. Because the radial-wire cell size is better matched to the varying hit density ($d^2N/drd\phi \sim 1/r$ in endcaps) the advantage grows to a factor of three.

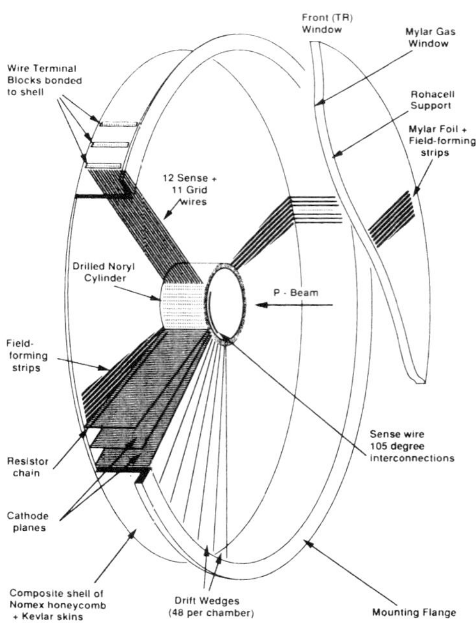

Figure 1 b) 48-sector radial chamber module used in the H1 experiment [5].

The equations of a track originating on the beam line at r = z = 0 in a uniform magnetic field are

$$\phi = \phi_o + {}^{eBz}/_{2p_z}$$
$$r = 2p_T . \sin({}^{eBz}/_{2p_z})/eB$$

Event associated tracks are thus exactly straight lines in the ϕ-z plane. This makes for easy track-finding within and between modules.

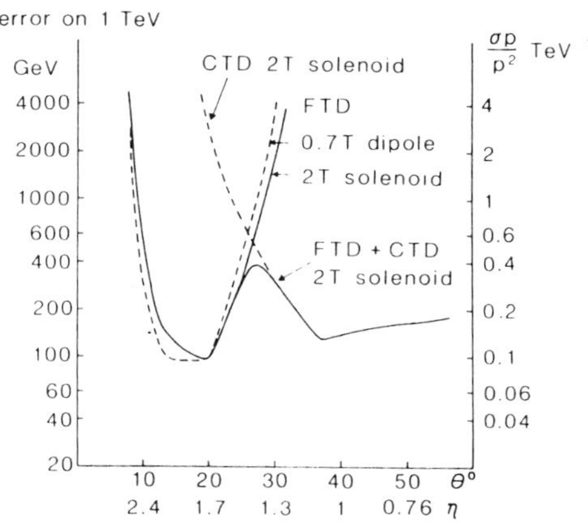

Figure 2: Comparative momentum resolutions for solenoidal and dipole intermediate angle tracking detectors (FTD) [4].

Detector Layout and Performance

The ITD in figure 1 has 6 (or 5) modules per end-cap with 300 sectors ($50 \leq r \leq 150$ cm). The maximum drift distance varies with radius from 5.2 to 15.7 mm. Each module has eight radial wire sense-layers giving a modest total of 14400 (or 12000) channels per end-cap. For readout, each sense wire is joined to another well separated in azimuth (105° in H1). Using resistive sense wire, charge division readout gives 2 cm position resolution in r. The whole detector is constructed of light composite materials. The z-space between modules is filled with suitable transition radiator (TR) material (polypropylene foils in H1). Traversing electrons emit ~1.4 collinear X-ray photons ($<E_X>$~6 keV) which convert in the chamber gas (30% Xe) in each module. Deposited charge ($^{dE}/_{dx}$ + TR) is used to distinguish e from hadrons.

With a gas gain of 2.10^4 the sense wire irradiation dose is 0.08 C cm^{-1} y^{-1} at $\mathcal{L} = 1.10^{33}$ cm^{-2} s^{-1} and the occupancy, (hit rate/cell) x (busy time), is 0.12. The busy time is equal to 2 x (two hit resolution)/(drift velocity). The H1 forward tracker has already achieved 1.5 mm two-hit resolution [5]. Given radial wire sense wire geometry and azimuthal electric drift fields in a solenoidal magnetic field, the component of drift velocity perpendicular to the wire cannot exceed

$$v_{perp} = {}^{E}/_{2kB}.\sin(2\theta_L)$$

where θ_L is the Lorentz angle of rotation of the electron drift relative to $\underline{E}$, and $k = {}^{E}/(vB \sin\theta_L)$. k ~ 1.1 for most gas mixtures. We therefore design our chamber for a fast gas with E = 3 - 4 kV cm^{-1}, B = 2T, $\theta_L = 45^o$, v_{perp}~ 100 μm ns^{-1}. (See [6] by J.M.Bailey for our work on gas mixtures).

The occupancy and lifetime figures indicate that the detector is viable up to $\mathcal{L} = 3.10^{33}$ cm^{-2} s^{-1}. Detector congestion and pattern recognition is discussed below. The major technical concern is the maintenance of the cathode-plane voltage gradient in the presence of the high anode current draw.

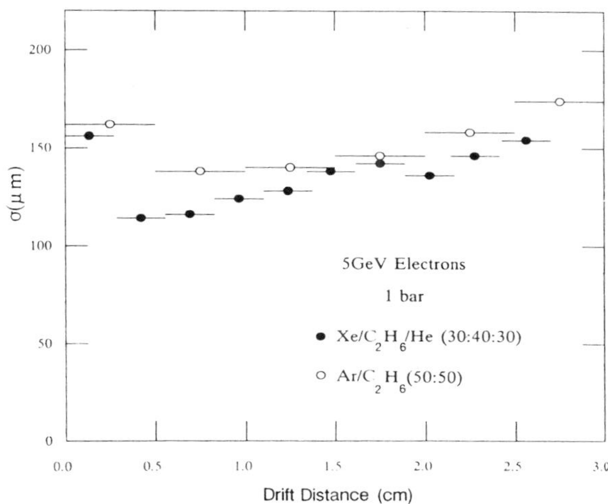

Figure 3 a) Position resolutions in ϕ.r as a function of drift distance in an H1 48 sector radial wire chamber [5]

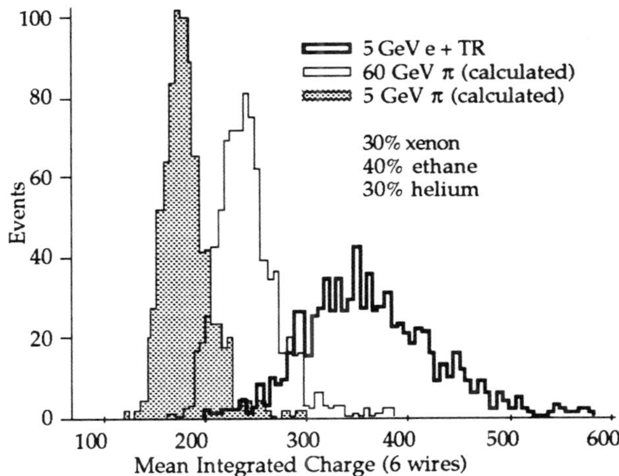

Figure 3 b) Integrated charge measurement for three 12 wire modules in the H1 configuration for 5 GeV e with preceding radiator; also shown are π spectra calculated from measurement of 5 GeV e without radiator [5].

The 48-sector H1 forward tracker [5] acts as a prototype for this device. Mechanical construction provides wire position accuracy of 38 μm rms. Pulse height measurement is independent of position along the sense wire to within 2% and is very uniform between sense wires. Figure 3(a) shows position resolution as a function of drift distance (everywhere better than 150 μm) and figure 3(b) the integrated charge measurement for electrons and pions. At 90% electron efficiency these provide a pion rejection of 200 at 5 GeV/c and 25 at 40 GeV/c. Similar precision is anticipated in each case for the SSC ITD as described. The ϕ.r and r measurements are naturally correlated by the readout (8-bit nonlinear FADC in this case). The device therefore provides three-dimensional space points with deposited charge information.

Crossing Tagging and Pattern Recognition

In a multi-hit device such as a drift chamber, one should distinguish between the busy time (related to the two-hit resolution) and the memory time (maximum drift time) from which the total number of hits read out with an event can be estimated. The number of read-out channels in the detector is kept below 30,000 (both end caps) by taking advantage of this distinction. We exchange time of drift for money (number of channels). The result is an accumulation of background hits due to out of time events. (The average memory time of a cell is 105 ns at v_{perp} = 100 μm ns^{-1}). To suppress them we propose the addition of an ancillary *crossing-tagger*. A conventional way to realise such a device is a 1.5 mm spacing MWPC with cathode readout strips arranged parallel to the drift chamber isochrones. These are read out (cheaply) via updating discriminators (3,000 channels per module at 1.5 mm spacing) and have a three-bunch-crossing latch time. An in-time ITD drift chamber hit thus has a crossing tagger latch exactly over it. Since the tagger strip matches the drift chamber isochrone geometry this can be done before any pattern recognition. The crossing tagger also trivially resolves the left-right ambiguity (the natural way to lay out the strips joins the isochrones of neighbouring cells), and provides prompt pattern signals that could be used in triggering. An exciting approach would be to use microstrip gas avalanche chambers with 30 μm position resolution and 125 μm two-track resolution [7]. These naturally have outstanding rate compatibility [1].

TABLE 1

ITD Hit rates for $H \rightarrow Z^oZ^o$ events

	without crossing trigger	with crossing trigger
Mean number of hits	73 (190) [108]	21 (31) [25]
Fraction of hits from Higgs	21% (8%) [14%]	73% (49%) [66%]
Fraction of hits from wrong bunch-crossing		1.5% (1.7%) [2.5%]
Fraction of Higgs hits lost (within 2 mm of earlier hit)	7% (14%) [9%]	
Luminosity	$\mathcal{L} = 1.10^{33}$ cm^{-2} s^{-1} ($\mathcal{L} = 3.10^{33}$ cm^{-2} s^{-1}) [maximal pair conversions]	

We have made pattern recognition studies using 20 + 20 TeV PYTHIA events [8] for $H \rightarrow Z^oZ^o$ superimposed on a background of minimum bias events. Table 1 gives the results for one detector plane. We conclude that the use of crossing triggers with the radial chambers provides a means of substantially eliminating hits from wrong bunch crossings. The remaining contamination comes almost exclusively from events in the same bunch crossing as the triggered event, and would of course be present in any detector with a time resolution equal to the bunch crossing interval. Such contamination must be removed by other means such as the use of the z-vertex, for which of course an accurate charged track detector is essential! The fraction of Higgs hits lost is small and is characteristic of the two-track resolution, better than in, for example, 4 mm diameter straw tubes.

Summary

The 300-wedge ITD is a powerful detection device. It gives 3D space points. Accurate ϕ.r measurement provides excellent momentum resolution and relatively straightforward track finding in a region of uniform axial (z) magnetic field. Electrons are identified in jets with pion rejection factors of 200 (5 GeV/c) to 25 (40 GeV/c). The geometry matches the physics - dense packed where needed, ϕ-z straight lines. The excellent resolution (150 μm in ϕ.r, < 2cm in r) has been proven in a large detector. Less than 30,000 full readout channels are needed. A crossing-tagger (30K hit channels) picks the wanted hits with only 1 - 2% background because it is precise in both time and space.

References

[1] D.H.Saxon, to appear in *Proceedings of the LHC Workshop, Aachen, F R Germany, October 1990.*

[2] SDC Expression of Interest, SSClab, May 1990.

[3] D.Blockus *et al.*, "SSC Detector Sub-system Proposal - Central and Forward Tracking with Wire Chambers", SSC-PC-024, October 1989.

[4] J.B.Dainton, "Forward Charged Track Detector at LHC/CLIC", in *Proceedings of the Workshop on Physics at Future Accelerators, La Thuile and Geneva, January 1987, ed. J.H.Mulvey*, CERN 87-07

[5] G.A.Beck *et al.*, Nucl.Instr.Meth. **A283** 471 (1989). H.Gräßler *et al.*, Nucl.Instr.Meth. **A283** 622 (1989). H.Gräßler *et al.*, "Precision Reconstruction of Charged Tracks with simultaneous Electron Identification using Transition Radiation", to appear in *Proceedings of the 2nd London conference on Position Sensitive Detectors, Imperial College, London, September 4-7 1990.*

[6] EHIT Collaboration, J.M.Bailey *et al.*, "Measurements of Electron drift in fast Gases with Crossed Electric and Magnetic Fields", presented at this conference by J.M.Bailey.

[7] A.Oed *et al.*, Nucl.Instr.Meth. **A263** 351 (1988). F.Angelini *et al.*, Nucl.Instr.Meth. **A283** 755 (1989), Particle World **1** 85 (1990), INFN PI/AE 90/6, R.Bellazzini, to appear in *Proceedings of the LHC Workshop, Aachen, F R Germany, October 1990.*

[8] H-U.Bengtsson, G.Ingelman and T.Sjostrand, "A manual to a Lund Monte Carlo for Hadronic Processes", PYTHIA version 5.3, CERN program writeup POOL W5044, November 1989

STUDY OF MWPCS WITH INTERPOLATING PAD READOUT FOR HIGH MULTIPLICITY CHARGED PARTICLE DETECTION*

R. Debbe, J. Fischer, D. Lissauer, T. W. Ludlam, D. Makowiecki, E. O'Brien, V. Radeka, S. Rescia, L. C. Rogers, G. C. Smith, D. Stephani, B. Yu

Brookhaven National Laboratory, Upton, New York 11973

S. V. Greene, T. K. Hemmick, J. T. Mitchell, B. Shivakumar

Yale University, New Haven, Connecticut 06511

E. Mathieson

Leicester University, Leicester LE1 7RH, England

Abstract

Two types of MWPC with interpolating cathode "pad" readout for two dimensional position sensitive detection have been studied. One type uses rectangular pads and resistive charge division. A detector with an active area of 26cm×16cm, over 1000 readout channels has been built. With readout spacing of 1 cm, a position resolution of $\sim 70\mu m$ (rms) for 5.4 keV X-rays and a differential non-linearity of ±6% have been achieved. Another type of detector uses "chevron" shaped pads for geometrical charge division. Position resolution $\sim 50\mu m$ (rms) for 5.4 keV X-rays has been observed in a test detector and results of an investigation to reduce differential non-linearity are reported.

Introduction

Two types of position sensitive detector have been studied at BNL. The "Pad Chamber" [1-2] is a detector with a cathode plane divided into a large number of rectangular pads (see Fig. 1). Anode wires are placed above the pads, and field wires with ground potential are placed between anode wires. All pads beneath one anode wire are interconnected by a resistive strip. Approximately every tenth pad is read out by a charge sensitive preamplifier. An avalanche created by a charged particle traversing the detector induces localized charge on several pads. The position of the avalanche can then be found by calculating the centroid of the charge collected from these pads [3]. Unambiguous pattern recognition is achieved because of this two dimensional segmented cathode layout.

The "Chevron Pad Detector", in which the rectangular pads and the resistive layers are replaced by rows of chevron shaped pads as the cathode (shown in Fig. 2), has also been studied. Each pad is read out by a charge sensitive preamplifier. In this scheme, position sensing is achieved by geometrical charge division [4-5].

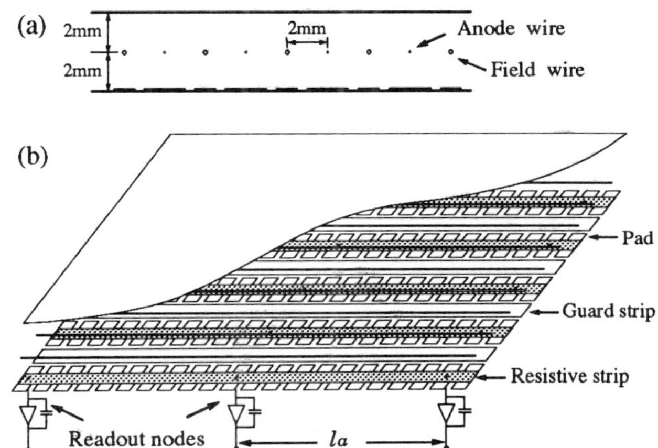

Figure 1: Schematic of a Pad Chamber. (a) Cell geometry viewed along the wire direction. (b) "3-D" view of the Pad Chamber.

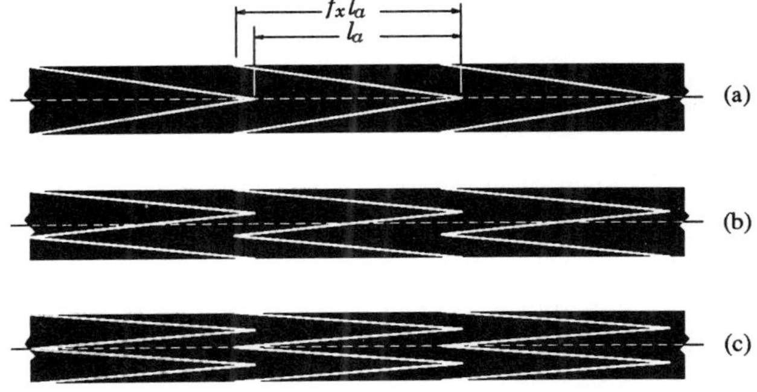

Figure 2: Three types of chevron pad cathode: (a) "Single chevron," (b) "One and a half chevron," (c) "Double chevron." (The dashed lines indicate anode wire positions.)

Pad Chamber with Resistive Charge Division

The design and construction of two versions of this detector are reported in Refs. [2] and [6]. The prototype detector [2] has an active area of 20 cm x 10 cm, with 525 readout channels. Monte Carlo studies have shown that this detector, with 1000 electrons rms noise and a gas gain of 10^4, can resolve a minimum ionizing particle track multiplicity of 20 with an efficiency of 96% [1]. The full-sized detector [6] has an active area of 26 cm x 16 cm,

* This research was supported in part by the U. S. Department of Energy: Contract No. DE-AC02-76CH00016.

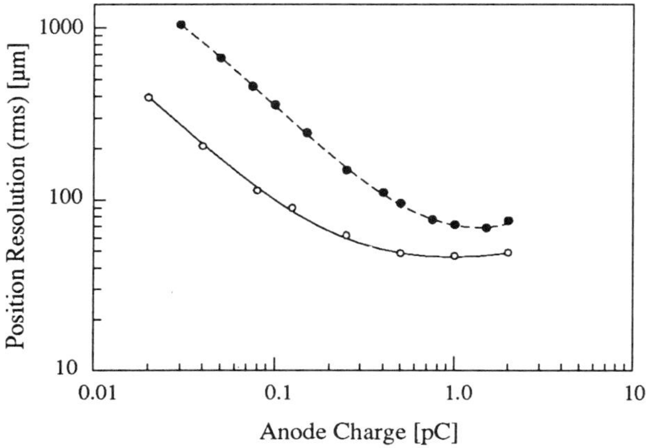

Figure 3: Position resolution as a function of anode charge for a pad chamber (solid circle, l_a = 10 mm, anode charge was measured with a 200 nsec time constant, 90% Ar / 10% CH_4 gas mixture), and a chevron pad chamber (hollow circle, l_a = 12 mm, single chevron, f_x = 1.05, anode charge was measured with a 1 μsec time constant, 50% Ar / 50% C_2H_6 gas mixture). Radiation was 5.4 keV X-rays.

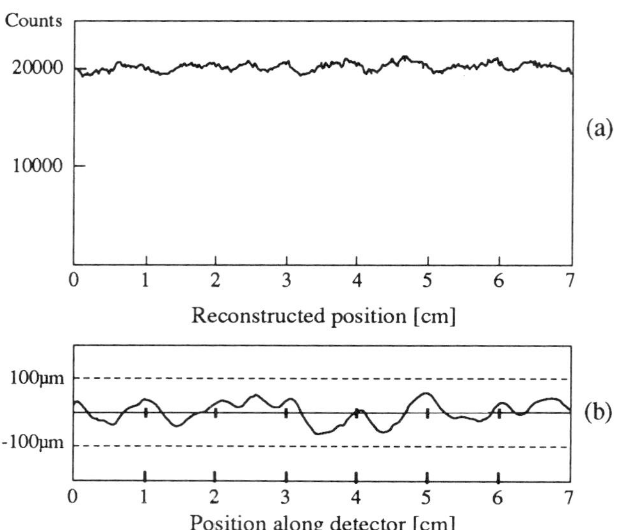

Figure 4: (a) Uniform Irradiation Response of a section of the prototype pad chamber. (b) Reconstructed position error derived from the spectrum in (a).

with 1016 readout channels. Both detectors use fiberglass honeycomb material on their thin cathode printed circuit boards to maintain their flatness and rigidity, which resulted in the cathode boards representing only 0.6% of a radiation length.

Absolute position resolution and linearity of the detectors have been measured by means of an electronic centroid finding system, in which the centroid of charge is determined by convolution of the sequentially switched outputs from several readout nodes with a linear centroid finding filter [7]. The test was performed with a 15 μm wide collimated X-ray beam and with a gas mixture of 90% Ar / 10% CH_4. The anode charge was measured with a 200 nsec shaping time. Figure 3 shows the behavior of the position resolution versus anode charge Q_A in the prototype chamber with a readout node spacing of 1 cm. For Q_A up to approximately 0.2 pC, the resolution improves according to a $1/Q_A$ relationship, since the electronic noise is the only significant perturbing factor. For Q_A above 0.2 pC, however, the position resolution improves more slowly than $1/Q_A$ because of the influence of phenomena such as electron range and diffusion. A minimum value of about 70 μm (rms) was attained for $Q_A \sim 1.5$ pC. Higher charge levels cause progressive deterioration in resolution because of spreading of the avalanche along the anode wire.

Differential non-linearity measurements have been made by obtaining a uniform irradiation response (UIR), with 5.4 keV X-rays, from a section of the anode wire. The preamplifier signals were fed to the centroid finding system and the resulting position signals analyzed with a pulse height analyzer. Figure 4a shows the UIR across eight readout nodes of the prototype pad chamber. The positions of the readout nodes, separated by 1 cm, correspond to the tick marks on the abscissa. Differential non-linearity is better than ±6%. Figure 4b shows the absolute position error (±60 μm), which was derived by evaluating integral non-linearity.

Pad Chamber with Geometrical Charge Division

A test chamber with interchangeable chevron cathode planes has been built to study their linearities and position resolutions. Its cell structure is identical to that of the pad chamber with resistive charge division. The testing of the chevron pad chamber was performed with the same centroid finding system as described earlier and 5.4 keV X-rays. The gas mixture used was 50% Ar and 50% C_2H_6. For a "single chevron" cathode (as shown in Fig. 2a) with l_a = 12 mm, a position resolution of $\sim 50\,\mu m$ has been observed at an anode charge level of 0.5 pC (measured with a 1 μsec time constant), as shown in Fig. 3. Experimental results indicate that for each type of geometry, the best linearity can be obtained with f_x (see Fig. 2) slightly greater than unity. A similar behavior has also been observed with zigzag strip cathodes [8].

One interesting phenomenon observed is the effect of the anode avalanche angular localization [9] on the position accuracy and linearity of the chamber. If a narrow beam of X-rays is incident on the detector from directly above one of the anode wires, X-rays are absorbed both above and below this wire. The induced charge on the chevron cathode pads for events initiating an avalanche on top of the wire (window side events) is different in time development and charge distribution from that for events from below the wire (pad side events). The sizes of induced charge from pad side events are smaller than that

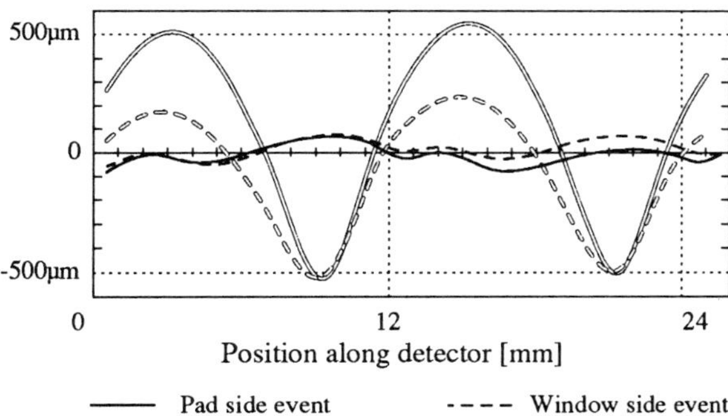

Figure 5: Comparison of reconstructed position errors between a single chevron cathode (shown in hollow lines) and a double chevron cathode (shown in solid lines). l_a = 12 mm and $f_x = 1.05$ for both cathodes.

of the window side events. If the width of the chevron is large compared with the size of the induced charge, it will result in poor sampling and, therefore, an error in the reconstructed position. The pad side events suffer more because of their smaller sizes of induced charge. This can result in a considerable difference in the reconstructed centroid position. A special filtering technique [9] allows us to select one type of event over another. The hollow lines in Fig. 5 show the reconstructed position errors from these two types of event obtained from two scans with collimated X-ray beam. The cathode pads are single chevrons, l_a = 12 mm, f_x = 1.05, and the gap width between chevrons is about 60 μm.

In order to minimize this position splitting, several other chevron patterns have been tested. Figure 2 (b) and (c) show two of them. These cathodes demonstrated significant improvements both in the reconstructed position linearity and the window side–pad side splitting. Figure 5 shows the experimental results of a double chevron and a single chevron. Figure 6 shows the measured differential non-linearities of some of the chevron patterns we tested. Details of this study will be reported elsewhere [10].

This splitting of centroid position for X-ray events is more pronounced with a long sampling time constant (1.4 μsec in our system) and a small anode–cathode spacing. For the case of a finely collimated beam of charged particles, a single position peak will be observed. Fluctuation of ionization clusters along the particle tracks will, however, cause a broadening of the position peak. A shorter time constant or a larger anode–cathode spacing will reduce the splitting and the broadening.

Conclusions

Both types of pad detector have been built and used in the E814 heavy ion experiment at the AGS in BNL [11]. Their ability to handle high particle multiplicity and provide unambiguous two-dimensional position information (1–3% of the readout node spacing along the wire direction) offers a very powerful detector technique. Pad

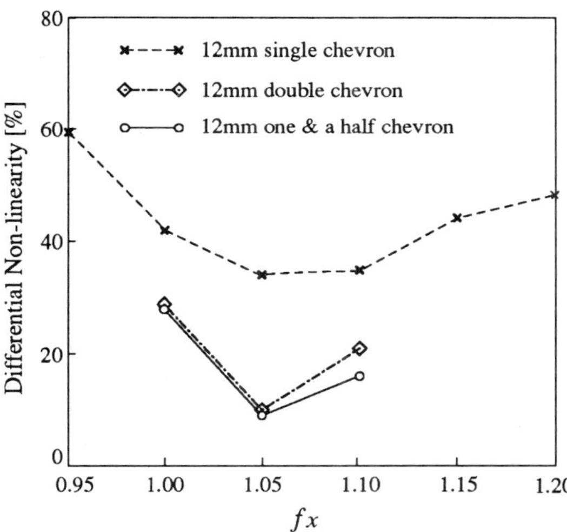

Figure 6: Differential non-linearity versus f_x for several chevron patterns

chambers with resistive charge division can provide very good position resolution and linearity. By using laser or abrasive trimming of resistors, one can expect to reduce the differential non-linearity down to the 2% level. However, resistor trimming on a detector with large active area becomes more difficult. Therefore they are most suited for small detectors or larger detectors whose cathodes can be divided into small modules. On the other hand, chevron pad cathodes which are relatively simple to fabricate, can be made in various sizes. They can provide exellent position resolution and good linearity with optimized chevron geometries. Such interpolating "gas pixel detector" with unambiguous two dimensional readout will be essential at the SSC as special layers in the barrel tracking detector and particularly in forward direction.

Acknowledgements

We would like to acknowledge the useful discussions with C. Woody and help from T. Willis during the testing of a detector. The production of all the printed circuit boards was done by A. De Libero, R. Angona and P. Borello. Various parts of the detectors were constructed by R. Chanda. E. Von Achen helped throughout the assembly and testing of the detectors.

References

1. T. Ludlam, R. Debbe and V. Radeka, "Many-track Chamber: A Design for T1," BNL E802 Note (Oct. 1985).
2. R. Debbe et al., "MWPC with Highly Segmented Cathode Pad Readout," *Nucl. Instrum. Methods* A283, 772-777 (1989).
3. J. L. Alberi and V. Radeka, "Position Sensing by Charge Division," *IEEE Trans. Nucl. Sci.* NS-23, 251-258 (1976).

4. C. Martin et al., "Wedge-and-Strip Anodes for Centroid-Finding Position-Sensitive Photon and Particle Detectors," *Rev. Sci. Instrum.* 52, 1067 (1981).

5. R. Chase, "Evaluation of the Linearity of Position Measurement with Chevron-Shaped Cathodes by Scale-Up Simulation," BNL Internal Note (Summer 1987).

6. R. Debbe et al., "A Study of Wire Chambers with Highly Segmented Cathode Pad Readout for High Multiplicity Charged Particle Detection," *IEEE Trans. Nucl. Sci.* NS-37, 88-94, (1990).

7. V. Radeka and R. Boie, "Centroid Finding Method for Position-Sensitive Detectors," *Nucl. Instrum. Methods* 178, 543-554 (1980).

8. E. Mathieson and G. C. Smith, "Reduction in Non-linearity in Position-Sensitive MWPCs," *IEEE Trans. Nucl. Sci.* NS-36, 305-310 (1989).

9. G. C. Smith, J. Fischer, and V. Radeka, "Photoelectron Range Limitations to the Spatial Resolution for X-rays in the Gas Proportional Chambers," *IEEE Trans. Nucl. Sci.* 31, 111-114 (1984).

10. B. Yu, G. C. Smith, V. Radeka and E. Mathieson, to be presented in IEEE 1990 Nuclear Science Symposium (Oct. 23-27, 1990).

11. J. Fischer et al., "A Many-Particle Tracking Detector with Drift Planes and Segmented Cathode Readout," *IEEE Trans. Nucl. Sci.* NS-37, 82-87, (1990).

THE MICROSTRIP CHAMBER ULTRAPRECISE TRACKING FOR SSC DETECTORS

E.F. Barasch, T.J.V. Bowcock, H.P. Demroff, M.M. Drew, S.M. Elliott, B. Lee, P.M. McIntyre, Y. Pang, K.J. Roller, D.D. Smith, J. Wahl
Department of Physics
Texas A&M University
College Station, Texas 77843

Abstract

A new technology for particle track detectors is being developed. Using standard IC fabrication techniques, a pattern of microscopic knife-edge electrodes is fabricated on a silicon substrate. The microstrip chamber uniquely offers attractive performance for the track chambers required for SSC detectors, for which no present technology is yet satisfactory. Its features include: excellent radiation hardness (10 Mrad), excellent spatial resolution (−20 μm), short live time (20 ns), and large pulse height (1 mV), and direct digital data flow.

Introduction

The Superconducting Super Collider presents formidable challenges to the detector technology required to instrument its experiments. The challenges are particularly distressing in the central magnetic spectrometer[1] which is the heart of most collider detectors. Figure 1 shows a simulation of a Higgs boson which decays into 2 Z bosons. The event contains 200 charged particles, half of which curl partly or fully within the chamber volume. On average there are 1.6 pp collisions in each bunch crossing, and the resolving time of most track chambers would indiscriminately record tracks from several consecutive crossings. In total some 20,000 signals must be processed from each event, and the (few) interesting tracks must be extracted from the (many) irrelevant backgrounds. The demand upon resolution, speed, and radiation hardness for tracking devices is one of the great challenges for SSC detector design.

The Accelerator Research Laboratory at Texas A&M University is developing a new technology for precision track chambers in which the anode plane of a multi-wire proportional chamber is replicated in miniature on a silicon substrate. The heart of the design is an array of microscopic knife-edge electrodes which can be fabricated in large arrays on a silicon substrate using conventional IC process techniques.[2] The knife-edge array can be configured with field-shaping electrodes and suitably biased to produce an extended region of high field near the knife-edge surface, as shown in Figure 2.

The knife-edge chamber has a number of advantageous properties for track detection at the SSC. A knife-edge spacing of 50 μm can be readily achieved. Operating in an appropriate gas (Xe/ethane/ethanol or Xe/CF_4), the knife-edge chamber can produce well-controlled gas gain ~ 10^4 and saturated drift velocity ~ 50 mm/μsec. With a 1 mm gas-filled sensitive region, the chamber should produce ~ 1 mV pulses and should be capable of a 99% efficiency, a spatial resolution of ~ 20 μm, and a resolving time of ~ 20 ns. The thickness of a complete chamber is only 0.1% radiation length, so that photon conversions and multiple scattering are minimized. The only elements of the device structure which are vulnerable to radiation damage are the gas medium and front-end readout electronics. The stability against gas aging effects should be

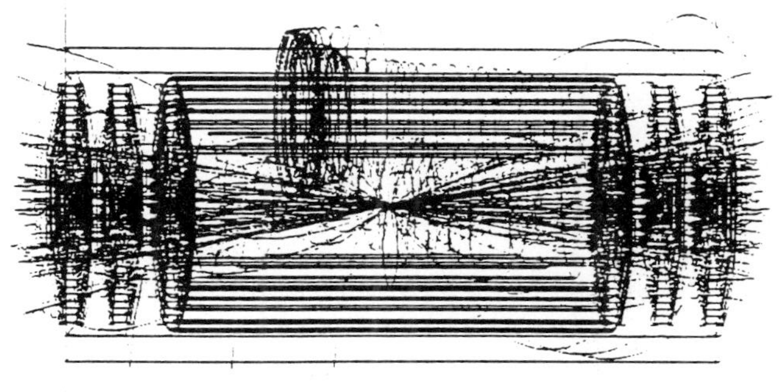

Figure 1. Simulation of a 400 GeV/c^2 Higgs boson in the SDC spectrometer.

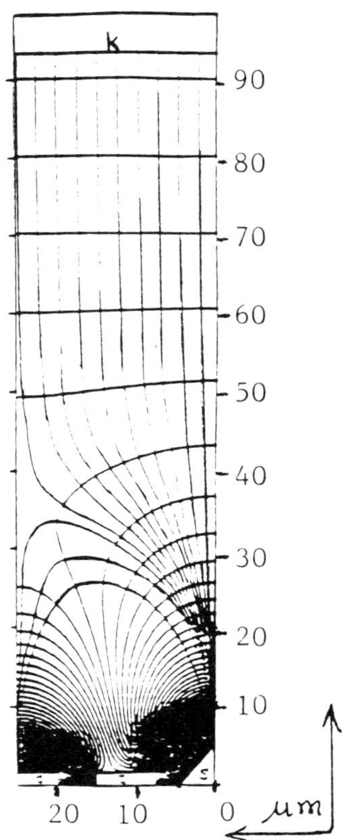

Figure 2. Calculated electric field geometry of the microstrip chamber.

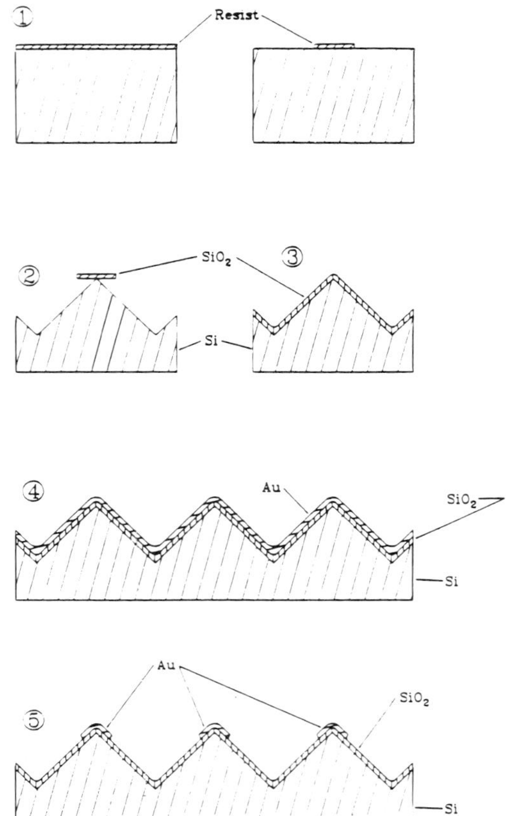

Figure 3. Fabrication sequence for microstrip chambers.

100 times better than in a straw tube chamber, because the charge collected per unit length of anode is 100 times less than in a straw tube for a given particle flux.[3] The stability against leakage currents from radiation damage of readout electronics should be 40 times better than that of silicon microstrip detectors, because the charge collected for each minumum-ionizing track is 40 times greater. This combination of properties would appear to make the knife-edge chamber an attractive technology for tracking at the SSC.

Fabrication of the Microstrip Chamber

The pattern of knife edges is fabricated on a silicon wafer using orientation-dependent etching (ODE) as shown in Figure 3. When a silicon surface is exposed to an appropriate alkaline wet-etch solution, the etch rate is $\sim$ 100 times faster in the 100 and 110 crystal directions than in the 111 direction. Knife-edge fabrication begins with a wafer of silicon (resistivity $\sim$ 0.01 Ω cm), cleaved on the 100 axis, with a thickness of 75 μm. The surface is coated with photoresist, and a pattern of lines is lithographically defined on the resist. The silicon surface is then etched in an ODE solution until the desired knife-edge pattern is produced. The resist pattern is then removed by a polishing etch. The silicon surface is oxidized at a high temperature to diffuse a SiO_2 insulating layer into a thickness of $\sim$ 2 μm. A layer of gold is then deposited. A layer of photoresist is deposited, and a second pattern of lines is exposed lithographically. The gold is etched to produce the signal and field-shaping electrodes. Figure 4 shows an electron micrograph of a knife-edge array just before the final etch step.

Electric Field Geometry

The knife-edge geometry provides the possibility to replicate on a microscopic scale the field geometry of a multiwire proportional chamber. Gas atoms are ionized by high energy charged particles traversing the chamber. The planar electric field between cathode and anode drifts the electrons towards the anode plane. The field-shaping electrodes are biased so as to produce a convergence of electric field upon each knife-edge. Figure 2 shows the electric field distribution. The field-shaping electrodes make it possible to locally control the gas gain of the chamber. A negative bias on the shaping eletrode can be used to enhance the electric field near the knife-edge anode and extend the high field region further from

Figure 4. Electron micrograph of chamber after process step 4.

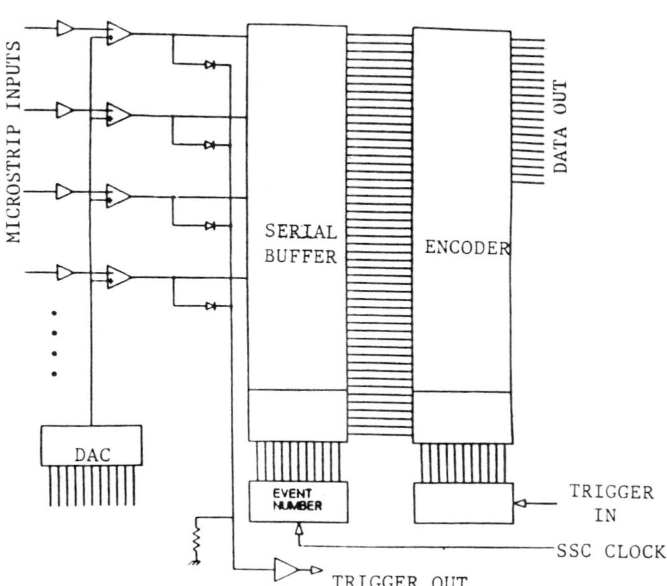

Figure 5. Functional schematic of readout electronics.

the knife-edge than would be obtained by simply miniaturizing a conventional MWPC. It is thereby possible to preserve a desired gas gain (10^4) while reducing the anode spacing to 50 μm.

Proportional Chamber Operation

The knife-edge chamber operates as a detector just as a microscopic multiwire proportional chamber. Indeed an entire channel would fit within the diameter of the very *wire* of a typical MWPC. For use as a track chamber in a collider detector, the chamber would be oriented face-on to the track direction. The chamber could be operated essentially as an ultraprecise hodoscope—no timing or pulse height digitization is required.

A response time of 20 nsec is compatible with limits from the maximum drift time for ionization electrons, the RLC response of the readout lines, and the response of the readout electronics. A spatial resolution of 20 μm is compatible with limits from transverse diffusion, track angle dispersion, and delta rays.

Readout Electronics

The readout electronics for each knife-edge chamber can be mounted on a single chip which is bonded to the readout electrode lines. Bonding can be accomplished using either wire bond or bump bonding.

Figure 5 shows the functional schematic of the readout electronics. All readout electronics is envisioned to be synchronous with the 64 MHz SSC clock. A front-end bifet amplifier buffers each input signal to a comparator. The comparator outputs are applied to a silicon gate array which serves two functions. First, outputs from manageable widths of the chamber may be OR'd to produce a fast hodoscope signal for use in constructing a trigger decision. Second, the pattern of hits is pipelined through CMOS registers for a number of beam crossings equal to the trigger decision time, e.g., 64 cycles = 1 μsec. In this way all high-speed, synchronous electronics are mounted locally on each knife-edge chamber; only trigger-selected, compacted data is communicated for higher level processing.

Collaborative Development

Related concepts for microstrip chambers are being developed at INFN (Pisa)[5] and KEK[6]. A collaboration[7] has been formed among investigators at Texas A&M, KEK, INFN, CERN, and Texas Instruments to fully develop and evaluate microstrip chambers for the Intermediate Angle Spectrometer in the Solenoidal Detector Collaboration (SDC). Figure 6 shows a design for one octant of this spectrometer consisting of three superlayers, each containing four layers of microstrip chambers.

This work is supported by the U.S. Department of Energy, contracts No. DE-AC01-85ER40236 and DE-AS05-81ER40039, and by the Texas Advanced Research Program.

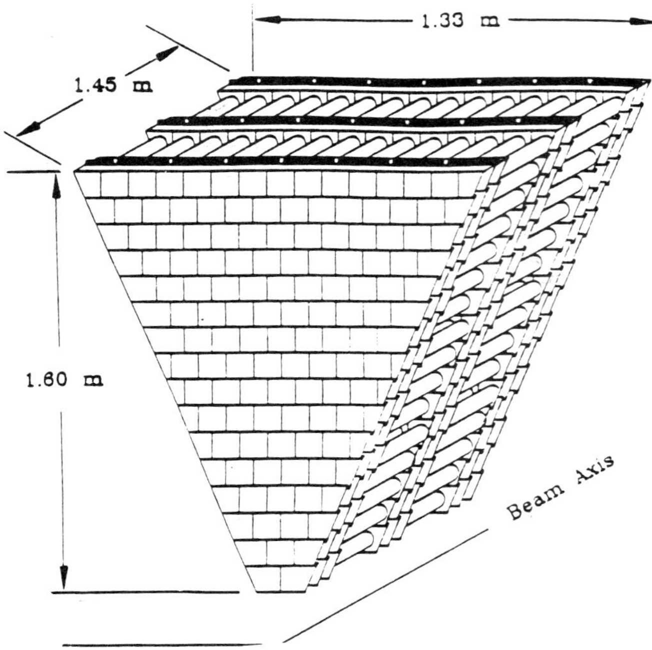

Figure 6. One octant of the SDC Intermediate Angle Spectrometer.

References

[1] G.G. Hanson, B.B. Niczyporuk, and A.P.T. Palounek, "Tracking Simulation and Wire Chamber Requirements for the SSC," Proc. DPF Summer Study: Snowmass '88, High Energy Physics in the 1990's, Snowmass, Colo., June 27–July 15, 1988.

[2] D.J. Campisi and H. Gray, "Microfabrication of field emission devices for vacuum integrated circuits using orientation dependent etching," in Proc. Mat. Res. Soc. Meeting, 1986.

[3] M. Atac, "Wire Chamber Aging and Wire Material," Vertex Detectors, F. Villa, ed., Plenum, New York, 1988.

[4] K. Kinoshita, Nucl. Inst. & Meth. **A276**, 242, (1989).

[5] F. Angelini et al., Nucl. Inst. & Meth. **A292**, 199 (1990). F. Angelini et al., "Test Beam Study of the Microstrip Gas Avalanche Chamber," INFN preprint, 1990.

[6] A. Maki, "A Large-Size Knife-Edge Chamber," Proc. 1990 Workshop on Solenoidal Detectors, Tsukuba, April 1990.

[7] D. Allen et al. "Microstrip Track Chambers," SSC Subsystem R&D Proposal, (1990).

STUDY OF SENSE WIRE STABILITY AND SUPPORT IN STRAW TUBE DETECTOR ELEMENTS

S. H. Oh and W. J. Robertson
Department of Physics
Duke University
Durham, N.C. 27706

Abstract

Electrostatic instability of one-mill sense wires in 4 mm-diameter straw tubes requires wire support in long detector elements. Our mechanism for wire support is currently being tested.

I. Electrostatic Stability of Straw Tubes

Our straw tube design for the central tracking chamber in a solenoidal detector at the Superconducting Super Collider (SSC) calls for 4 mm-diameter detector elements. The sense wire would be operated at 1.8–2.2 kV to achieve gas gains which give suitable signal pulses. The ideal case in which the sense wire is exactly centered within a perfectly cylindrical cathode is elctrostatically stable. Practically, however, the sense wire will not be exactly centered due to positioning error and gravitational sag and the straw tube cathode will be neither perfectly cylindrical nor perfectly straight. The electrostatic force due to offset of the sense wire in the straw tube is given by:

$$F_E = \frac{2\pi\epsilon_o V^2 \delta}{R^2(ln(R/r))^2}$$

where V is the sense wire potential, δ is the offset of the sense wire from the center of the straw tube, R is the radius of the straw tube cathode and r is the radius of the sense wire.

We have constructed a straw tube model of stainless-steel tubing. The tubing has an inner diameter of 3.9 mm and is placed in a horizontal orientation. A manifold on one end allows gas flow to purge air from the tube to eliminate molecular ionization currents which occur with oxygen and nitrogen at potentials in excess of 2 kV. We have used CH_4, CO_2 and CF_4 as stable gases. We find that the longest cell that is stable above 2.5 kV with a 25 μm sense wire centered in the tube under 50 grams tension is one meter. The tensile strength of 25 μm-diameter tungsten wire is 150–200 grams so the applied tension to the wire should be kept below 100 grams. Keeping the wire tension to a minimum will also reduce the mechanical load on the detector assembly.

Figure 1 shows the maximum stable operating voltage that a one-meter long cell filled with CH_4 can sustain when the sense wire is displaced from the center of the cathode for two different wire tensions of 50 and 100 grams. These tests show that wire offsets of up to 100 μm will not compromise cell operation.

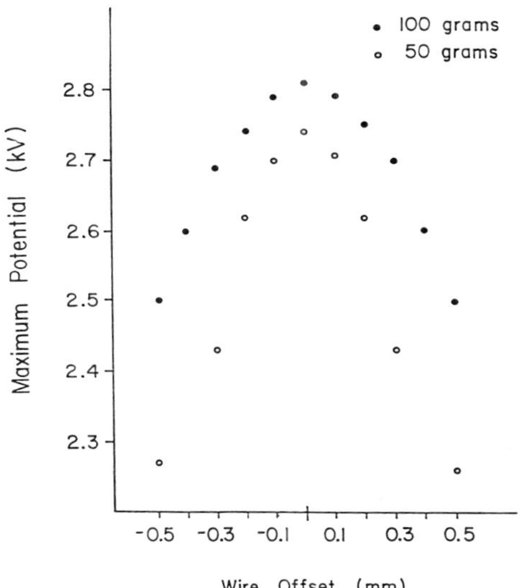

Figure 1

One can predict the motion of the sense wire under increasing electric potential. This has been studied by drilling a hole in the middle of the steel tube to observe the position of the sense wire with a travelling microscope as the wire potential is raised. The sense wire was positioned off-center by a known amount (δ). The sense wire potential was increased and the deflection of the midpoint of the wire (d) from its initial position was measured. This deflection can be calculated and is related to the potential on the wire to good approximation by the expressions:

$$d = \frac{L^2 F_E}{8T}$$

$$d = \frac{2\pi\epsilon_o V^2 L^2(\delta + d/2)}{R^2(ln(R/r))^2} \quad .$$

For a one-meter long wire the displacement is given by:

$$d = \frac{\delta(\frac{V}{V_o})^2}{(1 - \frac{1}{2}(\frac{V}{V_o})^2)}$$

where $V_o = 2700\sqrt{\frac{T}{50}}$; with T, the wire tension, in grams and V, the applied potential, in volts.

Figure 2 shows measurements of the wire deflection versus sense wire potential for tensions of 50 and 100 grams. The lower set of data in each case is with an initial offset of 100 μm and the upper set of data is for an initial offset of 200 μm . The solid curves are predictions from the formula for the wire deflection. The open circle data points are for initial offset in one direction while the closed circle data points are for initial offset in the opposite direction. The discrepancies between the data are indicative of some nonlinearity of the tube. The measurements agree quite well with the predictions of the above calculations.

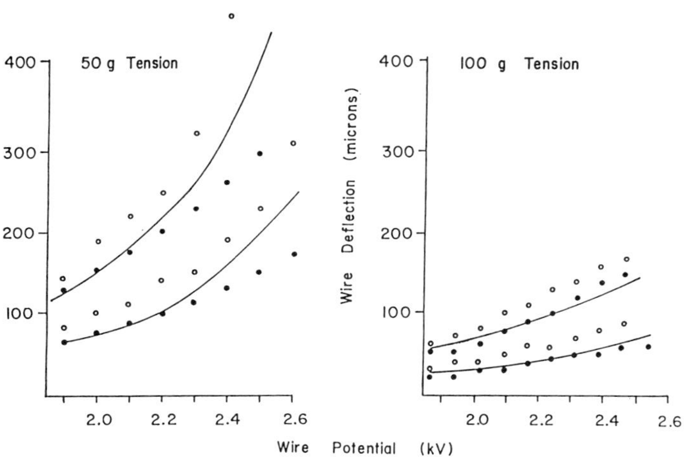

Figure 2

II. Sense Wire Support

To achieve adequate geometric acceptance for charged tracks the SSC central tracking chamber will have to be six meters long at an outer radius of 1.7 meters. Some designs call for a gap in the middle of the chamber in which case straw tube cells may be about three meters long. In either case it will be necessary to provide centering support for the sense wires in the straw tubes at about one-meter intervals to ensure electrostatic stability.

The wire supports must center the wire in the straw tube while, at the same time, not restrict gas flow through the tube. Our design for the wire support consists of a plastic cylinder with a helical groove which is a cylinder radius deep and makes at least one complete revolution around the cylinder in a length of 1–2 cm. Such a device will provide gravitational support regardless of the orientation of the tube. A schematic drawing of the wire-support design with cross-section views is shown in Figure 3. Production of an injection molding form is in progress and the second prototype version of the molded fixtures should be available for testing in the near future.

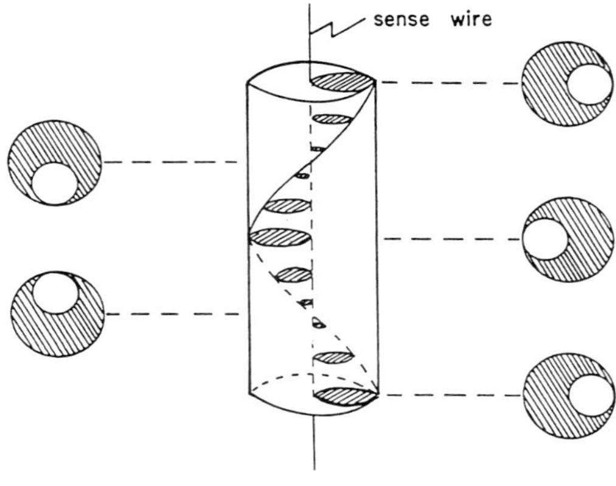

Figure 3

The wire supports will be attached to the straw tube. The wire will then be threaded through the tube. This has the advantage that if a wire should break during installation in a straw tube, it will be easy to remove the broken wire and restring the tube. We have used air flow through the tube to draw and guide the sense wire. We are confident that this procedure will also work with the wire-support pieces in place.

We have assembled eight layers of 2.7-meter long straw tubes on a machined base plate which has been aligned to be flat to better than 100 μm. The tubes are made up of three 0.9-meter long straws with the splices containing temporary wire support pieces which provide gravitational wire support. All of the tubes sustain voltages in excess of 2.2 kV and cosmic ray tracks have been observed using a dual scintillator coincidence to start CAMAC TDC's.

CONSTRUCTION AND TEST OF A PROTOTYPE STRAW CHAMBER DETECTOR OF LENGTH 2.7 meters

S.H. Oh, A.T. Goshaw, W.J. Robertson
Department of Physics
Duke University
Durham, NC 27706

Abstract

In this paper, we report the construction and test of a 2.7-meter long straw tube drift chamber consisting of 60 channels. The tubes are stacked in a pyramid shape with 8 tube layers. There are 11 tubes at the bottom layer and 4 tubes at the top layer. The tube radius is 2 mm. Twenty-eight channels out of 60 are instrumented for measurement of resolution, gain, and attenuation length.

I. Introduction

The design of the central tracking chamber in a solenoidal detector for the Superconducting Super Collider (SSC) calls for a long cylindrical tracking chamber with a length of about 6 meters. Using straw tubes with small radii is one of the options for the tracking chamber. Due to the bunch spacing and the occupancy rate, the drift space has to be less than 2 mm. For a feasibility study, we have constructed a 2.7 meter long, 60 channel prototype chamber with 2 mm radius straw tubes. By successfully constructing and operating the prototype, we have demonstrated that the basic design concept of the cylindrical chamber is sound and can be extended to a full-size cylindrical chamber. Specifically, we have shown the following. First, we showed that layers of tubes can be placed straight with an accuracy better than 100 microns over several meters once a flat base surface is provided. Second, an end plate designed by our group using two plates does work. The design provides a simple way to supply gas, tension the sense wire and mount electronics. Third, the wires can be supported every meter inside tubes with proper wire supports. (A contributed paper in this proceeding with the title "Study of sense wire stability and support in straw tube detector elements" describes this in detail). Fourth, a large number of channels can be made operational simultaneously without difficulty. This is important since the expected number of channels for the final tracking chamber is about 200000.

In the next section, we describe the construction of the prototype in detail. In the third section, we describe the tests performed with the prototype.

II. The Construction of the Prototype

We have successfully constructed a 2.7 m long prototype chamber with 8 layers of tubes. In this section, we present the construction of the prototype. The geometry of the tubes in the prototype is shown in Figure 1. The 60 tubes are arranged in a pyramid shape with 11 tubes at the bottom and 4 tubes at the top. The radius of each tube is 2 mm.

The construction of the prototype starts with a sturdy base. The base could be an optical table or a rail. It is important that the base has a flat surface that tubes can be placed on. Our base looks like the bottom third of a disk. It is 3.3 meters long with a 7 cm wide machined face. The surface was scanned with a survey telescope and found to be flat to better than 25 microns over the entire surface.

On the top of the base, four aluminum plates with machined grooves are placed. The grooves on the plates are to guide the tubes. The plates are aligned among themselves using the survey instrument. The plate simulates the cylindrical base to be needed in the final tracking chamber.

Straw tubes were assembled prior to placing them on the aluminum plate. The tubes used in the prototype have a wall thickness of 50 microns and were made of mylar strips. The strips were wound on top of 8- or 12-micron thick aluminum foil (cathode). Three tubes of 90 cm were joined. Before the tubes were put together using thin aluminum cylinders, wire supports were inserted inside the tubes as shown in Figure 3. Because we were not able to obtain the wire support described in an earlier paper in time for use in the prototype, we have used two V-shaped disks back to back. Although the disks were machined carefully, measurement using a microscope showed that they are about 75 microns in

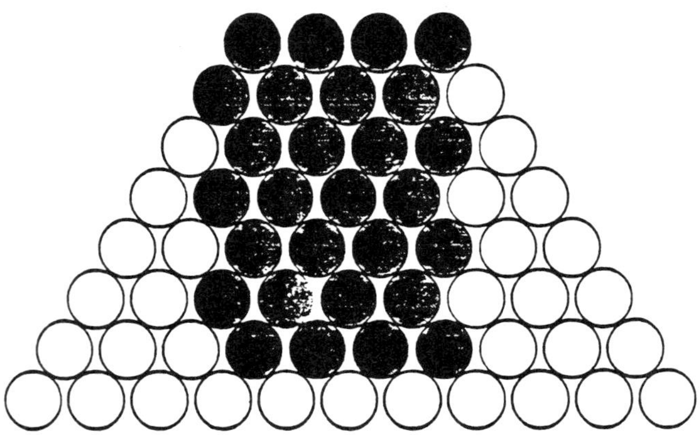

Figure 1: Cross-section of the straw tube super layer. The shaded tubes were instrumented.

tolerance. As we have discovered, joining tubes was not the best approach since the tube joints introduced errors along the vertical direction at the joint. The next prototype will not have joints and the wire supports will be inserted from the ends of the straws.

Once enough tubes are assembled for a layer, they are attached to the end plates (Figure 2) and placed in the grooves of the aluminum plate. In order to attach to end plates, the tubes are bent slightly. To obtain better placement accuracy, the tubes are pressed slightly using finger-shaped jigs from the top. After the tubes are properly placed, a small amount of fast-drying glue (cyanoacrylate) is used to glue the tubes to the plate. The gluing has two purposes. First, the fixed tubes act like the grooves on the aluminum plates, so that the next layer of tubes can be aligned accurately. Second, it straightens the tubes. The tubes from the factory are not only very flexible but also are slightly bent.

Each end plate consists of two walls. Tubes are attached to the inner plate by a thin wall cylinder and sense wires are tensioned from the other wall (Figure 2). The space between two walls serves as a plenum to provide gas flow to each tube.

As we place each layer, the vertical and horizontal position of the tubes is measured. Figure 4 shows that horizontal position of the tubes in a layer at different locations along the tube length. The figure shows that tubes can be positioned to an accuracy of 100 microns or better. Figure 5 shows the vertical position of the top layer (8th layer) measured along the tube length from one end. As we mentioned earlier, the tube joints (at 90 cm) are out of place by about 100-micron intervals.

After all tubes are placed and glued, sense wires are strung. To accomplish this, air at high pressure (we used 20 psi air from a tube with 3 mm diameter) is blown from one end of a tube to string a guide wire with 100 micron diameter. A 25-micron diameter sense wire is attached to the guide wire and pulled through.

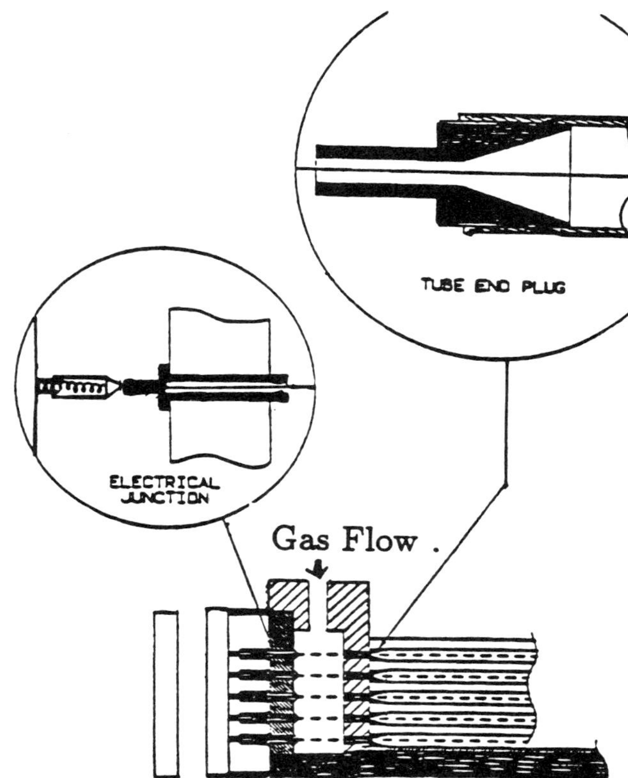

Figure 2: The end plate used in the prototype.

The sense wire is tensioned to 50 grams, pinned, and soldered.

Each cell is tested for high voltage. Out of 60 tubes, we only instrument 28 tubes (shaded tubes in Figure 1). Of these, we find that only one tube does not hold an operating voltage (1800 volts). For the rest of the tubes, we were able to raise the high voltage to at least 2500 volts. It is not clear why one cell is bad, but we think that the wire support inside the cell may not be positioned properly or was moved by high pressure air. In this one tube, we strung another wire with 100 grams tension, and were able to raise the voltage to 1900 volts.

III. The Tests of the Prototype

In this section, the tests performed with the prototype chamber are described. We studied the attenuation length, gain and resolution. Typical signals (using Sr^{90}) into 50 ohms from the chamber have a rise time of about 2–3 ns. About 30 ns later, the signal is reduced to about 15% of its maximum.

Figure 6 shows plateau curves obtained using a Lecroy 2735DC amplifier-discriminator with a 3 microamp threshold, using a Sr^{90} radiation source. One curve is taken with the source very near the readout end and the other with the source 250 cm from the readout end.

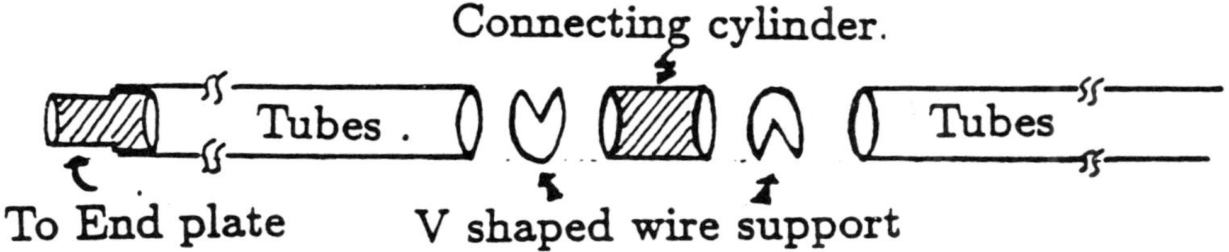

Figure 3: Tube assembly with tube connector and wire support.

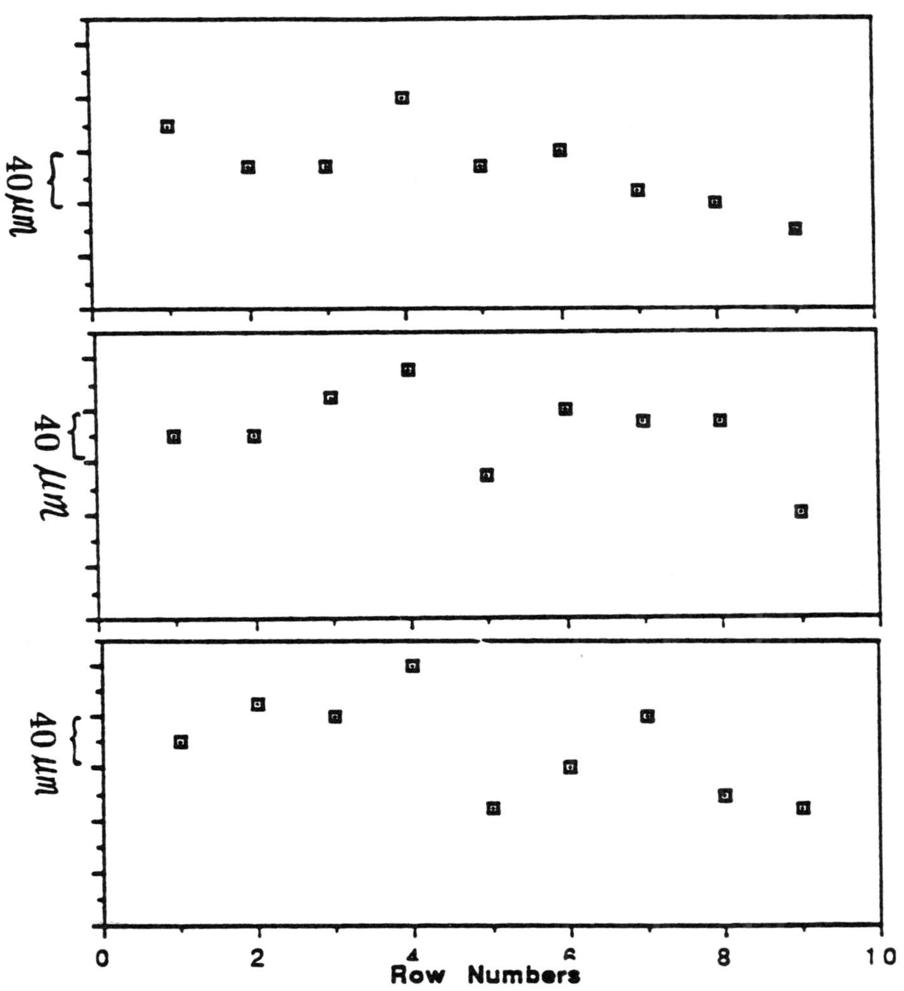

Figure 4: Horizontal position of straw tubes in a layer as a function of straw tube number.

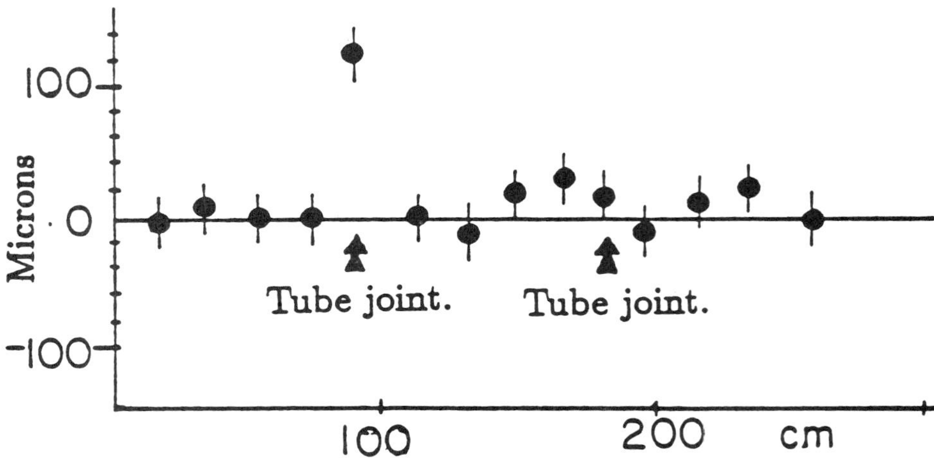

Figure 5: The vertical position of straw tubes on the top row as a function of distance from one end.

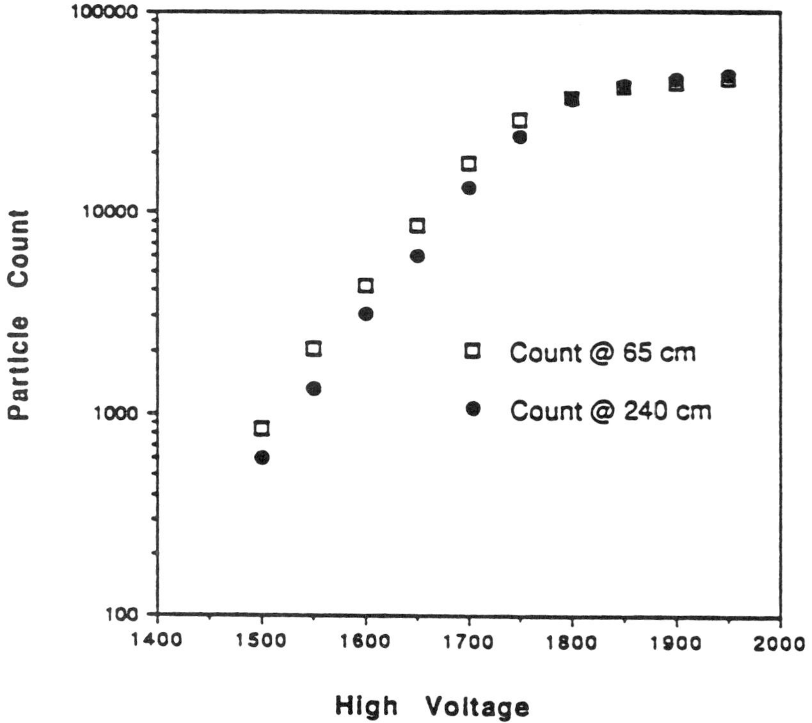

Figure 6: Plateau curves obtained using Sr^{90} source at two different positions from the end of tubes. The gas mixture is Ar-CF_4-Ethane (33-33-33).

The gas used is CF_4-Ar-Ethane (33-33-33) mixture. As expected, the knee of the plateau curve measured at the far end location moves compared to the other curve. The shape of the signal from the near end compared to the far end (with respect to the readout end) is very similar although the height is down by about half (i.e., the dispersion is small).

The attenuation length is measured using a Fe^{55} source. Because of the variation of gain near the straw tube joints, the peak of the signal is measured from both ends of the chamber. From the ratio of the two peaks as a function of distance from one end, the attenuation length is calculated. Figure 7 shows the plotted ratio for several channels. The overlapped curves are fits using an exponential function. Although there is some variation between tubes, an average attenuation length of 500 cm is obtained. (The attenuation length calculated this way is one-half of the true attenuation length. The 500 cm and the numbers in Figure 7 are already multiplied by a factor of 2.)

Figure 8 shows the gain measurement for several different gases using the Fe^{55} source. The output from a cell is connected to a charge amplifier and then to a multichannel analyzer. The absolute calibration is accomplished by inserting a known amount of charges into the amplifier. Due to the uncertainty in the calibration, we expect a 10–20 percent systematic error in the absolute gain values.

Twenty-eight straw tube channels were instrumented for the resolution study using a cosmic-ray trigger. The output from the chamber was connected to the Lecroy 2735DC amplifier-discriminator. The discriminated outputs were connected to Lecroy TDC's. Figure 9 shows a track traversing the superlayer. Using the fitted tracks, the residuals are calculated and plotted in Figure 10 (Ar-Ethane 50-50 mixture, operating voltage = 1700 volts). The residuals are calculated using tracks with 6 or more hits. A sigma of 110 microns is obtained by fitting the distribution of the residuals. We also have tried different gases. For CF_4-Ethane (50-50 mixture, operating voltage = 2100 volts), a sigma of 120 microns is obtained. For CF_4-Isobutane (80-20 mixture, operating voltage = 2200 volts), a sigma of 125 microns is obtained.

In conclusion, a 2.7 meter long prototype straw tube drift chamber has been successfully constructed and tested. The prototype consists of 8 layers of tubes with a total of 60 channels. Twenty-eight channels were instrumented with readout electronics. Using the prototype, we obtain about 500 cm for the attenuation length and about 120 microns for the resolution of the 2 mm radius straw tubes. The basic design concepts studied using the prototype can be extended to a full-size cylindrical straw tube superlayer.

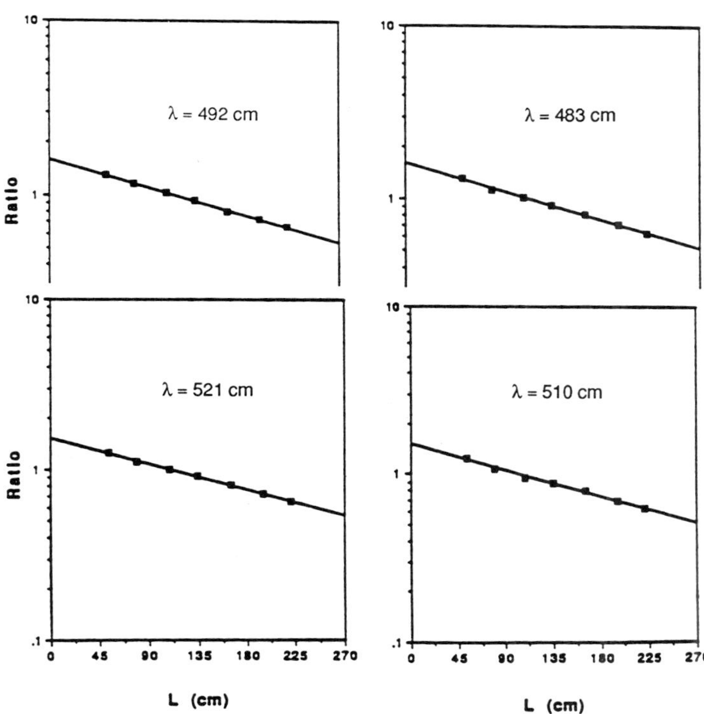

Figure 7: The ratio of amplitude from one end to the other end using Fe^{55} source. The attenuation length is obtained from the slope.

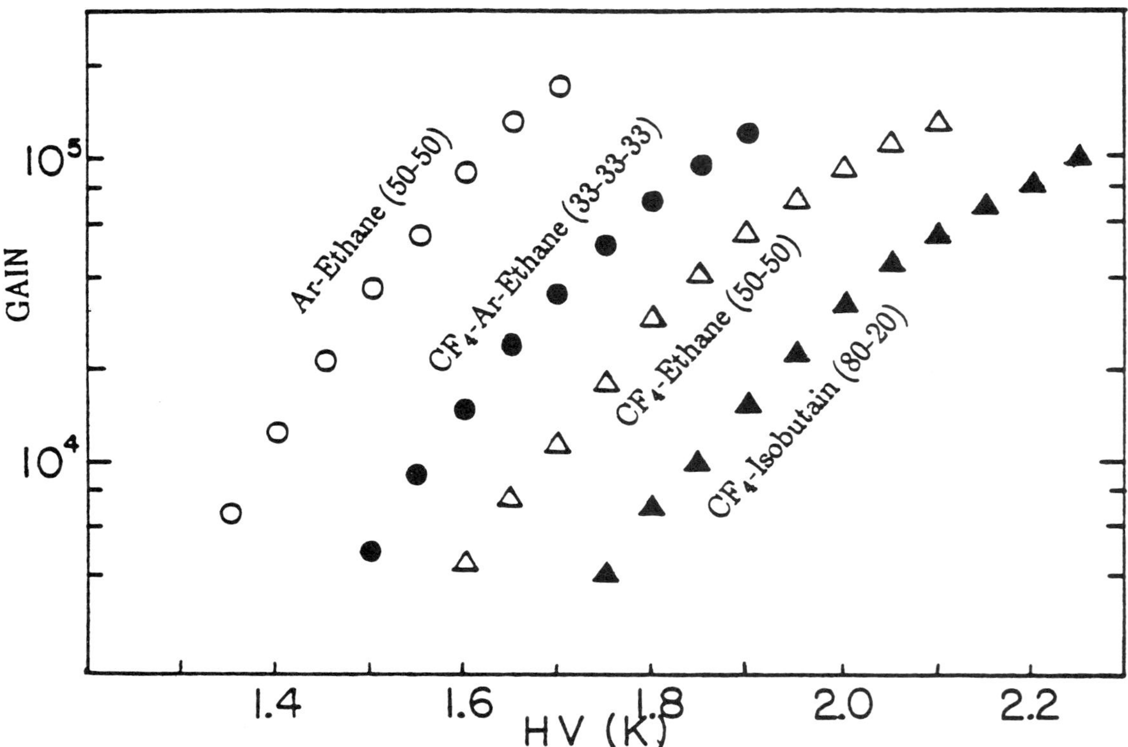

Figure 8: The gain as a function of high voltage for several gas mixtures.

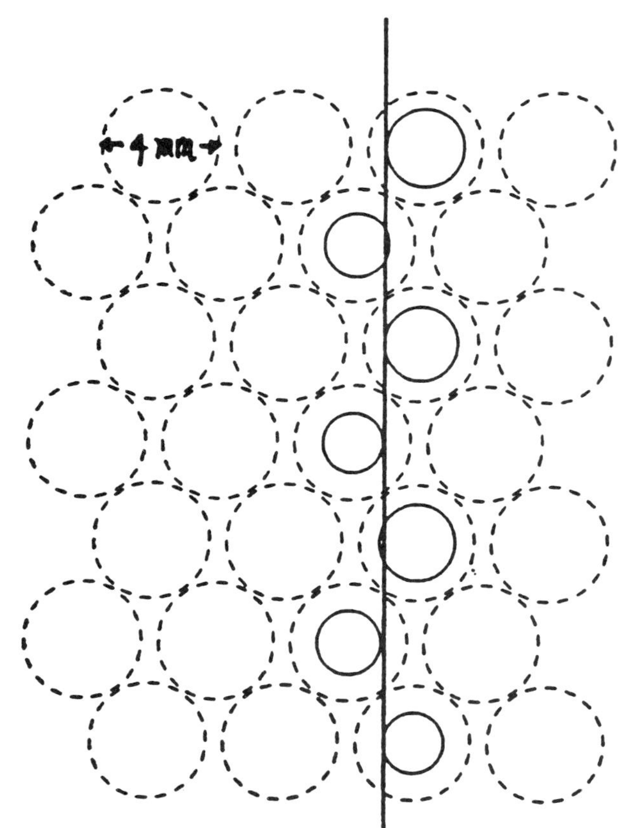

Figure 9: A triggered cosmic ray track traversing the superlayer.

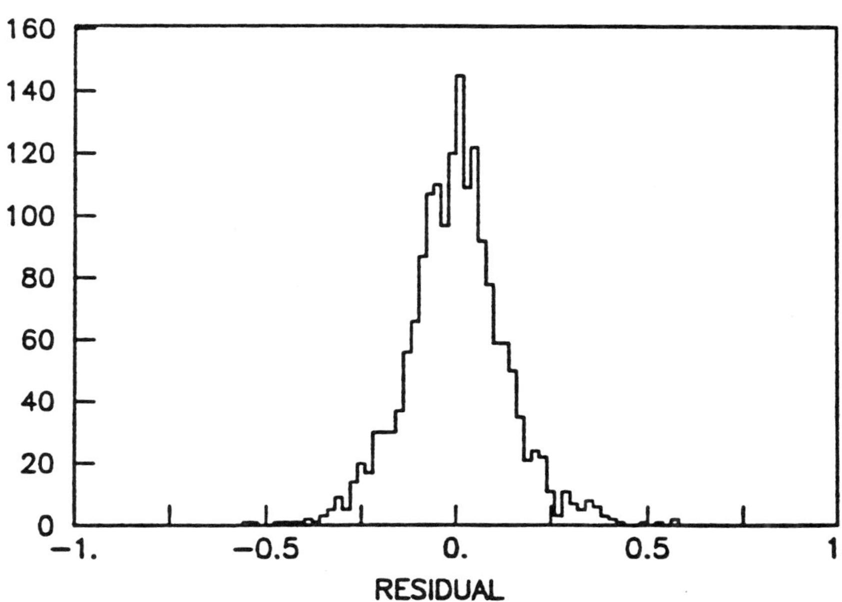

Figure 10: Residual in mm obtained using cosmic-ray tracks. Sigma of 110 microns is calculated from a gaussian fit to the curve.

INVESTIGATIONS OF SINGLE-ELECTRON AVALANCHES IN A PROPORTIONAL DRIFT TUBE

W.S. Anderson, J.C. Armitage,* P. Chevreau, J.G. Heinrich, C. Lu,
I. McDonald, K.T. McDonald, B. Miller, D. Secrest, and J. Weckel
Joseph Henry Laboratories, Princeton University, Princeton, NJ 08544

Abstract
Investigations of single-electron avalanches in a proportional drift tube have been carried out with a pulsed N_2 laser. The study consists of two aspects: timing properties, and fluctuations in the gas avalanche.

Detailed information on single-electron drift and avalanche behaviour has a basic interest in an investigation of gas-chamber performance. Its timing, avalanche distribution, attachment by the working gas mixtures, *etc.*, provide various criteria for choosing the best suitable gas mixture under a specific experimental circumstance.

Timing Studies with Single Electrons

A sketch of the experimental set-up is shown in Fig. 1.

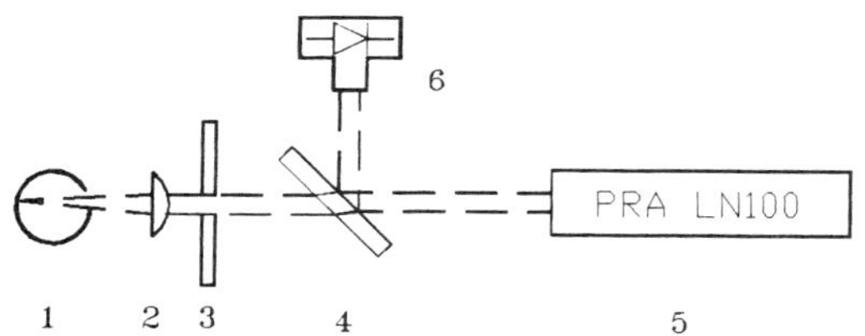

Figure 1: Sketch of the experimental set-up.

The N_2 laser generates pulsed 337-nm (3.67-eV) UV photons, with a pulse length of about 350 ps. The pulse energy is about 50 μJ, corresponding to about 10^{14} photons. The test drift tube was made of 7.67-mm-diameter aluminum tubing with a 25-μm-diameter gold-plated-tungsten anode wire.

The laser beam was focused onto the inner wall of aluminum tube after passing through a 1-mm-diameter hole in the wall. The hole is offset by 1 mm from the center line of the tube. Because the beam spot on the inner surface of the tube is quite small, we can neglect mechanical imperfections and consider that the drift distance of the photoelectrons was just the radius of the tube. A beam splitter was used to reflect part of the laser beam to a photodiode (RCA C30905E), which generates a fast, large electric pulse as a start signal for TDC system. A preamplifier was directly connected to the test drift tube through a high-voltage capacitor. The drift-tube signal was used as a stop signal.

Two different preamplifiers have been used in our measurements: a LeCroy TRA402 and U. Penn/AT&T preamplifier.[1] The U. Penn preamplifier was used for studying the performance of CF_4/Ar and CF_4/isobutane mixtures with different mixing ratios. The other gas mixtures were measured by LeCroy preamplifier.

Due to the very small quantum efficiency and large number of primary UV photons, the number of photoelectrons in each pulse should vary according to a Poisson distribution. We reduced the aperture of the iris diaphragm until only 1 in 10 laser pulses yielded any photoelectrons, and hence at least 90% of the recorded events were initiated by single photoelectrons.

Using simplified formulae to approximate the published curves on the drift velocity w *vs.* E/P,[2, 3] we are able to calculate the total drift time. The results are shown in Fig. 2. The agreement is impressive. Because our data are collected under various pressures, and the drift velocity curves used in our fitting were measured or calculated at 1 atm, our results further confirm the well-known fact: $w \propto f(E/P)$.

We have directly measured the time resolution σ_t, and converted to the spatial resolution σ_x using the x *vs.* t curve near the tube wall. Figure 3 shows our data as well as a fit of the form

$$\sigma_x = \sigma_0 + A/\sqrt{P}$$

for Ar/CO_2, Ar/CH_4 and Ar/C_2H_6.

We made a further investigation for gas mixtures containing CF_4. This "fast" gas is believed to be one of the most attractive candidates for the SSC environment.[2, 4]

We summarize the drift-time results of CF_4 gas mixtures in Fig. 4. The dependence of the drift time on the percentage of CF_4 and V/P is insensitive for Ar/CF_4.

*On sabbatical from Carleton U.

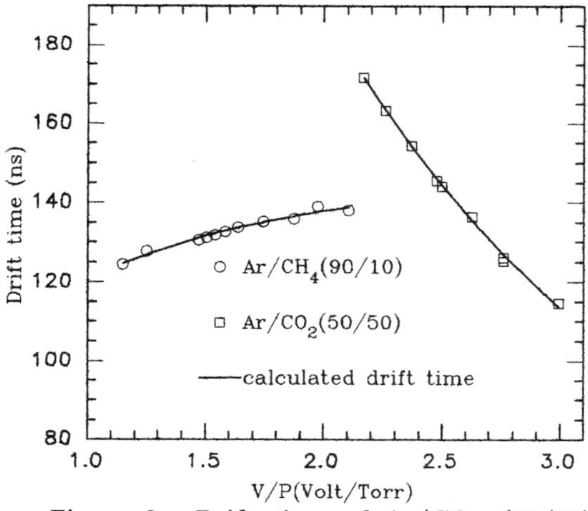

Figure 2: Drift time of Ar/CO_2 (50/50), Ar/CH_4 (90/10).

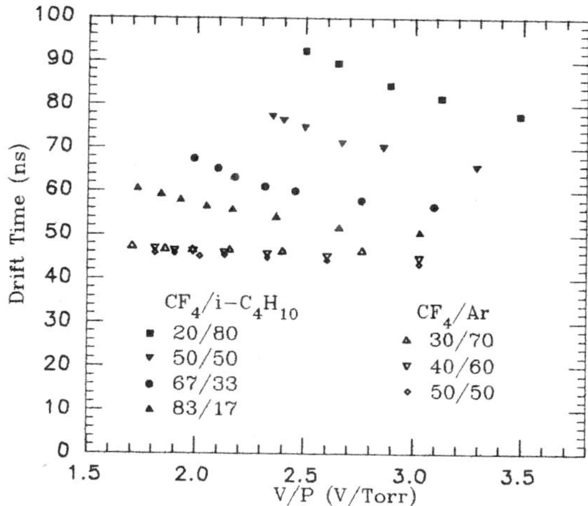

Figure 4: Total drift time of CF_4/Ar and CF_4/isobutane.

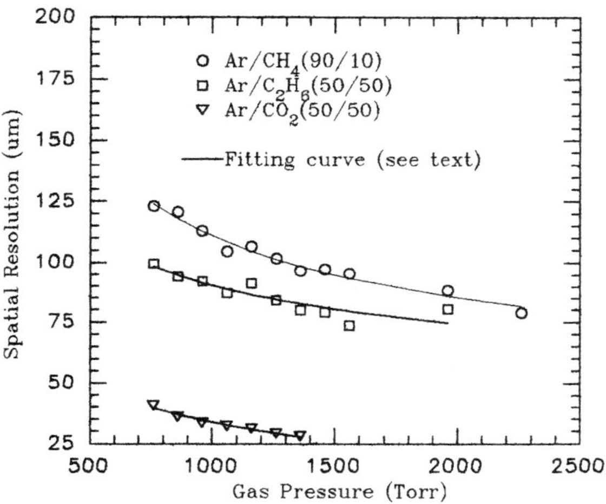

Figure 3: Spatial resolution of Ar/CO_2 (50/50), Ar/C_2H_6 and Ar/CH_4 (90/10).

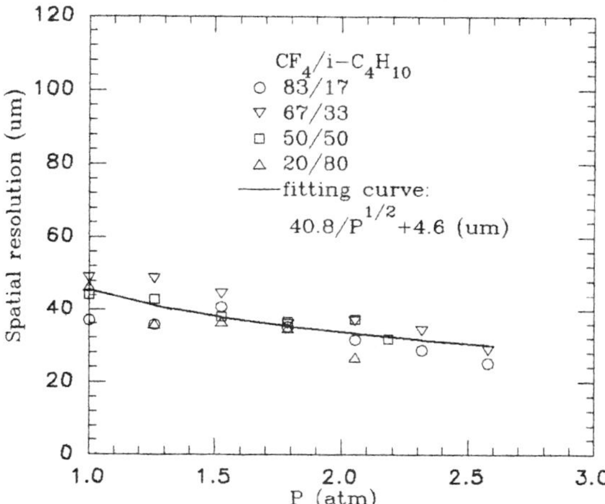

Figure 5: Spatial resolution of CF_4/isobutane.

The time resolutions are in the 0.55-0.75 ns range for Ar/CF_4 mixtures, and 0.55-1.0 ns for CF_4/isobutane mixtures. Using data from T. Yamashita *et al.*[5] we can infer the spatial resolution for Ar/isobutane near the tube wall. All of the resolutions are 25-50 μm, as shown in Fig. 5. Because there is no existing drift-velocity data for CF_4/isobutane (83/17), (67/33), we have adopted the (80/20), (70/30) data instead.

Fluctuations in Single-Electron Avalanches

We have studied this with the same experimental set-up as in the timing measurements. Here we use an EG&G ORTEC 142PC charge-sensitive preamplifier instead of the transresistance preamplifier used for timing measurements.

The single-electron-avalanche spectra from Ar/CF_4 and Ar/isobutane have been measured for several different mixing ratios. The single-electron-avalanche spectra were then fitted with the Polya distribution [6]

$$P(a) \propto (ba/\bar{a})^{b-1}e^{-ba/\bar{a}},$$

where a is the amplitude of an avalanche, $\bar{a}$ is the mean gas gain, and b is a measure of the fluctuation of gas gain: $(\sigma_a/a)^2 = 1/b$.

We are able to observe the presence of electron attachment in the chamber gas as follows. Charge spectra were also recorded for each gas mixture at several different high voltages using an Fe^{55} source. If the signal size observed with the Fe^{55} source is not $\sim$ 200 times that from a single photoelectron ejected by a laser pulse, then attachment has occurred. Comparisons of avalanches due to single photoelectrons with those from Fe^{55} x-rays are shown in Fig. 6.

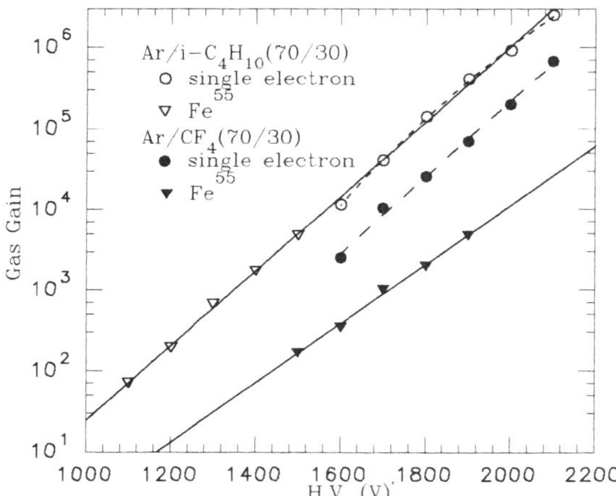

Figure 6: Gas gain from single photoelectrons and from Fe^{55} x-rays in Ar/isobutane and Ar/CF_4.

All three of the Ar/isobutane gas mixtures gave rather consistent results, both from single electrons as well as from Fe^{55} x-rays. Then when Ar/CF_4 gas mixtures were tested, the large discrepancy between single-electron and Fe^{55} measurements indicates that CF_4 may have serious electron attachment.[2] The gas gain from single electrons was about ten times higher than that from Fe^{55}.

Additional evidence for electron attachment is the poor energy resolution observed with Ar/CF_4 mixtures. Previous work[2] yielded energy resolutions with Ar/CF_4 (95/5) and (90/10) of $\sim$ 60% and $\sim$ 75%, respectively. We have measured the energy resolution for the mixing ratios (80/20), (70/30), (60/40), and (50/50) under different gas pressures. Typical values are summarized in Table 1.

A statistical model of the energy resolution of proportional counters has been well established,[7] which yields the relative variation of gas-gain A as following

$$\left(\frac{\sigma_A}{A}\right)^2 = \frac{1}{n_O}\left(F + \frac{1}{b}\right),$$

where F is the Fano factor and b is the parameter of Polya distribution.

Typical values of F and b are 0.05-0.2 and 1-2, respectively. Since n_O is the number of ion pairs created by the incident radiation, for Fe^{55}, $n_O \sim 200$. The Polya parameters b for these gas mixtures are greater than 1.

A rather surprising fact is that even under the worst case of F and b (but no electron attachment), the energy resolution of 5.9-keV x-rays still should be better than 18%. For Ar/CF_4 the measured energy resolution is worse than this by a factor of 3-5. If n_O has been reduced by a factor of 9-25 due to electron attachment this behavior would be explained. As noted previously, such a reduction is consistent with the low mean value of the gas gain observed for Ar/CF_4 mixtures with an Fe^{55} source .

Table 1: Energy resolution of Ar/CF_4 and Ar/isobutane using an Fe^{55} source.

Ar/CF_4	P (Torr)	Gas Gain (Fe^{55})	Resolution (Fe^{55})
80/20	760	98.8	48%
	1160	105.9	61%
	1960	107.2	82%
50/50	760	119.0	53%
	1160	92.2	63%
	1960	89.1	98%
Ar/isobutane			
80/20	760	206.0	16%
	1160	377.1	18.4%
	1960	243.8	17.3%
70/30	760	210.5	16.4%
29/71	760	994.9	16.7%
90/10	760	188.5	13.3%
	1160	2564.0	12.5%
	1960	1861.0	15.2%

References

[1] F.M. Newcomer *et al.*, *A Fast Monolithic Preamp and Shaper for High Rate Gas Tracking Datectors*, U. Pennsylvania, preprint.

[2] L.G. Christophrou *et al.*, *Fast Gas Mixtures for Gas-Filled Particle Detectors*, Nucl. Instr. and Meth. **163**, 141-149 (1979).

[3] A.Peisert and F. Sauli, *Drift and Diffusion of Electrons in Gases: A Compilation*, CERN 84-08, 13 July 1984.

[4] J. Fischer *et al.*, *Proportional Chambers for Very High Counting Rates Based on Gas Mixtures of CF_4 with Hydrocarbons*, Nucl. Instr. and Meth. A238, 249-264 (1985).

[5] T. Yamashita *et al.*, *Measurements of the Electron Drift Velocity and Positive Ion Mobility for the Gases Containing CF_4*, Preprint.

[6] J. Byrne, *Statistics of the Electron Multiplication Process in Proportional Counters*, Proc. Roy. Soc. Edin. **66A**, 33 (1962).

[7] G.F. Knoll, *Radiation Detection and Measurement*, 2nd ed., (John Wiley & Sons, New York, 1989).

ADVANCED FIELD SHAPING DRIFT CHAMBERS FOR SSC MUON TRACKING

A. F. Sill
Department of Physics and Astronomy
University of Rochester
Rochester, NY 14627-0011

Abstract

New construction methods are described which can be used to form high-precision drift chambers that are light and strong and have a linear voltage-to-distance relationship over large portions each of the drift cells. Use of resistive sheet film plastics is discussed. Examples are given with respect to a recent version of this sort of detector built for internal forward tracking the AMY experiment at TRISTAN.

Introduction

Muon tracking chambers for detectors at the SSC will have to meet some very stringent requirements for accuracy of alignment, physical size, and intrinsic resolution in order to provide useful input to triggering as well as offline data analysis. In addition they ought to be fairly robust, both in terms of relative immunity from rate-dependent effects and also physical rigidity and strength, and ought to be able to withstand broken wires and similar localized problems without greatly affecting the data-taking abilities of the entire detector subsystem.

These problems are not new, but take on added significance when compounded with the difficulties involved with handling chambers of the size typically needed for a large SSC detector. Spans of several meters in width and up to tens of meters in length are often discussed and have been seriously proposed. In order to meet these challenges it's desirable to have chambers that are light, strong, and rigid and that are constructed in a simple fashion which incorporates an appropriate method of electric field shaping directly into the mechanical design of the device.

We propose an option based on mechanical components that are stress-skin panels formed from a light core with a cladding of resistive or conductively-printed or etched metal-coated sheet plastic, then made into box girder layers by the addition of light side walls to give sets of closed vector- or single-wire drift cells. These can be made with recently available resistive sheet plastics to allow a construction that is light, strong, and economical to assemble.

An example based on AMY Forward Tracking

The AMY Forward Tracking Chambers[1] that were recently built by our group[2] for use at the e^+e^- collider TRISTAN were developed to meet a similar need. These chambers, which consist of approximately 2000 drift cells housed within the magnetic field of the detector, needed to be self-supporting with a minimum amount of material within the interior active volume of the detector and without much space devoted either to dead regions in the form of a support structure; hence they had to be essentially self-supporting, light, and strong. They were built to provide tracking in the endcap regions, approximately the 14° to 35° angular range with respect to the beam line.

Figure 1 shows the mechanical details of a 3-layer subset of the multi-layered construction for the Forward Tracking Chambers (FTCs) at AMY. They were formed from thin (80 μm) sheets of high resistance (10^{14} Ω per square) resistive-Kapton[3,4] plastic, preprinted with parallel lines of conductive epoxy paint, which were then laminated to 2 mm thick panels of nonconductive polymethacrylimide[5] foam.

The conductive traces, which were 2 mm in width, were printed at 1.5 cm intervals directly onto the high resistance Kapton. These corresponded to the centers and the sides of the 3 cm wide cells. Low-resistance side walls, made of a high conductivity (200 Ω per square) resistive Kapton laminated to 1 mm thick Rohacell foam, were used to complete the sides of each cell. The conductive traces in contact with the cell walls were placed at negative high voltage, and the center strips were placed at ground potential.

In this design, the high-resistance Kapton acted as a continuous, electrically noise-free sheet resistor between adjacent conductive traces to form a uniform potential gradient from the sides to the centers of the cells. Distortion of the equipotentials within the cell by the sense wires, which were placed at positive high voltage, was minimal, as shown in Figure 2. These cells consequently exhibited a uniform electric field throughout most of their width, which should lead in turn to an extremely linear time-to-distance relationship. Drift paths are shown for an example track in Figure 3. We built several prototypes using other methods of construction, including the multiple parallel strip method which involves the use of an external resistor chain. The resistive kapton method appears to be superior in terms of construction time and effort, as well as overall complexity (since it gets rid of the external voltage divider and field shaping traces or feed throughs), at roughly comparable cost.

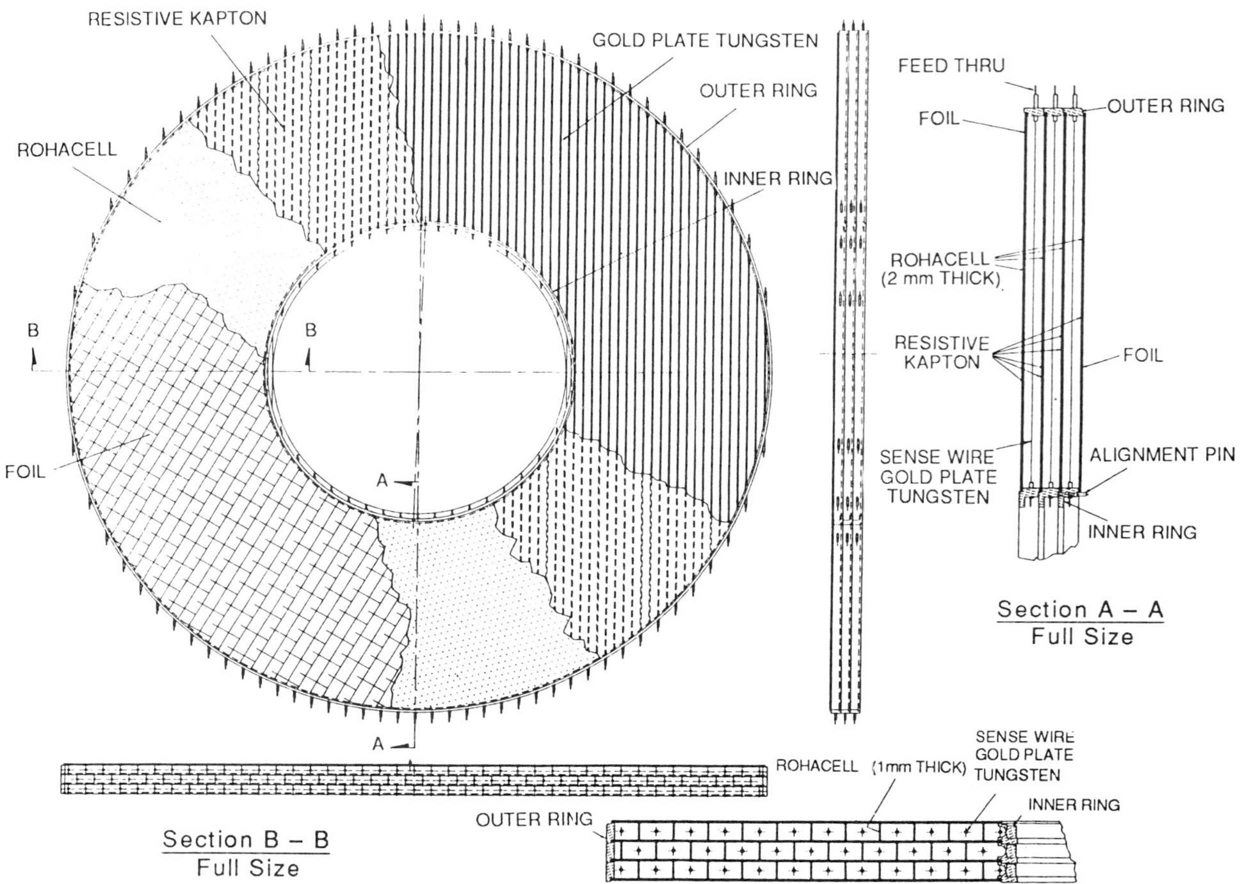

FIG. 1. Physical construction details for the AMY forward tracking chambers. The figure shows 3 of the 15 layers per endcap installed for the initial version of this detector.

Application to SSC Muon Tracking

The main features of the above design were its light weight, achieved by the use of polymethacrylimide foam and resistive sheet plastic, simplicity of field shaping, which allowed the use of comparatively large cells, and self-supporting semi-rigid construction. It is attractive to see whether with modest alterations in design these methods also could apply to the design of large-scale chambers for the SSC.

First, the resistive plastic would have to be chosen to allow for fast dissipation of excess accumulated charge from within the chamber at high rates. This would imply a larger conductivity (lower surface resistivity) than the 10^{14} Ω/square used in the top and bottom surfaces in the AMY chambers. As the conductivity is raised, the amount of power dissipated within the gas and carried away by the internal support structure will have to go up as well, leading to a situation in which a tradeoff exists between heat transfer and the intrinsic antistatic nature of the resistive plastic. The expression which gives this power dissipation is $P = 4V_F^2 N_{\text{cells}} l/(w\sigma)$ where l and w are the length and width of each cell, V_F is the field shaping difference in potential between the central and side conductive stripes, σ is the surface resistivity in terms of ohms per square, and the factor of 4 derives from the fact that there are four such long strip resistors per cell. In the case of AMY this gives approximately

$$P = \frac{4\ (2 \times 10^3\ \text{V})^2\ (2 \times 10^3)\ (50\ \text{cm})}{1.5\ \text{cm}\ \sigma} = \frac{10^{12}}{\sigma}\ \text{W}$$

or about 10 milliwatts for $\sigma = 10^{14}\Omega$/square. This is acceptable in the AMY case at low rates since the conductive central traces and low-resistance side walls should provide ion clearing sufficient to the task in the e^+e^- environment. It is substantially beyond the 10^5 to 10^6 Ω/square values usually needed to provide antistatic surfaces, however, and the above equation shows the difficulty of reaching such low values in any chamber with long thin cells. With hexcell panels of intrinsically higher heat conductivity than the extremely insulative foam used in AMY, it may be possible to allow up to several tens or even hundreds of watts to be dissipated within the entire support structure of a layer of cells, and surface resistivities approaching 10^8 Ω/square could be reached. The performance of such chambers in terms of ion clearing with respect to rate dependence will have to be checked by explicit tests.

Second, the mechanical properties of the support substrate and of the sidewalls will have to be improved to achieve such large spans. Possibilities include hexcell, graphite-reinforced composites, and extruded thin I-beams for the side walls. The solution could be a cross between the resistive plastic techniques used in the AMY chambers and those from several earlier I-beam-type chambers, such as those used by CDF and several other experments. The results would have the advantages of being relatively strong and light compared to extrusions, and more alignable and rigid than all-wire field shaping chambers, which would require massive end plates to achieve similar accuracy in placement.

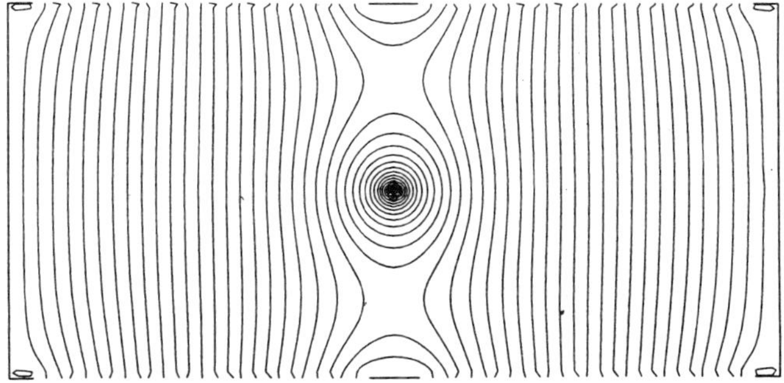

FIG. 2. Equipotentials for the example of the AMY drift cells described in the text.

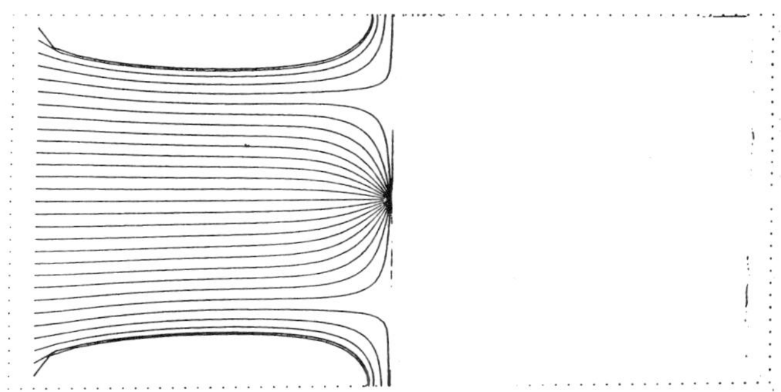

FIG. 3. Electron drift paths for the graded potential drift chamber cell shown in the previous figure.

References

1. A. Sill, Proceedings of the Meeting of the Division of Particles and Fields of the American Physical Society, Rice University, Houston, TX, Jan. 5-9, 1990.
2. The group involved physicists from Rochester, Louisiana State University, and KEK.
3. XC series resistive Kapton^{TM} polyimide film, DuPont Corporation, Wilmington DE 19880.
4. The desired value of resistivity in the AMY case had been two orders of magnitude lower, but DuPont was only able to supply this material within the time (and budget) limitations of the project.
5. Rohacell^{TM} polyimide/acrylic alloy foam, Röhm GmbH Chemische Fabrik, Darmstadt, W. Germany

DESIGN OF ELECTRONIC READOUT BOARD FOR STRAW TUBE ARRAYS*

Eric Erdos
Dept. of Physics, Campus Box 390
University of Colorado
Boulder, CO 80309

ABSTRACT

The electronics readout board for straw tube arrays is in the early design stage. At this time the emphasis is on developing techniques to produce a high density electronics board with reliable circuit connections. The choices of materials, techniques, and technologies are discussed.

Introduction

Wire drift arrays are suitable tracking detectors for SSC provided the drift distances and times are made small enough. The resulting large channel count puts a premium on the use of extensive on-board signal processing. The close packing of cells implies high densities of chips on the readout assemblies. We consider a model layout and techniques for its implementation.

The circuit designs have been described elsewhere[1]. Each drift wire is capacitively coupled to a preamp/-shaper/discriminator, whose differential output drives a time-to-voltage converter (TVC). Data collection chips (DCC) handle the digital data flow from some number of TVC's to the higher-level data stream. Information for triggering the experiment must flow both to and from the on-board circuits, which also have to supply high voltage to the wires.

A straw tube array such as that described by the Solenoidal Detector Collaboration[2] (SDC) comprises superlayers of 6-10 close-packed straw layers conforming to cylindrical surfaces, perhaps assembled from modules of a few cm width. For the present study we assume modules of trapezoidal cross section containing 228 straws, each 4 mm in diameter, in eight layers. Removable readout boards are placed perpendicular to the straw ends, respecting the 10-15 cm radial spacing between superlayers. In this paper we present the choices available to produce a board with such high density electronics.

Model Board Design

The model layout we have developed is shown in Fig. 1. The amplifiers, shapers, and discriminators for four wires are packaged into a single chip of dimensions 2 mm × 3 mm. Connections to each chip include four each of single-end wire input and differential pair output, one signal ground, and 2-3 DC voltage supply levels. The TVC is a single four channel chip 8 mm × 8 mm which has a pair of differential inputs per channel. Forty parallel buss lines handle data addressing and control, as well as power and ground. These connections terminate at the DCC which communicates between higher level DCC's.

Design Considerations

The main problem is making reliable connections from the IC die to the PC board and having enough room left over for test points and via connections to other layers. The technology available for making high density connections is not completely reliable. The physical properties of bonds are not easy to determine by visual inspection. Each board may have up to 5000 connections.

There are four different technologies that could be used:

- tape automated bonding (TAB)
- chip on board wire bonding (COB)
- flip chip
- thick film hybrid

FR4 Multilayer Circuit Board

A FR4 PC board is widely used, and it is fairly easy to find vendors capable of producing multilayers with traces of 4mil × 4mil. With multilayer boards vias are used to connect layers. There are three types of vias; conventional, blind and buried as illustrated in Fig. 2. To insure vias make connections to the other layers, pads have a relatively large size of .025", to provide for error in registration. Blind and buried vias can reduce the space taken up by the pads and allow more room for traces. However, these vias can trap contaminants thus affecting reliability. The cost of multilayer boards increases significantly with the more layers there are and the number of buried

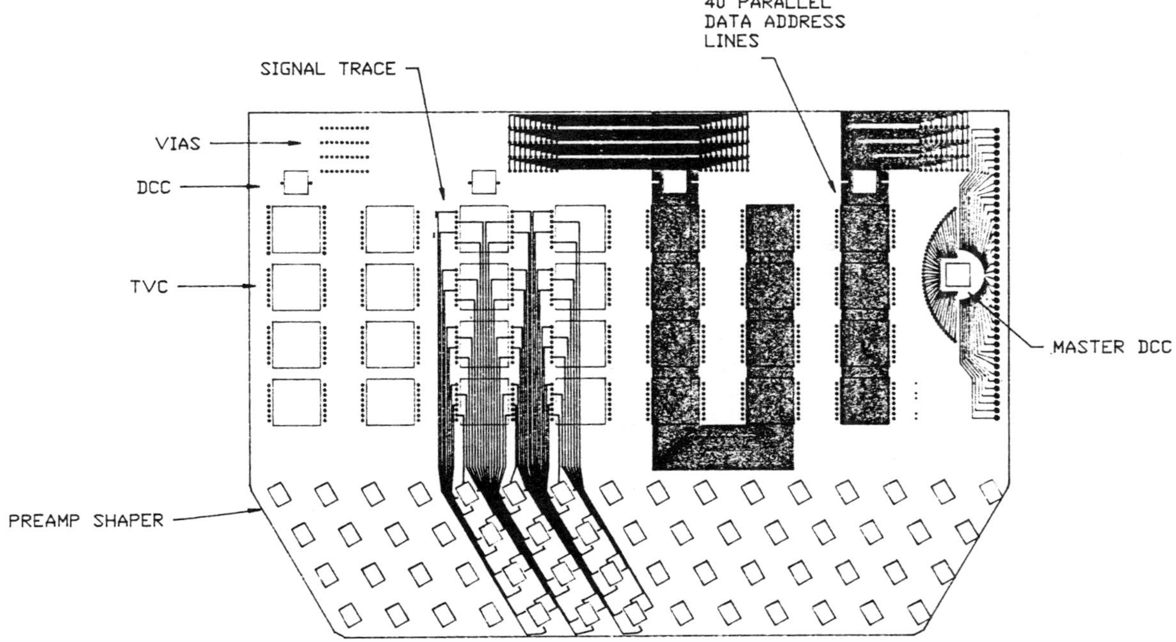

Fig. 1

and blind vias. Trace thickness also increases price when reducing trace and space sizes, due to lower yields.

A vacuum process is used to laminate the layers together. The board for this design will probably be about .062" thick. Fig 1 shows a layout suitable for this technique.

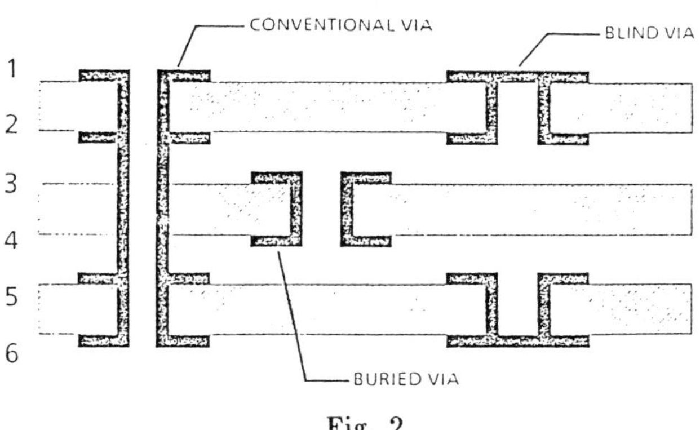

Fig. 2

FR4 Substrate TAB

Tape automated bonding uses an etched metal frame with inner leads that match the pattern of the die and outer leads that match the trace pattern and spacing on the PC board. The inner leads can be reliably connected in a variety of ways. The outer leads can be reflow soldered similarly to surface mount techniques. Problems arise with solder bridges between closely spaced traces. A solder mask cannot be used because of the closely spaced traces. However with a metal etched stencil, solder paste can be accurately printed through to the trace. Then the TAB is placed on the board and held in place by the tackiness of the solder paste. When all components are in place, the board can be heated to make the solder connection. If the solder paste is strictly controlled during application through the stencil, solder bridges can be kept to a minimum.

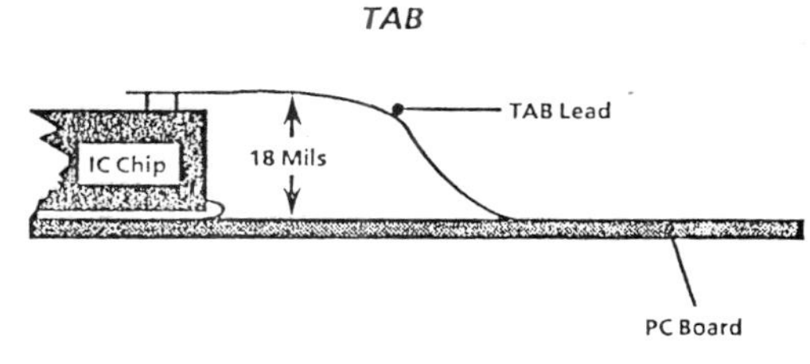

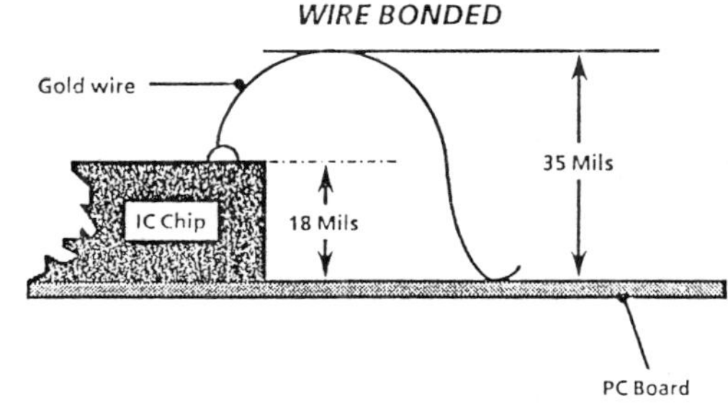

Fig. 3

Comparison of the COB and tape automated bonding (TAB) technologies

FR4 Chip on Board

With COB technology, the tab is done away with and the chip is mounted on the substrate. 25 micron gold wire is used to make the connections from the die to the trace, as illustrated in Fig. 3.

With FR4, heat is a problem. The stress generated by the differences of coefficent of thermal expansion (CTE) could cause the die to come off. Also, the temperature applied to the FR4 during bonding could cause it to melt, producing an unreliable connection. In order to bond to the trace, soft gold must be plated on the trace where bonds will be made. The purer the gold the lower the temperature of the bonding tool.

Flip Chip

Flip chip process eliminates the need for bond wires or tab mounting. The chip is turned over and mounted directly on the substrate. Solder on the die pads and substrate make the connections using surface mount soldering techniques. CTE is very important. To prevent solder joint cracking a substrate should have similar CTE to the die.

If a die is going to be used in a flip chip process, it needs to have solder bumps deposited on the pads. This is done in a vacuum deposition process while it is still in wafer form. The PC board also needs solder bumps. The solder bump size and diameter need to be tightly controlled. Before the chips are placed on the board a rosin flux is applied to the substrate to provide tackiness and to prevent chip movement until reflow soldering is done.

Ceramic Substrate Thick Film Hybrids

With polymer thick film technology it is possible to lay down layers of conductive traces, using screen printing techniques. Multilayers are easily formed by printing alternate layers of conductor and insulator. After each layer is screened, it is fired to cure the polymer. A board can have up to 20 layers per side. Interconnections between layers are made by printing through windows in the insulator layer. Printing line widths of 5mil with 5mil spaces are possible, however registration accuracy becomes a problem with smaller trace spacing. Any trace that might need to be soldered to can be screened with a solderable ink. Traces that need to be bonded to can be gold plated by using a platable ink base.

A ceramic substrate is the best choice for reliable wire bonding since the substrate is very stiff and can be heated. Thicknesses of ceramic are available as thin as .005", although the repeated firings of the polymer ink system tend to make ceramic brittle. A .025" thickness is commonly used.

High Voltage Input Board

Surface mount chip resistors could be used for current limiting, reducing board size. If a thick film process is used, the current limiting resistors would be screen printed on, further reducing the board mass.

The decoupling capacitors could stand on end (see Fig. 4) and be sandwiched between the HV input board and the readout board. Conductive epoxy could make the connection.

One possible way to make connections to the sense wire would be a conductive rubber tube from the HV board to the wire feedthrough.

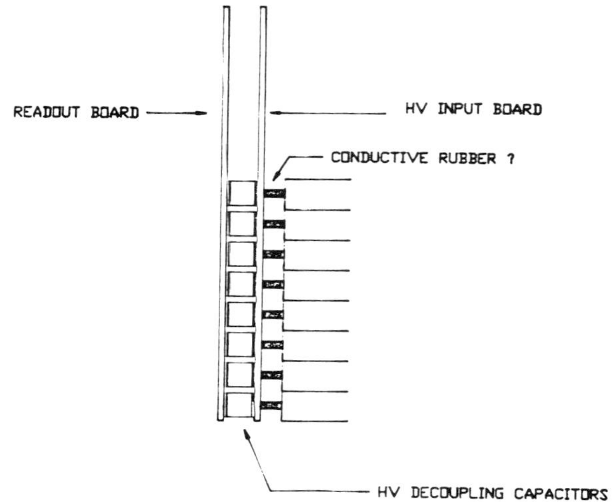

Fig. 4

*Work supported in part by the Department of Energy, contract DE-AC02-86ER40253, and by the Texas National Research Laboratory Commission.

[1] M. Newcomer, paper at this conference.

[2] Solenoidal Detector Collaboration, Expression of Interest (1990).

ETCHING OF ANODE WIRE DEPOSITS WITH CF_4 ISOBUTANE (80:20) AVALANCHES

R. Openshaw, R.S. Henderson[1)], W. Faszer and M. Salomon

TRIUMF, 4004 Wesbrook Mall, Vancouver, B.C., Canada V6T 2A3
[1]Also University of Melbourne, Parkville, Victoria, Australia 3052

Abstract

An ionization exposure of 0.5 coulombs per cm of wire in a gas mixture of CF_4/Isobutane (80:20) is shown to reverse anode wire damage in single-wire chambers. Several chambers aged in Argon/Ethane (50:50) and Argon/Ethane/Ethanol (50:50:0.2) and having pulse height reductions of 25-30% have recovered pulse heights and currents to greater than 98% of their initial values. Inspection of the anode wires indicates that the thick deposits caused by the exposure in Argon/Ethane have been removed. Auger Electron Spectroscopy reveals only a thin residual layer containing primarily carbon and oxygen. This etching ability of CF_4/Isobutane (80:20) avalanches may explain the extremely good ageing characteristics previously reported for this mixture.

Introduction

There is considerable interest in using CF_4 based gas mixtures for fast wire chambers. This is primarily due to the high drift velocities,[1,2)] high primary ionization,[2)] and low electron diffusion[3)] of CF_4 mixtures. Recent results,[4,5,6)] have shown that CF_4/Isobutane (80:20) does not produce any deposits on the anode wires, even with exposures up to 10 C/cm of wire. A 4% admixture of CF_4 has also been shown[5)] to inhibit anode damage in Argon/Ethane (50:50) for current density $\simeq$50 nA/cm-wire. The mechanism by which CF_4 prevents anode deposits has not been resolved.

Gas mixtures based on CF_4 are widely used in the semiconductor industry to etch silicon and silicon oxide substrates. While the conditions encountered in semiconductor plasma etching are not identical to those found in a typical MWPC avalanche, it is likely that similar chemical reactions take place.[7,8)] It may be reasonable then to expect that polymers found on the "aged" wires of a MWPC, containing carbon, silicon, oxygen and hydrogen, can be attacked by the radicals and negative ions produced from ionization in CF_4. The present work was undertaken primarily to determine if this was the reason for the resistance of CF_4 gas mixtures to ageing. The possibility of finding a non-invasive technique of "cleaning" MWPC's was also intriguing.

Experimental Procedure

The designs of the three types of test cells are shown in figure 1. The cells are placed in a stack separated by 3.2 mm thick lead collimators and exposed to the 10 mCi ^{90}Sr source through a 20 mm long aperture in the center of the 20 cm cell length. The amount of ionization occurring outside of the collimated area varies from 30% of the total ionization at the top of the test cell stack to 15% at the bottom. Within the stack were reference chambers for each gas mixture. These chambers had HV applied only during pulse height measurements with an ^{55}Fe source, and were used to monitor and correct for the combined effect of gas density and mixture variations.

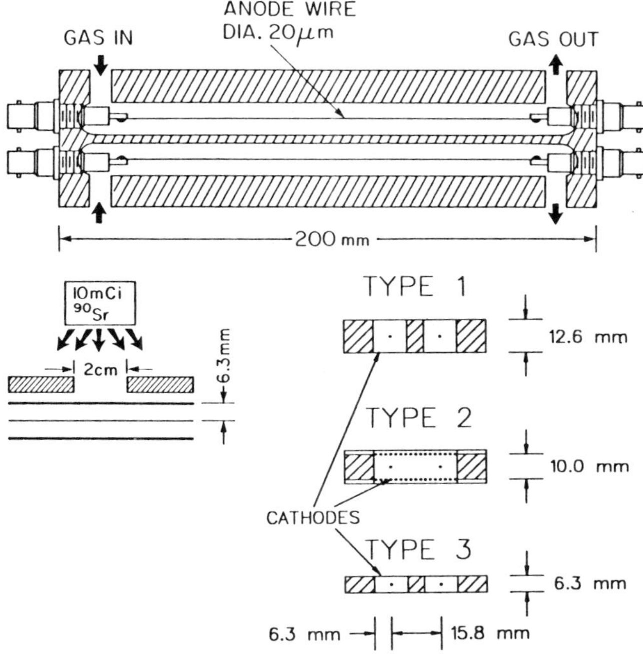

Figure 1: Schematic diagram of the single-wire test cells.

Five Single Wire Chambers (SWC's) which had previously been aged were selected for the CF_4/Isobutane recovery tests. Table 1 describes the conditions, total accumulated charge and the resulting damage from the exposure. Three of the chambers (E36, E47 and E48) were aged in Ar/Et (50:50) and E58 in Ar/Et/Ethanol (50:50:0.2), with typical pulse height reductions of 25% to 30% observed at the end of the ageing. The fifth chamber E34, had been aged in Ar/Et/CF_4 (48:48:4) at the high current density of 350 nA/cm.

After ageing but prior to exposure with CF_4/Iso, chambers E34 and E36 were opened and the anode wires observed with an optical microscope. Both of the SWC's had thick reddish/brown coatings covering the central 20 mm irradiated area and extending ±5–10 mm beyond. Also many small white lumps were observed scattered along the anode wires, becoming much more concentrated in the central irradiated regions. After observation of the anode wires, new cathode foils were installed.

For chamber E34 an improvement in pulse height in the damaged areas of about 4% was noticed when the gas was switched to CF_4/Iso. Switching back and forth between Ar/Et and CF_4/Iso yielded completely repeatable results. A similar improvement of $\simeq$8% was seen in chambers E47 and E48, but not with chambers E36 or E58. We have been unable to explain this effect. After ageing in Ar/Et and recovery in CF_4/Iso, chambers E34, E47 and E48 were again connected to the Ar/Et mixture and the pulse height profiles measured (see Table 2), demonstrating that the large improvement in performance is retained with operation in Ar/Et.

Recovery Results

Figure 2 shows plots of normalized pulse height and current as a function of accumulated charge and a selection of pulse height profiles for two of the five recovering test cells (E36 and E48). Table 2 shows the recovery parameters of the five chambers. The chambers were exposed to various current densities in order to observe the relationship between current density and recovery rate. The chambers damaged in Ar/Et (E36, E47, E48 and E58) were recovered with current densities ranging from 0.20 μA/cm to 1.15 μA/cm and showed uncorrelated recovery rates between 35 and 180 %/C/cm.

Chamber E34 was damaged in Ar/Et/CF_4 and "recovered" at the low current density of 40 nA/cm-wire. It displayed very slow pulse height recovery and very irregular values of current. This is partly due to the difficulty of accurately measuring the small currents used (typ. ±20%). Also the last current reading before removal of the strong 10 mCi ^{90}Sr source was consistently higher than the reading after its reinstalling some time later. This could be due to short lived chemical changes on the surface of deposits due to the strong source. The pulse height data for E34 had much smaller variation than the current, but was still quite erratic compared to the other chambers. Chamber E58 which recovered at the high rate of 180 %/C/cm also showed quite erratic currents and pulse heights during its recovery.

In all chambers except E34 the pulse heights and currents had recovered to >95% of their "clean wire" values by the time a total charge of 0.5 C/cm of wire had been accumulated. Chamber E34 showed only a slow recovery rate of $\simeq$10%/C/cm possibly because the low recovery current density of 40 nA/cm fell below some required energy density threshold or perhaps because the deposits formed in Ar/Et/CF_4 are more resistant to etching.

Chambers E36 at an accumulated recovery charge of Q=0.33 C/cm-wire (see Fig. 2) and E34 at Q=0.08 C/cm-wire, show a pronounced reduction in pulse height downstream of the irradiated area. Both downstream areas had greatly improved when the next pulse height profiles were taken five days later. Temporary condensation of low volatility reaction fragments moving from the central irradiated area with the gas flow could explain this effect. Another interesting effect was noted in chambers E48 (Fig. 2) and E47 where the first profiles taken after the commencement of recovery in CF_4/Iso, at Q1=0.091 C/cm-wire and Q=0.17 C/cm-wire respectively, actually showed reduced pulse heights in the irradiated area of $\simeq$6.3% and $\simeq$2.5% respectively. By the time the next profile was observed, both chambers had large improvements in pulse height in the irradiated area. It may be that during the initial stages of etching, the break up of large polymer chains into shorter fragments may loosen and expand the deposits thereby decreasing the field at the anode wire.

At the conclusion of the tests all of the test cells were disassembled and the anode wires observed by a variety of techniques including stereo microscopy, Secondary Electron Microscopy (SEM) and Auger Electron Spectroscopy (AES). Observation with the stereo microscope revealed that the white lumps and thick brown deposits seen on the anode wire of E36 before the exposure in CF_4/Iso were no longer visible. The anode wire of E36 still had a slight reddish tinge in the central area, while E47 and E48 appeared bright gold and clean with no deposits and no visible difference between irradiated and non-irradiated areas of the wires. Chamber E34 which had been removed from the ageing stack before complete recovery had a faint reddish discoloration in the central area along with a few of the white lumps. There were also some clear droplet-shaped lumps which are likely a product of the incomplete etching process, as they were not observed on the wire before the exposure in CF_4/Iso. The thickness and the area of the E34 anode wire covered by the various deposits were clearly less than that observed before the treatment with CF_4/Iso.

Using Auger Electron Spectroscopy (AES) the elemental composition of surface deposits on the anode wires of test cells which had recovered in CF_4/Iso were compared to those of typical cells damaged in Ar/Et. Since AES is only sensitive to surfaces to a depth of $\simeq$50Å the concentrations measured do not necessarily reflect the bulk composition of the deposit. In the AES spectra for E56, a cell which exhibited $\simeq$30% gain drop after 0.25 C/cm-wire ageing in Ar/Et, the 69 eV Au peak is absent and there are large Si (as SiO_2) and Oxygen peaks. In contrast AES from the "etched" E36 wire shows strong Au peaks, much less

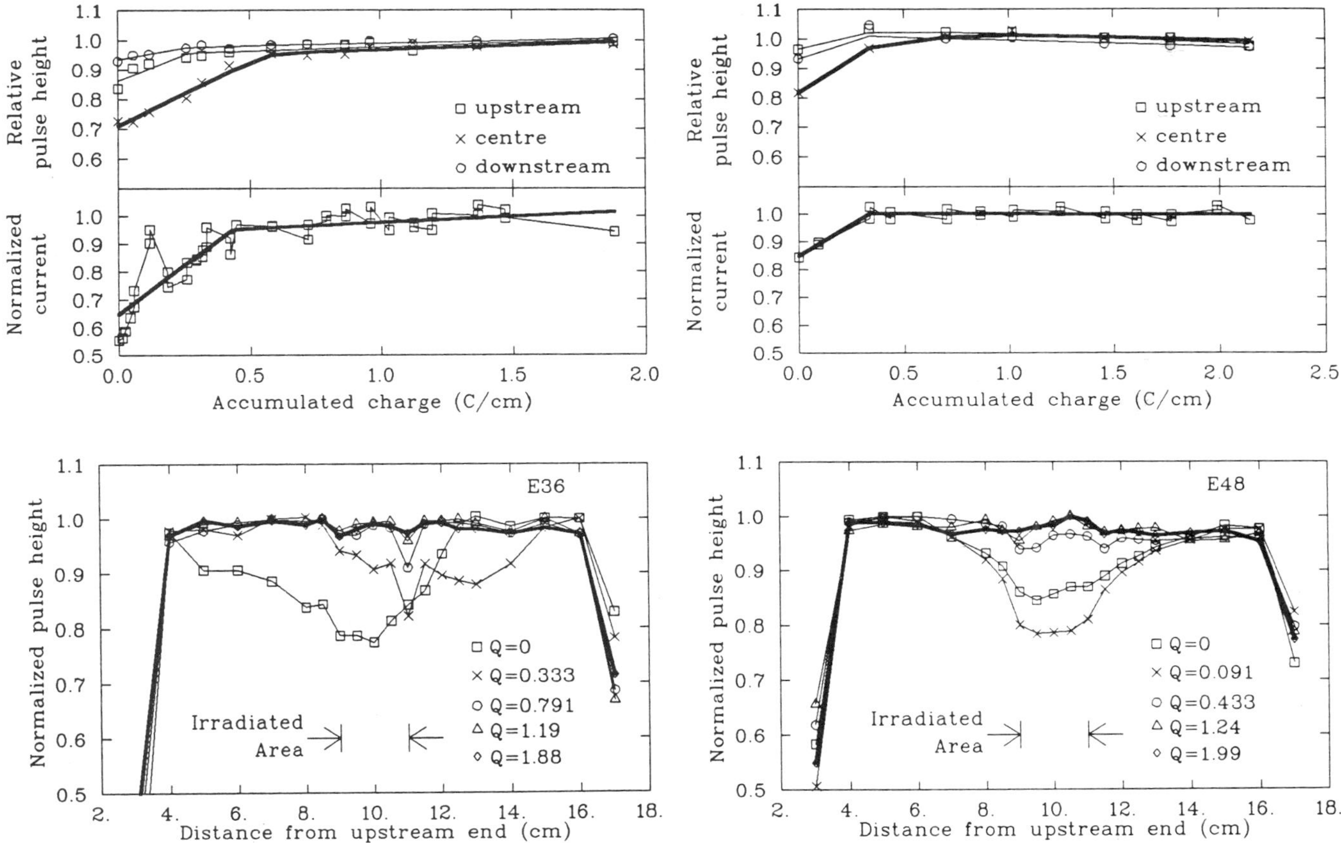

Figure 2: Plots of normalized pulse height and current versus accumulated charge for recovering chambers E36 (left) and E48 (right), also a selection of pulse height profiles at charge Q (C/cm-wire).

Oxygen, and no evidence of Si or SiO_2.

By comparing the peak heights and correcting for the Auger sensitivity to different elements it is possible to calculate relative atomic abundance. Table 3 lists the relative abundance of the four major elements (C,Si,S,O) found in anode wire deposits for: a fresh wire, three CF_4/Iso etched wires (E47,E36,E34) and an unetched Ar/Et damaged wire (E56). There is very little difference in elemental composition shown between the thin coatings on a fresh wire and those on the CF_4/Iso etched wires E36 and E47. The unetched wire E56 shows the largest relative silicon abundance while the the partially etched E34 show less and the other wires none at all, indicating that chemical processes in the avalanches in CF_4/Iso are capable of removing deposits containing silicon. The sulphur in the deposits could be from mercaptan impurities sometimes found in the ethane feedstock, however the sulphur on the fresh wire may also indicate a general environmental source. Fluorine was not observed in any of the deposits. It might be expected that if fluorine species were responsible for the etching then some fluorine should be bound to the deposits, particularly in the partially etched E34 deposits. The absence of fluorine may mean that any fluorinated products of the etching process are highly volatile and solid fluorinated products are not formed. This is found in RF plasma etching of SiO_2 with CF_4/O_2, where the products formed; SiF_4, CO_2 and COF_2 are stable and volatile[10]).

If the silicon in the deposits is in the form of SiO_2 it would be easily etched by CF_x radicals as it is in semiconductor plasma etching. The lower relative silicon concentration on the surface of E34 compared to E56 may even imply a preferential etching of silicon compounds over carbon compounds. Kadyk[11]) has repeated the etching experiment and seen recovery in CF_4/Isobutane (80:20) of single wire chambers prematurely aged by bubbling the Ar/Et mixture through silicon oil.

The deposit thickness for E36, E47 and the fresh wire were calculated by comparing the ratio of the intensities of the 69 eV and 2025 eV gold peaks to that of a standard, clean surfaced gold sample. For E56 the very large deposit thickness was estimated from SEM observations. On E34 even though the gold peaks were present, the deposit thickness was not calculated as the surface was clearly irregular, with some areas appearing relatively clean while others had clumpy deposits.

Chemistry

The chemical processes involved in plasma etching are

Table 1
Ageing Parameters of the five SWCs

Cell No.	Cell Type	Ageing Gas	Current Density (nA/cm)	Total Charge (C/cm)	Aged Pulse ht. (%)	Damage PH rate (%/C/cm)	Aged Current (%)
E34	1	$Ar/Et/CF_4$	350	4.48	62*	−8.8±1	73
E36	1	Ar/Et	210	2.82	75	−8.9±0.8	68
E47	2	Ar/Et	80	0.56	75	−39±5	74
E48	2	Ar/Et	100	0.64	70	−43±5	72
E58	3	Ar/Et/EtOH	50	0.20	74	−140±10	75

Gas gains $5x10^4$. * to 81% with replacement of cathode foil.

Table 2
Recovery Parameters in CF_4/Isobutane (80:20)

Cell No.	Gas Gain	Current Density (μA/cm)	Aged PH CF_4/Iso (%)	Recovered PH CF_4/Iso (%)	Total Charge (C/cm)	Initial Rec. Rate (%/C/cm)	Recovered PH Ar/Et (%)
E34	$5x10^4$	0.04	85±2	93±3	0.4	10	
E36	$5x10^4$	0.27	74±2	99±2	1.9	60	98±2
E47	$1x10^5$	1.15	84±2	98±2	3.9	35	98±2
E48	$5x10^4$	0.63	78±2	100±2	2.1	45	100±2
E58	$5x10^4$	0.20	72±3	98±3	0.91	180	

Table 3
Relative Elemental Abundance in Deposits

Cell	Description	Abundance relative to carbon C	Si	S	O	Deposit Thickness (Å)
	Fresh wire	1	0	0.19	0.14	11
E47	Ar/Et deposit etched by CF_4/Iso	1	0	0.08	0.11	6
E36	Ar/Et deposit etched by CF_4/Iso	1	0	0.06	0.14	33
E34	Ar/Et deposit partially etched	1	0.39	0.12	0.88	**
E56	Ar/Et deposit unetched	1	0.63	0.45	1.7	~10,000

** not measured

extremely complex. There are a large number of reactions between ions, electrons, free radicals, excited and stable molecules, which can occur in either the gas phase or at the gas/solid interface. In addition many of these reactions may be extremely sensitive to parameters such as flow rate, pressure, temperature, power density, concentration of reactants and minor impurities. Even a basic understanding of the chemistry involved would require as a minimum, detailed analysis of the solid and gas phase reaction products. Such an understanding is beyond the scope of this paper. Here we only suggest some possible chemical mechanisms for the removal of anode wire deposits by electron avalanches in CF_4/Isobutane. Plumb and Ryan[9] have shown that the most probable electron impact disassociation reactions in CF_4 are:

$$CF_4 + e \Rightarrow CF_3^{\bullet} + F^{\bullet} + e$$
$$\Rightarrow CF_2^{\bullet\bullet} + 2F^{\bullet} + e \qquad (1)$$

$$CF_4 + e \Rightarrow CF_3^{\bullet} + F^{-} \qquad (2)$$

$$CF_4 + e \Rightarrow CF_3^{-} + F^{\bullet} \qquad (3)$$

Reaction 1 occurs approximately twice as frequently as reactions 2–3 and involves the creation of short lived excited states of the radical $CF_3^{\bullet}$ which rapidly disassociate. In semiconductor plasma etching, the plasma is continuously pumped with RF energy, and the reactive species can remain in the plasma generator for 10's of milliseconds. During this time recombination and other gas phase reactions between the molecules, electrons, radicals and ions present can take place and substantially change the relative populations of the various chemical species. In a wire chamber avalanche the major production of electrons, ions and radicals occurs within a few mean free paths of the anode wire, and the avalanche lasts less than a nanosecond. It seems unlikely then that gas phase reactions would have time to significantly modify the initial disassociation products.

Due to the difference in mobility between ions and electrons, all surfaces in contact with an RF plasma assume a negative potential with respect to the plasma. As a result, positive ions are accelerated into surfaces. The process is distinctly different in an atmospheric pressure wire chamber, where moderate energy electrons (<10 eV) and low energy negative ions (<1 eV) bombard the anode and near thermal positive ions strike the cathode. The electron energies and densities are however similar[7]. In an avalanche in CF_4/Iso (80:20) perhaps 10^6 to 10^8 highly reactive F and CF_x radicals and ions are released over a small area ($\simeq$10x10 μm) of the anode wire. At precisely the same time energetic electrons (5–10 eV) strike the same area of the anode surface. The energies of the avalanche electrons span the range of disassociation energies of typical polymer deposits, and therefore the electrons are quite capable of breaking bonds between atoms of the polymer. It is likely that the fluorine containing radicals would bond to the reactive polymer fragments preventing recombination of the polymer bonds. If the resulting halocarbon reaction products are small enough to be volatile they would be carried out of the chamber by the gas flow. Heavier halocarbons would be subject to subsequent avalanches and eventually all polymers would be reduced to stable volatile products.

Conclusions

We have demonstrated that exposure to ionization in CF_4/Isobutane (80:20) reverses anode wire damage in single-wire chambers aged in Argon/Ethane (50:50) and Argon/Ethane/Ethanol (50:50:0.2). Four chambers with pulse height reductions of 25–30% have recovered pulse heights and currents to greater than 98% of their initial values, with most of the recovery occurring in the first 0.5 C/cm of exposure. Inspection of the anode wires indicates that the thick deposits caused by initial ageing have been removed. Auger Electron Spectroscopy reveals only thin (<40 Å) residual layers containing primarily carbon and oxygen. A fifth chamber aged in Argon/Ethane/CF_4 (48:48:4) has recovered to only 93%, probably because the recovery current density used was too low or possibly these deposits are more resistant to etching.

All of the recovery tests reported here have been done with the CF_4/Isobutane (80:20) gas mixture and it should be stressed that introducing different components or even changing the relative proportions of CF_4 and Isobutane could totally change the chemical reactions in the avalanche and on the surface of the anode wire.

A more complete understanding of the chemical processes involved could make it easier to customize gas mixtures to incorporate the excellent ageing characteristics of CF_4/Isobutane (80:20) for each application. As an example of empirically improved performance, a 4% admixture of CF_4 has also been shown[5] to inhibit anode damage in Argon/Ethane (50:50) for current densities of $\simeq$50 nA/cm-wire, well above the 5 to 10 nA/cm-wire expected in most high rate, large area applications. On the other hand, Kadyk[7] has seen an amazingly rapid ageing rate of $\simeq$123,000 %/C/cm with Ar/CF_4/O_2 (50:40:10).

The etching ability of CF_4/Isobutane (80:20) suggests the possibility of a non-invasive technique of "cleaning" wire chambers. For a chamber with an active area of 1 M^2 and an anode wire spacing of say 4 mm; a cleanup current of 0.5 μA/cm-wire would total 12.5 mA, and 23 days would be required to accumulate a recovery exposure of 1.0 C/cm. It would also need even and powerful illumination with ionizing radiation. Cleanup of small specific detector areas would of course be less extreme.

The demonstrated etching ability of CF_4/Isobutane (80:20) avalanches probably explains the extremely good ageing characteristics previously reported for this mixture. This ability is believed due to the highly reactive nature of the CF_x and F ions and radicals produced, and their tendency to form stable and volatile reaction products.

References

[1] Christophorou et al., "Fast gas mixtures for gas–filled particle detectors." Nucl. Instr. and Meth. **163**(1979)141.

[2] J. Fisher et al., "Proportional chambers for very high counting rates based on gas mixtures of CF_4 with hydrocarbons." Nucl. Instr. and Meth. **A238**(1987)249.

[3] B. Schmidt and S. Polenz, "Electron motion in counting gases – new answers and new questions." Nucl. Instr. and Meth. **A273**(1988)488.

[4] R. Henderson et al., "Wire chamber ageing with CF_4/Isobutane and Argon/Ethane mixtures." IEEE Trans. on Nuclear Science **35**(1988)477.

[5] R. Openshaw et al., "Tests of wire chamber ageing with CF_4/Isobutane (80:20), Argon/Ethane (50:50), and Argon/Ethane/CF_4 (48:48:4)." IEEE Trans. on Nuclear Science **36**(1989)567.

[6] J. Kadyk et al., to be published in IEEE Trans. on Nuclear Science **37**(1990).

[7] J. Va'vra, "Review of Wire Chamber Aging." Nucl. Instr. and Meth. **A252**(1986)547.

[8] D.W. Hess, "Plasma Chemistry in Wire Coatings." Proceedings of Workshop on Radiation Damage to Wire Chambers, p15-23. Lawrence Berkeley Laboratory 1986.

[9] I.C. Plumb and K.R. Ryan, "A model of the chemical processes occurring in CF_4/O_2 discharges used in plasma etching.", Plasma Chem. and Plasma Process, Vol.6 No.3(1986)205.

[10] K.R. Ryan and I.C. Plumb, "Gas-phase chemistry in the processing of materials for the semiconductor industry." CRC Crit. Revs. in Solid State and Mat. Sci., Vol.15 No.2(1988)153.

[11] J. Kadyk, private communication.

RECENT WORK ON RADIATION-HARD GASES AND STRAW TUBES*

J.A. Kadyk,[1] D.W. Hess,[2] J. Va'vra,[3] and J. Wise[1,2]

[1]Lawrence Berkeley Laboratory, University of California, Berkeley, CA 94720
[2]Department of Chemical Engineering, University of California, Berkeley, CA 94720
[3]Stanford Linear Accelerator Center, Stanford University, Stanford, CA 94309

Abstract

Studies have been made of radiation-hard gases having CF_4 as a major constituent, and aging rates based upon accelerated aging testing techniques are presented. A variety of aging tests have also been performed with straw tubes of 4 mm diameter. Using a straw tube with a carbon filament anode wire, the disposition of heat generated in the avalanche process was determined to be predominantly by conduction from the anode to the cathode tube wall along the radial direction. This eliminates the need for removal of the anticipated heat load at SSC experiments through exceptionally high gas flow rates.

I. Inhibition of Wire Aging Using CF_4-Based Gases

A. Introduction

Using techniques described previously [1] to perform accelerated aging tests, we have investigated several CF_4-based gas mixtures which have properties desirable for application in wire chambers to be used in high radiation environments. Among these properties are: 1) low diffusion [2], 2) high linear ionization density [3], 3) high drift velocity [2,4], and 4) low aging rates [5,1]. The aging properties of the gas mixture CF_4/isobutane(80/20) have been studied recently, and have exhibited an exceptionally low aging rates [5,1]. These low rates may be the result of the 'etching' action of radicals produced in the breakup of the CF_4 molecule in the avalanche process [6]. We report here on additional results obtained using CF_4/isobutane(80/20) as well as using other gas mixtures containing CF_4. These results can be interpreted in terms of the model of competitive ablation and polymerization [7].

In the subsequent discussion, the aging rate is defined as the rate of current decrease (gain decrease):

$$R = (1/I_o)(dI/dQ)\ \%/C/cm,$$

where: I = current, I_o = initial current, and Q = charge transfer. The length of the irradiated region on the wire is about 3 mm in most cases. The avalanche gains at beginning of tests are thought to be in the neighborhood of 20,000 to 50,000, but are not known with any precision.

B. Etching of Silicone-coated Wires Using CF_4/isobutane(80/20)

It has been recognized for some time now that the gas mixture CF4/isobutane(80/20) in wire chamber tests results in very low aging rates [5,1]. Additional tests have demonstrated that, in fact, previously coated wires can be restored by etching (i.e. chemical removal of deposits) using this gas mixture [8]. We have confirmed this observation by first coating anode wires using a previously developed technique [1], then restoring full gain in these wires using this gas mixture. Other types of tests were also performed to elucidate the mechanisms involved.

1. Aging Rate of CF4/isobutane(80/20)

We have done several tests of the CF_4/isobutane(80/20) mixture, and all give very low aging rates. A typical test is summarized in Table I. This confirms earlier results obtained by the TRIUMF group [5].

2. Silicone-Coating of Anode Wires

A technique for deliberately coating the anode wire with silicone polymer [1] was used to study the etching effects of CF_4/isobutane(80/20). By bubbling an argon/ethane(50/50) gas mixture through silicone diffusion pump oil (Dow 704) before the gas enters the wire chamber, a layer quickly coats the anode wire. The rate of wire aging during this process is as high as 300%/C/cm, being more rapid initially, then diminishing later as the gain decreases asymptotically toward a non-zero value. It was verified that the wire coating is not due to merely a film of oil mechanically deposited by the gas by performing a run as described above, with high voltage applied, but with no Fe^{55} source to induce avalanches. No aging was observed through either a gain decrease or a degradation in the Fe^{55} pulse height spectrum under these conditions. In contrast, a similar test, with a source applied for an equal length of time, resulted in significant gain decrease and a severely degraded pulse height spectrum. The avalanche is therefore necessary to produce the wire coating. The rate of coating seems remarkably high, since the vapor pressure of the silicone oil is only 1.4×10^{-8} Torr (at 20°C), corresponding to a concentration in the gas of only 0.018 ppb! An analysis of the surface of wires coated in this manner using

*This work was supported by U.S Department of Energy Contracts DE-AC03-76SF00098 (LBL) and DE-AC03-76SF00515 (SLAC).

FTIR revealed the presence of Si-O bonds, suggestive of a silicone polymer.

This technique of bubbling gas through silicone oil has been used repeatedly to create wire deposits, and is quite reproducible. Fig. 1 shows the current during a test in which a silicone coated wire was produced for the test described next.

3. Removal of Silicone Coating Using CF_4/isobutane(80/20): "Etching"

An anode wire coated in the manner described above in section IIB was subsequently irradiated while operating with a CF_4/isobutane(80/20) mixture (the Fe^{55} source location remained fixed relative to the wire between coating and etching runs). Fig. 2 shows the current (gain) during this run, and can be compared with Fig. 1. Initially, the current and gain are low, and the voltage was increased by ~5% (at "A" on Fig. 2) to accelerate the etching process. Subsequent to this point, there was a small drop in current, soon followed by an erratic increase in current up to the full-gain value, where the current becomes constant again, suggestive of complete removal of deposits in the irradiated area. Also associated with the process are frequent current surges suggestive of Malter-type breakdowns. This indicates that cathode as well as anode coatings were present. The cathode coatings also seem to be removed, by the etching action since they cease as the process of gain recovery proceeds.

4. Bubbling CF_4/isobutane(80/20) Through Silicone Oil

As an additional test of the resistance to aging, CF_4/isobutane(80/20) was bubbled through Dow 704 silicone oil before the gas entered the wire chamber, in the same manner used to coat wires using argon/ethane gas. There was a very small measured aging rate of 3%/C/cm. This test was performed with the same wire, source, and geometry used for tests described above in sections IIB and IIC.

5. Lack of Etching Using Argon/Ethane or DME

To establish that the removal of wire coating does indeed depend upon the use of the CF_4/isobutane gas mixture, two other gases, argon/ethane(50/50) and dimethyl ether (DME), were used instead, under conditions identical to those described in section IIC above, where etching was observed. In neither case is there any evidence that the current is increasing.

C. Other Results and Summary

The results of tests using several other CF_4-based gas mixtures are given in Table I. CF_4-based gases appear to be excellent candidates for many wire chamber applications, especially those involving high rates and radiation levels. The combination of low diffusion and high linear ionization density are important in obtaining good spatial resolution, and the relatively high drift velocity and very

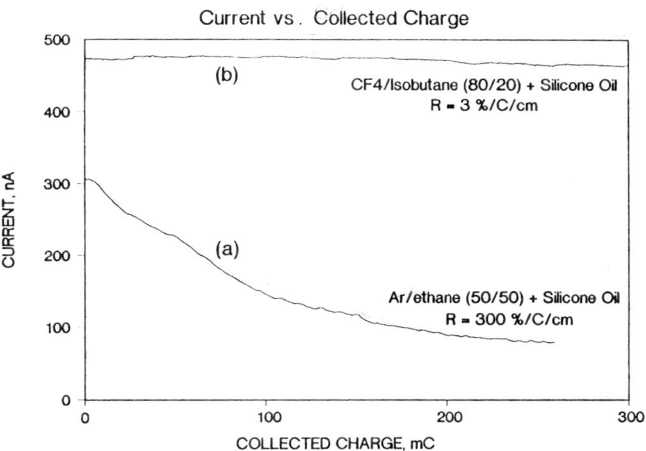

Fig. 1. (a) Aging run during which Ar/ethane(50/50) was bubbled through silicone oil (Dow 704) upstream of the wire chamber. Aging only occurs when high voltage and Fe^{55} source are both present. Note that current is asymptotic to a nonzero value. (b) Run similar to (a), except that CF_4/isobutane gas was bubbled through silicone oil.

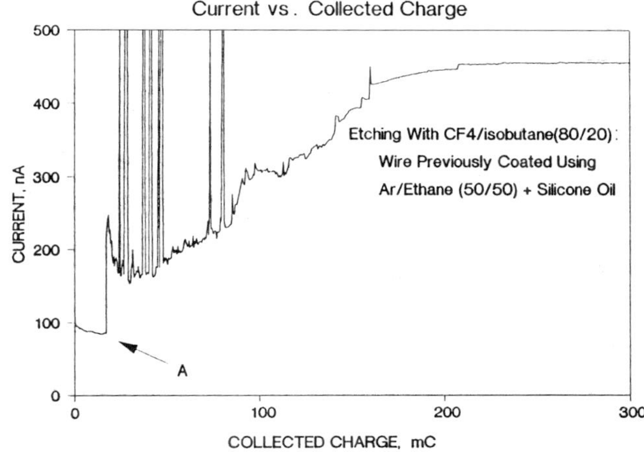

Fig. 2. Removal of Wire Deposits Using CF_4/isobutane(80/20) Gas. Avalanche-induced silicone deposits produced by bubbling Ar/ethane through silicone oil (Fig. 1) are subsequently removed by operating instead with CF_4/isobutane. All other variables were unchanged except for a 5% voltage increase to accelerate the etching process.

low aging rates are important for operation in a high radiation environment. Aging rates, using Fe^{55} sources to accelerate the aging process, have been measured for several gas mixtures containing CF_4.

The principal gas mixture investigated was CF_4/isobutane. The mixture CF_4/isobutane(80/20) was found to have a very low aging rate ($R < 2$), and to be capable of removing deposits from an existing wire coated with silicone polymer. It also exhibited a quite low aging rate ($R = 3$) when bubbled through silicone oil before entering the wire chamber, a procedure used deliberately to create the silicone wire coating using Ar/ethane(50/50) gas ($R \sim 100$). An additional proof of the distinctive etching properties of this mixture was the observation that two gases, Ar/ethane(50/50) or dimethyl ether did not result in a removal of deposits in the manner observed for the CF_4/isobutane(80/20).

By reducing the CF_4 content, aging could be induced: although the (50/50) mixture did not age, the (20/80) mixture did (R ~ 300). The mixtures CF_4/DME(90/10) and (20/80) and CF_4/CO_2(40/60) were all observed to have very low aging rates as well. This latter gas operated stably with little aging, while, for comparison, pure CO_2 appeared to be quite unstable, although not obviously as a result of recognizable aging. There may be some advantage in using CF_4/CO_2 in applications where background signals resulting from neutron-induced proton recoil particles may be experienced, such as at future SSC experiments. Use of gases with no hydrocarbon content should avoid such background processes occurring in the gas. There may, however, be recoil protons arising from wire chamber walls (e.g. straw tube walls), unless these are also made from non-hydrogenic materials.

A model of competitive ablation and polymerization processes has been proposed [7]. A non-zero aging rate and amount of deposition, which depends upon the relative strengths of the ablation and polymerization reactions, is implied by this model. The results obtained here may be explained in terms of this model.

Table I
Anode wire aging rates measured for CF_4-based gases and associated tests, using wire 50 μm diameter. The gas flow rates were 10-20 cc/min through a 3/8" diameter tube. Gains were 20,000-50,000. Wire is irradiated over ~3 mm, except for the first test listed. The silicone oil used was Dow 704 diffusion pump oil with a vapor pressure at 20°C of 1.4×10^{-8} torr.

Gas mixture	Aging rate, R %/C/cm	Current density, μA/cm	Charge, C/cm	Comments
Tests using gold-plated wire:				
CF_4/iC_4H_{10}				
(80/20)	<2	0.3	0.8	~6 mm of wire irradiated
+silicone oil	3	1.4	1.5	Bubbled through oil upstream of chamber
Ar/C_2H_6 (50/50) +silicone oil	300→50	0.9→0.3	0.6	Aging rate decreased during test (Fig. 2)
CF_4/iC_4H_{10}				
(20/80)	300	0.9	0.09	
(50/50)	-2	2.0	1.6	Same wire used in (20/80) test
CF_4/DME				
(90/10)	1	1.2	1.8	
(20/80)	0	1.0	0.5	
CF_4/CO_2				
(40/60)	7	1.2	0.8	
Tests using resistive wire:				
CF_4/iC_4H_{10} (80/20)				
	350	0.6	0.2	Nicotin
	110	0.6	0.2	Stablohm
	320	0.6	0.1	Stainless steel

II. Use of Straw Tubes in High-Radiation Environments

A. Introduction

There are two parts to this investigation: (1) study of aging effects in straw tubes, including straw tubes previously damaged by discharges and (2) use of straw tubes equipped with carbon wires to study the thermal effects of avalanches. The results of aging tests are contained in Table II, and will be discussed briefly in the summary. Heating effects in straw tubes are dicussed below and summarized in Table III.

B. Heating Effects

Possible limitations to the use of straw tubes in high radiation environments are gas temperature changes and the consequent gain changes. Indeed, temperature changes as small as 6°C, corresponding to ~10% gain variations, may be unacceptable in some applications. The potential difficulty arises in removal of the power dissipated as heat in the tubes, since a temperature rise of ~10^4 °C is obtained from an estimate based upon heat transport using the heat capacity of the gas alone, assuming laminar flow of 1 mm/sec, and 2 mW/m avalanche power dissipated. Various heat transport mechanisms will surely limit the temperature rise to a much smaller amount than this. However, from this analysis we recognize the possibility that unacceptably large increases in temperature may occur.

For the purpose of investigating the heat transfer mechanisms in detail, we used a straw tube of the abovementioned design with a 33 μm-diameter carbon anode wire through which current was passed to produce ohmic heating, and in this fashion to simulate the heat produced in a tube in a high-radiation environment. We believe this method simulates closely the gas heating near the surface of the wire and the associated heat flow through the tube.

A small dc current supplied by a battery was passed through the carbon wire, and the voltage across the wire and the current through the wire were measured simultaneously. These measurements determined directly both the resistance of the wire and the power dissipated. The resistance determination permitted an estimate of the temperature of the carbon wire, since the resistance of carbon is a function of temperature. To estimate the temperature, we used the coefficient for carbon resistivity as a function of temperature as measured by us, using a temperature-controlled oven (Varian model 3400 gas chromatograph).

A second method of estimating the temperature near the wire surface was measurement of the gain increase observed in the pulse-height spectrum of ^{55}Fe when the wire was heated. The dependence of gain on temperature has been well established from ambient pressure and temperature changes during many wire aging tests performed with this system.

A third method of determining the wire temperature follows from an analysis of the heat transfer. Estimates show that all except one of the potential channels for heat flow are small enough to be neglected: conduction through the gas along the radial direction between the anode and the cathode dominates the heat flow. With this simplification, an estimate of the temperature change is obtained directly from the conduction equation, using the known heat input to the wire.

Measurements of the temperature were made using all three methods, by varying the battery voltage up to about 100 V. Two specific cases, 9 V and 46 V, are summarized in Table III. The agreement between the three methods is good.

The level of agreement between these two cases gives some confidence that this procedure can be extrapolated to the still lower power levels appropriate to an SSC experiment. Although the power corresponding to the 9 V case is ~60 mW/m, perhaps 100 times the maximum expected in normal SSC operation, the temperature rise is nevertheless sufficiently small to be acceptable. Since the temperature rise is expected to scale linearly with power, the anticipated temperature rise in an SSC experiment would be very small indeed for an isolated tube.

As an additional verification of the conductive model of heat transfer, the gas flow rate was varied from 1.5 to 150 cm^3/min while monitoring the gain. No change of gain was observed for flow rates of 1.5 to 120 cm^3/min, indicating that the gas temperature in the avalanche region is largely unaffected by the gas flow rate, as expected if radial conduction is the dominant heat transfer mechanism. Some change in the gain was observed for flow rates in the 120 to 150 cm^3/min range, however. We believe that this was where the transport of heat by gas flow through the tube at higher flow rates began to compete with removal of heat by conduction radially.

Since conduction is the dominant heat transport mechanism, the gas flow rate is not dictated by heat removal, and can be as low as consistent with acceptable aging. We obtained good aging results with gas flows as low as ~1.0 cm^3/min, corresponding to an average linear gas velocity of only 1.3 mm/sec.

For isolated straw tubes, conduction alone is sufficient to remove the heat from the tube wall; some convection is expected to be necessary to remove heat from an array of tubes such as in a full-scale detector. If tubes are arranged in layers with a small number of tubes per layer, as in some current designs [9], there will still be only negligibly small temperature rises within each layer according to our estimate, with heat being carried away efficiently by the gas between layers.

Table II
Aging results in straw tubes. The aging rate, R, is defined as $R = -(1/I_0)(dI/dQ)$ where Q is the collected charge per length of wire and I is the current. All tests used gold-plated wires of 38 μm diameter except the last test listed, which used a carbon filament of 33 μm diameter. The total gain was ~5×10^4 for tests with gold-plated wires and ~2×10^4 for the test with the carbon filament. Nylon tubing was used for the gas plumbing in all tests.

gas	R %/C/cm	accumulated charge C/cm	anode voltage V	gas vel. mm/s
Ar/C_2H_6 (50/50)				
+1000 ppm H_2O	15	0.4	1600	53
CF_4/iC_4H_{10} (80/20)				
+1000 ppm H_2O	3	0.5	2240	53
w/out added H_2O	5	0.25	2240	53
CF_4/DME (90/10)	3	0.7	2200	26.5
Ar/C_2H_6 (50/50)	~0	0.8	1590	26.5

C. Summary

The simple prototype straw tube proportional counters described here were used for studies of aging and heating under conditions that are expected to be encountered at SSC experiments. Two major concerns investigated were the aging rates with several gases and the proportional gain fluctuations due to avalanche heating.

We have tested the aging of straw tubes having aluminum cathodes with several gases: argon/ethane (50/50) and CF_4/isobutane (80/20), with and without water added; and CF_4/dimethyl ether (90/10). With argon/ethane as the test gas, it was found that water vapor on the order of 1000 ppm was needed to prevent electrical discharges in a tube that had some previous damage from discharges. With the use of Nylon tubing, no direct water vapor addition was needed, presumably due to sufficient natural outgassing of water from or diffusion of water through the tube walls. In at least one case (CF_4/DME), however, use of stainless steel tubing resulted in immediate electrical breakdown, which we believe to have been due to the lack of water vapor. All of the gases tested were found to give satisfactory aging rates. In addition, argon/ethane gave satisfatory aging with gas flow rates as low as 1.0 cm^3/min.

The investigation of heating effects used a 33 μm-diameter carbon anode wire to model the avalanche gas heating by ohmic heating from small electric currents. It was found that the heat transfer from the wire is essentially all by conduction through the gas along the radial direction. The recognition of this may simplify considerably the design of large straw-tube detectors.

Table III

Measurements of heating effects in straw tube with 33 μm diameter carbon filament anode: 9 V and 46 V cases. Temperature dependence of carbon resistivity: $(1/R_0)\mathrm{d}R/\mathrm{d}T = -2.0\times10^{-4}/°\mathrm{C}$. The resistance of the filament was 9.22 kΩ at 21°C. The electrical length of the wire was 22.9 cm; the active length was 12.2 cm.

	9 V	46 V
Voltage across filament (V)	9.0	45.8
Current through filament (mA)	0.98	5.0
Resistance of filament (kΩ)	9.22	9.13
Power input (mW)	8.8	230
Power input to active region (mW)	4.7	123
Gas flow (cm^3/min)	20	18
^{55}Fe: Pulse-height of principal peak (channel, ped. subtr.)		
Base case* (0 V)	400	376
With voltage applied	423	769
Gain ratio G(V)/G(0):	1.06	2.04
Computation of temperature rise (°C) (estimated error)		
From filament resistance (±20%)	1.6	43
From gain change (±10%)	2.4	45
From power and conduction (±5%)	1.6	41

*The tests were made on different days, and the differences in the pulse heights of the base cases are due to barometric pressure changes.

References

[1] J. Kadyk *et al.*, *IEEE Trans. Nucl. Sci.*, **37**(2) (1990) 478.

[2] B. Schmidt, Internal Note HD-PY/86/06, Univ. of Heidelberg. Also given in: "Radiation Effects At The SSC", SSC-SR-1035, (1988) 189, published by SSC Central Design Group, LBL, M. Gilchriese, ed.

[3] J. Fehlmann and G. Viertel, Compliation of Data for Drift Chamber Operation, ETH Zurich (August 1983)

[4] L.G. Christophorou *et al.*, *Nucl. Instr. and Meth.*, **163** (1979) 141; M.S. Naidu and A.N. Prasad, *J. Phys. D: Appl. Phys.*, **5** (1972) 983; T. Yamashita *et al.*, *Nucl. Instr. and Meth.*, **A283** (1989) 709.

[5] R. Openshaw *et al.*, *IEEE Trans. Nucl. Sci.*, **36**(1) (1989) 567.

[6] D.W. Hess, Proceedings of the Workshop on Radiation Damage to Wire Chambers, Lawrence Berkeley Laboratory, LBL-21170 (1986) 15.

[7] H. Yasuda, "Plasma Polymerization" (Academic Press, New York, 1985) pp. 179,180.

[8] R. Openshaw *et al.*, TRIUMF, paper submitted to the 1990 IEEE Nucl. Sci. Symposium, and private communication.

[9] H. Ogren, Tracking Review for the SCD Collaboration, 1990 Snowmass DPF Meeting.

DESIGN OF A TRACKING SYSTEM FOR A SOLENOIDAL DETECTOR*

Gail G. Hanson
Indiana University
Bloomington, Indiana 47405

Abstract

The design of an integrated tracking system for a solenoidal detector will be presented. The tracking system consists of a silicon pixel and microstrip detector at smaller radii from the beam collision point and wire chambers at larger radii. The tracking system provides momentum measurements and a fast trigger for all charged particles with p_T above a few GeV/c and for $|\eta| \leq 2.5$. Research and development issues will be discussed.

Introduction

The goals of a tracking system for the SSC are to provide momentum measurement and a fast trigger for charged particles with p_T above a few GeV/c and $|\eta| \leq 2.5$. In addition, the tracking system should provide a precise vertex measurement in order to identify long-lived tracks, for example, from *B* decays, and detect separated vertices from multiple *p-p* interactions. Since the tracking system is only a part of the complete detector, it must provide these functions in an economical manner. The tracking system must operate in the high-rate environment of the SSC at and above the design luminosity.

We are engaged in detailed design studies of an integrated tracking system[1] for a solenoidal detector for the SSC. The tracking system consists of a silicon pixel and microstrip detector[2] at small radii from the beam collision point and wire chambers at larger radii. The R&D includes wire chamber detector design for straw tubes and intermediate angle tracking detectors with wires transverse to the beam direction, engineering R&D including support structure and alignment, front end and triggering electronics, and computer simulation.

Tracking System Design

A conceptual design of the tracking system, contained inside a 2 Tesla solenoidal magnetic field, is shown in Fig. 1. The inner radius of the coil is located at a radius of 1.7 m, and the tracking volume half-length is 4.0 m. In order to aid the pattern recognition in the high-rate SSC environment, the entire tracking system design, except for the pixel detector, is based on the concept of track segments, rather than individual track hits. Track segments in the outer tracking system (radius > 50 cm) are used in the first level trigger. Track segments in the superlayers are linked to form tracks.

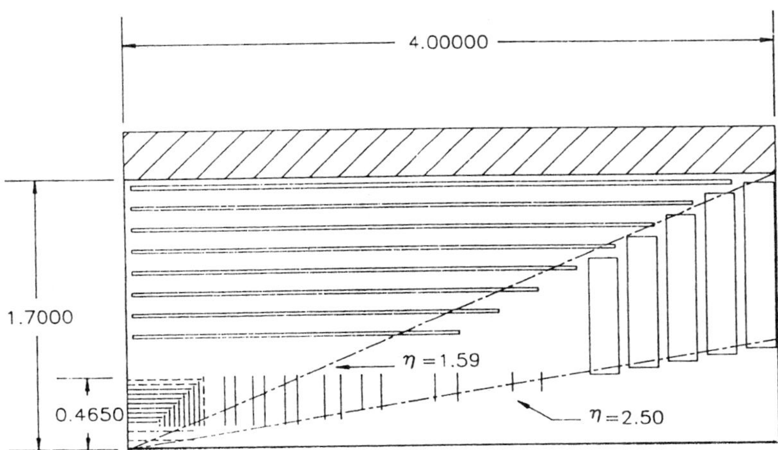

Fig. 1. Conceptual design for a silicon and wire chamber tracking system for a solenoidal detector.

The silicon tracking detector, inside 50 cm radius, consists of an array of high-resolution two-dimensional pixel detectors and segmented microstrip detectors; these provide vertex measurement and pattern recognition capability even in the cores of dense jets.[3] Some of the microstrip layers are double-sided silicon with axial strips on one side and stereo on the other. Track segments are found in superlayers consisting of two layers of silicon.

Wire chambers are used for tracking between 70 and about 163 cm radius. The central tracking system, covering the region $|\eta| \leq 1.5$, consists of superlayers of 4 mm diameter straw tubes[4], as shown in Fig. 2. In order to reduce radiation damage and current draw, the straws are made as small as practical and run at low gas gain, $\sim 2 \times 10^4$. Occupancy is reduced by the small straw diameter and the use of a fast gas, such as mixtures containing CF_4.[4,5] The half-cell stagger between layers permits resolution of left-right and crossing-time ambiguities. The coordinate along the wire is measured by means of small-angle (~3°) stereo. With a spatial resolution of 100-150 μm per wire in the r-ϕ projection, the expected resolution in z is 2-3 mm. The superlayers alternate axial and stereo. Local track segments are found in each superlayer.

The intermediate angle tracking system[6] covers the region of radius > 50 cm and $|\eta| > 1.5$. It is composed of drift chambers with wires transverse to the beam direction and "crossing taggers," as shown in Fig. 3. A crossing tagger is set of planes of gaseous proportional detector with cathode strip readout. Crossing taggers are used to associate a wire chamber hit to within 3 or 4 bunch crossings and could be used for the trigger. The wire chambers are presently envisaged to be radial wire chambers.

The performance of the tracking system is shown in Fig. 4.

Engineering R&D

The tracking system will be supported and aligned using a structure of graphite fiber composites.[7] There are inner and outer support cylinders for structural stability. The superlayers in the central tracking system are made up of modules of straws with an outer graphite composite shell to provide rigidity and positioning. The replaceable modules are positioned and supported by an open framework. The intermediate angle and silicon tracking systems are supported by the same structure.

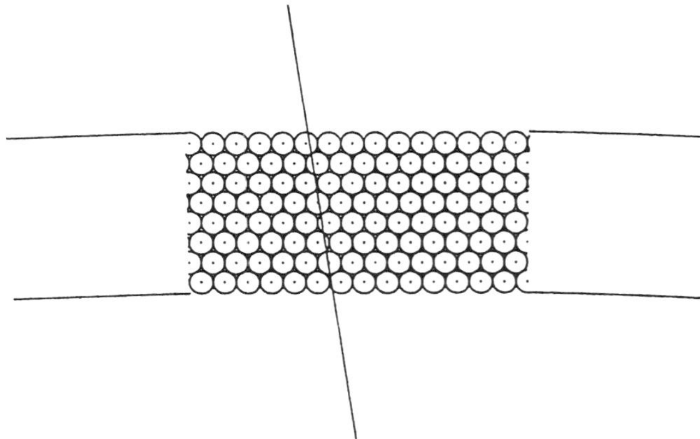

Fig. 2. Cross sectional view of a superlayer of straw tubes.

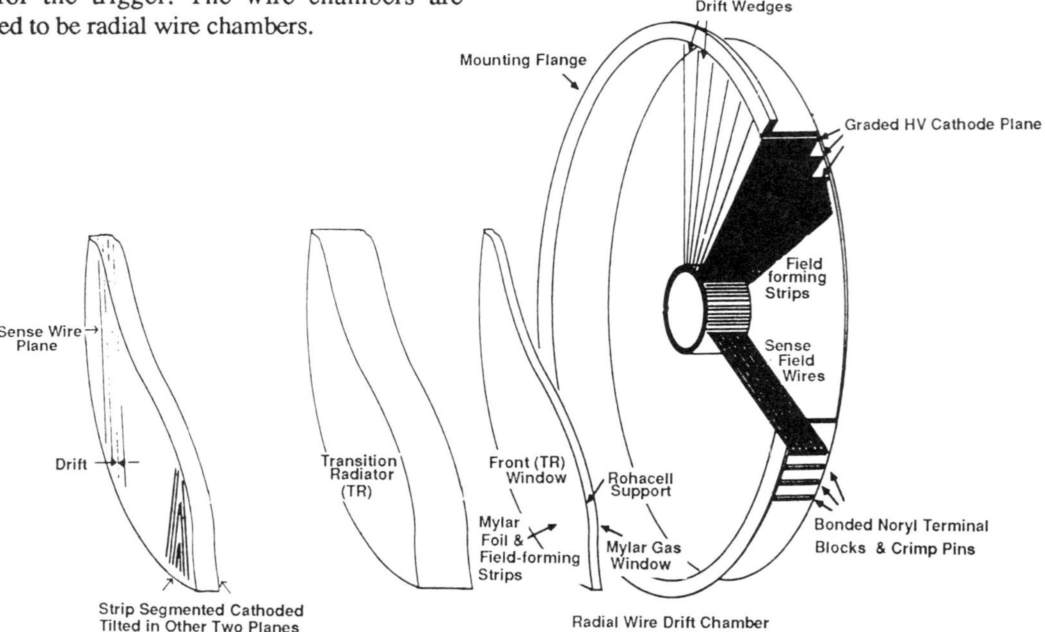

Fig. 3. Intermediate angle tracking supermodule layout.

Front End and Triggering Electronics

Custom integrated circuits for the front end and triggering electronics are being developed both as part of this subsystem effort and as part of separate electronics subsystem R&D projects. A low-power radiation-hard integrated preamp, shaper, discriminator and time-to-voltage converter are being developed as part of an R&D program centered at the University of Pennsylvania.[8] Similar developments are taking place in Japan. Circuits for triggering on track segments in superlayers using drift time (synchronizers) are being developed at the University of Michigan.

R&D Issues

The R&D issues involved in the design of the tracking system are being addressed by design and prototyping, as well as computer calculations and simulations. Some of these R&D issues are summarized below.

In the design of the tracking system, we must minimize material (primarily because of the problem of photon conversions) and cost. Materials issues include sufficient rigidity and structural support in the module walls and support structure, and connections to the wires at the endplates, including terminations, wire-holding devices, interface boards, electronics, and cabling. Finite element analyses are carried out to optimize the designs of the module walls and the support structure. Design and prototyping are being carried out to minimize material at the wire connections.

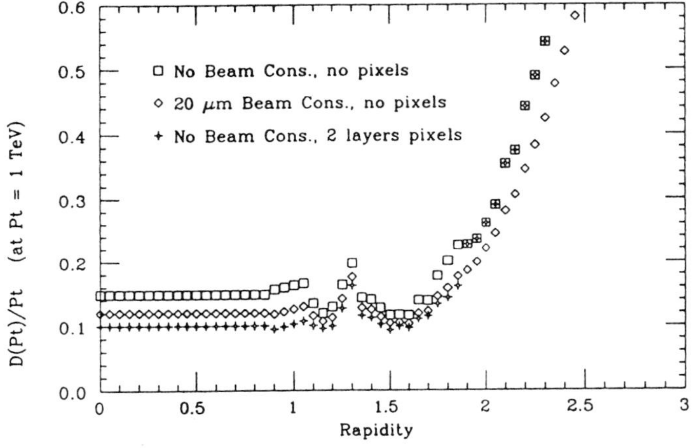

Fig. 4. Momentum resolution in a 2 Tesla magnetic field as a function of $|\eta|$ for the tracking system.

Wire chamber design and prototyping are being carried out for both the straws and the radial wire chambers. These studies include minimizing the resistance of the straws and determining the optimum technique for supplying graded high voltage to the cathodes for the radial wire chambers. Tests of suitable fast gases are being carried out. In addition, there are many construction and assembly problems to be solved.

We are building prototypes for all of the components of the design: straw modules, radial wire chamber sectors, connections, electronics, and support structure. These will be tested in a beam when possible.

Many crucial design issues are being addressed by computer simulation of the response of the detector to SSC background and physics events. With 16 ns between bunch crossings and 1.6 interactions per bunch crossing at the SSC design luminosity, high occupancy and its result, lost hits, are a major issue for wire chambers. We are using simulations to determine the best configuration for triggering on track segments in the outer superlayers of the central tracking and in the crossing taggers of the intermediate angle tracking. We need to continue to trigger on track segments as the luminosity increases beyond the design value, and we must do this with a sufficiently robust trigger using a minimum number of channels and at a minimum outer radius for the tracking system. We are also carrying out studies of pattern recognition for finding tracks.[9]

*Work supported by the Department of Energy, contract DE-AC02-84ER40125.

Presented at the Symposium on Detector Research and Development for the Superconducting Super Collider, Fort Worth, Texas, October 15 - 18, 1990.

REFERENCES

1. "SSC Detector Subsystem Summary Report and Proposal for FY 1991," Central and Forward Tracking Collaboration, G. Hanson, contact person.
2. "Subsystem R&D Proposal to Develop a Silicon Tracking System," A. Seiden, contact person.
3. A. Seiden, "Silicon Tracking," in these Proceedings.
4. H. Ogren, "Wire Chamber Tracking," in these Proceedings.
5. J.A. Kadyk *et al.*, in these Proceedings.
6. J.B. Dainton *et al.*, in these Proceedings; D.H. Saxon *et al.*, in these Proceedings.
7. R. Swensrud, in these Proceedings.
8. L. Callewaert *et al.*, in these Proceedings.
9. W.T. Ford, "Pattern Recognition in Tracking Systems with Small Drift Cell Arrays," in these Proceedings.

TRACK RECONSTRUCTION IN STRAW SUPERLAYERS*

W. T. Ford and M. Lohner
Department of Physics, University of Colorado
Boulder, Colorado 80309-0390

Abstract

We have developed a program to reconstruct track segments in dense wire drift chamber arrays. We compare our ISAJET simulations of signal and background events for SSC parameters with previous studies and report measurements of the pattern recognition efficiency for a large solenoid detector geometry. We find that the tracking is quite reliable for superlayers placed more than 50 cm from the beam axis. Without accounting for detection inefficiencies we find comparable performance for either six or eight layers per superlayer.

Introduction

We have begun a study of the reliability of track reconstruction in a combined silicon strip and wire drift chamber detector with solenoidal magnetic field under design for operation at the SSC[1]. Previous studies[2,3,4] have delineated some of the parameters for such a system. To make the pattern recognition manageable, cylindrical layers of drift cells are organized into groups (superlayers). A preliminary phase of data reduction is performed to identify strings of consecutive measurements along a track (segments), resolve right/left ambiguities, and fit for local position and direction. High particle fluxes pose the potential problems of information loss caused by pileup and confusion of the pattern recognition. Qualitative evidence that the information of interest can be extracted has been presented previously[4]. Here we discuss a quantitative study of pattern recognition at the segment finding level. One of the objectives is to establish the appropriate number of layers required in a superlayer, bearing in mind that less is better from the perspective of minimizing cost and the amount of material in the particle path.

The following sections of this paper describe the model used for simulation, the pattern recognition scheme, results and conclusions.

Event simulation and occupancy estimate

To make sure that our performance tests will be sufficiently realistic we first made some simple calculations of occupancies with the available physics process simulation programs, putting in the appropriate detector geometry but omitting detailed tracing of particle progress through the detector. Some of the omitted effects are accounted for by hand after comparison of our results with previous studies.

We use the ISAJET package to generate signal events. In the present work, these are production of Higgs bosons of mass 800 GeV/c^2 and their subsequent decay to leptons through Z^0 pairs. For the background, we ran ISAJET with the TWOJET option and a minimum transverse momentum ($p_\perp$) of 4.5 GeV/c, the value that gives a two-jet cross section equal to the total inelastic cross section extrapolated from data (about 95 mbarns). The charged particle yield is about 10 per unit of rapidity.

The detector geometry is that of the barrel straw tracker described in the SDC expression of interest[1], namely, eight superlayers at radii between 0.73 and 1.80 m and maximum half length 3.0 m, in a magnetic field of 2 T. To measure occupancy in this detector we compute the intersections of helical trajectories with the cylindrical detector surfaces. The number of crossings of one track may be zero (looper never reaches the layer), one (track exits through the coil and does not return), a few (looper traverses the superlayer, then reenters), or many (looper passes through the superlayer tangentially). We take the number of loops made before a track leaves the detector through the ends to be two, a rough average inferred from event pictures.

The integration time is taken to be 40 ns, based upon a drift cell radius of 2 mm, drift velocity of 100 μm/ns, and allowance for time-of-flight plus signal propagation delays and pulse duration. This implies that the detector integrates over 2.6 bunch crossings on average. At the design luminosity of 10^{33} cm^{-2}s^{-1} there are 1.6 interactions per crossing, so the detector sees 4.2 events per trigger. The occupancy we find from this calculation is shown by the dashed curve in Fig. 1.

An earlier study by Makoto Asai[5] can be compared with this one, if we interpolate to the approximate rapidity acceptance of SDC and correct for the slightly different straw radius. Asai's calculation used the PYTHIA event generator. The present result at 1 m superlayer radius is about ten percent higher than Asai's for primary particles only. He finds that with secondary interactions the occupancy is about two times as large. Therefore we have scaled our result to get the solid curve in Fig. 1.

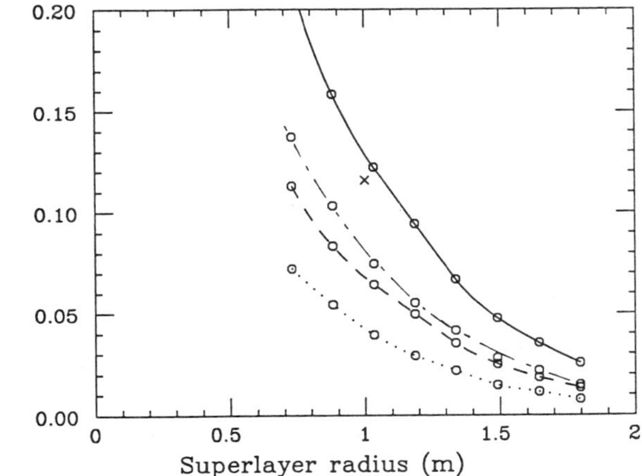

Figure 1. Occupancy vs radius in the SDC barrel straw tube tracker. The curves are: (dashed, dotted) primary charged tracks from (ISAJET,PYTHIA); (solid, dot-dashed) scaled by the factor 1.9 to account for secondary interactions.

Our calculation repeated with the PYTHIA event generator shows that the yield is about 60-70% of the ISAJET value (6.5 per unit of rapidity). The corresponding occupancy curves are included also in Fig. 1.

For the pattern recognition study discussed below we used a file of background events generated with ISAJET as described above, selecting the number of interactions from a Poisson distribution with mean of 4.2. Secondary interactions within the straw detector are included, but none in the silicon detector. The net occupancy is within the range of our present best estimates.

Pattern recognition

Each superlayer contains six or eight layers of straw tubes in a close-packed configuration. The time measurement can be converted into a simple circular contour of distance from the wire, to which the trajectory is tangent. This is true provided the time of arrival of the signal at the readout electronics is given solely by the drift time. In fact the delay from particle time of flight to the detector and propagation of the signal pulse along the wire to its end is not completely negligible: it amounts to the time equivalent of roughly 100 microns of track displacement, which is comparable to the measurement resolution from electron diffusion in the gas. Systematic variation of the pulse height may aggrevate this effect. Of course this delay is not random, and can be corrected for once a track is fully reconstructed and the axial position of the track is known. This information is not available, however, to the segment finder. It may be necessary to treat the track crossing time as a parameter to be inferred from the segment fit, along with the azimuth and direction of the track. In the present study, we have just set the resolution to 150 μm for the segment reconstruction to account for this effect.

The segment finder works as follows: one superlayer is considered at a time. A search over layers begins with the outer one (layer 8 if that is how many we have and we number from the interaction point outward). Each hit in that layer is a starting point for a candidate segment. The search continues in layers beginning with layer 1 for a hit with azimuth not too different from that of the first hit. This pair of hits is a seed for the segment; a segment is quite well determined for each of the four choices of right/left ambiguity for the seed hits.

The track is interpolated to layer 2 (or layer 3 if the second hit of the seed was found in layer 2, etc.). That layer is searched for a hit within a window of the interpolated track (both of the hit's ambiguity signs being considered). When one is found, the three hits are fed to a least-squares fit. The fit is subjected to a chi-squared test and, if passed, provides an updated set of segment parameters. The program is written so that this fit can be readily expanded to include additional parameters, such as the track crossing time, if needed. The search continues through all of the internal layers, with a chi-squared test and parameter update at each stage. When the layers and hits are exhausted, we ask are there enough hits; if so, the candidate is saved temporarily until all four right/left choices for the seed have been considered. Then the (up to four) condidates are compared. The test variable is: $(N_{layers} - N_{hits})^2 + \chi^2/\text{dof}$. The candidate with smallest test variable value is saved as a segment. Hits on the segment are excluded from further consideration as the search for more segments proceeds. The results below were obtained with the requirement of at least four hits per segment.

Performance results

For the evaluation studies we have used GEANT, with a geometry package written by Hanson, Palounek, et al.[4,6], which we shall refer to as the Large Solenoid Detector (LSD) design. There are thirteen superlayers of eight layers each, with radii between 0.57 and 1.6 m. The lengths are stepped in three groups (superlayers 1-3, 4-7, and 8-13). The straw diameters are graded from 4 mm at the inner radius to about 6 mm at the outermost. The straws run the full length of the detector.

The question we address is: given a reasonably stiff track ($p_\perp > 1$ GeV/c), which has crossed a superlayer, is it found, in the sense that we have a segment all of whose hits came from that track. The results are given in the plot of Fig. 2. For each superlayer there are four histogram bins: from left to right, these are the number of track crossings, the number of segments matching a generated track perfectly, the number containing one hit that does not belong, and the number with more than one wrongly assigned hit. For the inner superlayers, the efficiency is around 90%, if we take the strict criterion for success. If we accept one misassigned hit, the number is about 98%. Efficiency is higher for the outer superlayers, as expected from the decrease of occupancy with radius.

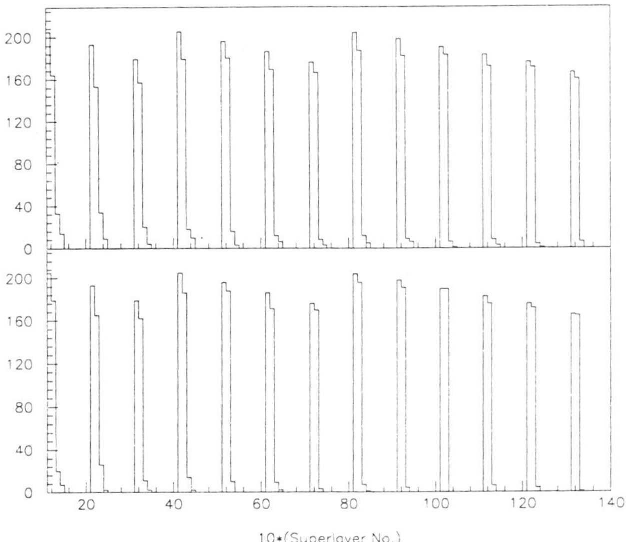

Figure 2. Segment finding efficiency and accuracy for (above) 8-layer and (below) 6-layer superlayers. For each superlayer (1-13) are plotted the reconstruction statistics described in the text.

Comparing the two histograms in Fig. 2, we find the efficiency is just as good if we have only six layers per superlayer as with eight. It should be noted here that we have not accounted for inefficiency of the cells.

A different accounting is illustrated in Fig. 3. The open histograms show the number of hits in each layer (identified with stiff tracks, as defined above); singly hatched histograms are those that have been correctly assigned to segments; doubly hatched histograms are the hits for which the correct choice of ambiguity sign was made. The performance with eight layers per superlayer is slightly better in terms of ambiguity resolution than with six.

Conclusions

We have developed a segment finding program that is quite efficient according to simulations for SSC conditions at design luminosity. Superlayers with either six or eight layers perform quite well, given that all cells are perfectly efficient (a limitation of the evaluation that will be removed in future work). These results are rather encouraging in view of the high occupancies at the smaller radii.

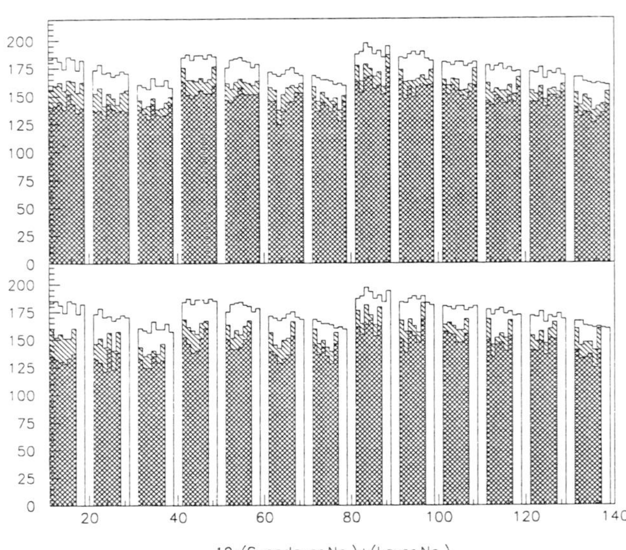

Figure 3. Linking accuracy by layer for (above) 8-layer and (below) 6-layer superlayers. Open histogram is number of layer crossings by stiff tracks; single hatched is hits correctly detected in the layer; double hatched is hits assigned the correct ambiguity sign.

*Work supported in part by the Department of Energy, contract DE-AC02-86ER40253, and by the Texas National Research Laboratory Commission.

[1] Solenoidal Detector Collaboration, Expression of Interest (1990).

[2] D. G. Cassel, et al., "Report of the Central Tracking Group", in *Proceedings of the 1986 Summer Study on the Physics of the Superconducting Super Collider*, Snowmass, CO, June 23-July 11, 1986.

[3] R. Donaldson and M. G. D. Gilchriese, *Experiments, Detectors, and Experimental Areas for the Supercollider*, Berkeley, CA, July 7-17, 1987.

[4] G. G. Hanson, B. B. Niczyporuk, and A. P. T. Palounek, "Wire chamber requirements and tracking simulation studies for tracking systems at the superconducting super collider", *Proceedings of the Wire Chamber Conference*, Vienna, Austria, February 13-17, 1989 (SLAC-PUB-4860).

[5] M. Asai, "Hit rates of the straw chamber tracker", SDC Note SSC-SDE-28 (Hiroshima Institute of Technology preprint 9001, 1990).

[6] A. P. T. Palounek, "Simulating a central drift chamber for a large solenoid detector at the SSC, using GEANT3", SLAC-PUB-4787(1988).

PROPERTIES OF A STRAW TUBE / SCINTILLATING FIBER HYBRID CENTRAL TRACKING CHAMBER FROM COMPUTER SIMULATION

Martyn Corden

Supercomputer Computations Research Institute,
Florida State University.

Abstract

A detailed simulation of the performance of a hybrid central tracking chamber in a realistic SSC environment has been carried out. Drift time spectra and cell occupancies are presented for the design luminosity of 10^{33} $cm^{-2}s^{-1}$, along with preliminary estimates for 10^{34} $cm^{-2}s^{-1}$.

Detector

The detector as simulated, (before downscoping), consists of two cylindrical chambers, each 3 metres long. These contain 8 superlayers equally spaced in radius between 70 and 180 cm. Each superlayer consists of 8 axial layers of 4mm diameter aluminized mylar straw tubes supported by a carbon fiber mandrel. The straws in the 3 outer superlayers are 3 m long, falling to 2 m for the inner superlayer. The 3 outer superlayers also contain 0.5 mm scintillating fibers, in staggered pairs of axial, +5 and -5 degree, helically wound stereo layers. The total amount of material in the tracker is 9% of a radiation length (9% X0) at 90 degrees, including the support structure. An additional 6% X0 from a silicon inner tracker is assumed. The whole detector is immersed in a uniform, solenoidal magnetic field of 2 Tesla. The straw tube gas is assumed to have an effective radial drift velocity of 66 μ/ns in the magnetic field, (equivalent to 100 μ/ns without field), giving a maximum drift time of 30 ns. The position resolution is 100 μ, increasing to 150 μ near the wire. The thickness of the aluminized mylar straw walls is 50 μ.

Simulation

The SSCSIM application of the GEANT 313 simulation code is used. The detector digitization and treatment of pileup are specific to this application. The event generators ISAJET and PYTHIA are used as sources of Higgs (trigger) and minimum bias (background) events respectively.

The program explicitly simulates 6 consecutive beam crossings, 3 before and 2 after that containing the triggering event. Particles are tracked through the detector for up to 1 μs. Digitizings from background events from these beam crossings, which fall within a time gate delayed by a multiple of 96 ns (6 beam crossings), are used to represent the in-time digitizings from beam crossings up to 1 μs before the trigger event. Such digitizings arise from slow or looping tracks.

The simulation includes particle time-of-flight, signal propagation time along the straw or fiber, and the straw drift time and its fluctuations. For straws, the first hit to occur within a 40ns time gate is digitized. All scintillating fibers giving light within a 16ns gate are recorded. The basic simulation assumes an SSC luminosity of 10^{33} $cm^{-2}s^{-1}$, corresponding to a mean of 1.6 background events per beam crossing. The straw and fiber material is approximated by concentric cylindrical layers of π x wall thickness (straws) or $\pi/4$ times diameter (fibers). The true circular cross section is used for the digitization of both straws and fibers.

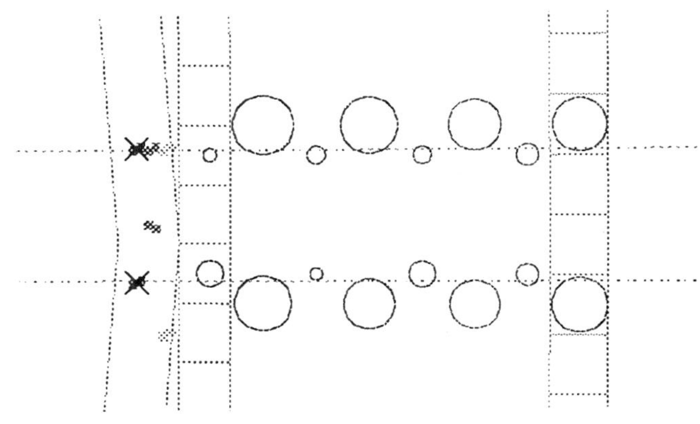

Drift Spectra

The above figure shows a simulation of two tracks traversing a single superlayer containing both straw and fiber layers. The drift times of struck straw tubes are represented by circles, and the struck fibers by solid squares as if at $z=0$. The offset of the stereo fiber layers corresponding to a 4cm offset in z can be seen, and the 3d point reconstructed from the stereo fibers is represented by a cross. Track segments are found initially from the stack straw numbers only, and then refit using the drift times. The segments are then combined into tracks, and an overall fit is performed [1]. A time gate of 40 ns is used for digitizing straw tube hits. Approximately 27% of the hits from a triggering event fall outside this gate, mostly due to slow or looping particles.

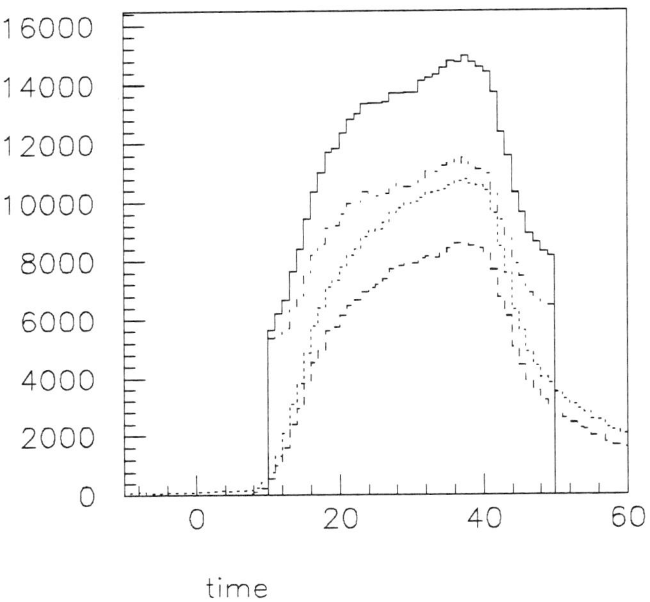

The figure shows superposed drift time spectra for: hits from trigger events only (lower dashed line), hits from all events from triggering beam crossing (next dotted line), hits within the 40 ns time window from all events, and the digitized time spectrum for all events limited to one hit (the first) per straw tube. At a luminosity of 10^{33} cm^{-2}s^{-1}, approximately half of the hits within the time gate come from the triggering event, and half from the minimum bias background. About 20% of these hits are not recorded due to an prior hit on the same straw. The following figure shows the same quantities obtained from a preliminary, low statistics simulation at a luminosity of 10^{34} cm^{-2}s^{-1}. Approximately 10% of all hits in the time gate come from the trigger event, as expected. Roughly 40% of these hits are not recorded due to an earlier hit on the same straw.

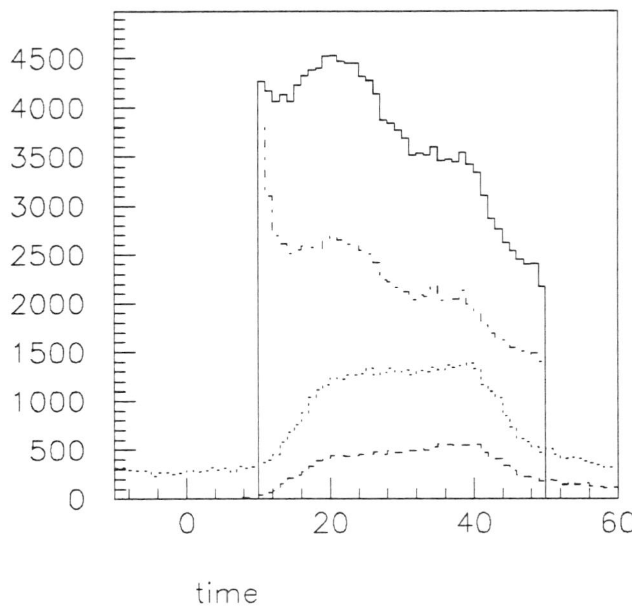

Occupancies

The occupancies of the various detector elements have been determined for a luminosity of 10^{33} cm^{-2}s^{-1} under the above conditions, and are tabulated below:

straws	L=10^{33}	L=10^{34}	superlayer inner radius
layer 1	17%	~ 28%	70 cm
layer 3	11%	~ 25%	100 cm
layer 5	7%	~ 15%	130 cm
layer 8	3%	~ 8%	175 cm
fibers			
layer 6	0.4%	~ 1%	145 cm
layer 8	0.2%	~0.5%	175 cm

The straw tube occupancy varies between 3 and 17%. The inner layers are quite active, but at a level that still permits local track segment finding. The fiber occupancies are an order of magnitude lower than the straw tube occupancies at similar radius, as expected from the smaller cell size and shorter sensitive time. The fiber occupancies are consistent with independent estimates of the occupancy of a fiber-only tracker at a similar radius [2].

Preliminary estimates have been made of occupancies at a luminosity of 10^{34} $cm^{-2}s^{-1}$. They show the effect of the factor of ten increase in background tracks. At this luminosity, the inner straw tube layers are becoming congested, and it will likely be necessary to concentrate on global methods of finding fast tracks rather than looking for local track segments.

An example of a simulated Higgs event at a luminosity of 10^{33} $cm^{-2}s^{-1}$ is shown below. Reconstructed, high p_t tracks are represented by dotted lines. The decrease of occupancy with increasing radius can be seen. Much of the background is due to low momentum secondary and out of time tracks.

Conclusions

A detailed simulation of a hybrid central tracking chamber has been carried out. Studies of straw tube drift time spectra and occupancies indicate that the system should work satisfactorily at the design luminosity of 10^{33} $cm^{-2}s^{-1}$.

References

[1] M.Mermikides,"A track finding strategy for superlayered drift chamber design studies", FSU-SCRI-90T-145 and paper submitted to this conference.

[2] R.A.Lewis, "Simulation studies for a scintillating fiber tracker", PSU LEPS 90/14 and paper submitted to this conference.

TRACK FINDING IN A HYBRID CENTRAL TRACKING CHAMBER

Michael E.Mermikides
Supercomputer Computations Research Institute
Florida State University
Tallahassee, FL 32306, USA

Abstract
A track finding strategy has been developed as an aid to design studies of a hybrid tracking chamber. The method combines local and global search methods to link the superlayer track elements into full tracks, and incorporates a fit to reconstruct their momenta.

HCTC geometry and readout

The prototype HCTC design[1] consists of groups of straw chambers (superlayers) arranged in concentric cylindrical layers surrounding the interaction point. Each superlayer is composed of eight axial straw tube layers in a close-packed formation (fig 1) on a light carbon fiber support. The tubes thus provide up to eight drift time measurements of a track crossing the superlayer. z measurement is derived from scintillating fiber stereo layers embedded in the carbon support structure. In the present conceptual design only the three outermost superlayers are equipped with scintillating fiber readout.

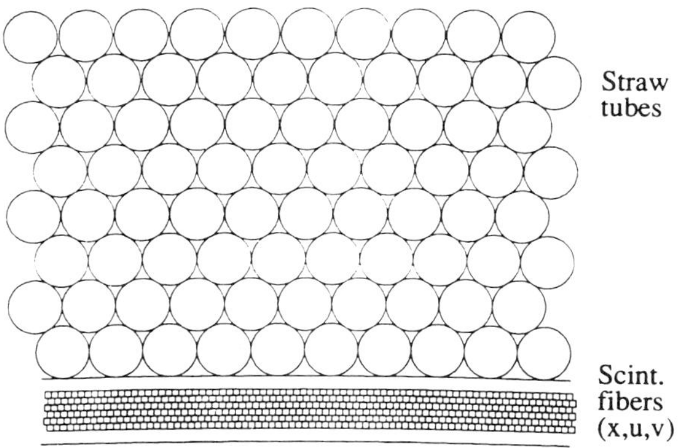

Figure 1. Superlayer geometry

The track segments are found locally starting from the pattern of hit tubes, and then refined by a fit to the drift times, taking into account the effects of signal propagation and time of flight[2]. The fit serves to define the track crossing azimuthal position and direction and their errors at a nominal radius in the superlayer, and the z coordinate is obtained by associating the track with local scintillating fiber information.

Track Finding Procedure

Pattern recognition is performed in the $(x - y)$ plane where the resolution is highest. The search starts by connecting a pair of track segments (referred to a "hits" in what follows), one from each of the outermost superlayers, which potentially belong to the same track (a link). The process is repeated by forming a new link between the inner hit and one from the next superlayer thus building a chain of connected hits. For chains that have not been followed over their full potential length, we enter a second (global) phase where the track found so far is used to define a road within which we look for further hits. This combined strategy has a number of advantages over purely local or global methods. By starting from the outside, where the population of hits is lowest, we have a better chance of forming unambiguous links initially. As the chain grows, the improved track parameters enable us to select further hits in the inner congested region with progressively increased precision. The relative emphasis given to the local and global approaches can be adjusted to suit a wide range of detector characteristics and operating conditions. There is no a-priori constraint that the tracks come from a central vertex, and the method lends itself to parallelization since the search for each link and each chain can proceed independently.

Local search (link finding)

Starting from a hit on the outer layer we form a link with each hit in the inner layer within an azimuthal window $\Delta\phi$. This procedure makes use of the ϕ binning for fast lookup in the region of interest. The track projection (circular in a uniform solenoidal field) is described by three parameters. In the interest of execution speed, we use a parabolic track model locally (this can be shown to be a very good approximation for our purposes). With two spatial and direction measurements, we can use the redundant information to perform a consistency test for accepting the link. The parameters of the parabola $v = a + bu + cu^2$ are obtained from two coordinates and one angle (β_1), in the local reference frame (u, v), (fig 2):

$$c = \frac{v_2 - \tan\beta_1(u_2 - u_1)}{(u_2 - u_1)^2} \quad (1)$$

$$b = \tan\beta_1 - 2cu_1 \quad (2)$$

$$a = -bu_1 - cu_1^2 \quad (3)$$

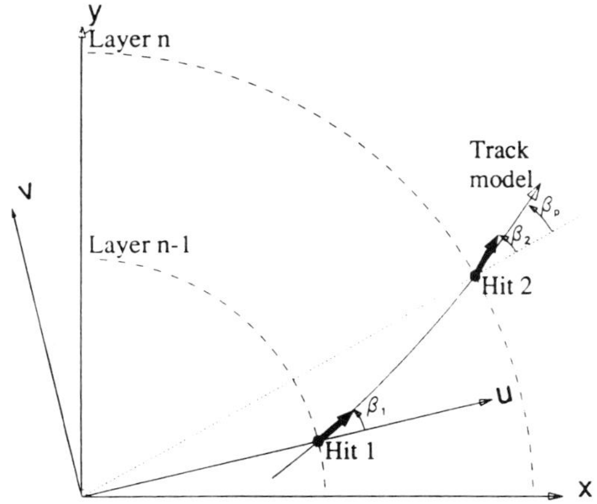

Figure 2. Link definition

The consistency test is based on the comparison of the predicted angle at the outer point,

$$\beta_p = \tan^{-1}[b + 2c(u_2 - u_1)] - \phi_2 + \phi_1 \quad (4)$$

to the measured value β_2, expressed as

$$\chi_L^2 = (\tan\beta_p - \tan\beta_2)^2/(\delta^2\tan\beta_1 + \delta^2\tan\beta_2) \quad (5)$$

where $\delta\tan\beta_{1,2}$ are the errors of the measured tangents at the two superlayers. A link is accepted if the value of χ_L^2 is below some cut χ_{Lmax}^2. In the case where more than one hit from the inner layer passes the test, we keep the one the gives the smallest χ^2. If no link can be found among the hits in the inner superlayer, the procedure is repeated using the next layer in, skipping up to a predefined maximum number of layers N_g. The valid links found from the current starting layer are saved in a common array for the subsequent steps of the track finding.

Chain building

The next stage is to propagate each initial link, starting from the stiffest one (this is not strictly necessary, but useful for getting the fast tracks first). The procedure entails the formation of a new link between the inner hit of the existing one and a hit from the next superlayer in. A link is acceptable if it passes the χ^2 test described above and is also consistent within errors in curvature with the parent link. This process is repeated until one of the following either the innermost superlayer is reached, or more than N_g layers without acceptable hits are encountered, or more than one acceptable link is found from a given parent. (Ambiguities are deliberately avoided at this stage)

Once a chain is completed it is validated with the aid of a fast circle fit in the $x - y$ plane, followed by a line fit in the $s - z$ projection if z information is available to complete the measurement of the helix parameters. If the fit passes a chi-squared test, the track parameters are stored and the associated hits are logically removed from so they are not considered in ensuing track searches. A fit that fails the χ^2 test is repeated after removing the hit with the worst residual. If the fit is still unsuccessful the chain is abandoned.

Chain extrapolation

Due to chamber inefficiencies and effects of multiple tracks in a given cell (the straw chambers operate on

single-hit electronics), there will inevitably be missing hits resulting in chains which do not span the full length of the track. As mentioned above, chain propagation will also be interrupted when ambiguities are encountered. The global method uses the fitted chain parameters to define a circular "road" within which more hits are sought for. The length over which the chain can be reliably projected is a parameter to be adjusted to the charateristics of the detector. For the present HCTC design we extrapolate the chain over half of its original length, along both directions of the track (the upsteam projection is useful for picking up hits which were missed in the local search). In the case of low p_t tracks which are trapped within the chamber volume, the projection is limited to the furthest point on the track circle; spiralling tracks will therefore be found as semi-circular pieces to be connected later.
New hits which are found within the road are added to the chain if they pass the test:

$$\chi_R^2 = \frac{1}{2}[(\phi_p - \phi)^2/\delta^2\phi + (\beta_p - \beta)^2/\delta^2\beta] < \chi_{Rmax}^2 \quad (6)$$

where the subscript p refers to the predicted quantities, defined by the intersection of the chain with a given superlayer, and the unsubscripted quantities refer to the measured values. In the case where there are two or more hit candidates at a given superlayer, we choose the one with the smallest χ_R^2. If more hits are found, the chain is refitted to improve its parameters and the process is repeated until the full potential track length is covered. If at any stage the fit fails, even after rejecting the worst hit, the process is stopped and we retain the parameters of the last succesful fit.
After all links from the starting (outermost) superlayer are propagated, the entire cycle is repeated starting the link search from the next layer in until all layers are covered.

Implementation

The code is being used to reconstruct realistically simulated events in the SSC environment using a GEANT framework [2],[3]. Fig 3 shows a display of the readout of one endplate of the HCTC arising from a Higgs decay at a machine luminosity of $10^{33} cm^{-2} sec^{-1}$. The simulation covers a period from three beam crossings before to two after the triggering event to account for pileup during the total drift and signal propagation time ($\approx \pm 2 \times 50$ nsec). Background is also added to simulate late arrival of hits from slow particles up to 1 μsec before the trigger. The display shows the drift time circles in individual straws and the reconstructed tracks; the segments which successfully match the scintillating fiber hits (and thus provide z-coordinates) are indicated with a cross.
The average reconstruction time, starting from the track segments, is approximately 5 msec/track on a VAXstation 3100 workstation. In the scalar implementation, the reconstruction time is found to vary approximately linearly with track multiplicity.

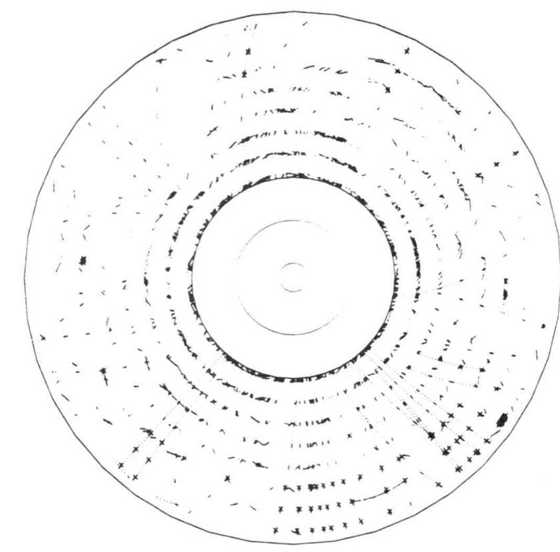

Figure 3. Reconstruction in simulated SSC environment using detailed GEANT detector simulation.

References

[1] A. T. Goshaw et al, " A proposal for the development of a prototype central tracking chamber" submitted to SSC Laboratory

[2] M.J.Corden, "Properties of a straw tube/ scintillating fiber hybrid central tracking chamber from computer simulation", presented at the Symposium on Detector R & D for the SSC, Fort Worth, TX, Oct 15-18.

[3] S.L. Linn, "SSCSIM USERS GUIDE", SCRI Technical Report FSU-SCRI-90-1

MECHANICAL CONCERNS FOR LONG STRAW-TUBE ARRAYS

W.S. Anderson, J.C. Armitage,* H. Edwards, C. Lu,
K.T. McDonald, B. Miller, W. Sands, and J. Weckel
Joseph Henry Laboratories, Princeton University, Princeton, NJ 08544
M. O'Neill
Carleton University, Ottawa K1S 5B6, Canada

Abstract

A calculation of the force required to buckle an array of straw tubes is presented together with a description of an experiment being carried out to test this. A way of safely handling long straws during the assembly process is discussed which incorporates quality control on the straw integrity. A new method is presented for fixing the wire in a straw in such a way that a specified tension of the wire is assured. Finally, a rapid method of measuring the tension of wires in long tubes is discussed.

Introduction

This study of techniques of constructing straw-tube-chamber arrays was carried out under the auspices of the Central and Forward Tracking Subsystems Group, and was motivated in part a the proposal to study B-decays at the SSC.[1] The latter consideration leads to design parameters that are somewhat different than those of our collaborators who were motivated by experiments such as the Solenoidal Detector (SDC):

- We are concerned with self-supporting arrays of glued straws. The strength of the walls of the array should be sufficient to resist buckling under the tension of the wires, rather than the necessary rigidity being provided by an exoskeleton.

- A B-physics experiment would need straw tubes of lengths up to only 2 m, compared to 4-6 m for the SDC, and these straws would be in a vertical, or near vertical orientation. Hence we are pursuing a straw-tube-chamber design with a 2-m-long unsupported wire.

To get some immediate experience of the constructional problems, an Ohio State U. design[2] has been used for the straw-tube end plugs. This has resulted in our working with straws of a larger diameter (7 mm) than might be optimal at the SSC, but has given us valuable experience that will be needed for work with smaller straws. Indeed, after assembling a hundred or so straws to that design, our technical staff found that there were improvements that could be made to facilitate the attainment of correct wire tension. Also, after working with the O.S.U. method of straw alignment, it was not found to be suitable for extension to longer straws, and a different method has been developed.

*On sabbatical from Carleton U.

A proper wire tension is very important in minimizing the gravitational sag of the wire and determines the ultimate high voltage that can be applied to the straw. We anticipate having to measure and remeasure the wire tensions at several stages during the fabrication, and therefore need a fast efficient method for wire tension measurement. We have built two instruments using differing techniques, and from the experience gained with these we are able to identify a better technique.

This paper describes results from a buckling calculation and the work that is being done to check those results. It discusses improvements to the O.S.U. State method of straw assembly, and describes wire-tension measurement techniques suited to long wires.

Buckling of Straw-Tube Arrays

The buckling of single tubes is described by the Euler criterion

$$F = \frac{\pi^2 EI}{L^2}, \tag{1}$$

where F is the load, E is Young's modulus for the straw material, I is the bending moment, and L the length. This can be extended to $N \times N$ arrays of straws using the parallel-axis theorem giving

$$\frac{F}{N^2} = \pi^3(N^2+1)E\frac{R^3 T}{2L^2}, \tag{2}$$

where R is the radius of the straw, and T is the wall thickness. For convenience, the force required to buckle an array has been divided by the total number of tubes in the array, N^2, giving a measure of the maximum wire tension that may safely be applied. A plot of F/N^2 is shown in Fig. 1as a function of the length of the straw. The buckling load has been calculated for two wall thicknesses: 50.8, and 25.4 μm. The plot is for a 7-mm straw radius and assumes free (unrestrained) ends.

If the only force on the straw-tube array is the compressive force due to the combined wire tensions, then it

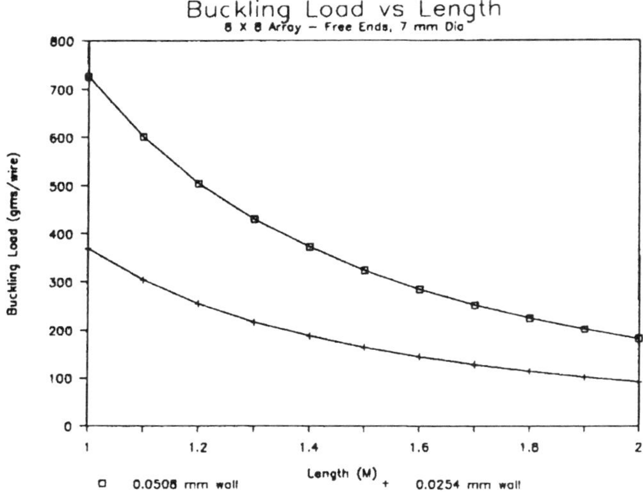

Figure 1: The buckling force per straw in an $N \times N$ array of straw tubes as a function of the length of the straws. The tubes are glued together to provide a large bending moment.

can be seen that, for a 2-m-long array constructed from straws with 25.4-μm wall thickness, the maximum load before buckling is 100 gms, only a factor of two larger than the force required to keep a 20-μm-diameter gold-plated-tungsten wire taut. The figures, that these buckling formulae predict, can vary by large factors from the figures obtained by experiment. This discrepancy is due partly to the theoretical model used: as there are many different modes of buckling and there are many different axes along which buckling can occur, and partly due to the non-symmetrical nature of the actual straws: they are after all not a true cykinder and the material properties may be non-uniform. Usually, all these effects combine to reduce the actual observed buckling force so that the factor of two is not a large enough safety factor.

The prediction (2) is being checked with an array of tubes fabricated at Princeton U. Once glued, the array will be fixed to a granite surface plate and restrained from buckling along the diagonal. A force meter is then used in conjunction with a screw, to apply and measure the longitudinal force.

An improvement in the strength of the structure, whilst still maintaining a minimum of material, can be obtained through the use of an epoxy glue filled with microspheres.[3] Because the glue has a very low density, all the interstitial gaps can be filled, resulting in a much stronger structure. Alternately, the number of straws in an array can be increased.

Improvements in Array Assembly

The Ohio State U. method of assembly of straw-tube arrays[2, 4] has been used to great effect for short (typically 50-cm-long) arrays. This method aligns the straws accurately by inserting a precision-ground, stainless-steel rod into each straw and then positioning the rod with dowel pins located in a precisely drilled end caps. This method becomes quite cumbersome when 2-m-long straws are anticipated. Instead of locating the straw by using its internal surface, it is more desirable to use the external surface. However, straws with thin walls are very fragile and deform easily under any force perpendicular to the axis of the straw.

One solution that we are pursuing is to pressurize the straw to 15 p.s.i. before placing it in a precisely machined positioning jig. This has the added advantage of performing a quality control on the straw, as punctured tubes are easily detected. Pressurization can be conveniently carried out by attaching bicycle-tire valves temporarily to the end plugs of the tubes After inflation, the straws can be left overnight and checked for leaks the following day. While still inflated, the straw can be handled safely with little danger of buckling. This speeds up the gluing and positioning process.

After gluing the assembly, the next step is the stringing of the anode wires. When using the Ohio State feedthrough, Fig. 2a, we encountered difficulty in attaining the correct wire tension. The problem is that the wire is secured by clamping it between a conical brass pin and the internal cylindrical surface of the plastic feedthrough. The pin is inserted in a direction towards the centre of the straw and opposite to the tensioning force. Any friction between the pin and the wire tends to push the wire back into the straw and reduces the wire tension.

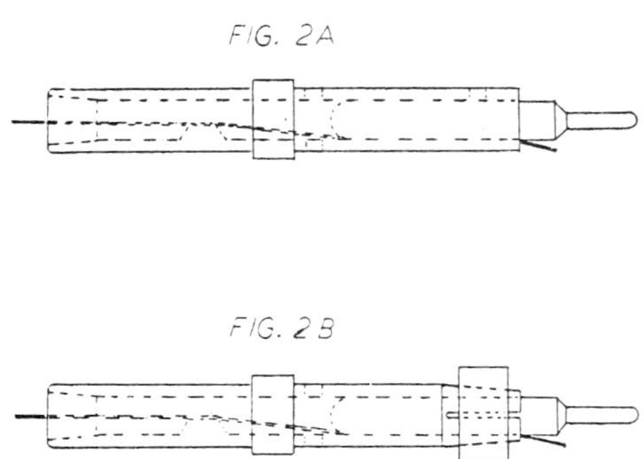

Figure 2: **a)**: The Ohio State U. design of a plastic feedthrough for positioning and clamping the anode wire of a straw-tube chamber. **b)**: A modified feedthrough with a slotted end and a clamping ring that squeezes the feedthrough to the brass taper pin without causing longitudinal forces that change the wire tension.

We have had some success with a modified design that takes the original plastic feedthrough and puts a longitudinal cut, Fig. 2b, into the end of it, allowing the ends to splay out. The pin is inserted without much force and the wire is clamped by pushing a ring over the splayed ends, forcing them together. In this design there are no extra longitudinal forces introduced on the wire during the clamping process.

Measurement of Wire Tension

Most wire-tension measurements make use of voltages induced by the movement of a current-carrying wire in a magnetic field. There are three general techniques for exploiting this effect:

1. The induced voltage can be amplified and measured. This is a maximum at resonance. A sweep of the frequencies around the suspected resonant frequency is made and terminated when the induced voltage is a maximum. A meter using this principle has been developed at Fermilab,[5] and has also been built at our lab.

2. The phase difference between the driving and driven voltages can be utilized.[6] Again a frequency sweep around the suspected resonant frequency is made, and the frequency at which the phase difference drops to zero, is recorded. An example of this type of device has been built at Princeton U. and is capable of very precise measurements to 0.25 Hz.A plot of the phase shifts measured on a 50cm long wire is shown in Fig. 3

3. The wire can be subjected to a short current pulse and the subsequent ringing-frequency recorded.[7]

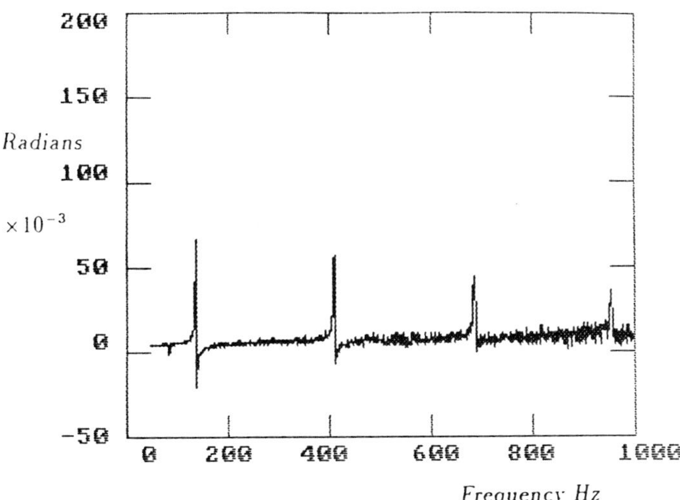

Figure 3: Phase shift versus Frequency showing four clear resonances

With 2-m-long straws the resonant frequency is very low ($\sim$ 60 Hz), though higher harmonics can be used to good effect. These mechanical systems have a fairly high Q so that the resonance curve is quite narrow, typically 2 to 3 Hz, at these frequencies. They also seem to have quite a lot of "inertia" in that changing from a driving frequency of say 60 Hz to 62 Hz, requires a wait of several seconds for the wire to stabilize its oscillations at the new frequency. A long time is required to achieve an accurate measure with the frequency-scanning techniques as the possible range of frequencies has to be spanned in a set of small increments with pauses at each step. We are now pursuing the third option, which offers the possibility of making an instantaneous measurement – crucial for large arrays of tubes.

Conclusion

The force required to buckle an array of straw tubes has been calculated. These calculations have a large uncertainty due to material imperfections and nonuniformities in assembly – for example, the nonuniform application of glue. They are to be compared with experimental results, and if they do indeed describe the buckling in a reasonable manner, they can be easily extended to describe straw arrays with different diameters, lengths, and numbers of straws.

Practical experience is being gained in the assembly of large straw-tube arrays, which has led to new methods of assembly that are better suited to long straws. Improvements have been made in the manner of clamping the wire so as to leave the wire tension unaffected.

Experience with two techniques for measuring wire tensions has shown that they are very time consuming when working with long wires. A third method that utilizes the ringing frequency is being exploited which has the potential for making rapid measurements. This is essential when large arrays of tens of thousands of straws are being contemplated.

References

[1] BCD Collaboration, *Expression of Interest for a Bottom Collider Detector at the SSC*, SSC EOI0008 (May 25, 1990).

[2] M. Frautschi *et al.*, *The AMY Inner tracking Chamber* Ohio State U. preprint, (submitted to Nucl. Instr. and Meth.).

[3] H. Kagan, private communication.

[4] R. Fulton *et al.*, *Proposal to Add Precision Tracking Layers to CLEO II*, Ohio State U. preprint (1990).

[5] T. Regan, *A Sensitive Instrument for Measuring Wire Tensions in Multiwire Proportional and Drift Chambers*, Nucl. Instr. and Meth. **219**, 100 (1984).

[6] Y. Asano *et al.*, *Tension Measurement of 7.6-m Drift Tubes of the Venus Muon Detector*, Nucl. Instr. and Meth. **A254**, 35 (1987).

[7] Wire Tension Meter Model NE-660 A, manufactured by K.F.K.I. H-1525 Budapest, Hungary P.F.49.

MEASUREMENTS OF ELECTRON DRIFT IN FAST GASES WITH CROSSED ELECTRIC AND MAGNETIC FIELDS†

EHIT Collaboration

J.M.Bailey[(a)], G.A.Beck[(a)], S.F.Biagi[(a)], P.J.Bussey[(b)], P.Colrain[(b)], J.B.Dainton[(a)], E.Gabathuler[(a)], J.C.Hart[(c)], J.W.Hiddleston[(c)], S.J.Maxfield[(a)], N.A.McCubbin[(c)], J.M.Morton[(a)], B.T.Payne[(c)], C.Raine[(b)], E.J.Romans[(c)], D.P.C.Sankey[(a)], D.H.Saxon[(b)], I.O.Skillicorn[(b)], D.J.White[(c)]

†Supported by the UK SERC and the US DoE through SSClab R & D project SSC–PC–024

(a) Department of Physics, University of Liverpool, P O Box 147, Liverpool L69 3BX, England
(b) Department of Physics and Astronomy, University of Glasgow, Glasgow G12 8QQ, Scotland
(c) Rutherford Appleton Laboratory, Chilton, Didcot, Oxon. OX11 0QX, England

Abstract

Results are presented from theoretical and experimental studies of electron drift in gas mixtures which aim for high drift speed and good X-ray detection. Implications for the design of an Intermediate Track Detector at a supercollider are briefly considered.

Introduction

This paper reports progress in establishing the optimum gas mixture for an intermediate angle gaseous chamber system at an experiment at a future proton supercollider (SSC or LHC). The detector (Intermediate Track Detector ITD) will be designed to reconstruct charged tracks, and possibly also to identify electrons by means of transition radiation (TR) detection in the pseudo-rapidity range $1.2 < \eta < 2.2$ [1]. The foreseen technique follows closely that of the forward track detector at the H1 experiment at HERA [2,3,4,5]. An essential ingredient for the proper design of an ITD is an understanding of the electron drift properties of potentially suitable gas mixtures.

Preliminary Considerations

Theoretical and experimental studies of various gas mixtures are under way. Interaction rates at a proposed luminosity (10^{33} cm^{-2} s^{-1}) of future supercolliders together with the nature and configuration of drift cells in an ITD at a "magnetic detector" (e g the SDC experiment at SSC [6]) pose the following minimum set of requirements on choice of gas mixture if optimal drift chamber operation is to be possible:

- high drift velocity for minimal cell occupancy with minimal drift Lorentz angle;
- good spatial accuracy in proportional mode;
- minimal depreciation with radiation exposure;
- manageable physical and chemical properties;
- X-ray sensitivity for efficient TR detection.

Figure 1 shows a compilation of measurements of the electron drift velocity as a function of drift field for the six fastest known pure gases [7]. When mixed with noble gases for comfortable HV operation (Ar for cheapness, He for low density, Kr and Xe for TR X-ray absorption), high drift speed is still to be expected. Other components may also be included as quenchers, ionization increasers, or fillers.

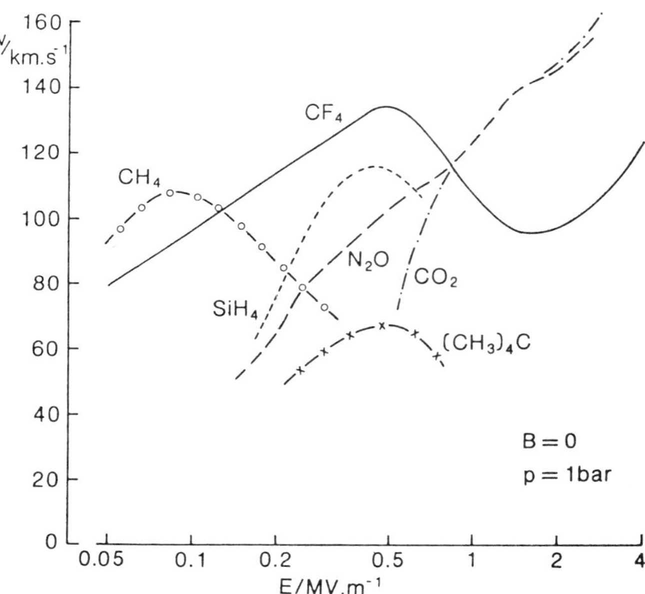

Fig 1: Electron drift velocities in fast polyatomic gases.

Admixtures of the gases of figure 1 to the noble gases also cool electron drift by virtue of their (inelastic) vibrational degrees of freedom. Nitrous oxide (N_2O) and neo-pentane ($(CH_3)_4C$) have further good properties. For example N_2O contains no C atoms, so can never polymerize, while $(CH_3)_4C$ absorbs the UV emission of Kr or Xe better than iso-C_4H_{10} or other quenchers. Kr or

In choosing a particular mix we try of course also to minimize the toxicity and explosion risk. In what follows, we present first\ measurements of mixtures containing the cheapest noble gas (Ar) and the fastest pure gas (CF_4).

Experiment

A multi-wire drift cell designed and constructed specifically for the purpose is used to measure drift velocity and Lorentz angle. The chamber follows closely the design of Atac et al [8]. Drift time in a constant electric field is measured over perpendicular distances of a 16, 32 or 48 mm from a sense wire. Drift direction is measured using the known position of irradiation by a collimated ^{90}Sr source (β end-point energies 0.55 MeV ^{90}Sr, 2.23 MeV ^{90}Y) and the position of gas avalanche along the sense wire by means of the induced signal on adjacent cathode strips. The chamber itself is operated in the uniform magnetic field of a C-magnet ($B \leq 2$ T) with the plane of ionisation drift and sense wire perpendicular to the magnetic field direction.

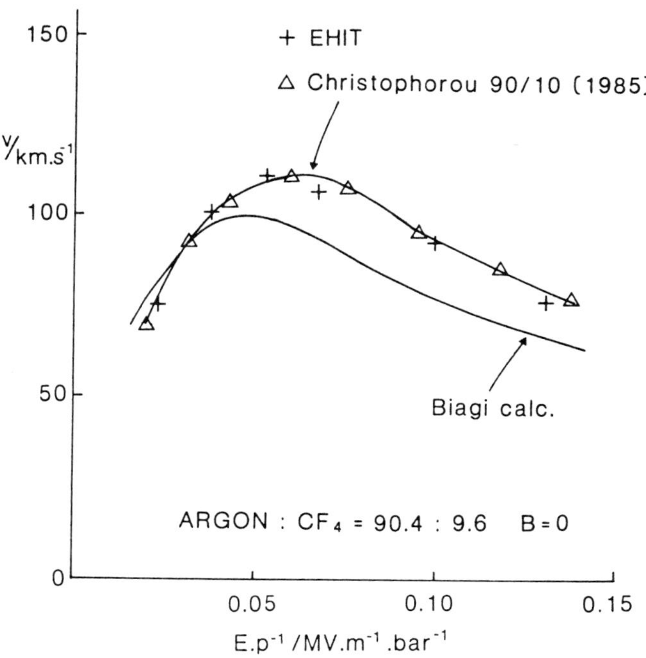

Fig 2: Drift velocity measurement in an Ar:CF_4 90:10 mixture with B=0; previous measurements and calculation are also shown [9,11,12].

Results of drift velocity measurements for CF_4:Ar mixtures in zero magnetic field are shown in figure 2 together with preliminary calculations [9,10]. In zero field our drift velocity measurements agree with previous measurements [11,12]. Systematic errors of measurement are of order 10%, and may be reduced after further study.

In finite applied magnetic fields (examples are shown in figures 3 and 4), where our measurements are unique, there is however some discrepancy of drift speed and angle with calculation. These calculations involving CF_4, and the parametrisations of input cross sections which they use, are new so the discrepancies should not yet be regarded as established.

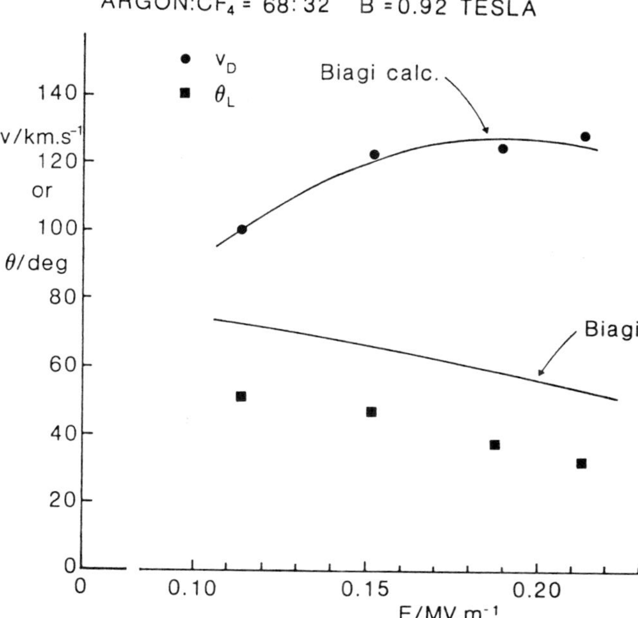

Fig 3: Drift velocity and angle measurements in an Ar:CF_4 68:32 mixture and crossed B field of 0.92 T; theoretical calculation is also shown [9].

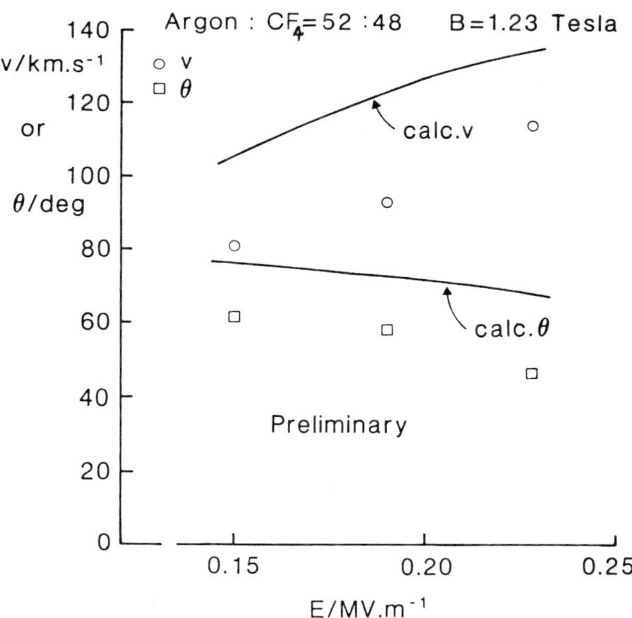

Fig 4: Drift velocity and angle measurements in an Ar:CF_4 52:48 mixture and crossed B field of 1.23 T; theoretical calculation is also shown [9].

It is already clear however that the use of CF_4 as a major component of drift chamber gas for the purpose of increasing ionisation drift velocity to at least 100 μm ns^{-1} is confirmed and for the first time demonstrated in substantial crossed magnetic field. Nevertheless its use in a magnetic field of ~1.8 T (likely in the SSC experiment SDC) with a suitable drift electric field may involve Lorentz angles substantially greater than 45° unless the chamber can be operated with a high drift field ≥ ~2 kV cm^{-1}.

A major problem associated with the use of CF_4 is the chemical effect it has on materials familiar to us in the construction of low mass drift chambers, e g foam, phenolic and epoxy based composites. At least one of our insulating materials, probably phenolic-bonded fabric ("Tufnol"), and perhaps also epoxy-bonded fabric ("G10"), had shrunk significantly due to absorbing CF_4.

We are therefore now carrying out measurements of mixtures with N_2O, which on paper looks also to be a suitable gas for drift velocities of 100 μm ns^{-1}, as quencher while we prepare a new prototype using CF_4 resistant materials. To date it is not yet clear that we can achieve adequate avalanche gas gain in the externally applied magnetic field for the hitherto possible electric drift fields.

Implications for ITD Chamber Design

The requirement of high drift field and thus acceptable Lorentz angle (~ 45°) is easily achieved with modest applied voltage for the small chamber aperture which is essential at high rate supercollider experiments. Any significant Lorentz angle increases the memory time of a drift cell and thus increases the occupancy. One possible way of reducing this effect is to string the ITD sense wires to follow as closely as possible to equiangular (logarithmic) spirals (figure 5).

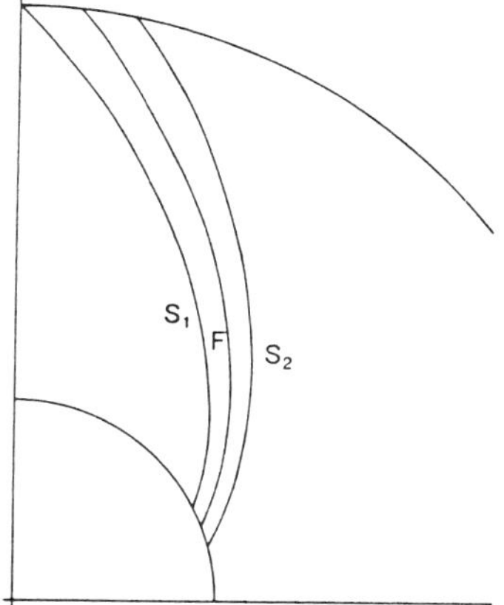

Fig 5: Equiangular spirals for 45°.

The problems experienced with the deformation, chemical and/or physical, of the construction materials used dictate the following:

- pre-shrinkage by long immersion in CF_4 gas, or perhaps other CFCs
- identification and use of low mass, inert materials which also have the necessary mechanical properties, e g Teflon (PTFE)

and they are being pursued.

Conclusions

First measurements have been made of electron drift characteristics in substantial crossed magnetic fields with gas mixtures which appear suitable theoretically for gaseous drift chambers at high rate supercolliders. It is possible to achieve acceptable drift velocities (>~100 km s^{-1}) in Ar:CF_4 mixtures, but with the disadvantage of substantial drift angles for values of drift electric fields used hitherto (~0.12 MV m^{-1}). At higher drift field (> ~0.3 MV m^{-1}) the drift angle can probably be reduced to 45° or less. Present theoretical calculations of drift velocity and angle for these mixtures are not yet of adequate precision for chamber design.

A serious drawback of the mixtures so far investigated is their deleterious effects on the materials traditionally used for low mass construction.

References

1 EHIT collaboration, J.M.Bailey *et al.*, "Forward Tracking with Enhanced Electron Identification", presented at this conference by D.H.Saxon.

2 G.A.Beck *et al.*, Nucl.Inst.Meth **A283** 471 (1989).

3 H.Gräßler *et al.*, Nucl.Inst.Meth **A283** 622 (1989).

4 H.Gräßler *et al.*, "Precision Reconstruction of Charged Tracks with Simultaneous Electron Identification in a Gaseous Detector using Transition Radiation", to be published in *Proceedings of the 2nd London conference on Position Sensitive Detectors, Imperial College of Science Technology and Medicine, London*, September 4-7 1990.

5 J.B.Dainton, in *Proceedings of the Workshop on Tracking Systems for the SSC, Triumf Laboratory, Vancouver, Canada*, ed W Frisken, July 24-28 1989.

6 SDC Expression of Interest, SSClab May 1990.

7 J.M.Bailey, "Compilation of Electron Drift Velocity Measurements", in preparation and references therein.

8 M.Atac *et al.*, Nucl Inst Meth **A249** 265 (1986).

9 S.F.Biagi, Nucl Inst Meth **A283** 716 (1989).

10 M.Yousfi, private communication.

11 L.G.Christophorou *et al.*, Nucl Inst Meth **163** 141 (1979).

12 L.G.Christophorou *et al.*, J Appl Phys **58** 3001 (1985).

3 Scintillating Fiber Tracking

SCINTILLATING FIBER TRACKING USING VISIBLE LIGHT PHOTON COUNTERS

M. Atac*
Fermi National Accelerator Laboratory, Batavia, IL 60510
University of California, Los Angeles, CA 90024-1547

Michael D. Petroff
Science Center
Rockwell International Corporation
Anaheim, CA 92803

Abstract

Work on scintillating fiber tracking using recently developed Solid State Photomultipliers (SSPM) is presented. Recent tests show that more than 4 photoelectrons are obtainable from a 0.5 mm thick 3HF-PTP doped and PMMA clad scintillating plastic fiber using 1 MeV electrons of a collimated Bi 207 source at a distance of 60 cm. Possible development of a visible light optimized version of the SSPM is presented; some test results and future plans are summarized.

Introduction

High luminosity operation of collider detectors create the need for fast tracking systems. The SSC is designed for a luminosity of 10^{33} cm^{-2} sec^{-1} and is considered to be increased to 10^{34}. The CDF and D0 experiments at Fermilab may be upgraded to operate at luminosities approaching $5 \cdot 10^{31}$ cm^{-2}sec^{-1}. LHC at CERN is being advertised to run at 10^{34} and beyond. If tracking is to work well at these luminosities, it needs to produce a signal faster than any previous tracking technology. We believe that the future collider detectors should utilize light transmission speeds within the tracking volume. The tracker should also be finely segmented to minimize occupancy rates per tracking element. All of this can be achieved using scintillating fibers connected to Visible Light Photon Counters (VLPCs), recently developed by Rockwell International Science Center.

The first suggestion to use scintillating fibers for charged particle tracking [1] were made in 1980, but to date results have not been totally successful. The problem has been the lack of an adequate photon detector having single photon sensitivity. Until now avalanche photodiodes (APDs) were the most promising candidates but their operation tends to be unstable. Gain variations when operated in the avalanche mode are large. Typical gains for 200-300 in this mode made single photon counting impractical. Operating the ADP in the Geiger mode generates gains as large as 10^7, but the deadtime of the device is too long for high rate applications. The problem occurs because there are trapped charges that can take up to a second to remove [2].

One successful application of scintillating fibers to charged particle tracking was achieved by the UA2 experiment at CERN [3]. The success of the tracker was somewhat limited by the quantum efficiency of their photon detectors. Typically bialcali photocathodes have quantum efficiencies of ~ 15%- 20%. Nonetheless, using fibers of 1mm diameter, tracking efficiencies of 80% were obtained for detecting minimum ionizing particles.

Collaborative work between Rockwell International Science Center (M. Petroff) and UCLA (M. Atac) has produced very encouraging experimental results using the newly developed Solid State Photomultipliers (SSPMs) [4]. These devices overcome the past shortcomings of other photodetectors. They have a quantum efficiency of between 60% and 80% and can be operated up to a gain of $5 \cdot 10^4$. Furthermore, they have a rate capability of several times 10^9 photoelectrons per cm^2 per second.

Presently Rockwell and UCLA are working on visible light-optimized version of the SSPMs. These are the Visible Light Photon Counters (VLPCs). In this operation, dark pulse count rate will be reduced, the count rate capability will be increased, time reso-

*Talk given at the Dallas SSC Detector R&D Workshop, October 13-15, 1990. This work is supported by the U.S. DOE under Contract No. DE-AMO3-76SF00010, Project Agreement No. DE-ATO3-883440384 with UCLA.

lution and the quantum efficiency will be improved.

The VLPCs and SSPMs need to be cryogenically cooled. The operating temperature is between 7°K and 10°K, otherwise the thermal electron pulse rate would be intolerably high. Repeated tests have shown that thermal cycling is not a problem for the thin plastic fibers that transmit the photons to the VLPCs.

The success of the work at UCLA and Rockwell led to a larger collaborative effort: the Fiber Tracking Group (FTG) [6]. Its focus is to fully develop the fiber tracking concept for the SSC. The work reported here is the result of this research and development effort which is supported by SSCL and DOE.

Central Tracking for the SDC

The SDC is designed to be a general purpose large solenoidal detector to do physics at the SSC (see SDC EoI). It is designed to identify and measure momenta precisely for isolated leptons, to use in first level trigger, to determine transverse momenta for electrons and muons, and to identify secondary and multiple vertices, especially for multi-lepton events. To do this physics at the SSC, we need a tracking system with low occupancy for each track element, with very good time resolution (< 16 nsec), providing a powerful trigger to select high momentum tracks, and that can operate up to a luminosity of 10^{34}. The tracking system should have an uncomplicated readout. We are convinced that scintillating fiber tracking using VLPC readout will satisfy the above requirements. A simple latch to identify a hit fiber is sufficient for the readout.

Figure 1 shows a cross section view of the SDC. The tracking for the SDC needs to be an integrated system for obtaining excellent momentum resolution with triggering capability to point charged particle tracks to calorimeters with precision. The best combination is to use silicon-strip tracking around the beam pipe up to a radius of 50 cm and follow up with scintillating fiber tracker up to the solenoid using superlayers of fibers. An enlarged view of the central tracking system is shown in Fig. 2. Since our subject is mainly the readout of the scintillating fiber system, we will not go into great detail about the structure of the superlayers here.

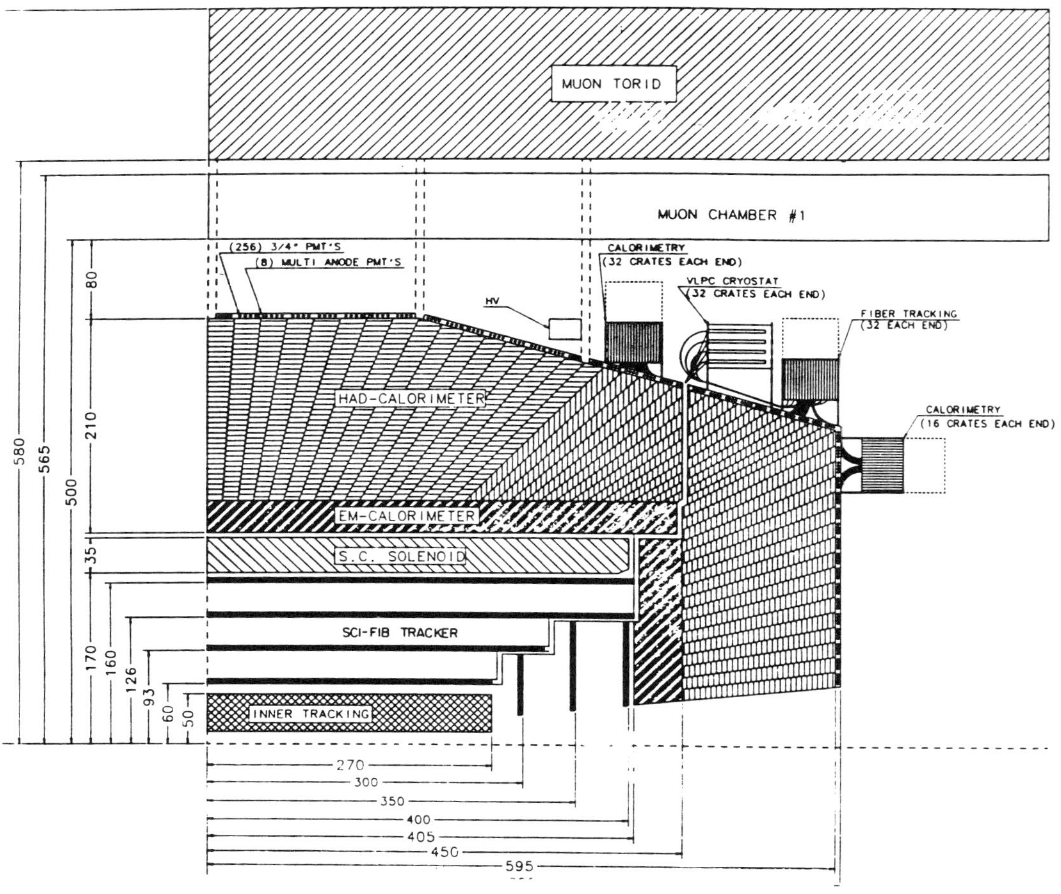

Fig. 1. Schematic view of quarter cross section of central part of the SDC.

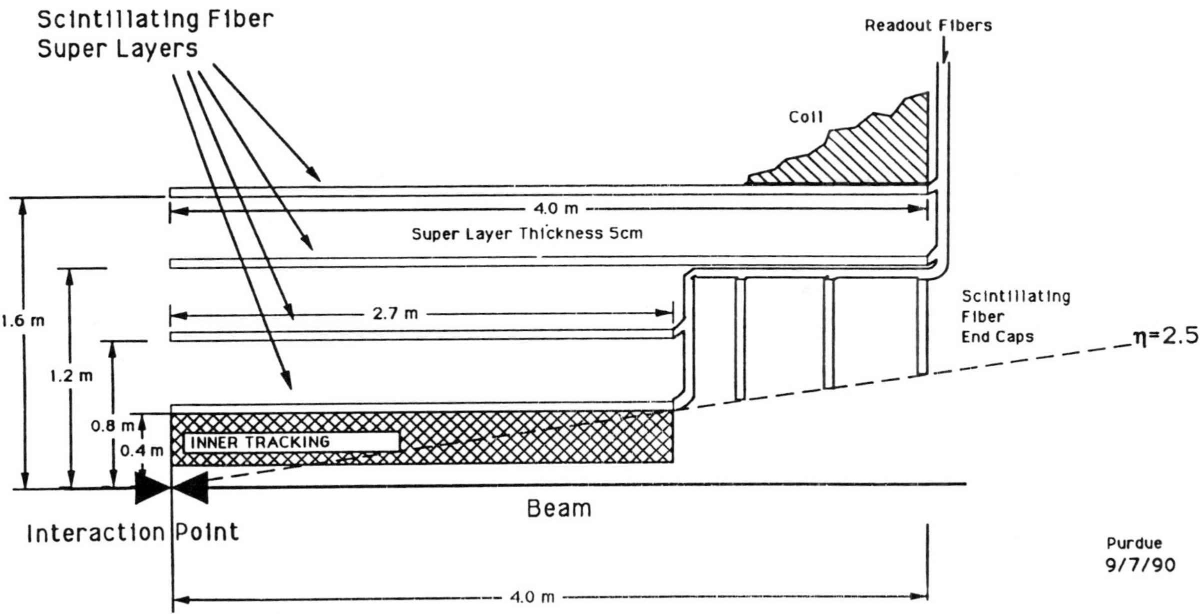

Fig. 2. An enlarged view of the inner tracking of the SDC.

In general each scintillating fiber is spliced to a nonscintillating optical fiber guide to carry the photons onto a VLPC element. This is schematically shown in Fig. 3.

Studies of Fibers with VLPC Readout

First experiments using scintillating fibers and SSPMs were carried out at Rockwell International Science Center, Anaheim [4]. Figure 4 shows the experimental arrangement. Photons produced from two 180 micron thick scintillating fibers with square cross section were detected by two SSPMs in coincidence. One of the fibers was used for triggering a pulse height analyzer for obtaining the spectra in Fig. 5. A collimated beta source was held at two different positions, resulting in an attenuation length of 180 cm in the thin scintillating fiber. This was the first time the SSPMs were used in the Visible Photon Counting (VLPC) mode using scintillating fibers.

Later more tests were done at UCLA and Fermilab. Recent results obtained from 0.5 mm diameter scintillating fiber, polystyrene core doped with 3HF+PTP and PMMA clad (produced by BICRON Corp.) show photoelectron yield using a collimated Bi^{207}–1 MeV electrons at distance of 60 cm from the SSPM. The average number of photoelectrons is ~ 4. Figure 6 shows the incredible resolution obtainable up to 10 photoelectron peaks. They are well resolved. Corresponding oscilloscope picture is shown in Fig. 7. The pulse shape is mainly determined by an integrating FET preamplifier. Faster rise time, about 20 ns, is obtainable from a faster preamplifier (Fig. 8).

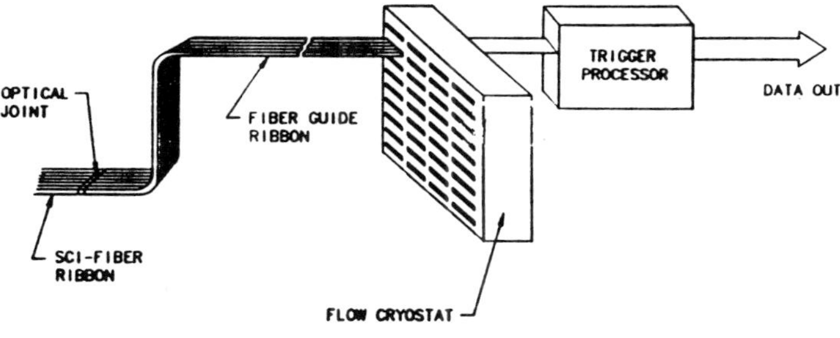

Fig. 3.

For the above results, a simple dipstick cryostat was used (Fig. 9). The temperature of the SSPM housing was monitored and controlled by simple Allen-Bradley carbon resistors. The temperature needs to be controlled within 1°K around 7°K. It was controlled by a simple operational amplifier feedback circuit.

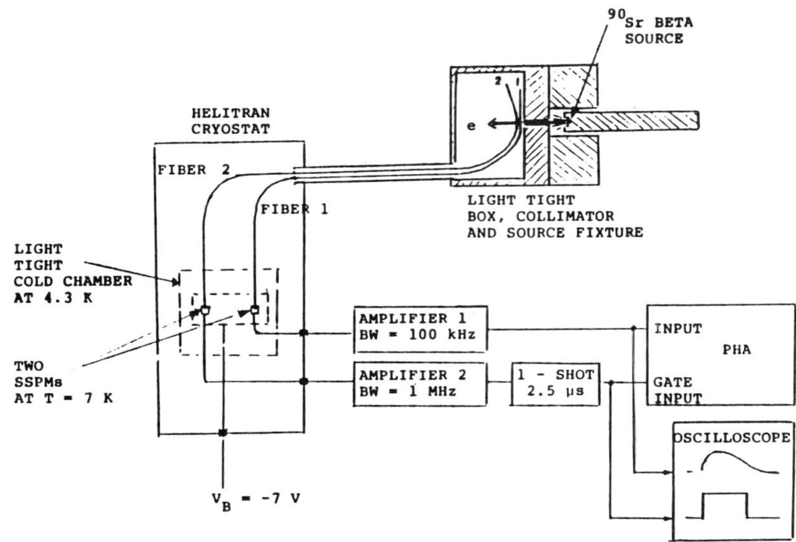

Experimental Arrangement for Detection Of Fiber Scintillations

Fig. 4.

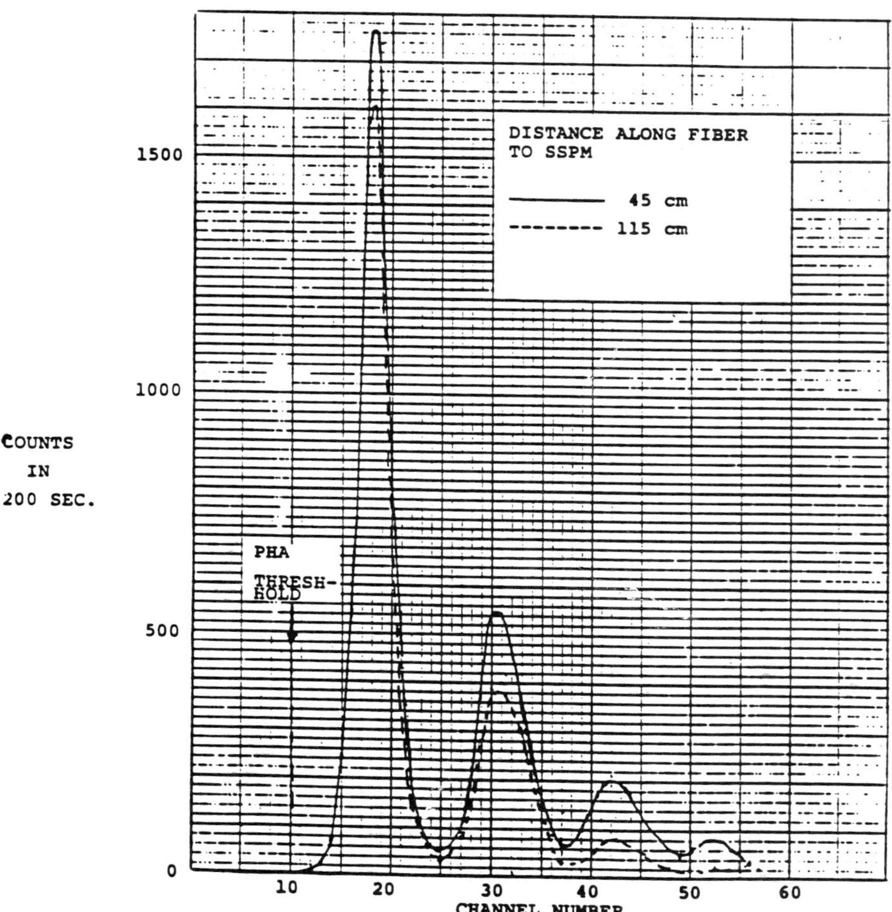

SSPM Pulse Height Distributions for Fiber Scintillations at two Positions of Beta Beam

Fig. 5.

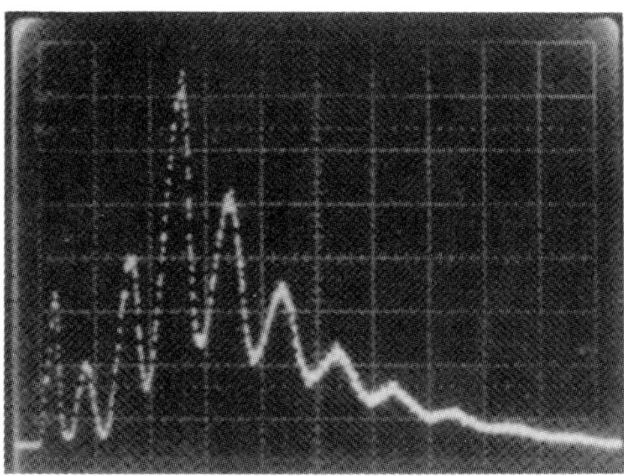

Fig. 6. Frequency distribution of simultaneous multiple photoelectrons detected by the SSPM from a 0.5 mm 3HF fiber.

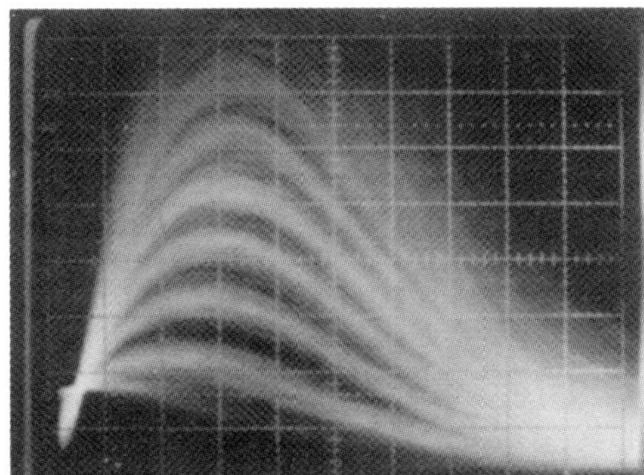

Fig. 7. Oscilloscope picture of the multiple photoelectron pulses in well resolved bands detected by the SSPM.

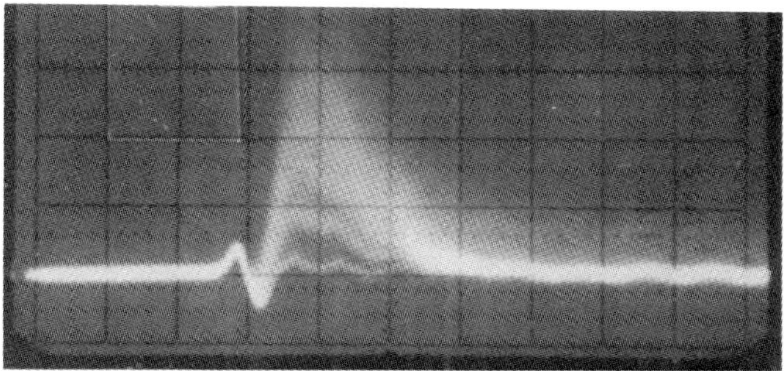

Fig. 8. Oscilloscope picture of the multiple photoelectron pulses with a FET charge integrating amplifier showing that a risetime of about 20 ns is obtainable.

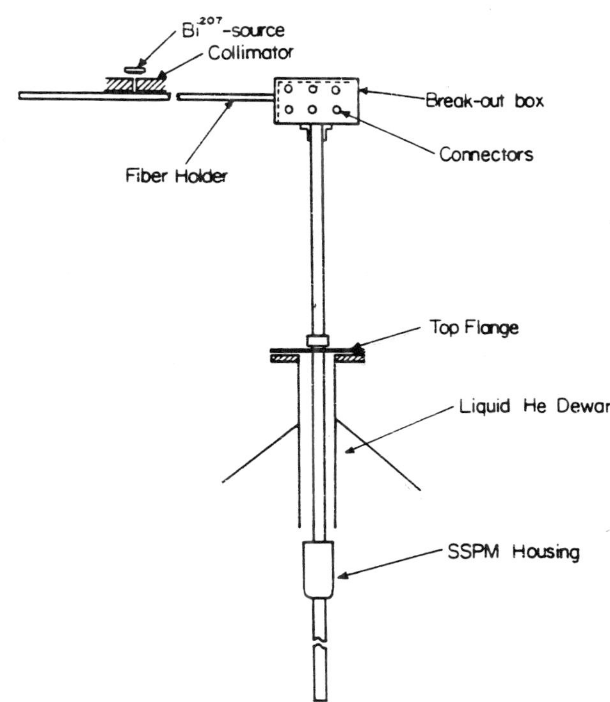

Fig. 9. The dipstick cryostat that is used to obtain the above results.

Visible Light Photon Counters (VLPC) and Future Experiments

Rockwell International Science Center under a contract from UCLA is working jointly to produce VLPCs. The work is supported by Superconducting Supercollider, Department of Energy. This is a part of the Detector Subsystem R&D for the SSC. The VLPC differs from the SSPM in spectral response, rate capability, and the risetime. We expect high quantum efficiency (around 80%) for visible photons, wavelengths between 500 to 600 nanometers, higher rate capability than that is obtainable from the SSPM, and the pulse risetime better than 10 ns. We believe that the quantum efficiency of the VLPCs for wavelengths about 1 micron will be very poor.

The Fiber Tracking Group is preparing a beam test using 256 channel VLPC-fiber system to study operation characteristics of the VLPCs together with fibers. In this we plan to have four groups of 64 fibers which are spaced apart to study tracking. We hope to determine photon yield per fiber as a function of the fiber length, track angle, and tracking accuracy, double track resolution, and tracking efficiency.

Figure 10 shows VLPC socket assembly. With this prototype arrangement each VLPC element is wire microbonded to a socket pin to carry the signal out individually. Groups of fiber ribbons are directly held across the VLPCs at a distance of about 75 microns with a precision of 25 microns. The optical fiber ribbons are clamped and potted in their precise positions. The VLPC socket assembly with the non-scintillating optical fiber ribbon is kept in a liquid helium flow cryostat at helium atmosphere.

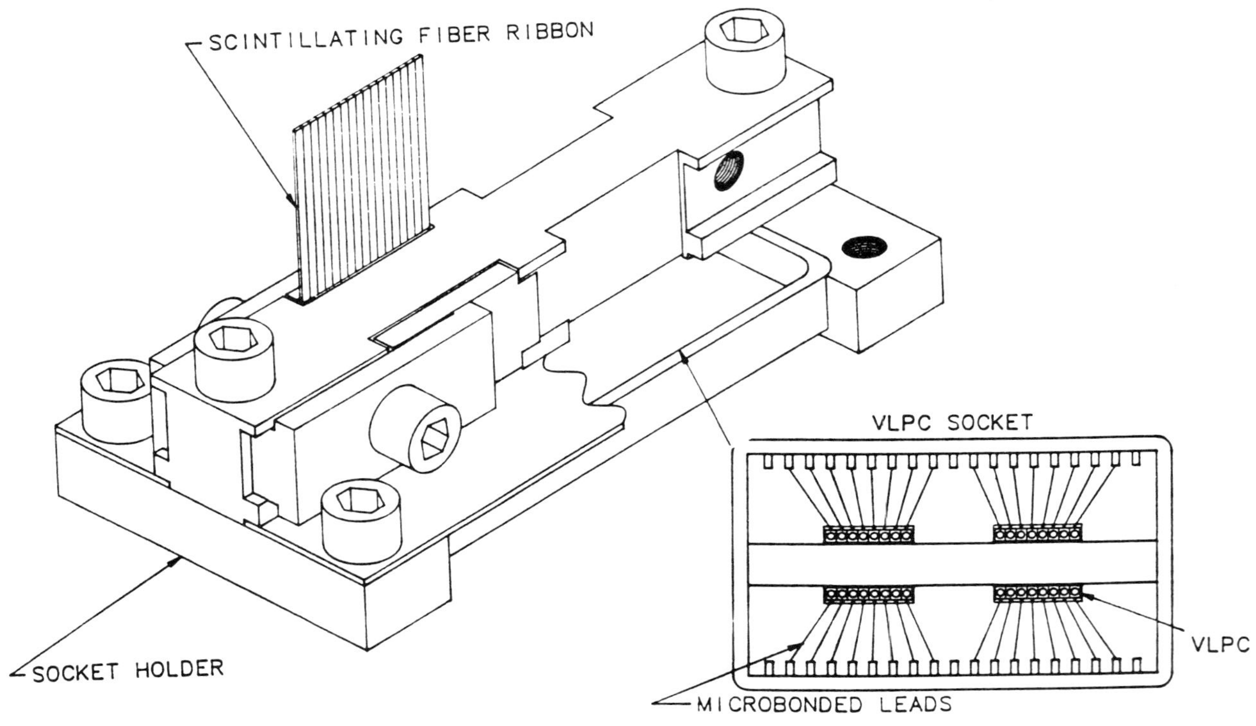

Fig. 10. The VLPC-socket assembly with fiber ribbon position arrangement for multiple fibers and the VLPCs.

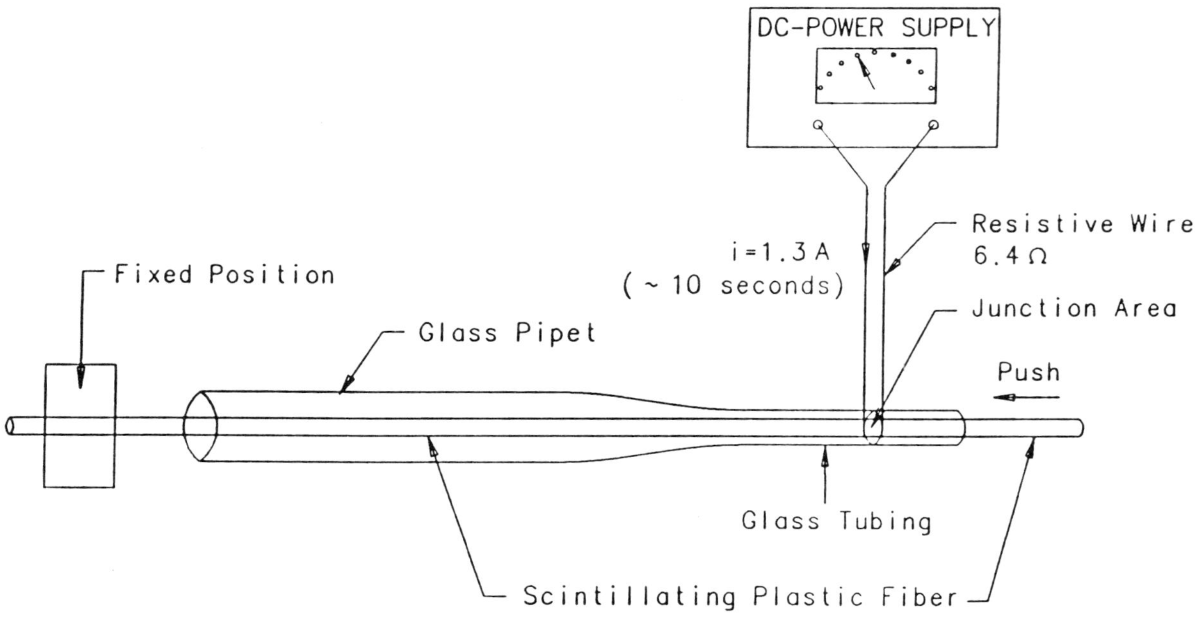

Fig. 11. Describes a simple method for splicing plastic scintillating fibers.

For the beam test each scintillating fiber of a ribbon is spliced to a nonscintillating optical fiber using a simple technique that was developed by M. Atac and W. Foster at Fermilab [5]. Figure 11 describes how the splicing is accomplished. Using this technique, 95-98% optical transmission efficiency can be accomplished through the spliced junction. Figure 12 shows that the transmission loss is within the experimental error.

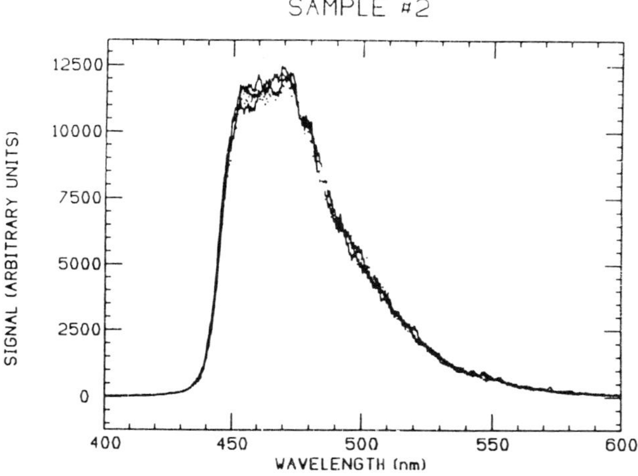

Fig. 12. Shows the transmission efficiency through a spliced joint. Dotted curves are before and solid curves are after the spliced joint when the fiber was excited by a UV-light source.

The Fiber Tracking Group is planning to do the beam tests in conjunction with the scintillating tile calorimeter tests prepared by J. Freeman and W. Foster of the CDF using the pipelined trigger system (see W. Foster's contribution to this conference).

Summary

The scintillating fiber tracking with VLPC readout looks very promising for the high luminosity operation of the SSC due to high rate capability, very small occupancy per fiber (less than 0.5%) at the design luminosity of 10^{33}, simplicity of the digital readout, fast triggering capability, the fiber and the readout being immuned to magnetic fields, present availability of radiation hard scintillating fibers around 500-600 mm wavelength of photons where the quantum efficiency of the VLPCs expected to peak around 80%.

We believe that we have some hard work ahead of us to prove the capability of the scintillating fiber tracking with the VLPC readout, realizing the need and importance of it.

References

[1] S. R. Borenstein, R. B. Palmer, and R. C. Strand, Physica Scripta 23, 550 (1981).

[2] S. Reucroft and M. Salomon; two separate contributions to this conference proceedings.

[3] Proceedings of the Workshop on Scintillating Fiber Detector Development for the SSC, Vol. 1&2 (Nov. 1988).

[4] M. D. Petroff and M. G. Stapelbroek, IEEE Trans. Nucl. Sci. N.S.36, 158 (1989); M. D. Petroff and M. Atac, IEEE Trans. Nucl. Sci. N.S.36, 163 (1989).

[5] M. Atac, W. Foster, and M. Lundin, Fermilab Report FN-537 (March 1990).

[6] Fiber Tracking Group: B. Abbott[h], D. Adams[i], E. Anderson[j], M. Atac[a,b], C. Anway[a], A. Baumbaugh[b,d], P. Berge[b], P. Besser[j], R. Bharat[j], M. Binkley[b], J. Bishop[b], N. Biswas[d], A.D. Bross[b], C. Buchanan[a], N. Cason[d], R. Chaney[k], D. Chrisman[a], A. Clark[b], D. Cline[a], H. Cohn[e], M. Corcoran[i], R. Davis[h], J. Elias[b], E. Fenyves[k], D. Finley[b], G.W. Foster[b], J.-M. Gaillard[b], J. Godfrey[d], H. Goldberg[c], H. Hammack[k], X. Huang[d], J. Jacques[d], V. Kenney[d], R. Kephart[b], D. Koltick[h], J. Kolonko[a], K. Kondo[ℓ], R.A. Lewis[g], J. Marchant[d], R. McIlwain[h], S. Margulies[c], H. Miettenen[i], R. Moore[g], R. J. Mountain[d], T. Okusawa[f], M. Petroff[j], J. Piekarz[d], A. Pla-Dalmau[b], R. Ruchti[d], R. Scalise[g], W. Shephard[d], E. Shibata[h], G.A. Smith[g], J. Solomon[c], K. Takikawa[ℓ], S. Tkaczyk[b], and R. Wagner[b]

[a] University of California at Los Angeles
[b] Fermi National Accelerator Laboratory
[c] University of Illinois at Chicago
[d] University of Notre Dame
[e] Oak Ridge National Laboratory
[f] Osaka City University
[g] Pennsylvania State University
[h] Purdue University
[i] Rice University
[j] Rockwell International
[k] University of Texas at Dallas
[ℓ] Tsukuba University

GEANT SIMULATIONS WITH SHELL AND MODULE GEOMETRY CODE AND SCINTILLATING FIBER CENTRAL TRACKING

Carol Anway*
Department of Physics
University of California
Los Angeles, CA 90024-1547

Abstract

This paper describes two GEANT geometry subroutines. One is a new geometry definition coordinating subroutine for use in GEANT simulations of the SDC, modular in design, and allows linking time replacement of subsystem geometry defintion routines. The other is a scintillating fiber central tracking geometry with four superlayers, three disk end caps, defined fiber layers, and a simple method for changing a vector position to layer number and fiber number.

Introduction

At 40 TeV center of mass energy, luminosity 10^{33}, and a bunch spacing of 16 nanoseconds, the SSC provides us with a substantial challenge: to design a 4π detector that will survive and produce accurate data, but keep costs to a minimum. To design this detector it is essential to use data from accurate simulation programs. Almost every working group has some sort of simulation program for that group's section of the proposed detector. We can get substantial results from these limited simulations, but it is time to have available more detailed detector simulations, in which a researcher studying muon detector response (for example) may take into account the effect of each internal structure on the muons, as well as coordinate the muon chamber data with that from the central tracker. This information is invaluable in creating detailed track fitting programs, accurate ideas of background radiation, etc.

Researchers at LBL and FNAL are working in conjunction with various working groups to bring together existing geometry definition routines, translate them into a format appropriate for use in GEANT, and make that package available to researchers who want an accurate detector geometry. This means the creation of a system of modular geometry subroutines, internal to GEANT, and flexible enough to use existing code in spite of its various different styles and sources. In particular, this code should have the following characteristics:

1. GEANT-based
2. Optional fast geometry
3. Accurate and up to date precision geometry
4. Useful for different approaches to the same subsystem (e.g., straw tubes versus scintillating fiber central tracking)
5. Choice of subsystem version at linking time
6. Only slight changes required of existing GEANT subsystem code for inclusion
7. Simplified GEANT volume number, material number systems for easy reference and debugging

To meet these demands, we need one geometry coordinating subroutine which then calls subsystem subroutines to collect (in common blocks) the information needed to define every volume. These subsystem subroutines can be exchanged easily, as long as the replacement routine has the same name. The internal structure of these subroutines is irrelevant as long as the common blocks are filled properly, which allows existing code to be adapted easily to the modular format. Finally, the common blocks may be useful later in tracking or digitizations.

Proposed Modular Approach

In the code written at UCLA and submitted for approval, this one geometry coordinating routine is called GEOMETRY. It is called from UGINIT, and coordinates calls to all the modular subsystem subroutines. GEOMETRY performs the following six main tasks: initialization of the common blocks, creation of a materials table, filling the common blocks in the subsystem routines, definition and positioning of each volume, set definitions, and then calls to GSORD and GGCLOS.

*This work supported in part by the DOE Contract Number DE-AM03-76SF0010, Project Agreement Number DE-AT03-88ER40384. This research was conducted by the Fiber Tracking Group, a list of authors can be found in "Simulation Studies for a Scintillating Fiber Tracker", Raymond Lewis, these proceedings.

The creation of a materials table takes place in the subroutine MATERIALS. It starts with the GEANT table created by a call to GMATE. Then it adds the straw tube mixture, scintillating fiber material, and other useful materials. This table can easily be extended to include any other materials required by subsystem subroutines. There is some discussion over whether there should be a table set up such that each material is listed only once or whether each volume should have its own entry in that table. Right now there is code to do either. Magnetic fields and tracking parameters are also set (with GSTMED) at this time, so there may be several definitions for the same material, depending on the magnetic field.

Every other task that GEOMETRY performs requires use of the common block CGEOM. This common block holds all the defining information for each volume in arrays. The arrays are set up so that for each value of the pointer, I, the name of the volume is NAME(I), the position of the volume is X(I), Y(I), Z(I), the number of divisions on the volume is NDIV(I), etc. Most importantly, the assignment of the volumes is carefully set up so that the internal number that GEANT uses to identify the volume matches the pointer value, I, for that volume; something that certainly doesn't happen automatically.

These variables are initialized to default values in the subroutine INITIALIZE. Most variables are just set to zero, but not all. Two of the more interesting assignments are NMAT(I) = 16 and MOTHER(I) = 'SDCV'. NMAT is the material number from the table set up in MATERIALS. It is set up so that if the user doesn't specify otherwise, the material within a defined volume is vacuum without a magnetic field. MOTHER(I) is the volume in which volume I is to be positioned. This assignment places it in the main, overall volume, SDCV.

The next step is to fill the common blocks by calling various subsystem subroutines. The pointer, I, which keeps track of the volume number in the arrays being filled, is passed as a parameter. The first subroutine to be called is GENERAL. This routine defines SDCV, the volume which houses all other volumes. It also defines "shadow" volumes, which outline places for each subsystem. This may change, as it is difficult to reconcile such a rigid requirement–that the entire subsystem fit inside some simple but artificial shape–with the necessity of programming volumes of cables and the like, which snake between volumes and may cause problems with overlapping volumes in GEANT.

Then GEOMETRY calls a routine for each subsystem, which are at present MVTX, CENTRAL, COIL, ENDPLUGCAL, BARRELCAL, FORWARDCAL, and MUON (See Figure 1). The subsystems may change as people consider where the natural breaking points are in the structure of the SDC detector. These routines are the ones we hope to have at least two versions of, one fast but simple, and the other detailed if slow. The fast version is a block of matter with density, radiation length, and absorption length equal to the over-all average value within that area of the SDC detector. That is, rather than programming each layer of a calorimeter into the subroutine, the subroutine would simply define a few, or even only one, volume shapes containing a uniform material with properties an average of those of the constituent parts.

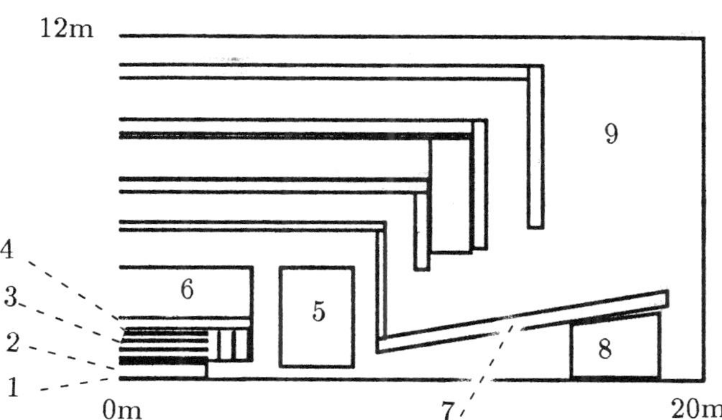

Figure 1: Full SDC geometry including scintillating fiber central tracking and muon chamber subsystems. 1: BEAM PIPE, 2: MVTX, 3: CENTRAL, 4: COIL, 5: ENDPLUGCAL, 6: BARRELCAL, 7: ABSORBER, 8: FORWARDCAL, 9: MUON

Detailed subroutines with the same names as the fast ones can replace them at compilation time. These versions don't take the shortcuts of the fast type, but define individual elements of each subsystem, with particulars such as fiber layers, support structures, readout cable bundles, etc. Rather than duplicate work that other researchers have done in this area, it is preferred to adapt existing work on these subsystem geometries to this format. The modifications required to existing GEANT code are simple: add the pointer, I, and replace the calls to GSVOLU, etc, with simple assignments to the common block variables (leaving GEOMETRY to handle all the calls to GSVOLU, etc, in a coordinated way). For code other than GEANT code, it may be easier to start from the beginning, using only the same figures that were used in the non-GEANT code. If, however, the geometrical parameters are in a CAD file, there may be more hope. Andrea Palounek (LBL) is checking into the possibility of a translation routine, capable of reading an output file from CAD and putting the information into GEANT language.

After all the common block variables are full, GEOMETRY loops through the pointer I again, calling the subroutines GSVOLU, GSPOS, and GSDVN. This defines, positions, copies and divides each volume in turn. Then GEOMETRY finished up with the requisite calls to GSORD and GGCLOS.

Scintillating Fiber Central Tracking Subsystem

The subsystem subroutine CENTRAL uses scintillating fiber technology in layers, stereo and axial, to create

Figure 2: Central Tracker with superlayers and disks.

an effective tracking volume. The general construction of the scintillating fiber central tracker is simple, owing to the convenient properties of the fibers. The fibers are laid in parallel layers with four layers all in the same direction making a tier. These tiers fit together with the help of a tubular support structure to make a superlayer. Since fibers can be twisted and bent, they may be placed in layers such that a fiber in the layer describes a helical shape on the surface of a tube. The fibers in layers with fiber axis parallel to the beam pipe are called Z fibers, and the fibers in layers with fiber axis at a small angle to the Z axis are called stereo fibers, axial fibers, or U and V fibers.

The central tracker described here has four tubular superlayers and three disks of fibers on each end (See Figure 2). The tubular layers are each divided at the center of the tracker, and readout takes place at the outer ends. The inner two superlayers are constructed the same way. They each have a tier of four 0.5 mm Z fiber layers on the inside, a 5 cm spacer, and then another tier of Z fibers on the outside. The outer two superlayers make use of the U and V layers possible with fibers. They each have a tier of four Z fiber layers, then of U fibers, then a 5 cm spacer, V fibers, and Z fibers. The end disks each have a unique geometry that also utilizes the bending capabilities of the fibers. These layers are stacked together with more spacer material for support, and arranged as three disks on each end of the detector.

In the GEANT user code, each tier of fibers (2 mm thick) is a volume, and each single layer (0.5 mm) is a division of that tier. It is then straightforward to calculate a layer number corresponding to a hit fiber. The program simply counts through the active fiber layers, right and left, using some counting convention, and assigns the layer number for that hit. It is slightly more complicated to assign a fiber number (within the layer) to the hit. What the subroutine does is to calculate the phi angle of the hit, adjust it for the stereo angle (in the case of U or V fibers), divide by the diameter of the fibers, and truncate that answer to the next lowest integer. This effectively numbers the fibers at the Z = 0 end, starting at phi = 0 and going in the +phi direction. The process can be reversed to obtain an X, Y coordinate from a layer number, fiber number combination.

Using this set-up, the next step is to run ISAJET events through the detector to estimate an expected occupancy. The plan is to approximate data from a typical Higgs event as consisting of 16 events, one Higgs and 15 minimum bias. Then one may count the number of fiber number pairs corresponding to hits of charged particles, and calculate the occupancy. If one uses the coordinating geometry routine to include at least the solenoidal magnet and fast microvertex chamber, then we can also take into consideration the extra particles coming from decays and bounce-back.

Future Work

The most important logical step to take with the geometry coordinating routine is to finalize the module spaces. In the code described here there are seven spaces, MVTX, CENTRAL, COIL, ENDPLUGCAL, BARRELCAL, FORWARDCAL, and MUON. Using this structure places rigid constraints on the modular code: the modular geometries must fit in the spaces provided. On the other hand, the spaces are natural ones, given the working groups that we have. A scheme proposed by Dr. Kunori of FNAL[1] has only three spaces to fill (CENTRAL, CALOR, and MUON). It allows more freedom within each module, but presents problems merging different working groups' code into each module. Another option is to do away with proscribed spaces altogether. The modules could then be written and called in the most natural manner, according to the code already written. There would also be space to include cables and support structure, a major difficulty with the other two schemes. But then the burden would fall to the user to ensure that nothing overlaps. Note: errors arise in GEANT when volumes overlap. There is a portion of code written in Florida by Dr. Saul Youssef (SCRI) that checks for overlapping volumes. Perhaps this could be used here.

Future work should also include finding a continuing person to be in charge of maintaining and updating the coordinating geometry code, as well as creating and maintaining a library of modules for use by researchers. In addition, we need a naming scheme to avoid future confusion. There are several in use. Also a CAD translating subroutine is in the works, as mentioned earlier. Finally, this geometry structure needs to be incorporated into the SHELL program being developed for the SDC collaboration by Kunori et al.

Acknowlegements

Many thanks to Dr. M. Atac, for his support, and to Dr. Andrea Palounek for her kind and thorough answers to my many GEANT questions.

[1]SDC communication, unnumbered.

SIMULATION OF A SCINTILLATING FIBER TRACKER FOR THE SSC

D. Adams, S. Panwar, J. Skeens, J. Wilson
Rice University
Houston TX, 77251

Fiber Tracking Group*

Abstract

A simulation and analysis code for an SSC scintillating fiber tracking system is described. Initial results regarding occupancy, pattern recognition and momentum resolution are presented.

Introduction

The development of a tracking system is a difficult but essential task in the construction of a general purpose SSC detector. An ideal tracker provides imaging, measures global and jet multiplicities, flags charged tracks and associates each track with the correct primary or secondary vertex. The system becomes even more powerful with the addition of a solenoidal magnetic field providing measurement of the momentum and charge sign. All of the above are useful for both off-line analysis and on-line triggering.

Approaching this ideal is difficult because of the high multiplicities expected in the SSC environment. One expects hundreds of charged tracks in a single crossing at the design luminosity and would like the detector to continue to function as the luminosity is increased by a factor of ten or more. On the order of a million channels are required to sample in tens of layers and keep the occupancy around or below one percent. The tracking system must also have a very fast recovery time to minimize the pileup of events from other crossings.

We belive that a scintillating fiber tracking system is the best way to address these issues. With submillimeter fibers, one can obtain low occupancy and good precision (100 μm in a superlayer) along with time response and recovery within a single crossing period (16 ns). The important and difficult issues of readout and radiation damage are being actively studied and we are optimistic they can be resolved. Once these and other generic issues are settled, there are two major tasks required to confidently design a system. The first is the the construction of a series of increasingly complex prototypes to demonstrate feasibility and measure the efficiency, precision and other properties of single fibers, layers and superlayers. The second is the development of detailed simulation and analysis programs in which the design is varied to evaluate the capabilities of the system in the SSC environment.

SFTSIM is intended to address the latter. It provides a detailed description of the response of a scintillating fiber tracking system to events at the SSC. Routines for clustering hits, reconstructing tracks and simulating trigger response are also included. Here we describe the inner workings of the code and the user interface and give some initial results.

Inner Workings

After initialization, physics events are assumed to be provided by an external event generator and then tracked in the usual way by GEANT. An SFTSIM routine is called each time a track ends a step inside a tracking volume. The particle type, the position of the particle, its energy loss in the last step, and some GEANT flags are recorded in the GEANT hit bank. After the GEANT tracking is completed for all tracks, the hits are digitized and clustered. Next, tracks are reconstructed and trigger simulation is performed. Most of the features can be switched on or off at run time using an FFREAD control card. For instance, if the user is only interested in looking at cluster multiplicities, he can turn off the track reconstruction and triggering.

The bulk of the initialization involves determining the geometry of the detector which is defined at run time using an input file. The user defines volumes which GEANT assumes are filled with scintillator material (CH with density 1.0 g/cm^3). The user also defines the size, shape and positions of fibers within these volumes. Three types of volumes are available and all of them are described by the GEANT cylindrical shell shape TUBE. The difference lies in the arrangement of fibers. For the cylinder, the fibers are at constant radius and for the disk and spiral, the fibers are at constant z. In the disk, the fibers are straight and all run in the same direction (x, y, u, ...). In the spiral, the fibers run

*See paper by R. Lewis in these proceedings

from inner to outer radius with increasing curvature so that their spacing remains constant. Each volume may contain zero, one or more layers of fibers. The user can also define inactive silicon volumes to include material effects of an inner tracker.

The SFTSIM digitization routines extract adjacent pairs of hits from the GEANT hit bank and do a linear interpolation to calculate the path length overlap with individual fibers. This overlap is used to estimate the amount of energy deposited in the fiber which is used to calculate the number of photons produced. This number is then attenuated to account for the fiber acceptance, losses in the scintillating fiber, optical fiber and couplings and the quantum efficiency of the photodetector. The detected number of photons is then chosen from a Poisson distribution and if this number is greater than zero, a signal of fixed width is produced. The signal is sampled at specified intervals and if present, a bit is set. The parameters defining the fiber and readout characteristics can all be set in the geometry input file. In our present runs, we sample once and assign the signal a width of 16 ns.

Instructions for assigning layers to superlayers are contained in another file. This file also contains the map for combining the superclusters from each superlayer to reconstruct tracks.

The clustering algorithms are fairly simple. Adjacent hits in a layer are assigned to the same cluster and then overlapping clusters within a superlayer are grouped into a supercluster. The position, direction and corresponding uncertainties for each supercluster are calculated using a least squares fit treating each fiber as an independent measurement. Clearly this is not an ideal solution for large clusters or those coming from overlapping tracks and it is intended these algorithms will eventually be improved. One avenue is to develop a lookup table possibly using experimental data as input. The assignment of layers to superlayers is taken from another input file.

This input file also contains instructions for the track reconstruction. The reconstruction begins by starting tracks in a specified superlayer (one track/supercluster) and then extrapolating them to a second superlayer where an attempt is made to add each nearby supercluster to the track assuming it came from the origin. All tracks with reasonable fits are kept and the extrapolation is repeated for each of the superlayers listed in the map. To date, only the code for dealing with cylindrical fibers parallel to the axis has been developed. Stereo layers and disk and spiral volumes are important areas for future development.

The code for the trigger simulation has not yet been completed although the framework has been defined and simple examples written. This is another high priority area for future work. We would like to implement trigger strategies like those under hardware development.

User Interface

SFTSIM is intended to be inserted in a GEANT-based simulation program and is also available as a stand-alone code imbedded in a old version of the Argonne SSC simulation package ANLSIM. Here we describe the user interface to that version. It is currently running under VAX/VMS and on an IBM R6000. This version is a batch program in the sense that the input and output are all through files except for some messages that are repeated on the screen and the option of directing graphical output to the screen.

There are five input files. The first two are the geometry and track reconstruction files described above. Another contains instructions for graphics. Many different views can be defined and each can be directed to the screen, a metafile or to a ZEBRA bank for later plotting with PAW. Each view can contain different components including the detector, GEANT tracks, superclusters, reconstructed tracks or any combination of these. The fourth input file is read using FFREAD and contains cards used by GEANT and ANLSIM as well SFTSIM. The last contains the trigger map. The logical unit numbers and our local file names are:

```
81  sf.dat
82  trf.dat
83  drw.dat
84  anlsim.dat
85  trg.dat
```

There are 17 output files containing errors, output from the GEANT routines GPXXXX, histograms and lists of tracks, hits, digitizations, clusters, etc. The logical unit numbers and contents of each file are listed below:

```
61  SFTSIM error messages
62  GEANT geometry and other setup output
63  Pythia messages
64  SFTSIM listing of each GEANT track
65  GEANT hit listing
66  GEANT vertex listing
67  histograms filled each event (event.hst)
68  histograms filled each track (track.hst)
69  SFTSIM listing of the GEANT hits
70  GEANT digitization listing
71  SFTSIM listing of GEANT digitizations
72  SFTSIM list of clusters
73  SFTSIM list of superclusters
74  SFTSIM list of all reconstructed tracks
75  SFTSIM list of final reconstructed tracks
76  SFTSIM triggering output
77  not yet used
78  graphics ZEBRA file -- picture.dat
79  not used
80  not used (often the graphics metafile)
```

An FFREAD card can be used to limit the number of events for which most of these files are written. Typi-

superlayer numbers	radius (cm)	# fibers/ superlayer	mean # hit fibers			total occupancy	
			1 bg	int bg	$H \to ZZ$	10^{33}	10^{34}
1 - 4	70	36k	125	400	240	1.8 %	12 %
5 - 8	100	51k	80	260	180	0.9	5
9 - 12	130	66k	55	180	130	0.5	3
13 - 16	160	81k	32	80	100	0.22	1.4

Table 1. Radius, fiber count, number of hit fibers per superlayer and total occupancy for each superlayer. The number of hit fibers includes columns for a single minimum bias event, 1.6 minimum bias events/crossing integrated over all crossings and one $H \to ZZ \to 4\mu$ event. The total occupancy is given for luminosities of 10^{33} and $10^{34}/cm^2/sec$.

cally only the first one or few events are written to most of the files.

For our current production runs, the most interesting information is in four n-tuples. The first (501) is intended for studying our track reconstruction capability. It is filled for the subset of tracks that reach the outer layer of the tracker and contains the initial kinematics and the reconstructed track parameters. The second (502) contains the kinematics for each event generated by Pythia. The third n-tuple (503) is filled as often as every step and is intended for looking at track multiplicities. The last (504) is intended for studying occupancy and is filled each event for each superlayer. The number of superclusters, hit fibers, and entering tracks are recorded. Each of the n-tuples may be independently activated or deactivated with an FFREAD card.

Results

In parallel with the development of SFTSIM, we have used the program to study some of the issues relevant to a scintillating fiber tracking system. Here we present results in the areas of occupancy, pattern recognition and momentum and vertex resolution.

Occupancy was studied by generating mininum bias (low p_T) events with Pythia. Figure 1 shows the pseudorapidity and transverse momentum (p_T) distributions calculated from 100 such events. There were an average of 150 particles per event most of which were pions distributed evenly between the three charge states. The detector used for these studies is split along a plane perpendicular to the beam axis at the intersection point and contains eight superlayers in each half. Each superlayer contains four layers of fibers all parallel to the beam axis. The superlayers are in pairs separated by 2 cm at inner radii of 70, 100, 130 and 160 cm. The fibers are 0.5 mm wide and 4 m long. The total fiber count is 930,000. The detector is shown in figure 2.

To look a the contributions of out-of-time crossings we moved our digitization interval in 16 ns increments. We found that the out-of-time crossings increase the multiplicity by a factor of 2.0 at 70, 100 and 130 cm and by 1.5 at 160 cm.

The total occupancy at design luminosity is calculated by assuming 1.6 minimum bias interac-

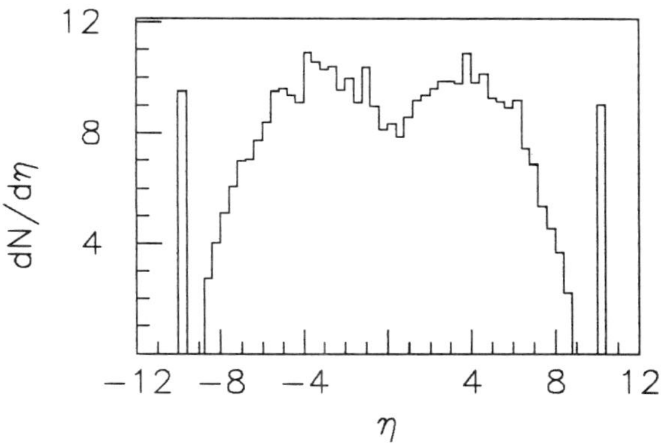

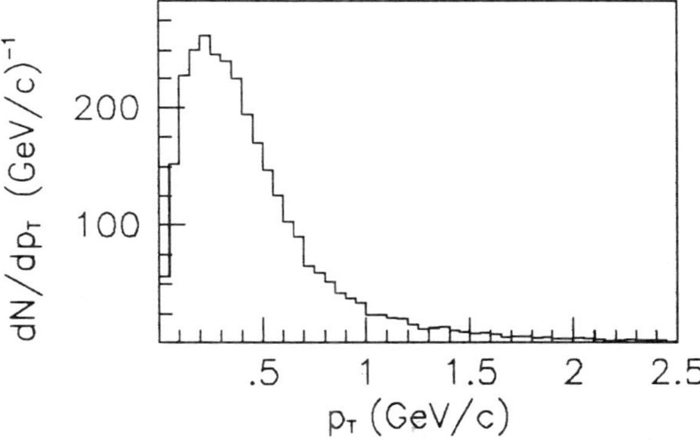

Figure 1: Pseudorapidity and transverse momentum distributions for Pythia mininum bias events normalized to one event.

tions/crossing and tossing in one "interesting" event. The latter is a 500 GeV $H \to ZZ \to 4\mu$ with Q_{Tmin} set to 200 GeV in Pythia. Occupancy at ten times design luminosity is calculated by including 16 minimum bias events/crossing. The results are shown in table 1. The design is not intended to be a final one but the results are quite encouraging. One would probably like to break up (or give up) the inner layer at least to run at the higher luminosities.

We have just begun to assess our pattern recognition capabilities. Using single mininum bias events and the tracker described above, we attempted to reconstruct all the tracks which reached the outer layer of the detector. This corresponds to a p_T threshold of

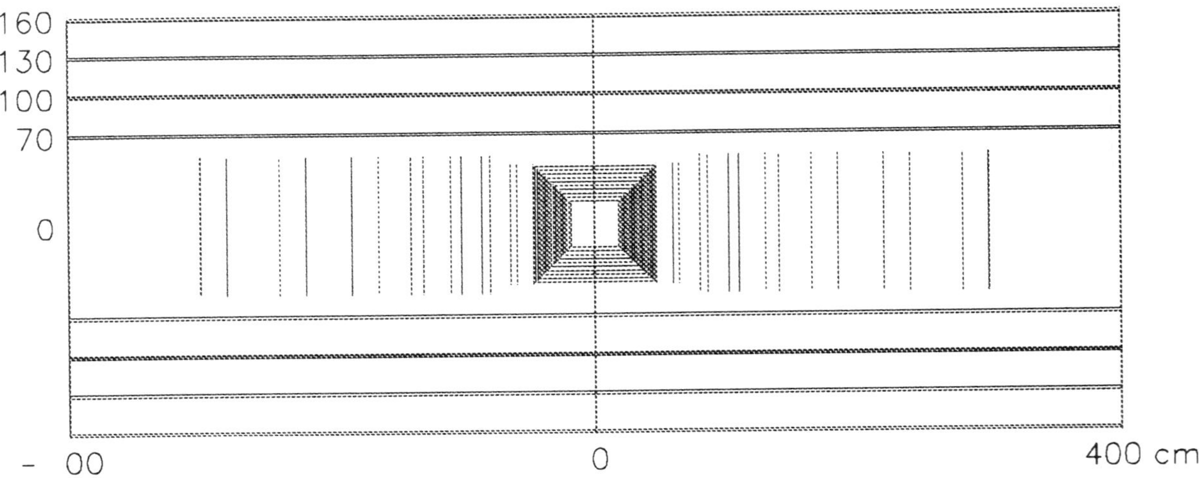

Figure 2: Side view of the detector used in the occupancy and pattern recognition studies. The outer four double superlayers are scintillating fibers. The inner cylinders and disks are silicon.

true	type I			type II		
p_T	p_T	σ	σ/p_T^2	p_T	σ	σ/p_T^2
20	20.0	0.31	0.78	20.0	0.16	0.41
50	50.1	1.81	0.72	50.1	0.81	0.32
100	100.	7.8	0.78	100.	3.5	0.35
200	204.	34.	0.83	201.	14.6	0.37
500				520.	131.	0.53

Table 2. Mean reconstructed momenta and root-mean-square deviations for type I and type II trackers for a variety of momenta.

0.5 GeV/c. We compared the reconstructed and initial transverse momenta and found that 98% of the tracks were within 0.1 GeV/c and 95% within 0.03 GeV/c.

Finally we present some early results in the areas of momentum and vertex resolution. These studies were done using a different tracker with double layers at 50, 100 and 150 cm. In the standard configuration, two of the four layers in a superlayer are offset by 0.5 fiber spacings to improve efficiency and spatial resolution. We refer to this as the type I arrangement. In the type II configuration, the fibers have quarter spacing offsets in the following sequence: 0.0, 0.5, 0.25, 0.75. Results obtained with single muon tracks at zero rapidity and a variety of momenta are presented in table 2. We find that the resolution is a factor of two better for the type II tracker. We also found that the type I tracker gave the wrong charge sign for 0.4% of the 500 GeV/c events while type II was always correct. The radial vertex reconstruction showed similar behavior and had a root-mean-square deviation of 0.02 cm for the type I tracker.

Future

It is clear that we have just begun to address the simulation issues facing a scintillating fiber tracking system. These occupancy results are encouraging but we need to demonstrate the pattern recognition and momentum resolution capabilities in an environment including both interesting and background events. The geometry of the detector needs to be considered very seriously to minimize the channel count (proportional to cost) without seriously compromising the physics capabilities.

The problem of forward tracking, i.e., for rapidities above 1.6 needs serious attention. It is not clear whether this can be carried out with barrel detectors of the type described here or if one must orient the fibers perpendicular to the beam. If so, one is probably forced to put in full stereo as part of the pattern recognition increasing the channel count and the complexity of the algorithms. In the simple trackers described above we did not include any stereo in our barrel detectors. We need to evaluate the physics gain as a function of channel count increase in both cases.

Another important issue is the integration of this simulation into a full detector package. Perhaps most important is the addition of active inner silicon layers so that the problems of pattern recognition, momentum resolution, etc. can be addressed from an integrated point of view. To do full physics capability simulations one will also have to include descriptions of the calorimeter and muon detector.

The support of the TNRLC, the SSCL and the DOE are gratefully acknowledged.

SIMULATION STUDIES FOR A SCINTILLATING FIBER TRACKER*

B. Abbott[h], D. Adams[i], M. Atac[a,b], C. Anway[a], A.E. Baumbaugh[b], B. Baumbaugh[d], P. Berge[b], M. Binkley[b], J. Bishop[d], N. Biswas[d], A.D. Bross[b], C. Buchanan[a], N. Cason[d], R. Chaney[j], D. Chrisman[a], A. Clark[b], D. Cline[a], H. Cohn[e], M. Corcoran[i], R. Davies[h], J. Elias[b], E. Fenyves[j], D. Finley[b], G. Foster[b], J.M. Gaillard[b], J. Godfrey[d], H. Goldberg[c], H. Hammack[j], J. Jaques[d], V. Kenney[d], R. Kephart[b], D. Koltick[h], J. Kolonko[a], K. Kondo[k], R.A. Lewis[g], J. Marchant[d], R. McIlwain[h], S. Margulies[c], H. Miettenen[i], R. Moore[g], R.J. Mountain[d], T. Okusawa[f], J. Piekarz[d], R. Ruchti[d], R. Scalise[g], W. Shephard[d], E. Shibata[h], G.A. Smith[g], J. Solomon[c], K. Takikawa[k], S. Tkaczyk[b] and R. Wagner[b]

University of California at Los Angeles[a]
Fermi National Accelerator Laboratory[b]
University of Illinois at Chicago[c]
University of Notre Dame[d]
Oak Ridge National Laboratory[e]
Osaka City University[f]
Pennsylvania State University[g]
Purdue University[h]
Rice University[i]
University of Texas at Dallas[j]
Tsukuba University[k]

Presented by
Raymond A. Lewis
Pennsylvania State University

Abstract

High p_t leptons will be used to trigger on candidate events for physics studies (e.g. Higgs decay) at the SSC. Rates for producing high p_t leptons are low, compared with the maximum 1000 MHz pp interaction rate. Unusual combinations of background processes can imitate features of the high p_t signal, requiring a considerable degree of event reconstruction at the trigger level. We are presently addressing the problem of recognizing high p_t tracks in the SciFi central tracker proposed for SDC. Minimum bias events generated by Pythia are superposed, and hit patterns are analyzed to find candidates for high p_t tracks from unassociated hits. The fiber tracker is compared with a straw tube tracker, showing selectivity for various choices of redundancy as a function of luminosity.

A. Introduction

A major challenge is to construct a tracking detector which is useful for triggering at luminosities up to $10^{34}/\text{cm}^2$–sec. The fiber tracker solves a potential problem of occupancy by providing a large number of detector elements over which the hits are spread.

The layout of the SciFi tracker proposed for SDC is shown in Figure 1. The central tracker consists of 4 superlayers of 650 micron scintillating fibers, with two sublayers of axial fibers in each superlayer. The intermediate tracker is comprised of 3 superlayers of 1 mm diameter fibers.

B. Backgrounds from Pileup

Particles from 40 TeV pp interactions generated by Pythia [1] were swum through the 2T magnetic field, simulating the effects of gamma pair conversions, Coulomb scattering, and ionization losses. Hits due to particles which loop for more than 16 nsec, the time between bunch crossings, were superposed on events

*Work supported in part by the U.S. Dept. of Energy and the Texas National Research Laboratory Commission.

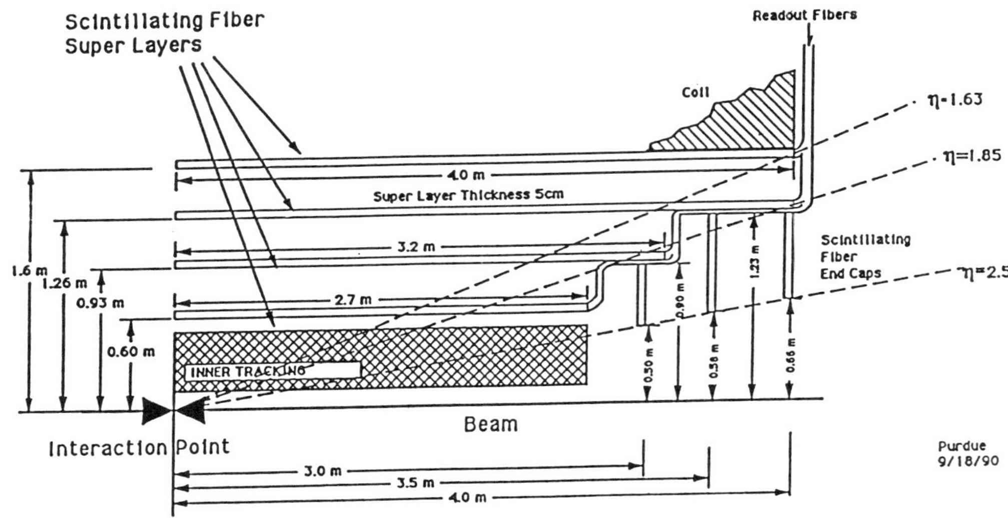

Figure 1 – Central and Intermediate Scintillating Fiber Tracker Layout.

occurring in up to 10 subsequent beam crossings. Table 1 shows the average multiplicity of hits/beam crossing in each of the superlayers, assuming one pp interaction/beam crossing.

Table 1 - Occupancies in SciFi Tracker.

Superlayer	Hits/event
(Central Tracker)	
1	63.6
2	42.2
3	31.5
4	19.0
(Intermediate Tracker)	
1	7.1
2	9.0
3	7.7

A trigger based on the central tracker uses two hits to define the azimuth and p_t of a high p_t track candidate, and the other 6 hits as a redundancy check. Figures 2 and 3 illustrate the effects of occupancy, showing the probability of finding a high p_t (> 10 GeV/c) track among non-associated hits. At a luminosity of 10^{33}/cm^2–sec or greater, hits in 8 layers of 4 mm straw tubes (Fig. 2), consistent with high p_t track, can be found at a rate of at least 0.3/bunch crossing. With the fiber tracker (Fig. 3), the fake high p_t rate of 10^{-8}/bunch crossing is well below the rate of

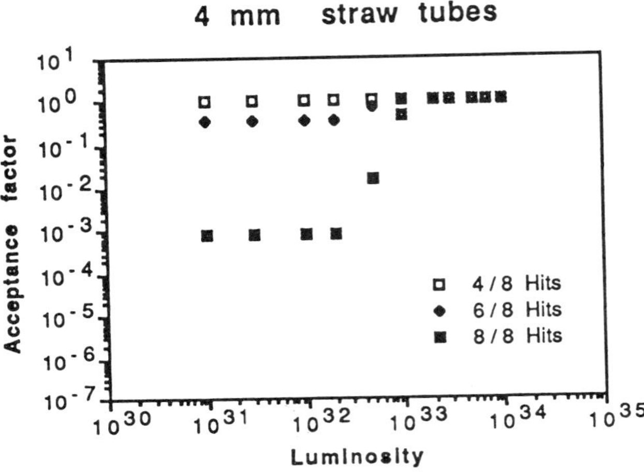

Figure 2 – Simulated acceptance for fake high p_t (> 1 GeV/c) tracks in a 4 mm straw tube tracker.

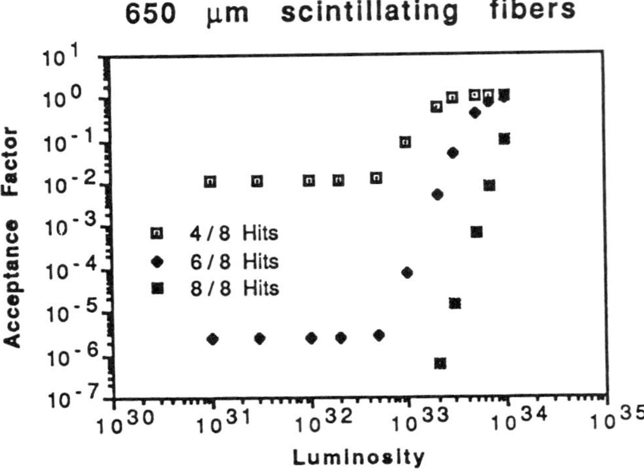

Figure 3 – Simulated acceptance for fake high p_t (> 1 GeV/c) tracks in a 650 μm scintillating fiber tracker.

0.02/bunch crossing at which high p_t pions are created. By contrast, the straw tracker will yield a noise-to-signal of 15 at the same luminosity. At $10^{34}/cm^2$-sec, the fiber tracker provides a noise-to-signal of 0.5, still satisfactory as a first level trigger. At this luminosity, a straw tube tracker is clearly swamped with background.

A similar high p_t algorithm applied to the intermediate tracker results in a fake trigger rate of ~ 5/bunch crossing, at L = $10^{34}/cm^2$-sec. Smaller diameter fibers of 500 microns are desirable, if intermediate tracking information is to be used at an early stage of the trigger logic.

C. High p_t Resolution

The p_t resolution of the central tracker is important both at the trigger level, and in extracting physics at a later stage of analysis. The resolution is studied by simulating the effects of fiber granularity, efficiency, finite target size, delta rays, and pileup. At the trigger level, information about the vertex is ignored. The p_t resolution expected, assuming that the beam crossing is within a 1000 micron radius of the nominal beam centerline, is 60% r.m.s. at p_t = 1 TeV/c. However, the vertex can be defined more accurately at a later stage of the analysis by using more complete information. The p_t resolution versus p_t shown in Figure 4 is calculated assuming that the vertex is known within 20 microns. In this case, the p_t resolution is 25% r.m.s. at 1 TeV/c.

D. Neutrons in the Central Tracker

Neutrons from hadronic interactions in nearby calorimeters produce hits in the fiber tracker from np scattering. The neutrons have typically 1 MeV energies, whereas a 3 MeV proton will range out in a fiber radius of 250 microns. Thus neutrons produce hits in single fibers which can be rejected in the electronics.

At a luminosity of $10^{34}/cm^2$-sec, neutron hits are expected at a rate of about 76 per monolayer of 15,000 fibers, at a radius of 160 cm. Neutron hits within 1 fiber diameter in any two of the

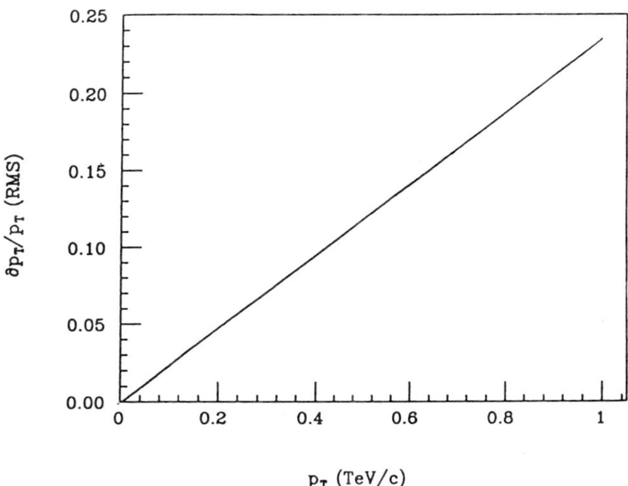

Figure 4 — Simulated momentum resolution versus p_t with a 20 μm vertex constraint imposed.

4 layers of fibers will occur at a rate of 2.7/proton bunch crossing. This rate is small compared with about 200 hits from pions and electrons.

E. Conclusions

Background rates due to pileup faking a high p_t trigger are used as a criterion for evaluating the performance of a central tracker design. At a luminosity of $10^{33}/cm^2$-sec, background rates with 650 micron fibers are small compared with the production of pions with p_t > 10 GeV/c. At a luminosity of 10^{34}, the fiber tracker is still able to cope with all conceivable backgrounds. However, a 4 mm straw tube tracker starts to have serious difficulties (noise/signal = 15) with background at $10^{33}/cm^2$-sec, and at $10^{34}/cm^2$-sec is overwhelmed with backgrounds.

[1] H. Bengtsson and T. Sjöstrand, "The Lund Monte Carlo for Jet Fragmentation and e^+e^- Physics — JETSET Version 6.3 — An Update," Comp. Phys. Commun. 43, 367 (1987).

MONTE CARLO SIMULATIONS OF SCINTILLATING FIBER OPTIC WAVEGUIDES

Raymond J. Mountain
Department of Physics, University of Notre Dame, Notre Dame, IN 46556

Representing the Fiber Tracking Group*

University of California at Los Angeles
Fermi National Accelerator Laboratory
University of Illinois at Chicago
University of Notre Dame
Oak Ridge National Laboratory
Osaka City University
The Pennsylvania State University
Purdue University
Rice University
University of Texas at Dallas
Tsukuba University

ABSTRACT

Preliminary results from model calculations of a single scintillating fiber optic waveguide are discussed.

INTRODUCTION

We have been studying the light transmission properties of single scintillating fiber optics waveguides through detailed Monte Carlo simulation and experimental measurements. In particular, we have been attempting to probe the nature of the single-photon detection problem in this context as part of the effort to develop a scintillating fiber tracking subsystem for use in SSC experiments.

As currently envisioned, a single channel for such a tracking detector will consist of an active scintillating fiber waveguide optically coupled to a passive waveguide and then to a highly efficient photodetector. Refer to Figure 1. From the standard lore one may determine five distinct stages to this single-photon detection problem, each with its own concomitant processes and problems:

(1) passage of an ionizing particle through the scintillating fiber deposits a certain amount of energy, causing a series of excitations, non-radiative and radiative transfers dependent on the detailed bulk composition and molecular structure, and resulting in the emission of a photon;
(2) propagation of light along the fiber with a certain complex light-collection efficiency determined by the geometry of the waveguide, the degree of structural imperfection, and the amount of (self-)absorption of the light by the scintillator material itself;
(3) transmission across a physical discontinuity, a splice from the scintillator waveguide to a non-scintillator ("clear") waveguide for transport of light out of the fiducial volume of the detector;
(4) negotiation of one or several bends in the clear waveguide light-pipe;
(5) conversion of single photons to electronic signals by an optically-coupled photodetector with high quantum efficiency.

The first of these involves development of new efficient bulk scintillators, which is currently under investigation [1,2]. The last of these involves the development of SSPMs and VLPCs [1,3]. Thusfar, the focus of the present effort has been on the second of these items, and herein we shall concentrate on certain preliminary estimates from this study. We stress that it is extremely important to understand in the most precise possible terms the detailed behavior and response of a single waveguide. The systematic study of such a system is our ultimate goal.

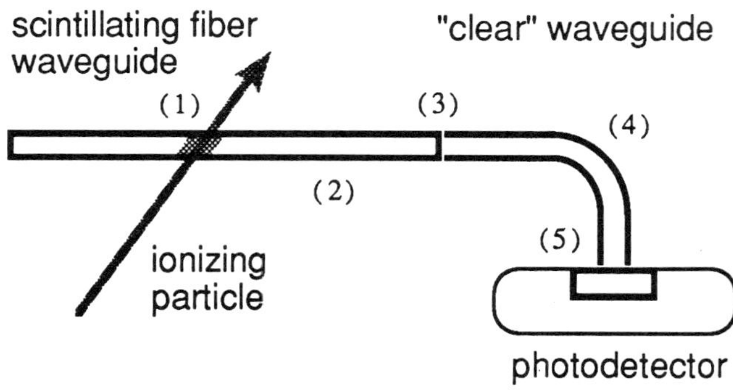

Figure 1. Schematic of a single fiber channel.

* The full author list for the Fiber Tracking Group may be found in Reference 1.

SCINTILLATOR PHOTON YIELD

The raw photon yield from a ionizing charged particle passing through the bulk scintillator is dependent on the molecular properties of the material. For our present use, it may be expressed as

$$I_0 = \left(\frac{dE}{dx}\right) \Delta x \frac{Q}{E_\gamma} \tag{1}$$

where dE/dx is the mean energy loss associated with the ionizing radiation, Δx is the given path length traversed by the particle in the fiber waveguide, Q is the quantum efficiency of the core material, and E_γ is the "effective" energy required to liberate a single photon. Typical numbers for these quantities give $I_0/\Delta x \approx$ (2 MeV/cm)(0.05)/(4.7eV) $\approx$ 2000 photons/mm of track length.

Thus we have a nearly-constant number of photons produced per unit track length for such materials. The photodetection problem then becomes one of geometrically determining the appropriate path length of the particle across the fiber, and tracing the propagation of the light produced along this path through the waveguide in order to estimate the number of photons reaching the detector-end of the fiber. In practice, the above-quoted number will be reduced by a factor on the order of 100.

LIGHT PROPAGATION

In principle, the parameters relevant to the detailed understanding of a scintillating fiber waveguide and to the evaluation of its efficiency are quite large in number [4]. In attempting to systematically study this parameter space, we have been developing a detailed waveguide simulation code OPTICKS [5] which has thusfar given good qualitative agreement with observation. Our code adopts an approach that is principally one of geometric optics, with modifications from macroscopic electrodynamics.

Define the *survival fraction* as that fraction of photons produced in the waveguide which propagate without escaping through the clad or being absorbed in the core and thus exit the waveguide at the detector end. What are the major effects involved? They are:

(0) the geometry and sizes of the waveguide (cross-sectional shape, core, clad and coat thicknesses, length);
(1) structural imperfections, specifically due to mechanical imperfections of the interface between the core and clad, producing reflection coefficients less than unity;
(2) self-absorption of the scintillation light by the core material itself, depending on the overlap of the emission and absorption spectra and the wavelength of the propagating light.

Effect (1) yields power-law absorption; whilst effect (2) yields exponential absorption. The combination reduces the initial intensity, Equation (1) by the factor

$$g = R^N \cdot \exp(-D/\lambda_B) \tag{2}$$

where R is the reflectance, N is the number of times the photon encounters the interface, D is the total path length of the photon in the fiber, and λ_B is the bulk attenuation length of the material (related inversely to the concentration times the molar extinction coefficient).

Efficiency vs. Radial Position of Source

Consider the case of a cylindrical waveguide in which a point-source is placed on the optical axis and then successively moved radially away from the axis. We have calculated the survival fraction as a function of the position of the point-source for the specific configuration of: PS/PMMA core/clad materials, fiber length of 3m, fiber diameter of 500μm, effectively *no* bulk absorption in the core, and 95% absorbing "black" coat exterior to the clad. This simple case is to be seen as almost pure geometry. Refer to Figure 2.

With the point-source on the optical axis, the only modes of the waveguide which are excited involve meridional-rays (i.e. modes for which the propagation vector crosses the optical axis). The survival fraction in this case is ~0.02. As the point-source is moved further away from the optical axis, an increasing number of skew-ray modes are excited (i.e., the nonmeridional-rays or "winding" modes). Many more channels will then open up for light propagation. There is an increasingly-significant sampling of the relative contribution of the skew-ray modes, so that the survival fraction grows quadratically with radial distance to ~0.07.

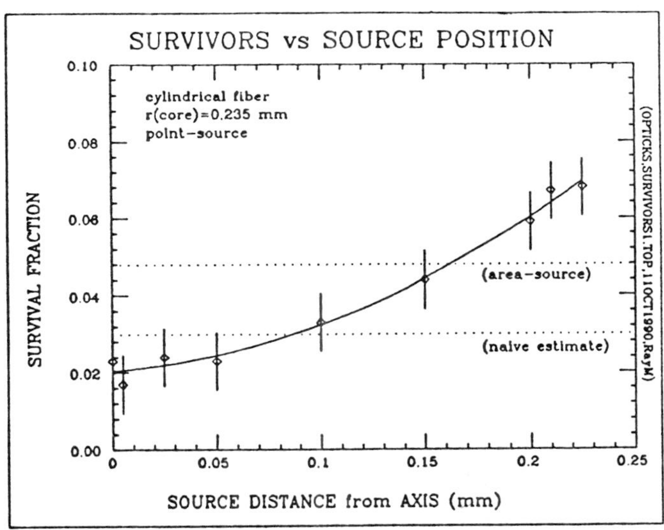

Figure 2. Variation of the survival fraction from a point-source with radial distance from the optical axis.

The naive efficiency estimate for meridional rays *only* is the well-known result $\varepsilon_0 = 0.5\ (1\text{-n(clad)/n(core)}) \approx 0.03$ and is indicated in Figure 2 by the lower dotted line. Although this estimate is often quoted as the de facto efficiency, it falls short by greater than a factor of two for many modes (in even this simple case). It is clear from this that there is a significant contribution to the light collection efficiency made by skew rays. A more appealing mean behavior may be inferred from a random sampling of the full cross-sectional area of the cylindrical fiber. As is reasonable, the resulting value of 0.048 properly weights the more distant areas. The upper dotted line in Figure 2 indicates the survival fraction so obtained. This is a reliable value for the mean fraction of photons trapped in the direction of the detector.

Forster Transfer Mean Free Path

For these scintillators, there is a mean free path (inversely related to dye concentration) between the sites of ionization energy deposition and scintillation photon emission. Our geometric studies indicate that the survival fraction drops to 80% linearly with this mean free path up to the fiber diameter. This implies that for efficient fiber production, one must maximize the dye concentration in order to minimize the mean free path and increase the geometric efficiency. However, self-absorption effects will compete with this gain.

End-face Distributions

We have investigated the intensity distribution of photons emerging from the end-face of the waveguide for the case of different sources of light. From a point-source on axis (meridional-rays) we obtain an areal distribution peaked in the center of the end-face disk; from a point-source near the core-clad interface (skew-rays) we obtain a distribution which is depleted in the central region but more heavily populated on the periphery; and from an area-source (sampling) we obtain a distribution which is fairly uniform over the end-face disk. Figure 3 shows the skew-ray distribution, which contributes the long time-component of fiber traversal times ("afterglow").

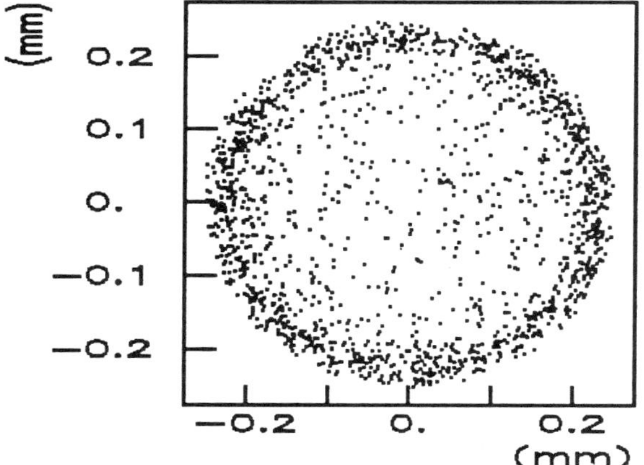

Figure 3. Intensity distribution at end-face of fiber from point-source near interface (mostly skew-rays).

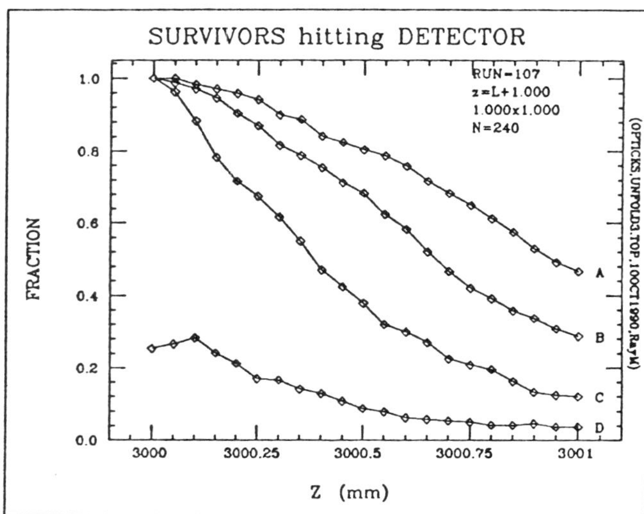

Figure 4. Fraction of photons emerging from the end of the waveguide and accepted by the photodetector.

As for the angular distributions of emergent photons, the general feature is that skew-rays contribute larger angles to the distribution, while meridional-rays tend to contribute more paraxial modes. The mean polar angle is in fact the value given by the numerical aperture. This (and the previous) result is in agreement with intuition and experimental observation of light emerging from the end of a fiber.

Additionally, we have modelled the geometric acceptance of a single-element photodetector by calculating the variation of the collection efficiency of emergent photons with the distance from the end of the fiber to the detector. Perfect quantum efficiency is assumed. Figure 4 shows the results for an area-source of light in the waveguide and for different photosensitive sizes. Detector "A" corresponds to $(1\text{mm})^2$ active area, "B" to $(0.750\text{mm})^2$, "C" to $(0.500\text{mm})^2$, and "D" to $(0.250\text{mm})^2$. Detector "B" is the nominal size of a SSPM active region. Considering the area-source to be a good indicator of average behavior, over 90% efficiency may be maintained out to almost 250μm from the fiber end-face.

CONCLUSIONS

We have begun a systematic study of single scintillating fiber waveguides and have developed a detailed Monte Carlo simulation for this purpose. Preliminary results have been presented. We shall continue the study of bulk absorption effects, time-dispersion, reflective coatings, and bend losses.

REFERENCES

[1] R. Ruchti, et al., these proceedings.
[2] J. Piekarz, et al., these proceedings.
[3] M. Atac, et al., these proceedings.
[4] M. Born and E. Wolf, *Principles of Opticks* (3rd rev ed), Pergamon Press, London, 1965.
[5] RJ Mountain and R Ruchti, "Model of a Scintillating Fiber Optic Waveguide", (in preparation).

MEASUREMENTS OF FLUORESCENCE EFFICIENCY OF SCINTILLATORS BEFORE AND AFTER IRRADIATION*

J.Piekarz
University of Notre Dame

October 22, 1990

Abstract

Various polystyrene-based scintillators were prepared in the Chemistry Lab. of Fermi National Accelerator Laboratory and cut into discs of 1.2 cm dia. and 1.0 cm thick. They were irradiated with a ^{60}Co source to 10 Mrad dose. The scintillation light output and fluorescence and transmittance spectra were measured. After annealing for 10 days in oxygen and 3 days in the air the light output of the best scintillators was 90% of the light output of the original samples.

1 Introduction

This study is part of the effort to develop a high-rate tracking subsystem appropriate for SSC experiments using scintillating fiber technology*. Our goal is to develop highly-efficient and fast scintillators that are resistant to radiation damage.

Common commercial plastic scintillators consist of a polymer base and wavelength shifting fluors.Ionizing particles excite polymer molecules which normally emit in UV region. In the presence of the fluors,however, the energy is transfered to the primary,and often to a secondary fluor, which emits at longer wavelengths.For standard scintillators the emission spectrum peaks in the blue violet region and coincides with the wavelength region most strongly affected by the radiation,thus causing a decrease in light output[1,2,3]. The radiation problem is alleviated in plastic scintillators doped with 3HF(3-hydroxyflavone) emitting green light at 530 nm since at these long wavelengths the radiation induced damage is minimal. One obtains still brighter scintillators using 3HF derivatives[4].

In this study we manufactured various polystyrene-based scintillators using the fluors listed in Table 1. BBQ and K27 emit green light and therefore could lead to radiation hard and possibly bright scintillators. They were coupled to a polystyrene base through the following primary fluors:MOPOM,OLIGO408 and OLIGO415A. The latter two primaries were expected to be radiation hard[5] in polystyrene. p-Terphenyl and DMPOPOP are familiar dyes used for the production of standard scintillators. DAT was tested as a replacement for p-terphenyl because of its high solubility and expected radiation hardness[6].

Table 1. Fluors used in this study.

Fluor[a]	Description
MOPOM(420)	syntesized by J.Kauffman
O415(415)	OLIGO415A[b,c]
O408(410)	OLIGO408[c,d]
BBQ[e](470)	Isoquinoline derivative[f]
K27[e](500)	Benzoxanthene derivative[g]
DAT(375)	di-t-amyl-p-terphenyl[d]
DMPOPOP[e](430)	dimethyl-POPOP[h]
3HF[e](530)	3-hydroxyflavone[h]
pT(360)	p-terphenyl [d]

[a]Wavelength of emission peak in brackets.[b]Exciton,Inc. product Exalite 416.[c]Bridged oligophenylenes. Obtained from [d]Bicron Corp.,[f]Sigma Chem. Co., [g]Hoechst Chem. Co.,[h]Aldrich Chem. Co. [e]Purified by recrystalization.

2 Preparation of scintillators

The styrene was first deinhibited through a column and then purified by vacuum distillation.The glass polymerization tubes were cleaned with nitric and hydrochloric acids,rinsed in distilled water and treated for about 4 h in a 30% solution of dimethyldichlorosilane in chloroform,and finally rinsed with chloroform,methanol,and de-ionized water. This treatment builds a hydrophobic Langmuir layer on the walls of the tube which helps in the removal of the plastic after polymerization. Styrene and fluors were placed in the tubes and degassed with repeated freeze-pump-thaw cycles.The solutions were polymerized in an oil bath at 110°C-24h,125°C-48h,140°C-12h and ramped back down to

[*]This research was conducted by the Fiber Tracking Group,a list of authors can be found in "Scintillating Fiber Tracking for SSC", R.Ruchti,et. al.,these proceedings.

90°C at a rate of 10°C/h. Test tubes were quenched in liquid nitrogen for fast release of the plastic rods and the rods were cut and polished into discs 1.2 cm diameter and 1 cm thick.

At first we prepared scintillators with one fluor only(binary scintillators), varying the amount of the fluor from 0.05% to 2% of the total weight. In Figure 1 we plot the scintillation light output resulting from the excitation of the samples by 1 MeV electrons from a ^{207}Bi source.(In the case of DAT the amount of DAT was varied in the ternary scintillator DAT+0.01%DMPOPOP. DMPOPOP was added to DAT to render the fluorescence visible to a standard phototube). The saturation occurs at 1 percent concentration by weight.This minimal concentration was chosen for primary fluors in ternary scintillators to minimize possible absorption from the fluor.

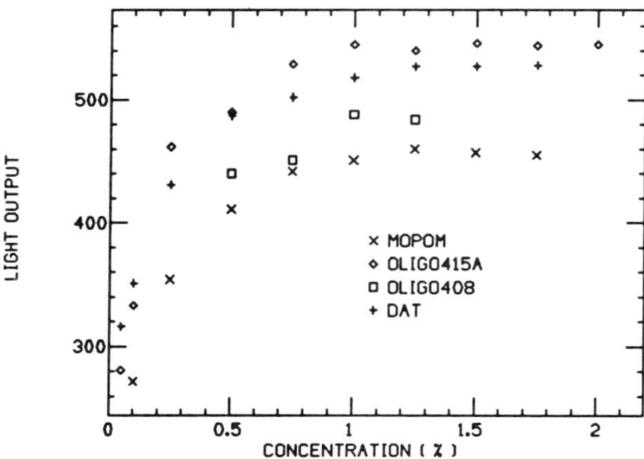

Figure 1.Scintillation light output as a function of weight concentration of the fluor in percent.

Next,the samples with two fluors were prepared with the concentrations of secondary fluors 0.01% and 0.02%. The light output was the same at both concentrations.

The composition of scintillators used for radiation studies is given in Table 2.

3 Measurements and Results

Samples were irradiated in nitrogen gas with a ^{60}Co source at the Nuclear Reactor Laboratory of the University of Michigan at a rate of 1 Mrad/h,to a total dose of 10 Mrad.

To evaluate the scintillators the measurements were performed with both a spectrophotometer and radiation source. The fluorescence and transmittance spectra were recorded with a Hewlett Packard model 8451A diode array spectrophotometer.Fluorescence spectra presented in this paper are from back surface excitation. They were obtained by illuminating one side of the sample with 254 nm light and then measuring the emission spectrum from the other side(samples are 1 cm thick). For the transmission spectra,the absorption of undoped,un-irradiated polystyrene was subtracted out. The fluorescence spectra from back and front surface irradiation with 313 nm light were also taken[7].

Tests with a radiation source used ^{207}Bi conversion electrons and Hamamatsu photomultoplier R669(our photomultiplier has quantum efficiency which is almost constant from 430 nm to 600 nm). The scintillator discs were placed directly on the photomultiplier with immersion oil providing optical contact. The source was put directly on the top of the discs and pulse height spectra were registered with a LeCroy qVt multi-channel analyzer.

The fluorescence and transmittance spectra for scintillators with OLIGO408+K27 and DAT+3HF are presented in Figures 2 and 3.

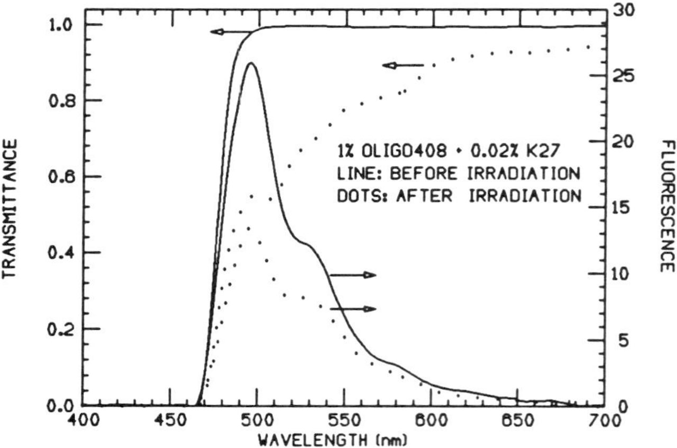

Figure 2. Transmittance and flurescence spectra for the scintillator with OLIGO408+K27.

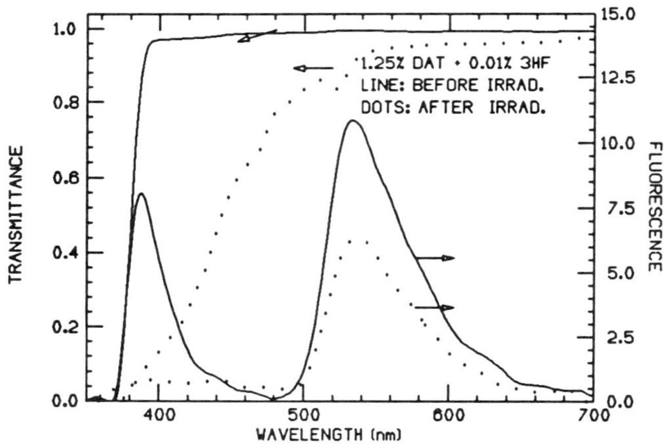

Figure 3. Transmittance and fluorescence spectra for the scintillator with DAT+3HF.

After irradiation ,transmittance in the peak of emission of OLIGO408+K27 and DAT+3HF is 0.5 and 0.9 respectively,which causes a decrease in light output.The decrease in DAT+3HF is however worse than what we would expect from the deterioration of transmittance alone.This effect has been observed before[3]. Similar data were taken for all studied samples.

Both the total area under the fluorescence spectra,and the peak position in the pulse height spectra from ^{207}Bi excitation measure light output of the scintillators. Both measurements agreed to within a few percent except of the annealed samples OLIGO+BBQ(or K27) for which the fluorescence spectra measure 15% more light than the photomultiplier. In Table 2 we give the values of light output measured by the photomultiplier viewing the scintillator samples excited by 1 MeV electrons from ^{207}Bi source. The light outputs from the scintillators before irradiation,after irradiation, and after annealing for 10 days in oxygen and 3 days in air are given in Table 2(in arbitrary units). In the last row we give the light output from the standard scintillator p- terphenyl+DMPOPOP manufactured in our Laboratory and cut to the same size as all studied samples.

Table 2. Light output before and after 10 Mrad radiation dose from ^{60}Co source in arbitrary units.

Scintillator^a	λ^b (nm)	Light output Before	After	After 13 d^c
MOPOM	420	0.79	.04	.42
MOPOM+.01%BBQ	470	0.95	.13	.63
MOPOM+.02%BBQ	470	0.96	.16	.69
MOPOM+.01%K27	500	0.95	.27	.75
MOPOM+.02%K27	500	0.95	.27	.75
O415A	415	0.92	.49	.79
O415A+.01%BBQ	470	1.01	.42	.86
O415A+.02%BBQ	470	1.02	.34	.89
O415A+.01%K27	500	0.97	.57	.87
O415A+.02%K27	500	0.99	.53	.88
O408	410	0.88	.48	.76
O408+.02%BBQ	470	1.02	.34	.87
O408+.02%K27	500	0.97	.57	.89
DAT	375	0.73	.16	.46
DAT+DMPOPOP	430	0.86	.31	.70
DAT+.01%3HF	530	0.85	.40	.63
pT	360	0.84	.21	.41
pT+DMPOPOP	430	0.88		

^{a}Fluor concentrations are by weight,the concentrations of primaries are 1%(1.25% for DAT).^{b}Wavelength of peak emission. ^{c}Annealed 10 days in oxygen and 3 days in air.

The light output of non-irradiated samples is given in the column "Before" of Table 2. All new fluors used in this study produced bright scintillators(DAT emitting at 375 nm is the only exception).The brightest ones are OLIGO+BBQ and OLIGO+K27 and their output is ten percent higher than that from pT+DMPOPOP. Light output after irradiation is given in the column "After".Here we note big differences between various scintillators,OLIGO+K27 being least effected .After irradiation the light output of OLIGO415A+K27 and OLIGO408+K27 is 0.55 and 0.57 compared to 0.88 for non-irradiated pT+DMPOPOP.After annealing for 10 days in oxygen and 3 days in air the above scintillators and OLIGO+BBQ are the brightest ones with average light output 0.88.The next brightest green scintillator is MOPOM+K27 with light output 0.75.We note here that after irradiation OLIGO+BBQ and OLIGO+K27 exhibited phosphorescence in room light with the time constant of minutes. This phenomenon disappears in the annealed samples. By dividing the numbers in the column "After 13 d" by the numbers in the column "Before" ,we compute the ratio of the light output of the annealed scintillators to the light output of the original samples.This ratio is as high as 90 percent for OLIGO+K27 scintillators.

4 Conclusions

We have identified several candidates for radiation hard scintillators :OLIGO415A(or OLIGO408)+K27(or BBQ). Their fluorescence emission peaks at 500 nm (470 nm). After irradiation with a 10 Mrad dose from a ^{60}Co source followed by 13 days of annealing,the light output was observed to be the same as of the non-irradiated standard scintillator pT+DMPOPOP.

Acknowledgements

We thank C.Hurlbut of Bicron Corporation and J.Kauffman of the Philadelphia College of Pharmacy and Science for helpful discussion and for providing some of the fluors used in this work.

References

[1]A.Bross,Proceedings of the Workshop on Scintillating Fiber Detector Development for the SSC,Fermi National Accelerator Laboratory,Batavia,Il., Nov. 14-16,1988,pp.741.

[2]S.Majewski et al.,Nucl. Instr. and Meth. A281 (1989) 500.

[3]A.D.Bross and A.Pla,Polymer Preprints,vol.31,no.2 (1990) p.355.

[4]A.Pla-Dalmau, "Design of Fluorescent Compounds for Scintillation Detection", Northern Illinois University, Dekalb,Il,Ph.D. dissertation,1990.

[5]J.Kauffman,private communication.

[6]J.Kauffman,Proceedings of the Workshop on Scintillating Fiber Detector Development for the SSC,Fermi National Accelerator Laboratory,Batavia,Il., Nov. 14-16,1988,pp.675.

[7]A.Pla ,1990 Nuclear Science Symposium.

FABRICATION OF A STEP-INDEX SCINTILLATING OPTICAL FIBER BY A COEXTRUSION PROCESS

W.-S. Lee and C.-W. Park, Department of Chemical Engineering
&
J. P. Harmon and J. K. Walker, Department of Physics
University of Florida
Gainesville, FL 32611

Abstract

A bicomponent, coextrusion fiber spinning process has been studied theoretically and experimentally. The theory was to model the mechanics of the two-phase flow as affected by the rheological property differences between the two materials which are used for core and cladding, respectively. A small scale experiment has been also performed to confirm the viability of the coextrusion process and to fabricate scintillating fibers for the SSC using various materials.

Introduction

Bicomponent optical fibers have been coextruded for about fifteen years in Japan, Europe and the USA. Attenuation lengths of better than 30 db per kilometer have been routinely achieved. At the SSC, attenuation lengths of 6 db per 3 meters is probably adequate. Thus, it is quite clear that the drawing or spinning methods are both capable of meeting the requirements of the SSC. Up to now, plastic optical fibers have been made with thermoplastic materials which are typically polystyrene (PS) and poly(methyl methacrylate) (PMMA) for core and cladding, respectively. Such conventional plastic scintillators, however, may not be adequate for the SSC due to their radiation damage. For this reason, we initiated a research program to develop a radiation hard scintillator and identified polysiloxane as a base polymer for scintillating plastic which is highly resistant to radiation damage. We evaluated a variety of siloxane polymers based on the monomers of dimethyl, methyl phenyl, and diphenyl siloxane and found that polymethylphenylsiloxane is the most desirable polymer for the SSC.

From the processing point of view, the siloxane polymers we evaluated have various rheological properties. Depending upon the monomer type and the degree of polymerization some are elastomeric with their glass transition temperatures (T_g) lower than the room temperature (T_R), whereas others have a very low viscosity. Since they are not in solid phase at room temperature, a drawing process (from a rod or a preform) cannot be applied to make fibers from these materials. The only alternative may be a coextrusion process, and the major objective of our work reported here is to develop a process of spinning optical fiber whose core material has $T_g < T_R$ or has a very low viscosity.

Through a rigorous modeling study, we have developed for the first time the theory of bicomponent fiber spinning flow. We have shown that the flow mechanics are determined by the characteristics of the cladding material even though it is very thin. The core can have $T_g < T_R$ and/or have a low viscosity. Various predictions for fiber radius profiles and stress distributions have been made.

We also carried out small scale experiments and successfully produced scintillating optical fibers using two different siloxane polymers. Our study showed both theoretically and experimentally that the resultant fiber can be made mechanically stable if the core of the fiber is left as a sufficiently viscous thermoplastic linear molecule. Alternatively, the core polymer can be crosslinked after spinning and again constitute a stable fiber.

Our theory on bicomponent fiber spinning is discussed first in the following section. Then, a brief description follows for the experiment including the equipment set-up and the various polymeric materials we used.

Theoretical Development

The mathematical modeling of a bicomponent fiber spinning process has been already published [1] and here we give only a qualitative description. The objective of the study is to investigate the effects of rheological property interactions on the mechanics of the fiber spinning flow. The cladding materials being developed under our current project include commercial fluoropolymers, PMMA, their blends, and a novel copolymer synthesized in our lab. They are extrusion-grade viscoelastic polymers which are spinnable as a single component. The core materials, on the other hand, are various polyorganosiloxanes with different comonomers (dimethyl, diphenyl, or methylphenyl siloxane). Depending upon their molecular weights and monomer type, some of them behave like Newtonian fluids and not all of them are readily spinnable as a single component. In general their T_g is lower than room temperature T_R. When they are coextruded with a viscoelastic cladding material, however, it may be possible to fabricate optical fibers. The theoretical study addresses how the spinning characteristics of core materials are affected by those of cladding materials and vice versa.

We adopted a simplified model for the study in which core and cladding materials were assumed to be a Newtonian fluid and an upper-convected Maxwell fluid, respectively (Figure 1). The upper-convected Maxwell model is a linear viscoelastic model which provides a first approximation to elastic phenomena in fluids. This modeling study provides a predictive model for the radius and interface profiles (R(z) &

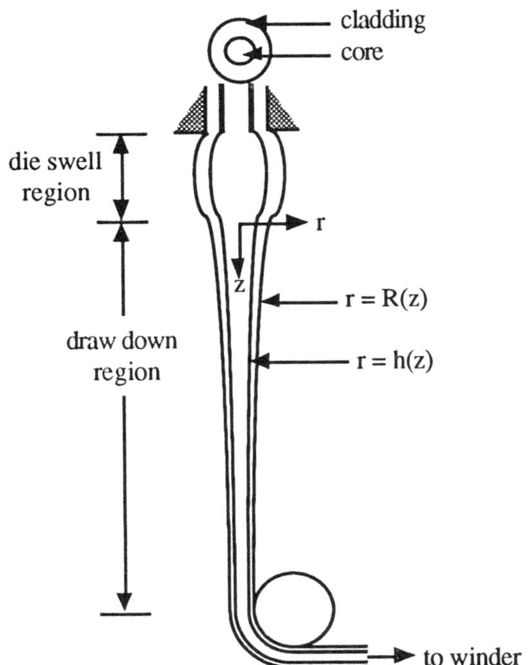

Figure 1. Schematic of a coextrusion fiber spinning

h(z) in Figure 1) as well as the melt stress distribution as a function of the flow rate ratio of the two fluids.

While there have been many studies on the flow mechanics of a single-component fiber spinning [2, 3], a bicomponent fiber spinning flow was never studied before and ours has been the first attempt for such a configuration that is relevant to optical fibers [1].

When a Newtonian fluid is spun as a single component, the velocity profile (or the radius profile, R(z)) is exponential [2]. An upper-convected Maxwell fluid, on the other hand, shows a near linear velocity profile and requires much larger stress (or draw force) to be spun than a Newtonian fluid [3]. When the two materials are coextruded, our results indicate that the mechanics of the flow is dictated by the viscoelastic component (i.e., cladding material) and, more importantly, is relatively insensitive to the flow rate ratio of the two materials. Thus, for typical spinning conditions the flow mechanics is determined by the characteristics of the cladding material although its thickness may be very thin.

Stability of the flow is also a very important factor since radius uniformity is crucial for optical fibers. Newtonian fluids are known to show "draw resonance" when the draw ratio (which is defined as the ratio of the die radius to the fiber radius) is greater than about 20. Draw resonance is an instability phenomenon in which the radius of the fiber fluctuates periodically. Viscoelastic materials, on the other hand, tend to be less prone to draw resonance due to the stabilizing effect of viscoelasticity. As we showed that the flow mechanics of the two-phase flow is dictated by the viscoelastic cladding material, we may speculate that the stability of the flow will also be dictated by the viscoelastic material.

Although the theory is based on a simplified model, the study indicates that radiation hard polyorganosiloxane can be spun into optical fibers with an appropriate cladding material whether the core material has a $T_g < T_R$ or a low viscosity. As will be discussed in the following section, optical fibers in fact could be fabricated using elastomeric as well as a low viscosity siloxane polymers as a core material.

Fiber Spinning Experiments

Equipment

A schematic of our coextrusion equipment is given in Figure 2. The piston-cylinder and the lead screw unit have been custom-made according to our design. Total volume of the cylinder (100 cm^3) is small enough to handle experimental samples that are available only in limited quantities yet large enough to perform pilot scale fiber spinning. The lead screw which drives the piston is driven by a motor which can provide a precise feed volume control of core materials from 4 cm^3/min to 40 cm^3/min.

Coextrusion die design is given schematically in Figure 3. The most important feature of this design is a short thermal contact time between core and cladding materials. While the typical processing temperature of cladding materials is about 170°C, it is desirable to maintain the temperature of the core material below 100°C due to the thermal stability of fluors incorporated into the core material for scintillating fibers. Since the capillary diameter is small (3.2 mm) and its length is short (7.5 cm), the residence time of the core material is small and the material can be kept cool except for a thin region in the immediate vicinity of the capillary wall.

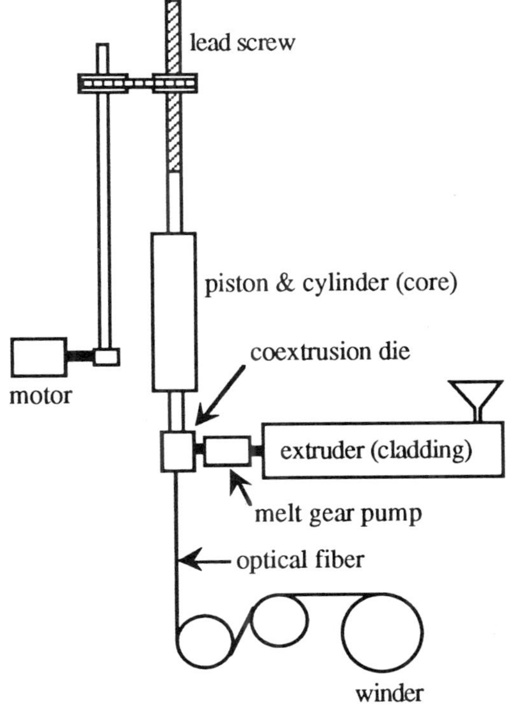

Figure 2. Schematic of a coextrusion equipment

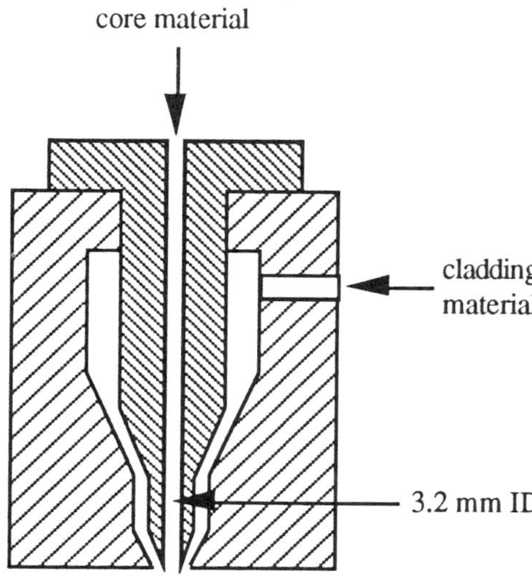

Figure 3. Schematic of a coextrusion die

Optical Fiber Spinning

Small scale fiber spinning was performed using four different core materials and three cladding materials. A PS-PMMA (polystyrene-poly(methyl methacrylate)) fiber was fabricated using a 15 melt-flow PS and a 29 melt-flow PMMA as a core and cladding material, respectively. This experiment was simply to demonstrate that a coextrusion process is also capable of producing the conventional PS-PMMA optical fibers. A 0.8 mm fiber with a 0.1 mm cladding thickness was made and its optical properties are currently under evaluation.

Poly(dimethyl diphenylsiloxane), poly(methylphenyl diphenylsiloxane), and poly(methylphenylsiloxane) were then evaluated with a PMMA or a fluorinated blend as a cladding material. The PMMA we used was a 16.5 melt-flow material and had adequate melt strength and extrudability. Since PMMA is hygroscopic, it was treated in a vacuum oven (12 hours at 90°C) prior to any extrusion to avoid the formation of moisture bubbles during fiber spinning.

The poly(dimethyl diphenylsiloxane) contained 20% diphenyl and was elastomeric. While there were no extrusion difficulties with this material, it showed several streaks of small bubbles at the contact surface with the cladding material upon exiting the coextrusion die. It was due to the volatile residuals (e.g., residual monomers) in the material which evaporated at the process temperature of the cladding materials (180°C). Since this material has a vinyl end group and its viscosity is insensitive to temperature, a vacuum treatment at an elevated temperature is not appropriate. For this type of material a vented extrusion is the most efficient way of devolatilization, and it will be reevaluated with a vented extruder once the equipment is acquired.

The poly(methylphenyl diphenylsiloxane) we used was rather glassy at room temperature. Its viscosity, however, decreased rapidly to a very low value at about 50°C. Nevertheless, this material could be fabricated into defect-free optical fibers. As a next step, we incorporated a fluor into this material in order to investigate if the coextrusion process can have any adverse thermal effect on the efficiency of the fluor. For this experiment 3-HF was added at a 0.4% concentration by weight and a good quality scintillating fiber was produced without any difficulty.

The viscosity of the poly(methylphenylsiloxane) was very low even at the room temperature. While a higher molecular weight (thus, a higher viscosity) material is currently being developed, we tried to fabricate optical fibers using this low viscosity material. Due to its very low viscosity a liquid gear pump was used instead of the cylinder-piston unit described in Figure 2. As was the case with the poly(methylphenyl diphenylsiloxane), defect-free fibers could be made without too much difficulty.

Uniformity of Fiber Radius

Uniformity of fiber radius was investigated using a micrometer which has an accuracy of 0.01 mm. For a one-meter long fiber the outermost diameter was measured at 10 different locations which are 10 cm apart. For an average diameter of 0.88 mm, the standard deviation was 0.02 mm. While this level of uniformity may be tolerable, it should be pointed out that a melt gear pump (which was not yet available for the reported experiment) will significantly improve the radius uniformity.

Summary

Our study on a coextrusion fiber spinning process has shown that a variety of core materials with various rheological properties can be spun into optical fibers with a proper choice of a cladding material. As our theory predicted, it is due to the fact that the mechanics of the bicomponent spinning flow is controlled by the viscoelastic cladding material under typical spinning conditions. While the conventional coextrusion process is for extruding only thermoplastic materials, our research has opened the way to explore a variety of core material options for optical fibers whose long term radiation and mechanical stability will be optimized for use at the SSC.

Acknowledgement

This work was supported by the Office of Superconducting Super Collider of the U.S. Department of Energy under Contract No. DE-FG05-90ER40547.

References

[1] C.-W. Park., "Extensional Flow of a Two-Phase Fiber," *AIChE J.*, **36**, 197-206 (1990).

[2] M. A. Matovich and J. R. A. Pearson, "Spinning a Molten Threadline, Steady-State Isothermal Viscous Flows," *Ind. Eng. Chem. Fundam.*, **8**, 512-520 (1969).

[3] M. M. Denn, C. J. S. Petrie, and P. Avenas, "Mechanics of Steady Spinning of a Viscoelastic Liquid," *AIChE J.*, **21**, 791-799 (1975).

AVALANCHE PHOTODIODES AS PLASTIC SCINTILLATING FIBER READOUT DEVICES

S. Reucroft, H. Fenker*, A. Grimes, M. Hulbert and I.D. Leedom
Northeastern University, Boston MA

H. Dautet and R. McIntyre
General Electric Canada Inc., Vaudreuil

W. L. Dunn and J. D. Simpkins
Quantum Research Services Inc., Durham NC

Abstract

We present the first results of tests of the Avalanche Photodiode as a readout device for plastic scintillating fibers.

Introduction

The development of Plastic Scintillating Fiber (PSF) oriented detectors at the SSC requires research into two principal areas, namely the properties of the fibers themselves, and the method by which they are to be read out. The low cost, high speed and good spatial resolution of PSF are qualities well suited to the requirements of tracking detectors; the problem is then to find a readout device able to perform to a similarly high standard without becoming prohibitively expensive when produced on the grand scale dictated by the SSC.

Our initial work in the field of scintillating fibers has resulted in the successful installation and operation of detectors in the L3 experiment at CERN and in E774 at Fermilab, with the arrays of fibers being read out by multiple-anode devices: microchannel-plate multi-anode tubes at L3, multi-anode photomultiplier tubes at E774. Whilst effective, these devices rely on a bi-alkali photocathode, which has the disadvantage of a relatively low quantum efficiency (typically 20%), thus limiting the overall efficiency of the system. Bi-alkali photocathodes also suffer from a relatively narrow spectral response. Experience with these devices has lead us to investigate the avalanche photodiode, which with its improved efficiency (silicon devices can have quantum efficiency of up to $\approx$80%), broader spectral response and compact size, presents an inexpensive solution to the problem of selecting a scintillating fiber readout alternative.

APD Operation and Characteristics

In evaluating the performance of the avalanche photodiode (APD) we used as our test device the RCA C30921S, which has sensitive area 0.5mm in diameter packaged in a $\approx$5mm by $\approx$5mm cylindrical can. Depending upon the operating voltage, the diode can function in two distinct modes. When biased below its breakdown voltage, the diode yields an output approximately proportional to the intensity of incident light. However, since the magnitude of the signal in this 'linear mode' is relatively small, it is advantageous to bias the diode a few volts above its breakdown voltage. When operated in this 'Geiger mode' light of arbitrary intensity triggers an avalanche within the diode allowing current to flow, resulting in a pulse of much greater magnitude. While sacrificing linearity, this mode is still quite suitable for reading out arrays of fibers as part of a tracking system (wherein a single bit is adequate to record a hit or miss fiber).

We utilized the APD in its Geiger mode, the advantage being that, as hinted above, since the ouput signal is relatively large (enough to drive a 50Ω cable and subsequent processing electronics) there is no need for any amplification at the diode itself. The APD was tested using a simple 'passive quenching' circuit. Here, the diode is placed in series with a load resistance, actually two separate resistors, of (say) 500kΩ and 5kΩ, respectively, which enables it to quench via an RC circuit after each avalanche. This arrangement is then biased above the APD breakdown voltage (around 220V for these particular models) and the output signal measured across the smaller resistor. The output is very similar to that produced by a photomultiplier tube (PMT) and depends

* Superconducting Super Collider Laboratory, Dallas, Texas. Operated by Universities Research Association, for the U.S. Department of Energy under Contract No. DE–AC02–89ER40486.

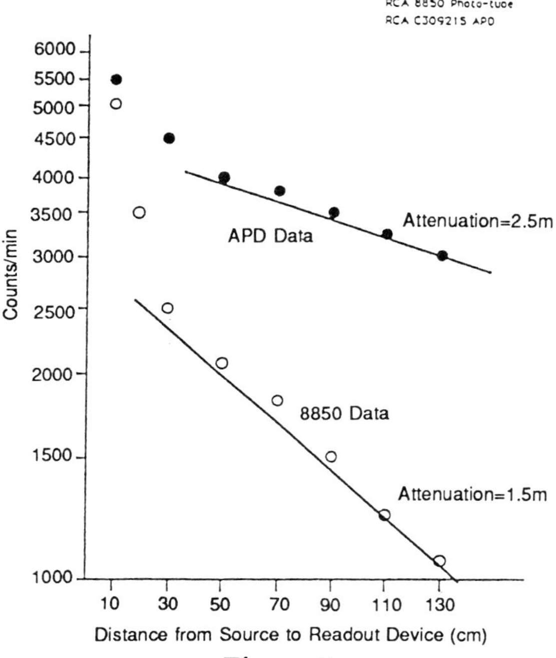

Figure 1

upon the operating voltage; biased at 10V above breakdown, for example, the pulse height is typically around 25mV.

Light Curve

The APD was mounted in one end of a tube-like structure, and the end of a scintillating fiber (set into a cylindrical 'core') pressed up against it from the other; optical grease was used to minimize the light-loss at this APD-fiber interface. As a fundamental measure of the APD performance, we compared its light curve, that is the rate of events triggered by a radioactive source placed above the fiber, as a function of the distance between source and detector, to that of a standard PMT under identical conditions.

A 2mCi Ru^{106} source was placed above the 0.5mm Kyowa 4B test-fiber, 30cm from the interface between fiber and APD. Two further PMT's were used to read out a pair of trigger fibers placed directly above and below the test fiber; only coincident events from all three fibers were counted so as to eliminate any signals not initiated by radiation from the source. After aligning the test fiber with the face of the APD (see below), the source was moved in 10cm increments away from the detector, and the count rate, the number of coincidences per minute, was recorded. By plotting this against source distance (see figure) the distance in which the rate dropped by a factor of 1/e was determined. This attenuation length is found to be around 250cm when the APD is used to read out the test fiber, but only 150cm when the procedure is repeated using an RCA 8850 PMT in place of the APD. The improved light curve of the APD is a direct result of its excellent spectral response. After an event, blue light traveling down the fiber undergoes self-absorption, followed by re-emission at a longer wavelength; this process continues until the light reaches the detector. Moving the source away from the detector increases the shift towards the red end of the spectrum, and the count-rate from the PMT falls off quite rapidly, since the spectral response of the 8850 is virtually zero above $\approx$650nm. The attenuation seen by the APD is less severe, since the APD is relatively efficient over the entire visible spectrum and beyond, and is thus able to detect the red-shifted events.

Efficiency and Alignment

The efficiency of the APD in detecting light from a plastic scintillating fiber was then compared with that of the PMT. Initially the test fiber was read out by the RCA 8850 PMT, and the count rate recorded. When the 8850 was replaced with an APD, the count rate was found to vary considerably as the fiber was rotated with respect to the face of the APD, since the two were not centered precisely. The APD was found to be between 55% and 108% efficient in detecting events seen by the 8850 PMT depending upon fiber/diode alignment - i.e. upon the amount of light from the 0.5mm diameter fiber falling on the sensitive area of the APD, also 0.5mm in diameter. This efficiency could clearly be improved with better centering.

Timing Characteristics

Inherent to the so-called passive-quenching circuit employed here is a certain dead-time following each avalanche, during which the APD is blind to further incoming photons. The large load resistance, which limits the current through the diode, ensures that the APD ceases to conduct immediately after breakdown; it is thereby quenched. However the circuit does take a certain time to recharge, approximately equal to its RC time constant. Taking $C = 2pF$ and $R = 500k\Omega$, we expect this dead time to be around $1\mu s$.

An experimental measure of the dead-time was found by sending a pair of pulses, separated by a variable time interval T, to an LED adjacent to the face of the diode. This procedure was repeated a large number of times (with at least $100\mu s$ between one pair and the next so as to allow the APD to fully recover) and the number of corresponding signals registered by the APD was recorded. For large T (several μs) almost 100% of the pulses were detected. As T is decreased, one expects this percentage to fall, since when the LED flashes the second time, the APD is still recovering from the initial discharge; consequently the likelihood of detecting the second event in each pair is reduced. This was borne out by our results which showed the percentage of events detected dropping to around 50% at around $T = 1\mu s$. By reducing the load resistance, and thereby the RC time constant, this measure of the dead-time was decreased to around 500ns. This time may still be too long for some tracking purposes; one possible remedy is to employ an 'active quenching' circuit (reset pulse) to accelerate the recovery process.

Dark Current: Temperature Effects and Radiation Hardness

The APD dark count rate, approximately 2000Hz at 22°C, decreases by an order of magnitude for each 20° drop in temperature.[1] Irradiation with neutrons pushes up the dark count. For example, after a fast neutron fluence of $\approx 10^7 \text{cm}^{-2}$, the count rate increases by almost an order of magnitude.[2] Though the APD exhibits a slow recovery (about 10% per month), this increase in noise probably prohibits the installation of APD's too close to the SSC beam pipe (for example, within the calorimeter of an SSC experiment).

Future Considerations

Bearing in mind the large number of fibers necessary for PSF systems, the logical extension of using individual APD's to read out single fibers is to create an array of APD's capable of servicing a fiber ribbon (i.e. a flat strip of precisely spaced fibers cemented together.) G.E. has recently constructed such an array, comprising four RCA C30902S diodes in a single unit. It is forseeable that arrays of 32 or 64 APD's could follow later. The present device, in addition to the bias line and the four output lines, has a reset line to facilitate active quenching of the diodes, enabling their dead-times to be reduced. Also undergoing constant development is the design of the APD itself, in particular to improve its efficiency and increase the diameter of its sensitive area or incorporate a tapered light-guide to allow the readout of 1mm diameter fibers.

Conclusions

We started to test the suitability of existing avalanche photodiodes for plastic scintillating fiber readout purposes. Though aware that better devices are coming, and that the present ones have not yet been adequately optimized, we have nevertheless found the initial results to be very encouraging.

When considering the applicability to the SSC, all present difficulties seem readily solvable. By active quenching, the dead-time of these devices may be reduced from hundreds to tens of ns. Since the APD is sensitive to neutron damage, it must be positioned in a region where the neutron fluence is low ($< 10^8 \text{cm}^{-2}$) and thereby keep the dark current at an acceptable level. Other problems, such as the delicate ones of alignment, the splicing together of fibers, and packaging costs, are all part of a broader set of PSF considerations. However, the actual detectors should cost no more than a few dollars per channel.

References

[1] R. Brown et al., Applied Optics **25**, 4122, (1986).

[2] W. L. Dunn et al., "Radiation Damage Studies of Straw Tube and Scintillating Fiber Elements," in this Proceeding, (1990).

PRELIMINARY RESULTS ON THE USE OF AVALANCHE PHOTODIODES FOR SCINTILLATING FIBER READOUT

M. Salomon and J. Kitching
TRIUMF, 4004 Wesbrook Mall, Vancouver, B.C., Canada V6T 2A3

Abstract

We have measured the properties of avalanche photodiodes for possible use in conjunction with scintillating fibers as charged particle detectors.

The quantities measured so far have been the gain of the APD's, the amplitude distribution of pulses from particles through the fibers, and the efficiency of detecting these particles.

The preliminary results obtained indicate the importance of the choice of associated electronics and good optical coupling between fiber and APD. The results obtained so far have been measured at room temperature and $-100°$ C.

Introduction

Scintillating Fibers offer a very attractive method to detect charged particles because of their speed and their small granularity, allowing very good position information. They have already been used in several large experiments[1] and in many small ones as well. The fibers are now being manufactured by several commercial companies, and their properties are very well known. Although the idea is rather old[2] their development and applications in high energy physics are relatively recent.

A major reason for this delay, and still a problem nowadays, is the difficulty of providing a readout method that shares the advantage of the fibers themselves, i.e. to develop a photodetector that is fast, cheap, with very low noise, insensitive to magnetic fields and with very good spacial resolution. Such a device is still elusive. Several candidates have been tried[1] among them Image Intensifiers coupled to CCD's[3], large arrays of photomultipliers[4], position sensitive photomultipliers[5], *etc.* None of these satisfy all the criteria mentioned above, and for this reason we have started our studies of Avalanche Photodiodes (APD's), which in principle have the possibility of doing so. We should mention here that another device that has great promise is the Solid State Photomultiplier (SSPM)[6] but it has the disadvantage of operating near LHe temperatures. The major interest in these detectors is the need to satisfy the difficult operating conditions of detectors at the SSC for which the scintillating fibers offer a very attractive solution.

The use of APD's to detect scintillation light is not new. They have been used recently with inorganic scintillators at room temperature with very good results[7]. What makes the use of APD's with fibers difficult is the low level of light to be detected, typically 10 to 50 photons. A very good analysis of this problem as well as measurements and possible design improvements in the APD's have been described[8]. Previous work on the coupling of scintillating fibers to APD's was done by Borenstein *et al.*[9]

Our preliminary work in this field was done to determine the capabilities of Avalanche Photodiodes, their properties and possibilities for use in scintillating fiber detection systems. This work is not exhaustive and represents only a first step in our study of APD's coupled to scintillating fibers. A comprehensive discussion of the properties and operation of APD's has been published by one of the manufacturers[10]. Further work related to several properties of APD's, such as operation in the avalanche mode, enhancement of response at short wave lengths, single photon detection and different structures and arrays, can be found in Ref. 11.

Experimental setup

Two types of APD were used, the C30817 manufactured by RCA and the S2383 SPL manufactured by Hamamatsu. The results presented in this paper were obtained from the latter because its mechanical construction allows closer proximity of the fiber to the diode.

The S2383 has the following properties. The active surface is 1 mm diameter, the photoefficiency at 420 nm is 15 to 20%. The dark current is 0.5 nA at room temperature, the junction capacitance 7 pF and a gain of up to 100, although it could be operated at a gain of 200 before the onset of the avalanche mode. In this diode the shot noise increases faster than the signal gain at gains of about 100. Consequently there is an optimal bias at which to operate the device.

A scintillating fiber of square cross section 1×1 mm^2 and 30 cm long (Kyowa SCSF-81) was coupled to the APD on one end and to a photomultiplier (Phillips XP2252) at the other end as shown in Fig. 1.

A second photomultiplier coupled to a small ($1\times1\times0.5$ cm^3) scintillator was used in coincidence with the fiber signals. The source used was ^{106}Ru providing electrons from beta decay with a maximum energy of 3.1 MeV. There was also a pulsed light source (LED) used initially to measure the gain of the APD, which in turn was thermally coupled to a cold finger which was immersed in liquid nitrogen for the low temperature measurements. A small resistor near the APD was used to stabilize the temperature at various values between LN_2 and room temperature.

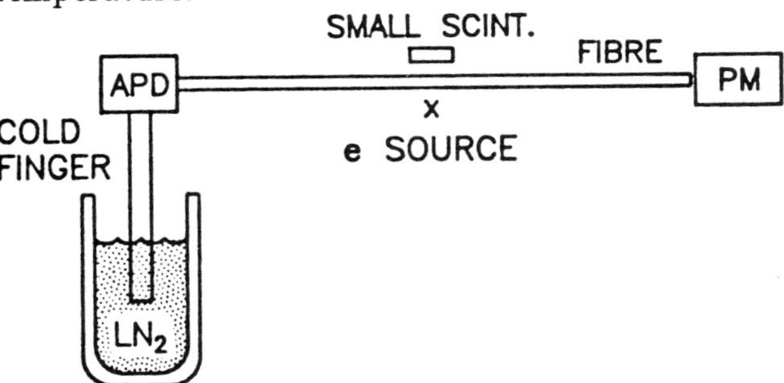

Figure 1. A 30 cm scintillating fiber of 1 mm square cross-section is viewed at one end by an APD and at the other by a photomultiplier (PM). The electron source (Ru 106) produces minimum ionizing electrons traversing the fiber and stopping in another scintillator also viewed by a PM. The APD could be cooled to LN_2 temperatures by means of a cold finger.

The signals from the APD were amplified first by a charge integrating preamplifier (Ortec 142-A) and then by a linear amplifier (Canberra 2010).

We started by measuring the number of photons emitted by the fiber for minimum ionizing particles (mip). Using the photomultiplier signals and the single electron peak as calibration an average of 6 photoelectrons was obtained for mip's, compatible with previous results[12,3]. Using a photoefficiency of 25% in the PM, we obtain an average of 24 photons at a distance of 15 cm. The attenuation of light in a fiber is the result of several different processes and the number of photons at the end of a fiber as a function of the distance to the source of light is a superposition of several exponentials. In our fiber the intensity reduced to half in approximately 20 cm.

The numbers mentioned above are consistent with a scintillator efficiency of 1 photon/120 eV and a light trapping efficiency of 3.5%.

Gain Measurements

Light pulses of fixed amplitude were injected by a light emitting diode (LED), and by varying the bias of APD we determined the gain at several temperatures. The results are shown in Fig. 2. Beyond gains of approximately 200 the diode started to operate in the avalanche or geiger mode, that is the

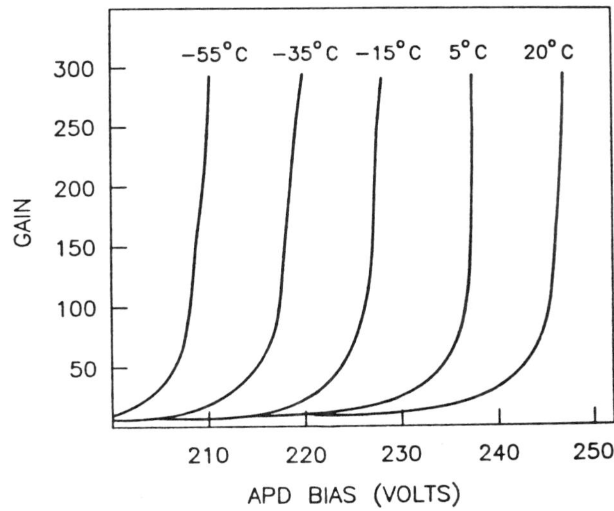

Figure 2. The gain in the APD measured with an LED shining light pulses at the fiber. The bias dependance for several temperatures is shown.

device was not linear and the signals increased in size by a factor of 100 or more. These results indicate one area of difficulty. If one wants to have a fixed gain it is imperative to have very good temperature control and bias voltage stability as the gain curve is very steep. This problem can be reduced by using a compensating current source to bias the APD.

The signals from the APD had a rise time of 3 to 5 ns and a decay time of about 20 ns. When the gain increased and the APD started operating in the avalanche mode, the pulses became much longer, and decay times fluctuated about several microseconds.

Amplitude Spectra

We measured the amplitude distribution of the signals of the APD after amplification. The best shaping time was found to be 250 ns, for largest signal to noise ratio. The signals were collected in a MCA gated by the coincidence of both PM's. The result is shown in Fig. 3. Also included in the spectrum is the superimposed peak of a pulser injecting signals in the preamplifier. The width of the peak indicates the magnitude of the electronic noise. In the same figure we include the spectrum of the noise signals, obtained as a singles spectrum in a separate measurement at the same bias and temperature, and normalized to have the same counts per channel at the cut-off point.

The shape of the coincidence spectrum in Fig. 3 is similar to previous calculations and experimental observations[8] performed to study the response of APD to a low number of photons. These distributions are found to deviate significantly from a guassian shape.

We compared the shape and number of noise pulses above the threshold of the MCA and found that for equal gains of 100 at low temperature (-105°C) and at room temperature (25°C), there is a difference of only 34% and that the amplitude distributions are similar.

Efficiency Measurements

For practical applications of the APD it is necessary to determine its efficiency. This was done by comparing the number of triple coincidence APD.SC1.SC2 and the double coincidences between both PM's SC1.SC2. Random coincidences were measured with the delay method and subtracted. This was done at two temperatures +22° C and -105° C as a function of the bias applied to the APD. The results are shown in Fig. 4. Clearly the best results are obtained at the lower temperature, but

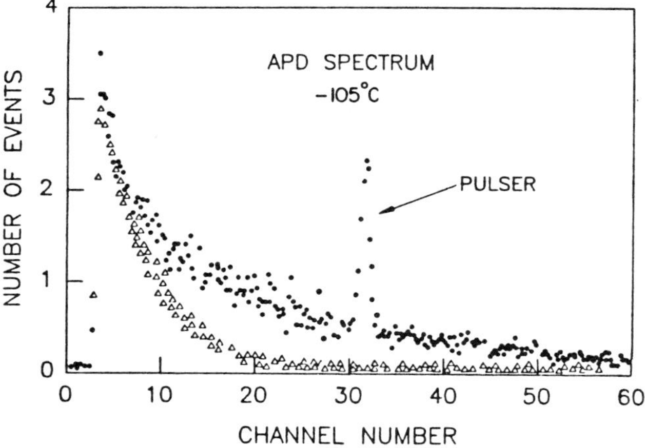

Figure 3. The pulse amplitude spectrum of the APD after integration (250 ns) and amplification. The signals were collected in a MCA gated by the coincidence signals from both PM's. The peak is superimposed by injecting pulses in the preamplifier. Its width measures the electronic noise of the system. The triangles are the noise spectrum obtained as singles at the same temperature and bias.

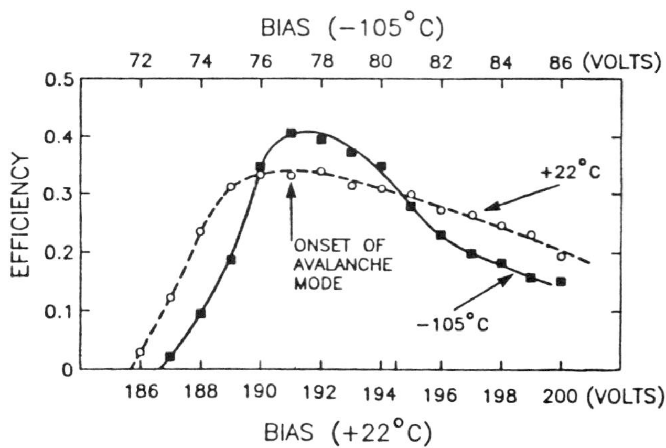

Figure 4. The efficiency obtained from the ratio of coincidences between. APD and both PM's and the PM's rate, after subtraction of accidentals. These ratios are plotted for several biases of the APD, and at two temperatures. The arrow indicates the transition to avalanche mode at both temperatures.

the best value of the efficiency (40%) is still rather low.

The reason for the efficiency decrease at larger biases is due to the onset of avalanche mode operation. The number of carriers is so large that the electric field in the APD is distorted and will only recover with the characteristic RC of the circuit[13]. This recovery time could be reduced by active quenching circuits.

It was suggested[13] that when APD's are operated in the avalanche mode, they also produce light pulses. We tested this by looking at coincidences between the APD operating in the geiger mode and the photomultiplier in the absence of the electron source. We couldn't find any coincidences down to the single photoelectron level of the photomultiplier.

The values of efficiency obtained here with a 1 mm fiber are much lower than the value of 80% obtained by Borenstein[9] with a 2 mm fiber and operating the APD in the geiger mode at –70° C.

Summary

We describe above our preliminary results of the use of commercial APD's to detect the light from scintillating fibers. The device tested seems to be marginally usable, although the efficiency obtained is still too low. Several improvements should be possible: It is important to minimize photon loss. This can be done by decreasing the distance between the fiber and the diode and by increasing somewhat the diode area.

The electronics should be tailored to the APD. Here, the preamplifier was designed for larger capacitance diodes. Cooling the input FET should also be considered.

Methods to increase the photoefficiency at short wavelengths[8] should be attempted, or fibers with longer wavelength emission and more photons per MeV should be manufactured. We have tried several commercial types with peak emissions from 500 to 650 nm but the number of photons emitted by the fibers was much too low.

References

[1] Workshop on Scintillating Fiber Detector Development for the SSC (Fermilab, Nov. 1988).

[2] G.T. Reynolds, IEEE Trans. Nucl. Science **NS-7**, (1960) 115
H.C. Borrows, Nucl. Electronics **1** (1962) 153.

[3] R.E. Ansorge, *et al.*, Nucl. Instr. and Methods, **A265**, (1988) 33.
R. Ruchti *et al.*, IEEE Trans. Nucl. Science **NS-35** (1988) 441.
A. Konaka, Nucl. Instr. and Methods, **A256** (1987) 70.

[4] J.S. Frank, R.C. Strand, Ref. 1, p 359, vol I

[5] H. Akbari *et al.*, Ref. 1 p1061, Vol II.
K. Kuroda *et al.*, Ref. 1, p 429, Vol I.
M. Salomon, *et al.*, Ref. 1, p 409, Vol. I and IEEE Trans. Nucl. Science **NS-36**, (1989) 94.

[6] M.D. Petroff and M. Atac, IEEE Trans. in Nucl. Science **NS-36** (1989) 163.
M. Atac, Ref. 1 p 841 Vol II.

[7] A. W. Lightstone *et al.*, IEEE Trans. in Nucl. Science **NS-33** (1986) 456.
G. Entine *et al.*, IEEE Trans. in Nucl. Science, **NS-30** (1983) 431.
S.E. Derenzo *et al.*, IEEE Trans. in Nucl. Science, **NS-30** (1983) 665.

[8] R.J. McIntyre, IEEE Trans. on Electron Devices, **ED19** (1972) 703.
J. Conradi, IEEE Trans. on Electron Devices, **ED19** (1972) 713.

[9] S.R. Borenstein *et al.*, Invited talk at the Int. Conf. on Experiments at LEP (Uppsala, Sweden, June 1980) IEEE Trans. in Nucl. Science, **NS-29** (1982) 402, and IEEE Trans. Nucl. Science, **NS-31** (1984) 396.

[10] P.P. Webb, R.J. McIntyre and J. Conradi, RCA Review **35** (1974) 234.

[11] R.J. McIntyre, IEEE Trans. on Electron Devices, **ED20** (1973) 637.
R.J. McIntyre, Measurement **3** (1985) 146.
M. Trakalo *et al.*, Applied Optics **26** (1987) 3594.
A.W. Lightstone *et al.*, NASA Laser Light Scattering Workshop, Cleveland, Ohio, 1988.

[12] C.M. Hawkes *et al.*, Internal Report CALTECH (CALT-68-1598, November 1989) and Nucl. Instr. and Methods **A292** (1990) 329.
T.O. White, Nucl. Instr. and Methods, **A273** (1988) 820.

[13] A.W. Lightstone and R.J. McIntyre, OSA Proc. on Photon Correlation Techniques and Applications, (Washington, D.C., May 1988).

REVIEW OF DETECTOR DEVELOPMENT WORK AT THE UNIVERSITY OF FLORIDA

J.K. Walker, J.P. Harmon, C.W. Park, S. Li
The University of Florida
Gainesville, Florida 32611-2085

Abstract

The dominant mechanism for optical polymer degradation due to radiation has been identified. As a result of this understanding new optical polymers have been synthesized which are extremely resistant to degradation due to radiation. Fast, good light output plastic scintillator, resistant to 20 MRad has been developed. Clad scintillating fibers and plates have been produced. A novel avalanche photo diode with ten times the gain, one-tenth the noise and at least ten times the radiation resistance has been designed, fabricated and tested. These developments open up new vistas at the Superconducting Super Collider (SSC).

Introduction

We briefly describe the achievements of the University of Florida detector development group. Major breakthroughs have been made in optical polymers, material processing and Avalanche Photo Diode (APD) development. It is anticipated that these breakthroughs have the potential to transform the way in which scintillating plastic detectors are constructed and readout at the SSC. This paper summarizes three talks to be given by Drs. J.P. Harmon, C.W. Park and S. Li.

Polymer Development

The scintillating plastic technique for particle detection at the SSC is attracting greater attention due to its fast time response. At a luminosity of $10^{33}\,cm^{-2}sec^{-1}$, the radiation levels at the SSC are high and the anticipation of even higher luminosity means even greater demands upon long-term reliable operation of scintillator detectors. Consequently, there is an acute need for improving the stability of the optical transparency of polymers which are subject to a large, long-term exposure to ionizing radiation.

An optical polymer, such as polystyrene, suffers damage to its transparency in two ways—Type I color centers are generated in the plastic which very slowly disappear (anneal) with time. These centers are dominant at the wavelength range of interest in plastic scintillators. Type II color centers appear to be permanent, or at least much longer-lived, and are dominant only in the u.v. portion of the spectrum.

We ahve identified the physical mechanism for the generation of Type I centers. Ionizing radiation produces scission and cross-linking in the polymer matrix. The free volume associated with this new matrix is redistributed in a way which introduces density fluctuations and induces stress in the glassy polymer. We have demonstrated that elimination of this stress totally eliminates the existence of Type I color center production. Elastomeric optical polymers meet the necessary condition for elimination of this stress. We have synthesized some polymers of this type which exhibit extreme optical stability to radiation degradation.

We have speculated on the nature of color center annealing associated with radiation-induced stress in the polymer. Color centers may be trapped in regions of stress-induced voids. These centers will be quenched when stressed segments relax and the centers come in contact with oxygen. Alternatively, there is a second possible explanation. The induced density fluctuations constitute a non-uniform electron density distribution in the polymer. Light traversing this polymer is scattered in a manner described by Rayleigh scattering from the non-uniform electron distribution. This may account for the approximate $1/\lambda^4$ behavior of light absorption in damaged plastics.

Stress, and associated non-uniform density distribution, in the polymer is slowly annealed with time. Higher temperature speeds the recovery process in a manner described by Arrhenius behavior. A discontinuity in the recovery occurs at the polymer glass transition temperature. At this temperature, the polymer backbone motion and segmental motions are rapid and transparency is recovered essentially instantaneously. This is a key demonstration of the validity of the proposed mechanism of damage to optical polymers. Much more work is required to confirm the specific mechanisms of optical absorption.

As a consequence of the above measurements and identification of the polymer damage mechanism, we modified the chemical structure of polystyrene and PMMA to lower their glass transition temperatures below room temperature. These new polymers exhibit no damage in the visible range for a 10 MRad radiation exposure. The mechanical properties of these polymers can be controlled by their degree of cross-linking. They are flexible cross-linked elastomers which do not develop internal stress and hence do not develop Type I centers. Three years ago, we discovered that siloxane rubber had extreme optical resistance to radiation damage[1]—we now understand the physical basis for this characteristic.

These new radiation stable optical elastomeric polymers open the way to a new era of stable scintillation counters at the SSC.[2] There is a second reason why elastomers are preferable to glassy polymers for use at the SSC. We have found that the radiation-induced stress in glassy polymers can produce the small surface cracks known as crazing. This may be particularly true and critical for fibers which are sometimes used in situations with pre-existing stress due to small radii of curvature.

For both of the above reasons, namely optical and mechanical stability, it appears that elastomeric polymers offer advantages for scintillation counters at the SSC. There is one consideration which must be mentioned. The mobility of dyes in an elastomer is approximately two orders of magnitude greater than in a glassy polymer. In this situation, careful measurements of dye concentration must be performed to ensure adequate stability of performance. If required, the dye can be covalently bound to the polymer matrix to ensure stability. This has been done in the past and does not represent a problem.

We have made stable elastomeric scintillator with light output constant to $\pm 10\%$ at radiation exposures up to 20 MRad. Light output is comparable (within 20%) to that of polystyrene-based scintillator. The measured fluorescence time of the scintillation is about one nanosecond and is described by a single exponential over three decades in light intensity.

Fibers and Plates

We have developed for the first time the theory of bicomponent fiber extrusion. The theory modeled the mechanics of the two-phase flow as affected by the rheological property differences between the two materials which are used for the core and the cladding. We were particularly interested in the theoretical model results for core materials whose glass transition temperatures were below room temperature due to their demonstrated optical and mechanical stability. We have shown that even in this case the flow mechanics are determined by the characteristics of the thermoplastic cladding material even though it may be very thin. After extrusion, the core material can be cross-linked.

In recent months, we have put into operation a bicomponent extruder for optical fibers to test the new core and cladding materials and our theory of the process. Preliminary fibers have been produced. The conditions for 1) adequate polymer purity, 2) dust-free air, 3) extruder cleanliness, 4) fiber diameter uniformity, and 5) bubble-free core, etc. are beginning to be understood. Most importantly, fiber production with the new stable elastomeric polymer core material and cladding material follows our theoretical predictions rather well and we see no fundamental difficulties in fiber production. We have made both 1 mm diameter scintillating fiber and clear read out fiber. At present, due to the variety of experimental parameters identified above we are limited to optical attenuation lengths of about one-half meter. We expect this to substantially improve in the coming months. We have also succeeded in casting good quality elastomeric plates with embedded wavelength shifting fibers. This technology is of great interest to the major detectors being considered for operation at the SSC.

Avalanche Photo Diode (APD) Development

Reading our fibers from either tracking detectors, pre-radiators or the wavelength shifting fibers from calorimeter plates is a challenging problem at the SSC. There are various candidate technologies. In the last year, we have developed a novel ultra high gain, low noise and radiation hard device. It is based on an AlGaAs/GaAs Separated-Absorption Multiplication (SAM) avalanche photodiode. It incorporates an effective mass filtering superlattice absorption layer to enhance the value of electron-to-hole ionization rate ratio to a value of 25 at an electric field strength of 1.25×10^6 volt/cm. Under these conditions, the maximum gain factor is about 6,000. The measured quantum efficiency is in excess of 70%. The peak in quantum efficiency response can be tailored anywhere in the range 400–800 nm. The high gain factor of the new APD structure permits the use of low-cost pre-amplifiers which are not at the frontier of low-noise operation.

This novel APD[3], compared to commercial devices, has: 1) a factor of ten more gain, 2) one-tenth the noise, and 3) a factor of ten-to-one-hundred the radiation resistance. We intend to develop these devices in the form of 1) arrays for pre-radiator electromagnetic calorimetry and possibly tracking purposes, and 2) larger sensitive areas for read-out of multi-fiber bundles from calorimeter towers. The device immunity to magnetic field, extremely large linear dynamic range ($> 3 \times 10^4$), small space requirements and lack of need of cryogenic fluids suggest they will find an important application at the SSC.

In summary, we have made major progress in developing plastic scintillator with extreme optical and mechanical stability. We have produced scintillating fibers and plates from this elastomeric plastic. A novel avalanche photodiode has been developed to detect light from scintillating fibers.

References

Note: Technical references are provided in the three contributions to this Symposium of which this contribution is a summary.

[1–3] The University of Florida has filed for patents for this technology.

CRYOGENIC PREAMPLIFIERS FOR SSPM/VLPC DEVICES FOR SCINTILLATING FIBER TRACKING*

Roy C. Chaney
University of Texas at Dallas, P. O. Box 830688
Richardson, TX 75083-0688

Abstract

The application of scintillating fiber technology for particle tracking [1] requires the development of solid state photodetectors, such as the Solid State Photomultiplier (SSPM) and the Visible Light Photon Counter (VLPC) under development at Rockwell International Science Center [2]. These devices are operated at about 7° K, and therefore the design of front-end electronics operating at cryogenic temperatures is needed. We report on the development of six alternative conceptual designs for the cryogenic front-end electronics using traditional JFET technology and GaAs FETs currently under development.

Introduction

Six alternative conceptual designs for the cryogenic front-end electronics of the Scintillating Fiber Tracking Subsystem for the SDC detector [3] were developed. We first studied the various design possibilities to determine which ones are compatible with the following electronic and thermal constraints of the system:

(a) The preamplifiers must operate at or below 77° K.

(b) The pulse width must be less than 16 nsec, and the rise time 5 nsec or less.

After detailed discussions with manufacturers of fast cryogenic electronic devices we developed six alternative conceptual designs (Fig. 1). The first two of them use traditional silicon JFET technology, while the other four are based upon Gallium Arsenide FETs currently available in prototype form.

We have also identified manufacturers to provide cables for cryogenic temperatures. We propose to use cables made from 2 mil SS304 stainless steel wires laminated in Teflon for the interconnection between the SSPM's and the wall of the cold He gas chamber. Teflon has very good characteristics at liquid helium temperatures and does not bond to the stainless steel wire. This eliminates stress produced by the different thermal expansion of the two materials. The stainless steel-Teflon ribbon will have a relatively low thermal conductivity. An alternative approach to the stainless steel-Teflon ribbons might be stainless steel-silicon-rubber ribbons. We have used copper silicon-rubber ribbons before in cold hostile environments, in the Apollo moon flight program.

* This research was conducted by the Fiber Tracking Group, a list of authors can be found in "Scintillating Fiber Tracking for SSC", et. al., these proceedings.

We propose to use flexible copper-Kapton multilayer ribbon for the region from the cold He gas chamber out to the room temperature region. With copper trace widths as low as 4 mils, minimum heat transfer can be assured between devices placed in the cryogenic compartments. Surface mount preamplifier/amplifier devices can be directly applied to the multilayer ribbon cable.

Comparison of the Conceptual Designs

The various conceptual designs to be evaluated and tested are shown in Fig. 1. The first uses a single silicon JFET surface mounted on the copper-Kapton cable. It has the advantage of minimum heat transfer dissipation, and is the simplest device to implement, using mainly off-the-shelf components with minimum design cost. The third design is very similar to the first one, but uses GaAs FETs rather than silicon. The advantage of the third design over the first is the significantly faster rise time of the GaAs device. The second conceptual design shown in Fig. 1 has the advantage of a single smaller preamplifier/amplifier running at cryogenic temperature. This results in a lower noise level and more compact packaging. It does, however, restrict the design to the use of components which can operate at cryogenic temperatures. Placing the amplifier inside will also, of course, produce additional heat load for the cryogenic system. These same remarks also apply to the fourth design, which differs only in the use of GaAs devices instead of silicon JFETs. The most promising design will most likely be either design five or six. They use GaAs devices operating at helium temperature and directly coupled to the SSPM/VLPC devices. These GaAs devices have been tested below 10° K and have the additional advantage of eliminating the capacitance of the interconnecting cable. This fact will result in a system rise time limited only by the impedance of the SSPM. We are, however, concerned about the additional heat load produced by placing the amplifier system in the helium chamber. The difference between conceptual designs five and six addresses this concern.

The above two devices have the additional important characteristics of consuming relatively low power and have low impedance output (less than 100 ohm).

It is our opinion that these devices will enable the fiber tracking detector prototype to be constructed and tested at cryogenic temperatures and 60 MHz operation. The preamplifiers have additional important applications at room temperatures in the silicon tracker readout system and in some of the calorimetric devices to be used in the SDC detector.

Acknowledgments

This work was supported by the Department of Energy and the Texas National Research Laboratory Commission.

References

[1] *Proc. of the Workshop on Scintillating Fiber Detector Development for the SSC*, Vol. 1 & 2 (Nov. 1988)

[2] M. D. Petroff and M. G. Stapelbroek, IEEE Trans. Nucl. Sci. NS-36, 158 (1989); M. D. Petroff and M. Atac, IEEE Trans. Nucl. Sci. NS-36, pp. 163 (1989)

[3] Solenoidal Detector Collaboration, Expression of Interest, 24 May 1990

DESIGN ALTERNATIVES

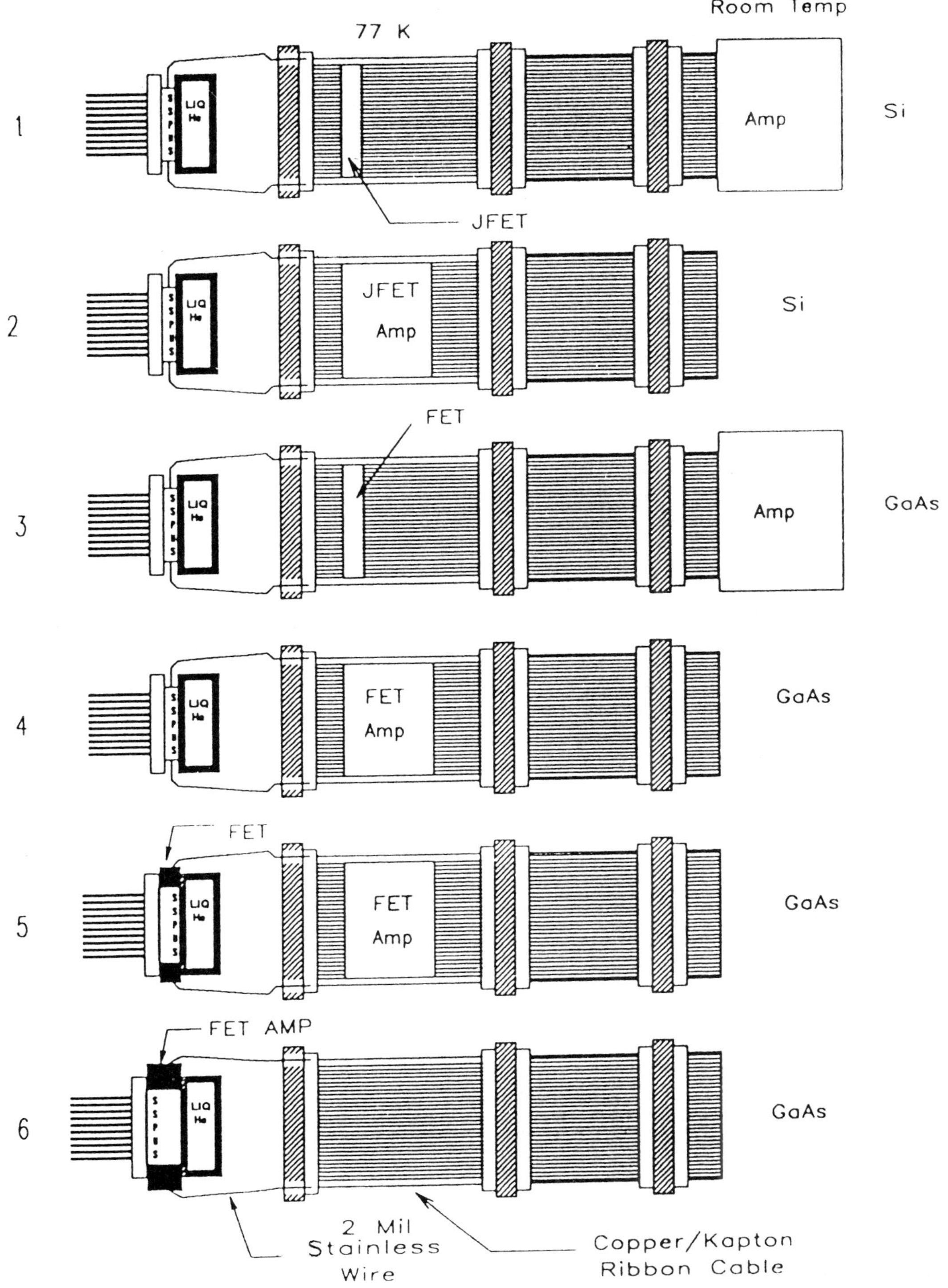

Figure 1. Various conceptual design alternatives for cryogenic preamps to be used in conjunction with VLPC/SSPM devices: 1-2 use silicon JFET technology; 3-6 are based on GaAs FETs currently available in prototype form.

CRYOGENIC ISSUES FOR SOLID-STATE PHOTOMULTIPLIER OPERATION*

H. Goldberg, S. Margulies, and J. Solomon
Department of Physics
University of Illinois at Chicago
Chicago, IL 60680

Abstract

A tracking system for the SSC employing some 10^6 scintillating fibers has been proposed. This system is made possible by the recently developed solid-state photomultiplier, a multichannel device that is extremely small and can detect single photons. However, since the device must be operated near 7°K, a cryogenics system is required. This paper presents a simple conceptual environment for the cold region, and estimates the resulting heat loads.

Introduction

The solid-state photomultiplier, developed by Rockwell International Science Center, may revolutionize photon detection in high-energy physics. Capable of detecting single photons, a multichannel array can have individual sensitive areas a fraction of a millimeter on a side.[1] A version called the visible light photon counter (VLPC) currently being optimized for high-energy-physics applications will have a quantum efficiency of 80% and a gain of 5×10^4. Based on this technology, the Solenoid Detector Collaboration has proposed, as one option, a tracking system employing some 10^6 scintillating fibers each of which is read out by a VLPC channel.[2]

Along with its promise, the VLPC presents challenges. Foremost, to keep noise to a reasonable level, the device must be maintained near 7°K. For a large system, several liquid-helium-temperature cryostats are required, each involving the penetration of thousands of entering fibers and exiting signal leads. Also, cold preamps located near the VLPCs are desirable.

This paper presents a simple conceptual environment for the cold VLPC volume and estimates the heat loads presented to the liquid-helium cryogenics. The estimates can form the basis for an initial design of the cryosystem.

Conceptual Design

A conceptual design for the cold volume is shown in Fig. 1. Wave-guide fibers enter a vacuum region and are thermally grounded at liquid nitrogen temperature (77°K). They continue through a second vacuum region into the 7°K volume and are coupled to 96-channel VLPC arrays. The VLPC output signals pass through a vacuum region to preamps operating at 77°K. Finally, amplified signals emerge into a room temperature environment.

The fibers have a 750-μm-dia. polystyrene core and a thin acrylic cladding, and extend 5 cm between the 77°K and 7°K regions. The VLPC outputs are sent on 15-cm-long ribbon cables made of 2-mil-dia. SS304 wires spaced at 25 mils and embedded in 10-mil-thick Teflon. A similar two-conductor cable 10 cm long supplies a 7V bias to each photodetector array.

Heat Load Estimates

These per-channel estimates consider only heat in the 7°K VLPC region. Neither heat transferred to the liquid nitrogen nor static cryostat heat loads are considered here.

---

*Work performed in collaboration with the Fiber Tracking Group; a list of members may be found in "Scintillating Fiber Tracking for SSC", R. Ruchti *et al.*, these proceedings. Supported in part by the Superconducting Super Collider Laboratory/ U.S. Department of Energy and the Texas National Laboratory Research Commission.

Conduction

Neglecting heat losses through their sides, the steady-state rate of conductive heat transfer through the fibers and electrical cables is determined by $P_c = -k(T)A\ dT/dx$, where A is the cross-sectional area and k(T) is the thermal conductivity at temperature T. Integration over a length L gives $P_c = AK/L$, where K is the integral of k(T) dT between 7°K and 77°K.

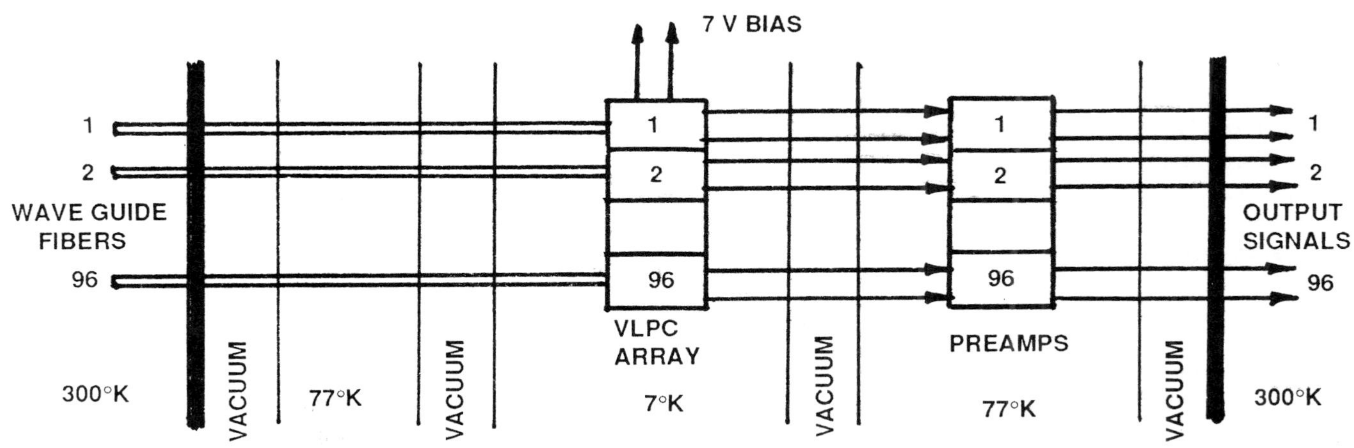

Figure 1. Schematic diagram of the cryostat system.

The values of K used were calculated using data from Ref. [3]. For a polystyrene fiber, where K = 2.34×10^{-2} W/cm, there follows $P_c = 2.1\times10^{-5}$ W. For the signal and d.c. bias cables, the contribution from the wires (K = 3.14 W/cm for SS304) is 8.6×10^{-6} W and that from the Teflon (K = 0.13 W/cm) is 2.8×10^{-5} W. The average cable rate is then 3.7×10^{-5} W per channel.

Radiation

Visible photons, produced by hits on the scintillating sections of the fibers, have typical energies of 3.3×10^{-19} J. Because a maximum ionizing hit due to a neutron-induced recoil proton should produce about 2.6×10^{3} photons at the VLPC, the energy transfer is negligible.

Heat from infrared photons is difficult to calculate. Basically, each section of a fiber is in thermal equilibrium with black-body radiation of the corresponding temperature, and photons may be trapped and propagate. As an extreme case, if 3% of the 300°K radiation was trapped and propagated without loss, the heat load would be 2.7×10^{-4} W. However, although the infrared absorption spectrum of polystyrene is complicated, little transmission is expected over the distances involved.

VLPC Heat

Electron multiplication in a VLPC produces heat. As a worst case, consider scintillation light from a 2.7-m-long fiber in the innermost tracker superlayer located 0.6m from the beam. A minimum-ionizing particle should produce about 10 photons at the VLPC, and a recoil proton-- from a hit by an albedo neutron -- will yield about 2.6×10^{3} photons. With the stated VLPC parameters, because the avalanche charge is collected in about 1 ns,[1] the corresponding signal currents are 6.4×10^{-6} A and 1.7×10^{-2} A.

A layer at 0.6m radius contains 5.0×10^{3} fibers. At SSC operation at $\mathbf{L} = 10^{33}$ $cm^{-2}s^{-1}$, 1.6 interactions are expected per crossing, with each interaction producing some 150 charged particles. Doubling this value to include spiraling particles and conversion electrons, the probability that a fiber is hit by a minimum-ionizing particle is 9.6×10^{-2}/crossing. Assuming a 1-ns-wide VLPC current pulse, the 16 ns crossing period results in a duty factor of 1/16. These values yield a time-averaged signal current of 3.8×10^{-7} A/channel.

At the above **L**, the fluence of the neutron albedo in the tracking volume will be $0.5 \times 10^{12}/cm^2$ in one year,[4] producing an inner fiber hit rate of $3.2 \times 10^5\ s^{-1}$. The np elastic cross-section at 1 MeV, the peak of the neutron spectrum, is 6 barns. Thus, the probability that a neutron hit will produce scintillation via a recoil proton is 2.0×10^{-2}. These values result in a time-averaged signal current of 1.1×10^{-7} A/channel.

Because the quiescent current is 10^{-8} A,[1] the total average VLPC current is thus 5.0×10^{-7} A/channel which, with a bias potential of 7V, corresponds to 3.5×10^{-6} W of heat. At $\mathbf{L} = 10^{34}\ cm^{-2}s^{-1}$, both the interactions per crossing and the neutron fluence increase tenfold, and the heat rate is 3.4×10^{-5} W.

Joule Heating

Joule heating, $P_j = RI^2_{total}$, is produced in the output signal and bias cables. At 7°K, the resistance of a pair of output leads is 73Ω; that of the two conductor bias cable is 49Ω.

The output current is the sum of signals due to minimum-ionizing particles and neutrons. For estimation, assume the pulses to be 1 ns wide and have amplitudes and frequencies as described above. Because the two signals are uncorrelated, their cross terms vanish when the time average of I^2_{total} is calculated. At $\mathbf{L} = 10^{33}\ cm^{-2}s^{-1}$, this average is $1.1 \times 10^{-9}\ A^2$/channel, due principally to neutrons. The resulting joule heating is 8.1×10^{-8} W. At $\mathbf{L} = 10^{34}\ cm^{-2}s^{-1}$, this value increases to 8.8×10^{-7} W.

One bias cable supplies all 96 channels in a VLPC array. The current consists of pulses due to minimum-ionizing particles and neutrons superimposed on the quiescent current. The signal pulses are assumed to be as described above. Again, in calculating the time average of I^2_{total} their cross terms vanish; however, their cross terms with the quiescent current contribute. At $\mathbf{L} = 10^{33}\ cm^{-2}s^{-1}$, the per-channel value of the average of I^2_{total} is $1.1 \times 10^{-7}\ A^2$, due principally to neutrons. The corresponding joule heating is 5.4×10^{-6} W. This value increases tenfold at $\mathbf{L} = 10^{34}\ cm^{-2}s^{-1}$.

Conclusion

The individual heat loads estimated for the assumed configuration are summarized in Table 1, as are their sums. For a tracking system of some 10^6 fibers, the total load at a luminosity of $10^{34}\ cm^{-2}\ s^{-1}$ is about 150 W. This is quite manageable since, for comparison, one Fermilab superconducting ring magnet requires about 10 W of refrigeration.

Table 1. Summary of heat loads.

Mechanism	Head Load (W/channel) $\mathbf{L}=10^{33}cm^{-2}s^{-1}$	Head Load (W/channel) $\mathbf{L}=10^{34}cm^{-2}s^{-1}$
Conduction		
Fibers	2.1×10^{-5}	2.1×10^{-5}
Cables	3.7×10^{-5}	3.7×10^{-5}
Radiation		
Visible	negligible	negligible
Infrared	negligible	negligible
VLPC Heat	3.5×10^{-6}	3.4×10^{-5}
Joule Heating		
Signal Cable	8.1×10^{-8}	8.8×10^{-7}
Bias Cable	5.4×10^{-6}	5.4×10^{-5}
Total	7.0×10^{-5}	1.5×10^{-4}

References

[1] M.D. Petroff et al., "Detection of individual 0.4-28 μm wavelength photons via impurity-impact ionization in a solid-state photomultiplier", Appl. Phys. Lett. 51, 406-408, 1987; M.D. Petroff and M.G. Stapelbroek, "Photon-counting solid-state photomultiplier", IEEE Trans. Nucl. Sci. 36, 158-162 (1989).

[2] "Expression of Interest by the Solenoid Detector Collaboration", 24 May 1990, pp. 22-31, (unpublished).

[3] Y.S. Touloukian, R.W. Powell, C.Y. Ho, and P.G. Klemens, "Thermophysical Properties of Matter". New York: IFI/Plenum Press, Vol. 1, pp. 1160-1162; Vol. 2, pp. 963-969.

[4] Ref. [2], p. 19.

OPTICAL SUBSYSTEMS FOR SCINTILLATING FIBER READOUT

James K. Woosley, Alphonsus J. Fennelly, and Melinda B. Johnson
Technology Applications Department
Teledyne Brown Engineering, MS-50
300 Sparkman Drive, NW
Huntsville, AL 35807-7007

Abstract

Scintillating fiber elements suffer high signal loss rates due primarily to their high intrinsic absorbtivity. The utility of scintillating fibers for certain applications can be increased if the output signal strength is increased. Two approaches are presented which have the potential for increasing the output signal strength significantly: (a) inclusion of optical gain media within the fiber; and, (b) use of nonlinear optical systems with gain, such as the bistable Fabry-Perot etalon. These approaches are discussed conceptually, with a description of development requirements and concerns.

Introduction

The use of scintillating fiber elements in particle detectors has become standard in high energy physics in the past decade (see, for example, [1], [2], and [3]). However, scintillating fibers suffer from two intrinsic signal loss mechanisms: emissions outside the critical angle, as shown in Figure 1, are lost to absorption or transmission through the cladding[1]; and, scintillations on-axis suffer high absorbtivity, particularly after absorption of radiation dosages of order 10 MRad (note, however, several reports elsewhere in these proceedings which demonstrate that the absorbtivity problem has been at least partially solved). These attenuation problems are such that UA2 [1] reports only 23 photons for minimum-ionizing particles (MIPs) transiting their fibers 1.3 m from the detector end of their 2.4 m tracking fibers.

While such photon counts are adequate as a hit indicator, the loss mechanisms limit the full potential utility of scintillating fiber tracking systems. For example: (a) the use of long fibers is restricted by signal attenuation; the use of thicker fibers with intrinsically poorer position resolution becomes necessary in applications in which fiber length is important (such as the longer tracking elements in a SSC tracking detector); (b) the low signal strength substantially eliminates any capability toward axial position determination through timing and dispersion in long fibers; (c) stronger signals simplify the porting of signal to radiation-sensitive detector elements; (d) significantly larger signal strengths may permit the use of optical processing [4] in the tracking trigger.

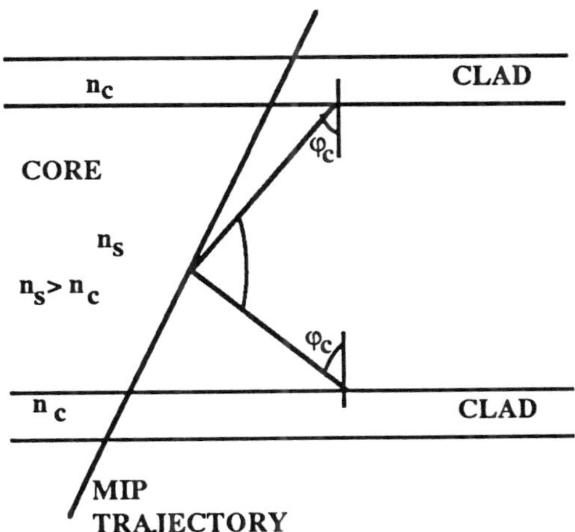

Figure 1. Acceptance angle of ordinary scintillating fiber

The present report describes two conceptual approaches to increasing the output signal strength of scintillating fibers. The first concept involves the use of embedded optical gain media, such as erbium-doped fibers or semiconductor diode lasers (SDLs), perhaps in conjuction with frequency shifters, to increase the optical signal on output from the fiber. The second concept involves the use of the Fabry-Perot etalon as an optical amplifier. These concepts are described in the succeeding sections.

---

[1] Consider a standard scintillating fiber cross-section; due to transmission by and/or absorption in the cladding material, light outside the forward cone represented by total internal reflection is lost to the cladding. The backward cone may be temporarily ignored due to the high intrinsic absorbtivity of the medium. The fractional forward transmission is thus

$$f = \frac{n_s - n_c}{n_s}$$

where n_c (n_s) is the index of refraction of the cladding material (scintillating material). For typical plastic scintillating fibers [2], this leads to f = 2.5%, and an effective source strength of order 50 photons per millimeter of fiber diameter.

Embedded Gain Media

Due to the fact that optical fibers for communications systems have attenuation lengths short compared to system length, considerable attention has been devoted to the development of repeater systems to compensate for this attenuation. All repeater systems currently in operation make use of an intermediate electrical step; however, two approaches have been developed which may eliminate (or alleviate) the need for an electrical step.

Semiconductor Diode Laser

Since a laser intrinsically represents an optical gain medium (see Figure 2), the use of an embedded semiconductor diode laser (SDL) represents the most obvious approach to introducing gain into scintillating fibers. Such approaches are described in the literature with respect to communications systems [5], where travelling-wave mode signal gains of 20 dB at a bandwidth of 20 THz are believed to be possible. These capabilities have been shown at 1500 nm wavelength, however; extrapolation to typical scintillation fluor wavelengths (shorter than about 650 nm) has not been performed to date.

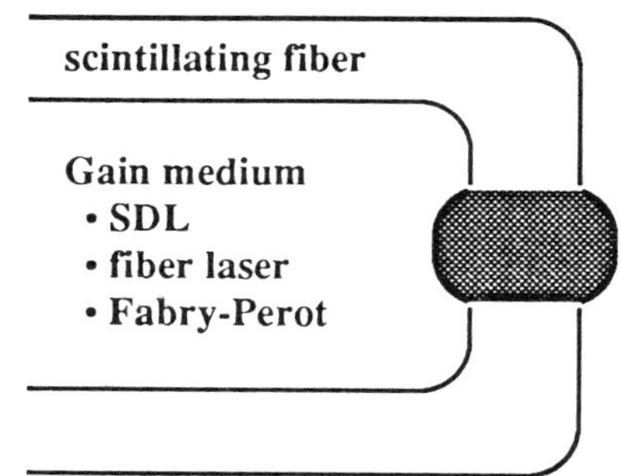

Figure 2. Conceptual view of scintillating fiber containing gain media

Other issues regarding the use of SDLs for signal amplification include:

• Size. Can a small SDL amplifier be constructed; for tracking applications, the SDL radial and azimuthal profile must match the fiber within reasonably close tolerance in order to decrease dead space within the tracking system.

• Power. SDL operation will require incorporation of an electrical power source into the design.

• Cost. The single most significant factor; this approach will require at least one SDL amplifier per channel, perhaps more. While this approach may achieve the desired gain, the increased channel costs may be prohibitive for SSC applications.

On the whole, the SDL approach is believed to be of high risk at this time; this statement is primarily driven by cost considerations.

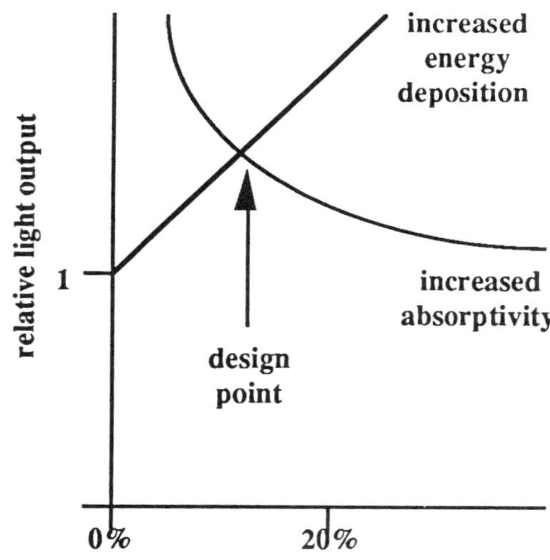

Figure 3. Increased light output due to high-Z dopant

Fiber Amplifier and Fiber Laser

Considerable attention [6,7] has been devoted of late to the potential use of fiber amplifiers and fiber lasers as self-contained communications systems with a gain factor (from the amplifying or lasing fluor) to compensate the attenuation within the fiber. Most popular attention has focused on triply-ionized erbium (Er^{3+}) and neodymium (Nd^{3+}). However, terbium (Tb^{3+}) is particularly interesting for scintillation applications; it has three broad lines in the region 520-620 nm, potentially compatible with amplification of the signal from the new low-loss scintillating fibers with fluors in this region.

The main issues involved in the development of scintillating fiber lasers are chemistry and fiber architecture. No work has been performed on the integration of scintillating and lasing fluors in the same constitutive media (silica, plastic, ...). While there is no evident *a priori* reason why this approach should not work, laboratory confirmation of chemical compatability is required. It may also be necessary to develop a specialized fiber architecture, consisting of a series of embedded lasers within the fiber, in the same form as the embedded etalons discussed below and elsewhere in these proceedings.

Another conceptual approach to in-fiber amplification bears mention: doping of the fiber with a high-Z material should increase energy deposition within the fiber, and hence, potentially, light output at the source. This approach is somewhat limited; as the fraction of high-Z dopant increases, the fiber transmittivity will in general be decreased, limiting the utility of the approach (Figure 3). However, with a doubling of energy deposition in a fiber with 20% by mass of a high-Z material, the potential gains of attempting high-Z doping are clear. It is also worth mention that a high-Z dopant may increase the effectiveness of scintillating fiber calorimetry. This suggestion makes lead the obvious first choice for a high-Z dopant; however, concerns about the radiation softness of lead glasses may require the use of another dopant (bismuth, tungsten, and depleted uranium are the most obvious alternates).

The costs of a scintillating fiber doped with multiple materials – terbium, bismuth, ... – will clearly be larger per unit length than a fiber not so doped. However, in large system costing, the advantages of potentially being able to reduce the total number of, for example, tracking readouts by a factor of two to four while maintaining adequate azimuthal segmentation must be factored into the development cost.

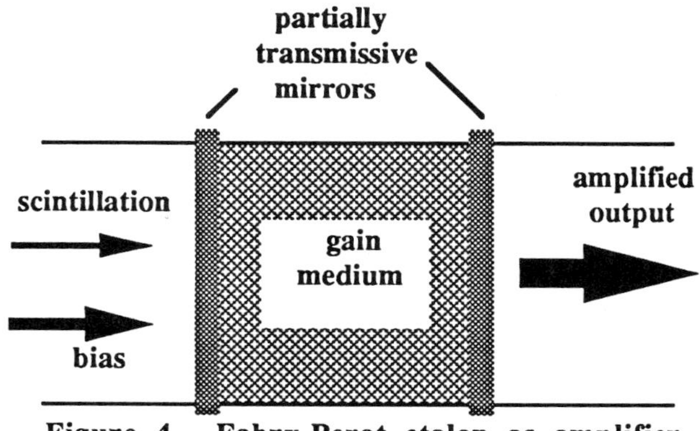

Figure 4. Fabry-Perot etalon as amplifier

Nonlinear Gain

Another, not unrelated, approach to scintillating fiber signal amplification is the use of embedded Fabry-Perot etalons as amplifiers. Operating in this mode, the etalons (Figure 4) consists of a pair of partially transmissive mirrors separated by a nonlinear optical medium, such that [7]

$$n = n_0 + n_2 I_C.$$

Consequently, the etalon is optically bistable; that is, for a certain range of input light intensities, there are two allowable output light intensities, permitting (for example) a gain profile such as the one shown in Figure 5. Operation requires that a subthreshold bias photocurrent be provided to the etalon; the increased signal represented by the

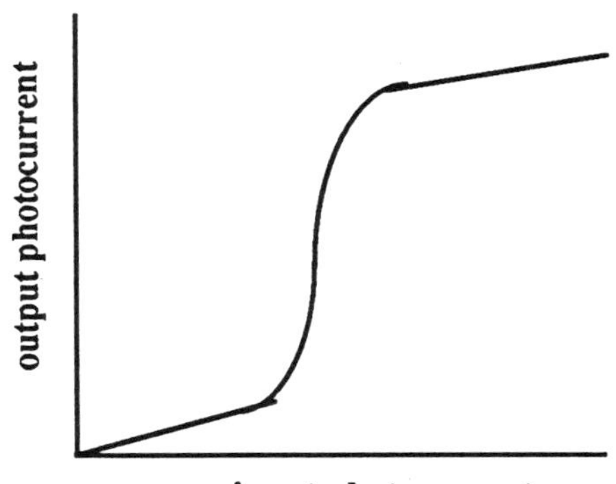

Figure 5. Gain from Fabry-Perot etalon in amplifier mode

scintillation light should be sufficient to trigger the transition to the high-output mode of the etalon, with consequent signal amplification. Fabry-Perot etalons offer high speed operation (sub-picosecond), but require large amounts of energy for large-scale operation.

Two uses of the Fabry-Perot etalon are possible in scintillation tracking systems. In the first approach, as represented above in Figure 2, the etalon is embedded within the fiber and serves as a repeater; spacing of etalons will be determined by experimental circumstances in order to ensure adequate gain. The energy requirements in this serial mode are somewhat mitigated by the fact that, to within fiber absorbtivity, the bias current is linearly additive. This approach also has the consequence of introducing dead space into the detector system. In the second approach, the etalon is placed on the output and is configured to permit high optical gain in the output. Depending upon desired signal output strength, short strings of etalons may be concatenated.

Conclusions

A variety of approaches which may be used to amply the output signal of scintillating fibers have been presented. The lowest cost risk is believed to be associated with the development of in-fiber amplification techniques involving the use of terbium doping and doping with a high-Z material. The use of optically bistable amplifying devices such as the Fabry-Perot etalon, embedded within the fiber, also has the potential of increasing signal strength with minimal increase in channel cost. It must always be realized that the cost should be viewed in the context of minimizing total detector system cost, by using fibers with higher output signal strength in conjunction with a reduced number of detector readout channels at constant azimuthal segmentation.

References

1. J. Alitti *et al.*, Nuc. Inst. and Meth. A283 (1989), 646.
2. R. Ruchti, *et al.*, IEEE Trans. Nuc. Sci. 35 (1988), 441.
3. S. Dye, *et al.*, A Calorimeter-Based High-Rate Detector for the SSC (TEXAS), Boston University, May, 1990.
4. M. B. Johnson, *et al.*, these proceedings.
5. T. Kimura, in T. Suematsu (ed.), JARECT 11 (1984),57.
6. "Erbium-Doped Fibers: Tommorrow's Repeaters – An Interview with Emmanuel Desurvire," Lasers and Optotronics, May, 1990.
7. I. Andonovic and D. Uttamchandani (ed.), *Principles of Modern Optical Systems*, Artech House, 1989.

Liquid Ionization and Silicon Calorimetry

4

RECENT DEVELOPMENTS IN SILICON CALORIMETRY†

James E. Brau
Department of Physics, University of Oregon, Eugene, OR 97403

Abstract

We present a survey of some of the recent calorimeter applications of silicon detectors. The numerous attractive features of silicon detectors are summarized, with an emphasis on those aspects important to calorimetry. Several of the uses of this technology are summarized and referenced. We consider applications for electromagnetic calorimetry, hadronic calorimetry, and proposals for the SSC.

We present a review of the applications of silicon detectors to calorimetry in high energy physics experiments. The use of silicon detectors in high energy physics is an expanding activity with new applications emerging frequently. Emphasis will be on recent applications and developments. The presentation is subdivided into three topics: 1.) electromagnetic calorimetry, 2.) hadronic calorimetry, and 3.) the proposals for applications at the SSC as presented in the Expressions of Interest.

Introduction

Experiments at the Superconducting Super Collider (SSC) will drive calorimetry to new levels of performance. As the dominant element of the central detection system, the calorimeter faces the exceptional experimental conditions created by a very high luminosity, high multiplicities, and the large variety of physics phenomena to study. The characteristics demanded of these calorimeters will include: compact construction, flexibility, fine segmentation, fast charge collection, easy calibration, good resistance to radiation damage, good energy resolution, and a compensating (e/h=1) or nearly compensating response. With respect to these issues of performance, silicon detectors offer attractive capabilities either in stand-alone systems or as complementary components with other calorimeters, as in the present applications of plug calorimeters, small angle calorimeters, and electron/hadron separators.[1]

Excellent signal to noise ratios for single minimum ionizing particles (mip) can be achieved with an active layer thickness of less than 500 μm. Since thin sampling gaps are sufficient, silicon instrumented calorimeters can be made very compact. In fact they may be the only choice where limited space does not permit other solutions. Additionally, these detectors can be used in normal laboratory environments and at room temperature. No special provision, such as a vacuum tank or a cryostat vessel, is necessary. Therefore, the construction of calorimeters with nearly perfect hermeticity is feasible. The technology of silicon detector fabrication enables a tailored geometry with a fine lateral and longitudinal segmentation. Another advantage is the overall modular design of calorimetric devices possible with this technique. Thus the sampling geometry can be easily changed and optimized in test setups. Dead cells resulting from radiation damage or electronic failures are in principle replaceable. Due to their small thickness of less than 500 μm and the high ohmic resistivity of about 5-10 kohm-cm, silicon detectors can be operated at comparatively low bias voltages. In most cases total depletion, that is the extension of the electric field throughout the detector thickness, is achieved at or below 100 V. Like other active media based on ionization sampling, silicon detectors exhibit a direct charge readout. Since their response is practically nonsaturating, independent of particle type, and exhibits perfect charge collection, a direct and absolute calibration of the deposited visible energy is possible and can be readily accomplished using, for example, an alpha source. The Hall effect is not expected to play a significant role especially for thin, large area detectors. Therefore, the signal response is nearly insensitive to magnetic fields. The high value of the average electric field which is established in these thin detectors at a low operating voltage, together with the comparatively high mobilities of the charge carriers (electrons and holes) leads to very short charge collection times of less than 20 nanoseconds. Finally, a significant advantage of semiconductor detectors is the potential for a direct integration of the front-end electronics into the detector structure. In summary, it is the combination of these outstanding properties, like compactness and flexibility of the overall geometry, single mip resolution, direct charge readout with a non-saturating absolute energy calibration, and high speed, which make silicon detectors most promising for future applications.

The many advantages of silicon (as they apply to various applications) have resulted in the rapid growth of their use. Though the first applications of silicon detectors in high energy physics experiments were largely motivated by the high resolution possible with microstrip devices in vertex detectors, the demand for large area pad detectors is increasing. This application has now become quite active, especially for the instrumentation of calorimeters and other related equipment. C. Goessling presented an earlier survey of the applications of pad detectors in high energy physics.[2] Total areas between 0.1 and 50 m^2 are already employed or under design. The largest, the SICAPO calorimeter, is the first and presently the only project for a full hadronic test calorimeter. Other projects are mainly designed for either dE/dx-measurements, electron hadron separation, luminosity monitoring, or plug calorimeters. The prospects of SSC applications were studied at the Workshop on Calorimetry for the SSC,[1] where the above outlined issues were discussed in detail.

The UA2 detector at CERN was the first experiment to employ a large area of pad detectors.[3] For the inner cylindrical layer each detector contains 16 pads of 2 × 16 mm^2. The detectors of the outer layer have 7 pads of 9 × 40 mm^2. The total area covered is 0.1 m^2 for the inner and 1 m^2 for the outer layer. The main purpose of the installation is a fast and simple vertex reconstruction as well as a rejection of otherwise unresolved tracks. Since the UA2 experience many groups have adopted the silicon technique. We will discuss many of these below.

1. Electromagnetic Calorimetry

SLD LMSAT/MASC - Small angle calorimetry.

Motivated by earlier successful work with silicon high energy electromagnetic calorimeters[4], a set of calorimeters employing silicon diode sampling has been constructed to provide the small angle electromagnetic calorimetry for SLD at SLAC.[5] The luminosity monitor and small angle tagger (LMSAT) and the medium angle silicon calorimeter (MASC) measure the direction and energy of electrons and photons in the region of 23-190 milliradians from the SLD beam axis. They provide a precision measurement ($\leq$ 3%) of the integrated luminosity by precise tagging of small-angle Bhabha scattering (which is practically free of electroweak interference effects) and extend electromagnetic coverage of SLD to small angles.

The SLD will make sensitive tests of the standard model of the electroweak interaction through precision measurements of the decays of highly polarized Z^0s.

The fine transverse segmentation of the luminosity monitor permits track directions to be determined by the calorimeter itself, rather than by external tracking chambers. This is important, since synchrotron radiation background from the final focus quadrupoles and the interaction region makes it difficult to use conventional PWC or drift chambers, particularly when the inner radius of the vertex detector is small.

The LMSAT covers the angles from 23 to 68 milliradians, with its front face at a distance of 100 centimeters along the beam axis from the interaction point. The angles from 68 to 190 milli-

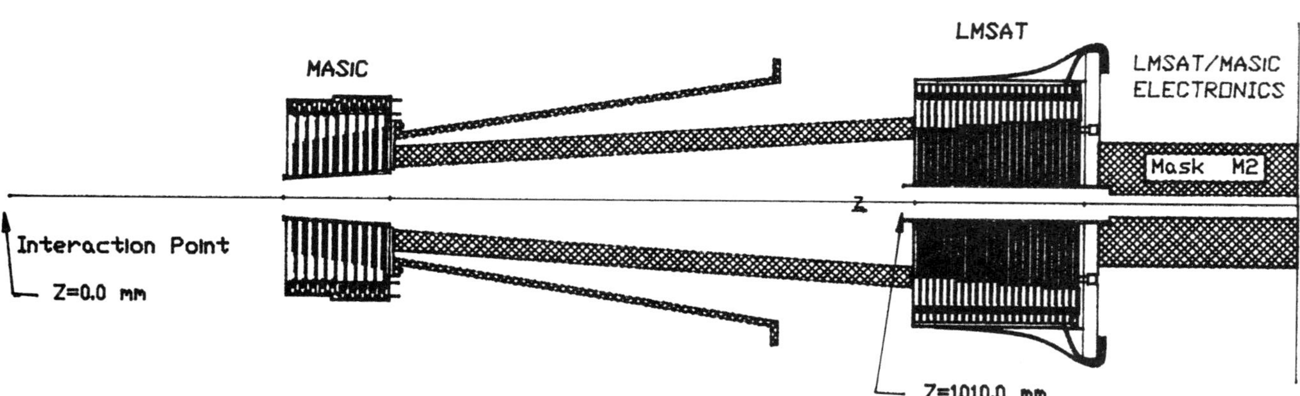

Figure 1. Side view of the small angle calorimeters: the LMSAT and the MASC

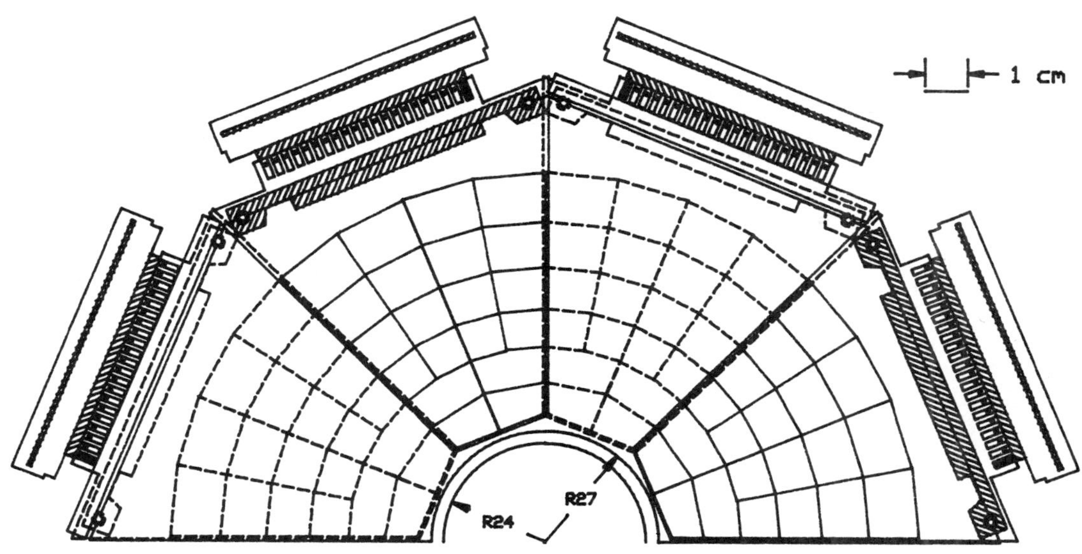

Figure 2. Cell layout at the first layer of the LMSAT.

radians are covered by the MASC whose front face is located 31 centimeters along the beam axis from the interaction point (see Figure 1). This layout (with the split at 68 milliradians) resulted from extensive background studies designed to minimize the effects of beam associated radiation on all components of SLD.

Each LMSAT and each MASC (one of each is mounted on either side of the interaction point) are constructed in two halves for easy disassembly. Figure 1 represents a simplified side view of the LMSAT and MASC showing their respective positions within SLD. The LMSAT consists of 23 heavy met (90% tungsten with copper and nickel binders) radiator plates, placed normal to the beam axis, each approximately one radiation length (3.5 mm) thick. The plates are spaced every 7.0 mm creating a gap of 3.5 mm between plates. This spacing allows an easy removable or detectors on each layer and broadens the transverse spread of the showers for position measurements. The silicon detectors, which provide the readout, are mounted directly onto printed circuit boards. Electromagnetic showers, which develop in the tungsten, liberate electrons in the fully depleted silicon detectors which are then collected by charge sensitive preamplifiers. This configuration was chosen from a set of designs studied with the electromagnetic shower simulation code EGS. The expected energy resolution for 50 GeV electromagnetic showers is about 3%, being dominated by sampling and leakage fluctuations. The LMSAT consists of transversely segmented detector arrays, with 20 cells per array. The sizes of the arrays were determined by the division of a planar circular region at a known distance from the interaction point into 8 equal radial pieces. Figure 2 shows the layout of arrays on the first layer of the LMSAT. The angular resolution of the monitor is determined by this segmentation. Ideally, each of the 23 layers would have cells which increased in size, so that each one would subtend the same angle as its predecessor, creating projective towers from the interaction point. This would have led to 23 unique detector array designs. To reduce costs, only 4 unique designs per calorimeter type were created. Thus,the first 6 layers, the second 6 layers, the third 6 layers, and the last 5 layers use the same respective array designs. For readout purposes, nearly projective towers are formed by electrically connecting successive layers of detectors in parallel. The first 6 layers combine to form the signal EM1 and the remaining 17 layers combine to form the signal EM2. Combined with the effects of array segmentation, this configuration results in the formation of 160 towers each in EM1 and EM2 for the LMSAT.

The MASC consists of ten 6.6 mm thick heavy met plates. As in the LMSAT, silicon detectors reside between the plates. The silicon detectors are transversely segmented into 12 cells per array with eight arrays in the full azimuth. With signals EM1 (EM2) from the first three (last seven) layers, there are 192 signals from each MASC, north and south.

The entire system, which consists of four LMSAT halves and four MASC halves, produces a total of 1024 signals to the preamps. The system contains 736 silicon detector arrays. A total of 888 arrays were produced to allow for spares.

The detector arrays are constructed by Hamamatsu Corporation from 300 (± 10) micron thick high resistivity n-type silicon wafers with p-type ion-implantation from the front surface to form the junction diode. The detectors fully deplete at between 30 and 50 VDC. One operating voltage for the entire set of detector arrays was specified by the manufacturer to be 75 VDC. The detectors are, of course, operated in the reverse bias mode at this voltage to increase the depletion region to the full volume of the device. The typical leakage current for this type of operation is much less than 10 nA/cm^2.

One LMSAT module was tested at the A2 test beam at the Brookhaven National Laboratory Alternating Gradient Synchrotron.[6] The A2 test beam delivers hadrons with a few percent electrons at 2 GeV/c, falling to less than one percent at 7 GeV/c. The calorimeter was calibrated both with muons and electrons, as described in Reference 6. Figure 3 shows the electron energy distribution for electron showers centered on any one of 12 towers at 4 and 5.4 GeV/c and 9 towers at 2 GeV. The fitted Gaussians to these figures yields measured resolutions (σ_E/E) of 0.166, 0.120, and 0.109 at 2.0, 4.0, and 5.42 GeV/c, respectively.

The position resolution was also verified in the beam test. The position of an electromagnetic shower is determined by the first moment of the tower pulse heights:

$$<x,y>= \sum_{\substack{i \\ towers}} (x_i, y_i)E_i / \sum_{\substack{i \\ towers}} E_i$$

This estimator is corrected in a well known way for the bias toward the edges of the towers.[6] Since beam position measurements were not available in the A2 beam line, the position resolution of the calorimeter was experimentally verified by comparing the shower position measured in EM1 with that measured in EM2. The fluctuations in these two measurements about the real position are only weakly correlated, as the EGS Monte Carlo result indicates. Figure 4 shows the difference between the measured position in EM1 and EM2. The curve on the figure is the expectation from the EGS Monte Carlo calculation. The Gaussian fit to this distribution yields a standard deviation of 0.04 radians, confirming the azimuthal resolution of 1.0 mm for the shower centroid.

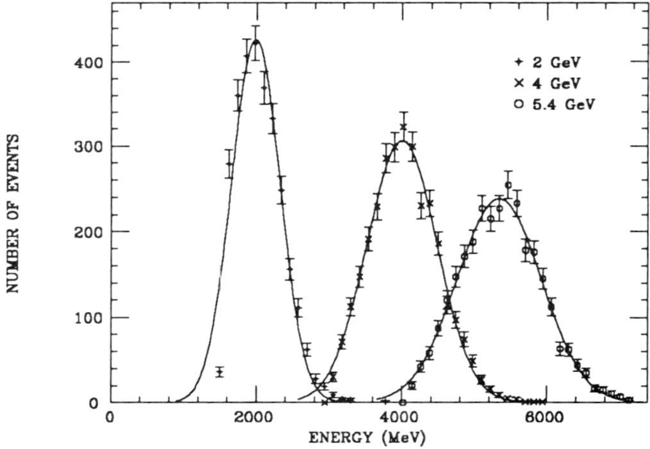

Figure 3. Electron energy distribution based on gains determined from 4 GeV electron showers.

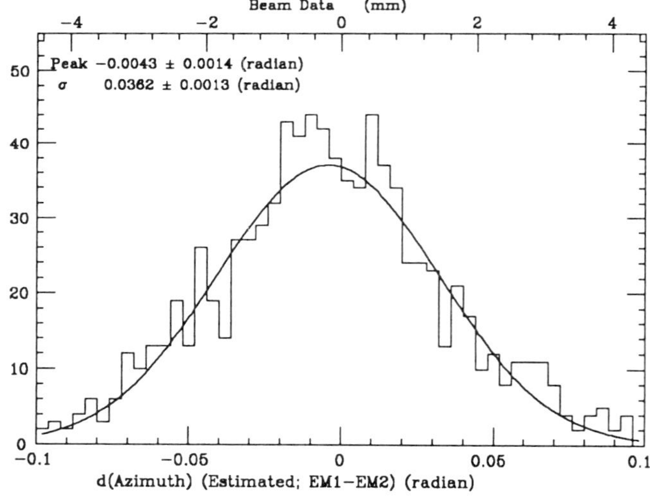

Figure 4. Distribution of differences between positions measured in EM1 and EM2. The curve is the EGS Monte Carlo prediction.

In summary, the test beam performance of the SLD luminosity monitor module was found to be close to expectation. Table 1 presents the measured values and extrapolates them to the expected performance for 45.5 GeV/c electrons at SLD.

measured variable	resolution determined at 4 GeV/c	expected resolution at 45.5 GeV/c
energy	0.116	0.034
shower centroid	1.0 mm	0.3 mm
azimuth	21.9 mrad	6.5 mrad
theta		0.3 mrad

Table 1. Summary of test results and extrapolation to 45.5 GeV/c at SLD.

Moscow test calorimeter.

The Moscow State University group has recently constructed and tested with 10 and 20 GeV electron beams, two silicon-lead electromagnetic calorimeter modules.[7] These tests resulted in an energy resolution measurement of $\sigma/E = 23.0 \pm 0.7\%\sqrt{\tau/E}$, where τ is the radiator thickness in radiation lengths. The calorimeters sampled the showers every one radiation length in one configuration and every two in a second configuration. Longitudinal and transverse shower profiles have been measured and presented.[7]

DELPHI: VSAT and MSAT.

The DELPHI Collaboration at LEP has been operating a silicon sandwich calorimeter (the VSAT) as part of their luminosity monitor since July, 1990.[8] This system consists of 900 cm^2 of silicon diodes. Based on the successful operation of this device, and motivated by the possibility that polarized beams could play an important physics role at LEP in the future, DELPHI is proposing the construction of the Mid Small Angle Tagger (MSAT).[8] Operation with polarized beams could produce measurements of $sin^2\theta_W$ with an error of 0.00035, but would require very precise measurements of the luminosity, with excellent point-to-point reproducibility and high statistics. DELPHI proposes this new luminosity monitor based on silicon technology to provide this capability. The DELPHI concept provides a systematic uncertainty in luminosity of less than 1% and a point-to-point relative uncertainty of 0.3 %. This is achieved with the MSAT system that covers the Bhabha scattering between 30 and 60 mrad with 1.3 m^2 of silicon detectors.

ALEPH: SICAL proposed.

The ALEPH Collaboration has also recently proposed complementing their existing ALEPH luminosity calorimeter with a compact silicon-tungsten calorimeter (SICAL) which extends the low-angle acceptance for Bhabhas from 55 mrad down to 28 mrad.[9] As stated above, this proposal is motivated by the possibility for polarized beams at LEP. The proposed ALEPH system employs 1.5 m^2 of silicon detectors.

ZEUS e/h separator.

The ZEUS Collaboration is preparing to install silicon diodes within their uranium-scintillator calorimeter to separate electron showers from hadrons. This is based on the principle of measuring the charged particle flux near the maximum of electron initiated electromagnetic showers. Such measurements have been shown to provide very powerful discrimination between electrons and hadrons when silicon diodes are so employed.[10] There are a number of factors contributing to the selection of silicon for this application, many of them the same that led other groups to choose silicon for full calorimeters. First, silicon provides excellent dE/dx resolution and the large dynamic range is limited in practice only by the readout electronics. Secondly, the silicon detectors introduce very little material, being compact and flexible in design, with almost any desired size and shape possible without bulky frames or enclosures. Thirdly, the detectors are operated at low power levels with very safe voltages, at room temperature, and are not influenced by magnetic fields. Finally, the silicon detectors do not require the use of any flammable gases.

The beam tests of the ZEUS Collaboration have achieved hadron misidentification levels of 2% at an electron efficiency of 90% with just one plane of diodes at 3, 5, and 9 GeV. Addition of a second plane reduced the hadron misidentification to 0.3% with an electron efficiency of 90% .[10]

LHC preshower proposal.

A proposal has recently been presented at CERN to study the track stub/preshower technique of electron identification at the highest operating luminosity of the Large Hadron Collider (LHC).[11] This proposal strives for 1.) fast response of better than 15 nanoseconds in order to tag the bunch crossing, and 2.) high granularity in order to achieve excellent pattern recognition. This group has proposed the use of silicon, motivated, at least in part, by the UA2 experience with a scintillating fiber pre-radiator in which the silicon pad layer was found to be essential in resolving ambiguities inherent in the stereo triplet geometry of the fibers. The proposal calls for pad sizes of $3 \times 3\ mm^2$ based on a compromise of radiation resistance, detector capacitance and speed, pattern recognition, mechanical and electrical considerations, and cost. While many details of this proposal remain undetermined and are in fact the subject of the proposed study, they propose to employ strips as well as pads, with sets of pads and strips on each side of the converter plane. The proposed preshower scheme would employ one preamplifier per pad, a charge sampling integrator based on the HARP system, data compression in both time and space, with radiation hardened electronics being developed for the final application.

LEP luminosity monitors.

Mention must be made of the successful implementation of silicon calorimeters in the monitoring of Bhabha scattering at the four LEP interaction regions.[12] These calorimeters are designed to provide a valuable measurement of the luminosity in each interaction region, as well as the synchrotron radiation background.

2. Hadron Calorimetry

Compensation.

The underlying mechanisms of compensating calorimetry are now well understood.[13] The fundamental question that needs to be addressed is that of the relative response of a calorimeter to electrons and hadrons. One property of silicon that plays an important role in this is its extreme linearity up to very large stopping powers: i.e., silicon exhibits very little saturation. It has been demonstrated that saturation limits compensation in currently designed uranium-liquid argon calorimeters, so saturation could be a very important factor in silicon calorimeters as well. With little saturation, silicon will provide very good response to some of the low energy components of hadronic showers.

It is useful to consider the differences between the response of silicon and liquid argon since, except for the saturation in liquid argon, full compensation could possibly be achieved.[13] The three critical factors which determine the fraction of the neutron energy which is detected in the active medium are neutron cross sections, the kinematic limitation on energy transfer, and saturation. First the neutron cross section in argon decreases for energies below a few MeV, while the cross section in silicon remains large and even increases below 500 keV. This will result in a larger energy deposition by low energy neutrons in silicon than in liquid argon. Secondly, the effect of the kinematic limitation on neutron energy transfer restricts the transfer in liquid argon to ten percent, while in silicon, due to the smaller size of the atom, the transfer is limited to about 13 percent. This should lead to an additional increase in the energy transferred to the silicon. Finally, the very limited saturation which occurs in silicon will enable these increased neutron depositions to be more observable.

The potential for compensation with silicon certainly seems better than for liquid argon. To make quantitative predictions it is necessary to make use of a Monte Carlo code system which takes into account all of the basic processes and examines the details of the shower development. The Oak Ridge system CALOR has been applied to this problem and the results indicate that one should expect compensation in uranium-silicon calorimeters.[13]

In addition to these basic considerations for compensation in silicon hadron calorimeters, other means of achieving compensation should these fail have been considered. These include the introduction of hydrogenous material such as polyethylene coatings to enhance the neutron signal through the detection of knock-on protons,[14] and the introduction of low-Z materials within the calorimeter which will decrease the response of the calorimeter to electrons relative to its response to hadrons.[15]

SICAPO studies - "filtering".

The most aggressive application of silicon detectors is the work of the SICAPO Collaboration in its studies of hadronic calorimetry.[15] The SICAPO Collaboration has employed large area silicon detectors in their beam test calorimeters. These detectors have a standard area of 28 cm^2 and can be assembled into mosaics with an active area of up to several thousands of cm^2. The test done to date involve mosaics of 250-524 cm^2 active areas.

One approach to compensation the SICAPO Collaboration has been pursuing is based on the fact that the lower the critical energy of the absorber, the softer the energy spectrum of the electromagnetic shower. By mixing low-Z and high-Z absorbers, the energy distribution of the shower is transformed and the response of the calorimeter to the incoming shower is modified. This is particularly dramatic when entering a low-Z (iron) absorber and exiting from a high-Z (lead) absorber. The degree of the effect obviously depends on the fraction of lead in an absorber made of a combination of iron and lead. In a series of measurements, the SICAPO Collaboration studied the relative electron and hadron responses as the mix of iron and lead was varied and showed that some degree of tuning of the e/h ratio was possible.[15] They also studied the insertion of G10 (a low-Z medium inserted to alter the soft electromagnetic energy content). This resulted in a smaller fraction of lead being needed in the absorber to achieve the compensation condition.[16]

In a continuation of these silicon hadron calorimeter studies, the SICAPO Collaboration is now building a set of 50 planes of $40 \times 40\ cm^2$ active area built from individual $2 \times 2\ cm^2$, $400\mu m$ thick detectors.[17] When complete, this system will consist of 20,000 detectors. Presently it is 1/3 built, another 1/3 will be completed by December, and the full calorimeter is scheduled for completion at the start of next year. The front-end electronics and programmable amplifiers are operational.[17]

H1 Plug Calorimeter.

The H1 Collaboration designed and has under development a silicon instrumented hadronic plug calorimeter. It will fill the gap in the very forward direction between the liquid argon calorimeter and the beam pipe. The hermeticity so achieved is very important in measuring the missing energy of the 800 GeV proton collisions with the 30 GeV electrons at HERA. The requirements for this calorimeter are about 5 mrad in angular resolution demanding good lateral segmentation, but only about $200\%/\sqrt{E}$ energy resolution. To achieve these goals the H1 group samples roughly twice per interaction length. The absorber material is copper since the ratio of interaction length to radiation length is relatively low, resulting in reasonable sampling of electromagnetic showers even with this coarse construction. Planes of 84 detectors, mostly $5 \times 5\ cm^2$, are inserted in each detector module. Eight planes make up the full eight interaction length depth of the calorimeter. The detectors are being manufactured in the facilities of Hamburg University and are described in more detail in the presentation of G. Lindstroem to this symposium.[18]

In concert with the H1 work, the Hamburg group has recently collaborated with ITEP on measurements of the hadronic shower development in various calorimeter stacks by moving a silicon mosaic plane from gap to gap within the stack. Early results of these measurements were also presented by Lindstroem at this symposium.[19]

U.S. Development of 28 cm^2 detectors.

As noted above, experiments at the Superconducting Super Collider (SSC) will challenge calorimeter performance with the exceptional experimental conditions created by a very high luminosity, high multiplicities, and the large variety of physics phenomena to study. These challenges are motivating further development of silicon detectors, particularly low cost, large area devices. It should be noted that a large area device can be subdivided electrically to reduce capacitance, if desired.

Driven by the potential applications at the SSC, large area silicon detectors for calorimetry applications have been fabricated at United Detector Technology, Hawthorne, CA, using solid state diffusion technology.[20] The big advantage of diffusion technology is its lower cost of doping compared with the ion implantation process. This motivated the adoption of the diffusion technology for the fabrication of these silicon detectors. Trapezoidal detectors with 28 cm^2 active area were fabricated on $> 2500\ \Omega cm$, 4 inch diameter n-type silicon wafers. The low-cost, high volume solid state diffusion technology along with phosphosilicate-glass and TCA gettering was adopted for boron and phosphorus doping. Typically the diode dark current achieved was 15 μA @ 100 volts. Excellent response to ionizing radiation has been observed.

To meet the price goal of $\$2/cm^2$ for large area silicon detectors needed in the SSC project, the finished detector yield should be over 80%. For the best performance the detector is used in a fully depleted mode needing breakdown voltage in excess of 100 volts. Fabrication of large area fully depleted detectors with over 80% yield is extremely challenging because of the defect generation during processing which reduces the diode breakdown voltage and/or increases the leakage current. Defects in semiconductor devices may be broadly divided into three groups. (i) crystallographic defects, (ii) metallurgical defects and (iii) pattern defects. The efforts to reduce these defects during manufacturing of 28 cm^2 active area trapezoidal detectors were described at this symposium.[20]

3. Super Collider Expressions of Interest

L* EOI.

The L* Collaboration has proposed building a full silicon hadron calorimeter for the SSC.[21] This calorimeter is estimated to consist of about 5000 m^2 of silicon detector diodes which they estimate will take 4-5 years of Soviet production if produced in the Ukraine. The calorimeter would be cooled to -20 degrees centigrade to withstand the intense radiation environment and its effects on the detectors. They have proposed two options, one with a high resolution electromagnetic calorimeter (either barium fluoride or liquid xenon) in front of the hadron calorimeter, and the second a full silicon calorimeter for electromagnetic and hadronic calorimetry. In the first option the hadron resolution is unavoidably compromised, with an estimated resolution for single hadrons of $50\%/\sqrt{E} + 4\%$.

EMPACT EOI.

The EMPACT Collaboration has presented as one of its calorimetry options, a scintillating fiber lead hadron calorimeter with a silicon electromagnetic calorimeter.[22]

The motivation for selecting silicon calorimetry as one of the options rests on the same issues which have been enumerated in the introduction of this paper. The fast charge collection, ease of segmentation, and complementarity to scintillating fiber calorimetry are the major attractive features.

The issues of construction and operation (for example cooling and power dissipation) are being addressed as part of the SSC subsystem research and development project on silicon calorimetry by the Silicon Electromagnetic Calorimeter Collaboration.[23] This subsystem work will apply the recent developments of the ORNL electronics group for fast, current compensated preamplifiers based on ASIC technology.[24] This work will also address some of the issues of the L* proposal.

While there are a number of advantages in using silicon detectors in calorimetry, the radiation hardness will be the main problem to be overcome at the SSC. In fact, survivability and radiation resistance is one of the most critical issues for any detector at the SSC. The most hostile radiation in the calorimeter comes from low energy (MeV) neutrons. These levels have been studied and documented by the Task Force on Radiation Levels in the SSC Interaction Regions.[25] Many studies have recently been done to investigate the effect of different types of radiation on silicon detectors. A particularly complete set of investigations have been carried out by the Hamburg group.[26] Figure 5 comes from this work. Here one sees that the damage from 1.8 MeV electrons is particularly weak, while the neutron damage constants are large. This coupled with the large numbers of neutrons produced, makes neutrons the number one radiation problem to overcome.

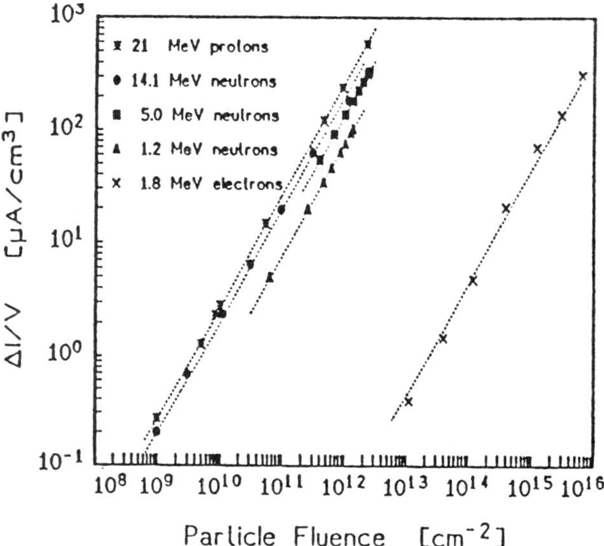

Figure 5. Increase of measured bulk generation current as function of particle fluence from reference 26.

Monte Carlo calculations have shown that a silicon calorimeter can be significantly hardened by adding thin polyethylene sheets in each of the sampling gaps.[27] The hydrogenous material would be expected to suppress neutron fluence effectively since hydrogen has larger cross sections and larger kinematic energy transfer limits for neutrons than heavier nuclei. The EMPACT electromagnetic calorimeter, with 24 radiation lengths depth, has substantial neutron fluxes. A large fraction of the neutrons in the EM section actually comes from the hadron section of the calorimeter, as is seen from observing the flux reduction effect of adding a 10 *cm* polyethylene shield between the EM and hadron sections. Calculations show that the neutron 'gas' produced originally in the EM section can also be suppressed by adding polyethylene sheets in each sampling gap.[28] These results emphasize that inserting both thin polyethylene moderators ($\sim 2mm$) in the sampling gaps and a thick ($\sim 2cm$) shield between EM and hadron sections is necessary to reduce the neutron fluence. A preliminary neutron flux measurement result employing an activation detector technique at the University of Oregon 5 MeV Van de Graaff accelerator has been presented at this symposium and is consistent with these expectations, although more sensitive measurements are still needed and will be pursued.[28]

Conclusions

Silicon detectors are finding numerous calorimetric applications in high energy physics experiments. We have reviewed some of the more significant recent applications.

Acknowledgements

I would like to thank my collaborators for their contributions to the work reported in this survey.

†Supported in part by U.S. Department of Energy.

References

[1] C. Bertrand et al., "Silicon Calorimetry for the SSC", Proceedings of the Workshop on Calorimetry for the SSC, p. 489 (1990); and references cited therein.

[2] C. Goessling, "Large Area Silicon Detectors", Proceedings of the XXIV International Conference on High Energy Physics, Munich, 1988, p. 1208.

[3] K. Borer et al., Nucl. Instr. and Methods A257, 591 (1987).

[4] G. Barbiellini et al., Nucl. Inst. and Methods A235, 55 (1985); G. Barbiellini et al., Nucl. Inst. and Methods A236, 316 (1985). M. Bormann et al., Nucl. Inst. and Methods A240, 63 (1985); and A. Nakamoto et al., Nucl. Inst. and Methods A251, 275 (1985).

[5] S. Berridge et al., IEEE Trans. Nucl. Sci. NS-36, 339 (1989).

[6] S. Berridge et al., IEEE Trans. Nucl. Sci. NS-37, 1191 (1990).

[7] A.A. Adodzero et al, "Silicon Pad Detectors and Integrated Electronics for Collider Calorimetry," presented at the Workshop on Physics at UNK, Protvino, September, 1990.

[8] "The Delphi MSAT detector," September 3, 1990, unpublished.

[9] The ALEPH Collaboration, "A Proposal for a Low-Angle Solid-State Luminosity Calorimeter for ALEPH: SICAL," CERN/LEPC/90-3, January 26, 1990.

[10] A. Dwurazny et al., Nucl. Inst. and Methods A277, 176 (1989).

[11] D.J. Munday et al., "A Proposal to Study a Tracking/Preshower Detector for the LHC," CERN/DRDC 90-27, August 2, 1990.

[12] J.Y. Hemery. F. Lemeilleur, and G. von Holtey, CERN/LEP-BI/86-5.

[13] J.E. Brau and T.A. Gabriel, "Theoretical studies of hadronic calorimeters for high-luminosity high-energy colliders," presented by J.E. Brau at the International Conference on Advanced Technology and Particle Physics, Como, 13-16 June 1988, Nucl. Instr. and Methods A279, 40 (1989).

[14] James S. Russ, "Neutron Energy Deposition Mechanisms in Silicon Calorimetry," Proceedings of the Workshop on Calorimetry for the SSC, p. 531 (1990).

[15] E. Borchi et al., "Electromagnetic Shower Energy Filtering Effect. A Way to Achieve the Compensation Condition ($e/\pi=1$) in Hadronic Calorimetry," Phys. Lett. B222, 525 (1989).

[16] E. Borchi et al., "Evidence for the Compensation Condition in Si/U Hadronic Calorimetry by the Local Hardening Effect," Phys. Lett. B242, 293 (1990).

[17] P.G. Rancoita, private communication.

[18] E. Fretwurst et al., "The Silicon Instrumented Plug-Calorimeter for H1 at HERA," Proceedings of the ECFA Study Week on Instrumentation for High-Luminosity Hadron Colliders, CERN 89-10, ECFA 89-124, p.319 (1989).

[19] G. Lindstroem, "Application of Silicon Detectors for Calorimetry and Monte Carlo Studies," presented at the Ft. Worth Symposium on Detector Research and Development for the Superconducting Super Collider, October, 1990.

[20] R. Korde, K. Furuno, H. Hwang, J.E. Brau, and W.M. Bugg, "Low Cost, Large Area Silicon Detectors for Calorimetry," presented at the Ft. Worth Symposium by R. Korde, OREXP-90-1002, October, 1990.

[21] L* Collaboration, Expression of Interest, May 25, 1990.

[22] EMPACT Collaboration, Expression of Interest, May 25, 1990.

[23] Silicon Electromagnetic Calorimeter Collaboration, "The Design and Construction of an SSC Silicon Electromagnetic Calorimeter Prototype," presented at the Ft. Worth Symposium by S. Berridge.

[24] C. Britton, "Bipolar Monolithic Preamplifiers for SSC Silicon Calorimetry," presented at the Ft. Worth Symposium on Detector Research and Development for the Superconducting Super Collider, October, 1990.

[25] Radiation Levels in the SSC Interaction Regions, SSC-SR-1033, edited by D.E.Groom (1988).

[26] R. Wunstorf et al., "Radiation Damage of

PILEUP AND THERMAL NOISE IN A LIQUID IONIZATION CALORIMETER*

W. E. Cleland, E. A. Scholle, and E. G. Stern
Department of Physics and Astronomy
University of Pittsburgh
Pittsburgh, PA 15260

Abstract

A study of the effects of thermal and pileup noise in liquid ionization calorimeters operating in a high luminosity environment is presented. The method of optimal filtering of multiply-sampled signals is examined, and results for the dependence on the resolution of signal parameters on rapidity, luminosity, and the number of data samples are given.

Introduction

We describe here techniques for the determination of the amplitude and timing information contained in a shaped signal from a liquid ionization calorimeter of the type typically used in high energy physics. In particular we wish to concentrate on the effects of distortion of the signal due to thermal noise and to real signals from uncorrelated events which are close in time to the event of interest, the so-called pileup noise. Because these two sources of noise have different characteristics, it is possible to devise a method which maximizes the signal/noise ratio for determination of the time origin and/or the amplitude of the signal. We show how this can be done for the case of a signal which is sampled several times over the waveform.

The motivation for this study is the need to understand such questions in designing signal processing electronics for liquid ionization calorimeters at high luminosity colliders, such as SSC and LHC. Practical considerations limit the peaking time of shaped signals to 50 ns or greater, and the entire waveform will dwell at least four times that value The colliders in question will have collisions at intervals much shorter than the dwell time and hence pileup effects will occur. Because of the possibility of a collision occurring at each beam crossing (16 ns for the SSC), it is necessary to sample the waveform at least at the frequency of beam crossings. Samples may be discarded, if the trigger logic in the experiment indicates that they are associated with an uninteresting event. Even for interesting events, it may not be necessary to keep all samples; one only needs to preserve enough samples to accurately reconstruct the amplitude and timing information.

There are two principal sources of noise which must be considered for liquid ionization calorimeters running in a high luminosity environment: (1) thermal noise, which has a white spectrum and whose amplitude depends only upon the detector characteristics in an optimized system; and (2) pileup or physics noise, which is due to the constant bombardment of calorimeter cells by products of collisions reactions which high cross sections (mainly minimum bias and two jet events). These signals are generated at the beam crossing frequency, and the amplitude spectrum is given by a Poisson distribution, whose mean value is the average number of events produced at each crossing. The mean number of events/crossing n_i for event type i is related to the cross section σ_i, the bunch spacing T, and the collider luminosity $\mathcal{L}$ by the simple formula

$$\overline{n_i} = \mathcal{L} T \sigma_i \tag{1}$$

Thus it is clear that the level of pileup noise depends on the luminosity $\mathcal{L}$ of the machine.

Since we are dealing with a system of limited bandwidth, the limitation being introduced by the shaping amplifier, data samples taken in a time short compared to the dwell time of the shaped signal, there are necessarily correlations introduced between the samples. It is necessary, when considering the treatment of highly correlated data, to understand the autocorrelation function for the system, from which we can obtain the covariance matrix for the data samples.

We assume for this study that the impulse response for the system is a bipolar pulse with zero area, of the form shown in Figure 1. We characterize the time base of the signal by using τ_p, the peaking time, which is the time to reach the peak of the first lobe. For these studies we have compared pulses of two different peaking times, 48 ns and 96 ns, which are near values commonly cited as reasonable for use at the SSC. Because of the 16 ns spacings between beam crossings and the high luminosity, which leads to a collision rate of about 1.6/crossing, the ability to correlate signals in time will depend critically on the ability to extract the timing information from the

*Supported in part by the U.S. Department of Energy under Contract No. DE-AC02-80ER10667 under the SSC Generic Detector R&D Program

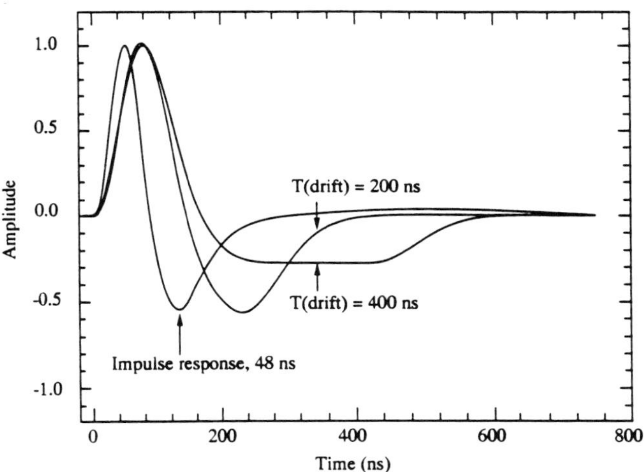

Figure 1: Shapes of the bipolar pulses used in the pileup calculations for $\tau_p = 48$ ns . Convolution of the impulse response with drift times of 200 ns and 400 ns are shown.

signal itself. Intuitively, one expects that an improvement in timing resolution and a reduction in the effect of event pileup would be realized as the peaking time is decreased. However, such a reduction is accompanied by two effects which lead to a deterioration in the signal to noise ratio:

1. The thermal noise increases as the bandwidth of the system is increased so σ_S, the series noise, is proportional to $1/\sqrt{\tau_p}$.
2. As τ_p is decreased, a smaller fraction of the charged deposited in the calorimeter is utilized in the signal processing, which leads to a dependence of the signal height on the drift time. The charge induced on the electrodes is given by

$$Q_s = N_e[\frac{t_m}{t_d} - \frac{1}{2}(\frac{t_m}{t_d})^2], \qquad (2)$$

in which t_m, the measurement time, is approximately equal to the peaking time τ_p, t_d is the drift time, and N_e is the number of electrons liberated by the charged particle.

Thus τ_p cannot be reduced to arbitrarily small values, due to the above effects, and it cannot be increased to arbitrarily large values due to the effects of event pileup and degradation of the timing resolution. It is the interplay between the effects of shaping time and drift time on the values of the thermal and pileup noise that is the subject of this study.

Pileup Noise

The problem of the effects of overlapping events at a high luminosity collider has received considerable attention at workshops for the SSC and LHC. The previous estimates have been concerned more with the random noise level seen in individual calorimeter cells than with effects on sums over larger areas as would be used in the trigger logic. The assumption may be made that the variance of pileup noise, like thermal noise, should scale with the area, which would be correct for uncorrelated energy deposition. However, we know that a significant fraction of the pileup will come from two jet events, since the cross section for their production is about half of the total cross section. In such events, the energy in neighboring cells will be correlated, due to the energy flow characteristics of jet events. In addition, various approximations to the shaping functions have been used in the previous calculations, with little or no attention given to the effects of different types of filtering. We have therefore undertaken a detailed calculation of pileup effects, permitting the use of a shaping function defined numerically, and convoluting it with the drift current from the ionization chamber.

Method of Pileup Calculations

To estimate the effects of pileup, we simulate the time structure of the machine, by permitting collisions at points in time separated by the bunch crossing time of 16 ns, and we use Poisson statistics to determine the number of collisions of each type of events at each crossing. We consider only two jet (with transverse energy of the jet above 5 GeV) and minimum bias events, as generated in ISAJET version 6.21. The two types of events occur in approximately equal proportions, and together they give rise to a mean collision rate of 1.6 collisions/crossing.

Energy deposition in the calorimeter is simulated by the program QFL, in which a spherical calorimeter of inner radius 3 m is assumed. QFL permits only two depth segmentations, so we choose an electromagnetic section of thickness 25 radiation lengths, and a hadronic section of 10 absorption lengths. Transverse spreading in the calorimeter is parameterized in a simple way, but comparisons with experimental data indicate that the width is correct to an accuracy of about 10%. In this calculation, we deal with sums of transverse energy, i.e. we assume that $\sin\theta$ weighting is made immediately after the preamplifier.

After the transverse energy deposition for each calorimeter cell is determined from QFL, a sum for the cell over all events in a given crossing is made, and we then initiate the shaping function of an amplitude corresponding to this sum. This procedure is carried out for 12 cells, electromagnetic and hadronic, spaced in six equal increments in rapidity over the interval $0 < \eta < 5$. We then step to the next crossing, continuing the shaping functions from all previous crossings, and add the contribution from that crossing to the simulated shaper output. In this way we simulate the signal waveform produced by the shaping and summing circuitry under the assumed conditions. We then calculate the mean value and mean square value, from which (after 1000 crossings) a variance is evaluated for each of the 12 cells.

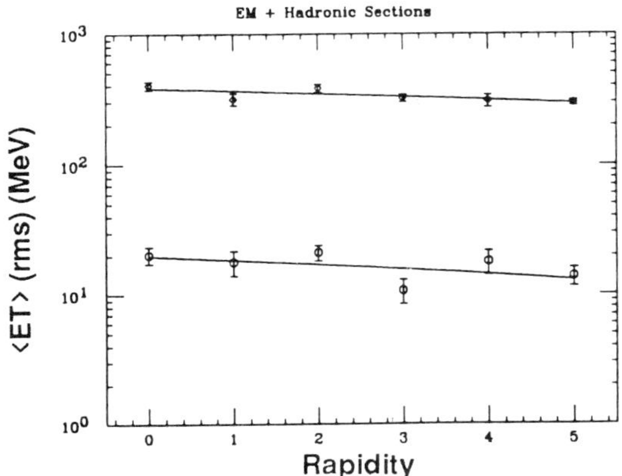

Figure 2: Values of σ_E vs. rapidity for the EM + HAD sections for minimum bias and two jet events. Diamonds:towers of size 0.32 × 0.32; circles: cells of size 0.04 × 0.04. The lines shown are the least squares fits to the linear dependence on η.

Evaluation of pileup variance

In the approximation that the process is continuously distributed in time, Campbell's theorem may be used to find the pileup variance. However, at the SSC, the signals are generated at intervals of 16 ns, and it can easily be shown that the proper expression for the pileup variance in this case is:

$$\sigma_P^2 = \sigma_E^2 \int_{-\infty}^{\infty} w^2(t)dt = \sigma_E^2 I_{2p} \tag{3}$$

in which σ_E^2 is the variance of the energy deposition in the calorimeter and $w(t)$ is the signal waveform (see Equation 8) below. Due to the use of bipolar shaping, the mean value of the shaped signal is zero, so the pileup variance is simply the average of the square of the shaped signal. As is seen from Equation 3, the pileup variance separates into two conceptually independent factors: σ_E, which depends only on the rate of collisions and the distribution of deposited energy, and the pileup integral I_{2p}, which depends on the parameters of the calorimeter and signal processing. We calculate this first factor by evaluating the quantity

$$\sigma_E^2 = \frac{\sum_i E_i^2}{n} - \frac{(\sum_i E_i)^2}{n^2}, \tag{4}$$

which is a characteristic of the sample of events contributing to the pileup. We show results for two sizes of calorimeter cells: $\Delta\eta \times \Delta\phi$= (0.04 × 0.04) and (0.32 × 0.32), the latter being a reasonable value for a "trigger tower". In Figure 2 we show the rapidity dependence of σ_E. It is, to a good approximation approximation, rapidity independent, although this is less true for the EM section than for the total transverse energy. One also notes that the ratio of transverse energy deposition for the two cell sizes is larger than the factor 8 which would be expected from scaling the area of the cells. The factor is closer to a factor of 15-20, presumably due to the effect of jets as mentioned above. We fit our results to a linear expression in η:

$$\sigma_E = A + B|\eta|, \tag{5}$$

Table 1: Values of the constants A and B in the expression for σ_E the two calorimeter sections and the two tower sizes used in the study.

Section	Tower Size	A(MeV)	B(MeV)
EM	0.04	15.2±1.8	-1.9±0.6
TOT	0.04	19.9±2.2	-1.4±0.7
EM	0.32	282±16	-29±4
TOT	0.32	384±21	-18±5

Table 2: Values of thermal noise calculated by S. Rescia for the seven-layer liquid argon calorimeter (for t_m = 50 ns and t_d = 400 ns and for a tower size of 0.32 × 0.32) discussed at the Alabama Calorimetry workshop, April 1989.

Section	σ_S(GeV)
EM	1.17
HAD	4.65
TOT	4.79

and the results are shown in Table 1. We now consider the variance of the shaped signal. Using a bipolar signal (see Figure 1) of peaking time 96 ns and a drift time of 200 ns, we carry out the variance calculation described above. As expected from the plots of σ_E, this quantity is also essentially independent of rapidity. From these results, we can test Equation 3 by calculating the integral I_{2p} numerically. We find, for this particular case the value I_{2p}=158 ns. To compare this value with that found numerically in the Monte Carlo, we extract a value of the pileup integral numerically by calculating the ratio of σ_P to σ_E. At $\eta = 0$, for the 0.32 × 0.32 towers, we find for I_{2p} the values 135±15 ns and 149±16 ns for the EM and TOT sections, respectively, in agreement with the value calculated from the waveform integral.

Thermal Noise

The thermal noise σ_S depends upon the signal processing parameters through the ratio of the integral I_1, defined as:

$$I_1 = \int_{-\infty}^{\infty} (\frac{dh}{dt})^2 dt, \tag{6}$$

to the square of the signal charge Q_s, given in Equation 2, as follows:

$$\sigma_S^2 = \frac{kI_1}{Q_s^2}. \tag{7}$$

The proportionality constant k depends upon the parameters of the the detector. We use, for our calculations, the values calculated by S. Rescia shown in Table 2. These values may be scaled for different choices of the drift times and shaping function peaking time. We do not include the effect of signal loss due to the addition of additives or choice of a medium different from pure liquid argon in these calculations.

Effects of Noise on Resolution

We have carried out a detailed study of the effect of noise on the amplitude and timing information of a sampled signal from the calorimeter. We assume that at the SSC, the shaped signals from the calorimeter will be sampled with the clock of the accelerator, i.e. at each 16 ns interval. While there will very likely be a circuit which will extract timing information from the signal, this study concentrates on the use of the digitized information for the extraction of the timing, as would be done in an offline analysis. In this case, one uses the entire pulse shape, and, if the resolution of the digitization is sufficient, these results represent an upper limit to the timing resolution obtainable in hardware. Similarly, whereas discriminators will be used to establish whether the amplitude of the signals are above certain threshold levels, in an offline analysis, one would use the entire waveform to obtain this information, and the resolution obtained in this way should represent an upper limit to the discriminator threshold resolution. We also have studied the case where a number of samples are eliminated, and we obtain results for the accuracy of both timing and amplitude parameters as a function of the number of samples.

We approach the problem of determining the amplitude and timing information from a set of measured samples by looking first at the traditional method for extracting information from an overdetermined data set: the χ^2 method. We then turn to a solution from signal processing theory, optimal filtering(the optimality criterion is the minimization of the least square error), and show that under certain conditions, the two methods are equivalent. We then discuss some results using the method of optimal filtering through the use of the system weighting function. Finally, we have carried out some Monte Carlo simulations of the noise signals, and these are compared with the results of the optimal filtering method.

We assume that the form of the signal at the output of the shaping amplifier is known, except for its amplitude A and time origin τ (the deviation from the assumed crossing time). The expression for this signal is

$$S_i = Aw(t_i - \tau), \tag{8}$$

in which w is the convolution of the impulse response $h(t)$ of the shaping amplifier with the drift current $i(t)$:

$$w(t) = K \int_{-\infty}^{t} h(u)\, i(t-u) du. \tag{9}$$

The normalization constant K is chosen to set the maximum value of the first lobe of the shaping function to be unity. We also assume that the waveform will be sampled many times, giving a set of measurements $S_1, \ldots, S_n$. We wish to determine the parameters A and τ from the data set S_i.

χ^2 approach

We define the χ^2 function as follows :

$$\chi^2(A,\tau) = \sum_{ij} (S_i - Aw(t_i-\tau))V_{ij}(S_j - Aw(t_j - \tau)) \tag{10}$$

in which V_{ij} is an element of the weight matrix (the inverse of the covariance matrix) for the measured points S_i. The dependence on τ is linearized by the use of a Taylor expansion:

$$w(t_i - \tau) = w(t_i) + \tau w'(t_i), \tag{11}$$

in which $w' = dw/dt$. Defining the two parameters in the fit to be $\alpha_1 = A$ and $\alpha_2 = A\tau$, we have

$$\chi^2 = \sum_{ij} (S_i - \alpha_1 w_i - \alpha_2 w_i')V_{ij}(S_j - \alpha_1 w_j - \alpha_2 w_j'). \tag{12}$$

We now form the following sums, expressed also in the equivalent matrix notation:

$$Q_1 = \sum_{ij} w_i V_{ij} w_j = \vec{w}^{+}\mathbf{V}\vec{w} \tag{13}$$

$$Q_2 = \sum_{ij} w_i' V_{ij} w_j' = \vec{w}'^{+}\mathbf{V}\vec{w}'$$

$$Q_3 = \sum_{ij} w_i' V_{ij} w_j = \vec{w}'^{+}\mathbf{V}\vec{w}$$

$$Q_4 = \sum_{ij} w_i V_{ij} S_j = \vec{w}^{+}\mathbf{V}\vec{S}$$

$$Q_5 = \sum_{ij} w_i' V_{ij} S_j = \vec{w}'^{+}\mathbf{V}\vec{S}$$

In these equations $\vec{w}$ is the line vector of the unit-amplitude shaping function samples, $\vec{w}^{+}$ is its transpose, and $\vec{w}'$ is its time derivative. The line vector $\vec{S}$ is the set of measured data points. We set $\partial\chi^2/\partial\alpha_i = 0$, and solve the resulting linear equations for α_i. The results are:

$$\alpha_1 = A = \frac{1}{\Delta}[Q_2Q_4 - Q_5Q_3] \tag{14}$$

$$\alpha_2 = A\tau = \frac{1}{\Delta}[Q_1Q_5 - Q_3Q_4], \tag{15}$$

in which

$$\Delta = Q_1Q_2 - Q_3^2. \tag{16}$$

The covariance matrix elements

$$\epsilon_{ij} = \frac{1}{2}\frac{\partial^2\chi^2}{\partial\alpha_i\partial\alpha_j} \tag{17}$$

for the parameters α_1 and α_2 are:

$$\epsilon_{11} = \sigma_A^2 = \frac{Q_2}{\Delta} \tag{18}$$

$$\epsilon_{22} = A^2\sigma_\tau^2 = \frac{Q_1}{\Delta}$$

$$\epsilon_{12} = \frac{-Q_3}{\Delta}.$$

In using this method, one normally solves the above equations, then recalculates the function with the new time parameter, and iterates the problem. This avoids errors due to the truncation of the Taylor expansion for w at the first term.

Method of optimal filtering

In this method, we search for a linear combination of the data samples to yield values for the parameters A and τ, which are determined independently. The coefficients are found by minimizing the variance of the parameters. For the amplitude parameter, we define

$$v = \sum_i a_i S_i = A \sum_i a_i w_i. \tag{19}$$

Solving for A, we find

$$A \equiv \beta_1 = \frac{\sum_i a_i S_i}{\sum_i a_i w_i.} \tag{20}$$

To solve for the parameter τ, let

$$u = \sum_i b_i S_i = A \sum_i b_i w_i. \tag{21}$$

If we displace the waveform in time by an amount τ, we find

$$\Delta u = \tau \frac{du}{dt} = A\tau \sum_i b_i w_i'. \tag{22}$$

Since $\Delta u = u(\tau) - u(0)$, we have

$$\Delta u = \sum_i b_i S_i - A \sum_i b_i w_i = A\tau \sum_i b_i w_i'. \tag{23}$$

Solving for $A\tau$, we find

$$A\tau \equiv \beta_2 = \frac{\sum_i b_i S_i}{\sum_i b_i w_i'} - A \frac{\sum_i b_i w_i}{\sum_i b_i w_i'}. \tag{24}$$

We now calculate the variances of the parameters β_1 and β_2:

$$\xi_1^2 = \sigma_A^2 = \frac{\left\langle \sum_i a_i n_i \right\rangle^2}{(\sum_i a_i w_i)^2} = \frac{\sum_{ij} a_i a_j \langle n_i n_j \rangle}{(\sum_i a_i w_i)^2} \tag{25}$$

$$\xi_2^2 = a^2 \sigma_\tau^2 \frac{\left\langle \sum_i b_i n_i \right\rangle^2}{(\sum_i b_i w_i')^2} = \frac{\sum_{ij} b_i b_j \langle n_i n_j \rangle}{(\sum_i b_i w_i')^2}, \tag{26}$$

in which n_i is the noise for the sample S_i. The expectation value $\langle n_i, n_j \rangle$ is the autocorrelation function evaluated at the time $t_i - t_j$, which can be represented as the sum of two autocorrelation functions R_S and R_P:

$$\langle n_i, n_j \rangle = R(t_i - t_j) \equiv R_{ij} = R_S(t_i - t_j) + R_P(t_i - t_j). \tag{27}$$

R_{ij} is an element of the covariance matrix for the samples S_i. The function R_S is the series (thermal noise) autocorrelation function, given by

$$R_S(t) = \frac{\sigma_S^2}{\int_{-\infty}^{\infty} h'^2(t)dt} \int_{-\infty}^{\infty} h'(t+u)\, h'(u) du, \tag{28}$$

in which $h' = dh/dt$ and σ_S is defined in Equation 7. The function R_P is the autocorrelation function for pileup noise:

$$R_P(t) = \sigma_E^2 \int w(t+u)\, w(u) du. \tag{29}$$

Note that $R(0) = \sigma_S^2 + \sigma_P^2 = \sigma_{sample}^2$, in which σ_{sample}^2 is the variance of an individual sample S_i. σ_P is given by Equation 3 and σ_S is given by Equation 7.

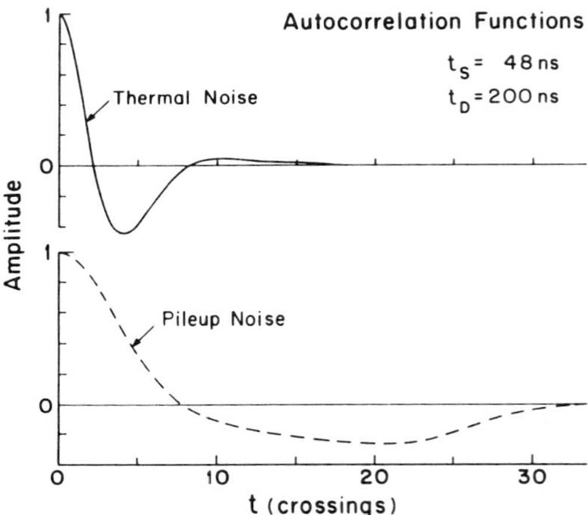

Figure 3: Autocorrelation functions for the case $\tau_p = 48$ ns and $t_d = 200$ ns. The upper curve is the function R_S as a function of time, in units of 16 ns, and the lower curve is R_P. Both curves are normalized to unit amplitude.

In Figure 3 we show, as an illustration of the two types of noise, an example of $R_S(t)$ and $R_P(t)$ for one case studied (t_S=48 ns, t_d=200 ns). Note that R_P is a much wider function than R_S, which is due primarily to the fact that the function w(t) is considerably wider than h(t) because of the convolution with i(t) (see Equation 35).

In order to simplify slightly the equations, we introduce two new sets of coefficients c_i and d_i and a constant C:

$$A = \sum_i c_i S_i \tag{30}$$

$$\tau = \frac{1}{A} \sum_i d_i S_i - C. \tag{31}$$

In the optimal filtering method, the solution for the parameters c_i and d_i is found by minimizing the variances of A and τ. For the case of uncorrelated variables ($\epsilon_{12} = 0$), one finds a solution identical to that of the χ^2 method. For the correlated case, inspection of the χ^2 solution gives directly the result:

$$c_i = (Q_2 \sum_j w_j V_{ij} - Q_3 \sum_j w_j' V_{ij})/\Delta \tag{32}$$

$$d_i = (Q_1 \sum_j w'_j V_{ij} - Q_3 \sum_j w_j V_{ij})/\Delta \tag{33}$$

$$C = 0.$$

The weight matrix $\mathbf{V}$ is obtained by inverting the covariance matrix $\mathbf{R}$.

Weighting functions

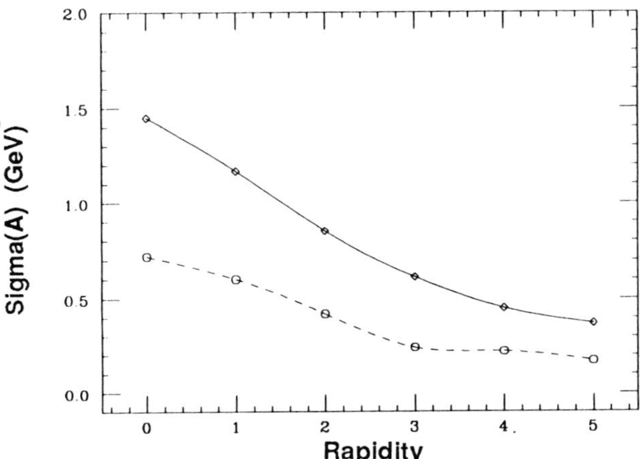

Figure 4: Rapidity dependence of the amplitude parameter, using the method of optimal filtering. Solid curve: EM+HAD sections, dashed curve: EM section.

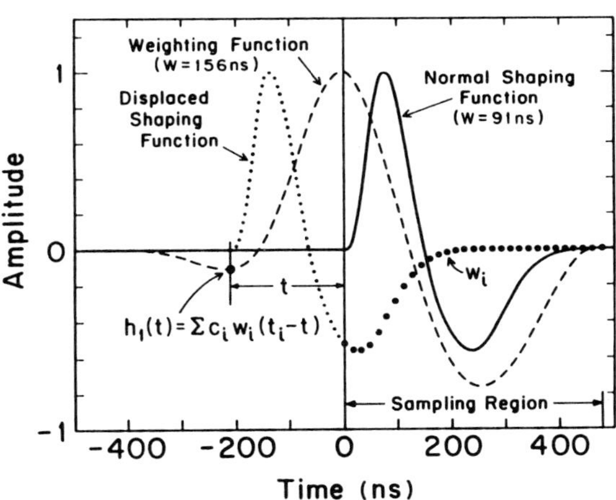

Figure 5: Illustration of the concept of the weighting function. The unit-amplitude shaping function is displaced by t, and the sampled points w_i (heavy dots) are multiplied by the optimal filter constants a_i to yield the value of $h_1(t)$.

In order to understand the implications of the solutions found above, we describe here the weighting functions appropriate to the problem, presented in Equations 30-33 above. Because these equations are simply a linear function of the measured signal (no convergence criterion is needed for their solution, as is the case for an iterative χ^2 solution), it is possible to define a response function for the system as a function of its time origin. To do this, we replace S_i by the unit amplitude response, shifted in time by an amount t:

$$h_1(t) = \sum_i c_i w(t_i - t). \tag{34}$$

This function is the amplitude that is obtained by the optimal filtering method for a signal of unit amplitude displaced in time. It can be viewed as a measure of the response of the system to out-of-time signals. By definition $h_1(0)$ is unity. The concept of the weighting function is illustrated in Figure 5. The effective width of the weighting function is defined analogously to bandwidth, in the time domain. It is the width of a rectangle of unit height which has the area of the integral of the square of $h_1(t)$:

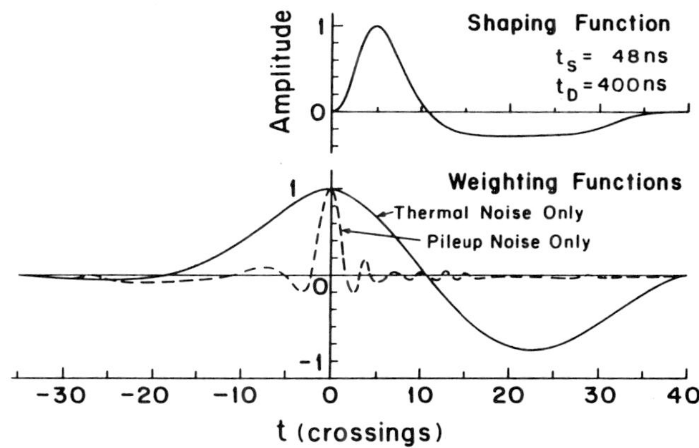

Figure 6: Weighting function for two extreme cases in which(1) thermal noise = 0, and (2) pileup noise = 0. In the case where the thermal noise is negligible, the method minimizes the width of the weighting function in order to reduce the effects of pileup noise.

$$W = \int_{-\infty}^{\infty} h_1^2(t)dt = \int_{-\infty}^{\infty} [\sum_i c_i w(t_i - t)]^2 dt \tag{35}$$

$$= \sum_{ij} c_i c_j \int_{-\infty}^{\infty} w(u)\, w(t_j - t_i + u) du$$

$$= \frac{1}{\sigma_E^2} \sum_{ij} c_i c_j R_P(t_i - t_j),$$

where the latter substitution comes from Equation 29. Recall (Equations 25 and 27) that

$$\sigma_A^2 = \sum_{ij} c_i c_j R_{ij} \tag{36}$$

$$= \sum_{ij} c_i c_j R_S(t_i - t_j) + \sum_{ij} c_i c_j R_P(t_i - t_j)$$

$$= \xi_S^2 + \xi_P^2$$

$$= \xi_S^2 + \sigma_E^2 W, \tag{37}$$

in which ξ_S^2 and ξ_P^2 are the components of the variance of A due to thermal and pileup noise, respectively. By comparison with Equation 3, we see that the effective width of the weighting function W plays the same role as the integral I_2p plays for the variance of a single data point.

In Figure 6 we illustrate the weighting function for two extreme cases, to demonstrate its properties. For this example we have chosen a shaping function of $\tau_s = 48$ ns and a drift time of $t_D = 400$ ns, giving the waveform shown in Figure 1. For illustrative purposes, we calculate the weighting function for two extreme cases: (1) the case for thermal noise only (i.e. $\mathcal{L}$=0) and (2) the case for pileup noise only ($\sigma_S = 0$). In the first case, the weighting function maximizes its width, since there is no penalty for pileup, and the effective measurement time is made as long as the shaping function permits, thereby minimizing the thermal noise. In the second case, only pileup is important, so the effective width becomes as small as possible, to minimize its effects.

We see, therefore that the shape of the weighting function for a solution for the c_i and d_i gives an indication of the relative importance of thermal and pileup noise. The weighting function for a solution in which pileup is the dominant component of the noise will be narrower than one in which thermal noise dominates. As is shown by Equation 37, the contribution of pileup to the variance of the amplitude A is simply the effective width W times σ_E^2.

We also calculate a weighting function for the measurement of τ, defined analogously to h_1:

$$h_2(t) = \sum_i d_i w(t_i - t). \tag{38}$$

This weighting function has the property that $h_2(0)$ =0, and it is linear in t in the region near t =0. The width of this linear region indicates the range over which the optimal filtering method will produce useful results for the timing correction. An example of both types of weighting functions is given in Figure 8.

Results

We now turn to the results obtained by using the optimal filter formalism discussed above, along with the thermal noise values of Table 2 and the pileup noise values of Table 1. We have studied the following specific questions

1. Rapidity dependence of the resolution of a liquid ionization calorimeter operating at a luminosity of 10^{33}cm^{-2}sec^{-1}.
2. The way in which the amplitude and timing resolution depend on the thermal and pileup noise contributions, for different shaping function.
3. Dependence of the resolution on luminosity.
4. Dependence of the resolution on the number of data samples recorded.

In general the examples given here are for the larger "trigger towers" ($\Delta\eta \times \Delta\phi = 0.32 \times 0.32$), since it is for this case that the two types of noise are comparable. Each of the studies is described in detail below.

Rapidity dependence of resolution

Since we are working with a spherical calorimeter simulation, we assume that the angular dependence of the thermal noise is proportional to $\sin\theta$ (recall that we work with E_T):

$$\sigma_S(\theta) = \sigma_S(90^\circ)\sin\theta. \tag{39}$$

We combine this with the pseudorapidity dependence of the pileup noise, which is given by Equation 5 and determine optimal values of the c_i and d_i at each value of η.

The result for a particular case (τ_p = 48 ns, t_d = 200 ns, $\mathcal{L} = 10^{33}$) is shown in Figure 4. We note the following properties:

- $\eta = 0$ is the worst case for both amplitude and timing resolution;
- the timing resolution improves significantly faster than the amplitude resolution as one moves away from η =0;
- The rapidity dependence of the resolution will become weaker at higher luminosity, as the pileup noise depends only weakly on the rapidity.

Dependence on shaping time and drift time

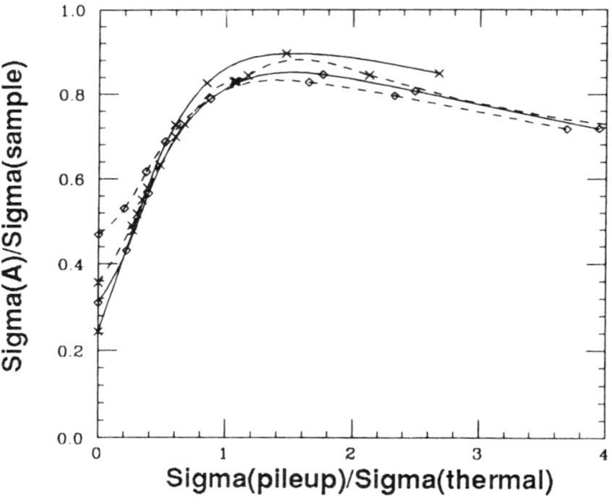

Figure 7: Values of the ratio $R = \sigma_A/\sigma_{sample}$ versus the ratio of σ_P to σ_S for bipolar shaping functions. Crosses: τ_p = 48 ns, diamonds: τ_p = 96 ns, solid lines: t_d = 400 ns, dashed lines: t_d = 200 ns.

For a variety of cases, we have calculated the reduction in the amplitude resolution over that for a single sample achieved by optimal filtering. We calculate the ratio of the resolution of A (Equation 25) divided by the uncertainty on a single data sample:

$$R = \sigma_A/\sigma_{sample} \tag{40}$$

We plot R as a function of the ratio of σ_P/σ_S, the contributions to the single sample variance due to pileup and thermal noise. The results for a variety of bipolar functions are shown in Figure 7. We note that there is a certain universal behavior in these curves, except near the origin, where $\sigma_P \approx 0$. The lesson one learns from this study is that, for cases where $\sigma_P/\sigma_S \geq 0.5$, the improvement in resolution using optimal filtering, is essentially independent of the choice of shaping function.

Luminosity dependence

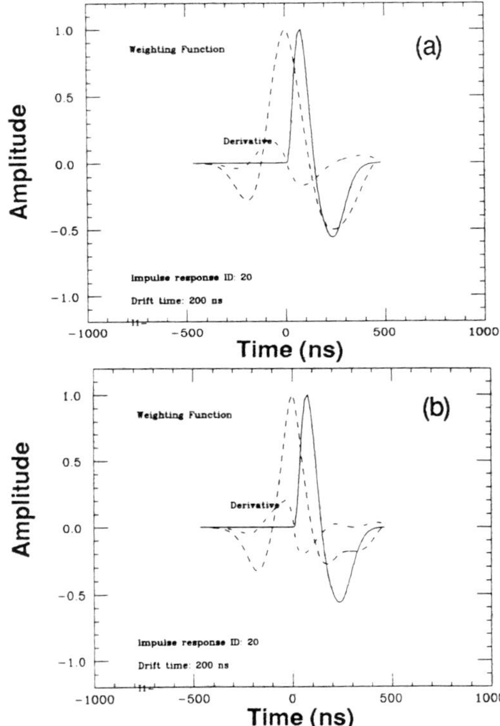

Figure 8: Weighting functions (dashed lines) for both time and amplitude at $\eta = 0$. (a) $\mathcal{L} = 10^{33}$; (b) $\mathcal{L} = 10^{34}$. Note that the amplitude weighting function $h_1(t)$ becomes narrower and the timing weighting function $h_2(t)$, labeled "Derivative", becomes steeper as the luminosity increases.

In this exercise we study the dependence of the amplitude resolution as a function of machine luminosity. Clearly the thermal noise contribution is luminosity independent, while the pileup variance is proportional to luminosity:

$$\sigma_P^2 = \frac{\mathcal{L}}{\mathcal{L}_0}\sigma_P^2(\mathcal{L}_0). \tag{41}$$

We choose optimal values for the coefficients c_i and d_i for each value of the luminosity. The results are shown in Table 3 and, for one choice of peaking time τ_p and drift time t_d, in Figure 9. We also show, in Figure 8 the weighting functions for both amplitude and timing information at $\eta = 0$ for two values of the luminosity: 10^{33} and 10^{34}. Note from this figure how the width of the amplitude weighting function decreases as the luminosity increases, in accordance with expectations. Note also that the weighting function for the timing information has a steeper slope and narrower useful range (the linear portion of the curve) at the higher luminosity. The particular case we have chosen

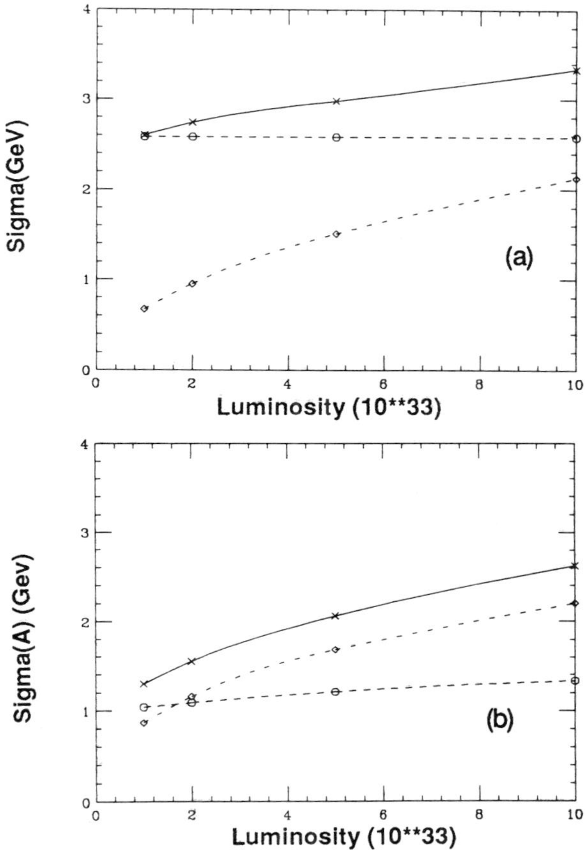

Figure 9: Illustration of the combination of pileup and thermal noise for 0.32 ×0.32 trigger towers (TOT deposition) using the method of optimal filtering. (a) Noise contribution versus luminosity for single samples. (b) Value of sigma for the amplitude function obtained using optimal filtering.

for this study is $\tau_s = 48$ ns, $t_D = 200$ ns, trigger sum = 0.32 × .32 (EM + hadronic). The value of σ_P used was an average over all rapidities. From the behavior seen in the curves, we see that, approximately

$$\sigma_A \propto \mathcal{L}^{0.30} \tag{42}$$

$$A\sigma_\tau \propto \mathcal{L}^{0.12}. \tag{43}$$

Table 3: Values of resolution at two luminosities for a 0.32 × 0.32 cell size at $\eta = 0$.

			$\mathcal{L} = 10^{33}\text{cm}^{-2}\text{sec}^{-1}$		$\mathcal{L} = 10^{34}\text{cm}^{-2}\text{sec}^{-1}$	
τ_p (ns)	t_d (ns)	Section	σ_A (GeV)	$A\sigma_\tau$ (GeV-ns)	σ_A (GeV)	$A\sigma_\tau$ (GeV-ns)
48	200	EM	0.8	23	1.6	34
48	200	TOT	1.5	72	3.1	94
48	400	EM	0.9	43	2.0	53
48	400	TOT	1.9	134	3.8	174
96	200	EM	0.7	27	1.4	39
96	200	TOT	1.3	80	2.8	111
96	400	EM	0.9	46	1.8	62
96	400	TOT	1.6	139	3.6	187

Dependence on the number of samples

From the χ^2 solution for the covariance matrix for the parameters, we see that for the case in which $Q_3^2 << Q_1Q_2$ (the uncorrelated case), the variances of the fitted parameters are:

$$\sigma_A^2 = \frac{1}{Q_1} = \frac{1}{\sum_i jw_j V_i jw_i} \tag{44}$$

$$A^2\sigma_\tau^2 = \frac{1}{Q_2}\frac{1}{\sum_i jw_j' V_i jw_i'}. \tag{45}$$

We see from these expressions that terms in the sums which are small contribute little to reducing the variance. We therefore eliminate such terms from the sample, one by one (independently for σ_A and σ_τ), reevaluate the coefficients, and find the new values for σ_A and σ_τ. This process is carried out until only two samples are left. The results indicate that, with 4 or 5 data points, approximately 80 % of the ultimate resolution can be obtained for either σ_A or σ_τ. However, the data points chosen by this method are not the same for the two parameters.

We have considered another method, in which data points have been chosen by an algorithm similar to the one above, but involving simultaneously terms from both sums. In this way we obtain samples which are suitable for determination of either A or τ. The results of this study are shown in Figure 10, and for the case of 5 data points, the resulting coefficients are shown in Figure 11. Although this method requires more study, it seems clear from the one case studied that the resolution in both A and τ is improved by using this common sample.

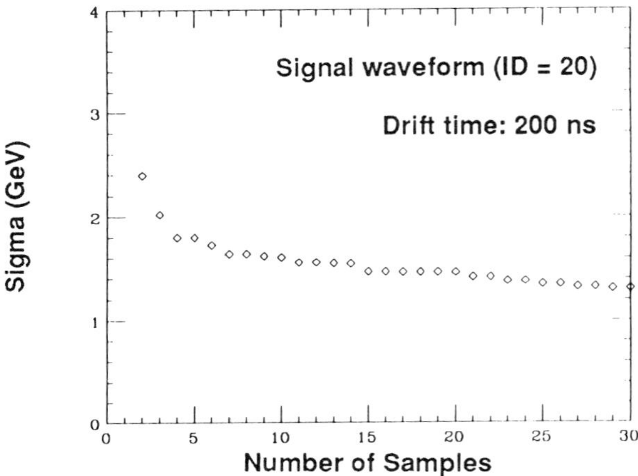

Figure 10: Amplitude resolution using optimal filtering method versus the number of data samples. For this study, $\tau_p = 50$ ns, $t_d = 200$ ns, and $\mathcal{L} = 10^{33}$.

Effects of Noise on Triggering Quantities

It is instructive to examine some of the distributions of possible utility in the trigger to obtain a qualitative feel for the effects of noise and the benefits of optimal filtering.

Total E_t is calculated by summing the transverse energy for all calorimeter cells. The single sample noise contribution to total E_t is about 35 GeV from thermal sources, and about 24 GeV from pileup at luminosity 10^{33}. Adding these two quantities in quadrature gives 43 GeV of noise. Our studies show that the total E_t distributions are sufficiently broad that the effects of noise are visible but do not change qualitatively the total E_t distribution.

Noise has a more significant effect on $\not{p}_t$, the magnitude of transverse momentum imbalance. Figure 12 shows $\not{p}_t$ distributions of minimum bias and two jet events for various levels of noise. Curve (a) is $\not{p}_t$ with no additional calorimeter noise and shows the level of momentum imbalance from neutrinos and particles escaping down the beam hole. Curves (b) and (c) are the level of momentum imbalance at $\mathcal{L} = 10^{33}$ from single sample noise and optimally filtered noise, respectively.

As shown in the figures, the addition of noise has a drastic effect on $\not{p}_t$. With no noise, the missing p_t distribution is peaked at 0. Adding noise shifts the peak of the missing p_t distribution to larger values, and increases the width. The effect of noise on $\not{p}_t$ is ameliorated by optimal filtering but is still an important effect.

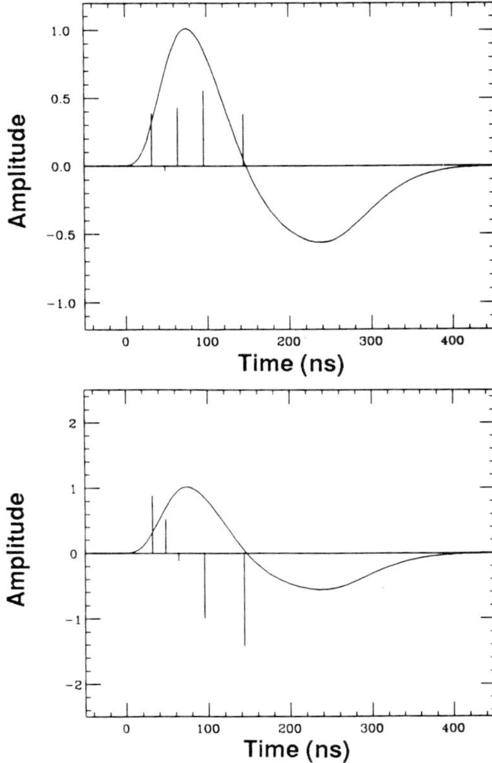

Figure 11: Shaping functions showing the position of five samples and the value of the coefficients (indicated by the height of the vertical bars). The upper graph show the coefficients c_i for the amplitude measurement and the lower graph shows the coefficients d_i for the timing measurement.

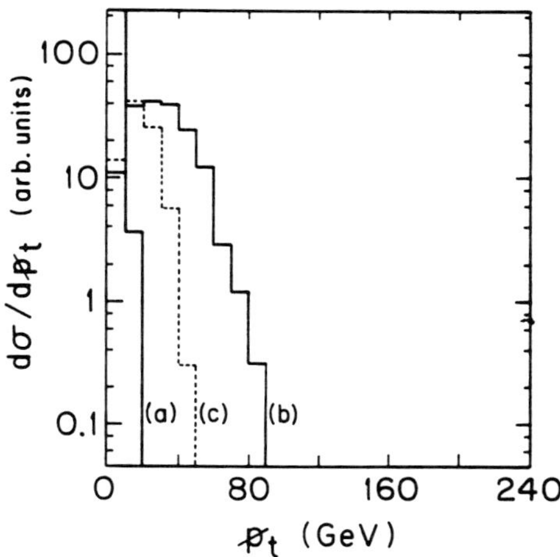

Figure 12: The momentum imbalance $\not{p}_t$ as measured in the calorimeter for minimum bias and two jet events. (a): no noise; (b): single sample noise and (c): optimally filtered noise for $\mathcal{L} = 10^{33}$.

Summary

In this study we have shown that it is possible to obtain values of A and τ from a multiply sampled waveform, with a statistical accuracy significantly better than that of a single sample. The use of optimal filtering permits a rapid calculation of the parameters with coefficients that can be determined from a knowledge of the thermal noise of the calorimeter. This technique can be used to quantitatively evaluate the dependence of resolution on rapidity, luminosity, and the number of data samples.

Acknowledgements

It is a pleasure to acknowledge many useful discussions with Veljko Radeka and Sergio Rescia.

THE DESIGN AND CONSTRUCTION OF AN SSC SILICON ELECTROMAGNETIC CALORIMETER PROTOTYPE

Brookhaven National Laboratory, Carnegie-Mellon University, University of Colorado, University of Mississippi, Oak Ridge National Laboratory, University of Oregon, University of Tennessee, The Silicon Electromagnetic Calorimeter Collaboration
presented by S. Berridge

Abstract

The SECC will design, construct and beam test a silicon electromagnetic calorimeter of cross sectional area 1000 cm^2, employing about three square meters of silicon diodes. The 28 layer device will consist of one radiation length tungsten radiators each instrumented with 300um silicon detectors. The design goal for the response of the electronics is one SSC beam crossing (16 nanoseconds). Signals will be summed at the periphery of the radiator structure, for a total channel count of 1520. The development of this subsystem module builds on the SSC generic research and development work of the collaboration, silicon diode fabrication, current compensated preamplification, and radiation effects. This work is supported in part by the Department of Energy.

INTRODUCTION

The advantages and versatility of silicon calorimetry have been described in detail within the literature. [Ref. 1,2,3] For these reasons silicon based calorimeters are being considered for a number of experimental applications within the Superconducting Super Collider. The SECC submitted a proposal to the Department of Energy to construct a prototype end cap calorimeter suitable for SSC R&D. The project was scaled back yet will still provide us with an excellent test bed for virtually all of the new design problems which have to be addressed prior to full scale SSC calorimeter construction. This paper describes these obstacles and explores the currently envisioned solutions.

SECC Calorimeter Design

Configuration

Much consideration has been given to the configuration of the SECC silicon calorimeter layout in order to produce the most versatile and cost effective test calorimeter possible. The primary design considerations revolved around pad sizes, sampling fraction, number of interaction lengths, and longitudinal segmentation.

We have decided on three pad sizes for the silicon detectors. The smallest cell size will be 1.067cm x 1.067cm to enable us to test hadron background rejection of an EMPACT type electromagnetic calorimeter. The largest pad size will be 3.2cm x 3.2cm designed to correspond to the basic pad size for an L* hadron calorimeter. The third size was chosen in to be one half the largest pad size and is 1.6cm x 1.6cm. Nominal choices for pad sizes would have been 1 x 1, 2 x 2 and 3 x 3cm but the actual values were chosen to better utilize the available silicon area on 4" wafers. Hamamatsu has been chosen to produce the 300μm silicon detector arrays from N-type silicon with P$^+$ junctions. Each of the arrays will have the same size active area, namely 41cm^2, but that area will be divided into different numbers of pads. The arrays resulting from the chosen pad sizes are 6 x 6, 4 x 4 and 2 x 2. With these arrays, we produce calorimeter layers each with a square active region of 32cm on a side. Of course, there are dead areas between the arrays which are spaced 0.5mm apart. These dead areas are negligible when compared to the shower diameters. The array configuration is shown in Figure 1. The smallest pads are placed in the first 12 radiation lengths, where the EM shower dimensions are only a few millimeters. Deeper in the shower, where the Molieré radius is an appropriate scale, the pad sizes are increased. For hadron rejection studies, this configuration allows us to study the effects of shower development measurements in both transverse and longitudinal dimensions. The hadrons can be introduced at varying distances from the electron axis, up to a few transverse interaction length separation. This is sufficient to give a good parameterization of the e/h separation over a wide range of energies of e and h. The 28 layers of the calorimeter will require about $3m^2$ of silicon detectors.

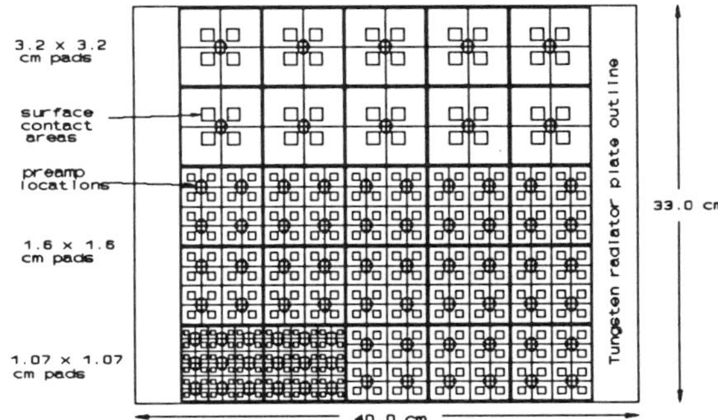

Figure 1. Layers 1-12

The active silicon planes will be placed onto tungsten alloy radiators which are 3.5mm thick. When assembled they will provide $1X_0$ sampling throughout the calorimeter. The radiators will be 40cm x 33cm which provides space for mechanical supports, etc. The total depth of the calorimeter will be 28 radiation lengths with sampling at each one. This depth corresponds to about one interaction length. Longitudinal segmentation will generate 5 signals from layer summations of $4X_0$, $4X_0$, $4X_0$, $8X_0$, and $8X_0$. Figure 2 is an oblique block cutaway view of the calorimeter.

Detector Mounting and Interconnection

The calorimeter will be constructed with 0.3cm spaces between the plates. This space will house the silicon detectors, preamplifiers and associated circuitry. Refer to Figure 3. The tungsten radiator plates will serve as a rigid mounting surface for the assembly. The assembly of the individual planes of the

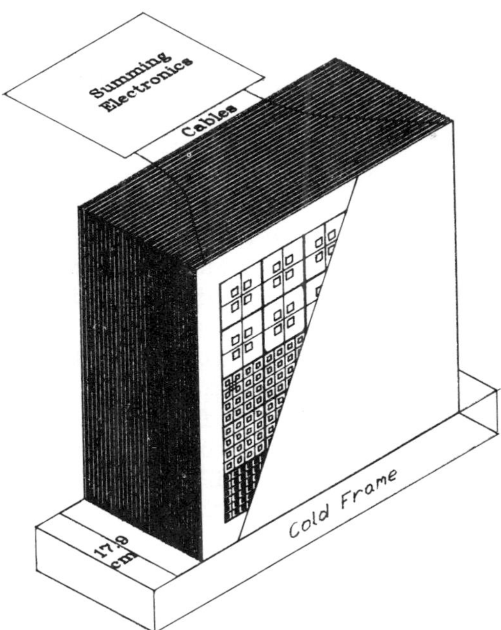

Figure 2. Oblique Block Cutaway of SECC Calorimeter

calorimeter is planned to be accomplished as follows. A kapton type circuit will be permanently attached to the rigid radiator plate. A reverse bias voltage of +40 to +150 VDC will ultimately be applied to the detectors via this circuit. The detectors will then be mounted to this circuit providing both mechanical support and electrical contact to the N^- side of the detector arrays. This contact during mounting will be made using conductive epoxy previously proven to be harmless to the silicon detectors.

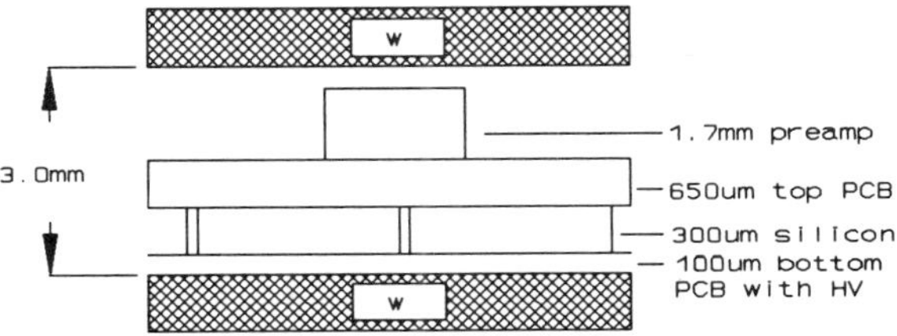

Figure 3. Sampling Layer Detail (no scale)

Since the calorimeter may operate over a range from room temperature to -20°C, it is imperative that these contacts as well as all interfaces be able to withstand such temperature changes. Testing to verify these tolerances has begun with the testing of existing devices in an environmental chamber. The interconnection layer, a thin multilayer circuit, will be applied to the top (P^+ side) of the detectors.

This circuit will be applied with the quad preamps already mounted onto it. Connection will be made from this circuit to the aluminized top surface of the silicon. The options for this connection which are currently being evaluated are (primarily) Al wire bonding and (secondarily) conductive epoxy. This interconnection circuit will also provide power distribution for the preamps and traces running to one edge for connection to the summing amplifiers. The completed planes will then be stacked into a final assembly. Since the preamps are located directly in the calorimeter it will be necessary to provide cooling. Within the calorimeter, virtually all of the heat generated comes from the preamps. Their typical dissipation is about 20mW per channel. There is one channel for each pad. Heat from leakage current in silicon detectors is negligible.

Signal Routing and Summing Amplifiers

The signal routing occurs on the interconnection circuit which will need to carry signals from the detector pads to the inputs of the preamps and signals from the preamp outputs to the edge of the calorimeter and to the summing amplifiers. In order for these lines to be able to carry very fast signals with low distortion, attenuation and crosstalk, they will be designed as striplines. Work has begun to analyze this configuration in a thin multilayer circuit board. These signals will terminate into the summing amplifiers which will be developed from off the shelf components. Digitization and data collection beyond the summing stage will be done with standard commercial equipment.

Tungsten Radiators

The radiators will be sintered from a nonmagnetic tungsten powder alloy. We have used this material previously in other detectors and have found it to have reasonable machinability and structural integrity. The rough sintered plates will be machined to accommodate structural reinforcing rods and then ground flat to a thickness of 3.5mm (±0.050mm). This level of control will insure uniform shower development throughout the calorimeter. Once assembled the plates will become an integral part of the mechanical structure. The assembly will contact a cold plate attached to the top face of the calorimeter for cooling.

Cooling and Heat Loads

Cooling for the calorimeter will be provided by a standard commercial refrigeration system. The operating point for the system will be adjustable from room temperature to -20°C. For all practical purposes, all of the heat load comes from the preamps. Each quad preamp package will dissipate about 80mW of power. This heat will be convected from the preamps to the radiator plates which are physically separated by approximately 0.25mm. The temperature of the system will be maintained to within ±2°C of the operating point. Analysis of this heat transfer has shown that the total heat load for the 28 layer calorimeter will be about 600 BTUs/hr or 170 watts. Calculations also show that this amount of heat can easily be conducted to the heat sink via the tungsten radiators within the temperature tolerance.

References

[1] S. Berridge et al., "The Small Angle E-M Calorimeter at SLD: a $2m^2$ Application of Silicon Detector Diodes. *Trans. on Nuc. Sci.*, vol. 36, No. 1 (1988), 339.

[2] L* Collaboration, "Expression of Interest to the Superconducting Super Collider Laboratory," (1990), Sec. V, 46.

[3] EMPACT Collaboration, "Expression of Interest," (1990).

RADIATION DAMAGE EFFECTS IN SILICON ELECTROMAGNETIC CALORIMETRY AT THE SSC

J.S. Russ and Yang-Ling Zhang
Carnegie Mellon University
Pittsburgh, PA 15213

Abstract

Active silicon devices like transistors or capacitors exposed to ionizing radiation exhibit a saturation of the radiation damage effects after several hundred kilorads of dose. In the Superconducting Super Collider (SSC) environment the peak annual dose at a shower maximum in the forward electromagnetic calorimeter due to *ionizing* radiation is estimated to be 1 Mrad or less for a calorimeter at 3.5 m from the crossing, set at $\eta = 3$. We report experiments on the response of silicon p-i-n diodes to electron and gamma dose at the 1 Mrad level. The increase in leakage current saturates, rising initially by a factor of order 10 and decaying with a room temperature relaxation time of two weeks. No measurable change in the electron charge transport is seen, as determined by energy resolution measurements with α-sources of several energies.

1. Introduction

The major effects of ionizing radiation on semiconductor devices center in charge buildup in the oxide layer(s).[1] In MOS devices the major consequences are shifts in the pre-irradiation values of R, C, or threshold. For silicon p-i-n diodes used as sampling layers for electromagnetic calorimetry the concern about radiation damage centers on two areas: increase of leakage current, which can paralyze the preamplifier and/or produce excessive shot noise, and introduction of charge traps, thereby changing the calibration of the detector.

Normal operation of silicon diodes for charge collection is done at "low" electric fields, typically 3 kV/cm for a 300 μm detector. This voltage level may be increased to operate with short charge collection times at the SSC, but it is unlikely to move into the "high field" regime (100 kV/cm) in which charge amplification and hence dramatic increase in leakage currents can be observed.[2] *If* silicon diodes behave in an oxide-limited fashion, then one should expect saturation of the leakage current at high dose. This effect has been reported for silicon diodes exposed to 20 keV X-rays.[3] Our interest is to verify this saturation in the mixed radiation field to be expected in a sampling electromagnetic calorimeter. Most of the dose in such circumstances is delivered by photons and electrons near the critical energy of such a device, i.e., typically 10 MeV, not 20 keV. The radiation damage literature claims an equivalence of total damage for ionizing dose delivered by 20 keV X-rays, Co^{60} photons, and "linac" electrons, i.e., electrons delivering a dose equivalent to minimum-ionizing particles.[4] The purpose of this work is to evaluate this equivalence specifically for sampling diodes and to begin the study of the charge trapping problem.

2. Detector Characteristics

The detectors used in this study were Hamamatsu 25 cm^2 diodes, 300 μm thick. The leakage currents before irradiation ranged from 0.3 μA/detector to 1.6 μA/detector (12–60 nA/cm^2). The detectors performed identically when exposed to α-sources of different energies. Spectra for four different α-sources seen by one detector are displayed in Figure 1. The spectra were taken with standard Tennelec nuclear spectroscopy electronics. Two different integration times were used: 1 μsec and 16 μsec. No variations were noted in the detector performance for the two integration times, either before or after irradiation. The ability to locate the peak of the pulse height distribution is obviously good. The skewed shape is due to the Landau tail of the energy loss in an air gap of 500 μm and a resin protective coating of thickness 2 μm. These data were taken with the sources on the negative-bias side and so reflect the transport of electrons across the thickness of the detector. The resolution is 1.6% (σ), measured from peak to high energy side, for Am^{241}. For U^{238} the peak is ill-defined, but the half-maximum point on the high energy side is very well defined and serves to look for peak shifts after irradiation.

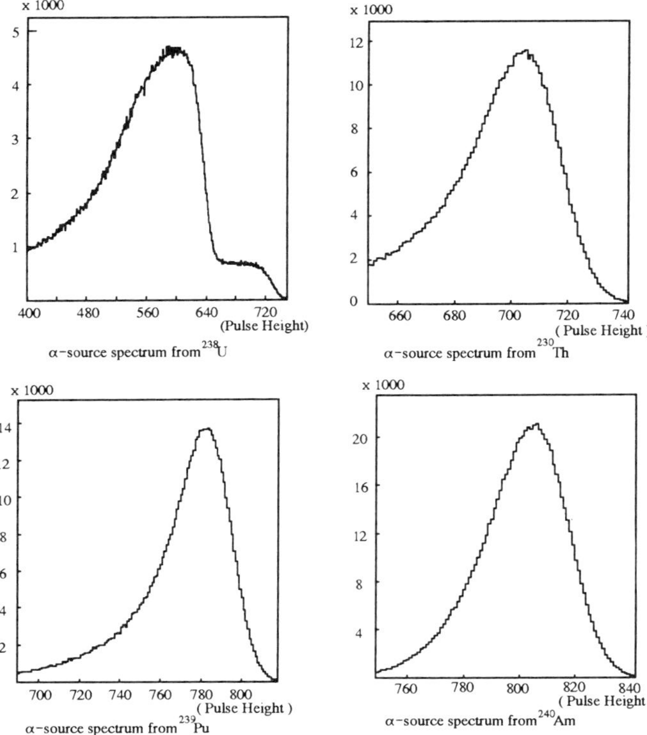

Figure 1: α particle spectra before irradiation for four different α emitters: Am^{241} (5.44 MeV), Pu^{239} (5.1 MeV), Th^{230} (4.62 MeV), and U^{238} (4.15 MeV).

3. Electron Dose Effects

The 21 MeV linac at the Argonne National Laboratory's Chemistry Department is available to outside users, and these irradiations were conducted there. An extracted beam is available, and it was collimated to an area of 5 cm^2. The beam current delivered was 1.3 nA/cm^2, and the exposure time was 90 minutes. Total dose is 1.2 Mrad. Two detectors were irradiated, one powered and one unpowered. No differences in the gross effects were seen. The leakage currents were measured immediately before and after the exposure. The increase for each detector was from approximately 50 nA/cm^2 to approximately 500 nA/cm^2. Immediately after the irradiation there was a short-time decay to around 400 nA/cm^2, followed by a slow decrease in current under room temperature storage conditions. The decrease is exponential with a half-life of 85 days. It is not affected by whether the detector was powered or not at the fields used here (1.5 kV/cm).

Source spectra of the sort shown in Figure 1 were collected after the irradiation. Using various analytical functions to fit the peak shape, we were unable to see any statistically-meaningful systematic change in the shape or peak position after the electron beam irradiation. The overall shifts in the peak position or half-maximum position on the high side were less than 1%. One detector moved up, the other down. Therefore, there is no evidence of increased noise or of charge loss at the 1% level after a dose of 1.2 Mrad in an electron beam.

4. Gamma Irradiation

To check the equivalence of electron and gama irradiations, one detector from the electron-beam run and another unexposed detector were placed in the Co^{60} irradiation facility of the Argonne National Laboratory Biology Department. In contrast to the electron beam exposure, which applied dose over a well-defined region in the center of the detector, the gamma exposure irradiated the entire detector. The dose was 1.5 Mrad, approximately uniform over the entire detector surface. Again, one of the detectors was powered, the other not.

In contrast to the electron beam case, quite different behaviors were observed in the two cases here. The powered detector, which had previously been exposed to the electron beam, exhibited a short-term decrease in the leakage current before settling into a steady but slow long-term decay. This behavior is similar to its behavior after the electron beam dose. Following the short-term behavior, the leakage current for this previously-irradiated detector was only 20% higher than it had been prior to the gamma dose. This is very similar to the behavior of voltage threshold shifts in MOS devices and is the phenomenon labelled "saturation". The current in this detector has continued to decrease since the irradiation with the same slow time constant that we see in the detector exposed only to the electron beam. The α-source resolution was the same as before. There was no evidence of a peak shift.

The unpowered detector, which had not been irradiated previously, displayed instabilities in its leakage current after the gamma dose. Prior to the exposure, it was indistinguishable from the other detectors. Afterward, α-source spectra were significantly broadened, but not significantly shifted. Oshugi, et al., reported similar behavior for Hamamatsu silicon p-i-n photodiodes.[5] At the dose levels reported here, they observe erratic behavior of the diodes. At some larger dose level it recovered to stable operation. We note that this effect did *not* occur for the detector which had been exposed first to electrons. It suggests that a mixed radiation field of electrons and photons, as would be found in an electromagnetic calorimeter, would not produce the same kind of instability. We plan to explore this in future work.

5. Summary

The exposure of silicon p-i-n diodes to Mrad-scale doses of ionizing radiation produces only limited increases in the overall leakage current, similar to radiation dose effects seen in other semiconductor systems. This means that the leakage current increase will *not* be a limiting factor in the application of silicon sampling electromagnetic calorimetry in the SSC. There are small and erratic shifts in the electron charge transport efficiency ($< 1\%$) at these dose levels, as measured with α-source spectra. This says that the absolute energy scale is unaffected by dose at this level, unless there are effects in the hole transport efficiency. We will measure the hole transport and also explore further the instability seen in one detector exposed only to photon radiation.

As part of the Silicon Electromagnetic Calorimeter Collaboration activity, we plan to repeat these measurements on a large sample of detectors to establish a baseline for system behavior under irradiation.

6. Acknowledgements

We wish to thank A. Stevens, W. Haberichter and the staffs of the ANL linac and Co^{60} irradiation facilities for their help in carrying out the exposures.

References

[1] H.E. Boesch, Jr., F.B. McLean, J.M. Benedetto, J.M. McGarrity, and W.E. Bailey, "Saturation of Threshold Voltage Shift in MOSFETs at High Total Dose," IEEE Trans. on Nucl. Sci. NS-33, No. 6, 1191 (1986).

[2] P.W. Marshall, C.J. Dale, E.A. Burke, G.P. Summers, and G.E. Bender, "Displacement Damage Extremes in Silicon Depeletion Regions," IEEE Trans. on Nucl. Sci. NS-36, No. 6, 1831 (1989).

[3] G. Lindström, E. Fretwurst, H. Herdan, M. Rollwagen, P. Thomsen, and R. Wunstorf, "Radiation Damage Effects in Silicon Detectors," *Proceeding of the Workshop on Calorimetry for the Supercollider*, R. Donaldson and M.G.D. Gilchriese, eds., World Scientific (1990).

[4] J.M. Benedetto and H.E. Boesch, Jr., "The Relationship between Co^{60} and 10 keV X-ray Damage in MOS Structures," IEEE Trans. on Nucl. Sci. NS-33, No. 6, 1318 (1986).

[5] T. Oshugi, H. Kojama, H. Himeno, T. Kondo, M. Noguchi, M. Hasegawa, and S. Mori, "Radiation Damage of PIN Photodiodes by Gamma Rays," *Proceedings of the Workshop on Calorimetry for the Supercollider*, R. Donaldson and M.G.D. Gilchriese, eds., World Scientific (1990).

LOW COST, LARGE AREA SILICON DETECTORS FOR CALORIMETRY†

R. Korde[a], K. Furuno[b], H. Hwang[b], J.E. Brau[b], and W.M. Bugg[c]

[a]International Radiation Detectors, Torrance, CA 90509-2806
[b]University of Oregon, Eugene, OR 97403
[c]University of Tennessee, Knoxville, TN 37996

Abstract

Trapezoidal detectors with 28 cm^2 active area have been fabricated on > 2500 Ωcm, 4 *inch* diameter n-type silicon wafers. Instead of the commonly used ion implantation method, low-cost, high volume solid state diffusion technology along with phosphosilicate-glass and TCA gettering was adopted for boron and phosphorus doping. Typically the diode dark current was 15 μA @ 100 volts. Efforts are being made to obtain a finished device yield of 80% to meet the $\$2/cm^2$ price goal of SSC semiconductor detector group.

Introduction

Experiments at the Superconducting Supercollider (SSC) will challenge calorimeter performance. As the core element of the central detection system, the calorimeter must face the exceptional experimental conditions created by a very high luminosity, high multiplicities, and the large variety of physics phenomena to study. The following features will be asked of the hadron calorimeter: compact construction, easy calibration, good resistance to radiation damages, good energy resolution and a compensating (e/h=1) response. With respect to these issues of performance, silicon detectors offer many attractive features for SSC calorimeters [1]. Futhermore, they can be employed as complementary components with other calorimeters[2], as the present applications of plug calorimeters and small angle calorimeters[3]. Silicon detectors have been employed in sampling calorimeters for quite some time now[4], and there is strong motivation for pursuing low cost, large area devices. It should be noted that a large area device can easily be electrically subdivided to reduce capacitance, if desired.

Large area silicon detectors for calorimetry applications have been fabricated by us at United Detector Technology, Hawthorne, CA, using solid state diffusion technology. The first advantage of the diffusion technology is its lower cost of doping [5] compared with the ion implantation process. Secondly, diodes without any front dead region have been manufactured,so far, by adopting solid state diffusion technology [6][7]. These advantages motivated us to adopt the diffusion technology for the fabrication of the silicon detectors reported here.

To meet the price goal of $\$2/cm^2$ for large area silicon detectors needed in the SSC project,[8] the finished detector yield should be over 80%. For the best performance, the detector is used in a fully depleted mode needing breakdown voltage in excess of 100 volts. Fabrication of large area fully depleted detector with over 80% yield is really challenging because of the defect generation during processing which reduces the diode breakdown voltage and/or increases the leakage current [9]. Defects in semiconductor devices may be broadly divided into three groups. (i) crystallographic defects, (ii) metallurgical defects and (iii) pattern defects. This paper describes our efforts to reduce these defects during manufacturing of 28 cm^2 active area trapezoidal detectors.

Starting Wafers

Float zone, four inch diameter, n-type, one side mirror polished silicon wafers with bulk resistivity greater than 2500 Ωcm were used to fabricate the detectors. The starting wafers had thickness ranging from 400-525 μm with thickness variation on a given wafer limited to ± 15 μm. The dislocation density measured on a few representative wafers by Van Sirtl etch was less than $25/cm^2$. Two $28cm^2$ detectors were produced from each wafer following the trapezoidal design of the SICAPO collaboration [10].

The bulk quality of the high resistivity float zone starting wafers is usually excellent with excess carrier lifetime over 1 ms. However, it is known that if the surface of the silicon wafer is abraded or damaged (which results into lattice strain or stress) during the starting wafer preparation operations, stacking faults are generated during thermal oxidation [11]. To verify the surface quality of the starting wafers X-ray topographs on a few representative wafers were obtained from North Carolina State University. Wafer perfection depicted by the X-ray topograph was found to be satisfactory for the large active area detector fabrication.

Device Fabrication

Fig.1 shows the process sequence used to fabricate detectors reported in this work. Most of the oxidations and diffusions shown in Fig.1 were carried at around 1000°C. Phosphorus and boron diffusions were performed by phosphorus-oxy-chloride and boron nitride sources respectively. All the high temperature processing except boron deposition was carried out on 200 wafer run. Boron deposition was performed on 50 wafers at a time. To avoid the slip and dislocation generation during high-temperature processing, suggestions reported [12][13][14] were adopted. TCA ambient was used during all the oxidation and drive-in diffusion steps. TCA is known to reduce the metallic contamination of wafers during processing and is helpful in stacking fault suppression [15].

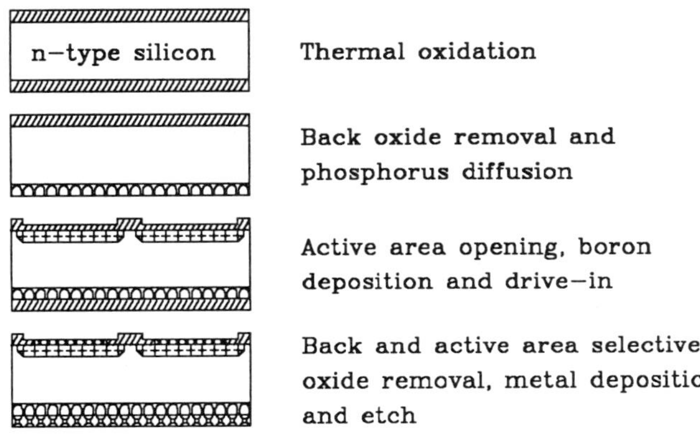

Figure 1. Process flow schematic.

Several gettering schemes may be utilized during silicon wafer processing to remove harmful impurities and defects from the critical device regions [16]. In addition to TCA gettering, heavy phosphorus doping, with phosphorus concentration greater than 10^{20} $atoms/cm^3$, has been used by us on the back of the wafers for gettering. Highly phosphorus doped layers are known to be the most effective gettering agent during silicon wafer processing [17]. Recently, Steve Holland has reported the use of phosphorus doped poly-silicon layer on back of the silicon wafers to reduce leakage currents in detectors [18]. Although the leakage currents of 1 nA/cm^2 obtained were excellent, the fabricated diodes had only 1 mm radius. In our opinion, the possibility of realizing a large area detector with proportionately low leakage currents using the above gettering method is dubious.

To reduce pattern defects, photoplates with defect size less than 1 μm were used. Absence of oxide islands within the pre-diffusion active area opening is extremely important during large area detector fabrication. If these islands are allowed, they will impede the boron diffusion leaving n-type pipes in the boron diffused region. These pipes may also be generated if pinholes exist in the front oxide before the back phosphorus diffusion. The pipes occur because the high phosphorus concentration in these regions will not be compensated during subsequent lower concentration boron diffusion. The pipe formation was mitigated by critical visual inspection of all the wafers before phosphorus and boron depositions. Wafers with questionable quality were subjected to rework.

Present 25% finished device yield is caused by failure due to large leakage currents and low breakdown voltages. Large leakage currents are a direct result of excessive pipes (or less probably a single large pipe) and localized metallic impurity contamination. Micro or meso plasmas in the wafer bulk originating at the impurity segregates or at the partical contaminates are known to be the causes for abnormally low breakdown voltages [19].

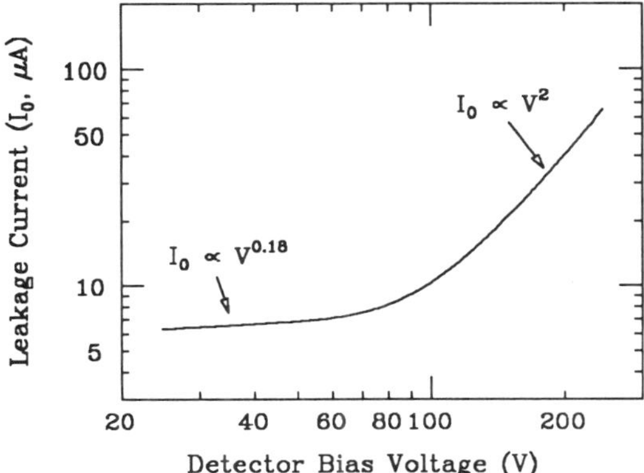

Figure 2. Leakage current observed.

Results and Discussion

Figure 2 shows the dependence of leakage current (I_0) on the reversed bias voltage, V. I_0 was found to be proportional to $V^{0.18}$ for voltages less than 80 volts. Generation currents proportional to $V^{0.5}$ are dominant leakage currents in majority of small area silicon detectors. Currents associated with resistive shunt paths, pipes, (these shunt paths get pinched-off at few volts) and diffusion current components are predominant in the detectors fabricated in this work. Our preliminary calculations show that these voltage independent current components have a total value of 4.2 μA for detector of Fig.2.

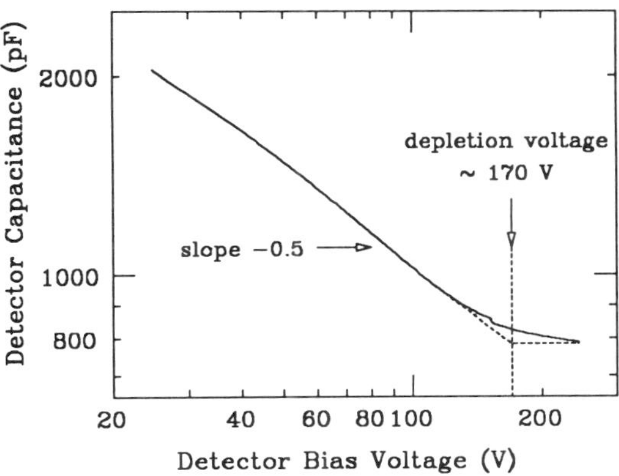

Figure 3. Capacitance observed.

The large current increase seen in Fig.2 for voltages in excess of 100 volts may be explained by the leakage currents associated with decorated microdefects at the back of the wafer. At 100 volts, the depletion region width in the detector is 300 μm, only 100 μm away from the back where large number of impurities and defects exist. As shown in

Fig.2, this current is proportional to V^2. Much stronger voltage dependence of the currents associated with defects have been reported by other workers [20].

Detector capacitance shown in Fig.3 has a slope of -0.5 as per the abrupt $p-n$ junction theory. The detector depletes fully at about 170 volts which is consistent with the bulk resistivity of 3100 Ωcm and thickness of 385 μm. The curvature observed around 170 volts is due to large lateral depletion region widths which are changing the effective area of the detector.

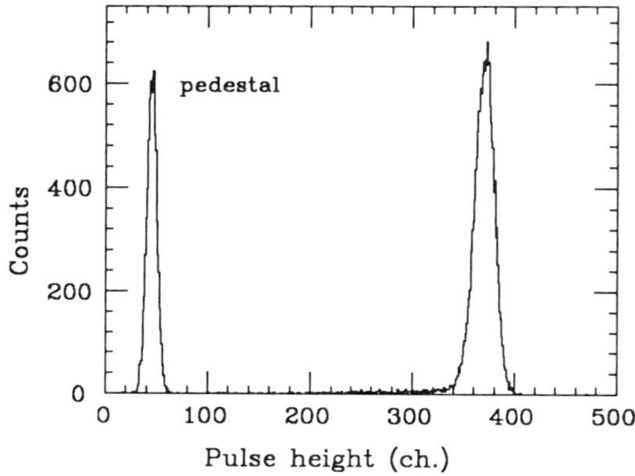

Figure 4. α (^{241}Am) signal

A sample of observed ^{241}Am α signal (E_α= 5.5 MeV) pulse height distribution detected by the detector described above is shown in Fig.4. The α ray was collimated to avoid path length fluctuations. The pedestal is also shown in the figure. The S/N ratio of 67 was observed in this particular measurement electronics.

Conclusions

Motivated by calorimetry application at the SSC, 28 cm^2 active area silicon detectors have been successfully fabricated using the conventional less expensive diffusion technology. To decrease the leakage current at voltages higher than 100 volts additional gettering steps on the back of the wafer are necessary. Use of a double sided mirror polished wafer may be helpful to reduce this leakage current. Efforts to increase the finished detector yield to 80% should concentrate on reducing the pipe formation in the P-diffused region and stricter particulate control during processing.

† Supported in part by U.S. Department of Energy

References

[1] J.E. Brau (ed.), Proceedings for Workshop on Calorimetry for the SSC, World Scientific, (1990), 489.

[2] EMPACT Collaboration, "EMPACT - An Expression of Interest for an Experiment to be Performed at the Superconducting Super Collider," May 25, 1990.

[3] G. Lindstroem et al, DESY-90-109 (1990); S. Berridge et al, IEEE Trans. Nucl. Sci. 36, 339 (1989); IEEE Trans. Nucl. Sci. 37,1191 (1990).

[4] G. Barbiellini et al, Nucl. Inst. and Methods A235, 55(1985); G. Barbiellini et al, Nucl. Inst. and Methods A236, 316 (1985); M. Bormann et al, Nucl. Inst. and Methods A240, 63 (1985); and A. Nakamoto et al, Nucl. Inst. and Methods A251, 275 (1985)

[5] K. T. Robinson and J. L. Tworek "Effectiveness of Planar Diffusion Sources", Semiconductor International, **10**, 88-91, (June 1987).

[6] R. Korde and J. Geist, "Stable High Quantum Efficient UV-enhanced Silicon Photodiodes by Arsenic Diffusion", Solid State Electronics, **30**, 89-92 (1987).

[7] R. Korde and J. Geist, "Internal Quantum Efficiency Stability of Silicon Photodiodes", Applied Optics, **26**, 5284-5290 (1987).

[8] J. Kirkby et al., Proceedings of the Workshop on Experiments Detectors and Experimental Areas for the Supercollider, Editors R. Donaldson and M.G.D. Gilchriese (World Scientific Publishing, 1988), 388.

[9] L. Jastrebski, "Origin and Control of Material Defects in Silicon VLSI Technologies : An Overview", IEEE Trans., Electron Devices, **ED-29**, 475-487 (1982).

[10] E. Borchi et al, Nucl. Inst. and Methods A279, 57 (1989).

[11] C. M. Chevlin and L. J. Demer, "The Nucleation Mechanisms for Oxidation-Induced Stacking Faults in Silicon Crystals Containing Surface Damage" Philosophical Magazine,**A40**, 685-699 (1979).

[12] Fairfield J. M. and Schwuttke G. H., "The Influence of Crystallographic Defects on Device Performance", J. Electrochemical Soc., **113**, 1229-1231 (1966)

[13] Porter W. A. et al, "Thermal Conditions When Dislocations are Generated in Silicon", J. Appl. Physics, **43**, 1477-1479 (1972).

[14] A. E. Widmer and W. Rehwald, "Thermoplastic Deformation of Silicon Wafers", J. Electrochem. Soc., **133**, 2403-2409 (1986).

[15] E. J. Janssen and G. J. Declerck, "The use of 1,1,1-Trichloroethane as an Optimized Additive to Improve the Silicon Thermal oxidation Technology", J. Electrochem. Soc., **125**, 1696-1703 (1978).

[16] For example, see : S. M. Sze, VLSI Technology, McGraw-Hill, New York, 1988, ch. 1, pp. 45-47.

[17] E. R. Weber and D. Gilles, "Transition Metals in Silicon : Fundamentals and Gettering Mechanisms", in Semiconductor Silicon 1990, Electrochemical Soc. Proc., **90-7**, 585-600 (1990) and reference cited there.

[18] S. Holland, "Fabrication of Detectors and Transisters on High Resistivity Silicon", Nucl. Inst. and Methods in Phys. Res., **A275**, 537-541 (1989).

[19] R. Korde and M. Tyagi, "Breakdown Voltage Reduction of Silicon Pin Diodes by a Process Induced Linear Defect", Physica Status Solidi (a), **64**, k89-k91 (1981).

[20] Varker C. J. and Ravi K. V.,"Oxidation Induced Stacking Faults in Silicon Electrical Effects in p-n Diodes", Applied Phys., **45**, 272-287 (1974).

DIAMOND FILM IONIZATION DETECTORS FOR CALORIMETRY

C. P. Beetz, B. Lincoln
ATM, Inc., New Milford, CT

D. R. Winn, K. Segall, M. Vasas, D. Wall
Fairfield University, Fairfield, CT

Abstract

Synthetic diamond film diodes have been fabricated and tested for the SSC calorimetry. A discussion of the potential for diamond films applied to SSC calorimeters is presented.

Introduction

Compensated sampling calorimetry is a fundamental method for energy measurements in high energy physics[1,2,3,4,5,6] The use of solid state semiconductor materials such as Si are being considered for calorimeters[7] and tracking detectors at the SSC. Si has been the medium of choice, because of its availability, low electron affinity, good mobility and sufficient carrier lifetimes. One of the major shortcomings of the Si based detectors is their high susceptibility to damage when placed in radiation intensive environments.[8] Also due to the relatively small bandgap of Si, exponential thermal generation of carriers severely restricts use of Si devices to temperatures below ~150°C. Diamond[9] has several advantages with respect to Si including: (1) much lower thermal noise (5.5 eV bandgap compared with 1.3 eV for Si); (2) a dielectric constant less than half that of Si, and therefore lower capacitance per unit thickness, (3) higher radiation damage threshold for neutrons and gammas by at least 1 order of magnitude; (4) higher carrier mobilities (electron and hole mobilities are nearly equal); (5) higher saturated drift velocity, by a factor of 2.7; and (6) a much higher thermal conductivity, by a factor of 20. The Johnson (Keyes) Figure of Merit is 8,200 (32) times that of Si, respectively, for the operation of high bandwidth electronic devices. Diamond (especially natural diamond) was one of the earliest materials considered for use as an extremely fast X-ray or particle detector[10] or indeed for ionization detectors in general.[11,12,13,14,15,16,17,18,19] Diamond was shown to be radiation hard in high neutron and gamma fluxes, maintaining a ~0.5-1% energy resolution at 1 MeV[20,21] after neutron fluxes up to 10^{14} neutrons/cm^2.

The potential disadvantages of diamond compared with Si include a lower number of electron hole pairs per KeV lost, by a factor of ~4-5, due to the large intrinsic bandgap (5.5 eV), yielding 34 electrons per micron per MIP, with 18 eV as the pair energy. A very short carrier lifetime, of order ~1 ns, severely limits the sampling collection volume to ~200 μ, even with $v_{sat}=2.7\times10^7$cm/s. (Note that because of the lower applied field, Si takes ~20 ns to drift 200μ.) Nevertheless, for some applications where radiation hardness, speed, and stability are absolutely crucial to performance, diamond films may be a medium of choice for a sampling calorimeters, especially in high radiation regions.

Even if the carrier lifetime is ~5 ns, under complete charge collection conditions, an analysis indicates the maximum depth of the active region of a diamond detector is ~300 μm[10]. This thickness is within the capability of the diamond CVD growth techniques presently in use, but defines the limits of calorimetric yields to about ~10,000 electrons per sampling layer per normal incidence MIP. In a tungsten electromagnetic calorimeter with diamond film (chips of ~1 cm^2 area) sampling every radiation length, for example, this could yield ~1,200,000 electrons per GeV deposited at a noise level of order ~20,000 electrons (typical input capacitances with a noise filter time of ~50 ns), implying an energy resolution of ~15%/$\sqrt{E}$.

Fabrication

Due to recent advancements in diamond film growth, it is now possible to reproducibly synthesize high resistivity (>10^{10} Ω-cm) monolithic and polycrystalline diamond films that can be used for detector evaluation. We synthesize diamond films by either hot filament assisted or microwave assisted chemical vapor deposition using a modified Perkin Elmer UHV chamber, attached to a Perkin Elmer MBE-400 Molecular Beam Epitaxial growth system. We were able to introduce selected donor/acceptor impurity atoms (nitrogen and boron) into the diamond film directly during the growth process to produce near midgap impurity states that contributed to the photocurrent and gave rise to a peak in the photoconductivity excitation spectrum centered at ~4600Å. The diamond films were grown on pseudo-lattice matched W, SiO_2, Si, or Si_3N_4 substrates; the thermal expansion coefficient of the latter substrates are close to those of diamond over the range of temperatures used for film growth and can be chemically etched so to form a window. Typical films consist of crystallites of sizes of order of the film thickness - 10- 100 μ; however in thinner films, crystallinity has been observed to be as large as a few mm. A cross-section through a ~10-100 μ thick detector is shown in Figure 1.

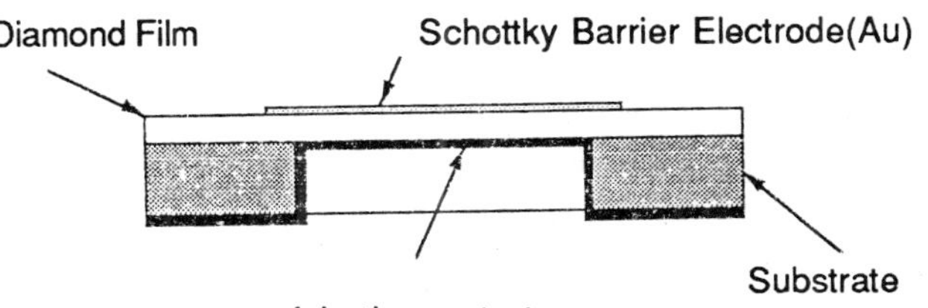

Figure 1. Cross section view of diamond film detector

The quality of the diamond films are judged by measuring the sub-bandgap DC photoconductivity in the vicinity of the band edge, and by the Raman and photoluminescence spectra. The absence of photoconductivity excited at 2535Å is the main selection criteria, as used by Kozlov et al.[21] on natural diamond. The absence of photoluminescence at ~2.1 eV (band-A, radiative recombination across an indirect gap of separately trapped electrons and holes, with the donor center an isolated substitutional nitrogen atom[22]) and at 1.67 eV (GR1 neutral vacancy defect) was also used as a guide to aid in film selection. Films that fulfilled the selection criteria were be fabricated into detectors using a Schottky barrier electrode on the front surface of the diamond with sputtered Al (200 A thick). Ohmic rear injection contacts were fabricated with alternating layers of Ti/Mo/Au, with 800° post heat treatment.

Fast Ionization Detector Results

Sub-bandgap Detectors

We have demonstrated the use of selected polycrystalline diamond films as fast photoconductive light detectors for both the UV (above bandgap) and visible (sub-bandgap) regions[23]. The pulsed sub-bandgap photoconductivity consisted of a short lifetime (<5ns, limited by the lead inductance) and a longer lifetime component (~10 ms), when excited with a sub-nanosecond laser pulse. Figure 2 shows a diode film mounted in front of a laser, connected to a digital oscilloscope (400 MHz bandwidth limit). Figure 3 shows a pulse initiated by a 100 KW nitrogen laser pulse ~1ns wide, at a film bias voltage of 20 V, with an internal field of 20 kV/cm (10 μ thick, 0.4 cm square).

Figure 2. Pulsed laser diamond diode test stand.

The long photoluminescent tail at ~mV scales is not shown; 32 pulses were averaged to obtain Fig. 3. The signal corresponds to about 4×10^{-5} electrons per sub-bandgap photon (3.7 eV) on 50 Ohms. It is obvious that the recombination dynamics of the carriers in the films must be extremely fast. Other reports of photoconductivity in diamond films using above bandgap excitation[24,25], have yielded similar promising results, demonstrating efficient switching with fields as large as 1MV/cm and carrier recombination times ~ 600 ps. The origin of the fast recombination dynamics is most likely the presence of impurity atoms such as hydrogen and nitrogen which have been shown to be present in CVD diamond films[26,27]. The origin of the slower lifetime component is most likely due to the excitation of carrier pairs in the vicinity of the Schottky barrier electrodes.

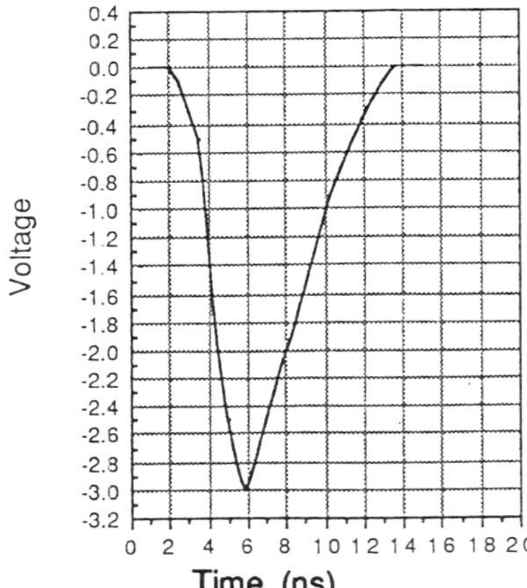

Figure 3. Photoconductive diamond film response to a 1-ns nitrogen laser (32-pulse average - 20 V bias)

Low Energy X-Rays

We have evaluated one of our nitrogen doped photoconductive detectors as a detector for X-rays from a 30 KV Cu target low power (50 μA) X-ray tube. The detector was ~15 mm^2 x 10μm thick, placed 1 cm from the x-ray target. Using a bias voltage of 10V we were able to generate 30 nA current in the detector while the dark current was ~10^{-2} pA. Using 18 eV to form an electron hole pair in diamond, an average energy from the x-ray flux of the Cu K and L lines, and the absorption probability in the thin film, we conservatively estimate an electron collection efficiency of at least 25% and possibly higher. We are encouraged that much better detectors can be fabricated, since this detector had a low resistivity ~10^6 Ω-cm and exhibited below bandgap photoconductivity. Figure 4 shows the Ohmic excess x-ray induced photocurrent vs bias voltage.

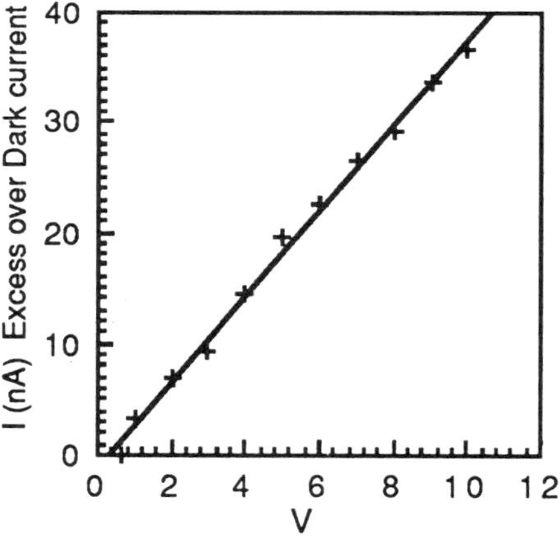

Figure 4. I (excess over I_{dark}) vs. V in a diamond ionization chamber under 30 KV endpoint X-rays.

Figure 5 shows some diamond films mounted in TO-5 cans as X-ray detector (left) and as an optically coupled high voltage switch.

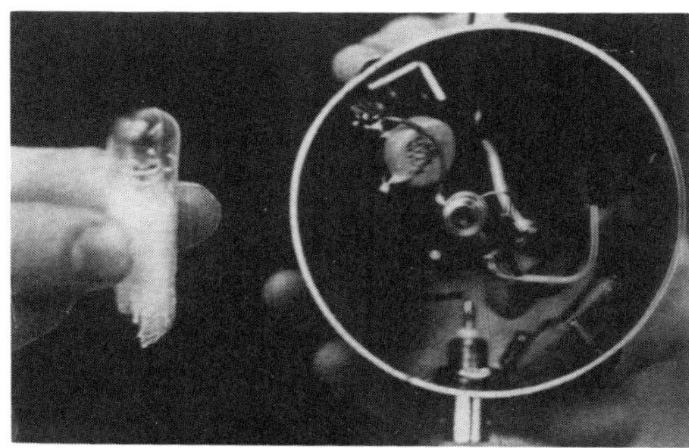

Figure 5. Diamond Films packaged as X-ray detector (L), and as a fast high voltage optically coupled switch (R).

Minimum Ionizing Particles
We have preliminary results showing sensitivity to beta particles, giving ~200 electron pulses from a 10μ thick film, at the edge of detectability with a an ultra-low noise battery operated pre-amp, used inside a shielded box. These yields are roughly a factor of 2 lower than expected.

Conclusions

We have shown that drifted ionization properties and yields in synthetic diamond film diodes within a factor of 2-4 of predicted values are achievable with minimal effort in films with thicknesses less than 50 μ. Thicker films await demonstration as reliable diode detectors. While synthetic high resistivity diamond film ionization detector technology has interesting features for SSC calorimetry radiation and readout speed, it also carries a high cost per area (at present), and has modest predicted calorimetric performance, relative to other techniques. We also caution about the difficulties of: (1) handling free-standing films, and of substrate thermal and mechanical mismatches if the films are left as grown on substrates; (2) high quality Ohmic contact technology; (3) thickness, growth and dopant uniformity over large areas; and (4) maintaining electronic quality at larger thicknesses.

We thank the DoD SBIR program for support.

[1] C. Fabjan and T. Ludlam, "Calorimetry in High Energy Physics", Ann. Rev. Nucl. Part. Sci. Vol 32 (1982)

[2] U. Amaldi, "Fluctuations in Calorimetry Measurements", publ. in *Expt.Tech.in High Energy Physics*, A-W (1987) T. Ferbel, Ed.

[3] K. Kleinknecht, *Detectors for Particle Radiation* (1986) Cambridge U.P., ch. 6.

[4] R.C. Fernow, *Intro. to Exptl. Part. Physics* (1986) Cambridge U.P., ch. 11

[5] W. R. Leo, *Techniques of Nuclear and Particle Physics Experiments*, Springer-Verlag (1987)

[6] R. Wigmans, " On the Energy Resolution of U and other Hadron Calorimeters", CERN/EF 86-18, (Sept, 1986), also publ. in N.I.M. (1987)

[7] G. Lindstrom, M. Eberle, I. Fedder, E. Fretwurst, V. Riech and M. Seidel, Workshop on Calorimetry for the SSC, Tuscaloosa, Alabama, March (1989), DESY 89-104.

[8] G. Lindstrom, E. Fretwurst, H. Herdan, M. Rollwagen, P. Thomsen and R. Wunstorf, Workshop on Calorimetry for the SSC, Tuscaloosa, Alabama, March (1989), DESY 89-105.

[9] DIAMAS SSC Collaboration, SSC EOI-9

[10] E. A. Konorova and S. F. Kozlov, Soviet Phys. Semicond. 4, 1600 (1971).

[11] P. J. Van Heerden, Thesis Utrecht (1945).

[12] H. Friedman, L. S. Birks, H. P. Gauvin, Phys. Rev. 73, 186 (1948).

[13] R. Hofstadter, Phys. Rev. 73, 631 (1948).

[14] D. E. Woolridge, A. J. Ahearn and J. A. Burton, Phys. Rev. 73, 913 (1948).

[15] A. J. Ahearn, Phys. Rev. 73, 1113 (1948).

[16] P. J. Kennedy, Proc. Roy. Soc., London, A253, 37 (1959).

[17] P. J. Dean and J. C. Male, J. Phys. Chem. Solids, 25, 311 (1964).

[18] H. H. Goldsmith, Nucleonics 4, 142 May (1949).

[19] S. P. Kozlov, R. Stuck, M. Hage-Ali, and P. Siffert, IEEE Trans. Nucl. Sci. NS-22, 160 (1975).

[20] S. P. Kozlov, E. A. Konorova, A. L. Barinov and V. P. Jarkov, IEEE Trans. Nucl. Sci. NS-22, 171 (1975).

[21] S. F. Kozlov, A. V. Bachurin, S. S. Petrusev and Y. P. Fedorovsky, IEEE Trans. Nucl. Sci. NS-24, 240 (1977).

[22] H. B. Dyer, F. A. Raal, L. DuPreez and J. H. N. Loubser, Phil. Mag. 11, 763 (1965).

[23] C. P. Beetz, Jr. and B. Lincoln, "Ultrafast Diamond Film Photoconductive Detector," Final Report, Contract #SDIO-DASG60-89-C-0066, December (1989).

[24] J. Curry, S. T. Feng, P. T. Ho, and J. Goldhar, SDIO/IST-ONR Diamond Initiative Symposium, Arlington, VA, July 1988.

[25] L. S. Pan, P. Pianetta, D. R. Kania, O. L. Landen and K. V. Ravi, Extended Abstracts Electrochemical Society Spring Mtg. Los Angeles, CA 89-1, 159,1989.

[26] C. P. Beetz, Jr. and T. A. Perry, "Elastic Recoil Detection of Hydrogen Atoms in CVD Grown Diamond Films," , paper presented at the Fall Meeting of the New England Section of the American Physical Society, 7-8 October, Amherst, MA, 1988.

[27] T. A. Perry, C. P. Beetz, "Photoluminescence Studies of Filament Assisted CVD Diamond Films," in *Raman and Luminescence Spectroscopic Instrumentation in Technology*, eds. F. Adar and J.E. Griffiths, SPIE Proc. 1055, 152 (1989).

A PRECISION BARIUM FLUORIDE CRYSTAL CALORIMETER FOR THE SUPERCONDUCTING SUPER COLLIDER

Ren-yuan Zhu*
Lauritsen Laboratory
California Institute of Technology
Pasadena, CA 91125

Abstract

A high speed and highly radiation-resistant barium fluoride crystal array is under development by the Caltech group in collaboration with the Brookhaven National Lab, the Shanghai Institute of Ceramics in China, the KEK in Japan and the AccSys Technology Inc. in California. The barium fluoride array will serve as a prototype for a high precision electromagnetic calorimeter at the SSC, which has a unique capability in detecting the Higgs particles in the mass region between 80 to 180 GeV.

INTRODUCTION

This report present a design study of a very high precision electromagnetic calorimeter composed of large size barium fluoride (BaF_2) crystals for the Superconducting Super Collider (SSC). The speed (gating in a single beam croosing), granularity (typically $\Delta\eta \times \Delta\phi = 0.04 \times 0.04$ per crystal), radiation hardness ($\geq 10^5$ Gy), high scintillation light output and the large acceptance ($|\ \eta\ |\leq 2.87$) of BaF_2 calorimeter will provide an excellent precision detector for the SSC to probe new physics in the high intensity SSC environment.

The BaF_2 electromagnetic calorimeter will have energy resolution of $1.3\%/\sqrt{E} + 0.5\%$, comparable with what has been achieved by the L3 BGO electromagnetic calorimeter [1]. The physics motivation for a precision EM calorimeter is to make very precise energy measurement for photons and electrons. The unique capability of this detector is shown, for example, in the ability to detect Higgs particles in a mass region between 80 to 180 GeV [2].

It is well known that there exists a gap between the upper limit for Higgs detection at LEP Phase II (80 GeV) [3] and the lower limit for Higgs detection at the SSC using the four lepton final state (140 GeV) [4]. The precision EM detector will cover this gap by measuring the Higgs in its $\gamma\gamma$ decay channel.

The high resolution calorimeter also will be used to search for new massive gauge bosons through their decays into e^+e^-. This will allow us to explore the energy regime where high energy symmetries, such as superstring-inspired E6, break down to form the gauge structure observed at low energies.

*Work supported in part by U.S. Department of Energy Contract No.DE-AC03-81-ER40050.

PHYSICS OF PHOTONS AND ELECTRONS

The most important discoveries at the SSC could result from the detection of entirely new phenomena at the 1 TeV scale, outside of the Standard Model or its simplest extensions. These discoveries could appear as new multilepton final-states, or as new configurations of photons and jets. The ability to distinguish and precisely measure electrons and photons, as well as muons, at the SSC will allow isolate the new physics signals from the backgrounds of the *"Standard"* physics. Should the Standard Model survive in the SSC era, then the highest mass resolution, and good acceptance will be required to detect the Higgs, through its distinctive decays into multileptons or photons.

$H^0 \rightarrow \gamma\gamma$ Search for $M_{H^0} \in (80, 160)$ GeV

Figure 1 shows a lego plot of transverse energies in the $\eta - \phi$ plane for a 150 GeV Higgs which decays into two photons together with associated QCD jets. The distinctive feature of isolated photons in the Higgs decay can be used to reject the QCD background.

The main background in the $H^0 \rightarrow \gamma\gamma$ search is direct photon production: $q\bar{q} \rightarrow \gamma\gamma$ (59 pb) and $gg \rightarrow \gamma\gamma$ via a box diagram (493 pb)[7]. This so-called *"irredu-*

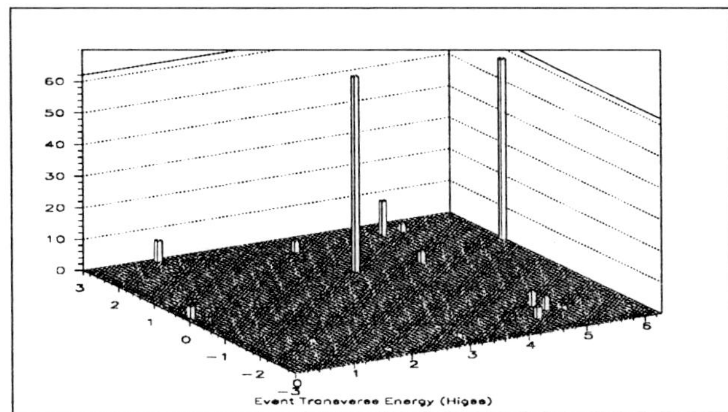

Figure 1: Transverse energy distribution of a $H^0 \to \gamma\gamma$ event in the $\eta - \phi$ plane.

ciable background" has to be reduced by event selection cuts:

- $|\eta_\gamma| < 2.8$;
- $E_T^\gamma > 20$ GeV;
- $|\eta_{\gamma\gamma}| < 3$ ($\Rightarrow$ to reduce qq$\to \gamma\gamma$).
- $|\cos\theta_\gamma^*| < 0.8$ ($\Rightarrow$ to reduce gg$\to \gamma\gamma$).

The background cross sections after event selection cuts are reduced to 91 pb for gg $\to \gamma\gamma$ and 14 pb for $q\bar{q} \to \gamma\gamma$. Since the Higgs width in this mass region is very narrow (~10 MeV) a very high resolution electromagnetic calorimeter, such as the BaF_2, is needed to detect a typical 0.2 pb signal [5] from $H^0 \to \gamma\gamma$.

There are copious π^0's, and thus photons, produced at the SSC. The QCD two jet cross section is 1.2 mb for $\hat{P}_T > 20$ GeV and $|\eta| < 2.8$. There are also processes producing one real photon and a QCD jet, such as qg $\to q\gamma$ (119 nb) and qq $\to g\gamma$ (6 nb), which will fake a two-isolated-photon final state with a higher probability. An isolation cut is used to reject photons produced from fluctuations in jet fragmentation. The cut requires that the sum of the transverse energy in a cone of radius R=0.6 ($R=\sqrt{\Delta\eta^2 + \Delta\phi^2}$), excluding the transverse energy of the photon itself, is less than 10% of the transverse energy of the photon plus 5 GeV.

By using this isolation cut, the rate of one isolated photon from a jet is reduced by a factor of 10^{-4}. The rate of two isolated photons from two jets is thus reduced by a factor of 10^{-8}, assuming that the two jets are statistically independent. Figure 2a) shows the signal/background ratio of 5, 9, 15, 14 standard deviations for Higgs mass of 80, 100, 120 and 150 GeV, respectively, with BaF_2 calorimeter. For comparison, the same signal is shown in 2b) for a conventional electromagnetic calorimeter with energy resolution of $15\%/\sqrt{E} + 1\%$. The Higgs peaks after background subtraction is shown in figure 2 c) and d), respectively. The detector with a conventional resolution will certainly not be able to do this physics.

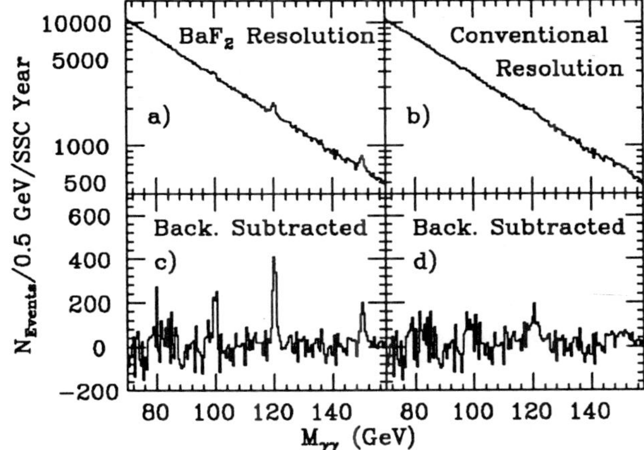

Figure 2: $H^0 \to \gamma\gamma$ Detection for electromagnetic calorimeters with a) and c) BaF_2 EM resolution, and b) and d) conventional resolution.

$H^0 \to ZZ/ZZ^* \to e^+e^-e^+e^-$ Search for $M_{H^0} \in (140, 800)$ GeV

The cleanest signatures for the heavy Higgs search in this mass region is four isolated electrons or muons in its final state: $H^0 \to ZZ/ZZ^* \to e^+e^-e^+e^-$, where Z^* represents a virtual Z with mass $< (M_{H^0} - M_{Z^0})$ for M_H less than $2M_Z$. The proposed BaF_2 calorimeter has excellent capability in detecting 4 isolated electron final state. All of the QCD background, including contributions from charm, bottom and even top decays (computed with top mass = 150 GeV), is rejected with the isolation cut described before (with a reduced radius R=0.3).

One of the remaining backgrounds is pp $\to ZZ/ZZ^* \to e^+e^-e^+e^-$ which has a cross section of 40 femtobarns (fb). Another major background is the production of one real Z^0 and a pair of heavy quarks, such as top: pp $\to Zt\bar{t} \to e^+e^-e^+e^-X$ [6].

Figure 3a) shows the mass spectrum of a 150 GeV Higgs signal superimposed with the sum of all background processes described above in a period of one SSC year. As a comparison, Fig. 3b) shows the same spectrum observed by an electromagnetic calorimeter with conventional energy resolution: $15\%/\sqrt{E} + 1\%$.

BaF_2 E-M CALORIMETER

The proposed barium fluoride electromagnetic calorimeter consists of 18,044 large size (50 cm long, 24.5 radiation lengths) crystals. The calorimeter has two parts:

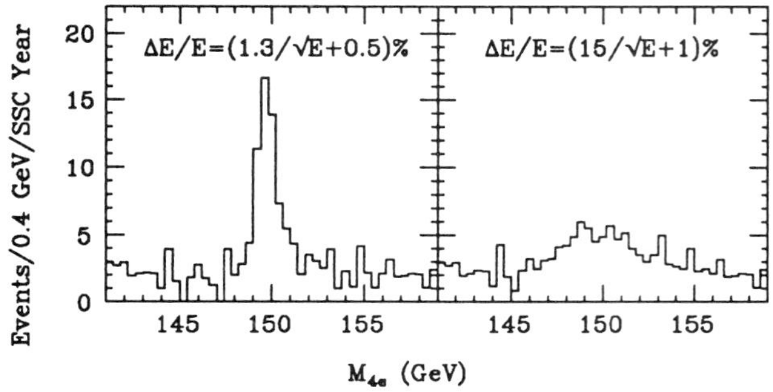

Figure 3: $H^0 \rightarrow ZZ^* \rightarrow e^+e^-e^+e^-$ detection for electromagnetic calorimeters with BaF_2 resolution and conventional resolution in one SSC year.

- A central barrel calorimeter has an inner radius of 75 cm and an outer radius of 140 cm. It covers the rapidity of ± 1.45.
- Two endcaps, located at z=± 150 cm, cover the rapidity of $1.45 \leq |\eta| \leq 2.87$.

Figure 4 is a schematic showing the central BaF_2 calorimeter consisting of the barrel and the end caps. The crystals are approximately square in cross section, with sizes ranging from 2×2 cm^2 in the front for the smallest crystal, to 5.2×5.2 cm^2 in the back of the largest crystal. The typical crystal covers a $\Delta\eta \times \Delta\phi$ interval of 0.04 × 0.04. The total volume of crystals is 11.7 m^3 with 57.1 tons.

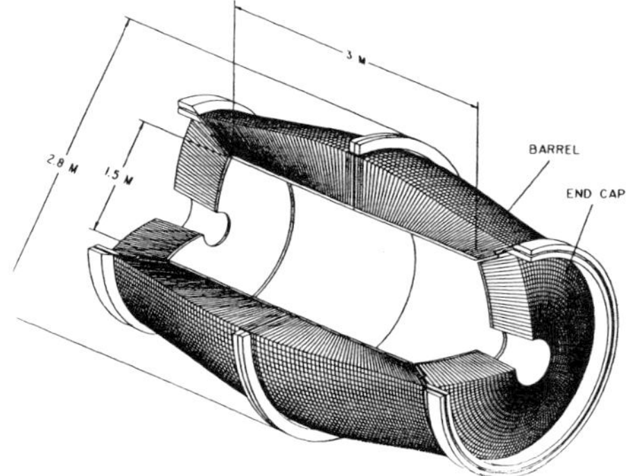

Figure 4: A schematic showing central BaF_2 calorimeter.

The proposed BaF_2 calorimeter has following features:

- The scintillation light readout has less than 10 nsec peaking time, i.e. gating in a single beam crossing;
- Bipolar signal output with residual tail of less than 10^{-4} after 35 nsec;
- Radiation resistance up to at least 10^5 Gy to survive the SSC environment;
- Energy resolution of a $1.3\%/\sqrt{E} + 0.5\%$.
- π/e suppression ratio is in an order of 10^{-4}.

Readout of the Fast Scintillation Components

Table 1 lists the properties of some candidate fast scintillation materials: pure CsI, BaF_2, CeF_3 and liquid xenon.

Table 1: Fast Scintillators

	P. CsI	BaF_2	CeF_3	L. Xe
Density (g/cm^3)	4.51	4.88	6.16	3.05
X_{rad} (cm)	1.85	2.10	1.7	2.77
X_{int} (cm)	37.0	29.9	26.2	55
X_{rad}/X_{int}	0.051	0.068	0.065	0.050
$R_{Moliere}$ (cm)	3.5	4.4	2.6	5.6
Ref. Index	1.80	1.49	1.62	N/A
Hygroscopic	No	No	No	N/A
Lumin. (nm)	300	320	340	170
		210	310	
τ_{Decay} (nsec)	10	630	27	45
		0.9	≈2	
Light Output	<2000	5000	700	40,000
(photons/MeV)		1000	?	

Among all these scintillation crystals, BaF_2 is unique in its fast decay components at 195 and 220 nm with decay time constants of 0.87 and 0.88 nsec [8]. Because of the high speed of the fast component and the evidence of high radiation resistance [9], BaF_2 has gained vast interest in recent years [10]. There is, however, a stronger scintillation component at 310 nm which has a decay time of about 600 nsec. Because of the 16 nsec beam crossing time, the slow component could lead to pile-up and thus could degrade the high resolution at the SSC. The suppression of the slow component thus is a key issue in designing a high energy-resolution BaF_2 calorimeter for the SSC.

Slow Component Suppression

Previous measurements show that the total light output of BaF_2 is one fifth of Na(Tl) [8]. The intensity ratio of fast components (195 and 220 nm) to the slow component (310 nm) has been measured to be about 1:5 [8]. Studies by a group at Delft University [11] and C. Woody [12] indicated that an improvement of the fast to slow ratio of more than a factor of 5 may be obtained by using lanthanum doping.

Figure 5 shows the emission spectra for a sample of pure BaF_2 and a BaF_2 sample doped with 1% of lanthanum. The peak intensity of the slow component is reduced by a factor of about five, with little change to the fast components. It is also known that the intensity of the fast components have no temperature dependence, while the slow component increases with decreasing temperature, at a rate of -2.4%/°C [13].

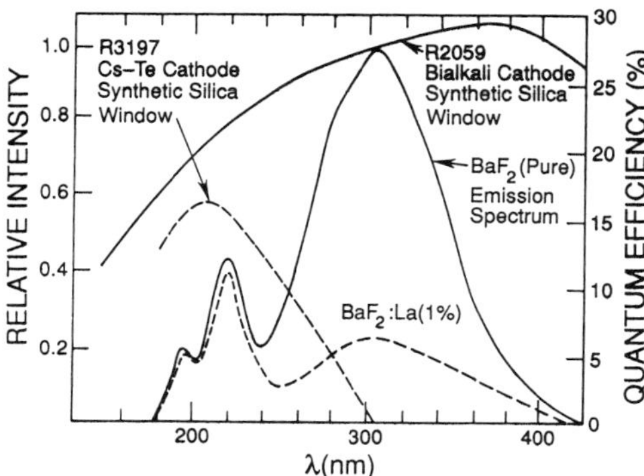

Figure 5: BaF_2 scintillation spectra and PMT quantum efficiencies.

Since the fast components and the slow component of BaF_2 are at different wavelengths, they are spectroscopically separable. Figure 5 also shows the quantum efficiencies of two photomultipliers (PMT): a special UV-sensitive, solar-blind PMT with a Cesium Telluride (Cs-Te) photocathode (Hamamatsu R3197) and a PMT with a bialkali photocathode (Hamamatsu R2059). Both PMT's have a synthetic silica (quartz) windows. The Cs-Te photocathode has a quantum efficiency of 12% around 220 nm, while the bialkali photocathode has a quantum efficiency of about 18% around 220 nm. It is clear that the solar-blind photocathode (Cs-Te) is mainly sensitive to the fast scintillation light.

Figure 6 shows pictures of the scintillation light pulses recorded on an HP54111D digital scope at Caltech using a bialkali photocathode (a and d), and a Cs-Te photocathode (b and e). Figure 6 c and f show the scinillation pulse obtained with a Cs-Te photocathode, together with a BaF_2 crystal doped with 1% lanthanum. The rise time of the scintillation light pulse in these pictures was completely dominated by the 2.3 nsec rise time of the PMT's. The optical suppression factors (F/S) for the slow component, defined as the number of photoelectrons in the fast components (F) divided by the number of photoelectrons in the slow component (S), are also shown in the figure. With lanthanum doping, a fast to slow ratio of 5 is obtained.

An additional optical suppression of the slow component can be achieved by using a new photocathode (K-Cs-Te) recently developed by the Hamamatsu Corporation [14]. Figure 7 shows a comparison of the quantum efficiency responses of this new solar-blind (K-Cs-Te) photocathode and a conventional solar-blind photocathode (Cs-Te). The K-Cs-Te photocathode has 10 times better suppression power, expressed in terms of Q.E.(220 nm)/Q.E.(310 nm), compared to the Cs-Te photocathode. It is thus expected that a F/S = 10 may be obtained by using this K-Cs-Te photocathode for pure BaF_2. The residual effect on the measured energy of the slow component can then be suppressed to below the level of 10^{-4} by a fast shaping of the signal without using lathanum doping.

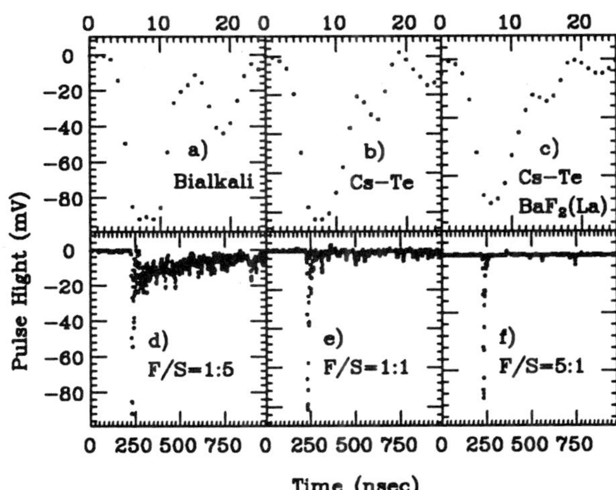

Figure 6: BaF_2 scintillation Light pulse observed by a) and d) a Hamamatsu R2059 PMT; b) and e) a Hamamatsu R3197 PMT; and c) and f) a Hamamatsu R3197 PMT and a La-doped BaF_2 crystal.

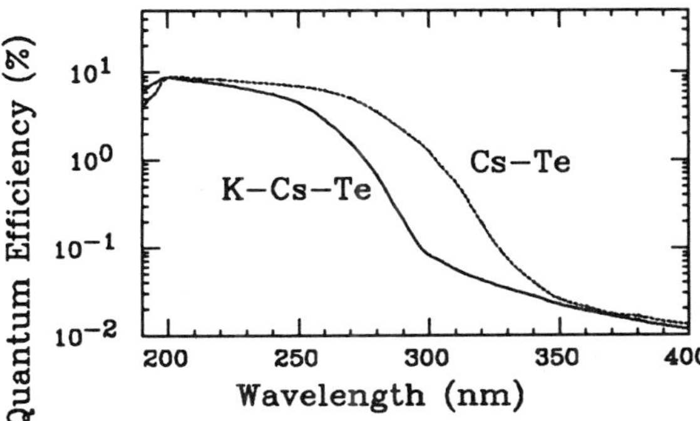

Figure 7: Quantum efficiency of a Hamamatsu K-Cs-Te photocathode comparing with a conventional Cs-Te photocathode.

Readout Design

The principal readout components of BaF_2 are designed as follows:

- A vacuum photodiode with K-Cs-Te photocathode and Quartz window as the photosensitive device. This device will suppress the slow component of BaF_2 by a factor of 50. The device also works in a magnetic field.
- A preamplifier, mounted on the base of vacuum phototube.
- A fast shaper with 3 – 10 nsec peaking time. The residual signal of the bipolar out put after 35 nsec is less than 10^{-4}.

Figure 8 shows the input and output pulse shapes simulated with the PSPICE program for the designed circuit. The input signal, which has a 3 nsec rise time and a Fast/Slow ratio equal to one, is an average of the digitized BaF_2 scintillation pulses obtained from a solar-blind photocathode (Cs-Te).

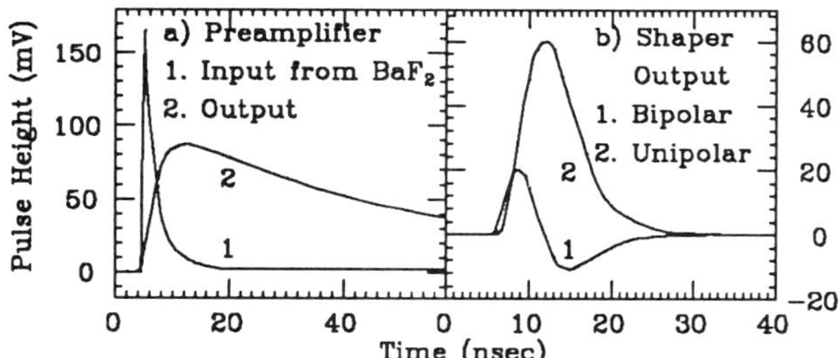

Figure 8: BaF_2 readout pulse shape simulated with a PSPICE program: a) input and after preamplifier and b) outputs.

Also shown in Fig. 8a) is the output from the preamplifier. Both bipolar and unipolar output pulses are shown in Fig. 8b). This very fast readout system will enable us to operate the detector in a single beam crossing, thus reduce the pile-up to the minimum.

BaF_2 Radiation Resistance

Early work done by S. Majewski and D. Anderson [9] showed that no color centers were formed in BaF_2 up to a dose of 1.3×10^5 Gy in an 800 GeV proton beam. Many other works [9] confirmed this early observation for irradiations from either charged particles or photons. Figure 9 shows transmittance of a 8 cm long BaF_2 crystal measured by Charpak *et al.* before and after 0.5 MGy dose [15].

It is also known that impurities in the crystal will increase its susceptibility to radiation damage. An absorption band around 205 nm was identified as originating from Pb contamination [18], and this was correlated with the susceptibility to radiation damage.

Some commercially available BaF_2 crystals, however, show different level of susceptibility to radiation damage. We have tested the radiation resistance of BaF_2 crystals provided by several suppliers, and have found that BaF_2 crystals are among the most radiation resistant crystals known. The main observations on the BaF_2 radiation damage measurements at Caltech, BNL and IHEP (Beijing) can be summarized as follows:

- Some BaF_2 samples show very high radiation resistance. Measurements by IHEP (Beijing) on a BaF_2 crystal sample from BGRI and by Woody on a BaF_2 sample from Optivac (Merck) found compatible radiation resistance with what has been observed by Charpak *et al.*[15].
- The radiation damage caused by γ-ray or neutron irradiation is recoverable by annealing the crystal at 500°C for 3 hours [16]. This indicates that neutrons, as well as photons, do not cause permanent damage in BaF_2 crystals. We thus assume that the BaF_2 damage mechanism from neutrons is the same as that from photons.
- Most samples suffer a small initial damage, during the first small dose (up to 1 kGy), followed by no further change for up to extremely high doses (0.1 MGy). This indicates that the BaF_2 radiation damage is not caused by an intrinsic color center in bulk material of crystal, such as O^{--} vacancies in BGO [17], rather by an externally introduced impurity. By controlling the level of impurities in BaF_2, one thus would be able to improve the radiation hardness of the crystal.

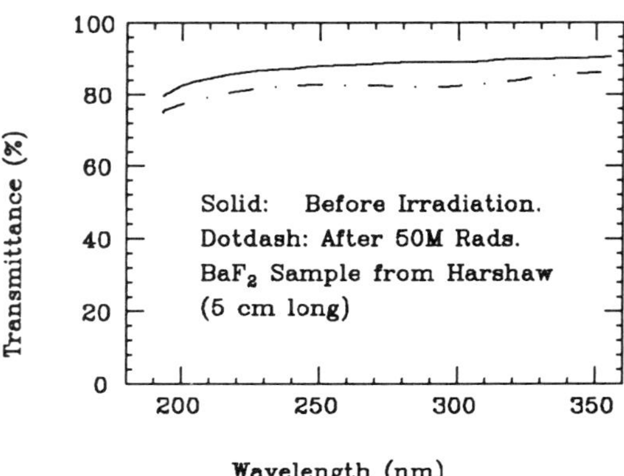

Figure 9: The transmittance as a function of wave length before and after 0.5 MGy dose measured by Charpak *et al.* for a 8 cm long BaF_2 crystal.

We conclude that one can readily obtain radiation hard crystals in small quantity. A further investigation on the correlations between the impurities in the crystal and the small initial damage will be carried out. It is expected that large size, highly radiation resistant BaF_2 crystals will be produced following this investigation.

Linearity

The scintillation light linearity of BaF_2 is known to be as good as BGO. A test carried out at CERN by E. Lorenz *et al.* [19], showed that the linearity of the light output of BaF_2 is excellent up to 50 GeV. We expect the light output linearity of BaF_2 will remain good for at least five orders of magnitude.

Energy Resolution

The energy resolution of a total absorption calorimeter, such as BaF_2 calorimeter, can be factorized as:

$$(\frac{\sigma}{E})^2 = (\frac{a_0}{\sqrt{E}})^2 + (\frac{a_1}{E})^2 + c^2$$

The contributions of each term to the resolution are:

- a_0 is the contribution from the photoelectron statistics.
- a_1 is the contribution from electrical noise, sum over a few Moliere radii.
- c is a constant term, $c^2 = \sigma_G^2 + \sigma_C^2 + \sigma_n^2$:
  - σ_G (geometry effect): shower leak (front, side and rear) and inactive material between cells;
  - σ_C (intercalibration error);
  - σ_n (physics noise): fluctuation of shower and light yield etc.

Light Yield

Figure 10 shows the measured photoelectron yield per MeV as a function of the integration time obtained from a R3197 PMT for six BaF_2 crystals. The data points in the figure were fitted to a sum of contributions from two fast components (F) and from the slow component (S) with an exponential decay time constant τ: y = F + S [(1 - exp(-t/τ)]. The numerical results of the fit are also shown in the figure. Around 80 fast photoelectrons/MeV and F/S $\sim$ 1 were observed with Cs-Te photocathode.

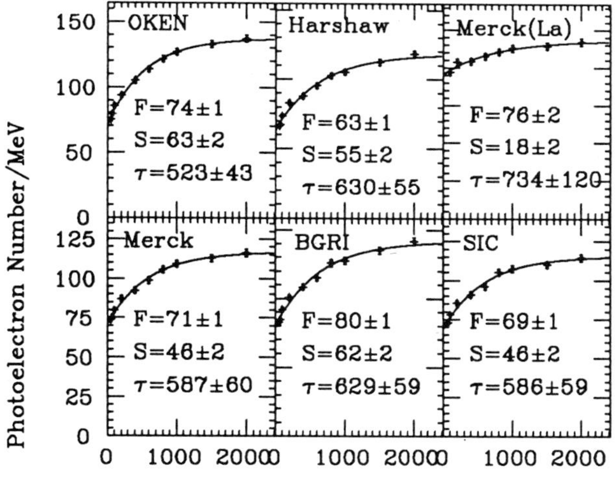

Figure 10: Photoelectron yields measured with a Hamamatsu R3197 PMT as a function of the integration time for six BaF_2 crystals.

This test result was obtained using BaF_2 samples with a typical length of 2.5 cm. A test with longer BaF_2 samples, such as a 25 cm long BaF_2 crystals recently delivered by Merck and SIC, show no apparent decrease in the light yield. The fast component measured with 25 cm long crystals is around 60 to 70 photoelectrons/MeV, which would contribute a term of about $0.4\%/\sqrt{E}$ to the overall energy resolution.

Electrical Noise

The contribution of the electronic noise to the total energy resolution is small with our design. In the construction of the BaF_2 calorimeter we will implement, and carefully check the electronic isolation between the readout channels, to reduce the correlated noise to a minimum.

Summary of Resolution

A GEANT simulation was carried out to calculate the intrinsic resolution of the BaF_2 calorimeter. The simulation was calculated for an array composed of 121 (11 × 11) BaF_2 crystals with the proposed size: (3 cm×3 cm) in front, by (5 cm×5 cm). Effects included in the simulation are:

- 250 μm carbon fiber wall between crystals;
- shower leakage because of summing a limited number of crystals;
- 0.30 radiation lengths of dead material, representing the beam pipe, tracker, and carbon fiber mechanical support, at the front of the BaF_2 array.

Table 2 lists the resulting BaF_2 resolution, including the intrinsic contribution, and the other contributions listed above. Assuming a 0.4% precision calibration, achieved, for example, in our RFQ test in 1988 [20]), it is clear that the BaF_2 calorimeter will be able to reach the design goal of $1.3\%/\sqrt{E} + 0.5\%$ resolution for electrons and photons.

Table 2: Energy Resolution (%)

E (GeV)	5	10	100	500
Photoelectrons	0.2	0.1	0.04	0.02
Noise	0.8	0.40	0.04	0.008
Calibration	0.40	0.40	0.40	0.40
GEANT	0.67	0.60	0.42	0.36
Total	1.1	0.83	0.58	0.54

π/e Separation

Because of the high uniformity and fine granularity of the proposed BaF_2 calorimeter, one can use the detailed transverse shower information to distinguish an electron from a charged pion. The ability to match the momentum measured in the central track detector, or the energy measured in the hadron calorimeter, with the energy measured in the electromagnetic calorimeter provides an additional powerful rejection.

The results of a detailed GEANT simulation are summarized in Table 3. As seen in the table, the complete detector will be able to separate isolated electrons and pions with a rejection ratio of $\sim 10^{-4}$.

Table 3: Acceptance of pions & electrons.

E (GeV)	10	100	1000
After	Lateral	Shower	Cut:
π (10^{-2})	32.6 ± 0.4	34.7 ± 0.5	39.4 ± 0.7
e (%)	99.3	99.5	99.5
After	E_{BaF_2}	and P_{trk}	match:
π (10^{-3})	< 0.1	12.3 ± 0.9	388 ± 7
e (%)	97.3	99.2	99.5
After	HCAL	Veto:	
π (10^{-4})	< 1	2.0 ± 1.4	< 2.2
e (%)	97.3	96.4	95.0

Calibration

Precise, frequent calibration *in situ* is vital in maintaining the high resolution of the precision. As shown in table 2, the dominant contribution to the resolution of BaF_2 calorimeter is the uncertainty of the calibration for high energy electrons and photons.

RFQ Calibration

Our primary calibration will use a novel technique based on a Radio Frequency Quadrupole accelerator (RFQ) [20]. This technique has been developed by Caltech and AccSys in past six years for calibrating the L3 BGO electromagnetic calorimeter. The first calibration RFQ will be installed in L3 soon.

Depending on the type of target, this calibration scheme can be run in two different modes:

- Lithium target: this reaction — ^{7}Li(p,γ)^{8}Be — produces a flux of 17.6 MeV photons which can be used to calibrate the thousands of electromagnetic calorimeter cells at once, with an absolute accuracy of better than 1% in 1–2 hours. The feasibility of this calibration mode has been proven in an experimental test of a 4 × 5 L3 BGO crystal array at the RFQ facility at AccSys Technology, Inc. in November, 1987 [20]. An absolute calibration precision of 0.7% was achieved in the test. This low energy photon calibration, however, is not directly relevant to an SSC electromagnetic calorimeter.
- Fluoride target: this reaction — ^{19}F(p,α)^{16}O* [22] and the subsequent decay of the excited oxygen nucleus ^{16}O* — produces hundreds to thousands of 6 MeV photons per millisteradian per beam pulse. These photons, functioning as a synchronized "equivalent high energy photon" of up to 30 GeV per calorimeter cell, would serve as a calibration source for SSC electromagnetic calorimeters. With a proper normalization, this technique provides a relative calibration with precision of 0.4% in a few minutes. The feasibility of this calibration mode has been proven in another experimental test with 4 BaF_2 counters and a 7 × 7 L3 BGO crystal array at the same RFQ facility at AccSys in September, 1988 [20]. A calibration precision of 0.4% was achieved in the test in two minutes.

Figure 11 shows the normalized ADC distributions of a crystal for runs with different beam intensities. The perfect Gaussian distribution and the correlation between the r.m.s. width and the total energy deposited in the crystal is clearly seen.

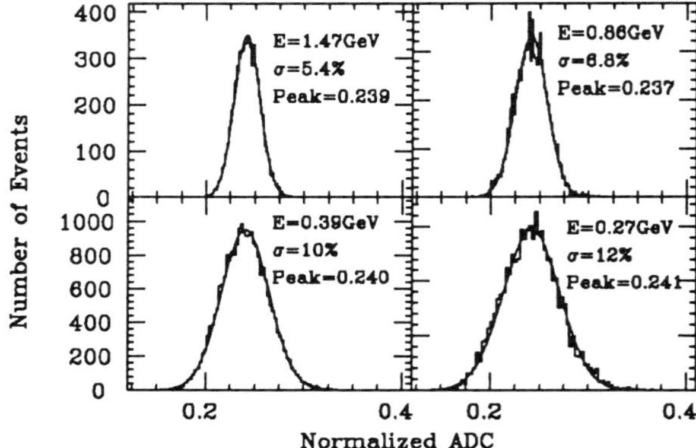

Figure 11: Normalized ADC distribution of a crystal at different energies.

Figure 12 shows that the relative deviation of the peaks of 49 BGO distributions, normalized to the sum of BGO energy in the 7 × 7 matrix, for several runs. This distribution has a gaussian shape with a standard deviation of 0.34%. It is therefore evident that a stability of 0.4% may be achieved.

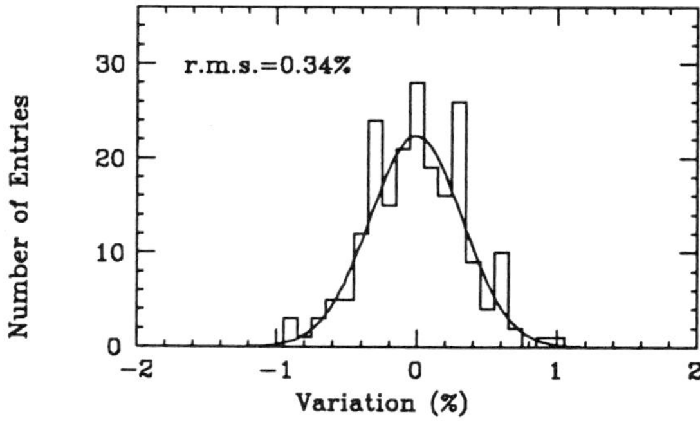

Figure 12: Variation of the normalized peak positions from BGO crystals.

The equivalent photon energy (EPE), defined as the sum of the energies of photons from one beam pulse hit-ting one crystal detector, was measured at AccSys with a LiF target. Up to 2.38 GeV/0.1μCoulomb/1.6msrad was observed for a 1.92 MeV beam. There are much stronger fluorine resonances between 2.0 and 4.0 MeV [22], beyond the 1.92 MeV beam energy at AccSys. By using a 3.85 MeV RFQ and a CaF_2 target, which would have no neutron production as a by-product below 4.05 MeV [22], an EPE of 30 GeV/0.1μCoulomb/1.6msrad or more is expected.

Table 4 lists EPE's measured with a CaF_2 target bombarded with a proton beam from a Van de Graaff at Kellogg Lab at Caltech. The expected equivalent photon energies, calculated with an integration of the resonances [22], are also listed in the table. It is clear that with a 3.85 MeV proton beam up to 40 GeV/1.6 msrad/0.1 μCoulomb is achievable.

Table 4: Measured and Calculated Equivalent Photon Energy.

Proton Energy (MeV)	2.0	2.5	3.0	3.5	3.85
EPE_{meas} (GeV)	1.5	13	22	30	37
EPE_{cal} (GeV)	2.6	14	24	33	38

MIP Calibration

Complementary to the RFQ calibration, we will also use the high flux of minimum ionizing particles (MIP) produced by the collider as a calibration souce. A MIP passing through a BaF_2 crystal longitudinally would deposit 0.33 GeV energy in the crystal which would be readout by the crystal with a few percent resolution. With a statistics of a few hundreds tracks, the peak position or the calibration point can be determined to 0.4%. The multiplicity of charged hadrons at SSC is large enough to provide a calibration within 12 hours during SSC running.

The daily calibrations with these techniques will be confirmed by the measurement and invariant mass reconstruction of inclusive Z^0, Υ and J production, where the vector bosons decay into e^+e^-. With a p_T cut of 10 GeV on the vector boson, a typical crystal in the central calorimeter will see 20 e^+ or e^- from Z^0decay, and 200 from Υ decay. [21] for each week of running.

R&D PROGRAM

The goal of the BaF_2 R&D prorram is to build a high precision electromagnetic calorimeter prototype, composed of an array of 144 large BaF_2 crystals, and to test and calibrate the array using high energy electron and pion beams, together with a hadron calorimeter prototype, at either BNL, CERN or Serpukhov. The array will also be tested at an RFQ calibration test stand, which will be setup at Caltech or AccSys. The research and development program will extend over two years. It will focus on the following six aspects:

- BaF_2 crystal mass production: to produce large size BaF_2 crystals at low cost, which meet the mechanical and transparency specifications and are radiation hard up to 10^5 Gy. We want to ensure that the cost of the crystals is at or below the goal of $ 2.5/cc;
- BaF_2 crystal quality control and radiation hardness tests: to establish the means of verifying that each crystal meets the specifications; to study UV-reflective crystal coatings, and other surface treatments, to provide uniform light output and high light yield; to study the correlation between radiation resistance of BaF_2 crystals and impurities in the crystal;
- BaF_2 readout development: to develop a selective readout which is sensitive only to the fast scintillation component of BaF_2 (0.9 nsec decay time), while fully suppressing the slow scintillation component; to construct prototype preamplifier, and fast shaper cricuits. We will also study the design of the higher-level digital readout.
- A BaF_2 prototype crystal array: to construct a BaF_2 array consisting of 144 BaF_2 crystals. The following related engineering problems will be studied:
  - Mechanical design and construction of a prototype section of the carbon fiber mechanical support.
  - Thermal control for stable operation of the readout.
  - Shielding to reduce electronic noise.
- An accurate (RFQ) calibration facility: to build a RadioFrequency Quadrupole (RFQ) accelerator facility which will produce a pulsed, low energy proton beam to bombard a fluoride target. The photons produced by the target will be used to produce an accurate calibration [20].
- Test beam and cosmic ray (MIP) bench: to test the prototype BaF_2 crystal matrix with high energy electron and pion beams, to verify the resolution and radiation hardness; a cosmic ray test bench also will be used, to test the feasibility of calibration with minimum ionizing particles.

ACKNOWLEDGEMENTS

I am grateful to Drs. F. Takasaki and D. Scigocki and to Merck, Inc. who provided BaF_2 crystal samples for this test. Mr. M. Chen and D. Kirkby at Caltech carried out part of the Physics Monte Carlo simulations described in this report. Dr. Z.Y. Wei tested all BaF_2 samples at Caltech. Dr. H. Ma carried out the analog readout design for the BaF_2 readout at BNL. Mr. M. LeBeau, M. Rennich and R. Barber provided many valuable informations about the mechanical structure design for the BaF_2 calorimeter. Many useful discussions with Drs. A. Gurtu, M. Chen, J. Gunion, R. Hamm, H. Newman, T. Sjostrand, S. Ting, S .Suzuki, C. Woody, Z.W. Ying and B. Zhou are greatly appreciated. The assistance of S. Sondergaard, J. Hanson, and many members of the Caltech technical staff is also acknowledged.

References

[1] L3 Collaboration, *L3 Technical Proposal*, May 1983
J. Bakken *et al.*, *Nucl. Instr. and Meth.* **A254** (1987) 535 and **A228** (1985) 294.

[2] J. Gunion et al., *The Higgs Hunter's Guide*, UCD Preprint, 89-4 (1989).

[3] CERN Green Book, *ECFA Workshop on LEP 200*, Aachen, September, 1986.

[4] *Snowmass Summer Study*, Snowmass, July, 1990; *ECFA Workshop on LHC*, Aachen, October, 1990.

[5] H. Benson and T. Sjostrand, PYTHIA version 5.4, June, 1990.

[6] F. Paige and S.D. Protopopescu, ISAJET version 6.27, April, 1990.

[7] D. Dicus and S. Willenbrock, Phys. Rev. **D37** (1988) 1801.

[8] M. Lavel *et al.*, *Nucl. Instr. and Meth.* **A206** (1983) 169;
P. Schotanus *et al.*, *Nucl. Instr. and Meth.* **A259** (1987) 586; and *IEEE-NS* **34** (1987) 76.

[9] S. Majewski and D. Anderson *et al.*, *Nucl. Instr. and Meth.* **A241** (1985) 76;
A. J. Caffrey *et al.*, IEEE Trans. Nucl. Sci. **NS-33** (1986) 230;
S. Majewski *et al.*, *Nucl. Instr. and Meth.* **A260** (1987) 373.

[10] D.F. Anderson *et al.*, *Nucl. Instr. and Meth.* **A228** (1984) 33.

[11] P. Schotanus *et al.*, *IEEE-NS* **34** (1987) 272; and *Nucl. Instr. and Meth.* **A281** (1989) 162.

[12] C.L. Woody *et al.*, *IEEE-NS* **36** (1989) 536.

[13] P. Schotanus *et al.*, *Nucl. Instr. and Meth.* **A238** (1985) 564;
Kobayashi *et al.*, *Nucl. Instr. and Meth.* **A270** (1988) 106.

[14] S. Suzuki, private communication.

[15] G. Charpak *et al.*, CERN-EP/90-41.

[16] Z. Wei *et al.*, **CALT-68-1659**, 1990, submitted to *Nucl. Instr. and Meth.*

[17] Z. Wei *et al.*, **CALT-68-1628**, 1990, accepted by *Nucl. Instr. and Meth.*;
R.Y. Zhu *et al.*, **CALT-68-1392**, accepted by *Nucl. Instr. and Meth.*;
T.Q. Zhou *et al.*, *Nucl. Instr. and Meth.* **A258** (1987) 464.

[18] P. Schotanus *et al.*, *Nucl. Instr. and Meth.* **A272** (1987) 917; and *Technical Univ. at Delft Preprint* **88-1**.

[19] E. Lorenz *et al.*, *Nucl. Instr. and Meth.* **A249** (1986) 235;
J. Giehl *et al.*, *Nucl. Instr. and Meth.* **A263** (1988) 392.

[20] R.Y. Zhu *et al.*, *Supper Collider I*, Plenum Press, edited by McAshan, (1989) 587;
H. Ma *et al.*, *Nucl. Instr. and Meth.* **A274** (1989) 113; and *Nucl. Instr. and Meth.* **A281** (1989) 469.

[21] H. Bergstroem *et al.*, Penn State Preprint PSU/TH/63, March, 1990.

[22] F. Ajzenberg-Selove, *Nucl. Phys.* **A475** (1987) 1;
H. B. Willard *et al.*, *Phys. Rev.* **85** (1952) 849.

CRYOGENIC LIQUID MIXTURES FOR SSC CALORIMETRY*

D.L. McCorkle, L.G. Christophorou, and H. Faidas
Department of Physics and Astronomy
The University of Tennessee, Knoxville
Knoxville, Tennessee 37996-1200
U.S.A.

and

Atomic and High Voltage Physics Group
Health and Safety Research Division
Oak Ridge National Laboratory
Oak Ridge, Tennessee 37831
U.S.A.

Abstract

The solubilities of a number of potential molecular additives to liquid Ar and to liquid Xe were calculated using the Scatchard-Hildebrand theory.

Introduction

A potentially important component of the SSC detector array is cryogenic liquid-filled calorimeters. Pure cryogenic liquids (Ar or Xe) offer the advantages that they are resistant to radiation damage and they can be purified relatively easily. However, the response of pure cryogenic liquids may not be fast enough for the high fluxes generated. Preliminary studies[1] have indicated the performance of cryogenic liquid-filled detectors can be enhanced by one of two types of additives to the basic cryogenic liquid (Ar or Xe) in a binary or ternary liquid system. These include (i) molecular additives in concentrations of 0.5 to 5% that increase the electron drift velocity, w, especially at high electric fields and (ii) low-ionization threshold additives[2] in parts per million (ppm) concentration to improve the free electron yield and thus the detector sensitivity. It would of course be advantageous to use a single molecular additive which performs both functions, i.e. increases the electron drift velocity and the free electron yield, G_{fe}.

It would be advantageous if the molecular additive to the cryogenic liquid also has a negative electron ground state energy, V_0. The spherically symmetric molecules neopentane [TMC], tetramethylsilane [TMS], tetramethylgermanium [TMG], tetramethyltin [TMT], and 2,2,4,4-tetramethylpentane [TMP] all have negative V_0, and therefore, we have focused our search for ultrafast and ultrasensitive liquid media on mixtures of these components with liquid Ar and Xe.

Results

The solubilities of solid phase TMC, TMS, TMG, TMT, and TMP in liquid Ar (87–100K) and liquid Xe (165–178K) were estimated using the Scatchard-Hildebrand theory.[3] It is assumed that a solid-liquid equilibrium of a saturated solution exists between the pure solute and the solvent. This theory yields a set of three thermodynamic equations which must be solved simultaneously. Thermodynamic critical constants for both the solute and solvent must be known. The results are given in Table 1.

Table 1. Solubilities of molecular additives in liquid argon and xenon.

Additive	Argon (87–100K)	Xenon (165–178K)
TMC	3–4%	40–50%
TMS	0.2–0.3%	75–100%
TMG	0.03–0.1%	45–75%
TMT	0.03–0.07%	25–35%
TMP	0.02–0.06%	25–40%

Discussion

The magnitude of w can be considerably increased, especially at high electric fields, E, by the addition of

*Research sponsored by the U.S. Department of Energy, under Contract No. DE-AC05-84OR21400 with Martin Marietta Energy Systems, Inc., and under Contract No. DE-AS05-76ER03956 with the University of Tennessee, Knoxville.

suitable molecular additives to liquid Ar and Xe. Electron drift velocities approaching 10^7 cm s^{-1} at E values of $\sim 10^5$ V cm^{-1} have been measured in less than percent mixtures of methane in Ar.[4] Similar enhancements in w can be expected for other molecular additives if they are soluble in liquid Ar in sufficient concentrations. Our solubility calculations suggest that TMC and perhaps TMS will dissolve in liquid Ar in concentrations large enough to enhance w significantly. All compounds studied will dissolve in liquid Xe to several percent and therefore can be used in this liquid to enhance w.

In order for a single molecular additive such as TMC (or TMS) at a few percent concentration in liquid Ar to enhance G_{fe} (as well as w) it must have an ionization threshold in liquid lower than the energy of the recombination luminescence photons in liquid Ar [mean energy of 9.5 eV (1300 Å)].[2] For all compounds studied here the ionization threshold of the neat liquid is less than 9.5 eV.[1] Anderson has shown that low ionization potential dopants in ppm concentrations can increase G_{fe} by a factor of 2 to 7 with respect to pure Ar. We would expect therefore that a single molecular additive (TMC or TMS) in percent concentrations in liquid Ar would enhance both w and G_{fe}.

If indeed it is necessary to use a second additive in ppm concentration to enhance G_{fe}, a potential problem could be that the more abundant additive might absorb the Ar recombination luminescence photons before they ionize the photosensitive dopant. For TMC and TMS the photo-absorption extinction coefficients at 1300 Å are 33,000 and 10,000 mole l^{-1} cm^{-1}, respectively.[5,6] A 5% TMC/Ar mixture would therefore absorb 50% of the 1300 Å photons within $\sim$ 100 Å. The mean distance between dopant molecules at 100 ppm in liquid Ar is $\sim$ 60 Å. Since this value is comparable to the attenuation length of a 5% TMC/Ar mixture it is likely that appreciable recombination luminescence photons would not be absorbed by the more abundant additive before they could ionize the photosensitive dopant.

References

[1] L.G. Christophorou, H. Faidas, and D.L. McCorkle. J.W. Gallagher et al., eds. *Non-Equilibrium Effects in Ion and Electron Transport.* Plenum, New York (in press).

[2] D.F. Anderson, Nucl. Instr. Meth. Phys. Res. **A245**, 361 (1986).

[3] G.T. Preston and J.M. Prausnitz, Ind. Eng. Chem. Process Des. Develop. **9**, 264 (1970).

[4] K. Yoshino, U. Sowada, and W.F. Schmidt, Phys. Rev. **A14**, 438 (1976).

[5] B.A. Lombos, P. Sauvageau, and C. Sandorfy, Chem. Phys. Lett. **1**, 221 (1967).

[6] A.G. Alexander, O.P. Strausz, R. Pottier, and G.P. Semeluk, Chem. Phys. Lett. **13**, 608 (1972).

TEST AND MODELING OF HEAT TRANSPORT IN LIQUID ARGON

T.S. Adams
Martin Marietta Astronautics
Denver, CO 80201

For the EMPACT Collaboration

Abstract

The results of tests and analyses to simulate the thermal characteristics of preamplifiers in a liquid argon calorimeter are presented. The test apparatus, consisting of discrete heat sources mounted on parallel plates, is instrumented with thermocouples and submerged in liquid argon. Bubble formation and temperature profiles are recorded as a function of heat input and test assembly orientation. The analyses model the behavior of the test apparatus using a 3D coupled fluid and thermal finite-element-method

Introduction

Several of the proposed experiments at the Superconducting Super Collider (SSC) are considering liquid argon calorimeter (LAC) technology [1,2]. Although LAC technology is widely used in high energy physics and is generally considered well understood there remain many difficult engineering issues unique to the SSC. One of the more challenging issues involves the uncontrolled formation of argon gas within sensing elements where performance is degraded by the resulting density fluctuations and high voltage arcing. This is not a new problem for LAr but it is complicated by the extraordinary size of an SSC LAC. A typical SSC LAC, depicted in Figure 1, will be 7 meters in diameter and 14 meters long with 270,000 data channels [3]. Signal quality and transfer speed considerations for such a large instrument may require a preamplifier for each channel [4] placed near the submerged charge collection plates perhaps within the module support structure. However, preamplifiers are active devices that produce heat and even at the modest power level of 50 milliwatts per channel they can dissipate upwards of 15 kilowatts of heat within the LAC cryostats. Techniques exist for preventing the evolution of gas within critical areas but such solutions tend to increase design complexity and degrade hermeticity.

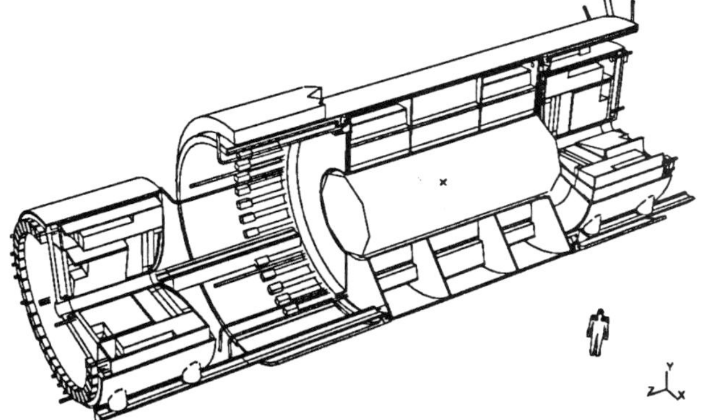

Figure 1. Cutaway View of an SSC LAC Design, Showing the Barrel and Endcaps and Revealing the Internal Modules.

New solutions, beyond the traditional approach of using conservative gross thermal effect calculations and prototyping and testing are needed for the SSC. The ability to quickly and inexpensively investigate new engineering solutions is absolutely essential to producing an optimized design. To this end we have undertaken to develop an analysis capability that will allow confident exploration of alternative design solutions with minimal dependence on tests and prototypes. This report presents the results of our analysis and corroborative test conducted on a simulated LAC module electronics section.

Simulated LAC Electronics Test

Objectives and Overview

The main objective of the test is to obtain data on conductive and convective heat transfer from a group of discrete heat sources to a surrounding bath of liquid argon. These data are used for benchmark verification of two and three dimensional analytical models developed for use with a coupled thermal and fluid analysis computer code. The validated steady state and transient modeling and analysis approaches can then be used in the detector design process with an established level of confidence. A second objective is to obtain relevant data on LAC preamplifier thermal conditions. To this end every attempt has been made to remain faithful to the predicted envelope of critical parameter values for a SSC LAC such as power density, flow channel geometry and LAr depth.

Test Description

The test approach is typical of classic pool boiling experiments with perhaps the exceptions of using many discrete heat sources and allowing convective flow through a narrow channel. The simulated preamplifiers consist of 2 sets of 10 closely matched strain gauge heaters each mounted to a 1.83 x 1.02 cm x 1.07 mm thick piece of perforated fiberglass phenolic circuit board. The two sets differ only by the manner in which the strain gauges are mounted to the fiberglass phenolic. One set is bonded directly to the fiberglass phenolic board while the other set is mounted parallel to and slightly above the circuit board substrate via the solder tabs. These two mounting configurations are intended to represent direct and I/C type mounting respectively.

The circuit board/strain gauge assemblies are in turn bonded to a 0.635 x 10.16 x 10.16 cm aluminum back plate in the pattern shown in Figure 2. The aluminum plate is fitted with a Teflon backing to simulate an insulative interface with module substructure. Heater power is controlled and measured via the four parallel circuits indicated in Figure 2 thus allowing for independent control of the two different configurations. A plexiglass cover is mounted above the aluminum plate and heaters to simulate the narrow channels within a module strongback and to allow direct observation of the heaters. The support assembly base is constructed of two thick parallel aluminum plates to provide stability during orientation adjustments. The base plate is isolated thermally from the rest of the fixture by Teflon screws to minimize conduction from the dewar floor. The assembly can be rotated about the horizontal axis so that the direction in which the heater plane faces can be varied. The strain gauge heaters, each representing one channel/preamplifier, can be operated between 50 and 200 milliwatts. This power range in conjunction with the spatial distribution of heaters is representative of a 6 channel detector tower with a 4 cm x 4 cm area allocated to electronics. The strain gauges can be observed through the LAr and recorded by video camera. The strain gauge thermal input is measured as a function of voltage and current.

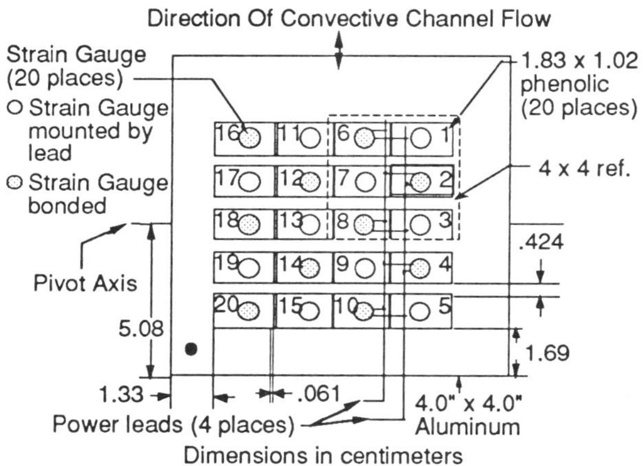

Figure 2. Simulated Preamplifier Layout

The test dewar is a triple wall container manufactured by Andonian Cryogenics Inc. The instrumented test article is placed inside the dewar as illustrated in Figure 3.

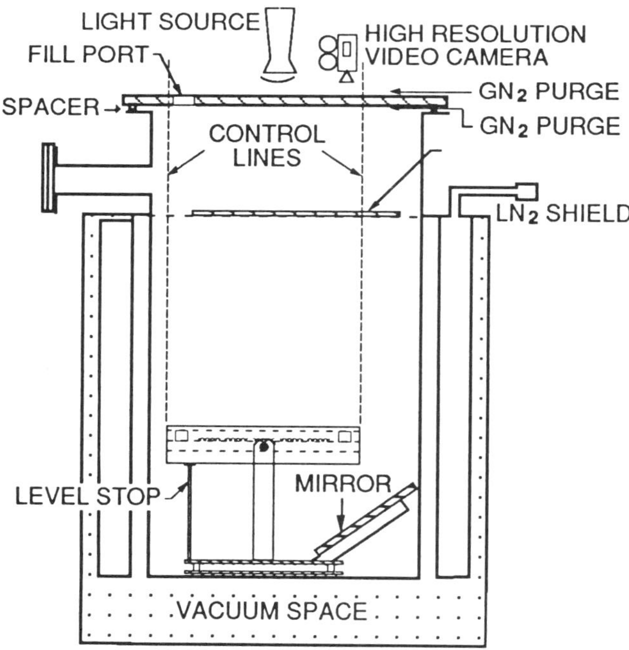

Figure 3. Test Fixture Configuration

Instrumentation

Temperature measurements are made with 21 standard Type E thermocouples and a single calibrated four wire platinum resistive thermometer (PRT) mounted at locations selected to obtain thermal gradient data for the substrate, cover plate, and LAr in areas considered critical for later comparison with analytical predictions. The calculated "at depth" saturation temperature (denoted Tsat) is calculated by summing the barometric and liquid head pressures. Temperatures exceeding Tsat indicate boiling. Subsequent thermocouple data taken during the tests are adjusted by the difference between the initial PRT data and each thermocouple value to calibrate the thermocouple response to the PRT response. Although our intent here is to demonstrate the capability to predict the <u>onset</u> of boiling, we take this opportunity to pass along an interesting observation made during testing. That is, once initiated, boiling tends to persist even as power is reduced to much lower settings (< 10 mW). This behavior, which has been observed in other fluids, and described variously as hysteretic, thermal over-shoot, and/or metastable is strongly influenced by the microscopic charcteristics of individual nucleation sites. Not surprisingly this behavior was first observed from the non bonded guages at relatively low power levels (< 20 mW/strain gauge). By comparison, boiling from bonded guages began at much higher power settings (> 50 mW) usually reached at the end of tests as power was increased in rapid succession. Clearly, future designs should incorporate direct conductive mounting of active electronic elements to a supporting heat sink as a means to mitigate boiling.

Analysis

Objectives and Overview

The main objectives of the analysis are, a) to model the behavior of the test article using a commercially available three dimensional coupled thermal and fluid finite element computer code and, b) to establish confidence in the analysis process through correlations with the test data. A simple hand calculation reveals that dewer radiative heat losses are several orders of magnitude lower than the energy dissipated by the heaters and it makes sense to characterize the problem as adiabatic. Although it first appears deceptively simple, the problem becomes complicated because of the size of the Rayleigh number, which is on the order of $1x10^{10}$ for these conditions. The Rayleigh number is a measure of the relative magnitude of bouyancy and viscous forces within the fluid and is used to indicate the level of fluid turbulence. Turbulence can begin to contribute significantly to heat transfer at Rayleigh numbers of the order of 10^{10}. For this analysis, however, we assume that turbulence effects are confined to thin layers near the dewer walls and as such have a negligible effect on temperatures near the test article located at the dewer centerline.

Analytical Computer Code

The computer program selected for the analysis is called FIDAP (Fluid Dynamics Analysis Package, version 5.0), marketed by Fluid Dynamics International. This particular code was chosen because of its ability to perform conjugate fluid and heat transfer analyses of arbitrary 2-D and 3-D geometries, and because it is able to interface with the SDRC I-DEAS finite element mesh generator, Supertab [5]. The latter requirement is derived from the desire to maintain a single modeling data base accessible by a suite of analysis codes spanning all of the major engineering analysis disciplines [3]. FIDAP is attractive in this respect because it is a finite element method (FEM) code and promises to simplify the task of data sharing with mechanical codes.

Analysis and Modeling Approach

Scale of the problem easily ranges over an order of magnitude with the upper and lower bounds defined by dewer and channel gap dimensions respectively. This suggests dividing the problem into two more manageable models we call "macro" and "micro". As its name implies, the macro model encompasses both the dewar and the test article and is meshed to the extent necessary to obtain large scale effects within the dewar and around the test article. The results obtained from the macro model are, in turn, used to define the boundary conditions for the micro model. The micro model focuses on the test article itself and contains more detail of the heated plate geometry and flow channel. Both models are two dimensional representations of the test setup.

Results

Figures 4 and 5 provide graphical comparisons of analysis results with test measurements for the heater plane in the horizontal position and the flow channel gap set at 10mm.

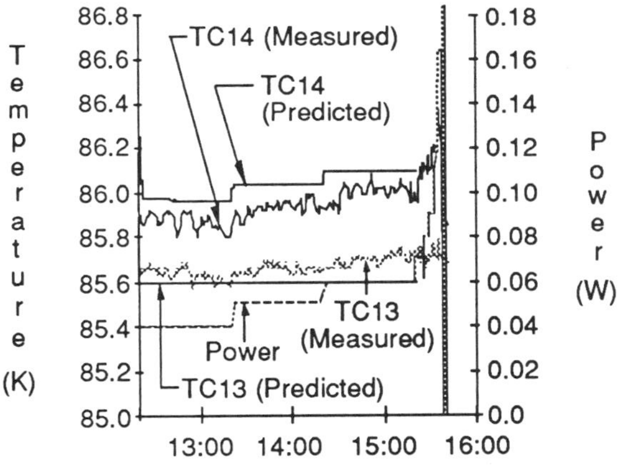

Figure 4. Comparison of Calculations With Test Data for the Test 31-32 Configuration.

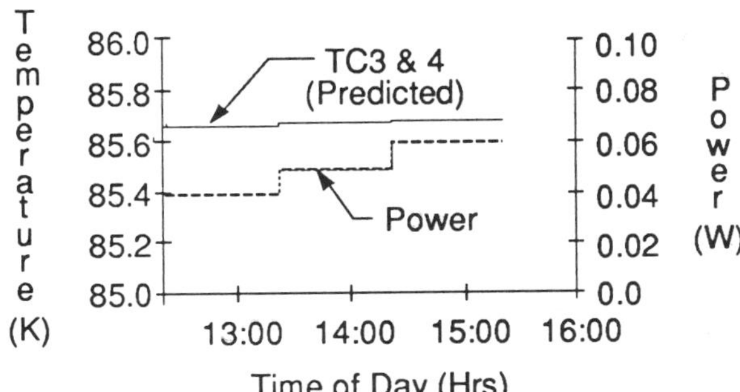

Figure 5. Calculated Responses of Thermocouple Locations 3 & 4 for the Test 31-32 Configuration.

During the last three hours of the test the power per bonded guage is increased hourly by 10 mW increments begining at 40 mW and ending at 60 mW. No power is applied to the non bonded circuit. Data are shown for thermocouple locations 13 & 14 corresponding to the bottom and top of the plexiglass cover plate respectively and 3 & 4 which are surface mounted at opposite ends of the aluminum plate. The calculated responses for locations 13 and 14 are overlaid on the corresponding test data in Figure 4 to aid comparison. As can be seen, the analysis and test data correlate quite well. The predicted temperature delta between positions 13 and 14 for the first hour is 0.37 K as compared to ~ 0.31 K measured in the test. Good correlation between predicted and measured responses of thermocouples 13 & 14 are especially important because they depend on heat transfer via both conduction and advection within the gap and as such represent a difficult modeling challenge. Similar good agreement is seen for the responses of thermocouple locations 3 & 4. The test data indicate that no appreciable gradient exists along the length of the aluminum surface and indeed, the predictions for 3 & 4 are indistinguishable in Figure 5. The temperature increases predicted for locations 3 & 4 also compare well against measurements with 0.08 K predicted versus ~0.1 K measured.

Summary

Comparisons of test data with initial analyses are quite promising as we are able to predict thermal responses well for those conditions in which turbulence plays a relatively minor role. We recognize, however, that turbulence within a LAC cannot be ignored and our future development plans include exploration of available techniques to model these effects. We also plan to conduct three dimensional analyses in order to gain experience in identifying when the benefits of such calculations outweigh the inevitable increase in model complexity.

Clearly, there is room for improvement but this first study serves to demonstrate the advantages of performing high fidelity integrated analyses early in the design phase.

References

[1] M. D. Marx *et al.*, "EMPACT: Electrons Muons Partons with Air Core Toroids, An expression of Interest for an Experiment to be Performed at the Superconducting Super Collider", SSC-EOI0006, (May 25, 1990)

[2] G. H. Trilling *et al.*, "Expression of Interest by the Solenodial Detector Collaboration to construct and operate a detector at the Superconducting Super Collider", SSC-EOI0003, (24 May 1990).

[3] T. S. Adams *et al.*, "An Engineering Design Study of a Hermetic Liquid Argon Calorimeter for the SSC", Submitted to NIM, (1990).

[4] V. Radeka and S. Rescia, Nucl. Instrum. Methods. A265,228 (1988).

[5] I-DEAS, Structural Dynamics Research Corporation, Milford, Ohio, (1987).

A MEASUREMENT OF E/π FOR A FAST LEAD LIQUID ARGON CALORIMETER

D. Makowiecki, [2] G.S. Abrams,[8] K. Amako,[7] A.R. Baden,[10] T. Bowen,[1] T. Burnett,[17] V. Cook,[17] L. Cremaldi,[12] R. Davisson,[17] N. DiGiacomo,[9] P. Draper,[14] J. Faust,[6] T. Ferbel,[14] D. Fong,[10] G. Forden,[1] T.A. Gabriel,[13] H.A. Gordon,[2] D.E. Groom,[8] N. Hadley,[10] V. Hagopian,[5] T. Handler,[15] J. Hauptman,[6] D. Hitlin,[3] O. Inaba,[7] E. Jenkins,[1] K. Johns,[1] F. Kirsten,[8] T. Kondo,[7] S. Kunori,[10] M. Levi,[8] F. Lobkowicz,[14] H. Ma,[2] P. Mockett,[17] G.T. Mulholland,[4] M. Murtagh,[2] M. Pang,[6] V. Radeka,[2] D. Rahm,[2] J. Reidy,[12] S. Rescia,[2] J. Rothberg,[17] J. Rutherfoord,[1] M. Shupe,[1] J. Siegrist,[8] A. Skuja,[10] D. Summers,[12] Y. Watanabe,[16] H. Weerts,[11] R.W. Williams,[17] J. Womersley[5]

[1] University of Arizona, Tucson, AZ 85721
[2] Brookhaven National Laboratory, Upton, NY 11973
[3] California Institute of Technology, Pasadena, CA 91125
[4] Fermilab, Batavia, IL 60510
[5] Florida State University, Tallahassee, FL 32306-3016
[6] Iowa State University, Ames, IA 50011
[7] KEK, Tsukuba-shi, Ibaraki-Ken, Japan
[8] Lawrence Berkeley Laboratory, Berkeley, CA 94720
[9] Martin Marietta Astronautics, Denver, CO 80201
[10] University of Maryland, College Park, MD 20742
[11] Michigan State University, East Lansing, MI 48824-1116
[12] University of Mississippi, University, MS 38677
[13] Oak Ridge National Laboratory, Oak Ridge, TN 37831-6369
[14] University of Rochester, Rochester, NY 14627
[15] University of Tennessee, Knoxville, TN 37996-1200
[16] Tokyo Institute of Technology, Tokyo 152, JAPAN
[17] University of Washington, Seattle, WA 98195

Abstract

The NA34 (HELIOS) calorimeter has measured $e/\pi \cong 1.1$ in a uranium / liquid argon calorimeter with a shaping time of 135 nsec. Lead may be a viable alternative, but e/π must first be measured at fast shaping times in lead. We are preparing to measure e/π at momenta ranging from 0.5 to 20 GeV/c and with shaping times of 50, 100 and 150 nsec.

Introduction

The use of uranium as an absorber in an SSC detector may prove to costly. It would be useful to make an accurate measurement of the performance of a liquid argon calorimeter with lead as an absorber. Some EGS studies have suggested that e/π is more nearly equal to one if thick lead plates are used. Described is a lead liquid argon calorimeter that closely follows the HELIOS design.[1] Our motivation for building this calorimeter has been to study the issue of compensation in lead. Where we could improve on the HELIOS design we have, when we could study other issues that might be applicable to a general SSC detector we have, but the major emphasis has been to study the e/π ratio. That has driven the cost and the time line.

Lead Stack

Some EGS studies by our collaborators from the University of Washington [2] have suggested that e/π will more nearly equal one if 12 mm lead plate is used instead of 6mm plate. We have designed a stack approximately 100 cm X 100cm with a depth of 169 cm (7.5 λ). It is segmented into three longitudinal sections, each weighing five tons. Lead that is 12 mm thick, with 0.06% calcium and 1.3% tin added for hardness, will be used as an absorber. The signal boards are three layer printed circuit boards with the signal strip on the inner layer and high voltage strips on the outside. Figure 1. The strips are produced by machining copper clad G-10 printed circuit material and laminating them together. An offset is

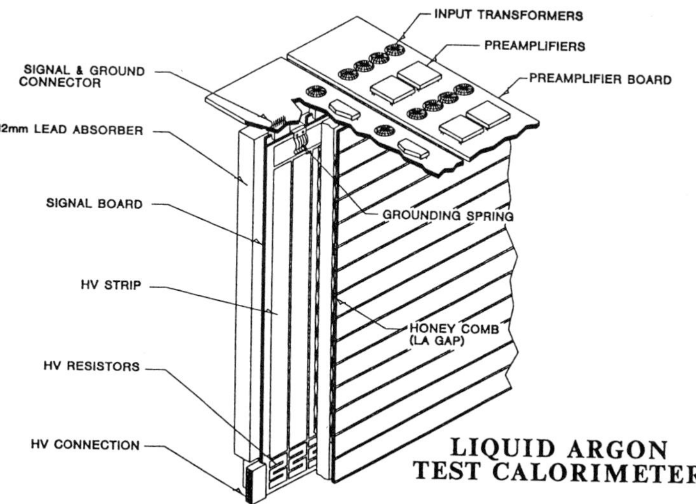

Fig. 1: A conceptual drawing of a liquid argon calorimeter (not to scale)

provided at the edge of each board to make electrical connections to the inner signal electrodes. The outer or high voltage electrodes are connected to a high voltage bus using epoxy carbon silk screen resistors. Charge deposited on the outer electrodes is capacitively coupled to the signal strips. A low inductance ground connection to the lead absorber is made with a specially designed copper-berylium spring. The input transformers and preamplifiers are located in the liquid argon in close proximity to the signal boards. The connection scheme follows closely the HELIOS method of achieving low induction charge transfer to the preamplifiers [3]. Each longitudinal section contains 32 cells, each cell consists of a lead absorber, a 2 mm argon gap, a 1.6 mm G-10 printed circuit signal board and another 2mm argon gap. The signal boards each have forty 2.5 cm strips. These strips alternate in X and Y in depth. The interleaving of the X and Y strips provide information about the transverse profile of the shower and a measurement of the sampling fluctuations. Figure 2. shows breakdown

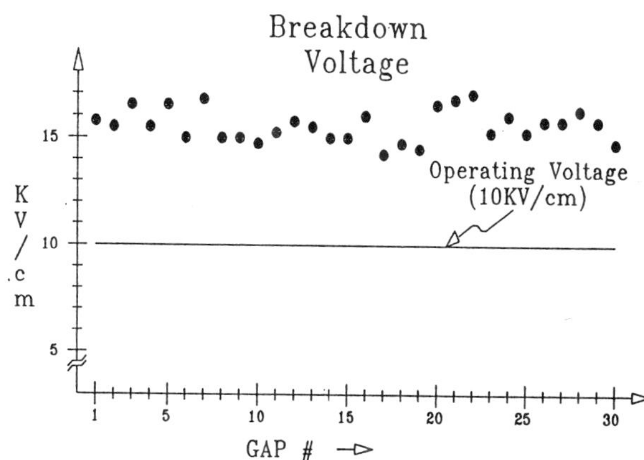

Fig. 2: Breakdown voltage for individual gaps in kV/cm)

voltage for the first 30 gaps in kV/cm. A large safety margin exists between operating voltage and breakdown voltage.

External Electronics

The external electronics consists of an intermediate amplifier stage a shaping and a summing amplifier. The intermediate amplifier stage is attached to the outside wall of the calorimeters cryogenics containment vessel, very close to the signal feedthroughs. A good electrical and mechanical connection between the vessel and the intermediate amplifier housing defines a faraday cage containing the stack, preamplifiers and intermediate amplifiers. See Figure 3. Besides driving the signal cables, the intermediate amplifier provides two fixed gains. This minimizes the dynamic range required for succeeding electronics and allows the same electronics chain to be used for tests at both the AGS and Fermilab energies. Signal cables are transformer-coupled at the output of the intermediate amplifier to provide electrical isolation and to break ground loops.

Following the intermediate amplifier is a Fastbus - sized printed circuit board which contains 48 channels of variable gain amplifiers, summing electronics and shaping amplifiers. The variable gain amplifier stages are controlled by an onboard microprocessor. This microprocessor accepts gain information for the individual channels via a RS 232 port. Gain is adjustable over a $\pm 25\%$ range. Individual channels are normalized to a fraction of a percent by injecting a known charge via precision resistors on the preamplifier boards located on the stack. One can also turn off individual channels in the variable gain amplifier. This aids in normalizing the gain per channel or in suppressing noisy channels. The signal from the variable gain amplifier is routed to two circuits. One circuit is a shaping amplifier with a fixed shaping time of 100 nsec. The output from these circuits is read

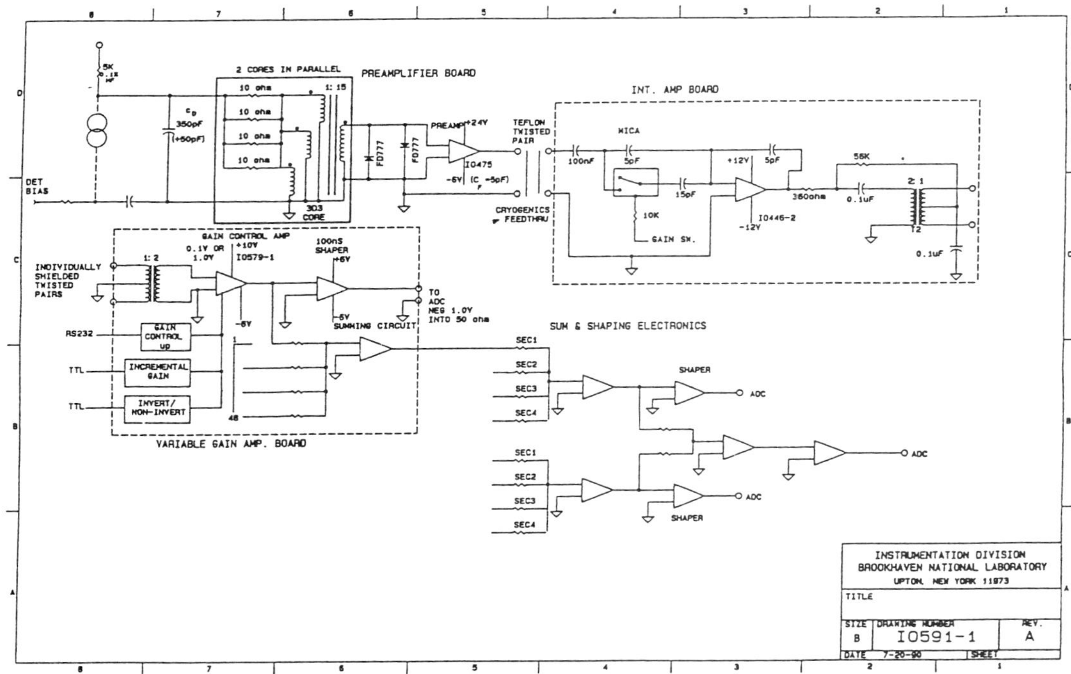

Fig. 3: Electronics Chain

out via a CAMAC based ADC system. The second circuit is an onboard summing amplifier. The 48 signal channels are summed at this point. It is important to note that the fast shape of the output signal from the stack preamplifiers is preserved throughout this chain. The output sums of the 48 channels are then summed again in X and Y. The X,Y and total sums go to individual shaping amplifiers with changeable shaping times between 50 and 200 nsec. Figure 4. is a photograph of the various moduled which constitute the external electronics chain. We plan to test this calorimeter in the spring of 1991 in the AGS beam at Brookhaven. The beam will have two Cerenkov counters to identify electrons. We will vary the momentum between 0.5 GeV and 20 GeV. The intensity of the beam can be as high as 10^7 particles / sec. This will allow the study of pileup effects directly.

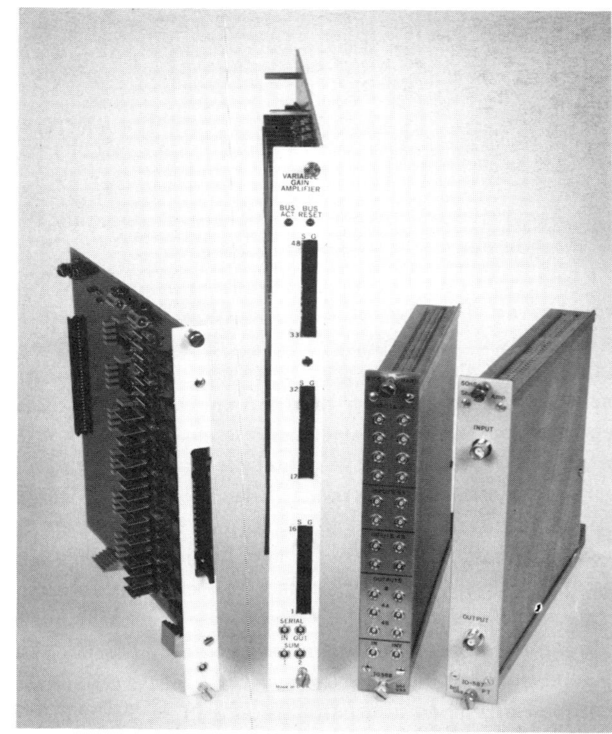

Fig. 4: Various modules which constitute the external electronic chain

References

1. The Calorimeter Team of the HELIOS Collaboration, CERN Report 0775C, August 1988.
2. Paul Mockett, unpublished.
3. V. Radeka and S. Rescia, Nucl. Inst. and Methods. A265 (1988) 228.

DEVELOPMENTS IN INTERPOLATIVE PAD READ OUT OF LIQUID ARGON SAMPLING CALORIMETERS *

P. M. Mockett, Vic Cook
University of Washington
Seattle, Washington 98195

Abstract

We report here recent developments on interpolative pad read out, dubbed BRITE Pads, of liquid argon sampling calorimeters. We have constructed model devices and made bench tests of their performance under various conditions. We report on the performance of a two element linear BRITE PAD system and a four element four moment BRITE PAD system. The results from the linear system are from a model electromagnetic calorimeter tower. To duplicate the fast pulse of an electromagnetic shower, we developed a fast laser, optical fiber and photodiode system; the photodiode fits within the 2 mm gap. The fast signals, about 10 pC discharged in 10 ns, were read out with BNL designed amplifiers followed by 50 ns RCCR shaping.

Introduction

Interpolative pad read out, or BRITE PADS, is designed to preserve the accuracy of a very fine transverse segmentation but with a much reduced channel count.[1] It is possible that towers with very small transverse dimensions, on the order on the order of 1 cm × 1 cm will be required to obtain the shower position resolution required for physics at the SSC in a jet environment.[2] Such segmentation would require far too many channels to be feasible. Modest size towers could be segmented into centimeter wide strips but this too would lead to an unwieldy number of channels. However the important information on the strips can be obtained by recording only the first few moments of the charge distribution . BRITE PADS present an economical way of achieving this. Fig. 1 shows a two element linear system that provides the zeroth and first moments of the charge distribution. The strips, outlined as dashed lines, capacitively couple to the pads to provide capacitive charge division. More complicated designs have been devised that provide the first four charge moments; from these moments not only the position but also the width and even the individual energies of two displaced showers within the same tower can be calculated.[1]

We have constructed a test model of an electromagnetic calorimeter tower with which the BRITE PAD technique can be tested. Fig. 2 shows the apparatus including the laser and fiber optic link used in this study. The diode laser is Mitsubishi ML4402 and the photodiode is Siemens SFH250. The laser system can discharge

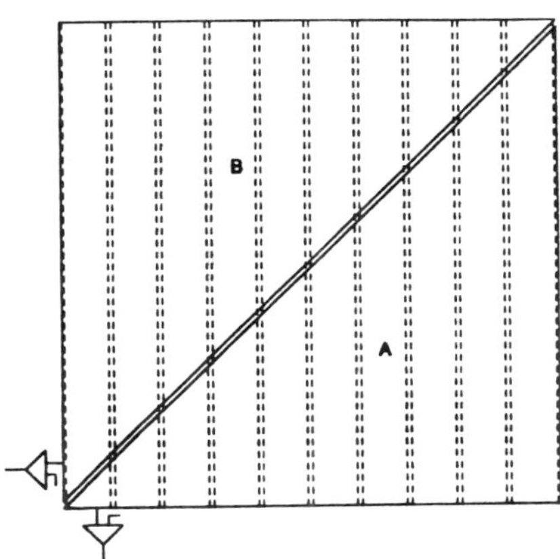

Figure 1. A schematic of a two element BRITE PAD. The charge induced on triangles A and B are amplified. The strips are outlined with a dashed line.

about 10 pC in 10 ns when the photodiode is biased with about 20 volts. The tower pads are segmented into x and y strips in alternate gaps. The strips are connected together at their ends and are coupled to the amplifiers through the two element BRITE PAD described above. An example of a two element BRITE PAD is shown in the foreground.

Fig. 3 shows a schematic of the strip connections giving more detail on the construction.

The signals from the two triangular elements are amplified by BNL-Helios type amplifiers shown on top of the tower. The amplifier signals are shaped by 50 ns RCCR shapers. The shaped signals are then recorded by a fast storage scope. Typical signals are shown in Fig. 4.

*Work supported by the Department of Energy, grant DE-FG06-90ER40541.

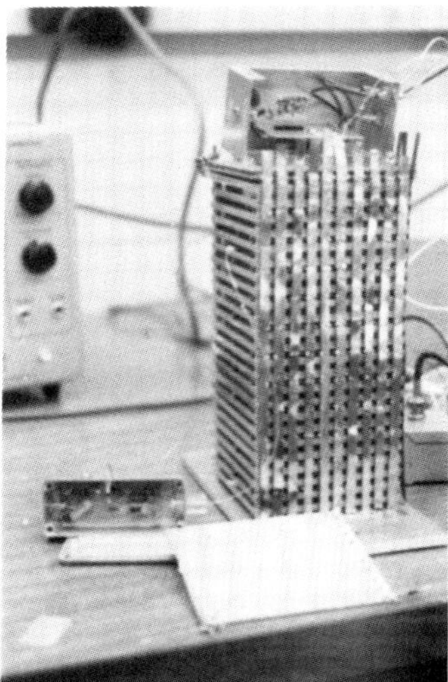

Figure 2. Photograph of the test assembly. The strips up the side connect the strips in the gaps to the BRITE PAD located at the top of the tower. The laser and fiber optic link are shown to the left of the tower. The amplifier is on the top. In the foreground an extra two element BRITE PAD is shown. The black diagonal line indicates boundary of the two elements.

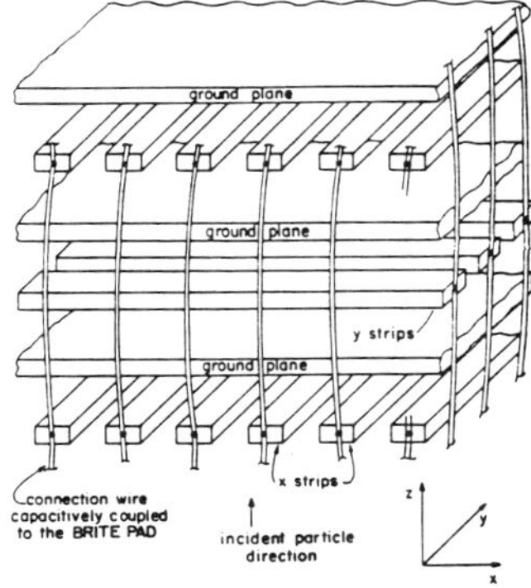

Figure 3. Schematic of the x- and y-strips forming a BRITE PAD tower. The strips are connected together projectively and then coupled capacitively to the BRITE PAD elements.

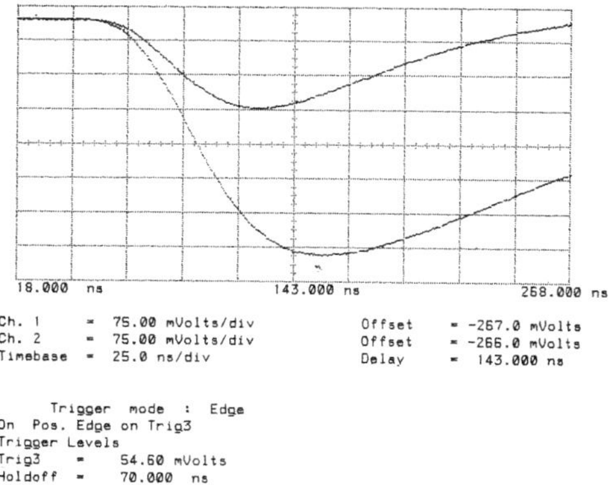

Figure 4. Photodiode signals induced on the two elements of the BRITE PAD, amplified and shaped.

BRITE PAD TESTS

We have tested a two element BRITE PAD connected to the model electromagnetic tower and a four element BRITE PAD separate form the tower. The position of the charge deposition is determined from the equation:

$$X = \frac{A}{A+B} \times L$$

where A and B are the charges on elements A and B and L is the length of the array. Because of the fast signal shaping, there is a fixed cross talk between the elements that must must be removed before calculating the position. The cross talk does not affect the sum of A and B but only the values. The transformation is a simple one given by the equation

$$\begin{pmatrix} A' \\ B' \end{pmatrix} = \begin{pmatrix} 1-f & f \\ f & 1-f \end{pmatrix} \begin{pmatrix} A \\ B \end{pmatrix}$$

where A and B are the true charge and A' and B' are the observed charge; f is the fraction of the signal on A that appears on B. f is independent of where the signal arrives on the BRITE PAD. Fig. 5 shows the results of measurements of the two element system on the electromagnetic tower. The abscissa is the position and the ordinate the fraction of charge on A divided by the sum with and without the cross talk correction. Analysis finds that f is about 0.22. The deviation in the data points from a straight line is attributed wholely to a lack of precision in the BRITE PAD construction. Future versions will use photo-etched double sided Kapton PC material where tolerances can be much more tightly controlled. In the above case, sheared strips of brass were stuck to sheared brass triangles using double stick tape. The results are highly encouraging that fast signals can be collected from the tower and processed through the BRITE PAD to give good positional accuracy.

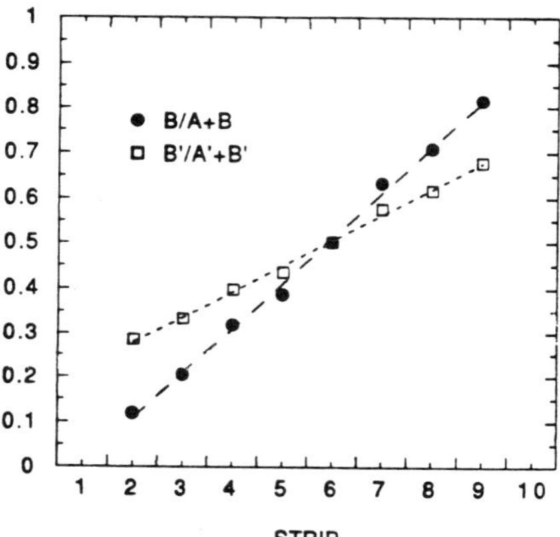

Figure 5. The fractional position of the charge deposition determined by the two element BRITE PAD with (squares) and without (dots) the cross talk.

We have also tested a four element BRITE PAD. The design of the elements is shown in Fig. 6. The four pads determine the zeroth through third moment of the charge distribution. With these moments the total charge, the charge position and width can be determined; if two showers are present the charge on each can also be determined.[1]

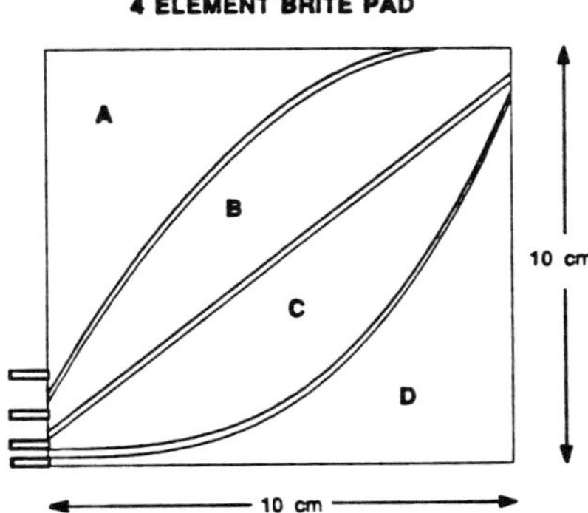

Figure 6. Schematic of a four element BRITE PAD.

Fig. 7 shows the results of measurements with the four element BRITE PAD. A cross talk uncorrected plot of the charge on A+B divided by the sum is shown. Again a good linear fit is obtained but with a shallower slope because of the cross talk present.

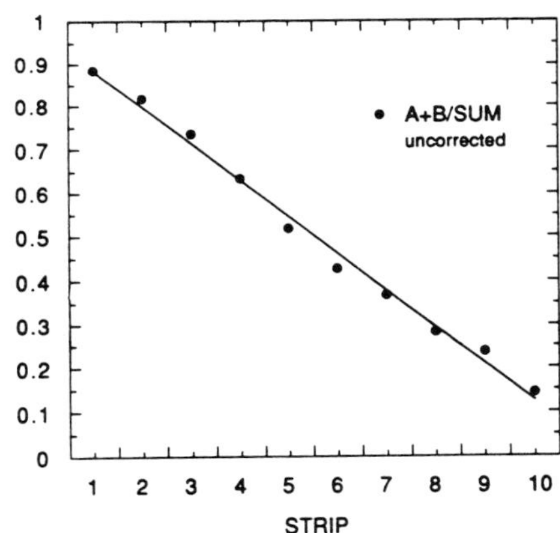

Figure 7. A four element BRITE PAD measurement of the charge position.

Fig. 8 shows the quadratic measurement and a quadratic fit. The shape agrees very well with the shape of the elements.

Conclusions

We have constructed two and four element BRITE PADs and tested them with fast signals. The two element BRITE PAD has been tested on a model electromagnetic tower. A laser photodiode system was developed to duplicate fast shower signals. We have demonstrated that fast signals can be successfully processed with a BRITE PAD system. Such a system makes fine segmentation possible with a modest number of electronic channels. The next step is to fabricate and test BRITE PADS made with photo-etched double sided Kapton PC material where much higher tolerances can be obtained for the element shapes, their positioning and capacitive coupling to the strips.

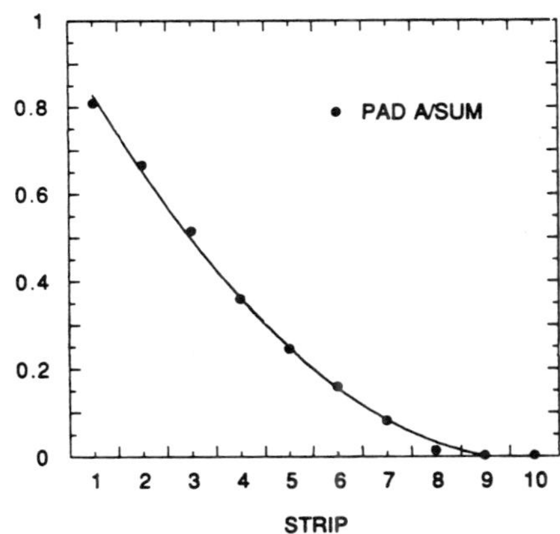

Figure 8. Quadratic measurement from a four element BRITE PAD. The shape follows the contour of element A very well.

References

1. P. M. Mockett, "A Novel Segmentation for Charge Sampling Calorimeters," Nucl. Instr. and Meth. **A289**, 139–145, (1990).

2. H. Gordon, NA34, private communication.

LIQUID ARGON CALORIMETER READOUT WITH AN ELECTROSTATIC TRANSFORMER AND EXTERNAL PREAMPS *

C. Hearty[a], G. S. Abrams[a,b], M. E. Levi[a], J. L. Siegrist[b]

[a]Lawrence Berkeley Laboratory
One Cyclotron Road
Berkeley, CA 94720

[b]Superconducting Super Collider Laboratory
2550 Beckleymeade Avenue
Dallas, TX 75237

Abstract

Progress is reported on research underway to verify that fast, low-noise readout of liquid-argon calorimeters can be achieved via the electrostatic transformer with preamps external to the cryostat.

Introduction — Liquid Argon Readout

Liquid argon calorimeters are not generally considered to be fast, despite an inherently fast signal risetime. There are several factors that can slow the readout:

1. The deposited charge has a long drift time — 200 ns/mm. Shaping the preamp output can produce a short peaking time, at the cost of a reduced signal (the ballistic deficit).
2. The calorimeter towers (stacks) have large capacitance, so it can take a long time to transfer the charge from the stack to the preamp.
3. The electronic noise (σ_N) becomes worse for short shaping times. For series dominated noise [1],

$$\sigma_N \propto \sqrt{\frac{R_s T}{\Lambda}}(C_d + C_{cable} + C_{in}), \quad (1)$$

where R_s is the preamp equivalent series resistance, T the preamp temperature, Λ the shaping time, and C_d, C_{cable} and C_{in} the detector, cable and preamp input capacitances.

Points 1 and 2 indicate that the signal-to-noise a decreases rapidly with peaking time t_p ($t_p = \Lambda/4$ for bipolar shaping). For $t_p \leq$ drift time (t_d),

$$S/N \propto \left(\frac{t_p}{t_d}\right)^{\frac{3}{2}} \left(1 - \frac{t_p}{2t_d}\right). \quad (2)$$

For $t_d = 400$ ns, S/N decreases by 4.6× as the shaping time is reduced from 150 ns to 50 ns. This equation does not include the time to transfer the charge from the stack to the preamp, which can significantly reduce the signal at short peaking times.

An Existing Solution: HELIOS

The HELIOS [2] experiment has developed a LAr calorimeter with fast readout and low noise. A ferrite core transformer (FCT) is used between the tower and the preamp and preamp itself is placed on the stack. A transformer ratio of n reduces the capacitance seen by the preamp by n^2, while reducing the signal by only n. The electronic noise is then

$$\sigma_N \propto \sqrt{\frac{R_s T}{\Lambda}} \left(\frac{C_d}{n} + n(C_{cable} + C_{in})\right). \quad (3)$$

In this case, $C_{cable} = 0$. The optimal value for n is $n^2 = C_d/C_{in}$. The characteristic risetime is $\tau \simeq \sqrt{LC_d}$, where L is the inductance between the stack and transformer and C_d is the untransformed detector capacitance; τ does not depend on n. The total risetime is approximately $3\tau = 45$ ns for HELIOS.

Difficulties with the HELIOS Solution

The disadvantages of the HELIOS solution for SSC experiments are:

- The preamps are inaccessible and, therefore, cannot be repaired or upgraded.
- Since the preamps are in the LAr, they must be low power and be cooled with liquid nitrogen

*This research was supported by a Fellowship from the Texas National Research Laboratory Commission and by the Director, Office of Energy Research, Office of High Energy and Nuclear Physics, Division of High Energy Physics of the U. S. Department of Energy under Contract Number DE-AC03-76SF00098.

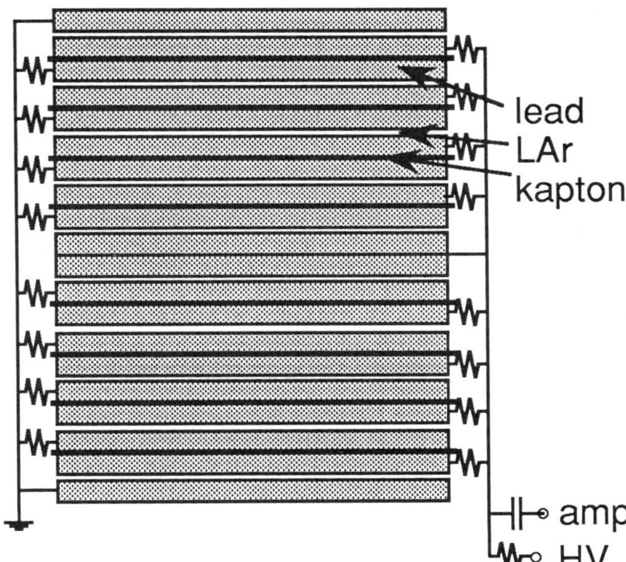

Figure 1: Ganging scheme used in the EST cosmic ray stack. Shown are two parallel sets of five gaps in series; the complete structure has four sets. The absorbers are divided in half by kapton to give both HV and ground planes. All charge drifts in the direction of the center absorber. The signal from other layers is capacitively coupled through the gaps.

to prevent bubbling and possible HV problems. With a nitrogen cooling system comes the risk of a leak contaminating the LAr.

- Preamps must be radiation hard to the level of at least several Mrad. Possible high luminosity running would place even stricter requirements.
- Ferrite core transformers, even with shielding, saturate in magnetic fields of 0.5–0.9 T.

Members of the LAr large subsystem group or the SDC collaboration are pursuing solutions to these problems that retain FCT's and preamps in the LAr. The alternative we are pursuing differs in two substantial respects — it uses an electrostatic transformer (EST) [3] and places the preamps outside of the cryostat. These options are not necessarily coupled. It may be possible, for example, to use FCT's with external preamps or EST's with internal preamps. The EST eliminates magnetic saturation of the transformer, while the external preamps simplify the cryogenics system and allow for repairs and upgrades.

The Electrostatic Transformer

The EST consists of combining longitudinal gaps in series rather than in parallel. The capacitance of n gaps in series is C_{gap}/n, versus $n \cdot C_{gap}$ for parallel ganging: the detector capacitance is reduced by n^2 while

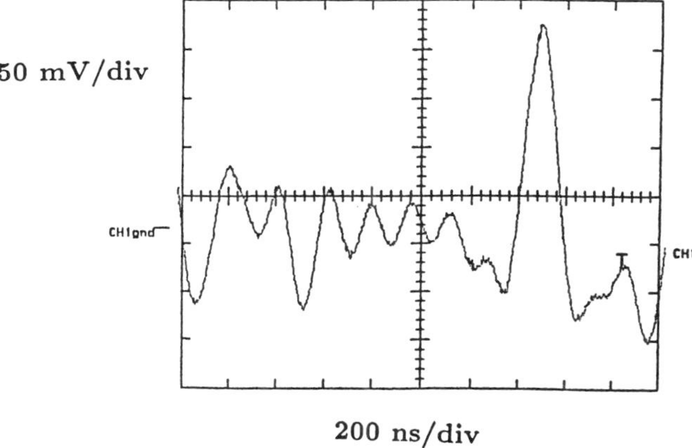

Figure 2: A cosmic ray observed in a 20-gap EST liquid argon calorimeter stack. The peaking time is 170 ns — the minimum for the Ortec 450 shaper. Signal-to-noise was 3:1.

the signal is reduced by only n. (A FCT of ratio n has the same effect — hence the name EST.) We have constructed a 20 gap EST stack (1.5 mm LAr, 12 mm lead) consisting of four parallel sets of five gaps in series (Figure 1). An Amptek A250 preamp located outside of the LAr cryostat was connected to the stack by a 3.6 m long 50Ω cable. Shaping was performed with an Ortec 450 research amplifier. The first tests with cosmic rays indicate that it is possible to obtain a fast risetime and good signal-to-noise for minimum ionizing particles (Figure 2). After corrections for the ballistic deficit, the minimum ionizing signal is constant to within 20% over the range of shaping times tested, and agrees with the expected value. Additional stacks will be installed to study crosstalk between towers.

Preamp Development

Equation 3 indicates that the additional cable capacitance associated with remote preamps will contribute to the electronic noise. However, the preamps can be operated at higher power and, therefore, with lower noise. One such method is to use multiple FET's at the input. For m FET's in parallel, the series equivalent resistance is reduced and the input capacitance increased: $R_s \to R_s/m$, $C_{in} \to m \cdot C_{in}$. The optimal value for m is $m = (C_{cable} + C_d/n^2)/C_{in}$. For example, for the cosmic ray stack, $C_{cable} + C_d/n^2 \simeq 400$ pF, $C_{in} \simeq 9$ pF and $m_{opt} = 44$. The achievable value, however, is lower. The input of the preamp appears as C_{in} in parallel with the electronically cooled resistance R_{in}. For proper termination of the cable, the input impedance must be approximately $R_{in} \simeq Z_{cable}$, which indicates that $C_{in} \leq 100$ pF. We are currently testing preamps using between five and ten 2SK152 JFET's. Each has $R_s \simeq 22\Omega$ (transconductance $g_m \simeq 30$ mA/V) and $C_{in} \simeq 9$ pF, so the total $R_s \simeq 2$–$5\ \Omega$. Other FET's are also being tested. The goal is to

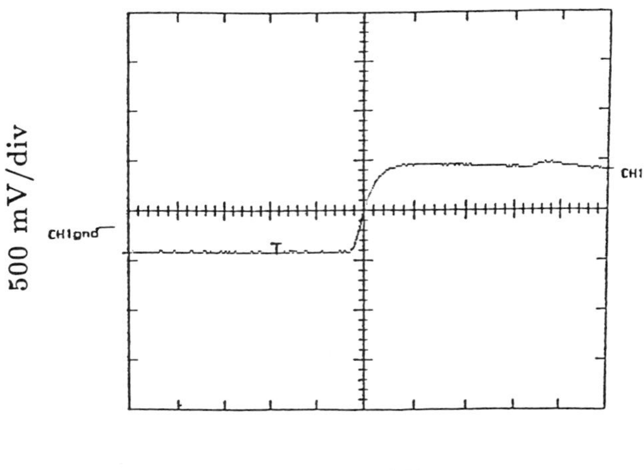

Figure 3: Risetime of a 390 pF load connected to the preamp via a 32 m long 50Ω cable. The expected risetime is $\tau \simeq 20$ ns, for a total risetime of approximately 60 ns.

achieve $R_s \leq 2\Omega$ with $C_{in} \leq 100$ pF, preferably using only one or two larger FET's. The preamp will not have to be radiation hard.

The characteristic charge transfer time is $\tau \simeq Z_{cable}C_d/n^2$, for $R_{in} = Z_{cable}$. Z_{cable} is chosen to give $\tau \leq 20$ ns. This relationship is illustrated in Figure 3, which shows the risetime of a signal from a 390 pF load connected to an A250 preamp by a 32 m long 50 Ω cable. The measured risetime is consistent with the expected value of 20 ns. This cable length is substantially longer than the maximum of 8 m indicated by preliminary SDC designs.

The noise performance of the EST with external preamps can be compared to the HELIOS solution using as an example the first 2λ-thick hadronic section, 0.15×0.15 in eta and phi. This section is a sum of nine towers, each consisting of forty 10 cm square 8.5 mm-thick lead plates with 1.5 mm LAr gaps. $C_d = 3730$ pF (untransformed). For internal preamps, $R_s = 22\Omega$, $T = 90$, $\Lambda = 300$ ns, $C_{in} = 9$ pF, and $n = 20$, giving a total noise — electronic and pileup [4] — of 220 MeV. For external preamps, $R_s = 2\Omega$, $T = 290$, $C_{in} = 100$, $n = 5$, $C_{cable} = 300$ pF (8 m of $Z_{cable} = 120\Omega$) and $\Lambda = 600$ ns, for total noise of 330 MeV. Λ is chosen to minimize the total noise, which occurs approximately when the contribution of electronic noise and pileup are equal. The longer shaping time in the EST case slightly decreases the timing resolution. To achieve $\sigma_{time} = 5$ ns requires an energy deposition of 10 GeV for the EST, 3.2 GeV for the FCT.

Larger System Tests

We are working on two larger systems tests, the goal of which is to test that adequately fast and low-noise solutions exist for liquid argon readout using an EST with external preamps. The first uses a 3 × 3 aluminum air stack (no LAr or HV). This stack has been previously used [3] by the Warm Liquid Calorimeter collaboration to test the EST concept. The intent is to verify the response of the EST with realistic cables, feedthroughs and preamps to injected charge. The second and larger test will be a 3 × 3 array of lead/LAr stacks 22 X_0 deep. Each tower will have 10 layers of 12 mm lead plates with 1.5 mm LAr gaps, ganged as in Figure 1. This array will be tested in spring 1991 at the Brookhaven National Laboratory e^- beam as part of the LAr large subsystem beam test.

Conclusion

A program of research and development is under way to study the feasibility of reading out liquid argon calorimeters using EST's and external preamps. Such a system may have noise performance compatible with SDC physics goals, but with the advantages of preamp accessibility, a simplified cryogenics system, and the absence of transformer magnetic saturation.

References

[1] V. Radeka, S. Rescia, "Speed and Noise Limits in Ionization Chamber Calorimeters," Nucl. Instum. Methods **A265**, 228–242, (1988).

[2] H. J. Specht, "The HELIOS Experiment," Nucl. Phys. **A447**, 387C–391C, (1986).

[3] G. Knies, D. Neuffer, "Development and Test of a Liquid Argon Shower Detector," Nucl. Instum. Methods **120**, 1–7, (1974); J. Colas, M. Pripstein, W. A. Wenzel, "The Electrostatic Transformer," LBL-28511, (1990); submitted to Nucl. Instrum. Methods A.

[4] G. O. Alverson, J. Huston, "Estimation of Background Noise in LAr Detectors Due to Pileup," in *Proceedings of the 1986 Summer Study on the Design and Utilization of the Superconducting Super Collider*, 1986, pp 368–370.

FREE ION YIELDS FOR SEVERAL SILICON, GERMANIUM AND TIN CONTAINING LIQUIDS

R. A. Holroyd
Chemistry Department, Brookhaven National Laboratory, Upton, NY 11973

S. Geer and F. Ptohos
High Energy Physics Laboratory, Harvard University, 42 Oxford Street, Cambridge, MA 02138

Abstract

A survey was made of compounds containing silicon, germanium or tin atoms to find new room temperature liquids for use in ionization chambers. The results show an interesting correlation of free ion yields with molecular structure. The data were also analyzed to obtain the average thermalization ranges of electrons formed in the ionization process.

Introduction

An effort is being made to find better room temperature liquids for use in ionization chambers. Such liquids must have both a high electron mobility as well as a large free ion yield (G_{fi}), where G_{fi} is the number of ion pairs detected per 100 eV absorbed. Liquids like neopentane and tetramethylsilane satisfy these requirements but have the disadvantage of low boiling points and flash points. We have measured the field dependence of the free ion yield for several higher boiling liquids containing silicon, germanium or tin atoms.[1,2]

Experimental

The details of the experimental method have been described.[1,2] In brief, the liquids were exposed in a parallel plate ionization chamber to x-rays produced by 1.95 MeV electrons incident on a tungsten target. The currents measured for each liquid were compared to the current measured with 2,2,4,4-tetramethylpentane (TMP) at an applied field of 9.2 kV/cm for which G_{fi} is 1.1 ion pairs per 100 eV[3]. The dose in each liquid was calculated with the EGS4 Monte Carlo code.[4]

Results and Discussion

The results show that tetraalkyl compounds of the same molecular structure have very similar free ion yields. Thus if the silicon atom in tetraethylsilane is replaced by germanium or tin the free ion yield remains unchanged (see Fig. 1). Similarly, tetraethylsilane (TES) and tetraethyltin (TESn) have comparable ion yields; the same is true for tetrabutylgermane (TBG) and tetrabutyltin (TBSn). The yields are dependent on the nature of the alkyl groups: the

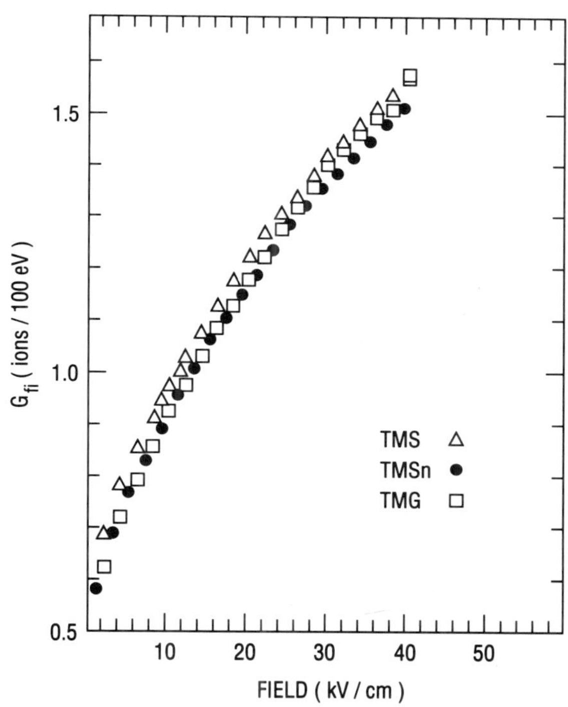

Fig. 1. Free ion yields shown as a function of applied electric field for tetramethylsilane (TMS), tetramethylgermane (TMG) and tetramethyltin (TMSn).

tetramethyl liquids have higher yields than the tetraethyl liquids which in turn have higher yields than the tetrabutyl liquids.

From an analysis of the data we derived the range (b) of epithermal electrons; that is, the distance traveled by the electron in separating from the positive ion. A distribution of separation distances of the form:

$$D(Y) = (3.84\ y^2/\pi^{1/2}b^3)\ \exp(-y^2/b^2) \qquad (1)$$

was assumed for $y < 2.4$ b; for larger distances a y^{-3} power tail was added. The free ion yield depends on D(y) and the probability of escape P(y,E). Thus

$$G_{fi}(E) = G_{tot} \int_0^\infty D(y)P(y,E)dy, \qquad (2)$$

where G_{tot} is the total yield of ions formed initially per 100 eV absorbed. The probability of escape according to Onsager theory is:

$$P(y,E) = \exp(-r_c/y)(1 + F(y,E)), \qquad (3)$$

where $r_c = e^2/\varepsilon kT$ and ε is the dielectric constant of the liquid. The best fit values of G_{tot} are in the neighborhood of 3 ion pairs per 100 eV deposited for many of the liquids studied; lower than the value of 4.3 ion pairs per 100 eV sometimes used. For G_{tot} fixed at 3.0 the resulting values of b, shown in Table 1, are near 160 Å for the tetramethyl compounds but much lower for the tetrabutyl compounds. When both absolute normalization and shape uncertainties in the $G_{fi}(E)$ measurements are taken into account the errors on b and $G_{fi}(0)$ are determined to be about ±20% and ±12% respectively.

Table 1[a]. Values of b, free ion yields and mobilities.

Liquid	G_{tot} = 3.0 per 100 eV			G_{fi} (0)[b]	μ
	b (Å)	b x ε	b x d	ions per 100 eV	$cm^2V^{-1}s^{-1}$
TMS	166	319	107	0.64	100
TMG	156	314	157	0.63	90
TMSn	145	326	190	0.65	70
TES	-	-	-	0.20	0.5
TESn	-	-	-	0.21	-
TEC	-	-	-	0.23	0.8
TBG	63	134	59	0.14	-
TBSn	62	138	66	0.15	-
BTMSM[c]	126	270	95	0.50	-
TMP	187	370	135	0.76	30

a) Data from Ref. 2.
b) Free ion yield at zero applied field.
c) Preliminary results.

The product b x ε is comparable for tetraalkyl compounds of similar molecular structure. However the product of b x d, where d is the density, is not constant as suggested by earlier studies.[4] The electron mobilities (μ) and free ion yields are also comparable for similar structures. Thus, $G_{fi}(0)$ is 0.64 ± .01 per 100 eV and μ is of the order of 100 cm^2/Vs for the tetramethyl compounds. In contrast $G_{fi}(0)$ is 0.20 per 100 eV and μ is near 0.6 cm^2/Vs for the tetraethyl compounds.

The observation, made here, that substitution of a central atom has little effect on the electron properties: G_{fi} and μ could also apply to other compounds but there is little data available as yet to test this. We show the results for tetramethylpentane (TMP) and the analogous silicon substituted compound bis-trimethylsilylmethane (BTMSM) in Table I. The ion yield is less for BTMSM than for TMP. The mobility is unknown but may be larger in BTMSM by analogy to the tetramethyl liquids where the mobility in TMS exceeds that of neopentane.

In summary, the best liquids found so far are the tetramethyl compounds which exhibit both high ion yields and electron mobilities. The nature of the central group IV atom has little effect on these properties. However, tetramethyltin is not recommended because of its toxicity and electrical breakdown phenomena.[6] BTMSM, which is of the same structure as TMP except two carbon atoms have been substituted by silicon, may be of interest but the mobility is as yet unmeasured..

Acknowledgments

S.G. and F.P. are supported under contract DE-AC02-76ER03064 with the Department of Energy. R.H. is supported by contract DE-AC02-76CH00016 with the Department of Energy and its Division of Chemical Science, Office of Basic Energy Sciences.

References

[1] S. Geer, R. A. Holroyd and F. Ptohos, *Nucl. Inst. and Meth. in Phys. Res.* **A287** 447 (1990).

[2] S. Geer, R. A. Holroyd and F. Ptohos, *Nucl. Inst. and Meth. in Phys. Res.*, submitted.

[3] T. G. Ryan and G. R. Freeman, *J. Chem. Phys.* **68**, 5144 (1978).

[4] The EGS4 Code System, W. R. Nelson et al. SLAC-265 (1985), unpublished.

[5] W. F. Schmidt and A. O. Allen, *J. Chem. Phys.* **52**, 2345 (1970).

[6] H. Faidas, et al., *Nucl. Inst. and Meth. in Phys. Res.*, submitted.

WARM LIQUIDS FOR SSC CALORIMETRY: ELECTRON TRANSPORT AND ELECTRICAL PROPERTIES*

H. Faidas, D.L. McCorkle and L.G. Christophorou
Department of Physics and Astronomy
The University of Tennessee
Knoxville, TN 37996-1200
U.S.A.

Atomic Molecular and High Voltage Physics Group
Health and Safety Research Division
Oak Ridge National Laboratory
Oak Ridge, TN 37831-6199
U.S.A.

Abstract

Electron drift velocity measurements and findings on the electrical properties for a number of likely warm calorimeter liquids are presented.

Introduction

Warm liquids appear as promising candidates for the Superconducting Super Collider calorimeter provided they meet certain requirements [1], such as, high electron drift velocity, w, (resulting in fast calorimeter response), high free electron yield, G_{fe}, (resulting in high sensitivity), good electrical properties and radiation strength. In this paper, we report on our w and mobility ($\mu = w/E$) measurements as a function of the uniform applied electric field, E, for a number of fast warm liquids and their mixtures. These include the tetramethyl liquids neopentane (TMC), tetramethylsilane (TMS), tetramethylgermanium (TMG), tetramethyltin (TMT), 2,2,4,4-tetramethylpentane (TMP) and mixtures of TMS with TMP (mole ratio M= 1.31/1) and n-pentane (n-Pt) (M= 102/1, 17/1 and 5.6/1). For these measurements, we have used our newly developed laser photoinjection technique which is especially suited for fast liquids [2–4]. We also report on certain findings concerning the electrical properties of these liquids, namely, their dielectric strength, the effects of the electrical breakdown, and the products produced by electrical breakdown (EB).

Results

The $w(E)$ and $\mu(E)$ data for the neat liquids are shown in Figures 1a and 1b and the data for the mixtures, in Figures 1c and 1d. The salient features of the $w(E)$ curves for both neat liquids and mixtures are: at low fields and up to a critical field E_c, w increases linearly with E, the electrons are thermal, and $\mu = w/E = \mu_{th}$. For $E > E_c$, the electrons become hot (they have energies above thermal) and the mobility decreases. At higher E values, $w(E)$ tends toward saturation. Table 1 lists M, E_c, μ_{th}, the maximum field reached without having an EB, E_{max}, and the corresponding maximum drift velocity measured w_{max}, for the liquids/mixtures studied. An interesting observation is (see Figures 1a and 1b) that TMG and TMT for $E > 2 \times 10^4$ and $6 \times 10^4\ \mathrm{V\,cm^{-1}}$, respectively, have higher w-values than TMS.

Electrical Properties

Table 2 shows the comparative properties of TMC, TMS, TMG, TMT, and TMP as detector liquids. In all liquids except TMT, EB occurred at $E > 10^5\ \mathrm{V\,cm^{-1}}$. The TMS/n-pentane mixtures appear to have relatively higher dielectric strengths. All liquids underwent various degrees of decomposition after an EB. In TMS, decomposition was minimal and did not affect noticeably the subsequent performance of the system or the measurements. In TMG, and especially in TMT, extensive decomposition occurred after an EB; the cell surfaces were covered with heavy layers of Ge and Sn, re-

*Research sponsored by the Office of Naval Research and the U.S. Department of Energy under respectively, Contract Nos. N00014-89-J-1990 and DE-AS05-76ER03956 with the University of Tennessee, and by the Office of Health and Environmental Research, U.S. Department of Energy, under Contract No. DE-AC05-84OR21400 with Martin Marietta Energy Systems, Inc.

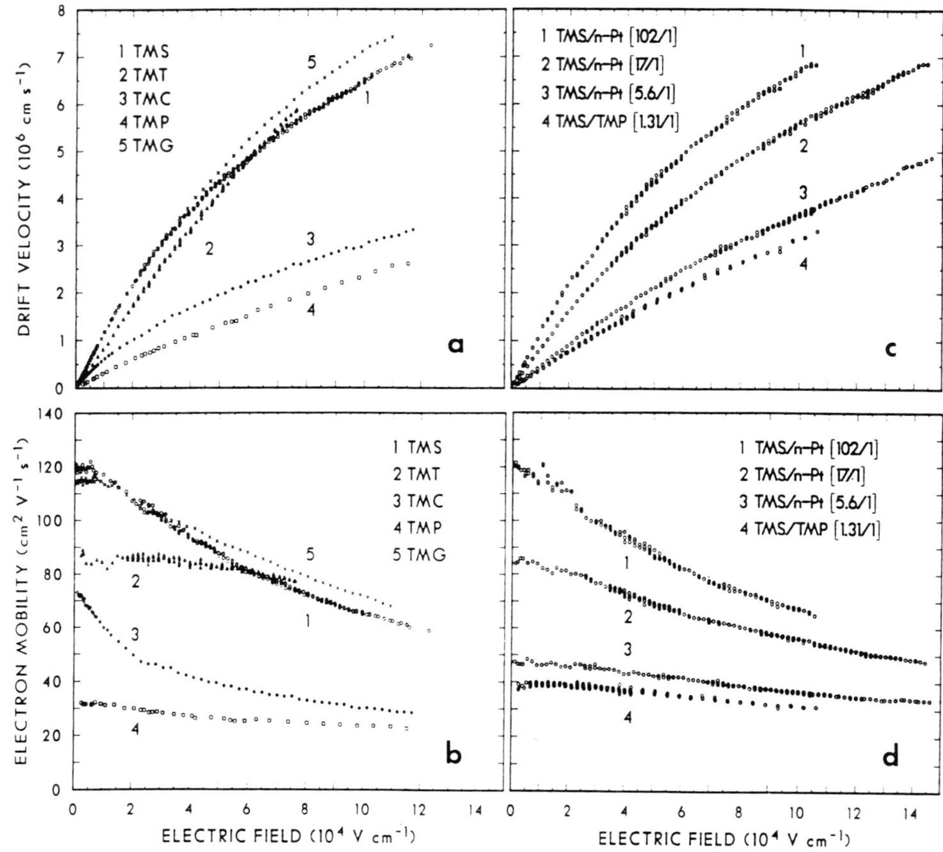

Figure 1: Electron drift velocities and electron mobilities in pure fast liquids and mixtures.

spectively, and the dielectric strength of the liquid was lowered to $\sim$ 20 to 30 kV cm^{-1}. In TMP and TMG, electronegative impurities were formed that reduced the electron lifetime to less than $\sim$ 100 ns and 10 ns, respectively. From our studies, it appears that unless electrical breakdown can be totally avoided, TMS is the most promising liquid for detectors.

References

[1] M.G. Albrow et al., "Performance of a Uranium/ tetramethylpentane Electromagnetic Calorimeter," Nucl. Instr. and Meth. **A265**, 303–318 (1988).

[2] H. Faidas, L.G. Christophorou and D.L. McCorkle, "Drift Velocities of Excess Electrons in TMP and TMS: A Fast Drift Technique," Chem. Phys. Lett. **163**, 495–498 (1989).

[3] H. Faidas, L.G. Christophorou, D.L. McCorkle and J.G. Carter, "Electron Drift Velocities and Electron Mobilities in Fast Room-temperature Dielectric Liquids and Their Corresponding Vapors," Nucl. Instr. and Meth. **A294**, 575–582 (1990).

[4] H. Faidas, L.G. Christophorou, D.L. McCorkle and J.G. Carter, "Electron Drift Velocities in Fast Dielectric Liquids and Their Vapors," in *Proceedings of the 6th Int. Symp. on Gaseous Dielectrics*, Knoxville, Tennessee, September 23–27, 1990.

Table 1. Mole Ratio, M; Critical Electric Field, E_c; Thermal Electron Mobility, μ_{th}; Maximum Electric Field, E_{max}; and Maximum Electron Drift Velocity, w_{max} for the Fast Neat Liquids and Mixtures Studied.

Liquid/Mixture	M	E_c (kV cm^{-1})	μ_{th} (cm^2 s^{-1} V^{-1})	E_{max} (kV cm^{-1})	w_{max} (10^6 cm s^{-1})
TMC	Neat	3.5	71.5	116[a]	3.3
TMS	Neat	7	119.3	125	7.2
TMG	Neat	15	114.7	109[b]	7.4
TMT	Neat	30	85.7	75	6.0
TMP	Neat	15	31.8	115	2.6
TMS/TMP	1.31/1	18	39.1	105	3.2
TMS/n-pentane	102/1	7	118	105[a]	6.8
TMS/n-pentane	17/1	8	85	145[a]	6.8
TMS/n-pentane	5.6/1	15	47.6	145	4.9

[a] E_{max} was not limited by electrical breakdown.
[b] Electrical breakdown occurred at 120 kV cm^{-1}.

Table 2. Comparative "Practical" Properties of TMC, TMS, TMG, TMT, and TMP

	Purification	Drift Velocity	Dielectric Strength	Effects of Electrical Breakdown
TMC	Easy/Quick	Intermediate	High	
TMS	Easy/Quick	Fast	High	No apparent liquid deterioration. Damage to electronics/pitting of electrodes
TMG	Easy/Quick	Fast	High	Extensive liquid decomposition rendered system unusable. Electronegative impurity(ies) formed. Damage to electronics/ pitting of electrodes
TMT	Difficult, reacts with H_2SO_4 and NaK	Fast	Low	Same as TMG
TMP	Slow	Slow	High	Electronegative impurity(ies) formed

MATERIAL COMPATIBILITY TEST FOR A WARM LIQUID CALORIMETER

H. Yuta, K. Abe, K. Hasegawa, F. Suekane, N. Kawamura, J. Katayama, H. Hanada, K. Neichi and Y. Hasegawa
Department of Physics, Tohoku University, Sendai, Japan

M. Higuchi, Y. Hoshi, and H. Iso
Department of Engineering, Tohoku Gakuin University, Tagajo, Japan

K. Masuda
Saitama College of Health, Urawa, Japan

R. Kikuchi
Department of Physics, Kyoto University, Kyoto, Japan

K. Miyano
Department of Physics, Niigata University, Niigata, Japan

K. Ishimaru
KEK, Oho,Japan

Abstract

We present preliminary results of compatibility tests of lead using a warm-liquid ionization chamber for the SSC calorimeter R+D work.

I. Introduction

Warm liquid calorimetry is one of the promising technologies for the SDC detector. The modules of the calorimeter are considered to be a type of swimming pool configuration for which absorber plates are arranged in a module filled with warm liquid. Because of possible e/h compensation with lead absorber in the warm liquid calorimeter, it is the material most likely to be used as the absorber. The total weight of lead absorber for the barrel and endcap calorimeters is approximately 5000tons. The surface contact rate of lead to liquid is $10000 cm^2/\ell$ for a 2mm gap. Because of the stringent requirements for liquid purity, we need to study extensively the compatibility of the materials used for the module. In particular, lead is one of the most important materials to be tested for the warm liquid calorimeter. The status of the material compatibility tests is summarized in ref. [1].

In this report, we present preliminary results of compatibility tests of lead in an ionization chamber. In this study, we have used liquid tetramethylgermanium (TMG) obtained from Wiley Organics Co. with a purity specification of 99.9%. TMG has a molecular structure similar to tetramethylsilane(TMS) and tetramethylpentane(TMP). It has a high mobility, 90cm/kVs, like TMS and is less flammable than TMS. From these properties, TMG is more suitable liquid for the SSC calorimeter compared to TMS, though the liquid is expensive.

II. Experimental procedure

Fig. 1 shows the test ionization chamber used for this test experiment. The chamber vessel is made of stainless steel with inside dimensions of 100mm diameter and 50mm height. The two circular plates, anode and cathode, are placed in parallel by a gap of 7mm with pyrex glass spacers and installed in the chamber. About 0.01μC of the isotope ^{207}Bi is deposited in the center of the cathode plate for the signal readout of 976keV conversion electrons from the isotope. A total volume of liquid is about 260cc to fill the chamber.

Two test piece of lead plates were inserted in a lead container of about 200cc. The lead used for this test contains 0.065% Calcium and 1.65% Tin. The lead container is connected to the chamber through a metal valve so that TMG in the chamber can come in contact with any material inserted in the container.

The size of the Pb plates is 100×30×5mm giving a surface contact rate of 550cm²/ℓ. The surfaces of the lead plates were machined without using oil. Then the plates were cleaned in an acetone ultrasound bath and kept in acetone until the plates were inserted into the container.

The TMG having an initial purity of 99.9% was purified further. Fig. 2 shows the purification system, which consists of a vacuum distillation system and a 3 liter vessel containing molecular sieves of 4A. The vacuum distillation was performed by transferring the liquid from one vessel to other in a vacuum of $\sim 5\times10^{-5}$ Torr. After two distillations, the liquid was transferred to the molecular sieve vessel where the liquid was kept for two days. This series of steps was repeated twice, and the liquid was then transferred to the test chamber for measurements of free ion yields (G_{fi}) and free electron lifetimes (τ).

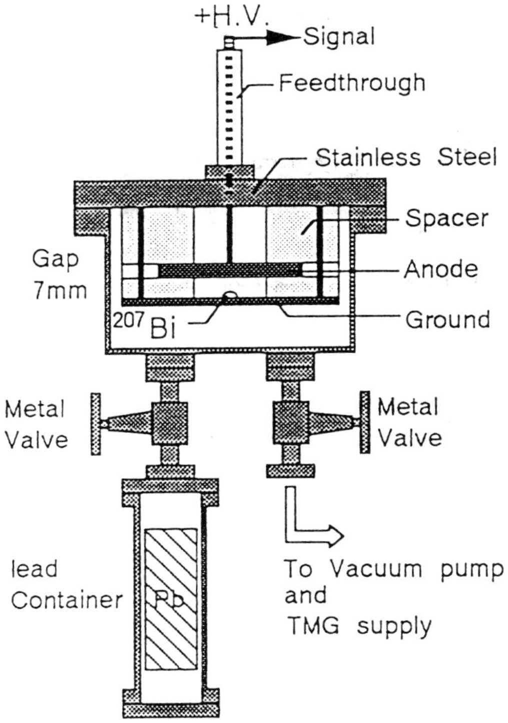

Fig. 1 Test chamber with a lead container

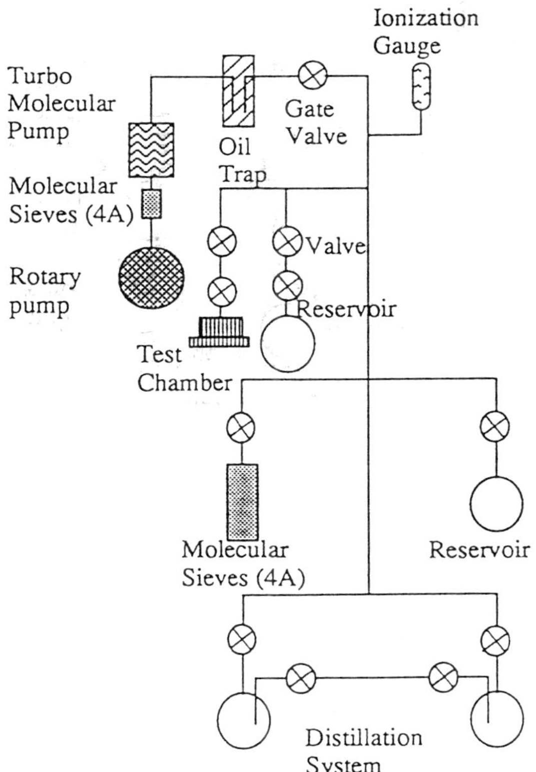

Fig. 2 Purification system used for this test

The electron lifetimes and the free electron yields, G_{fi}, are obtained from the measurements of charges induced by the 976keV conversion electrons from ^{207}Bi using the Onsager method [2]. The lifetimes and the G_{fi} were measured to be τ=2.5±0.4μsec and G_{fi}=0.49±0.02 immediately after the chamber was filled with liquid.

III. Results

The material compatibility tests were performed by measuring the free ion yields at several electric fields. The deterioration of the liquid reduces the ion yields which directly affect the energy calibration for the calorimeter.

Figs. 3(a) and (b) show the free ion yield, G_{fi}, at an electric field E=5keV/cm and the electron lifetime as a function of days. The data were taken for a period of three months. To confirm first that the chamber was operating stably, the G_{fi}'s and the lifetimes were measured keeping the liquid in the chamber for the first three weeks. Subsequently, the liquid was transferred to and left in the lead container. The liquid was transferred to the chamber only when the G_{fi} and the lifetimes were measured. The lambda shape marks in Fig. 3 indicate the date when the immersion of lead in the liquid was started.

Assuming that the G_{fi} varies exponentially as a function of days, we fitted the G_{fi}'s over the whole period. The curve in Fig. 3(a) shows the result of the best fit, having the form of

$$G_{fi} \propto 10**(-0.000037\pm0.000233)\cdot t(\text{day}) \quad \cdots\cdots(1)$$

Extrapolating this fit to one year later, we found that the change of G_{fi}, equivalent to the signal change ΔQ, is

ΔG_{fi}=0.03±0.19 at t=360days.

This result seems to indicate that the liquid pollution due to out-gassing from the lead surface is insignificant,

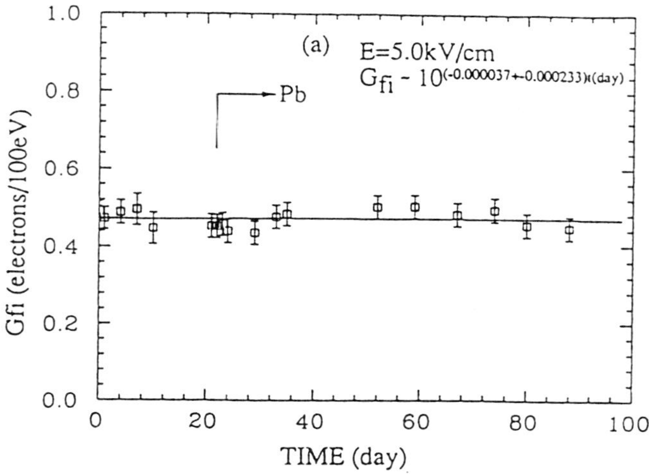

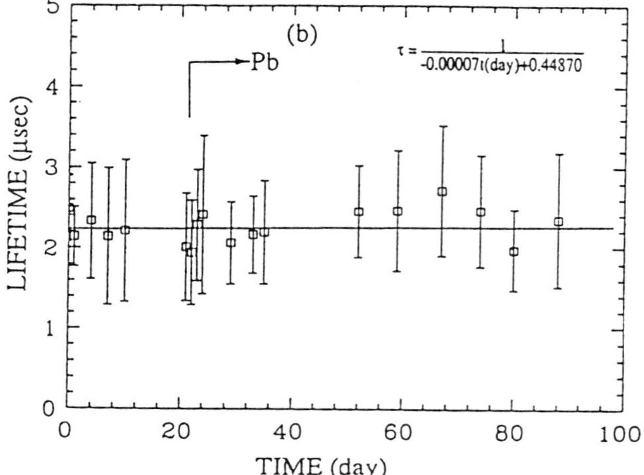

Fig. 3(a) Free ion yield, G_{fi}, and (b) electron lifetime, τ, as a function of days.

but the error is large and we need more long-term measurements.

References

[1] M. Pripstein et al., "Status Report on warm Liquid Calorimetry and Preliminary Result from E795 at Fermilab", in Proceedings of the International Workshop on Solenoidal Detector for the SSC, held at KEK, April 23-25, 1990, P-336-341.

[2] M. I. Lopes et al., "Estimation of the electron lifetime in tetramethylsilave", NIM A271, 464-470 (1988).

RADIATION DAMAGE TO A WARM LIQUID CALORIMETER

H. Yuta, K. Abe, K. Hasegawa, F. Suekane, N. Kawamura, J. Katayama, H. Hanada, K. Neichi and Y. Hasegawa
Department of Physics, Tohoku University, Sendai, Japan

M. Higuchi, Y. Hoshi, and H. Iso
Department of Engineering, Tohoku Gakuin University, Tagajo, Japan

K. Masuda
Saitama College of Health, Urawa, Japan

R. Kikuchi
Department of Physics, Kyoto University, Kyoto, Japan

K. Miyano
Department of Physics, Niigata University, Niigata, Japan

K. Ishimaru
KEK, Oho,Japan

Abstract

We present preliminary results of radiation damage to a warm-liquid ionization chamber. The electron lifetime is reduced substantially at 1.2×10^3Grays for the chamber with a gap of 7mm. However, a calculation for a 2mm gap chamber indicates no significant changes in free ion yields using the lifetimes measured.

I. Introduction

One of the most critical issues for the SSC calorimeter is the radiation hardness of the detector. The radiation level is expected to be in a range from 10 grays/year at the central region to 10^6grays/year at the forward region for the beam luminosity of $10^{33}/cm^2sec$[1]. For the warm-liquid calorimeter of "swimming-pool type", it is possible to circulate or exchange the warm liquid if the signal or the collected charge changes significantly due to the radiation damage. However, it is important to study the effect of the radiation damage to the liquid prior to the engineering design of the SDC calorimeter.

In this report, we present the preliminary results of our test on the effects of radiation damage to tetramethylgermanium (TMG) exposing an ionization chamber to the ^{60}Co radiation source. In this test, we have used TMG obtained from Wiley Organics Co. with a the purity specification of 99.9% because the liquid with such initial purity requires only a few cycles of purification in our system to obtain a reasonable electron lifetime.

Holroyd reported the effects of radiation damage to tetramethylsilane (TMS), 2, 2, 4, 4-tetramethylpentane (TMP), and TMG and discussed pressure buildup due to radiolysis products of H_2 and CH_4, and changes of the electron drift velocity and charge collection [2]. However, the changes of charge collection were measured indirectly from the lifetime measurements. In this test, we present the results of the measurement of the charge collection from the ionization chamber, which directly affect on the energy calibration for the calorimeter.

II. Experimental procedure

We have used the ^{60}Co source of 500 C (Sept. 10, 1990) at the engineering school, Tohoku University for this exposure. This source gives about 2 grays for one hour exposure of the chamber at 150m far from the center of the source. The dose rates were varied by changing the chamber position. Fig. 1 shows the dose rates and the exposure times used for this test.

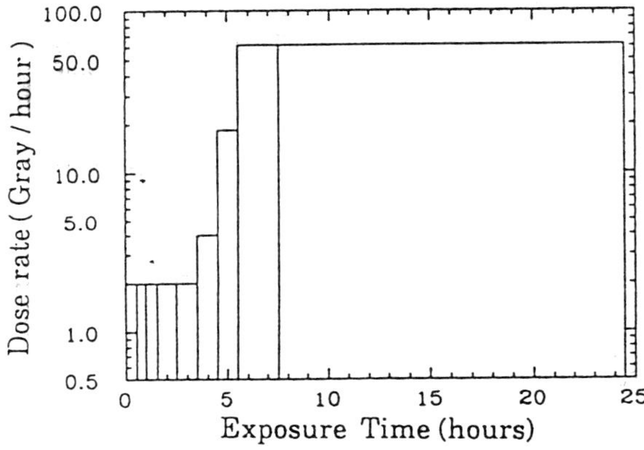

Fig. 1 Dose rate as a function of exposure time.

Fig. 2 shows the test ionization chamber used for this test experiment. The chamber vessel is made of stainless steel with inside dimensions of 100mm diameter and 50mm height. The two circular plates, anode and cathode, are placed in parallel by a gap of 7mm with pyrex glass spacers and installed in the chamber. About 0.01μC of the isotope ^{207}Bi is deposited in the center of the cathode plate for the signal readout of 976keV conversion electrons from the isotope. A total volume of liquid is about 300cc to fill the chamber.

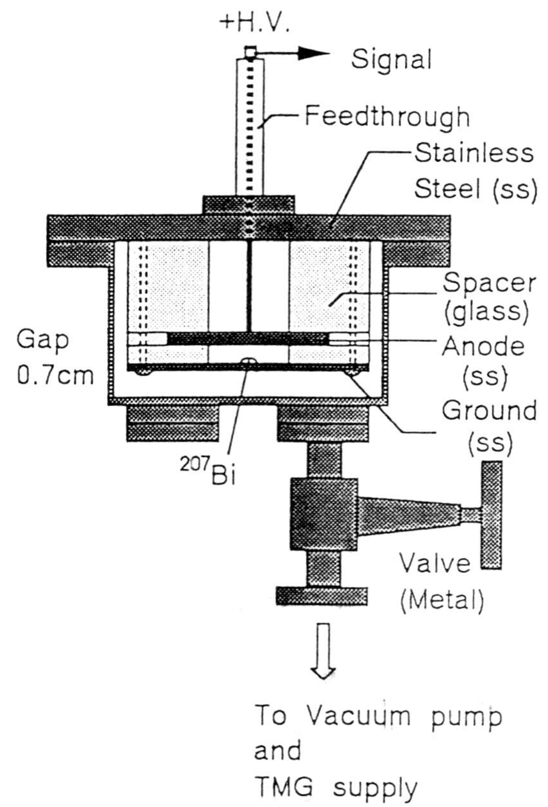

Fig. 2 Chamber used for the radiation damage tests.

The liquid, TMG, used for this test was obtained by purifying the liquid through the purification system [3]. The purification procedure is the same as described in ref.[3]. The chamber was filled with TMG initially having the free ion yield G_{fi}=0.71±0.05 and the electron lifetime τ=16±7μsec and was left for a month to confirm that the chamber was working stably before the exposure.

III. Results

The exposures of the chamber was done in the dose rate shown in Fig. 1. After each exposure, the signals (G_{fi}) from ^{207}Bi source were measured at several electric fields. With these measured G_{fi}'s, the electron lifetime was obtained by fitting the formula based on the Onsager's method [4].

Figs. 3(a) and (b) show the free ion yields G_{fi}, at E=5keV/cm and the electron lifetimes, τ, as a function of radiation dose. The curves in Figs. 3(a) and (b) are drawn to guide eyes. The G_{fi}'s and the lifetimes first increase slightly at low does. This improvement of G_{fi} and τ in small doses was observed previously [5,6]. At higher doses, the G_{fi} and the τ start decreasing. The lifetime decreased to 0.9 μsec at 1.2×10^3 grays.

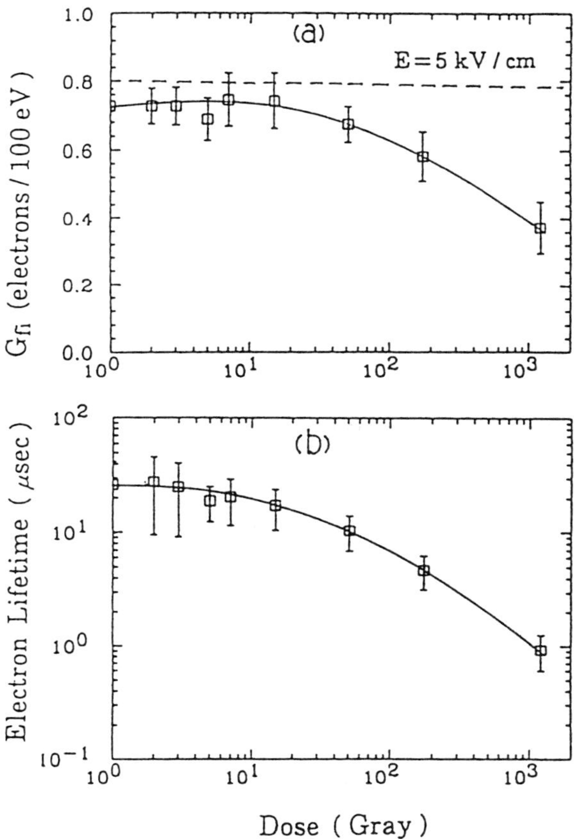

Fig. 3(a) Free ion yield and (b) electron lifetime as a function of dose. The dashed line in Fig. 3(a) is obtained from a calculation for a gap of 2mm.

The G_{fi} corresponds to the collected charge which is used to measure the deposited energy in a calorimeter. Thus, the change in G_{fi} is directly related with the change of energy calibration. The G_{fi} is reduced by about 20% at 170 grays and decreases to 50% at 1.2×10^3 grays for this chamber with the gap of 7mm. The dashed curve in Fig. 3(a) shows the G_{if} (not normalized) calculated from the Onsager method for a 2mm gap chamber with the lifetimes in Fig. 3(b). For this gap, the G_{if} changes by only 2% at 1.2×10^3 grays, indicating that the radiation damage gives insignificant effects on G_{fi}.

References

[1] Donalyd E. Groom, "Radiation Levels in SSC Detectors", NIM A279, 1-6, (1989).

[2] Richard Holroyd, "Effects of Radiation Damage to Ionization Chamber Liquids", IEEE 37, 513-521, (1990)

[3] H. Yuta et al., "Material Compatibility Test for a Warm Liquid Calorimeter", in Proceedings of the Symposium on Detector Research and Development for the SSC, Fort Worth, TX, Oct. 15-18, (1990)

[4] M. I. Lopes et al., "Estimation of the electron lifetime in tetramethylsilane", NIM A271, 464-470 (1988)

[5] R. A. Holroyd et al., "Transport of Electrons in Nonpolar Liquids', Annu. Rev. Phys. Chem. 40, 439-468 (1989)

[6] A. Gonidec et al., "Calorimetry with Room Temperature Liquids and Uranium", UTN-85-45.

A Room-Temperature Liquid Calorimeter Prototype for the SSC

G.W. Brandenburg, A. Daw, S. Geer, J. Oliver, E. Sadowski
Harvard Univ. High Energy Physics Laboratory, 42 Oxford Street, Cambridge, MA02138

D. Theriot
Fermi National Accelerator Laboratory, Batavia, IL 60510

In the last few years there has been considerable interest in using room temperature liquids as the active medium in the next generation of total absorption calorimeters at present and future high energy hadron colliders. The motivations for using this technology at the SSC include good control over the absolute calibration, uniformity of response, radiation hardness, moderately fast readout times, and the possibility of implementing complicated geometries. Tests with prototype calorimeter modules containing 2,2,4,4-tetramethylpentane (TMP) and tetramethylsilane (TMS) have given encouraging results [1]. Furthermore, searches for new liquids for use in ionization detectors indicate that there are other candidates [2] that might also be of interest for future calorimeters.

In this paper we report on the construction of a prototype electromagnetic calorimeter module in which tungsten radiator plates and stainless-steel electrode planes are supported by ceramic spacers in a stainless steel tank containing TMP ("swimming pool" geometry). This calorimeter module is being constructed to (i) find simple and reliable modular construction techniques, (ii) understand how to make the electrical connections to obtain fast readout, low cross-talk, and minimal dead spaces, and (iii) identify materials and construction techniques which are compatible with maintaining the purity of the liquid. The mechanical design of the module is illustrated in figure 1, and relevent parameters summarised in table 1. The stainless steel tank has a removable lid which is sealed with a copper gasket. The lid also has (not shown in the figure) two fill tubes for inserting and removing TMP, and a cylindrical expansion reservoir for safety. Immersed in the TMP inside the tank is an assembly consisting of two adjacent stacks, each stack having 25 tunsten radiator plates, 25 electrode planes,

Table 1: Parameters of the prototype module.

Tank	9.5"x8.75"x8"
	304L Stainless Steel 3/16" thick
	Holds 4.5 Liters of TMP
Radiators	50 Tungsten plates 4mm thick
	(ρ=19.3 g/cm^3)
Electrodes	400 2"x2"x0.020" stainless 304L
Spacers	1200 Al_2O_3 (96%)

and 50 active 1.5mm TMP gaps. The stainless steel electrode planes are supported by ceramic spacers in the middle of the TMP region between the radiator plates. The electrode planes in each stack are segmented into eight 5cm x 5cm pads. Thus there are 16 electrode towers in the calorimeter module. Each tower is read out in two longitudinal segments, the first segment consisting of the front five planes, and the second segment consisting of the back 20 planes. The gaps can be run up to a potential difference of 7.5KV corressponding to a drift field in the TMP of 50 KV/cm and a maximum electron drift time of approximately 100ns. The electrode towers are read out by transmission lines consisting of one wire at high voltage connected to the radiators, and a second wire at ground connected to the electrode pads. This configuration is necessary to reduce tower-tower cross talk and minimize signal ringing. Each transmission line exits the tank via a ceramic feedthrough which is laser welded to the tank lid.

The module is assembled by putting together each of the two stacks separately. Each stack is assembled by constructing an electrode plane from the eight pads which are held in place with cereamic spacers and stacking these planes between the radiatiors. The stacks are fastened to compression plates with tie bars to form rigid structures which can be fastened to the lid. After the electrical connections have been made, the lid plus stack assembly is lowered into the tank.

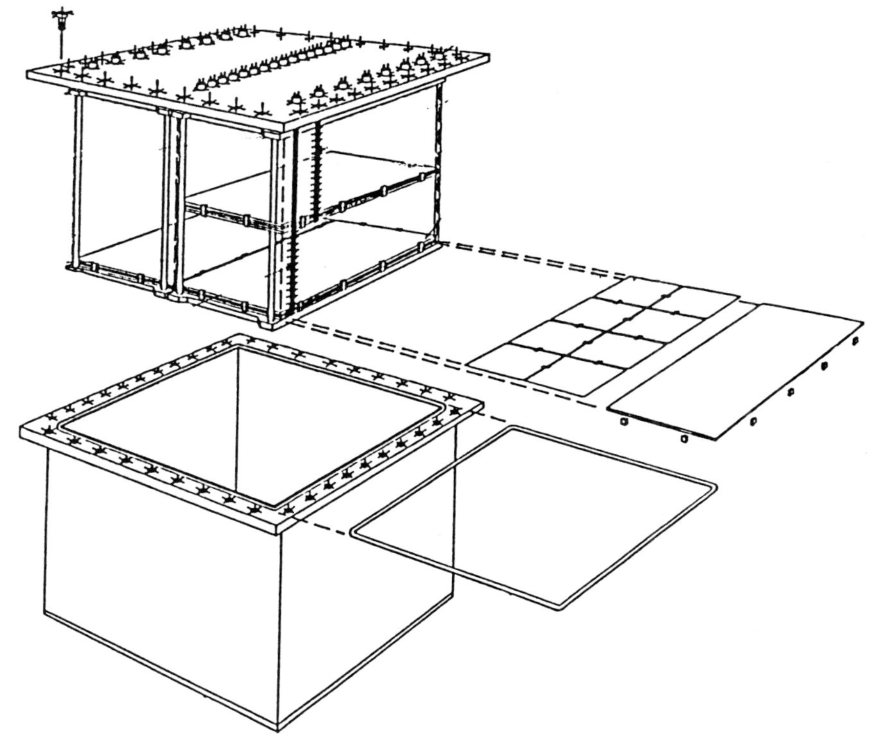

Figure 1: Schematic of the prototype Tungsten-TMP electromagnetic calorimeter module.

The cleaning proceedures are a critical part of preparing the module. Before assembly the radiators are cleaned by (i) degreasing in the laboratory cleaning liquid "micro" at $60^{o}C$ for 10 minutes using ultrasonics, (ii) rinsing in 18 MΩ water for 5 minutes, (iii) cleaning in 18 MΩ water for 10 minutes using ultrasonics, (iv) rinsing in 18 MΩ water for 5 minutes, (v) rinsing in acetone, (vi) drying in argon in an oven at $150^{o}C$ for 30 minutes, (vii) degassing in vacuum at $300^{o}C$ for 4 hours, and (viii) cooling down in argon over approximately 8 hours. Once assembled the radiator and electrode structure is again washed in 18 MΩ water, dried in hot argon, baked out at $300^{o}C$ for 4 hours, and cooled in argon over an 8 hour period.

Readout of the module is via preamplifier cards mounted on the outside of the tank. The preamps have an input impedance of 15Ω and see 300 pf (1200 pf) detector capacitance for the 5 layer (20 layer) tower segments. The preamps provide a differential output which will be shaped in the counting room with a variety of optional shaping times.

Assembly of the prototype is currently in progress, and the completed module should be ready for filling with TMP later this year. Once we have observed signals from cosmic rays we plan to transport the module to a Fermilab testbeam and measure energy resolution, linearity, and signal characteristics obtained when exposed to electrons in the energy range 10-200 GeV.

This work is supported in part by U.S. D.O.E. Contract DE-AC02-76ER03064.

References

[1] M. Albrow et al. (UA1 Collab) Nucl. Instr. and Meth. A265 (1988) 303.
K. Ankoviak et al; Nucl. Instr. and Meth. A279 (1989) 83.
J. Engler and H. Heim, Nucl. Instr. and Meth. 233 (1984) 47.
WALIC Collab; see contribution to this symposium by M. Pripstein

[2] S. Geer, R.A. Holroyd, F. Ptohos, Nucl. Instr. and Meth. A287(1990)447; and BNL 44709, June 1990, submitted to NIM; See also contribution to this symposium.

HIGH-PRESSURE IONIZATION CHAMBERS FOR CALORIMETRY IN HIGH-ENERGY PHYSICS

S. Cihangir, D. F. Anderson and B. J. Kross
Fermi National Accelerator Laboratory, Batavia, IL 60510, USA

N.D. Giokaris
The Rockefeller University, New York, NY 10021, USA

D. Carlsmith
University of Wisconsin, Madison, WI 53706, USA

Abstract

The size and speed of signals produced by alpha particles in pure argon gas at pressures up to 160 atm have been studied. At 100 atm the charge collection efficiency is 80% with much of the loss attributable to recombination. With the addition of 1% methane and an electric field of 1 kV/mm the electron collection time is 60 nsec/mm. The signal size and speed at 100 atm is not affected by O_2 for concentrations of up to 80 ppm. A tile constructed to utilize the high pressure gas calorimetry technique is described.

1. Introduction

There is now a great deal of interest in unity gain readout techniques for calorimetry. The most notable ones are liquid argon (LAr) and the warm liquids TMS and TMP. The major advantage of these liquids is that they lend themselves to instruments with good granularity and fine longitudinal segmentation. The major disadvantage of LAr is that it requires a cryostat; 4π detectors using this technique have considerable numbers of cracks, or regions that are blind to passing particles.

The main disadvantage of warm liquids is that they are very sensitive to impurities on the ppb level. This problem has been solved by enclosing the liquid in many sealed boxes. These boxes are very expensive and have a substantial fraction of their area that is insensitive. Also, warm liquids have been shown to have about the same sensitivity to radiation damage as plastic scintillators[1].

An additional problem of LAr and warm liquids is that the collection time for the electrons is long. At 1.5 kV/mm the collection time for both liquids is about 250 nsec/mm. This, of course, is not a problem for many experiments, but for the SSC and for many of the high-rate fixed-target experiments this is a real concern. Their high capacitance makes fast charge collection even more difficult.

An additional means of readout for calorimetry that is now receiving a great deal of attention is the use of scintillation fibers. This technique solves the problem of the liquids, but does not lend itself to longitudinal segmentation, which is important for particle identification. Also, used with photo multiplier tubes the readout is not unity gain and thus requires constant calibration.

The technique we have been studying is that of high-pressure ionization chambers. After we started our work, we found out that similar work was done earlier in U. S. S. R[2] and at CERN[3]. We have studied argon gas at pressures of up to 160 atm and have also added small amounts of methane to increase the drift velocity. The results reported in this work were all obtained with an ^{241}Am alpha source of strength 20 nCi in a planar geometry.

2. Experiment and Results

We designed and built a high-pressure vessel with a gas volume of about one liter. Details of the experimental setup, the procedure and the results are published earlier[4]. In this article we emphasize some of the results.

The dependence of the signal size on the pressure is shown in Fig.1 for pure argon and for argon mixed with 100 ppm O_2. Almost all of the charge liberated in the gas is collected at pressures of up to about 50 atm. At 100 atm, about 80% of the total charge is collected. Alpha-particles ionize the argon gas heavily creating highly dense positive charges. Most of the loss in charge is due to the recombination of charges in this environment. We expect to collect a larger fraction of the charges liberated by a minimum ionizing particle. The signal loss due to the 100-ppm O_2 contamination is very small (about 10%). We estimate that the signal produced by a minimum ionizing particle in 100 atm of pure argon would be at least 10% of what would be collected in the same thickness of liquid argon.

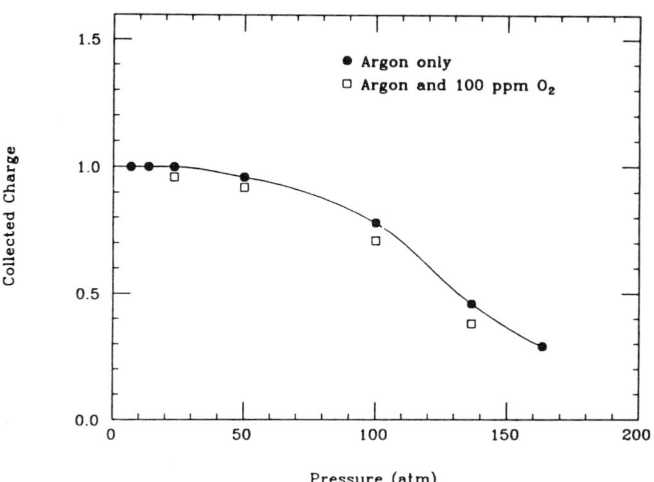

Figure 1) Signal size as a function of pressure.

Furthermore, our calculations show that in a calorimeter with a sampling medium of argon gas at 100 atm plus ~1% CH_4 and lead absorber, there is enough primary ionization charge produced so that particle energies down to 100 MeV to 200 MeV could be easily measured. [1]

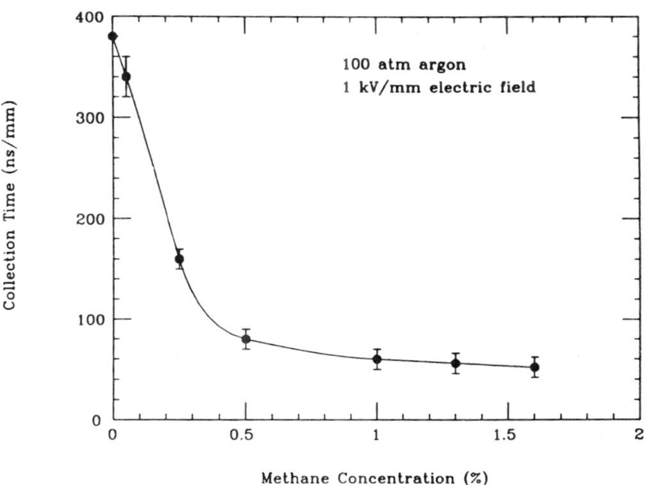

Figure 2) Signal Speed as a function of methane concentration.

The time needed to collect all of the charge at 100-atm argon with 1 kV/mm electric field is about 1 μsec corresponding to an electron collection time of 400 nsec/mm. This is longer than one would like for high rate applications. However, we were able to improve the charge collection time by adding a small amount of methane (CH_4). The price paid is the loss of some of the signal. The dependence of the speed of the signal on the CH_4 concentration, for a total pressure of 100 atm and an electric field of 1 kV/mm, is shown in Fig.2. With the addition of 1.3% CH_4 there is about a 25% signal loss relative to the pure argon case. However, the electron collection time drops to ~60 nsec/mm.

Fig.3 shows the collected charge as a function of O_2 concentration for argon and 1.3% CH_4 mixture at 100 atm. For O_2 concentrations of up to 50 ppm in argon with 1.3% CH_4 there is only a small loss of signal due to O_2.

[1] To do this calculation, a sampling medium to absorber volume ratio of 1:3 was assumed. The electron to muon and the electron to pion ratios were taken to be 1.5 and 1.15, respectively. The maximum source capacitance was assumed to be 1 nF and the noise level for a standard calorimeter preamp was taken to be 1 fC, or 6000 electrons per 1 nF of source capacitance.

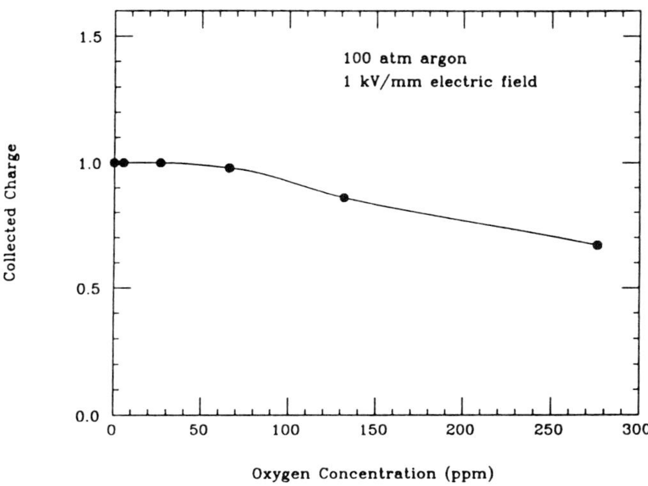

Figure 3) Collected charge as a function of O_2 concentration.

Such a calorimeter could be extremely radiation hard, and it could be an attractive alternative to existing technologies for high-luminosity colliders (e.g., SSC, LHC) especially at the very forward regions. This technique has unity gain, does not require a cryostat as does LAr, is less sensitive to impurities by several orders of magnitude than warm liquids, and also solves the problems of neutron-induced pulses and glow mode (localized and self sustained discharge) encountered at high rates in calorimeters using 1 atm gas as a sampling medium.

As a next step, we designed a tile utilizing the technolgy we described and had it built by Ability Engineering Technology[5]. The details of the tile are shown in Fig.4. Two sets of holes, 4 mm in diameter, are drilled across the stainless steel block. Wires with 1 mm diameter are extended through the holes and cross welded to form a continuous wire with one end floating and the other end connected to electronics and high voltage. We are setting up a test stand to operate it. We expect to present some results in a future publication.

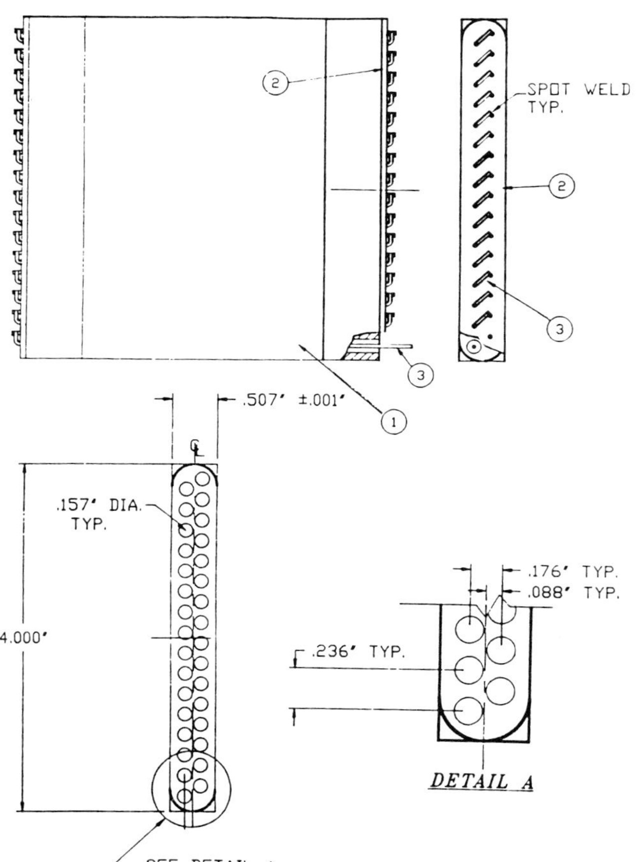

Figure 4) High presure gas calorimetry tile.

Acknowledgments

We are grateful to J. Zimmerman of Fermilab who contributed to the construction of the tile and is helping to setup the tile for our test.

References

[1] R. A. Holroyd, Effects of Radiation Damage to Ionization Chamber Liquids, to be published in IEEE Trans. Nucl. Sci. NS-37 (1990).

[2] V. I. Baskakov et al., Nucl. Instr. and Methods 159 (1979) 83.

[3] M. Barranco-Luque et al., BNL31100, OG 624.

[4] N. D. Giokaris et al., Nucl. Instr. and Methods A291 (1990) 552.

[5] Ability Engineering Technology, Inc. South Holland, IL 60473.

NEUTRON FLUX SUPPRESSION WITH POLYETHYLENE MODERATORS IN THE EMPACT SILICON EM CALORIMETER†

Koichiro Furuno, James E. Brau, Raymond E. Frey
William Hough, Hyun Hwang, Harlan Lefevre and Hwanbae Park

Department of Physics, University of Oregon, Eugene, OR 97403

Abstract

While there are a number of advantages in using silicon detectors in calorimetry, the radiation hardness will be the main issue to be solved at the SSC. The most hostile radiation in the calorimeter comes from low energy (MeV) neutrons. We report on work based on the EMPACT option of the calorimeter. The Monte Carlo calculations show that the silicon EM calorimeter can be significantly hardened by adding thin polyethylene sheets in each of the sampling gaps and inserting a thicker shield between EM and HAD sections. We present a preliminary neutron flux measurement result employing an activation detector technique at the University of Oregon 5 MeV Van de Graaff accelerator.

One approach to calorimetry that shows promise for a number of reasons, including its fast response time, is the use of silicon diode arrays as the active readout. The use of silicon pads in a large calorimeter would build on the rapid expansion of silicon pads in high energy physics experiments [1]. The numerous advantages of silicon have been discussed [2]. Among the few problems that need solutions is that of the radiation hardness since the radiation level in the calorimeter at the SSC will be intense due to the high event rate [3] and the influence of the cascade particles. The most hostile radiation in the calorimeter comes from the low energy (MeV) neutrons.

The hydrogenous material would be expected to suppress neutron fluence effectively since hydrogen has larger cross sections and larger kinematic energy transfer limits for neutrons than heavy nuclei [2]. Detailed Monte Carlo studies of this idea employing scintillator [4] and polyethylene [5] have shown excellent suppression of neutron fluences, especially in low energy spectra. The EMPACT electromagnetic calorimeter [6], with 24 radiation lengths depth, has substantial neutron fluxes while a technique such as this can suppress it. In this study, we consider a hadron calorimeter consisting of a lead-scintillator(PVX) sandwich configuration instead of the EMPACT spaghetti for simplicity of calculations, with a tungsten-silicon electromagnetic calorimeter. We also calculate an extreme case, a full tungsten-silicon calorimeter without a polyethylene moderator. The calculations have been performed with the Oak Ridge Monte Carlo, CALOR89 [7]. The hadron showers initiated by 10 GeV π^-'s have been described by HETC88, followed by MORSE which modeled the detailed low-energy neutron propagation below 20 MeV. A cut-off energy of 25 keV for neutron transport was used in this study, which is below the silicon damage threshold of 160 keV [8]. The 100×100 cm^2 cross section of calorimeters was used.

The results of calculations are shown in Figures 1-4. The calorimeter configurations used are shown in Table 1. All neutron fluxes have been calculated as the summation of neutrons over the cross section of calorimeters, weighted by the factor $1/|\cos\theta|$ called the boundary crossing estimator, where θ is the boundary crossing angle. The longitudinal profiles were obtained by integrating the energy spectra between 25 keV and 20 MeV.

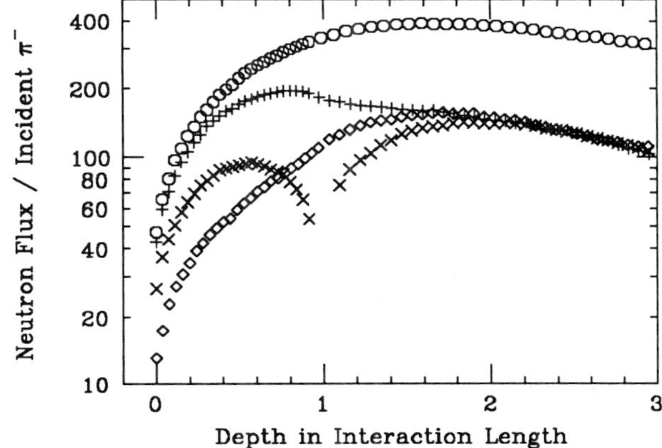

Figure 1. Neutron Longitudinal Profiles

Fig.1 shows the longitudinal profiles calculated for four different calorimeter configurations of Table 1. The effect of the scintillator in suppressing the neutron flux is clearly seen. A large fraction of the neutrons in the EM section is actually coming from the HAD section, as can be seen from observing the flux reduction effect of adding a 10 cm polyethylene shield between EM and HAD sections. The neutron 'gas' produced originally in the EM section can be suppressed by adding polyethylene sheets in each sampling gap, as is clearly seen in the profile resulted from the configuration with two 1 mm polyethylene sheets in each gap of EM section and no shield between EM and HAD sections. From these results, it can be emphasized that inserting both thinner polyethylene moderators in the sampling gaps and thicker shield between EM and HAD sections is necessary to reduce neutron fluence in the EM section.

The energy dependence of the silicon damage from low energy neutrons below 20 MeV is well known [8]. In Figures 2 and 3, we show the longitudinal profile obtained by integrating the neutron flux weighted by the energy-dependent factor taken from Ref.8 over neutron energies between 25 keV and 20 MeV. The differences between Fig.1 and Fig.2 mainly come from very low damage effect of neutrons with their energy below 160 keV. This can be clearly seen in Fig. 4, which shows the neutron energy spectra unweighted (symbols with dashed line) and weighted (symbols with solid line) by the damage factor at the flux maxima in the longitudinal profiles of the EM section for some calorimeter configurations.

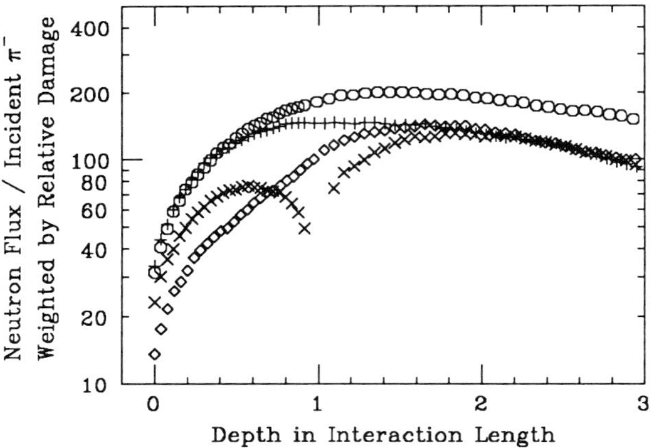

Figure 2. Neutron Longitudinal Profiles

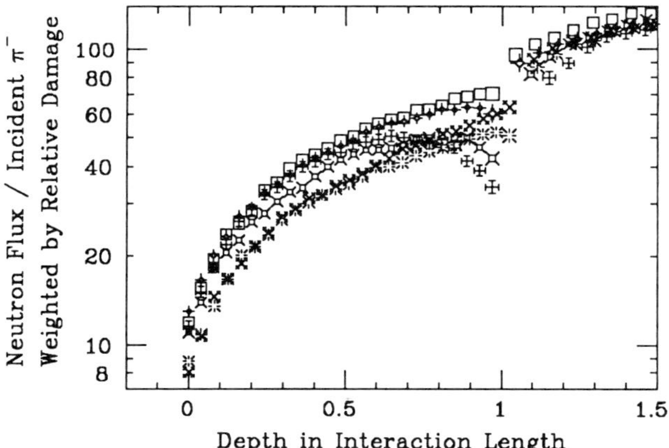

Figure 3. Neutron Longitudinal Profiles

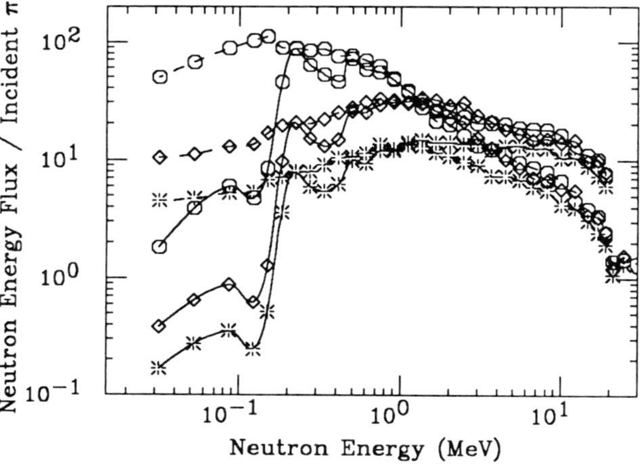

Figure 4. Neutron Energy Spectra $(1/N_{\pi^-})dN_n/d\ln E_n$

The neutron suppression effect in terms of silicon damage should be discussed with Figures 2 and 3, where a factor of about 3 reduction is observed for the configuration of two 1 *mm* polyethylene sheets and a 2 *cm* shield, compared to the 'bare' silicon case, and a factor of about 2.4 reduction compared to the 'bare' silicon-EM and lead-scintillator case.

Table 1. Calorimeter Configurations in the Figures 1-4.

	EM	Shield	HAD
○	W/Si/G10	None	W/Si/G10
+	W/Si/G10	None	Pb/Sci
×	W/Si/G10	Polye 10cm	Pb/Sci
◇	W/P1/Si/G10/P1	None	Pb/Sci
□	W/P1/Si/G10/P1	Polye 1cm	Pb/Sci
✧	W/P1/Si/G10/P1	Polye 2cm	Pb/Sci
⌘	W/P1/Si/G10/P1	Polye 5cm	Pb/Sci
⊞	W/P1/Si/G10/P1	Polye 10cm	Pb/Sci
⊠	W/P2/Si/G10/P2	Polye 1cm	Pb/Sci
⁂	W/P2/Si/G10/P2	Polye 2cm	Pb/Sci

Si 0.03 cm, Sci 0.25 cm, G10 0.05 cm
W 0.35 cm(EM), 0.70 cm(HAD), Pb 1.00 cm(HAD)
P1 0.10 cm, P2 0.20 cm (polyethylene)
EM ; 24 samplings(24Xo)
HAD ; 12 λint

With this hardening approach in hand, we have estimated the effects of radiation on the EMPACT electromagnetic calorimeter. Table 2 presents the angular dependence of the radiation exposure. The maximum dosage is primarily from electromagnetic showers, peaking at EM shower maximum. We assume a damage constant of $10^{10}s/cm^2$ for electromagnetic-initiated energy deposition.

Table 2.

Summary of dosage and leakage current increases for one year at a luminosity of $10^{33}/cm^2/s$. (all current increases in nA/cm^2)

θ	η	em dose (Gy)	dI_{em}	had dose (Gy)	dI_{had}	effective n fluence	dI_n
90	0	290	3.8	8	11	1.6×10^{11}	98
40.4	1	459	6.2	16	21	3.2×10^{11}	192
23.8	1.6	745	10.0	32	43	5.4×10^{11}	324
15.4	2	2832	38.0	115	154	1.8×10^{12}	1092
9.4	2.5	12640	169	490	657	9.6×10^{12}	5800

The relative insensitivity of the silicon to electromagnetic dosage means the hadronic charged particle dosage, while smaller, has a large overall effect. For it we assume a damage constant of $10^8s/cm^2$. The neutron fluence peaks at the back of the silicon EM calorimeter and will be suppressed with the polyethylene insertions. For the neutron damage, we assume a damage constant of $10^7s/cm^2$ with a factor of 18 reduction in the resulting current from a combination of the hydrogenous material insertions, the damage threshold effect and annealing, which will reduce the effect of the integrated exposure [2]. The appendix shows how the neutron flux and leakage current were estimated.

We conclude from this Monte Carlo study that the EMPACT silicon electromagnetic calorimeter can be significantly hardened against neutron damage by adding thin sheets of polyethylene in each of the sampling gaps and inserting a 2 *cm* polyethylene shield between EM and HAD sections.

In order to investigate this reduction prediction of low energy neutron fluence experimentally, we are presently engaged in the experiment at the University of Oregon 5 MeV Van de Graaff accelerator producing 3 MeV deuteron beam incident on a thick carbon target yielding the large flux of low energy neutrons [10]. A preliminary measurement of the neutron fluences has been performed in a calorimeter-like stack which consists of 3/4 inch iron plates separated by 9/32 inch air gaps. Thin aluminium plates with the Indium activation detectors attached where inserted in the air gaps. The neutron fluences were estimated by measuring the gamma activities from disintegrations of ^{115m}In, which were produced during the neutron irradiation through the reaction ^{115}In(n,n')^{115m}In, in the detectors using a NaI spectrometer. The reaction cross section has a threshold at about 350 keV and is about 250 *m*barn at 2 MeV [11]. ^{115m}In has a half-life of 4.486 hours and decays mainly by an isomeric transition leading to the emission of conversion electrons and 336 keV photons, and the abundance of these photons is about 47%.

A result from these measurements is presented in Fig. 5 with statistical error bars. Indium detectors were located at four different radial distances from the neutron beam center and various depths in the iron stacks, as shown in the figure. All measured values in the figure are normalized at the time of the end of the irradiation and as described in the figure caption. This result shows the feasibility of low-energy neutron measurement and the promise of observing the reduction effect by polyethylene moderator.

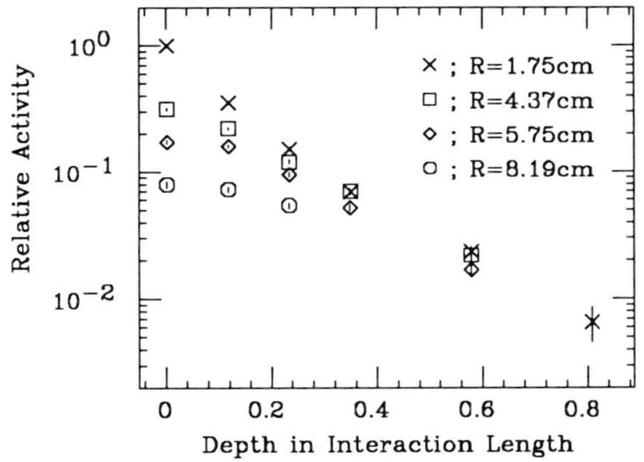

Figure 5. Measured activity of ^{115m}In relative to the detector at R=1.75*cm* in the first slot.

The exact comparison of this result and the MORSE Monte Carlo calculation in an absolute manner is under investigation. The experiment observing the moderator effect will be performed in the near future. Those experiments will provide a determination of the reliability of the Monte Carlo calculations of low-energy neutron transportation through materials and the prediction of the suppression effect.

Appendix

The neutron fluences in the tungsten-silicon calorimeter with polyethylene moderator were extrapolated from the lead-silicon fluences presented in Ref.3 from the following formulae,

$f_n(\text{Lead/Si}) = 3.8 \times 10^{11} \times \cosh^{2.67}(\eta) \times (200/R)^2 \times 1.67$

$f_n(\text{W/Si with Polyethylene}) =$

$$f_n(\text{Lead/Si})/F_{SUPPR} \ (/cm^2/year)$$

where $F_{SUPPR} = 6$ as obtained from our calculation for two 1 *mm* polyethylene sheets in each gap and 2 *cm* shield assuming the same neutron fluence between Lead/Si and W/Si calorimeters, including the damage threshold effect.

Then the increments of leakage current are calculated as follows [9],

$dI_n = 1.2 \times 10^{-9} \times f_n \times W/K/F_{ANNEAL} \ (A/cm^2)$,
where $W = 0.03cm$, $K = 10^7/cm^2$ and $F_{ANNEAL} = 3$.

†Supported in part by U.S. Department of Energy.

References

[1] C. Goessling, "Large Area Silicon Detectors", Proceedings of the XXIV International Conference on High Energy Physics, Munich, 1988, p. 1208.

[2] J.E. Brau et al, "Silicon Calorimetry for the SSC", Proceedings of the Workshop on Calorimetry for the Supercollider, p. 489 (1990).

[3] Radiation Levels in the SSC Interaction Regions, SSC-SR-1033, edited by D.E.Groom (1988).

[4] J.E. Brau and T.A. Gabriel, "Calculations of Neutron Flux in Iron and Uranium Calorimeters*", Radiation Levels in the SSC Interaction Regions, SSC-SR-1033, edited by D.E.Groom (1988). *The spectra in figures 7-10 are a factor of two high due to plotting error. All other figures are correct.

[5] K. Furuno, J.E.Brau and H.Hwang, "Neutron Flux Suppression with Polyethylene Moderators in Silicon Hadron Calorimeters", CERN 89-10, ECFA 89-124, pp.325-335 (1985).

[6] EMPACT Collaboration, EOI, May 25, 1990.

[7] T.A.Gabriel et al, ORNL/TM-11185, unpublished.

[8] V.A.J. van Lint, T.M. Flanagan, R.E. Leadon, J.A. Nabor, and V.C. Rogers, Mechanisms of Radiation Effects in Electronic Materials, vol. 1, Wiley, New York, 1980.

[9] Radiation Effects at the SSC, SSC-SR-1035, edited by M.G.D. Gilchriese (1988).

[10] R. Bahnsen, C. A. Burke, H. W. Lefevre and W. R. Wylie, appeared in chapter I, Progress Report For the Year Ending May 31, 1969, on Nuclear Physics Research under U.S. Atomic Energy Commission Contract AT(45-1)-1925, RLO-1925-20, Univ. of Oregon, Dept of Physics.

[11] W. L. Zijp, H. J. Nolthenius and G. C.H.M. Verhaarg, Cross-Section Library DOSCROS84, Netherland Energy Research Foundation Report ECN-160 (1984).

ELECTROMAGNETIC SHOWER SAMPLING EFFICIENCY IN LIQUID ARGON SAMPLING CALORIMETERS AND THE SAMPLING EFFICIENCY DEPENDENCE ON LOW-Z CLADDING OF HIGH-Z ABSORBER PLATES *

P. M. Mockett, M. Boulware †
University of Washington Seattle, Washington 98195

Abstract

The electron shower sampling efficiency in liquid argon ionization sampling calorimeters is studied with the EGS Monte Carlo program. A formula is derived to describe the dependence of this efficiency on the absorber plate thickness and on the sampling gap thickness. This formula is used to calculate the effect on the sampling efficiency of cladding lead plates with iron. In addition, a model is developed and used to predict the e/π response for iron-clad lead absorber plates.

Introduction

Liquid Argon (LAr) ionization sampling calorimeters have been studied studied for some time.[1] It has long been recognized that the fraction of the electron shower energy deposited in the sampling gaps is not in general the same as the fraction deposited by a muon.[2,3] For iron absorber plates the fractions are nearly the same while for lead absorber plates the electron fraction is substantially less than the muon fraction. The ratio of the electron fraction to the muon fraction is called the sampling efficiency. The e/π response ratio is strongly affected by this sampling efficiency.

The electron shower energy deposition can be described by three components; there are base energy depositions in the absorber plate and in the LAr and a transition energy deposition (TED) near the absorber-LAr interface. These components were studied with the aid of the EGS Monte Carlo shower code for a lead-LAr calorimeter. The TED was found to fall off exponentially in the LAr. From these three distributions we compute the electromagnetic sampling efficiency for different configurations of absorber plates and argon gaps. We compare all metal systems with those employing G10 sampling planes. We also investigate the extent to which the TED can be shielded from the LAr gap by cladding the lead plates with iron.

In the last section we develop a model for the hadron shower sampling and use it to determine e/π for a liquid argon sampling calorimeter employing iron-clad lead plates for the absorber.

EGS Study of the Electromagnetic Shower

We tune the cutoffs in EGS so that the ionization density sampled by an infinitely thin argon gap in a block of lead is the same as that in the lead except for the photo electron ionization energy. This photo ionization loss of about 6 % of the shower energy is not counted because it is purely local. Good continuity was obtained with the electron cutoff at 0.55 MeV, the photon cutoff at 2.0 KeV and ESTEPE at 0.5 %. If the lead absorber plate is thicker than a couple of millimeters, the EGS Monte Carlo shows there is a fixed ratio of the energy density in the lead and in the LAr except near the boundary between the two materials. Fig. 1 shows the net energy deposition in the LAr after the base energy deposition has been subtracted; the units used for this TED are in equivalent millimeters of the base energy in LAr. Both the front and back interface contributions are included. The TED begins at 30% of the LAr base energy deposition and has an attenuation length of 2.6 mm of LAr.

The TED has the exponential form:

$$dt/dx = .30e^{x/2.6mm}. \tag{1}$$

The base energy deposition in the LAr is defined as 1 in these units and it is found to be 10.65 in the lead. For a minimum ionizing muon the lead base energy value is calculated to be 6.08. Using these parameters, we obtain the following expression for the sampling efficiency, ϵ_e:

$$\epsilon_e = \frac{6.08x_{Pb} + x_{Ar}}{10.65x_{Pb} + x_{Ar}} \frac{x_{Ar} + t_G}{x_{Ar}} \tag{2}$$

*Work supported by the National Science Foundation, grant PHY-8822028.

†Present Address: Stanford Univ., Stanford, CA

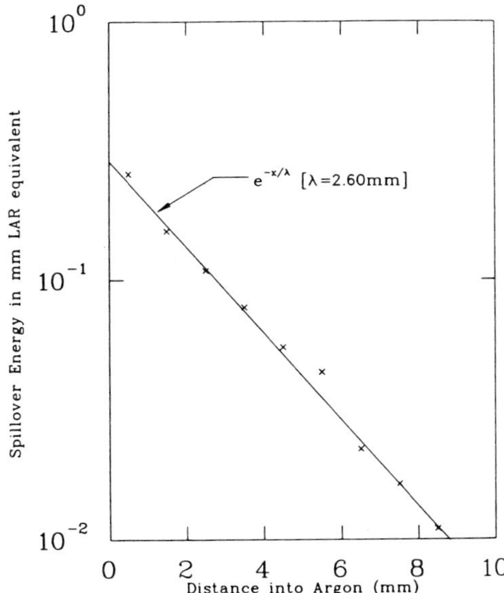

Figure 1: The transition energy deposition in millimeters of LAr equivalent.

where t_G is the transition energy in the LAr gap obtained by integrating equation 1 over the appropriate interval. The value of t_G depends upon the details of the gap.

In Fig. 2 we show the dependence of ϵ_e on the thickness of the absorber plates and the thickness of the sampling gap as obtained from equation 2. The general trend is for the sampling efficiency to decrease slightly with thicker absorber plates and thicker sampling gaps.

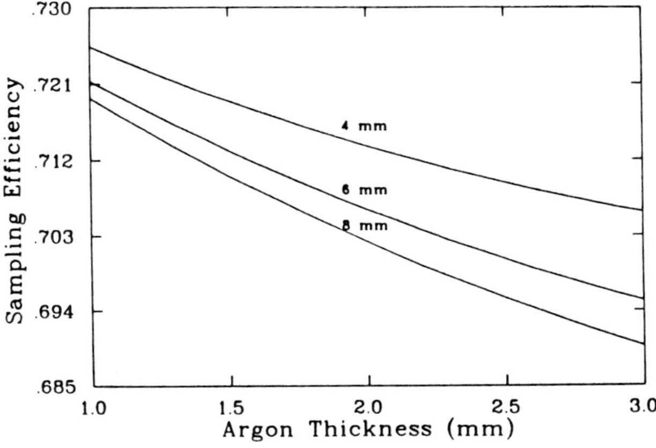

Figure 2: Sampling efficiency for various absorber thicknesses as a function of the LAr thickness.

For G10 in the middle of the LAr gap, equation 2 can be modified to include the additional material. Letting g_{eq} designate an amount of LAr equivalent to the G10, equation 2 becomes:

$$\epsilon_e = \frac{6.08x_{Pb} + x_{Ar} + g_{eq}}{10.65x_{Pb} + x_{Ar} + g_{eq}} \frac{x_{Ar} + t_G}{x_{Ar}}. \quad (3)$$

The integration of equation 1 to obtain t_G must exclude the region of the G10. Fig. 3 shows the sampling efficiency for a 1 mm G10 sheet at the gap center where the abscissa designates the thickness of the LAr on each side of the G10.

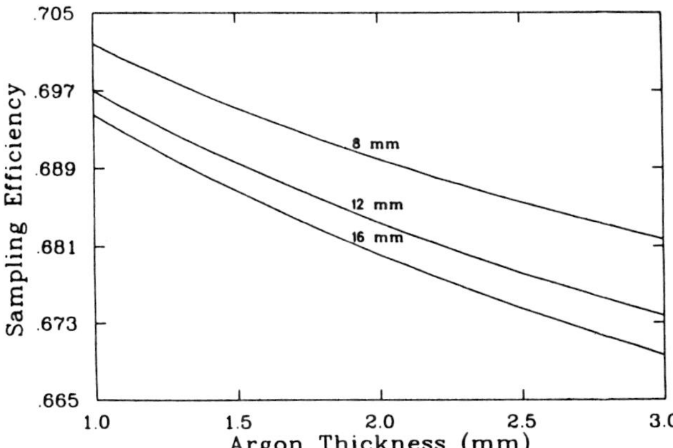

Figure 3: Sampling efficiency for various absorber thicknesses as a function of the argon thickness (1 mm G10 sheet at the center of the sampling cell).

Fe Cladding of the Absorber Plates

Since the TED decreases exponentially in the LAr gap, it is possible to partially shield the LAr from this energy by cladding the absorber plates with a low-z material. Iron is ideal for this since it has about the same absorption length as lead but about the same ionization characteristics as argon. Letting f_{eq} designate an amount of LAr equivalent to the iron cladding, equation 3 becomes:

$$\epsilon_e = \frac{6.08x_{Pb} + x_{Ar} + f_{eq} + g_{eq}}{10.65x_{Pb} + x_{Ar} + f_{eq} + g_{eq}} \frac{x_{Ar} + t_G}{x_{Ar}}. \quad (4)$$

Fig. 4 shows the sampling efficiency for various cladding thicknesses and a 2 mm argon sampling gap on each side of a 1 mm G10 sheet. The overall thickness of the sandwich is kept fixed as labelled on the figure. Note that the minimum in the efficiency is quite broad so that the amount of iron can be increased substantially over the value at minimum efficiency without increasing the sampling efficiency appreciably.

e/π Response with Iron-Clad Absorber Plates

The introduction of iron into the calorimeter also alters the pion response. The nuclear losses are reduced which in principle could boost the pion response. We have constructed a simple model of the hadron response in an attempt to get a qualitative picture of how the

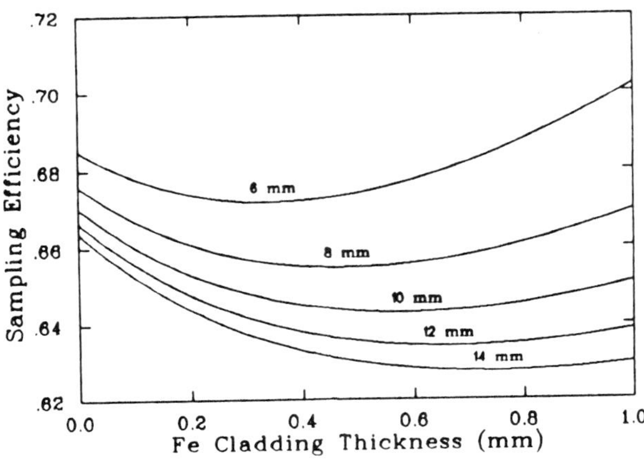

Figure 4: Sampling efficiency for various absorber thicknesses as a function of the iron cladding thickness. A 2 mm argon gap is used on each side of a 1 mm G10 sheet.

addition of iron might affect the pion response. The following table gives the shower components of a 10 GeV proton (kinetic energy) in iron and in lead according to the HETC calculation of Gabriel, et al..[4] The shower components of lead are thought to be very similar to those given for uranium without fission.

Table 1. Hadronic Shower Components

Shower components	Fe	Pb
hadronic	3.854	3.230
gamma	3.186	2.531
neutron	0.848	1.330
not sampled	2.110	2.912

From the measured value[1] of e/π for a iron LAr sampling calorimeter we obtain the neutron sampling efficiency, ϵ_n, for the neutron spectrum emitted in iron nucleus collisions. After correcting for leakage we obtain an e/π at 10 GeV of 1.31. The first two elements under Fe in table 1. are sampled with unit efficiency in an iron calorimeter. Hence we can solve for the neutron sampling efficiency from the equation for e/π.

$$e/\pi = \frac{10.}{3.854 + 3.186 + \epsilon_n .848} = 1.31. \quad (5)$$

We find ϵ_n is 0.70. The average neutron energy is about 8 MeV. We note that some fraction of the energy labeled not sampled may indeed actually be sampled. However this energy and the neutron energy are proportional so we choose to lump any such energy into an effective neutron sampling efficiency. For the lead-LAr calorimeter case the uncertainty in the e/π is large.[5] We assume a value of 1.2 in line with Monte Carlo studies.[6] From equation 4 we get ϵ_e of 0.667 for 12 mm thick lead plates and 1 mm G10 between 2 mm LAr gaps.

$$e/\pi = \frac{.667 \times 10.}{3.230 + .667 \times 2.531 + \epsilon_n 1.33} = 1.20. \quad (6)$$

For the approximately 5 MeV mean energy neutrons in lead, the value of ϵ_n obtained is 0.48.

We now use these neutron efficiencies to predict what the e/π response might be in a composite calorimeter. Starting from the value of 1.2 we find e/π at a minimum value of 1.13 with 1.0 mm Fe cladding. However iron as thick as 1.5 mm changes the e/π very little. Starting with a larger value of e/π lowers the calculated neutron sampling efficiency and raises very slightly the optimal thickness for the iron cladding.

The iron cladding not only lowers e/π but improves the calorimeter in other ways. Iron increases the sampling fraction, it decreases the radiation length of the sampling plates and decreases the neutron flux released into the tracking volume. Cladding the plates with iron also provides structural advantages, iron being both stronger and lighter than lead.

Conclusions

The EGS Monte Carlo was used to determine the form of the transition energy deposition in the LAr gap. This energy has an exponential decay length of 2.6 mm and contributes to the sampling efficiency. Together with the base energy depositions, the sampling efficiency for various gap and plate configurations, including iron-clad lead plates, were determined. A model was developed to optimize the e/π response in iron-clad lead calorimeters; an optimum cladding thickness of about 1.25 mm was found. Beam tests of lead-LAr calorimeters early next year will help test this model.

References

1. C. W. Fabjan et al., "Iron Liquid-Argon and Uranium Liquid-Argon Calorimeters for Hadron Energy Measurement," NIM **141**, 61–80, (1977).

2. P. M. Mockett, "A Review of the Physics and Technology of High-Energy Calorimeter Devices," 1983 SLAC Summer Institute, 335–393, (1983).

3. C. W. Fabjan et al., "Calorimetry in High-Energy Physics," CERN/EP 85-54, 1–57, (1985).

4. T. A. Gabriel et al., "The Physics of Compensating Calorimetry and the New CALOR89 Code System," IEEE Transactions on Nuclear Science, **36**, 14–22, (1989).

5. R. Dubois et al., "SLD Liquid Argon Calorimeter Prototype Test Results," IEEE Transactions on Nuclear Science, **33**, 194–196, (1986).

6. J. Brau, "Monte Carlo Studies with HETC," Workshop on Compensated Calorimetry, CALT-68-1305, (1985).

JET RESPONSE OF A HOMOGENOUS CALORIMETER*

D. E. Groom and E. M. Wang
Lawrence Berkeley Laboratory, Berkeley CA 94720, USA

Abstract

The response of a large calorimeter with uniform structure to an incident hadron with energy E, relative to that for an incident electron or photon with the same energy, is adequately described by $R_h/R_e = [1-(1-\epsilon_h/\epsilon_e)(E/E_0)^{m-1}]$. Here $m \approx 0.86$, $E_0 \approx 1$ GeV, and ϵ_e/ϵ_h is essentially the "intrinsic e/h" introduced by Wigmans. In a jet with energy E_J the hadron spectrum is described by a fragmentation function $D(z)$, where (for particle energy $E \approx p_{\|}$) $E = zE_J$. The calorimeter response to a jet may be found by integrating the single-hadron response over this distribution:

$$R_J/\epsilon_e = E_J\left[1-(1-\epsilon_h/\epsilon_e)(E_J/E_0)^{m-1}\int_0^1 z^m D(z)dz\right]$$

The response to a jet is the same as that to a single hadron only if the integral is unity. For $D(z)$ obtained from published DELPHI and CDF data and for ISAJET jets at $\sqrt{s} = 40$ TeV, the integral lies between 0.92 and 1.06, with significant uncertainties concerning low-z behavior and the nonelectromagnetic energy fraction. Even for a low-energy jet in a very noncompensating calorimeter ($\epsilon_e/\epsilon_h = 1.3$), the error in jet energy determination is less than 2%. Such errors are likely to be small compared with those resulting from the low-z cutoff of a possible solenoidal magnetic field and the effects of dead material in front of the calorimeter.

1. Introduction

Not all jets are created equal. Calorimeter response, measured in a test beam with incident hadrons, can be thought of as the response to the "jet" produced in the first collision of the hadron with a heavy nucleus in the calorimeter—or in a convertor out front, as is often done in test beam simulations of jet response. However, the incident jet might consist of just a few energetic particles, for example from the decay of a high-energy W or Z. At the other extreme, a high-energy jet from the primary collision at the SSC contains a large number of very soft particles. If the calorimeter is noncompensating, there is no guarantee that the response to these different kinds of jets with same energy will be the same.

We report here a study of the effect of differences in jet composition, as given by a range of fragmentation functions, on the response of an ideal calorimeter. By "ideal" we mean a calorimeter is large, in the sense that leakage effects can be ignored, and "homogenous," in the sense that the structure is the same throughout—if there is a plate structure, it is the same front-to-back.

2. Response to a single incident hadron

The response of a calorimeter to an incident hadron with energy E may be written as[1]

$$R_h = \epsilon_e E\left[1-(1-\epsilon_h/\epsilon_e)(E/E_0)^{m-1}\right] , \quad (1)$$

where R_h is the response of the calorimeter to an incident hadron, ϵ_e is the efficiency with which electromagnetic energy deposition is converted to output signal, ϵ_h is the efficiency with which energy deposition by low-energy hadrons is converted to output signal, E_0 is a scale energy (very close to 1 GeV), and $m \approx 0.85$–0.87. In practice, only the product $(1-\epsilon_h/\epsilon_e)E_0^{1-m}$ can be measured. The ratio ϵ_e/ϵ_h is very close to Wigmans"s "intrinsic e/h."

The response to an electron or photon is taken to be linear:

$$R_e = \epsilon_e E \quad (2)$$

The ratio R_e/R_h ("e/π") is asymptotically unity, and the "linearity" R_h/E is proportional to its reciprocal.

3. Response to an ensemble of particles

A jet with energy E_J consists of photons from π^0 decay and "stable" hadrons. (in the present analysis, we ignore energy which might be carried away by leptons.) We need only to sum the responses of all of these particles to obtain the response to a jet. If R_{ej} is the response to the jth π^0 in the jet (with energy $E_{\pi^0 j}$) and R_{hk} the response to the kth stable hadron (with energy E_{hk}), then the response to the jet is given by

$$R_J = \sum_{j=1}^{N_{\pi^0}} R_{ej} + \sum_{k=1}^{N_{\rm had}} R_{hk} . \quad (3)$$

Using Eqs. (1) and (2) to evaluate R_{hk} and R_{ej}, this reduces to

$$R_J = \epsilon_e E_J\Big[1-(1-\epsilon_h/\epsilon_e)(E_J/E_0)^{m-1} \times \sum_{k=1}^{N_{\rm had}}(E_{hk}/E_J)^m\Big] . \quad (4)$$

Alternatively, the spectrum of stable hadrons can

* This work was supported by the U. S. Department of Energy under Contract No. DE-AC03-76SF00098.

be described by a fragmentation function $D(z)$, where z is the hadron's momentum parallel to the jet direction, scaled by the jet's momentum. In the present study we treat z as the fractional energy, *i.e.* $z \approx E_{\rm had}/E_J$. When the arguments leading to Eq. (4) are repeated, one obtains

$$R_J = \epsilon_e E_J \Big[1-(1-\epsilon_h/\epsilon_e)(E_J/E_0)^{m-1} \times \int_0^1 z^m D(z)dz\Big] , \tag{5}$$

where $D(z)$ describes the spectrum of all hadrons except for π^0's.

If the summation in Eq. (4) or the integral in Eq. (5) is evaluated with for $m = 0$, the stable hadron multiplicity ($N_{\rm had}$) is obtained. If $m = 1$, the result is the nonelectromagnetic fraction of the jet's energy ($F_{\rm had}$). The desired integral, with $m \approx 0.86$, is in some sense an interpolation between the two.

If the correction factor given by the sum or integral is unity, then the calorimeter has the same response for a jet as for an incident hadron. Imposition of this requirement for the "jet" from the first hadronic collision in the calorimeter results in recovery of the relationship between m, the mean multiplicity, and the average hadronic energy fraction in a single collision which was given in Ref. 1. If the factor is not unity, then the compensation of the calorimeter appears to be different for jets than for single hadrons.

In using either experimental or Monte Carlo distributions to evaluate the sum or integral, special treatment of the very low-z region is necessary, as is normalization to an appropriate $F_{\rm had}$.

4. Examples

We evaluate the integral in Eq. (6) for four representative cases:

Two experimental results, both with jet energies at or near $M_Z/2$. Since the measurements are for *charged* hadrons, the distributions must be renormalized to include the contributions of such particles as Λ's and K_L's.

1. Jets from Z decay, as measured by the DELPHI collaboration at LEP[2]. The published fragmentation function is for the entire event, so we have normalized the function downward by a factor of two to describe the individual jets. Data were read from their Fig. 3(b) and extrapolated to $z = 0$.
2. CDF charged fragmentation function at $\sqrt{s} = 1800$ GeV[3]. $z dN_{\rm ch}/dz$ was extrapolated to $z = 0$ to force $\langle F_{\rm ch}\rangle = 0.65$, their reported value. (Since some of the energy is carried by neutrals, this value is probably too high for consistency with isospin conservation.)

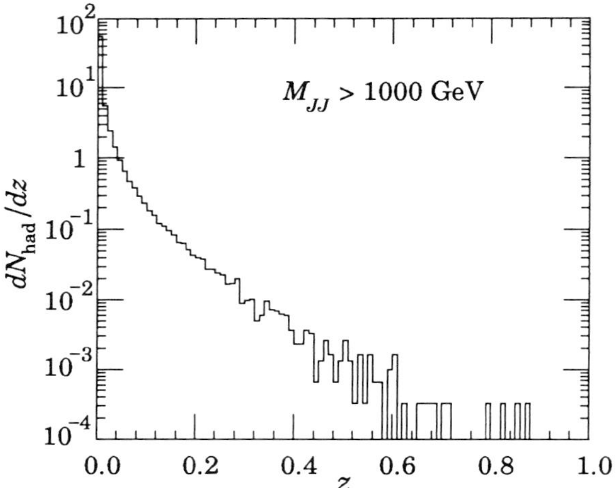

FIG. 1. Distribution in z for ISAJET TWOJET events at $\sqrt{s} = 40$ TeV, for 1000 GeV $< M_{JJ} <$ 2000 GeV.

Two samples of TWOJET ISAJET events at $\sqrt{s} =$ 40 TeV. In both cases, all hadrons other than π^0's are used:

3. 3226 events with p_t (hard scatter) > 40 GeV/c, and 100 GeV $< M_{JJ} <$ 200 GeV. The mean jet momentum is 73 GeV/c, and the mean non-π^0 hadronic multiplicity is 26.
4. 3042 events with p_t (hard scatter) > 400 GeV/c, and 1000 GeV $< M_{JJ} <$ 2000 GeV. The mean jet momentum is 677 GeV/c, and the mean non-π^0 hadronic multiplicity is 70. The z distribution for these events is shown in Fig. 1.

The results are summarized in Table 1. In all cases, there is ambiguity because of uncertainty in $F_{\rm had}$ (or $F_{\rm ch}$, in the the experimental results). For $I = 1$ pion production one would expect $F_{\rm had} = 2/3$, and the value should be lowered by the presence of $I = 1/2$ states. Some of the bias can probably be removed by normalizing $\int zD(z)dz$ to 0.67 (entries in parentheses). As can be seen, the integral is slightly less than unity for the similar low-energy LEP and Tevatron functions, and is slightly greater than unity for simulated high-energy SSC jets. Values lie between 0.84 and 1.15 before normalization, and 0.92 to 1.06 after normalization. The integrals also change by about 0.03 if m is changed by 0.01, introducing an additional uncertainty of perhaps 5%. Given the various uncertainties, we conclude that the correction factor for jets at the SSC will be between 0.85 and 1.20, and will depend upon the kind of jet. In the following discussion, we treat the limits as 0.9 and 1.2, in part to avoid confusion between the lower limit and the value of the exponent m.

6. Implications for calibration and compensation

The only reasonable way to calibrate a calorimeter is in terms of equivalent electron response, so for purposes of this section we take the first factor on the right of Eq. (1) as unity, and measure the response in GeV. For example, if $\epsilon_e/\epsilon_h = 1.30$ and $E_0 \approx 1$ GeV, then a 1000 GeV pion produces a response of 912.3 GeV. Conversely, if a 912.3 GeV signal is observed for an incident hadron, Eq. (1) can be inverted to find $E_h = 1000$ GeV. For an incident jet, Eq. (4) or (5) must be used instead. If the correction factor is 0.90 and if the energy scale has been calibrated in a hadron test beam, then a 1000 GeV jet will appear to have 1009.5 GeV. The percentage errors for the two limiting cases 0.9 and 1.2 are plotted in Fig. 2 for calorimeters with $\epsilon_e/\epsilon_h = 1.30$ and $\epsilon_e/\epsilon_h = 1.15$. (A lead–liquid argon calorimeter with fast readout might have $\epsilon_e/\epsilon_h = 1.30$, and 1.15 might be obtained for a poorly-designed metal-scintillator calorimeter.) The error is 4% in the worst case at 10 GeV, and decreases with energy and with the degree of compensation.

Table 1

Integrals over representative fragmentation functions. Numbers in parentheses are calculated for the nonelectromagnetic energy fraction normalized to 0.67. In the case of the DELPHI and CDF results, the unrenormalized energy fraction is for charged hadrons only.

Source	Process	$\int_0^1 D(z)dz$	$\int_0^1 z^{0.86}D(z)dz$	$\int_0^1 zD(z)dz$
DELPHI	$Z \to jet\ jet$	11.0(12.1)	0.84(0.92)	0.61(0.67)
CDF	$\sqrt{s} = 1.8$ TeV	17.8*(19.9)	0.94(0.97)	0.65(0.67)
ISAJET	40 TeV, $\langle p_J \rangle = 73$ GeV/c	26.2(25.2)	1.04(1.00)	0.69(0.67)
ISAJET	40 TeV, $\langle p_J \rangle = 677$ GeV/c	69.8(64.7)	1.15(1.06)	0.72(0.67)

*The extrapolated low-momentum part of the function contributes 10 to this total.

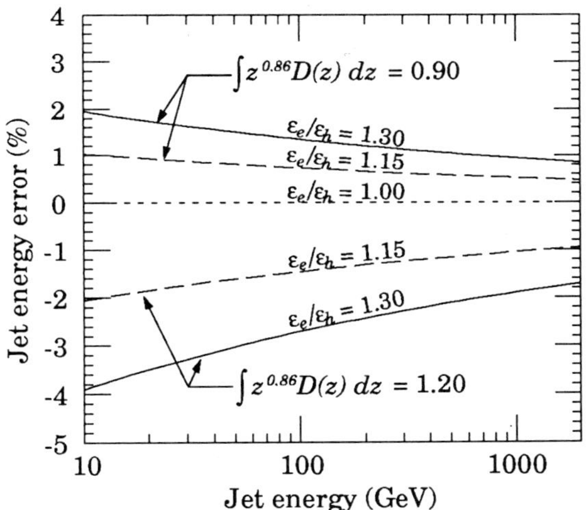

FIG. 2. Energy determination error as a function of jet energy for representative values of ϵ_e/ϵ_h, for the two extreme case of the correction factor: 0.90 (top curves, for hard, low-multiplicity jets) and 1.20 (bottom curves, for soft, high-multiplicity jets).

Such errors are likely to be small in comparison with errors associated with a possible solenoidal field and dead matter in front of the calorimeter. These are important because much of a jet's energy is carried by very low-energy particles. For example, in our 100 GeV ISAJET sample 4% of the energy is carried by particles with energies below 600 MeV ($z = 0.006$), and 8% by particles below 1 GeV. In addition, the hadrons which reach the calorimeter tend to interact earlier than those in a single-hadron cascade, and in a real calorimeter the early part (electromagnetic section) might be different than the rest. Further studies of these effects are planned.

References

1. D. E. Groom, "Energy Scaling of Low-Energy Neutron Yield, the e/π Ratio, and Hadronic Response in a Calorimeter," in *Proc. of the Workshop on Calorimetry for the Supercollider,* Tuscaloosa, Alabama, 13–17 March 1989, ed. R. Donaldson and M. G. D. Gilchriese, World Scientific (June 1990), 59–75.
2. P. Aarnio *et al.*, Phys. Lett. **240B**, 271 (1990).
3. F. Abe *et al.*, Phys. Rev. Lett. **65**, 968 (1990).
4. F. E. Paige and S. D. Protopopescu, in *Physics of the Superconducting Supercollider*, ed. by R. Donaldson and J. Marx (Snowmass CO, 1986), 320.

Scintillation Calorimetry

BEAMTESTS OF LARGE LEAD AND SCINTILLATING FIBER PROTOTYPE CALORIMETERS

Hans P. Paar
Physics Department
University of California at San Diego
La Jolla, CA 92093

Abstract

We report preliminary results of beamtests of 1.5 ton and 13 ton lead and scintillating fiber prototype calorimeters. The tests were done by the SPACAL collaboration* at CERN in three different periods over the past year with electrons and pions in the energy range 5 to 150 GeV. After a discussion of the geometry and the readout of the prototypes and of the testbeam we give results on lateral energy distributions, e/π-ratio, hadronic lineshape, and hadronic energy resolution for single pions and 'jets'. Also results on electron-pion discrimination are given for isolated electrons as well as for electrons near pions.

Introduction

Within the context of the CERN/LAA program the SPAghetti CALorimeter (SPACAL) collaboration has been engaged in the development of a lead and scintillating fiber calorimeter technique since 1987 [1]. The principal objective of the R & D effort is to provide the high energy physics community with more choice in the selection of a calorimeter technique for an experiment at the LHC or SSC. The technique might also be of use in heavy ion collider experiments or in stationary target experiments.

Results on the electromagnetic response of a small number of prototype towers have been obtained and published [2]. In view of the performance of these modules, the SPACAL collaboration decided to construct and test two larger prototype calorimeters to study the hadronic response. We report below on the first preliminary results obtained with these two prototypes.

Prototypes and Testbeam

Both prototypes have parallel (as opposed to projective) geometry and are of identical construction. They consist of lead and scintillating fibers in the volume ratio 4:1. The combination has a radiation length $L_{\text{rad}} = 7.5\,\text{mm}$ and a nuclear interaction length [3] $\lambda_I = 21\,\text{cm}$. Each consists of towers with a hexagonal cross section, 86 mm apex-to-apex, and 2 m ($9.5\,\lambda_I$) long. One, the '1.5 ton prototype', consists of 20 towers and has 1.5 ton of instrumented lead. The other '13 ton prototype' consists of 155 towers, has 13 ton of instrumented lead, and is 1.1m ($5\,\lambda_I$) in diameter. We show in Fig. 1 the organization of the towers of the 13 ton prototype into rings.

Figure 1: Geometry of the 13 ton prototype.

The towers consist of 2.2 m long scintillating fibers of 1 mm diameter, embedded in a lead matrix in an equi-

*37 physicists from Cagliari, UCSD, CERN, CERN (LAA), CERN (World Lab), COPPE/EE/UFRJ (Rio de Janeiro), Cornell, Heidelberg, Lisbon, LPNHE (Paris VI & VII), Max Planck (Heidelberg), Napoli, NIKHEF, Pavia, Palermo.

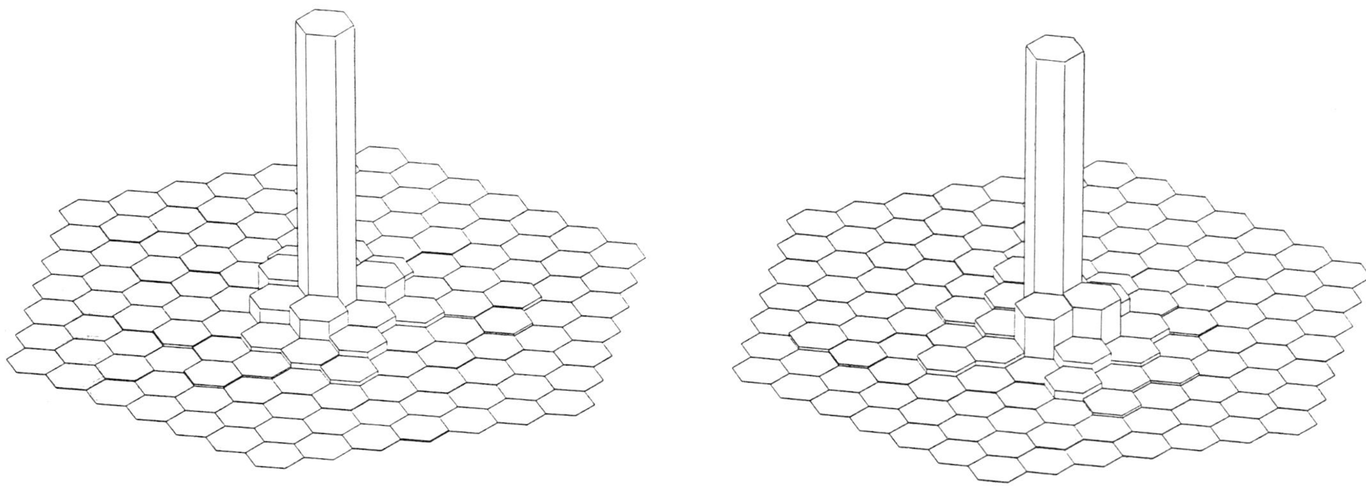

Figure 2: Event display for an 80 GeV pion and 'jet' (target at 44 cm).

lateral pattern, 2.21 mm center to center. The SCSF-38 fibers were purchased from Kyowa [4]. The front faces of the fibers were mirrored (85% reflectivity) by aluminum sputtering. This doubles the effective attenuation length of the fibers to about 7 m and thereby flattens the response to hadron showers produced at different depths. The lead matrix is formed by stacking 1.91 mm thick, 48.6 mm wide, and 2 m long grooved lead strips with fibers embedded in the grooves.

For each tower the 20 cm long ends of the 1141 fibers protruding from the lead (the 'spaghetti') are bunched into a close-packed hexagonal pattern, their ends machined and polished, and coupled by means of a hexagonal lightpipe (42 mm apex-to-apex and 79 mm long) to an 8 stage (7 dynodes) photomultiplier (Philips XP2282). The photomultiplier bases are cooled by forced air for gain stability. The photomultipliers are connected to ADC's with 10 m long Lemo cables followed by a 70 m long RG58 cables. There are active splitters for each signal to obtain high gain (×10) and low gain (×1) readout. The signals were digitized with a LeCroy 2280 series charge-sensitive ADC system with a gate length of 400 ns. The 12-bit ADC's gain was 4 counts per pC for a 1000 pC full scale. Sparse data readout was enabled: a digitizer output had to be four or more counts over pedestal in order to be written to tape. The high voltages of the photomultipliers were set to obtain low gain signals of 4 pC per GeV in the 1.5 ton prototype and in the central region of the 13 ton prototype and 20 pC per GeV in the outer rings of the 13 ton prototype. Thus the high gain readout threshold is about 5 MeV for the outer rings of the 13 ton prototype.

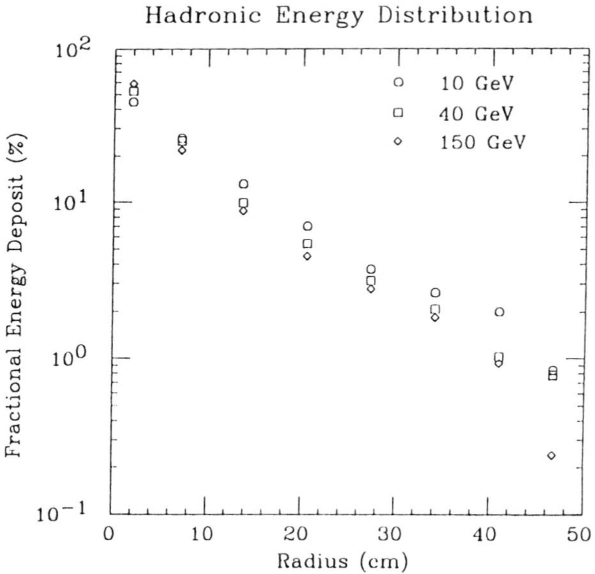

Figure 3: Lateral energy distribution of pions.

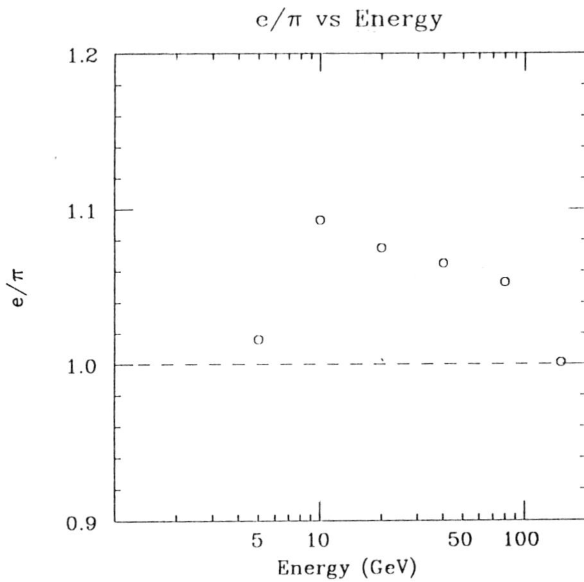

Figure 4: The e/π-ratio as function of energy.

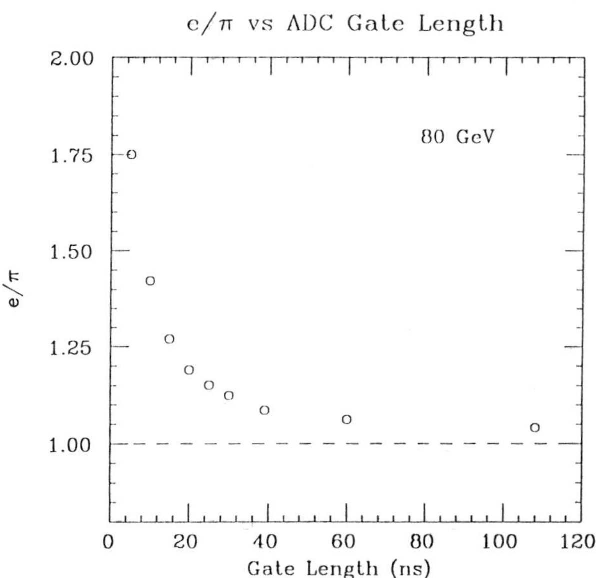

Figure 5: The e/π-ratio as function of ADC gate length for 80 GeV particles.

The 1.5 ton prototype was tested in the H2 beam line of the SPS at CERN in December 1989 and April 1990 while the 13 ton prototype, with the 1.5 ton prototype positioned transversely behind it, was tested in June 1990. The H2 beam was set to deliver relatively pure beams of negatively charged particles (electrons, pions, and muons) in the energy range 20 to 150 GeV. The electron contamination in the pion beam was measured (see the section on the pre-shower detector) and found to be 1 : 500 at 40 GeV and 1 : 1700 at 150 GeV. Special runs were made in June 1990 at 5 and 10 GeV, at these low energies the beam contains a mixture of electrons and pions. A pre-shower detector was used in the trigger for the 5 and 10 GeV runs to select either electrons or pions. The response to 'jets' was simulated by placing a 10 cm long paraffin target in the beam, either 44 or 147 cm upstream of the calorimeter. For these runs the trigger required pulseheight over minimum ionizing in a scintillator counter that was placed between the target and the calorimeter. Obviously this leaves much to be desired as a simulation of a 'jet'. Using the counter's pulseheight offline may improve that in the future. We show in Fig. 2 event displays for an 80 GeV pion and for an 80 GeV 'jet' (target at 44 cm).

The results for the 13 ton prototype that are discussed below are preliminary because they were obtained online. No cuts were made, except those that result from effects of the pre-shower detector in the trigger for the 5 and 10 GeV points. In particular no calibration has been applied beyond a careful adjustment of the high voltages and no cuts were made on the starting point of the showers or on the containment (no cuts on the energy in the backing calorimeter).

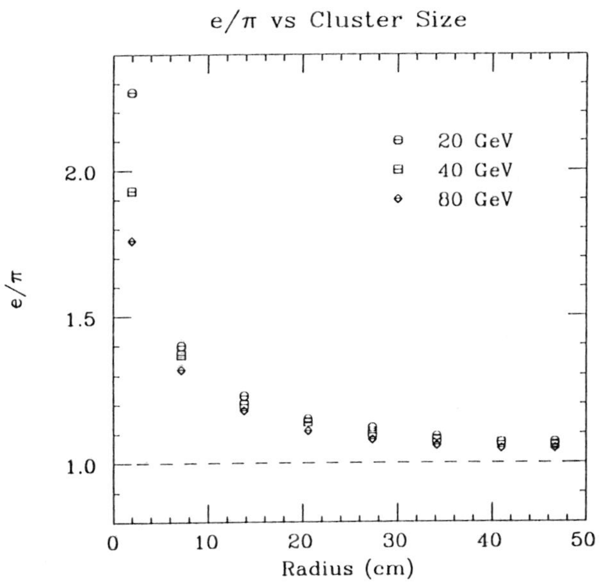

Figure 6: The e/π-ratio as function of cluster size for 80 GeV particles.

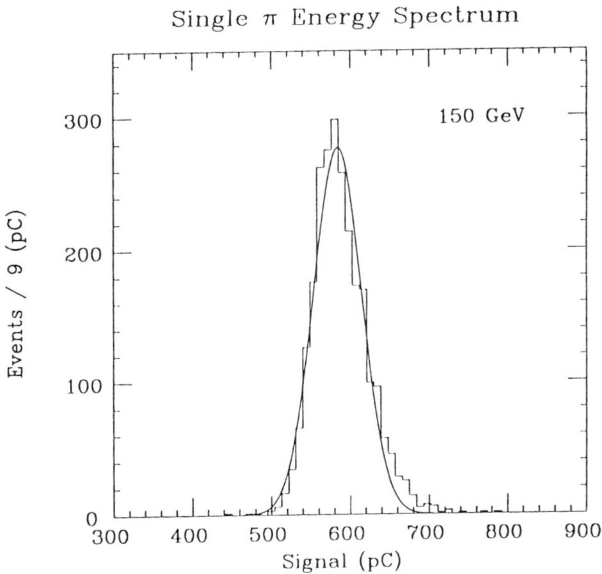

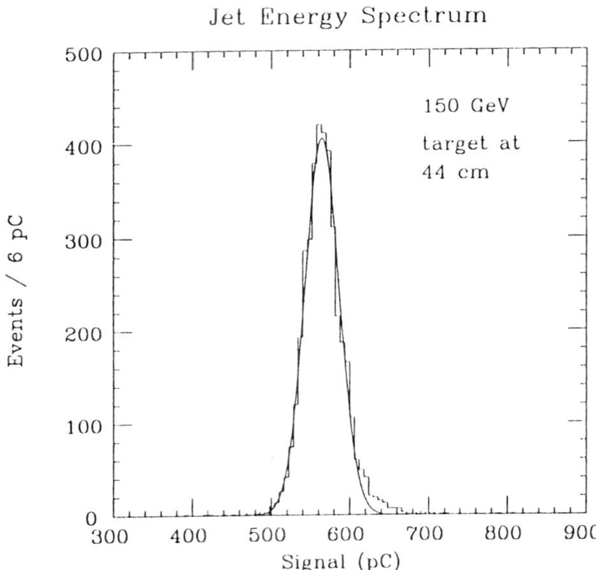

Figure 7: Energy spectra for 150 GeV pions and 'jets'.

As an example of the wealth of detailed information that is available from the beamtests we show the lateral energy distribution in the 13 ton prototype for 10, 40, and 150 GeV pions in Fig. 3. The points at the largest value of the radius are affected by limitations in the online readout.

e/π Ratio

The e/π-ratio has been measured as function of energy, see Fig. 4. The 5 GeV point is suspect because of impurity of the electron and pion samples. The preliminary value of e/π-ratio and its energy dependence might indicate that the calorimeter is not completely compensating. The intrinsic e/π-ratio is approximately equal [5] to the observed e/π-ratio at 10 GeV, about 1.1 in the present case.

In Fig. 5 we show the e/π-ratio as function of ADC gate length. It is seen that the final value for the e/π-ratio is essentially obtained with gate lengths of 40 ns. Analysis shows that this result is determined to a large extend by the use of the long signal cables. If the electronics are mounted near the calorimeter a still shorter gate length might give acceptable results.

In Fig. 6 we show the e/π-ratio as function of the cluster size. It is seen that summing over a cluster with a radius of 35 cm gives essentially the final value.

One would like the intrinsic e/π-ratio to be in the range 0.95 to 1.08 in order to keep the constant term in the energy resolution to 1% [5]. If the final analysis results should show that the e/π-ratio is too high, this could be corrected by increasing the volume ratio of lead to scintillating fiber, for example from 4:1 to 5:1. It is advantageous to aim for too low a value for the e/π-ratio because the effects of both a short ADC gate and a small cluster size are to raise the e/π-ratio slightly.

In summary we expect that short ADC gates and small cluster sizes, both important for high luminosity applications, will give adequate values for the the e/π-ratio.

Hadronic Lineshape.

The energy spectra for 150 GeV pions and 'jets' are shown in Fig. 7. The curves are Gaussians, fitted from -2σ to $+2\sigma$ from the fitted mean. A high energy tail is evident, the tail is larger for single pions than for 'jets'. The tail is observed to be smaller at lower energies. There is no such tail in the energy spectra for electrons [2] as long as the incident angle is larger than 2 deg. Analysis shows that for pions the tail is mainly due to the finite attenuation length of the scintillating fibers. Showers that develop deeper into the calorimeter, for example because they start later, will give rise to larger pulseheight. A 'jet' has many independently developing showers, leading to a reduction of the tail. In addition, part of the tail might be due to the the calorimeter not being entirely compensating. The ever improving quality of commercially available fibers will hopefully lead to further increases in the attenuation length thereby reducing the high energy tail.

Hadronic Energy Resolution.

In Fig. 8 the hadronic energy resolution σ_E/E is shown as function of $1/\sqrt{E}$ for single pions incident at 0 deg

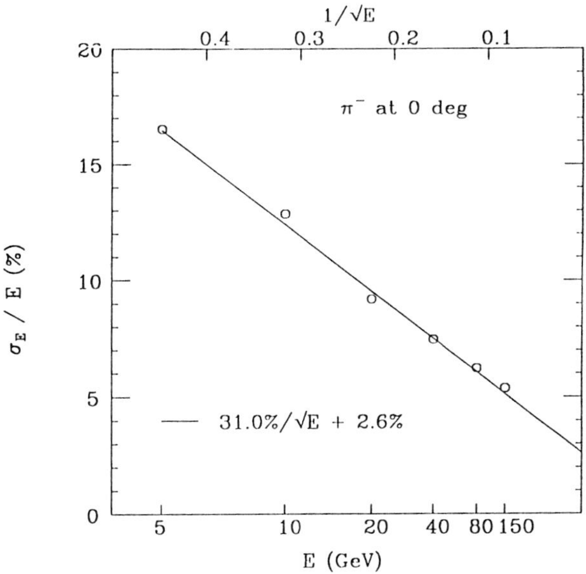

Figure 8: Hadronic energy resolution for single pions incident at 0 deg.

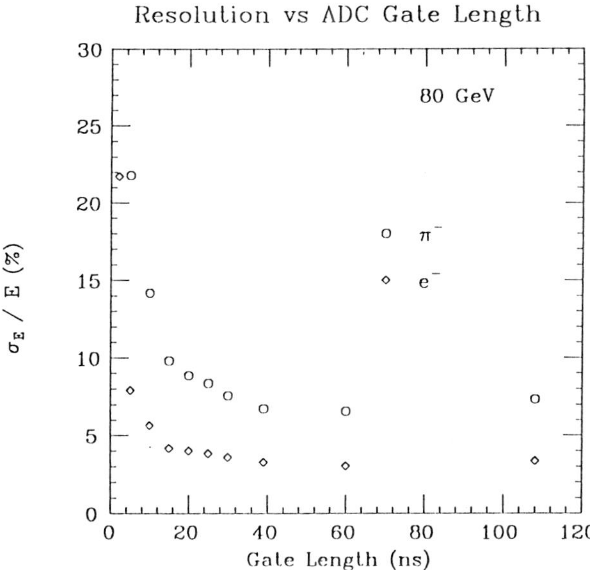

Figure 9: The energy resolution as function of the ADC gate length for 80 GeV electrons and pions.

where σ_E is taken from the Gaussian fit described above. The line is a fit to the data over the energy range 5 to 150 GeV and is parameterized by

$$\frac{\sigma_E}{E} = \frac{31.0\%}{\sqrt{E}} + 2.6\% \tag{1}$$

The values of the stochastic and the constant term have a large negative correlation. The data are equally well fitted by adding the constant term in quadrature instead of linearly, the difference becomes only apparent at energies greater than 150 GeV for which we have no measurements (yet). Obviously the given parameterization should only be used in the energy range 5 to 150 GeV where it has been measured. It is seen that the stochastic term in the resolution is similar to the one obtained with a compensating uranium and scintillator calorimeter. For 'jets' the constant term is measured to be considerably smaller than for single pions. The challenge is to reduce the constant term below 1% for jets.

The dependences of the electromagnetic and hadronic energy resolution upon the ADC gate length are shown in Fig. 9 for 80 GeV electrons and pions. It is seen that the final hadronic resolution is obtained for gate lengths as short as 40 ns.

In Fig. 10 the hadronic energy resolution is shown as function of the cluster size. It is seen that summing over a cluster with a radius of 35 cm gives essentially the final value. The minimum required ADC gate length and hadronic cluster size found here are similar to those required for the e/π-ratio; the values are very promising for high luminosity applications.

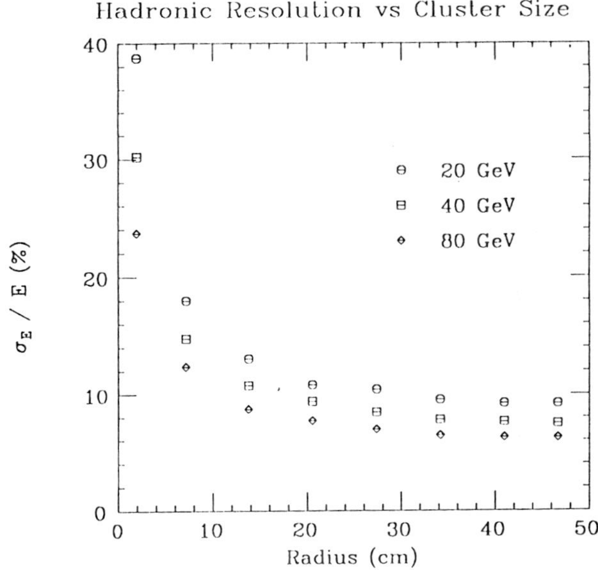

Figure 10: The hadronic energy resolution as function of the cluster size.

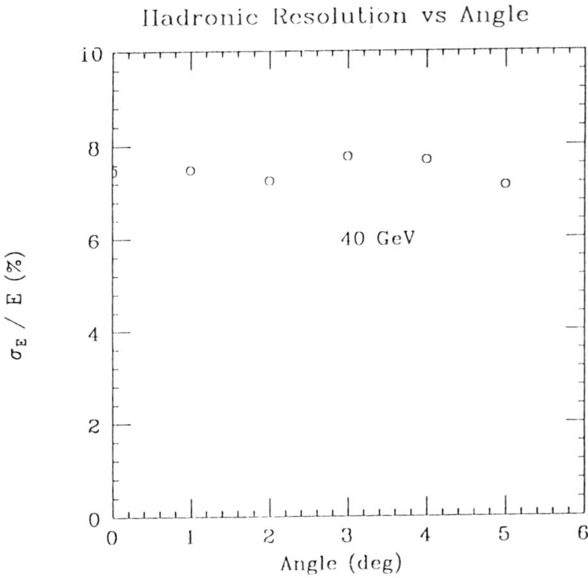

Figure 11: The hadronic energy resolution as function of the incident angle for 40 GeV pions.

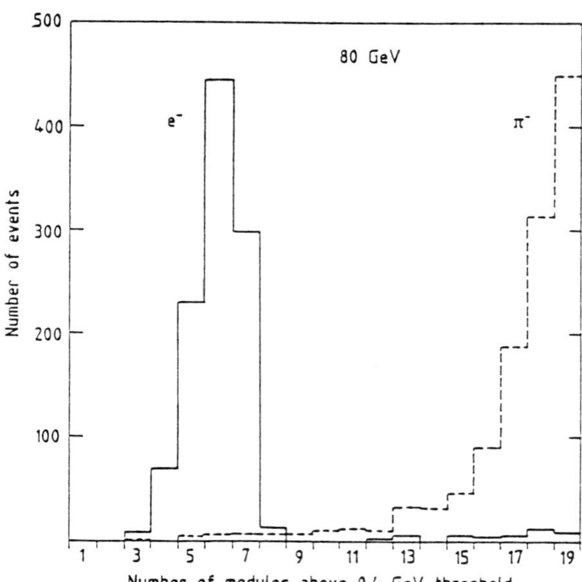

Figure 12: Frequency distribution of the number of towers with energy over $E_{\text{thr}} = 0.4\,\text{GeV}$ for 80 GeV electrons and pions.

In Fig. 11 we show the dependence of the hadronic resolution upon the angle of incidence. It is seen that the resolution is independent of the angle, as opposed to the case for the electromagnetic response which has a strong angular dependence [2] for angles in the range 0 to 2 degrees. Should a calorimeter have separate electromagnetic and hadronic sections then the fibers in the hadronic section may point exactly to the interaction point while the fibers in the electromagnetic section should be offset by at least 2 deg.

Electron-Pion Discrimination.

General

Electron-pion discrimination is traditionally achieved by longitudinal segmentation of the calorimeter, each segment with its own readout. The resulting information on the longitudinal shower development is then used to identify a given energy deposit as either electromagnetic or hadronic in nature. Longitudinal segmentation is possible for the SPACAL technique too but it would be awkward to say the least. It is much more elegant and mechanically sound to construct towers out of single mechanical units. As longitudinal and lateral shower development are correlated (showers that are small longitudinally are in general small laterally), we studied to what extend pion rejection can be achieved in the SPACAL technique without longitudinally segmented readout. Several methods have been developed [6] to identify electrons that are either isolated or near a hadron. These methods were developed in offline analyses of the 1.5 ton prototype. Most are simple enough to lend themselves to application in triggering. We discuss some representative methods and refer to reader to Ref. 6 for details on other methods.

Tower counting

Electromagnetic showers are much smaller laterally than hadronic showers. One can count the number of towers in a cluster with energy over a certain energy threshold E_{thr}. In Fig. 12 we show the frequency distribution of that number for 80 GeV electrons and pions incident near the center of a tower and for $E_{\text{thr}} = 0.4\,\text{GeV}$. The number of towers is clearly smaller for electrons than for pions. A small pion contamination in the electron beam is clearly visible. A pion rejection factor of 100 was found for a cut at seven towers, that is a particle was called an electron if seven or fewer towers in the cluster were over threshold and it was called a pion otherwise. The electron efficiency for passing this cut was 98.8%. We have studied the pion rejection factor as function of E_{thr} and the cut on the number of towers, and as function of incident energy. Better pion rejection factors were obtained for higher values of E_{thr} and for a cut at three towers instead of seven but in the latter case E_{thr} must be varied according to impact point to maintain constant electron efficiency. For simplicity of an application to triggering we prefer a cut at seven towers. Inspection of the tower geometry shown in Fig. 1 should make clear why the cuts at seven towers (central one plus six surrounding ones) or three (adjacent) towers are logical choices.

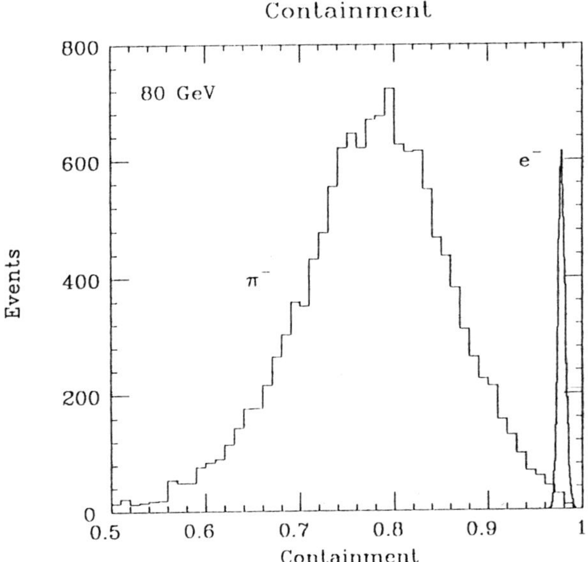

Figure 13: The containment $\mathcal{C}$ for 80 GeV pions and electrons.

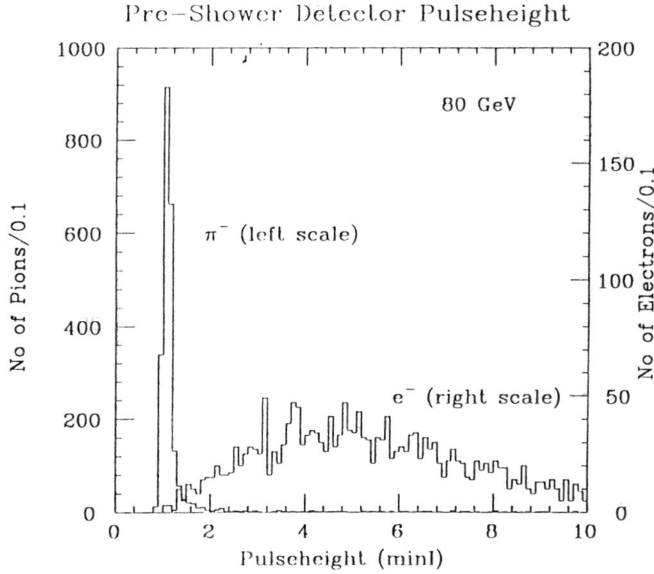

Figure 14: The pulseheight in the pre-shower detector for 80 GeV pions (left hand scale) and electrons (right hand scale). The abscissa is in units of minimum ionizing (minI) pulseheight.

Containment

Another, more precise way, to parameterize the lateral shower size is to use the containment $\mathcal{C}$ defined as:

$$\mathcal{C} = \frac{\sum_{i\in\{3\}} E_i}{\sum_{i\in\{12\}} E_i} \tag{2}$$

$\mathcal{C}$ is defined independent of knowledge of impact point as follows, see Fig. 1. The triplet $\{3\}$ of three towers is defined as the three adjacent towers with the largest energy sum. The triplet is surrounded by nine neighboring towers. The cluster $\{12\}$ consists of these nine and the triplet. Because electromagnetic showers are are well contained in at most three neighboring towers while hadronic showers spread out over many more towers (see Fig. 2 and 3), one may expect $\mathcal{C}$ to be near 1.0 for electrons and significantly smaller than 1.0 for pions. We show in Fig. 13 the containment $\mathcal{C}$ for 80 GeV pions and electrons, incident near the center of a tower. A cut at 0.970 results in a pion rejection factor of 150 with an electron efficiency of 98%. The pion rejection factor is measured to be larger for other impact points because for off-center incidence electromagnetic showers are better contained in a triplet ($\mathcal{C}$ is larger) while the containment for hadronic showers is essentially independent of impact point. The energy dependences of the pion rejection factor and the corresponding electron efficiency are measured to be negligible in the range 40 to 150 GeV.

Even though the cut on $\mathcal{C}$ is specified in three decimal places, precise calibration constants are not required. For values of $\mathcal{C}$ near the cut at 0.970 the numerator and denominator in (2) are almost equal. This means that the sum over the cluster $\{12\}$ is essentially over the same three towers as in the triplet $\{3\}$ of the numerator. Therefore variations in the corresponding calibration constants should leave the value of $\mathcal{C}$ almost unchanged. This was tested by varying the calibration constants in a random way by 10%. The pion rejection factor was found to be unchanged while the electron efficiency dropped by 0.5%. Thus the possible absence of precise calibration constants at the time of data taking does not present an obstacle to the use of the containment in the trigger. A possible hardware implementation is described in Ref. 7.

Pre-Shower Detector

It is well known that electron-pion discrimination can be obtained with a pre-shower detector. We used an improvised pre-shower detector that consisted of a thin (two radiation lengths) lead and tungsten radiator followed by a 1 cm thick scintillation counter. The radiator thickness corresponds to 5% of a pion nuclear interaction length [3]. The pulseheight spectra for 80 GeV electrons and pions are shown in Fig. 14. The electron and pion pulseheights are well separated from each other. A cut at 1.75 minI pulseheight gives a pion rejection factor of 13.5, approximately what one would expect when the extra material upstream of the pre-shower detector and the Landau tail of non-interacting pions is taken into account. The electron efficiency is 98%. The energy dependences of the pion rejection factor and the corre-

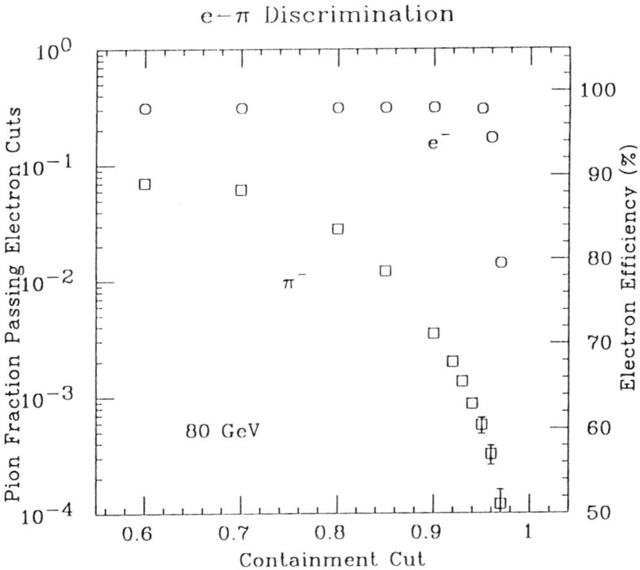

Figure 15: The fraction of pions (left scale) passing combined electron selection cuts and the electron efficiency (right scale) as function of the containment cut for 80 GeV electrons and pions.

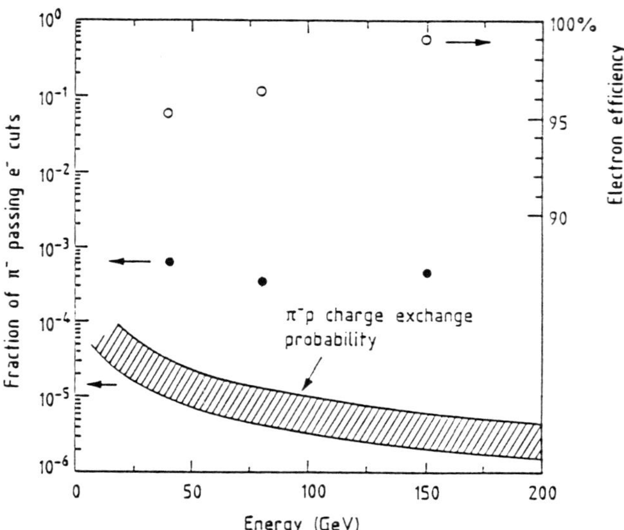

Figure 16: The fraction of pions (left scale) passing combined electron selection cuts and the electron efficiency (right scale) as function of energy. An estimate of the charge exchange contribution is also shown.

sponding electron efficiencies were measured and found to be small in the energy range 40 to 150 GeV.

The pion fraction passing the combined pre-shower detector pulseheight cut (at 1.75 minI) and containment cut are shown in Fig. 15 as function of the containment cut together with the corresponding electron efficiency for 80 GeV particles. It is seen that pion rejection factors of several thousand are obtained with good electron efficiency. The energy dependence of the pion fraction passing the combined electron selection cuts and the corresponding electron efficiency are shown in Fig. 16 together with an estimate of the contribution of the charge exchange reaction $\pi^- p \longrightarrow \pi^0 n$ in the radiator. The calculation shows it to be at least a factor 20 lower than the pion fraction passing electron selection cuts. For this measurement the small electron contamination in the pion beam was removed by placing a 25 mm thick lead sheet in the pion beam 80 m upstream of the calorimeter. A brief run with electrons with the lead sheet in place verified that the electron contamination in the pion beam was reduced by a factor 200 to less than $1 : 10^5$, allowing the measurement of these large pion rejection factors. A comparison of the fraction of pions passing the combined electron selection cuts with and without the lead sheet in place gave the estimate of the electron contamination in the pion beam stated in the section on 'Prototypes and Testbeam'.

Timing

It has been shown [8] that the time dependence of the photomultiplier signals is very different for electrons than for pions. This is due to the fact that hadronic showers are large longitudinally while electromagnetic showers are short. Direct light generated in the fibers by particles deep in a hadronic shower not only reaches the photomultiplier earlier than light from electromagnetic showers but these particles will also send light towards the front of the fibers where it is reflected by the mirrored fiber ends, and arrives extra late at the photomultiplier. Also the longitudinal development of hadronic showers has much larger spatial fluctuations (of order of a pion nuclear interaction length: 30 cm [3]) than electromagnetic showers (of order of a radiation length: 7.5 mm). As a result the photomultiplier signals that result from hadronic showers are wider and show larger variation in their pulseshape than the signals from electromagnetic showers. In Fig. 17 we show the distribution of the Full Width at Fifth Maximum (FWFM) for 80 GeV electrons and pions. The mean value and the variation of the FWFM are smaller for electrons than for pions. A cut at 14 ns gives a pion rejection factor of 800 with an electron efficiency of 99%. The FWFM distributions are largely independent of energy and impact point. Therefore a fixed cut on the FWFM can be used in a trigger application.

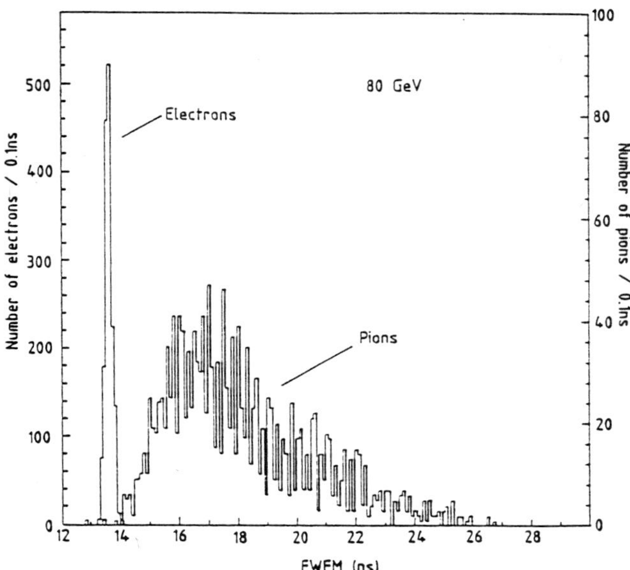

Figure 17: The distribution of Full Width at Fifth Maximum (FWFM) for 80 GeV electrons and pions.

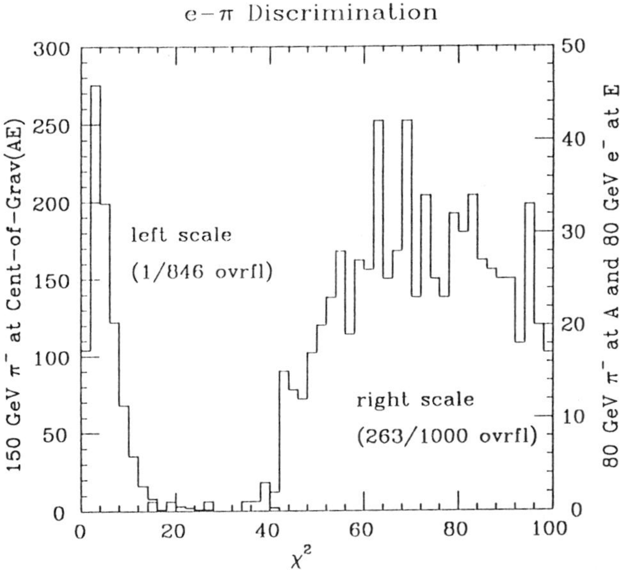

Figure 18: The χ^2 distribution for a 150 GeV pion-hypothesis applied to 150 GeV pions (left side) and to mixed electron-pion events of 80 GeV each (right side).

Covariance Matrix Method

This method is still under development and is intended to recognize electrons near or in jets. It is based upon the H-matrix method of D0 [9] and is defined in analogy with the covariance and error matrices of error analysis. As an example, the method is used to define a χ^2 to test the hypothesis that a given cluster of energy is consistent with being generated by a single pion. If the χ^2 is large we reject the hypothesis and assume that the cluster is instead generated by a pion and a nearby electron. Following Ref. 9 we define the correlation matrix:

$$H_{ij} = \overline{(E_i - \overline{E_i})(E_j - \overline{E_j})} \quad (3)$$

where the average is over events, i and j are tower labels. The H-matrix is defined by means of a sample of pion events of energy E, impacting the calorimeter at a given point. The H-matrix can be inverted to define the χ^2 for a single event as:

$$\chi^2 = (E_i - \overline{E_i})H_{ij}^{-1}(E_j - \overline{E_j}) \quad (4)$$

We have chosen an eight-dimensional H-matrix, element 1 is the tower with the largest energy deposit, elements 2-7 are its six neighbors, and element 8 is the sum of all remaining towers. When the χ^2 is evaluated for the same sample of pion events as was used to define the H-matrix, the distribution shown in Fig. 18 (left side) results. Its mean is 8, the number of degrees of freedom. Next we mix an 80 GeV pion event with an 80 GeV electron event whose impact points are 74 mm from each other and are chosen such that the center of gravity of their combined energy cluster coincides with the impact point of the single 150 GeV pions used to calculate the H-matrix. The χ^2 is calculated as above. This represents a test of a 150 GeV single pion hypothesis applied to the mixed events. The resulting χ^2 distribution is also shown in Fig. 18 (right side). The two distributions are clearly separated from each other. A cut at $\chi^2 = 50$ is 93% efficient for recognizing the mixed events as mixed events while misidentifying only 1 event out of 846 pions as a mixed event while it was not a mixed event. We have found a simple scaling law with energy (this allowed us to use the H-matrix derived from 150 GeV pions at 160 GeV). Thus H-matrices need to be generated at only a small number of energies. The method is currently being extended to jets. As jets have a large lateral size it is hoped that the impact point dependence will be slight and that the H-matrices therefore need to be generated at only a small number of impact points. The method is expected to work better the finer the lateral segmentation of the calorimeter. The present 85 mm tower size is larger than is foreseen for most LHC or SSC calorimeters. When a finely segmented pre-shower detector is present (with for example 1 cm x 1 cm square silicon pad readout), its information can be incorporated in the H-matrix, improving the effectiveness of the method.

Summary

Calorimeter requirements for the LHC and SSC are extremely demanding. The physics requirements include speed of response, radiation resistance, compensation, linearity, good energy resolution for electromagnetic and hadronic showers, uniformity of response as function of impact point and angle, hermeticity, ease of lateral and longitudinal segmentation, spatial resolution, low noise, sensitivity to minimum ionizing particles, and electron-hadron discrimination. Engineering requirements having to do with construction and cost are equally important.

The SPACAL collaboration have shown in Ref. 2 for electromagnetic response and with the present preliminary results for hadronic response that the scintillating fiber and lead combination can satisfy many of the physics requirements.

There are major outstanding issues that require further R & D. Among these are radiation resistance and the engineering of a several thousand ton calorimeter. A good beginning has been made in addressing the engineering issues [10, 11, 12]. The SPACAL collaboration has been approved and partially funded in September 1990 for a second phase of R & D in which the remaining issues will hopefully be addressed.

[1] P. Jenni, P. Sonderegger, H.P. Paar, R. Wigmans, "The High Resolution Spaghetti Hadron Calorimeter", NIKHEF report NIKHEF-H/87-7, unpublished.

[2] D. Acosta *et al.*, "Results of prototype Studies for a Spaghetti Calorimeter", NIM **A294**, 193-210, (1990).

[3] The nuclear interaction length λ_I is defined as the mean free path for protons between inelastic interactions, as defined by the Particle Data Group in Physics Letters **B239** (1990), page III5 and III6. It is a factor 1.5 larger for pions.

[4] Kyowa Gas Chemical Industry Co, LTD, Tokyo, Japan.

[5] R. Wigmans, "High Resolution Hadron Calorimetry", NIM **A265**, 273-290 (1988).

[6] D. Acosta *et al.*, "Electron-Pion Discrimination with a Scintillating Fiber Calorimeter", CERN preprint CERN-PPE/90-172 (1990), submitted to NIM.

]7] R.K. Bock, V. Buzuloiu, W. Krischer, CERN-LAA-RT/90-01 (May 1990), unpublished.

[8] R. DeSalvo *et al.*, "A Novel Way of Electron Identification in Calorimeters", NIM **A279**, 467-472 (1989).

[9] R. Engelmann *et al.*, "Response of a Highly Segmented Extruded Lead Glas Calorimeter to Electrons and Pions between 15 and 45 GeV", NIM **216**, 45-56, (1983).

[10] "Electrons, Muons, Partons with Aircore Toroids" (EMPACT collaboration Expression of Interest), SSC report SSC-EoI-0006 (1990).

[11] "Totally Hermetic EXperiment at the Supercollider", (TEXAS collaboration Expression cf Interest), SSC report SSC-EoI-0011 (1990).

[10] "EMPACT/TEXAS Letter of Intent to write a Proposal for an Experiment at the Superconducting Super Collider", (EMPACT/TEXAS collaboration Letter of Intent), unpublished (November 1990)

SCINTILLATING LIQUID XENON CALORIMETER FOR PRECISE ELECTRON/PHOTON/JET PHYSICS AT HIGH ENERGY HIGH LUMINOSITY HADRON COLLIDERS

M. Chen et al.[1]

Abstract

We use αs as well as e, π, p, d and heavy ion beams to test prototype scintillating liquid xenon detectors, with large UV photodiodes and fast amplifiers submersed directly in liquid xenon. The data show very large photoelectron yields (10^7/GeV) and high energy resolution ($\sigma(E)/E < 0.7\%$ for $E > 1.6$ GeV). The α spectra are stable over long term and can be used to calibrate the detectors. Full size liquid xenon detectors have been constructed, to study cosmic μ's and heavy ions. We report the progress on the design and construction of the 5×5 and 11×11 cell liquid xenon detectors which will be tested in high energy beams to determine the e/π ratio. We describe the design and the unique properties of the proposed scintillating LXe calorimeter for the SSC.

Xenon Calorimeter for Precision e/γ and "jet" Physics

Introduction

The liquid xenon (LXe) calorimeter uses multiple layers of UV-photodiodes and fast amplifiers, developed and manufactured by MIT and Waseda, submersed directly in LXe to measure the scintillation light from liquid xenon. The goal is to detect photons and electrons with an EM energy resolution of $< 0.7\%$; as well as a vertex determination of ~ 0.7 cm and a spatial resolution of of their impact points to ~ 1.5 mm by measuring their three dimensional shower profiles. Our recent test results using full size prototype detectors show the following[2]:

- large size UV photodiodes and fast (10 ns peaking time, defined as the time interval for pulse height to rise from 0 to 100%, corresponding to about twice of the value of the rise time) amplifiers work well inside commercial liquid xenon (even without any purification) with a typical photoelectron yield of 10^7/GeV;
- calibration can be performed using 5.5 MeV α sources, both situated inside liquid xenon and on the diodes;
- $\delta E/E < 0.7\%$ for heavy ions at $E > 1.6$ GeV with fast amplifiers; and
- scintillation is 3 orders of magnitude less vulnerable to impurity than ionization.

Description of the Detector

The proposed detector shown in Figure 1 consists of a hollow ellipse filled with $\approx 3 \times 10^4$ LXe cells. The full detector is composed of two independent half shells bolted to a vertical plate located at the midplane. The overall dimensions are 5 m in length and 2.8 m in diameter. The depth of the active volume is 24.5 X_o (radiation length) long for most parts and 22 X_o for $-0.5 < \eta < 0.5$. Liquid temperature is monitored by both sensors and the vapor pressure. The detector can be serviced in situ by first pumping out the liquid and then cutting

[1]M. Chen, D. Luckey, D. Pelly, S. Shotkin, K. Sumorok, B. Wadsworth, X.J. Yan, C. You, X. Zhang, Laboratory for Nuclear Science, MIT, Cambridge, MA, USA; E.G. Chen, M.P.J. Gaudreau, D.B. Montgomery, J.D. Sullivan, Plasma Fusion Center, MIT, Cambridge, MA USA; A. Bolozdynya, V. Chernyshev, P. Goritchev, V. Khovansky, A. Kouchenkov, A. Kovalenko, V. Lebedenko, V.A. Vinogradov, V. Epstein, S. Zeldovich, ITEP, Moscow, USSR; R. Krasnokutsky, R. Shuvalov, IHEP, Serpuchov, USSR; E. Aprile, R. Mukherjee, M. Suzuki, M. Moulsen, Dept. of Physics, Columbia University, New York, NY, USA; S. Sugimoto, K. Okada, Dept. of Physics and OULNS, Osaka University, Osaka 560, Japan; T. Fujino, T. Matsuda, M. Miyajima, Nat. Lab. for HEP, KEK, Tsukuba, Ibaraki 305, Japan; T. Doke, J. Kikuchi, A. Hitachi, T. Kashiwagi, Y. Nagasawa, H. Ichinose, N. Ishida, T. Nakasugi, T. Ito, Science and Eng. Res. Lab., Waseda University, Tokyo 162, Japan; K. Masuda, E. Shibamura, Saitama College of Health, Saitama 338, Japan; T. Takahashi, Riken, Saitama 351, Japan; W. Wallraff, I. Physikalisches Institut, RWTH Aachen, D-51 Aachen, Germany; M. Vivargent, Laboratoire de Physique des Particules, LAPP, Annecy, France; Manfred Mutterer, Dept. of Physics, Technische Hochschule, Darmstadt, Germany; H.S. Chen, H.W. Tang, K.L. Tung, Institute of High Energy Physics, Beijing, China; H.L. Ding, Institute of Atomic Energy, Beijing, China.

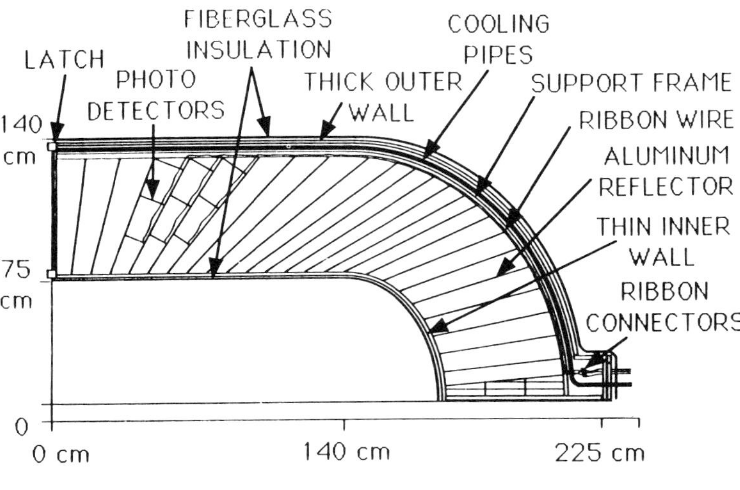

Figure 1: Schematic of the full cross section of two half-hemispheres of the liquid xenon "Olive" calorimeter, bolted together at the mid-plane, showing cells pointing toward the intersecting point, with 3 diodes/cell at large angles and many layers of diodes (with degraders) for forward regions (crossed).

two thinly welded seals and recessing the outer shell and cooling radiator; all photodiodes and cabling are then directly accessible.

A typical cell of the xenon EM shower counter contains photodiodes, UV reflectors, and amplifiers with cables.

UV photodiodes

Because phototubes, including triodes and (even narrow gap) vacumm diodes, do not work when magnetic field is not parallel to the axis of the tube, we have developed large UV sensitive silicon solid-state photodiodes covered by gold meshes, which are insensitive to magnetic fields and have a quantum efficiency an order of magnitude greater than that of vacuum phototubes. The diodes are so thin that < 1% of the signal of an EM shower is due to particles passing through them. This makes hermetical calorimeters with 3-D shower measurement possible. Using these diodes directly in LXe also eliminates UV-transparent windows and glue which simplifies the construction of the detector and increase its reliability.

The small effect of 1% of signal originated in the diodes can be eliminated by using inacitive pads of a thickness of about 1.5 mm to make the effective W value of the diode-pad system to be the same as the LXe. This could be useful when many diodes are used in one single cell. By inserting absorbers behind each photosensor, one can reduce the effective radiation length of the EM shower counters for the forward regions. The leakage current from the diodes is so small at -110 °C that even if it increases by three orders of magnitude after heavy doses of radiation, the noise due to leakage current is still negligible. We are now ready to produce The first 100 diodes for the beam test of the 5 x 5 cell LXe calorimeter. Three Chinese Institutes are preparing for mass production.

Fast amplifiers

Fast amplifiers developed at MIT and beam tested at KEK and the Riken heavy ion accelerator operate in LXe and has a peaking time of 10 ns for the largest size diode (5 cm diameter) planned for the LXe calorimeter. The detector yields < 0.54% energy resolution for 2.47 GeV ^{27}Al ions. These tests show that our present diodes have already satisfied the criteria for the LXe calorimeter. The fast amplifiers are situated behind the last (largest) diode, attached to the N_2 cooling lines outside the active xenon region so that the heat generated can be directly absorbed by the cooling system.

The present MIT amplifier driving a 50 Ω cable to the outside of the cryostat, has a low gain channel which is linear up to 10^9 electrons or one hundred GeV and has a high gain shaper to give sufficient sensitivity at the lower end. We therefore require 2 FADC channels per photodiode. The measured noise is 1.5×10^4 electrons.

Our aim is to have 100 amplifier channels (200 FADC channels) ready in 1991 for the 5 × 5 cell prototype and 400 amplifier channels (800 ADC channels) ready for the 11 × 11 cell prototype in 1992. In order to cover the full dynamical range required at the SSC at reasonable cost, we just start to study the relative merits of using two diodes (one large and one small) versus two amplifiers (one high gain and one low gain), etc. Further R&D on electronics is still needed.

UV reflecting cell walls

The cell walls are used to reflect UV light and do not mechanically support the weight of the liquid. Consequently, we have made reflectors of thin (100 μm) UV reflecting aluminum foils, welded flat piece by piece using electron guns and then expanded into proper shape. Small grooves in the foil allow liquid xenon to flow from cell to cell, causing practically no dead space and no deterioration in energy measurement. These metal UV reflectors also serve as Faraday shields for individual cells to eliminate cross talk and noise, which often limits the energy resolution for crystal cells.

Assuming a reflectivity of 0.88 at 170 nm, which has been achieved by 300 RICH counter Al mirrors used by DELPHI at LEP, Monte Carlo studies for a 30 cm long section (the longest section of a full size, 66 cm, truncated pyramidal shaped cell in the "olive') show excellent uniformity for the following two configurations:

1. Using two photodiodes (one at either end), the uniform composite signal ($\frac{cS_1S_2}{S_1+cS_2}$) corresponds to $\sim 60 \pm 1.5\%$ light collection. Uniformity is insensitive to reflectivity or attenuation length in this case.
2. Using a single photodiode and darkening the nearest ~ 1 cm of the reflector, the signal corresponds to 40% light collection.

Monte Carlo Results for these cases are shown in Figure 2. We are scheduled for heavy ion beam tests using full size LXe detectors starting Nov. 17, 1990 as will be described later.

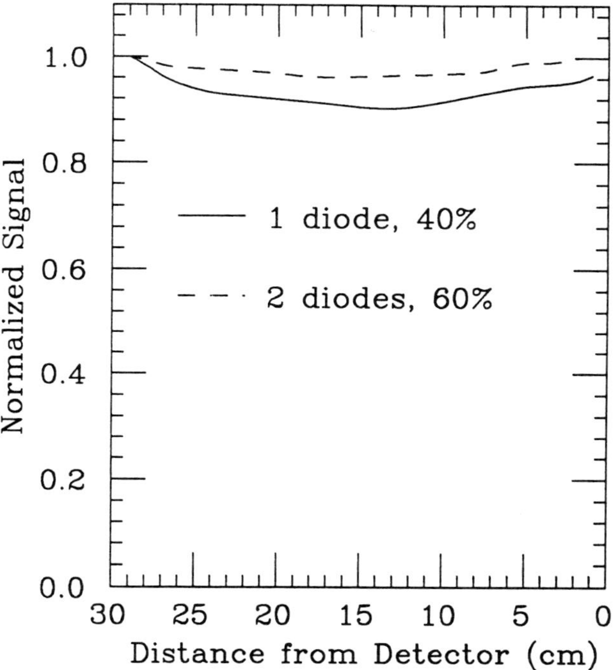

Figure 2: Monte Carlo studies of the signal uniformity as a function of distance from the large end of a 30 cm long section of a 66 cm long, truncated pyramid reflector cell with a normal reflectivity of 0.88.

Physical properties of liquid xenon detectors

Some physical properties of liquid xenon detectors are summarized in Tables 1 and 2.

Two unique properties of the detector

The purpose of a Xenon calorimeter is to achieve an EM energy resolution of $\sim 0.5\%$ with $e/\pi = 1$. The energy resolution depends on calibration.

Calibration: Due to the very large number of photoelectrons produced (3×10^7/GeV) in liquid xenon, we have observed a large signal-to-noise ratio for 5.5 MeV α from ^{241}Am, which serves as a precision calibration of each of the proposed 10^4 cells.

$e/\pi = 1 \pm 0.1$*:* Monte Carlo results show that scintillating liquid xenon detectors can achieve compensation by using fast signals from low velocity nuclear spallation fragments, such as protons.

We describe both properties in more detail in the following sections.

Table 1: Physical properties of liquid xenon

triple point	
temperature	161.36 K
pressure	0.805 atm
liquid density @165 K	3.057 g cm^3
radiation length	2.8 cm
Moliere radius	4.06 cm
absorption length	55 cm
fast component	5 ns
slow component	20 ns
attenuation length	$\gg 7$ cm measured by us and $\gg 20$ cm by T. Ypsilantis at CERN
index of refraction @ 170 nm	1.5
photon yield	$4 - 7 \times 10^7$/ GeV

Table 2: Physical properties of liquid xenon with our diode/amplifiers

photoelectrons/GeV	10^7, including Q.E. and acceptance
calibration	5.49 MeV α
longitudinal sampling	3 to 10 layers
saturation	100 times higher than NaI
$\sigma(E)/E$	$< 0.7\%$ for $E > 1.6$ GeV

Scintillation Mechanism in Liquid Xenon and e/π

The overall scintillation in liquid xenon consists of the excitation luminescence and the recombination luminescence. The liquid xenon scintillation is dependent on the dE/dx of the incident particles, because the recombination time is a function of the electron-ion density in the ionization tracks produced by the incident ionizing particles. It is known that the intensity of the fast component is drastically enhanced and the output signals shortened when liquid xenon is irradiated by high

dE/dx particles such as alpha particles, heavy ions, or, most important for jet physics, nuclear spallation fragments from hadron-nucleus interaction when compared with irradiation by low dE/dx particles such as energetic electrons.

Thus when a hadron loses energy in nuclear inelastic interaction due to nuclear binding energy loss, it also gain energy due to higher dL/dE of nuclear spallation fragments from the same inelastic interaction. The amount of energy gained from nuclear fragments increases for shorter gate time, since the signals from nuclear fragments are faster than that of an electron. These effects compensate well the nuclear binding energy loss due to hadron nuclear inelastic scattering and makes the e/π of liquid xenon close to 1. We checked this by using the Geant Monte Carlo code to compute the excess energy deposited (within 15 ns) by slow nuclear spallation fragments produced from 10, 50 and 100 GeV π's incident in a 60 cm long, 5×5 cell, liquid xenon detector. We plan to experimentally confirm this by beam testing a prototype detector together with a compensating calorimeter (e.g. scintillator-U calorimeter of ZEUS, TMP calorimeter of UA1, or scintillating fibre calorimeters of SPACAL/LAA); thus experimentally verifying that scintillating liquid xenon has the property, unique among all homogenous calorimeters, that the e/π ratio can be adjusted close to unity as has been predicted by Monte Carlo calculations.

Major Test Results in 1990

From the investigations on 5000 cm^3 liquid xenon detectors[3] with heavy ions and radioactive sources, we conclude that the energy loss in liquid xenon is transferred up to extremely high ionization densities with high efficiency and linearity into UV scintillation light producing large currents in UV photodiodes. In particular, the diodes and amplifiers made by us, can detect small signals (5 MeV) with high resolution and fast peaking time (10 ns) and have already met the requirements in speed and signal/noise ratio for the SSC liquid xenon calorimeter proposed by us and are ready for mass production.

Calibration using α Particles in situ

We determined the photoelectron yield and study calibration methods for LXe detectors using 5.49 MeV α's. The details can be found in the Xenon Prototype Detector Subsystem Proposal[1]. The measured pulse height of photoelectrons corresponds to 4×10^4 electrons with a resolution, dominated by the electronic noise of the amplifier, of 6.6% (slow) and 17% (fast). The temperature dependence of the scintillation yield, in the temperature region between $-110°$C and $-75°$C, was about $-0.4\%/°$C, three times smaller than that of the BGO. Temperature variation for liquids is much less important than solid where time constant for heat conduction can be many hours and temperature must be deduced from surface measurements.

When α's are allowed to stop directly in the diode, the measured output pulse height equals that theoretically expected value. The half width of the leading edge (high energy side) of the spectrum is only limited by electronic noise whereas the half width for the low energy side is increased by the energy levels of the ^{241}Am source and the straggling of the α's through the material. The width the pulse height distribution is 0.5% with 3 μs gate time and 1.5% with 20 ns[4]. This high resolution demonstrates the excellent uniformity of the photodiodes, both home made and from Hamamatsu. Both diodes and amplifiers work well at low temperature. The α spectra are very stable and, therefore, can serve as a very sensitive and reliable calibration for the experiments, which will be described in the next section.

One can use α, cosmic μ or RFQ for calibration, after the detectors have been calibrated in beams Figure 3. Here we describe the calibration scheme using two

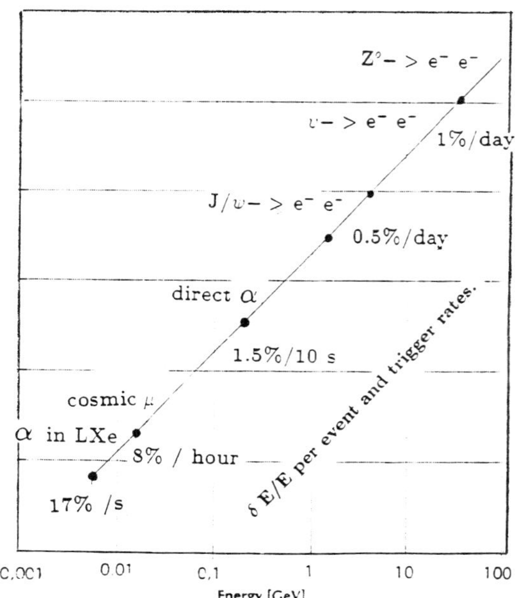

Figure 3: Calibration of liquid xenon detector using α particle in LXe and direct on the diodes, cosmic μ's, and monitored by J and $Z \rightarrow ee$ and the RFQ.

α sources, one situated in LXe to calibrate the complete detector system using LXe scintillation light, and one situated on the front surface of the diode to calibrate the electronics system using the direct hits of α's.

1. The gains of the electronics (ADC and amplifiers) can be readily calibrated using direct α hits on the diodes, which produce the equivalent of 150 MeV energy in liquid xenon. Our data shows the peak

position can be measured to an accuracy of 1.5% with a fast amplifier. Therefore a few hundred hits/diode yields 0.15% accuracy.

2. The light yield of liquid xenon due to different amount of impurity can be monitored by an ionization chamber to better than 0.2%. As noted earlier, the temperature dependence of the scintillation yield, is −0.4%/°C, three times smaller than that of the BGO. The heat leakage is $< 10^{-4}$ of the heat capacitance of the xenon and thus takes hours to change LXe temperature by 1°C even if the heat leakage is doubled without being noticed.. With the additional advantage of liquid over solid in shorter time constant to reach thermal equilibrium, it should be possible to control the temperature of LXe to better than ± 0.5°C, corresponding to < 0.2% in the variation of light yield.

3. We can calibrate the overall effect, including electronics-LXe-reflectivity, using UV light produced by α's in liquid xenon. Using fast amplifiers we need 10000 events to determine the peak position to $17\%/\sqrt{10000}$ <0.2%. With slow amplifiers, we need 1000 events/diode to determine the peak position to $6\%/\sqrt{1000}$ <0.2%. The systematic errors of a large system still need to be studied.

Attenuation length measurements

We measure the attenuation length of LXe for the 170 nm scintillation light from α's by varying the distance between the α source and the photosensor. There is no indication of any light attenuation in liquid xenon over a distance between 1 and 9.5 cm. The data are consistent with the variation expected from light acceptance and 82.5% random reflection effects[5]. An independent measurement by T. Ypsilantis et al. at CERN, using an evaporation method to change the depth of LXe between the light source and the photosensor, without having to move the diode, has determined the attenuation length to be much larger than 20 cm[6]. We plan to determine the attenuation length to much higher accuracy when we measure the uniformity of full size LXe detectors in heavy ion beams starting Nov. 17, 1990.

dE/dx of e, π, p and d for Particle Identification

We measured the dE/dx response of various particles between 700 and 1500 MeV/c in a 55 mm LXe cell at KEK, to study the particle identification capability of scintillating liquid xenon detectors[7]. Some typical results are shown in Figure 4. The difference in the pulse height of π's and p's is due to the difference in dE/dx or beta. The electron distribution is higher and broader due to shower development and leakage, which could have been picked up by the subsequent cells in the proposed Xenon Olive Detector to yield a precision total energy measurement. The observed narrow peaks due to pions and protons show the intrinsic good resolution of scintillating liquid xenon detectors and their potential use as particle identification devices.

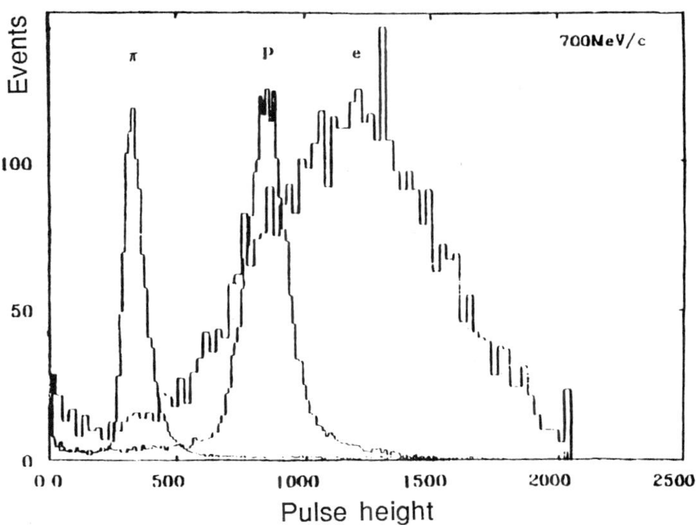

Figure 4: The dE/dx pulse height distributions for e, π and p for 700 MeV/c momentum in a 55 mm LXe detector. The small peak at the left of the electron curve is due to the non-interacting electrons. Similar data exist up to 1500 MeV/c, including d also.

Energy Resolution measured with Heavy Ion Beams

We determine the intrinsic energy resolution of a scintillating LXe detector equipped with full size (5cm diameter) silicon photodiodes using ^{14}N, ^{40}Ar and ^{27}Al ion beams from the Ring Cyclotron at Riken, Japan. The energy of ^{14}N ions, after passage through the front materials and the window, was 1.64 GeV. The observed energy resolution was 0.6% RMS. The electronic noise contributes about 0.1% and the energy loss straggling due to known front materials another 0.15%. The other 0.57% of the resolution could come from beam energy spread and scattering off the collimator wall[8].

The energy of ^{40}Ar (beam energy of 80 MeV/n) ions entering liquid xenon, after passing through material in front, was estimated to be 2.65 GeV. The fast amplifier installed just behind the photodiode to minimize the inductance of the readout cable and rise time of the signals, operated at -75°C with a shaping time of the amplifier of 10 ns. The peaking time of the observed signals was less than 20 ns. Since then, the peaking time of the system has been further improved to be 10 ns (see the section on amplifiers). The charge observed is 2.91×10^7 electrons (expected 2.35×10^7). The measured energy resolution without corrections is 0.68%.

To test the reliability of the detectors, in particular the UV photodiodes, we baked the same diode used for previous measurements to 100°C until the quantum efficiency of the diode, determined using α's as described earlier, has dropped to about 50% of its previous value. The diode was then installed into the same liquid xenon detector and we repeated the energy measurement using 2.47 GeV ^{27}Al ions 2 months after the previous tests. The charge observed is reduced to 1.36×10^7 electrons, yet the measured energy resolution without corrections is 0.48% with 200 ns gate and 0.54% with 40 ns gate. These results show that the measured energy resolution is mainly due to beam energy spread, not electronics or intrinsic photon statistics: the actual intrinsic energy resolution of LXe detectors is still much better than what are quoted above.

Figure 5 shows the pulse shape and height distributions of LXe detector for the above heavy ion tests. Figure 6 shows the energy resolution of scintillating LXe detector together with that of the L3 BGO EM calorimeter[9]. LXe has better resolution owing to the much larger number of photo-electrons generated.

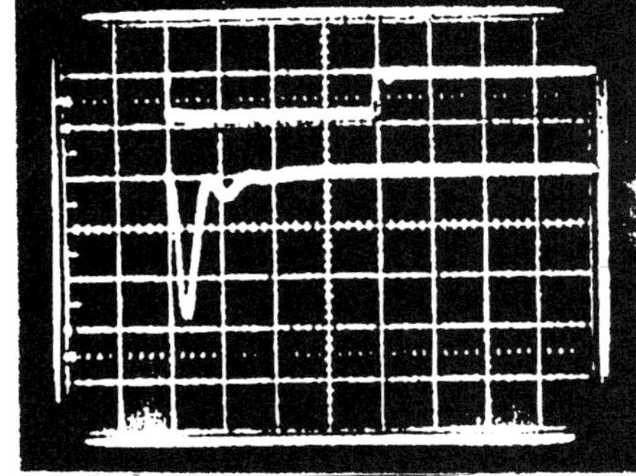

Figure 5: a) Pulse shape (50 ns/div); and b) pulse height distributions with 40 ns gate time for 2.47 GeV ^{27}Al ions in a prototype LXe detector.

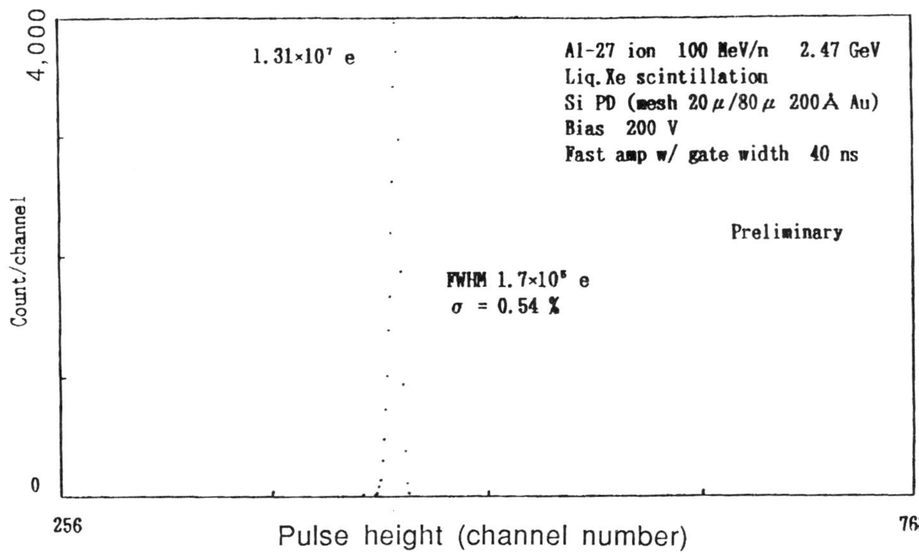

Tests of Full size (65cm long) LXe Detectors

Prototype Cell at MIT for Cosmic μ's

At MIT, a versatile xenon prototype liquid xenon chamber with foam insulation is completed (cf. Figure 7) to study uniformity (including impurity effects and UV reflectivity) in scintillating liquid xenon detector. It is being used to research many aspects of the "Olive" design. In order to simulate the actual cooling and insulation properties of the "Olive", the MIT chamber is being operated in all orientations; thus, issues of xenon flow to prevent localized boiling (at the amplifier) as well as freezing at the cooling coils can be addressed (note the viewports in Figure 8). Also, the practicality of foam versus vacuum insulation and of xenon purity and performance can be studied, and other design issues addressed as they arise.

Prototype Cell for Heavy Ion Tests

A full size 65 cm long prototype cell with vacuum insulation has also been constructed and tested at MIT. Figure 9 shows a schematic. This chamber has 7 beam windows, spaced 10 cm apart; 2 of the 7 windows are transparent so that we can visually check for bubbling in the liquid xenon or freezing. These are never observed after filly is completed. Inside the chamber, tapered reflector plates are installed with a diode at each end. These will be used in coincidence to measure the uniformity over a distance of 65 cm under realistic condition using heavy ion beams at Riken (to enter each of the 7 thin windows) at the end of Oct., 1990.

Measured Energy Resolution for

- BGO;
- LXe using α's and heavy ions (including beam energy and other uncertainties).

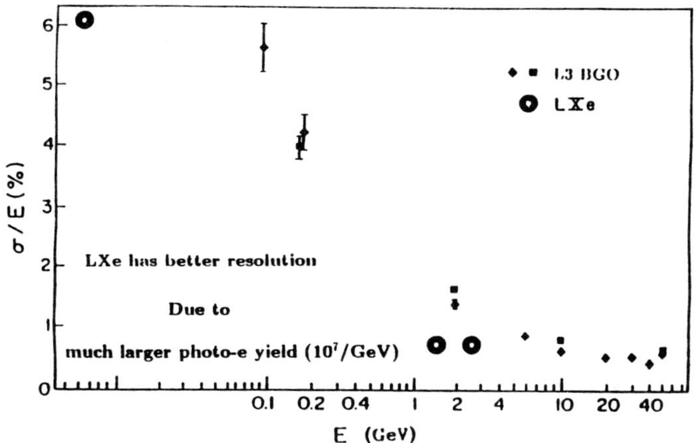

Figure 6: Measured energy resolution versus incident beam energy for the L3 BGO EM calorimeter (squares) and scintillating liquid xenon detectors (circles) measured with α's and heavy ions as described in this report. The LXe resolution includes uncertainties such as the beam energy spread, etc..

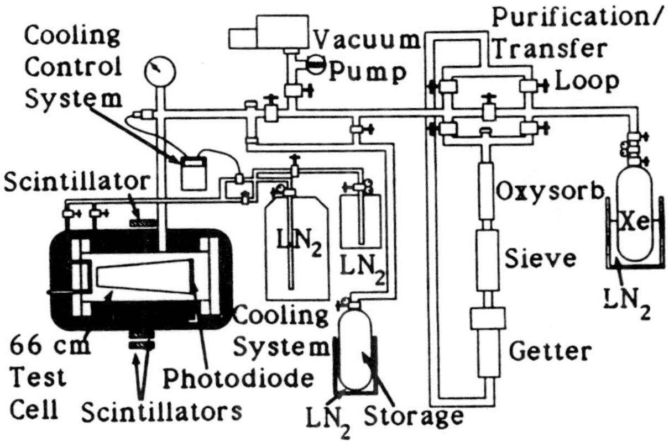

Figure 7: Schematic of the full size (65 cm long) prototype liquid xenon system being tested at MIT with cosmic μ's.

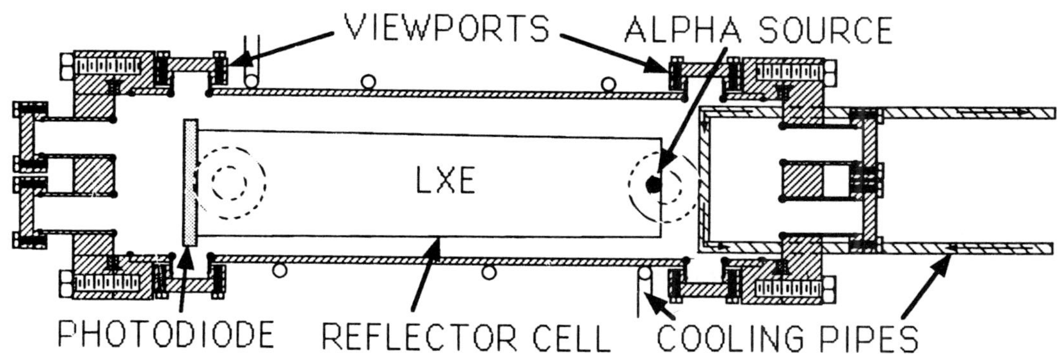

Figure 8: Schematic of the full size prototype liquid xenon cell being tested at MIT with α's and cosmic μ's.

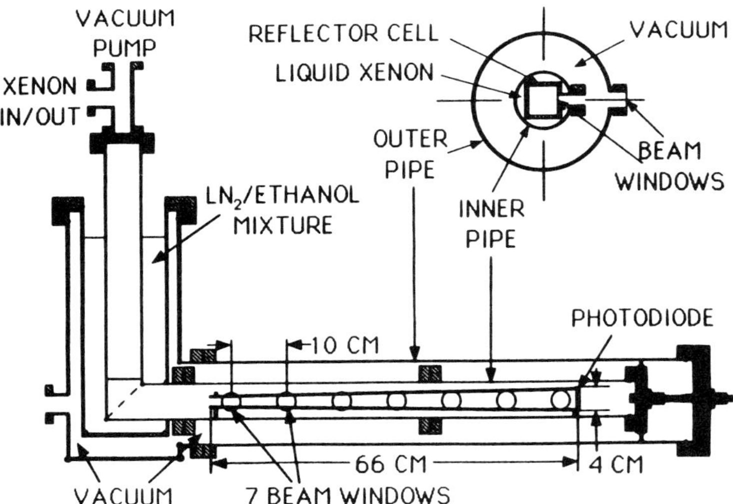

Figure 9: Schematic of a full size prototype liquid xenon cell to measure the uniformity using heavy ions, which can enter any of the seven thin beam windows.

Purification Tests

Purification for large systems has been a major task for ionization type of calorimeters. For LXe, a Ba-Ti getter purifier gives an attenuation length of electrons more than 1 m in an ionization chamber[10]. It has been shown that for liquid xenon with impurity of a few ppm O_2 equivalent, the attenuation length of electrons is reduced by a factor of a thousand to about 1 mm, while the scintillation light yield remains stable at a level of a few times 10^7/GeV%[5]. This shows scintillation is more than 3 orders of magnitude less sensitive to impurity than ionization.

At MIT, we have studied the effect of impurity by filling up two full size scintillating liquid xenon cells with commercial grade xenon and found that scintillation works well even without any purification. These tests show that our presently designed purification system are certainly good enough for scintillating xenon detectors. We will use ionization as a sensitive monitor for the purity of LXe, in addition to α's in LXe.

R&D

Milestones

Having achieved $\sigma(E)/E < 0.7\%$ for heavy ion beams at E > 1.6 GeV and two working full size cells with different types of insulating technique, the present development program is aiming at the measurements of fully contained 50 GeV electron showers with high precision and reproducibility using a 5×5 matrix of full size liquid xenon cells in 1991, and (with a hadron calorimeter behind) fully contained 50 GeV pion showers with high precision and reproducibility using an 11×11 matrix of full size liquid xenon cells in 1992. The engineering design work on the 5×5 cells have been finished. We have set up and UV spectrometer to measure the reflectivity of mirrors systematically for the purpose of mass production of reflectors.

The major milestones in our planned R&D program are as follows:

- Fourth Quarter 1990 :
  1. Start the construction of 5×5 cells
  2. Radiation hardness tests of photodiodes at liquid xenon temperatures
  3. UV reflector development
- Third Quarter 1991 : Beam test of the 5×5 prototype
- Fourth Quarter 1991 :
  1. Start the construction of the 11×11 cells as one real sector of the EM calorimeter
  2. Start engineering design of the "Olive"
- Third Quarter 1992 : Beam test of the 11×11 prototype together with a compensated hadron calorimeter to measure e/π.

Xenon Availability

Four commercial companies: Air Liquide, Matheson, Spectra Gases and Union Carbide, have submitted letters that each of them can produce up to 15 m^3 of new LXe by 1999 at a price of about 2.5 M$/$m^3$ of LXe, if there is sufficient guarantee/encouragement from SSC/DOE to induce any one of them to start new production of xenon in 1991. The current procurement plan for xenon is shown in Figure 10. There is also a recent Japanese patent on how to produce xenon directly from liquid air using absorbers; the effect of this on production and cost is still being evaluated.

Summary

Our test data on prototype scintillating liquid xenon detectors, have shown that scintillating xenon detectors have the following properties important for an SSC precision EM calorimeter:

- α and heavy ion tests with full size photodiode/amp. submersed in LXe, show:
  1. quantum efficiency > 50%
  2. 10 ns peaking time

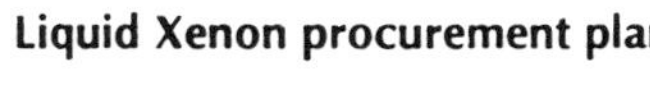
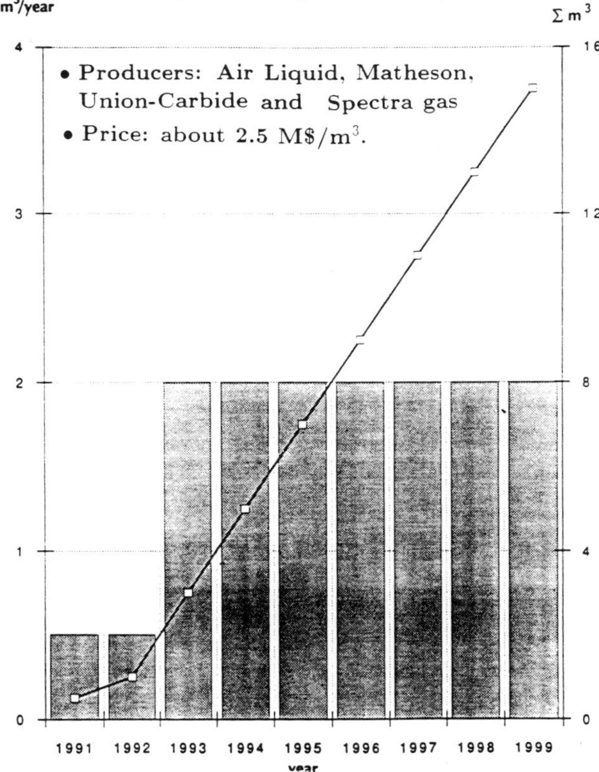

Figure 10: Xenon procurement plan 1991–1999. Xenon procurement in m^3 per year (left scale) and total to date (right scale).

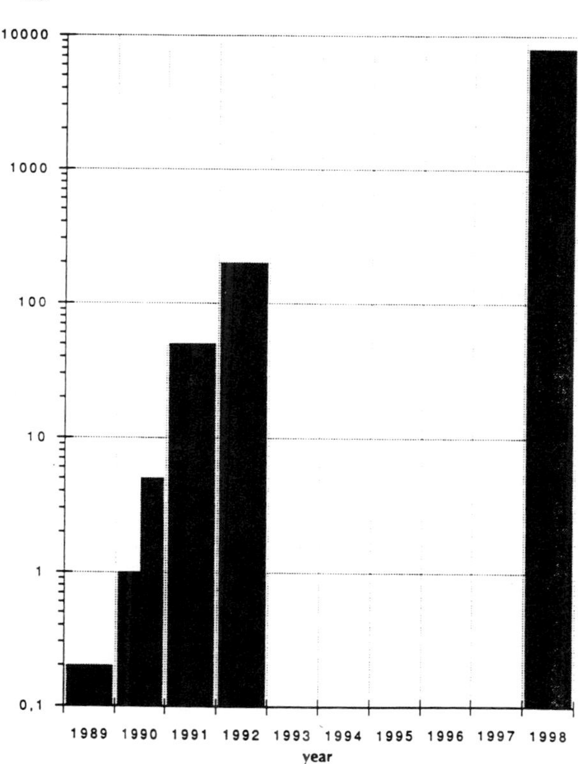

Figure 11: Liquid xenon detector development plan 1989–1998. The detector volume is shown as a function of time.

3. $\sigma(E)/E < 0.7\%$ at $E > 1.6$ GeV.

Therefore 3-D shower measurement possible.

- First 100 diode/amp. under fabrication at MIT/Waseda, 75% usable.
- LXe detectors have large output ($> 10^7$ photo-e/GeV) and thus can be calibrated using α's.
- Two full size cells, with reflector/diode/amp. are under test at MIT and work well even with commercial grade xenon unpurified.
- Scintillation in LXe is $10^3\times$ less sensitive to impurities than ionization.
- Engineering design of the 5×5 LXe cells for e/π beam tests is complete and quotes exist for construction.
- Each of four companies is ready to commit producing the required 15 m^3 of LXe by 1999 for about 2.5 M$/$m^3$.

Finally, the development plan for the liquid xenon EM calorimeter is summarized in Figure 11.

---

[2] M. Chen et al., The Xenon Prototype Detector Subsystem Proposal, 1989 and 1990

[3] M. Gaudreau et al., MIT-LNS Tech. Rep. no. 174, 1990; T. Doke et al., MIT-LNS Tech. Rep. no. 175, 1990; and S. Sugimoto et al., OULNS 90-2, Osaka Univ., May 1990.

[4] M. Gaudreau et al., OULNS Rep. 90-01, May 1990., S. Sugimoto et al., OULNS Rep. 90-02, May 1990, T. Doke et al., private communication.

[5] M. Miyajima et al., Japanese Physical Society Meeting at Osaka, March, 1990.

[6] J. Seguinot, T. Ypsilantis et al., LAA Progress Report, CERN 90/1

[7] S.Sugimoto, M.Miyajima, S.Sasaki, and H.Tawara, Osaka Univ. OULNS 90-8, June 30, 1990.

[8] Note: $0.6 = \sqrt{0.1^2 + 0.15^2 + 0.57^2}$.

[9] L3 collaboration, B. Adeva et al., Nucl. Instr. and Meth. A 289 (1990) 35.

[10] Masuda et al., NIM **188** (1981) 629.

SIMULATION STUDIES FOR DESIGN OPTIMIZATION OF A SCINTILLATOR PLATE CALORIMETER*

J. Proudfoot, P.K. Job, H.-J. Trost
Argonne National Laboratory

T. Handler
University of Tennessee

T. Gabriel
Oak Ridge National Laboratory

Presented by James Proudfoot
High Energy Physics Division
Argonne National Laboratory
9700 South Cass Avenue
Argonne, Illinois 60439

Abstract

Results on simulations studies relating to the optimization of a sampling scintillator plate calorimeter for an SSC detector system are presented. These studies show that whereas a compensating sampling geometry can be obtained using a variety of configurations using either lead or depleted uranium as the principal absorber, no configuration based on a pure iron absorber is compensating. Unlike in a lead system, delayed energy release from long-lived shower products produced in a uranium system pose a serious pile-up problem. Therefore we advocate the use of lead as the principal absorber in this calorimeter. Work on optimization of the mechanical structure is in progress and results are presented on issues such as structural support, tolerances and on the degradation in response due to other detector material within the volume of the calorimeter.

Introduction

The detector designers of experiments at the Superconducting Super Collider (SSC) have the task of determining the most effective means of achieving high performance physics measurements in the distinctly hostile radiation field from the hadron collisions themselves. To further increase the difficulty of this task such a detector must be designed in the next few years, fabricated in a timely fashion to allow installation by beam turn on at the end of the decade and then yield physics measurements for many years thereafter. For a detector component such as the calorimeter, major refurbishment is unimaginable and therefore this detector must be designed for the maximum luminosity lifetime anticipated. This varies from a lifetime of ten years at a luminosity of $10^{33}\,cm^{-2}s^{-1}$ to $10^{34}\,cm^{-2}s^{-1}$.

The use of sophisticated simulation programs to evaluate calorimeter performance is now commonplace. Over the last ten years, the models and approximations embedded in such programs have been well tested and the use of EGS[1] to simulate electromagnetic response, CALOR[2,3] to simulate hadronic response and GEANT[4] to simulate the physics response is expected to give reliable predictions of detector performance. The talks presented by T. Handler, and L.E. Price at this conference present the current state of the art with these programs and give an insight into the physics processes of relevance. The central issue for this presentation is the reliability of the CALOR codes to simulate hadronic and electromagnetic showers, including energy deposition from neutrons produced in the cascade. That these ideas do indeed yield reliable results is shown later when

*Work supported in part by the U.S. Department of Energy, Division of High Energy Physics, Contract W-31-109-ENG-38.

the results from CALOR are compared with experimental results for specific geometries.

The studies described here were carried out in the context of the SSC subsystem collaboration of Spinka et al. whose objective is the "Development of a Compensating Scintillator Plate Calorimeter for the SSC." The key issues in this proposal are: compensation—namely a calorimeter having equal response to electrons and pions of the same energy; scintillator—the use of scintillating plastic as the sensitive material in order to take advantage of its intrinsic characteristics of low noise and fast response time; plate—the use of plate scintillators in a sampling geometry in order to build on the many years of development of such calorimeters.

General Specifications and Performance

The performance goals for this calorimeter design are:

- e/h < 1.05 in 16 nsec (ratio of electron-to-hadron response)
- $\sigma_E/\sqrt{E} < 18\%/\sqrt{E}$ with $\lesssim 1\%$ constant term for electromagnetic showers
- $\sigma_E/\sqrt{E} < 50\%/\sqrt{E}$ with $\lesssim 2\%$ constant term for hadronic showers
- Maximum hermeticity in $|\eta| < 3$
- < 10% degradation in performance over a 10 year lifetime in the barrel section at a luminosity of $10^{34}\,cm^{-2}s^{-1}$.

The CALOR, EGS, and GEANT simulation codes were used initially to evaluate the potential of three absorbers (depleted uranium, lead, and iron) against each other and against the above specification without the need for fabrication and testing of actual prototypes in a particle beam.

Firstly, a GEANT simulation was used to determine the appropriate depth for the longitudinal energy containment of single pions in the nominally compensating geometries associated with depleted uranium (DU), lead and iron/scintillator. This is shown in Figure 1 for 10 GeV pions, from which we determined that all subsequent calculations should use a total absorber depth of 8 interaction lengths.

Thereafter, the CALOR and EGS codes were used to address the choice of absorber and its associated sampling geometry required for equalization of electromagnetic and hadronic responses (e/h = 1). The parameter space considered was:

- Absorber composition (*pure* lead, iron, and depleted uranium)
- Absorber thickness
- Scintillator thickness

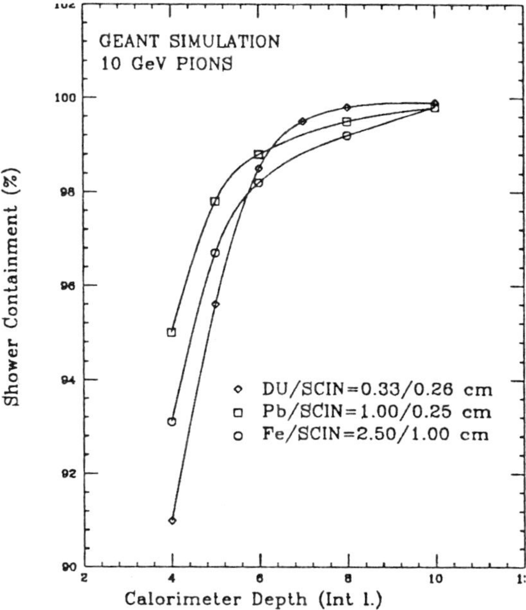

Figure 1: Shower containment as a function of calorimeter depth for 10 GeV pions in depleted uranium, lead and iron scintillator stacks.

- Signal integration gate
- Energy deposition time profile

For these studies the scintillator light saturation constant (Birk's Constant) was assumed to be 0.0131, which is typical of scintillators under consideration.

The e/h ratio as as a function of absorber thickness for a constant scintillator thickness of 0.25 cm and an integration gate of 48 nsec is shown in Figure 2. Both lead and uranium systems allow a geometry yielding e/h = 1. This is not the case for the iron system. A measure of the systematic uncertainty in the model predictions can be obtained by varying the energy cut (E_{spect}) used to determine whether a cascade scaling model or a fragmentation model is used to describe the hadronic cascade. Reasonable bounds on this cut are 5 and 15 GeV. The difference in e/h predictions for these cut values is also shown in Figure 2, indicating a model uncertainty of about 8%. The relative difference as a function of absorber is, however, much smaller. The calculations are seen to be in moderate agreement with the available experimental data (References 5–7).

Figure 3 shows e/h as a function of pseudorapidity (η) under the assumption of constant plate and scintillator thickness in the compensating uranium/scintillator geometry of 0.33 : 0.26 cm. Only a modest increase of e/h of around 5% is seen which is emminently acceptable with respect to degradation of the constant term in the resolution. The slope of e/h versus absorber thickness for lead/scintillator is flatter than that for depleted uranium/scintillator (Figure 2). Therefore, the rapidity

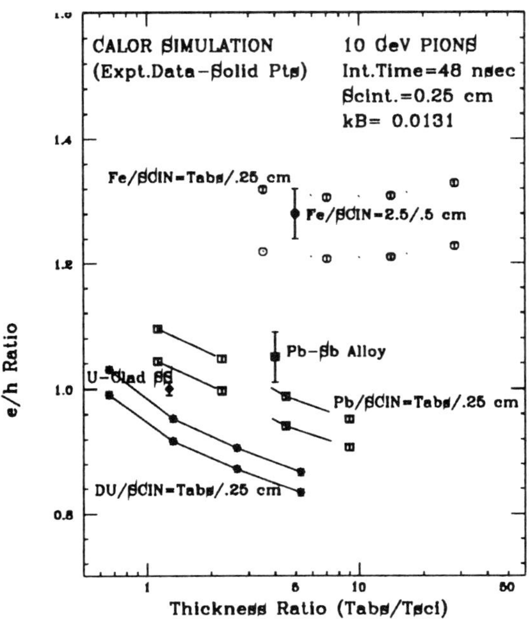

Figure 2: Relative response of electrons to pions (e/h) as a function of absorber thickness (Tab) and composition for a fixed scintillator thickness of 0.25 cm. Lower (upper) sets of points are for $E_{\text{spect}} = 5$ (15) GeV as described in text. Also shown are experimental data (References 5–7).

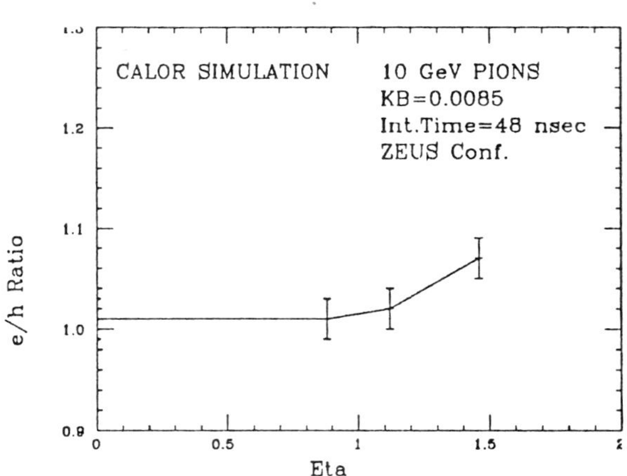

Figure 3: e/h as a function of pseudorapidity (η) for the ZEUS compensating uranium geometry having a scintillator thickness of 0.26 cm.

variation in the lead system will be comparable to or less than that measured in the uranium/scintillator system. Clearly, no rapidity variation is expected in the iron/scintillator system as e/h is computed to be independent of absorber thickness.

The variation of e/h for fixed absorber thickness is shown in Figure 4 (again for an integration gate of 48 nsec). This predicts that for a sampling frequency in lead of 1 radiation length, which is characteristic of that required for an electromagnetic calorimeter, a scintillator thickness of 0.25 should yield compensation. This is a key issue in the use of lead as the passive absorber, where fine sampling is required for good resolution and where the nominal 4 : 1 ratio of absorber-to-scintillator thickness would seriously compromise the performance of the electromagnetic calorimeter.

Finally, the dependence of e/h with integration time is shown in Figure 5 for the nominal compensating iron, lead and depleted uranium systems. Both lead and depleted uranium systems rapidly achieve e/h $\sim$ 1. However, in the case of depleted uranium, energy from long-lived products of the hadron shower continue to be measured up to 500 nsec after the primary shower. These contribute a significant amount of energy ($\sim$ 20%) and pose a serious problem with regard to pile-up.

Our conclusion from this analysis is that firstly, pure iron should be removed from consideration as it

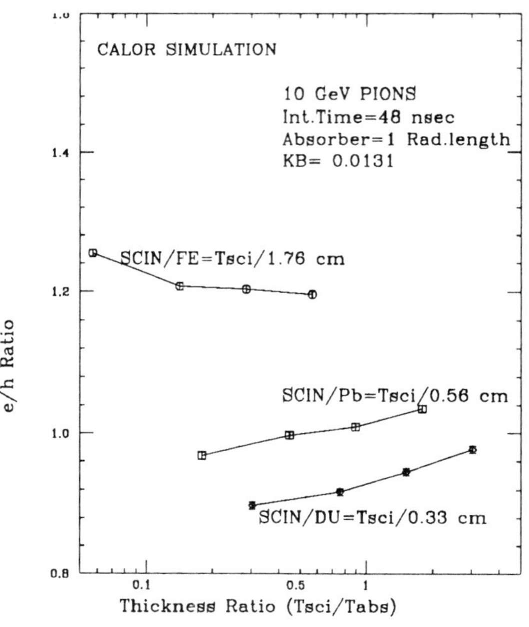

Figure 4: e/h as a function of scintillator thickness for a fixed absorber thickness of 1 radiation length.

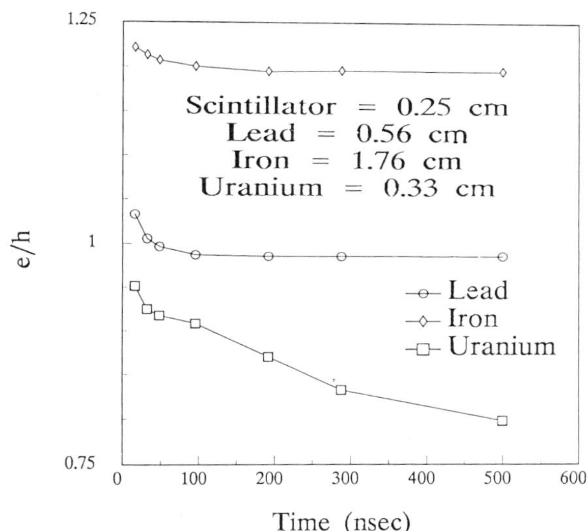

Figure 5: e/h for lead, iron and uranium calorimeters with nominal compensating geometries as a function of signal integration time.

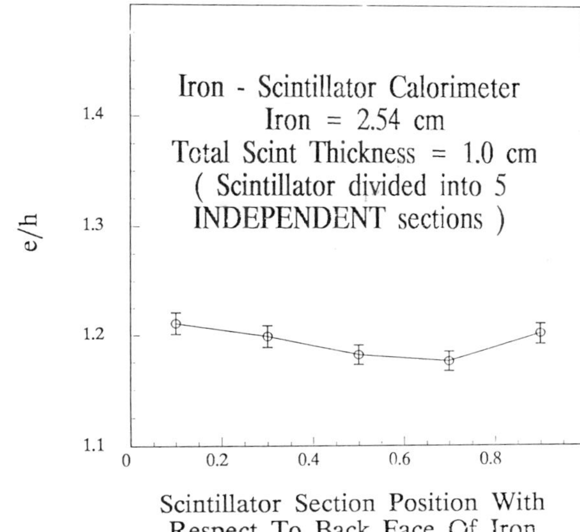

Figure 6: e/h as a function of scintillator section position.

does not yield compensation in any configuration. However, this result for iron is a topic of considerable debate and we are currently pursuing studies to determine the possibility of obtaining compensation in a composite iron structure. These results are discussed in more detail in the talk of T. Handler at this conference. An initial study used 2.54 cm-thick iron absorber and 1 cm of scintillator segmented into five independent sections. Figure 6 shows e/h as a function of scintillator segment depth in this geometry. Little improvement in e/h is observed. These studies are continuing.

Both lead and depleted uranium can yield identical levels of compensation in comparable sampling geometries. As there are finite limits on scintillator plate thickness, this results in calorimeters with essentially identical resolutions. Both are relatively insensitive to the rapidity non-uniformity intrinsic to the sampling plate geometry. However, the significant delayed energy release predicted in the uranium system relative to the lead system indicates a potentially serious pile-up problem. (This result must be verified by experiment.) This, in addition to other factors such as safety, materials, and fabrication costs, results in our advocating the use of lead as the principal absorber in the calorimeter (despite the advantages of calibration and reduced radius associated with the use of depleted uranium). We are therefore proposing to further evaluate the mechanical design of a lead/scintillator calorimeter in FY91.

Detailed Design Optimization

To proceed from the general design outline above which advocates the use of a pure lead calorimeter with 0.5 cm lead plates and 0.25 cm scintillator plates, one must include additional constraints. From the mechanical perspective these include:

- Scintillator composition (especially as is required for radhard characteristics).
- Location, size and orientation of structural supports.
- Gap, absorber, and scintillator tolerances and their effect on response uniformity.
- Global mechanical design issues such as flat versus staggered plate geometries.

Furthermore, from the perspective of the calorimeter as a component in a full detector system, these include:

- Transverse cell size
- Longitudinal sampling frequency (and its resulting effect on electromagnetic and hadronic energy resolution)
- Depth segmentation (and its effect on e-π separation)
- Impact and correction for detector systems within the inner volume of the solenoid.
- Cost
- Radiation damage effects on uniformity and stability

I will briefly illustrate studies into some of these issues.

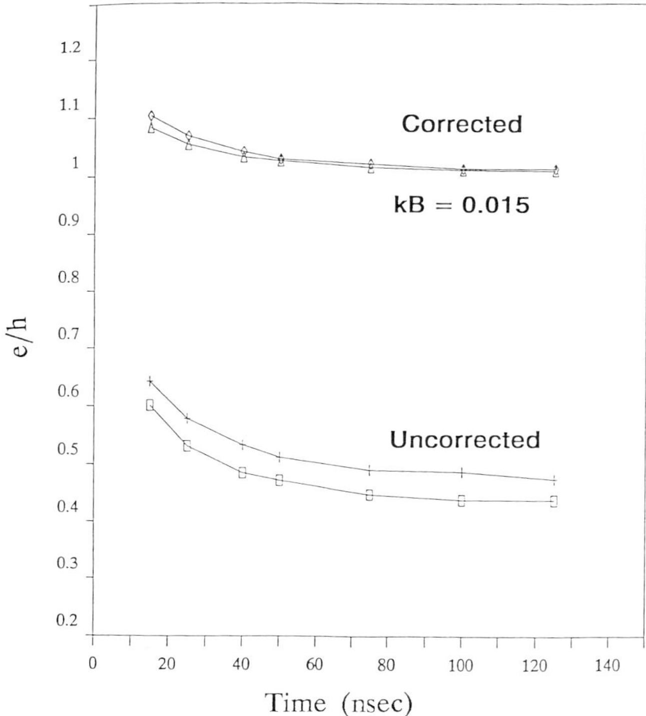

Figure 7: e/h in a Pb scintillator calorimeter with 0.56 : 0.25 cm sampling as a function of neutron cutoff time, assuming kB = 0 and kB = 0.015.

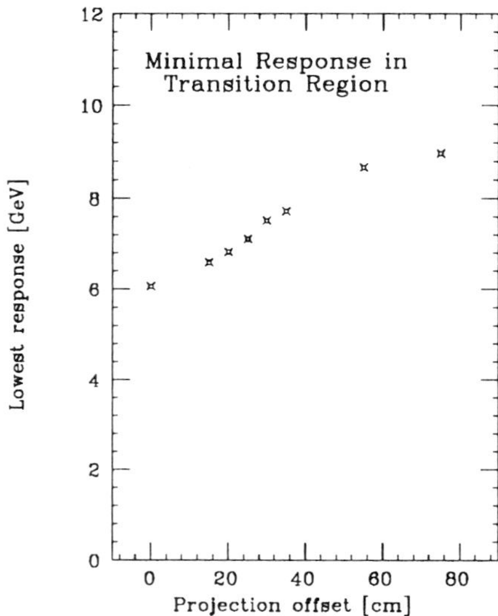

Figure 8: Minimal response to 10 GeV π^- in the barrel-endplug transition region as a function of projective offset, assuming a 1″ aluminum support ring at the boundary.

Mechanical Issues

A commonly neglected variable in compensation studies is the value of Birk's constant (kB), which describes scintillator saturation effects. This is discussed in more detail in the presentation of T. Handler at this conference. The difference between corrected and uncorrected light yields is shown in Figure 7 for a lead scintillator system with Pb : scintillator ratio of 0.56 : 0.25 cm. The effect of saturation of the light yield from low-energy particles (including the important proton recoils) is seen to be enormous. At a more realistic level the difference between kB = 0.015 and kB = 0.008, which are values for actual acrylic scintillators, corresponds to a 10% effect in e/h for equivalent systems. The present optimization assumes kB = 0.0131. Once data on the actual radhard scintillator to be used is available, the design optimization will be reiterated.

The support structure for the calorimeter and detectors within are an area of concern as they can result in projective cracks and loss of heremeticity. As an alternative to a strictly radial cut for these supports we have studied the feasibility of using a tower structure which is pseudoprojective in η to improve the measurement of signal in the barrel-endplug transition region. This is achieved by projecting the towers at large η to a point offset by some distance in z from the true beam crossing point. The minimal response in the barrel-endcap transition region for 10 GeV π^- incident as a function of this offset is shown in Figure 8 (where we have assumed a 1″ aluminum support ring). It is possible to cover this gap by using a rather modest offset of 60 cm. However, resolution is degraded in this region and the scale of this degradation is sensitive to the size of the physical gap required for cables and support services. As a result, our collaboration decided to retain a true endcap design as the more conservative choice for a sampling plate calorimeter (the alternative is still a viable option, however, for this or other technologies).

Another issue for the mechanical design, in particular for a lead calorimeter, is that of thickness tolerance. Lead is well-known to creep, to be difficult to roll flat, and in the option of a cast lead fabrication, liable to shrink during molding. This was judged to be principally an issue for the electromagnetic calorimeter and was studied using EGS in a detector whose nominal sampling was 30 layers of 0.5 cm Pb and 0.25 cm scintillator. The absorber thicknesses were varied randomly in a window of 0, ±5, 10, and 20% from the nominal thickness and the resulting calorimeters surveyed with electrons of energy 1, 2.5, 5.0, 10.0, and 25 GeV. The samples were chosen to yield approximately identical statistical errors on the mean responses for all energies. The overall performance was computed as the average response in MeV (deposited)/GeV (incident energy) over the five beam energies. The resulting comparison of average deposited energy with respect to a perfect calorimeter as a function of absorber thickness tolerance is shown in Figure 9. A systematic change in response is observed due to the dispersion of the EM shower profile. For a simple system in which plates are flat but of varying thickness,

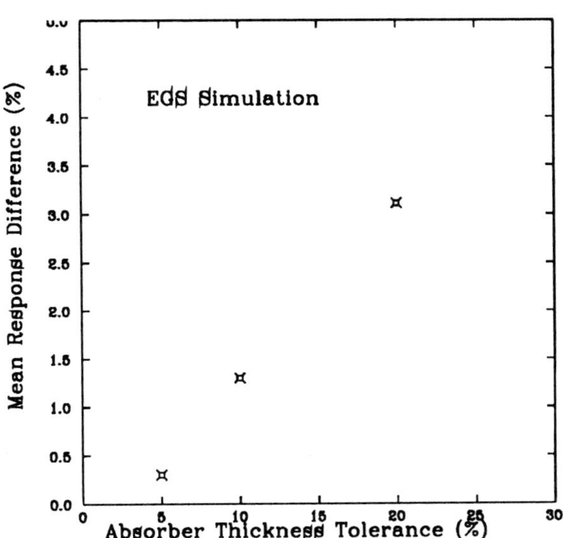

Figure 9: Difference in scintillator energy deposition averaged over incident electron energies of 1, 2.5, 5.0, 10.0, 25.0 GeV with respect to perfect absorber thickness of 0.5 cm for 0.5 : 0.25 cm ratio Pb : Scint.

in situ calibration can correct this mismatch for electron showers. This, however, is not wholly the case for either non-flat plates or for hadron showers. The data show that to maintain a constant term in the resolution of $\leq 1\%$ the absorber thickness tolerance must be maintained to within 8% for 0.5 cm nominal absorber sampling. This is not anticipated to be a limitation on the casting approach currently advocated by the scintillator plate collaboration for lead electromagnetic and hadronic calorimeters.

Other issues for the mechanical design such as the difference between staggered and flat plate geometries (as advocated for an iron calorimeter and as may be demanded for mechanical reasons in a lead-composite calorimeter) are currently under study. Preliminary results show that although the staggered plate geometry may yield superior hadron resolution it may also contribute to a significant constant term. Our current position is to proceed with a flat plate design for both electromagnetic and hadronic calorimeters.

Detector Issues

The detector within whose context this calorimeter design is being carried out is shown in Figure 10. Topics of hot debate in this design are those of transverse and longitudinal granularity. A GEANT calculation of the lateral shower profile for 10 GeV pions is shown in Figure 11. The significant characteristic is that 65% of the shower is contained within a projective tower subtending $10 \times 10\,\text{cm}^2$ at a radius of 2 m from the interaction point (note that this conclusion is approximately independent of absorber medium). Therefore, transverse granularity finer than $10 \times 10\,\text{cm}^2$ only increases the sensitivity to the single particle shower profile. As this is of no import in a jet environment we advocated (and adopted) the angular cell size of $10 \times 10\,\text{cm}^2$ at a 2 m radius and 90° to the beam.

As is indicated in Figure 10, three depth measurements of the energy deposition are envisaged. The first segment is for use in identification of electrons and photons by longitudinal shower development. Current knowledge states that to maximize e/π separation this should be roughly $25X_o$ by maximizing the containment of electromagnetic showers for a minimum number of hadron interaction lengths. GEANT simulations of the optimum geometry and the effect of increased absorber with η are in progress. In the cast lead fabrication approach there is reduced flexibility in the choice of absorber geometry at high η coming from mechanical limitations and these studies will be used to determine the optimal working point.

The outer calorimeter compartment is currently considered to be required to return magnetic flux and is therefore iron (and hence, allowing less than perfect compensation). In this design, therefore, the intermediate section comprises the high-resolution compensating hadron calorimeter. The depth of this section is therefore crucial to the overall linearity of the calorimeter. The energy deposition from neutrons in a compensating Pb/scintillator calorimeter is shown in Figure 12 for 10 GeV π^- incident. Ninety percent of the neutron energy is contained with 4.3 interaction lengths of absorber. This result can be extrapolated to high-energy hadron showers by appealing to existing experimental data (Reference 8). These data indicate that hadron shower maximum moves approximately 0.6 interaction lengths deeper into a Pb stack as the incident pion energy increases from 10 to 100 GeV. Therefore, a minimum of 5 interaction lengths is required to contain 90% of the neutron energy deposition from a 100 GeV pion. In the above configuration, the iron calorimeter would thus mismeasure the pion energy by 0.2% due to mismeasurement of the neutron component. This is an acceptable compromise with respect to the constraints of outer radius and magnetic flux return.

The final issue I shall discuss in this presentation is that of the coil required to generate the solenoidal field in this detector. This contributes a thickness of aluminum of around $1.1X_o$ distributed at lumped radii over 35 cm in total radius. It, therefore, smears electromagnetic showers by absorption of low-energy electrons and by allowing showers initiated in the coil to propagate in vacuum. This effect is largest at maximum angle of incidence, which is roughly 25° to the beam in the detector under consideration. This is illustrated in Figure 13, where a Pb calorimeter with its first layer being 0.5 cm Pb (and no coil) is compared with one having its first layer being 0.25 cm scintillator with a simulated coil and front plate as the first absorber. This cell geometry is sketched in Figure 14. Considerable non-linearity is

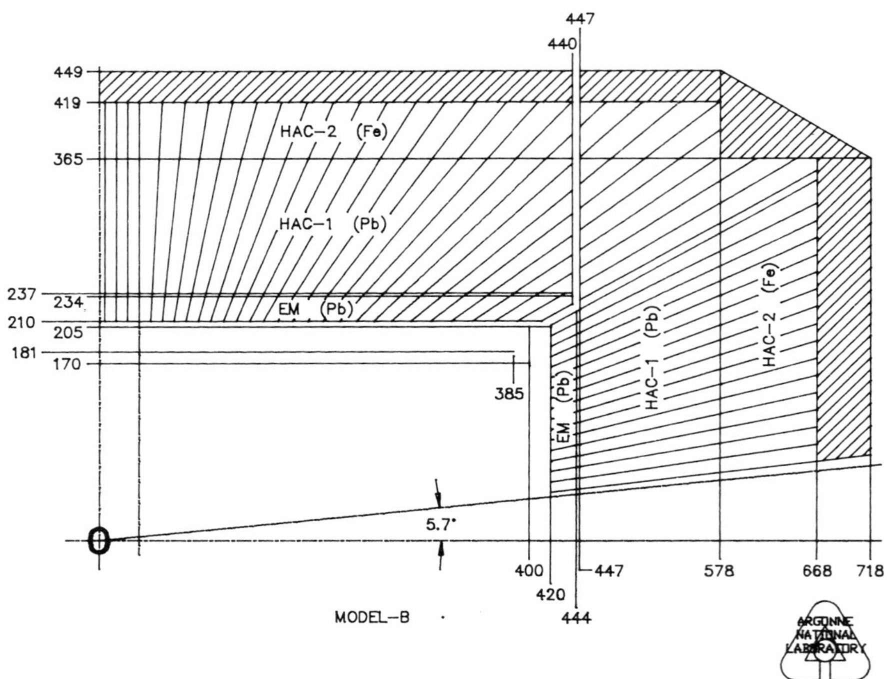

Figure 10: Conceptual design layout for the SDC detector showing one electromagnetic, and two hadronic calorimeter compartments.

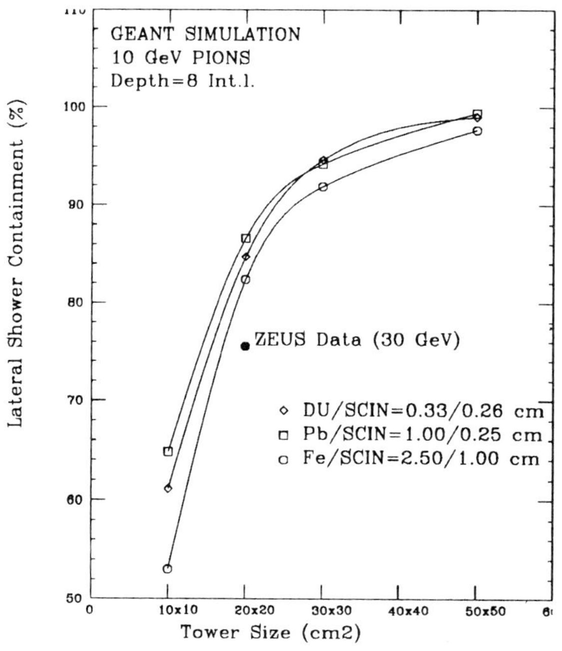

Figure 11: Lateral shower profile for 10 GeV pions incident on projective calorimeter towers.

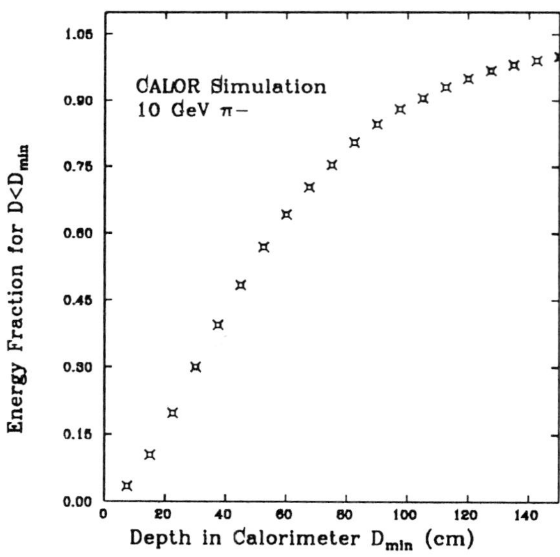

Figure 12: Fraction of energy deposited in scintillator by neutrons as a function of depth in a lead calorimeter with unit cell 0.56 : 0.25 Pb : Scintillator.

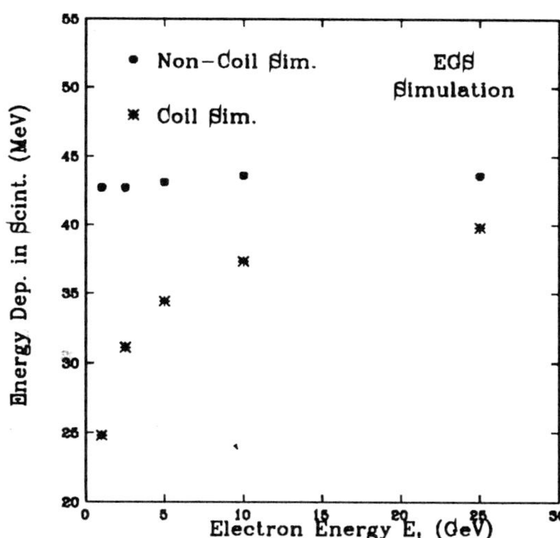

Figure 13: Calorimeter response at 25° incident angle as a function of energy for the case of a perfect calorimeter and one including material to simulate a solenoid coil.

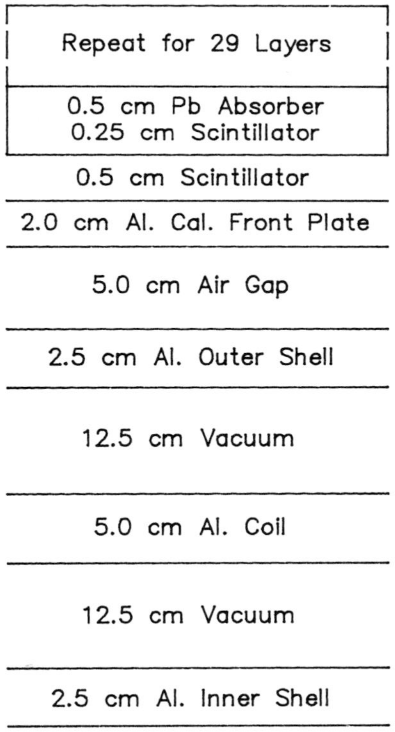

Repeat for 29 Layers
0.5 cm Pb Absorber 0.25 cm Scintillator
0.5 cm Scintillator
2.0 cm Al. Cal. Front Plate
5.0 cm Air Gap
2.5 cm Al. Outer Shell
12.5 cm Vacuum
5.0 cm Al. Coil
12.5 cm Vacuum
2.5 cm Al. Inner Shell

Figure 14: Material distribution for coil simulation studies on calorimeter performance for configuration using double sampling in the first scintillator layer.

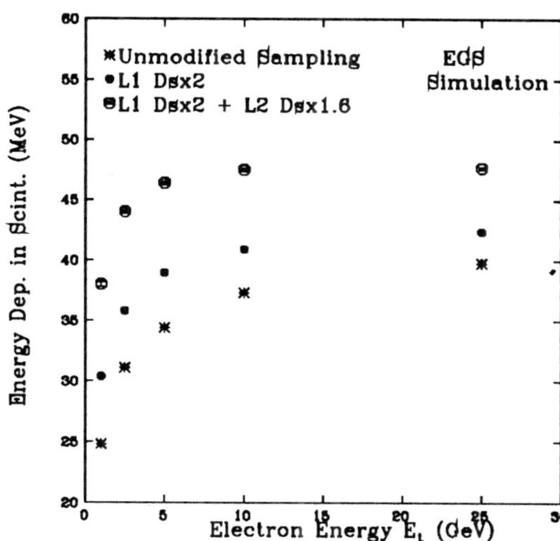

Figure 15: Calorimeter response at 25° incident angle with coil simulation and modified sampling distributions as described in text.

predicted (with corresponding detrimental effects being seen in the resolution). Studies are now in progress to utilize modification of the scintillator thickness and sampling distribution to reduce this effect. These are shown in Figure 15 for the unmodified sampling configuration and two initial candidate geometries:

a) L1 Ds × 2, in which the first scintillator layer thickness is doubled to 0.5 cm.

b) L1 Ds × 2 + L2 Ds × 1.6, in which in addition to doubling the thickness of the first scintillator layer, the thickness of the second layer is increased by 1.6.

Considerable improvement is observed in the linearity and in the energy resolution, which is shown in Figure 16. However, much work remains to be done to optimize the overall geometry for electron energy measurement within the constraint of retaining e/h close to 1.

Conclusions

We have shown that a high-resolution compensating calorimeter can be built using either lead or depleted uranium as the principal absorber. Studies to date indicate that this is not possible, even with some composite tuning, in an iron-based calorimeter. Energy release from long-lived shower products contributes a significant fraction of the shower energy in a uranium calorimeter for times up to 500 nsec after the collision. We judge, for this reason, that lead should be used as the principal absorber.

Detailed mechanical design studies of a lead-composite calorimeter have begun. Radial and pseudo-projective barrel-endcap transitions in which the sup-

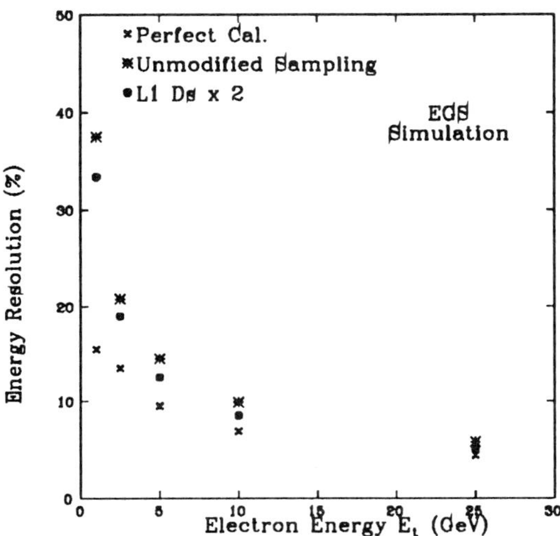

Figure 16: Calorimeter resolution at 25° incident angle for perfect, coil simulation and modified sampling distributions as described in the text.

port structure for the barrel calorimeter and coil is located have been evaluated. Both are viable. However, it is our opinion that the more conservative design to use is that of a radial cut. The casting technology anticipated for the fabrication of this calorimeter can be anticipated to yield non-uniform absorber thickness due to shrinkage. Our studies show that this predominantly affects the constant term in the resolution of the electromagnetic calorimeter. This can be held to $< 1\%$ by maintaining an absorber thickness tolerance of better than $\pm 8\%$, which is far worse than the expected tolerance. The detector design requires magnetic flux return and, therefore, an iron calorimeter segment at the outer radius (to minimize the overall radius). We have studied the longitudinal energy deposition from neutrons in a lead calorimeter system and conclude that this section should have a depth of at least 6.9 absorption lengths to yield an acceptable mismeasurement of 100 GeV pions. Much more work, however, is needed to evaluate the effect of the non-uniform sampling distribution on the constant term in the resolution for hadron showers. Other issues requiring further study include material alloys and internal support structure.

Finally, the coil of the solenoid required in the detector is a necessary evil. As is now well-known, "massless gaps" can compensate for energy loss and smearing in the coil and thereby improve linearity and resolution. Studies are in progress and show severe degradation in performance at the outer ends of the barrel calorimeter. Much more work is needed in this area to properly compensate this degradation.

References

[1] W.R. Nelson, H. Hirayama and D.W.O. Rogers, *EGS4 Code System*, SLAC-265, (1985).

[2] T.W. Armstrong and K.C. Chandler, *HETC—A High Energy Transport Code*, Nucl. Sci. and Eng. **49**, 110 (1972).

[3] *Monte Carlo High Energy Nucleon-Meson Transport Code*, ORNL RSIC Computer Code Collection, CCC-178.

[4] R. Brun, F. Brayant, M. Maire, A.C. McPherson and P. Zanarini, *GEANT3 Users Guide*, CERN Report DD/EE/84-1 (1986).

[5] H. Abrahamowicz et al., *The Response and Resolution of an Iron-Scintillator Calorimeter for Hadronic and Electromagnetic Showers Between 10 GeV and 140 GeV*, Nucl. Inst. Meth. **180**, 429 (1980).

[6] E. Bernadi et al., Nucl. Inst. Meth. **A262**, 229 (1987).

[7] E. Ros and T. Tsurugai, ZEUS Note 88-086 (1988).

[8] W.J. Womersley et al., Nucl. Inst. Meth. **A267**, 49 (1988).

PROPERTIES OF PHOTODETECTORS FOR SSC SCINTILLATING CALORIMETERS

D. R. Winn, K. Segall, D. Wall
Dept. of Physics, Fairfield University, Fairfield CT

P. Cushman, D. Grigoriev
Dept. of Physics, Yale University, New Haven, CT

Abstract

Properties of photodetectors for SSC scintillation calorimetry are reviewed. Data on the hysteresis of photomultiplier output pulses and a novel optical pulse train test device for SSC photodetectors are presented, together with implications for SSC calorimeters. Photodetector development issues are discussed.

Introduction

SSC scintillation calorimeters place unprecedented demands on the ability of optical transducers to convert an optical pulse train into electrical signals with sufficient fidelity and gain to extract physics signals of interest. For example, an electromagnetic calorimeter might be designed to detect both H->γγ with high resolution, and Z'->ee with a mass of 2 TeV, at high luminosity. A hadron calorimeter needs jet energy response linear up to 10 TeV for Compositeness searches. The photodetector must therefore be linear from a deposition of ~20-100 MeV (~10-20% of a MIP for physics & calibration) up to 1-2 TeV, a dynamic range of 10^4 -10^5, and accurate in pulse amplitude to 0.1- 1%. For a least count of a few hundred photoelectrons (20% quantum efficiency), the gain must be sufficiently high (>1,000) to overcome electronic noise. A photodetector might need a dynamic range of peak output voltages from 20 μV to 2 V on 50 Ohms.

The amplitude properties necessary for a photodetector are complicated by the temporal properties needed at the SSC. At the highest luminosities, in the forward region, cell occupancy may approach 10-100% for depositions at the level of 1-0.1 GeV. Thus the amplitude conditions must be satisfied for arbitrary activity in the photodetector within ~16 ns of the pulse to be measured. The change in gain, risetime and afterpulsing of a typical photomultiplier (PMT) with both overall rate, avarage anode current, and with previous pulse amplitude is well-known[1]. A typical small-diameter (<5 cm) PMT's gain is lowered by a factor of ~2(10) at repetition rates of ~10(50) MHz[2] with high current active bases[3]. The typical gain variation with average anode current is ~10% for a modest average anode current change between 10-100 μA[4]. In the ~100 PC peak pulse regime (modest pulses for the SSC), many high gain PMT show a recovery time to stable gain of many tens of seconds when cycling from a 10 kHz pulse train to a 100 Hz pulse train, with a 50% gain shift. Another possible serious effect in some triggering applications are risetime effects with anode current.

While we note that wide variations exist between PMT types, it is nevertheless useful to study in detail the pulse history effects on the performance of a few fast photomultipliers which might be used in the SSC. Unfortunately, much of the data to date has been in terms of averages, such as a steady rate or averaged anode current. Furthermore, much of the rate dependent data is base-dependent in a way that is not separable from intrinsic PMT performance. We have therefore begun a study of pulse-to-pulse variation of PMT properties, using a novel optical delay line pulse train and waveform digitization.

Optical Pulse Train Generator

The basic idea for a low-cost optical pulse train generator capable of producing pulses a few ns to hundreds of ns apart to simulate SSC signals on photodetectors is sketched in Figure 1.

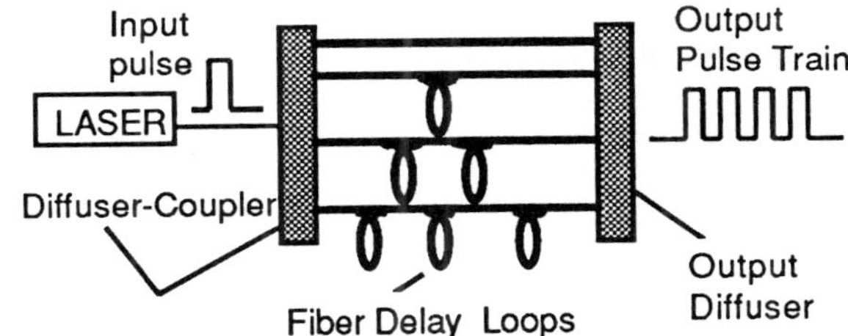

Figure 1. Schematic of an Optical Pulse Train Generator

A laser with a short pulse is used to drive a set of optical fibers of different lengths through a diffuser or dye shifter cell. The fiber outputs are then recombined to illuminate a photodetector. The result is an optical pulse train with the number of pulses in the train equal to the number of fibers, the pulse width roughly equal to the width of the laser or dye pulse (up to modal dispersion in the fibers), but with the pulse-pulse separation arbitrarily variable by adjusting the fiber lengths, and the intensity of each individual pulse variable with the fiber coupling to the laser, through a variety of attenuation methods (coupling angle, neutral density filters, index matchers, fiber numerical aperture, etc). The data acquisition is triggered by a small fast silicon photodiode intercepting a fraction of the laser pulse. Figure 2 shows a photograph of the pulse train generator with sealed N_2 laser totally self-contained in a dark box (the laser is on the lower left and a reel of fiber delays on the upper left). The fibers are first adjusted individually.When the fibers are then recombined into the pulse train, it is possible to study the effect of one pulse on the properties of another pulse. For example, the effect of a small leading pulse on a large pulse or vice-versa can be studied as a function of delay, or the effect of a train of several pulses can be studied. One virtue of this pulser is that intrinsic high rep-rate tube properties can be studied at low average base current because it operates in burst-mode. A fuller simulation of the SSC optical calorimeter pulses could be obained with a 62 MHz ring laser modulated according to MC data.

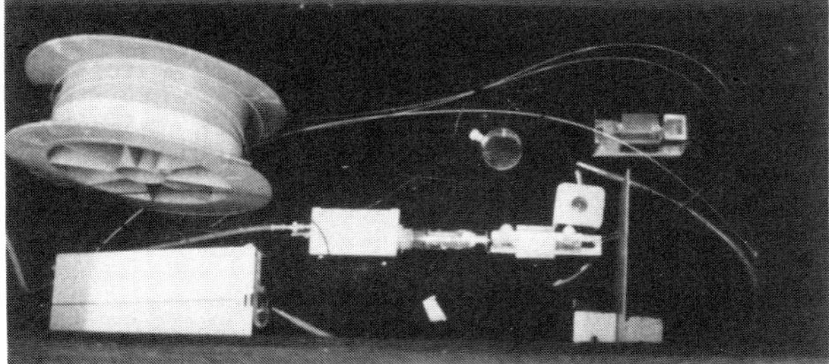

Figure 2. Photograph of a compact optical pulser.

PMT Pulse Train Test Results

The data shown here are mainly on double pulse behavior. Figure 3 shows the effect of applying 2 ~identical optical pulses separated by 20 ns to a high gain 1.25" PMT with a coaxial output (0.7 ns ristime)[5], with the pulses digitized by a 1.3 GHz digitizer[6]. The fibers were trimmed using a TDC, and the couplings were adjusted to give identical pulse peak at the 100 photoelectron level on each fiber. The gain shifts by ~ 15% and the time of the peak by ~2 ns. Pulse trains with 3-4 pulses show progressively worse pulse fidelity.

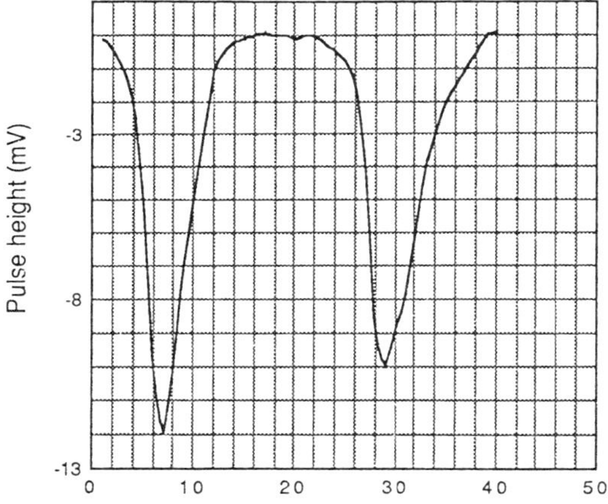

Figure 3. Double optical pulse fast PMT response.

Figure 4 shows a summary of typical non-linear temporal behavior for double pulse separations below ~50-60 ns by plotting the measured second pulse peak time from the digital oscilloscope vs the delay from the trimmed fibers. The points are connected by line segments to guide the eye, and the error bars are estimated from the trimming and measuring process. On the other end of the time/energy scale, Figure 5 shows the time shift of a second pulse at a fixed 350 ns delay with an amplitude equivalent to about 1-2 GeV equivalent, when varying the amplitude of a large leading pulse from about 1 GeV up to 1 TeV equivalent (assume 1000 p.e./GeV).

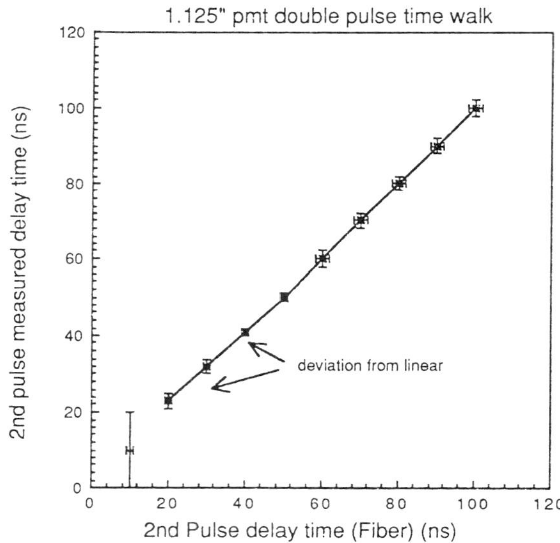

Figure 4. Measured pulse delay vs Fiber-timed Delay

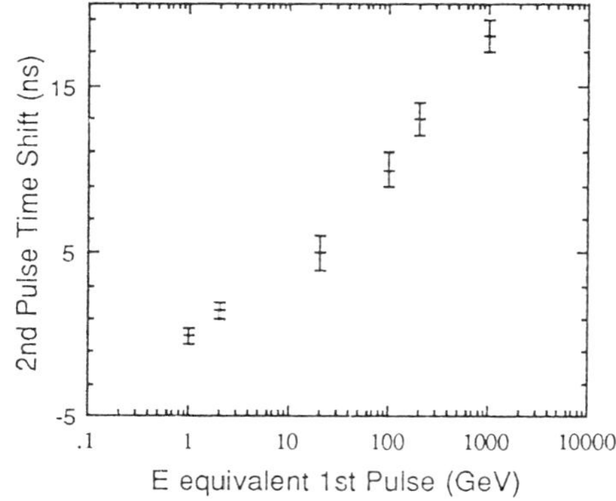

Figure 5. Time shift of 2nd pulse vs Equivalent Energy of the 1st pulse for a fixed 350 ns delay.

These initial hysteresis results point out the difficulties for some high rate SSC calorimeter applications, but are somewhat preliminary. The PMT's tested so far were chosen for best risetime and high gain, with focussed dynode structures to eliminate transit time spread, and not for the best high pulse currents. Furthermore, the photocathodes were bi-alkali, a higher resistivity material which affects the ability of the photocathode to recover. Also, the bases were standard resistive bases without Zener or active components.

SSC Photodetector Development

Secondary Emission Gain Photomultipliers

Much development work is forseen that could improve linearity and pulse recovery for standard PMT. PMT for SSC calorimeters may not need pulse timing jitter or risetime better than a few ns, and therefore non-focused dynode structures with low inter-dynode capacitance, such as meshes or blinds, may be used, lowering space charge effects at the anode, potentially raising the linearity intrinsic to the PMT[7]. The cathode resistivities should be as low as possible for pulse-mode operation[8], either by photocathode material (S-11 is about 100 times less resistive than bialkali cathodes[9]) or by subtrate materials, such as cathode window embedded meshes. Multiple anodes could be a method to lower capacitance and charging effects. Lower gains enable higher dynamic range, but at a premium on the electronics. A dual system of a low gain photo-diode or -triode together with a high gain PMT solves the dynamic range problem but at cost/complexity. The non-linearities in the dynode voltage supply extraneous to the tube itself also limit PMT performance. PMT gain is a very sensitive function of voltage (exponential in or a high power of the tube voltage), leading to supply voltage stability requirements in the 10-100 ppm range to reach a signal accuracy of 0.1-1% for high precision calorimetry, and is difficult to maintain under pulse operation. Lastly, the large vacuum electron currents inPMT lead to ion-feed-back after-pulsing

EBS Gain Hybrid Phototubes

A potential solution to some problems forseen in dynode PMT for precision SSC calorimetry is to replace the dynodes by a semiconductor diode, the top electrode of which serves as the phototube anode. Photoelectrons are accelerated to high voltage, focussed onto the diode, creating ion pairs, which when collected serve as the gain mechanism (electron-bombarded semiconductors or EBS)[10,11]. Figure 6 shows a schematic of an EBS/Hybrid Tube developed in the mid-1960's[12,13,14]. With a Si diode, gains are about 2,000 per 10 KV lost; up to ~40 KV at 0.1 μA of photocathode current has been demonstrated. Proposals and tests of these tubes have recently been made by R. DeSalvo[15].

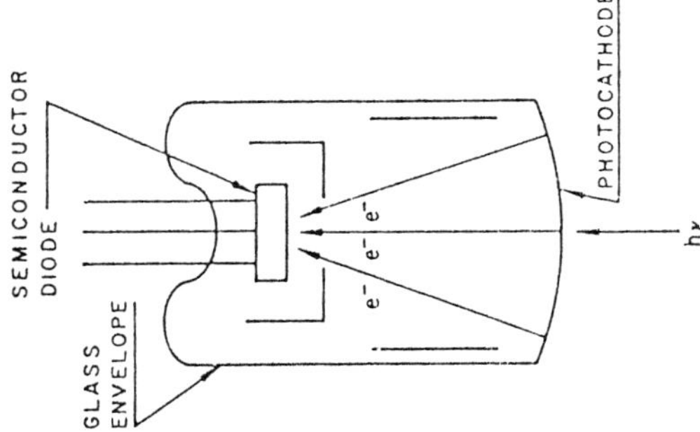

Figure 6: Schematic of an EBS Hybrid Tube (ref. 12)

The virtues of EBST are :

(1) High linearity dynamic range - up to 10 A peak pulse have been drawn from the diode of these tubes;

(2) Gain stability because the gain is linear in the voltage. For a tube operating at 20 KV, a ~5 V ripple corresponds to ~1 electron of ripple in the output;

(3) Highly compact;

(4) Low high voltage power.

As an example of the latter, a 2 TeV equivalent pulse at 1,000 photoelectrons per GeV would correspond to 0.3 pC in the pulse drawn from a 25 KV HV supply on an EBST, whereas a similar gain of 5,000 (~6 stage) PMT would draw 1,500 pC from a 1 KV supply. A 1 GeV tower deposition on every beam crossing would draw only 0.01 μA from an EBST 25 KV supply, but 5 mA from a 1 KV PMT supply.

The potential problems with EBS hybrid tubes for calorimetry are: (1) the fabrication process necessary to "passivate" the surface of the diode to protect it from the cesiation process; (2) diode aging effects. Recent CERN tests on prototypes made by DEP indicate that these problems can be solved[15]. The EBST is potentially a cost-competitive alternative to PMT for calorimeter applications. Furthermore, EBST techniques can easily be adapted to fiber pixel detectors by substituting an array of diode anodes for a single anode.

Conclusions and Future Directions

A novel optical pulse train generator has been constructed for SSC photodetector testing, demonstrating the difficulties of SSC precision calorimetry at high rates. Improvements in standard PMT for high linearity and pulse recovery appear to be possible and essential, and EBS Hybrid Tube development is promising. A more exact simulation of the optical pulses generated in SSC calorimeters could be developed with laser technology for photodetector tests, and on-line calibration in experiments.

---

[1] R. Engstrom, RCA PMT Handbook, 44-53 (1980)
[2] C. Ohmori et al., Test of Phototube Bases for High count Rates", NIM A256, 361 (1987)
[3] C. R. Kerns, A High Rate Phototube Base, IEEE TNS V NS-24, 353 (1977)
[4] M. De Vincenzi et al., Exptl. Study of Non-linear effects on photomultiplier Gain, NIM 255, 104 (1984)
[5] Hammamatsu PMT R2083.
[6] The overall pulses were normalized using a small diode intercepting the laser beam, and 32 pulses were dithered and averaged in order to capture the rise-times. The laser pulse-pulse variation was about 5%. The graph was drawn with an interpolater connecting the points.
[7] For a complete discussion, see references 1 and 9
[8] B. R. C. Garfield et al., Improvements to Photocathodes for Pulse Operation, Adv. in Electronics and Electron Physics V 20, 375 (1964) Academic Press
[9] J. Boutot et al, Recents Trends in Photomultipliers, Adv. in Electronics and electron Physics V 60, 223 (1983)
[10] D. J. Bates er al., "Electron Bombarded Semiconductor Devices", in Adv. in Electronics and Electron Physics", V 44,221 (1977), Academic Press
[11] J. Abraham et al. "Application of Solid State Elements to Photoemissive Devices", in Adv. in Electronics and Electron Physics", V 22, 671 (1965), AP
[12] R. Kalibjian, "A phototube using a semiconductor diode as multiplier element",IEEE TNS NS-13,54 (1966)
[13] J Fertin et al., Reverse epitaxial Si diode for hybrid photomultiplier tubes, IEEE TNS NS-13(3), 179 (1966)
[14] L. Wolfgang et al, "Hybrid Photomultiplier tubes with solid state elements",IEEE TNS NS-13(3), 54 (1966)
[15] R. DeSalvo, (1) "Hybrid Photodiode Tube", Cornell Preprint CLNS 87-92, *and* (2) CERN LAA/HC/89-004, *and* presentation/communications @ Snowmass 1990

PRESHOWER DETECTORS FOR ELECTRON IDENTIFICATION

P. Cushman, V. Singh

Yale University, New Haven

P. Melese, R. Rusack

The Rockefeller University, New York

Abstract

We have built a combination scintillating fiber tracker/preshower detector to study electron and pion separation using differences in the early development of a shower. Data were collected in a test beam at Fermilab and results of our analysis is presented. In addition for some part of the data collection we placed a high field magnet upstream of our detector to study how synchrotron radiation might be used to further improve the separation.

Design of the Prototype Preshower Detector

We have built and tested a fine-grained scintillating fiber preshower detector designed to study the early development of electron showers and the feasibility of tagging electrons by observing the synchrotron radiation produced as the electrons pass through a high magnetic field. A schematic of the preshower detector is shown in Fig [1]. The 9600 fibers (nominally 500 μm dia) were arranged in 8 sampling layers (SL) separated by 1.4 mm lead sheets, making a total thickness of 1.75 radiation lengths (R.L.). Each SL was formed from 6 fiber ribbons arranged in pairs with each pair placed at one of three stereo angles, 0° and $\pm 15^o$. The two ribbons in each stereo projection were offset from each other by the width of half a fiber with a pitch of 550 μm.

The fibers were drawn by Bicron from a preform supplied by Alan Bross at FNAL. The core was 420 μm in diameter with a 20 μm thick cladding and an additional 15 μm of white reflective paint. The fiber ribbons were made in pairs by winding the fiber onto a grooved drum with thin adhesive tape placed on it to form the first ribbon. The drum had a circumference of 1 m and was mounted on a lathe which advanced 550 μm each revolution. The same lathe was used to cut the groove in the drum. Once the first ribbon was wound the second ribbon of the pair was wound directly on top of the grooves formed by the first, and the set was then painted with ULTRA-SEAL glue. *

The fibers were read out at both ends. At one end the light was amplified by a chain of two image intensifiers (II), manufactured by DEP, providing a demagnification of 80/18 and an optical gain of 10^4. The output of the second II was divided into four quadrants and the image from each quadrant was transferred to a CCD (Thompson TH7852) by a fiber optic light guide manufactured by Collimated Holes. The electro-optic chain is illustrated in Fig. (2). Two such chains were used to read out 4 SL's each from our detector. The CCD pixel information was processed by Fastbus digitizers which were built following the UA2 design [1] and provided a factor of 10^4 data sparsification. The other end of the fibers were bundled according to SL and were connected to 8 PMT's read out with CAMAC ADC's. Selected fibers were not connected to the PMT and went instead to a LED pulser system to provide the fiducials necessary to map the fibers to the pixels of the CCD ($\sim$ 10 pixels per fiber).

Test Setup

The detector was placed in a test beam at FNAL and data were collected during April 1990 with incident electrons and charged pions. The setup in the test beam is shown in Fig [3]. The particle beam passed through a magnet placed directly upstream of the detector. The maximum magnetic field at which the magnet was operated was 2.5 T. Electrons bending in the magnetic field emitted synchrotron radiation (SR) which was observed in our detector. Two 2.5 cm^2 scintillator paddles placed in front of the preshower detector and a block of lead glass positioned behind it provided the trigger. Data for roughly 25,000 electrons were collected at each energy setting: 10, 25, 50, 100, 150, and 200 GeV, and for magnetic fields of B=0.0, 2.0, and 2.5 Tesla. The

*ETI Fields Landing CA 95537

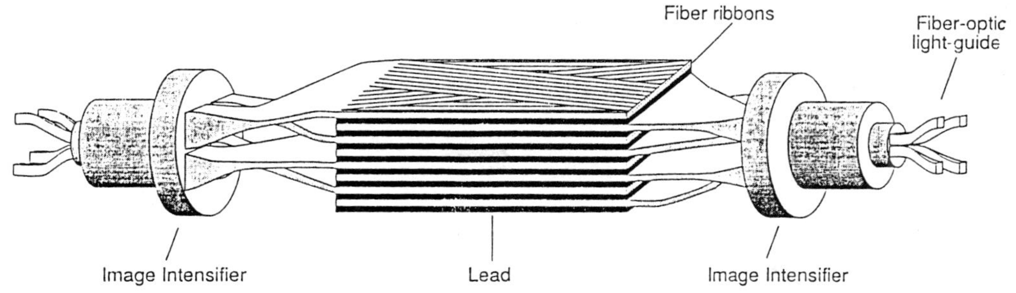

Figure 1: Schematic of the Prototype Imaging preshower Detector.

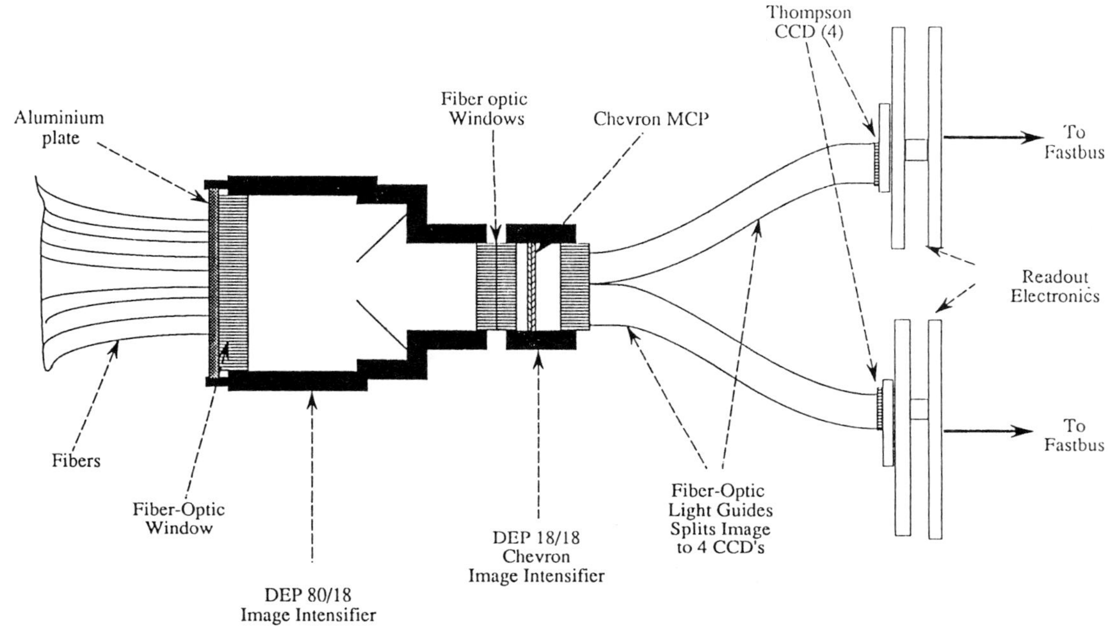

Figure 2: Electro-optic readout of the detector.

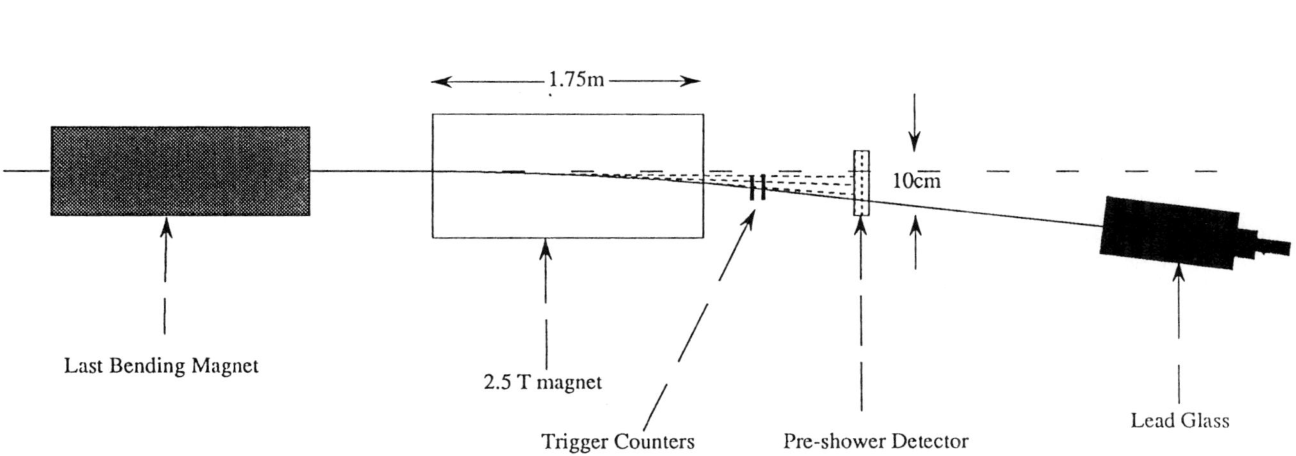

Figure 3: The Test Beam Setup

test beam was reconfigured to deliver pions for a short while, during which time we collected 20,000 π's.

Reconstruction and Tracking Efficiency

Tracks or showers are identified off line in each of the three views individually and then combined to form a 3-dimensional trajectory. Each fiber ribbon (one layer of fibers) had an efficiency of 39% for detecting minimum ionizing particles. This is partly due to geometrical acceptance resulting from the thickness of the cladding and the cracks between the fibers and partly due to light detection inefficiencies. With our Monte Carlo calculation we have determined that that geometrical effects are responsible for a 27% decrease in detection efficiency. If 4 SL's (3 SL's) are required to find a track, then 71% (88.1%) of the pion tracks are reconstructed. Electron "tracks" are found more than 99% of the time. The resolution per view is determined by removing SL 4 from the track finding algorithm and then plotting the residual between the energy weighted mean of SL 4 and position of the reconstructed track. The sigma of this distribution is presently 200 μm, 60 μm of which results from uncertainties in the fiber alignment according to our Monte Carlo calculations.

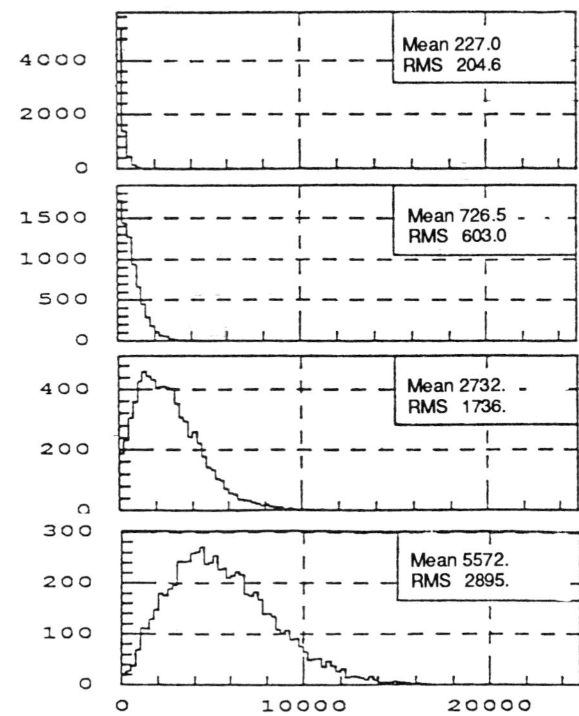

Figure 4: Energy deposited by a 100 GeV electron in superlayers 2,4,6 and 8 (.25,.75,1.25, and 1.75 RL's of Pb) starting from the top. The energy deposited by a pion in all the superlayers is similar to the plot for an electron in superlayer 2.

Shower Energy and Width

In each view, a cone is defined about the trajectory and the energy within its boundaries is summed to form the shower energy. The cone's opening angle is increased until a further angle increment results in less than a 5% change in the shower energy. In Fig [4] the longitudinal development of a 100 GeV electron shower as it passes through 1.75 R.L.'s of lead is shown as a series of snapshots of the energy in every other SL. Although the π's and e's resemble each other early in the detector, by the 8th SL the response is quite different. We define the average energy deposited by the π in our detector (1600 ADC counts overall, 200 counts in SL8) as 1 minimum ionizing particle (MIP). The mean of the total energy deposited in the detector in MIP's and the RMS fluctuations around the mean are plotted in Fig [5] as a function of the incident energy of the particle. There is a nonlinear energy dependence without much analyzing power due to the large shower fluctuations present at this depth. The widths of the showers, on the other hand, are not sensitive to the incident particle energy. The standard deviation of the energy weighted shower width in SL 8 has a mean value of 1.15 mm. It is plotted in Fig [6] for different incident energies and particle type.

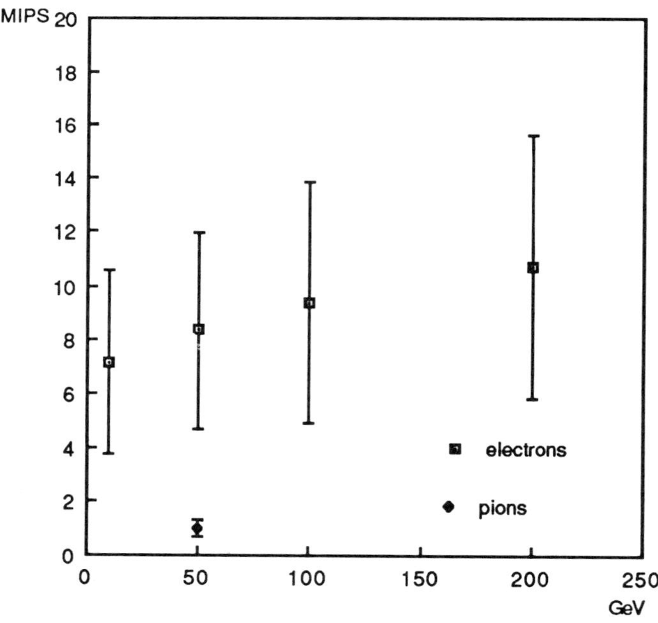

Figure 5: Mean of the total energy deposited in the detector. The error bars represent the RMS fluctuations about the mean.

Electron Pion Discrimination

Cuts on both the energy and width can be used to obtain good e/π discrimination. They are highly correlated with each other and, in general, the energy cuts

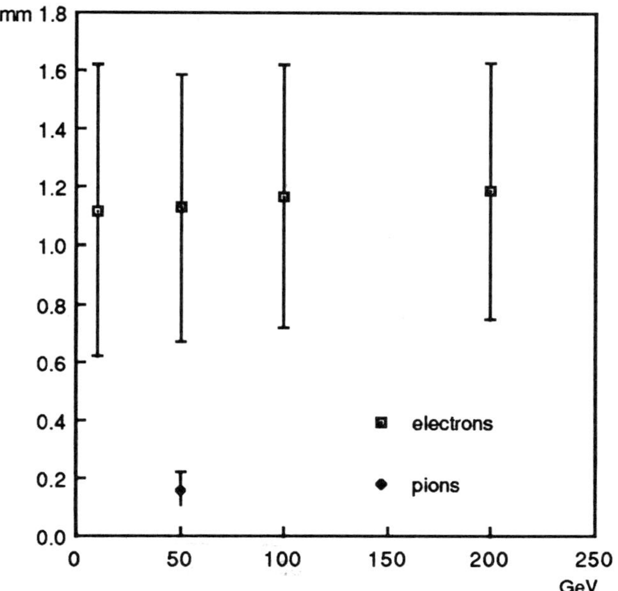

Figure 6: Shower widths in SL 8 versus incident energy. The errors bars represent the RMS fluctuations about the mean.

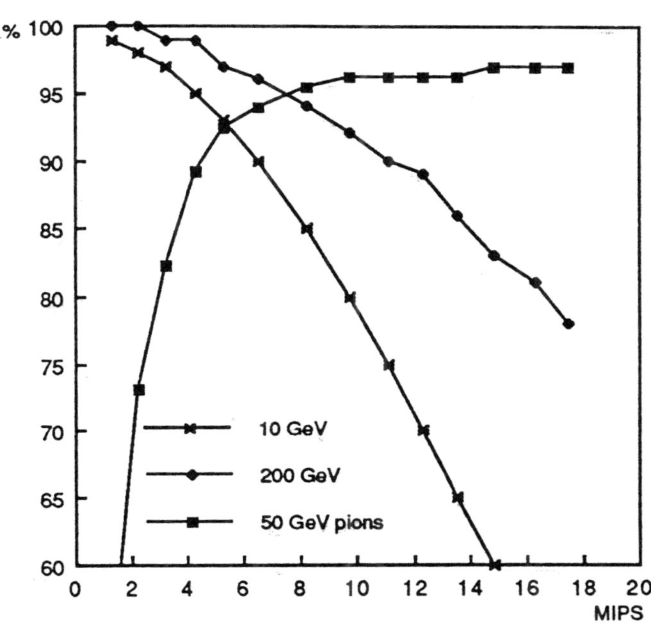

Figure 7: Electron efficiency and pion rejection efficiency using energy cuts in superlayer 8

do better than width cuts. Figures [7] and [8] show the electron efficiency and π rejection obtained for 50 GeV π's, 10 and 200 GeV e's as a function of the energy deposited (ESL8) and the width of the shower (WSL8) in the last SL. A cut of ESL8 > 8.5 MIP's gives a π rejection of 96% with 93% efficiency for 100-200 GeV electrons (the efficiency begins to drop off for 50 GeV electrons). A cut of WSL8 > 0.56 mm gives only 93% π rejection and 91% electron efficiency, but is less sensitive to the electron energy. The systematic error on these numbers is less than 1%. The greatest contribution to this error comes from the misidentification of pions in the lead glass which was estimated with our Monte Carlo to be less than 0.5%.

For all these efficiency curves, it is important to note that recent results from the SPACAL collaboration [2] have shown that shower spreading cuts on $\pi's$ in the calorimeter are not highly correlated with pulse height cuts in the preconverter. Thus 93% pion rejection in their preshower detector alone gives $\sim 2\text{x}10^{-4}$ e/π separation when used in combination with a spaghetti calorimeter.

We have found that another variable which can be used to descriminate between e's and π's is the total number of hits in a shower. This is due to the fine-grained nature of our preshower detector and suggests that fine segmentation may be able to take the place of pulse height information. Efficiency curves for cuts on the total number of hits in a shower are shown in Fig [9]. Making a cut on the number of hits in SL 8 alone degrades the e/π separation only slightly. The number of hits is highly correlated with the total energy deposited in the preshower detector as can be seen in Fig [10]. This implies that ignoring the pulse height

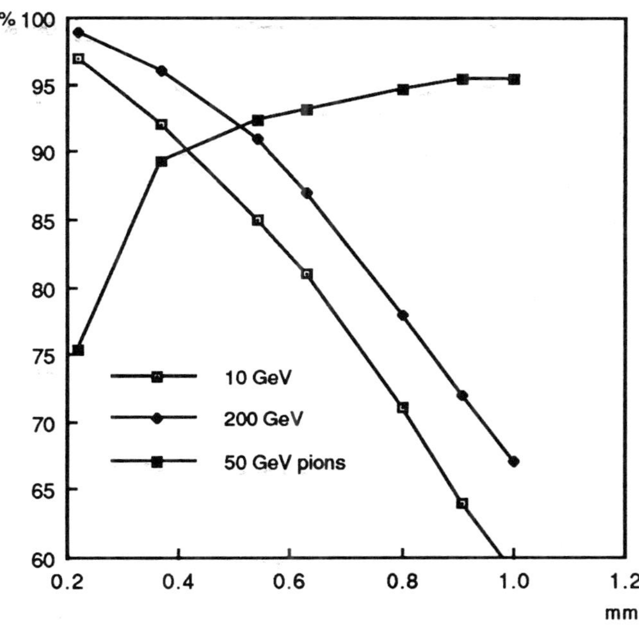

Figure 8: Electron efficiency and pion rejection efficiency using shower width cuts in superlayer 8

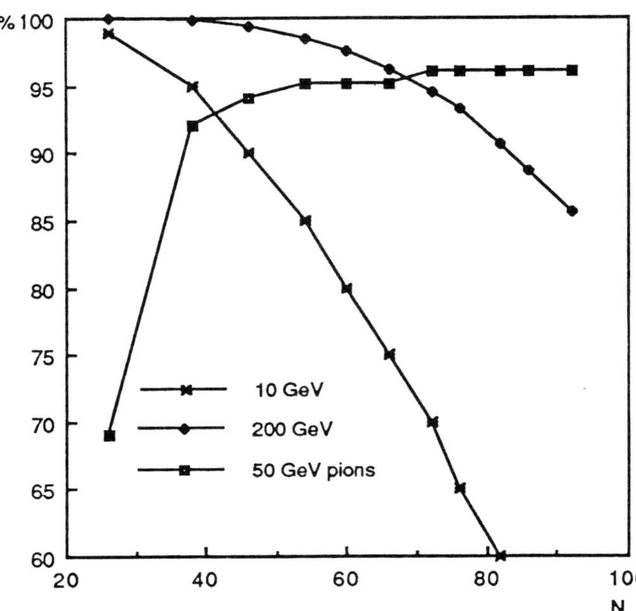

Figure 9: Electron efficiency and pion rejection efficiency using the total number of hits in the shower

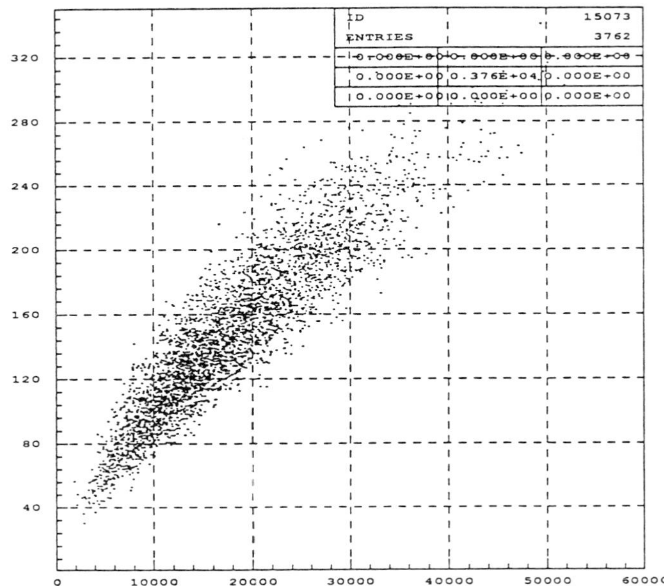

Figure 10: Number of hits in a shower versus the shower energy.

information would result in little or no loss of energy information. This is important if one is planning to sum the energy in the preshower detector with that deposited in the calorimeter, in order to maintain good energy resolution.

Synchrotron Radiation Tagging

In a magnetic detector, an imaging preradiator can use the synchrotron radiation (SR) produced by electrons as they bend in the magnetic field of the solenoid. The SR is emitted tangentially to and in the same plane as the electron orbit. Thus, it is deposited in a narrow strip in the bend plane of the magnet. The length of the strip is proportional to the bend of the electron, varying inversely with energy (see Table 1). If the energy of a reasonable fraction of the SR photons is above the pair production threshold then they can be detected in a lead/scintillator sandwich. The characteristic pattern allows easy recognition by a preshower detector with position resolution of the order of a millimeter. This method of identifying electrons is uncorrelated with shower development characteristics in either a calorimeter or a preshower detector.

The SR spectrum can be characterized by the total emitted energy I and a critical photon energy $\hbar\omega_c$ above which the intensity becomes negligible [3, 4, 5]. For electrons: $I = 1.27E^2B^2L$ [KeV] and $\hbar\omega_c = 1.33E^2B$ [KeV], where E is the electron energy in GeV, B is in Tesla and the path length, L, is in meters. Since the mean photon energy is $\hbar\omega_c/4$, the average number of photons can be estimated as $N_{av} = 3.8BL$ [6]. Values for $\hbar\omega_c$ and strip length, δ, are shown in Table 1 for

Table 1: Properties of Synchrotron Radiation

Field	E_e (GeV)	50	100	150	200
5 T - 1 m	$\hbar\omega_c$ (MeV)	16.6	66.5	150	266
(N_{av}=18)	δ (mm)	24	12	8	6
2.5 T - 1.8 m	$\hbar\omega_c$ (MeV)	8	33	75	133
(N_{av}=17)	δ (mm)	58	29	20	15

electron energies between 50 and 250 GeV for our test beam configuration and a possible high field compact solenoid with roughly the same number of SR γ's, but twice the field.

Due to the low energy of the electrons produced in our detector by the SR and the multiple scattering they undergo while traversing the lead radiator it is not possible to track them between several layers of our detector. The most sensitive parameter turns out to be the extra energy to the bend side of the shower (we assume that a magnetic detector will know which side to look for the SR γ's). The E(left) parameter which is plotted in Fig [11] and [12] is the energy deposited in a band 60 fibers wide (3 cm) to the left of the shower definition cone, but not including the hits inside the shower. In Fig [11] this parameter is plotted for pions, 100 GeV and 200 GeV electrons passing through a 2.5T field and 200 GeV electrons with the magnet off. The E(left) distribution for magnet-off electrons is wider than that of pions because there is some additional background out-

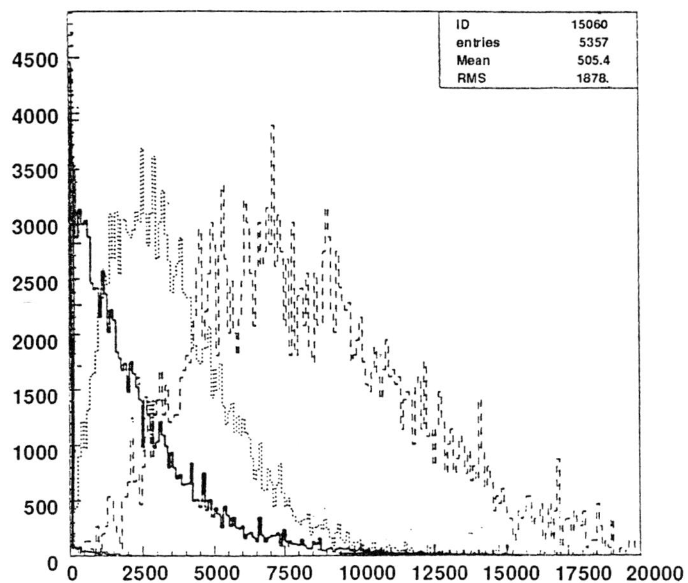

Figure 11: Energy deposited to the left of the shower for 100 GeV (dotted) and 200 Gev (dashed) electrons with B=2.5T Also plotted are $\pi's$ (narrow solid) and 200 GeV electrons with the magnet off (wider solid).

side the shower cone (on both sides) for electron showers that is not present in the pions. In Fig [12] the electron efficiency and π rejection is plotted with respect to a cut on E(left). For E(left) $>$ 1.4 MIP's, we achieve 96% π rejection with 98% efficiency for 200 GeV electrons. Detection of 100 GeV electrons is only 68% efficient here using the E(left) criterion alone, since they have less energetic SR γ's. However, since we are using the SR to enhance the purity of electron identification for higher energy electrons, we would in any case want to combine the E(left) cut with a general shower or width cut for all energy electrons. Such a double cut of E(left) $>$ 1.4 MIP's + E(total) $>$ 2.9 MIP's gives 95% and 100% efficiency for 100 and 200 GeV electrons with 96% pion rejection.

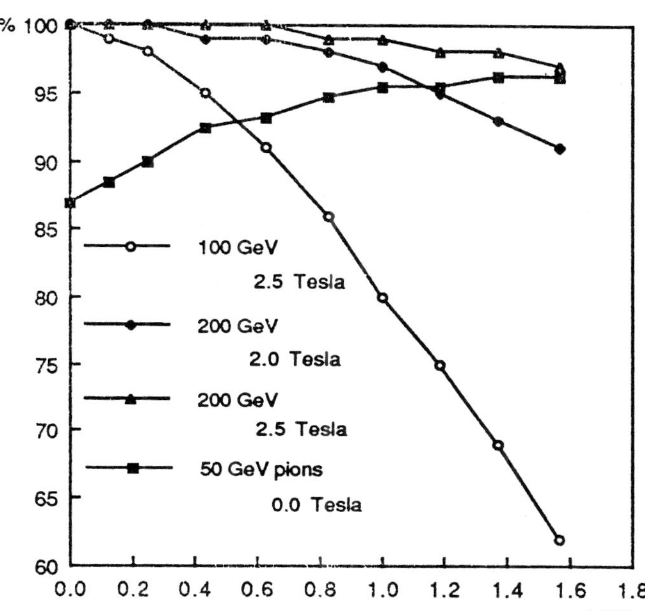

Figure 12: Electron efficiency and pion rejection efficiency as a function of the E(left) cut.

References

[1] R. Ansorge *et al.*, "Performance of a Scintillating Fiber Detector in the UA2 Upgrade," NIM **A265**, 33, (1988)

[2] D. Acosta *et al.*, "Results of Prototype Studies for a Spaghetti Calorimeter," CERN EP/90-37, 1990.

[3] J. D. Jackson, *Classical Electrodynamics.* New York: Wiley, 1975

[4] B. Merkel *et al.*, "Transition and Synchrotron Radiation Produced by Electrons and Particle Discrimination," NIM **138**, 625, (1976).

[5] H. Harada *et al.*, "Electron Identification using Synchrotron Radiation",NIM **A265**, 141, (1988).

[6] J. S. Dworkin *et al.*, "Electron Identification Using a Synchrotron Radiation Detector", NIM **A247**, 412, (1986).

CAST LEAD-EUTECTIC SOLID AND LIQUID SCINTILLATING FIBER SHOWER CALORIMETERS

W. Worstell, J. Miller, L. Roberts, L. Sulak
Department of Physics, Boston University

D. R. Winn
Department of Physics, Fairfield University

P. Cushman, S. Dhawan, V.W. Hughes
Department of Physics, Yale University

Abstract

Compact cast lead-eutectic scintillating fiber and scintillating liquid capillary electromagnetic calorimeter modules have been constructed. Construction methods are described and results from lab bench tests and beam test results are presented, with energy resolutions about 15-19%/√E. These calorimeters are prototypes for a muon g-2 experiment, and precision compensated calorimetry at the SSC.

Introduction

Scintillating fiber calorimeters show much promise for calorimetry in medium and high energy physics experiments [1-4]. The longitudinal fiber technique is a particularly efficient way to obtain a relatively homogeneous mixture of absorber and scintillator, avoiding dead regions from readout apparatus and offering the potential for reducing sampling fluctuations to a minimum.The technique has demonstrated the ability to be compensated (e/h ~ 1.0), with energy resolution better than 30%/√E+1% for hadrons. Furthermore, by increasing the density of fibers[3], or by using sufficiently fine (0.5 mm diameter) fibers maintaining a compensated volumetric ratio[4], an electromagnetic resolution of 7%-8%/√E may be obtained.

Among several draw-backs with the technique, however, is the method to embed the fibers in a heavy absorber matrix, usually lead, in an efficient, precise and cost-effective manner. Typically (lead) absorber sheets are grooved and the scintillating fibers are inserted or laid into the grooves, a labor-intensive and slow technique.We have recently developed a casting technique whereby the fibers are stretched in a mold and an absorber is cast in place around the fibers.

Technique

To date, this technique has been demonstrated on small prototypes (electromagnetic calorimeters), and shows promise for large-scale production, and for larger sizes. We make use of low-melting point Pb-eutectic alloys (the so-called "Wood's metals"). Our eutectic uses bismuth,tin, and cadmium alloyed with lead to obtain a low-melting point absorber with a density of 9.5 g/cc [5]. The melting temperature of 70° C does not soften the plastic fibers or affect the cladding during a casting process.(~70° C), since fibers typically soften at a 10-20° C higher temperaturess.The eutectic together with the longitudinal fibers forms a homogeneous sensitive material with a radiation length of ~0.9 cm and an interaction length of ~22 cm, when used in a compensated ratio by volume of ~4:1 eutectic to plastic (this is the same as the transverse area ratio for a longitudinal fiber calorimeter) .

To start, we designed 4x4x12 cm electromagnetic calorimeter modules for the g-2 experiment at BNL (maximum electron energy about 3 GeV), but in a transverse size appropriate for a large SSC electromagnetic calorimeter (Fig 1).

Figure 1. Nine calorimeter prototypes with a clear fiber light guide and matching plate on the left.

The fiber matrix was assembled and pre-stretched before insertion in a mold with precision machined alignment plates at both ends of the mold for matching to a fiber guide. Two or 3 thin aluminum spacers were placed in the middle of the fiber matrix for alignment inside the mold, as shown in Figure 2, before casting. The plates had machined holes for the fibers, either on 3.3 mm centers, or on 1.91 mm centers, for 1 mm fibers or for

1 mm teflon capillary tubes. The fibers were strung by hand through the plates and stretched with clamping jigs and weights, an operation that typically required about half an hour.

Figure 2. Assembled fiber matrix before casting.

The mold was then filled with the low melting-point metal When cooled and extracted from the the mold, the 2 precision (2 μ accuracy) machined aluminum endplates with protruding fibers are then machined with a diamond-tooled end mill to a high polish. These end plates have alignment pin-holes to couple a flexible clear fiber-bundle light guide to a matching end plate on one end of the calorimeter as shown on the lower left of Fig. 1. The light pipe coupling efficiency was shown to be about 85%, and was highly uniform over the bundle. To date more than a dozen prototype modules have been produced (Fig. 3).

Test Results

The electromagnetic shower counter prototypes were tested at BNL in the A-2 1-4 GeV test beam, using Cerenkov-tagged electrons. (A troublesome background of in-time pions and muons was nearly always present.) Results indicate an energy resolution scaling of about 15-20%/√E, with a 20 ns integration time (Fig. 3) extrapolated from the data from electrons between 1.5-4 GeV. The light yield using a bi-alkali photocathode PMT was about 600 p.e./GeV.

Liquid Capillary Fiber Calorimeters

A novel extension of this technique is to use low refractive-index capillary tubes instead of fibers. The low-index capillaries are filled with liquid scintillator. The use of liquid scintillators might be necessary to avoid radiation damage in some regions of SSC experiments or in other high dosage experiments, where damage to plastic scintillating fibers may prohibit their use. (We note in passing that the casting technique offers a way to seal out oxygen from the fibers, although the benefits of this for radiation damage are controversial at present).

To test this concept, we designed a simple vacuum-pumpable stainless tank with pyrex windows on the front and back, slightly larger in interior dimensions than the calorimeter module dimensions, to hold one of the cast modules and contain the liquid (Fig. 4).This was done in order to be able to easily fill the capillaries with liquid without generating air bubbles, and to read them out without sealing difficulties for a test prototype.

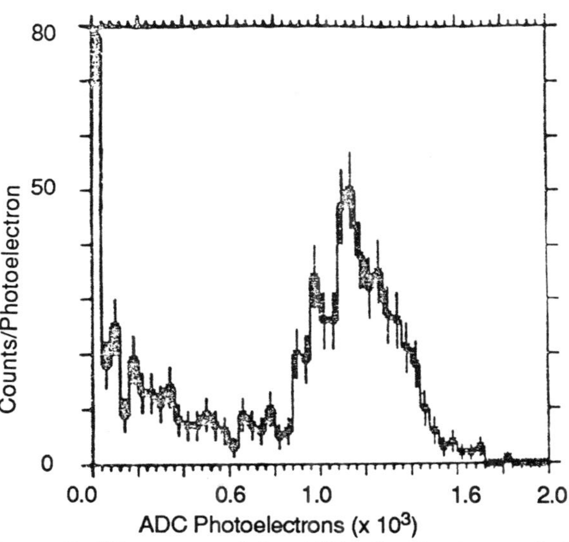

Figure 3. Histogram of the ADC distribution scaled to photoelectrons for 2 GeV electrons incident on a fiber calorimeter prototype

Figure 4. Tank holding a cast calorimeter module with teflon capillaries, filled with non-hydrogeneous liquid scintillator.

For convenience, we cast teflon capillary spaghetti (non-rad-hard) into the shower counter blocks. We used both hydrogeneous and non-hydrogeneous (perfluorobenzene-based) liquid scintillators to fill the capillaries. The non-hydrogeneous scintillator makes a unique low-cost neutron-insensitive shower counter. The light output of the liquid capillary scintillating fiber calorimeter was similar or superior to those of standard plastic fibers. Radiation-hard low refractive index material is commercially available and can be used to line small tubes, but was not tested here. We are attempting to obtain thin-wall stainless-steel tubes lined with a thin film (~few microns thick) optical coating of sodium (n~1.34) or magnesium fluoride as a low index total internal reflection rad-hard lining. These are available in large areas, and are used in commercial optics as anti-reflection coatings.

Extensions of the Technique

We have studied methods to automate the fiber stringing (weaving), in preparation for use in a mold. One proposed technique uses ribbons of fiber arranged in the appropriate side-to-side spacing. The fiber ribbon is pulled from a reel as a layer, and right-angle spacer combs are inserted and used to hold the layers apart. The resulting fiber matrix may then be cast-in-place with the low-volume steel combs in place. We have shown in detail the 9 elementary steps necessary to achieve this automation that could be easily implemented with existing robotics and robotic programming languages. Figure 5 shows schematically the last step in this process, producing a fiber matrix on the right side, similar to the photograph of Fig. 2.

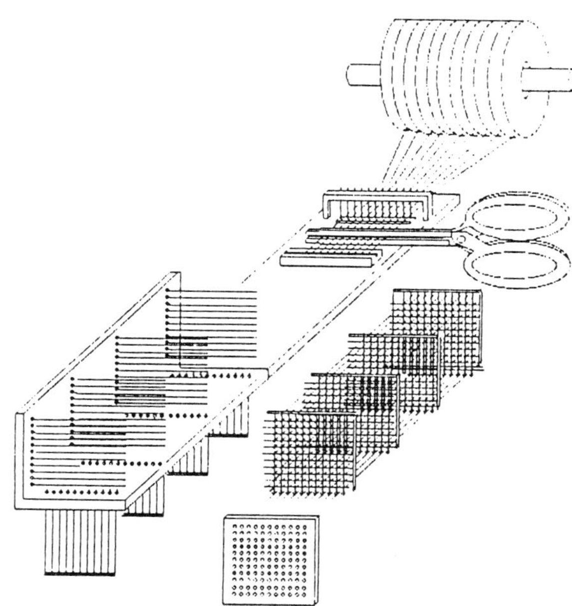

Figure 5. Last stage of fiber weaving automation scenario using inserted metal combs to generate a finished free-standing fiber matrix, as on the right of the picture (compare with Fig. 2). The prewound reel of parallel fibers lays down one fiber layer when pulled into place and cut by the "scissors".

Extending the casting technique to the SSC is trivial for electromagnetic calorimeters, but needs demonstration for a full-size hadron calorimeter (about 2.5 m deep for 11 absorption lengths). The buoyancy of the fibers must be shown to be controlled when the mold is filled with the melt during casting. A large mold is being constructed at Boston University for this purpose. Possible materials besides the lead-eutectic for casting are metal powder-filled polymers. Recently a low-expansion polymer has been shown to accept up to 70% by volume loading of metal powders [7].

Another difficulty with a full-size hadron fiber calorimeter is to maintain compensation in a projective tower along its length. In this case, two designs are being studied, one with parallel fibers, most of which terminate on the projective side panels of the mold, and one with conical tapered fibers which all terminate on the front and back faces, oriented fully radially from the interaction region. Fibers with a taper ratio exceeding the requirements of the SSC (0.5 mm to 1.2 mm over 50 cm has been achieved) have been produced by Optectron [5].

Conclusion

The cast-in-place technique using lead-based low melting point absorbers may offer significant time and cost savings in the construction of scintillating fiber calorimeters, while aiding the very good uniformity of the fiber installation which is necessary for the best energy resolution in such a system. We plan to use this technique in the g-2 experiment at BNL[8], and have proposed it for the EMPACT/TEXAS [9] collaboration to be proposed as an experiment at the SSC.

This development work is part of the DOE supported SSCINTCAL SSC R&D collaboration.

[1] R. Wigmans, NIM A259, 389 (1987), and NIM A265, 273 (1988)
[2] H. Paar, "Status of Scintillating Fiber calorimetry" in Proc.of Barcelona ECFA Meeting, Sept. 1989, CERN Report (1990)
[3] D. Hertzog, "Scintillating Fiber Calorimetry", in Proc.of Tallahassee Workshop on Radiation Damage to Scintillator, (1990), K. Johnson, ed.
[4] R. DeSalvo, "Results from the SPACAL Collaboration", presented at Snowmass 1990 (tbp)
[5] Ostalloy 158, Arconium Specialty Alloys, Providence, RI
[5] Fiber are obtained from Optectron, Inc, 91946 Les Ulis, France; *and* Moritex, USA, San Diego CA; *and* Bicron, Inc., Newbury, OH
[6] General Electric Corp Materials Group
[7] "A New High Precision Measurement of the Muon G-2 Value at the Level of 0.35 ppm", Proposal to the DOE, Dec. 1986, and accepted as BNL E821.
[8] EMPACT and TEXAS Expressions Of Interest were submitted to the SSCL in May, 1990, and are now the combined EMPACT/TEXAS Collaboration.

The Manufacturing Engineering of a Hermetic Cast Fiber Calorimeter

T. Coan, D. Higby, L. Sulak, and W. Worstell
Physics Department, Boston University, Boston, MA 02215

D. Winn
Physics Department, Fairfield University, Fairfield, CT 06430

F. Ayer, C. Elder, and D. Sullivan
The Charles Stark Draper Laboratory, Cambridge, MA 02138

Abstract

We have made the first pass at designing and engineering a cast lead calorimeter with a rapidity coverage to $\eta = 5.5$. The design preserves detector hermeticity. Plastic scintillating fibers provide a fast, hadronically compensated,high-resolution device. A lead–eutectic, which melts below the softening point of the plastic, provides an easily manufactured high Z absorber. This calorimeter, designed with the TEXAS SSC detector [1] as a baseline, is easily scaled in size and in segmentation without major design changes.

Introduction

An economically attractive method of calorimeter construction is to cast scintillating fibers in a low-melting point lead alloy,[2]as has been described in an accompanying paper and an associated R&D effort [3]. Small electromagnetic calorimeters constructed by this method have yielded $\leq 13\%\sqrt{E}$ resolution with an electron/hadron compensating packing fraction of 1:5. Casting minimizes construction costs by eliminating the need to machine tower surfaces and handle laminated plates. A keystoned structure of cast towers has been optimized by finite-element analysis. It is self-supporting and withstands creep even for modules formed entirely of the eutectic material. Alternative construction methods using grooved or punched lead, as pioneered by SPACAL,[4] are also amenable to automation. However, in comparisons made by both Martin Marietta Aerospace and by the C.S. Draper Laboratory, the cost economies achieved by casting are estimated at 30–40 % .

The manufacturing technique combines the electromagnetic and hadronic calorimeter sections into a monolithic tower. To preserve compensation, the fiber to lead ratio is held constant. Several possible realizations using parallel fibers of varying lengths in keystoned towers are presented in the SSCintCal Subsystem proposal.[5] These designs provide natural longitudinal segmentation.

In the design described here, the photomultipliers are distributed on the outer surface of the device with no dead structural elements within the body of the calorimeter. Detailed drawings of the device and an engineering conceptual design for its mechanical support appear in reference 6. Projective towers of scintillating fibers and lead alloy are combined into two hermetically sealed halves at $\eta = 0$ which are smoothly integrated with forward liquid scintillator–based calorimetry. A narrow gap at $\eta = 0$ allows for readout of the central tracking and imaging preradiator systems.

Calorimetry performance improves with larger radius and increased transverse segmentation. For this study, we use an inner radius of 1.5 to 2 meters and segmentation of $0.05\eta \times 0.05\phi$ in the hadronic and $0.025\eta \times 0.025\phi$ in the electromagnetic compartment.

The calorimeter uses projective geometry throughout, with the front faces of the calorimeter $\geq 60°$ relative to the normal from the interaction region. To avoid feedthrough to adjacent towers and to minimize albedo to avoid channeling of electromagnetic showers in the fibers, an aprojective angle of 3° in ϕ is employed. The thickness of the calorimeter is dictated by the requirement of 99% containment. This implies a minimum of 11 absorption lengths at 90° growing to 14 interaction lengths at $\eta = 3.0$.

Calorimeter Structural Design

We have designed, modeled, analyzed and evaluated an external structural system capable of maintaining the towers of the calorimeter within the dimensional tolerances required by physics and by assembly ease (see Figure 1). It introduces negligible dead material in the active portion of the device. The structural support for the calorimeter consists of two identical halves (mirror images) that move away from each other to gain access

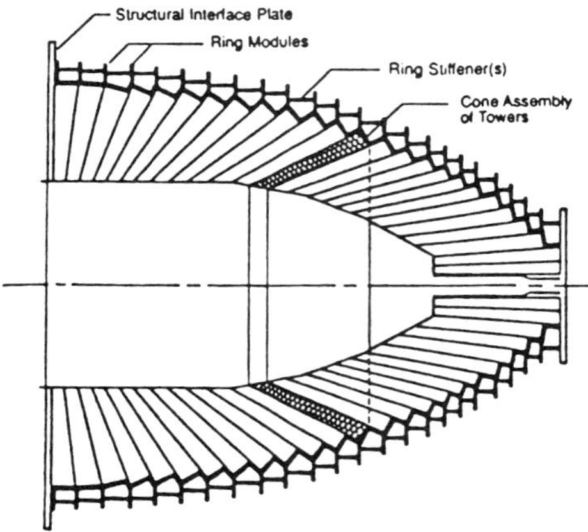

Figure 1: Modular ring calorimeter enclosure with projective

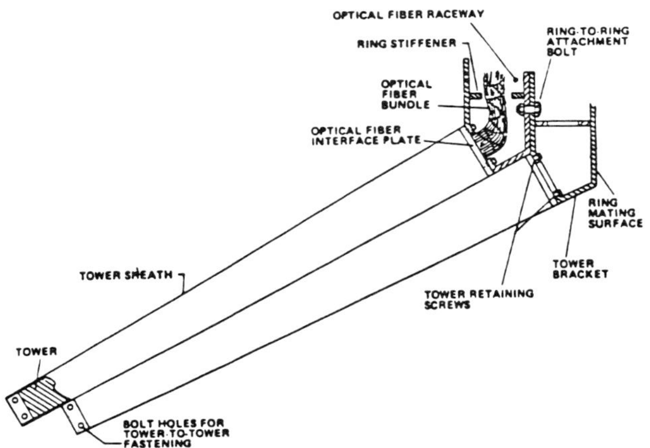

Figure 2: Two towers attached, including the fiber bundle exiting from the top tower

to the inside tracker and the front face of the calorimeter. A finite element model was used to identify the deformations of the structure under the 4.4 kt mass of the lead-eutectic, establish stress levels throughout the support structure, and provide force and moment components necessary for the evaluation of structural stability. The support system carries the weight of the calorimeter with acceptable deformations (< 2 mm) and tolerable stress levels. Global and local stability checks exhibit generous safety factors. The total weight of the entire structure is 300 tons of steel.

The structure supports (with no struts, fins, etc.) the very heavy and dense mass of the supertowers (see Figure 2) with a ring stiffened ellipsoidal shell. The shell behaves as a very deep beam. Its overall depth (9.4 m diameter maximum at the center) offers the stiffness necessary to deliver reaction loads to the end support frames, approximately 10 m apart. The center line of the calorimeter is 10 m above ground, placing special demands on the end frames designed to insure adequate

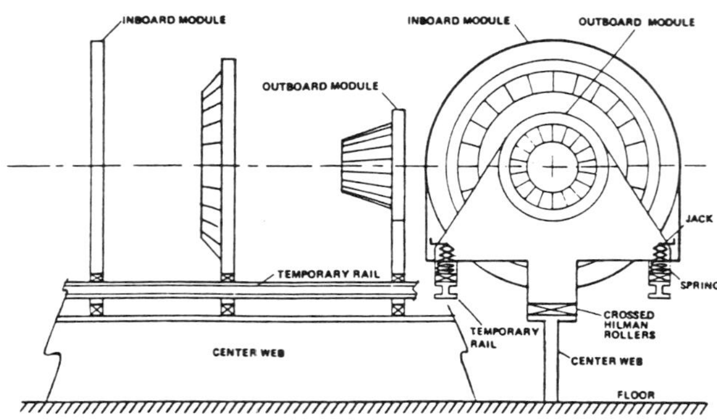

Figure 3: Side (left) and beams eye (right) view of tower assembly procedure. Three stages are shown

stability. A longitudinal central wall provides the necessary bracing. It also offers additional vertical support to the ellipsoidal shell. The gravity load picked up by this wall is a function of the relative stiffness between the shell and the wall. It does not carry a disproportionate portion of the mass since it could distort the calorimeter circularity. The design of the frames satisfies the physics requirement that they be no thicker than 25 cm while providing the necessary stiffness to insure acceptable deformations, stress levels and stability. The shell plate is pierced by one hole per tower to accomodate the passage of optical fibers (see Figure 2). The complexity of the structure is limited to minimize interference with the pmts, fibers, and electronics network. Each half of the calorimeter consists of 7,680 pyramidal supertowers. Rings of 128 towers are arranged to form the shape of a hollow truncated cone. Sixty of these hollow cones nest to form a stack that completes the calorimeter half.

Calorimeter Assembly Procedure

The assembly of the structural system and the supertowers is a critical issue; it affects not only delicate construction procedures with tight tolerances but has a decisive influence on the design concept. The towers are cast in thin sheet metal sheaths for strength, protection, and for joining the towers together at their small ends. Threaded steel inserts are cast into the large end of the towers to fix that end to the tower support structure. Two methods of assembly of the calorimeter have been developed. We describe the "modular ring" method, where each cone (made up of 128 towers) is attached to a separate steel support ring. The 60 rings are then bolted, surface-to-surface, to form the nested stack of cones which completes the calorimeter and structural assembly (see Figure 3). The first ring assembled is the largest diameter ring. With it horizontal, the towers are installed to form the cone. After alignment of the towers, their inner ends are fastened together to make the cone rigid. The ring is lowered into the hall and bolted, axis horizontal, to the end support frames. The next ring in the stack is assembled, aligned, lowered into the hall and bolted to the first ring. Temporary rails support the rings until the calorimeter assembly is finished. When the calorimeter stacking is completed, the outboard end support frame is attached and the temporary rails for ring support are removed.

Finite Element Model and Analysis Results

The main structural system is the ring stiffened ellipsoidal shell. The shell is truncated 10.17 m from the center with an opening of 3.14 m diameter to accomodate the calorimeter end plug. The 2 cm thick stell shell is reinforced by 10 outside rings 1 meter on center. Eight reinforcements run the length of the ellipsoid 45° apart around the circumference. Longitudinal reinforcements are added every 22.5° for the first 4 m of the shell near the center where additional stiffness is required. The center and end frames are disposed in a truss like configuration for material efficiency. These frames are configured to preserve shell circularity in areas of large reaction loads. The widths of all frames are minimized to maintain hermiticity. The material at the shell and the frames is a high strength steel of F_y=50 ksi (yield stress).

The finite-element model contains 528 shell elements and 618 beams for a total of 4326 degrees of freedom. The mass of the supertowers is assumed to be distributed uniformly along the circumference at each station of the shell axis. No structural rigidity is assumed for the eutectic. Figure 4 shows views of the undeformed and deformed (3 mm maximum, grossly exaggerated) structural system for comparison. Most of the deformations occur in the highly stressed supporting frames; the shell itself remains within 1 mm of its original configuration but undergoes rigid body motions. Shell stresses remain under 5 ksi. Local bending stresses are significant. Conservative assumptions have been used for the stability check. The analysis shows generous factors of safety (between 1.5 and 4.0) above those included in the critical allowable load.

Tower Production

We discuss one concept for the production of the cast towers. The essential idea is that the fibers are automatically loaded from the top into a stack of horizontal meshes inside a steel sheath. The meshes are aligned vertically with each other by a tie-rod to provide accurate positioning of the fibers. The meshes and their captured fibers are then inverted, and the liquid euctetic is injected into the entire assembly. The large bouyant force of the fibers in the high density eutectic precisely aligns the fibers vertically. Upon freezing, the fibers, meshes, sheath and eutectic are all bonded together.

The tower production process takes place in a simple assembly line. The first production step cuts the

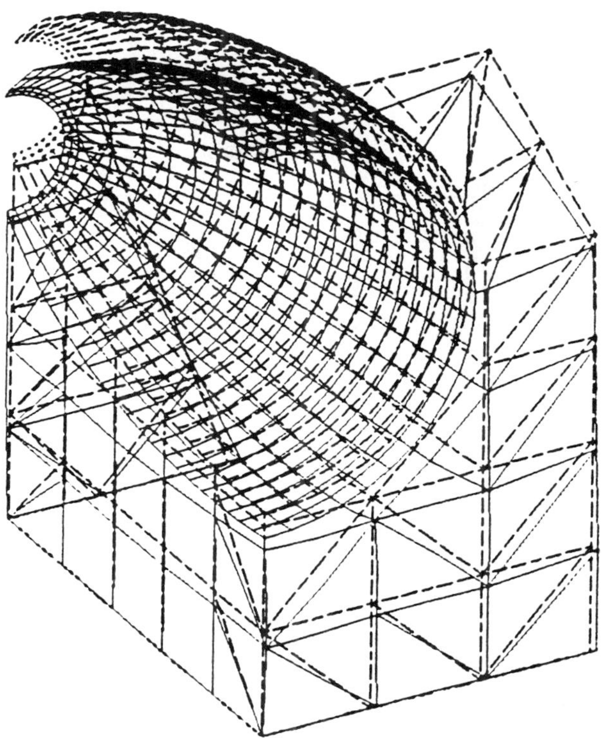

Figure 4: Finite-element deformations of the calorimeter exoskeleton and support frames under gravity load.

scintillating fibers for a particular tower to their appropriate lengths. As they are cut with an IR laser, the fibers are "headed" to provide vertical registration and constraint inside the stack of meshes, especially for fibers that are shorter than the full calorimeter height. A conventional pick and place type arm grasps a fiber with a vacuum V–groove, holds it vertically, and places it over the stack of meshes. The pick and place device needs only one degree of freedom and closed loop control is not required.

Each mesh is composed of two horizontal planes of round steel rods, 1 mm in diameter(see Figure 5). The rods in each plane are parallel to one another with

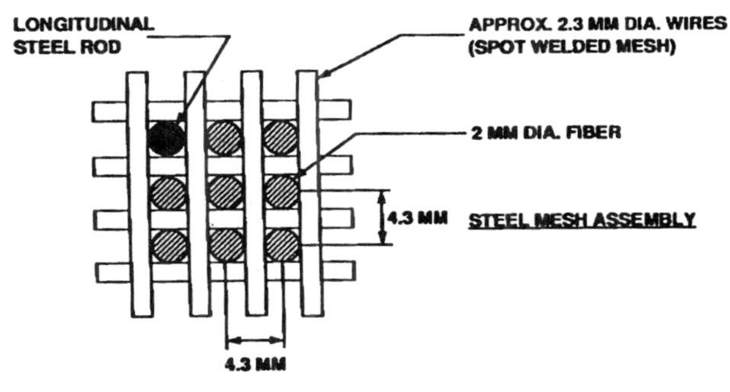

Figure 5: Top view of steel rods with fibers inserted

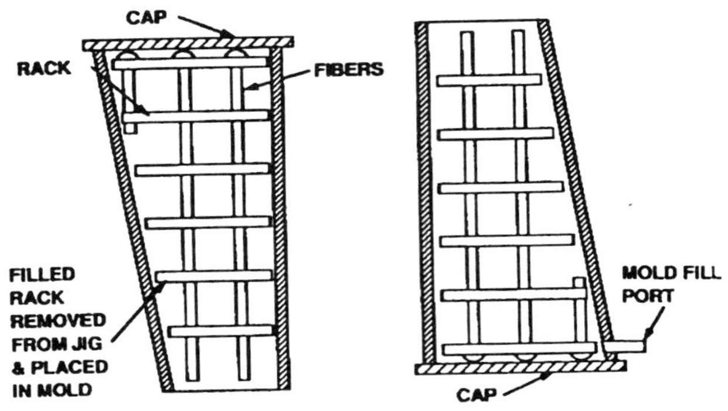

Figure 6: Side view of rack of stainless steel rods after filling and capping (left) and ready for injection (right)

a pitch of 4.3 mm and the rods in one plane are perpindicular to those the other. The rods are spot welded to each other. The spaces between the rods locate the scintillating fibers to an accuracy of 0.010 inches. The meshes are held apart vertically by three vertical round rods per tower that all fit through the mesh holes. This stack of meshes, or rack, is positioned all inside a precision jig which aligns the holes of a given mesh with those above and beneath it. This rack jig is placed on an x-y translation stage and awaits the fiber from the pick and place device.

The advantages of the wire mesh are significant: i) The round rods provide a tapered hole to help guide the scintillating fiber into the hole; ii) The rods are smooth sided which avoids damage to the fibers; iii) The liquid euctectic flows easily around the mesh; iv) The construction technique is cheap and thoroughly conventional.

The pick and place device holds the prepared fiber vertically and positions it to a fixed position over the rack jig. The x-y table locates the rack jig so that the proper mesh hole is positioned directly beneath the fiber. The pick and place device then drops the fiber through the rack jig. The headed end of the fiber is caught by the top-most mesh which prevents it from falling completely through the rack.

This process is repeated until an entire rack is filled. Typical machines of this nature require no more than 10 seconds to load each fiber; presumably rows of fibers could also be accommodated by this technique.

After a rack is filled with fibers, the precision mold is then capped at the end closest to the headed fibers (see figure 6). This mold defines the shape of a tower and is constructed to the same accuracy as the rack jig. The mold is inverted and filled from the bottom (the capped end) with the liquid eutectic. Due to the significant difference in specific gravity between the liquid eutectic and the plastic fibers (9.4:1), the fibers are accurately held vertically by a large bouyant force.

This development work is part of the DOE supported SSCINTCAL SSC R&D Collaboration. It was also aided by support from the Superconducting Supercollider Laboratory. We thank R. DeSalvo, H. Paar, A. Para, T. Thuston, J.S. Whitaker, R. Wigmans for many suggestion.

[1.] EMPACT and TEXAS Expressions of Interest paper were submitted to the SSCL in May 1990 and are now the combined EMPACT/TEXAS Collaboration.This paper describes one of the two calorimeter options advocated by the collaboration. It is described in detail in the SSC Expression in Interest, TEXAS collaboration, May 25, 1990.

[2.] Ostalloy 158, Arconium Specialty Alloys, Providence, RI.

[3.] "Cast Lead-Eutectic Solid and Liquid Scintillating Fiber Shower Calorimeters", W. Worstell, J. Miller, B.L. Roberts, L. Sulak, D.R. Winn, P. Cushman, S. Dhawan, V.W. Hughes, Pros. of the Conference for Research and Development for the SSC, Ft. Worth, October 1990; "A New High Precision Measurement of theMuon G-2 Value at the Level of 0.35 ppm", Proposal to the DOE, Dec. 1986, and accepted as BNL E821.

[4.]R. DeSalvo, "Results from the SPACAL Collaboration", presented at Snowmass 1990; R. Wigmans, NIM A259, 389 (1987), and NIM A265, 273 (1988); H. Paar, "Status of Scintillating Fiber Calorimetry" in Proc. of Barcelona ECFA Meeting, Sept 1989, CERN Report (1990).

[5.]"Scintillating Calorimetry for the SSC", SSC Research and Development Subsystem Proposals, The SSCintCal Collaboration, A. Para, L. Sulak, et.al. September 27,1989 and September 19, 1990.

[6.]"Scintillator Calorimeter Support Structural Design and Evaluation", SSCL-N-727, F. Ayer and C. Elder, Draper Laboratories, 31 May 1990.

SCINTILLATING PLATE CALORIMETER OPTICAL DESIGN

R. McNeil, A. Fazely, R. Gunasingha, R. Imlay, J. Lim
Louisiana State University

Abstract

Progress on designs for the optical readout of a compensating scintillator plate calorimeter are presented. These designs include readout of the scintillator plates via wavelength shifter plates or fiber readout. Results from radiation damage studies of the optical components are presented.

Introduction

A major technical challenge facing the builder of a general purpose detector for the SSC is to achieve an optimum design for the calorimeter. The calorimeter must measure jets as well as electrons from heavy boson decays. Ideally, a calorimeter should have good energy resolution, fine segmentation, and a high degree of hermiticity. In addition, the SSC imposes constraints related to the interaction rate of 62 MHz and the anticipated radiation dose of more than one Mrad per year in the forward region. These requirements place severe constraints on the technologies that can be used for the calorimeter.

Because of its fast response and good energy resolution, scintillating plate sampling calorimeters should be considered as a possible technology option. The work of the Scintillating Plate Calorimeter Collaboration is focused on compensating plate calorimeters. Based on experimental and simulation studies [1], it is expected that a sampling calorimeter with alternating layers of high-Z absorber (Pb, W, DU, etc.) and plastic scintillator can be made compensating (e/h = 1.00) by suitable choice of the ratio of absorber/ scintillator thickness. Two conceptual designs have been pursued by this subsystem collaboration. One is based on lead as the absorber, with readout of the scintillator plates via wavelength shifter fibers. The other design is based on depleted uranium as the absorber with wavelength shifter (WLS) plate readout. The guidelines for these two options are given in another paper contributed to this conference [2].

This report is on the progress of the measurements made to evaluate the two optical system design options. Recent progress [3] by Bicron Corp. on radiation tolerent scintillator have resulted in a 2.5mm thick plate which has good surface quality and thickness tolerance. This scintillator (Type B/PS) is one in a series of blue-emitting materials being developed by Bicron for rad-resistance and sheet production for SSC use. This particular material is based on polystyrene and contains selected fluors and stabilizing additives known to enhance radiation resistance. The scintillation efficinecy is high, about 2.4% for MIPs, similar to that of most commercial high efficiency products. Also tested with the scintillator tiles are 1mm diameter Wavelength Shifter (WLS) fibers based on Bicron's RH-1 material doped with Y7. The radiation tolerance of these fibers have been measured [4].

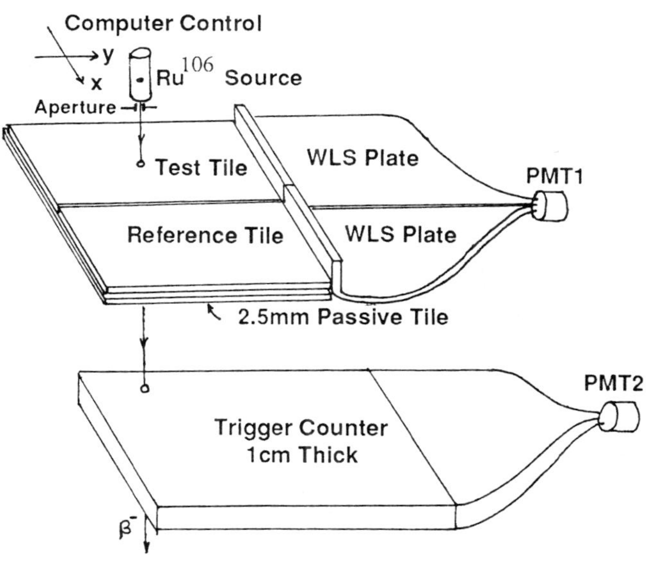

Figure 1. Experimental Setup for testing scintillator using the WLS Plate Option

Experimental Setup

The measurements of optical performance of the scintillator tiles were performed at LSU using a computer controlled X-Y scan table designed and built at Ohio State University and used for measurements of ZEUS EMC wavelength shifter plates. The scintillator test scheme is shown in figure 1. We used a β^- source of Ruthenium 106. The output of the source consists of an energetic beta (E_{max} = 3.5MeV) and softer photons. We select an energetic beta by requiring a large pulse height in a 1cm thick trigger scintillator counter underneath the 2.5mm thick test tile and a 2.5 mm thick passive scintillator tile. The beta electrons are collimated to about a 3mm wide beam by means of a source holder with thick aluminum aperture.

The tiles received from Bicron are wrapped with Al foil and were readout by a 4mm thick WLS plate or WLS fiber doped with Y7 dye. In both cases the light from the light guide was coupled to the PMT via a transition piece. A reference tile is used to monitor PMT gain shifts. The trigger tile and test tile PMT signal are readout via CAMAC ADC. Each measurement of the the output pulseheight at a position on the tile is the average from 6,000 triggers (about 1/2 minute). It was found that the single point measurements showed no significant variation with time over a 1 hour period with a measurement uncertainty of less than 1%. Repeatability studies performed by removing the test tile and placing it back into contact with the waveshifter plate resulted in a overall (including the above 1%) uncertainty in the measurements of 1.3% .

It was found that 4% of the accepted triggers are due to cosmic ray muons and source photons which do not result in a signal in the test tile and hence are in the pedestal. To obtain the light output measurement from a point, the contribution from the cosmics/photons to the pedestal bin is removed and the spectrum is shifted so that the remaining pedestal is at channel zero. It was estimated that 1.5% of the accepted triggers are from cosmic ray muons which do make a minI signal in the test tile. We do not correct the light output measurement for these events.

Results

A total of 18 10cm × 10cm × 2.5mm tiles were measured at LSU. The thickness of the tiles were measured by micrometer prior to wrapping. Figure 3 compares the light output at a particular position on the tile (9cm from the WLS plate) from the 18 tiles to the thickness measurement. Overall there is good correlation between the two measurements indicating that the surface quality and material quality of the scintillator is uniform and the output variations are due to thickness.

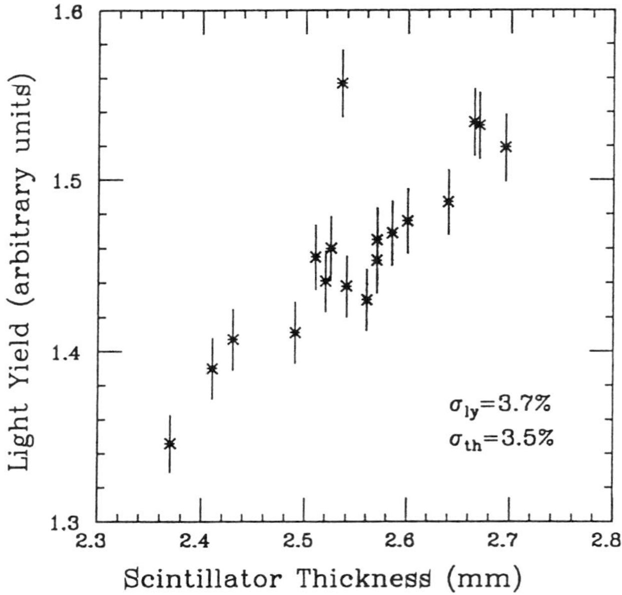

Figure 2. Light output vs scintillator thickness

A study was performed to compared the uniformity of the 10cm × 10cm tile readout on a single side by WLS plate with the same tile readout via a embedded WLS fiber. A groove was machined in the tile in the shape of a "U" deep enough so the fiber would be entirely embedded in the tile. In the first attempt to make a groove we chose a U-shaped groove pattern as shown in figure 3b. It was found that the absolute light yield was sensitive to the light coupling between the fiber and the transition piece and between the fiber and tile within the grove. A quantitative comparison of light yields as well as repeatability between WLS Plate and Fiber option is forthcoming. In figure 3a is shown the light yield as a function of position on the tile for the WLS plate option where correction for the thickness of the plate have been performed. The light yield have been normalized to the center of the tile. In figure 3b is shown the light yield for the same tile now with a groove machined into it as shown and a WLS fiber embedded into the tile using optical grease. As can be seen the WLS plate option has a monotonic decrease in the light yield as a function of the distance from the WLS plate with a maximum decrease from 1cm to 9cm away of 16%. This can be corrected using reflective masking on the tiles. The WLS fiber option has comparable uniformity variations as the WLS plate with the region of lowest relative output being at the corners opposite to where the fibers exit the tile. This nonuniformity can presummably be minimized by altering the groove path.

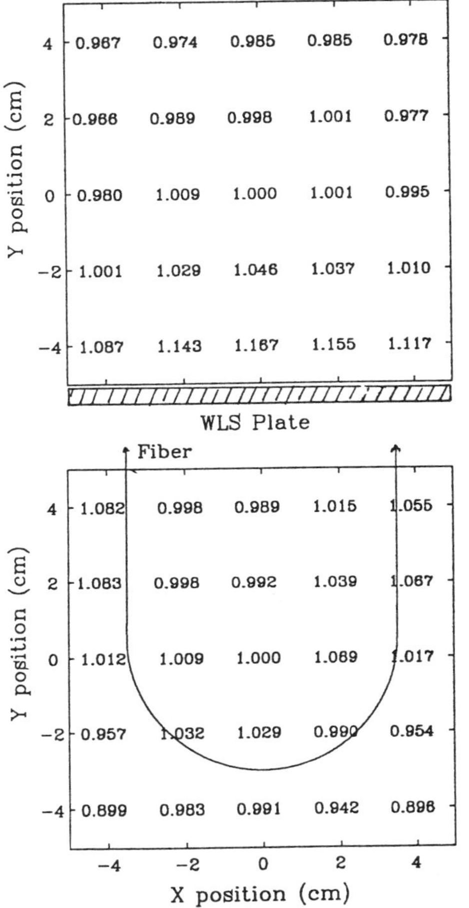

Figure 3. Light Yield vs position on tile for a) WLS Plate and b) WLS Fiber option

Two of the 18 tiles were selected for radiation exposure at LSU with a Co^{60} source. From preliminary calculations, the dose rate was estimated to be 140krad/hr uniformly spread over the surface of the tile. The two tiles were exposed to 1Mrad. Following exposure the tiles appeared yellow in color. This color diminished becoming clear after only 24hours. Figure 4 shows the light output of one of the tiles as function of distance from the waveshifter plate (down the middle of the tile). The recovery was found to be very fast, reaching 70% of the initial light (at the mid-point) in only 30 hours. However, the recovery then either has stopped (with a residual damage) or is continuing at a greatly reduced rate. We will continue to monitor these tiles to see if they eventually recover.

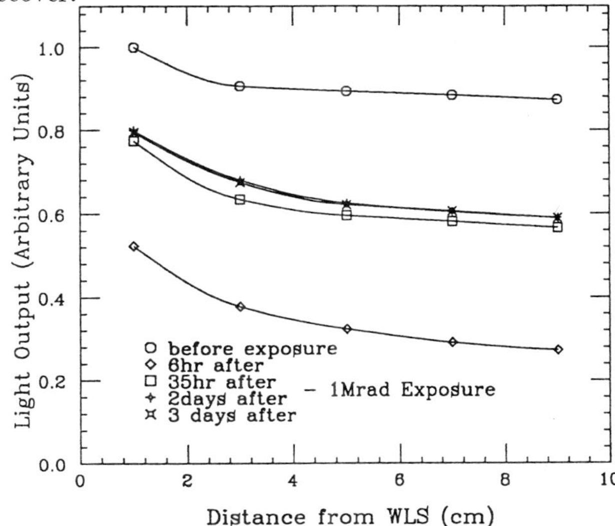

Figure 4. Light output vs distance from WLS before and after radiation exposure.

In conclusion, LSU has an experimental setup for measuring the uniformity of plastic scintillator. The new scintillator materials from Bicron Corp. are of good surface quality and material uniformity. Preliminary measurements from a tile with readout via an embedded WLS fiber shows the uniformity over the surface of the tile to be comparable to that from one-sided WLS plate readout. Work will continue to find the optimum fiber placement in the tile. Preliminary results from a 1Mrad exposure show that the new material recovers quickly but not fully. We would like to thank the Nuclear Science Center at LSU and J.C. Courtney for helping us expose the scintillator tiles. This work is supported by the U.S. Department of Energy under Contract No. DE-AS05-77ER05490.

References

[1] See for example J. Proudfoot et al. , Simulation Studies for a Scintillator Plate Calorimeter, submitted to these proceedings.

[2] H. Spinka et al., Scintillating Plate Calorimeter Mechanical Design, submitted to these proceedings.

[3] C. Hurbut (Bicron Corp.), Private Comm.

[4] V. Hagopian et al., 1990 Snowmass Summer Study, and these proceedings.

DESIGN OF READOUT ELECTRONICS FOR A SCINTILLATING PLATE CALORIMETER

H. B. Crawley[a], W. T. Meyer[a], E. I. Rosenberg[a], W. D. Thomas[a], R. E. Blair[b], A. Buehring[b], J. Dawson[b], N. Hill[b], R. Noland[b], E. Petereit[b], L. E. Price[b], J. Proudfoot[b], H. Spinka[b], R. Talaga[b], H. J. Trost[b], D. Underwood[b], A. B. Wickland[b], C. Hurlbut[c], V. Hagopian[d], K. Johnson[d], R. Imlay[e], R. McNeil[e], W. Metcalf[e], L. Bolen[f], L. Cremaldi[f], J. Reidy[f], D. Summers[f], P. Fu[g], T. Gabriel[g], T. Handler[h], J. R. Ficenec[i], B. Lu[i], L. Mo[i], L. E. Piilonen[i], T. Nunamaker[i], M. Burke[j], D. T. Hackworth[j], T. F. Porter[j], R. J. Ravas[j], D. Scherbarth[j], R. Swensrud[j], D. Carlsmith[k], C. Foudas[k], J. Lackey[k], D. Loveless[k], D. Reeder[k], P. Robb[k], and W. H. Smith[k]

Abstract

We describe the progress made on the design of readout electronics for a compensating scintillator plate calorimeter.

INTRODUCTION

A scintillator calorimeter produces unique problems for the designer of readout electronics. On the one hand the narrow time structure of scintillator pulses, ~10 nsec, is well matched to the rf structure of the SSC and gives hope of isolating information from individual beam crossings. On the other hand, the compensation mechanism and the need to broaden the pulse shape for use with analog signal sampling devices gives a somewhat wider time structure, ~50-100 nsec. Furthermore the granularity of such a device implies that the full energy of an electromagnetic shower may be totally contained within one readout channel. If the resolution of the electronics is not to compromise the intrinsic resolution of the calorimeter, assumed to be $\sigma/E \approx 15\%/\surd E + 1\%$ (E in Gev), coverage of the full dynamic range (40,000:1) requires at least two 12-bit devices with 7 bits of overlap for a linear front-end electronics chain.

The positioning of the electronics also is a critical issue. At luminosities of 10^{33} $cm^{-2}sec^{-1}$, electronics placed on the calorimeter must withstand doses of at least 10^{10} neutrons/cm^2 and 2000 Rad per year at 90°.

In the past year, the scintillating calorimeter collaboration has begun studying these and related issues. Among the work reported below is: a study related to remote location of the calorimeter electronics, a comprehensive program to evaluate the properties of FADCs capable of operation at 60-80 MHz, design of a analog memory unit and development of a "benchmark system" to help evaluate components under development both within and outside our collaboration.

REMOTING OF ELECTRONICS

The advantages of locating a substantial fraction of the front-end electronics away from the calorimeter include: accessibility and hence ease of maintenance and future upgrades, significantly reduced exposure to radiation and more space for cooling. The systems evaluated involve either using fiber optics for transmitting signals from phototransducers on the calorimeter to remote electronics, or bringing the scintillator light signals along fiber optic paths to remote phototransducers and electronics.

An investigation was made of the performance and cost of fiber optic links for sending analog data signals (and a possibly digitized trigger signal) from and digital control and clock signals to a multiplexed set of phototransducers. Analog bandwidths above 200 MHz and digital rates of up to 2 Gigabits/sec exist for such systems. The dominant cost is in the fiber optic links, ≥$300/channel.

The dominant cost in the alternative scheme of bringing the scintillator light away from the surface of the calorimeter is in the fiber itself. Long, low attenuation fibers are required and the total fiber length is in excess of 10,000 km. Low loss glass fibers are available, but the cost is prohibitive ($3000/km). There also may be difficulty in bonding glass fibers to the wavelength shifter fibers of the calorimeter. Clear acrylic fibers are currently being evaluated as an alternative approach.

FADC EVALUATION

A systematic study of commercially available Flash A/D converters capable of 100 Megasample per second operation has begun [1]. Such FADCs could be used to digitize the phototransducer output for trigger purposes or could be used with a non-linear preamplifier as part of the data collection stream. (Development of such an ASIC preamplifier is part of a Fermilab effort.)

We have built a CAMAC based test bench controlled by a VAXstation 3100 for the testing of FADCs. Communication with the test bench components is via both SCSI and GPIB buses. Two special purpose programmable CAMAC boards have been built: one provides clock and trigger signals, the other interfaces to various evaluation boards. The clock/trigger board allows ten discrete frequencies in the range 20-140 MHz to be supplied to the FADC under test. The FADC board contains 1 kbyte of 12-bit memory to store the digitized data. FADCs with dynamic ranges up to 12-bits can be tested. (Minor modification of the clock/trigger board can extend the range of available clock frequencies to lower values.) The test procedures follow the guidelines given in IEEE Std-1057, "IEEE Trial-use Standard for Digitizing Waveform Recorders." Among the properties measured are gain, offset, differential and integral non-linearity, maximum static error, monotonicity, hysteresis, effective number of digitizing bits, signal-to-noise ratio, peak error, aperture uncertainty, settling times, etc. Table 1 gives preliminary results of some of these studies for three of the FADCs currently under test.

Flash ADC	AD9002	CXA1176	CX20116
Manufacturer	Analog Devices	Sony	Sony
Advertised Maximum Rate (MSPS)	125	300	110
Clock Rate for Measurements(MHz)	30.0	80.0	20.0
Gain			
Static	0.995	0.994	0.996
Dynamic	0.983	0.989	0.995
Differential Non-Linearity			
Sine Wave	0.300	1.000	0.800
Triangle Wave	0.297	not measured	0.559
Integral Non-Linearity	0.576	0.638	0.333
Signal-to-Noise Ratio	168.1	156.0	202.5
Effective Number of Bits	7.271	6.988	7.405

Table 1 -- Preliminary Results from FADC Testing

ANALOG MEMORY

One possible scheme for collecting data from a scintillating calorimeter is given in Figure 1. In this scheme A/D conversion is done after two levels of triggering. This reduces the speed at which the conversion must be accomplished, but requires the data be stored in analog form for periods of up to 50 μsec. In addition if the storage is to be done on the outer surface of the calorimeter the analog memory units must be radiation hard. A VLSI chip to store analog signals from the calorimeter has been designed and is scheduled for submission to the MOSIS foundry in early November, 1990. The analog phototransducer output signals are

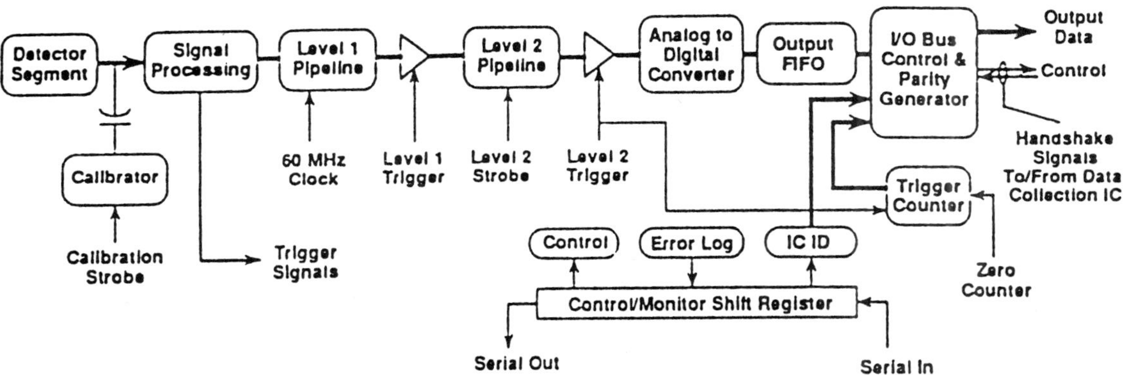

Figure 1 -- Possible Readout Architecture

integrated for 16 nsec and stored on capacitors. Integration, storage and digitization are performed on a single chip upon receipt of a second level trigger signal. The architecture stresses the simplicity of the logic and has the potential for deadtimeless operation. Figure 2 is a schematic diagram of one of the eight capacitor arrays to be located on a single chip. Figure 3 shows the output circuit comparator for digitization.

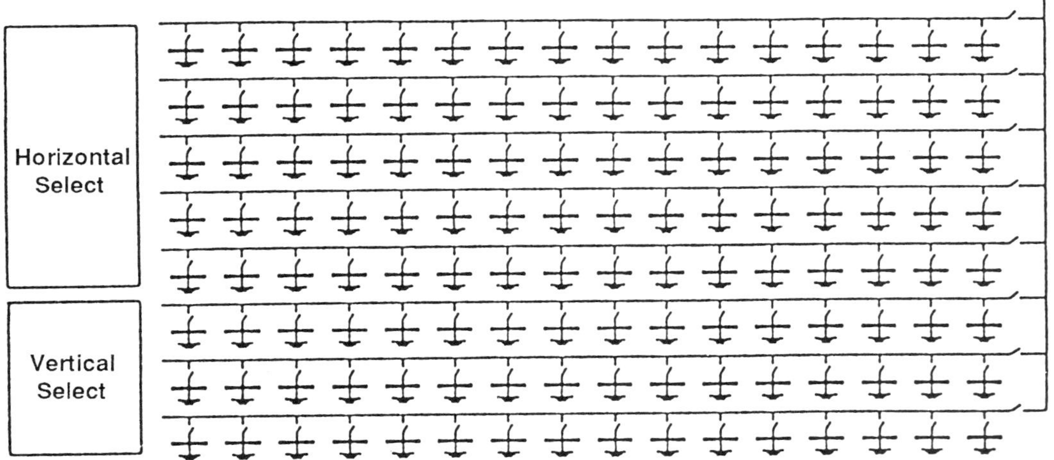

Figure 2 -- 8x16 Capacitor Array with Shift Registers Shown

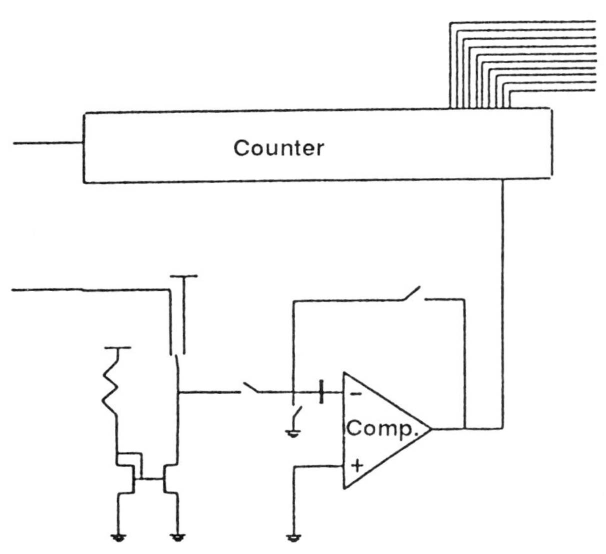

Figure 3 -- Output Stage of VLSI Chip

Discussions have begun with industry on creating a radiation hardened version of this chip.

BENCHMARK SYSTEM

We also are developing a "benchmark system" to use in a Fermilab test beam with calorimeter prototype modules which will allow the testing of new components as they become available.

Notes and References

a Ames Laboratory and Department of Physics, Iowa State University, Ames, Iowa

b High Energy Physics Division, Argonne National Laboratory, Argonne, Illinois

c Bicron Corporation, Newbury, Ohio

d Florida State University, Tallahassee, Florida

e Louisiana State University, Baton Rouge, Louisiana

f University of Mississippi, University, Mississippi

g Oak Ridge National Laboratory, Oak Ridge, Tennessee

h University of Tennessee, Knoxville, Tennessee

i Virginia Polytechnic Institute, Blacksburg, Virginia

j Science and Technology Center, Westinghouse Electric Corporation, Pittsburgh, Pennsylvania

k University of Wisconsin, Madison, Wisconsin

[1] H. B. Crawley et al, "Performance of Flash ADCs in the 100 MHz Range I. Test Bench and Preliminary Results," submitted to this symposium and to the 1990 Nuclear Science Symposium of the IEEE.

CALIBRATION OF A SCINTILLATING FIBER CALORIMETER PROTOTYPE USING A MOVING RADIOACTIVE SOURCE

Virgil E. Barnes and Alvin T. Laasanen

Purdue University
West Lafayette, Indiana 47907

Abstract

A full-depth iron/lead/scintillating fiber calorimeter prototype consisting of 30 projective towers was instrumented with a set of 45 tubes oriented longitudinally and transversely to the tower axes. A remotely activated source driver permits selection of any tube and insertion of a 12-foot long 0.71 mm OD hypodermic tube carrying a pointlike 8.8 mCi Cs-137 gamma source. Preliminary data on the photomultiplier current responses of the towers are presented, bearing on the uniformity of the towers, the optical attenuation lengths of the installed scintillating fibers, and the reproducibility of the source response

Introduction

Moving Cs-137 gamma sources have proved very useful in characterizing and calibrating scintillator calorimeters, for example in construction and use of the central and endwall calorimeters of the Collider Detector at Fermilab (CDF), and have proved capable of maintaining test beam calibrations at better than the 1% level [1].

We have developed a device capable of inserting at constant speed a very long source carrier into any one of a large number of fine (1.80 mm OD) tubes which in principle permits us to expose many towers using one source. These tubes can be imbedded in a wide variety of tile or fiber calorimeters, and in many positions and orientations within the towers.

The new source driver and tube system were installed in the scintillating fiber calorimeter prototype discussed by Adam Para earlier today [2]. The prototype was then moved immediately to the test beam, in August, 1990, where access to the data acquisition system was very limited. Source response data were obtained separated in time from the test beam exposures. After the end of the test beam period, a complete set of source runs was made in a period of about 10 hours. We will summarize these data.

The Hardware

The heart of the driver is a 7-inch diameter polycarbonate disc with a storage groove 0.1 inch wide radially and 0.2 inches deep, rotating snugly against a teflon sheet backed by a metal plate. The source carrier is attached to the outer wall of the groove, and passes through a hole in the plate. It is pulled by the inner wall of the groove or pushed by the outer wall of the groove. The reel is driven by a reversible DC motor. The carrier wire passes over a wheel attached to a reversible quad pulse generator, with a 0.1 mm least count.

The carrier wire passes through a nozzle which can recoil as the wire hits the closed end of a tube, actuating an extension limiting microswitch. The reel and nozzle are moved by a linear indexer with 1/8 inch increments over a 30 inch long Delrin block, into which are plugged 45 stainless steel tubes (1.30 mm ID, 1.80 mm OD hypodermic tubing). A series of raised plates alongside the "active" holes are sensed by a microswitch so that the indexer can avoid unused locations. Various electromechanical interlocks prevent indexing unless the wire is fully retracted and prevent extension of the wire unless the indexer is lined up with an active hole. The driver can be operated either manually or remotely via CAMAC or PC. The indexing position is also digitized.

Functionality of the System

The source wire can be extended more than ten feet (this is limited at present by the 12 foot length of the wire, which could be up to 20 feet long). It can pass at least four 90° bends, of radii three inches or greater. It was essential to coat the interior of each tube copiously with "dry" graphite film. This dramatically reduces friction and scuffing. It is believed that wear of the source carrier will not be a problem, but extended wear tests must be done.

A typical source run was a full extension into one tube, a brief pause, and a full retraction into the lead shielding garage near the reel. The source speed was roughly constant at 2.4 cm/s extending and 1.9 cm/s retracting (this asymmetry will be corrected electronically).

Source Tube Layout and Geometry

As described in Adam Para's talk [2], the 30 projective calorimeters are sandwiches of lead and steel plates, with parallel scintillating fibers running longitudinally in the lead. Source tubes are placed in gaps in the steel plates, as shown in Fig. 1. The source activity is .38 mm in diameter and 1.1 cm in length, compared to a 2.5 mm fiber repeat distance.

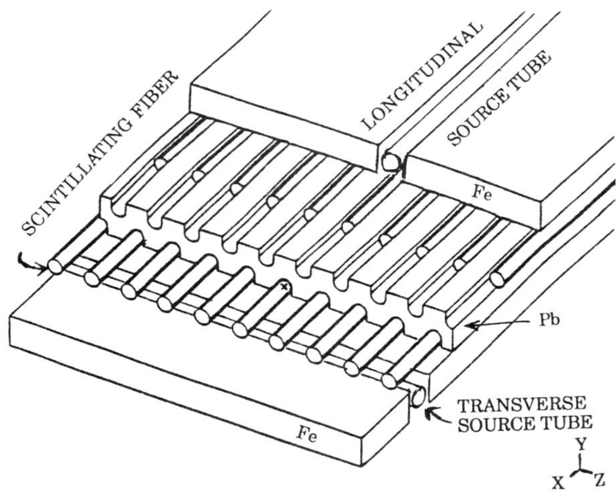

Fig. 1 Source tube geometry.

One tube is imbedded longitudinally near the center of each tower, terminating at the front end of the tower. Groups of six towers have transverse tubes imbedded in the midplane steel sheet at 10, 100, and 190 cm from the front end, and labelled T1, T2, and T3. The T tubes also run longitudinally along the edge of a group of towers, and both T and Longitudinal tubes pass near the rear fiber bundles and the phototubes.

Fibers starting in depth are bundled separately into side towers, while the "core" fibers which go within 30 cm of the front are bundled into center towers. Transverse T2 and T3 scans will cause center tower peaks and also pairs of side tower peaks. T1 scans do not cause useful side peaks, but do light up the edge tower while in transit.

Data Acquisition and Analysis

The DC current channels of the CDF Central Electromagnetic Calorimeter Rabbit electronics were used. Current readings were taken at about 15 Hz. The gains are uniformly 11.62 pA/count, with a sigma of about 0.1% [3]. Peaks range in height from 30 to 150 nA. The associated automatic peak fitting software was adapted for the more complex and more spiky patterns of peaks seen in the present calorimeter.

Longitudinal Source Scan Responses

A typical longitudinal source run is shown in Fig. 2. The structure is real, and is seen in mirror image in the extension (EXT) and retraction (RET) portions of the run, which consists of several thousand data points. The structure is presumed to reflect the texture of the 2.5 mm fiber spacing.

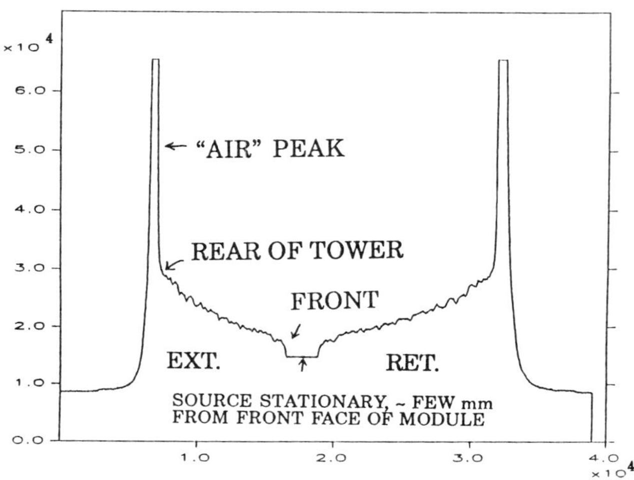

Fig. 2 A Longitudinal scan.

The large peak is with the source in air. The midpoint near bin 1700 is with the source at rest and within a few mm of the front end of the tower, where it subtends close to 2π steradians. The optical attenuation length of the fibers can be estimated at about 2 m, and is increasing from the back to the front of the towers.

Transverse Scan Responses

A T2 scan is shown for the second tower in a group of six, in Figs. 3a (side) and b (center). The first and last peaks are the broad "air" peaks. In Fig. 3 a one sees four small side peaks (EXT and RET).

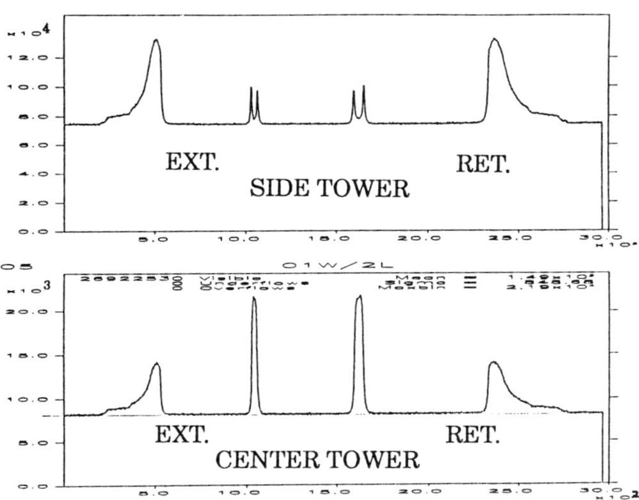

Fig. 3 a,b T2 scans: side, center.

The two side peaks of a tower are typically unequal, presumably due to some combination of fiber non-uniformity and geometrical nonuniformity. The fitted peak 2/peak 1 ratios have a mean of 1.00 and a Gaussian sigma of 12%.

Defining "peak" as the maximum bin in a peak, the ratio Fit/Peak has a mean of 0.998 and sigma of 1.0% for side towers, and mean of 0.994 and sigma of 0.5% for center towers. This is surprisingly good in the absence of real plateaus, and indicates a good signal to noise ratio and robust peak fitting.

The RET/EXT ratios are 1.00 with sigmas of 1.3% and 0.8%, respectively, for side and center, indicating the reproducibility of the measurements. Slower source motion and/or a stronger source might improve the reproducibility.

One group of towers was measured at the beginning and near the end of the 10 hour complete set of measurements. The Late/Early ratio averages 0.981 with a 2.3% sigma, indicating some drift of the phototubes. The phototube high voltages were on for approximately 14 hours prior to the start of data taking, and were on for even shorter periods before taking test beam data because the supplies were shared with other groups. Phototube and tower temperatures were not controlled.

For the center towers, response ratios at different depths are an almost direct measurement of the average optical attenuation lengths of the fibers, and are also an indication of the uniformity and homogeneity of the towers.

RATIO:	T2/T3	T1/T3	T1/T2
Mean:	0.593	0.403	0.685
σ:	0.038	0.025	0.025
σ (%)	6.4%	6.2%	3.7%
λ_{opt}:	1.72m	1.98m	2.38m

The values of λ_{opt} involving T1 have not been corrected for the 10% core taper between 30cm depth (shortest core fibers) and 10cm depth (T1 location). The variance is an indication of the nonuniformity within a tower, due to some combination of fiber to fiber variations in light output, light transmission, and optical coupling to the phototube. This variance indicates how well we can equalize the tower responses prior to a test beam exposure. It also raises questions about the accuracy with which source responses could substitute for test beam exposures as the primary calibration of towers. Direct measurement of e/source and π/source ratios vs. tower, position, and time, will be done in the early 1991 test beam.

The 0.8% reproducibility seen in the RET/EXT ratios in the Electromagnetic compartment (center tower), which we believe can be improved, indicates the accuracy with which electron test beam calibrations can be maintained in a system with reasonable gain stability.

References

[1] S. M. Bertolucci et al., NIM A267,(1988) 301, S. R. Hahn et al., NIM A267 (1988) 351.
[2] A. Para, these proceedings.
[3] S. Hahn, private communication.

A LIQUID SCINTILLATOR SPAGHETTI CALORIMETER EM MODULE FOR FORWARD CALORIMETRY AT THE SSC*

P. Datte, G. Dudgeon, R. Huson, L. Peters, A. Sanzgiri, R. Webb, J. White and S. Zaman
Texas A&M University
College Station, TX 77843
and
A. Para and G. Yasuda
Fermilab
Batavia, IL 60510

Abstract

In the forward ($\eta > 3$) regions of SSC experiments, the radiation resistance, speed and granularity requirements for calorimetry may best be satisfied by "spaghetti" style calorimetry using liquid scintillating fibers in place of solid fibers. Here are described the design, construction and first results of a test of a module using 1 mm borosilicate glass tubes (n=1.47) containing a high refractive index liquid scintillator (n=1.57) embedded in a lead matrix.

Introduction

Scintillating fiber calorimetry in a "spaghetti" geometry is one of the technologies being investigated for use at the SSC.[1] Some of its more attractive features include fast response, easily adjustable granularity, excellent resolution for both hadrons and electrons and very good hermiticity and uniformity.

One property that must be carefully considered is the ability of such a device to survive high radiation levels. Existing fibers are believed to be able to withstand doses up to about 10 Megarads per year. The doses in regions with $\eta > 3$ will far exceed this value even at a large distance from the interaction region. Although it is possible that even more robust fibers will be developed before it is necessary to begin construction, another solution that does not depend on future developments is to switch to liquid fibers in these regions. Liquid scintillators are attractive because they are inherently more radiation resistant than solids, and, besides, can be recirculated if damaged.

In the following sections we describe the design, construction and a first look at a prototype liquid fiber calorimeter employing glass tubes containing a high refractive index liquid. The dimensions for this particular prototype were chosen so that beam test results can be directly compared with prototypes using 1 mm plastic scintillating fibers.

*Work supported in part by the DOE contracts DE-AS05-87ER40356 and DE-AS05-81ER40039 and through SSCintCal.

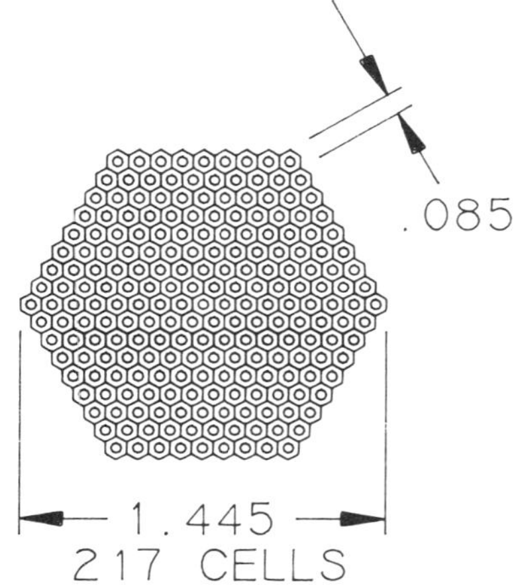

Figure 1: Components of a cell - borosilicate glass tubing and hexagonal lead tubing. Dimensions are in inches.

Components

For test beam studies, the prototype will be composed of seven modules each read out separately. A module is composed of 217 cells of hexagonal lead tubing with a glass tube in the center. The lead tubes are 280 mm in length, 2.16 mm face-to-face and have a 1.07 mm hole down the center. The glass tubes have an outer diameter of 1.02 mm and have an 85% open area. The tolerances in both cases were specified to be +/- 0.05 mm, but, the observed variations were larger by at least a factor of two in both cases. The dimensions of the components of a single cell are shown in Figure 1. In

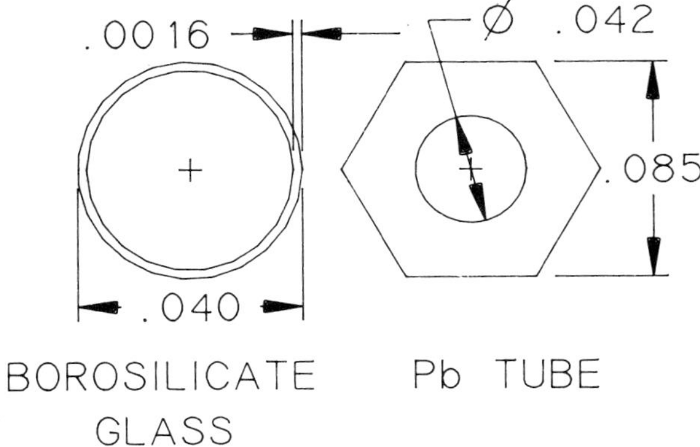

Figure 2: End view of a 217-cell module. Dimensions are in inches.

all drawings, the dimensions are labeled in inches. The lead was purchased from Vulcan Lead Company in Illinois, and the glass from Collimated Holes in California. These dimensions were chosen to correspond to 1 mm fibers in an absorber to fiber ratio of 4:1. Considering only the lead at 4/5 normal density, the module depth is about 40 radiation lengths. Figure 2 shows an end view of a module.

The glass tubes are made of a borosilicate glass with an index of refraction of 1.47. The liquid scintillators used in this device must have high refractive indices. The liquid used for this study was obtained from National Diagnostics and had a refractive index of 1.57. This combination yields a total-internal-reflection angle of about 20 degrees.

Plumbing and Light Extraction

The plumbing was designed to guarantee a predictable fluid flow even at small pressures. This is a crucial part of the design since a primary advantage to using liquids is the recirculability of damaged scintillator. Bubbles or stagnant regions could severely degrade the performance of the device. The desired reliability is achieved by forcing the liquid to flow through the tubes in series instead of in parallel. The ends of tubes are connected in pairs through slots in end plates as illustrated in Figure 3. The slots are alternated on the two ends so that one continuous tube is formed. Filling takes only a few minutes for this 217 tube design.

The light coupling is achieved by making the end plates out of a clear plastic. Plexiglas is used in these modules, but a higher index plastic or a glass may be used in the future. Once the light is through the end plates, it is transported to the phototube through light guides made of optical fiber bundles composed of 0.5 mm OD fibers. The bundles cover the entire rear face of the module. For a realistic detector, a plausible scheme to reduce the area will be required. Before reaching the phototube, the light goes through a 100 mm hexagonal plexiglas light mixer for even illumination of the photocathode.

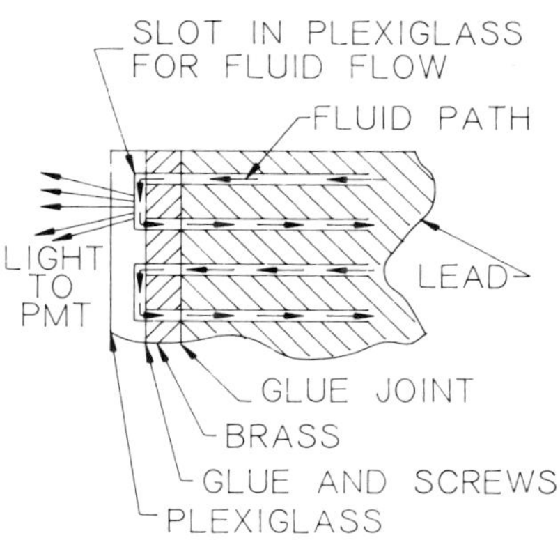

Figure 3: Serial pluming scheme using slotted end caps.

Assembly

The modules are assembled in several steps. The lead tubes are first clamped and glued together. The glass tubes are then inserted through the holes in the lead and through matching holes in brass pieces. The brass, glass and lead are then glued together bonding the brass to the lead and the glass to the brass. The plexiglas end caps with slots are then attached to the brass with both glue and fine screws. The fluid enters and exits from one end of the module and light is collected from the other end. A side view of a complete module is illustrated in Figure 4.

First Look at Performance

In order to begin performance studies before availibility of a test beam, a system was set up at Fermilab to measure the signals from cosmic ray muons. The system consisted of two scintillator paddles (20 by 25 mm) placed above and below the module (about 30 mm spacing). The signal from the phototube attached to the module was digitized and read out by a PC through CAMAC upon a coincidence from the paddles. For this test, the fiber bundle light guide was not used, so the light propagated from the module to the phototube through the 100 mm light mixer only. No glue or grease was used for coupling.

Unfortunately, the module was damaged during shipment from Texas A&M to Fermilab. It was slightly bent with a deflection of about 5 mm in the center. The

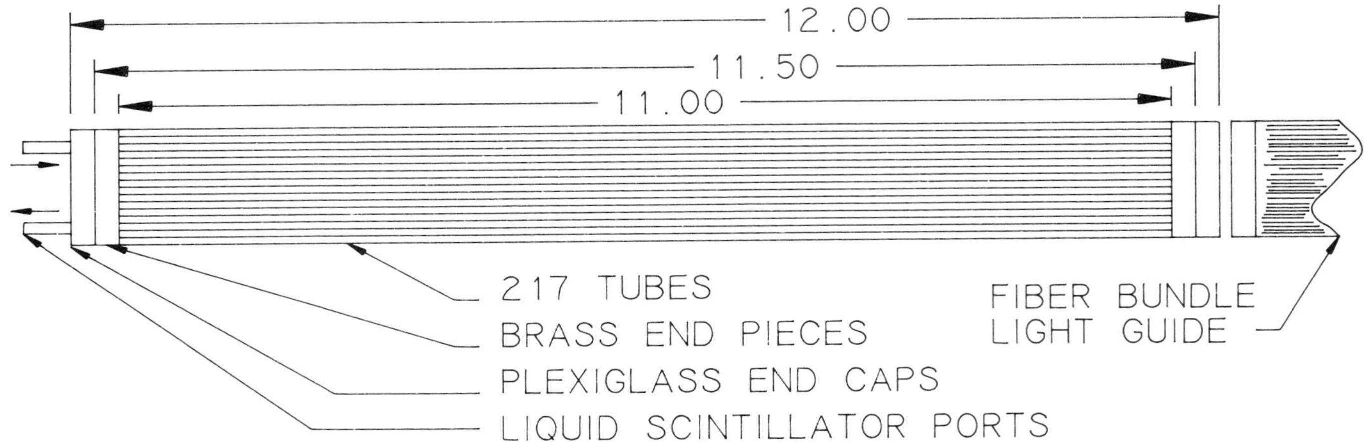

Figure 4: Side view of a module. Dimensions are in inches.

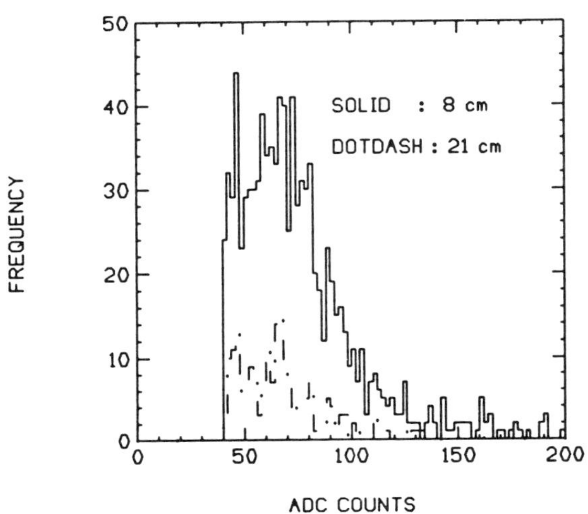

Figure 5: Cosmic ray muon signal 8 cm from light mixer (solid curve) and 21 cm from mixer (dotdash).

glass tubes were almost certainly snapped (could not actually see this), but we decided to test the module anyway. We measured the pulse height from the cosmic muons at two different positions along the module: at 80 mm and 210 mm from the phototube. Figure 5 shows the spectra for the two positions. The solid curve represents 1000 events with the paddles at the 80 mm position and the dotdash curve is the signal for 200 events at the 210 mm position. Time did not permit a systematic study of voltage settings or high statistic runs. However, the curves do show that that light actually emerges from the device and that output from the two positions does not differ greatly.

Future Plans

The seven modules are being assembled and testing for uniformity using sources and cosmic rays is underway. The prototype will be studied in a test beam when the Fermilab fixed target program resumes this fall. Two of the modules will contain solid fibers for comparison. A number of different liquids and slight variations in module construction will be tested.

Acknowledgements

We would like to thank the Texas Accelerator Center for use of its facilities and help in constructing these devices and National Diagnostics for supplying us with a sample of high refractive index liquid scintillator. We also wish to thank Karen Kephart and others at Fermilab for valuable assistance in this project.

[1] See, for example, the proceedings of this conference.

FULL-SIZED MODULES FOR THE ZEUS HIGH RESOLUTION FORWARD AND REAR CALORIMETERS

Presented by Bill Frisken, York University
for the ZEUS Calorimeter Group

Abstract

Production of the ZEUS high resolution forward and rear calorimeter (FCAL and RCAL) modules was undertaken in the Netherlands and in Canada. The Barrel Calorimeter is of the same conceptual design, but will not be discussed here. Most of the massive modules were shipped directly to DESY, but several went first to CERN for beam calibration tests. This paper discusses special features of the design and construction technique, and test beam performance.

Introduction

ZEUS is an electron-quark collider experiment for HERA. It must measure deep inelastic scattering events both for neutral current and charged current processes. The latter have a scattered neutrino in the final state, and must be measured by the quark jet alone. Other physics measurements accessible at HERA require detection of missing transverse energy, and/or extra electrons within the quark jet. The ZEUS calorimeter is designed to have the best possible capability for the measurement of jets, with a particular view to hermeticity, energy resolution, angular resolution and hadron electron separation. The design process proceeded hand in hand with an vigorous testbeam program.[1,2)]

Design Features of the ZEUS FCAL/RCAL

Energy Resolution. In tests with preprototype Uranium-Scintillator calorimeters, the ZEUS Calorimeter Group had obtained hadron energy resolution equal to 35%/sqrt{E} using 3.3mm thick (1 radiation length) plates of Uranium, sampled by 2.6mm thick tiles of SCSN-38 plastic scintillator (see reference 2 and previous work cited therein). This layer modularity became the basis of the ZEUS design.
Angular Resolution. FCAL/RCAL hadronic calorimeter (HAC) towers are 20cm square, transverse to the beam direction, and the electromagnetic (EMC) towers are 5cm high by 20cm wide (10x20 in RCAL). The towers are read out by wavelength shifter plates (WLS) at both sides, which allows the EMC shower centroids to be located to within $5x5cm^2$. The front plate of FCAL is located at 2.2m from the interaction point and that of RCAL at 1.5m.
Prototypes: Four prototype modules, each comprising four full HAC towers and 16 EMC towers, were designed and assembled at York University, and shipped to CERN late in 1987. Some of the parts and components were provided by other members of the ZEUS collaboration. These modules followed the intended final design very closely, both physically, and mechanically. The resulting four module array of sixteen complete HAC towers and 64 EMC towers has been subjected to detailed test beam investigation at CERN for more than two years.[1,2)] The excellent performance of these prototypes not only proved that the basic layer modularity was correct, but also gave us a detailed prediction of the behaviour to be expected of the ZEUS calorimeter. This led to changes in the design of the final modules (lightguides changed from clear PMMA to wavelength shifter material, Uranium plate spacers in the EMC changed from WC to the lower Z material, TiC), and to the introduction of a 3mm sheet of Pb between modules when assembling them to form the ZEUS calorimeter.
It taught us a great deal about fabrication logistics, and it also proved that the unusual mechanical design could withstand the rigors and hazards of transatlantic shipping.
Hermeticity: The hermeticity requirement was concerned with loss of energy due to particles escaping through gaps in the calorimeter, loss of energy due to jet punchthrough, and also loss of energy deposited in inert structural members of the calorimeter.
-Minimal Intermodule Cracks. The number of boundaries is minimised by making the modules large, typically comprising 20 supertowers ($20x20cm^2$). They were also required to be geometrically precise, so that they fit well together, and to be strong enough and stiff enough that this precision survives installation into ZEUS. The tower geometry is intentionally rectangular rather than pointing, and our design allows us to have the most forward intermodule crack aimed just 50mr from the interaction point.
-Punchthrough. The calorimeter is deep enough (7 lambda deep in the FCAL) for 95% containment for 90% of the showers. Events with excessive punchthrough are detected by a backing calorimeter formed by instrumenting the iron of the magnet return yoke, which surrounds the ZEUS calorimeter.
-Inert structural elements are kept outside the active volume of the calorimeter. A stretched steel skin binds the module to the outside steel structure.
Hadron Electron Separation: The sampling structure of the ZEUS calorimeter is uniform in depth, and consists of 185 one radiation length layers for the deepest FCAL modules. Hadron electron separation to the 1% level is achieved by segmenting the readout longitudinally into a 25 layer EMC, followed by two Hadronic Calorimeter segments (HAC1 and HAC2) of up to 80 layers per section. The separating power is enhanced by two orders of magnitude through the introduction of two special layers near shower max in the EMC (only one in RCAL) tiled with an array of $3x3cm^2$ Silicon Diodes in addition to their normal scintillators.
Built-in Calibration/Monitoring: Uranium-Scintillator calorimeters can be self calibrated, and self monitored by the

Uranium's radioactivity. When the photomultiplier (PM) tube gains of ZEUS calorimeter modules are set so that all PM's draw the expected amount of "Uranium Noise" current, then they will have the same response to a test beam to about 1% (see below). To check long term behaviour of light generation and transmission, PM gain and linearity, and functioning of electronics, we have added the ability to remotely probe each scintillator plate of each tower with a strong, cable driven Co60 source, each PM with pulses of laser light (via light fibres), and each module of front end electronics by charge injection.

Some Design Details of the ZEUS FCAL/RCAL

General Configuration: The resulting modules[2,3] are very tall and narrow, having the width of only one supertower (20cm), but the full height of the calorimeter (4.6m). The Uranium plates run the full height of the module. The figure shows a cutaway view of one of the largest FCAL modules. Each module has an external C-shaped steel structure, to which its stack of plates is bound by its own highly stressed stainless steel external skin.

Towers and Supertowers: The towers are nonpointing, and rectangular. They are defined optically rather than mechanically. The HAC scintillator tiles have the width and height of one HAC tower, 20x20cm^2, while the EMC scintillator tiles have the width and height of one EMC tower, 5x20cm^2. The HAC scintillation light is collected and integrated in depth in two longitudinal segments by 20cm high wavelength shifter (WLS) plates attached to the sides of the module. The EMC scintillation light is collected by 5cm high WLS plates.

An FCAL supertower comprises 4 EMC towers, one HAC1 tower, and one HAC2 tower (RCAL supertowers have no HAC2 section). The WLS plates and their attached graded reflector system are preassembled into cassettes for speedy and reliable installation, each cassette containing the WLS array to read out one side of a supertower.

Structural: An F1 module will be described. It is the largest, and the others are structurally very similar. The Uranium plates are 20cm wide by up to 4.6m high, and have similar structural properties to steel plates . The stack of 185 such plates is strapped to a strong steel T-beam backspine member by the module's stretched steel skin. The machined steel end pieces (see below) of the Uranium plates are then clamped to a second structural member at the bottom, and to a third structural member at the top. These two members (called C-arms) are then attached solidly to the vertical T-beam back-spine to complete the C-shaped structure which surrounds the module on three sides, but leaves its front face bare to the particles from the interaction point. This outside steel C-structure is important whenever movement of a module occurs, and holds the module in its stationary operating position.

The Stretched Steel Skin: The module is covered by a 250 micron thick stretched stainless steel skin, which provides the structural integrity of the module, and helps to form the light seal. The skin is actually applied in 20cm wide straps, so that each strap covers the wavelength shifter cassette for one super tower. Each of these steel skin straps is tensioned to 2 tons, the corresponding compressive force being transmitted from plate to plate by spacers, which keep the load off the scintillator. The front plate is only 1.5cm thick Aluminum, so that the first layer of scintillator is almost bare to the incoming particles.

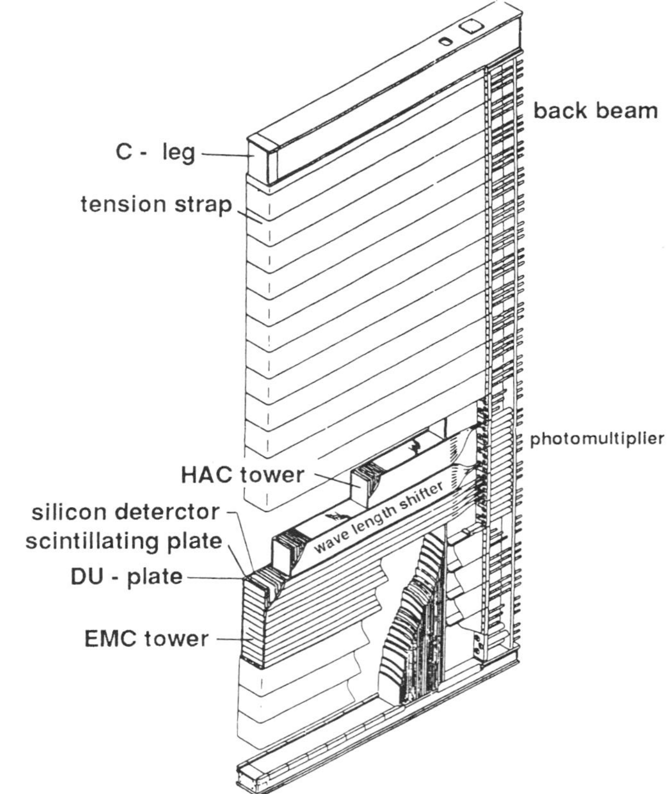

Cutaway view of a 5m high ZEUS FCAL module

Spacers: Spacers are placed between plates at the corners of supertowers, so that there are columns of spacers throughout the structure on a 20x20cm^2 square grid. In the EMC, where lateral shower extent is small, spacers have a cross sectional area only 5x6cm^2, and are made of Titanium Carbide. Elsewhere they are made of Tungsten Carbide, and in the bottom of HAC2, where maximum stresses are expected during shipping, they are 15x6cm^2.

Individually Clad Uranium Plates: The Uranium plates were rolled and then sheared to typical tolerances of 200 μm and 500 μm respectively. Since the maximum shearable length was just over 3 meters, plates for most modules were fabricated by e-beam butt-welding two shorter plates end-to-end. The plates were not otherwise machined, but each plate was then clad in a 400 μm thick stainless steel jacket (200mμ in EMC, to adjust the Uranium noise current), closed at each end by a protruding 3.3 mm thick stainless steel endpiece. These jackets were roller-formed to standard dimensions, and closed by laser welding the jacket material to itself along the longitudinal seams and to the endpieces at the ends. A computer driven plate profile measuring and laser-cutting table was then used to establish the longitudinal central axis of each plate, and cut ears on the endpieces centred on this symmetry axis. These precision-cut ears were used to locate plates during module stacking.

Module Stacking

A module was stacked by first placing its endbeam in the horizontal position, and then adding plates, scintillator tiles, and spacers in sequential horizontal layers. Thus in stacking terminology, the calorimeter depth becomes the stack height, and the height of a module becomes the length of the stack.

Open Stacking: A feature of a stack of Uranium plates of this thickness (3.3mm), is that, because of the very high density, it is difficult to move even the 3rd plate from the top of the stack. Our stacker design was based on this behaviour, and used a "rising form" technique to control precision location of the machined ears on the plate endpieces during and after delivery to the top of the stack. The technique also allowed us to constrain a mild case of "banana" distortion at the middle of a plate until it was covered by an additional 10 layers, and could safely be released. As the stacking of each layer of plate plus scintillator/spacers was completed, the online computer raised the 10 layer constraining rising form structure (RFS) by one complete layer thickness, before proceeding to stack the next layer. This RFS feature of the Markham stacker allowed us to accommodate plates with a banana distortion showing sagittas up to 2 mm.

Apart from the top 10 layers, the rest of the stack was completely free, and the alignment of its endpiece ears open to monitoring during the whole stacking process.

A Rigid Rectangular Coordinate System: Two essential requirements for such a system are that the RFS structure continue to move along fixed axes normal to the plane of the stacked plates, and that the stack's underlying base structure not deflect or tip progressively during stacking, as the weight of the stack increases. The stacker's external frame was of heavy steel construction, had a footprint $1x6m^2$, and its overall height was 5.5m. Thomson bearing arrays guided the 300 Kg RFS along two vertical 50mm diameter ground steel rods, placed 6m apart, one at each end of the stacker. The RFS was lifted by four 40 mm diameter ballscrews, whose rotation was computer driven and monitored, so that the height of this platform was always known to arbitrary accuracy.

Each module spent its early life on a strong, stiff "pallet" from the time its end beam was laid until the completed module had been reinforced and stiffened to its shipping condition. The pallets were 6m long 3 ton steel structures standing 0.8 m high, and capable of motion on wheels rolling on steel vee-rails or by flotation on a system of four 0.6 m diameter airpads. As stacking of a new module was begun, its backspine T-beam was laid using a system of levelling screws placed every 20cm along the upper surface of the pallet. The pallet (plus T-beam) was then rolled into the stacker, and picked up on jack points located on the corner columns of the stacker frame.

Granularity database: The thickness and width profiles of each Uranium plate were recorded at the rolling mill, and were entered into the plate database prior to stacking, as were weights of individual scintillator tiles. Module stacking was computer controlled, and the i.d. number and stack location of each Uranium plate, and of each scintillator tile were logged as part of the stacking process.

Stack Height Control - the Stack Compression System: Uranium plate rolling tolerances were ±150 μm for the 25 EMC plates, ±200 μm for the 80 HAC1 plates, ±250 μm for the 80 HAC2 plates, and ±50 μm for the associated 186 spacers. Variations in cladding thickness were negligible. The variations in the plate thickness reflected the dynamics of the rolling process, and were highly correlated plate to plate. Since the design of the wavelength shifter plate assemblies required the stack height to adhere to design values at the boundaries of EMC, HAC1, and HAC2, the stack height was controlled by using spacers of three different heights (3.8, 3.9, and 4.0 mm), as needed. To control variations in the stack height, the stack was compressed hydraulically several times during the stacking process, typically after stacking each 25 layers, but with special interest near the HAC1/HAC2 and HAC1/EMC boundaries. During compression, a force of 1 ton was exerted on each column of spacers simultaneously. In this condition, the stack height at each of the 48 spacer columns was sensed relative to the known height of the RFS structure under computer control. The heights of the 48 columns were automatically entered into the module's database, and the schedule of spacer thicknesses to be used at each column while stacking the next 25 layers was output to the stacker operator's terminal.

During the final compression, which was executed with the 15 mm thick Aluminum front plate in place, half of the stainless steel skin (every second 20cm wide strap) was installed and tensioned. The module, still supported solidly on its pallet, was sufficiently stable in this condition to be rolled out of the stacker, to make way for the next module.

Wavelength Shifter Installation: Since the removal and retensioning of an individual strap of the stretched skin is a field operation, installation of the WLS cassettes could be readily accommodated in a special area well away from the stacker. Typically the C-arms were also mounted in this area, whereafter the module (still on its pallet) proceeded to the shipping area for the installation of extra stiffening members in preparation for its transatlantic sea voyage.

Production Site, Schedule and Costs

Site: The Markham site was a 1000 m^2 floor area light industrial building leased for the project for 3 years. We used its 25% office space for our design office, small labs, conference room, and computer room. We used its 75% warehouse space, which featured 5.5 m height to roof trusses, and 2 loading bays at truck bed height, to build our module factory. This space easily accommodated the massive stacker, the WLS and shipping areas, a small mechanical shop, parts stock and supplies (including shielded storage for Uranium plate for up to 4 modules).

Schedule: In principle a module could be stacked, fitted with WLS and prepared for shipping in 3 weeks, and since those operations could be performed in parallel, we could (and did, briefly) produce a rate at one module per week. In practice, of the 26 modules produced, the first took 8 weeks from laying the endbeam to shipping day, and the remaining 25 took just 12 months, with our assembly line typically putting out one module every 2 weeks.

Costs: A current cost estimate of the ZEUS FCAL /RCAL cost about US$75 per Kg of Uranium content, including building, instrumenting, and delivery to DESY. The biggest cost is preparation of the Uranium plates, followed closely by electronics.

ZEUS FCAL/RCAL Cost Breakdown

Uranium billets	5%
Roll/shear Uranium plates	22%
Weld/clad Uranium Plates, cut ears	10%
Mechanical parts	8%
Plant rent, operations	13%
PM's, Bases	10%
Electronics to 1st Level Trig.	20%
Scintillator/WLS	10%
Monitoring	2%

Module Calibration and Performance

CERN test beam calibration work on the FCAL/RCAL modules just ended this summer, but preliminary analysis shows that the full-sized modules improve on the performance of the prototypes, with light output of 110 photoelectrons per PM, and 17.6%/sqrt{E} energy resolution for 15 GeV electrons in FCAL (140 p.e. and 17.4%/sqrt{E} in RCAL), plus better uniformity and better precision of calibration.

Uniformity (Tower to tower, and module to module): We have calibrated six of the largest FCAL modules and four of the largest RCAL modules in the X5 testbeam at the CERN SPS between November 1989 and July 31, 1990. Each tower was calibrated with electrons at 15 and 100 GeV/c, and with muons and hadrons at 100 Gev/c. Towers were selected by scanning each 14 tonne module across the beam on a computer controlled high precision heavy duty mechanical stage. Details of the set-up and of data analysis will be the subject of forthcoming NIM articles by the ZEUS Calorimeter Group, but preliminary results are shown in the following table. A module's PM gains are adjusted approximately during set-up by setting each one's measured Uranium noise current equal to within 2% of the expected value. Precise normalisation to Uranium noise current is done during analysis, yielding the results shown in the table below. The notation (H) and (L) appended to the energy of a particular scan, indicates whether the high gain (H) or low gain (L) ADC record was being used for analysis input.

Summary: Based on the electron, muon, and hadron runs with this sample (33% of all FCAL towers and 22% of all RCAL towers), this preliminary analysis shows the precision of the Uranium noise current calibration to be 1% in the EMC and HAC1 sections of FCAL, and 1.4% in HAC2. The spread in response of the RCAL EMC towers is more like 1.5%. Since most shower energy is deposited in EMC and HAC1, Uranium noise current alone will give a 1% overall calibration of FCAL, and a 1.5% calibration of RCAL.

This result has strong implications for future calorimeters, whose size and complexity makes the logistics of complete test beam calibration, module by module, tower by tower quite prohibitive.

Spread in Response of ZEUS FCAL and RCAL Towers when Calibrated by Uranium Noise Current (Preliminary Analysis)

Section Particle Module	HAC1 muons 100GeV(H)	HAC1 hadrons 100GeV(L)	HAC2 muons 100GeV(H)	HAC2 hadrons 100GeV(L)	EMC muons 100GeV(H)	EMC electrons 15GeV(H)	EMC electrons 15GeV(L)	EMC electrons 100GeV(L)
stat.error	0.6 %	0.5 %	0.6 %	1.7 %	1.7 %	0.3 %	0.3 %	0.1 %
F12•NL1	1.0 %	1.1 %	1.1 %	2.4 %	2.3 %	1.1 %	1.2 %	1.0 %
F12•NL3	0.9 %	1.3 %	1.5 %	3.3 %	1.8 %	0.7 %	1.1 %	1.0 %
F12•CD1	0.9 %	0.8 %	1.0 %	2.7 %	2.0 %	1.2 %	1.2 %	1.3 %
F12•NL2	0.8 %	1.4 %	0.9 %	3.4 %	2.0 %	0.8 %	0.8 %	1.0 %
F11•CD2	0.9 %	0.8 %	1.0 %	2.6 %	2.7 %	0.9 %	0.9 %	0.9 %
F12•NL4	1.0 %	1.1 %	1.2 %	1.2 %	2.4 %	1.0 %	1.3 %	1.1 %
all channels	1.0 %	1.7 %	1.4 %	3.4 %	2.6 %	1.1 %	1.2 %	1.1 %
R12•NL2	0.9 %	1.2 %	-	-	2.5 %	1.4 %	1.4 %	1.7 %
R11•CD1	0.8 %	1.5 %	-	-	1.7 %	1.1 %	1.1 %	0.9 %
R12•NL3	1.0 %	1.1 %	-	-	1.8 %	1.1 %	1.1 %	1.4 %
R11•CD2	1.1 %	1.0 %	-	-	1.6 %	1.5 %	1.5 %	1.2 %
all channels	1.1 %	1.5 %	-	-	2.0 %	1.5 %	1.5 %	1.6 %

References:

1. Response of a Uranium Scintillator Calorimeter to electron, pions and protons in the Momentum Range 0.5-10 GeV/c. ZEUS Calorimeter Group, NIM A290(1990) 95-108.

2. Test of the ZEUS Forward Calorimeter Prototype. ZEUS Calorimeter Group, NIM A289(1990) 115-138.

3. Experience with Uranium-Scintillator Calorimetry, D.K. Hasell and W.R. Frisken, Proc. Workshop On Calorimetry for the Supercollider, March 1989, Eds. R. Donaldson and M. Gilchriese, Published by World Scientific, June 1990.

Front End, Data Acquisition and Triggering Electronics

6

A PROGRESS REPORT OF THE SWITCH-BASED DATA ACQUISITION SYSTEM PROTOTYPE PROJECT & THE APPLICATION OF SWITCHES FROM INDUSTRY TO HIGH-ENERGY PHYSICS EVENT BUILDING

Worthington Hotel
Fort Worth, TX
October 15-18, 1990

Ed Barsotti, Alex Booth, Mark Bowden & Carl Swoboda
Data Acquisition Electronics Department
Computing Division
Fermilab*
P.O. Box 500
Batavia, IL 60510

Abstract

A prototype of a data acquisition system based on a new scalable, highly-parallel, open-system architecture is being developed at Fermilab. The major component of the new architecture, the parallel event builder, is based on a telecommunications industry technique used in the implementation of switching systems, a barrel-shift switch. The architecture is "scalable" both in the expandability of the number of input channels and in the throughput of the system. Because of its scalability, the system is well suited for low to high-rate experiments, test beams and all SSC detectors. The architecture is "open" in that as new technologies are developed and made into commercial products (e.g., arrays of processors and workstations and standard data links), these new products can be easily integrated into the system with minimal system modifications and no modifications to the system's basic architecture. Scalability and openness should guarantee that the data acquisition system does not become obsolete during the lifetime of the experiment. The paper first gives a description of the architecture and the prototype project and then details both the prototype project's software and hardware status including details of some architecture simulation studies. Suggestions for future R&D work on the new data acquisition system architecture are then described. The paper concludes by examining interconnection networks from industry and their application to event building and to other areas of high-energy physics data acquisition systems.

Introduction

A new data acquisition system architecture, called the Scalable Parallel Open Architecture Data Acquisition System[1] (SPOA DAQ), is being developed at Fermilab. The goal of the project is to test the new system's architecture by simulating then building a prototype data acquisition system whose central component is a switch-based self-routing parallel event builder[2]. In order to test the system, a crate of test modules emulating physics event data sources will be used to provide high-speed parallel inputs to the event builder and a crate of electronics emulating an online processor farm will be used to receive high-speed "built" events in parallel from the event builder.

* Operated by Universities Research Association, Inc. for the U.S. Department of Energy under Contract No. DE-AC02-76-CHO3000.

The architecture is scalable, firstly, in so much as it is well suited for data acquisition systems in low to high-rate experiments, test beams and all SSC detectors. Secondly, as both technology and physics needs change, the architecture "scales" for higher throughput (by adding more channels and/or processors) without modifying the fundamental structure of the system. This last feature is also implied in the "Open Architecture" part of the project's name. "Open Architecture" means that new technologies (e.g., new online processors from several companies, newer and faster data links, etc.) can be added to the system (or can replace existing elements) with little extra system development and no architecture modifications required.

Extensive behavioral modeling and simulations[3] have been undertaken with the goal of understanding not only individual components of the architecture but also how they interact in the system. Methods for diagnosing faults in the system and easily maintaining the system were of prime importance in the simulations. A number of diagnostic features were added to the prototype hardware as a result of the simulations.

Simulations are almost complete as are most all of the prototype project's hardware designs. System tests should be completed sometime in April, 1990 at which point there will be an extensive demonstration of the project's user interface, normal operations and diagnostic features.

Interconnection networks from both the telecommunications and computer industries have been extensively studied[4] before choosing the network used for event building in the prototype project. The dominant use of interconnection networks in the future probably will be in the broadband ISDN telecommunications area. Computer and computer peripheral interconnections will be a second major use of interconnection networks in the future. The application of commercially-available interconnection networks in high-energy physics data acquisition systems is discussed towards the end of this paper.

New Data Acquisition System Architecture

Late in 1988 work began at Fermilab to define a new data acquisition architecture which would meet and exceed the needs of high-energy physics detectors in the next decades. Some initial architecture goals are listed below:

- Scalable from low-rate experiments and test beams through very high-rate experiments
- Data rates from bytes per second through hundreds of gigabytes per second
- Easy expansion for more data sources and/or higher data rates without the need for architecture modifications
- Commercially-available hardware and software products and industry-supported standards used wherever practical and cost effective
- Simple control structure (i.e., very few messages to keep data flowing from the detector)

Shortly after this project began, several fundamental architectural concepts which would allow us to achieve the above goals and others were studied and decided upon. Some of these concepts are listed below:

- Parallel not sequential readout
- Push not pull data transfers
- Separate data and control links
- Decentralized control
- The use of simple switching techniques for parallel event building
- Synchronous control within the parallel event builder
- Event request link(s) from processors within the online farm to a trigger system/data acquisition system interface

Sometime later the SSC formed the Task Force on Electronics, Triggering and Data Acquisition for Experiments at the SSC. Fermilab's new architecture was presented at a Task Force meeting and enhanced during subsequent meetings. The name given to the new data acquisition system architecture is the Scalable Parallel Open Architecture Data Acquisition System.

Basic Architecture & Its Components

Figure 1 illustrates the main components of the new data acquisition system architecture. Brief descriptions of these components follow:

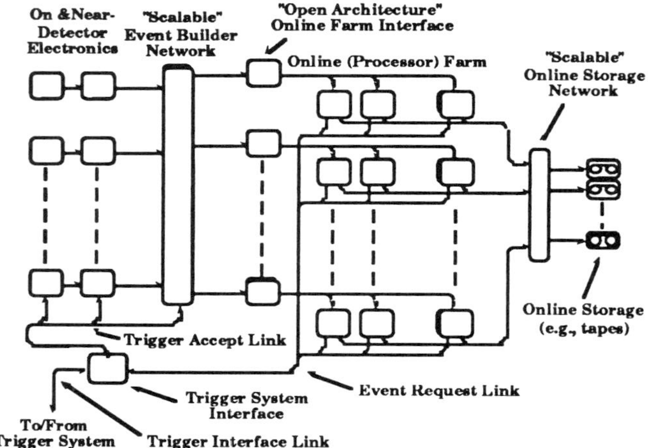

Figure 1.
Scalable Parallel Open Architecture
Data Acquisition System
Block Diagram

Trigger System Interface

The Trigger System Interface, as the name implies, interfaces trigger accept and, optionally, reject information from the experiment's trigger system to the data acquisition system. Two simple serial links, the Trigger Interface Link and Trigger Accept Link, connect this interface to the trigger and data acquisition subsystems, respectively. For each trigger, information contained on the Trigger Interface Link includes a Trigger Type, and optionally an Event ID, a Partition ID (of the received trigger type), and error detection and correction information.

From information transmitted from the online processor farm via the Event Request Link, the Trigger System Interface contains tables of available online farm processors. Tables of desired trigger types for each online farm processor, downloaded at system initialization, can also be included but are optional. When a (good) trigger is received from the trigger system, a match to an available processor is quickly made. The interface immediately sends, via the Trigger Accept Link, the received trigger information in addition to a Destination ID (which processor) and an Event ID to the Event Builder Network and optionally to the On and Near-Detector Electronics. Which information (e.g., Destination IC and Event ID) is sent where depends both on system implementation details and on where various levels of data integrity begin in the system. Simulations have shown the Trigger System Interface to be easily capable of handling tens of thousands of events per second.

Data from the Trigger System is interfaced to the data acquisition system in the same manner as that from front-end electronics.

On & Near-Detector Electronics

At various SSC meetings, emphasis has been placed on minimizing the number of interfaces between front-end electronics and the data acquisition system. Ideally, a single interface standard (protocol, signal levels, etc.) would be used to interface to all detector subsystems. Figure 2 illustrates a possible implementation of this standard where relatively slow (a few Megabytes per second) and simple serial links interface Front-End ICs and subsystems to Data Readout ICs. Another possible implementation of this interface would daisy-chain multiple Front-End ICs and a Data Readout IC together using a parallel path with a similar clock rate as in the serial link implementation. The first implementation is simpler to checkout and maintain but puts reduced limits on the amount of Front-End ICs which can be connected to a single Data Readout IC. Data Readout ICs can be multiplexed to other Data Readout ICs as shown in the figure. The degree of multiplexing depends upon predictions and tests of the amount of data anticipated from various Front-End ICs and subsystems. The Front-End ICs, Front-End subsystems & Data Readout ICs comprise what is titled "On-Detector Electronics" in Figure 1.

As the name implies, Near-Detector Electronics is situated near the detector but out of high-radiation areas. The area where this logic sits represents a "staging" area for experimental data sent to counting rooms. Any front-end or trigger electronics not on the detector is placed in this area along with the data acquisition electronics shown in the figure. The Near-Detector Electronics illustrated in Figure 2 assumes a data acquisition architecture with a parallel Event Building Network. For efficient use of a network and thus highest throughput, the network must both be preceded by input queueing logic and have roughly balanced amounts of data at each of its inputs averaged over time. Ideally, input queueing buffers fragments of equal sizes from different events and presents these fragments to the event builder network during a single time slot so that no two fragments are being transmitted to the same output during any given time slot. Realistically, there are differences in event fragment sizes even averaged over a significant amount of time (heavy buffering) such that, for long term differences, Data Balancing Logic as shown in Figure 2 is desirable to increase the throughput of an event building network. Data Balancing Logic is very simple. Groups of (serial) cables from the detector are multiplexed so sets of cables within a particular group can be connected to specific Data Buffering & Ordering Logic/(Fiber) Data Link Driver pairs. Since this logic is only included for long term changes in event fragment size distributions, there is no need for "real-time" interconnectability. When the system is down, system control computers would remotely change the Data Balancing Logic configurations.

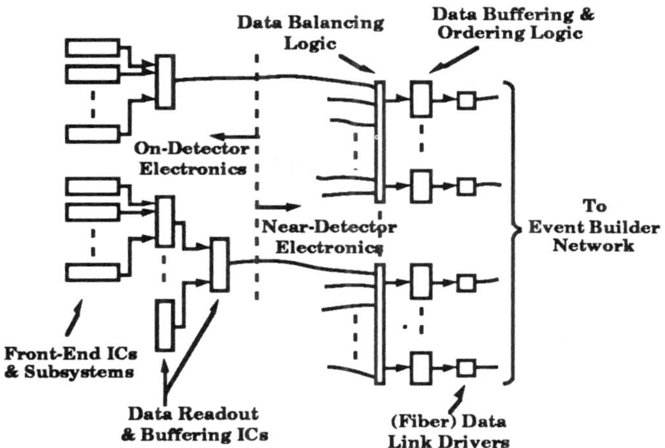

Figure 2.
On & Near-Detector Electronics
Block Diagram

Details of other Near-Detector Electronics is very subsystem implementation dependent. Data Buffering & Ordering Logic and (Fiber) Data Link Drivers are shown in Figure 2. The final implementations of front-end electronics, for example, could heavily buffer events such that (event) data ordering of event fragments is done within each front-end IC. In this implementation, Data Readout ICs could further buffer and event order these fragments such that, when data arrives at the near-detector electronics, it is already event ordered and no further event ordering is needed.

Another implementation lightly buffers event data in front-end ICs and transmits event fragments as quickly as possibly from these ICs. In this implementation, event fragment ordering, if required, is done in both the Data Readout ICs and the Near-Detector Data Ordering Logic.

The Near-Detector Electronics could also include the Event Builder Network input queueing logic mentioned earlier as an integral part of its required event data buffering.

Perhaps a thousand or so serial cables would be used to transmit event data from the detector to the Near-Detector Electronics. After multiplexing, etc., less than one hundred fiber-optic cables operating at medium cost-effective speeds (tens of Megabytes per second) transmit event data over a long distance (a few hundred meters) to counting room electronics racks and into the Event Builder Network.

Event Builder Network

The Event Builder Network would probably be switch-based and require input and output queueing networks to allow for throughputs in the Gigabytes per second range. Industry is working on such networks but are concentrating on specific markets such as broadband ISDN (telecommunication industry) and processor/disk etc. interconnections (computer industry). Since physics event building is substantially different than message or file transfers (event fragments from all network inputs get "built" into total event streams), it is possible that these "general purpose networks" from industry are overkill for high-energy physics event building. It presently appears that building a switch-based parallel event builder "in-house" based on switching techniques from industry is a better approach for high-energy physics. This subject will be covered in more detail later.

Online Storage Subsystem

For a typical SSC detector, the online storage subsystem will accept 10 to 100 (one Megabyte) events per second and be purchased from industry. Figure 1 includes an Online Storage Network (switch). This network would be used to allow specific type events to be sent to a specific storage medium to simplify data retrieval at a later time. This network would be substantially smaller in size and require much lower throughput than the Event Builder Network.

Data & Control Links

All data and control links, with the exception of data links from the detector to the Near-Detector electronics can be standard industry-supported data links.

Event Data Links

Throughput requirements for a typical SSC detector are in the few Gigabytes per second range. For a conservative design, this implies several tens of fiber-optic cables each with several tens of Megabytes per second data rate capability. Event data links into and out of an event building network need not have the same data rate. Data links which connect the event builder network to commercial banks of processors will have higher data rates in the future and will also be fiber-optic links. From the Near-Detector electronics through online storage, industry will provide event data links and interfaces (which use these links) to processors, storage media, etc.

Trigger Interface & Trigger Accept Links

Triggers occur every few microseconds in a typical SSC detector. Even if both good and bad triggers were sent over the Trigger Interface Link to the Trigger System Interface and even if the trigger message required as many as 16 32-bit words, it would require just one microsecond to transmit trigger information using a commercially-available 500 Megabit per second serial link. If only good triggers were transmitted to the Trigger System Interface, a much slower link could be used.

Event Request Link(s)

The maximum average rate that a request for an event originates from an online farm processor equals the good trigger rate to the Trigger System Interface. Since this rate is typically 1 KHz and not more than 10 KHz, a relatively slow commercially-available serial link could be used for the Event Request Link.

Data Acquisition System Prototype Project

As work progressed towards defining a new data acquisition system architecture, it soon became apparent that to thoroughly study and understand the new architecture, a simulation and prototype project should be defined which tests out the architectural concepts and doesn't require simulating and building an entire data acquisition system to do so. The project chosen divides the architecture into four manageable parts as follows enabling us to complete the project and prove (or disprove!) its concepts in a reasonable amount of time with a reasonable amount of effort:

- Detector data source "emulator"
- Trigger System Interface "emulator"
- Event Builder Network
- Online Processor Farm "emulator"

Figure 3 illustrates the components of the new data acquisition system architecture which have been simulated and are being built at Fermilab. They will be described in detail later in the paper. Besides simulating the architecture, the goal of the project is to produce a "scalable" 8-input x 8-output event builder subsystem modularly-expandable in a single crate to a 64 x 64 parallel event builder capable of a maximum 1.28 Gigabytes per second throughput (64 links x 20 Megabytes per second per link).

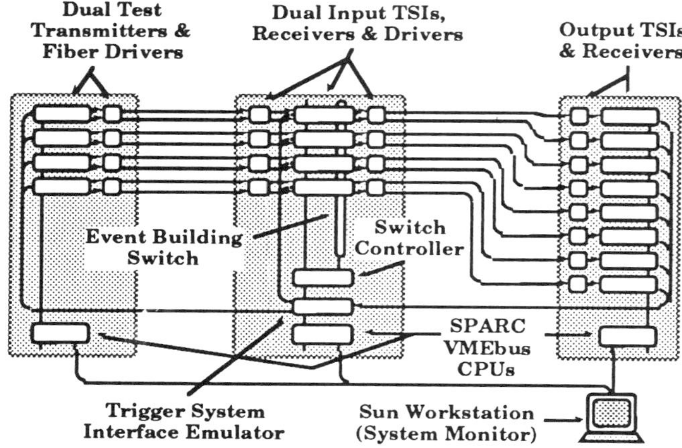

Figure 3
Prototype Data Acquisition System Project Block Diagram

Prototype Project Software

Three commercially-available software tools were chosen to aid implementation of the prototype project, Verilog[5] for modeling and simulations, DataViews[6] for the project's user interface, and NEXPERT[7], a expert systems tool, for diagnostics.

Simulations

Verilog-XL is a digital logic simulator based on the computer hardware description language Verilog-HDL. It provides advanced simulation capabilities designed to handle complex electronic designs and has many features which can be summarized as follows:

- Different levels of abstraction (i.e., system or architectural, behavioral or algorithmic or register-based, and transfer or gate/circuit levels).
- Mixed-level modeling.
- Stochastic analysis.
- Interactive debugging environment.
- Open design environment and library support.
- Programming language interface (e.g., C).

These features make Verilog very useful for modeling data acquisition systems, even those which include multiple ASIC's and complex VLSI devices.

Verilog provides three kinds of graphical display:

- A logic analyzer type display which is very useful for low level debugging.
- A register display where the programmer can display the value of any variable or register in his HDL code.
- A bar graph display for showing such things as queue occupancy and overflows in the stochastic/queue management type modeling.

User Interface

DataViews is a powerful graphical design environment for developing custom color displays for real-time monitoring and control. DataViews is written in C and runs on most 32-bit workstations. It comprises two main components; a powerful drawing editor called DVdraw, and a comprehensive set of utilities called DVtools. DVdraw enables users to create and modify color pictures. In the prototype project, these pictures were system diagrams of the project as well as menu driven displays for user interaction. DVtools allows the user to specify the dynamic interactions of all components on each screen and between screens, as well as to integrate the displays into user application programs. DataViews is also used to "front-end" Verilog so that the infrequent user of Verilog can easily set up parameters for a simulation run in a user friendly way, without having to know Verilog HDL code.

A state-of-the-art user interface was considered one of the most important aspects of the Scalable Parallel Open Architecture Data Acquisition System.

Diagnostics

Nexpert is a powerful "knowledge representation and reasoning" system. It includes a rule and "object" expert system shell and was particularly attractive to us for three reasons:

- It meets some goals we have in terms of diagnosing data acquisition systems.
- It includes a unified database bridge which interfaces to a variety of database packages.
- It has a software bridge to Dataviews

This last feature was very attractive because the bi-directional relationship between Nexpert and Dataviews means that by clicking on buttons on a Dataviews "view", rules fire which can cause for example, other programs to be invoked (e.g., reading status registers in the DAQ system), and even other "views" to be displayed (e.g., a lower level in the DAQ system). In the context of diagnosing a switch-based data acquisition system, we desire to narrow the problem down to a particular functional sub-unit as quickly as possible. One can imagine for example a "rule" which tests whether all nodes in the processor farm are "seeing" a particular problem, or whether nodes only on a particular channel are "seeing" the problem. If all nodes are "seeing" the problem, then the problem is probably pre-switch or on the input to the switch itself, whereas if only one channel is experiencing a problem, then the problem is post-switch or on the outputs of the switch itself. Either way, this simple "rule" allows the diagnostic software to perform the first "binary cut" on diagnosing the total DAQ system. At the present time, expert system software is the least developed part of the project, but still remains a positive area for investigation and development.

Prototype Project Hardware

The hardware part of the project consists of three crates, one crate housing the detector data source emulators, one crate housing the Event Builder Network, its input queueing network and the trigger system emulator, and one crate housing the online processor farm emulators. Detector data sources (On and Near-Detector Electronics) are being emulated using Test Transmitter modules. The Event Builder Network and its input queueing logic consists of a barrel-shift switch and time division multiplexors (Input TSIs), respectively. The Output TSIs serve two purposes, online processor farm emulators and output queueing logic for the Event Builder Network.

The project includes Trigger Accept and Event Request Links, as described earlier, to control event data flow.

Trigger System Interface Emulator (TRIF)

The Trigger System Interface is the only central control point in the overall data acquisition system architecture. Simulations have shown that this interface easily accommodates tens of thousands of events per second <u>today</u> so it will not be a "bottleneck" in the system. The Trigger System Interface module or emulator processes event requests from online farm processors (via the Output TSI in the prototype system) and matches these requests with triggers from the trigger system. If all processors on all output channels are busy, no trigger accept is generated. The destination ID of the selected processor is broadcast over a fast serial link to all Input TSI channels for use in routing the event.

The Trigger System Interface Emulator module in the prototype system includes a trigger simulator and a second serial broadcast link. The "event type" information from the simulated trigger is sent to the Test Transmitters to select one of up to 256 stored event patterns.

The module is built on a 9U X 400mm Eurocard and has a VMEbus interface. For diagnostic purposes, there are software "diagnostic FIFOs" at each of its outputs and inputs. A SPARC processor is included for embedded control and a 68008 processor is included to handle Event Request Link communication. This module would become a true interface between the trigger system and the event builder subsystem with the simple addition of a fast serial port to connect the module to the trigger system.

The Trigger System Interface Emulator module receives Event Requests from the individual Output TSIs and stores these requests in queues corresponding to each trigger type (there need only be one trigger type if desired). If all processors can accept all event types, then only one queue is necessary. When a trigger is generated, a processor destination ID is removed from the queue and transmitted to the Input TSIs for use in routing the event data. At the same time, a "trigger accept" message is transmitted to the front-end readout modules (or Test Transmitters) indicating the event type for selective readout.

The Trigger System Interface Emulator module uses a simple priority table to help ensure that the data flow is balanced. Sequential events are assigned to the same output channel only if no other output channel has a request pending. To avoid processor deadtime caused by the Event Request latency, processors typically support multiple event buffers and request several events ahead of the one currently being analyzed.

In the prototype system, the trigger is simulated by the Trigger System Interface Emulator module and all trigger/request matching is performed by software. In a real system, the matching function could be placed in hardware to reduce front-end buffer requirements.

Detector Data Source Emulators (Test Transmitters)

Each Test Transmitter channel emulates one high-speed multiplexed data channel from a detector front-end system. The prototype project consists of four Dual Test Transmitters. Data for up to 256 real or simulated "event fragments" can be placed in the Test Transmitter memory and selected on the basis of a trigger received over a fast serial input. Programmable headers and trailers can be added to these static data packets to form a dynamic event. For diagnostic purposes, the Test Transmitter can also generate errors in the header information.

The Test Transmitter can drive fiber-optic or differential serial outputs at up to 100 Megabytes per second to allow testing of high-speed data links at full bandwidth. The normal output rate is approximately 20 Megabytes per second when used in the prototype system.

A Dual Test Transmitter is built on a 9U X 400mm Eurocard and has a VMEbus interface. For diagnostic purposes, there is a "diagnostic FIFO" at each of its outputs. A SPARC processor is included for embedded control.

A Test Transmitter receives a "Trigger Accept" message containing the Trigger Type (and optional Destination ID, etc.) from the Trigger System Interface. One of up to 256 stored event fragments is then selected by the Test Transmitter. The processor constructs an event header and trailer which are written to the header/trailer FIFO. The event header contains a word count and source ID, plus the Trigger Type/Destination ID information received in the Trigger Accept message. Any of these fields may be altered under program control for test/diagnostic purposes.

A Test Transmitter sends the header from the header/trailer FIFO, followed by the event data and data checksum from the event memory, followed again by any trailer information from the header/trailer FIFO. Events are pipelined so that header/trailer information for the current trigger is assembled while previous events are being transmitted.

Input Time Slot Interchangers (ITSIs)

The Input TSIs package asynchronous event data fragments into fixed-sized packets, buffer and time division multiplex these packets through the Event Builder Network.

The prototype project consists of four Dual Input TSIs, one for each pair of detector data sources. An Input Time Slot Interchanger (ITSI) contains up to 64 logical buffers with one buffer for each Output TSI channel (although the prototype project consists of only an 8 X 8 event builder network, the crate which houses the prototype project can accommodate up to a 64 X 64 network). Incoming data is placed in one of these buffers based on either a destination header in the data packet or a message from the Trigger System Interface. Fixed length packets are removed from each buffer and transmitted through the switch in a predefined order (normally a standard barrel shift rotation).

In the prototype system, each of four Dual Input TSI modules contains two independent data channels and occupies one slot of the switch crate. The Dual Input TSI module also contains serial drivers for the switch output connection (to the Output TSI). The Input TSI logical buffers are managed by a SPARC embedded processor (software compatible with the System Monitor/Local Controllers). This implements a high-speed paged memory with 512 pages of 512 words each. The pages are allocated to individual buffers as needed.

A Dual Input TSI module is built on a 9U X 400mm Eurocard and has a VMEbus interface. For diagnostic purposes, there are "diagnostic FIFOs" at both the input and output of each of its two channels.

The Input TSI receives variable length event data packets from a Test Transmitter over the event data link. It also receives trigger messages from the Trigger System Interface module. Based on routing information contained in the event data packet header or the trigger message (or both), the packet is copied into the appropriate output buffer. Memory pages are dynamically allocated to each output buffer as needed.

Fixed length data packets are copied from each output buffer to switch synchronization FIFOs, in the same order as the switch rotation (normally sequential). In the prototype version, the Input TSI module contains the switch input and output synchronization FIFOs, and the Output Data Link driver.

Event Builder Network (Barrel-Shift Switch)

The Time Multiplexed Switch is implemented on the backplane of the switch crate. It is a fully synchronous packet based switch with up to 64 bytewide channels. The switch configuration is determined by the Switch Control Logic module which contains the data clocks and a configuration memory. The content of the configuration memory is static and does not change during system operation. Therefore, no external switch control is required. The switch configuration control registers are double-buffered, allowing full reconfiguration of the switch in a single clock cycle. The first stage control registers are loaded during the packet transmission time. Because all of this circuitry is integrated into the L64270, only eight ICs are required to implement a 64 channel (capable of greater than two Gigabytes per second bandwidth) data switch.

The switch backplane can accommodate up to 16 Quad Input TSIs. This allows 1 to 64 inputs and 1 to 64 outputs in a single event builder crate. The Switch Control Logic module is built on a 9U X 400mm Eurocard and has a VMEbus interface.

The Time Multiplexed Switch operates in a synchronous, fixed rotation mode. For each packet timeslot, the interconnect pattern is transferred from a static memory on the Switch Control Logic module. This memory is normally programmed to connect the switch as a simple N x N barrel shifter, but can also be used to implement multiple independent barrel shifters (static partitioning), N x M switches and switches with different input/output data rates or inactive channels. The switch control registers can also be loaded directly from VME for test purposes.

Barrel-Shift Switch Operation

The operation of the barrel-shift switch has been described in detail in another document[8]. Briefly, a barrel-shift switch of size N x N has N different states, state 1 connects input 1 to output 1, input 2 to output 2, etc. State 2 connects input 1 to output 2, input 2 to output 3, etc. This "shifting" continues until the state N which connects input 1 to output N, input 2 to output 1, etc. At this point the switch has "shifted" through all its N states and returns to state 1. If input queueing logic (time division multiplexing) is used and equal packets of event data are sent through the switch at each switch state, event building at all outputs occurs in parallel producing an truely parallel event builder simple but capable of very high data rates.

Output Time Slot Interchangers (OTSIs)

Each Output TSI emulates a string of many processors connected to a single output of the Event Builder Network. Since they are also standard VMEbus master/slave modules, these modules can also be used as interfaces to VMEbus CPUs or workstations with VMEbus interfaces. A main function of the Output TSI is to concatenate packets of event fragments into total events.

The Output Time Slot Interchanger contains up to 64 logical buffers with one buffer for each Input TSI channel. Fixed length packets from the switch are placed in buffers in a predefined order (normally a standard barrel shift rotation). Complete events are assembled on output by concatenating event fragments from each buffer.

In the prototype system, the Output TSI module contains a single data channel and can be installed in a standard VME processor system or a commercial workstation with a VMEbus interface. The Output TSI supports single VMEbus read/write operations both as a slave or master and 32 bit master/slave block transfers for data output from the TSI buffers. The Output TSI module also contains the driver for the Event Request Link which is accessed through a VME message from the processors.

An Output TSI module is built on a standard VMEbus 6U X 160mm Eurocard. For diagnostic purposes, there is a "diagnostic FIFO" at its input. A SPARC processor is included for embedded control.

The Output TSI receives fixed length event data packets from the Input TSI (switch) over the output event data link. It also sends Event Request messages to the Trigger Interface module. Data packets are copied to input buffers in the same order as the switch rotation (normally sequential). Memory pages are dynamically allocated to each input buffer as needed.

On output, data is copied from the first buffer until the end-of-event is reached, followed by data from the second buffer and so on. As data is output, the header information and checksums are verified. When a VMEbus processor or workstation processor with a VMEbus interface has completed event analysis it writes a new Event Request to the Output TSI and the message is forwarded over the Event Request Link to the Trigger System Interface (module). For test purposes in the prototype system, the I/O controller on the Output TSI module can simulate a bank of processors.

VMEbus CPU Local Controllers

SPARC 1E VMEbus CPUs from Sun Microsystems, Inc.[9] are used, one in each prototype project crate, to perform local monitoring and other control and diagnostic functions and to interface to the projects overall control computer.

Sun, Solbourne Workstations & Ethernet

Two workstations, one from Sun Microsystems, Inc. and one from Solbourne, Inc.[10] are used, interchangeably, for project monitoring and simulations. These two workstations, along with the VMEbus CPU Local Controllers, are interconnected via an Ethernet LAN.

System Operating Modes

There are four operating modes in the new data acquisition system architecture, two open-loop and two closed-loop. The difference between the open-loop and closed-loop mode of operation is that the Event Request Link is only used in closed-loop operating modes. Both open-loop operating modes do not use the Event Request Link; both closed-loop operating modes do use the Event Request Link.

In the first open-loop operating mode, the Input TSI modules automatically assign events to the next sequential destination. No trigger system "throttle" is used and events without ready destinations "fall off the end of the table". Statistics are kept which log information on all events including whether a processor accepted them. In the second open-loop operating mode, the Input TSI modules automatically assign events to destinations based on stored processor distribution information.

In the first closed-loop operating mode, the Event Request Link is used to indicate whether processors are ready to receive an event. The Trigger System Interface Emulator module assigns events to the next sequential ready processor. In the second closed-loop operating mode, the Event Request Link is again used to indicate whether processors are ready to receive an event. The Trigger System Emulator module assigns events to the next ready processor which is interested in a specific trigger type.

System Expansion

System expansion for either more channels into the event builder network or higher data rates is very simple. For example, if twice as many channels were needed, additional event builder network crates would be added as follows. A second "front-row" crate would would be added to interface to the second half of the data sources from the Near-Detector Electronics. Two "back-row" crates would then be added. Half the outputs of each of the "front-row" crates would be connected to the inputs of the first "back-row" crate and the other half of the outputs of each of the "front-row" crates would be connected to the inputs of the second "back-row" crate thus providing "event building" over the entire detector.

To provide the highest throughput for a given event builder network, the amount of data on each input should be relatively balanced when averaged over time. Additionally the input channel data rate times the number of inputs should be equal to the output channel data rate times the number of outputs. The most practical system might be one with more inputs than outputs so that the output channel data rate should be higher than the input channel data rate for highest event builder network efficiency. The cost effective approach would then use medium-speed input links and high-speed output links. This approach more easily allows higher event builder network throughput and more input channels without modifications to the online processor farm or its interfaces.

Software Status

The different software tasks included in the prototype project and the current status of each task are as follows:

Simulation & Modeling

Many models have been written and incorporated in a whole series of simulation experiments. Results include charts of system throughput and TSI buffer occupancy for different trigger rates, event sizes, number and online farm processors power, etc. A detailed report and a brief paper[4] describing simulation results are presently being written.

User Interface (Dataviews)

On the premise that everything a user would like to do to the actual data acquisition system, he would also like to do to his model of the data acquisition system, an elegant user interface has been developed which allows you to configure, initialize, monitor and diagnose the data acquisition system. A commercial software package, DataViews, was used to implement the user interface. Clicking on buttons with a mouse typically causes remote procedures on the SPARC 1E VMEbus CPU modules (Local Controllers), to read VMEbus locations and return the results, which are then displayed to the user. All software required for this user interface is coded and working.

Database (DataViews]

A small database has been developed under DataViews to handle an online processor farm. The database's design was significantly influenced by the information needed by CDF during its 1989 run. For example, the database contains the VMEbus information for each node, the name of the file containing the Level 3 algorithm, etc. The initial database, although coded and working, will need enhancements as we develop demonstration software for the project.

Remote Procedure Calls

These functions run in the Local Controllers and are responsible for configuring, initializing, monitoring and providing diagnostic information. They typically interact over VMEbus with a remote procedure running in one of the embedded SPARC processors in the Trigger System Interface Emulator, Test Transmitters, Input TSIs or Output TSIs. The design of the software is complete but there are still 50% of the functions to write.

Embedded Software

This software runs in the embedded SPARC processors in the Trigger System Interface Emulator, Test Transmitters, Input TSIs and Output TSIs. For example in the case of the input TSI, this software is responsible for low level functions such as looking at data header words, checking consistency with trigger information and reporting errors, etc. This software is almost all coded but needs testing.

System Diagnostics

We have designed the prototype project to be able to quickly and easily diagnose and locate all known potential failures. The code for these online diagnostics will be completed as part of the system demonstration due to be completed in April, 1991.

Low-Level Test & Diagnostic Routines

Tests of in-house designed hardware modules with embedded SPARC processors will be done using a commercial SPARC monitor/debugger connected to the module via RS-232. Tests of the first hardware module, the Trigger System Interface Emulator (TRIF) using this technique will begin in mid-November. Low-level test & diagnostic routings for this and other modules have begun and are 30% completed.

Hardware Status

The Trigger System Interface Emulator (TRIF), the Dual Test Transmitter and the Input TSI module designs and the barrel-shift switch backplane design and their printed circuit board layouts are all completed. The design and printed circuit board layout of a module used to control the switching of the barrel-shift switch, the Switch Local Control Logic module, will be completed in late November. Design of the last hardware module needed for the prototype project, the Output TSI, will begin by mid-November with an expected completion date in February, 1991.

System Testing Schedule

The following lists tentative completion dates of some of the planned system tests. All tests are closed-loop tests in that, under software control, it will be possible to check that all data and control messages got transmitted and were received correctly.

Late 12/90: Trigger System Interface Emulator sending event triggers over the Trigger Accept Link to a Test Transmitter; Test Transmitter transmitting previously-loaded event fragments.

Mid 1/91: Addition of an Input TSI receiving data from the Test Transmitter.

Early 2/91: Addition of the (switch) backplane and testing various barrel-shift switch functions.

Early 3/91: Addition of an Output TSI receiving data from the event builder network.

Late 3/91: Addition of an Event Request Link to "close" the system.

Late 4/91: Completion of the prototype project which includes four Dual Test Transmitters, four Dual Input TSIs, the switch (backplane) and its control logic, and eight Output TSIs.

Future R&D On The New Architecture

The (Scalable Parallel Open Architecture Data Acquisition System) prototype project should be completed sometime in April, 1990. A report on initial simulations of the new data acquisition system architecture is being written. Assuming a successful completion of the prototype, there are several "next steps" that should lead to this architecture being proven for future detector data acquisition systems at the SSC and elsewhere. Some of these "next steps" are described in the next paragraphs.

Interfacing To Workstations & VMEbus Processors

The Output TSI is being designed as a standard VMEbus module with master and slave block transfer capabilities. Several workstation and computer manufacturers provide VMEbus interfaces to their products. Since Fermilab supports both VMEbus-based online farm processors and VMEbus-interfaced Silicon Graphics, Inc.[11] workstations, it is desirable to interface both to a commercially-available VMEbus CPU and to a Silicon Graphics workstation. If time and available resources permit, a closed-loop demonstration of event building from Test Transmitters to VMEbus CPUs and to Silicon Graphics workstations will be part of the prototype project system tests.

Interfacing To Front-End Electronics

A draft of a very brief conceptual proposal (for discussion purposes only) for interfacing front-end ICs and subsystems to the Scalable Parallel Open Architecture Data Acquisition System prototype project has been written and distributed for comments. There has been several SSC and detector-specific meetings and documents written discussing requirements for the front-end/data acquisition interface. We recommend that a small collaboration develop data readout ICs, support electronics and software for interfacing front-end ICs and subsystems to the Scalable Parallel Open Architecture Data Acquisition System prototype project. This work could provide several institutions with a test bed for their R&D work on front-end electronics and exercise the new data acquisition system architecture as well. A partial list of institutions expressing interest in this work is Fermilab, University of Pennsylvania, KEK, Brown University, University of Toronto and Rutherford Appelton Laboratory.

Additional Simulations

The simulations completed to date have been extensive and detailed in the event builder network area but very broad and general in the detector to event builder network area. We need to integrate our simulations with ongoing and future simulation efforts from other institutions which actually model and simulate the detailed interworkings of front-end ICs, data readout ICs, etc., rather than detector data source emulators as in the prototype project. Before this occurs, simulations using actual expected (Monte Carlo) data in the detector data source emulators should be completed to test throughput variations with data which more closely represents that which might occur in an actual detector. To date we have only used simple common data distributions for data variations with channels and events.

Expert Systems Diagnostic Software

Knowledge system-based diagnostic software will not be completed as part of the prototype project. It would be very beneficial to use the prototype data acquisition system to test out the concept of using this type of software for online maintenance of high-energy physics data acquisition systems.

Integration & Repackaging

Steps were taken in the prototype project to shorten its development time (e.g., relatively slow fiber optic data rates, no ASIC development, sparse packaging, etc.). If this new data acquisition architecture is chosen as the "shell" of the data acquisition system for an SSC detector (hopefully for all SSC detectors), the prototype project should have several of its components integrated into ASICs and the system repackaged for higher density. If repackaged, Input TSIs may become an integral part of the Near-Detector Electronics, depending upon where various buffering,event ordering, etc. functions take place in the rest of the system.

Interconnection Networks From Industry & High-Energy Physics Event Building

Only a small fraction of the total data from each event is available for use in the initial trigger decision. The remaining data is scattered over many front-end buffers and must be collected in one place for detailed analysis. An "event builder" is the device in a data acquisition system which provides a physical connection between the individual data sources (detector front-end electronics) and the data destinations (high-level event processors or online data storage).

Regardless of the implementation, all event builders function as simple data multiplexers. If data rates are low, this multiplexing operation can take place over a single time-shared bus using software controlled selection of source and destination. This is the technique used in the majority of data acquisition systems to date. High-speed event builders have not been necessary because the data rates which could be supported by the sources and destinations were limited. This situation is changing rapidly. The ability to acquire, digitize and buffer data using VLSI front-end circuitry has increased allowable trigger rates by a factor of at least 1000. Similarly, the performance of high-level processors and the density of on-line data storage have both improved by a factor of almost 1000 over the last fifteen years. Unfortunately, the speeds of standard busses used for event building have improved by only a factor of ten in the same time period. The event builder has become the bottleneck.

There are two possible solutions to this problem. Either the trigger efficiency can be increased, limiting data rates to the bandwidth of the event builder, or the event builder bandwidth can be increased. Techniques for improving trigger efficiency are dependent on the experiment. Techniques for improving event builder bandwidth can be considered independently. The use of commercially-available Interconnection Networks from the computer and telecommunications industries for high-energy physics event building and/or other data acquisition system functions will be discussed in the following paragraphs.

Interconnection Networks...Types & Features

Figure 4 shows a generic Interconnection Network (IN) used in multiprocessor and telecommunications systems. The network is completely general in that any message can originate from any source and go to any destination and that all possible message paths are bidirectional. Additionally, a total "message" requires minimally one source and one destination.

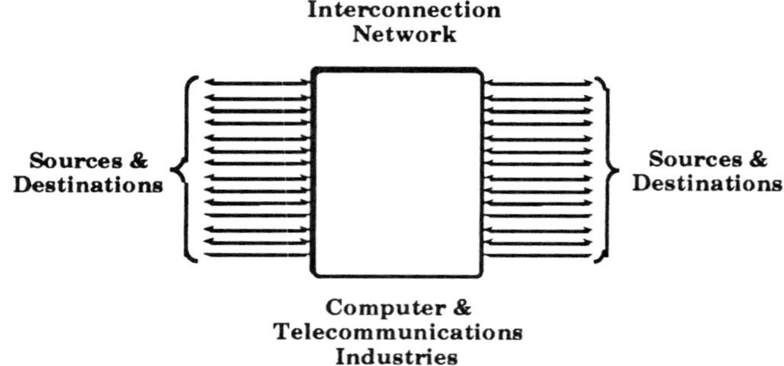

Figure 4.
General-Purpose
Interconnection Network

As shown in Figure 5., in high energy physics messages all originate from one side of the IN (from the detector) and all terminate at the other side of the IN (the online processor farm). A "message" in high-energy physics requires data from several, usually all, sources. The IN and its associated control is referred to as an "event builder". Because the pattern of data flow is well defined (unidirectional and evenly distributed over time), a general-purpose IN can often be simplified for use as an event builder.

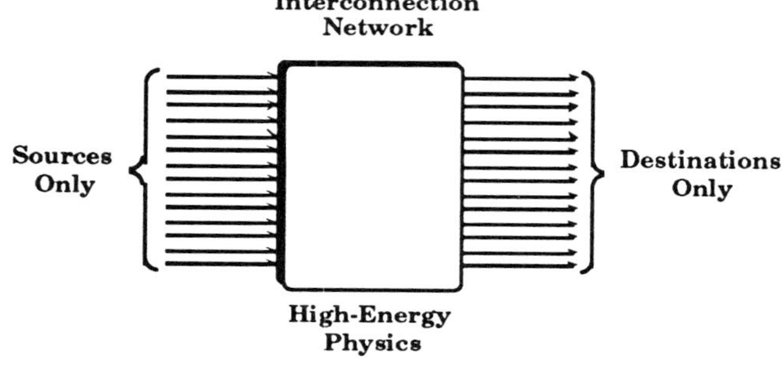

Figure 5.
High-Energy Physics
Interconnection Network

An enormous advantage is gained for high energy physics if we are able to draw from both the computer and telecommunications industries when designing data acquisition systems requiring parallel event builders with high-throughput.

Presently, only a few companies have INs available commercially. These initial INs are limited to a few tens of I/O channels and a few tens of Megabytes per second throughput if operated efficiently. The computer industry, because of the demand for multiprocessing, large processor and storage arrays, etc., and the telecommunications industry, because of satellite communications, high-density TV, other broadband ISDN applications, and the general glut the world has for information, are spending enormous resources to develop larger and faster INs.

The Need For Input & Output Queueing

Most all Interconnection Networks will be designed to be non-blocking in that if all its inputs had a message to be transmitted through the IN and all the messages were to be directed to different outputs, the IN would provide a distinct (non-blocked) path for each message, regardless of which combination of input or output any message had. For sure, applications within the computer and telecommunications industries rarely have such non-randomness. Because of the general nature of their industries, severe overhead in introduced (and substantial throughput reduction) is messages are not "queued" at the input and/or output of an IN. Input queueing logic buffers and sorts messages attempting to sent as many messages (to different outputs) through the switch at any given time. Output queueing logic buffers up outgoing messages until the subsequent destination(s) are ready to receive.

High-energy physics is quite different from general networks, not only for unidirectional data flow and "message" (event) fragmentation, but also the ease with which this application lends itself to simple input queueing logic. Figure 6 shows an IN with its integral input and output queueing logic. Note that, in the case of the Scalable Parallel Open Architecture Data Acquisition System prototype project, Input TSIs and Output TSIs function as input and output queueing logic, respectively.

Higher IN throughput is realized with properly chosen input and/or output queueing logic. Additional IN throughput is realized by somewhat balancing of the amount of data (event fragments) at all IN inputs averaged over time.

Interconnection Networks...Technology & The Future

Four technologies currently used or planned for implementing Interconnection Networks are as follows:

- VLSI Networks
- Electro-Optical Networks
- Integrated-Optical Networks
- Optical Networks

VLSI networks are available in small sizes now. In three to five years, up to 64 x 64 INs with tens to hundreds of Megabytes per second throughput will be commercially-available.

Electro-optical networks are non-integrated (GaAs primarily) switches. They are very expensive to build and will not be commercially-available for at least five years.

Integrated-optical networks are at least ten years away. No one has integrated GaAs onto fiber yet. Obviously, these networks are just beginning to be studied in R&D laboratories.

Very small totally optical networks are being build in R&D laboratories now. The SSC will have run for many years before suitable optical networks are commercially-available.

In the future, I/O data links to/from INs will use the HIPPI[12] fiber channel in the computer industry and the SONET[13] fiber standard in the telecommunications industry. Computer and computer peripheral manufacturer's will use the HIPPI fiber channel where SCSI or SCSI II does not suffice (e.g., higher speeds). Large computer systems will interconnect computers to disk and/or tape "farms" via an IN using HIPPI fiber channel data links. Broadband ISDN applications in the telecommunications industry will dominate the IN industry.

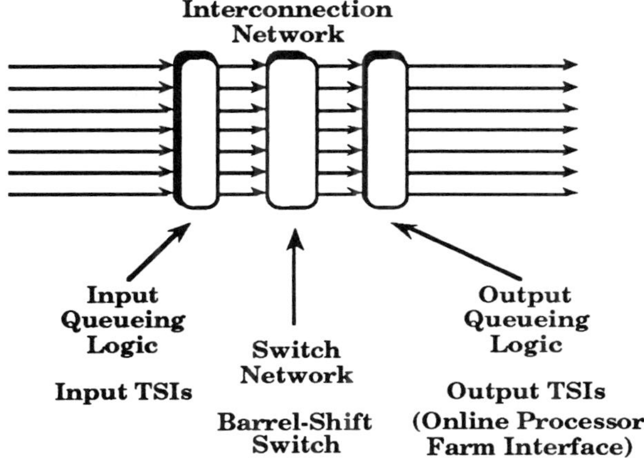

Figure 6.
SPOA DAQ Interconnection Network
With Input & Output Queueing

Event Building Techniques

Conventional thinking regarding SSC front-end electronics and data acquisition leads one to buffering event data on and/or near the detector until a total event's data has been collected then transmitting the event data efficiently to an event building network. Perhaps this is the correct approach, but other "non-conventional" approaches should not be overlooked.

Another (and an asynchronous) approach to event building differs from the "conventional" method in that event fragment data from single detector data sources are transmitted to the event builder network independent of what other detector data sources are doing. Less buffering is required on or near the detector but more is required after the event building network. This approach lends itself to the use of a more general-purpose (and commercially-avaliable) event building network. Realizable throughput is suspect with this approach because no queueing logic is used and multiple event buffering management and event "concatenation" is both done after the event builder network and are much harder tasks.

Two other approaches to event building which have been discussed at various SSC meetings are using mesh interconnects (switching nodes with processing power) for "staged" event building and transmitting data only when needed over a subsystem's individual network.

Event Builder Subsystem...Build Or Buy

As shown in Figure 6 and described in this paper, the (switch) network itself is very simple and straightforward in high-energy physics event building applications. The simple barrel-shift switch easily suffices. The dominant part of the effort is in the queueing logic. Both the computer and telecommunications industries require a general-purpose network in most all their applications. It appears that broadband ISDN applications in the telecommunications industry will dominate the IN sales for the next decade with computer interconnect applications a distant second. It presently appears that using a commercially-available IN might be overkill for high-energy physics event building applications and that building the event builder network "in-house" might be the cost-effective choice. Additionally, there are commercial switch ICs available today[14] and soon to be available[15] which operate at a maximum of 0.32 and 1.6 Gigabytes per second per IC, respectively. Only a few of these ICs are needed for a typical SSC experiment's event building (switch) network.

Because of its message randomness and modest size and data rate requirements, the Online Storage Network shown in Figure 1 should be a commercially-available product.

Summary

Unless major design problems occur, the Scalable Parallel Open Architecture Data Acquisition System prototype project should be successfully demonstrated within six months. Not only will a new, highly-parallel data acquisition system architecture be demonstrated but a Macintosh-like menu-driven user interface capable of being used for an experiment's run control program's user interface will also be demonstrated. The next obvious step would be to interface this project to the work of SSC front-end IC developers. Once this is finished, tests on front-end ICs and subsystems and their interfacing to a data acquisition system can be completed efficiently.

The semiconductor, telecommunications and computer industries will all provide an immense set of hardware and/or software tools and standards which should be used wherever possible in all SSC data acquisition systems. Interconnection networks surely could be used to interface a data acquisition system's online processor farm to its online storage subsystem. Whether general interconnects are practical for high-energy physics event building still remains to be seen.

Acknowledgements

We wish to thank all participants in the prototype project including Jeff Andresen, Al Baumbaugh, Dennis Black, Gustavo Cancelo, Betty Conrin, Derek Esterline, Kelly Knickerbocker, Rick Kwarciany, Gary Moore, Ray Reitz, Ken Treptow, Oscar Trevizo, John Urish, Rick VanConant and Don Walsh. We also would like to thank everyone else who gave us valuable input throughout this project.

References

[1] E. Barsotti et. al., A Proposed Scalable Parallel Open Architecture Data Acquisition System For Low to High-Rate Experiments, Test Beams and All SSC Detectors, IEEE Transactions On Nuclear Science, Vol. 37, No. 3, June 1990, pp. 1216-1221.

[2] M. Bowden et. al.,A High-Throughput Data Acquisition Architecture Based On Serial Interconnects, IEEE Transactions On Nuclear Science, Vol. 36, No. 1, February 1989, pp. 760-764.

[3] A. Booth et. al., Simulation And Modeling Of Data Acquisition Systems For Future HEP Experiments, to be published in the February 1991 issue of the IEEE Transactions On Nuclear Science.

[4] E. Barsotti et. al., Effects Of Various Event Building Techniques On Data Acquisition System Architectures, AIP Conference Proceedings, No. 209, 1990, pp. 82-101.

[5] Verilog-XL, Cadence (Gateway), Lowell, Massachusetts.

[6] DataViews, V.I. Corporation, Amherst, Massachusetts.

[7] Nexpert, Nueron Data, Inc., Palo Alto, California.

[8] M. Bowden, System Overview of the Scalable Parallel Open Architecture Data Acquisition System, not published, May 1990, pp. 18-25.

[9] Sun Microsystems, Inc., Mountain View, California.

[10] Solbourne Computer, Inc., Longmont, Colorado.

[11] Silicon Graphics, Inc., Mountain View, California.

[12] HIPPI Fiber Channel, X3T9.3 Committee of the American National Standard Institute

[13] SONET, Synchronous Optical Networks T1X1 Committee of the American National Standard Institute.

[14] LSI Logic Corporation, Milpitas, California, L64270 64 x 64 Crossbar IC.

[15] Vitesse Semiconductor Corporation, Camarillo, California, VSC864 64 x 64 Crossbar IC.

HARP: Hierarchical Analog Readout Processor with Analog Pipelining in CMOS

Francis Anghinolfi, Paul Aspell, Michael Campbell, Jheroen Dorenbosch(*), Erik Heijne, Pierre Jarron, Gerrit Meddeler
CERN, CH–1211 Geneva 23, Switzerland
Alf Olsen(**), Henrik von der Lippe
SI, Oslo, Norway

Abstract

Basic functional blocks have been designed and tested: Bipoltest, CSI, SAPE and PIPAD which are a front–end amplifier, a current sampling integrator, an analog pipeline and an analog/digital converter respectively. These functional blocks will eventually be combined in a single chip, and this processor will be the starting point for the HARP architecture. The signal amplifier peaking time for low capacitance detector elements is less than 15 ns. The sampling rate of the pipeline element in the 3 μm CMOS is measured to be 20 MHz with a dynamic range of ±1.2 pC of input charge which corresponds to 11 bits and the sampling rate in the 1.5 μm technology is over 60 MHz. The ADC operates at 1 MHz sampling rate with 12 mW for 11 bits precision.

Introduction

Experiments in the future supercolliders need a new approach to the organization of the detector readout and data acquisition, because of the large number of detector elements, 10^6 – 10^8, and even more so because of the increased event repetition rate, ≈ 70 MHz. The Italian-funded CERN–LAA R&D project was specifically conceived to initiate efforts on such new basic techniques. We presented earlier some preliminary results in a progress report [1].

In a conventional particle physics experiment as illustrated in fig. 1, the analog signals from the detector elements are amplified, shaped and transmitted via cable to a remote electronic system, located in the 'counting room'. After discrimination on a subset of the signals, followed by some logic operations one builds a trigger decision. Upon a positive trigger the analog to digital conversion (ADC) of all signals is executed, the event is tentatively reconstructed and if indeed of the desired type, all information for this event is stored in mass memory. It will be difficult to apply this scheme in the future high luminosity hadron colliders because of the cable volume and the cable characteristics needed for the 100 MHz transmission bandwidth for the millions of detector elements now envisaged. An unacceptably large volume would become insensitive, being occupied by these cables. Anyway, it appears that the power dissipation in the analog driver circuits can be equivalent to the power needed for local high speed ADC. An obvious alternative is then to install flash ADCs inside the experiment and to transmit compacted digital data to the counting room, as shown in fig. 2. For digital data transmission one may use high speed optical fiber links, which are more compact but still require high power drivers. Added to the relatively high power dissipation of FADC this leads to a probably unmanageable cooling problem, in particular in the inner regions of the detector system.

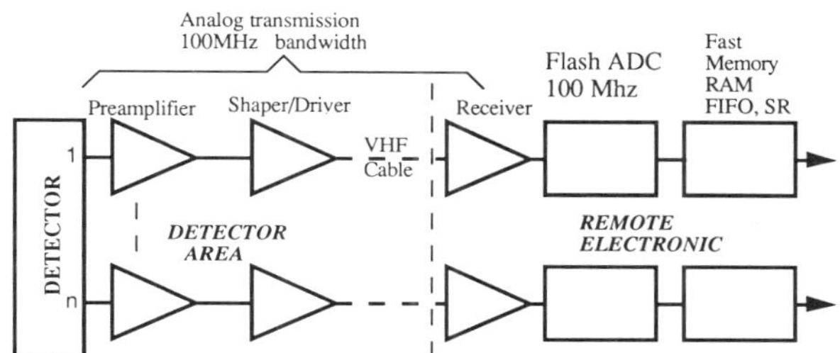

Figure 1. Conventional detector readout via analog signal transmission using cable. Often, the cable acts also as the signal delay element.

Here we propose an alternative approach (fig. 3) based on the high performance possibilities offered by Application Specific Integrated Circuits (ASIC). We have named this new system 'Hierarchical Analog Readout Pipeline Processor' (HARP). The heart of it will be an ASIC chip for which we develop presently different building blocks. The most essential one is the analog pipeline SAPE

(*) now at the SSC Laboratory, Dallas, Texas
(**) now at GeTec AS Oslo, Norway

and we will present here the operational results obtained on the first experimental circuits, in a 3 μm and in a 1.5 μm CMOS technology. SAPE stands for Simultaneous write/read Analog Pipeline Element, and one important new feature is the absence of the deadtime usually occurring during a readout cycle. This SAPE chip will be completed by front-end amplifiers, discriminators (comparators), an integrated ADC and the associated logic, needed for data selection upon an external trigger decision. These functional blocks have been developed and tested separately. An overall schematic of the future HARP chip is shown in fig. 4.

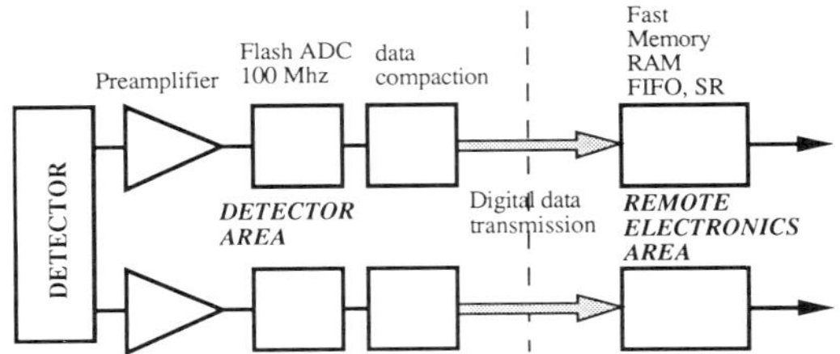

Figure 2. 'Brute Force' method consisting in digitization of all signals inside the detector, and transmission of compacted digital data. Delay in digital memory in the remote electronics area, the 'counting room'.

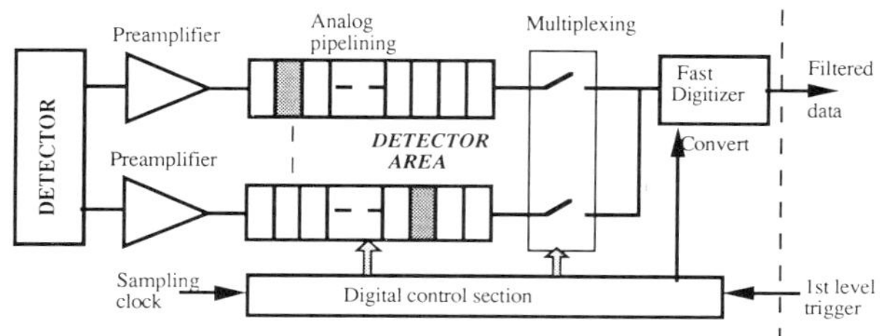

Figure 3. New approach uses analog pipeline inside the detector during the first level trigger decision time. Only data selected in this first level trigger are digitized, also inside the detector, and the reduced amount of data is transmitted in digital form.

The emphasis in this paper is on the design and performance of the two experimental SAPE circuits made to date. Apart from this, one section is devoted to the fast front-end amplifier which couples the detector element to the input of the SAPE circuit, and another section briefly describes a custom-designed low power ADC which could be integrated on the HARP chip. In the discussion some perspective will be given on the implementation of the HARP approach in the future detector subsystems.

Front-end amplifier 'Bipoltest'

The collider event repetition time of ≈15 ns imposes fast detectors and fast electronics. It is outside the scope of this article to discuss the detector options, but one may assume that some of the detector systems will use segmented silicon devices which at least offer adequate charge collection speed for both the generated electrons and the holes, in the order of 10 ns. In the by now canonical Si detector thickness of 300 μm a minimum ionizing particle (MIP) generates 24000 free e-h pairs, equivalent to a charge of 3.8 fC. This amount of charge is often quoted as a 1 MIP (Si) input signal.

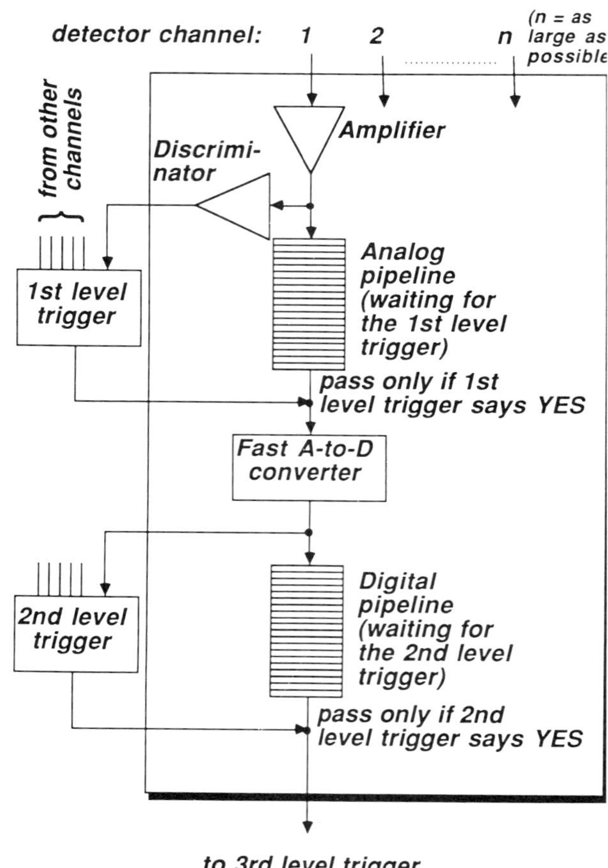

Figure 4. Schematic of the future HARP chip, including a number of building blocks, for n identical channels.

Regarding speed, one should prefer silicon bipolar devices over CMOS and good results have been already reported [2,3]. The better accessability of CMOS and the need for integration of complicated logic circuits led us to study alternatives in CMOS technology, in spite of the lesser performance. A true BiCMOS technology may ultimately prove the best compromise. For the moment we have used the lateral bipolar transistor structure [4] which has quite acceptable characteristics in the 3 μm SACMOS process(*). The p-well serves as the base of this npn lateral bipolar transistor. The collector n-type implant should closely surround a central n emitter, in order to have a short base width. We find a transition frequency f_T = 140 MHz.

The schematic diagram of the experimental amplifier 'bipoltest' is shown in fig. 5. The input transistor T1 in the common emitter configuration is of the lateral bipolar type.

(*) Proprietary process of Faselec AG, Zurich, a subsidiary company of Philips

The current gain is 170 for a standing collector current of 100 μA. A second stage with unity gain is formed by the n-channel transistor T2. Together, T1 and T2 form a current feedback pair, followed by the vertical bipolar transistor T3 as source follower. To overcome the large Miller capacitance of the lateral bipolar T1, which limits high-frequency operation, a neutralization capacitance C_{nf} is connected to the base of T1. The effect of C_{nf} is illustrated in the SPICE simulation in fig. 6. The sharp frequency cut-off filtering inherent to this circuit design produces a nearly Gaussian shaping of the output signal. The peaking time measured in fig. 7 is 15 ns. The conversion gain of the amplifier is 0.5 mV per fC or 2 mV per MIP. The circuit is designed to operate at 3.5 V and dissipates then 2 mW. The Equivalent Noise Charge (ENC) is 800 e^- r.m.s. with a slope of 60 e^- per pF. A still faster CMOS preamplifier and comparator block which does not use lateral bipolar transistors is under design.

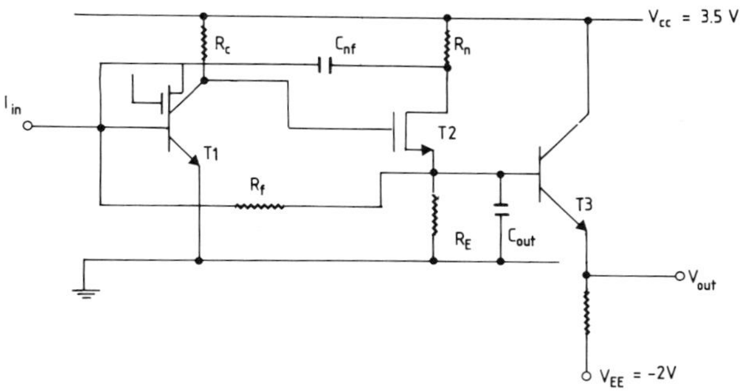

Figure 5. Schematic diagram of current sensitive amplifier 'bipoltest' using a lateral bipolar transistor T1 and a vertical bipolar transistor T3.

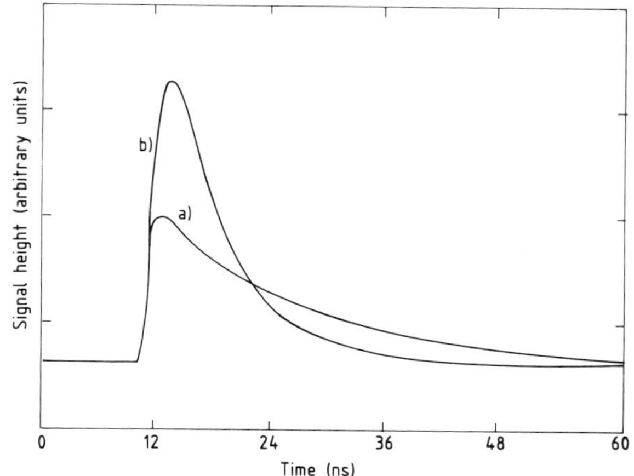

Figure 6. Simulated output signal of 'bipoltest': a) represents the signal without, and b) with a positive feedback neutralization capacitance of 0.2 pF

Principles of the SAPE circuit

The signals from the basic detector segments, amplified by a front-end circuit like 'bipoltest', pass into an analog memory

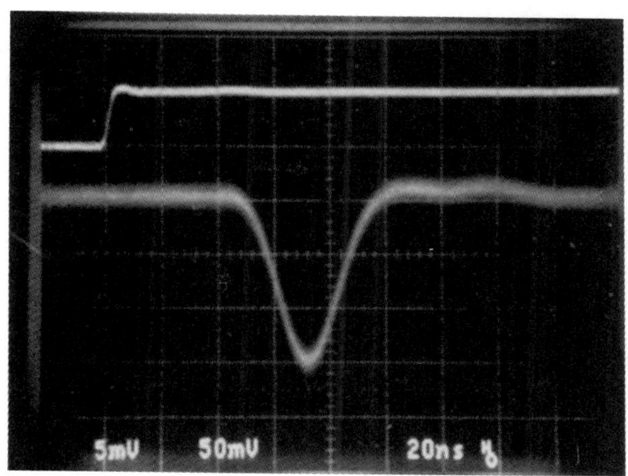

Figure 7. Measured output pulse of 'bipoltest'. The signal peaking time is about 15 ns.

which allows storage for the time T1 needed to build the first level trigger decision. The memory must be organized such that selected event data can be retrieved upon the first level trigger, while event writing continues in order to eliminate system deadtime. To our knowledge this simultaneous write and read operation is a novelty in analog pipelines which will be particularly needed under the high data rate conditions in the supercolliders. The principle is illustrated in fig. 8 for a matrix of memory cells with separate addressing of the write and read locations. The write and the read operations use separate bus lines, switches and amplifiers. The delay T1 between write and read ensures that they cannot address the same memory location and T1 can be adjusted to cover at least the latency period of the first level trigger. A simplified electrical scheme of the addressing of the memory cells is

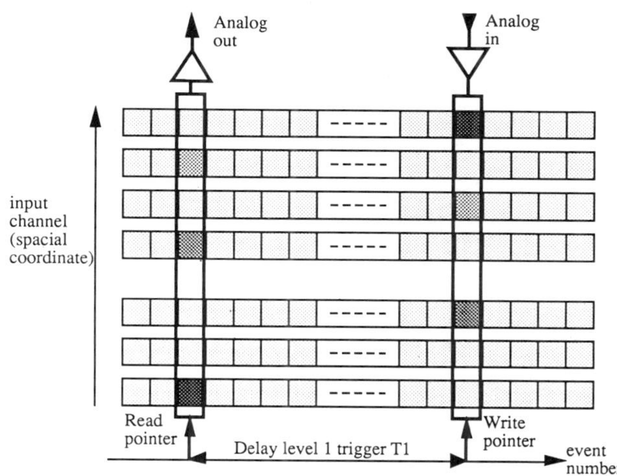

Figure 8. Matrix of analog signal storage elements. Each row contains the consecutive data from a single detector segment. The signals of the present event are being written in the column of elements indicated by the write pointer. At the same time data from an event, T1 seconds ago, indicated by the read pointer, can be read out if selected by the trigger.

shown in fig. 9 for a 3 cell array. Each cell consists of a capacitor C_S, a write switch and a read switch which are operated at frequencies f_w and f_r, respectively. While the write switch register ***WSR*** enables one capacitor to be charged the read switch register ***RSR*** enables a different capacitor to be read out. During this clock cycle all other capacitors remain disconnected from the bus lines. The ±3 V clock signals consist of adjustable, non-overlapping phases in order to avoid charge sharing between memory cells.

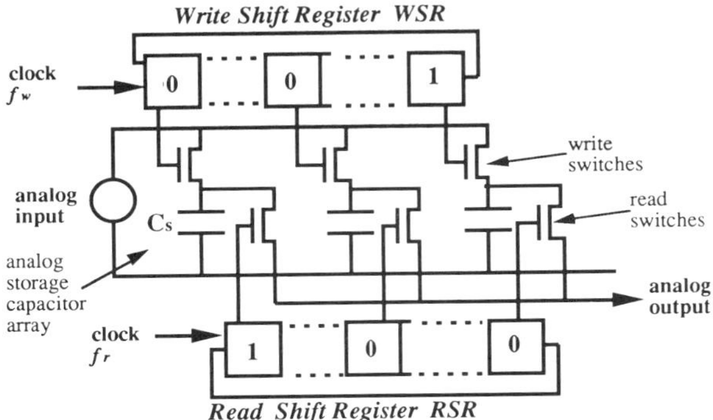

Figure 9. Schematic of a circuit which allows to write and read simultaeously in different storage capacitors C_S. The essential point is the need for separate bus lines and switches for writing and reading.

Voltage sampling and current sampling (CSI)

The input to the memory element can be configured as a voltage sampler or a current integrating sampler. These configurations are illustrated in figs 10 and 11, again for a symbolic 3-element capacitor array. The voltage sampling technique (fig. 10) is the most widely used. The write amplifier W_{amp} in the voltage follower configuration simply stores the instantaneous input voltage on capacitor C_{S1} while the write switch is closed. The voltage stored represents the input value available just before the write switch is opened again. In this example C_{Sn} is simultaneously being read by the read amplifier, and C_{Si} is disconnected.

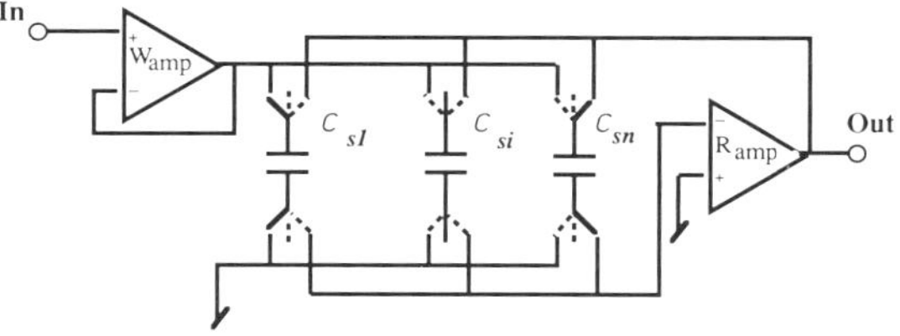

Figure 10. Voltage sampling input configuration.

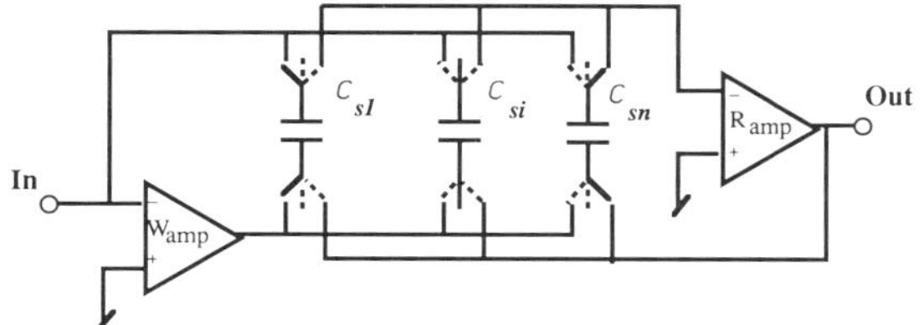

Figure 11. Current sampling input configuration.

In the charge sampling configuration of fig. 11 the capacitors C_{Si} act as feedback capacitor in the feedback loop of the write amplifier W_{amp} so that this circuit operates like a traditional charge amplifier. The resulting voltage V_C on the storage capacitor is proportional to the integral of the input signal current i(t) from the detector during the sampling interval $T_W = 1/f_W$, while the write switch is closed.

$$V_c = (C_{si})^{-1} \int_0^{T_w} i(t)\,dt$$

We call this input configuration the Current Sampling Integrator **CSI.**

In particle physics detector applications **CSI** has several advantages over voltage sampling, principally because the integrated charge is the direct output of most types of particle detector. In the supercolliders the detector signal may spread over several clock periods of 15 ns. It will be straightforward to reconstruct the detector output, just by summing consecutive signals. With voltage sampling a relatively complicated algorithm involving knowledge of the pulse shape would be needed to reconstruct this detector output charge. Moreover, for analog pulses of the order of 10 ns the individual shapes may well be sensitive to local variations. The detector geometry may lead to varying input load capacitances, the charge collection speed may vary, e.g. after radiation damage, and interconnect stray capacitances may be different.

The current sampling integrator can be implemented as a compact circuit, even with the extra switches and bus lines needed for the simultaneous W/R operations. It may even turn out to be superfluous to have a separate front-end amplifier between detector and pipeline circuit. The **CSI** itself has the capability for low noise operation, like any charge amplifier. Also, it acts as a time variant filter on each sampling, because the **CSI** circuit is equivalent to a gated integrator, followed by a time invariant filter consisting of the transfer functions of the write and read amplifiers. In principle, no further analog signal processing such as filtering should be needed to enhance the signal to noise ratio.

In order to reduce the clock feedthrough effects double write and read switches should be placed symmetrically around the storage capacitor C_S as illustrated in fig. 12. The Analog Memory Element (AME) is completed by a reset switch which discharges the capacitor, usually just after reading it out. The clock feedthrough via this switch is also cancelled by keeping all other switches around the capacitor open during reset. The injected charge is then equipartitioned on both capacitor electrodes. A more detailed discussion of the factors affecting the reproducibility of the analog memory cells, giving rise to pedestal non-uniformity, is given elsewhere [5].

The total equivalent noise charge of the **CSI** circuit is expected to have three principal contributions. First the equivalent noise charge of the gated integrator W_{amp}. This contribution can be calculated following the approach of Radeka for signal filtering for Ge detectors [6]. The kT/C noise voltage caused by the reset switch is 64 µV for C_s = 1 pF. The third contribution is the noise voltage of the read amplifier. Further analysis is expected to show that the gated integrating function is the dominant noise factor, at least in the ideal case.

One disadvantage of **CSI** is that the mismatch between capacitors in a single pipeline may cause a gain spread. If double polysilicon capacitors are used the spread may be 1–2%. If better precision is needed the voltage sampling configuration may be more appropriate.

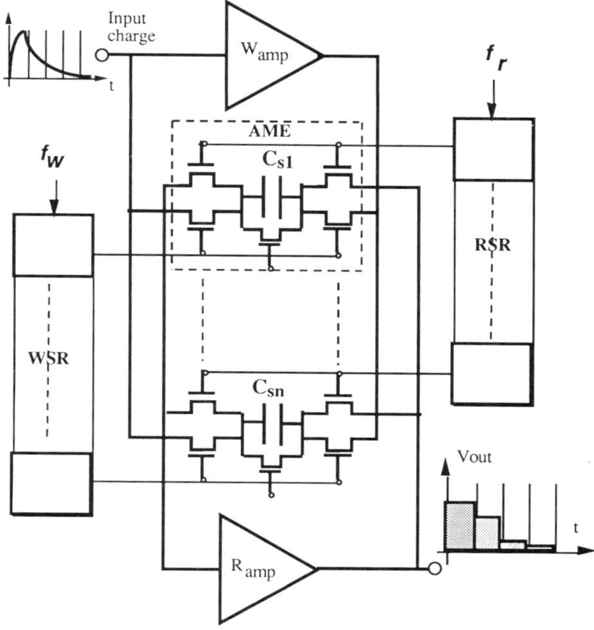

Figure 12. Schematic diagram of the analog memory elements (AME) in the current sampling integrator circuit. The symmetric write switch and read switch transistors are actuated by the respective shift registers WSR and RSR.

Design of SAPE in 3 µm CMOS

After the testing of an experimental 8-element pipeline [7] a more elaborate prototype with four parallel channels has been designed using a 3 µm n-well CMOS technology with double poly(*). The block diagram in fig. 13 shows the double pipeline structure with 64 memory cells in the first pipeline and 8 cells in the second. This is also clearly visible on the chip photograph in fig. 14. The shift registers WSR and RSR act as pointers to the 0.6 pF storage capacitors to be written or read. The size of each AME cell is 110 µm × 65 µm. The total length of each channel is 4140 µm. Two of the channels have voltage sampling inputs and the other two have a CSI configuration. The write amplifier is a folded cascode Operational Transconductance Amplifier with n-channel input transistor and appr. 0.5 pF input capacitance, open loop gain of 75 dB, unity gain at 117 MHz, a linear dynamic range of ±2 V and a risetime of 3 ns. The power dissipation of the OTA is 5 mW and the non-linearity in the range ±1.5 V is less than 0.2%.

The transfer of a signal from a cell in the first array to the second array is controlled by an external trigger signal and the addresses of the original cell and the destination cell are externally available also in order to allow pedestal correction and gain equalization, if needed. A 4-bit counter keeps track of the trigger number. The operation of this circuit has been visualised in fig. 15, where two turning wheels represent the pipelines containing the memory locations, which are written or read if they pass the respective pointers. The frequency in the second pipeline is designed to be 1/10th of that in the first. In fact, the second pipeline has room for 1/8th of the data. The transfer of a signal from the left wheel to the right wheel only becomes effective if an external trigger signal moves the write pointer in the second wheel to the following position. The cells in the right wheel contain 'good' data which can be read out at any desired moment. Therefore, this pipeline acts as a derandomizer element. The operation of the SAPE3 circuit as a delay line is illustrated in fig. 16 with a write clock frequency of 20 MHz in the first pipeline. The read clock frequency is 2 MHz so that 1/10th of the data can be read out.

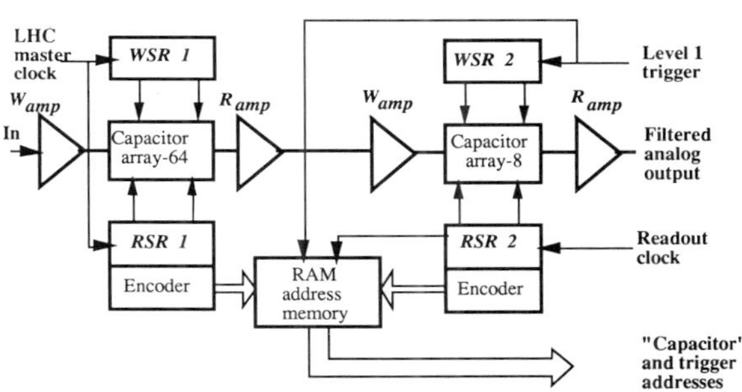

Figure 13. Block diagram of the double SAPE circuit made in a 3 µm CMOS technology

Experimental results

The SAPE circuit performs in an integrated manner several functions: low-noise amplification, analog signal storage and data filtering. One should like to characterize each of these functions as in the case of separate circuits: noise, sensitivity, dynamic range for the input amplifier, sampling speed, pedestal and gain uniformity for the analog memory function. The evaluation of such a system is complicated by various summing and interference effects and it is not evident to identify the origin of a performance limitation like noise, non-linearity or non-uniformity. Crosstalk between the digital and the analog circuitry obviously is a major concern.

(*) Mietec N.V. Oudenaarde, Belgium

Figure 14. Chip photograph of SAPE 3. The inputs to the 4 parallel pipelines are at the left. The second pipeline of 8 elements is at the right.The chip dimensions are 4 mm × 8 mm.

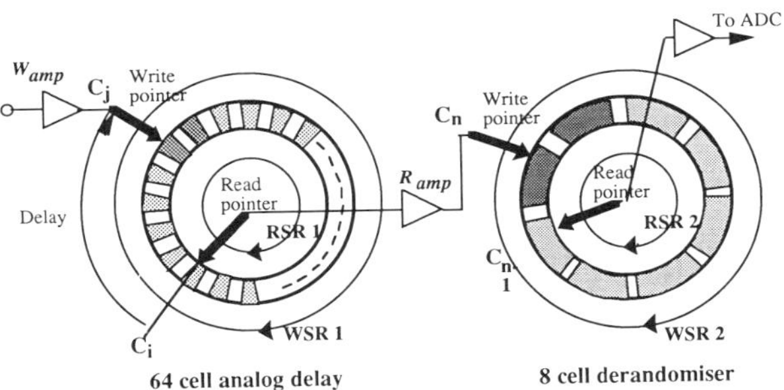

Figure 15. Visualization of the operation of the double pipeline circuit. The write pointer in the derandomizer array at the right will only move to a new cell after a 'good' event. The 'good' events can be read out of the second array at any desired moment.

The test set–up uses the commercially available programmable data aqcuisition system ASYST with LabMaster modules, running on a Olivetti M300 Personal Computer. Although this system provides ADC and DAC functions, another low noise DAC is added. The programmed clocking sequences are stored in a RAM, which is mounted in a test box, together with an address decoder and an output signal track/hold circuit.

Figure 17 shows an oscilloscope picture of a single pulse in voltage sampling in and out, at the 20 MHz write clock frequency. Several tests have been made to verify that one can store signals in successive capacitors and retrieve them correctly, as shown in fig. 18 for 5 pulses. The current sampling input is shown in fig. 19 where a current is presented at the input via a 100 kΩ resistor. This picture is one of a sequence in which the input current pulse was gradually moved from one clock cycle to the next. The charge is either integrated in exactly one memory cell or distributed the two storage capacitors proportionally to the time division.

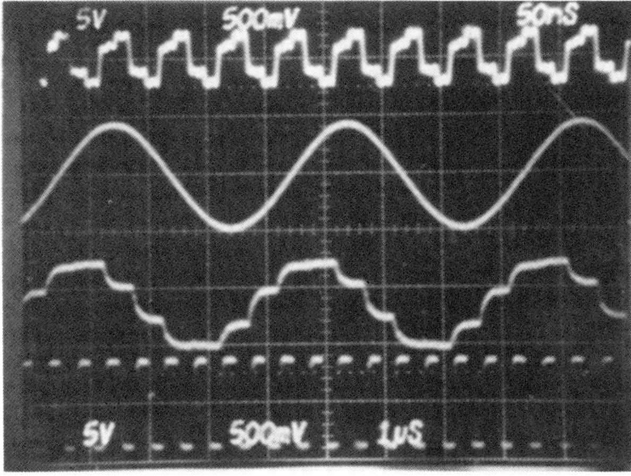

Figure 16. Operation of SAPE 3 as delay line. The top trace is the 20 MHz write clock and the bottom trace the 2 MHz read clock. The second trace is a 250 kHz sinusoidal input signal, which is reproduced at the output.

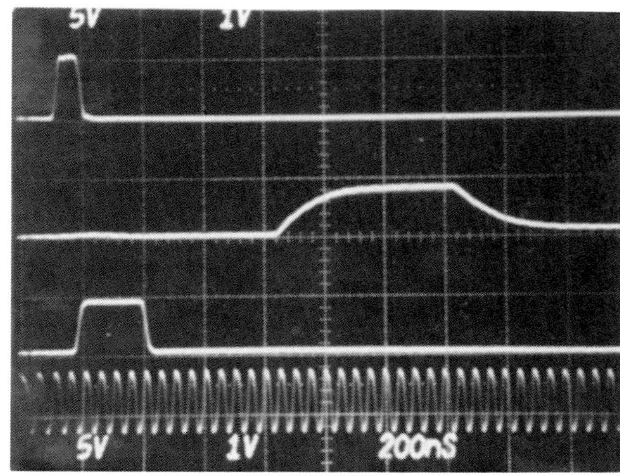

Figure 17. Top trace: input voltage signal of 1 V, second trace shows the corresponding output signal. The 3rd trace gives the trigger signal for readout and the 20 MHz write clock is shown at the bottom. The read clock is a ÷10 derivation of the write clock frequency.

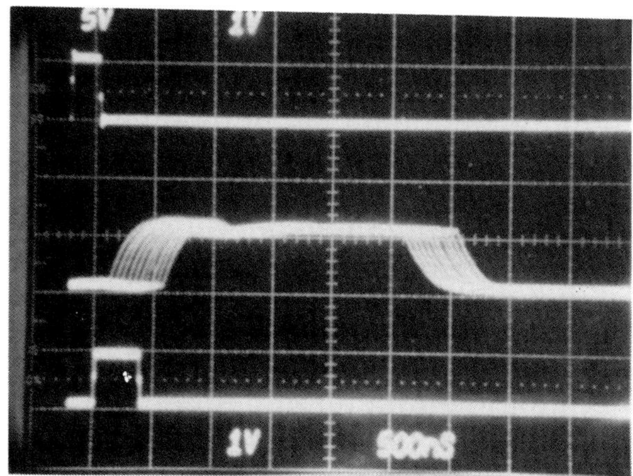

Figure 18. An input signal with a duration of 5 clock cycles at 10 MHz (top) in voltage sampling is stored in consecutive storage cells and read out (middle trace). The bottom trace is the trigger signal.

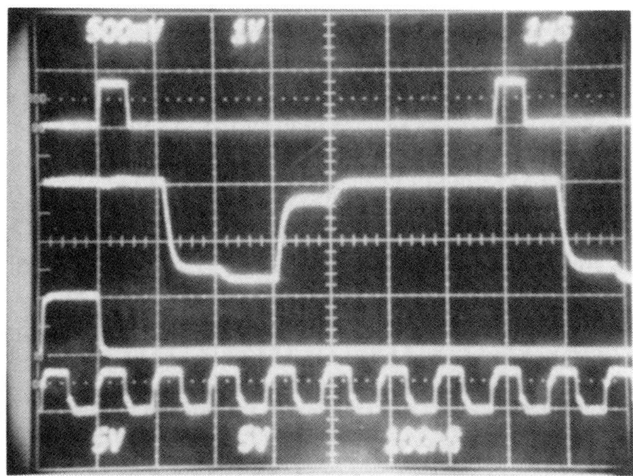

Figure 19. Current sampling of a signal, top, in a 100 kΩ resistor. A delay is applied in order to shift the 100 ns long signal across two read cycles, so that the integrated charge appears divided over two storage elements, second trace. Trigger signal and 10 MHz clock are as before.

The sensitivity and the linearity of the current input are shown in fig. 20 for positive input charge. A similar curve is obtained for negative signals. The sensitivity is 1.62 V/pC. The linear range is ±1.2 pC. Fluctuations with a variance of 1.1 mV or 58 dB have been measured on the simplified, 8-cell pipeline[7]. This dynamic range corresponds to 11 bits. Similar values will be shown for the SAPE 1.5. The writing speed of the SAPE 3 is limited by the digital circuitry, and the settling time of the output signal is limiting the readout frequency.

Cross talk problems appeared in the measurement setup, in particular digital clock feedthrough was serious at first. It was found that the main path to the sensitive inputs passes via the ceramic Dual-In-Line package, because the pedestal uniformity became nearly perfect after cutting the

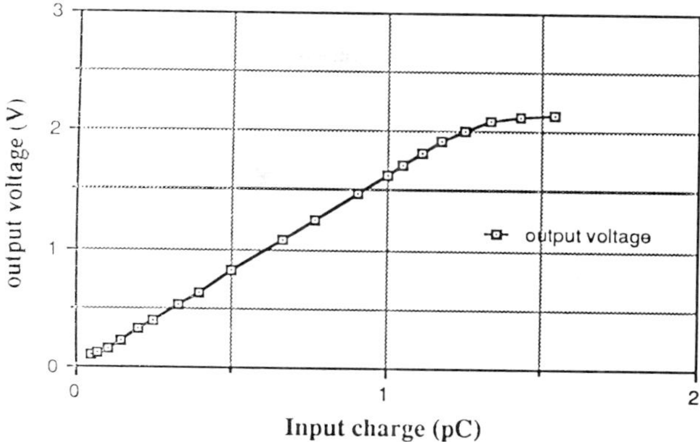

Figure 20. Sensitivity of the SAPE 3 current sampling input configuration.

input wire bonds. Disconnecting the pins from the board contacts did not help. One will have to use low-level, differential logic signals like ECL as well as sophisticated packaging for this type of devices.

Design of SAPE in 1.5 μm CMOS

A single pipeline simultaneous write/read circuit has been made in the Mietec 1.5 μm double-poly, double-metal CMOS process. It is similar to but much faster than the SAPE 3 circuit. The block diagram is shown in fig. 21 and the chip photograph in fig. 22. The derandomizer part has been omitted to save time during design, but also to obtain a more direct access to the analog memory elements for testing. The circuit consists again of 4 parallel channels with 2 voltage and 2 CSI inputs. The read amplifier and the write amplifier for each channel are identical folded cascode OTA. Unity gain is situated at 220 MHz with 0.8 pF load and the risetime is less than 5 ns. n-channel cascode transistors are used to obtain a settling time shorter than 10 ns. Dissipation is ≈ 5 mW. The value of the storage capacitor of 0.4 pF is a trade-off between accuracy and noise on the one hand and speed and area occupation on the other hand. Obviously the digital control circuitry is much simpler than for the SAPE 3.

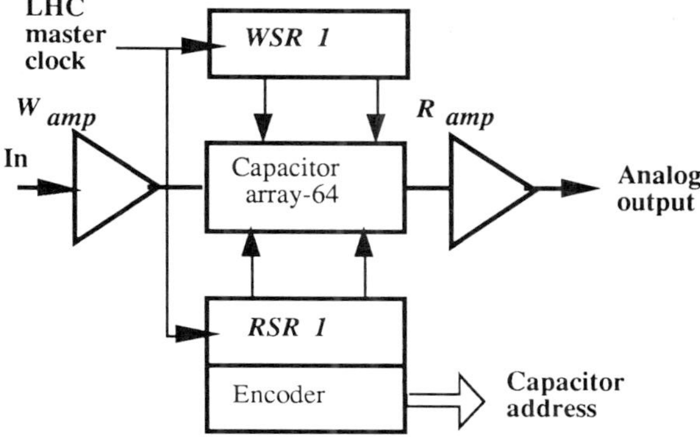

Figure 21. Block diagram of the SAPE 1.5 circuit.

Figure 22. Chip photograph of the SAPE 1.5 circuit. The chip dimensions are 4 mm × 2.5 mm.

The operation of the circuit as a simple delay line in voltage sampling mode is shown in fig. 23 and the maximum write/read clock frequency is found to be 70 MHz. However, presently our measurement setup does not yet allow digital data acquisition at this frequency and most measurements reported are therefore performed at 1 MHz, using the current sampling integrator channels. The sensitivity of the circuit (fig. 24) is 2.29 V/pC and the measured linearity extends from –0.7 pC to + 1 pC. A 2 MIP(Si) signal corresponding to 8 fC is visible in one of the cells in fig. 25. This proves that the circuit might operate even without an external front-end amplifier. The pedestal distribution of the 64 memory elements is plotted in fig. 26. The mean value of –1.28 V is determined by the threshold voltage of the output transistor in a source follower. This has been added as an output buffer to facilitate testing of the prototype circuit . The variance of this pedestal distribution is a few mV and also the variance of the output signals for a single memory element is in this range (2.7 mV). At present we are studying which are the limiting factors and it would be premature to regard these fluctuations as the noise performance of the circuit. Crosstalk between several channels is measured to be lower than 5%. Power dissipation has been measured as a function of the operating frequency and is illustrated in fig. 27.

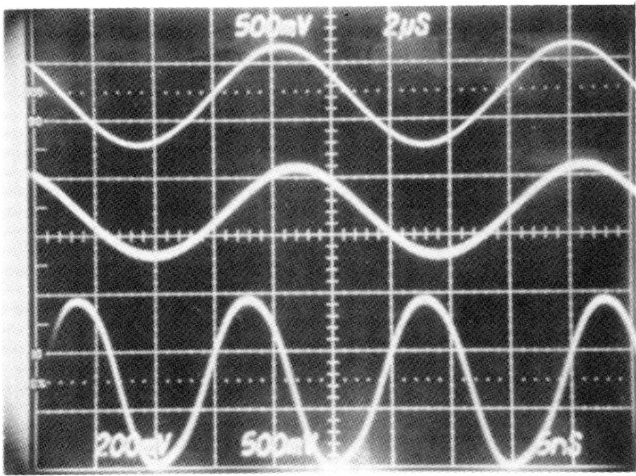

Figure 23. Oscilloscope picture of the operation of the SAPE 1.5 circuit as a delay line, operating at 66.6 MHz. The top trace is a 100 kHz sinusoidal input signal which is reproduced at the output, middle trace. The bottom trace is the clock frequency.

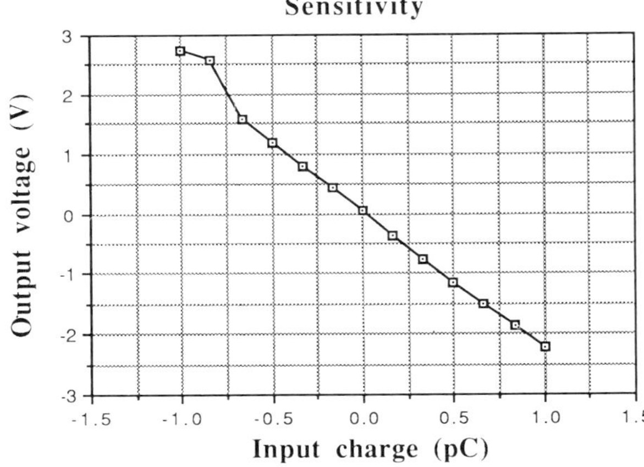

Figure 24. Graph of the output voltage as a function of the input charge for SAPE 1.5, measured at 1 MHz.

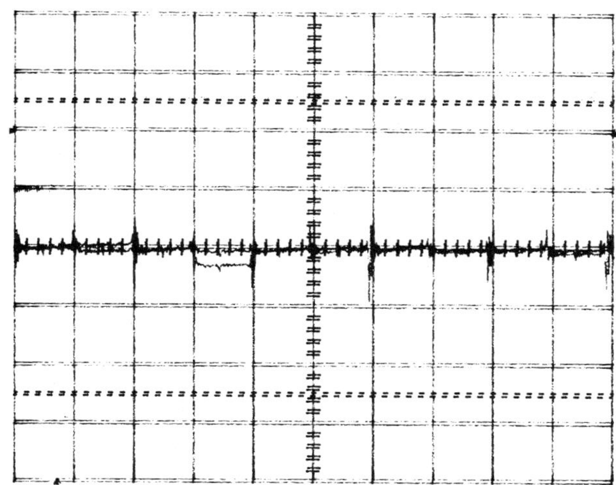

Figure 25. Two graphs from a digital oscilloscope, which are superimposed. The output from the cells without signal remain stable, but the 4th cell shows the signal above the pedestal. This signal corresponds to a charge of 2 MIP.

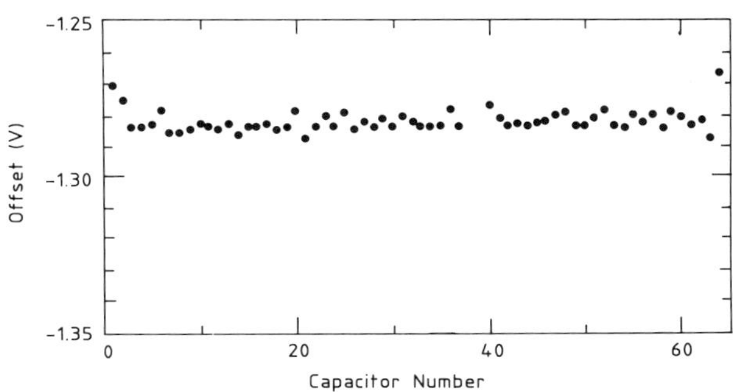

Figure 26. Plot of the pedestal values of the 64 memory elements of SAPE 1.5 in the current sampling integrator channels. The average value is –1.28 V.

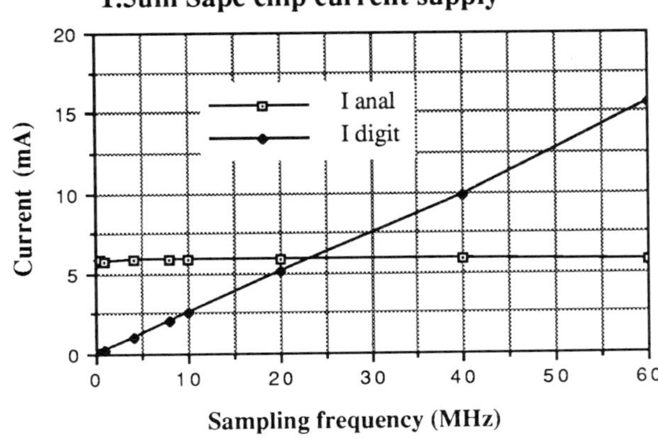

Figure 27. Power dissipation of the 4-channel SAPE 1.5 circuit as a function of operating frequency for the analog and the digital. The digital part is designed to serve many more channels.

Low power pipelined ADC

A low power 12 bit ADC has been made as a separate building block for the HARP architecture [8]. It avoids the massive power consumption of the usual Flash ADC (FADC) by using a pipelined structure. Successive signals are converted in 12 steps, as is shown in the block diagram in fig. 28. After the most significant bit has been determined the input voltage is multiplied by two, while the reference voltage is conditionally subtracted. The remainder signal undergoes the identical operation in the second stage, and so on.

The circuit operates well at 1 MHz sampling rate and achieves a resolution of 11 bits with values for the integral and differential non-linearities better than 0.75 times the value of the least significant bit. The power consumption of this device is only 12 mW at ±3 V supply voltage. The area is 5 mm^2 in the 3 μm SACMOS technology.

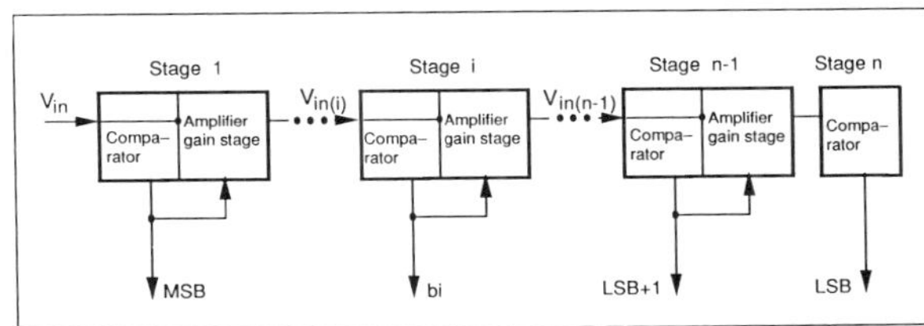

Figure 28. Block diagram of the pipelined A to D converter.

Discussion

The first and still preliminary results on the operation of functional building blocks for the HARP processor have been described. In particular, the SAPE building block made in a 1.5 μm CMOS technology performs already very satisfactorily up to 70 MHz, which shows promise for the use in the supercolliders. The essential features of the SAPE circuit are:

1. simultaneous write and read operation with no dead time

2. several triggers can be in the pipeline at once

3. current sampling integration allows a direct image of the detector output and eases information processing

4. the delay between write and read can be adjusted at any time

Although the measurement results at present are somewhat less favourable, calculation shows that the variance on the pedestal non-uniformity could be as low as 0.7 mV. For a dynamic range of 2.5 V this would correspond to a 12 bit precision.

The further development of the HARP architecture will be serving various applications in the future supercolliders, among which the silicon preshower prototype detector.

Acknowledgements

Most of the circuits described in this report were manufactured as multi-project chips in the framework of the IMEC-INVOMEC (Leuven, Belgium) MPC program. We appreciate the assistance in the testing of the devices by Stephane Bouvier, technical student at CERN, and by Jean-Claude Santiard.

References

[1] F. Anghinolfi, P. Aspell, M. Campbell, E.H.M. Heijne, P. Jarron, G. Meddeler, Ch.C. Enz, F. Krummenacher, L. Moult, P. Sharp and A. Olsen, Progress report 1988–1989, Development of integrated CMOS circuits and silicon pixel detectors in the CERN–LAA project, CERN/LAA/SD 89–11.

[2] H. Ikeda, N. Ujiie and Y. Akazawa, Monolithic preamplifier with bipolar SST for silicon strip readout, IEEE Trans. Nucl. Sci. NS–36 (1989) 502.

[3] T. Zimmerman, A high speed low noise ASIC preamplifier for silicon strip detectors, IEEE Trans. Nucl. Sci. NS–37 (1990) 439.

[4] E.A. Vittoz, MOS transistors operated in the lateral bipolar mode and their application in CMOS technology , IEEE J. Solid–State Circuits, SC–18 (1983) 273.

[5] P. Jarron, San Miniato Conference, to be published.

[6] V. Radeka, Trapezoidal filtering of signals from large germanium detectors at high rates, IEEE Trans. Nucl. Sci. NS–16 (1972).

[7] A. Olsen, H. von der Lippe, P. Jarron, F. Anghinolfi and E. Heijne, A switched capacitor analogue storage line in Proceedings of ECFA study week, CERN report 89–10, Vol. 2 (1989) 678.

[8]. F. Anghinolfi, M. Declercq, P. Deval, E.H.M. Heijne, P. Jarron, F. Krummenacher and V. Valencic, 1 MHz sampling rate 12-bit low-power analog-to-digital converter for data processing in particle detectors, presented at 2nd Int. Conf. on advanced technology and particle physics, Como, June 1990, to be published in Nuclear Physics B, Supplements Section.

HIGH SPEED NETWORK FOR REAL-TIME DATA ACQUISITION AND CONTROL WITH LOSSLESS AND BALANCED SELF-ROUTING

Yoram Ofek
IBM T. J. Watson Research Center
P.O.Box 704, Yorktown Heights, NY 10598
Tel. (914) 784-7085, e-mail: ofek at ibm.com
S. E. Hicks
IBM E/S National Support Center
1503 LBJ Freeway, Dallas, Texas 75234
Tel., (214) 406-5524, e-mail: hicks at dfwvm03.iinus1.ibm.com

Abstract

We present a gigabit network architecture for interconnecting the on-line and off-line data acquisition and processing farms for future detector facilities at the SSC (Superconducting Super Collider). In our solution we concentrate on how to interconnect the front-end (ADCs and level 1&2 triggers) with the on-line (real-time) computer farm. We propose that the on-line solution be extended into the off-line data processing and mass-storage. As a result, this solution provides a single uniform structure which will simplify the overall system.

1.0 Introduction

A collider detector facility can be viewed as having three basic parts: (1) a detector, which collects information about particle-particle interactions, (2) an on-line system, both analog and digital, which is responsible for "filtering" out in real-time the good, or interesting events from the background events, and (3) off-line processing, in which the interesting events are further analyzed. This structure is described schematically in Figure 1. The on-line part is divided into a front-end which includes ADCs and level 1&2 triggers, a parallel event builder, the on-line processing, and on-line storage. The on-line storage may also be connected to the off-line processing units.

Our networking proposal for the connection between the front-end and the off-line storage, is called *MetaNet* [4,5,6], and it includes the parallel event builder and the on-line processing, as shown in Figure 1. For the connection between the off-line storage and the outside world, we propose the *PARIS/MetaRing* architecture [1,2,3,7]. The machines which analyze the archived data will reside on this part of the network.

The *PARIS/MetaRing* is a gigabit/second architecture which is a wide area network (*PARIS*) with a native local area network (*MetaRing*). The local area is connected to the wide area network in a seamless fashion, which implies simplicity, small delays and high efficiency. We comment that the *PARIS/MetaRing* is currently under development as part of the CNRI/Aurora Gigabit Testbed [8].

In the rest of this manuscript, we first describe, in Section 2, the critical requirements of the on-line network which interconnects the front-end and the off-line storage, and then in Section 3, we describe the basic characteristics that the *MetaNet* can provide for satisfying the required functionality and performance.

2.0 The On-line Network Requirements

The on-line data acquisition function is performed in three basic steps: (1) Merging each event from the thousands of readout points to one on-line processing unit in a consistent way, i.e., such that it has not been contaminated with information from temporally nearby events, making useful physics analysis impossible. (2) After an event has been delivered to an on-line processing unit it is examined to determine if it contains 'interesting' physics data. (3) In the last step the events deemed 'interesting' are sent to the off-line storage. Our networking solution is primarily a solution for the merging step.

The on-line event merging/building should be performed in a distributed and balanced manner. The distributed operation is required for fault tolerance purposes. It will enable graceful degradation of the on-line network in the case of failures. Load balancing is required in order to assure that the traffic load on the network is equally distributed and that the events are equally distributed among the on-line processors.

To guarantee effective performance of the event merging the on-line network should have the following properties:

1. ***Lossless-ness***. Any message, either part of an event or control information, should not be lost in the on-line network as a result of congestion. If we assume that with today's technology the bit error rate is very low (10^{-9} to 10^{-12}). Then the packet loss probability will be very small, and therefore, the on-line network will not require error recovery protocol, which implies simplicity and high efficiency. We comment that networking technology which can suffer from congestions can result in unpredictable packet loss rate, which may require a special error recovery protocol.
2. ***Broadcast***. A broadcast capability is required to notify each node or collection point where to send information about a particular event. This broadcast message should include two basic parameters, the event name or its time stamp and the processing unit network identification where the event is to be analyzed. The broadcast operation should also be free of congestion to ensure that with a very high probability, all read-out points at the front-end will receive the message. We comment that a reliable broadcast mechanism is essential for many other control functions in the data acquisition system, e.g., the operation of the distributed merging scheduler, which determines what event is to be merged to each processing unit.
3. ***Self-routing***. When a node that has part of an event receives the broadcast message to where this event should be merged, it will attach the destination identification to the message and send it over the network. The on-line network routing nodes should ensure that the message will reach its destination with no loss. Self-routing implies that messages from all nodes that are destined to the same on-line processing unit will have the same message header. As a result, no route computation is required at the source nodes, which implies simplicity and fast response.

The on-line communication network should be able to provide any required aggregate throughput, in a way that is not constrained with a given transmission technology. Furthermore, the networking technology for such a large system should be flexible and extendable, so it can support the data acquisition system over the lifetime of the project.

3.0 The MetaNet Architecture

We propose the *MetaNet* architecture as the solution for the SSC on-line network. The *MetaNet* can be viewed as either a local area network (LAN) with an ***arbitrary topology*** or as a connectionless (datagram) architecture. The *MetaNet* has properties that are typically found in LANs.

The objective of the following discussion is to show that the *MetaNet* architecture properties can match the on-line network requirements. A more detailed description of the architecture can be found in [4,5,6]. The *MetaNet* architecture is comprised of two basic primitives: virtual ring embedding on an arbitrary topology and high speed hardware control signals, as shown in Figure 2.

1. ***Virtual Ring embedding***. The virtual embedded ring is a one-dimensional structure with linear order which implies a ***global sense of direction*** between any source and destination nodes in the network. Based on this sense of direction, we can derive other properties like self-routing, broadcast, and deflection. Deflection means that the packet does not travel on a predefined route and can be deflected as a result of the local traffic pattern. This principle is also known by the name "hot-potato" routing. Deflection together with self-routing provides the *MetaNet* with the desirable lossless property. See Figure 2.
2. ***Hardware control signals***. These signals are short (possibly one byte long) with local control information and are exchanged between neighboring nodes. These signals are sent without queuing delay and are inserted in an arbitrary position in the data packet. The information transferred with these signals is used for synchronization and regulation of the access to the network. The fairness mechanism and the global event clock are based on these signals, as shown in Figure 2.

Based on these two primitives the *MetaNet* architecture is designed with the following principles:

1. ***Lossless internal flow control*** for arbitrary and asynchronous traffic pattern. Such a design implies that overflows can occur only at the destination node. Lossless transport mechanism implies that the end-to-end protocol can be simplified (since the transfer is more reliable and less error-handling is expected); this is a major contribution to reducing the end-to-end total transfer delay.
2. ***Single buffer***. At the end of each link, there is a buffer that can contain one maximum size packet, which is the optimum for a packet switch network. The reduction of buffer management reduces cost (to the necessary minimum) and may imply increased speed at a node. We note that the single buffer condition implies that a packet has to be sent whenever both ends of a link are ready and without making sure that the entire packet has arrived. This is a "cut-through" method which implies that the routing algorithm is performed in an iterative forwarding fashion --- the routing mechanism constantly monitors the links, and packets are transmitted as soon as possible. This is a gen-

eralization of the register (buffer) insertion mode suggested for ring architecture. Reduced buffer management of traffic in transit simplifies the network protocols.

3. ***Fairness mechanism***. The network interface that controls the media access has a built-in fairness mechanism that is fault tolerant and self-stabilized, and makes sure that all nodes will have equal opportunity to regularly transmit packets into the network (equi-fairness). Unfair mechanism is unacceptable in a high speed system as it may cause a total denial of service and imbalanced event merging.

4. ***Deadlock avoidance***. The flow control mechanism embedded in the interface together with the fairness property ensures that deadlocks will not occur. This is crucial, since deadlock detection and prevention mechanisms are contradictory to the nature of high-speed systems as they generate unpredictable delays.

5. ***Dynamic self-routing***. It is possible to send packets such that their headers contain only certain destination identification when it leaves the source. This information is sufficient enough for reaching the destination, not necessarily on the same route each time (i.e., adaptive or dynamic). It is also robust and sufficient to overcome congestion and failures. The self-routing property ensures that the routing function at the intermediate nodes is easy to perform given the special-purpose hardware, a fact which contributes to the speed messages are transferred through the network interface.

6. ***Broadcast-with-feedback***. Broadcast is an important control mechanism. Two broadcast algorithms are implemented on the *MetaNet*, one is a simple broadcast and the other is combined with feedback (or echo) to the source. The objective of the second algorithm is to provide fast acknowledgement to the source of the broadcast message. The algorithm starts with a single broadcast message, that can be duplicated into multiple messages along the way in order to speed-up the operation. The broadcast satisfies the following conditions:

    Each physical node on the embedded ring should be reached at least once.

    The broadcast message should be copied once into each physical node on this embedded ring, regardless of the number of times it was traversed by this broadcast.

    Under light load condition, the broadcast should reach all the nodes in time proportional to the propagation delay along the diameter of the network.

    Each virtual node on the embedded ring should be traversed at most once.

Using the *MetaNet* architecture, it will be possible to solve effectively the above problems. We present a fairness and flow-control mechanism, that is an integral part of the architecture, which, together with a novel self-routing technique, enable the network to merge many events concurrently from 10's of thousands o nodes to multiple sink nodes. This control scheme is self-balanced, lossless and can dynamically adapt itsel to arbitrary and unpredictable traffic pattern. This mechanism imposes global control cycles on the data acquisition system and can be used to generate a global logical event clock.

The arbitrary topology of the *MetaNet* architecture will make it possible to satisfy any throughput requirements, by simply adding more high speed links or nodes, according to the needs and geographic constraints.

References

1. I. Cidon and I. S. Gopal, "PARIS: An Approach to Integrated High-Speed Private Networks," *International Journal of Digital & Analog Cable Systems*, Vol. 1, No. 2, pp. 77-86, April-June 1988.
2. I. Cidon and Y. Ofek, "Distributed Fairness Algorithm for Local Area Networks with Concurrent Transmissions," the 3rd International Workshop on Distributed Algorithms, Nice, September 1989, IBM Research Report RC 15051, October 1989.
3. I. Cidon and Y. Ofek, "MetaRing - A Full-Duplex Ring with Fairness and Spatial Reuse," IBM Research Report RC 14961, September 1989, also INFOCOM'90.
4. Y. Ofek and M. Yung, "Principles for High Speed Network Control: loss-less and deadlock-freeness, self-routing and a single buffer per link," *AC PODC'90*, pp. 161-175.
5. Y. Ofek and M. Yung, "Efficient Mechanism for Fairness and Deadlock-Avoidance in High-Speed Networks," *The 4th International Workshop on Distributed Algorithms*, Sept. 1990.
6. Y. Ofek and M. Yung, "Lossless Asynchronous Broadcast-and-Feedback on the MetaNet Architecture," submitted to INFOCOM'91.
7. Y. Ofek, "Integration of Multi-ring on the MetaRing Architecture," IBM Research Report RC 16058, 1990.
8. "Gigabit Connection," *Scientific American*, October 1990, pp. 118-120.

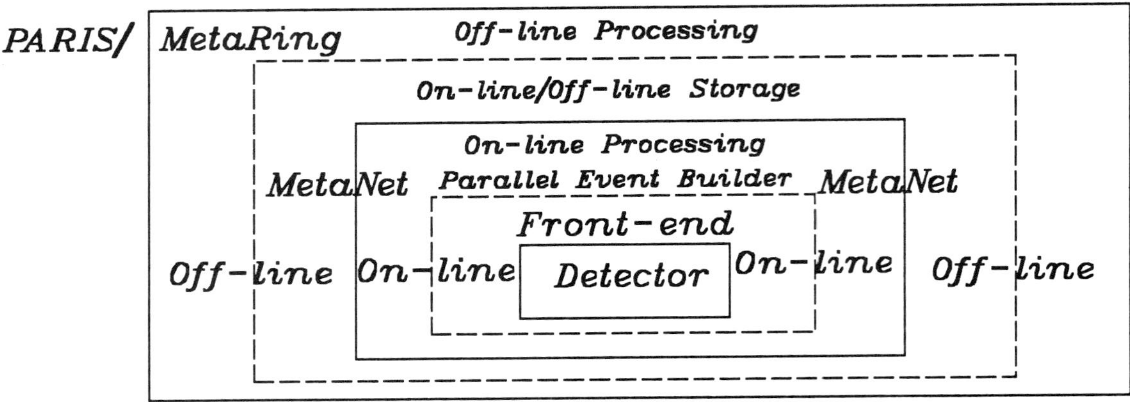

Figure 1. General Structure

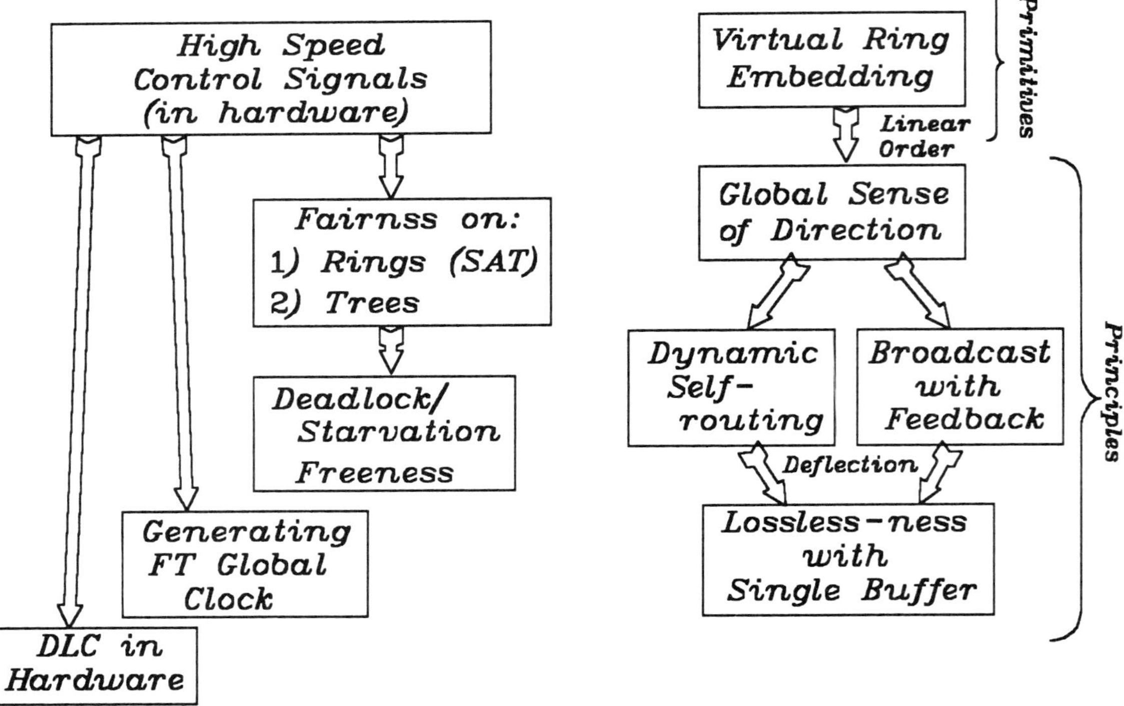

Figure 2. The *MetaNet* Principles

DATA PROCESSING AT THE SSC WITH STRUCTURED NEURAL NETS*

Klaus S. Lackner, Vernon D. Sandberg and David H. Sharp
Los Alamos National Laboratory
Los Alamos, New Mexico 87545

Abstract

SSC detectors will place extreme demands on data processing systems. One must reduce the data flux to manageable proportions at the earliest possible stage. This observation has led us to emphasize low-level data processing and track reconstruction. We report progress in three areas: A network compiler has been designed which generates programmable and trainable tree-structured nets; algorithms for track reconstruction have been designed and implemented in such nets; exploratory studies have been made on the use of such nets to carry out low-level processing on hardware components of a central detector.

1 Introduction

The work reported here has as its goal the development of a comprehensive approach to low-level data processing at the SSC. The enormous data flux ($> 10^{14}$ bit/s) anticipated in SSC experiments will require new approaches to data collection, event filtering, and on-line analysis. The highest data flux occurs at the lowest, or earliest, levels of the detector system. Thus emphasis must be placed on low-level data processing.

Algorithms are being developed for track reconstruction and other pattern recognition tasks. They lead to substantial data compression and $\log N$ performance in the input size N. Structured neural nets in the form of multiresolution pyramids have proven useful in developing these algorithms.

Software development must be conducted with hardware constraints in mind. Particle detectors must retain all data until a trigger decision has been made. This is an opportunity—not a problem. The delay allows for preprocessing on individual detector units, e.g. the silicon wafers in the detector core, and in the data pipeline which leads to the detector surface. We note that data compression at the lowest levels leads to the maximum absolute reduction in data flux.

2 The Bottom-up Approach

Pattern recognition at the SSC can be organized into a hierarchy of tasks, from simple to complex, as follows:

Hardware oriented preprocessing: The lowest level of processing is closely tied to the basic detector units and their architecture.

Local Track Segements: Spatially related hits are combined into local track segments. These may be viewed as line segments that have associated with them an orientation and a position.

Track Reconstruction: Spatially related hits and track segments are combined into longer tracks. The net must incorporate prior knowledge concerning rotational and translational invariance and momentum conservation. Parameterizing the raw data in terms of tracks leads to a significant data compression.

Vertex Reconstruction: Beginning and end points of tracks must be identified. The number of tracks originating from or ending in a vertex must be determined. Momentum conservation can be used to recognize missing momentum.

Jet identification: Tracks with a common origin and momentum pointing in the same direction may be combined to form a jet, characterized by its four-momentum flow and origin.

Event Reconstruction: Particles generating the tracks must be identified. Calculating reduced masses of intermediate particles could be done with analog neurons. Feedback from higher layers of the net could reduce the amount of processing necessary.

Our program is to design algorithms to perform these tasks as efficiently as possible, starting with the simplest task.

3 Structured Neural Networks for Track Reconstruction

Low-level data processing for particle detectors, in common with many other pattern recognition problems, involves the parallel and repetitive application of local rules. For inherently parallel, but local, problems such as these, a massively parallel processing architecture with relatively weak processing nodes is called for. This suggests that neural nets may be useful in attacking this problem [1–4].

In this section we discuss neural nets, structured neural nets and genetic neural nets, which are a recursively defined implementation of structured neural nets. We conclude with an outline of our approach to track recognition using genetic neural nets.

*Work supported by the US Department of Energy

3.1 Neural Nets

Neural nets are collections of simple processors which may be connected in a complex way. Typically, the processors are threshold devices with two states depending on the sum of all inputs. For example, in a neural net with a discrete time step, the state s_i of the ith processor (neuron) at time $t+1$ may be given by

$$s_i(t+1) = g\Big(\sum_{j=1}^{N} T_{ij} s_j(t)\Big),$$

where g is the thresholding function and T_{ij} the connection matrix which defines the coupling strength between any two processors. During training, the connection matrix is modified to achieve the desired behavior of the net. For conventional neural nets, the T_{ij}'s are independent. Hence, for N processors there are N^2 optimization parameters.

Neural nets have a number of advantages. They can be trained to solve problems for which a good solution algorithm is unknown. They are robust to noise in the data, small changes in the data structure and internal (hardware) failures. By design they are massively parallel. Of course, neural nets have disadvantages as well. Training large nets is prohibitively expensive, because with increasing N the optimization task becomes extremely difficult. Known features of the solution cannot be incorporated into the net from the outset but must be learned, making the learning task unnecessarily complicated. Furthermore, the complexity of the net is tied to its input size. This is in contrast to actual tasks, like track reconstruction, which only become marginally more difficult for larger input arrays.

3.2 Structured Neural Nets

To take advantage of the strong points of neural nets while overcoming their disadvantages, we work with structured nets. The idea is to design the net architecture to express what is known about a problem. A structured net approach can lead to a great variety of architectures, depending on the problem, ranging from nets which are completely hardwired implementations of a given algorithm to those which restrict connections to local areas and special pathways, but otherwise remain trainable. A structured architecture implies that the T_{ij}'s are not independent but are instead determined by a smaller set of independent parameters. This reduces the size of the solution space and consequently the time required for training. Structured neural nets have two further advantages: The number of independent parameters as function of the input size can be matched to the growth in complexity of the problem. Secondly, the architecture can be programmed to reflect the known structure of the problem.

3.3 Genetic Neural Nets

Genetic neural nets are a very general implementation of structured neural nets [5]. A family of nets of increasing size is defined through a recursion relation:

$$T^{a,n+1}_{(i_1,i_2),(j_1,j_2)} = \sum_b P^{a,b}{}_{i_1,j_1} T^{b,n}_{i_2,j_2},$$

where (i_1, i_2) is a multi-index spanning the product space of the i_1 and i_2 index. a and b label all nets in a given generation, n indicates the generation number. P, which is refered to as the propagator, establishes the relationship between nets in successive generations. For simple cases P is independent of n, but in general P may vary with the generation number. Typically, we consider the case where $P^{a,b}$ is a 2×2 matrix in the i and j indices. In this case, the number of neurons grows by a factor of 4 from one generation to the next. In effect, the matrix elements T_{ij} are implicit functions of the propagators which in turn may again be parameterized. It is those parameters which are optimized in a training procedure.

Genetic Neural Networks have been shown to train well and to generate understandable solution algorithms. The recursive definition of nets generates highly repetitive substructures and subunits called netlets. By breaking up the task into subtasks which are implemented as netlets, it is possible to refine the performance of subtasks by optimizing individual netlets or by replacing them with programmed modules.

Our general approach represents a balance between programming and training. We program into the net what we know about a problem and leave to training what cannot readily be reduced to algorithmic form, either because it is too complex, or because the scientific principles are not well understood. In practice this means that one starts with an essentially hardwired net reflecting the status of knowledge about the system. As the net is refined, training becomes more and more important. Thus, our approach is at variance to the black-box approach to neural nets.

3.4 An Approach to Track Reconstruction

The fact that genetic neural nets can be decomposed into small units (netlets), suggests an implementation where individual netlets are identified with simple feature detectors. The recursive definition of a genetic neural net leads to tree-structured architectures which perform their task in $N \log N$ operations and $\log N$ time steps; N is the number of inputs. The feature detectors are repeated throughout the tree structure and operate at different levels of resolution.

For minimimal feedback between levels, the net acts as a data pipeline processing the data during transfer. A pipeline makes it possible to process, filter and buffer multiple events simultaneously. An event does not just sit waiting for a trigger condition; it is refined and processed and thus may contribute to the trigger.

4 Progress to Date

We have progress to report in three areas: A network compiler has been designed which generates programmable and trainable tree-structured nets; algorithms for track reconstruction have been designed and implemented in such nets; exploratory studies have been made on the use of such nets to carry out low-level processing on hardware components of a central detector.

4.1 The Network Compiler

The function of a compiler is to translate between a high-level programming language and machine language. At its current state of development our network compiler generates a tree-structured net architecture from a high level description. The net is composed of neurons combined into netlets which are either trained or programmed to perform specific functions. The compiler determines the position and function of each netlet and lists its trainable parameters.

4.2 Track Reconstruction

The hits in an array of detector units must be organized into tracks. In our first examples we use a two-dimensional array of detectors. The multi-resolution, multi-level architecture introduced above is well suited to the task. Line segments at each level are generated from line segments on the next lower level by searching for a cluster in the space of line parameters (Hough Transform). On the lowest level multiple moments of the pixel distribution are calculated and used in a voting algorithm which decides on the presence of an oriented line segment. Processing at this level is strictly local and encompasses a 3×3 neighborhood of each pixel.

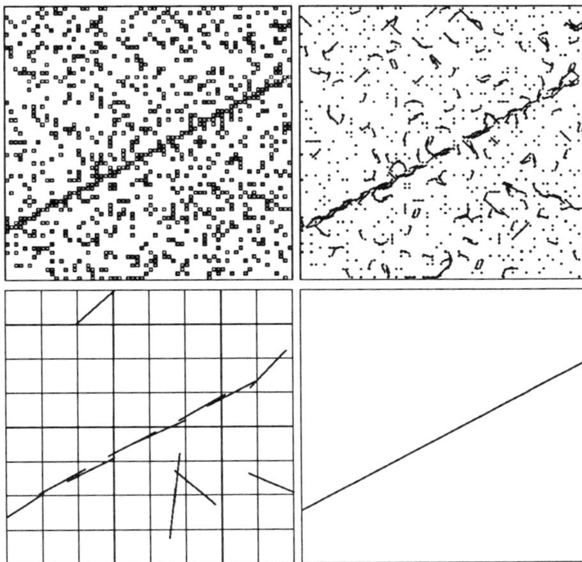

Fig. 1: Line reconstruction at high noise level from a 64×64 input array of pixels (upper left). The reconstruction of the track is shown at the bottom level (upper right, showing line segments and pixels), an intermediate level (lower left), and the top level (lower right) of a multi-resolution pyramid.

Hits which cannot be combined into line segments are discarded as noise. Those which form line segments are described in a concise way by specifying a set of parameters which define a line. This processing greatly reduces the data flux without loosing relevant information (Fig. 1).

4.3 Processing in the Central Detector

In collaboration with B. J. Cooke and C. L. Winter [6] we have outlined an approach to distributed data processing on the silicon wafers of a central detector. Hardware considerations will most likely prohibit inter-wafer processing inside the central detector.

The data collected by a wafer is compressed and separated into a prompt and delayed component. The prompt component is transmitted immediately and consists of crude statistical information characterizing the event, e.g. the number of hits, the centroid position and width of the hit distribution. This information will be available immediately and thus can be used in a trigger decision. The delayed channel completes the description of the event. Since complex events on a wafer are rare, we believe it is possible to send complete information on virtually all events to the detector surface obviating the need for information flow back into the detector with the associated synchronization problems.

5 Conclusion

A key question is how the pattern recognition methods described here should actually be used in real detectors. We believe that it is worthwhile to explore the possibility of carrying out a significant amount of the low-level processing on components such as silicon strips and optical fibers existing inside the detectors. Such a distributed data processing architecture will strongly influence trigger design philosophy.

References

[1] B. Denby, "Neural networks and cellular automata in experimental high energy physics", Comp. Phys. Comm. **49**, 429–448, (1988)

[2] B. Denby *et al.*. "Neural networks for triggering," IEEE Trans. Nucl. Science, in press.

[3] C. Peterson, "Track finding with neural networks," Nucl. Inst. Meth. **A279**, 537–448, (1989)

[4] L. Lönnblad, C. Peterson and T. Rögnvaldsson, "Finding gluon jets with a neural trigger," Phys. Rev. Lett. **65**, 1321–1324, (1990)

[5] E. Mjolsness, D. H. Sharp, B. K. Alpert, "Scaling, machine learning, and genetic neural nets, " Adv. Appl. Math. **10**, 137–163 (1989)

[6] B. J. Cooke, K. S. Lackner, V. D. Sandberg, D. H. Sharp and C. L. Winter, "Distributed data processing in the central detector," in preparation.

USING NEURAL NETWORKS TO IDENTIFY THE ORIGIN OF JETS

Carsten Peterson
Department of Theoretical Physics, University of Lund,
Sölvegatan 14A, S-223 62 Lund, Sweden
thepcap@seldc52 (bitnet)

Abstract

Recent progress using neural networks to identify the origin of hadronic jets is reviewed.

Introduction

When analyzing experimental data it is a standard procedure to make various cuts in order to single out desired features. A specific choice of cuts corresponds to a particular set of feature functions $y_i = F_i(x_k)$ in terms of the kinematical variables x_k. One would like to have an optimal choice of this function with respect to expected distributions. This is exactly what feature recognition neural networks (NN) aim at. A particular form of F_i is chosen

$$F_i(x_k) = g[\sum_i \omega_{ij} g(\sum_j \omega_{jk} x_k)] \tag{1}$$

where ω_{ij} and ω_{jk} are the parameters to be fitted to the expected distributions and $g(x) = 0.5[1 + \tanh(x)]$. Eq. (1) corresponds to the feed-forward architecture of fig. 1. The bottom layer (input) corresponds to the measured kinematical variables x_k and the top layer to the features y_i. The mission of the so-called hidden layer is to build up an internal representation of the observed data. Each unit or neuron has the threshold behaviour given by $g(x)$. Fitting to a given data set (or learning) takes place with gradient descent on e.g. a summed square error

$$E = \frac{1}{2}\sum_i (y_i - t_i)^2 \tag{2}$$

with respect to the weights ω_{ij} and ω_{jk}, where t_i are the desired feature values [1]. This is done by presenting all the **training patterns** over and over again with successive adjustments of the weights. Once this iterative learning has reached an acceptable level in terms of a low error E the weights are frozen and the NN is ready to be used on patterns it has never seen before. The capability of the network to correctly characterize these **test patterns** is called **generalization** performance.

Separating Quarks from Gluons

Being able to distinguish whether a jet of hadrons originates from a quark or a gluon is important from many perspectives. It can shed light on the hadronization mechanism. For example experimental studies of the so-called string effect needs identification of the gluon jet. Also a fairly precise identification of the gluon jet is required for establishing the existence of the 3-gluon coupling in e^+e^- -annihilation. To date the most of the gluon jet identification has been done by selecting the jet with smallest energy as the gluon. This procedure, which is based on the underlying perturbative QCD matrix element, typically yields $\sim$ 65% identification rate in e^+e^- reactions. In refs. [2][3][4] the neural network algorithm described above was used to do the separation both in e^+e^- reactions and in hadron induced large p_T processes.

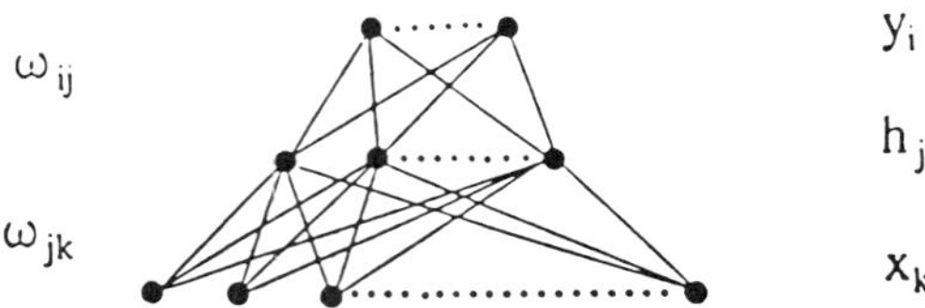

Figure 1: A one hidden layer feed-forward neural network architecture.

e^+e^- **reactions** [2][3]. In order not to reveal to much about the model dependent low momentum part of the jet four-momenta $(\vec{p}_k, E_k)$ of the four leading particles in the jet (k=1,..,4) were used as inputs to the network. One output node representing the quark/gluon option was used and 10 hidden units. Training and test sets were generated at 2 different energies (92 GeV and 29 GeV) with 3 different MC generators; JETSET [5], ARIADNE [6] and HERWIG [7]. The inputs were either single jets defined by the LUCLUS clustering algorithm in JETSET or entire 3-jet events. After training was completed the network was tested with a middle point succes criteria where > 0.5 for the output node is interpretted as a gluon jet and < 0.5 as a quark jet.

On the average the network was able to correctly classify 85% of the test set jets, which should be compared with 65% from just selecting the lowest energy jet as originating from the gluon. The theoretical upper limit of performance is set by the Bayesian limit, which is the minimal overlap between the two distributions. We have estimated this limit numerically and find it to be ~ 87%, which means that the neural network is only 2% away from the optimum. The model independence of the results were demonstrated by training on MC data generated by one model and tested on MC data from another. Almost no deterioration in performance was obeserved. Also runs where detector acceptance effects were included showed only $O(2\%)$ degradation.

Large p_T processes [4]. In this case the network gets no lead from matrix element information. A set of $p\bar{p}$ events at 630 GeV were generated with the PYTHIA MC [8]. The energy of the fragmentation products was mapped into a calorimeter with a granularity of $\Delta\eta = 0.20$ in pseudorapidity and $\Delta\phi = 0.26$ in azimuth. This calorimeter had a complete coverage in ϕ and extended down to $|\eta| \leq 2$. The set-up corresponds to the UA2 calorimeter. Jet transverse energies were collected in cones of radius 0.8 in η, ϕ space. The calorimeter information was presented to the network in two ways: **(a)**. The E_T's of the 7 ×7 cells around the jet center [3]. This case corresponds to 49 input nodes. **(b)**. Take the E_T of the leading cell in the 7×7 matrix and assign it to the first node x_1. Assign the η and ϕ coordinates relative to the center of the jet to x_2 and x_3 respectively. Then take the second leading cell and assign its E_T, η and ϕ to x_4, x_5 and x_6 and so on for the first 15 cells. This case corresponds to 45 input nodes.

After training the the network correctly classifies $70-72\%$ of the jets using the 0.5-critera as above. This number is consistent with the e^+e^- results above being ~ 20% above random choice; n e^+e^--reactions the matrix element information gives 65% and the remainding classification done by the network is just those 20%. Rather than having the 0.5 success criteria one might vary th **cut** of the output node and choose a value corresponding to an optimal efficiency and signal-to- background ratio. In fig. 2 the signal-to-background ratio as a function of the signal efficiency is shown. As can be seen the results are really impressive.

Heavy Quark Jet Tagging

The conventional way of identifying heavy quarks (c and b) in e^+e^--reactions is either through leptonic tagging or secondary vertex with efficiency/purity levels of approximately (10%/100%) and (25%/80%) respectively. We have designed a network identify heavy quarks [3], which is solely built upon hadronic information. As input variables we use the total jet energy and momentum along with the energy and direction for the 6 leading particles in the jet, giving a total of 20 input units.

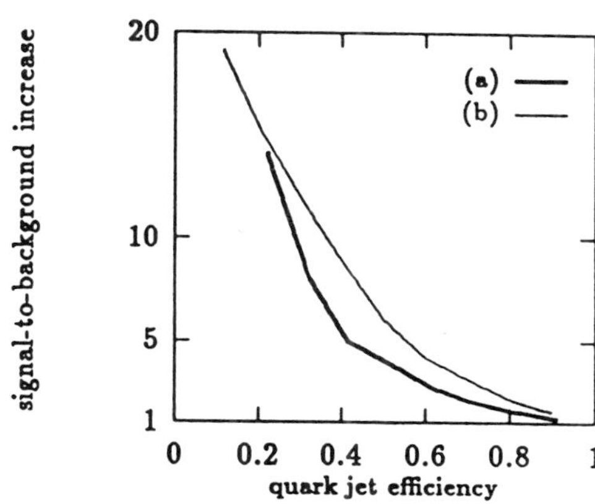

Figure 2: The increase in signal-to-backround ratio vs. the efficiency for a neural network quark trigger with options (**a**) and (b).

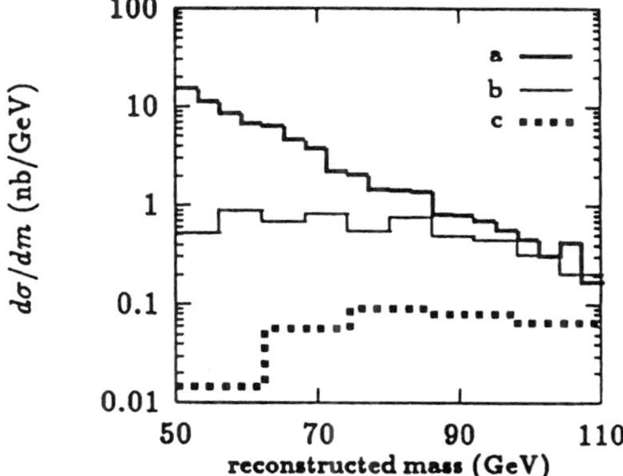

Figure 3: The two-jet invariant mass distribution for the QCD background (a) without trigger, (b) with NN trigger cut=0.50 and (c) cut=0.85.

Again one layer of 10 hidden units is used and a single output neuron (b-jet=1, non-b-jet=0). As in the large p_T application above the option of varying the cut on the output node can be used to select a desirable efficiency versus purity. The network is then able to produce efficiency/purity numbers comparable with what is expected from vertex detectors. This remarkable result implies that a general purpose hadronic detector could be very efficient also for heavy flavour tagging.

W/Z → 2 Jets

The networks described in the previous sections were designed to deal with individual jets. Next we want to use the entire event information without explicit jet reconstruction to distinguish hadronic decays of produced W in hadron-hadron collisions from the dominating QCD background [4]. The W events were generated, again using PYTHIA, with randomly chosen W masses between 50 and 150 GeV. Also an equal amount of QCD background events with a flat mass spectrum were generated. In both cases at least two jets were required in the rapidity interval $|\eta| \leq 0.7$. For each event the 30 highest transverse energy cells were presented to the NN using

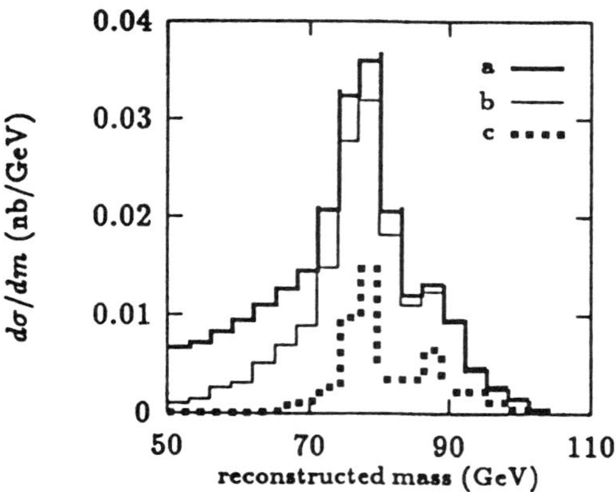

Figure 4: The two-jet invariant mass distribution for the W and Z signal (a) without trigger, (b) with NN trigger cut=0.50 and (c) cut=0.85.

a strategy similar to case (b) of the previous section. We used a NN with two hidden layers; 90 input nodes, 20 nodes in the first hidden layer and 4 in the second. After learning was completed the network was tested on natural mass distributions. In fig. 3 the invariant mass of the two largest jets of the QCD background is shown without and with the NN trigger using two different cuts on the output node. In fig. 4 the same quantity is shown for the W and Z signal. Looking at the W and Z peaks, it appears that for the weaker cut the efficiency is almost 1, while for the stronger cut the signal is reduced a factor 2. The QCD background in this region is however reduced by factors 3 and 20-50 respectively (due to the large rejection the statistics is very low for the strong cut on the QCD background). Note also that the W and Z peaks tend to get sharper when the NN trigger is applied.

Summary and Outlook

The neural network method is very efficient for extracting features in hadronic data. World record performance is obtained for quark/gluon separation. With respect to heavy quark tagging the results are in parity with those expected from a vertex detector. The network is able to reduce the QCD background to $W/Z \rightarrow jets$ by factors 20-30. All results reported here (refs.[2][3][4]) were obtained by using JETNET F77 software package [9].

The neural network approach is in general very noise- and damage-resistant. Hence it is suitable for high energy experiments where various parts of an detector may malfunction. Also with its inherent concurrency and simple structure fast execution custom made hardware could be an asset for real-time triggering.

The success in separating the QCD background from the W/Z hadronic decays is particularly encouraging if one at the SSC/LHC wants to look for Higgs bosons in $pp \rightarrow H \rightarrow WW$ where one W decays leptonically and the other into hadrons. Although the cross section for this process is much larger than the "gold-plated" channel where both W decay into leptons, it is plagued by severe backgounds from various QCD and heavy quark production processes. We are presently applying the neural network to this problem with encouraging results [10].

Neural network can also be designed to solve difficult optimization problems like track-finding in high energy physics [11].

References

[1] D.E. Rumelhart, G.E. Hinton and R.J. Williams, in *Parallel Distributed Processing: Explorations in the Microstructure of Cognition*, D.E.Rumelhart and J.L.McClelland (Eds.), (MIT Press, Cambridge, 1986), Vol. 1.

[2] L. Lönnblad, C. Peterson and T. Rögnvaldsson, "Finding Gluon Jets with a Neural trigger", *Physical Review Letters* **10**, 1321 (1990).

[3] L. Lönnblad, C. Peterson, and T. Rögnvaldsson, "Using neural Networks to identify Jets", LU TP 90-8 (to appear in *Nuclear Physics B*).

[4] P. Bhat, L. Lönnblad, K. Meier and K. Sugano, "Using Neural Networks to Identify Jets in Hadron-Hadron Collisions", LU TP 90-13 (to appear in *Proc. of the 1990 DPF Summer Study on High Energy Physics Research Directions for the Decade, Colorado, 1990*).

[5] T. Sjöstrand, JETSET 7.2 program and manual. See B. Bambah et al., QCD Generators for LEP, CERN-TH.5466/89.

[6] L. Lönnblad, "ARIADNE-3, A Monte Carlo for QCD Cascades in the Colour Dipole Formulation", Lund preprint LU TP 89-10.

[7] G. Marchesini and B. R. Webber, *Nucl. Phys.* **B310** (1988) 461; I. G. Knowles, *Nucl. Phys.* **B310** (1988) 571.

[8] H-U. Bengtsson, T. Sjöstrand, *Computer Phys. Comm.* **46** (1987)43.

[9] L. Lönnblad, C. Peterson and T. Rögnvaldsson, JETNET 1.1 program and manual [available via BITNET request].

[10] L. Lönnblad, C. Peterson and T. Rögnvaldsson, work in progress.

[11] B. Denby, "Neural Networks and Cellular Automata in Experimental High Energy Physics", *Comp. Phys. Comm.* **49**, 429(1988); C. Peterson, "Track Finding with Neural Networks", *Nuclear Instruments and Methods* **A279**, 537 (1989).

STUDIES OF TRIGGER REQUIREMENTS USING A RULE LEARNING PROGRAM

S. H. Clearwater, W. E. Cleland and E. G. Stern
University of Pittsburgh
Pittsburgh, PA 15260

Abstract

This paper describes a novel approach for studying trigger requirements in HEP experiments. We utilize a learning program that searches for rules subject to desired efficiencies and a set of pre-classified examples.

Motivation

The major physics topics to be studied at the SSC are well-known. Feasibility studies for separating signal from background events have been carried out for a variety of reactions. However, knowing that a particular type of event can in principle be separated from the background is not equivalent to knowing how best to do it. Of particular importance is the determination of a triggering strategy that will work in the presence of real-world problems such as calorimeter noise. We have chosen to use machine learning techniques as a generic study in trigger strategy selection.

Machine Learning

The learning program RL4[1-2] forms rules for classifying data into two data sets, one of which is considered the signal or positive examples, and the other which is background or negative examples. Its input is in the form of pre-classified training data as positive and negative examples, plus a vocabulary of attributes to be used in forming the classification rules.

This vocabulary takes form of a list of attributes thought to be useful in the classification of the data, and values that these attributes may take. The input of the physicist (in this case) to the problem is in the selection of useful attributes.

RL4 operates by finding rules that classify events as either signal or background using attributes as possible distinguishing characteristics. The space of possible rules is represented by a search tree with each node representing a potential rule. Deeper levels in the search tree correspond to more restrictive rules. By following a number of the best paths at each level, multiple satisfactory rules may be found, unlike other optimizing algorithms such as neural nets.

For inputs, RL4 takes a minimum desired signal threshold, a maximum allowable background threshold, or a minimum desired signal/background ratio and attempts to find separating rules that meet the specified thresholds. An inability to find rules that adequately classify the positive examples or to reject the negative examples indicates that the positive and negative data sets are too similar, or that no good attributes have been selected to discriminate between them, or the selection thresholds must be relaxed.

This approach has several advantages:

1) There is great flexibility in the selection of attributes used in the trigger.

2) The trigger strategy can be investigated under a wide variety of conditions, e.g. different luminosities, noise levels, etc. by changing the event sample that RL4 uses.

3) RL4 systematically identifies attributes individually or in conjunction that are useful for event selection and hence triggering purposes (attributes not found to be useful can be eliminated from consideration).

Triggers

The trigger we are modeling is a three stage trigger similar to what has been discussed in SSC literature. The quantities used in each trigger stage depend on the event rate at that stage; the data rate imposes restrictions on what processing may be performed to render a decision. Each stage in the trigger reduces the data rate to a level which can be handled by the following stage. As shown in Table 1, the first level of the trigger takes place in a fixed time, using quantities which are available after only minimal analog processing. The second level may take a variable time to reach a decision and use more sophisticated quantities that require processing, possibly in digital electronics. The third level is expected to utilize some form of general purpose microprocessor farm and will be able to handle complex calculations, using data from the entire detector.

Table 1. Assumed trigger scheme for trigger strategy studies.

Level	Decision time	Technique	Maximum rate (Hz)
1	1-2 μs	analog electronics	10^5
2	10-100 μs	digital arithmetic	10^3
3	1-10 ms	microprocessor farms	100

The simulation of this trigger involves three separate runs of the RL4 program, one run for each level of the trigger. Each run uses event attributes that are feasible to calculate at the corresponding trigger level. Each event of the positive and negative samples is assigned a weight proportional to the cross section for that event type. Acceptance thresholds for the positive and negative samples can then be specified in units of nanobarns (or Hertz). The thresholds for each run of RL4 (minimum acceptable signal, maximum acceptable background) are fixed by the desired rate at that trigger level. Since early levels of the trigger are dominated by background, the background threshold is just the allowable trigger rate at that level, while the signal threshold is set to give a reasonable acceptance. Events used in learning at one trigger level are filtered and used as input for learning the next level trigger rules. A flowchart of this methodology is shown in Figure 1.

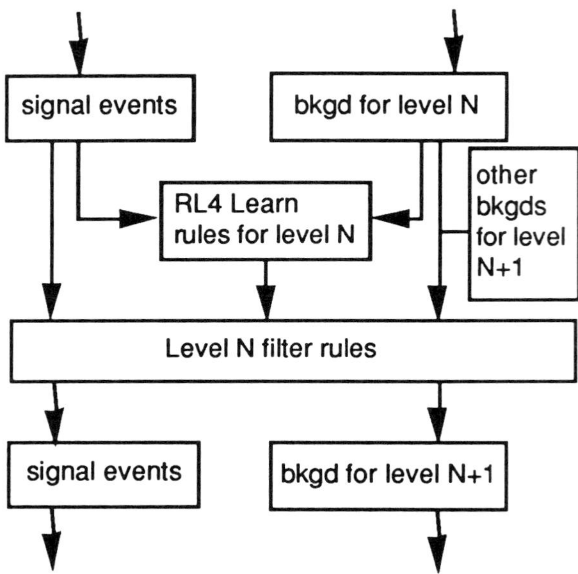

Figure 1. Flow diagram for using RL4 to learn trigger strategies at level N. Level N-1 is input to level N, which in turn is the input to level N+1.

Data Sample

RL4 uses preclassified signal and background event samples as input to identify rules that separate the two samples. For our purposes, these data samples can be efficiently generated by Monte Carlo programs. We use the ISAJET Monte Carlo (version 6.30) [3] to produce the events which make up the signal and background samples. The response of a calorimeter detector to these events is simulated by the QFL program. For each event, we calculate all attributes that could be used to form triggering rules.

For the positive and negative samples, we generate events which represent signal (top) and backgrounds likely to be of importance in any trigger that selects top quark events. Table 2 lists all of the event samples used in this study, along with their cross sections. Our positive sample consists of $q\bar{q} \to t\bar{t}$ and $gg \to t\bar{t}$ events.

Table 2. Event Types for Triggering Strategy Study for top

(At $\mathcal{L}=10^{33}$ cm^{-2}s^{-1}, 1 nb corresponds to a 1 Hz event rate.)

Process	pt range (GeV)	Cross section (nb)
	Positive sample (signal)	
t production	20-400	15.9,3.8(m_{top}=140,200)
	Negative sample (backgrounds)	
Minimum bias		1.37×10^8
QCD jets	5-1000	8.86×10^7
c, b production	20-1000	1.26×10^4
Drell-Yan W	20-240	38.0

The negative sample includes a variety of processes. Some of the reactions are not really expected to contribute events to the top candidate sample such as low p_t two jet events. However, they have a high cross section which form the bulk of the events that a low-level trigger will have to handle. They are included here so that RL4 will include rules for dealing with them at an early stage.

Results of Trigger Strategy Studies

We performed analyses using two values of the top mass and two values of luminosity (10^{33} cm^{-2}s^{-1} and 10^{34} cm^{-2}s^{-1}). The positive and negative event samples were used by RL4 to find rules appropriate for a trigger that finds top quark events. It formed rules using attributes chosen from those listed in Table 3. At the first level, RL4 had the task of finding rules that reduce the background to a level of 10^5 Hz. The rule chosen was to require that at least one jet sum contains more than 50 GeV of deposited transverse energy or a minimum total calorimeter E_t. Removing the events in this region eliminates all minimum-bias and two jet events of p_t<100 GeV while retaining most of the signal events, leaving 1×10^4 nb of background.

At the second level, the goal was to achieve a background level of 1×10^3 Hz. RL4 found several rules at this level. A common feature to all these rules was presence of either an electron or muon with a p_t of at least 10 GeV. Electron and muon cuts are expected to be disjoint, so we chose to require at least one lepton with at least 10 GeV as the second level rule. The higher luminosity samples required an additional condition in order to meet the desired trigger rate, further reducing the signal.

Table 3. Attributes Used in Forming Trigger Rules

Electrons and muons are assumed to be identified by non-calorimeter systems. A cluster is a group of locally correlated energy deposits which lie within a cone of radius $\Delta R<0.7$ in $\eta\times\phi$ space where $\Delta R^2=\Delta\eta^2+\Delta\phi^2$. At level 3, all calorimeter attributes are recalculated using optimal filtering to reduce noise. All energy units are in GeV. Higher trigger levels also contain lower level attributes.

Level 1
no. jet sums w/E_t > 10, 30, 50, total calorimeter E_t
Level 2
no. clusters w/E_t >10, 30, 50, no. leptons w/p_t>10,30, 50, 1st-, 2nd-, 3rd-largest jet sum E_t
Level 3
1st-, 2nd-, 3rd-largest cluster E_t, 1st-, 2nd-largest lepton p_t, 1st-, 2nd-largest E_t clusters & leptons, event circularity, $\not{p}_t$, E_t not clusters or leptons, M_t e-ν system, cos$\phi(e,\not{p}_t)$, cosϕ(e, event axis), isolated electrons E_t w/($\Delta R<x$, x = 0.2, 0.3,0.4, 0.5), corresponding muon rules.

The third trigger level must reduce the trigger rate to 100 Hz while preserving the signal. In one case found by RL4, using m_{top}=140 and $\mathcal{L}=10^{33}$ cm^{-2}s^{-1}, a cut of missing p_t ($\not{p}_t$)>50 GeV and transverse mass M_t of the electron-missing momentum system larger than 25 GeV gives a signal of 2.0 nb and a background contamination of 6.2 nb yielding a signal/background of about 0.3. RL4 also found that requiring $\not{p}_t$>75 GeV satisfies the background threshold for this run, accepting 25.9 nb of background events and 2.4 nb signal. The signal is marginally better than before, but the background contamination is approximately four times larger. Although the lower luminosity cuts have a larger signal acceptance, the signal rate is greater for the higher luminosity sample.

Table 4. Results of Trigger Strategy Studies for two values of top mass and luminosity

(Rates are in Hz and E_t , p_t M_t,$\not{p}_t$, etc. are in GeV.)

$\mathcal{L}$/top mass	Trig. Lev.	top rate	bkgd rate	Rule
10^{33} cm^{-2}s^{-1}/	1	11.6	1.3x10^4	num-jet sums (w/E_t > 50) ≥1
140 GeV	2	6.2	3.2x10^2	lepton p_t>18
	3	2.0	6.1	$\not{p}_t$ >50& M_t e-ν > 25
10^{34} cm^{-2}s^{-1}/	1	98.	1.0x10^5	E_t > 465
140 GeV	2	5.9	4.2x10^1	lepton p_t>10& $\not{p}_t$ > 160
	3	5.9	4.2x10^1	none needed
10^{33} cm^{-2}s^{-1}/	1	3.6	2.9x10^4	E_t > 370
200 GeV	2	2.2	4.3x10^2	lepton p_t>10
	3	0.9	8.1	M_t e-ν>65 & M_t μ-ν>40
10^{34} cm^{-2}s^{-1}/	1	24.	2.3x10^4	E_t > 600
200 GeV	2	6.0	5.6x10^2	lepton p_t>10& 2nd-largest-jet sum>90
	3	4.9	1.1	num-clusters (w/Et >10) ≤ 5

Note that RL4 was able to relax the requirement on $\not{p}_t$, with a cut on M_t, reducing the background while not significantly affecting the signal. Considering $\not{p}_t$, alone, it would appear that a severe cut is necessary to achieve a separation of signal from background. However, after the cut M_t>25 GeV, a much lower cut on $\not{p}_t$, gives a clean separation between signal and background.

Discussion

In a recent workshop, Barnett, Wang and Paige[4] presented the results of a study for discriminating top (assumed mass of 140 GeV) from the conjectured principal background of W's. To separate the top from the background, they utilized a number of cuts, selected by plotting a large number of distributions of top and background events.

Briefly, their results consist of cuts on an isolated lepton with high p_t, three jets and a total E_t cut. With these cuts they show that the top is separable from the background. However they did not consider the effects of noise which would have seriously impacted their isolated lepton requirement.

A paper by Hauptman[5] also discusses the recursive binary partitioning algorithm for separating *W -->jets* from a QCD background. This method could also be used to determine top trigger trigger requirements. Basically, the method relies on a measure of statistical distance between the top events and the background events for each potential cut variable. The statistical measure is used to choose the best cut variable to use for each iteration of the procedure. Each iteration relies on the cuts made at the previous levels. The procedure continues until the sample cannot be divided into statistically dissimilar samples.

However, the recursive binary partitioning can choose only one cut at a time, and sometimes chooses a bad cut at an early level. For instance, recursive binary partitioning is unable to find a cut such as the previously mentioned correlated cut RL4 found at the third level. The recursive binary algorithm instead finds the cut on missing p_t at 75 GeV and terminates its search. It does not continue to find that a lower $\not{p}_t$ cut combined with a transverse mass cut gives better efficiency and signal/background as RL4 did.

RL4 uses more general search strategies which allow it to follow multiple interesting paths. The number of interesting paths to follow can be set by the user. RL4 often finds many possible rules that satisfy the threshold criteria. It also returns the range over which the cuts may be varied while still satisfying the thresholds. The physicist can then choose among them using other criteria: ease of implementation, insensitivity to calibrations, absence of bias, etc.

Further work in this area will more naturally occur in the context of a particular SSC or HEP experiment. Any HEP experiment will have to make many design decisions and this methodology can be used in many of them as an adjunct to other more traditional methods to further the understanding of the possible choices.

This work was partially supported by DOE Contract No. DE-AC02-80ER10667 under the Generic Detector R&D Program. Support for the Monte Carlo calculations was provided by the Pittsburgh Supercomputing Center.

References

[1] S. H. Clearwater and F. J. Provost, "RL4: A Tool for Knowledge-Based Induction," in *Proceedings of Tools for Artificial Intelligence 90*, 1990, in press.

[2] S. H. Clearwater, W. E. Cleland, E. G. Stern, "Use of Learning Programs for SSC Trigger Strategy Studies," in *Proceedings of the International Workshop on Software Engineering, Artificial Intelligence and Expert Systems for High Energy and Nuclear Physics*, in press.

[3] F. E. Paige and S. D. Protopopescu, "ISAJET 5.30: A Monte Carlo Event Generator for pp and $\bar{p}$p Interactions in *Proceedings of the 1986 Snowmass Summer Study*, R. Donaldson and J. Marx, (eds.), 1986.

[4] R. M. Barnett, E. M. Wang, and F. E. Paige, "On top at SSC," in *Proceedings of the 4th Generation Workshop*, 1989.

[5] J. M. Hauptman, "Identification of W Boson and Top quark Hadronic Jets by Recursive Binary Partitioning," in *Proceedings of the 1984 Snowmass Summer Study*, pp. 121-123, 1984.

OPTICAL FIBER NETWORKS FOR THE SUPERCONDUCTING SUPER COLLIDER DETECTOR

Branko Leskovar
Electronics Engineering Department
Lawrence Berkeley Laboratory
One Cyclotron Road
Berkeley, CA 94720 U.S.A.

Abstract

The Superconducting Super Collider Detector System should be capable of operating with luminosity corresponding to an interaction rate of the order of 10^8 events per second. The detector system will involve more than 10^6 channels of readout electronics. Consequently, a new high performance data acquisition system, incorporating high-speed optical fiber networks, will be needed to process the large quantities of data from detector elements. Also, extensive online processing will be required to reduce the amount of data per interaction and data rates to amounts that can be handled by present storage techniques and computing capacity.

System Overview

A simplified block diagram of the new data acquisition system architecture incorporating optical fiber networks is given in Fig. 1. The architecture of the system is based on a number of earlier workshops on triggering and data acquisition organized by the SSC Central Design Group.[1,2] The system will consist of the following subsystems: detector front-end electronics based on custom integrated circuits, an efficient high speed data collection and transmission using fiber optics, a parallel event building network, special purpose processors for prompt triggers, calibration, and data compression, online processing units and a large data storage. Using an open system architecture, the data acquisition system will be capable to process data rates of several thousand Gbit/s from detector elements to the online processing subsystem.

Front-end integrated circuits accept signals from the detector elements and store the signals during the time necessary to process level 1 and level 2 triggers. Data collection integrated circuits collect the data from a set of front-end integrated circuits and prepare them for multiplexing. High speed data transmission will be implemented by optical fiber networks consisting of optical transmitters, receivers and associated electronics. The data balancer shifts multiplexed data to different fiber optic links and into the parallel event building network. This network simultaneously collects its input information from several detector channels. Also, the network simultaneously transmits totally built events to its multiple output ports.

It has been shown earlier that the application of optical data transmission in the SSC data acquisition system can significantly increase its data rate capability. Furthermore, it has also been shown that optical transmission will significantly reduce electromagnetic interference and eliminate ground currents with their attendant problems. Also, wide band optical transmission improves the maintainability, hermeticity and reliability because of the reduction of the cable plant size and simplification of the overall architecture of the detector systems.[3-5]

At previous workshops relating to the SSC data acquisition it was also shown that the system will be partially imbedded within the structure of the detector. Consequently, the power dissipation, system reliability, and radiation hardness will be of prime importance.

During the full operation of the data collection system of the detector front-end electronics, the digital data and fast timing signals will be present on the local data networks at the same time when the front-end electronics is receiving low-level analog signals. Consequently, there is a strong possibility for cross coupling between signal channels if electrical transmission is used. Therefore, the optical data transmission will be integrated in the data collection system to minimize cross-coupling effects. Also, the overall architecture of the collection system will be simplified by signal multiplexing and optical data transmission. Specifically, optical fiber networks will be applied in the following areas of the detector electronics subsystem:

Distribution of Timing Signals

Fast timing signals, such as the 60 MHz clock, the level 1 and level 2 triggers, as well as calibration strobes, will be distributed by fiber waveguides or guided-wave optical interconnections with a timing precision significantly better than 1 ns and the bit error rate smaller than 10^{-9}.

The optical interconnecting circuits would consist of a semiconductor-based guided-wave circuit, laser diodes and photodiodes. The laser diodes and photodiodes will be integrated with the guided-wave circuit. Such monolithic integration of electronic and photonic devices on a single chip has become the focus of considerable research effort during the last several years.[6]

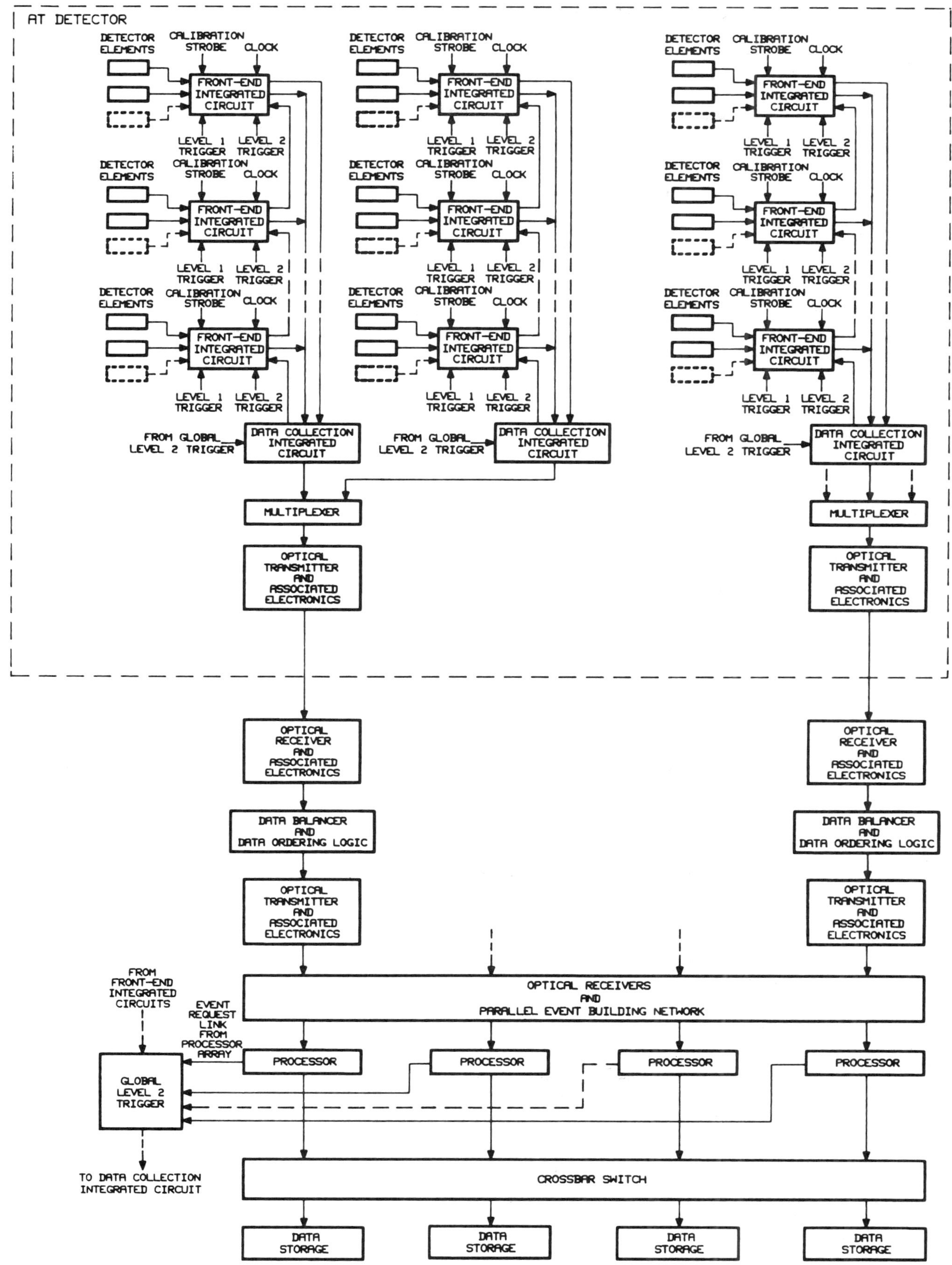

Fig. 1 DATA ACQUISITION ARCHITECTURE FOR THE SSC DETECTOR XBL 9010-3357

Transmission of Digital Data

The undirectional data from front-end IC's to digital data collection ICs, the handshake and interconnecting signals, as well as the output event data will be distributed, in a number of cases, by optical fibers. Furthermore, the handshake signals between front-end ICs and data collection IC's can be implemented by bidirectional data transmission systems using IC optical transmitters and receivers.[6]

Outside the detector, optical fiber networks incorporating subassemblies having low power dissipation will be applied in the following areas of the data acquisition system: high speed data transmission for point to point links, optoelectronic integrated circuits off the detector subsystem, high speed data transmission for data and control signals, interconnection of special purpose devices in trigger systems and high data rate transmission into an on line multiprocessor subsystem.

For example, in the system shown in Fig. 1 the data can be transmitted from multiplexers to data balancers by means of 1000 parallel high speed fiber links. At present, a single optical fiber data transmitting channel can operate with a typical data rate of 1.7 Gbit/s with a high reliability. Consequently, the combined data rate capability of the system is 1.7×10^3 Gbit/s.

Design Considerations

With respect to the detailed design of optical fiber networks for specific SSC detectors and the choice of technology used for their implementation, the issues requiring further effort are high-speed low power dissipation fiber optics subassemblies, wavelength division multiplexing and demultiplexing, component and subassembly radiation sensitivity and network reliability and failure modes.

High performance fiber optic subassemblies, such as time division multiplexers and demultiplexers, clock and data recovery circuits, multiple crosspoint switches, optical transmitters and receivers typically require a significant amount of power for their operation. Design studies must be made of these subassembly configurations and of integrated circuit technologies giving minimum power dissipation. In particular, studies are required on the design of these subassemblies based on the application of specific integrated circuits.

Multiplexing increases the information capacity of optical fibers by simultaneously transmitting two or more different signals at different wavelengths. The technique requires a separate source for each desired wavelength and multiplexing devices to couple all the signals into the fiber and to separate them at the other end of the fiber. Since the transmitting channels are completely independent, information with different formats and data rates can be transmitted on the same fiber waveguide without requiring any synchronization between the channels. Further studies should be carried out on the implementation of this technique for bidirectional data transmission using very low power dissipation IC optical transmitters, single mode fibers and wide-band IC optical receivers.

Most optical and mechanical characteristics of silica-based optical fibers are well understood. Other problems, such as radiation effects on optical sources, fiber waveguides, optical receivers and associated electronic circuits are presently being investigated in the Generic SSC R&D program.[4] However, further studies are needed on the reliability and failure modes of low power dissipation fiber optic networks.

Acknowledgments

This work was performed as part of the program of the Electronics Research and Development Group, Electronics Engineering Department of the Lawrence Berkeley Laboratory, University of California, Berkeley. The work was partially supported by the U.S. Department of Energy under Contract Number DE-AC03-76SF00098. Reference to a company or product name does not imply approval or recommendation of the product by the University of California or the U.S. Department of Energy to the exclusion of others that may be suitable.

References

[1] R. Donaldson (Ed.), Proceedings of the Workshop on Triggering and Data Acquisition for Experiments at the Supercollider, Workshop held at the University of Toronto, Ontario, Canada, 16-19 January 1989, SSC Central Design Group, LBL, Berkeley, California, April 1989.

[2] E. Barsotti, A. Booth, M. Bowden, C. Swoboda, N. Lockyer, and R. VanBerg, A Proposed Scalable Parallel Open Architecture Data Acquisition System for Low to High Rate Experiments, Test Beams and all SCC Detectors, IEEE Transactions on Nuclear Science, Vol. 37, No. 3, pp. 1216-1221, June 1990.

[3] B. Leskovar, Optical Receivers for Wide Band Data Transmission Systems, IEEE Transactions on Nuclear Science, Vol. 36, No. 1, pp. 787-793, Orlando, Florida, February 1989.

[4] B. Leskovar, Radiation Effects on Optical Data Transmission Systems, IEEE Transactions on Nuclear Science, Vol. 36, No. 1, pp. 543-551, Orlando, Florida, February 1989.

[5] B. Leskovar, Optical Data Transmission at the Superconducting Super Collider, IEEE Transactions on Nuclear Science, Vol. 37, No. 2, pp. 271-287, San Francisco, CA, April 1990.

[6] K.K. Wong, H.J. Wojtunik, S.T. Peng, M.A. Mentzer and L. McGoughan (Eds.), Integrated Optics and Optoelectronics, Proceedings SPIE - The International Society for Optical Engineering, Vol. 1177, Bellingham, Washington, 1990.

A STUDY OF ELECTRO-OPTICAL MODULATORS FOR TRANSFER OF DETECTOR SIGNALS BY OPTICAL FIBERS

T. Tsang, V. Radeka, T. Srinivasan-Rao, and W. J. Willis*

Brookhaven National Laboratory, Upton, NY 11973
*BNL and Columbia University, New York, NY 10027

ABSTRACT

Preliminary results of an investigation to minimize the local electronics on the detector by the use of electro-optical intensity modulators are reported. The electrical signal from the detector is converted into an optical signal by a Mach-Zehnder interferometer modulator based on the properties of electro-optical crystals such as $LiNbO_3$. The modulator is driven by the electrical signal from a detector element such as the gas proportional straw tube. The optical signal is brought in from a remote source and exits on an optical fiber to a data collection site. The results of the first tests of optical modulators with respect to their application to particle detectors will be presented. These include the optical power requirement, the linearity, the dynamic range, and the noise characteristics. Finally, the outlook for their application on a large scale is discussed.

I. Introduction

In high energy particle accelerators such as the Superconducting Super Collider, the readout of detectors with 10^5 to 10^6 signal channels is recognized as a major challenge. This is particularly the case where the electron identification information has to be obtained at the trigger level, since it requires a high degree of parallelism in signal transfer and processing. Electron identification can be performed by transition radiation detectors and calorimeters. In this study, we address this challenge by using a novel concept of integrated electrooptic modulators and optical fiber "cables" to read all channels in parallel. The use of the optical modulator with high input impedance can relieve the power problem of electronic readout circuits, give more freedom in preamplifier design, and dramatically reduce the mass and volume of the output "cables", due to the lower mass and volume of the optical fibers compared to electrical cables. This technique is of particular importance for transition radiation detectors. In a cryogenic detector such as the liquid argon calorimeter this technique simplifies the cold feedthroughs and lowers the heat loss due to copper cables.

The novel concept will utilize a system of electrooptic modulators, fabricated on a single substrate, and powered by a remote CW operated 1.3 μm laser source. The light is brought in remotely on single-mode polarization preserving fibers and the optical output signal is carried on fibers. The principle of operation is based on the Mach-Zehnder optical interferometer. A schematic of this type of intensity modulator is shown in Figure 1. A polarized light beam travelling in a single transverse mode optical waveguide is split by refraction into two beams, carried a suitable distance and then rejoined, giving rise to interference. The interference is constructive when the two beams are in phase, and the full power exits to the output waveguide. If the lengths of the two arms differ by half of a wavelength (optical phase of π), the interference is destructive, and the optical power is radiated into the substrate.

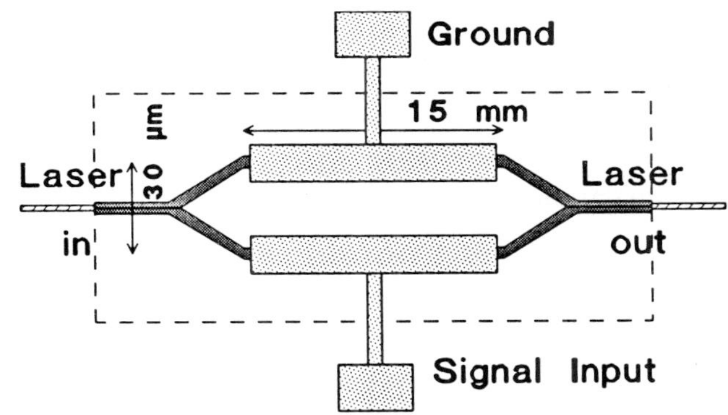

Figure 1. Mach-Zehnder Modulator

The single transverse mode optical waveguide is formed by Ti diffused into a planar $LiNbO_3$ substrate.

The principles of the device and the details of the state-of-the-art fabrication technology are given in review articles by Korotky [1] and Bulmer [2]. When a charge is injected on the electrode that is placed over one of the two arms of the electrooptic Ti : $LiNbO_3$ waveguides, the optical length of one arm becomes shorter than the other, changing the relative phase of the two beams at the end of the interferometer. The output power will then vary as the square of the cosine of the total phase shift, composed of a constant phase offset depending on the construction of the device, and a term due to the signal. The light signal then propagates on a low loss output fiber with very low time dispersion. The fiber can be one or more kilometers long, if a delay of that length is required. There it is received by a low noise semiconductor photodiode of high quantum efficiency and connected to a low noise preamplifier. The subsequent fast shaping and processing operations are done remote from the detector, so that the electronics on the detector can be minimized.

In this paper we intend to explore the operating parameters by studying a single modulator channel system first, and particularly, to examine closely the noise limitations, linearity, and the dynamic range.

II. Experimental Setup

The optical system comprises of a commercial diode-pumped Nd:YAG laser (Amoco), a Mach-Zehnder interferometric modulator (Crystal Technology) with x-cut y propagating Ti : $LiNbO_3$ electrooptic waveguides, low noise fast InGaAs photodiodes (Hamamatsu), low noise preamplifiers (BNL), and fast shaping amplifier (Ortec 579 and BNL). Figure 2 gives an overview of the experimental setup. The CW operated 1.3 μm infrared laser has a linearly polarized output with a power of 30 mW at the fiber output. It has a built-in 30 dB Faraday isolator, and a 50 meters long uncoated fiber pigtail. The output beam is recollimated and focused onto the optical fiber at the input end of the Mach-Zehnder modulator. The optical fiber ends were pigtailed to the input and the output ends of the waveguides. A half wave plate is employed to optimize the input polarization along the y-axis of the $LiNbO_3$ modulator to take advantage of the larger r_{33} electrooptic coefficient. A neutral density filter is used to adjust the optical power. After the output beam from the modulator is recollimated and focused onto the InGaAs photodiode, a neutral density filter is inserted until the photodiode achieved a photocurrent of $\sim$ 1 mA. Because the quantum efficiency of the InGaAs photodiode is 0.85 at 1.3 μm, this photocurrent corresponds to an incident laser power of 1.12 mW. The AC component of the photocurrent is then fed into a modified BNL IO-323 low noise preamplifier. The output of the preamplifier is then fed to an Ortec 579 fast timing filter amplifier which has integration and differentation times set at 20 ns. For faster pulse shaping time of $\sim$ 3 ns, a hybrid type filter amplifier made at BNL is employed. The data are collected on a 300 MHz Tektronic 2440 digital oscilloscope and stored on a computer for further analysis.

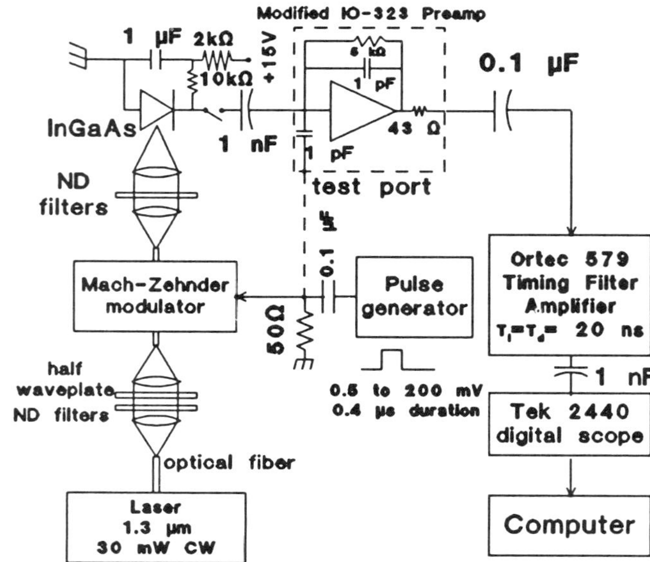

Figure 2. Experimental setup for noise measurements

The signal source used to investigate the transfer characteristics of the modulator is a $\pm$10 V ramp. To study the noise and the linearity of the modulated signal at the 20 ns shaping time, a voltage pulse with 6 ns risetime from a HP 8116A function generator is employed. For faster shaping time, a BNL pulser, BNL I0-391, with 2 ns risetime is used.

III. Results and Discussion

A. Transfer Characteristics

The particular Mach-Zehnder modulator employed has a -3 dB intensity modulation point close to the zero voltage bias point. The transfer characteristic is shown in Figure 3. The fringe intensity has an extinction ratio of 26 dB when the input polarization is optimized. The V_π voltage is 2.68 V. This low half wave voltage is achieved with an electrode length of 15 mm. An unoptimized input polarization of $\sim 2^\circ$ would shift up the visibility fringe pattern resulting in a change of the -3 dB intensity point by less than 0.01 dB. Therefore, the tolerance of the input polarization can be a few degrees.

B. Dynamic Range and Linearity

Given the output varying as the square of the cosine of the modulation voltage, the dynamic range will be determined by the requirement on linearity. In particle detector applications, two approaches are of interest. One is the case where a linear analog sum of many channels is needed, then linearity is required at some specified level in each channel in order to retain the accuracy of the sum. The other is where the nonlinear response is well defined and the nonlinearity can be corrected. The modulator response is well defined and very large signals could be analyzed by observing the output signal over multiple fringes.

For operation in the quasi-linear regime the nonlinearity can be determined from the modulator response function. For a 1% maximum deviation from a tangent through the -3 dB point, the maximum signal is $\frac{0.25V_\pi}{\pi}$ or ~ 200 mV in our case. If the noise is limited by the photon statistics as discussed in the next section, the dynamic range with respect to the noise is given by,

$$\frac{v_{\text{max}}}{\sigma_{\text{in}}(v)} = \frac{Q_{\text{max}}}{\sigma_{\text{in}}(q)} = 0.25\sqrt{\frac{t_m n_o}{a_F}}\,, \qquad (1)$$

where n_o is the photoelectron rate at the detector, t_m is the integration time, and a_F is a filter parameter, usually in the range of $\frac{4}{3}$ to 2.

Assumming a light intensity at the detector of 1 mW, the photon rate is $n_o = 5.6 \times 10^{15}$ sec^{-1}. The integration time for our slower filter was $t_m = 35$ ns and $a_F = 1.9$. The expected dynamic range is then, $\frac{v_{\text{max}}}{\sigma_{\text{in}}(v)} = 2.7 \times 10^3$ All measurements were done with an input laser power of ~ 1 mW on the detection system. However, after all optical losses have been considered in our experimental setup, the optical power incident on the modulator is ~ 14 mW. The insertion loss of the modulator is ≤ 7 dB.

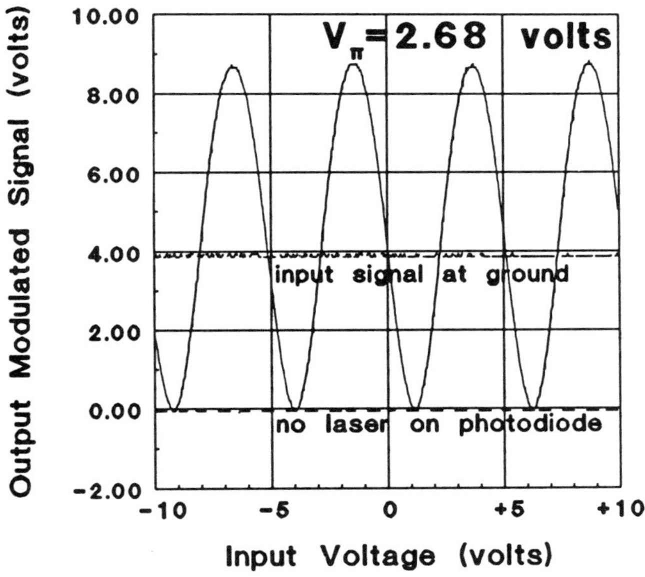

Figure 3. Transfer Characteristics

By measuring the voltage height of the modulator-receiver responses shown in Figure 4 with different input modulation voltages, a plot of the linearity curve is generated in Figure 5. The linearity of the modulated signal is quite good as expected, and the dynamic range approaches $3\frac{1}{2}$ decades. One possible improvement of the dynamic range is to operate the modulator at 0.2 V above or below the -3 dB point, but still in the linear regime, and inject an opposite polarity signal to the modulator. Thus, the bias point is very important for the dynamic range.

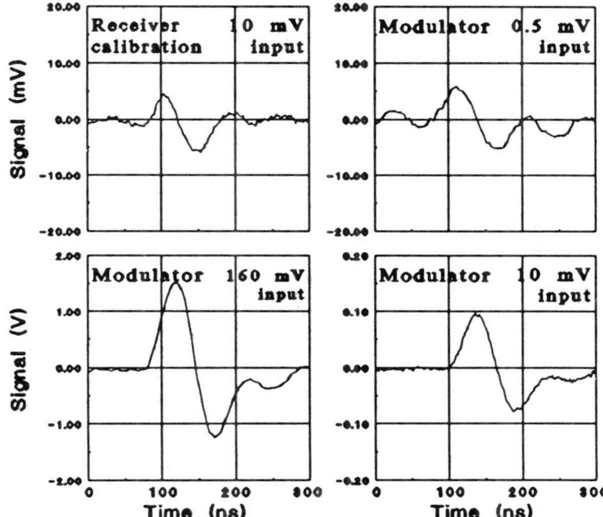

Figure 4. Modulator–Receiver Response at shaping time of $\tau_i=\tau_d=20$ ns (integration time $t_m=35$ ns) input equivalent noise $\sigma_{in}(v)=0.1$ mV

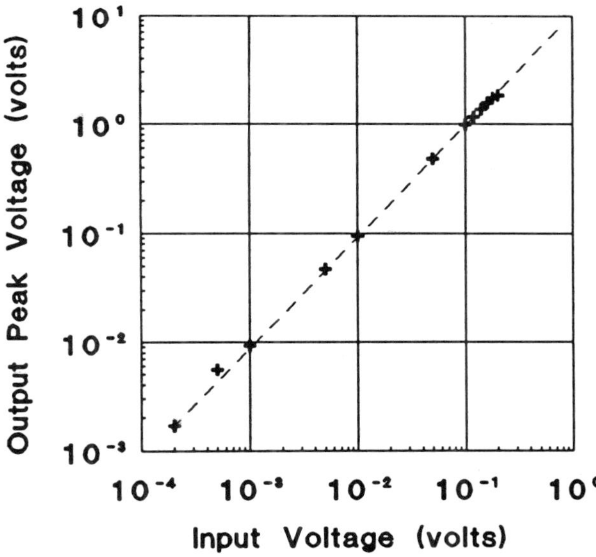

Figure 5. Linearity of the MZ Modulator

C. Noise Analysis

Five sources of noise may be considered in this experiment: (1) noise from the photodiode, (2) noise from the receiver system, (3) dielectric noise in the modulator (4) photon fluctations according to Poisson statistics, and (5) excess noise in the laser. We will discuss each of them individually in the following.

(1). Photodiodes

The Hamamatsu G3476-1 photodiode with infrared response has a junction capacitance of 0.8 pF and a measured dark current of ~ 1nA at a reverse bias of 15V. In a measurement time of 20 to 40 ns (peaking time of the pulse shaper), this photocurrent corresponds to an rms noise of 200 electrons or less, which is much less than the photon noise that we would anticipate.

(2). Receiver-Detector System

The noise of the receiver-detector system includes the photodiode, the preamplifier and the pulse shaper, see Figure 2. Based on a calibration run of the system (see Figure 4), the receiver-detector system has a rms of $\leq 3 \times 10^3$ electrons (~ 0.5 fC), for $t_m = 35$ nsec, referred to the input of the receiver.

(3). Dielectric noise

The dielectric noise is that due to the thermal fluctuations in the dielectric constituting the modulator [3].It can be evaluated from the dielectric loss factor for the $Ti : LiNbO_3$. Our estimate is that this will be less than 500 rms electrons, referred to the input of the modulator.

(4). Photon noise

The photoelectron statistics noise at the receiver is

$$\sigma(n) = \sqrt{a_F n_o t_m} \, , \qquad (2)$$

where a_F is the ratio of the effective measurement time for noise and the measurement time for the signal t_m. In the simplest case, the measurement is made on the modulator signal plus the baseline in one measurement interval, and in the other interval the baseline is measured and substracted from the first measurement. In this case $a_F = 2$. With pulse shaping of the signals from the receiver, lower values of a_F can be achieved (e.g., $a_F = \frac{4}{3}$ for a triangular bipolar shaping). The number of photoelectrons at the receiver produced during measurement time t_m by a step in the modulator input voltage is,

$$\Delta n \, t_m = \pi \frac{\Delta v_{in}}{V_\pi} t_m n_o \, , \qquad (3)$$

where Δn is the change in photoelectron rate caused by Δv_{in}. We obtain the equivalent input noise voltage from Eqs. (2) and (3) by substituting $\Delta n \, t_m = \sigma(n)$ and $\Delta v_{in} = \sigma_{in}(v)$,

$$\sigma_{in}(v) = \frac{V_\pi}{\pi} \sqrt{\frac{a_F}{t_m n_o}} \, . \qquad (4)$$

The equivalent input noise charge is,

$$\sigma_{in}(q) = (C_D + C_M) \, \sigma_{in}(v) \, ,$$

$$ENC_{in} = \sigma_{in}(q) = \frac{Q_\pi}{\pi} \left(1 + \frac{C_D}{C_M}\right) \sqrt{\frac{a_F}{t_m n_o}} \, , \qquad (5)$$

where $Q_\pi = C_M V_\pi$, C_M is the modulator electrode capacitance and C_D is the capacitance of the particle detector connected to the modulator. The parameter Q_π is independent of the modulator length ℓ, since $C_M \propto \ell$ and $V_\pi \propto 1/\ell$.

For 1.2 mW light input to the receiver diode, the average currrent is $I_o \simeq 1.07$ mA, and $n_o \simeq 6.7 \times 10^{15}$ sec^{-1}. For $t_m = 35$ ns, $a_F = 1.9$, the equivalent input noise voltage is $\sigma_{in}(v) = 84$ μV. For $C_D \simeq C_M = 10$ pF, the equivalent noise charge is $\sigma_{in}(q) = 1.68$ fC. The noise at the receiver input is obtained from Eq. (2), $\sigma(n) = 2.05 \times 10^4$ e^-. The noise of the photon statistics is clearly dominant over other noise sources. With a step signal of 10 mV at the modulator input, and charge calibration of the receiver, we measured a signal of 1.28×10^6 e^- and a signal to noise ratio of 57. This gives a noise of 2.25×10^4 rms electrons, which is about 11% higher than the calculated value. This difference is within the experimental accuracy of the noise measurement. To obtain a faster shaping time, a BNL made 2 ns risetime pulser and a hybrid circuit fast preamp and shaper with 3 ns peaking (integration) time were employed in the receiver-detector system. With 20 mV pulse input to the modulator, the response of this fast detection system is shown in Figure 6. Signal to noise ratio of 100 is obtained with an rms noise of ~ 4×10^3 electrons at the receiver. The expected dynamic range with this integration time is 7×10^2.

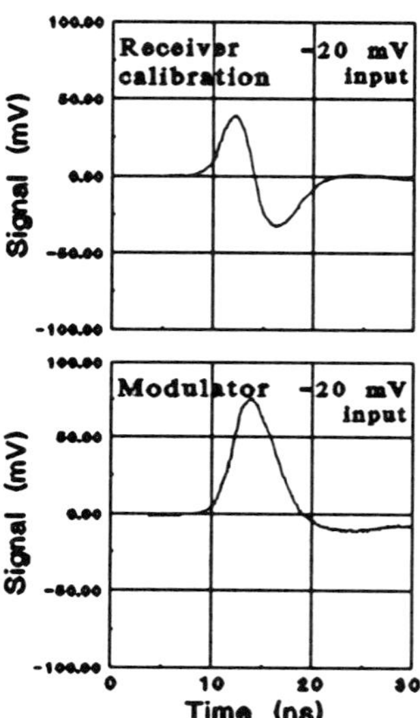

Figure 6. Modulator–Receiver Reponse at peaking time of t_m=3 ns

(5). Laser Excess Noise

Laser noise above the photon statistics was observed in the frequency range of 150-200 kHz. However this noise is not measureable after the pulse shaping filters used in our test.

IV. Application of Modulators for Transfer of Detector Signals

We are interested in two kinds of applications of modulators to particle detectors. One is where a low noise preamplifier is used, and the other is without any preamplifier when the detector is connected directly to the modulator. In the first case the low capacitance and high sensitivity of the modulator allow much lower power dissipation in the preamplifier than with electrical cables. In the case of a modulator replacing a preamplifier, sufficiently low noise is obtained for some applications. For low capacitance detectors ($C_D \simeq 10$ pF), such as gas proportional tubes of up to 1 meter length, the noise will be in the range of $\approx$2 fC at $t_m \approx 35$ μsec. The equivalent noise charge (shown in Eq. (5)) is inversely proportional to the square root of the measurement time and the photon rate (laser power).

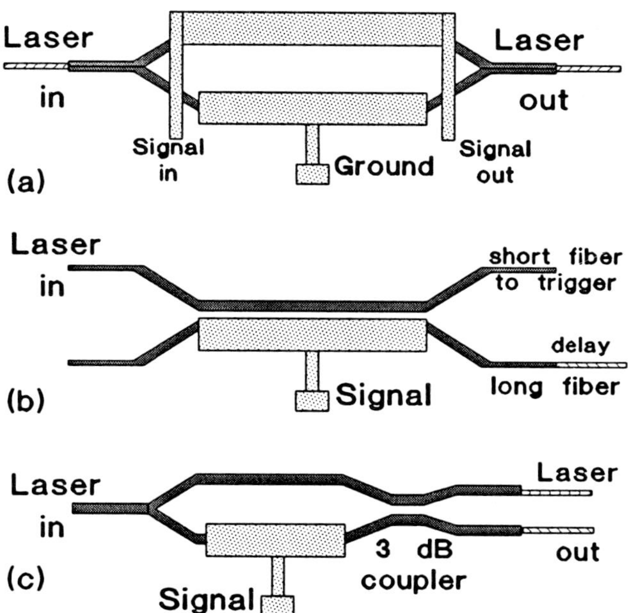

Figure 7. Schematics of (a) Mach-Zehnder type device (b) directional, and (c) 1x2 Y-junction couplers

Transfer of detector signals can be done in principle with various schemes using fiber optics communication technology [4]. Mach-Zehnder type device, 2×2 directional coupling, and 1 × 2 Y-junction splitting schemes (see Figure 7) have been used. In the last two schemes, a trigger signal can be obtained from the short fiber output of one signal arm while delaying the long fiber arm, to $1-5$ μs, for data collection. This scheme is particularly interesting because, one can perform some electronic processing or discrimination before storing the data. Finally, we should mention that there are several issues that are critical in realizing such a state-of-the-art detector. They are (a) the achievement of the optical power splitting among the channels on the device with adequate uniformity, (b) stability of the intrinsic and extrinsic bias point, (c) design and optimization of a full size modulator detector, (d) reliability of the fiber connectors, (e) power dissipation, (f) operation at low temperature, and (g) radiation damage to the modulators. These issues will be investigated in the next phase of this work.

Acknowledgements

We would like to acknowledge fruitful discussions with Drs. Steven Korotky of the AT&T and Catherine Bulmer of the Naval Research Laboratory. We appreciate the expert technical assistance of John Schill and Dmitri Stephani, and Barbara Kponou for the assistance with the manuscript.

This work has been supported under contract number DE-AC02-76CH00016 with the U.S. Department of Energy. Accordingly, the U.S. Government retains a nonexclusive, royalty-free license to publish or reproduce the published form of this contribution, or allow others to do so, for U.S. Government purposes.

References

1. S. K. Korotky and R. C. Alferness, *Optical Fiber Telecommunication II*, New York, Academic Press, Inc., 1988, ch. 11, pp. 421-465.
2. C. K. Bulmer and W. K. Burns, "Linear Interferometric Modulators in Ti : $LiNbO_3$," J. of Lightwave Technology, **LT-2**, pp. 512-521.
3. V. Radeka, Ann. Rev. Nucl. Part. Sci. **38**, pp. 217-277, 1988.
4. L. Thylen, "Integrated Optics in $LiNbO_3$: Recent Developments in Devices for Telecommunications," J. of Lightwave Technology, **LT-6**, pp. 847-861, 1988.

SSC HIGH-SPEED COMMUNICATION CHANNEL AND INTERCONNECTS

Brad Cooke, Clayton Smith and Ron Wagner
Los Alamos National Laboratory
Los Alamos, NM 87545

Abstract

This paper investigates potential metallic and optical technology applications in Superconducting Super Collider (SSC) high-speed interconnects and/or communication channels. Primary focus will be placed on evaluating several metallic and optical interconnect structures.

Introduction

The essential elements required for a typical communication channel are shown in Figure 1. They consist of an electronic driver circuit, a modulator (for optics), an interconnect channel (the transmission medium through which optical or electrical signals propagate) and a receiver. The encoder/decoder is optional. All of these elements, along with bandwidth, noise figure, thermal dissipation, volume and radiation length, must be considered, from a systems perspective, for optional performance.

The three types of interconnect technology available are the following:

(1) Metallic (the most mature of the technologies): Many of the simulation tools have been developed. However, metallic interconnects have relatively large bulk volume and mass (i.e., metal cables and transmission lines) that require high-powered terminated drivers and have low channel capacity, when compared with optics.

(2) Dielectric guided optics: Dielectric guides (fiber optics) have high channel capacity (1–10 THz). Although they are available commercially, technology is still at the development stage, and they require optical drivers/receivers made from materials that may not be compatible with silicon. The required simulation tools are still under development.

(3) Free-space optics: Free-space optics have very high channel capacity and small mass/volume requirements. Potential problems are the same as those of dielectric guides. In addition, one needs to consider optical alignment.

Metallic Interconnects

Metallic-interconnect transmission-line characteristics were calculated from first- and second-order electromagnetic field models, available in the literature,[1,2,3,4] for the structures shown in Figure 2. Parameters selected were those that provided an effective radiation length of approximately 0.4% to 1% and a characteristic impedance of 50 ohms (100 ohms for twin lead) for aluminum and kapton. Characteristic impedances under 50 ohms are undesirable because of the excessive loading of bipolar or metal oxide semiconductor (MOS) drivers. A summary of the parameters and simulation results follow.

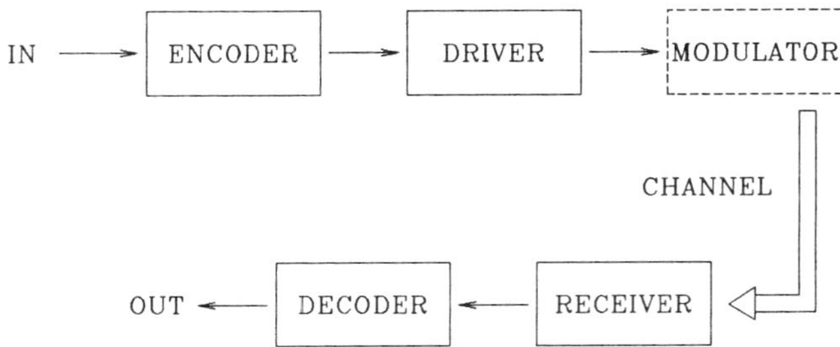

Figure 1. Essential Elements for a Communication Channel.

Aluminum: Resistivity: 2.65×10^{-4} ohm-m,
Radiation length: 0.089 m

Kapton: Dielectric constant: 3.55 F/m,
Radiation length: 0.2856 m

COAX: $\omega = 10$ Mrad/s, $a = 2 \times 10^{-4}$ m,
$b = 2.4 \times 10^{-4}$ m, $M_t = 1 \times 10^{-4}$ m,
$I_t = 1 \times 10^{-4}$ m

Characteristic impedance = 52.6 ohms
Attenuation = 4.92 dB/m
Radiation length = 0.5%.

TWIN LEAD: $\omega = 10$ Mrad/s, $s = 2 \times 10^{-3}$ m,
$d = 1.5 \times 10^{-3}$ m, I_{t1},
$I_{t2} = 1 \times 10^{-4}$ m

Characteristic impedance = 109 ohms
Attenuation = 1.0 dB/m
Radiation length = 0.52%.

MICROSTRIP: $\omega = 10$ Mrad/s, $w = 1 \times 10^{-3}$ m,
$h = 1 \times 10^{-4}$ m, $t = 1 \times 10^{-4}$ m,
$M_t = 1 \times 10^{-4}$ m, I_{t1},
$I_{t2} = 1 \times 10^{-4}$ m,

Characteristic impedance: 50.5 ohms
Attenuation: 3.2 dB/m
Radiation length: 0.4%.

STRIPLINE: $\omega = 10$ Mrad/s, $w = 1\ 10^{-3}$ m,
$b = 2.1\ 10^{-3}$ m, $t = 1\ 10^{-4}$ m,
$M_t = 1\ 10^{-4}$ m, I_{t1}, $I_{t2} = 1\ 10^{-4}$ m

Characteristic impedance = 49.3 ohms
Attenuation = 3 dB/m
Radiation length = 1.06%.

Because of the relatively small geometries of the transmission line structures that result in small surface areas, the frequency-dependent skin effect[5,6] dominates the attenuation at higher frequencies, limiting the effective transmission line bandwidth. Figure 3 shows the attenuation in decibels/meter for the four transmission line geometries over the frequency range of 1 to 100 Mrad/s. Note how structures with higher shielding (coax and stripline) have higher attenuation rates (this is most likely due to the higher confinement of field components). One could conclude that a structure similar to the twin lead transmission line (no ground planes) would provide the best radiation-length-to-bandwidth performance. However, the adjacent conductor coupling and shielding capabilities of such a structure would probably be unacceptable in a densely packed environment. In such a case, the stripline structure provides the best performance. Note that the conductor width, w, for both the stripline and the microstrip can be increased (the surface area can be increased without affecting the radiation length), thus decreasing the attenuation. However, increasing w will also decrease the characteristic impedance, potentially overloading the electronic drivers.

Optical Interconnects

Currently, the optical interconnect of interest is the single-mode, high-quality silica optical fiber. One can expect a bandwidth of greater than 1 THz (Terra = 10^{12}), and an attenuation[7] on the order of 10^{-4} dB/m for high-quality single-mode fibers 50 to 100 m long. A 125-μm silica fiber optic (core and cladding) has an effective radiation length of 10^{-5}% and can be made sufficiently radiation resistant for all SSC applications.

Conclusion

While several metallic and optical interconnect media have been discussed, one should bear in mind that any SSC high-speed communication channel implemented must be designed from a systems perspective, as discussed in the Introduction. A 1-THz optical fiber interconnect may not be cost effective or particularly desirable if the expected data rate or electronic driver capability is limited to 10 MHz. On the other hand, if the data transmission rate is expected to exceed 1 GHz, it is not realistic to consider a metallic stripline structure unless the radiation length limits are significantly increased.

The evaluation and design of electronic driver circuits (bipolar, MOS and GaAs), optical modulators/absorbers (lithium niobate and III-V compounds) and encoder/decoders is underway, along with the advanced modeling of metallic/optical interconnects.

References

[1] S. Ramo, J. R. Whinnery, and T. V. Duzer, *Fields and Waves in Communications Electronics.* John Wiley & Sons, 1965, ch. 6, Tbl. 8.09.

[2] L. J. Bahl and R. Garg, "Simple and Accurate Formulas for a Finite Strip Thickness," in Proceedings of IEEE, Vol. 65, No. 11 (November 1977), pp. 1611–1612.

[3] K. C. Gupta, R. Garg, and R. Chadha, "Computer Aided Design of Microwave Circuits," *Artech*, 1981, pp. 56–65.

[4] H. A. Wheeler, "Transmission-Line Properties of a Strip Line Between Parallel Planes," *IEEE Trans. Microwave Theory and Techniques*, MTT-26, No. 11 (November 1978), pp. 866–876.

[5] H. A. Wheeler, "Formulas for the Skin Effect," in Proceedings of the IRE, September 1942, pp. 412–424.

[6] R. A. Pucel, D. J. Masse, C. P. Hartwig, "Losses in Microstrip," *IEEE Trans. of Microwave Theory and Tech.*, Vol. MTT-16, No. 6 (June 1968), pp. 342–350.

[7] A. Yariv, *Optical Electronics.* Holt, Rinehart, and Winstron, Inc., 1985, Ch. 3.

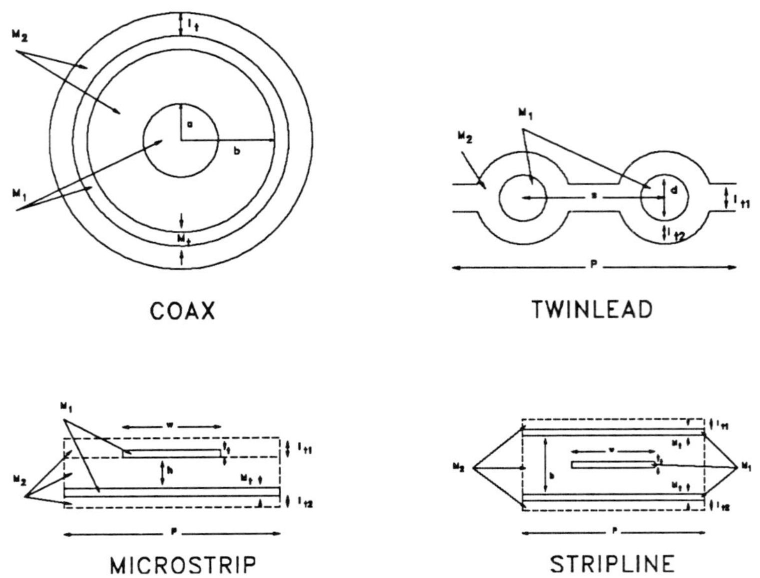

Figure 2. Metallic Transmission Line Structure Used in Evaluation

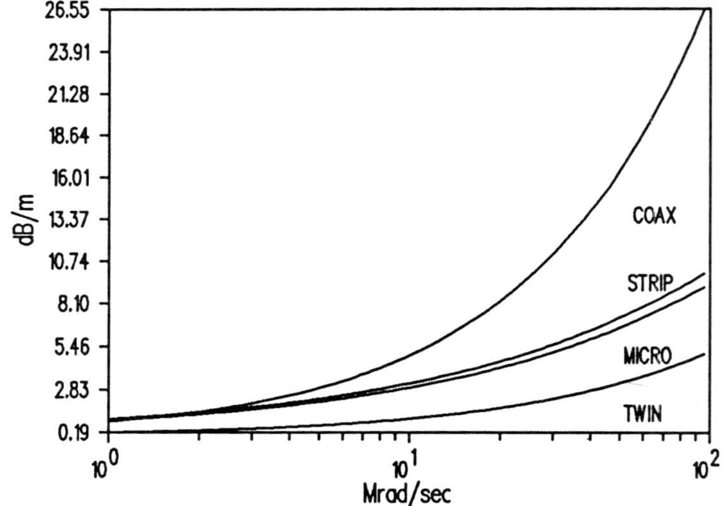

Figure 3. Transmission Line Attenuation as a Function of Frequency

FIRST LEVEL TRIGGERING USING SCINTILLATING FIBERS

Fiber Tracking Group (1)

Presented by John E. Elias
Fermi National Accelerator Laboratory
P.O. Box 500
Batavia, IL 60510

Abstract

A central tracking detector for SSC based on scintillating fibers arranged in azimuthal shells within a strong solenoidal magnetic field offers single crossing time resolution accompanied by minimal occupancy and simple binary readout electronics. These characteristics are exploited in the conceptual design of a first level trigger which performs full 2-dimensional track reconstruction for each crossing with momentum resolution of 0.004 p_t $(Gev/c)^{-1}$. Reconstruction is accomplished in a time of order 40 beam crossings, and the algorithm is easily vectorized. When implemented with piplined digital logic, the trigger is deatimeless producing lists of reconstructed tracks for each crossing at the 60 MHz crossing frequency after the pipeline delay. It is suggested that incorporating track segment decoding logic on the same ASIC as the required pipeline delay coupled with a separate, and programmable, segment linker may be an effective realization.

I. Introduction

Generation of a first level trigger for electrons and muons involves matching momentum information from tracking with electromagnetic energy depositions in calorimeters or with stiff penetrating tracks in the muon system. At the SSC, the design luminosity of 10^{33} cm^{-2} sec^{-1} leads to experimental conditions in the tracking volume where the occupancies will affect the trigger performance. Not only will there be inefficiencies due to pileup and deadtime losses, but chance correlations (combinatoric probablilites) will provide false high momentum signals from minimum bias fluctuations at substantial rate (2). A standard stiff track trigger from a single superlayer may not be viable at 10^{33} luminosity. For example, the radial hit pattern corresponding to 10 GeV/c or greater momentum in 8 layers of 4 mm straws at a radius of 1.6m occurs every third crossing from minimum bias alone. The false trigger problem is alleviated by using elements with the smallest product of size times sensitive time and by requiring high momentum patterns from several superlayers. The apporach described here uses scintillating fiber elements to minimize the sensitivity to occupancy and requires linking of track segments between all supperlayers for full pattern recognition.

Triggering on stiff tracks can be viewed as an image processing operation in which the clutter of low momentum tracks and spirals is suppressed to enhance the underlying picture of high momentum tracks. In imaging technology, pipelined processing in which a fixed series of operations is performed on a sequential set of data images has been found to be the most effective and highest performance technique. At SSC, a pipelined data acquisition structure is essentially mandatory because of the 16 nsec interval between beam crossings. The length of the pipeline is determined by the time needed to form a trigger decision for each crossing.

Continuing the imaging analogy, the fiber tracker trigger described here can be considered to constitute a pipelined, digital-filter image processor. Data images enter the pipeline at 60 MHz and exit, less than a microsecond later, reduced to their essential stiff track content. Moreover, the unique low occupancy and simple yes-no readout features of scintillating fiber structures allow for momentum reconstruction of all stiff tracks within the pipeline delay. The momentum resolution of 0.004 p_t $(GeV/c)^{-1}$ expected from adding sagitta determination to the process of linking superlayer track segments corresponds to a position rms resolution of one fiber diameter.

II. Method

The R-z view of a scintillating fiber tracking detector, depicted in Figure 1., shows four axial superlayers covering the pseudorapidity region $|\eta| < 1.85$ and ranging from R = 0.6 to 1.6 meters in radius. In detail, each fiber superlayer (ignoring stereo layers) is composed of a pair of axial sense layers separated by an internal gap in radius as shown in Figure 2. A sense layer is constructed as a closely packed set of fibers, four deep. A specific offset arrangement is used in packing the fibers four deep to both maximize the light output and the information content. The four individual layers are staggered by 0, 1/2, 1/4, and 3/4 diameters respectively. For perfect placement, the resolution of this structure is one fourth of the fiber diameter divided by root (12). For 0.75 mm fibers the resolution of such a four deep staggered package is 55 microns.

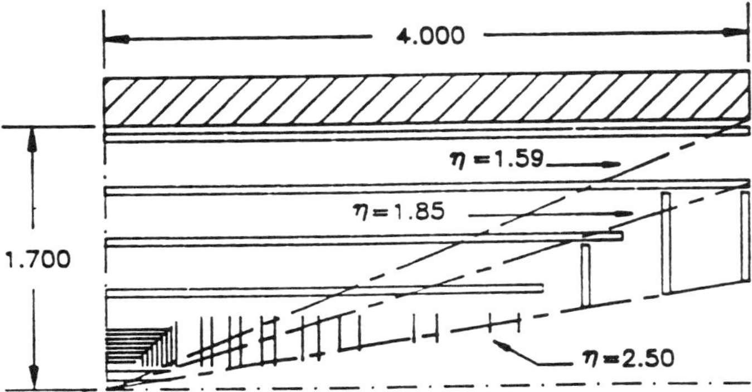

FIG. 1. Scintillating fiber and silicon microstrip tracking system for the SDC detector.

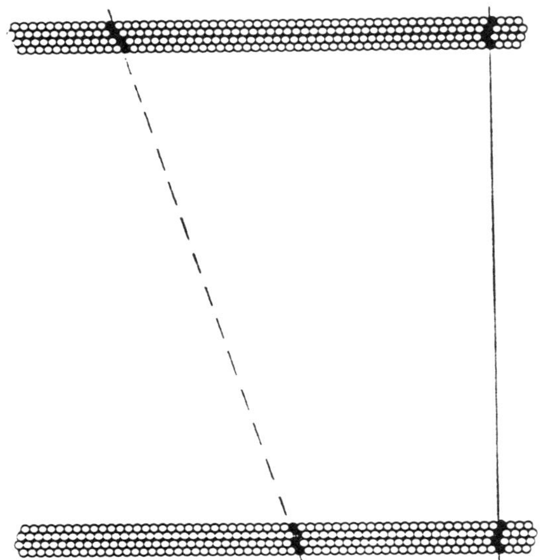

FIG. 2. Scintillating fiber superlayer with a radial track (solid line) and a low momentum track (dashed line). The occupancy is 3%.

The concept of tracking with radially separated superlayers containing highly correlated local information as opposed to uniform sampling was pioneered by CDF and shown to enable a fast momentum sensitive trigger (3). The fact that the orientation of sense layers intercepted radial tracks provided prompt hit information that was key in overcoming drift time limitations in the first level trigger. Scintillating fibers provide fine-grained information that is entriely prompt hits with a resolution set by the fiber diameter.

Trigger reconstruction of high momentum tracks using the fiber hit patterns proceeds through three logical steps:

- •Clustering - The hit patterns corresponding to radial tracks in sense layers are identified.
- •Connecting- Correlated pairs of clusters are connected across the superlayer internal gap.
- •Linking - Track segments are linked between superlayers to form tracks.

Cluster reconstruction in a four deep sense layer amounts to searching for clusters of hits that correspond to the patterns for radial trajectories. At present, only 3-out-of-4 and 4-out-of-4 patterns are being considered because of uncertainties in backgrounds to 2-out-of-4 patterns. All such valid clusters are kept at this stage; the occurrence of a dense clump of hits will trigger all valid radial clusters contained within. Hardwired AND/OR logic is the preferred implementation.

Connecting clusters across the superlayer internal gap is restricted to stiff tracks only, where stiff is taken to mean greater than 10 GeV/c momentum. The search road is thus restricted to only a small number of fiber diameters as indicated in Figure 2. Three classes of track segments are identified corresponding to positive, negative or zero slope along with the azimuthal coordinate for input to the linker.

Starting from the outer superlayer, linking proceeds back toward the origin and assumes the beam constraint. Again the search road is narrow as a 10 GeV/c track has only 2 cm of sagitta from the origin to the outer radius at 1.6 m. Reconstructed tracks and their quality factor (whether found in three or four superlayers) are then sent to the trigger decision logic. The azimuthal coordinate is partly encoded by the physical cable and partly by the data, and the momentum is encoded as the sagitta in units of the fiber diameter.

III. Realization

SSC conditions mandate a pipelined data acquisition structure to create a time window for trigger operations. For fiber tracking, this is a digital pipeline for the hit bits. Considerations of chip packaging imply one pipeline chip for every 128 fibers or 8000 chips overall. Full parallel processing for track segment finding can be achieved by allowing for a few chip to adjacent chip interconnects to share fibers and placing the AND/OR logic for clustering and connecting on the same silicon as the pipeline, shown in Figure 3. The narrow search road for stiff tracks makes this packaging possible. The present estimate is for 10 clock steps (or crossings) of delay including allowances for registering intermediate results due to settling times. Data for segments would be output as 30 bits, 10 phi bins by 3 slope bins (4).

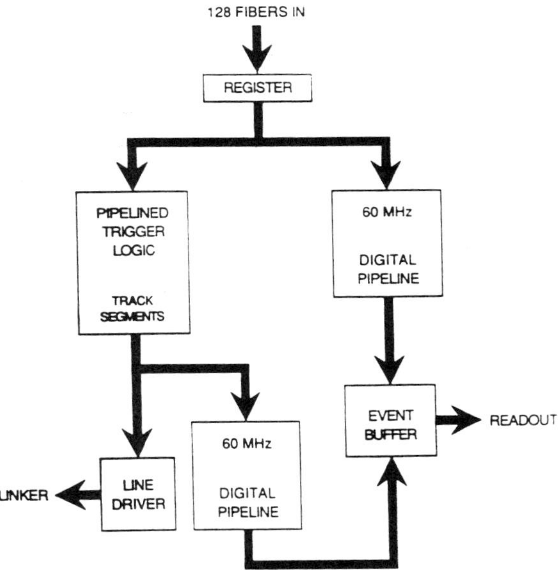

FIG. 3. Functional block diagram for digital ASIC combining pipeline delay with track segment finding for the trigger.

Linking, in contrast to segment finding, is thought of as a programmable task. Among the reasons are accomodating slow changes in superlayer registry, finding the beam off-center in the tracker, and dead or hot detector elements. For triggering, the goal is to search for the stiffest possible tracks consistent with the list of track segments. Only one number need be produced for each calorimeter tower for both electron and muon candidates, namely the highest momentum track interpretation of the track segments in front of that tower.. For a 10 GeV/c or 2 cm search road, about 30 clock steps (or crossings) of delay are estimated. (4).

IV. Summary

Track reconstruction in real time for high momentum particles is possible for the first level trigger using scintillating fiber tracking elements and simple hit-bit combinatorial logic. An implementation which integrates the pipeline delay with pipelined AND/OR logic results in time effective parallel processing for track segment finding. Linking track segments into tracks is also parallel pipelined processing, but in a stand-alone system due to the need for programmability. The resolution, at one fiber diameter precision in the track sagitta, is better than 0.004 p_t $(GeV/c)^{-1}$. The functionality of a momentum-determining charged track trigger, in particular the ability to apply variable threshold stiffness criteria, may be necessary for the first level trigger even at 10^{33} luminosity.

References

(1) R. Ruchti, et. al., these proceedings.

(2) R. A. Lewis, et. al., these proceedings.

(3) F. Bedeschi, et. al., NUCL. INST. and METH. A268 (1988) 50.

(4) G.W. Foster, et. al., these proceedings.

PERFORMANCE OF FLASH ADCs IN THE 100 MHz RANGE
I. TEST BENCH AND PRELIMINARY RESULTS

H. B. Crawley, R. McKay, W. T. Meyer, E. I. Rosenberg, and W. D. Thomas
Department of Physics and Ames Laboratory, USDOE
Iowa State University, Ames, Iowa 50011

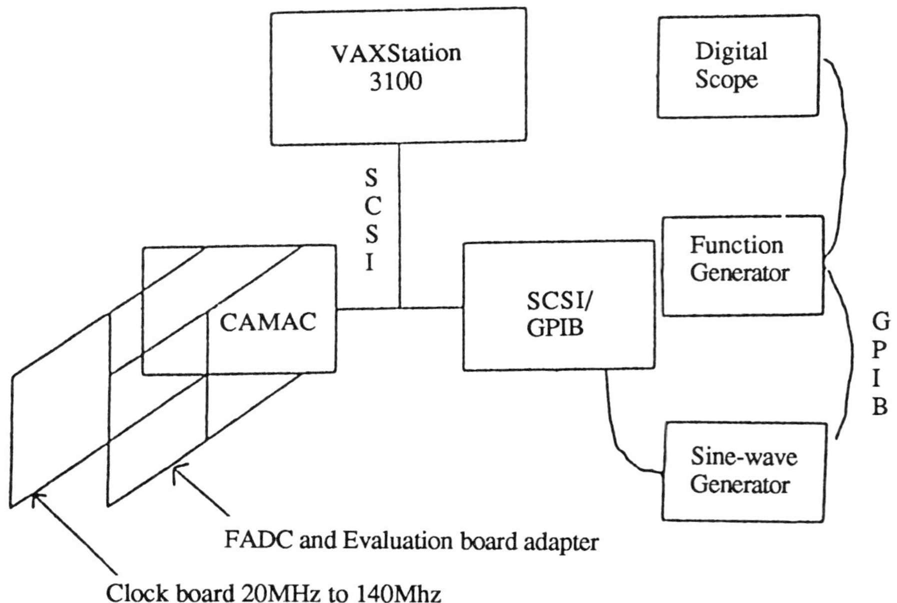

Figure 1 -- Test Bench Block Diagram

Abstract

We describe a systematic study of the performance of commercially available Flash ADCs in the 100 Megasample per second range, which might be suitable for use at the Superconducting Super Collider. Performance characteristics are measured using a CAMAC based test bench which is described. Among the FADC performance characteristics reported are linearity, differential linearity and the effective number of bits. This paper is the first in a series of reports to be presented within the next year as our tests continue.

INTRODUCTION

As presently designed, beams at the Superconducting Super Collider will collide once every 16 nsec. Thus it may be necessary to collect data at a comparable rate, i.e. 62.5 MHz. At the very least, trigger information will need to be gathered at this rate. In conjunction with the design of a compensating scintillator plate calorimeter, we have begun an investigation of devices capable of digitizing the analog signals produced by a detector at such rates. In particular, we report on commercially available Flash A/D converters in the 100 Megasample per second range. All of the devices we report on here use 8-bits to specify the digital value. However, it should be noted that for a given application dynamic ranges greater than 255:1 can be achieved using non-linear conversion curves.

In order to perform these investigations, we have developed a CAMAC-based test bench along with the appropriate software. Our testing procedure follows the guidelines given in IEEE Std 1057, "IEEE Trial-use Standard for Digitizing Waveform Recorders."

In the following sections we describe the test bench hardware and software and present preliminary results on three commercially available FADCs, the Analog Devices AD9002, the Sony CX20116 and the Sony CXA1176.

TEST BENCH DESCRIPTION

A. Hardware

A block diagram of our test bench hardware configuration is shown in figure 1. A VAXstation 3100 [1] running VMS controls and monitors the test bench. Communication between the computer and the test bench is via the SCSI port. A SCSI to CAMAC crate controller [2] provides communication with a single CAMAC crate containing two special purpose CAMAC boards. The bus converter [3] allows GPIB communication with a function generator [4], synthesized signal generator [5] and a digital oscilloscope [6]. The function and signal generators provide input signals to the FADC units.

A special clock generator board built for this test bench provides trigger and clock signals to the FADC unit. Ten discrete values of the clock signal provided to the FADC are available in the range 20 -140 MHz. Extension to lower values is possible. This allows us to test devices with dynamic ranges wider than those currently available at high sampling frequencies. Both TTL and ECL signals are available from this board.

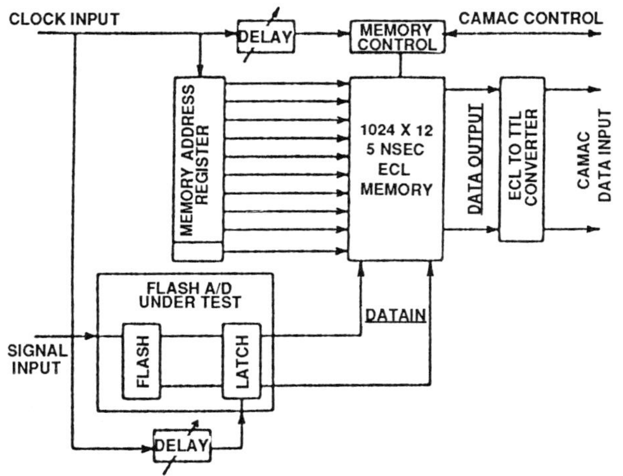

Figure 2 -- Evaluation Board Adapter

The second special board is designed to house an FADC evaluation unit. In some cases these are available from the chip manufacturers and in others we must design our own. This board also contains 1 kilobyte of fast 12 bit memory to store the FADC output prior to reading via CAMAC. A block diagram of this board is given in Figure 2.

B. Software

The test bench software consists of a set of program modules which perform the individual tests. The user interface uses the DEC screen management utility (SMG), permitting access with any standard DEC-supported terminals. Graphic routines are based on the GKS standard and support interactive printing of graphics output on a Postscript printer and interactive storage to GKS metafiles.

Software access to the bus devices attached to the GPIB bus is via a set of routines based on a standard library supplied with the IOtech SCSI/GPIB interface.

Readout and control of the CAMAC is done using routines which emulate a subset of the standard NIM/ESONE CAMAC library.

For a typical test, the user will have one window listing the options available, another listing parameters and the test results, and a third window to display graphical output. Information messages and alarms are displayed on top of these windows.

TEST PROCEDURES

As mentioned above, our tests are based on the IEEE trial standard Std 1057 and we refer the reader to that document for details. In what follows we give a brief description of these tests.

A. Gain and Offset

The gain and offset for a FADC can be measured independently by comparing the output signal to the input signal. We adjust the gain, a multiplicative factor applied to the input signal, and the offset, an additional additive factor applied to the input, so as to minimize the mean squared deviation between the two signals. For an ideal device the gain is 1.000 and there is no offset. A static determination of these quantities is made using a series of programmable dc levels. A dynamic determination is made by using about four cycles of a single sine wave input whose peak-to-peak amplitude covers the input range of the FADC under test.

B. Linearity

The differential and integral nonlinearity also are determined using the programmable dc-level inputs. The differential non-linearity is the normalized difference in the gain corrected measured range of input signals which yield a given output code and the expected, or "ideal", value of this range. The normalization is to the "ideal" value, or "code bin width". The same quantity is also determined using a triangular wave input asynchronous to the clock signal so that all input values are sampled. A perfect FADC has a differential non-linearity of 0.

The integral nonlinearity is computed as the maximum difference between the "ideal" and measured code transitions, after correcting for gain and offset, as a percentage of the full scale output value. We use the dc-level inputs to determine this quantity and a related quantity, the maximum static error. The maximum static error does not correct for gain and offset in comparing the transition levels and is not reported below.

C. Signal-to-noise and Effective Number of Bits

The sine wave data used to make a dynamic determination of the gain and offset also are used to determine other characteristics of the FADC. The rms noise is obtained from comparing the best fit of the sine wave to the measured data points. The signal-to-noise ratio is the rms signal (peak amplitude/$\sqrt{2}$) divided by the rms noise. The peak error is the largest term contributing to the rms noise divided by three times the sample standard deviation.

The effective number of bits also is determined from this data. This quantity is defined as the number of digitized bits less the $\log_2$ (actual rms error/the "ideal" quantization error). The actual rms error is determined for a given sine wave after gain and offset corrections are applied, The "ideal" quantization error assumes the rms error of a uniform distribution (bin width/$\sqrt{12}$). For an ideal device the number of digitized bits and the effective number of bits are identical.

PRELIMINARY RESULTS

In Table 1, we show the performance characteristics for the three devices mentioned previously. These preliminary results were taken at the one sampling frequency listed for each of the devices.

Flash ADC	AD9002	CXA1176	CX20116
Manufacturer	Analog Devices	Sony	Sony
Advertised Maximum Rate (MSPS)	125	300	110
Clock Rate for Measurements(MHz)	30.0	80.0	20.0
Gain			
Static	0.995	0.994	0.996
Dynamic	0.983	0.989	0.995
Differential Non-Linearity			
Sine Wave	0.300	1.000	0.800
Triangle Wave	0.297	not measured	0.559
Integral Non-Linearity	0.576	0.638	0.333
Signal-to-Noise Ratio	168.1	156.0	202.5
Effective Number of Bits	7.271	6.988	7.405

Table 1 -- Preliminary Results

FUTURE TESTS

In the future we will use multiple sine wave inputs to investigate time base errors and report on the aperture uncertainty and fixed errors in the settling time. Parameters such as analog input bandwidth, settling times, random noise, the word error rates, slew rates, and absolute overvoltage recovery will be given.

We will also present a more thorough summary of results as a function of sampling frequency for a large set of commercially available FADCs. As new devices under development are made available to us, we will test these as well.

Acknowledgements

This work was supported in part by a Research Development and Exploratory Fund grant by the director of the Ames Laboratory and by the U. S. Department of Energy under contract numbers W-7405-ENG-82-KA and W-7405-ENG-82-KS. Reference to a company or to product names does not imply approval or recommendation of the product by Iowa State University or the U. S. Department of Energy to the exclusion of others that may be suitable.

Notes and References

[1] DEC, SMG, VAXstation, and VMS are licensed trademarks of the Digital Equipment Corporation, Cambridge, Mass.

[2] Jorway Model 73 SCSI Bus Crate Controller.
Jorway Corporation, Westbury, New York.

[3] IOtech SCSI 488/D Bus Controller.
IOtech Inc., Cleveland, Ohio.

[4] Hewlett-Packard Model HP3314A Function Generator.
Hewlett-Packard Corp., Everett, Washington.

[5] Hewlett-Packard Model HP8656B Synthesized Signal Generator.
Hewlett-Packard, Ltd., West Lothian, Scotland.

[6] LeCroy Research Systems Model 9400.
LeCroy Research Systems, Chestnut Ridge, New York.

500 MHz TRANSIENT DIGITIZERS BASED ON GaAs CCDs

D. Bryman, J.V. Cresswell, M. LeNoble, R. Poutissou
TRIUMF, 4004 Wesbrook Mall, Vancouver, B.C., Canada V6T 2A3

Abstract

Due to high speed and radiation resistance, gallium arsenide devices may have important applications at the SSC. We report on the development of a 500 MHz transient digitizer based on a recently produced gallium arsenide charged coupled device which has 128 pixels and operates at 500 MHz. Initial testing of prototype digitizer modules is discussed.

Introduction

Gallium arsenide devices have potential benefits over their silicon counterparts due to the considerably higher mobility for electrons and ions and the significantly greater resistance to radiation. These potential attributes of GaAs structures may lead to SSC applications in detectors and readout systems where speed, low power and radiation hardness are to be premium characteristics. Wide bandwidth analog pipelines using GaAs charged coupled devices (CCD) or acoustic charge transport delay lines[1] are examples of potentially useful devices at the SSC. Much basic R&D is still necessary to realize the potential advantages of GaAs devices.

Over the past few years, a 500 MHz, 128-pixel GaAs CCD has been under development at TRIUMF. Its first application is in a high speed transient digitizer designed for use in the measurement of the rare decay $K^+ \to \pi^+ \nu \bar{\nu}$ (Experiment 787 at Brookhaven National Laboratory).[2] In the digitizer, the CCD is operated in a high frequency mode to capture a signal and then, on receipt of a trigger, its contents are shifted out at a lower frequency into an inexpensive flash-ADC and memory system. In this paper, we will discuss the GaAs CCD and its use in a prototype 500 MHz transient digitizer.

GaAs CCD based Transient Digitizer

The prototype digitizer module (see Fig. 1) is comprised of three primary sub-systems: the rf distribution system, the CCD transient digitizer modules and the FADC-processor system. The rf distribution system which provides rf signals to the CCD transient digitizer sub-modules, consists of a 1 GHz frequency synthesizer, a triggered frequency prescaler and a power distribution network. The frequency synthesizer followed by the triggered frequency prescaler provides the 500 MHz data acquisition clock signal source as well as the 4 MHz output clock. The high frequency output is normally active and is switched to the low frequency state for 128 clock cycles upon the application of an externally supplied trigger signal. An 'abort' input is used to asynchronously reset the prescaler to its high frequency output state.

The input signal is clocked into the CCD using the 500 MHz data acquisition clock. An externally generated trigger pulse disables the accumulation of the input signal and causes the 4 MHz low frequency pulse burst to be applied to the CCD. The captured signal is passed to the A/D converter and is transformed into a string of 128 8-bit words.

In the prototype digitizer, the CCD and its clock driver circuitry were located on a printed circuit board along with an A/D and a D/A converter. For readout, we are employing an 8-bit digitizer[3] developed at Iowa State University and Ames Laboratory for DELPHI. Each FASTBUS board contains 32 FADC channels and 256-deep memories capable of operating at frequencies up to 15 MHz and a programmable 16 MHz TS68000

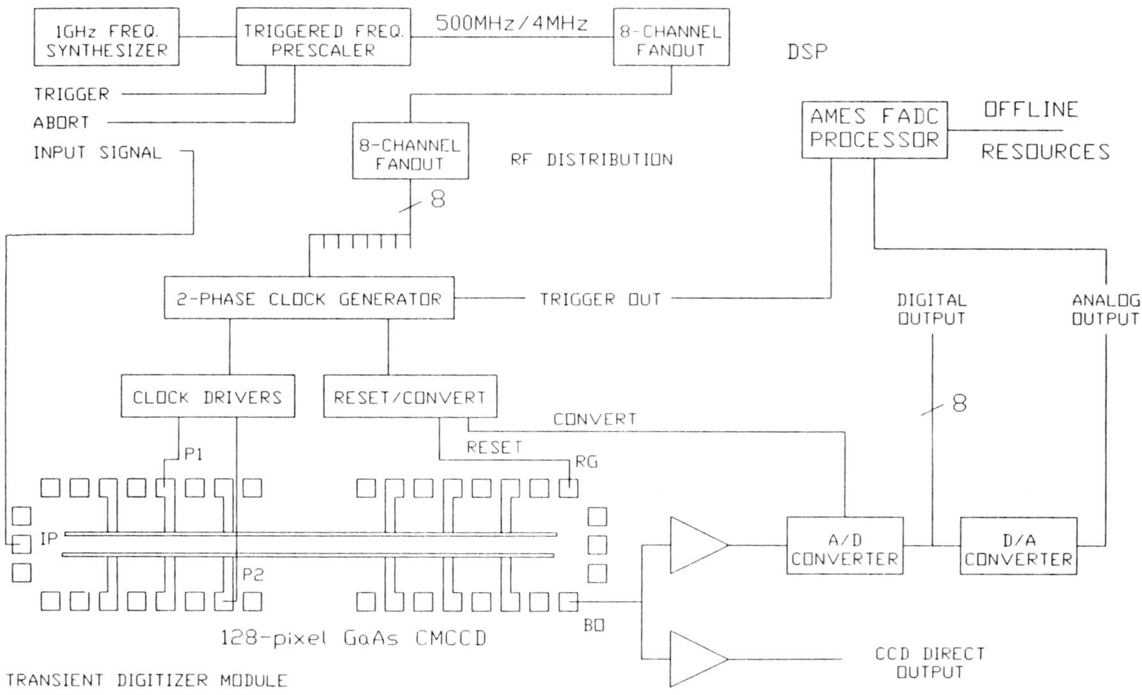

Figure 1. Block diagram of the GaAs CCD based transient recorder.

microprocessor which presently is used to repack over-threshold data for transmission to the data acquisition system. The CCD low frequency clock is used to synchronize the FADC's.

The response of the CCD digitizer system is demonstrated qualitatively in Fig. 2. A double Gaussian-like input signal covering a period of 20 ns was applied to the digitizer. In the figure, the corresponding D/A output spanning 2.5 μs is superimposed on the input signal.

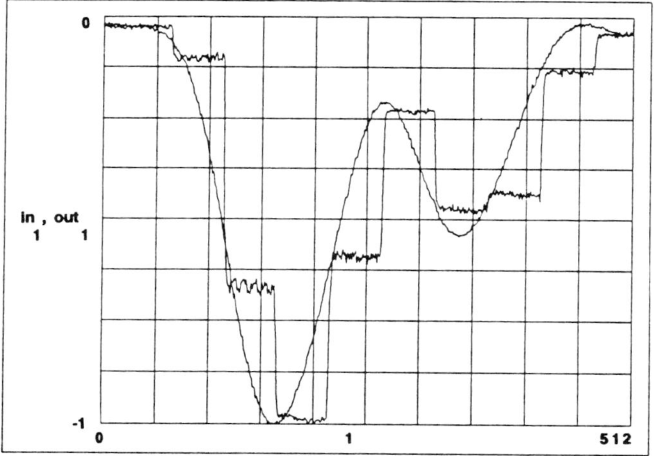

Figure 2. CCD response to a double pulse input signal (smooth curve). Each CCD step represents 2 ns.

Application to BNL E787

In E787, a 775 MeV/c kaon beam is slowed by a BeO degrader and stopped in a segmented scintillator fiber target[4] located at the center of the detector. The target consists of 3 m long, 2 mm diameter scintillating fibers grouped into 379 triangular clusters of six, each viewed by a 1 cm diameter photomultiplier tube. The target elements are used to track the incident kaons which travel along the fibers as well as to detect the signals from decay pions which cross the fibers transversely into the detector. After momentum analysis in a cylindrical central drift chamber in a 1 T solenoidal magnetic field, particles are stopped in a 15-layer range stack of 2 m long plastic scintillators viewed by phototubes at each end. Surrounding the drift chamber and range stack are calorimeters which cover almost 4π solid angle to detect photons.

Near the end of a recent data run, 18 channels of prototype CCD digitizers were deployed on the target. The signals from 21 non-adjacent target elements were multiplexed into each digitizer channel. Figure 3 shows an event display of the target elements hit in a sample event. The numbers within the target triangles correspond to the values recorded by integrating A/D converters. The arrows point to the signals digitized at 2 ns intervals for each element hit. CCD pedestals have been

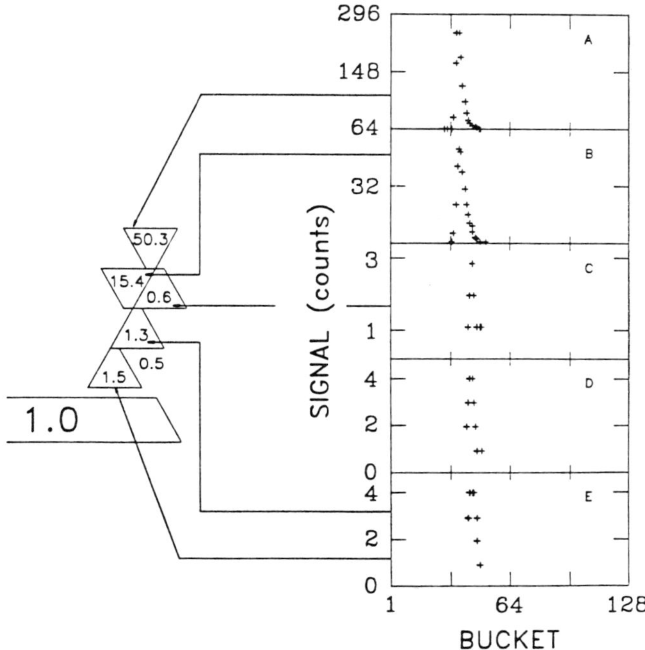

Figure 3. CCD digitized kaon decay event in the target. On the left are shown the target triangles hit along with the energy deposited as determined by integrating adc's. On the right are shown the digitized pulses as described in the text.

subtracted and a threshold of 3 counts has been required in software. Channels A and B were hit by the incident kaon which traversed the target elements longitudinally depositing 50.3 MeV and 15.4 MeV, respectively. The decay particle emerged transversely about 16 ns (8 CCD buckets) later striking channels C, D, and E leaving between 0.6 and 1.5 MeV in each element. The CCD calibration is roughly 3 counts/ MeV.

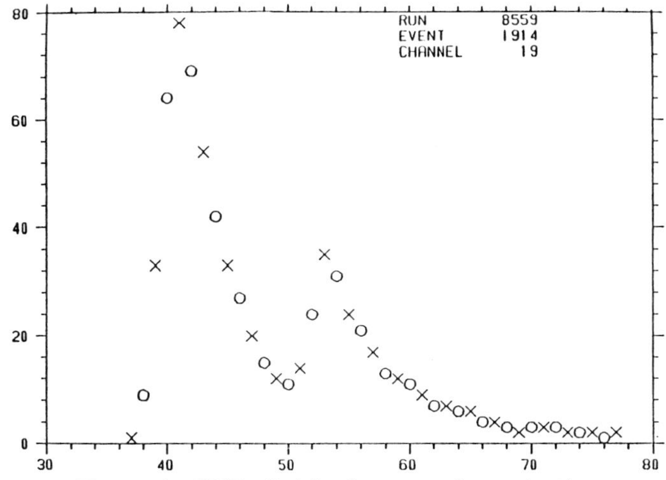

Figure 4. CCD digitized $\pi \rightarrow \mu$ decay in the range stack.

One range stack counter in the pion stopping region was also instrumented with a CCD digitizer. Figure 4 shows an example of a pion stop signal followed about 25 ns by the decay to a muon (approximately 4 MeV) as recorded by the CCD digitizer. The range stack scintillators are also instrumented with transient digitizers[5] (TD) based on the Tektronix TKAD508 A/D hybrid with two interleaved 250 Msample/s A/D converters. The CCD digitizers give results consistent with the TD's.

Conclusion

Prototype CCD digitizers which sample at 2 ns intervals for a period of 256 ns have been constructed and tested in Experiment 787 at BNL. The next step in the development of wide band CCD digitizers is to improve the layout by deploying the CCD chip and its driving circuitry on a separate daughter board and create a multi-element mother board. Further improvements to the on-chip CCD input circuit are also under development to attempt to increase the dynamic range. Other applications of GaAs technologies are being planned.

Acknowledgements

We would like to thank the E787 collaboration for their efforts and assistance in this work. We are particularly grateful to M. Constable, A. Daviel, S. Daviel, M. Burke and A. Turcot. In addition, we would like to thank the Ames group for their help in providing and using the FADC system.

References

[1] R. J. Davisson *et al.*, University of Washington preprint UWSEA PUB 90-6 (1990).

[2] M.S. Atiya *et al.*, Phys. Rev. Lett. **64** 21 (1990).

[3] H.B. Crawly *et al.*, IEEE Transactions on Nucl. Science **35**, 295 (1988).

[4] J.S. Frank and R.C. Strand, in *Proc.of the Workshop on Scintillating Fiber Detector Development for the SSC* (FNAL, Batavia, IL, 1988)

[5] M.S. Atiya *et al.*, Nucl. Instrum. Meth. **A 279**, 180 (1989).

CONSIDERATIONS ON THE DESIGN OF FRONT-END ELECTRONICS FOR SILICON CALORIMETRY FOR THE SSC*

A. L. Wintenberg, M. L. Bauer, C. L. Britton, Jr., E. J. Kennedy and R. A. Todd
Oak Ridge National Laboratory
P. O. Box 2008, Oak Ridge, Tennessee 37831-6006

S. C. Berridge and W. M. Bugg
The University of Tennessee, Knoxville
Knoxville, Tennessee 37996

Abstract

Some considerations are described for the design of a silicon-based sampling calorimetry detector for the Superconducting Super Collider (SSC). The use of silicon as the detection medium allows fast, accurate, and fine-grained energy measurements— but for optimal performance, the front-end electronics must be matched to the detector characteristics and have the speed required by the high SSC interaction rates. The relation between the signal-to-noise ratio of the calorimeter electronics and the charge collection time, the preamplifier power dissipation, detector capacitance and leakage, charge gain, and signal shaping and sampling was studied. The electrostatic transformer connection was analyzed and found to be unusable for a tightly arranged calorimeter because of stray capacitance effects. The method of deconvolutional sampling was developed as a means for pileup correction following synchronous sampling and analog storage.

Introduction

This paper presents a graphical analysis of the noise inherent in a silicon detector and the associated preamplifier and shaper in relation to the detector leakage current, capacitance and charge collection time, and preamplifier power consumption and dynamic range. It includes an examination of the effects of stray capacitances on the electrostatic transformer connection of detector elements and presents a method to correct for pulse pileup in a synchronous detector system.

System Noise Issues

In a well designed system, the noise level is mainly determined by the detector, preamplifier, and pulse-shaping function. Achieving a low noise level depends upon adjusting several parameters, but the values of those parameters needed may conflict with other requirements such as dynamic range or power dissipation. To better understand the interaction of various system parameters on the noise generated, we studied the simple model of a silicon detector and a bipolar preamplifier (shown in Fig. 1). The detector is represented by the current source i_{DET} and capacitance C_{DET}. The shot noise due to the silicon leakage is lumped together with that due to the preamp input transistor base current and is labelled i_n. The source e_n represents the equivalent noise voltage of the preamp. This model is a simplified version of one analyzed by Blalock[1].

A graphical study was made of the noise contours for bipolar preamplifiers at very short shaping time constants (10–100 ns). The nominal system for these studies was a detector with 35-pF capacitance and 10-µA leakage current, and a preamplifier with 5-pF feedback capacitance, 30-kΩ feedback resistance, and 100-Ω equivalent noise resistance. The noise current (i_n) was assumed to be predominantly due to detector leakage. Families of curves of equivalent noise charge vs shaping time were generated for various detector leakage currents (Fig. 2), detector capacitances (Fig. 3) and amplifier input noise as related to preamplifier first-stage emitter currents (Fig. 4). The equivalent noise charge is given in terms of rms electrons (rms e).

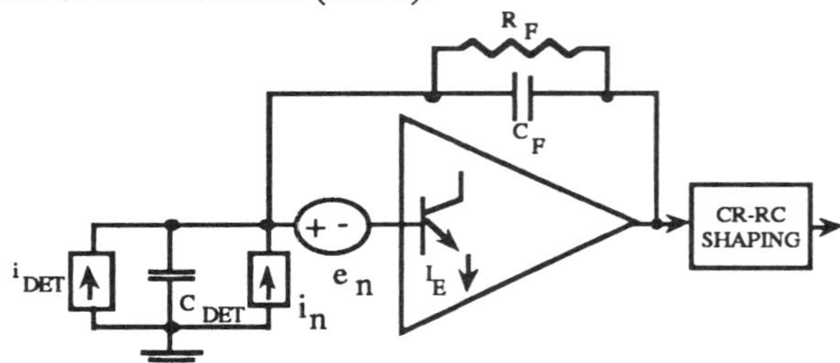

Figure 1. Silicon detector and preamplifier noise model.

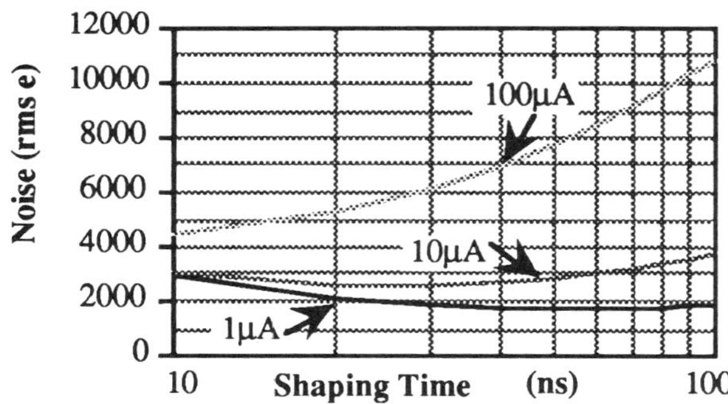

Figure 2. Amplifier noise as a function of shaping time and detector leakage current for bipolar preamplifier.

A point extremely important to SSC operation is that detector leakage is not a major noise source in the likely operating region (i.e., shaping times of 15 to 30 ns). The equivalent noise voltage due to the preamplifier is the dominant source for detector leakage currents under ~100 µA

*Research performed at Oak Ridge National Laboratory, operated for the U. S. Department of Energy under contract DE-AC05-84OR21400 with Martin Marietta Energy Systems, Inc.

(Fig. 2). The contribution due to the feedback resistor (30 kΩ) was negligible even though it is much smaller than the usual 100 MΩ or more. These results are consequences of the short shaping times.

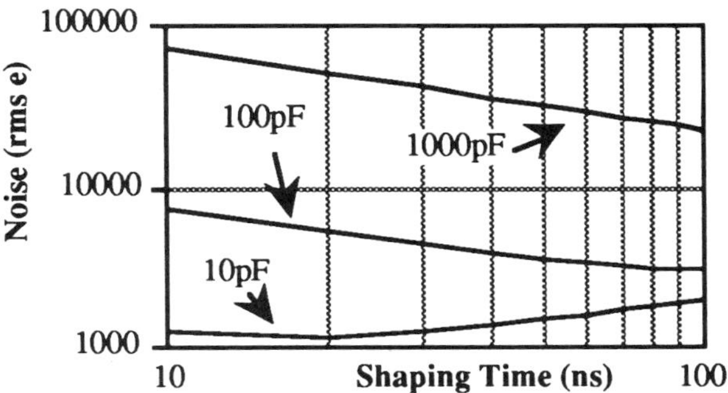

Figure 3. Amplifier noise as a function of shaping time and detector capacitance for bipolar preamplifier.

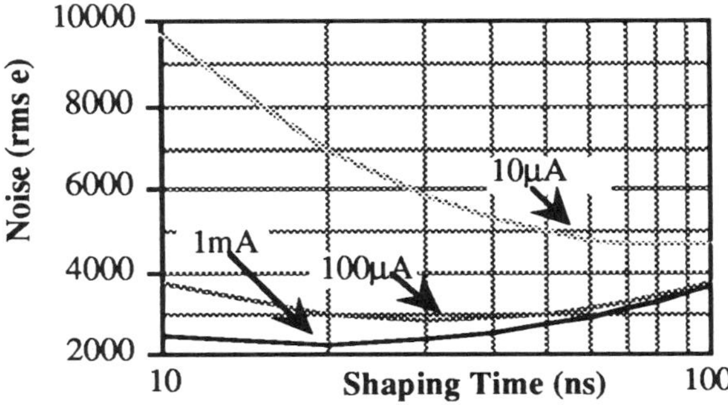

Figure 4. Amplifier noise as a function of shaping time and emitter current in the input device of the preamplifier. An input device with 50-Ω base resistance and 35-pF detector capacitor were assumed.

These studies also highlighted the tradeoffs between detector capacitance and noise. The silicon detector thickness for the electromagnetic calorimeter has largely been set at 400 μm, but the thickness for the hadronic calorimeter has varied between 400 and 500 μm. Several different pad sizes will probably be used. The capacitance is first-order inverse with thickness and the dominant noise mechanism is first-order with capacitance (Fig. 3), while the collected signal is proportional to the thickness for a particle that passes through. Thus, disregarding collection time, the thicker the detector, the better the signal-to-noise ratio is. At room temperature, with 100-V bias, the collection time is 11 ns for electrons and 34 ns for holes with a 400-μm detector. Since the collection time is proportional to the thickness squared, making the detector thinner improves the collection time but degrades the signal-to-noise ratio. Cooling the detector decreases the collection time without signal-to-noise ratio degradation.

Figure 4 illustrates the tradeoff between power consumption and noise. As the current in the input device is increased (and, consequently, power consumption is increased) the equivalent noise charge is decreased. A more subtle interaction occurs with the preamplifier feedback capacitor and the system gain, noise, and power consumption. For a given detector capacitance, the output due to preamp series noise is inversely proportional to the value of the feedback capacitor, as is the output signal (constant signal-to-noise ratio). Because of other noise sources, the total noise contribution increases with increasing capacitor (feedback and detector) values. The feedback capacitor value sets the preamp charge gain, while the feedback resistor and the input count rate affect pileup. The maximum linear signal amplitude output of the preamp is determined by the dc power supply voltage: more dynamic range costs power dissipation (even with devices operating at the same currents). Thus, the selection of the feedback capacitance is a tradeoff between the noise floor and the maximum signal charge for a given supply voltage.

Electrostatic Transformer Connection Analysis

One of the suggested solutions to the problem associated with high detector capacitance has been the use of the electrostatic (series) transformer interconnection of detector elements. Series connections in silicon calorimetry offer advantages of lower detector capacitance (presented to the preamplifier), fewer amplifiers per unit detector area, and fewer interconnect wires when compared with the use of a separate preamplifier for each detector segment and summing the signals from multiple preamplifiers in the same plane and/or tower. For *n* identical detector elements, the series capacitance is reduced by a factor of *n* over the single-element case. Likewise, the portion of charge seen by a low-impedance amplifier connected to the series arrangement is a factor of *n* less than in the single element-case, and this is illustrated in Fig. 5 for a case with five identical elements (C1, C2, C3, C4, and C5).

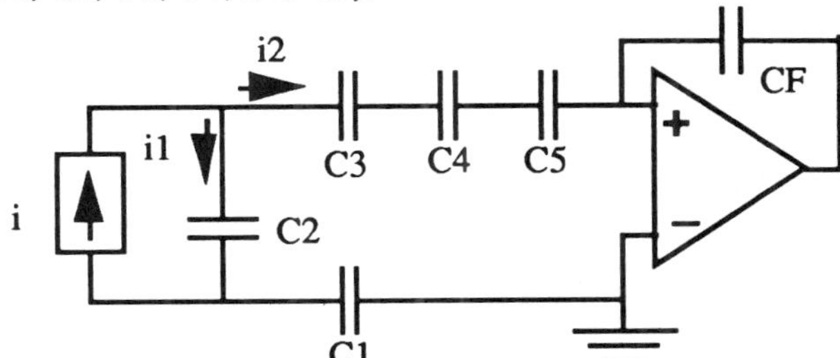

Figure 5. Equivalent circuit of series transformer charge collected from ionization track.

An example of the series transformer was shown in work published by the SICAPO collaboration[2]. With a 4-mm air gap on each side of the silicon, the parasitics were only 1.7% of the detector capacitance and had relatively little effect. Our current silicon calorimeter design has much smaller distances between the absorber plates, and with printed circuit boards serving as the dielectric, the stray capacitance jumps considerably. We plan to fill a 3-mm gap between absorber planes with 400- to 500-μm silicon, 2-mm thick electronics and 0.5-mm insulation; and the stray capacitance between each detector and the absorbers approachs 50% of the detector capacitance.

A SPICE simulation showed very nonuniform response due to differing charge collection as a function of the position of the detector in the series chain when stray capacitances were included. The model was run for a five-plane tower with stray capacitance equal to 5% of the segment capacitance, a value only 10% of what we expect. Even with this optimistic model, the detector farthest from the preamplifier had half the charge sensitivity of the closest. One might argue that nonuniformity along the position of the tower is inconsequential and can be compensated with a correction factor if the particle traverses all elements in the tower. However, for energy depositions of up to 80 to 90 MeV in a single detector element, the charge gain will depend on its location within the tower. Series transformer connections of silicon detector elements would also add series inductance, which would result in very long preamplifier settling times. Our analysis of signal degradation due to stray capacitances has shown the series transformer option to be untenable.

Shaping and Sampling Considerations

The high clock frequency of the SSC means that shaping times considerably less than 16 ns would be required to prevent pileup. This short shaping time is somewhat of a problem with silicon because the charge collection time is greater than a bunch crossing time. We decided to investigate this problem and see if a method to correct for pileup in a synchronous detector could be developed.

It is generally assumed that sampling followed by analog storage will be needed to save data for analog-to-digital conversion after a first-level trigger. Since earlier samples are available, they can be used in a pileup-correction scheme. For example, if the pulse shape presented to the sampler is a decaying exponential with time constant τ, the correction for the immediately preceding pulse tail is the preceding pulse height times $\exp(-T/\tau)$, where T is the bunch crossing time. This correction is made by subtracting a fraction of the previous sample from the sample that is to be converted, and this can be implemented very simply with a divider and a difference amplifier. This special case has been implemented before for a high-rate liquid argon calorimeter[3].

A more general approach is possible if the system of detector, preamplifier, shaper, sampler, and analog memory is viewed as a sampled data system. That view makes it simple to realize that the pileup due to all previous events can be corrected for if the response of the detector, preamplifier, and shaper is known at the sample times. If the deposition of energy in the detector is considered to be impulse-like and is denoted by $p(t)$ with $p(t) = p(nT)$ at bunch crossing n and zero otherwise, and the impulse response of the system is $h(t)$, then the output of the preamplifier is the convolution of $h(t)$ and $p(t)$ or

$$y(t) = h(t) * p(t) = \sum_{m=-\infty}^{\infty} h(t-mT)\, p(mT). \qquad (1)$$

If $y(t)$ is sampled at ΔT after each bunch crossing, then the stored values are

$$y(nT+\Delta T) = \sum_{m=-\infty}^{n} h(nT+\Delta T-mT)\, p(mT). \qquad (2)$$

This equation can be solved for $p(nT)$ in terms of $y(mT+\Delta T)$ and earlier samples of the output. The solution is in the form of a finite impulse response (FIR) digital filter, and for the case where $h(t)$ is a decaying function (the low feedback resistor guarantees this), only a small number of terms are needed to provide sufficient accuracy for even extreme pileup. This method can be implemented with dividers and a difference amplifier as illustrated in Fig. 6. We call this method of pileup correction "deconvolutional sampling." This method has some advantages over double correlated sampling. It requires only half the sample rate and can use longer shaping times that are more optimal for noise performance with the preamplifiers likely to be used. Further investigation of this method is planned.

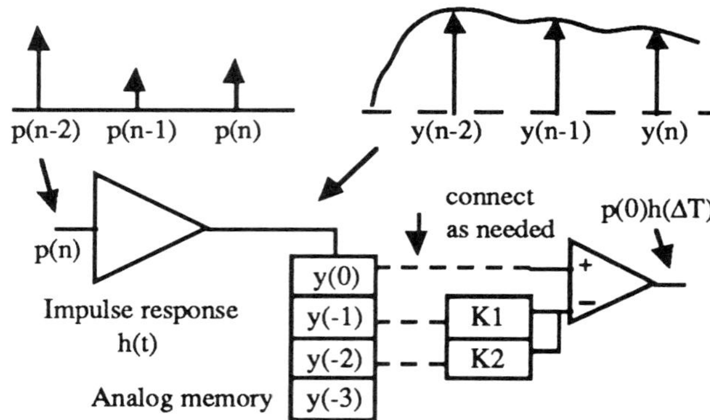

Figure 6. Deconvolutional sampling.

Conclusions

Noise issues specific to silicon calorimetry were explored, and it was found that for likely detector geometries, preamplifiers, and operating region, the noise would be dominated by the equivalent noise voltage of the preamplifier and not by the noise current due to detector leakage. We determined that the series transformer connection of detector elements was not a viable option for closely packed elements with electronics and insulation in between. The deconvolutional sampling method for pileup correction following synchronous sampling and analog storage was developed.

References

[1] T. V. Blalock, "A Low-Noise Charge Sensitive Preamplifier with a Field-Effect Transistor in the Input Stage," IEEE Trans. Nucl. Sci., **NS-11**, 365–72 (1964).

[2] C. Furetta et al., "Large-Area Sandwich Calorimeter for Hadronic Calorimetry," IEEE Trans. Nucl. Sci., **NS-35**, 446–50 (1988).

[3] T. F. Droege et al., "Design and Operating Experience with Electronic Systems for High Rate Liquid Argon Calorimeters," IEEE Trans. Nucl. Sci., **NS-27**, 64-67 (1980).

A TIME TO VOLTAGE CONVERTER AND ANALOG MEMORY UNIT FOR STRAW TRACKING DETECTORS

L. Callewaert*, W. Eyckmans*, A. Stevens, W. Sansen*,
J. Van der Spiegel, R. Van Berg, H.H. Williams, T.Y. Yau
University of Pennsylvania
Philadelphia, Pa. 19104

Abstract

A low power, sub-nanosecond accuracy, quick recovery, data-driven, multiple sample Time to Voltage Converter suitable for use on high rate straw tracking detectors is described. The described TVC includes "virtual" storage of analog data in both Level 1 and Level 2 queues and an on board ADC with first order correction for capacitance variations and non- linearities.

Introduction

In a high precision drift tube or straw tracking system, one measures the time of arrival of the first electron at the anode. While many possible schemes exist, our initial judgement was that an analog time measurement would offer both lower power and greater resolution than an equally complex digital system. In addition, we believe that it will be necessary to incorporate *all* of the system features such as connection to the trigger and DAQ systems in any usable design in order to keep the power, mass and complexity of the final system under control.

Design Goals

The minimum set of specifications necessary for a successful time measuring device would seem to be:

- $<$ 0.5ns time accuracy, $\sim$ 0.2ns time resolution - in order to ensure spatial precision of $< 100\mu m$ without extensive (preferably without any) calibration.
- Deadtimeless continuous multihit operation - second hit or double track resolution better than any conceivable chamber (i.e. $\leq 30ns$).
- High rate capability - at the SSC average rates for a straw chamber near the inner radius of survivability approach 10 MHz.
- Local storage of data during the trigger decision time(s) – to keep cabling and power from dominating a detector design, it will be necessary to move data off the detector only after the trigger has had time to reject most of the uninteresting events.

Time to Voltage Conversion

In previous work, we have fabricated and measured a Time to Voltage Converter [1] [2] in a 1.6 μm digital CMOS process that easily met the first criterion as can be seen in the differential nonlinearity distribution, Figure 1.

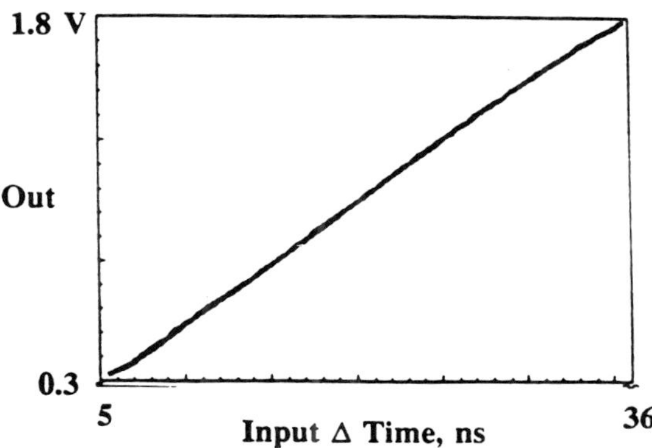

Figure 1: TVC Differential Non-Linearity from 7 to 31 ns. Note that 50 mV is equivalent to 1 ns.

This eight channel prototype TVC design depended upon careful matching of capacitance values from one sample to the next using common centroid layouts. It became obvious in this design that while the common centroid layout gave very good matching channel to channel, the cost in area and layout complexity for a much larger number of capacitors would be prohibitive. Because of this, we decided to pursue a charge–measurement scheme that would, to first order, be insensitive to capacitance values. In addition, we have attempted to include in the TVC/AMU all of the logic necessary for a full SSC compatible readout system. The block diagram in Figure 2 shows many of these features, the most salient of which will be discussed below. The first prototype of the TVC/AMU was also fabricated in a 1.6μm CMOS process, but a number of layout errors prevented full operation of the device. A second version of the TVC/AMU is now being readied for fabrication and measurement.

Any capacitor memory scheme capable of simultaneous reading and writing will have at least three switches around each individual capacitor – an input charging switch, an output discharge switch, and a reset switch. By placing charge on the capacitor at a constant input rate and then using the output switch to remove charge at a much

*Catholic University of Leuven, Belgium

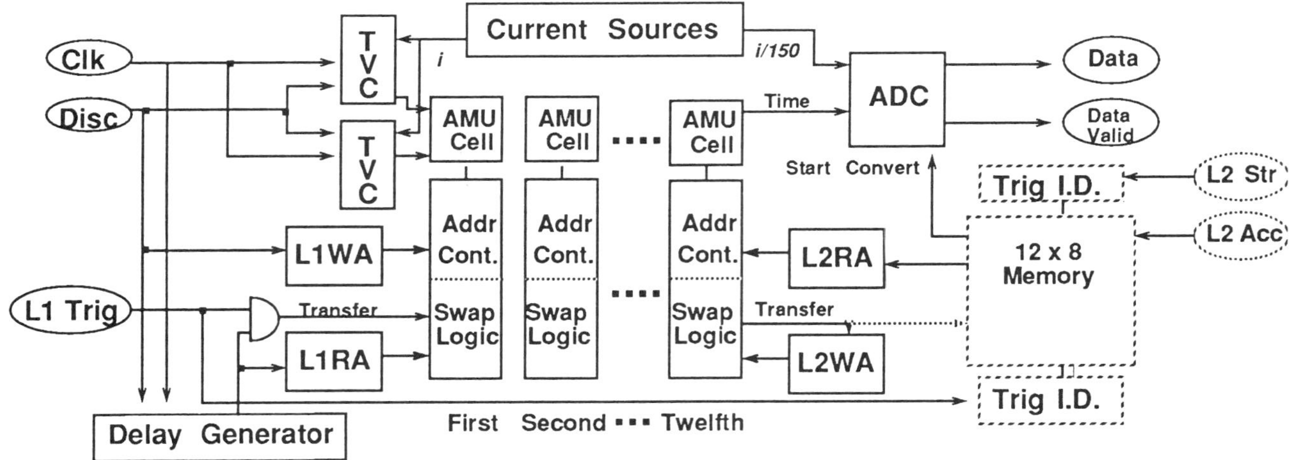

Figure 2: TVC/AMU Block Diagram. Dotted blocks associated with L2 Trigger identification are not implemented in the prototype version

lower rate, it is possible to measure the time *duration* of the input current to high precision with a relatively slow comparator. If the input and output currents are ratioed to each other, then the accuracy of the measurement can be kept high even with relatively large integrated circuit process and device variations since both the capacitance and the absolute values of the currents ratio out of the first order time equation, $\Delta t_{out} = \frac{\Delta t_{in} \times i_{in}}{i_{out}}$. We have chosen a ratio of $i_{in}/i_{out} \sim 150 :: 1$ in order to give $\sim 0.2ns$ resolution over a 16 ns (plus 8 ns pedestal) range.

Level 1 Storage

For a two to three μs Level 1 storage time it would be necessary to provide 128 to 192 storage locations for a simple synchronous system. However, by implementing a data driven system which uses a storage location only if new data is present, many fewer locations will suffice given the necessarily modest occupancies of any viable tracking system. Thus in the design of the prototype TVC/AMU, we have limited the number of Level 1 locations to eight which is more than enough for a 1 μs delay with a 5 MHz average input rate. This design may be expanded to 16 or more locations in Level 1 at the cost of additional silicon real estate.

Level 2 Storage

While a few microseconds of synchronous storage for Level 1 is conceivable, even if awkward, many tens of microseconds is clearly outside the realm of realizable circuits and thus Level 2 must also be implemented in a data driven fashion. For the TVC/AMU we have chosen to use a virtual Level 2 scheme in order to simplify the analog storage problems. The particular technique adopted is *address swapping* which is logically similar to minimum FIFO or pointer schemes, but minimizes the length of analog control signal lines - at the expense of longer control buses.

The Level 1 and Level 2 queues are implemented by using a write and read address counter for each queue. Thus the storage location next to be used is identified by the Level 1 Write Address (**L1WA**) counter which is incremented by the discriminator pulse, the **L1RA**, which is incremented by a delayed version of the discriminator pulse, identifies the next data to be tested for a valid Level 1 trigger. Similarly **L2WA** (incremented by a transfer from Level 1) and **L2RA** (incremented by an End of Convert) identify the beginning and end of the Level 2 queue. The output of these counters then serves to select the required capacitor by parallel comparison with the address contained in local address registers. The local address register is implemented as a five bit latch, the Level 1 counters (**L1WA, L1RA**) are implemented with four bits, and the Level 2 counters (**L2WA, L2RA**) are three bits wide, appropriate for a system with 16 Level 1 locations and 8 Level 2 locations even though the initial prototype TVC/AMU realizes only 8 and 4 locations respectively.

Time Measurement Logic

To accurately reconstruct tracks, it is necessary to know the time of arrival of the first electron at each straw anode relative to some common time that relates all of the data for a given event. The TVC/AMU operates as a common-stop time measuring device - starting on an anode discriminator pulse and stopping on the positive going clock edge after the next negative going clock edge.

This scheme leaves a guaranteed minimum of one-half of a clock cycle for any time measurement and for a nominal 16 ns clock means that the actual time ramp in the TVC goes from 8 to 24 ns. This minimum time allows the TVC response to be linear in the range of interest and avoids the possible race conditions associated with a system that allows a zero minimum measuring time.

Event Delay

In a data driven architecture, there is no one-to-one correspondence between physical storage location and event time. Thus there must be some provision in the TVC/AMU to allow synchronization of an event with a possible Level 1 Trigger signal. A simple 64 element shift register would allow about 1 μs of delay with a 16 ns clock period, but all 64 elements would be clocked at 60 MHz and significant power would be dissipated. In this design we have chosen to use a 64 element dynamic memory cell with read and write decoding done by a single decoder but with write select directed to the $(N)th$ element while read select is directed to the $(N+1)th$ element. One input line and one output line bussed to each of the elements then results in any input being delayed by 64 clock pulses before appearing on the output line.

Level 1 Trigger Interface

After a delayed data pulse exits from the delay generator, it is simply *AND*ed with the Level 1 Trigger signal and the resulting signal, **L1ok** is used to provide the transfer signal that logically moves data from Level 1 to Level 2. If the Level 1 Trigger signal remains high for more than one clock period, then the same simple logic will continue to transfer any existing Level 1 data to Level 2, thus providing for the case where the detector response time is greater than one clock period (in this case the electron drift in the straw detector). In order to differentiate data from different clock cycles within any one Level 1 Trigger time, a separate small counter must keep track of clock pulses.

Level 2 Trigger Interface

The Level 2 Trigger is asynchronous so that a delay line is not appropriate for providing the trigger-data synchronism. Since the Level 2 Trigger is monotonic, however, it is only necessary to provide two counters (**L2WA, L2RA**) to keep track of Level 2 inputs and outputs. Each new entry into Level 2 increments the **L2WA** counter and each read (or reset) of Level 2 increments the **L2RA** counter.

As the Level 2 Trigger is asynchronous, it must be accompanied by a timing signal. The Level 2 Strobe is used to clock a D-Flip Flop with the Level 2 Accept at the data input of the flip flop. For an accepted event, the ADC cycle is started, for a rejected event, the **L2RA** is simply incremented and the storage location is reset.

Event Identification

During normal SSC operation, individual detector elements and systems will be involved in simultaneous input processing, data storage, data conversion, and data output. Given the non-deterministic nature of multiple asynchronous devices, the data stream out of a detector will be disordered in time and it will be necessary to include in each data packet some time identifier.

The simplest time stamp, of course, would be a crossing counter - advancing every 16 ns. However, for Level 2 storage times of 50μs or more plus realistic DAQ pipeline delays of many tens of microseconds, a 13 or 14 bit crossing counter would be necessary in order to avoid ambiguity. We have chosen to count not the crossings at 60 MHz but the Level 1 Triggers at 1 to 100 KHz. Even at 100 KHz the eight bit Trigger I.D. Counter remains unambiguous for 2.5 ms.

Since the Level 1 Trigger may in many cases remain true for more than one crossing, to completely specify time it is necessary to have a small four bit Bunch Counter to keep track of bunches. This counter is enabled by Level 1 Trigger true and counts until Level 1 Trigger goes false.

Because the occupancy of a tracking system must be relatively low, most Level 1 and Level 2 Triggers will not have data in any given detector channel. For Level 1 Triggers, the Level 1 Trigger I.D. Counter advances on the L1 Accept, but the delayed data is *AND*ed with the L1 Accept and Level 2 data is stored only on a coincidence. At Level 2, it is necessary to tag each piece of data with the Level 1 Trigger I.D. and then, at each Level 2 Trigger, search the on chip memory for any relevant data.

The **L2WA** counter provides the write address for this memory and the **L2RA** provides the readout address. In order to skip over Level 2 Accepts not associated with stored data, the Trigger I.D. output of the present **L2RA** is compared with the value of the Level 2 Trigger I.D. Counter which is advanced by the Level 2 Strobe. If the memory-Level 2 Trigger I.D. Counter comparison is true and Level 2 Accept is true, the A/D converter is started and the **L2RA** is advanced on the end of the A/D conversion cycle. For events where Level 2 Accept is false, the **L2RA** is simply advanced on the Level 2 Strobe.

ADC

The Analog to Digital Converter is implemented as a Wilkinson run-down device with a capacitor (i_{out} discharge current ratioed (at 1/150) to the charging current (i_{in}). The capacitor voltage is then viewed by a comparator which trips when the voltage reaches the reset value. A 60 MHz counter that began counting when the discharge current began flowing is stopped when the comparator trips and the counter value is loaded into an output register. This value is then the relative time of the detector *hit* in units of about 0.2 ns per least significant bit.

Data Output

Data from the TVC/AMU prototype consists of the contents of the ADC register and the contents of the four bit Bunch Counter. The Data Present signal goes true at the beginning of the ADC cycle and the Data Valid line goes true at the end of the ADC cycle. For the production version, the data output must contain the ADC and Bunch Counter information, but also a geographic address of the *hit* wire (or element), and a temporal address in the form of the Trigger I.D. Number.

References

[1] A.E. Stevens et. al., *A Time to Voltage Converter for Colliding Beam Detectors*, IEEE Transactions on Nuclear Science, **36** No. 1, 517-21, (1989).

[2] A.E. Stevens et. al., *A Time-to-Voltage Converter and Analog Memory Unit for Colliding Beam Detectors*, Journal of Solid State Circuits, **24**, No. 6, 1748-51 (1989).

SIMULATION AND MODELLING STUDY OF AN SSC FRONT END ELECTRONICS ARCHITECTURE

K. J. Ragan
McGill University

P. K. Sinervo
University of Toronto

R. Van Berg and H. H. Williams
University of Pennsylvania

Abstract

An architecture and design for a front end electronics circuit intended for the readout of a wire tracking chamber at the SSC is described. A detailed high-level simulation has been developed and is being used to study the behavior of this circuit. Initial results from this simulation study are described.

Introduction

Proton-proton collisions at the SSC will occur at a rate of 60 MHz, but of this we estimate that the rate of "interesting" collisions will only be of order 100 to 1000 Hz. Thus, a detector will have to reduce the event rate by at least five orders of magnitude in order to obtain a reasonable readout rate of events. We also expect a typical SSC detector to have of order 10^7 channels of front end electronics, which means that these front end (FE) systems must have high channel density. In addition, they must have low power consumption and be resistant to radiation doses in excess of 1 MRad. These functional and technical requirements can only be met by electronics systems that employ the latest fabrication technologies and rather sophisticated control structures.

The behaviour of such systems is difficult to predict without detailed modelling and simulation. We have employed a high-level modelling language known as Verilog* to simulate the behaviour of a specific FE circuit designed for the readout of an SSC wire tracking detector (such as a "straw" chamber). Our simulation models the circuit by implementing each of its functional components and the control and logic signals required by each component (including signal delays). The simulation is driven by input signals from the wire chamber, the SSC beam clock and the Level 1 (L1) and Level 2 (L2) trigger system. We describe this model and report the preliminary results of a study using this simulation.

*Verilog is a product of Cadence Design Systems

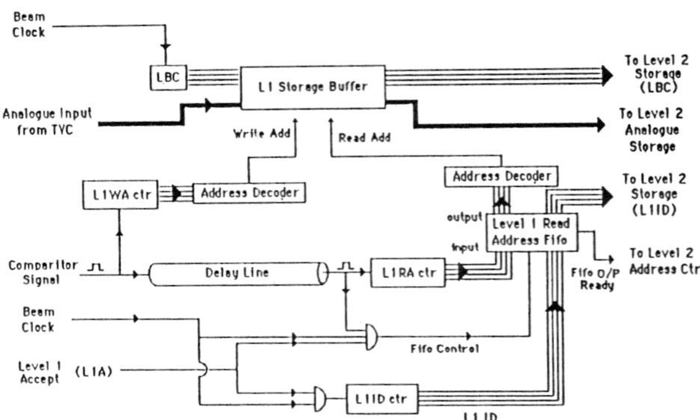

Figure 1: The block schematic of the portion of the FE circuit that manipulates the data prior to a L1 decision.

System Architecture

We show the schematic of the readout system in Figs. 1 and 2. Because of the inherently low channel occupancy of a wire tracking detector, we have taken a "data-driven" approach to buffering data on the FE circuit. The data consists of an analogue voltage signal produced by a time-to-voltage circuit (TVC) [1], which converts the time elapsed from the arrival of a hit on the wire chamber anode to the next crossing into a voltage signal. The edge of the discriminator pulse that starts the time-to-voltage conversion also signals that valid analogue data is available for storage. The circuit must store this data for a fixed delay time (on the order of several μs), during which time the L1 trigger has made a decision and a L1 Accept (L1A) signal has propogated to the FE circuit. Since the L1A signal is always delayed with respect to the interaction by a fixed number of crossings, this signal is synchronized to

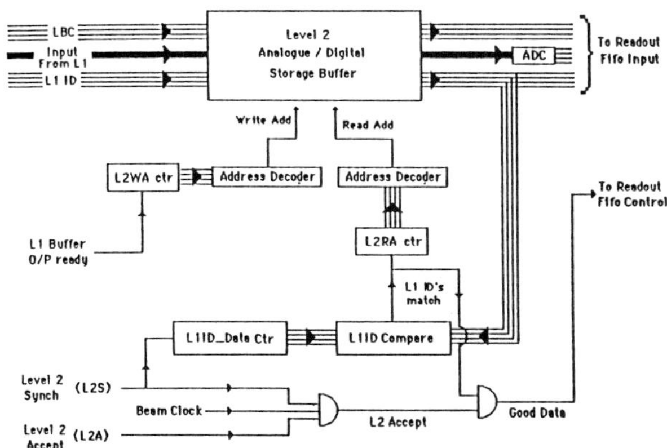

Figure 2: The block schematic of the portion of the FE circuit that manages the data prior to the L2 trigger decision.

the beam clock. This implies that only one signal is essential between the L1 trigger and the FE circuit (excluding the beam clock).

When the L1 trigger decides to accept a crossing, the FE circuit must store all data associated with the interaction that occurred on that crossing, which for a wire chamber system will be any data that is recorded by the circuit within some number of crossings of the interaction. The circuit then waits for a L2 trigger decision, which in our model may come at any time between a minimum and maximum time after the L1 trigger decision is made. When the L1 trigger rejects a crossing, any recorded data associated with that crossing are discarded by the circuit.

We expect the L2 trigger decision time to be of order 50 μs, and to vary considerably from event to event. We assume that the L2 trigger decisions will be made in the same order as the L1 trigger accepts, as otherwise the L2 trigger would have to identify to the FE circuit the L1 accept to which the L2 decision pertains; this would involve the distribution of a large number of signals from the L2 trigger to each FE circuit. The L2 trigger decision is communicated to the FE circuit using a pair of signals: the L2 Strobe signal (L2S) indicates to the circuit that a L2 trigger decision has been made, while the L2 Accept (L2A) signal is used to inform the circuit whether the L2 trigger accepted or rejected the L1 accept. The data associated with interactions rejected by the L2 trigger are then discarded by the circuit, while the data associated with a L2 accept are digitized and transferred to the data acquisition (DAQ) system.

Model for the FE Circuit Inputs

We have developed a Verilog module that provides as output a beam clock signal, a L1A signal, a L2S signal and a L2A signal. The behaviour of the L1 Accept signal is determined by two parameters: the L1 accept rate, and the L1 trigger delay. We define the L1 accept rate as the probability that the L1 trigger will generate a L1 accept for a crossing (a typical value would be 0.001). The L1 trigger delay is the number of crossings that is required before the L1 decision has propagated to the FE circuit (typically $\sim$ 120 crossings).

We specify the behaviour of the L2 trigger by (i) the fraction of those interactions passing the L1 trigger that are accepted by the L2 trigger; and (ii) the L2 trigger decision time, which in our model is a random variable that can have various distributions (we typically choose a Gaussian distributon with a mean and width of 40 μs and 25 μs, respectively. We also restrict the L2 decision time; typically we require that it be at least 5 μs and no more than 200 μs.

A separate Verilog module generates the input signal from the TVC circuit that prepares the analogue input and the hit signal. We characterize this signal by (i) the nominal hit rate, or wire occupancy, and (ii) the minimum time between hits, which is defined by the width of the signal pulse from the chamber. Typical values for these parameters are 0.05 hits/crossing and 30 ns, respectively. We note that we define the nominal wire occupancy to be the probability that a hit will occur in a given time, and does not reflect the deadtime experienced by the wire due to the finite pulse width (*i.e.*, the observed hit rate is less than the nominal rate).

Schematic of the Front End Circuit

The design of the FE circuit features (i) a L1 buffer, which stores all the analogue data during the time required for a L1 decision; (ii) a L2 buffer, which stores the analogue data associated with interactions accepted by the L1 trigger; (iii) a "buffered" L1-L2 transfer, in which the data transfer between the L1 and L2 buffers is performed by using an intermediate FIFO; and (iv) an output buffer, which holds the digitized data that will be transferred to the DAQ system. These components can be seen in the block diagrams shown in Figs. 1 and 2.

The L1 analogue buffer is implemented as a circular buffer, with a write address register (L1WA) that holds the address of the next L1 location into which data will be stored, and a read address register (L1RA) that points to the location that contains the oldest piece of data in the buffer. The TVC hit signal triggers the circuit to store the data in the next L1 buffer location and then increments the L1WA counter. The same hit signal is also fed into a delay line that stores the signal for the time required for the L1 decision.

The arrival of the L1 trigger decision is coincident with the arrival of the hit signal out of the delay line. Each hit signal increments the read address register (L1RA). If the L1 trigger accepts the crossing, each hit signal that comes out of the delay line during the time resolution of the detector is associated with the same L1 trigger and is transferred to the L2 buffer. In order to be able to associate each hit with a specific L1 trigger, it is necessary at this stage to tag each data with a L1 identification (L1ID). We do this by having a separate register that is

Parameter	Min	Max	Default
L1 Accept rate	0.0001	1.0	0.001
L1 Delay (ns)	480	3000	960
L2 Accept rate	0.001	1.0	0.02
L2 decision time (μs)	0.5	100	8
Hit rate (hits/crossing)	0.001	0.20	0.05
Minimum hit separation (ns)	20	50	30
Maximum readout time per event (ns)	16	500	50
Length of L1 buffer	8	32	16
Length of L2 buffer	2	32	4
Length of L1 Read Fifo	2	8	4
Time for writing into L2 (ns)	100	2000	100
Time for reading from L2 (ns)	100	5000	100

Table 1: The range of values used to specify the readout system.

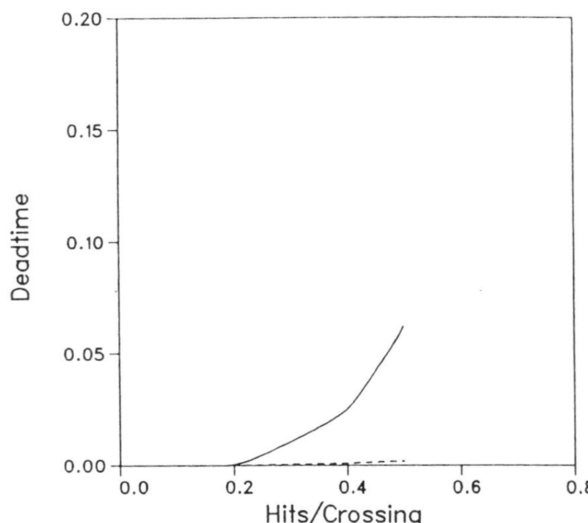

Figure 3: The deadtime of the the L1 buffer (solid line) and the L2 buffer (dashed line) in the as a function of the nominal wire occupancy (excluding the effects of the pulse width). Other system parameters are set to their default values.

incremented on each L1 accept. The data associated with a specific L1 accept is tagged with the L1ID value for that L1 trigger and this digital value is stored with each piece of analogue information that is transferred to the L2 buffer.

The data transfer from the L1 to L2 buffer is performed by storing the L1 address in the short address FIFO (L1RF). This can take place rapidly so several hits can be stored for the same interaction if necessary. The analogue data transfer to the L2 buffer can take place on a much slower time scale, even allowing for a low power analogue-to-digital conversion in the process if desired.

Simulation Results

We have begun a study of the performance of this buffered L1-L2 circuit using this simulation. We are varying the system parameters, as shown in Table 1, where we also indicate the default parameters for the readout system. Although the potential parameter space is large, it is significantly restricted by the demand that the overall L2 trigger rate be of order 10^3 Hz and that a significant fraction of the rejection be achieved in both trigger levels.

We show in Fig. 3 the measured "deadtime" of the circuit due to L1 and L2 buffer saturation as a function of the wire occupancy, keeping all other system parameters constant. We note that this deadtime (defined as the fraction of hits lost because the buffer was full) is small even for wire hit rates in excess of 0.20 hits/crossings. Since the actual wire efficiency is well below 0.90 in this case due to the width of the wire chamber pulse, the additional loss of efficiency because of the buffering on the circuit is acceptable.

The simulation executing on a SUN 4D/40 takes approximately 3 ms to perform a single 1 ns time-step (the time step we are currently using in the Verilog model). Thus, although it is possible to simulate a reasonable number of L1 and L2 triggers, the task of simulating 1 ms of actual operation requires approximately 1 hour of elapsed time.

Conclusions

We are continuing our simulation studies of this buffered L1-L2 data-driven circuit in order to fully characterize the behaviour of this circuit as a function of the salient system parameters such as input hit rate, L1 and L2 trigger rates and delays, and L1-L2 buffer transfer times. The simulation has already shown that the data-driven scheme explained here results in negligible excess deadtime for a sparsely occupied detector such as a wire tracking chamber.

We are also undertaking a simulation of the data collection network which will be responsible for collecting the data from each of the FE circuits after a L2 accept has been received. This requires us to take into account the correlations between nearby wire channels in a typical SSC event. We intend to do this by using as input to our hit generator the actual time structure of hits as predicted by a detailed simulation of the wire tracking detector in minimum-bias events.

References

[1] A. E. Stevens et al., *A Time-to-Voltage Converter and Analog Memory Unit for Colliding Beam Detectors*, IEEE Trans. Nucl. Sci., Vol. 36, No. 1 (1989).

FRONT-END ELECTRONICS FOR LONG STRAW TUBE SYSTEMS

John J. Paulos and Steven L. Blake

Department of Electrical and Computer Engineering
North Carolina State University
Campus Box 7911
Raleigh, North Carolina 27695-7911

Abstract

This paper examines the effects of signal attenuation and signal reflections due to improper termination in long, small diameter plastic straw tubes for central tracking subsystems.

Introduction

This paper addresses several critical issues in the readout of long, small diameter plastic straw tubes for central tracking subsystems. Of particular concern are signal attenuation in long straw tubes and signal reflections which arise from improper termination at the ends of the tube. This work is part of a 12 institution collaboration to design and validate a hybrid central tracking chamber (HCTC) utilizing both straw tube and scintillating fiber components. The HCTC design calls for 4 mm diameter plastic straw tubes spanning the entire central tracking region (6-8 m) with readout electronics at both ends. An electrical isolator may be used at the center of each wire to separate each tube into two electrically isolated regions so as to reduce occupancy by a factor of two. With this scheme, no track is farther than 4 m from the associated readout electronics. The HCTC collaboration includes the participation of researchers at the University of Pennsylvania who have contributed a preamplifier and shaper chip which is used in the simulations presented here. A more complete discussion of the HCTC design can be found in the paper by Dr. Alfred Goshaw.

This work supported in part by D.O.E. Contract No. DE-A505-90ER40576 and by Texas National Research Laboratory Commission Contract No. RFY9109.

Experimental Studies

An 8-layer stack of 4 mm straw tubes 2.7 m in length has been constructed and instrumented using commercial (LeCroy) readout electronics. The straw tube stack has been used to measure pulse height as a function of distance along the tube and to reconstruct tracks from cosmic rays. Pulses were recorded by ac coupling straw tubes directly into a 50 Ω scope input with a 250 Ω series resistor so as to maintain the proper termination of the straw tube transmission line (300 Ω in series with 60 pF). By varying the position of the ion source it was determined that pulse height falls by approximately 18 percent per meter, corresponding to an attenuation length of 5 m. Tracking studies from cosmic rays have been performed by Dr. Seog Oh which demonstrate tracking resolution on the order of 100-125 μm. A more complete discussion of this work can be found in the papers by Dr. Oh.

Simulation Studies

Lumped element, lossy transmission line simulations of the 2.7 m straw tubes were performed using the SPICE model of the University of Pennsylvania preamplifier/shaper chip for readout. A current pulse was calculated based on 35 primary electrons in a track at a radius of 1 mm, assuming a gas gain of 75,000 and an avalanche pulse shape of the form $t/(t+t_o)$ with t_o = 2 ns. Figure 1 shows the time-to-threshold (observed drift time) versus

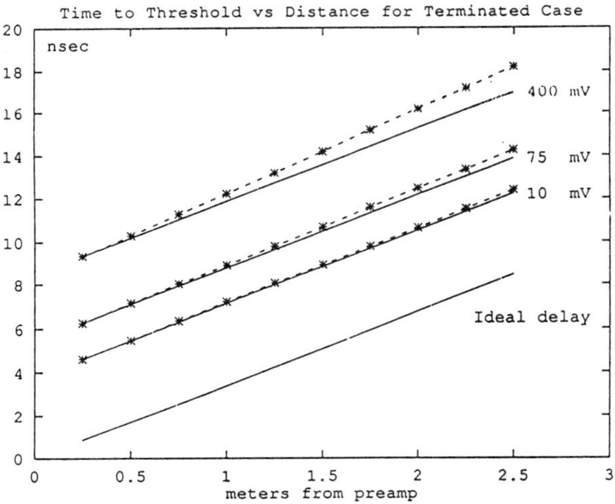

Figure 1. Time-to-threshold with termination.

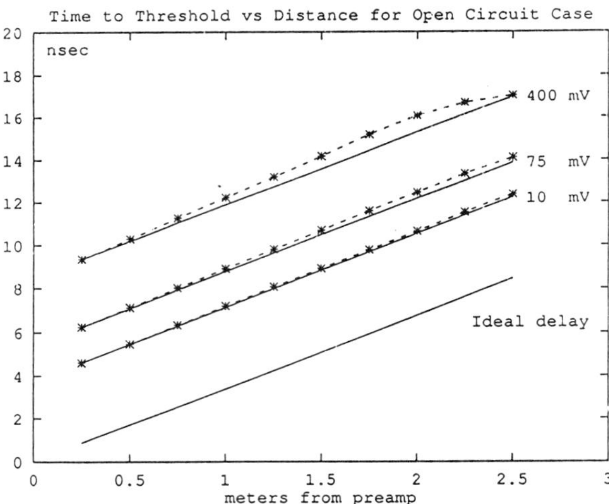

Figure 2. Time-to-threshold with open circuit.

distance from the preamplifier for a properly terminated (at both ends) straw tube for three different thresholds. The 10 mV threshold corresponds roughly to one half of the pulse height obtained for a single primary electron arriving close to the preamplifier. The 400 mV threshold is roughly one half of the total pulse height. The parallel solid lines represent the theoretical delay along the tube offset by the preamplifier/shaper delay for each threshold. The straight lines obtained at the lower thresholds indicate that there is little variation of the observed drift time as a result of signal attenuation.

Figure 2 shows a similar result for the case where the far end of the straw tube has an open circuit termination. Little change in time-to-threshold is observed relative to Figure 1 at the lower thresholds, which indicates that the reflections created by the open-circuit termination do not significantly affect the leading edge of the pulses seen by the preamplifier. Some change is seen at the high threshold for tracks near the far end of the tube where the reinforcing reflection arrives soon after the direct signal, thereby influencing pulse height. These results suggest that improper termination should have negligible effect on measured drift time for reasonable discriminator thresholds.

Double hit resolution was studied by comparing the time-to-threshold for two tracks at the same physical position separated by 40 ns in time. The results are shown in Figure 3. For the terminated case, a decrease in measured drift time of 1 to 1.5 ns is observed for a 400 mV threshold. For the open-circuit case, a threshold of 600 mV was required to keep the shift in measured drift time below 3 ns. Although these simulations included two stages of C-R shaping, we were unable to obtain reasonable double hit performance at the lower thresholds given t_o = 2 ns.) Although these simulations appear to be extremely pessimistic relative to available experimental data, they do establish the impact of improper termination on double hit resolution.

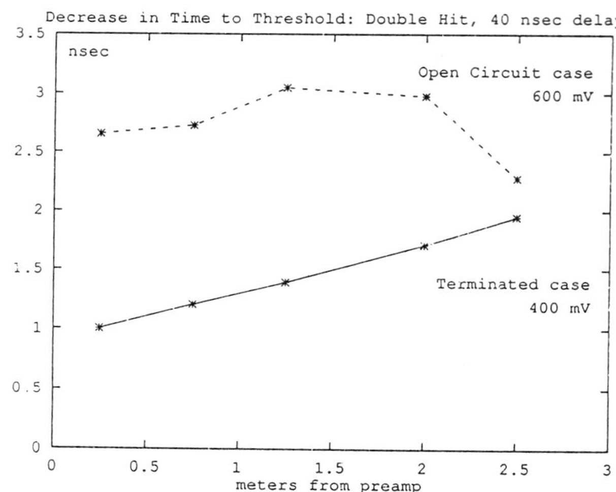

Figure 3. Double hit timing shifts.

Figure 4 shows a schematic diagram of the proposed HCTC straw tube configuration. For the outermost super layers, a continuous 8 m plastic straw tube would be strung with two 4 m gold-plated tungsten wires held together at the center with an isolating spacer. Simulations have been performed as shown in Figure 4 with the isolator modeled as a 0.1 pF capacitor. Figure 5 shows the results for a hit 1 m from preamp 1 followed 20 ns later by a hit 3 m from preamp 2. The resulting waveforms are nearly identical to those obtained when recorded singly, which indicates the absence of significant crosstalk. The distortions in pulse shape are the result of the open circuit termination implied by the 0.1 pF isolation capacitance.

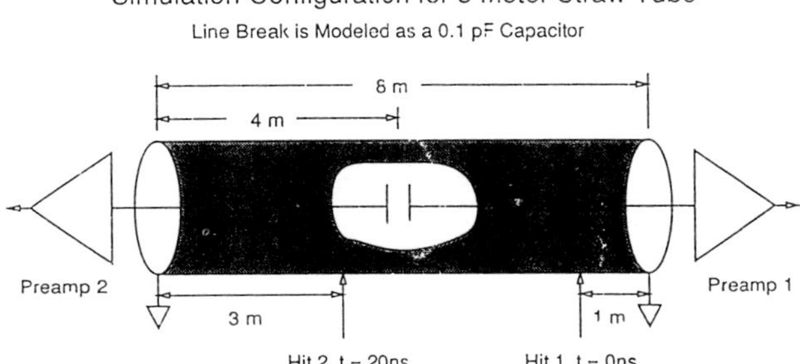

Figure 4. HCTC straw tube configuration.

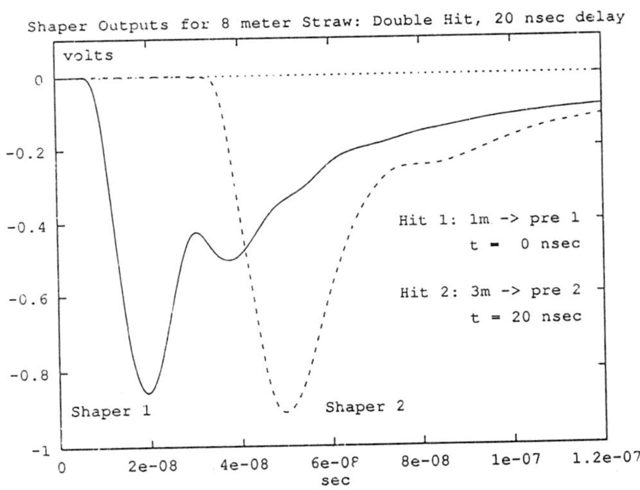

Figure 5. Shaper outputs for Figure 4.

Conclusions

Lumped element, lossy transmission line simulations have been performed which correlate at least roughly with experimental data. These simulations suggest that signal attenuation is not a significant concern in long (3-4 m) straw tube sections, and that improper termination has little effect on observed drift time. Improper termination is shown to be critical to double hit resolution. The use of 8 m straw tubes with preamplifiers at each end and with a central, isolating wire termination has been shown to be a viable approach, at least as far as signal attenuation and crosstalk are concerned.

TEST RESULTS OF A 90 MHZ INTEGRATED CIRCUIT SIXTEEN CHANNEL ANALOG PIPELINE FOR SSC DETECTOR CALORIMETRY

Stuart Kleinfelder, Michael Levi, Oren Milgrome
Lawrence Berkeley Laboratory*
1 Cyclotron Road
Berkeley, CA 94720

Abstract

A sixteen channel analog transient recorder with 128 cells per channel has been fabricated as an integrated circuit and tested at speeds of up to 90 MHz. The circuit uses a switched capacitor array technology to achieve a simultaneous read and write capability and twelve bit dynamic range. The high performance of this part should satisfy the demanding electronics requirements of calorimeter detectors at the SSC. The circuit parameters and test results are presented.

Introduction

The electronics for signal detection, event selection, and data recording for an SSC detector will be one of the most sophisticated electronics systems ever built. Signal processing must be performed on one trillion analog samples per second. The general task to be accomplished is the temporary storage, followed by slower readout, of brief but high speed analog transients. Acquisition with sampling frequencies up to the beam crossing rate of 62.5 MHz are needed. Usually, a trigger decision of a 1-2 microseconds latency is made to judge the value of the stored signals. These signals are then either discarded or digitized, processed, and transferred to computer systems. Thus, these systems usually require time delay via temporary storage and time stretching of the stored signal to match the bandwidth of the data processing system.

Analog Memories For Pipelined Storage

Traditional solutions to this problem have involved the use of flash analog to digital converters (FADC's) followed by digital memory, or use of charge coupled device (CCD) delay lines. The switched capacitor approach offers lower cost, lower power and higher density while providing fast sampling speeds, wider dynamic range and a higher degree of flexibility than these traditional techniques. The high density CMOS processes used are suitable for inclusion of greater signal processing power and integrated circuits (IC's) may ultimately include on-chip preamplification, signal shaping, analog to digital conversion and sophisticated addressing schemes including sparse or selective data readout and trigger buffering and selection.

The switched capacitor array (SCA) will allow for the retention of data during trigger decision. The length of the array will provide for 128 memory cells per channel of electronics, allowing up to 2 microseconds of analog information to be stored at the SSC. Three types of analog memory cells are being pursued. One of these, a single-ended single ported analog pipeline, has been previously reported.[1] A second type, with simultaneous read-write capability has recently been fabricated, the test results from this device are reported here. A third type, with fully differential inputs is in development but will not be discussed in this report.

Circuit Description

The SCA (Switched Capacitor Array) integrated circuit uses a two micron double metal CMOS process with two layers of polysilicon forming the high density capacitors prototyped through MOSIS. It has 16 parallel analog signal inputs and 16 parallel analog outputs. Two digital clock inputs, a reset control input, and two address busses are used to operate the circuit. Fig. 1 shows the device concept. The analog memory being developed permits simultaneous read and write operations providing deadtimeless operation of the memory.

The circuit contains 2048 sample and hold cells subdivided into 16 parallel channels of 128 cells per channel. Each of the 16 channels has a dedicated analog input which is connected to a bus distributing the input signal to 128 sample and hold cells. Each sample and hold cell consists of a complementary CMOS transmission gate (complementary switch) and a 1 pF double polysilicon capacitor. The capacitors use a high quality silicon-oxide dielectric of 70 nm thickness. An externally supplied reference voltage is applied to the bottom plate of all capacitors. The voltage stored on each sample and hold capacitor then corresponds to the difference between the input signal from a given channel at the time its sample and hold switches are opened and the applied reference voltage. The reference voltage can be adjusted to shift the baseline for better level matching.

An address decoder turns on and off sample and hold switches using a "break before make" action. "Break before make" insures that no charge sharing between subsequently engaged capacitors can occur. During the readout sequence each sample capacitor is placed in the feedback path of the reconstruction op-amp. There is one reconstruction op-amp per channel. Before readout of each sample, stray charge on the signal distribution busses are cleared by shorting each amplifiers inverting input nodes and output nodes through a reset switch. This prevents charge remaining on parasitic nodes from the previous samples from influencing the next read out sample.

* This work has been supported by the Director, Office of Energy Research, Office of High Energy and Nuclear Physics, Division of High Energy Physics, of the U.S. Department of Energy under Contract No. DE-AC03-76SF00098.

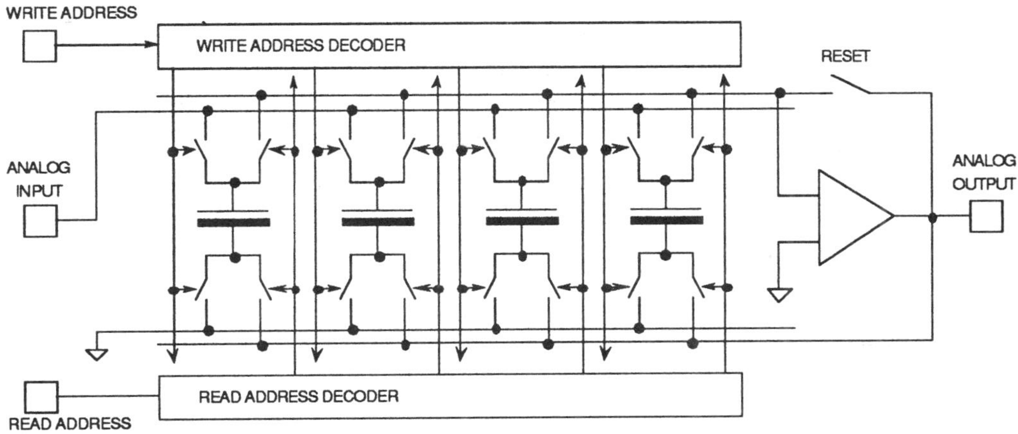

Fig. 1. A simplified block diagram of a single ended dual-ported analog memory cell structure as discussed in the text. The voltage presented at the input is stored as a charge on each capacitor in a time sequence. The selective reading and writing to and from a capacitor element is controlled by four CMOS switches. The reset switch is closed momentarily prior to each read operation. The input and output operations of the analog memory are completely separate and allow for full simultaneous read/write capability.

SCA Circuit Performance

Non-uniformities in the sample and hold elements limit dynamic range unless corrections are applied. The double polysilicon capacitors, although of excellent quality, are not likely to be uniform to better than 10 bits, but the SCA readout scheme is not sensitive to capacitor values to first order. Non-uniformity of the sample and hold switches can be important, as the charge injection from these switches (due to gate to drain capacitance) will then vary. These switches are complimentary, and therefore the charge injection is canceled to first order, but variations still exist. In addition, the rise and fall times of the clocks are very fast, and slight variations in clock skews (imperfect alignment of the complementary clocks) can also lead to imperfect cancellation.

In order to achieve 12 bit dynamic range the pedestal variation of each of the storage elements in a channel must be less than one part in 4096 of the full scale voltage range. A channel operating with a full scale range of 4 volts would then require less than 1 mV of pedestal variation per element.

To achieve 12 bit dynamic range 8.3 charging time constants must elapse in each 16 nsec write operation to charge the sample and hold capacitors, this fast input bandwidth requires a large input charging switch. For increasing switch sizes the amount of parasitic charge injection from the switch will also increase. In the version tested, with a 2 nsec charging time constant, the cell to cell pedestal variation at a 50 MHz acquisition rate was 0.4 mV rms. The cell-to-cell pedestal variation is shown for a single channel in Fig. 2.

The on-chip rail-to-rail readout operational amplifiers has been developed based on a published design.[2] The op-amp operates to within 100 mV of either voltage rail, thereby maximizing the useable dynamic range. The linearity of this op-amp, 3 mV out of a 6 volt range or 0.05%, substantially exceeds the intrinsic accuracy of any large calorimeter system under consideration. The amplifiers must drive the parasitic bottom plate capacitance of all sample capacitors (approximately 10 pF) which limits the readout rate. The overall linearity of the SCA has been limited by a layout error which will be rectified in subsequent designs.

The measured incoherent noise on any given single cell when operated with a 62.5 MHz acquisition rate is less than 0.6 mV rms. Thus, the total dynamic range (single cell) could exceed 8000:1, or 13 bits. In previous versions of the IC the clocking noise feedthrough was well below one millivolt; however, in the current version inadequate separation of the on chip digital and analog power lines have increased the clocking noise feedthrough to several millivolts. The clocking noise is absent in a differential manifestation of the IC and is expected to be absent in the next corrected version of the IC. The measured parameters of the SCA are summarized in table 1.

Table 1. Measured performance parameters of a simultaneous read-write switched capacitor array integrated circuit. Measurements were performed at a 50 MHz acquisition rate, but not utilizing the simultaneous read write capability.

Analog Store Parameters	Achieved Performance
No. channels/chip	16
No. of elements/channel	128
Power consumption/channel	23 mW
Non-linearity	+/- 0.9%
Cell to cell pedestal variation	0.3 mV rms
Charging time constant	2 ns
Channel dynamic range	4000
Maximum sample rate	90 MHz
Maximum readout rate	200 KHz
Noise floor	0.7 mV rms
Input voltage range	0.5-5.5 Volts
Capacitor droop rate	0.1 mV/msec
Output settling time (0.02%)	1.2 usec
Gain variation from unity	+/- 1.5%

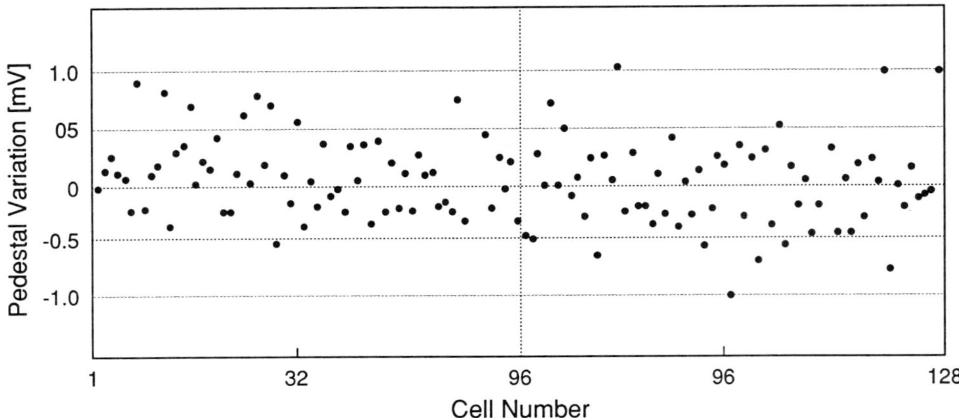

Fig. 2. Cell-to-cell pedestal variation as a function of cell number across a single channel. The small pedestal variation will permit large dynamic range operation of the circuit without corrections.

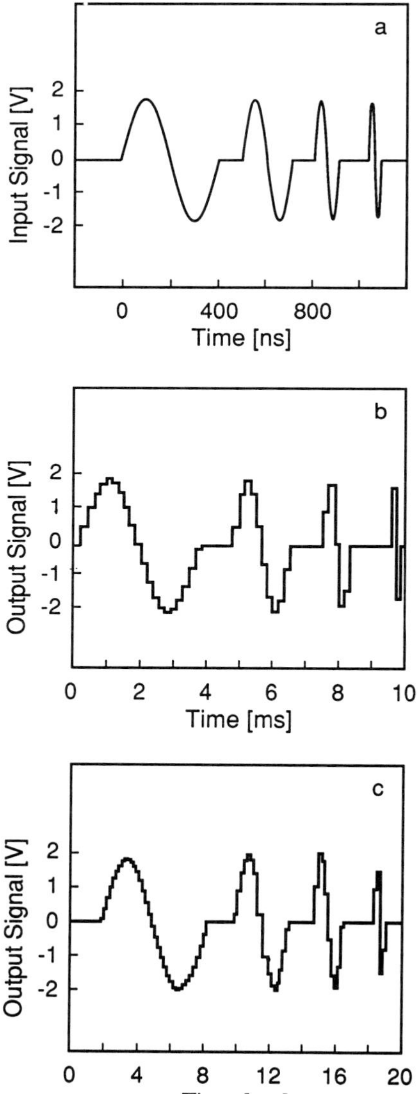

Fig. 3. In (a) is shown the input signal, 400, 200, 100, and 40 nsec period sinewaves of duration equal to the period. In (b) and (c) is shown the SCA output running at an acquisition frequency of 50 MHz and 90 MHz, respectively.

Operation of the integrated circuit has been verified on a sample of test pulses simulating the SSC calorimeter detector signals. Shown in Fig. 3 are a sequence of successively faster single cycle sinewaves, the pulses simulate the signals seen from the slowest to the fastest of the considered calorimeter detectors. Shown in (a) is the input pulse train, a 400 nsec, 200 nsec, 100 nsec and 40 nsec period sinewave. In (b) and (c) are the signals as sampled and recorded by the SCA at 50 MHz and 90 MHz acquisition rates, respectively. For illustrative purposes the readout rate has been slowed so that the comparatively short reset phase (1.6 usec) cannot be seen in these figures. The figures have been traced from AC coupled oscilloscope photographs.

Conclusions

A high accuracy analog pipeline has been implemented and is able to operate at higher speeds than those required for the SSC. Because of the low noise, pedestal variation, and rail-to-rail voltage range, 12-13 bit performance should be possible. Noise due to clock feedthrough should be corrected in the next version of the IC. At that point the chip will be fully capable of meeting the performance requirements for SSC calorimetry. Work is continuing in developing test circuits and procedures to study the chip while fully utilizing the simultaneous read-write capability of the device.

References

[1] S. Kleinfelder, IEEE Trans. Nucl. Sci **NS-37** (1990) 56. S. Kleinfelder, M. Levi, O. Milgrome, Toward a 62.5 MHz Analog Virtual Pipeline Integrated Data Acquisition System, LBL-29608, submitted to Nucl. Instr. Meth. W. Buttler, *et.al.*, Design and Performance of a 10 MHz CMOS Analog Pipeline, DESY 88-092, July 1988.

[2] J. Babanezhad, IEEE Journal of Solid State Circuits, **23** (1988) 1414.

BIPOLAR MONOLITHIC PREAMPLIFIERS FOR SSC SILICON CALORIMETRY*

C. L. Britton, Jr., R. A. Todd, and M. L. Bauer
Oak Ridge National Laboratory
P. O. Box 2008, Oak Ridge, TN 37831-6006

E. J. Kennedy
Dept. of Electrical and Computer Engineering
The University of Tennessee
Knoxville, TN 37996
and Oak Ridge National Laboratory

W. M. Bugg
Dept. of Physics
The University of Tennessee
Knoxville, TN 37996

Abstract

This paper describes preamplifiers designed specifically to address the requirements of silicon calorimetry for the Superconducting Super Collider (SSC). Eight different preamplifiers designed for detector capacitances ranging from 20 pF to 500 pF and operating temperatures from 25° C to -20° C are discussed. The preamplifiers were fabricated with two different high-frequency processes (one with the VTC, Inc. VJ900 process, seven with the Harris Semiconductor VHF Process). The different topologies and their features are discussed in addition to the design methodologies employed. The results for noise, power consumption, speed, and radiation damage effects as well as data for post-damage annealing are presented for the VTC process preamplifier. Simulations for the VHF Process circuits are presented. This work was funded through SSC Generic Detector funding, SSC Detector Subsystem funding, and the Oak Ridge National Laboratory (ORNL) Detector Center.

Introduction

Preamplifiers for SSC silicon calorimetry involve requirements that are common to other forms of calorimetry and some that are unique. Among the requirements are a need for low charge sensitivity, fast rise times, high slew rate, low noise, radiation hardness to > 5 Mrad, allowance for increasing detector leakage, low power consumption, and the fewest number of power supply connections possible. Developments undertaken at ORNL attempted to address these requirements with several preamplifier designs fabricated by two silicon foundries - VTC, Inc. and Harris Semiconductor. Both processes were dielectrically isolated complementary bipolar processes with transition frequencies in excess of 1 GHz. This paper describes the circuit topologies chosen, measured results of the circuit fabricated through VTC, and simulated results of the circuits being fabricated through Harris. All circuits presented have either been fabricated or are currently in fabrication. Both pre- and post-radiation measurements of the devices used in both the VTC and Harris processes are presented elsewhere in this conference[1].

Circuit Fabricated Through VTC

The initial design philosophy of this preamplifier was to develop a circuit for use with a silicon detector of ~ 100 pF capacitance and exhibit a noise of < 7500 rms electrons (erms) for a CR-RC peaking time of ~ 100 ns. The rise time was to be < 10 ns to allow the use of shorter peaking times if desired. The power dissipation was required to be < 10 mW, charge sensitivity was to be ~ 0.2 V/pC, and maximum charge input was to be 17 pC which corresponds to a 3000 MIP event. The power supply rails were chosen to be ± 3.5 V for reduced power dissipation.

Current Feedback Preamplifier Topology

The topology chosen for the first preamplifier development was that of current feedback (CF) described by Comlinear Corporation[2] and Kennedy[3]. Current feedback has many advantages when compared to more traditional topologies such as differential input[4] or grounded source (grounded emitter)[5,6,7]. Among the advantages are short settling time, excellent loop stability at low closed-loop gain, and high bidirectional slew rate for low static power dissipation. The major drawback is that the noise due to input bias current is larger than either of the other topologies by approximately the β of the input device. With the short peaking times necessary for the SSC, the current noise should be acceptable.

One of the design goals of this project was to allow the detector to be directly coupled to the preamplifier, thus precluding the need for a bulky external coupling capacitor. The current flowing from the detector into the input and, subsequently, the feedback resistor of the preamplifier shifts the preamplifier dc bias point as the detector current increases with progressive detector radiation damage. The shift in bias point causes two problems. The first and most apparent is that the output of the preamplifier will move closer to the negative supply rail. This reduces the

*Research performed at Oak Ridge National Laboratory, operated for the U. S. Department of Energy under contract DE-AC05-84OR21400 with Martin Marietta Energy Systems, Inc.

maximum output signal amplitude and, therefore, the dynamic range that can be tolerated prior to overload. The lower supply voltages required for low power dissipation only compound the problem. The second and less obvious is that as bias point changes, the preamplifier gain changes slightly because of the nonlinear nature of the transistors, resistors, and capacitors comprising the circuit. This condition is normally described as differential nonlinearity[8]. One desirable feature of a preamplifier would be to provide a method of restoring the dc operating point with minimal effect upon the impulse response, noise, or power dissipation of the circuit. The complete circuit, a block diagram of which is shown in Fig. 1, employs a gated rebalance amplifier (error amp) that samples the output for 10 μs to 50 μs every 40 ms and restores the preamplifier output to a value equal to the bandgap reference voltage allowing the preamplifier to continue to operate within its optimum range of performance, despite increases in detector leakage current. The particular implementation shown here will correct for up to 25 μA of detector leakage current. Table 1 summarizes the pre- and post-radiation performance results.

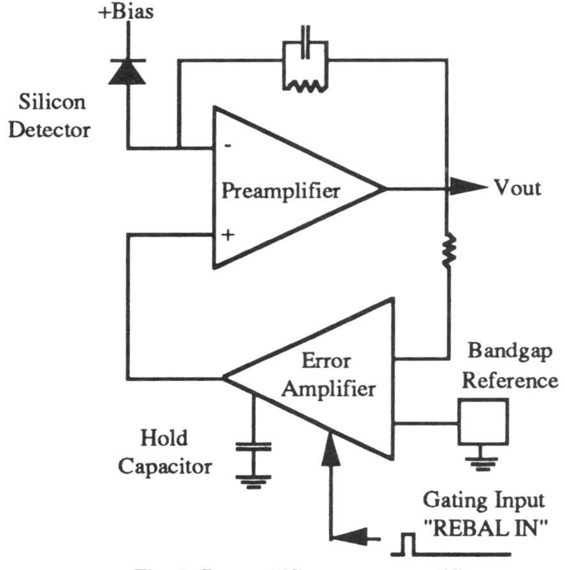

Fig. 1 Preamplifier - error amplifier Block Diagram

Table 1. Results of Tests (C_d=100pF)

Condition	Risetime	Power (+/-3.5 V)	ENC_e (60 ns Peaking)
Pre-rad	6.3 ns	7.5 mW	6300 erms
5.3 Mrad	20 ns	2.1 mW	6600 erms
5.3 Mrad 4 hrs. 100°C anneal	9 ns	5 mW	6800 erms

Circuits Fabricated Through Harris

Seven preamplifier designs for silicon calorimetry were implemented in a dielectrically isolated bipolar VHF Process using the Harris Semiconductor FASTRACK design system. The process used 1.2 GHz npns and 1.0 GHz pnps with breakdown voltages of ~ 20 V. All preamplifiers were charge-sensitive designs with feedback components on the chip. Each packaged chip housed four identical preamplifiers in a 16-pin, ceramic, leadless chip carrier. Three topologies were implemented with designs tailored for detector capacitances ranging from 20 pF to 500 pF. Power dissipation on all designs was typically < 20 mW for ±5 V power supplies. The designs featured low noise performance, radiation hardness, low power dissipation, wide dynamic range, minimum external parts count, and direct-coupled input connections. The dimensions and pad placements of all chips were physically and functionally identical to streamline wafer probe testing on the multi-project wafers.

Folded Cascode topology

Three quad preamplifier circuits were designed with a folded cascode input topology, having been optimized for detector sizes of 1 x 1, 2 x 2, and 3 x 3 cm^2 fully-depleted 400 μm silicon. The folded cascode topology exhibits the lowest input noise for given input device size and collector current of the three topologies. The output dc voltage was offset toward the positive supply voltage, V_{CC}, to maximize the useable dynamic range of the negative-going pulses using high-valued pinch resistors in the feedback network. A resistor was connected in series with the emitter of the input transistor in the case of the lowest detector capacitance (the 1 x 1 cm^2 design) to stabilize the circuit at the expense of additional noise. A feedforward path using a p-channel JFET was added from the input node to the dominant node to improve large signal response and to establish the amplifier's dc bias currents. The JFET was preferred over a pnp device because of its low gate leakage current when compared with the base current of the pnp after radiation exposure.

Current Feedback Topology

Two quad preamplifier circuits were designed with a current feedback topology, having been optimized for detector sizes of 1 x 1 and 2 x 2 cm^2 fully-depleted 400 μm silicon. The circuit was similar to the VTC chip without the bandgap reference circuit; the positive input node was connected to an internal bias voltage. Input device emitter currents were established using a p-channel JFET based current source with improved current mirrors to stabilize bias currents under post-radiation, low-beta conditions. Again, the output dc voltage was offset toward the positive supply voltage, V_{CC}, to maximize the useable dynamic range of the negative-going pulses. All feedback components are implemented on the chip with only external bypass capacitors at the positive input nodes required.

Differential Topology

Two quad preamplifier chip designs using the differential topology were implemented for detector sizes of 3 x 3 and 4 x 4 cm^2. For 400 μm detector thickness, these sizes correspond to 243 pF and 432 pF detector capacitances. Among the advantages of the differential topology are the

Table 2. Simulation results for the charge-sensitive bipolar preamplifiers using Harris' VHF process. C_d is detector capacitance for 400 um thickness, fully-depleted silicon; ENC_e (equivalent noise charge) is given in rms electrons at -20° C including ballistic deficit effects for a CR-RC peaking time of 50 ns; P_d (power dissipation) is given for ±5 V supplies; t_r (rise time) given for 10 mV output signal.

Preamp Type	Folded Cascode			Current Feedback		Differential	
Detector Size	1x1	2x2	3x3	1x1	2x2	3x3	4x4
C_d (pF)	27	108	243	27	108	243	432
ENC_e	4152	5001	7868	7619	9023	13,750	19,200
P_d (mW)	18.2	19.5	19	15	15	23.5	23.5
t_r (ns)	3.5	3.3	5.4	2.4	6.7	5.6	4.2
Ext. Cap. ?	NO	NO	NO	YES	YES	YES	YES

excellent dc bias resulting from a symmetrical design, and the high loop transmission possible with large detector capacitances. The series noise performance is dominated by the two input devices and as such is higher than in the folded cascode case for equivalent device sizes and emitter currents. Emitter resistors were added in the 3 x 3 cm^2 case for stability without degrading the large signal slew rate. The positive input node reference voltage is established from an internal bias network with an external 0.01 μF bypass capacitor to ground. The collectors of the input differential npns connect to pnp folded cascode stages with an improved current mirror in one path for bias stability under post-radiation conditions.

Performance Comparisons

Based on simulations using device models provided by Harris Semiconductor, Table 2 lists noise performance, transient response, and power dissipation results for the three circuit types as applied to the various detector sizes. The folded cascode topology is clearly the design of choice: the noise performance is superior, no external reference bypass capacitors are required, power dissipation and output dynamic range are comparable to the other designs, and component count (thereby required chip area) is lower indicating more circuits will be produced per wafer. While the speed of the current feedback design is superior, the response of the folded cascode design is more than adequate for the 7 ns and 22 ns respective electron and hole collection times anticipated for the 400 μm thick silicon detectors when operated at -20° C.

Conclusion

Eight preamplifiers were developed for silicon calorimetry in two different high-frequency bipolar processes. Three different topologies were implemented to evaluate the performance of each under the target conditions. The preamplifier fabricated through VTC was tested, while the preamplifiers currently being fabricated through Harris have been thoroughly simulated.

References

[1] E. J. Kennedy, et. al., "Radiation Effects On JFETS, MOSFETS, and Bipolar Transistors As Related To SSC Circuit Design", To be presented at this conference.

[2] Application Note, "Current-Feedback Amplifiers", 1989 Comlinear Databook, pp. 12-3 through 12-8.

[3] John Williams [Ed.], Analog Circuit Design: Art, Science, and Personalities, Butterworth Publications, 1991 (to be published).

[4] Paul R. Gray and Robert G. Meyer, Analysis and Design of Analog Integrated Circuits, pp. 194-211, John Wiley and Sons (1984).

[5] V. Radeka, "Signal, Noise, and Resolution in Position-Sensitive Detectors", IEEE Trans. Nucl. Sci., NS-21 (Feb. 1974), pp. 51-64.

[6] Neven Karlovac and Terry L. Mayhugh, "A Fast Low-Noise Charge Preamplifier", IEEE Trans. Nucl. Sci., NS-24, No. 1, Feb. 1977 (327-334).

[7] Charles L. Britton and Thomas H. Becker, "Design and Characterization of a Low Power Consumption, HPGe Spectrometer", IEEE Trans. Nucl. Sci., NS-32, No. 1, Feb. 1985 (36-40).

[8] Glenn F. Knoll, Radiation Detection and Measurement, p. 659, John Wiley (1989).

MONOLITHIC APPROACH FOR SOLID-STATE DETECTOR ELECTRONICS: DESIGN IMPLICATIONS

Guy Vanstraelen
Imec vzw, Kapeldreef 75, B-3001 Leuven, Belgium

Abstract

The monolithic integration is obtained using a p-well CMOS technology in which the p-channel devices are located in the high-resistivity silicon. The latter has enormous implications on their behaviour, as compared to standard devices. Based on an in depth evaluation and modeling of both their current-voltage and noise characteristics, some important design considerations are given.

Introduction

It is well known that transistors on high-resistivity silicon (HR-Si) suffer from punch-through, even for relative large gate lengths. A first consequence of this is a source-to-drain leakage current, which can only be eliminated by applying a large enough substrate bias [1]. A second result is a highly sloped I_{ds}-V_{ds} characteristic in the saturation region, indicating a low output resistance, even after total elimination of the punch-through current. This effect is modeled physically and analytically, and the implications for the performance of some often used circuits is investigated. A second important characteristic of these devices, their noise behaviour, was investigated in detail in both the off and the on region, and it leads to some important conclusions regarding the bias conditions under which a minimum noise is obtained.

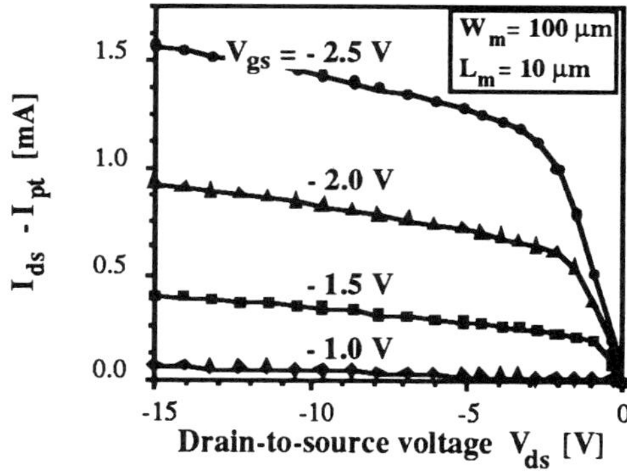

Figure 1. Pure surface current as a function of V_{ds}.

Sloped Saturation Region

Figure 1 shows the drain-to-source current I_{ds} corrected for the punch-through current I_{pt} as a function of V_{ds}. It can be noticed that the saturation region for this pure surface current is still characterized by a large slope. The basic reason is seen in the MINIMOS simulation result in figure 2, for a 3 μm length device, with V_{ds} = -3 V, V_{gs} = -1.5 V and V_T = -0.8 V.

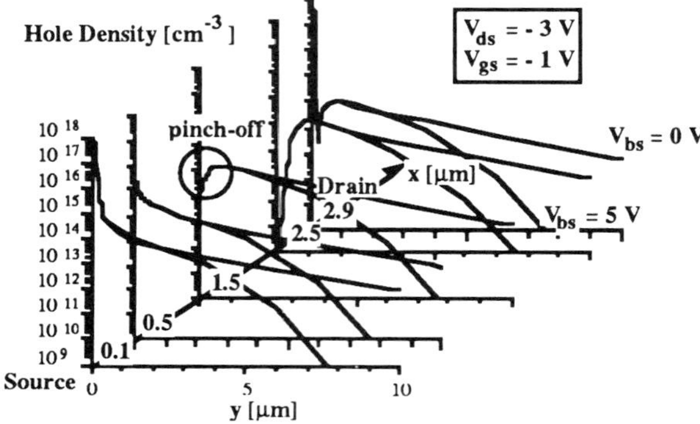

Figure 2. Carrier concentration along the channel for a p-channel device on HR-Si.

In a standard device the increase of I_{ds} with V_{ds} in the saturation region is due to a decrease in effective channel length due to the penetration of the drain depletion layer into the channel. Carriers originating from the inverted source side of the channel are swapped over this depleted part. For HR-Si, figure 2 indicates that the hole carriers leave the surface at the channel pinch-off point, and join the bulk punch-through channel, which remains present, even with application of a large V_{bs}, due to the 2-D situation. Because of the existence of this low resistive alternative current path, the channel will never be pinched off completely, but the surface part of the channel keeps operating in a quasi-linear mode, which can be described by [2]:

$$I_{ds} = I_{ql}(1 - \lambda_{ql} V_{ds}) \tag{1}$$

It is found that the parameters in this formula are a function of the device length L and V_{gs}. In figure 3 is depicted the output resistance $r_o = |\partial I_{ds}/\partial V_{ds}|$ calculated from (1), as a function of L. Also shown is $1/g_m$ for these p-channel devices and r_o for the n-channel devices resulting from the same process. Important conclusions are: i) for e.g. a 5 μm device, r_o is reduced by a factor of 10^5 for the pMOS as compared to the nMOS ii) $1/g_m$ is always larger than r_o iii) r_o changes more than proportional with L, meaning that e.g. a 10 μm device has a larger r_o that the equivalent r_o of two 5 μm devices connected in series.

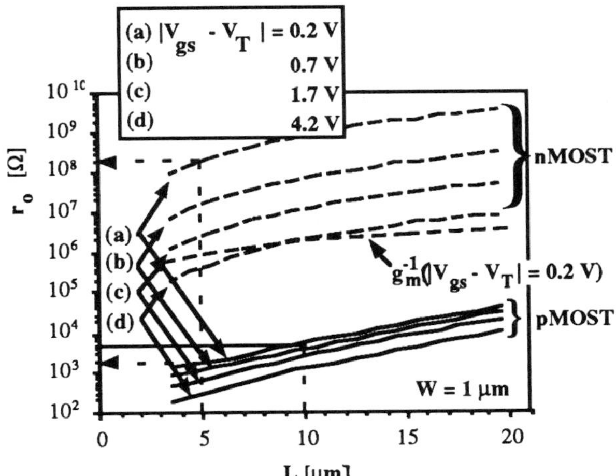

Figure 3. Output resistance as a function of device length L.

Implication for Circuits

Due the fact that $1/g_m > r_o$, the cascode amplifiers shown in figure 4a, do not benefit from the presence of the common gate device. Instead, output resistance and amplification are determined by the low resistive p-channel devices.

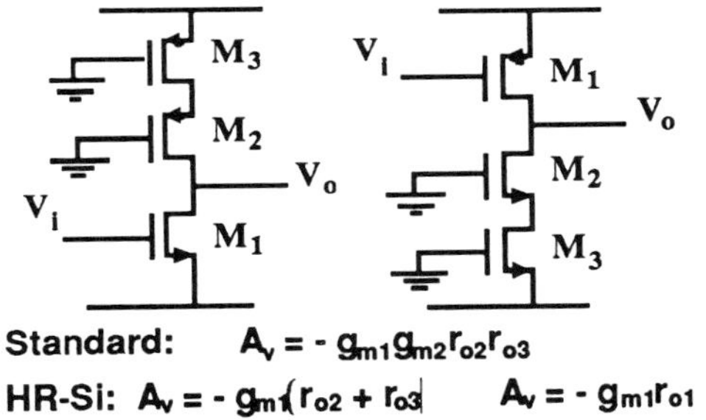

Figure 4a. Implication of the low r_o for cascode amplifiers.

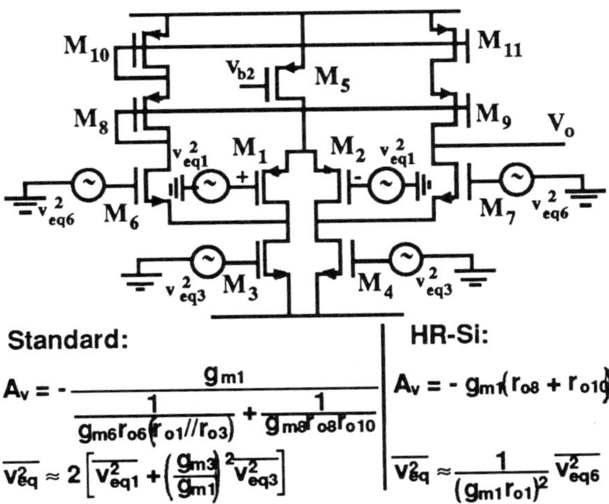

Figure 4. Implication of the low r_o for amplification and noise of a folded cascode.

A similar effect is found for the folded cascode of figure 4b. Also the noise performance is affected strongly. Whereas normally the equivalent input noise voltage of the folded cascode is determined mainly by the low noise p-type input devices M1-2, and load devices M3-M4, on HR-Si, the n-channel devices M6-7 become determining, and their noise is amplified towards the input by the factor $1/(g_{m1} r_{o1})$.

P-channel Noise Performance

Noise model

The small signal circuit equivalent for the p-channel devices is shown in figure 5. Beside the transconductances $g_{ms,g} = |\partial(I_{ds}-I_{pt})/\partial V_{gs}|$ and $g_{mpt,b} = |\partial I_{pt}/\partial V_{bs}|$, resistances $r_{o,s}$ and $r_{o,b}$, corresponding to respectively the surface and the punch-through channel, and the capacitances, it contains a current source representing thermal noise due to the channel and substrate resistance [3]:

$$\overline{i_t^2} = 4kTG_{eq}df \tag{2}$$

$$G_{eq} = \frac{2}{3} g_{ms,g} + R_{sub} g_{mpt,b}^2$$

and a current source representing the 1/f-noise contribution of both conduction channels, which can be considered to be independent. The surface part can be modeled as an equivalent input noise voltage, depending on the oxide capacitance C_{ox}, the gate area WL, and a technology factor K_f [4]:

$$\overline{v_{eq}^2} = \frac{K_f}{WLC_{ox}^2} \frac{df}{f} \tag{3}$$

In most cases an additional current dependence ($\sim I_{ds}^a$, $0.5<a<2$) has to be taken into account.

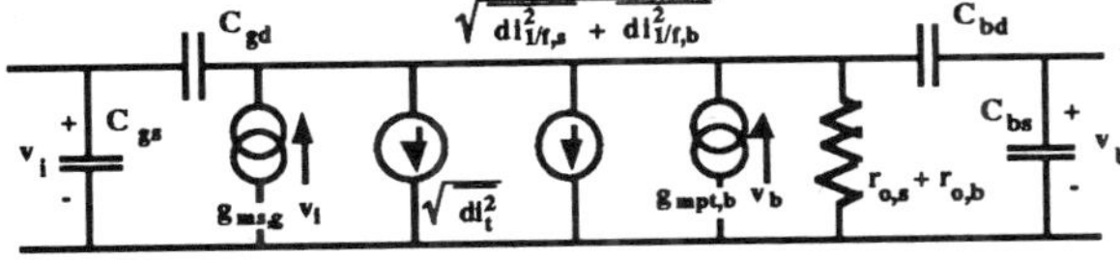

Figure 5. Small signal equivalent for the pMOST.

Experiments

The noise voltage spectral density at the drain S_{vo} [V^2/Hz] was measured for devices with W/L = 2000/5 and 2875/5, as a function of V_{ds} and I_{ds}, in both the off and the on region. The results for both device types were found to be in agreement. Figure 6a and b show the 1/f noise as a function of I_{ds}. In this figure, S_{vo} and S_{io} are the drain noise voltage and noise current spectral density, while S_{vi} is the equivalent input noise voltage spectral density. A comparison of figures 6a, b leads to the following conclusions: i) Both the off region and the on region at large currents are characterized by a low S_{io}. Use as low noise variable resistor is possible. ii) A low S_{vi}, making the device suited as amplifier input transistor is found for large currents, when the surface current is much larger than the punch-through current. iii) S_{vo} is independent of the operation mode, hereby indicating that the origin of the noise is in both cases the same.

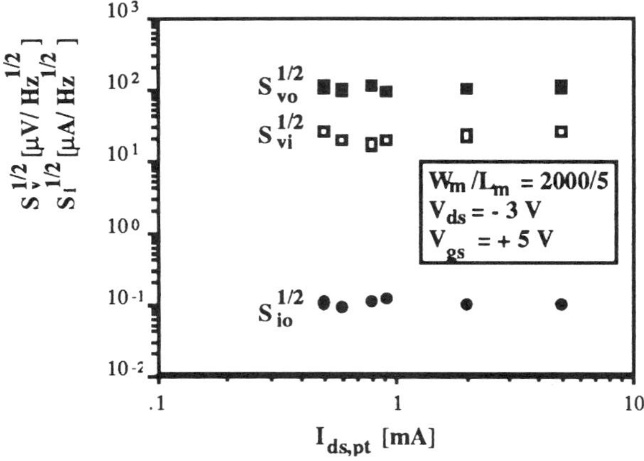

Figure 6a. 1/f noise characteristics for the pMOST in the off region.

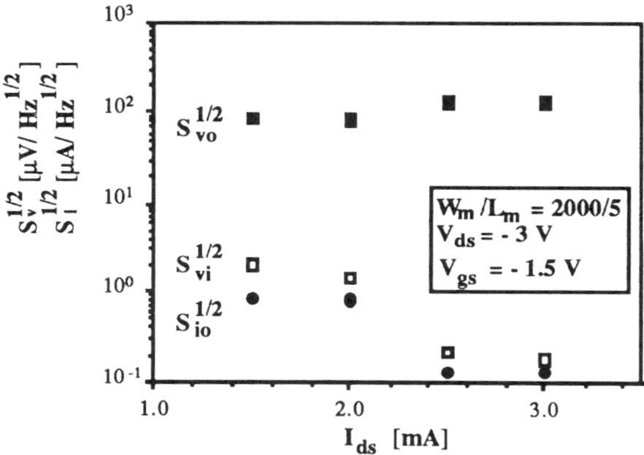

Figure 6b. 1/f noise characteristics for the pMOST in the on region.

I_{ds} [mA]	K_f [fV^{2}cm^2]	ENC [e$^-$ rms]
1.5	5.1x10^{-30}	140
2.0	2.5x10^{-30}	100
2.5	6.6x10^{-32}	19
3.0	7.7x10^{-32}	20
5.0	11.5x10^{-32}	24

Table 1. 1/f noise characteristics for the on mode; V_{ds} = -3 V, V_{bs} = 0 V.

Table 1 resumes the 1/f noise characteristics for the on region by giving the K_f factor and ENC.

For frequencies in excess of 100 Hz, the noise is found to be white. For both off and on mode the same S_{vo}-values are found. Comparing the corresponding S_{vi} of 1 µV/Hz$^{1/2}$ with the calculated thermal channel noise of ≈ 3 nV/Hz$^{1/2}$ allows the conclusion that the high value measured can be only due to the substrate resistance.

Conclusions

In this paper it is shown for pMOS transistors on HR-Si that a real saturation operation is never reached, but that they keep operating in a quasi-linear mode, due to the 2-D nature of the potential profile. Based on an analytical current model for the quasi-linear mode, it is shown that if the p-channel length is not carefully chosen, the active operation of the device in analog circuits is lost completely due to the low output resistance.

The second topic investigated is the p-channel noise behaviour. Experiments lead to the conclusion that a low 1/f noise in on mode is only guarantied if the surface current is much larger than the punch-through current. In off mode the device can be used as a low noise resistor. Furthermore it is found that the white noise is due to the substrate resistance, instead of the channel resistance.

References

[1] G. Vanstraelen, I. Debusschere, C. Claeys and G. Declerck, "New Concepts for Integrated Solid State Detector Electronics", Nucl. Instr. and Meth., A273, 625-629, (1988).

[2] G. Vanstraelen, E. Simoen, C. Claeys and G. Declerck, "Short Channel pMOST's in a High-Resistivity Si Substrate: I - Analytical Model", to be published in IEEE Trans. Electron Devices.

[3] E. Beuville, "Analyse du Bruit des Transistor a Effet de Champ en Technologie Complementaire MOS et Application a la Realisation d'un Amplificateur Integré Sensible au Courant", Thesis Univ. Louis Pasteur, August , 1982.

[4] W. Sansen, "Integrated Low-Noise Amplifiers in CMOS Technology", Nucl. Instr. and Methods, A253, 427-433, (1987).

INVESTIGATION OF ACOUSTIC CHARGE TRANSPORT DEVICES

H. J. Lubatti
Department of Physics FM-15
University of Washington
Seattle, Washington 98195
Representing the ACT delay line group*

Abstract

Acoustic Charge Transport (ACT) devices used as delay lines and/or transversal filters have been evaluated for use in front-end electronics of SSC Detectors. The device is capable of providing delays up to 2 μsec with a dynamic range of 12 bits (72 dB) and power consumption of < 100 mW per data channel. It is radiation hard to 10^{14} n/cm^2. The ACT used as a transversal filter can provide pulse shaping for calorimeter signals.

The basic principles of the ACT devices have been described in an earlier publication [1] and only a very brief description will be given here. The fundamental device is a delay line (Fig. 1).

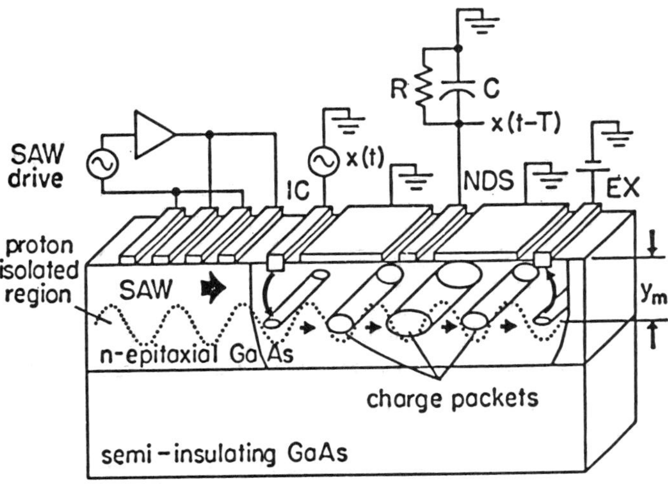

Figure 1: ACT device configuration.

The device exploits both the piezoelectric and semiconductor properties of GaAs. The piezoelectric properties are used to produce a surface acoustic wave (SAW) which travels in the <110> direction with a velocity of 2864 m/sec. The n-type GaAs is processed to produce a charge transport channel beneath the acoustic beam. The electric fields from the acoustic waves in combination with those of the transport channel produce a moving train of potential minima. Charge is injected into those wells, or buckets, as they pass the injection electrode. The amount of charge carried away in a bucket depends on, and so is a measure of, the potential of the signal electrode when the well moves out from under the electrode. The signal is thus sampled at the SAW frequency. The effective sampling aperture is only about 10% of the SAW period so there is negligible high frequency attenuation. The full Nyquist bandwidth is supported. Downstream of the injection electrode is a moving train of charge packets. These travel with an effective charge transfer efficiency up to 0.9999.

Capacitive readout electrodes downstream on the surface above the transport channel sense the delayed signal but do not alter the charge. These non-destructive sensing (NDS) electrodes, arranged in an equally spaced array, allow for repeated sampling of the signal at monotonically increasing time intervals. The technology for weighting the output of the NDS electrodes (NDS taps) exists so that the ACT may be operated as a transversal filter.

Two technologies exist: i) the buried channel ACT device developed by Electronic Decisions Incorporated (EDI) based on the work of Hoskins and Hunsinger [2]; ii) the Heterojunction ACT (HACT) developed at the United Technologies Research Center [3]. The essential difference is in the vertical charge confinement mechanism. In the ACT device the transport channel depth is determined by an externally applied DC potential whereas in the HACT the channel is formed by a minimum in the heterojunction band structure [3]. The ACT device is commercially available while the HACT device is still in the development stage. In this paper we report results of tests conducted with the ACT device. We are now beginning a detailed evaluation of the HACT.

The SAW frequency of the devices tested was 360 MHz and they could support the full 180 MHz Nyquist bandwidth (Fig. 2). This bandwidth is considerably larger than is needed for most HEP applications. Thus for SSC detectors, devices operating at lower SAW frequencies with reduced power dissipation are possible.

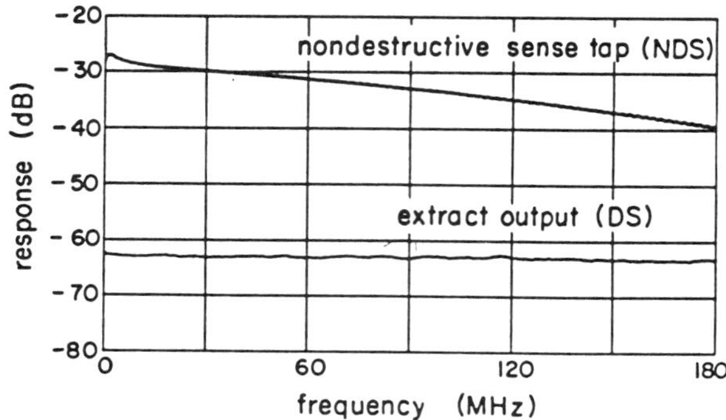

Figure 2: Frequency response of an ACT from 300 kHz to 300 MHz.

In a previous paper [1] we have discussed the ability to process signals by using the NDS tap-weighting features of the ACT technology and have given several examples. Here we only discuss some of our most recent work. Providing the input signal can be represented as a finite sum of exponentials with different time constants, it is possible to find a set of NDS tap weights which can shape the pulse and restore the base line to zero using a finite number of taps. Fig. 3 gives an example of the shaping technique for the input pulse given by

$$f(t) = \sum_{i=1}^{3} c_i e^{-t/\tau_i}, \tag{1}$$

with $c_1 = 1$, $c_2 = -1.5$, $c_3 = 0.5$, $\tau_1 = 40$ ns, $\tau_2 = 20$ ns and $\tau_3 = 10$ ns.

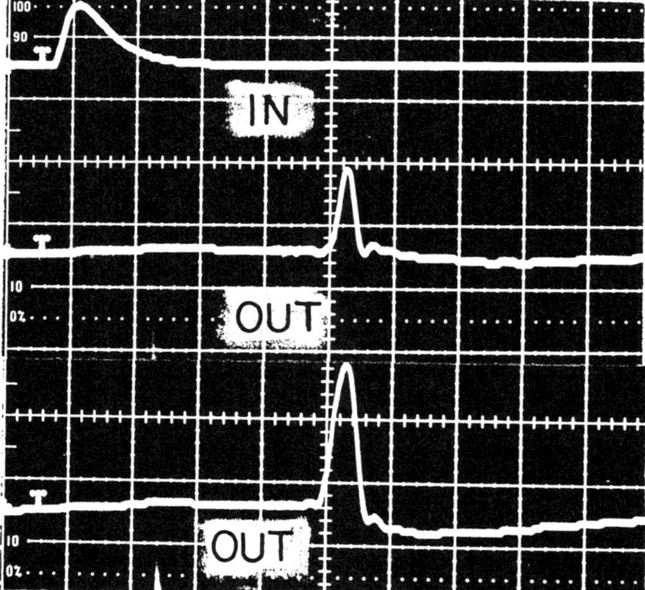

Figure 3: Pulse shaping, Scale: Input 500 mV/div, 100 ns/div; output 100 mV/div, 100 ns/div.

In order to test the pulse shaping capability of the ACT's we used a commercially available programable transversal filter (PTF) ACT with 128 NDS taps spaced 5.6 ns apart. The tap weight resolution is 5 bits. For a specific application which deals with fixed shape input signals a fixed tap weight ACT could be custom designed. Such a device would have 7 bit tap weight resolution, and be less expensive due to the absence of the on-chip tap weight circuitry.

In the examples shown in Fig. 3 we have used both 5.6 ns and 11.2 ns (skipping every other tap) tap spacings. The shaped signals have widths of approximately 30 and 60 ns for the 5.6 and 11.2 ns tap spacings, respectively. In both examples a total of 4 NDS taps with non-zero weights are used for the shaping and base line restoration. Using this technique a fast signal for pretriggering may be formed. The caveat is that when one does this shaping in the time domain it is possible to significantly distort the frequency spectrum and degrade the signal to noise. In general, for calorimeter signals the precise information will be in the low frequencies while for timing we emphasize the high frequencies. The signal is not affected by the operations with the NDS electrodes and it proceeds down the transport channel with the full analog information.

Proprietary software has been developed by EDI which will calculate the NDS tap weights necessary to approach a desired output pulse for an arbitrary input signal. This software also allows one to predict reliably the output given the design parameters of the ACT device.

Using the full power of the transversal filter it is possible to obtain an output which isolates the signals from different crossings by making the response zero at all the other crossings. Assuming the signal represented by Equation 1, the calculated output is shown in Fig. 4 for tap weight resolutions of 5 and 7 bits. Fig. 5 shows the response of the PTF which has 5 bit NDS tap weight resolution. The similarity to the predicted results is evident. We are convinced by this and other examples that the simulation software developed by EDI can accurately predict the response of the ACT. Thus we can rely on this software for the design of application specific devices.

A further example of the power of the ACT is shown in Fig. 6. The input pulse is a simulation of the output of the BNL Ar calorimeter preamplifier (the impulse response convoluted with the 400 ns current signal from the calorimeter). The tap weights are chosen to provide a shaped output with shaping time of 100 ns. The figure also shows the filter output when a second calorimeter signal appears 200 ns after the first.

We have investigated the neutron radiation damage sensitivity of ACT devices. Four GaAs wafer segments, containing ACT devices from EDI, were irradiated with neutrons at the University of Washington Nuclear Physics Laboratory. The neutrons were produced by the ^{9}Be(d,n) reaction using 9 MeV deuterons incident on a beryllium target. The four samples were exposed to fluxes of 5×10^{11}, 2×10^{12}, 1×10^{13}, and 1×10^{14} neutrons/cm^2, respectively [4]. The neutron energy spectrum was not directly measured, but may be in-

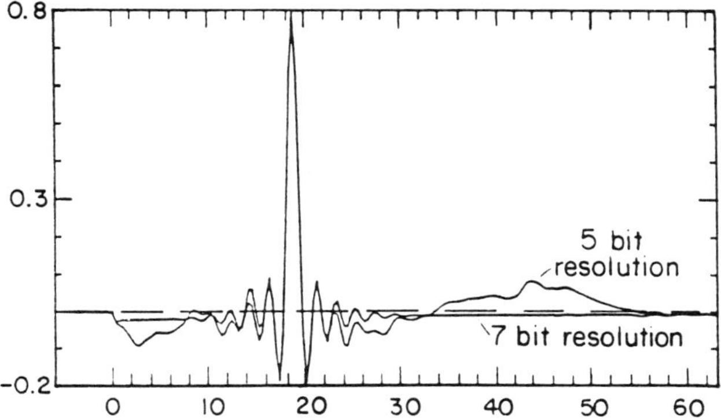

Figure 4: Calculated filter response to signal represented by Equation 1. Horizontal scale in beam crossings (16 ns units).

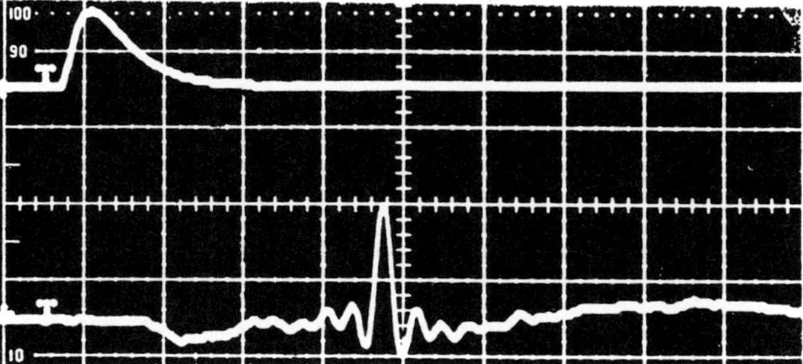

Figure 5: Filter response to signal represented by Equation 1 (sweep 100 ns/div).

ferred from the measurements of Weaver *et al.* [5]. The neutron energy distribution has a peak at about 4 MeV and drops almost linearly to 8 MeV.

The ACT device and active and passive ACTIC elements were measured before and after irradiation. Only at the greatest dose (10^{14} n/cm^2) did the device parameters show any measurable change, and then only minimal changes (less than 10% in FETS and p resistors). The ACT operating characteristics were not adversely affected. Thus we conclude that the device is radiation hard to 10^{14} n/cm^2. We plan to do more studies using ^{60}Co gammas and go to 5×10^{14} n/cm^2 before the end of the year.

Our current effort is focused on designing an ACT tailored to SSC applications. As an exercise M. Hoskins of EDI has used their software to design an ACT device with a single transducer driving 10 ACT channels each with 2 μsec delay. He has optimized the ACT for the signal represented by Equation 1. Each channel has one trigger output and one main output at the end of the delay line. He has tuned the device parameters, NDS electrode width, buffer amplifiers etc. to the frequency spectrum of the input signal with the aim of obtaining optimal signal to noise response. The estimated power per channel is approximately 150 mW with a dynamic range of 60 to 70 dB. It should be possible to further reduce the power by lowering the SAW frequency, since the bandwidth of the signal is presently far below

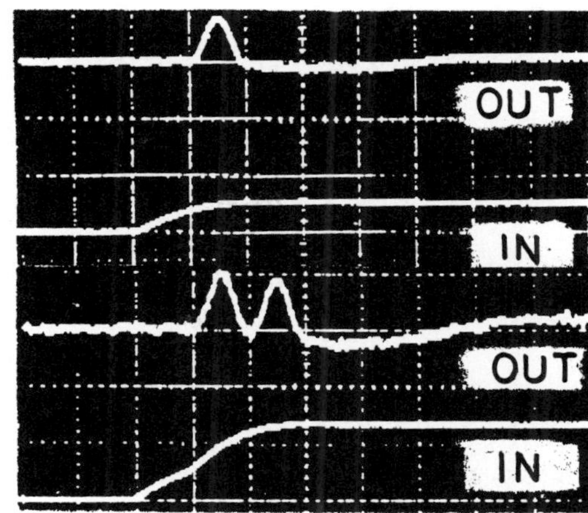

Figure 6: Pulse shaping of Ar calorimeter signal (sweep 200 ns/div).

the Nyquist limit. Power dissipation less than 100 mW seems possible.

Conclusions

The technology exists for manufacturing ACT devices with on-chip SAW frequency generator and NDS tap-weighting circuitry. The physics of the device is well understood and EDI modeling software can be used reliably to design application specific ACT's with predictable outputs.

During the remainder of the year we will complete the radiation damage tests and finish the design of an ACT tailored to the Ar calorimeter output. We also will continue to evaluate the HACT. We hope to be able to fabricate a prototype of a device tailored to a specific detector next year.

References

* D. Fleisch, M. Hoskins, R. Kansey, G. Pieters, Electronics Decision Incorporated (EDI), R. Davisson, W. Dougherty, H. Lubatti, W. Weitkamp, R. Wilkes, University of Washington, S. Merritt, W. Tanski, D. Cullen, United Technologies Research Center.

[1] R. J. Davisson *et al.*, "Evaluation of Acoustic Charge Transport Delay Lines", to be published in Nucl. Inst. and Meth.

[2] M. J. Hoskins *et al.*, Appl. Phys. Lett. 41(1982)4; M. J. Hoskins and B. J. Hunsinger, 1986 IEEE Ultrasonics Symposium Proceedings, p. 439.

[3] W. J. Tanski *et al.*, Appl. Phys. Lett. 52(1988)18

[4] W. Weitkamp, VTL-PHY-128, 1990

[5] K. S. Weaver *et al.*, Nucl. Sci. and Eng. 52(1973)35

A RAD-HARD, HIGH-GAIN, AND LOW-NOISE ALGAAS/GAAS AVALANCHE PHOTODIODE USING SUPERLATTICE ABSORPTION LAYER

Sheng S. Li, Y.H. Wang, S.S. Yu, and J.K. Walker*
Department of Electrical Engineering
University of Florida
Gainesville, FL 32611

Abstract

We report a novel ultra high-gain, low-noise AlGaAs/GaAs separate-absorption-multiplication (SAM) Avalanche Photo-Diode (APD) using an effective mass filtering (EMF) superlattice absorption layer to enhance the values of multiplication gain and electron/hole impact ionization rate ratio (α/β). The results yield a maximum multiplication gain factor of 6,000 and α/β equal to 25 at an electric field strength of 1.25×10^6 V/cm. The APD has a peak quantum efficiency of 72 % and responsivity of 0.45 A/W at 0.8 μm wavelength.

INTRODUCTION

Avalanche photodiodes (APDs) can provide unique features such as high-speed, high-sensitivity and large gain-bandwidth product for photon detection applications.[1] It is well known that the internal multiplication factor, bandwidth, and excess noise of an APD depends critically on the ionization rate ratio of electrons (α) and holes (β) in the APDs. Thus, enhancing the ratio of α to β is a key factor for achieving high-gain and low-noise operation. Unfortunately, for most III–V semiconductors, the electron and hole ionization rates are roughly equal, and as a result excess noise in the APDs formed by III-V compounds is expected to be high. Chin et al.[2] first proposed the use of band gap engineering in compound semiconductors to enhance the electron ionization rate over the hole ionization rate using the effective mass filtering (EMF) superlattice structure. Due to the nonlinear dependence of the ionization rate on the electric field, the ionization rate of electrons in the superlattice can be greatly enhanced over its bulk counterpart. It has been shown that by properly choosing the width and barrier height of the superlattice, the net electron ionization rate in a superlattice structure can be greatly enhanced over the corresponding electron ionization rate in the bulk III-V compounds [3-6]. Thus, it is important to study the effect of such a structure on the performance of an AlGaAs/GaAs APD for 0.4 to 0.85 μm detection.

In this paper we report the successful fabrication of an ultra-high gain, low-noise AlGaAs/GaAs Separate-Absorption-Multiplication (SAM) APD using an EMF superlattice absorption layer [7] to enhance the value of the electron/hole ionization rate ratio and hence enables to reduce the excess shot noise and to increase the multiplication gain in the APD.

EXPERIMENTAL

Figure 1 shows the cross sectional view of a planar AlGaAs/GaAs SAM APD with superlattice absorption layer. The AlGaAs/GaAs epilayers and the EMF superlattice structure were grown on a semi-insulating (S.I.) GaAs substrate using Varian Gen II molecular beam epitaxy (MBE) system. The absorption region consists of an undoped 165 periods of $Al_{0.3}Ga_{0.7}As$ (20 Å)/GaAs (56 Å) superlattice layers. The multiplication region consists of a 1 μm thick lightly-doped wide-bandgap p^-–$Al_{0.6}Ga_{0.4}As$ layer with a carrier concentration of 2.3×10^{14} cm^{-3} grown on top of the absorption layer. A heavily-doped n^+–$Al_{0.6}Ga_{0.4}As$ window layer with a carrier concentration of 10^{18} cm^{-3} and a 1000 Å n^+-GaAs layer with a doping concentration of 2×10^{18} cm^{-3} were grown sequentially on top of the p^--AlGaAs layer to serve as the n^+–emitter and the ohmic contact layer, respectively. Finally, the bottom ohmic contact was made on the p^+-GaAs layer with a carrier concentration of 5×10^{18} cm^{-3} which was grown the S.I. GaAs substrate. The mesa etched active area for this top illuminated APD is equal to 1.77×10^{-4} cm^2. A 100 μm diameter hole was etched through the n^+-GaAs layer by using a 1 : 1 : 4 mixture of H_3PO_4 : H_2O_2 : methanol etching solution. The wire contact pads were formed on the outer edge of the mesa etched active area to maintain the maximum sensitivity of the APD. The p-type ohmic contact was attained by using Cr/Au (300/2000 Å) and annealed at 480 °C for 30 seconds, whereas Ge-Au/Ni/Au was used for n-type ohmic contact and annealed at 450 °C for 2 minutes.

RESULTS AND DISCUSSION

Figure 2 shows the dark current-voltage (I-V) characteristics and the internal multiplication gain factors of electrons and holes for the APD under reverse bias

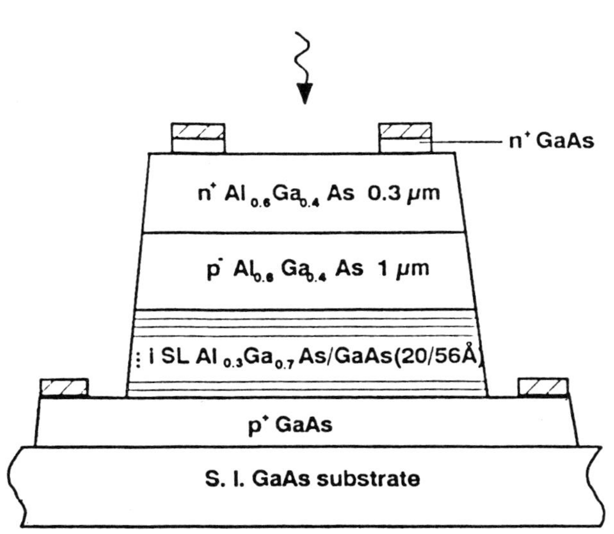

Fig. 1. Cross sectional view of a novel AlGaAs/GaAs SAM APD with an effective mass filtering superlattice structure in the absorption region.

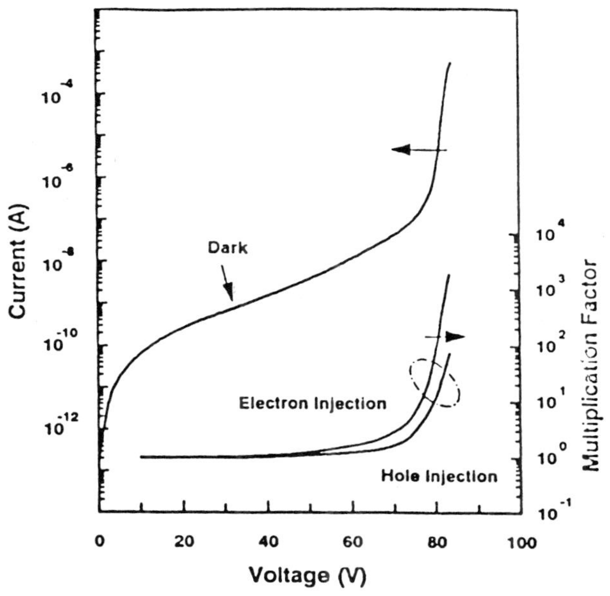

Current (A)
Dark
Electron Injection
Hole Injection
Multiplication Factor
Voltage (V)

Fig. 2. The dark I-V Characteristics and multiplication factors of electron and hole in an ALGaAs/GaAs SAM APD.

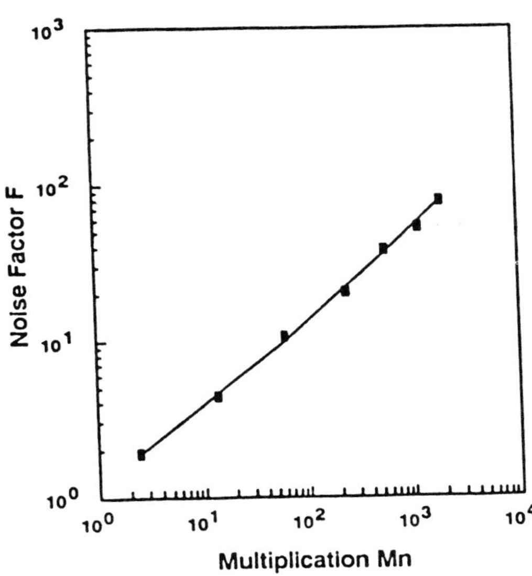

Fig. 3. Excess noise factor for electron injection in an AlGaAs/GaAs SAM APD.

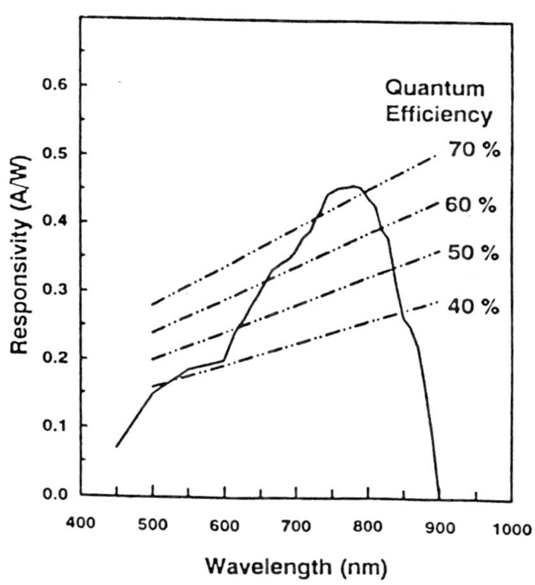

Fig. 4. The responsivity and quantum efficiency vs. wavelength for an AlGaAs/GaAs SAM APD.

condition. An excellent avalanche breakdown characteristics (V_B) was observed at $\sim$ 82 V for this APD. The results show that at 90 % breakdown voltage the dark current density is equal to 2.3×10^{-4} A/cm^2. From the photocurrent versus reverse bias plot, it is found that the p^--AlGaAs multiplication region of the APD becomes fully depleted at -3 V. An internal multiplication gain of 2,000 was obtained at -80 V when the initial photocurrent I_{po} was measured at100 nA. The multiplication gain factor for photogenerated carriers can be calculated by using the expression [8]

$$M = \frac{I - I_{MD}}{I_P - I_D} \qquad (1)$$

where I is the total multiplied current, I_P is the total primary current, and I_D and I_{MD} are the primary and multiplied dark currents, respectively. The maximum avalanche multiplication gain factor calculated for the APD shown in Fig. 1 is 6050. As shown in Fig.2, the results show that both the multiplication factor and the ionization rate for electrons is much larger than that of holes. The electron and hole ionization rates were determined from the bias dependent multiplication factors shown in Fig. 2. The relationships between the multiplication factors and the impact ionization rates of electrons and holes derived for bulk materials are assumed valid for our APD structure. Thus, we can write [9]

$$\alpha(E) = \frac{1}{W}\left(\frac{M_n - 1}{M_n - M_p}\right)\ln\left(\frac{M_n}{M_p}\right) \qquad (2)$$

$$\beta(E) = \frac{1}{W}\left(\frac{M_p - 1}{M_p - M_n}\right)\ln\left(\frac{M_p}{M_n}\right) \qquad (3)$$

where W is the width of the p^--region, and M_n and M_p are the multiplication factors for electrons and holes, respectively. We calculated the ionization rates for both electrons and holes by using a punch-through structure and assuming that the electric field varies linearly with the applied bias voltage in the avalanche region of the APD operation. The result yielded a maximum value of α/β equal to 25 at an electric filed strength of 1.25×10^6 V/cm. Both the maximum multiplication gain factor (M $\sim$ 6000) and the electron/hole ionization rate ratio ($\alpha/\beta \sim 25$) obtained for our APD are believed to be the highest ever reported for an AlGaAs/GaAs SAM APD.

For electron injection, the excess noise factor is given by [10]

$$F = \frac{M_n}{k'} + (2 - 1/M_n)(1 - \frac{1}{k'}) \qquad (4)$$

where k' = α/β, is the ratio of electron/hole ionization rates. Figure 3 shows the calculated excess noise factor as a function of the multiplication factor. The slow increase of the excess noise factor at higher multiplication factor seen in our APD is attributed to the fact that the value of α/β is enhanced by using the effective mass filtering superlattice structure in the absorption layer.

Figure 4 shows the spectral response for our APD. Note that both the responsivity and the quantum efficiency were measured at a reverse bias voltages between 2 and 20 V, in which no avalanche multiplication is occured. Using both the D.C. and A.C. light illumination with I_o = 75 pW, the detector has a peak responsivity of 0.45 A/W and a quantum efficiency greater than 70 % at 0.8 μm wavelength at V_R = 20 V. It is worth noting that no photoconductivity gain due to the EMF superlattice structure [7] in the absorption region was seen in this APD at low light level and under low bias condition, as evidenced by the responsivity and quantum efficiency vs. wavelength plot shown in Fig.4.

A simulation of the relative photoresponse vs. frequency for our APD using the measured values of C_J = 1 pF, R_s = 100 ohm, and R_D = 10 M ohms yielded a 3 dB cutoff frequency of around 1 GHz for this APD.

In conclusion, we have demonstrated in this work that by using an EMF superlattice structure in the absorption region of an AlGaAs/GaAs SAM APD, we have achieved a maximum multiplication gain factor of $\sim$ 6,000, a large value of α/β ($\geq$ 25), and a low excess noise factor for our APDs. The EMF superlattice and SAM structure used in this APD offers a new quantum-type photoconductivity scheme, which is used mainly to increase the electron/hole ionization rate ratio and hence enables to reduce the excess shot noise and to increase the multiplication gain in the APD. Finally, a 3 dB cutoff frequency of 1 GHz was obtained for our SAM APD.

REFERENCES

1. S. R. Forrest, J. Lightwave Tech. LT-3, 347 (1985).
2. R. Chin, N. Holonyak, Jr., G. E. Stillman, J. Y. Tang, and K. Hess, Electron. Lett. 16, 467 (1980).
3. K. Brennan and Y. Wang, J. Phys. (Paris), C5, suppl. 11, 277 (1987); IEEE Trans. Electron Devices, 35, 634 (1988).
4. F. Capasso, W. T. Tsang, A. L. Hutchinson, and G. F. Williams, Appl. Phys. Lett. 40, 38 (1982).
5. F. Osaka, T. Mikawa, and O. Wada, LEEE QE-22, 1986 (1986).
6. T. Kagawa, K. Mogi, H. Iwamura, and O. Mikami, IOOC'89 Technical Digest, 2, 122 (1989).
7. F. Capasso, in Semiconductors and Semimetals, Vol. 24, 319 (1987).
8. S. M. Sze, "Physics of Semiconductor Devices", Wiley, 2nd ed., (1981).
9. F. Y. Juang, U. Das, Y. Nashimoto, and P. K. Bhattacharya, Appl. Phys. Lett. 47, 972 (1985).
10. R. J. McIntyre, IEEE Trans. Electron Devices, ED-13, 164 (1966).

11. * With the Department of Physics, and Nanoptics Inc.

12. This work was supported by the DARPA under US Navy Grant No. MDA972-88-J-1006.

APPLICATIONS OF SUPERCONDUCTING ELECTRONICS FOR THE SUPERCONDUCTING SUPER COLLIDER

Edwin E. Stebbins
Hollis L. Caswell
HYPRES Inc., 175 Clearbrook Rd., Elmsford, NY 10523

John P. Wikswo, Jr.
Dept. of Physics and Astronomy, Vanderbilt Univ., Box 1807 Station B, Nashville, TN 37235

Abstract

Superconducting microelectronics based on low-temperature superconductors Niobium and Niobium Nitride have superior high speed and low power characteristics suitable for the demanding instrumentation and processing requirements of the SSC. HYPRES has been fabricating superconducting microelectronics for over four years, and is currently working in several areas which may be applicable to SSC, including 200GHz bandwidth comparators, 200GHz count rate counters, 20GS/s flash analog to digital converters, digital logic with sub 10ps gate delays, RISC processors with a 2GHz clock rate, and a 128 x 128 crossbar switch.

Introduction

Superconducting microelectronics based on Josephson junctions can be used to fabricate a wide variety of analog and digital circuits ranging from highly sensitive pulse detectors to complete digital processing systems. While many instrument and computer manufacturers are hesitant to use cryogenic electronics, the presence of liquid helium at the SSC may make it more effective to place high-performance superconducting electronics in close proximity to the SSC vertex detectors than to use conventional semiconductor electronics. Superconducting microelectronics fabricated from niobium or niobium nitride have several strategic properties, including high speed, low power, and radiation hardness. It is possible to fabricate high density transmission lines with low cross-talk, and to integrate digital, analog, and sensors on a single substrate. Because superconducting electronics offer low power consumption, they provide the potential for fabrication of large cooled instrumentation arrays. For processing applications, the most important feature is the low power - speed product. Intrinsic gate switching speeds of < 2 ps have been demonstrated[1] even at relatively large feature sizes. The extremely low power of the switching devices means very low system heat density and high circuit density. Superconducting microelectronics theoretically have superior radiation hardness due to the very small volume (2.5 micron diameter and 20 Angstroms thickness) of the active region in a Josephson Junction (JJ).

Superconducting Microelectronics

HYPRES currently designs and fabricates superconducting microelectronics in Niobium (Nb) or Niobium Nitride (NbN). Niobium becomes supercoducting at 9.2K, and is normally operated in liquid Helium (He) at 4.2K. Niobium Nitride has a T_c of around 17K and can be operated in a simple Closed Cycle Refrigerator (CCR). HYPRES has demonstrated the operation of a 70GHz NbN Time Domain Reflectometer IC[2].

The process flow for both Nb and NbN consist of ten layers. The HYPRES fabrication process makes use of refractory materials throughout, with the exception of Ti/Pd/Au contact metallization. Niobium is used due to its high critical temperature, electrical and thermal stability, and thermal cyclability. $Nb/Al_2O_3/Nb$ junctions are made by depositing an in-situ trilayer across the entire wafer, and subsequently defining junctions areas through etching and isolation using deposited Silicon Dioxide (SiO_2). HYPRES attains process uniformity of junction parameters of +/-5% across an active circuit area. The JJs are interconnected through two Nb wiring layers, a superconducting ground plane, and the resistive layers. The resistive metallizations provide resistors of 1.0Ω/ square and 0.01Ω/square. Deposited SiO_2 is used to isolate the junctions, the ground plane, and metallizations[3].

The process utilizes standard equipment from semiconductor production processing. The current design rules are limited to 2.5μm junctions and linewidths by photolithography equipment. HYPRES currently runs 12 3" wafers per week with an average process yield of 90%.

Superconducting Counters

The Single Flux Quantum (SFQ) Josephson binary counter was first demonstrated by Hamilton[4] in 1982 using lead alloy technology, after having been proposed by Hurrell and Silver[5]. This binary counter was able to toggle at rates close to 100GHz, and dissipated only microwatts. The first demonstration of an all-Nb SFQ counter was made at HYPRES[6].

HYPRES has recently fabricated and tested a 12-stage SFQ counter. This counter can be enabled on and off and reset by internal or external logic. The current counter value can transferred to a readout register by a read clock. This counter has been tested at 1GHz[7] count rate and is expected to be capable of a 200GHz count frequency. Testing is currently limited by the room-temperature test system. Using 4μm junctions and $1000A/cm^2$ current density, the counter dissipates less than 25uW and requires approximately 1.0mm x 0.7mm. The compact size of this circuit would allow 64 channels of 200GHz counter to be fabricated in $1cm^2$ IC, including multiplexing of the output data to a single I/O port.

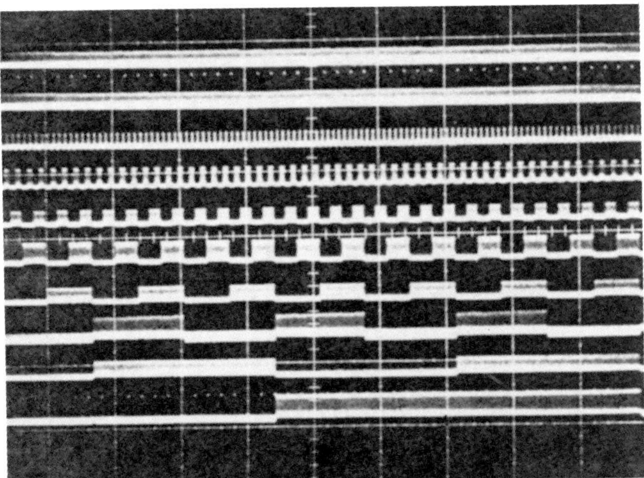

Figure 1 10bit Nb Counter Operation

Superconducting Analog to Digital Convertors

Two main types of Analog to Digital Convertors (ADC) are under development. One type, aimed at high speed / low resolution applications, uses parallel or flash conversion. The other type is based on the aforementioned binary counter, and is aimed at medium speed / high resolution applications.

The Flux-Quantizing (FQ) ADC utilizes a superconducting interferometer flux quantizer to generate pulses as flux is introduced by a magnetically-coupled analog signal[8]. The tracking type of FQ ADC utilizes two separate counters to count increasing and decreasing input signal changes. A single counter can be used for pulse height acquisition. HYPRES has designed and tested a 5-bit dual counter FQ type ADC. The ADC performance is determined by the maximum input signal slew rate, which is a function of the maximum counter rate. A 200GHz counter can track 10bits of resolution at a 200MS/s rate. The entire ADC requires approximately $1mm^2$ area and dissipates less than 50μW.

Superconducting interferometers (SQUIDs) exhibit a periodic transfer function[9]. This allows an N-bit ADC to be constructed from only N comparators, which has several advantages over semiconductor flash ADCs utilizing 2^N-1 comparators. Ko[10,11] has proposed using single junction SQUIDs. This design avoids the dynamic disturbances generated during flux state transitions that occur in previous two junction SQUID designs. This design has been improved and implemented at HYPRES by Bradley[12].

HYPRES has fabricated and is testing a 6 bit flash ADC, which is estimated to have a 20GS/s sample rate. The ADC is expected have 4 effective bits at a 10GHz input bandwidth and 6 effective bits at 2.5GHz. The ADC can be reduced to a $1mm^2$ active area, and can be interfaced to MVTL superconducting digital logic.

Figure 2 Microphotograph of the 12bit Flux Quantizer ADC

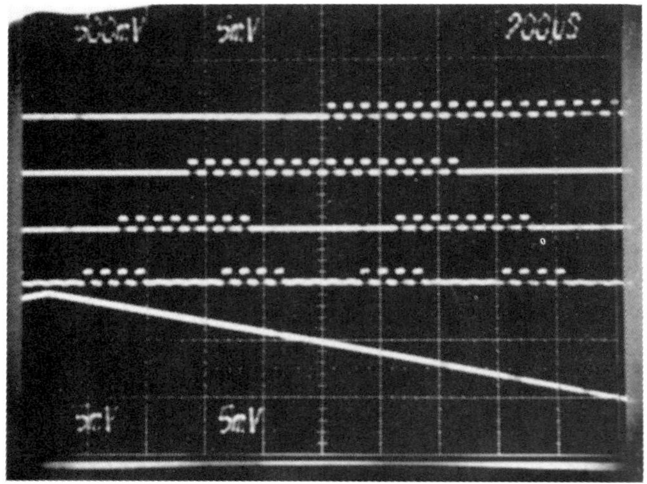

Figure 3 Display of 4bit Flash Operation with Ramp Input

Superconducting Digital Logic

HYPRES is using Modified Variable Threshold Logic (MVTL) for implementation of digital systems. MVTL logic, first developed by Kotani[13] of Fujitsu, is latching and synchronous; the gates are powered by the system clock signals and must be reset before each cycle. MVTL has fan-in and fan-outs of 3-4, and implements all required logic functions including inversion.

HYPRES currently produces MVTL gates which exhibit 18ps gate delays and 10μW per gate dissipation utilizing 4.0um geometries. This logic design is extendable to less than 0.5ps gate delays and less than 2μW per gate power dissipation using 1.0um geometries. HYPRES has fabricated and tested shift register chains of 1500 gates and 32bit shift regsiters have been tested at 500MHz. This circuit is being used as a complexity and yield driver. The figure below shows 288 stages operating correctly, the top trace is the input, bottom tree are stages 286,287, and 288.

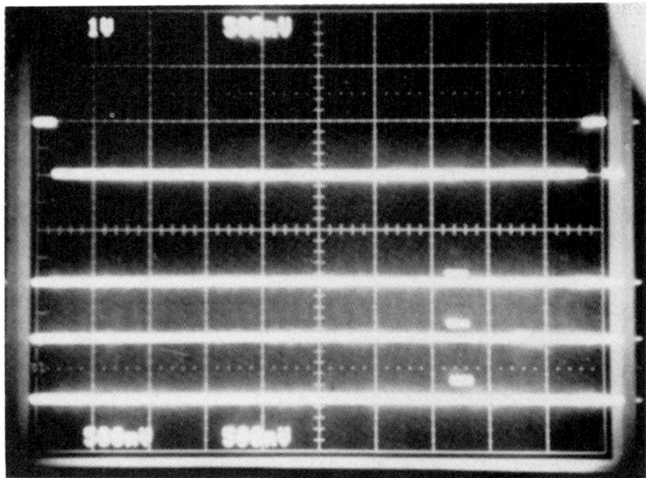

Figure 4 288 Stages of Shift Register Operation

HYPRES has designed and fabricated a 32 bit Arithmetic Logic Unit (ALU) that is constructed from 1512 superconducting MVTL logic gates, and is expected to have a carry-in to carry-out delay of less than 500ps, and consume less than 28mW of power. At Fujitsu, a partially functional Digital Signal Processor constructed from 6600 MVTL gates using 1.0um design rules has demonstrated 7ps loaded gate delays and consumes less than 2μW per gate[14]. Current HYPRES gate designs require 50μm by 100μm using 4μm lines and junctions. Use of high resolution process requipment and a redesign would yield 20μm x 20μm gate structures, thus allowing 25000 gates per 1.0cm^2 IC.

Pulse Characterization Applications

The devices mentioned above can be used to implement several pulse measurements including direct counting up to 200GHz; 5ps resolution measurements of time of arrival, delay, and pulse width; pulse height measurement with 1μA sensitivity; and complete pulse digitizing up to a 10GHz bandwidth.

Trigger Detectors

A pulse event can be detected by a superconducting comparator. Single junction comparators and SQUID comparators have input bandwidths ofover 100GHz, 1μA sensitivity, programmable polarity, hysterisis, and are latching or non-latching. The detector thresholds are externally controlled. These comparators interface directly to MVTL logic, which can implement complex logic-based trigger conditions.

Time of Arrival, Pulse Width, and Delay Measurements

Two superconducting latching comparators can detect both pulse start and end. The TOA measurement causes the value of a master clock counter to be latched in a register for subsequent read-out. The delay measurement is made relative to a common start event, and the PW counter measures the elapsed time from pulse start to end. The measurement resolution is determined by the maximum count rate. For a fixed 200GHz master clock, resolution would be 5ps. Hundreds of channels can use the same master clock.

The master clock source can be generated from a Josephson junction used as a Voltage to Frequency Convertor (VFC). Long-term stability of the master clock can be achieved by implementing a Phase Locked Loop (PLL) using a superconducting VFC and prescaler with an external reference and phase error comparator.

The counter can be interfaced to a shift register memory. Thus, multiple pulses can be acquired with a deadtime of less than 200ps. Using a current HYPRES memory design, each pulse parameter described above could be stored 30 times as a 16 bit value before read-out was required.

Real-time Vertex Dectector Angle of Arrival Trigger

A counter can be configured to generate a pulse after a programmable time delay, with 5ps of resolution and 5ps of jitter using a 200GHz master clock. After a common start event generated by logically "ORing" detectors, the delay pulses from each channel can be logically "ANDed". The resulting trigger pulse will occur if each detector is hit during its programmed time. This design can be used as a zero latency trigger detector for an event arriving from a specific angle relative to multiple detectors.

Pulse Height Measurements

The aforementioned flux-quantum ADC utilizes a superconducting interferometer that can detect a change of less than 1uA from an input signal. A flux-quantizer and single counter can measure the peak amplitude of a pulse. The counter counts the number of 1uA steps from an inter-pulse baseline level to the top of the pulse. The negative side of the input pulse can be detected and used to reset the counter and store the value into a shift register memory. The resolution of the pulse height measurement is a function of the maximum signal slew rate. If a 200GHz counter is used and the flux-quantizer has 1.0uA sensitivity, then the signal slew rate is limited to 200μA/ns.

Pulse Digital Acquisition

A complete monolithic 40GS/s pulse digitizer can be fabricated from the flash ADC described above and a shift register memory. The digitizer would operate at 40GS/s with 6 bits of resolution, utilizing two 20GS/s interleaved ADC's. The memory would consist of a 1024 x 12 bit shift register running at 20Gbps. At full speed, the on-chip memory can acquire a total 25ns of signal as one acquisition or as multiple detected pulses. The output port of the shift register is accessed by room-temperature electronics at a slower rate. The memory can operate in a pre-trigger mode, allowing the trigger detection of the pulse to occur after the signal has arrived. The single chip system will be 8mm x 8mm in size and dissipate less than 20mW. A multi-chip system can also be implemented utilizing an ADC and several memories mounted or fabricated on a single substrate. For example, a twenty-four 64Kb chip system could acquire 256K samples or 6.5μs of signal before read was required.

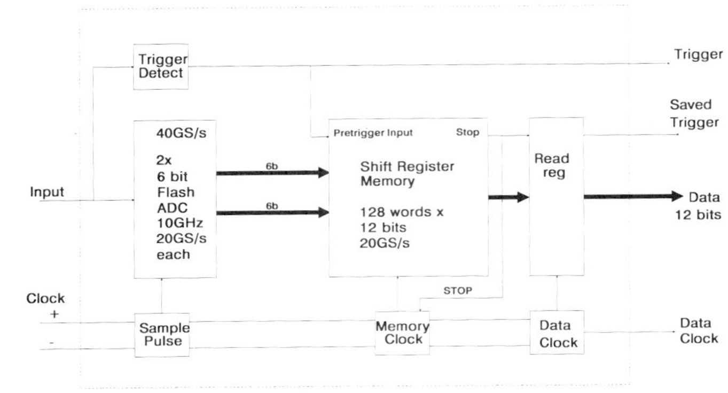

Figure 5 Monolithic 40GS/s Pulse Digitizer Block Diagram

Processing Applications

Superconducting RISC Processors

HYPRES is currently designing a complete Reduced Instruction Set Computer (RISC). The extraordinary speed and power characteristics of superconducting microelectronics would allow HYPRES to construct a 32bit RISC processor with a 2GHz clock rate. The RISC would include a 32bit integer math unit and separate pipelined 32bit floating point ALU and multiplier. The RISC would also include two separate 32K byte data and instruction superconducting 500ps memory caches. The RISC would be fabricated from 28 superconducting ICs mounted to a single 4" silicon wafer with superconducting interconnects, and would dissipate 500mW.

The expected peak performance would be 2000MFLOP/s and 1500MIPs. The sustained performance would depend on the memory hierarchy and data organization. The proposed processor would implement a standard scalar RISC specification and therefore use standard software at greatly enhanced performance levels.The low heat density of the single 4" processor substrate would allow compact multiple processor construction. For example, a parallel supercomputer could be constructed from 64 of the processor / memory substrates. The resulting product would exhibit 128GFLOP/s performance, while occupying only 4" x 8" volume and dissipating around 32W.

Superconducting Crossbar Switch

HYPRES is midway in a three year contract development to produce a 128 x 128 superconducting crossbar switch. The switch is designed to have a serial data rate of 2Gbps (Giga bits per second). Each channel of the crossbar has complete circuitry for address decoding, switch control, contention detection, and non-blocking arbitration. The latency from channel request to a busy or ready acknowledgement is dominated by wire delays and is expected to be less than 8ns. The crossbar is fabricated from superconducting integrated circuits which are each 32 x 32 crossbar switches. Sixteen switch circuits will be configured on a supercoducting substrate to form a complete 128 x 128 crossbar. Interface circuits which handle 64 I/O channels on the master and slave sides are also designed. The entire 128 x 128 crossbar is physically implemented as thin-film hybrid, with 20 1.2cm square superconducting ICs mounted on a single 4" diameter silicon wafer using superconducting transmission-line interconnects. The entire wafer is expected to dissipate 40mW. A 1024 x 1024 configuration could use 64 of the 128 x 128 substrates mounted in a 4" diameter by 8" volume.

Conclusions

The SSC has projected instrumentation and processing requirements that exceed the capablities of current technology. Superconducting microelectronics have superior speed, power, and size characteristics for many of the SSC applications. In an enviroment where superconductors are a way of life, the use of superconducting instrumentation and processing systems is appropriate

[1] S. Kotani, "1.5ps OR Gate" Tech. Digest Intl. Electron Devices Mtg., pp 884-885, 1988

[2] G.K.G. Hohenwarter, et al, "Demonstration of a Two-Stage Refrigerator Based Superconducting Time Domain Reflectometer", presented at Applied Superconductivity Conf. , Sept. 1990

[3] E. Hill et al, "HYPRES Niobium Design Rules", Hypres Doc. 22-80601, Proprietary, Rev. 3, p 2

[4] C.A. Hamilton and F.L. Lloyd, "100GHz Binary Counter Based on DC SQUIDs", IEEE Electron Devices Lett., Vol EDL-3, No. 11, pp 335-338, Nov. 1982

[5] J.P. Hurrel and A.H. Silver, "SQUID Digital Electronics", in AIP Conf. Proc., VOl. 44, pp 437-447, 1978

[6] F. Kuo, Hypres Internal Report, March 1989

[7] F. Kuo, "Superconducting A/D Convertors Based on Josephson Binary Counters", presented at Applied Superconductivity Conf., Sept. 1990

[8] J.P. Hurrel, D.C. Pridmore-Brown, and A.H. Silver, "Analog to Digital Conversion with unlatched SQUIDs", IEEE Trans. Electron Device, ED-27, pp 1887-1896, 1980

[9] K.K Likharev, *Dynamics of Josephson Junctions and Circuits*, pp153-163, New York: Gordon and Breach Science Publishers, 1986

[10] H. Ko and T. Van Duzer, "A New High-Speed Periodic Threshold Comparator for use in a Josephson A/D Convertor, "IEEE J. *Solid State Circuits*, Vol 23, No. 4, pp 1017-1021, Aug. 1988

[11] H. Ko, "A Flash A/D Constructed with One Junction SQUIDs", IEEE *Trans. Magn.*, Vol MAG-25, No. 2, pp 826-829, Mar. 1989

[12] P. Bradley, "Quasi-One Junction SQUIDs as Comparators for Analog to Digital Conversion", presented at Applied Superconductivity Conf., Sept. 1990

[13] S. Kotani et al, "Ultrahigh-Speed Logic Family with Nb/Al-AlOx/Nb Josephson Junctions", IEEE Trans. Electron Devices, Vol ED-33, No. 3, p 379, Mar. 1986

[14] S. Kotani et al. "A 1 GOPS 8b Josephson Digital Signal Processor", ISSCC Tech.Dig., p 148, 1990

USE OF NON-VOLATILE MEMORIES FOR SSC DETECTOR READOUT

A.J. Fennelly, J.K. Woosley, and M.B. Johnson
Technology Applications Department, MS 50
Teledyne Brown Engineering, P.O. Box 07007, Huntsville AL 35807-7007

Abstract

Use of non-volatile memory units at the end of each fiber optic bunch/strand would substantially increase information available from experiments by providing a complete event history, in addition to easing real time processing requirements. This may be an alternative to enhancing technology to optical computing techniques. Available and low-risk projected technologies will be surveyed, with costing addressed Some discussion will be given to conversion of optical signals, to electronic information, concepts for providing timing pulses to the memory units, and to the magnetoresistive (MRAM) and ferroelectric (FERAM) random access memory technologies that may be utilized in the prototype system.

Introduction

In this paper we describe the conceptual use of radiation-hardened (radhard), non-volatile memories for on-line data acquisition in a proposed SSC detector. The requirement for radhard memory is obvious, particularly when we want to put elements of the memories inside the detector up against the fiber-optic bunches/strands. The need of non-volatility is less obvious, but that eliminates the requirement for constant power supply to a true random-access memory (RAM) which may be rewritten when desired or necessary. There are available read-only memories (ROM) which are electrically erasable (EEPROM), but they cannot be written and rewritten like a RAM can. Also, up to the present, the development of a sufficiently radhard EEPROM has proven quite challenging. A non-volatile RAM, (NVRAM) would allow storage of information to activate detector triggers immediately upon initial reception of data, as well as storage of data as back-up in case power losses, voltage surges, etc. caused any loss of data in the main memories and data reduction systems. A radhard NVMEM would therefore be most useful in any SSC detector, if it were sufficiently small, cheap, and possess sufficiently fast read/write and access times.

As a historical reminder, Figure 1 exhibits approximate volumes and weights of selected memory technologies. Figure 2 further lists costs vs. speeds for a fairly comprehensive selection of NVMEM technologies which are also characterized by the nature of their readout, destructive (DRO) or nondestructive (NDRO) and by their achievabilities of radhardness. As a result of further available process and design technologies, there are now some programs to extend those capabilities into extremely dense, small, cheap, and fast, radhard NVMEM/NVRAM elements. Those include magnetoinductive (MI), magnetoresistive (MR), and ferroelectric (FE) memories. The first (MI) has been pursued by NASA and the US Air Force; the second two are under development by the US Army. We shall concentrate here on the MRAM and FERAM devices in this paper.

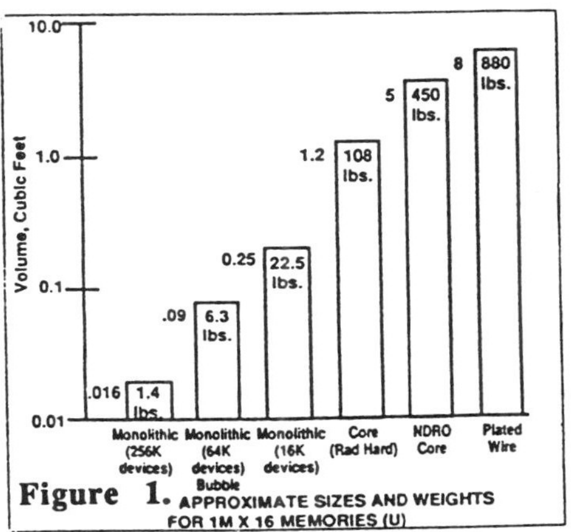

Figure 1. APPROXIMATE SIZES AND WEIGHTS FOR 1M X 16 MEMORIES (U)

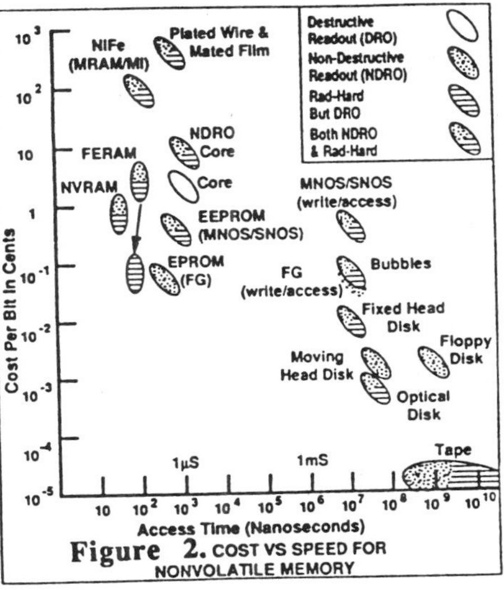

Figure 2. COST VS SPEED FOR NONVOLATILE MEMORY

Magnetoresistive RAM

The MRAM (short for MR/RAM) is under development by Honeywell SSEC in Plymouth MN. The basic bit is shown in Figure 3. It consists of two

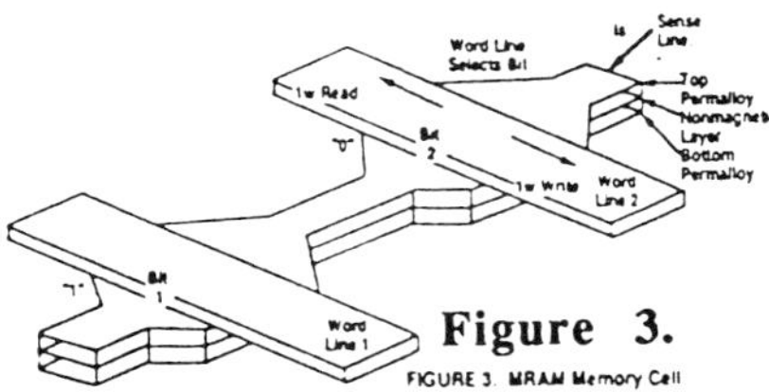

Figure 3.
FIGURE 3. MRAM Memory Cell

Cobalt Permalloy layers, separated by an exchange barrier, which is a layer of nonmagnetic alloy, with wordlines overlaid as shown in the figure. The word current has two values: (1) write and (2) readout (opposite direction). When the sense current opposes the stored flux of a bit, the magnetization rotation is large; when it aids the stored-flux direction, that rotation is small. The anisotropic magnetoresistance of the cell is thus larger in the former case, and small in the latter. Support electronics to the memory cell are on radhard bulk CMOS. The specification for the high-density version of the MRAM is given in Table I, both

	REQUIREMENT	GOAL
CAPACITY	8K x 8 BITS	(3) 64K x 1 or 8K x 8 BITS
READ-OUT	NONDESTRUCTIVE	NONDESTRUCTIVE
READ ACCESS	1 μs ((2) 300 ns)	700 ns ((2) 300 ns)
READ CYCLE	1 μs ((2) 300 ns)	800 ns ((2) 300 ns)
WRITE CYCLE	250 ns	200 ns
ACTIVE POWER	300 mW	250 mW
STANDBY POWER	75 mW	35 mW
VOLTAGE	5 volts ± 10%	5 volts ± 10%
READ ENDURANCE	1E12 cycles	1E15 cycles
WRITE ENDURANCE	1E12 cycles	1E15 cycles
(1) RETENTION	1 year	10 years
OPERATING TEMPERATURE (FULL SPEC COMPLIANCE)	0 to 80° C (case)	0 to 80° C (case)

(1) AFTER MAXIMUM ENDURANCE CYCLES
(2) EFFECTIVE CYCLE IN SEQUENTIAL BLOCK TRANSFER
(3) OTHER PARAMETERS IN TABLE BASED ON 64 K x1 CONFIGURATION

TABLE 1. HIGH DENSITY NONVOLATILE MEMORY SPECIFICATION FOR MRAM

requirements and ultimate goals. Intrinsic radhard properties have been demonstrated for prototype MRAM devices beyond 5 MRAD. The number of read/write cycles that an MRAM can endure is effectively infinite, and has been so demonstrated in several prototype devices. Also the readout is NDRO, which is another desirable property. However, the MRAM would be rather slow in its access times, which is a disadvantage.

Ferroelectric RAM

The FERAM is under development by a team le d by Raytheon EDL at Sudbury MA, and including Raytheon Research of Wayland MA and Raytheon Semiconductor from Mountain View CA. The team also includes a group of government laboratories and universities in a consortium to develop to ferroelectric materials for the memory capacitors.

The structure of the memory capacitor is notionally shown in Figure 4, along with the hysteresis curve which characterizes the ferroelectric response of the FE material to an applied voltage or electric field. In analogy with ferromagnetism, a remnant polarization field is induced in the material after its rise to a saturation polarization field. As long as the material is below its ferroelectric Curie temperature, it will exhibit that behaviour. The polarization field is reversed in direction by application of a reversed threshold electric

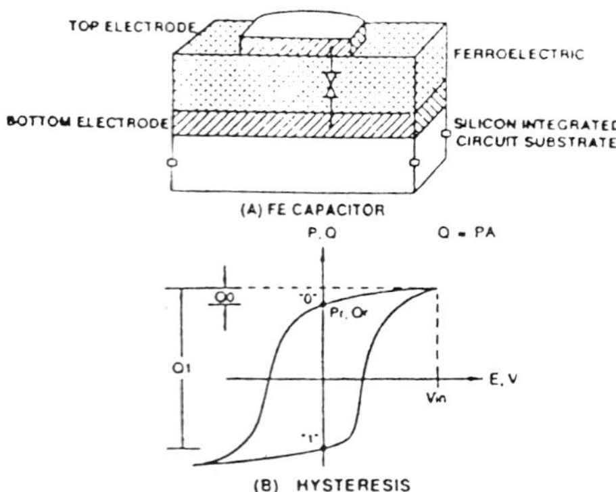

Figure 4. FERROELECTRIC CAPACITOR PHENOMENON

field or voltage. Thus, zero and one are represented by opposite polarization states of the remnant field. A representative FERAM Schottky diode memory cell is shown in Figure 5.

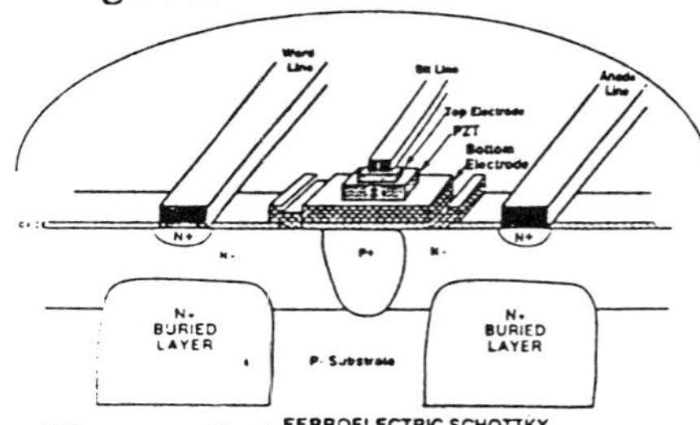

Figure 5. FERROELECTRIC SCHOTTKY DIODE MEMORY CELL

Open-literature studies of the radhard properties of FERAM devices are illustrated in Figure 6. A recent set of remnant polarization (i.e., retained memory) vs. radiation dose are given in Figure 7 (from Arizona State University) for a representative ferroelectric material.

RADIATION STRESS	DISPLACEMENT DAMAGE, n/cm²	IONIZING DOSE, Rads (Si)	IONIZING DOSE, Rads (Si)/s	FISSION FRAGMENTS Cf-252, Particles/cm²
THEORETICAL THRESHOLD (ESTIMATED LEVEL THRESHOLD)	5×10^{16}	$>10^{8}$	2.5×10^{13}	PROBABLE NO DAMAGE, NO SEU AT ANY LEVEL
EXPERIMENTAL DATA (MEASURED LEVEL AT WHICH NO DAMAGE IS OBSERVED)	10^{14}	10^{7}	1.2×10^{11}	AT 6×10^{6} Particles/cm², NO DAMAGE AND NO SEU

Figure 6. COMPARISON OF THEORETICAL AND EXPERIMENTAL RADIATION DATA

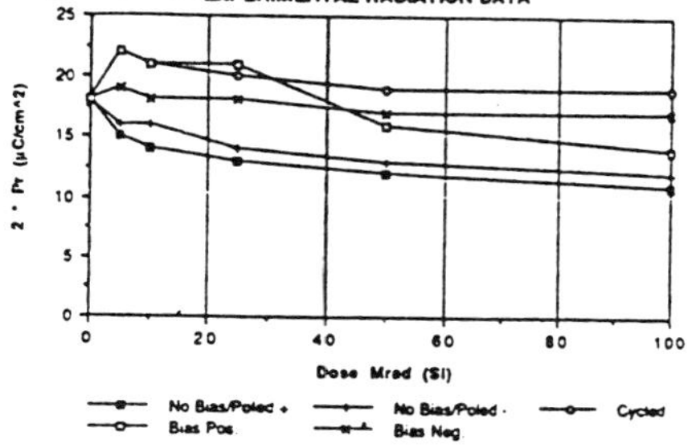

Figure 7. POLARIZATION VS DOSE

The question of fatigue has been a serious consideration for FE materials and is now being resolved in their favor, as Figure 8 shows. An SSC detector's NVMEM, used right at the scintillating fibers, photodetectors, etc. to implement trigger decisions, can be expected to experience many read/write cycles. The representative data in Figure 8 shown adequate endurance for the task in a presently available sample FE material.

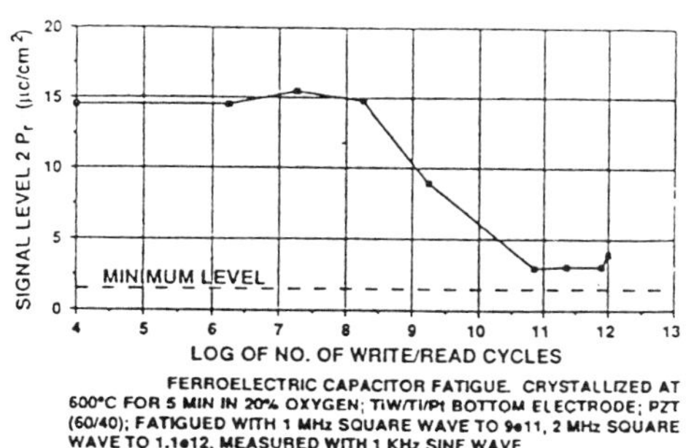

FERROELECTRIC CAPACITOR FATIGUE. CRYSTALLIZED AT 600°C FOR 5 MIN IN 20% OXYGEN; TiW/Ti/Pt BOTTOM ELECTRODE; PZT (60/40); FATIGUED WITH 1 MHz SQUARE WAVE TO 9e11, 2 MHz SQUARE WAVE TO 1.1e12, MEASURED WITH 1 KHz SINE WAVE.

Figure 8. FATIGUE TO 1 X 10^{12}

We show the specification for the high-density NVMEM in the case of the FERAM in Table II, Again, the basic requirements are listed with the ultimate program goals. It is clear that greater densities and faster read/write cycles and access times are expected with the FERAM than with the MRAM. The support electronics for the FERAM are based on a standard silicon bipolar process, which have a greater power consumption than equivalent CMOS support electronics would have.

	REQUIREMENT	GOAL
CAPACITY	64K x 1 BITS	(3) 256K x 1 or 32K x 8
READ-OUT	TBD	NONDESTRUCTIVE
READ ACCESS	200 ns	150 ns
READ CYCLE	250 ns ((2) 50 ns)	200 ns ((2) 50 ns)
WRITE CYCLE	250 ns	200 ns
ACTIVE POWER	600 mW	1000 mW
(1) STANDBY POWER	250 mW	350 mW
VOLTAGE	5 volts ± 10%	5 volts ± 10%
READ ENDURANCE	1E12 cycles	1E15 cycles
WRITE ENDURANCE	1E12 cycles	1E15 cycles
(1) RETENTION	1 year	10 years
OPERATING TEMPERATURE (FULL SPEC COMPLIANCE)	0 to 80° C (case)	0 to 80° C (case)

(1) AFTER MAXIMUM ENDURANCE CYCLES
(2) EFFECTIVE CYCLE IN SEQUENTIAL ACCESS MODE
(3) THE OTHER PERFORMANCE PARAMETERS IN THE GOAL COLUMN ARE BASED ON 256K X 1 CONFIGURATION

TABLE 2. HIGH DENSITY NONVOLATILE MEMORY SPECIFICATION FOR FERAM

SSC Implementation

Our concept for SSC detector implementation is shown schematically in Figure 9. In it, a bundle of scintillating fibers deliver their signals to a photodetector which sends signals to its contiguous NVMEM (NVM). Also shown are network interconnects of the fibers themselves with each other (with readouts to the photodetectors in the photodetector network), of the photodetectors (PD) themselves with each other, and of the NVM devices themselves with each other. Each network is furthermore network interconnected with the data/signal processing layer above and below it in the signal, data, processing, computing, discrimination, decision, etc. sequence. Those network interconnects allow for feed-forward and feed-backward signal propagation which enable the signals/data to actually process themselves, as well as letting the PD and NVM devices to communicate with each other to implement very fast real-time decisions for triggers and recognition of new events. Direct primary outputs from the NVM devices activate the radhard triggers line. Secondary outputs give a real-time write/non-real-time readout to allow greater selectivity in the

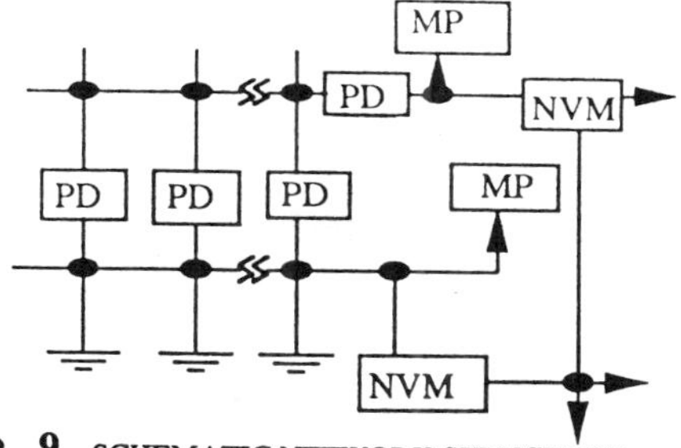

Figure 9. SCHEMATIC NETWORK STRUCTURE

trigger. Those line also network the NVM devices with the main data processing and computation systems for real-time network communication in both directions. Asynchronous timing with sequential readout is used by necessity, not choice. Then time records keep up with the active bits, with some dependence on access times.

The major issues are the densities of the current NVM devices (64K for the MRAM and 256K for the FERAM), and the cost per bit (at about $250 per 256K for the FERAM). Obviously, higher densities are desirable. As for the cost considerations, more implementation analysis will be required. Size is a consideration too, so that hermiticity of the detector is obstructed as little as possible.

Conclusion

The NVM devices discussed here offer considerable opportunities to enhance the operation of any SSC detector, and especially of one based on scintillating fibers (or any scintillation/photodetection systems). They are intrinsically very radiation hard. They will provide a fast delay line for the detector trigger. They will allow 100% storage in real time.

Their non-volatility assures 100% read-out when desired. They can be combined into networks for real-time computation with their signals and form a interactive network layer that cooperates with detector and fiber layers at the signal/data acquisition level, and with data-processing and computational network layers at higher levels in any SSC detector system.. All things considered, they will be a most useful system layer in any SSC detector, including the especially difficult case of detecting the Higgs (H) particle vs. the top quark (t), when we are as yet unsure of the relation of their masses, m(H) vs. m(t). Good triggers, with high discrimination capabilities, and fast real-time memories and processing capabilities, will be necessary to enhance those detections. NVM devices provide just those capabilities.

Acknowledgements

The authors thank A. C. Kuehl, USASDC, for reference to his October 1990 GOMAC paper. Figures 1 and 2 are by R. Fedorak, NADC.; Figure 3 by Honeywell SSEC; Figures 4-6 and 9, by Raytheon EDL.;Figure 7 by J. Benedetto , USALABCOM. For radhardness of FERAM materials see: J. F. Scott et al., "Radiation Effects on Ferroelectric Thin-Film Memories, **J. Appl. Phys. 66,** 1444-1453 (1989).

ELECTRO-OPTIC TRANSIENT IMAGING INSTRUMENTATION DEVELOPMENT AT LAWRENCE LIVERMORE NATIONAL LABORATORY: IMPLICATIONS FOR SSC INSTRUMENTATION DEVELOPMENT*

Mark Lowry, Barry Jacoby, and Hal Schulte
Lawrence Livermore National Laboratory
Livermore, California 94550

Introduction

Over the last decade, the underground weapons physics laboratories fielded by LLNL's Nuclear Test and Experimental Sciences (NTES) program have experienced marked change. This change is characterized by a phenomenal growth in the amount of data returned per event. These techniques have been developed as a result of the severe demands placed upon transient instrumentation by the physics requirements of our underground nuclear laboratories. The detector front-ends must quickly detect, process and transmit a large volume of data to recording stations located approximately 1 km from the event. In a recent event, the detector front-ends successfully handled data at a prompt rate of approximately *13 Terabits/sec.* Largely, this advance can be attributed directly to the increased use of electro-optic techniques.

These highly-parallel high-bandwidth imaging instrumentation systems developed for the test program may have a lot to offer the high-energy physics community tackling the challenge of the unprecedented luminosity and fidelity demands at the SSC. In what follows, we discuss details of a few of our prompt instrumentation techniques and compare these capabilities to the detector requirements for the challenging physics at the SSC.

Underground Weapons Physics Instrumentation

The power of these instrumentation developments is perhaps best illustrated by the x-ray imaging/spectroscopy experiment illustrated in Figure 1. Here the x-ray output from a nuclear event is spectroscopically dispersed by the bent-crystal spectrometer onto a fast fluor material (CdS, WL1201, etc.). The fluorescence is then coupled via coherent fiber arrays to streak cameras approximately 10 meters away. The streak camera is then read out by a very high-speed CCD camera.[1] Spatial sampling of the x-ray output through this system then yields an image of the event that is time resolved, spectrally resolved, and spatially resolved. In effect, a spectrally resolved x-ray motion picture of the event results.

To gain an appreciation for the information content of such techniques, let us first reflect on the number of independent data channels such a system yields. The limit on the number of spatial pixels is usually determined by the number of CCD pixel elements in the spatial direction on the streak camera (typically 380), the spatial resolution of system, and sampling considerations; the resulting number of independent channels is typically 100. The number of resolvable time samples for each of these channels is another dimension to the total data content. Here the time resolution is limited by several system considerations: the number of CCD pixels in the time direction (usually 122), often times the fluor lifetime, dispersion in the optical fiber, and sampling considerations. This time resolution is typically 1 ns, and the experiments are usually configured such that there would be 122 time samples, for a record length of 30 ns. The final dimension of the total data return is then the dynamic range (number of readable bits), and our current systems are capable of a dynamic range of 8 bits (presently going to 10 bits), with a precision of approximately 1%. Finally, we may calculate the prompt data rate as (N n b t)/T. Where N is the number of streak camera systems used, n is the number of channels per camera used, b is the dynamic range of each in bits, t is the number of resolvable time bins, and T is the temporal record length. In a recent event, where sixteen streak cameras were fielded, this piece of arithmetic yields the astonishing value of approximately 13 Terabits/sec for the prompt data capture rate.[2]

As powerful as these fluor-based imaging techniques are, we have found severe trade-offs between sensitivity and bandwidth that renders these techniques unsuitable for some applications that require good temporal response and high dynamic range. Not only does increased fluor speed usually result in less efficiency--through poisoning or short-circuiting of the fluorescent transitions with faster non-radiative channels, the optical output of the fluor is not well-matched to the propagation characteristics of optical fiber. The output of fluors is almost always isotropic and the spectrum is quite broad. These charac-

* This work was performed under the auspices of the U.S. Department of Energy by the Lawrence Livermore National Laboratory under contract number W-7405-ENG-48.

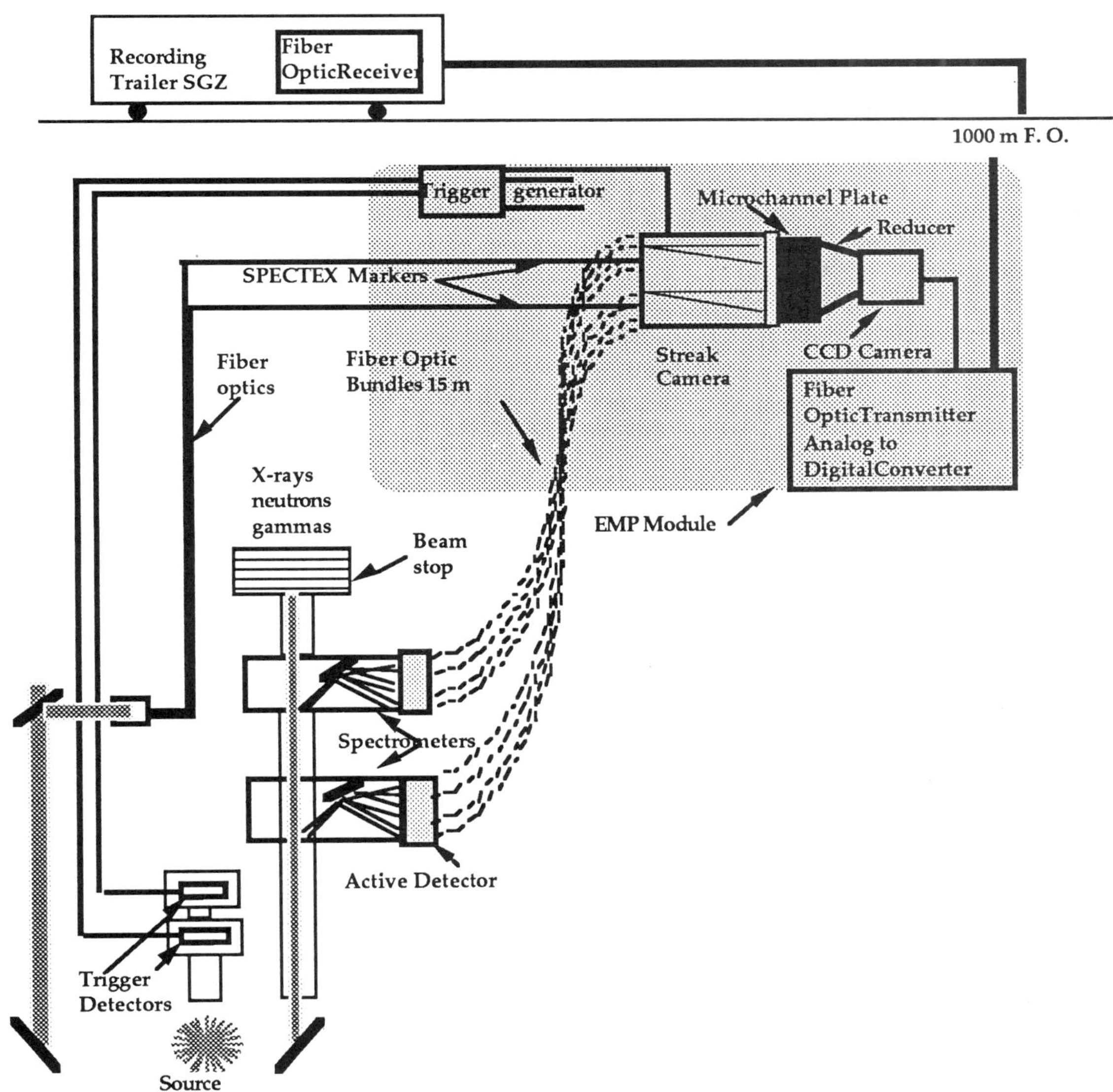

Figure 1. An example of LLNL's transient imaging instrumentation--time-resolved remote spectroscopy.

teristics lead to modal and chromatic dispersion, respectively, and can severely limit the bandwidth of such systems. Modal dispersion in 200 micron core fiber for instance is approximately 80 picoseconds/meter. The only cure is to limit the extent of the spectrum used (or use complicated spectral equalization techniques)[3] and limit the numerical aperture of the system; both measures lead to decreased sensitivity. Thus, for system temporal responses better than 1 ns with good sensitivity, other approaches are sought.

To better taylor the optical source to optical fiber, lasers are the obvious source of choice. However, encoding an electrical signal directly onto the gain mechanism of a laser is very difficult to do well at high frequency: the lasing mechanism is fundamentally very nonlinear in a dynamic sense,[4] therefore, the modulation is often chaotic at frequencies near or beyond the relaxation oscillation frequency. Since 1985, we have been pursuing development of integrated optical modulators to fill the role of encoding electrical signals onto an optical carrier in a very high bandwidth instrumentation system. By separating the functions of light generation and signal encoding, it is much easier to produce a more useful optical carrier while also allowing great flexibility in the characteristics of the signal encoding, i.e. the mapping of voltage or charge to optical power/energy space. For many electronic systems, linearity, not flexibility, of system response is the goal. However, nonlinearities as long as they are well behaved (not dynamic) can be successfully incorporated into instrumentation systems as we show below, and can be a tremendous sensitivity "lever" as discussed in the following paper.[5]

As we have hinted, there are many types of modulator schemes. A fundamentally important design is illustrated in Figure 2. The Mach-Zehnder modulator operates as follows. The optical carrier is injected into the single-mode waveguide on the lithium niobate substrate from a remotely located laser, via a single-mode optical fiber. After the y-split, the carrier is divided into two single-mode waveguide legs. The single-mode nature of the waveguides insures high correlation between the optical wavefronts as they propagate. As they come together again, if the optical path length of the two legs is identical, the two beams are exactly in phase. The in-phase beams couple into the symmetric mode of the single-mode output waveguide--this mode is guided. In the case of a π phase shift, the asymmetric mode of the output waveguide is excited--this mode is not guided and the light is scattered into the substrate. If a voltage is applied to the electrode

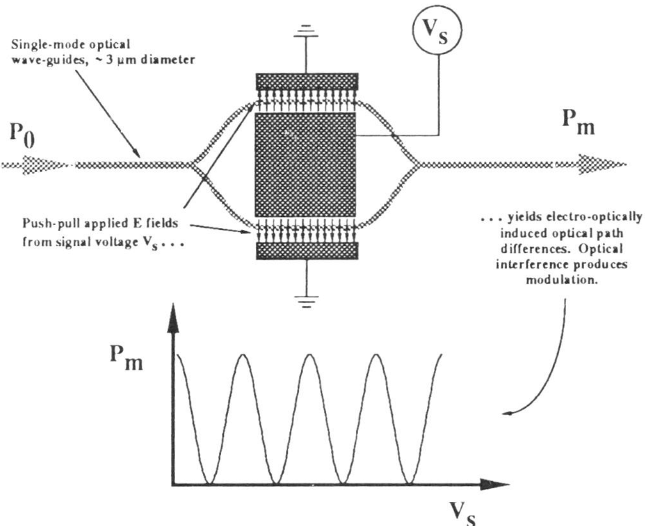

Figure 2. Schematic of an integrated optical Mach-Zehnder interferometer. See text for details of operation.

structure of Figure 2, the optical index is altered via the linear electro-optic effect. In the illustrated design, the two legs are phase shifted in a push-pull fashion. This relative phase shift then leads to a sinusoidal interferometric optical response as a function of the applied voltage. This electro-optic response can be expressed as,

$$P=\eta\left(1-\varepsilon f(V_s)\right)P_{in}\sin^2\left(\frac{\pi V_s}{2V_\pi}+\phi\right)+\eta\varepsilon f(V_s)P_{in} \quad \text{eq. 1}$$

where P is the modulated optical power output, P_{in} is the input optical power, V_S is the applied signal voltage, V_π is the voltage required to modulate from a minimum to maximum (1/2 of an interference fringe), η is the device insertion loss, $\in$ is the low voltage modulator extinction, ϕ is the quiescent phase shift of the device, and f is a phenomenological function that describes the behavior of the modulator extinction at high modulation voltages "f" is very nearly 1 except at voltages much larger than V_π; as V_S grows large, the guiding characteristics of the waveguides are altered such that one will become more than the other, this leads to incomplete interference and reduced extinction at large values of V_S.

It should be stressed here that this response function remains the same over a very large modulation frequency--as high as 20 GHz for some structures. Work to date indicates that as the frequency response is exceeded, the effective modulation voltage decreases, but the functional form of the response remains the same. The non-linearity is thus well-behaved.

Results from a working system are illustrated in Figure 3. These calibration data were taken from a field-system in place (buried over a thousand feet underground) with the data being transmitted over a distance of 1 Km over single-mode optical fiber. The plot compares a calibration waveform, measured with a sampling scope and averaged many times, to a single-shot measurement of the modulator output recorded on a streak camera. The noise in the streak camera measurement is nearly all shot-noise from the low-power laser that was used for this experiment (we are currently developing higher power sources to improve the signal-to-noise of the system). It should further be noted that the calibration waveform was of sufficiently high voltage to drive the modulator response through more than one fringe, i.e. over the peak of the sinusoid of eq. 1. This occurs at the peak value of the signal in Figure 3. The first spike in the signal used a timing fiducial. The sampling scope record of the electrical waveform was mapped through the independently measured interferometric response, eq. 1, on a computer, and that mapping is what is compared to the streak camera record in Figure 3. The good agreement between the two renditions of this rather complicated waveform demonstrates the high degree of fidelity obtainable using these modulators through the extremely nonlinear regions of the modulator response. Therefore, via eq. 1 the modulator response is said to be linearizable and well-behaved.

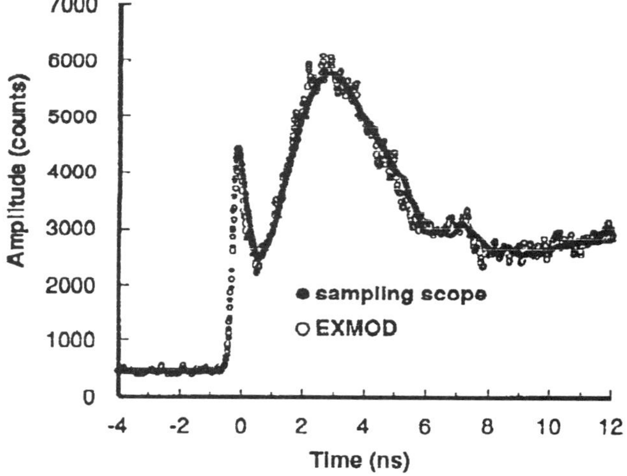

Figure 3. Comparison of a single-shot streak camera recording (open circles) of integrated-optic EXternal MODulator output driven by a calibration waveform and a sampling scope record--averaged many times--of the calibration waveform (filled in dots). See text for details of the comparison.

Conclusions

The instrumentation systems discussed here are in many ways capable of exceeding the requirements of the SSC, especially in terms of temporal response, data-density and remote recording; adaptations of these techniques may prove to be exceedingly well matched to SSC detector requirements.

We see similarities between the changes experienced by our own NTES program over the last decade, and the changes in technology that appear necessary to meet the challenge of high-energy physics (HEP) instrumentation for the SSC and into the next century. While the optimum solution to HEP instrumentation will not be identically what we have reviewed here, it does appear inevitable that the techniques will move in this direction.

Acknowledgements

Thanks are due to Richard Bionta and Mark Eckart of LLNL for recognizing the potential applicability of LLNL/NTES instrumentation techniques to the SSC. We also thank Dick Lear and Dick Fortner of LLNL for their support and encouragement. Several individuals who participated in the first fielding of LLNL integrated optical modulators for the HORNITOS event are recognized for their achievements and contributions to the instrumentation development: Greg Lancaster (LLNL), Dan Nelson (LLNL), and Phil Watts (EG&G).

References

1 M. E. Lowry and B. A. Jacoby, *Ultra High-Speed Single-Shot Electronic Imaging*, invited paper in the Proceeding of the Applications of Electronic Imaging Conference, January 1989, Los Angeles, SPIE Vol. 1082.

2 It should be noted that the single-transient nature of the event allows us to stretch out the transmission time of the bit stream without the pile-up difficulties associated with collider instrumentation. Nonetheless, as we show in the following paper, with appropriate triggering similar techniques may be applied to SSC instrumentation.

3 R. P. Reedy, F. Roeske, Jr., and D. E. Smith, *Fiber-Optic Spectral-Streak Equalizer*, Rev. Sci. Instr. Vol. 56, 7, pp. 1356-1367.

4 See for instance pp. 954-975 in A. E. Siegman, *Lasers*, University Science Books, Mill Valley, California (1986).

5 M. E. Lowry, E. Ables, R. M. Bionta, R. E. Haigh, K. F. Hugenberg, R. Kalibjian, C. F. McConaghy, D. Milton, M. D. Rotter, and H. F. Schulte, *An Electro-Optical Imaging Approach to the Prompt Signal Processing Problem of Mega-Channel SSC Detector Arrays*, the following paper in these proceedings.

AN ELECTRO-OPTICAL IMAGING APPROACH TO THE PROMPT SIGNAL PROCESSING PROBLEM OF MEGA-CHANNEL SSC DETECTOR ARRAYS

Mark Lowry, Eldon Ables, Richard Bionta, Ron Haigh, Keith Hugenberg,
Ralph Kalibjian, Charles McConaghy, Darin Milton, Mark Rotter, and Hal Schulte
Lawrence Livermore National Laboratory
Livermore, California 94550

Introduction

The physics demands of high luminosity at the SSC and the sometimes subtle measurements required to elucidate new physics will undoubtedly tax existing instrumentation. As is the case with most experimental fields, new physics follows from better measurement concepts and technologies. We expect this to be the case with the SSC as well. In what follows, we offer a glimpse of what may be possible using some of the recent results from the emerging technologies in the field of electro-optics.

A Proposed Electro-Optic System Architecture

The large number of channels required for SSC instrumentation and the temporal resolution required lead quite naturally to the consideration of electro-optical imaging techniques.[1] This is particularly true with the need to establish reliable and prompt event triggers that are based upon total energy and event topology, as we shall see with the system outlined in Figure 1.

Here we start with more-or-less conventional HEP detectors that yield charge as the measurement output. This charge is then used to modulate an externally provided optical carrier signal (amplifiers and pulse shaping may not even be required, as will be seen in what follows). The basic ideas regarding the optical modulator are discussed in ref. 1 . The optical carrier may be supplied by some, hopefully small, number of fibers that provide optical power to a chip containing many of the modulators discussed in ref. 1. The optical power division is accomplished on a chip as well (perhaps monolithically integrated with the modulator chip). This large

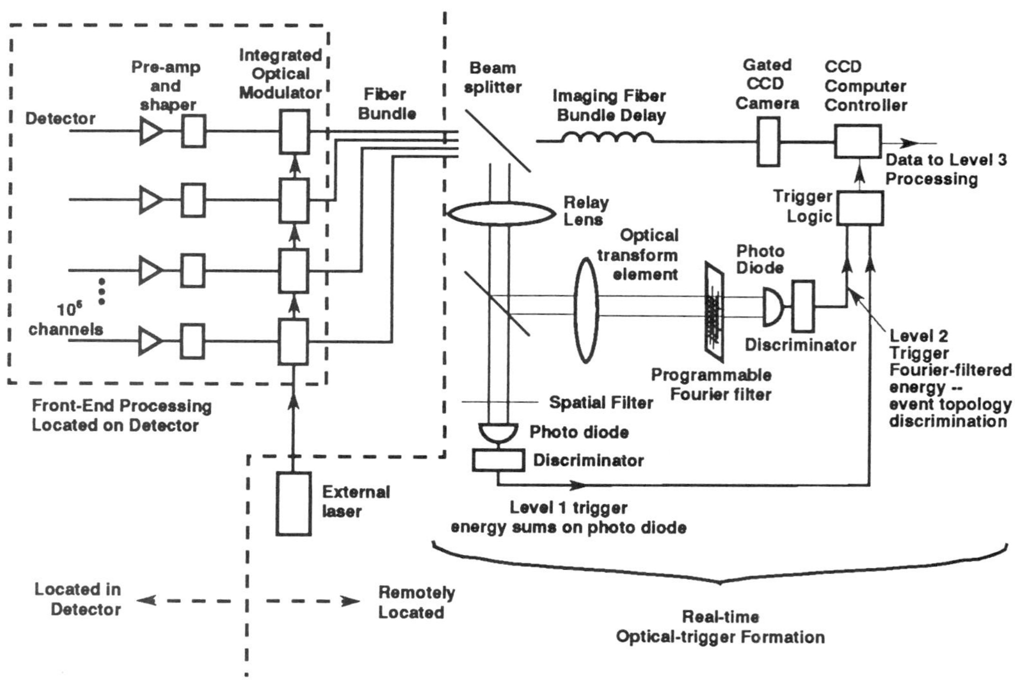

Figure 1. Proposed electro-optic system architecture for high luminosity data acquisition at the SSC. Some key features are the use of ultra-sensitive integrated-optical modulators and the use of optical pattern recognition for trigger formation.

* This work was performed under the auspices of the U.S. Department of Energy by the Lawrence Livermore National Laboratory under contract number W-7405-ENG-48.

number of independent optical channels (perhaps as many as 100 are possible per chip) are brought off chip in individual optical fibers. These optical fibers are arranged in an imaging bundle—such that image information is preserved. In the event of a hit, this bundle is then the conduit of a transient image that propagates to the level-one trigger.

The first step in the trigger generation process is to pick-off a portion of the image power through use of a conventional optical beam splitter. Let us call these beams, after the splitter, the trigger image and the major portion the record image. The trigger image can be split into two beams. In the case of data from a calorimeter, one leg of the trigger beam can be spatially filtered with a filter whose optical density is graded as, $\sin(\theta)$, where θ is the angle with respect to the beam direction for that pixel element. As this filtered image is relayed optically to a large area photodetector, the output charge from the photodiode is then a prompt measure of the total transverse energy—the delay for this part of the trigger formation is just the response time of the photodetector; a few nanoseconds is easy.

The second portion of the trigger image can be used to perform relatively complex event topology discrimination through the use of well-established optical pattern recognition techniques. The most straight-forward technique applies again to the case of calorimetry. By applying the optical Fourier transform to the second trigger image—realized conveniently with the use of simple optical lenses[2]—one can perform spatial frequency filtering operations by simply placing the appropriately patterned spatial transmission filter in the Fourier plane (by using the new technologies of spatial light modulation,[3] the characteristics of this frequency filter could be conveniently controlled electronically). The output of the spatial filter is then relayed to another large area photodetector. The prompt optical sum performed by this photodetector is a measure of the energy content only of features of the event that have a frequency content that is matched to the spatial filter in the Fourier plane. This then allows us to discriminate promptly (a few nanoseconds) on the basis of event topology, as well as total event energy. It should be noted that optical processing has at least one precedent[4] in HEP instrumentation.

Perhaps this optical pattern recognition may be extended to the case of tracking detectors. It may be possible to use the optical Hough transformation to detect the existence of stiff tracks. The Hough transformation maps straight lines into points—one coordinate of the point being the line's angle with respect to the abscissa, and the other is its distance from the origin. Using techniques similar to the optical Fourier transformation, the Hough transformation may be accomplished optically.[5, 6] It may prove possible to optically correlate this trigger with the calorimeter information to provide an elegant and comprehensive trigger, promptly, and with little or no electronics in difficult environs. It must be pointed out that all the optical Hough transformation work that we have seen to date is accomplished with a mechanically rotating optical element, and thus would be far too slow for HEP instrumentation. However, there may be a way to realize the transform in optical fashion instantly, and we are currently exploring that possibility.

We have thus far painted a simple picture of Fourier optics techniques. Our laboratory measurements to date—and some reflection on the Fourier optics theory—would suggest that performing the Fourier transform of the image from an optical fiber bundle may be very difficult. Basic Fourier optics work is done with the use of simple apertures, and the optical wavefronts employed are coherent. In our case, the individual image pixels have probably suffered large relative optical phase shifts. These phase shifts lead to severe interference noise in the Fourier plane. We can suggest three possible ways around this relative phase shift problem.

First, the fact that our optical carrier was derived from a single laser (or perhaps a phase locked set of lasers) may allow us to adjust these relative phases such that when we arrive at the Fourier plane the relative phase shifts have been compensated. Further, the use of integrated optical devices as our signal encoding mechanism may provide a convenient solution. The most fundamental form of optical modulation is the phase shifter. In fact, the Mach-Zehnder modulator discussed in ref. 1 is composed of two such devices. A phase shifter could be employed in each channel to compensate these phase shifts in response to the input from a feedback mechanism. Another phase compensation technique would be the use of a holographic element in the image plane after the bundle, this element could be used to compensate the optical phases of the individual image pixels by focusing the image onto this holographic phase plate.[7] We are exploring both of these phase compensation techniques, and hope to carry-out experiments soon. Finally, the partially coherent image from the fiber bundle could be transformed into an incoherent image. This may be achieved by imaging the fiber output unto an image intensifier. The image intensifier output is then incoherent light from the phosphor screen. Then the incoherent image could be processed using incoherent processing techniques. This offers the added advantage of allowing the image information to be gated; however, to retain the speed of the optical processing fast phosphors must be used—so, the gain of the intensifier may then become an issue.

The remaining beam, the record image, can be refocused onto another imaging optical fiber bundle, and be allowed to propagate for however long is required for complete trigger formation. After trigger formation, the image is captured by a gated ultra-fast readout CCD imager. Image readout times of 10 μs for 10,000 pixels with a dynamic ranges of 1,000 are attainable with today's technology. Extended dynamic ranges could be attained by using a slightly higher power optical carrier and splitting the record image with a 10:90 beam splitter and recording both images.

The advantages of using this electro-optical imaging approach are several:

- The transmission media (small diameter optical fibers with approximately $1cm^2$ cross-sectional area per 10,000 fibers) offers low radiation scattering, and a mechanically manageable cable plan.

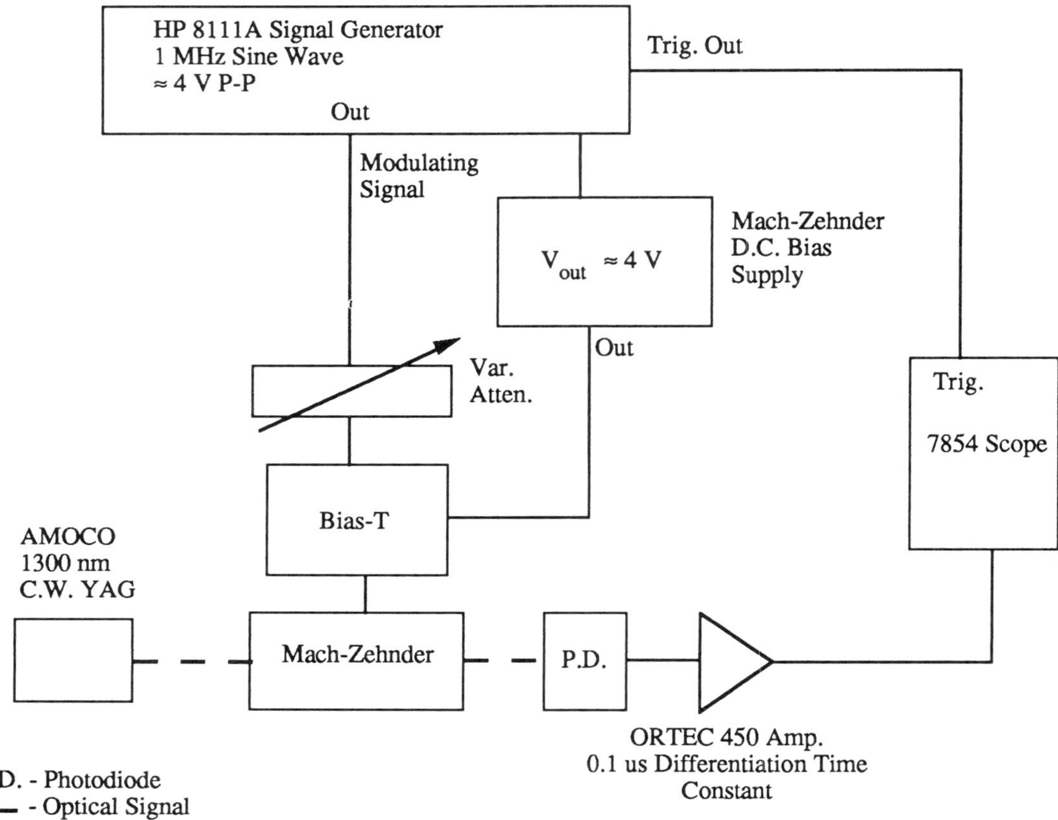

Figure 2. Experimental layout for Mach-Zehnder dynamic range measurements.

- Since all channels are brought outside in parallel, triggering architectures become very flexible and amenable to modification in the future.
- Ground-loops and other forms of EMI become less of a problem.
- The modulators and fiber medium easily support the required temporal response.
- The power consumption can be much less than 1 mW per channel for the power-dissipation critical inner-tracker applications.
- The fiber medium conveniently provides signal delay to accommodate the trigger formation time.
- The potential for operation without amplifiers (discussed in detail below) adds greatly to system reliability—particularly for the high radiation environments.
- Optical pattern recognition and optical summing techniques can be applied for prompt trigger formation.

While this total system architecture is quite attractive in many respects, we should not overlook the value of any one of the roughly three major subsystems: the optical modulators, the image processing techniques, and the fast CCD data capture. We will spend the rest of the paper addressing the optical modulators.

Integrated Optical Modulator Results

While we have been pursuing instrumentation development with modulators for several years, we have only recently considered them in the context of HEP experiments. To that end, we set out to understand how they might best be used in HEP experiments. A potential application of the technology is in calorimetry experiments, where exceptional dynamic range is required. We have made measurements of dynamic range at 800 nm, using 2x2 modulators, and measurements at 1300 nm using Mach-Zehnder modulators. Due to the limited space, we will only discuss the Mach-Zehnder measurements here.

The experimental setup is outlined in Figure 2. We used a sine-wave electrical input for convenience, but the conclusions should apply to pulse operation as well. The bandwidth of the system was limited by the amplifier at 1.5 MHz. The bias applied to the modulator was adjusted by observing the signal-to-noise of a very weak signal; at the point of maximum signal-to-noise, the bias voltage level was fixed. Figure 3 shows the result of three different levels of electrical attenuation (on the sinusoidal input signal) ranging from 0 dB to 80 dB (voltage). Even with 80 dB attenuation (x10,000), the output signal from the modulator is easily seen on the scope trace. For the 0 dB trace, the negative going portion of the sinusoidal input is distorted as it passes through the minimum in the sinusoidal response function (see ref. 1), however the positive going portion does not go through the response peak (indicating that the modulator was not biased precisely at the half-power point). Concentrating on the positive going portion of the sinusoid, we see that the input signal goes through a dynamic range of at least 10,000. At the lowest input signal level (80 dB attenuation), the signal is still much above the noise. By analyzing the rms noise of our system with the light beam blocked, and with the beam unblocked but no modulation signal, we find that at the attenuation value of 10,000 the ratio of the peak signal to the noise due to just the laser/modulator combination is ~ 4. However, these noise measurements were made from scope traces of the same time scale as Figure 3; here the noise bandwidth is determined by the amplifier roll-off at 1.5 MHz, and the record length at the low end; we find this to be 1.3 MHz. Extrapolating to the case of a 10 MHz bandwidth (more suitable for most SSC applications), we find that this increase in bandwidth should lead to a noise increase of a factor of ≈ 2.77. Thus operating with a bandwidth of 10 MHz, we would expect to find a dynamic range of ≈ (4/2.77)(10,000) or slightly more than 14,000. Further, since we know that this signal can be linearized, we can establish a linear relationship between the optical power from the modulator and the input voltage to the modulator.

At this point, it is convenient to discuss the relationship of the charge (or voltage sensitivity) to the parameters of the carrier modulator system. The optical power output of most modulators can be conveniently described by

$$P=P_0F(V) \qquad \text{eq. 1}$$

where $P_0 = \eta P_{in}$ is the maximum carrier power out of the modulator—when F(V)=1, η is the insertion loss, P_{in} is the input optical power, and F(V) is some modulation function of charge or voltage. The uncertainty of the measured optical throughput, ∂P, is made up of three contributions: carrier power instabilities, detector noise, and amplifier noise. We may lump these uncertainties together into the total uncertainty in the measured power, ∂P. The voltage uncertainty due to fluctuations in the measured optical power is given by

$$\delta V=(\delta P/P_0)/(dF/dV) \qquad \text{eq. 2}$$

assuming that ∂P is small compared to P_0, and dF/dV is not close to zero. Any fluctuations from the modulator itself[8] are then added in quadrature with this uncertainty to arrive at the total uncertainty in the measured voltage.

For the case of the Mach-Zehnder, reproducing eq. 1 from ref. 1 we note that

$$P=\eta\left(1-\varepsilon f(V_s)\right)P_{in}\sin^2\left(\frac{\pi V_s}{2V_\pi}+\phi\right)+\eta\varepsilon f(V_s)P_{in} \qquad \text{eq. 3}$$

where P is the modulated optical power output, P_{in} is the input optical power, V_s is the applied signal voltage, V_π is the voltage required to modulate from a minimum to maximum (1/2 of an interference fringe), η is the device insertion loss, ε is the low voltage modulator extinction and usually $\varepsilon < 0.01$, ϕ is

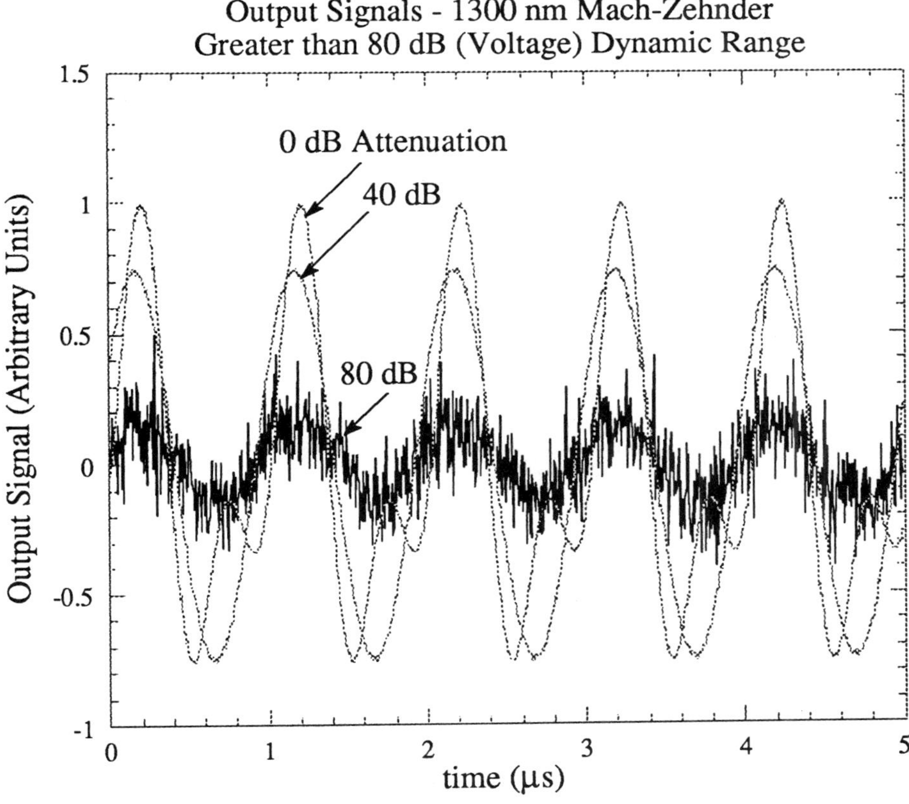

Figure 3. Results of Mach-Zehnder dynamic range measurements. A dynamic range well in excess of 10,000 is demonstrated.

the quiescent phase shift of the device, and f is a phenomenological function that describes the behavior of the modulator extinction at high modulation voltages "f" is very nearly 1 except at voltages much larger than V_π[9]. With dF/dV at its maximum (i.e., at the half-power point, in the Mach-Zehnder case), we find,

$$\frac{\delta Q}{Q_\pi} = \frac{\delta V}{V_\pi} = (\delta P/P_0)\frac{2}{\pi(1-\varepsilon)} \qquad \text{eq. 4}$$

so the voltage (charge) uncertainty is proportional to the relative optical power uncertainty, scaled by $V_\pi(Q_\pi)$.

For the measurements shown in Figure 3 extrapolated to 10 MHz, $\delta v/v_\pi$= 1/(dynamic range) = 1/14,000 = 7.1×10^{-5} Equivalently, $\delta Q/Q_\pi = 7.1\times10^{-5}$ Alternatively one may extract the relative optical power uncertainty using eq. 4, $\partial P/P_0=1.1=10^{-4}$. The modulator used for the measurements in Figure 3 was an in-house designed and fabricated high-bandwidth traveling wave Mach-Zehnder. In general, for SSC applications it appears that lumped element electrode structures may be preferred; to understand the charge sensitivity possible for SSC applications let's consider the lumped element case. Becker[10] discusses a figure-of-merit for lumped-element modulators: $M=V_\pi L$, where L is the length of the electrode structure. Becker finds that M= 14 V-mm and that the capacitance/electrode length ρ =0.85 pF/mm . Thus, one can easily calculate the amount of charge deposited on the electrode structure to yield a voltage of V_π; $Q_\pi=M\rho$=12 pC. Applying this to our dynamic range measurements above yields the result that the minimum charge that we are capable of detecting would be 12 pC/14,000 = 0.86 femtocoulombs. This assumes our value of the relative optical power uncertainty extrapolated to 10 MHz bandwidth, and the Q_π of a lumped element modulator structure. We are further making the tacit assumption that the electrode termination resistance is large enough (or actively gated) so that the charge will not dissipate on time scales consistent with the desired bandwidth.

It is useful to consider the shot noise (counting statistics) limitations on the dynamic range. For our optical power level of ~ 5 mW—at the Mach-Zehnder bias point, a projected bandwidth of 10 MHz, at a wavelength of 1300 nm, and a detector quantum efficiency of 90%, the shot noise is easily calculated using the usual Poisson distribution assumption. Given these circumstances we find that the shot noise fluctuations in the measured optical power would be 1.8×10^{-5} of the average power, i.e. $\partial P/P=1.8\times10^{-5}$, where ∂P is the fluctuations in the measured power (assuming only shot noise) and P is the average measured power. So we are a factor of 6.1 away from realizing a shot noise limited system. For the ideal optical shot-noise limited system, operating, with an optical power of 5 mW at the receiver, the charge sensitivity could be as good as 0.14 femtocoulombs. However, at these low charge levels, we are probably approaching the level of the thermal charge fluctuations in the modulator capacitance: at room temperature with a capacitance of 5 pF the thermal noise would be of order 1000 electrons or 0.16 fC (see ref. 8).

The dependence of the sensitivity on the slope of the modulation function , eq. 2, has led us to consider resonant modulator structures. These structures exhibit a modulation function that consists of resonances, i.e. the device is characterized by a "Q"—or finesse in the optical vernacular. These kinds of devices may be of interest where extremely high charge sensitivity is required but dynamic range is not. Most tracking detectors, by virtue of their necessarily small volumes, would seem to fall into this category.

This kind of modulator is very simple in design. It consists of a single waveguide with dielectric mirrors deposited on the input and output facets of the chip to form an optical cavity. The resonances in transmission are a function of the ratio of the optical length of the cavity and the optical free-space wavelength. For a Fabry-Perot modulator, the modulation function, in analogy with eq. 2 can be derived by following Born and Wolf[11] but including a loss factor,

$$G(\delta)=\frac{(1-\mathbf{R})^2 e^{-\alpha L}}{(1-\mathbf{R}e^{-\alpha L})^2+4\mathbf{R}e^{-\alpha L}\sin^2(\delta/2)} \qquad \text{eq. 5}$$

where α is the optical loss coefficient, R is the reflectivity of the cavity ends, δ is the optical phase difference; $\delta=4\pi nL/\lambda$, where n is the refractive index. Electrodes are then deposited on either side of the single waveguide. The electrodes are used to modulate the index of refraction and hence the optical length of the cavity. Thus, this structure can be used as a tuneable monochromater or a very sensitive modulator that responds to applied voltage or charge.

Figure 4 shows data taken at LLNL from an integrated optical Fabry-Perot modulator of LLNL design and fabrication. Here the resonance is clearly seen. In this figure, the finesse, *F*—the ratio of the free spectral range (distance between peaks) to the FWHM of the resonance—is approximately 12. This data was acquired by scanning the optical spectrum; which is equivalent to applying a voltage or charge. One can easily relate the finesse of this device to the maximum slope of the modulation function. An analytic expression is possible from eq. 5, however, the essential point can be made with far less algebra, by approximating the resonant peaks with triangular spikes. If we constrain the peak of the spike to correspond to the peak of the resonance and also force it to pass through the FWHM of the resonance, we arrive at a very simple approximation for the slope of the resonant peak, especially for a finesse that is large,

$$\frac{dG}{dV} \approx F/(2V_\pi) \qquad \text{eq. 6}$$

where *F*- is the finesse. Applying eq. 2, we may calculate the voltage and charge uncertainties (hence, theoretical sensitivities) for the Fabry-Perot modulator as,

$$\frac{\delta Q}{Q_\pi} = \frac{\delta V}{V_\pi} = (\delta P/P_0)\frac{2}{F} \qquad \text{eq. 7}$$

Naturally, this sensitivity scales with the inverse of the finesse which can in principal be made quite high. Given the same optical noise parameters of our MZ measurements and using the finesse value from Figure 4, we would estimate under these conditions that the charge sensitivity of the FP modulator would be 0.3 femtocoulombs or ~ 2000 electrons.

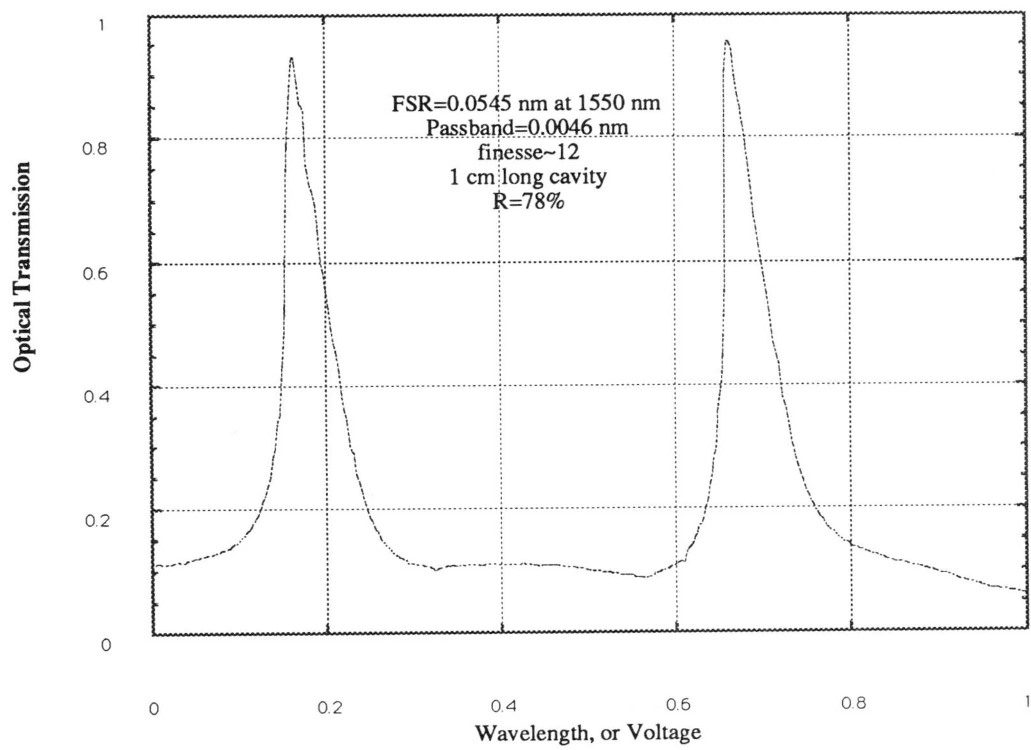

Figure 4. The modulation shape function for a Fabry-Perot structure. Note the resonant peaks in the response.

The sensitivity leverage of the resonant Fabry-Perot modulator is clear. We could presumably extract even better performance from a FP with a higher finesse and/or shot-noise limited optical detection, however we are again approaching the level of the thermal noise in the electrode structure itself (ref. 8).

A word of caution about such resonant structures is in order: Their excellent sensitivity may come at the price of significant instability, with respect to temperature in particular. However, it may be possible to compensate these instabilities with a feedback mechanism or conceive device designs that are self compensating, yet resonant—the Mach-Zehnder is an example of a self-compensating structure, unfortunately it is not resonant. The Mach-Zehnder's two parallel paths pass through the crystal so closely that any temperature-induced optical path change is experienced identically by both. Perhaps it is possible to design a resonant structure that also has this feature.

Modulator Power Dissipation

For many SSC applications, the power dissipated by subsystem elements is critical. For the case of the modulators, the dissipated power is from two sources the optical carrier and the charge driving the modulator. As we shall see, the most significant component is the optical power.

The optical power dissipation will be approximately equal to $(3/2)P_0$—on the order of 1/2 of the input optical power will probably be lost due to inefficiencies in optical coupling, and operation at the half power point will then account for an additional $1/2P_0$.

The energy dissipated due to the modulating charge is calculated as $q^2/2C$. Since the MZ will be the least sensitive of the two devices discussed here, we will consider it as the worst case. Taking C=2.55 pF and assuming $q=Q_\pi=12$ pC, we find that there will be 28 pJ of energy dissipated per "hit," maximum. Then, if every pixel registered a hit every beam crossing, the dissipated power from the charge would be 28pJ/ 1.6 nS = 1.8 mW. However, since the detector elements that *are not* hit far outnumber those that *are*, on average, this power dissipation term will be reduced by a couple orders of magnitude. So we would expect the charge dissipation power to be of order 10 microWatts, or less, on average. It should be pointed out that no matter how the charge is readout the energy must be dissipated.

While on the subject of power dissipation it should be noted that given the ultimate shot-noise limited system, where the optical noise goes as the $P_0^{1/2}$, there will be a trade-off between noise--hence, sensitivity—and optical power dissipation. This tradeoff should be considered when determining the optimum modulator design for a particular SSC application.

System Costs

There is not space here for a comprehensive discussion of projected costs. However, it is clear that without significant advances in integration density (number of modulators per chip) and automated packaging techniques the per channel costs will be prohibitive. We hope to pursue, with industrial partners, both of these critical areas.

Radiation Sensitivity

While some encouraging data on the radiation sensitivity of these integrated optical modulators does exist,[12] there is much that remains to be done at SSC dose-rate levels. And we have begun a program of investigation for this important issue for modulators at the SSC.

Conclusions

We have presented dynamic range, and sensitivity data for two distinct classes of integrated optical modulator. These measurements demonstrate that these devices may prove very useful for both calorimetry and tracking. Further, we have outlined a comprehensive electro-optical imaging system architecture that includes a novel and elegant optical approach to trigger formation and the means of capturing the triggered event data.

As high energy physics moves into the twenty first century with the completion of the Superconducting Super-Collider, we believe that only by incorporating the best and most sophisticated technology available will the physics move forward as efficiently as possible. We hope that we have demonstrated, in a substantive fashion, that electro-optics has a very large role to play.

Acknowledgements

We would like to thank Dan Marlow of Princeton University for many helpful discussions and much encouragement. Thanks are also due to Maged Atiya of Brookhaven for his encouragement and critical evaluation of some of these ideas. We also appreciate the time and consideration given by several researchers to some aspects of this work during a recent visit to Brookhaven: Veljko Radeka, Paul O'Connor, Venetios Polychronakos, Howard Gordon, William Willis, and Wilfred Cleland. David Nygren from Lawrence Berkeley Laboratory has also provided useful discussion and encouragement. As relative newcomers to the field, we gratefully acknowledge the help of many HEP researchers, too numerous to list, who attended the '90 Snowmass meeting and played an important role in shaping some of these ideas through their interest and constructive criticism.

We thank Richard Becker of Integrated Optics Circuit Consultants for useful discussions on modulator sensitivity.

At LLNL, we would like to thank Hye-Sook Park for her suggestion of the optical Hough transformation, Stan Thomas for useful discussions about current image intensifier technology, Wolfgang Stoeffl and Craig Wuest for their interest, and Orrin Fackler, Chip Britt, and Tony Chargin for their interest and encouragement.

References

1 M. E. Lowry, B. A. Jacoby, and H. F. Schulte, *Electro-Optic Transient Imaging Instrumentation Development at Lawrence Livermore National Laboratory: Implications for SSC Instrumentation Development,* in these proceedings (previous paper).

2 J. W. Goodman, *Introduction to Fourier Optics*, McGraw-Hill, San Francisco (1968)

3 W. P. Bleha, L. T. Lipton, E. Wienen-Auner, J. Grinberg, P. G. Reif, D. Casasent, H. B. Brown, and B. V. Markevitah, *Application of the Liquid Crystal Light Value to Real-Time Optical Data Processing,* Optical Engineering, Vol. 17, #4, pp. 371-384 (1978); note: Commercial products are now available from Hughes.

4 W. E. Cleland, D. E. Kraus, and J. A. Thompson, *Optical Trigger Processor for High Energy Physics,* Nuclear Instruments and Methods, Vol 216, #3, pp. 405-414, New-Holland, Amsterdam (1983).

5 W.H.Steier and R.K.Shori, *Optical Hough Transformation,* Applied Optics, vol. 25, # 16, pp. 2734-38, 1986

6 G.Eichmann and B.Z.Dong, *Coherent Optical Production of the Hough Transformation*, Applied Optics, vol. 22, #6, pp 830-34, 1983

7 Reynolds, Devetis, Pamert, and Thompson, *The New Physical Optics Notebook: Tutorials in Fourier Optics,* pp. 323-325, SPIE Optical Engineering Press, Bellingham, Washington (1989).

8 At least two potential sources of noise from within the modulator must be considered: The thermal noise of the electrode capacitance $\partial Q=(CkT)^{1/2}$ (integrated over all frequencies) and the ferro-electric noise from within the crystal itself. For a discussion of these two sources see D.A. Bell, *Noise and the Solid State*, Wiley (1985) At room temperature the thermal noise appears to dominate.

9 As V_S grows large, the guiding characteristics of the waveguides are altered such that one will become more lossy than the other—this leads to incomplete interference and reduced extinction at large values of V_S.

10 R. A. Becker, *Multigigahertz Lumped-Element Electro-optic Modulators*, IEEE J. Quant. El., QE-21, 8, pp 1144-1146 (1985).

11 See for instance p. 327 Born and Wolf, *Principles of Optics*, Pergamon Press, Oxford (1986) for a discussion of Fabry-Perot interferometers.

12 F. Roeske, Jr., D. R. Jander, G. D. Lancaster, M. E. Lowry, G. McWright, R. T. Peterson, and W. E. Tindall, *Preliminary Radiation Hardness Testing of Lithium Niobate Optical Directional Coupler Modulators Operating at 810 nm*, Proc. SPIE Technical Symposium Southeast on Optics, Electrooptics, and Sensors, Orlando, FL (May 1987).

A LOW POWER TIMING DISCRIMINATOR FOR SSC APPLICATIONS

F.M. Newcomer, S. Tedja
University of Pennsylvania
Philadelphia, Pa., 19104

Abstract

A bipolar monolithic discriminator with programmable threshold and hysteresis is being developed for proportional drift tube tracking sensors. Design goals include low power(4-8mW), sub-nanosecond cluster detection accuracy, and uniform channel to channel response. Technology, circuit design and expected performance are discussed.

Introduction

In a proportional wire tracking detector each sensor is instrumented with a signal amplifier, shaper and discriminator. The discriminator serves the dual purpose of providing a logic pulse to indicate that an element has been triggered and a timing edge to determine the closest point of approach of the track to the wire.

Critical design parameters for an SSC wire chamber discriminator are power, time slewing and chip to chip threshold matching.

Technology

Silicon bipolar technologies offer the best choice of trade-offs in speed, power, matching and reliability. Bipolar transistors have the highest transconductance, g_m, per unit standing current of any available technology. The typical, on chip, matching of the base emitter, or controlling, voltage is $1mV$ or less. This low offset voltage allows the use of a simple emitter coupled pair as a discriminator input stage without trimming or the use of other offset cancellation techniques. It also may lead to a reduced the gain requirement in earlier signal processing stages which will directly affect the power.

Low values of collector substrate capacitance, C_{cs}, and collector base capacitance, C_{cb}, in the advanced bipolar processes allow the designer to depend more heavily on resistive elements for gain. In the Tektronix *SHPi* process C_{cb} is 39fF and C_{cs} is 24fF for a minimum size transistor. These values are the same or smaller than the expected interconnect capacitance. A direct benefit of low transistor capacitance is high bandwidth at low power. A minimum size *SHPi* transistor has an expected unity gain bandwidth of $3GHz$ at $100\mu A$ of collector current.

Design Considerations

Power

The fixed logic output swing and requirement for a nearly constant internal delay requires a circuit with both high gain and bandwidth. Typical commercial comparators with ECL outputs and good delay versus input overdrive characteristics, consume several hundred milliwatts per channel and require off chip components to provide threshold and hysteresis feedback. The power budget for an entire channel of SSC wire chamber electronics is less than $25mW$.

Power depends on technology, output logic levels, drive load capacity, input range requirements and the details of the circuit design.

Output Signal

The discriminator couples directly to a time conversion unit located a few cm away. Differential outputs will minimize feedback to the preamplifier. Since the electronics is expected to be fully custom, the output logic levels can be set to any convenient magnitude.

A bipolar differential pair, or CMOS line receiver with an internal gain stage, can be fully switched with an input voltage difference of $150mV$. We have chosen this as the minimum acceptable logic swing for modeling purposes. Open collector outputs in this design allow larger logic levels, at the expense of output rise time.

Gain

Given the $150mV$ logic levels described above, a conventional analog amplifier would require a gain-bandwidth of more than 50GHz to switch its outputs (10 to 90%) in 1ns with a $1mV$ input. High speed bipolar transistors, operating at low current, rarely have a unity gain

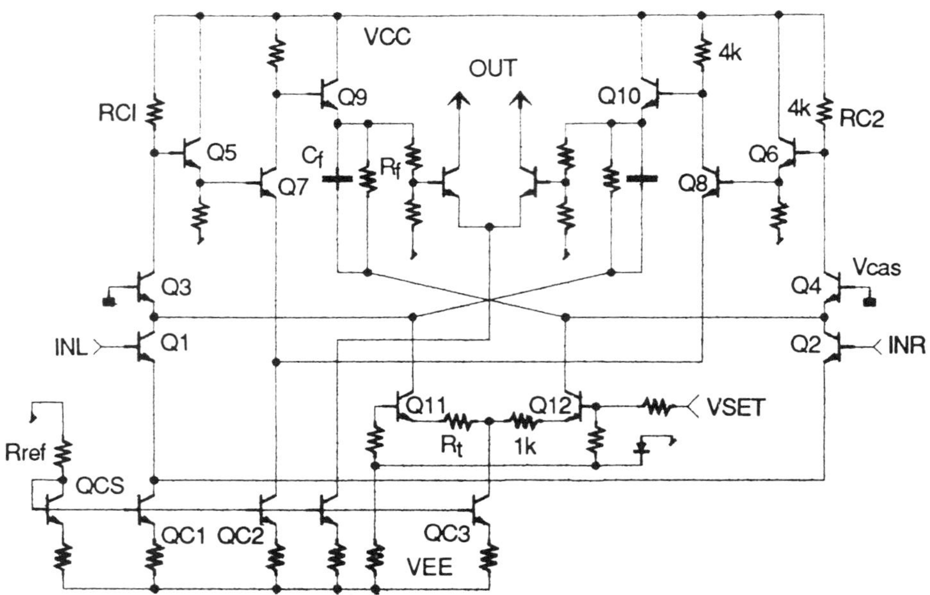

Figure 1: Prototype Wire Chamber Discriminator Circuit.

bandwidth, f_t, greater than 5GHz, so it is necessary to employ multiple gain stages to achieve a comparable level of performance. Since the objective is a digital output, positive feedback can be employed to boost the near crossing point gain.

Circuit Description

We have designed a prototype timing discriminator based on requirements for an intermediate tracking, proportional drift tube system at SSC. General characteristics based on known process variations and HSPICE simulations using model parameters for the Tektronix *SHPi* process are given below.

Performance Specification	
Power Dissipation	5mW
Time Slewing	Less than 1ns/decade of overdrive
Input offset	$1mV$ or less
Internal Threshold	Greater than $50mV$
Threshold Uniformity (chip to chip)	Better than 10% of setting

A simplified schematic is shown in figure 1. This design is similar to one previously developed for a fixed target silicon strip detector using the Tektronix *SH3* process. The major performance differences are in power and overdrive response [1].

Referring to the schematic, INL and INR are the differential inputs. If the outputs of the shaping stage are balanced to within an acceptable threshold error, they may be directly coupled to INL and INR. The DC level at the input may be between $\pm$ $600mV$.

Gain Stages

The input transistors, Q1 and Q2, operate at a quiescent current of $100\mu A$. They perform a voltage to current conversion with a gain, A_i, equal to the g_m of Q1 or Q2. This may be defined:

$$A_i \equiv \frac{i_o}{V_{in}} = \frac{I_c}{v_t}. \tag{1}$$

I_c is the $100\mu A$ collector current and v_t is a constant equal to $26mV$ at room temperature.

Q1 and Q2, together with cascode transistors, Q3 and Q4, form the first stage amplifier. Similarly Q7 and Q8 with a quiescent current of $50\mu A$ each, form the second stage amplifier.

The small signal voltage gain of a differential pair with load resistors R_l is:

$$A_v = g_m R_l; \quad (g_m = I_c/v_t). \tag{2}$$

In a monolithic design, semiconductor resistors offer the lowest capacitance choice for R_l, but process variations of up to $\pm30\%$ will directly affect A_v unless some form of compensation is applied.

The DC gain is stabilized by making the reference current in QCS depend on a resistance of the same type. This current is ratioed among QC1, QC2, and QC3, which contribute to the gain of the first two stages.

Assuming a small voltage drop across the emitter resistor of QCR and setting the current in QC1 to four times the reference current in QCR, the transconductance above may be expressed as follows:

$$g_m = 2\frac{VEE - v_{be}}{v_t Rref} \tag{3}$$

effectively placing the same process dependent resistance in the denominator of A_v.

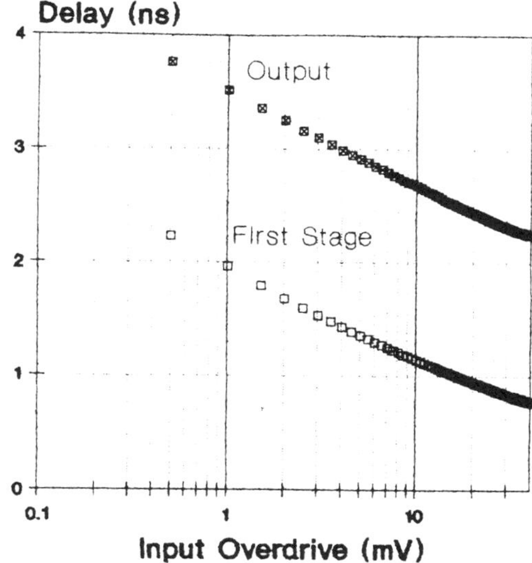

Figure 2: The plot above shows the delay from input to an output logic swing of 75% and the delay in the first stage from input to a $20mV$ cross over of the collector nodes.

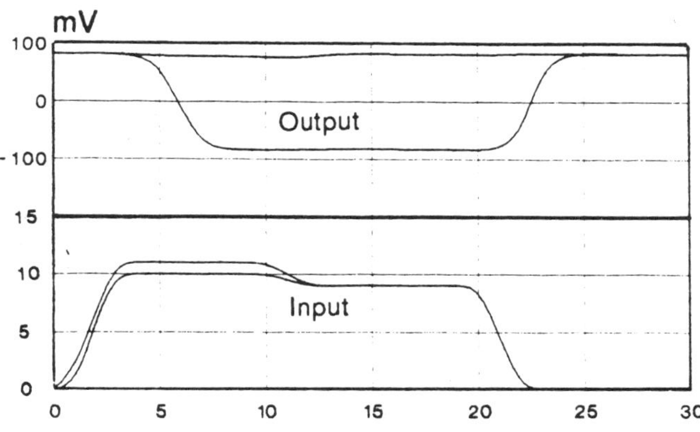

Figure 3: HSPICE simulation showing the effect of hysteresis.

The first and second stage gain is set to 15 and 7, respectively, by using 4kΩ collector resistors. Since the collector node time constant is nearly the same for both stages and the gain of the first stage is much larger than the second, the first stage dominates the overdrive delay. Figure 2 is a simulation result showing the elapsed time from input to a $20mV$ cross over in the first stage and from input to a 75% output transition, as a function of input overdrive. This plot clearly shows that the output overdrive characteristic is determined in the first stage.

Threshold and Hysteresis

The threshold and hysteresis circuits operate by supplying offset currents into resistors RC1 and RC2, shifting the differential voltage across the collector nodes.

The threshold input, VSET, is attenuated by a factor of ten and used to set a voltage difference across the transistor pair QTL and QTR. In the linear operating region, the current difference between QTL and QTR is approximately $VSET/10R_e$, where R_e is the sum of the resistor, Rt, and the emitter node impedance of one of the threshold setting transistors. Using A_i to relate this to an equivalent input voltage, V_{in}, defined in 1 and replacing the output signal, i_0, with the current difference in QTL and QTR, the effective threshold is:

$$V_{th} = \frac{v_t}{20}\frac{Rref}{R_e}\left(\frac{VSET}{VEE - v_{be}}\right) \tag{4}$$

where V_{in} has been re-labeled V_{th}. The threshold dependends on ratios of voltages, ratios of resistors and v_t . The inherent temperature of dependence v_t is partially compensated by the shift in v_{be}, resulting in an expected threshold shift of $.13\%/^{\circ}C$. Resistor ratioing errors are typically 1% or smaller for this process and should not contribute significantly to the chip to chip threshold variations.

Assuming an input offset error of $1mV$ or less, the gain of the shaper can be set so that a typical minimum signal will be 10 to $20mV$. The discriminator threshold setting would then be in the range of 5 to $15mV$ depending on amplifier and system noise. The prototype design maintains a linear response to the threshold reference, VSET, up to $60mV$ of equivalent input threshold.

A standard technique to eliminate output oscillation due to noise on near threshold signals, is to add a small amount of positive feedback, or hysteresis. This positive feedback may be tuned to greatly improve the gain of the input stage and therefore reduce the output delay for near threshold signals.

In this design, a resistor and capacitor pair, Rf and Cf, connect cross coupled from the emitter outputs of Q9 and Q10 to the cascode input nodes at Q3 and Q4. The capacitor and resistor are matched (within expected process variations) to provide the same level of positive feedback. Rf is chosen to provide a current difference equivalent to a few mV input offset. This value, added to V_{th}, determines the effective input threshold. When the output changes state, the hysteresis offset is reversed, lowering the threshold. Figure 3 shows an example this threshold shifting behavior. The two input pulses differ by only $1mV$ in their initial magnitude. After 10ns they both revert to a value $2mV$ below threshold. The output is triggered only once and remains *on* for the duration of the input pulse due to the lowering of the effective threshold.

This design is presently being prepared for fabrication in the Tektronix *SHPi* process.

References

[1] D. Christian et al., "The Development of two ASIC's for a Fast Silicon Strip Detector ", IEEE **NS36**, 507-511, (1989).

Superconducting Magnets

CONCEPTUAL DESIGN OF A SUPERCONDUCTING TOROIDAL MAGNET FOR THE SDC DETECTOR

John Purcell and Lew Creedon
Advanced Cryo Magnetics, Inc.*
7390 Trade Street
San Diego, CA 92121

Abstract

Advanced Cryo Magnetics, Inc. has completed a conceptual design and preliminary cost estimate for the toroidal magnets required for the Solenoidal Detector of the Superconducting Super Collider. The objectives of the design were to fulfill the physics requirements, minimize the cost, and keep the fabrication as simple as possible. In addition, new technology requiring extensive development was avoided.

Introduction

A conceptual design of the toroidal magnet for the solenoid detector for the SSC is presented. The major physics parameters are the thickness of the aluminum end plates and the magnetic field times the particle path length. This design provides a field times length of 8 tesla meters. The average end plate thickness is almost entirely aluminum. The design is simple and will give excellent reliability. Due to the difficult geometry, considerable effort was spent on an assembly procedure to assure the magnet could be built.

Conductor

After considering the various methods of stabilization, a simple bath cooled concept was chosen. Indirectly cooled conductors are not at all suitable for this magnet because a large amount of conductor motion is unavoidable. Indirect cooling cannot provide the necessary stabilization for large conductor motion. Force cooled conductor could probably be made to work in this magnet but it is much more expensive and may require costly development. Also the reliability of force cooled magnets is more dependant on a continuous source of refrigeration than bath cooled. Bath cooled magnets can easily tolerate an interruption in liquid supply as long as the level in the magnet helium vessel is maintained.

Monolithic, aluminum-stabilized NbTi, such as has been used in the "thin" solenoid detectors, is the chosen conductor. The aluminum minimizes the interaction with the high energy particles and provides low electrical resistance for good stablization. The conductor is wound on a former that will become the inner wall of the helium vessel. The total winding consists of 90 bundles of conductor, with 54 turns in each bundle. As the magnet is energized the conductor will be forced against the outer wall of the helium vessel. This will result in a considerable inelastic motion which means that the conductor must be very stable. An additional requirement on the conductor is that it can tolerate a small radius bend, less than 1/2 inch, at the corners of the magnet. This is an important cost feature because shaping a radius of even 1 inch at the corners of the structure is an expensive forming job. The bath cooled, monolithic, aluminum-stabilized conductor, with good helium ventilation, meets all the conductor requirements very nicely.

Stability

The monolithic, aluminum-stabilized conductor is 2×0.39 cm and operates at $7,000$ amperes. High purity aluminum, with some degree of work hardening, has a resistance ratio of about 400 to 1 at the operating field. This gives a resistance of 7.7×10^{-9} ohm per cm length. A current of $7,000$ amperes in the normal conductor produces 0.377 watts of heating per cm length; with 50% of the surface exposed to the liquid helium, the required heat transfer is 0.188 watts/square cm. With such a low heat transfer rate the conductor is completely stable. It is imperative the conductor in this magnet be very stable because as the magnet is charged, the conductor will move to transfer the magnetic load to the outside wall of the helium vessel and thus to the support structure.

The coil is well ventilated with a large volume of liquid available to absorb any energy release due to conductor motion. Turn-to-turn insulation will consist of a

*This work supported by Argonne National Laboratory

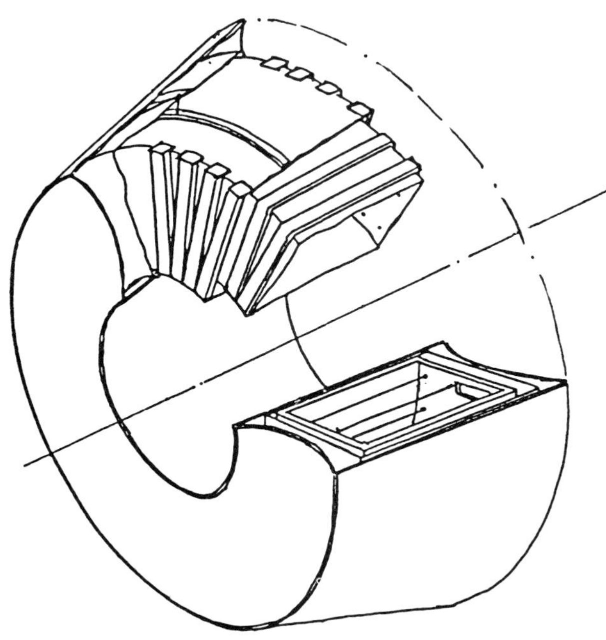

Figure 1. Cutaway view of the torus. The total end thickness is 20 centimeters.

1/16 inch thick strip in linen-filled phenolic with slots machined in the surface to provide good conductor support and expose 50% of the aluminum surface to the liquid helium. Thick phenolic will be packed around the conductor bundle to space it well away from the metal trays that hold it in position. The spacers will provide electrical insulation and still allow good helium circulation.

Quench Considerations

The total stored energy of about 300 megajoules in the coil is enough to warm about 15% of the conductor to room temperature. Helium vessel vent size and position need to be designed to ensure that during a quench, the liquid level is lowered enough to expose at least 15% of the conductor. The normal region will then propagate throughout the exposed conductor volume and distribute the energy, resulting in a safe temperature rise for the hot spot. This was the technique used to protect the FNAL 15 foot bubble chamber magnet, and it survived a quench with no damage. The FNAL magnet had stored energy of 400 megajoules.

It was beyond the scope of this study to design a vent system that would provide complete safety in the event of a quench; however, it is not a difficult task and will be part of the final design.

PARAMETER LIST

1. Minimum radius 2.3 meter
2. Maximum radius 5.95 meter
3. Length 3 meter
4. Peak field 3 Tesla
5. Stored energy, approx. 300MJ
6. Operating current 7,000 amps
7. Conducting A1 stablilized
8. Current density 9,000 A/cm^2
9. Ampere turns 34 × 106
10. Conductor length 51,000 m
11. Total weight 172,900 lbs
12. Liquid helium 15,000 liters

Structural Design

The basic structure consists of a double skinned torus 12 meters maximum diameter. Its cross section is a quadrilateral giving it flat, washerlike ends and conical inner and outer surfaces. The winding surface, made from 1/2" aluminum, Figure 2, is built in six pieces. The conductor, wound in approximately 15 bundles, each 2" × 6", around the minor toroidal axis of each pie-shaped piece, is fitted into a retainer channel. The six pieces are joined and the outer skin added, boxing in the conductor between the two shells. The closed core of the torus can be accessed through various ports connecting the inner and outer skins, the walls of which pass between the conductor bundles. The space between the two skins is the helium vessel. The conductor bundles are immersed in liquid helium.

The inner and outer conical surfaces of this body are both supported against the conductor forces by thick shells. The peak magnetic pressure is over 500 psi. The flat ends are also subject to this force and are supported by radial beams running along with each conductor bundle. The beam reactions are provided by ties which pass through the core of the torus from end to end via connecting ports through the helium vessel. These ties are made of carbon fiber and are pre-loaded to a small percentage of their working load.

An aluminum heat shield is built entirely around this torus at a 1 1/2" nominal distance from the helium vessel. It is maintained at this distance by 22 struts around the outer equator attached to 11 points on the vessel outer cone and 22 points, suitably reinforced, on the heat shield. These struts are 22" long. A further set of 22 struts, identical to the first, attaches to the heat

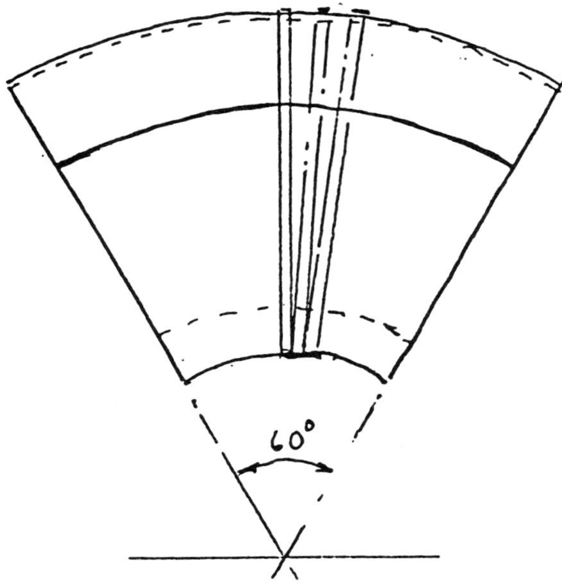

Figure 2. Winding former. Six sectors are wound and then assembled.

shield at the same reinforced points but on the outside of the shield, serving to connect the cold mass to 11 points on the warm vacuum vessel shell.

The vacuum vessel consists of inner and outer conical shells connected at each end by semi-toroidal concave heads. The considerable tension in these heads is reacted by compression rings at their inner and outer edges. The outer conical shell is strengthened, in its buckling mode, by the heavy ring required for the gravity support.

Except for the carbon fiber ties and the G-10 fiberglass epoxy struts, the entire structure is 6061 T-6 aluminum. The outer helium vessel wall is T-0 temper and all the rest is T-6. The welds are arranged in noncritical directions to avoid the loss of strength associated with annealing at the weld. At this stage of the design, 0.62 times yield has been used as the design stress due to magnetic forces and a factor of 3 against failure used for general weight carrying loads, which are somewhat less predictable. It is apparent that we are not pressing the capabilities of these materials in relation to physics requirements; and even if during detail design, the loads become somewhat larger, the necessary metal thickness will be within the physics requirements.

Design Issues

Cryogenic Considerations

The calculated heat leak to the helium bath is 15 watts which will result in an evaporation rate of 21 liters of liquid helium per hour. This is a good match to the 7,000 ampere current leads; they will take all of this gas flow for cooling. An additional heat load will occur during charge and discharge of the magnet due to eddy currents generated in the aluminum structure. Charging the magnet at 4.8 volts will produce a full field in 5 hours and dissipate 6.8 watts in the structure. The heat load on the 80 K shield will be 675 watts which will evaporate 15 liters of liquid nitrogen per hour. Rather than have a large nitrogen reservoir at the magnet, a better method is to pump the liquid nitrogen from a remote dewar through tubing on the 80 K shield and return to the dewar. By running nitrogen at a pressure of 3 atmospheres until it has left the magnet, it will stay liquid and absorb the heat with a small temperature rise. Keeping the flow single phase within the magnet minimizes the pressure drop and smaller piping can be used on the shield, thus saving on the valuable experimental space.

Structural

An examination was made of the shape and structure types to determine which configurations would offer the least material in the regions of experimental interest. It was clear that the outer and inner cones should have a simple, structurally efficient geometry since their material content is large.

Helium Vessel

A practical problem which may arise is the achievement of a good fit between the helium vessel and the heavy conical shells which butt against its inner and outer cones. We are uncertain whether these shells can be easily fitted as "thin" cones to the helium vessel and have a sufficiently accurate matching of cone angles to support the thin helium vessel wall and the conductor behind it. A "lack of fit" of approximately 0.030" would lead to undesirable local deflections and possible stick/slip thermal deposition at the conductor. Should further investigation show this to be a problem, then epoxy flooding of the helium vessel to load carrying shell interface appears to be feasible. Such epoxy fill crazes during cooldown, but is in confined compression of only 500 psi maximum during operation and is completely effective even after crazing.

The torus ends, however, need to be minimal; and many types of structure are possible. We considered support and have chosen multiple tie points.

Since this configuration has the simplest conceivable load path, i.e. a straight line from applied load to reaction, it obviously has the least possible material. Geometric considerations dictate that the number of straps must be limited. We have chosen to place the tie points along the line of each conductor bundle with a variable pitch of 25–40 inches. This span of continuous beam does not call for very much material on the average, only 4.5 cm at the bore edge of the "washer" and 1/3 of that of the outer edge. The ties between front and rear washers pass through ports in the helium vessel and do not involve the vessel in the load path at all. To minimize the deflection and the material content of the ties, carbon is used at a fairly low stress. Its specific stiffness (strength divided by density) is 7 times that of steel or aluminum.

Vacuum Vessel

The vacuum vessel is comparatively simple having an outer (buckling designed) shell, an inner tension shell and concave balloon ends. Due to the low pressure, these membranes are comparatively thin (< 1/4 in) and the support rings are not excessive.

Support Systems

The suspension of the 135,400 lb. cold mass is by means of a "Warren girder," curved and folded into cylinders between the outer 3 elements of the magnet. This was selected solely on the basis of experience and the design presented is very near final.

The only competitive suspension we considered was a similar Warren girder installed at the small end of the magnet. We thought that the loss of shell stabilizing and the magnet translation along axis during cooldown was not justified by the few tenths of a watt potential heat leak savings.

Cost Estimate

The cost for the toroidal coils of the SSC detector has been estimated based on the conceptual design by Advanced Cryo Magnetics, Inc. Labor cost is assumed to be $80.00 per hour. The cost is shown below:

1.	Engineering	$500,000
2.	First of a kind costs	$500,000
	Tooling	
	Manufacturing engineering	
	Shop procedures	
	Inspection procedures	
3.	Materials*	
	Toroid former, 24 klb.	$504,000
	Helium vessel, 12 klb.	$230,000
	Inner/outer cone and end beams, 39 klb.	$541,000
	Vacuum can, 20 klb.	$370,000
	Rings, 10 klb.	$180,000
	Insulation, 300 m^2	$120,000
	Carbon loops, 2 klb.	$500,000
	Miscellaneous	$489,000
	Subtotal	$2,934,000
	Contingency, 25%	$733,000
4.	On Site Labor	
	Vessel Forming	$420,000
	Welding	$400,000
	Coil Winding	$1,020,000
	Testing	$90,000
	Rework	$366,000
	Subtotal	$2,196,000
	Contingency, 20%	$439,000
5.	Conductor†	
	51,000 M at 7 kamp	$2,700,000
6.	Helium Liquefier	$500,000
	GRAND TOTAL	$10,502,000

*Material costs include off-site fabrication labor.

†Conductor costs are based on a Siemens quote of $52/meter. IGC has quoted $31/meter.

AN OVERVIEW OF A MUON SPECTROMETER BASED ON SUPERCONDUCTING AIR CORE TOROIDS

T. Brown[a], J. A. Pusateri[a].
P. Bonanos[d.], S. Gottesman[a.], J.N. Luton[c.], M. Marx[e.], J. Mueller[a.], L. Osborne[b.],L. Rosenson[b.], P. Spampinato[a.]
a. Grumman Aerospace Corporation
b. Massachusetts Institute of Technology
c. Oak Ridge Laboratory
d. Princeton Plasma Physics Laboratory
e. State University of New York at Stony Brook

Abstract

Superconducting air core toroids show great promise for use in a muon spectrometer for the SSC [1,2]. The design of a muon spectrometer based on superconducting air core toroids for the Superconducting Super Collider is now in the conceptual design stage. Central and end toroids are described in detail, including support of and interfaces for the calorimeters and muon chambers., Analysis of major elements is included. The toroid spectrometer consists of one central and two end toroid assemblies. System level studies on support structure, assembly, and material selection are provided. Details of the aluminum segment structure enclosing the low mass superconducting windings are shown. Access for the central tracker and calorimeters which are supported in the bore of the central toroid, along with eight inner central barrel muon chambers is discussed. The two layers of external muon chambers surrounding the toroids, and plug muon chambers covering the open ends are detailed. Resulting design is then compared to the scientific performance and benefits achieved.

Introduction

Although large superconducting air core toroid magnets have not been previously used in high energy physics experiments, they have been successfullly applied in the plasmal confinement fusion research program. The conceptual design of the air core toroid assembly shown in Figure 1 is based on a feasibility study commisioned by the SSCL [3] and scientific and engineering collaboration of a design selected for continued refinement [4]. This design places detectors external to the magnetic field. Access to muon measurements inside the magnetic volume is not required, thus producing a less complex toroid and magnet system and a simplified precision muon measuring system.

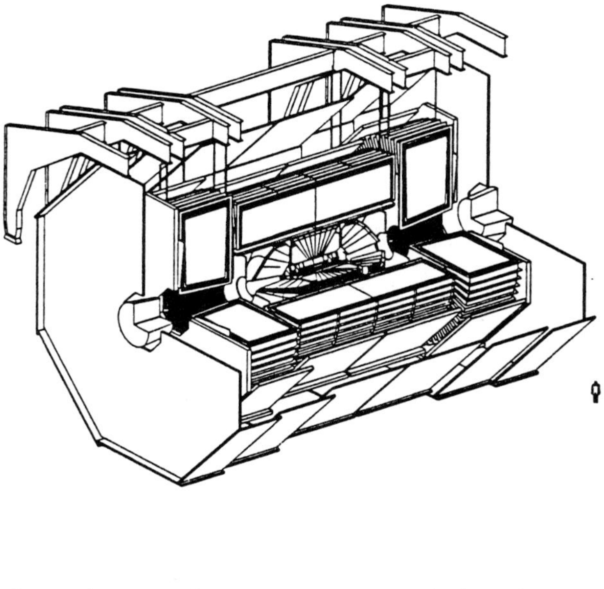

Figure 1. Muon Spectrometer Assembly Based on Superconducting Air Core Toroids

Muon Detectors

The current design measures muon momenta with detectors before and after the bending field of the toroidal magnets. The system has three point measurements (point-line) in the central region, and four point measurements (line-line or angle-angle) in the end region where the momenta are higher. Key dimensions of the muon chambers and toroids are shown in Fig. 2. Momentum resolution has been optimized in this design by maximizing bending power while minimizing material thickness (radiation length). Preliminary structural analysis has been carried out to optimize these material thicknesses and is the basis of the model utilized in this analysis. Momentum resolution also assumes the grouping of multilayer devices and a crude representation of the material at smaller radii than the muon system was made by defining the presence of a spherical "calorimeter" with a thickness equal to 450 radiation lengths.

Accuracy of the muon momentum measurement is dependent on knowing the relative positions of the muon detector elements. All the modules are referenced and aligned with an electro-optical system which provides continouous monitoring with a goal of surface location within 25 μm and spatial measurement of 200 μm. Momentum resolution as a function of θ for this design is shown in Fig. 3.

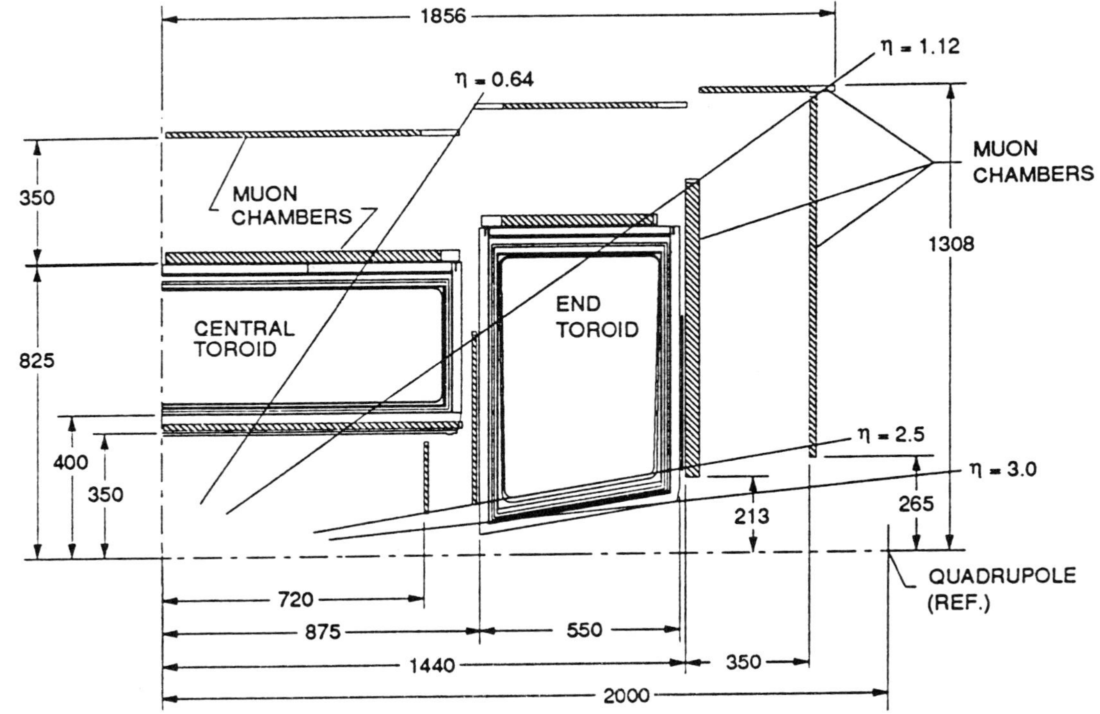

Figure 2. Key Dimensions of Muon Chambers and Toroids

Air Core Toroid Design and Analysis

Design is based on a set of initial parameters using rectangular cross section toroids with peak surface magnetic fields of 1.5T and 4.5T for the central and end toroids. All other physical parameters are being chosen to optimize design, i.e. maximize field integral, minimize radiation length, minimize cost and schedule for fabrication and assembly. Study of superconductors has led to the preference of a stabilized, forced cooled NbTi superconductor in an aluminum conduit with most of the stabilizer being aluminum. See Fig. 4.

A primary design issue for toroids of this size has been structural stability under both operating and fault

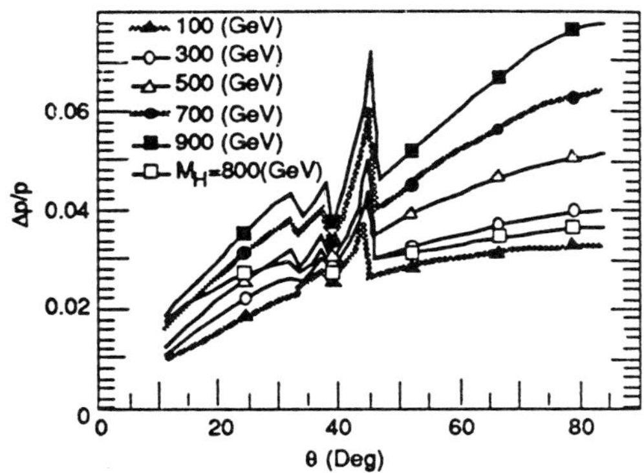

Figure 3. Momentum Resolution as a Function of θ For the Baseline Design

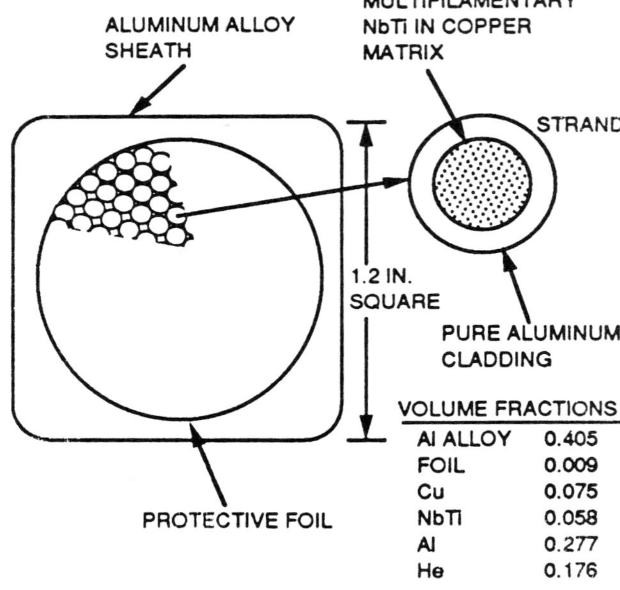

VOLUME FRACTIONS	
Al ALLOY	0.405
FOIL	0.009
Cu	0.075
NbTi	0.058
Al	0.277
He	0.176

Figure 4. Superconductor Details

conditions rather than stress analysis alone. In addition, consideration has been given to the size, weight, and transportability of the major elements. Three dimensional preliminary analysis of the 5 degree magnet segments (shown in Fig. 5) under the magnetic pressure $B^2/2\mu_0$, the dewar/calorimeter support cylinder under combined calorimeter, gravity, and pressure loads, and the detector supports under gravity loads has been completed to date (See Fig. 6, 7). This work, together with two-dimensional stability calculations has provided the confidence necessary to proceed with the design.

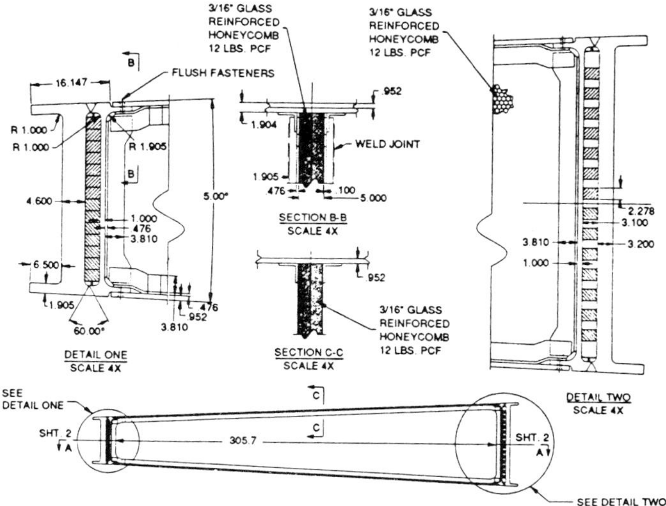

Figure 5. 5 Degree Magnet Section Details

Central Dewar FEM - Fins at 5 Deg

Loadcase 1 Shell Surface Top
Frame of Ref: Global
Stress - Von Mises Min 1.31E+01 Max 2.82E+04

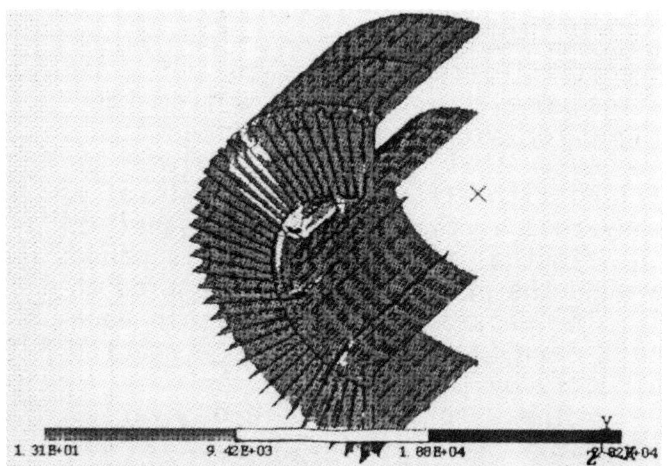

Figure 6. Finite Element Model of Dewar

Systems Considerations

Overhead crane and moving supports are required in the interaction hall in order to provide access to the central tracker and calorimeter which are in the bore of the central toroid. Completely assembled end toroids are capable of being moved into final position when central assembly of all elements are completed. A conceptual drawing showing an end toroid moved out of position to provide this access is seen in Figure 8.

Five Degree Segment FEM of Baseline Design

Loadcase 2 Shell Surface Top
Frame of Ref: Global
Stress - Von Mises Min 1.25E+02 Max 2.24E+04

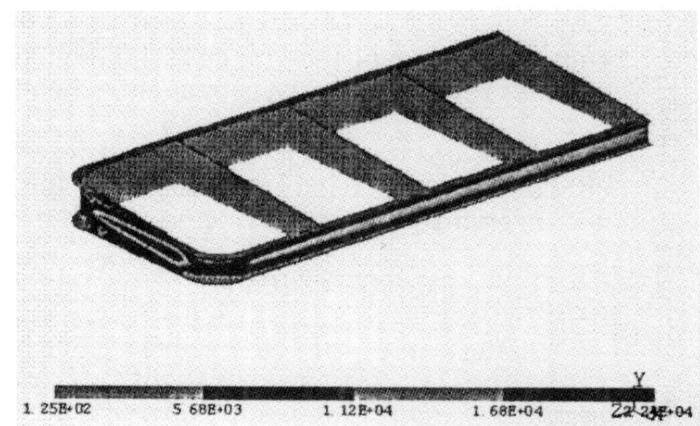

Figure 7. Finite Element Model of Five Degree Segment

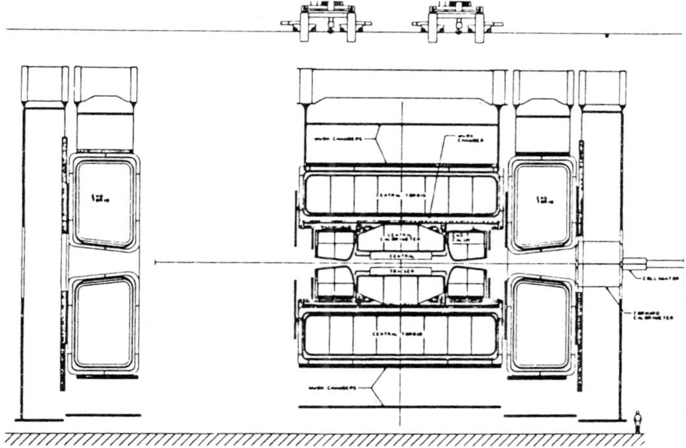

Figure 8. End Toroid Moved For Central Toroid Access

References

1. M. Marx, "EMPACT - An Alternative Approach to a High Pt SSC Experiment", SSC Report SSC-219, 1989.

2. T. Fields, et al., "Proposal For Development of Superconducting Air Core Toroids as a Precision Muon Spectrometer For SSC Experiments", 1989.

3. James N. Luton, Jr.,and Peter Bonanos, "Toroidal Detector Scoping Study", SSC Report SSCL-N-697, 1990.

4. Peter Bonanos, "Toroid Conceptual Design", EMPACT Note 183, 1990.

DESIGN OF RADIATION-THIN AIR CORE TOROIDS for HIGH RESOLUTION MEASUREMENT OF MUON MOMENTA

Peter Bonanos
Princeton University Plasma Physics Laboratory
P.O.Box 451, Princeton N.J.,08543

ABSTRACT

The general problem is defined and performance criteria established. A variety of possible designs is presented and the basis for each concept is outlined.

Toroidal coils which enclose central calorimetric detectors are described; end coils may be used in conjunction with solenoidal detectors for improved measurements in the forward, high rapidity range.

INRODUCTION

The use of air-core toroidal field coils for measurement of muon momenta is outlined and several design problems are discussed. The self-resolution provided by the magnet improves directly with bending length and decreases with the square root of the number radiation lengths encountered (i.e multiple scattering) along a charged particle trajectory. Air-core, as used here, means the absence of iron; it does not mean the absence of structures, which must be minimized, both within and without the field volume. The methods discussed include a variety of structural arrangements and the magnet shape itself.

CONFIGURATION,FORCES AND FAULTS

The general arrangement assumed (Fig.1) encloses a central calorimeter/detector. A central toroid used with two end toroids permits coverage to high rapidity (>2) and the gaps between magnets provide an exit path for detector and magnet utilities.

The coil windings are, to the scale of the magnet, thin and, if closely wound, produce toroidal field with little leakage. The force exerted on the structure is a magnetic

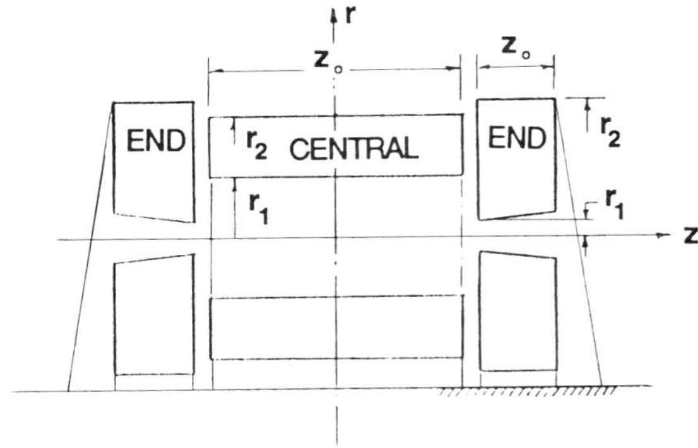

Fig.1 General Arrangemnt

pressure proportional to B^2 and directed everywhere outward from the magnetic volume. This immediately suggests that the field volume be filled with structure. Each coil of a many-coil array may contain a panel in an azimuthal plane. The coil is wound around the panel periphery and tied to it. The separating forces within each coil are then, in part, reacted directly by this vane.

The toroidal field varies inversely with radius hence, a net centering pressure is produced which compresses the structure tangentially and may be supported in several ways.

(1) A central support cylinder at the inner radius.(Fig.2) This leaves a magnetic volume which is

essentially hollow over most of the azimuth but must, as described later, include some tangential structure.

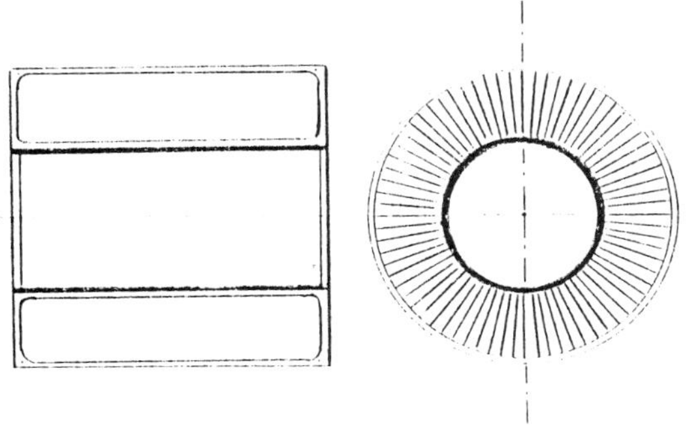

Fig.2 Vane/Cylinder Arrgmnt

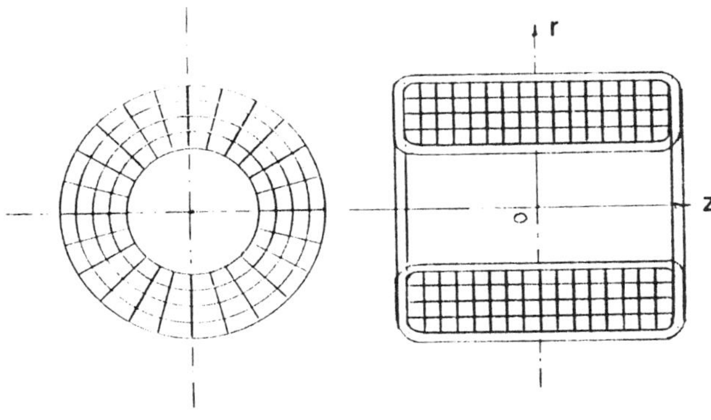

Fig.3 Disk/Vane Arrgmnt

(2) A series of annular disks spaced axially within the volume is combined with nested cylinders as well as vanes (Fig.3) to form a "3-D egg crate".

(3) A massive superstructure placed outside the outer radius and, perhaps, outside the muon chambers. The coils are then hung from this framing.

In all cases the supporting structure must be stable against buckling. The conditions promoting buckling are aggravated by fault loads. Each coil of each symmetric toroidal array is in unstable equilibrium. A coil is attracted to every other coil in the array; the net force on a coil is a radial centering force with no net tangential component. This is true for symmetric currents in all coils and perfect symmetry in the position of each coil. If one coil of the array is displaced tangentially so that the the toroidal gap angles between it and its two immediate neighbors are not equal then a tangential force is applied to the displaced coil and all other coils as well. The force tends to further open the larger gap and makes the centering forces unequal.

The coils are series connected and a "perfect" quench is dissipated in an external dump resistor; all coil currents decay at equal rates and the force pattern remains symmetric. Asymmetric quench-dumps or electrical shorts/grounds which destroy the symmetry of coil currents temporally also produce tangential forces and unequal centering forces.

The structure must therfore be capable of safely supporting asymmetric centering pressure and inter-coil tangential loads. The determination of likely faults and consequent forces requires extensive analysis as does the buckling behavior of the structure.

ONE DESIGN

A design using the azimuthal vane with central support cylinder was prepared for the EMPACT collaboration. It serves as a starting point for the necessary fault analysis which has not yet been done. The design is based on the following conditions:

	Central Toroid	End Toroid
r1 (m)	4	1
r2 (m)	8	9
z (m)	16	5
Bmax (T)	1.5	4.5
J (kA/sq cm)	4	2.2

Using symmetric buckling loads and reasonable end face thicknesses a 72 coil array was considered and evaluated. The space occupied by the vanes, considered "lost" to high resolution measurements, occupies less than 10% of the azimuth. Allowance for costruction tolerances and nitrogen shields reduces the "magnetic length" to about 3/4 of the dewar length. The self-resolution is better than 3% at 90 deg and is worse for rays which pass through both the central and end toroids.

An interesting variation is obtained if the magnetic length is adjusted to provide constant self-resolution over the entire range of rapidity. The shape of the magnetic volume and an enclosing coil is shown as Fig.4. The portions of the coil with straight conductors may be easily be curved and still achieve the identical result.

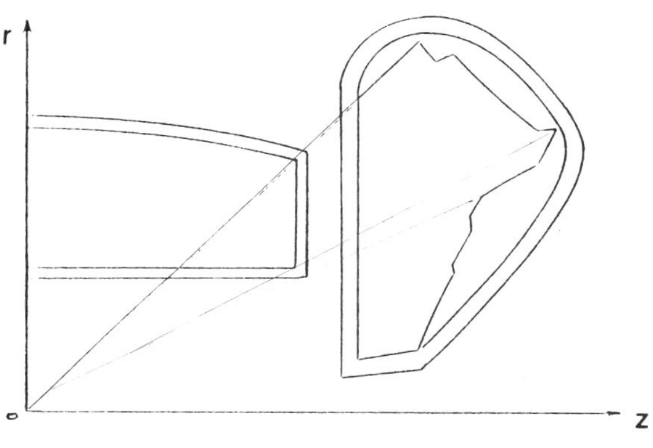

Fig.4 Magnet Shape for Constant Resolution

FUTURE WORK

Designs of the form of Fig.3 are suggested for further evaluation. They are inherently safer against buckling but not as radiation-thin. It is expected that higher fields will be possible and a smaller physical volume as well. The net effect on resolving power may be favorable.

A COMMENT ON FEASIBILITY

The question of feasibility is raised when a new technology is applied to conventional methods. There are presently three operating fusion research machines (tokamaks) utilizing superconducting toroids. Komarek (Ref.2) provides a convenient summary. At the 16th SOFT conference several papers describing the development of high current superconductors, as required here, were presented and are cited below.

CONCLUSIONS

Air-core toroids offer the prospect of high resolution measurement of muon momenta over a large rapidity range. Several structural arrangements and magnet shapes should be considered. Detailed design including fault analysis is necessary to determine the resolving power.

REFERENCES

(1) P.Bonanos,"EMPACT Technical Note No.183", presented at the EMPACT workshop, March, 1990, SSC Lab, Dallas Tx.

(2) P.Komarek, "Present Achievements and Prospects for S.C. Tokamaks in the World", presented at MT-11, Proc. of the 11th Int'l Conf. on Magnet Technology, Tsukuba, Japan (1989), Vol.1,pp 18-26, Elsevier Applied Science (1989)

(3) 16th Symposium on Fusion Technology, London, U.K.,Sept.1990, see A.Torossian et.al. and P.Blasio et.al. for 40kA conductor designs.(to be published)

THE FEASIBILITY OF LOW-MASS CONDUCTORS FOR TOROIDAL SUPERCONDUCTING MAGNETS FOR SSC DETECTORS*

J. N. Luton
Fusion Energy Division
Oak Ridge National Laboratory
Post Office Box 2009
Oak Ridge, TN 37831

Abstract

An earlier study by Luton and Bonanos (SSCL-N-697) concluded that the design and fabrication of superconducting toroidal bending magnets would require a major effort but would be feasible. In that study, the interaction length and thus the mass of the magnet were considered to be of secondary importance. This study is an extension to examine the feasibility of low-mass conductors for such use. It included a literature search, consultations with conductor manufacturers, and design calculations, but no experimental work. An unoptimized sample design that used a residual resistivity ratio for aluminum of 1360 and a current density of 3.5 kA/cm^2 over the uninsulated conductor for a 4.5-T toroid with 1 GJ of stored energy obtained a hot-spot temperature of 120 K with a maximum dump voltage of 3.6 kV and 24% of the initial current inductively transferred into the shorted aluminum structure. The stability margin was 200 mJ/cm^3 of cable space. Limiting the quench pressure to 360 atm to give conservative stresses in the sheath and assuming that the whole flow path quenched immediately resulted in helium taps that could be a kilometer apart if the flow friction factor were the same as that experienced in the Westinghouse (W) Large Coil Task (LCT) coil. This indicates that the 530-m conductor length of each of the 72 individual coil segments of a toroid would be a single flow path. If some practical uncertainties can be favorably resolved by producing and testing sample conductors, the use of a conductor with clad-aluminum stabilizer and extruded aluminum-alloy sheath should be feasible and economical.

Introduction

Particle detectors whose bending magnets are toroidal arrays of individual coils are of interest for the Superconducting Super Collider (SSC). In some of these cases, the windings and structure of the magnet must be designed so that they do not interfere with the passage of the particles, so low-density metals are required. This study investigated the feasibility of substituting aluminum or aluminum alloy for most of the copper and steel in a conductor. This was done by considering the configuration shown in Fig. 1, the present "external" arrangement of the EMPACT detector. The attempt was to find a feasible solution, not an optimized one.

Body

One winding configuration used in previous detector magnets, which are solenoidal and smaller than the proposed magnets, is an array of monolithic conductors epoxy potted together and cooled by conduction through the potting of the interfaces to the directly cooled structure. This arrangement was not considered because of a conviction, based on the experimental results of the LCT [1], that in a large toroid such as Figure 1, epoxy disbonding would be likely and the heat transfer path would be interrupted. However, pool boiling and cable-in-conduit conductors performed satisfactorily in the LCT, and the latter arrangement was selected for this study (and for an earlier study in which low mass was not a requirement [2]).

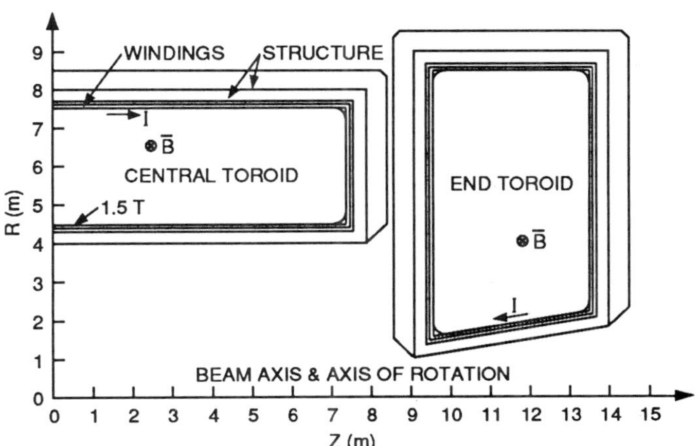

Figure 1. Vertical quarter-section of EMPACT magnet.

The conductor arrived at by this study is shown in Figure 2. Scaling trends on a plot of current density vs stored energy indicate that the conductor of Figure 2 has higher performance than would be expected by extrapolating the data for the existing indirectly cooled detector magnets to the 1-GJ level. The strands within the area of the cable labeled D_{MF} have cores that are 50 mil in diameter. These cores have

*Research sponsored jointly by the State University of New York at Stony Brook and the Office of Fusion Energy, U.S. Department of Energy, under contract DE-AC05-84OR21400 with Martin Marietta Energy Systems, Inc.

multifilamentary NbTi superconductor embedded in a copper matrix and can be fabricated with standard industrial techniques. Aluminum was not substituted for this matrix copper because earlier attempts to develop this technology were unsuccessful. The copper-to-superconductor ratio is 1.3, to match that of the SSC inner-layer strand. Around these cores is a layer of high-purity aluminum cladding that is used as a stabilizer, bringing the strand diameter to 75 mil. This cladding has been done on individual strands on an experimental basis by industry [3] and on a larger scale in cladding flattened cables for use in earlier detectors. Around the central part of the cable is an annular layer of pure aluminum strands. These provide additional aluminum for quench protection, at lower cost than it could be provided in the cladding, especially since it is suspected that fabrication difficulty would increase greatly if the core diameter were made smaller or the cladding thicker. The annulus also provides protection to the superconducting strands if helium penetrations are to be made through the conduit wall. The square conduit that surrounds the cable provides leak-tight helium confinement for the supercritical helium in which the cable lies. In a multiturn winding, it also is a structural member that carries transverse compressive load applied by adjacent turns. Therefore, the corners of the conduit are rather sharp to avoid bending stresses in the side wall, which acts as a column. The sheath can also support longitudinal tensile loads. The foil between the cable and conduit is for protection during manufacturing.

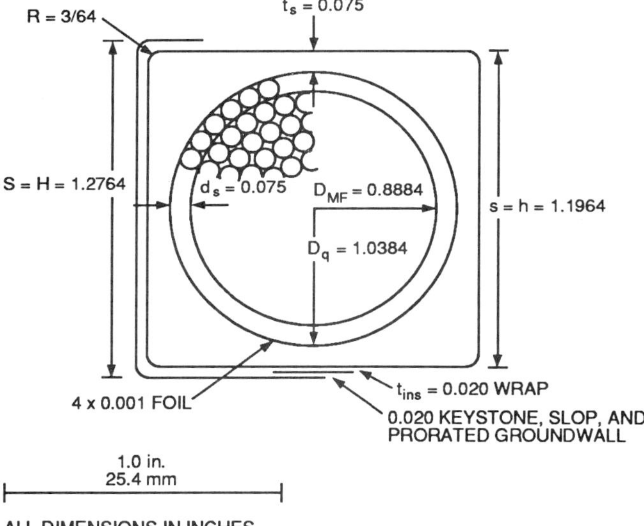

Figure 2. Conductor cross section.
Volume fractions: Al alloy, 0.405; foil, 0.009; Cu, 0.075; NbTi, 0.058; Al, 0.277; He, 0.176.

On the basis of the existence of aluminum-sheathed cable in the electrical industry (although the requirements are different) and on conversations with the Wendelstein VII group [4], extrusion of the sheath is believed to be feasible and economical. Two alternatives do exist: produce one or two sheath welds the full 59-mile length of the conductor, or pull the cable into the 530-m straight lengths of sheath and then compact it to the proper void fraction. The first technology was established during the LCT program for a steel sheath but not for aluminum, which must also be a strong alloy suitable for cryogenic service. The second has been done for a 400-m length of 40-kA cable, again in a steel sheath [5].

The circular shape of the cable in Fig. 1 is convenient but not necessarily optimal. For example, if the cable were squarish, the size of the sheath could be reduced somewhat, increasing the current density of the winding and decreasing its radiation thickness.

Five literature examples of practical experience with the resistivity of high-purity clad aluminum were examined. The post-extrusion residual resistivity ratio lay between 1360 and 2000. A resistivity ratio of 1360 was assumed for this study because it appears to be practicable and because Thoner [6] has provided convenient data for the resistivity as a function of field. His resistivity values of 5.0 and 5.1 nΩ•cm for fields of 1.5 and 4.5 T are for unstrained annealed material. However, at the spots where strands cross one another, some amount of plastic deformation is expected because of cabling, and the deformation may well be increased by the sheathing operation or by the magnetic forces, since the yield strength of pure annealed aluminum is only 2 to 3 ksi even at cryogenic temperatures. Hartwig [7] indicates that a residual plastic strain of 4 to 6% would add another 4 nΩ•cm to the resistivity at the affected regions.

The enthalpy of aluminum, although somewhat larger than that of copper on a weight basis, is lower on a volumetric basis by a factor of 2.5 at 100 K. This has a direct impact on the hot-spot temperature reached as a result of quench, as indicated in Figure 3. Here γ is the density, C_p the mass specific heat, ρ the resistivity, j_0 the current density at the start of the dump (quench), and τ_d the dump time constant.

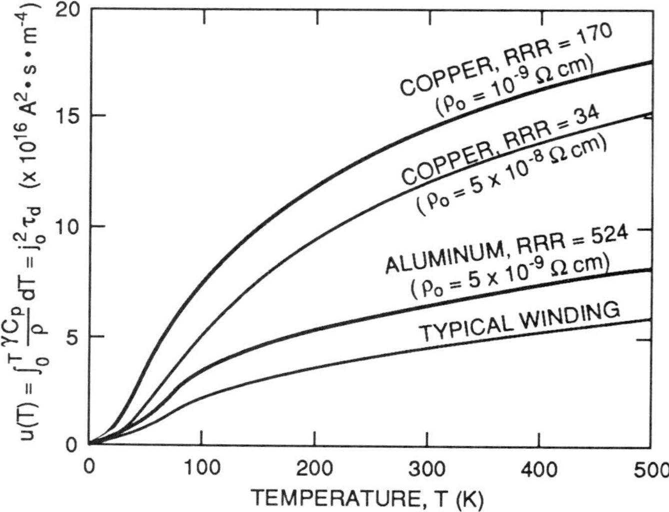

Figure 3. Hot-spot integral vs hot-spot temperature for copper, aluminum, and a small magnet containing 47% Cu, 23% NbTi, and 30% resin (adapted from Ref. [8]).

Not only is the aluminum worse than even the relatively impure copper, but also the integral goes up relatively slowly after the temperature reaches about 150 K. Thus, it is more profitable to try to decrease τ_d than to let the hot-spot temperature rise without bound. In turn, τ_d depends on the stored energy, the conductor current by way of inductance, and the dump voltage that the winding can withstand.

The 2.9 GJ of magnetic stored energy would not be a problem if it were evenly distributed over the windings. However, with a large coil that is stable this is difficult to achieve. An alternative is to provide an external dump resistor to absorb the energy. The enthalpy of the sheath is as large as that of the strands, and both are considered, since the experience with the W LCT coil was that the sheath was immediately heated. Experience with that coil also showed that at the start of a dump about 24% of the current in the conductor was immediately inductively transferred to the shorted aluminum structure and that this transfer was calculable if the coupling coefficient and the time constants of the two circuits were known [9]. With a hot-spot temperature of 120 K, the required dump time constant is 19.5 s, and the toroid voltage is 3.6 kV. This is an acceptable value for a forced-flow coil since the electrical insulation is solid and completely encloses the conductor.

For forced-flow conductors, the pressure reached within the sheath when a quench occurs must be considered. For a fixed field of 4.5 T and the flow friction factor observed in the W LCT coil, if we assume that an entire flow path quenches in a time much shorter than τ_d, that the current can transfer from the multifilamentary strands to the aluminum overwrap strands in a time short compared to τ_d, and that manufacturing limitations prevent a strand diameter smaller than 75 mil, then with structural considerations limiting the quench pressure to 5.3 ksi (360 atm), the allowed length of individual flow paths is $\approx$1000 m. A length of 530 m was chosen to give some margin and to match the length of conductor in an individual coil. The limit on quench pressure came from assuming that the stress in the sheath walls due to that internal pressure is added directly to the magnetic load in the walls, a very conservative assumption. This sum was allowed to be 48 ksi, which is two-thirds of the presumed yield strength of the aluminum sheath material when cold. The Wendelstein VII Group [4] is considering an alloy similar to the US 6082 aluminum and is anticipating a yield strength of at least 72 ksi.

The stability margin is defined as the maximum heat per unit volume of cable space that can be introduced into the conductor and still have it recover. For a conductor such as ours, with ample cooling surface and flow induced in the helium by the I^2R losses in the stabilizer, the stability margin can be approximated by the enthalpy available in the helium. For this calculation, we assume 0.5 K of headroom over the 4.5 K operating temperature, giving a helium temperature of 5 K. The field and current density of the NbTi are such that the current-sharing temperature is 5.88 K. To raise all components of the conductor to this temperature would require about 200 MJ per cubic centimeter of cable space. Reference [2] gives the rationale for believing that 150 MJ per cubic centimeter would be acceptable. The ratio of design current to critical current of the superconductor is 0.50. The ratio of limiting current to design current is greater than 3.

Conclusions

The conductor proposed for a radiation-thin toroid for an SSC toroidal detector is substantially and qualitatively different from the aluminum monolithic conductors used in existing detectors and from the forced-flow conductors of the fusion program. Conductor development is needed before designing the toroid to avoid schedule delays and interruptions of coil or conductor fabrication, to avoid increased cost and decreased performance as increased conservatism is required by the absence of needed data, and to reduce the risk of failure due to surprises. With successful completion of a development program, forced-flow conductor with pure aluminum stabilizer and aluminum-alloy sheath should be both feasible and economical for use in ironless toroidal magnets for SSC detectors.

References

[1] P. N. Haubenreich, P. Komarek, S. Shimamoto, G. Vécsey, L. Dresner, W. A. Fietz, T. Kato, M. S. Lubell, J. W. Lue, J. N. Luton, W. Maurer, K. Okuno, S. W. Schwenterly, Y. Takahashi, A. R. Ulbricht and J. A. Zichy, "The IEA Large Coil Task," Fusion Eng. Des. **7**, 1–230 (1988).

[2] J. N. Luton, Jr., and P. Bonanos, *Toroidal Detector Scoping Study,* SSCL-N-697, Superconducting Super Collider Laboratory, Dallas, Texas

[3] H. Kanithi, D. Phillips, C. King and B. Zeitlin, "Development and characterization of aluminum clad superconductors," IEEE Trans. Magn. **MAG-24** (2), 1029–1032 (1988).

[4] J. Sapper, private communication.

[5] B. Kazimierzak, private communication via P. Bonanos and J. W. Lue.

[6] M. Thoner, H. Krauth, J. Randolph, and A. Szulczyk, "Aluminum stabilized Nb_3Sn superconductors," in *Proceedings of the 7th International Cryogenic Materials Conference (ICMC7),* 1987, pp. 507–514.

[7] K. T. Hartwig, G. S. Yuan, and P. Leehmann, "Strain resistivity at 4.2 K in pure aluminum," in *Advances in Cryogenic Materials,* Vol. 32, R. P. Reed and A. F. Clark, eds. New York: Plenum Press, 1986.

[8] Martin N. Wilson, *Superconducting Magnets.* Oxford: Clarendon Press, 1983.

[9] J. N. Luton, S. S. Shen, T. J. McManamy, C. T. Wilson, J. F. Ellis, L. Dresner, and R. E. Wintenberg, "A study of eddy-current effects in the structure of the Westinghouse LCT coil," IEEE Trans. Magn. **MAG-23** (2), 1497–1500 (1987).

MONTE CARLO STUDY OF A SUPERCONDUCTING AIR CORE TOROID MUON SPECTROMETER FOR THE SSC

S.R. Gottesman(2), P. Bonanos(9), T.G. Brown(2), P. Burrows(6), W. Busza(6), I.-H. Chiang(1), J.J. Connelly(2), V. Fonseca(10), J. Frank(1), J.I. Friedman(6), M. Gundy(10), J. Haggerty(1), T. Hansl-Kozanecka(6), D. Hungerford(3), C.-K. Jung(10), H.W. Kendall(6), J. Klafin(2), K. Lau(3), L. Littenberg(1), J. Luton(8), T. Lyons(6), T.R. Marshall(4), B. Mayes(3), M.D. Marx(10), R.R. McNeil(5), W. Morse(1), J.R. Mueller(2), L.S. Osborne(6), F.E. Paige(1), D. Parks(3), L. Pinsky(3), B.G. Pope(7), J. Pusateri(2), J. Pyrlik(3), L. Rosenson(6), P.T. Spampinato(2), R. Strand(1), M. Tannenbaum(1), F. Taylor(6), R. Verdier(6), R. Weinstein(3), D. Weygand(1).

(1) Brookhaven National Laboratory, Upton, NY 11973
(2) Grumman Corporation, Bethpage, NY 11714
(3) University of Houston, Houston, TX 77204
(4) Indiana University, Bloomington, IN 47405
(5) Louisiana State University, Baton Rouge, LA 70803
(6) Massachusetts Institute of Technology, Cambridge, MA 02139
(7) Michigan State University, East Lansing, MI 48824
(8) Oak Ridge National Laboratory, Oak Ridge, TN 37831
(9) Princeton Plasma Physics Laboratory, Princeton, NJ 08544
(10) State University of New York at Stony Brook, Stony Brook, NY 11794.

Abstract

Substantial progress has been made during the past year in the conceptual design of a superconducting air core toroidal muon spectrometer (EMPACT/TEXAS) for the SSC. A Monte Carlo study of the EMPACT/TEXAS muon system shows the design will have excellent resolution and acceptance over a large range of rapidity, and will be capable of reconstructing Z^0s close to their natural width. There currently appears to be no major engineering obstacles to the development of such a system.

Introduction

The development of superconducting air core toroids for a muon spectrometer system offers substantial advantages over conventional magnet designs for the detection and characterization of muons. The toroidal field, always perpendicular to the particle trajectory, is a highly efficient configuration. In addition, the elimination of an iron yoke (for flux return) substantially reduces overall weight, leading to a design where the entire detector can be suspended from overhead supports, permitting maximal solid angle coverage. Other advantages stem from the reduction in multiple scattering muons experience while traversing the magnet, and the fact that particle tracking can now be optimized in a nonmagnetic environment.

Large superconducting air core toroids have been developed and successfully applied in the plasma fusion energy research program. The feasibility of using large superconducting air core toroids for high energy physics applications has been verified in an earlier study [1] commissioned by the SSC Lab which concluded that superconducting air core toroids for high energy physics are feasible, and furthermore, are likely to be built with a conservative design, despite the fact that they would be the world's largest superconducting air core toroids ever constructed.

System Configuration

The baseline concept is shown in Fig. 1. The muon spectrometer consists of a single central toroid of inner radius 4m and outer radius 8m, with two end toroids each having an inner radius of 1m (dewar corner) and outer radius of 9m. The maximum magnetic field strength B_{max} is 1.5T in the central toroid and 4.5T in the end toroids. This spectrometer has a magnetic field integral ($\int B \cdot dl$) of 3.6Tm at $\eta=0$ ($\theta=90$ deg.) which increases to a maximum of 18.6Tm at $\eta=2.5$ ($\theta=9.4$ deg.). The stored energy in each of the three magnets (one central and two end toroids) will be approximately 1 GJ. Additional design specification and performance detail may be found in references [2, 3, 4].

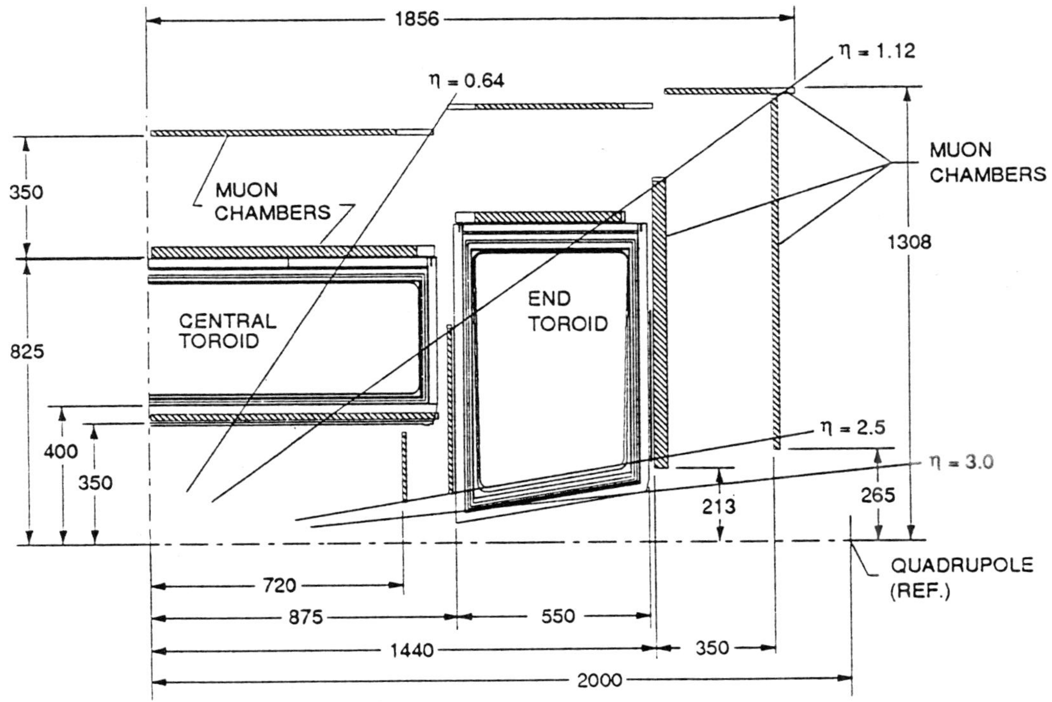

Figure 1. One quadrant of muon spectrometer, baseline concept. Maximum field strength is 1.5 Tesla in central toroid, 4.5 Tesla in end toroids.

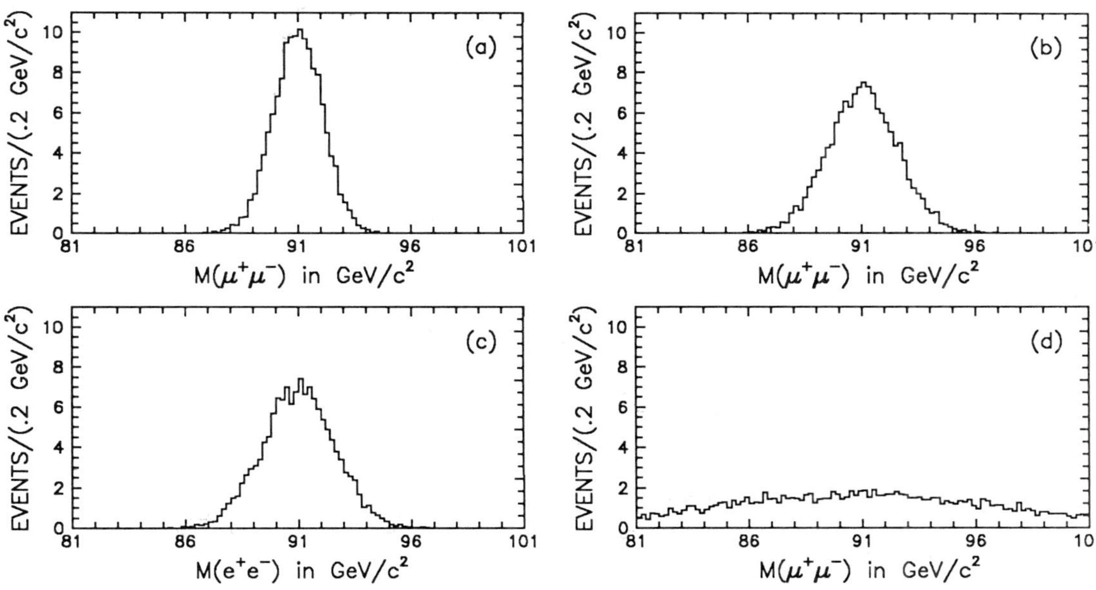

Figure 3. Reconstructed Z^0's from 400 GeV/c^2 Higgs decay for (a) perfect momentum resolution, (b) superconducting air core toroid, (c) calorimeter with $\Delta E/E = .15/\sqrt{E} + .01$ resolution, and (d) iron toroid, $\alpha = 10\%$.

The toroid design is predicated on the judicious use of aluminum, instead of steel and copper, in the toroid support structure and superconducting cable, in order to optimize momentum resolution by maximizing muon bending power while minimizing material thickness (radiation lengths) along any given muon trajectory. Initial results indicate that this approach is both promising and effective [5, 6].

System Performance

In order to assess the performance of the baseline design of Fig. 1, rays were projected from the interaction region through the spectrometer system for different angles (θ) with respect to the beamline, and for each ray the ratio of the RMS multiple scattering angle to magnetic bend angle predicted as a function of polar angle theta (θ). The results show that multiple scattering is minimal ($\approx$2.5%) for muons with trajectories perpendicular to the beam axis, and decreases even more (to $\approx$1.0%) for forward trajectories of θ=10 degrees. The multiple scattering function does show a spiked feature reaching a maximum of 4% at θ=35 degrees in the region where a muon trajectory intersects both the central and end toroids, but this condition exists only over a small range in pseudorapidity η.

A momentum resolution function (Δp/p) has been calculated for this design [7] under the assumption that the spatial measurement error will be smaller than 200μm for each detector plane. Detector planes were grouped into layers with an assumed effective multilayer resolution of 50 μm for each set of measurements. It was also assumed that an external alignment system would monitor the locations of the muon panels to an accuracy of 25 μm. The conceptual design of such a system has been developed and is described in detail in reference [8]. The momentum resolution function for such a muon system is shown in Fig. 2 for different muon momenta. The curve denoted by boxes (□) corresponds to the average resolution for muons from an 800 GeV/c^2 Higgs decay $H \rightarrow Z^0Z^0 \rightarrow \mu^+\mu^-\mu^+\mu^-$.

Assuming the resolution stated above, the reconstructed Z^0 mass distribution for a Higgs mass of 400 GeV/c^2 has been calculated and is presented in Fig. 3. In (3a) the Z^0 is shown at its natural width, as it would be measured with perfect (infinite) resolution. The second plot (3b) shows what can be expected from a toroid having the baseline design with momentum resolution of Fig. 2. For comparison purposes we present in (3c) the Z^0 reconstructed from a calorimeter measurement of $Z^0 \rightarrow e^+e^-$ with resolution $\Delta E/E = .15/\sqrt{E} + 0.01$, and in (3d) the results for an iron toroid with Δp/p=10% momentum resolution.

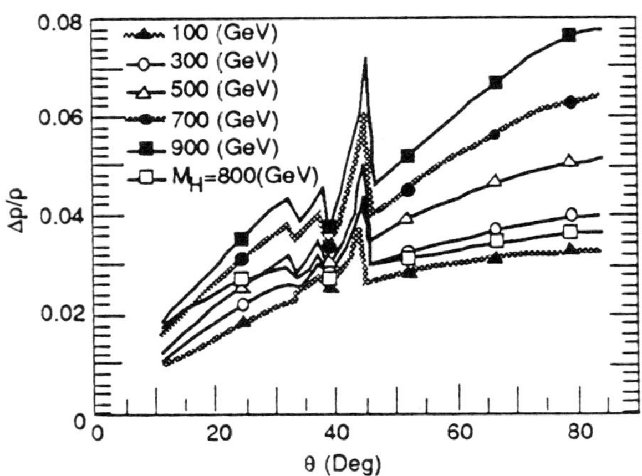

Figure 2. Momentum resolution of baseline design. Curve marked (□) is average momentum resolution for muons from 800 GeV/c^2 Higgs decay.

References

[1] P. Bonanos and J. Luton, "Toroidal Detector Scoping Study," SSCL-N-697 (1990).

[2] J.A. Pusateri et al., "The SSC Superconducting Air Core Toroid Design Development," Proceedings of the 2nd International Industrial Symposium on the Super Collider (IISSC), Miami Beach, Florida, March 1990.

[3] S.R. Gottesman et al., "Superconducting Air Core Toroids for precision Muon Measurements at the SSC," Proceedings of the 2nd International Industrial Symposium on the Super Collider (IISSC), Miami Beach, Florida, March 1990.

[4] M. D. Marx, et al., "EMPACT, Electrons Muons Partons with Air Core Toroids, An Expression of Interest for an Experiment to be Performed at the Superconducting Super Collider," submitted to SSCL May 25, 1990.

[5] P. Bonanos, "Toroid Conceptual Design," EMPACT Note 183 (1990).

[6] J. Luton, "The Feasibility of Low Mass Conductors for Toroidal Super-conducting Magnets for the Superconducting Super Collider (SSC)," EMPACT Note 202 (1990).

[7] L. Rosenson, "Momentum Resolution and Track Reconstruction Quality for the EMPACT Detector," EMPACT Note 224 (1990).

[8] J.J. Connelly, "A System Concept for Aligning the EMPACT Muon Detector Chambers for the Super Conducting Super Collider," EMPACT Note 225 (1990).

L-STAR MAGNET SYSTEM DESIGN STUDIES

L-Star Magnet Group

Presented by P.G. Marston
MIT/Plasma Fusion Center

Abstract

The paper presents a brief summary of the many magnet designs considered for the L-Star Detector. Options included uniform field toroids, solenoids, magnetic "bottles", concentric solenoids, cusps and transverse field dipoles. Fringe field shielding using a ferromagnetic "return frame" and/or superconducting windings have been considered.

Summary

As part of the L-Star development effort, MIT analyzed and evaluated more than fifty conceptual designs for the detector magnet system. The effort included the development of codes to calculate $\int Bdl$ along particle trajectories and ray tracing codes (including effects of multiple scattering) to evaluate the performance of the different designs.

The level of design effort for the various concepts varied from electromagnetic design only to a full conceptual design including stress analysis, structural design, conductor and winding design and drawings. In all instances the level of effort is adequate to establish credibility of performance, construction and compatibility with the intended use.

Concepts considered included:

- Toroids
  - R^{-1} Field Distribution
  - Uniform Field Distribution
- Solenoids
  - Uniform Field
  - "Bottles"
  - Concentric Solenoids
  - Superconducting Shields
  - Ferromagnetic Shields
- Cusps
- Transverse Field Dipoles

The purpose of this review is to present concepts only; several of which, although not considered for L-Star, may be appropriate for other or future Detector systems.

Toroids

Toroidal systems for the entire detector and also for the forward/backward (low angle) physics only were considered. The obvious advantage of the toroidal array is that the field is, in general, orthogonal to the particle trajectories at all (θ) angles. There are, however, a number of disadvantages particularly for conventional systems having R^{-1} field distributions and severely "lumped" windings at the outer diameter. These include:

- Loss of solid angle (particularly at low θ angles)
- High ratio of peak field in the winding to central (useful) field
- Severely non-uniform field distribution near the coil windings. This combined with potentially large motion of these conductors under magnetic load can complicate track reconstruction and reduce precision.
- The design and assembly procedures are comparatively complex and limit access to internal detector components.
- Although for some designs the R^{-1} field distribution is acceptable, for designs which impose reasonable limits on the field strength at small radii and yet require large bending power at large radii, the outer diameter of a toroidal system can become excessively large.

The field distribution in a toroid can, however, be controlled with suitable radial and azimuthal distribution of the windings. Several such systems have been considered.

Figures 2.1 through 2.4 show the winding layout and central field distribution for a "uniform field" toroid

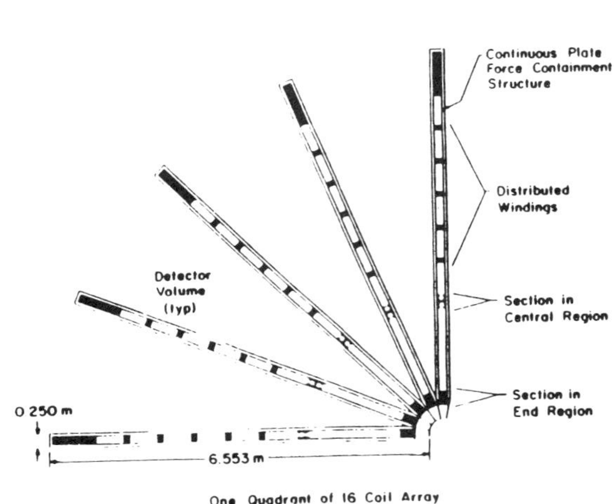

Figure 2.1

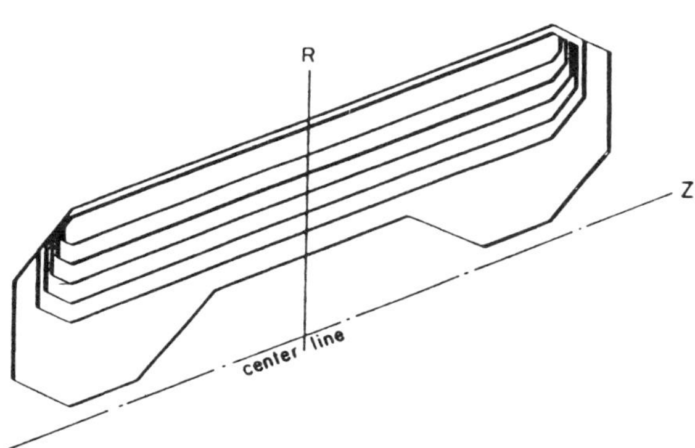

Figure 2.2

considered for the entire detector. The overall length of the windings is approximately 30 m.

Figure 2.5 shows one end of a winding concept wherein in addition to the radial distribution of Figures 2.1–2.4 the windings are distributed azimuthally as well. This will further improve uniformity and yet permit access at either end of the system.

Figure 2.6 shows an array in which the windings are also distributed over the ends of the system—thus providing a potentially perfectly uniform field over the entire inside volume of the toroid.

Figures 2.7 through 2.10 represent the implementation of such an array within the L-Star solenoid for $2° < \theta < 10°$. The muon chambers would be located inside the toroid. A similar system was considered located external to (at either end of) the central solenoid.

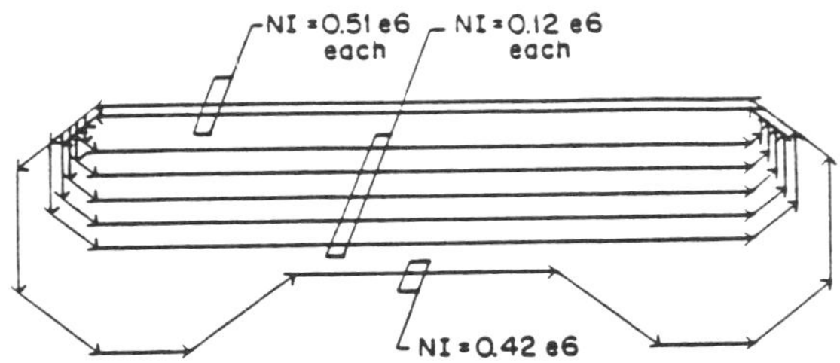

Figure 2.3. Current Filament Stick Model

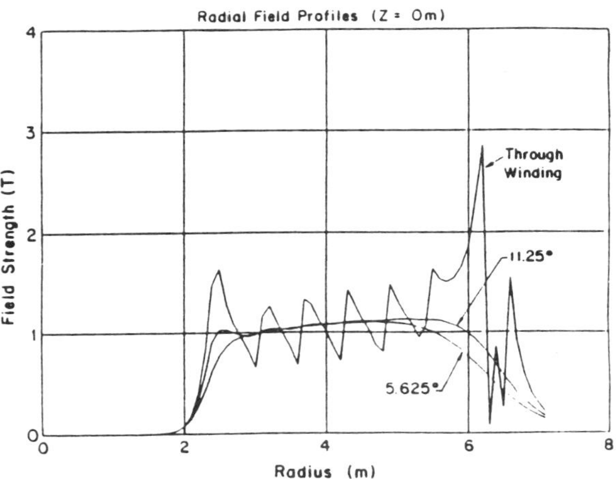

Figure 2.4

Solenoids

Although uniform field solenoids lose bending power rapidly at low angles, performance is acceptable down to $\theta \approx 10°$ (7.5° in the case of L-Star). For fully superconducting systems conical end windings can provide inexpensive additional length at low angles and thus be quite reasonable down to $\theta \approx 2.5°$. (Ref. Figure 3.5).

Low angle performance can also be improved with non-uniform fields wherein the field strength increases in the axial direction. The field curvature in these "magnetic bottle" configurations can improve low angle bending power by a factor of 2. Although the mirror ratio for the configurations studied is rather low, there could be some trapping of low energy particles. The degree of such particle containment has not been studied for these devices.

The solenoid designs considered "fringe field" shielding with both a ferromagnetic "return frame" or with an outer superconducting winding.

Concentric, reverse polarity, solenoids consisting of a relatively small diameter high field strength solenoid

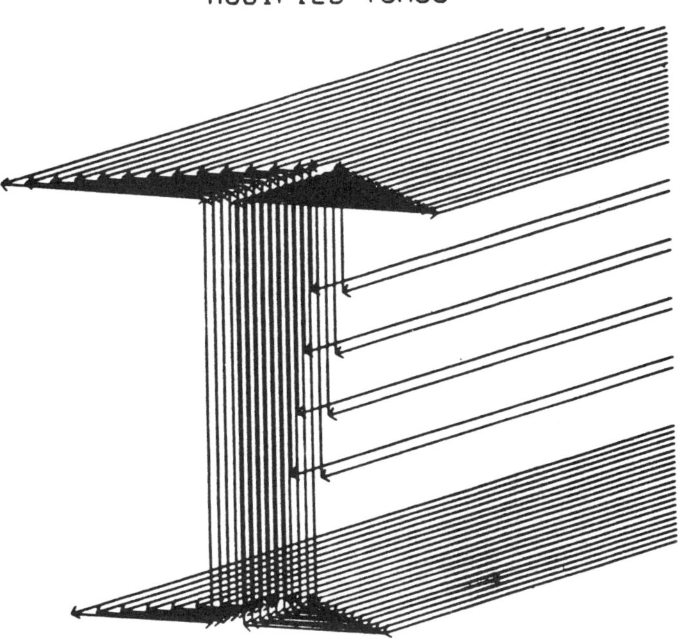

Figure 2.5

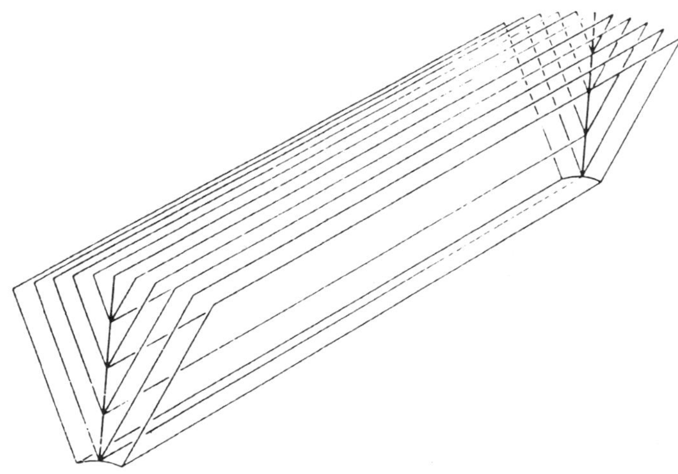

Figure 2.6. Isometric View of Current Filaments for Uniform Field Toroid Coil Module.

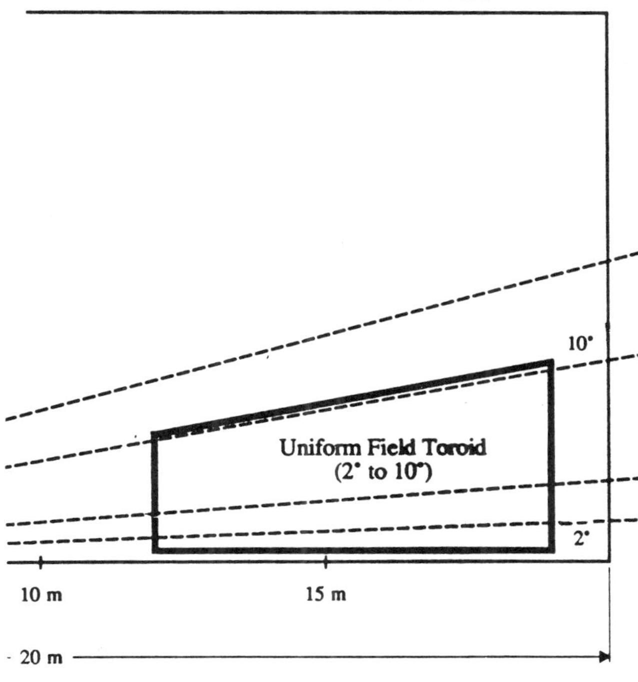

Figure 2.7

within a relatively large diameter low field strength solenoid create a field distribution in which the field is generally orthogonal to the particle trajectories (comparable to that of a toroid except that the field is radial rather than azimuthal). Such a device is inherently shielded having stray fields which are low enough that counting rooms, instruments, and other field sensitive components can easily be shielded locally if required. The high central field also benefits magnetic central trackers.

Figure 3.1 shows the general overall assembly for an iron bound solenoid. The coils can be of either conventional or superconducting designs.

Figure 3.2 shows the flux lines for such a magnet with very thin superconducting windings. In the case shown, the winding is made from six individual modules. The perturbations in the field resulting from such a discontinuous winding are clearly visible.

Figure 3.3 shows the elevation view of a fully superconducting solenoid with superconducting end windings and superconducting shields. The magnet shown has adequate bending power down to 7.5°. The dipole magnets at either end of the central solenoid cover the range of 7.5° to the beamline. These end dipoles are also solenoidal and represent a design wherein the solenoidal winding is shielded at its periphery by a superconducting shield—but instead of superconducting end windings, it has iron poles.

Figure 3.4 shows the plan view of this same system.

Figure 3.5 compares the previous design with one in which the central solenoid is identical but the end wind-

UNIFORM FIELD TOROID
(one octant / dimensional layout)

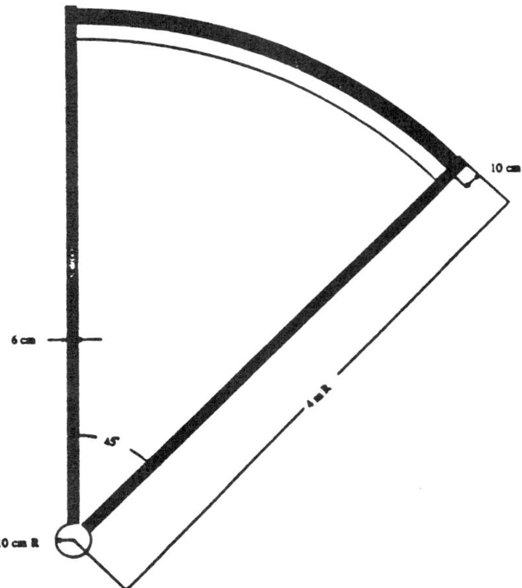

Figure 2.8

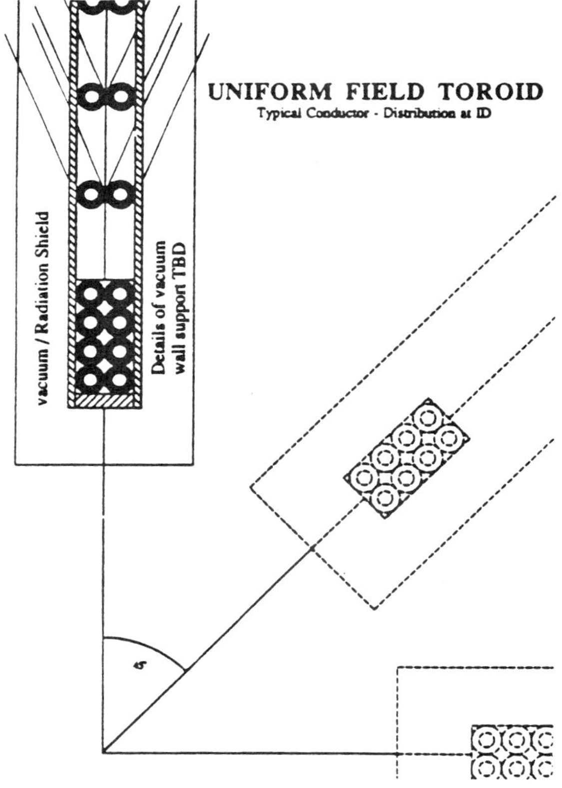

Figure 2.9

UNIFORM FIELD TOROID
Typical Conductor - Distribution at OD

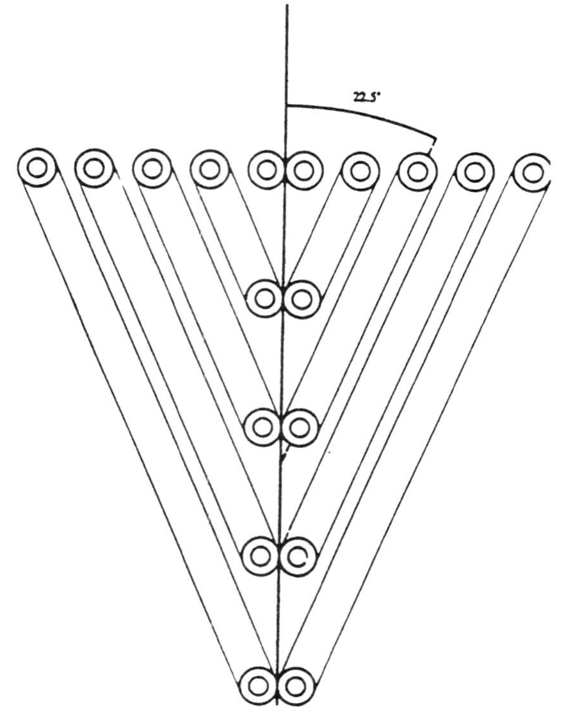

Figure 2.10

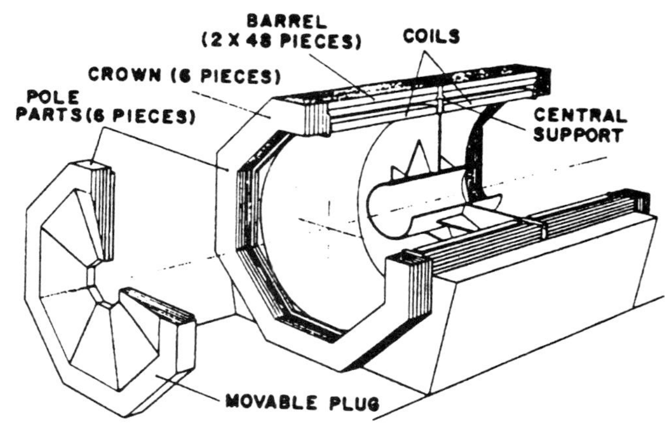

Figure 3.1

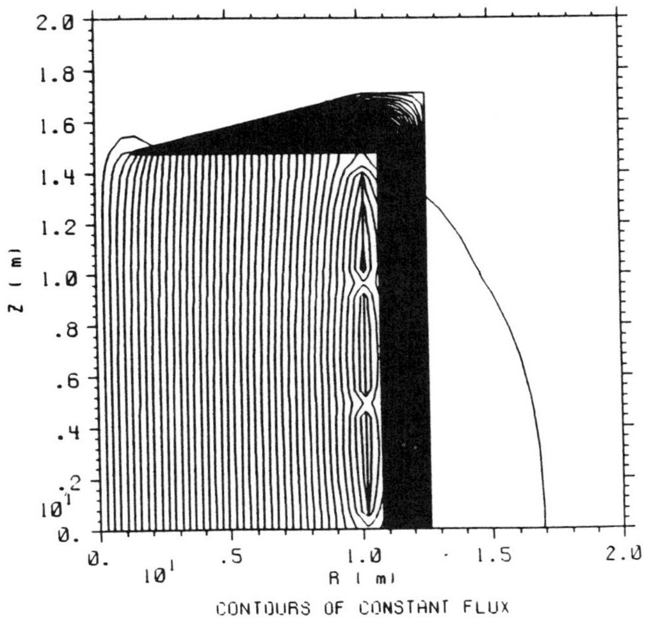

Figure 3.2

ings are conical, thus providing additional (low angle) length and bending power (Bl^2) with little additional cost compared to the square end shown on the left hand side of the drawing. This design has roughly equivalent performance and eliminates the end magnets.

Figure 3.6 shows the field distribution for such a system.

Figure 3.7 shows the field distribution for the end magnets (Ref. Figures 3.3 and 3.4). This concept represents a very simple design having considerable design flexibility and will be considered for the central solenoid as well as the end magnets.

Figure 3.8 shows the flux lines for an iron shielded "magnetic bottle". In the magnet shown, the field varies from 0.4 T at the midplane to 0.8 T peak field near the axis at the magnet end.

Figure 3.9 shows an air core version of a higher field "bottle". The superconducting shield is not shown.

Figure 3.10 shows a "concentric solenoid" configuration. The flux density in the inner solenoid is approximately 6 T and in the annulus between the inner and outer solenoids is 0.75 T max. Bending power is good down to about 10°. Below that, one could take advantage of a long moment arm with very low angle detectors ($\theta = 2.5$) located 30 m from the vertex.

The overall system length would thus be comparable to that of the "uniform field solenoid with end magnet" configuration but would be much simpler and less costly.

Figure 3.11 shows a concentric solenoid configuration with end windings and a small diameter high field solenoid located near the end of the system. The latter increases the strength of the radial field at low angles and provides adequate bending power within a total winding length of 30 m.

Cusps

This concept, dubbed the "fluxhorn", began as a simple technique to provide a strong radial field at the end (pole) of the iron shielded magnet. The concept proved interesting and was studied for a large number of superconducting variations as well. It is capable of enormous bending power at low angles in simple solenoidal systems but, in general, requires field strengths above the muon chamber performance specification of 0.75 T.

Figure 4.1 shows the flux lines for an iron bound air core solenoid with a reverse field, horn-shaped, coaxial solenoid added to the end of the system. This "cusp" configuration produces large radial field components proximate to the inside diameter of the magnet pole (in the hole) which can provide a large bending "impulse" to low angle particles. Detectors would be located inside of and outside of the pole.

Figure 4.2 shows a similar system except that an end winding has been added to the outside "horn-shaped" solenoid. This adds significant additional bending power in the region.

In both Figures 4.1 and 4.2, however, it can be seen that the external "fluxhorn" solenoid has a serious adverse effect on the field distribution inside the central iron bound air core solenoid.

Figure 4.3 shows that with the proper radial current distribution in a fully superconducting system (with superconducting end windings), this problem can be easily overcome.

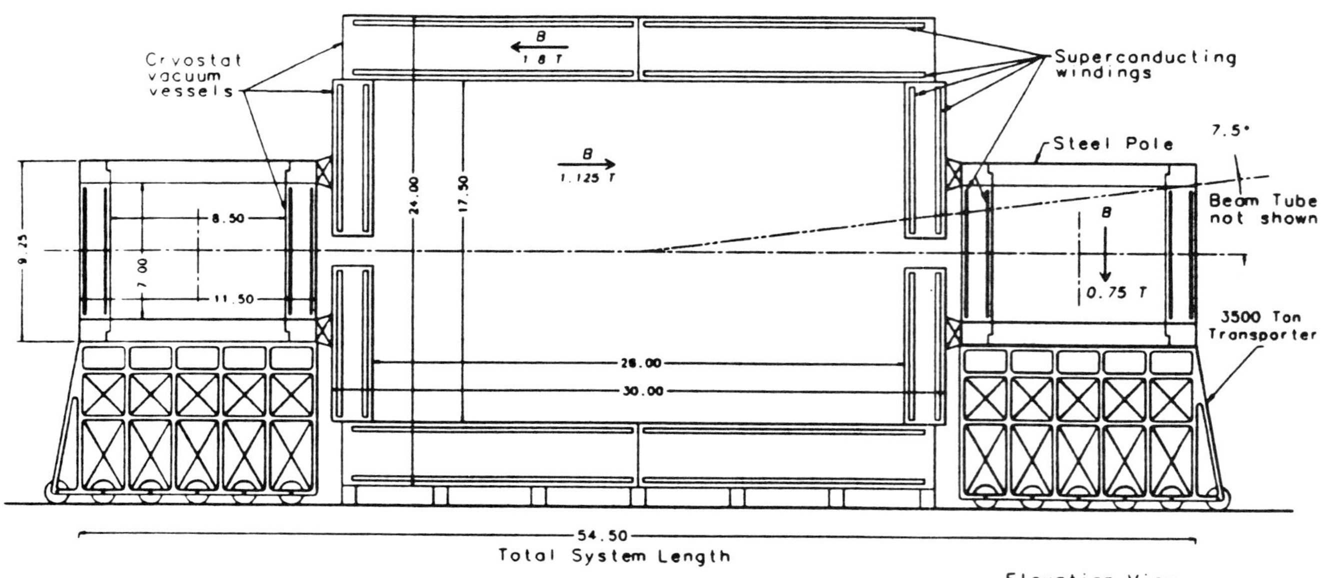

Figure 3.3

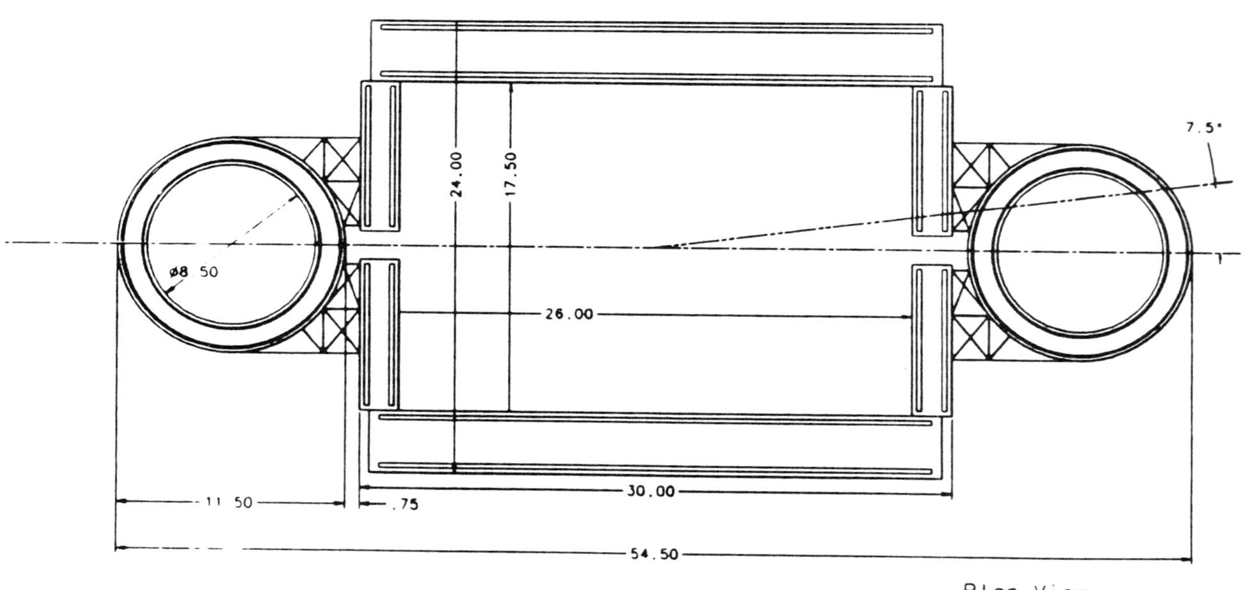

Figure 3.4

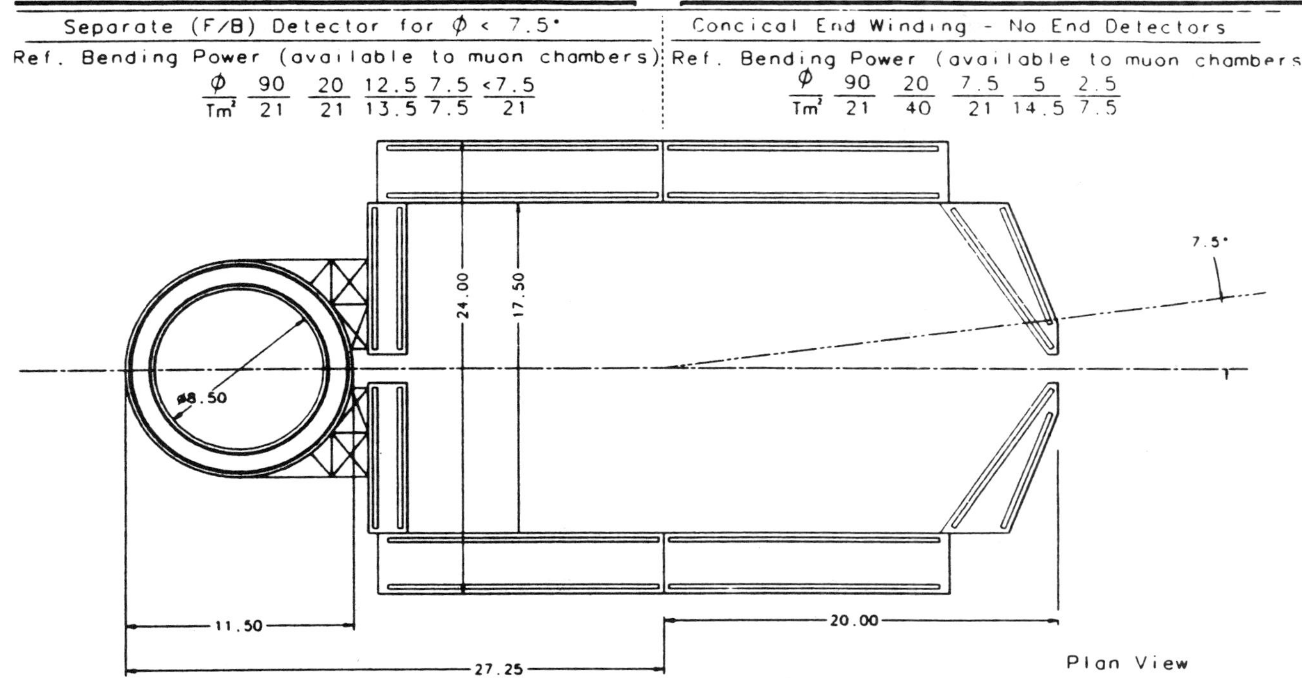

Figure 3.5

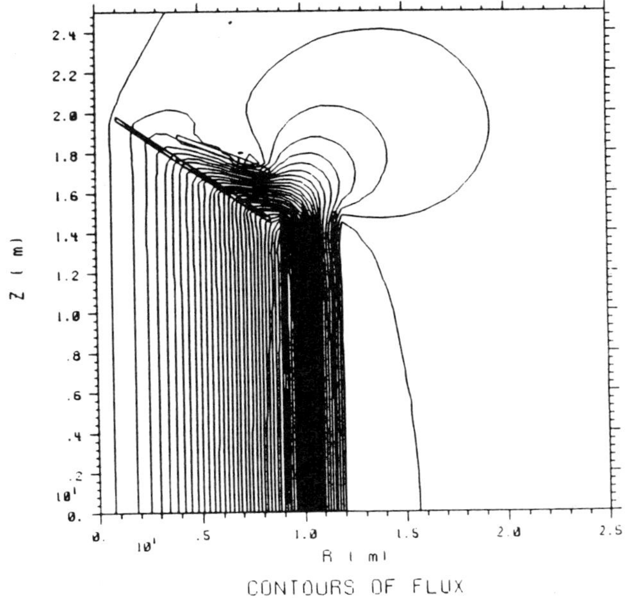

Figure 3.6

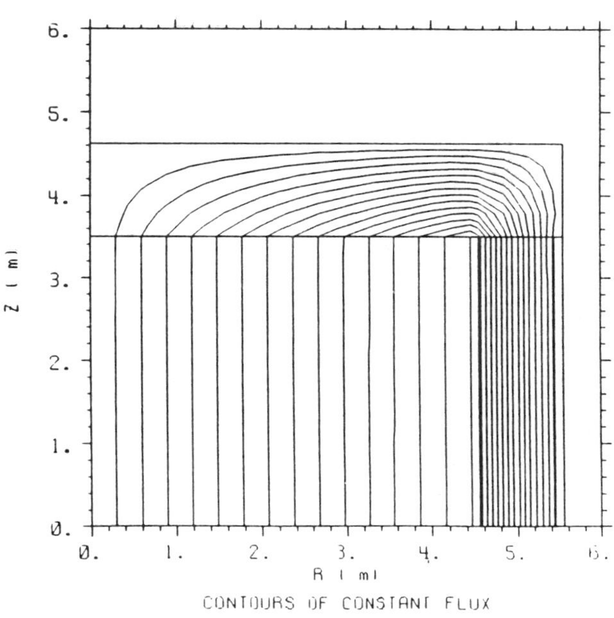

Figure 3.7

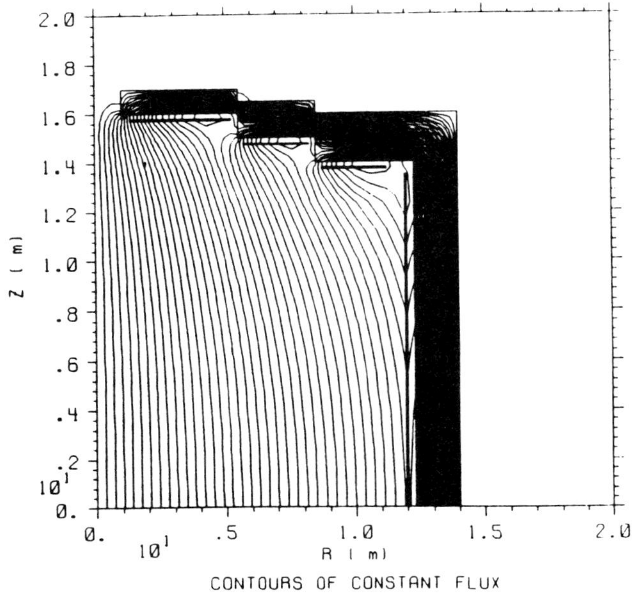

Figure 3.8

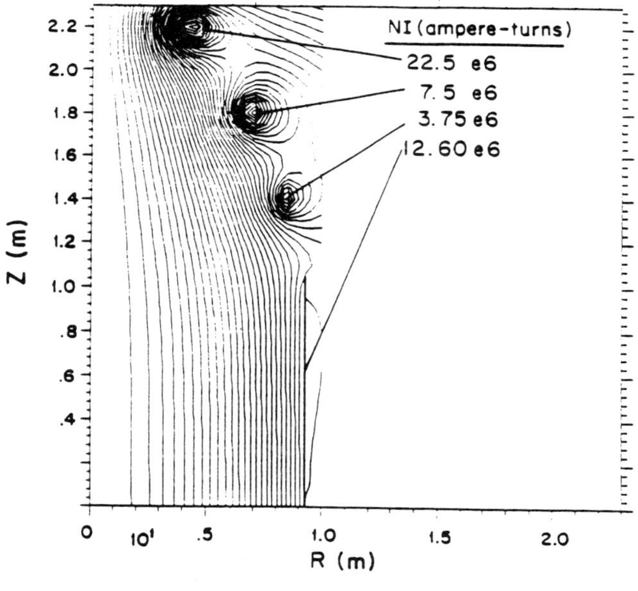

Figure 3.9

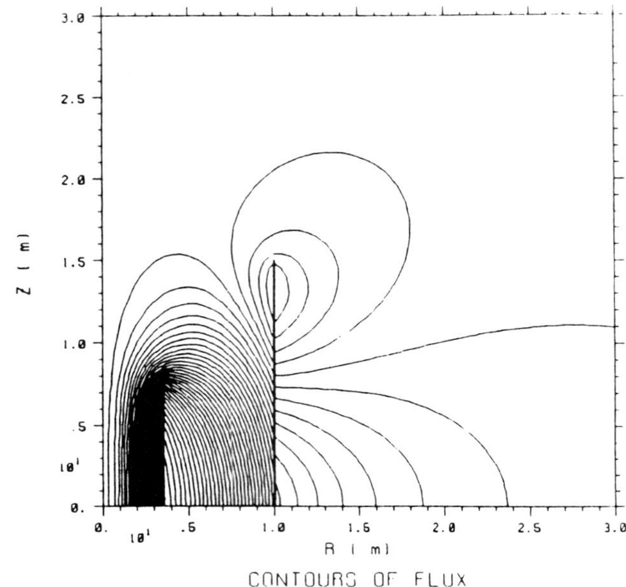

Figure 3.10

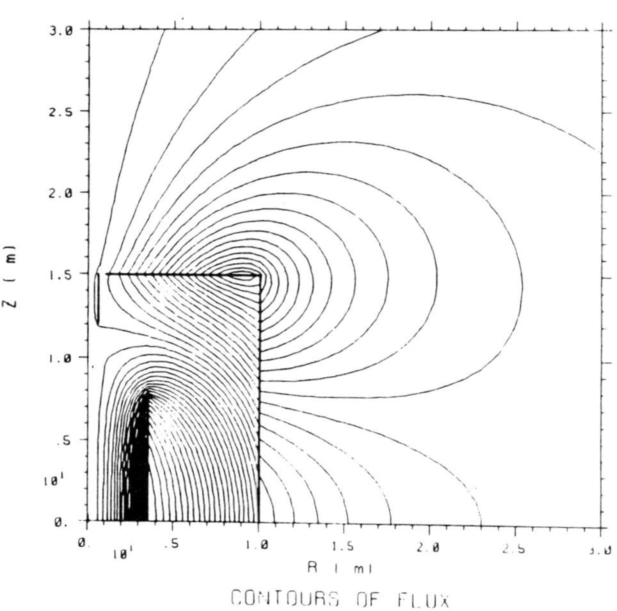

Figure 3.11

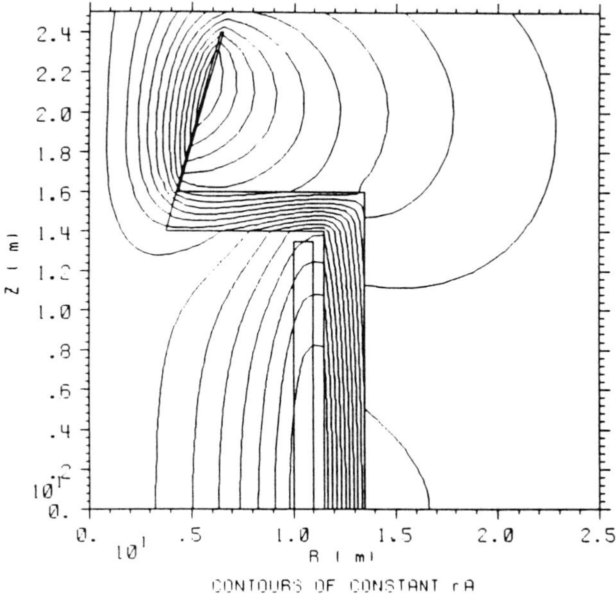

Figure 4.1

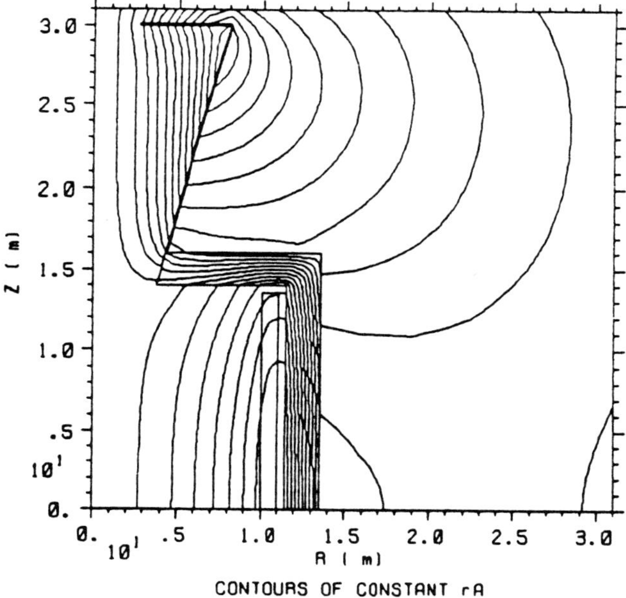

Figure 4.2

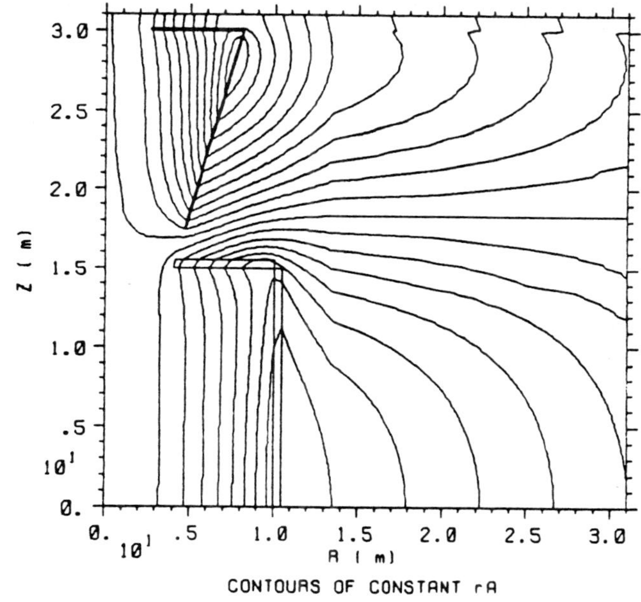

Figure 4.3

MATCHING FORWARD TOROIDS TO A CENTRAL SOLENOID

Thomas H. Fields
Argonne National Laboratory
Argonne, Illinois 60439

Abstract

Physics requirements and practical criteria for choosing superconducting toroid parameters for a solenoid detector system are described. Other approaches for momentum measurement of forward tracks are briefly outlined.

Introduction

The use of a superconducting toroid at forward angles to augment a central solenoid tracker was advocated by Jones[1] several years ago. At this conference, conceptual engineering designs of superconducting toroids for the SDC solenoid detector[2], as well as for the EMPACT all-toroid detector[3], have been described. The main purpose of this paper is to summarize the physics and other criteria for choosing the toroid parameters for a central solenoid detector. A second purpose is to briefly describe alternate approaches for achieving some of these physics capabilities at forward angles.

Momentum Accuracy

By using a toroid magnet in the forward angle region, momenta of forward-going muons can be measured more accurately than by using the solenoid central tracker alone. Such a toroid can substantially increase the momentum accuracy, detection efficiency, and Z° identification capability for rare multimuon events which are likely to be important in SSC experiments, such as Higgs→Z°Z°→$\mu^+\mu^-\mu^+\mu^-$. Fig. 1, which is adapted from the SDC Expression of Interest[2], shows the muon momentum accuracy as a function of angle (pseudorapidity) for both the solenoid and the toroid. Curves are shown for various values of P_T, and 100 GeV/c can be considered as a typical value for electroweak reactions. Note that the solenoid tracker and the toroid tracker both give fractional momentum measurement errors (solid lines) which are almost independent of angle at fixed P_T.

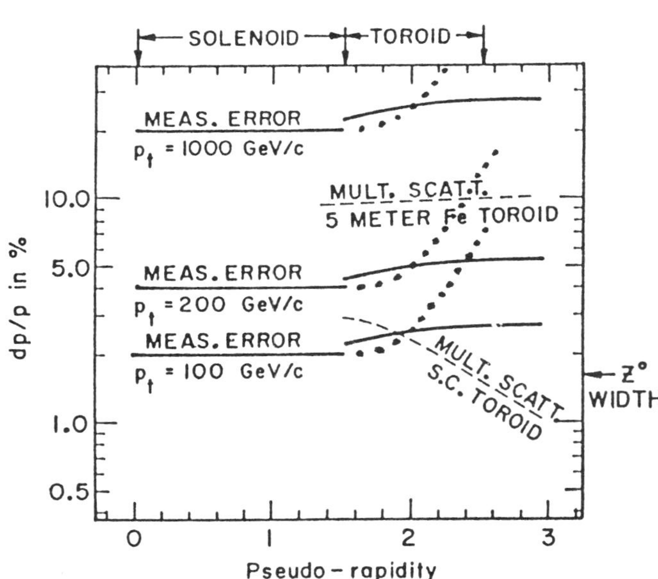

Fig. 1. Comparison of the muon momentum resolution of the SDC superconducting air-core toroid system with that of the central tracker. The solid lines show the resolutions dp_T/p_T^2 = 0.2 $(TeV/c)^{-1}$ for the central tracker and 0.25 $(TeV/c)^{-1}$ cosθ for the air-core toroids. The dashed lines show multiple-scattering contributions to the resolution of an air-core toroid and a toroid of magnetized iron. The dotted lines show the resolution which might be chosen for a small angle silicon strip tracking system in the solenoid region. Also shown is the resolution required to reconstruct the Z(→μμ) mass with an error equal to the natural width of the Z.

Thus it is possible to match the toroid system to the solenoid system, giving a fractional momentum error (and thus a fractional dimuon mass error) which is nearly independent of angle over a wide region.

As Fig. 1 shows, the actual SDC parameters have been chosen to achieve a $Z^\circ \rightarrow \mu^+\mu^-$ mass resolution which is comparable to the Z° natural width, for muon P_T up to about 100 GeV/c. This is an important goal for dimuon mass precision, since identifying Z°'s reliably in the presence of significant backgrounds is likely to be necessary for the study of rare processes at the SSC. The main parameters which determine the toroid curves shown in Fig. 1 are the toroid field integral (8 Tm at r=2.3m), the drift chamber precision (10o microns/ super-module, with 2-meter lever arms in front of and behind the toroid), and the muon entrance/exit wall thickness of the toroid (2-4 X_o). The latter determines the P_T-independent multiple scattering floor on the momentum accuracy of the superconducting toroid, shown as a dashed line in Fig. 1.

Toroid Design Goals

In addition to the momentum accuracy goals described above, there are important technical and practical criteria which need to be met by an engineering design for the super-conducting toroids. These criteria include reliability, safety, field uniformity, low leakage field, mobility, and minimum cost. Different engineering approaches to achieving these goals have been presented at this meeting and will require further evaluation before their technical tradeoffs and costs are accurately understood. Purcell has described a novel toroid design[4] which aims to achieve high reliability and low cost by using features, such as bath cooling of the superconductor, which have been proven technically conservative and highly cost effective in the performance of large superconducting magnets for bubble chambers.

Field uniformity and leakage field of the toroid are determined mainly by the number of discrete coils used. Fig.2 shows the leakage field for the SDC geometry as a function of the number of coils. In order to minimize problems with drift chamber operation and forces upon nearby iron structures, the number of coils should not be less than about 32. This arrangement will also yield good field uniformity.

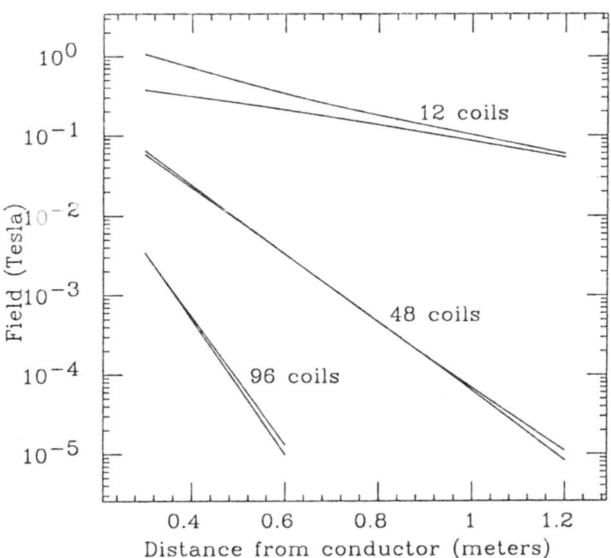

Fig. 2. The field outside the SDC toroid at maximum radius (6m), as a function of the number of (narrow) coils in the toroid. The distance is measured from the coil in the upstream axial (collider beam) direction. The field inside the toroid at this radius is about 1 Tesla. The two separate curves of each pair show the field within and between coil planes.

Other Methods For Measuring Forward Muons

For the SDC detector, several other methods can be used for making momentum measurements on forward angle muons. Choosing from among the available methods requires technical tradeoffs, accurate cost information, and relaxing the goal of closely matching the forward momentum accuracy to that of the central tracker. Brief descriptions of three possibilities are as follows:

1) Use a solid iron forward toroid rather than an air core superconducting toroid (ACT). This is expected to be less costly, and perhaps more reliable, than the ACT.

   The iron toroid has disadvantages in momentum accuracy (dp/p is limited to about 10%, see Fig. 1), mobility (4000 tons instead of 80 tons), and in producing more electromagnetic showers along the muon track. Mobility is important for achieving access to service the central tracker and calorimeter systems.

2. Make precise track measurements at intermediate angles inside the solenoid using new types of silicon strip chambers and/or radial wire chambers. The dotted curves in Fig. 2 show the momentum precision which might be chosen for the SDC. Here the advantages include the accurate measurement of other charged tracks as well as muons. The potential disadvantages include pattern recognition problems and radiation damage at high luminosity, lack of a good muon trigger signal, and rapid loss of momentum accuracy at small angles.

3) Use a superconducting air core toroid of lower magnetic field strength which gives, say, dp/p=10%. Its physics performance would be comparable to that of an iron toroid, it might cost less than an iron toroid, and it would have advantages in mobility and reduced shower production, as described above.

Conclusions

Superconducting toroids which match the momentum accuracy of the central solenoid system can offer valuable physics capabilities, and appear to be quite feasible technically. Present cost estimates for superconducting toroids range from comparable to iron to much more, and are still uncertain. It is therefore important to continue to develop and evaluate designs which are aimed at low cost and high reliability. Alternate devices of somewhat lower cost involve performance compromises which might turn out to be serious for physics experiments on rare processes.

Present planning for the SDC Letter of Intent is to use both alternates 1) and 2) as described above, instead of superconducting toroids, in order to achieve a significant cost reduction from the SDC ACT cost estimate used in the EOI[2].

1 L.W. Jones, Snowmass, 1986 Proceedings, p. 462.

2 Solenoidal Detector Collaboration (SDC), Expression of Intent, May, 1990.

3 EMPACT Detector Collaboration Expression of Intent, May 1990.

4 J. Purcell, talk at this conference.

Work supported by the U.S. Department of Energy, Division of High Energy Physics, Contract No. W-31-109-ENG-38.

Computing and Simulations

8

COMPUTING FACILITY AT SSC FOR DETECTORS

Phil Leibold
and
Brian Scipioni

Computation Group
Superconducting Super Collider Laboratory*
2550 Beckleymeade Avenue
Dallas, TX 75237

Abstract

A description of the RISC-based distributed computing facility for detector simulation being developed at the SSC Laboratory is discussed. The first phase of this facility is scheduled for completion in early 1991. Included is the status of the project, overview of the concepts used to model and define system architecture, networking capabilities for user access, plans for support of physics codes and related topics concerning the implementation of this facility.

Introduction

A Physics Detector Simulation Facility is being developed for the SSCL to support the design and development of experimental detector systems. The facility, when completed, will provide a minimum of 4000 MIPS of computing resource to scientists involved in the physics/detector effort for the SSC. Development is scheduled in three phases. The first phase, identified as the "500 MIPS Acquisition Project", will be made available to users in early 1991 with a total facility complete in 1992. The design philosophy for the detector simulation computing systems adheres to an overall SSCL computing strategy which stresses distributed architectures standardizing on "RISC/UNIX" environment of workstations, servers and compute engines.

The Physics Detector Simulation Project represents the first significant acquisition of computing resources at the SSCL providing off-line capabilities to Experimental Physics. Off-line computing encompasses Physics/Detector Simulation, Data Reduction, Analysis and Archiving.

The SSC Laboratory is in a start-up stage, developing plans, building organization and mustering resources. The Laboratory is in the enviable position of planning and developing computing resources without some of the traditional problems most Labs face, like maintaining existing systems and older technologies. It can (and must) take advantage of changing computer technology and position itself so that these technologies can be easily incorporated into its computer designs. The SSCL has adopted a policy that encourages and will foster the use of open environments for computing. The SSCL will require the use of open systems and industry standards where possible for operating systems, languages, utilities, and protocols. To meet this goal the Laboratory has decided to establish a computing environment which emphasizes the use of distributed, networked computing, graphics and peripherals from multiple vendor sources [1].

Organization

The Computation Group, Physics Research Division is responsible for the planning and development of off-line computing resources in support of Experimental Physics. Its tasks include Systems Development, Integration, Physics Software Support, and Data Analysis/Archiving. The group is organized into the following functions:

Systems Development - Develop computing systems supporting specific physics requirements where emphasis is placed on the use of commercial hardware using RISC technology and open software.

Systems Integration - Develop systems performance models, analysis and recommendation of software tools. Review of hardware and software technologies and provide focus to software engineering standards and development of design methodologies.

Physics Support - Maintain physics applications software and libraries, coordinate simulation code devel-

* Operated by Universities Research Association, Inc. for the U.S. Department of Energy under Contract No. DE-AC02-89ER40486.

opment, porting of simulation and analysis codes and testing, maintenance of software and development of user documentation.

Operations - Provide systems administration, backups, archival support and end user help, consulting facilities and training.

Communications - Support the planning and implementation of Local Area Networks (LANS) and Wide Area Networks (WANS) utilized by the SSCL, support Physics Research Division specific requirements. Plan and develop videoconferencing facilities and provide network user support.

In order to accomplish its mission and adhere to SSCL computing strategies the Computation Group must maintain a firm understanding of current technologies and their application to the physics environment. Continual interface and communication must be maintained between the scientist/user and the system developer/integrator so that the best in computing and networking can be matched with physics needs.

Basis for Strategies

Development of computing resources for Experimental Physics is not an activity which can be planned or developed by a single group; computing strategies for the SSCL are the result of recommendations of select committees and working groups made up of representatives of the High Energy Physics community. The establishment of initial SSCL Computing resources during startup which included LANs/WANs, and the development of the SSCL computing strategy for distributed systems, was the result of the recommendation of a 1988 Task Force on Computing for the SSC [2]. Shortly after the Laboratory was established a Computer Planning Committee was chartered to recommend specific direction for development of computing resources for Experimental Physics. This committee made the following recommendation which provided the basis for establishing the current Physics Detector Simulation Facility Project [3]:

- SSCL should be a major resource of computing in the SSC Physics program.
- Acquire and make available to users a minimum of 4000 MIPS of computing by 1992.
- Develop high volume storage and file service, batch job scheduling and other centralized services.
- Support robust wide area networking capabilities.
- Provide effective user support to both local and remote users.
- Provide software documentation support to physics codes and graphics.

Taking these recommendations the Laboratory formed a Computer Acquisition Working Group (CAWG) made up of members of the Computation Group, SSC Experimental Physicists and potential users. The CAWG was chartered for the following:

- Review the final recommendations of the Computer Planning Committee.
- Develop a plan for supporting detector simulation codes in a distributed environment.
- Establish a vendor review process.
- Coordinate activities with other SSCL Computing requirements.
- Develop acquisition plans (Specifications/RFP).
- Determine associated costs.

Based on the Computer Planning Committee recommendation the CAWG initiated the project to plan and acquire the Physics Detector Simulation Facility. This project will provide a minimum of 4000 MIPS of computing resources by 1992. The implementation of the Physics Simulation Facility will be accomplished in three phases as outlined in Table 1.

Table 1. Phased implementation of the Physics Detector Simulation Facility

	PHASE I 3/91	PHASE II 12/91	PHASE III 6/92
Computing Power	**500 MIPS**	**1000 MIPS**	**4000 MIPS**
On-Line Storage	**50 GB**	**100 GB**	**400 GB**
Tertiary Storage	**.75 TB**	**1.5 TB**	**6 TB**

The development of these systems has not followed the traditional process of anticipating user needs, procuring hardware, determining software requirements and implementation. A rigorous process which includes detailed analysis and design using accepted methodologies has been the key element driving the development of the Detector Simulation Facility.

Overview of the Development Process

The goal of the system development process is to provide traceability from user requirements analysis, to system design and implementation, through integration and test. This is intended to ensure that the facility provided is a faithful representation of users' needs, and that the degree to which these needs are satisfied can be ascertained. This also ensures that problems are easily identifiable so that they may be quickly fixed. In addition to traceability, the process, described as follows, is intended to reduce overall development time, provide for a generally higher quality product, and reduce cost and risk.

Requirements Analysis

The goal of the requirements analysis activity is to translate the users' needs into a form more amenable to (and as it turns out, more testable of) the system development effort. The first part of this activity is to acquire as much input as possible from prospective users. The second part is to apply standard techniques to this data to produce a structured specification.

Facility Utilization

The data concerning facility utilization comes from several reports [1,2,3,4], meetings, and discussions with individual users. These sources provided information concerning number of users, amount of computer time needed, data generation and storage capacities, and particular applications software requirements such as compilers, word processors and editors. They also provided a view of the individual activities performed by typical users of the system.

Structured Analysis

The technique of event partitioning was used to describe the independent transactions which will take place on the system. This is a list of the asynchronous activities performed by the interactive users which serves to identify the essential, technology independent data storage requirements, and also as a template for final integration testing of the facility. These essential activities are then broken down into the primary processes needed to perform the individual tasks, and the tasks are then assigned to the processors which service them.

System Design

Of all the requirements clearly the most demanding is that concerning GEANT detector simulation throughput. For SSC type simulations containing 1000 simulated events requiring 10^4 MIPS*sec/event, a single such run on a Vax 11/780 would take about 3.9 months. On the other hand, even if there were sufficient funds available for the multiple supercomputers necessary to meet the requirements, this is not the obvious solution for two reasons. First, the processing of a GEANT simulation event is essentially independent of the processing of other events and is therefore very parallelizable under just about any scheme. Second, most of the high throughput machines achieve their performance using vector processing, and in its current forms GEANT has a low average vector length.

Design Strategy

Concurrent with the need for high throughput is the challenge to effectively utilize new technology. In particular, both the low price/performance available in the RISC workstation environment, and the emerging multivendor, interoperable systems capability in the UNIX marketplace should be exploited. These considerations argue for a parallel network of RISC, UNIX workstations. However, cheap MIPS is not only an enabling technology for low cost supercomputing. It allows for system designs which were previously considered either to costly or too inefficient.

A principal design benchmark for multi-million dollar computers has been percent CPU utilization. This usually meant a time-sharing environment where interactive users battled each other and the batch processes for system resources. The frustration of this environment has been expressed to us by many of our clients. It therefore seems that an effective interactive working environment can be well worth the cost of lower CPU utilization. This conclusion, in fact, is nothing new in view of the rapid proliferation of PCs and workstations involved in desktop organizational computing. What is perhaps new is the suggestion that a single purpose interactive system can be built of many single user computers.

Also, it is generally good design practice to group like processes with each other and isolate dissimilar processes. This yields systems with high cohesion and low coupling. There are a host of benefits gleaned by designing this way, including faster, cheaper development and higher maintainability. Again the benefits reaped from good systems design are enabled by the availability of low cost computers.

Figure 1 shows a diagram of the conceptual evolution of a computer system having the above properties. First parallelism is exploited for batch GEANT job processing. Next, interactive processes are split off being only loosely coupled to the "back end" by job submission and retrieval. Then time sharing is eliminated by parallelizing the interactive "front end". Finally, a "dataless" client/server configuration is chosen. Dataless refers to the fact that all the data generated and user files are stored centrally on a file server/tape system. This is economically efficient, however the primary purpose lies in superior operational support and maintainability. Dataless is not diskless though. Each of the front end computers has enough local disk to support the more tightly coupled processes associated with the operating system and swapping. This has the advantage of greatly reducing network traffic.

System Simulation

Several candidate design solutions of this type were simulated using a Petri net model of the system. Although there would be too much effort using this method for a packet by packet analysis of ethernet traffic, for example, it gave a good picture of the gross capacities going through the system. Stochastic simulations including user logins, job throughput file transfer and storage quantities, and number and type of processors helped define the conceptual design. This design concept was then the basis of the RFP. The resulting vendor proposals are currently being evaluated.

Conceptual Design

Figure 2 shows a pared down version of the functional diagram of the system. The most salient features are a front end, a file server, and a "batch ranch", which is the Texas version of the familiar "batch processor farm" concept. The front end is composed of a bank of shelf mounted, monitor-less workstations for interactive use by remote users and a local front end of workstations for local users.

When a remote user accesses this system from the wide area network, a process resident on the file server assigns the user a vacant workstation - one per user. The users' personal files are stored on the file server but appear local because they are NFS (Network File System) mounted over the ethernet. The ethernet also provides connectivity to the batch ranch for batch job submission and status checking. All interactive tasks are performed on the front end, for example, running physics generators, editing GEANT routines, compiling, performing test runs, sending and answering mail, etc. When a user decides to debug on batch ranch architecture or submit a lengthy job, the NQS (Network Queueing System) provides access to the batch services. At the physics code level CPS (Cooperative Processes Software) provides the library support to run jobs in parallel over all the batch ranch processors.

As a job running on the ranch accumulates data, it is buffered to disk. When a job completes, all associated data is collated into data sets and archived onto tape by means of the tape robot subsystem.

When it is necessary to analyze the data, a batch request copies the data onto the appropriate file server disk where it appears for the user. Analysis directly from tape will also be supported as a batch process.

This architecture has the advantage of being easily scalable. All one needs to do is add more batch processors and more front end/file server combinations isolated by bridges or routers.

Systems Integration and Test

In order to close the loop on this development process, it is necessary to first test the system before release to the user community, and then to assess its effectiveness in actual operations.

The testing process is one of assessing the working condition of the individual isolated units comprising the system. Next, the units are combined into subsystems (batch ranch, for example) and then tested for subsystem functionality. Finally, full integration testing consists of performing the activities defined in the event list to verify that the planned functionality exists.

After the system becomes operational large amounts of data must be gathered concerning actual operational utility. Currently a systems integration database is being designed to accomodate this data. The goal is to become as proactive as possible in repairing the problems that all systems experience. In particular, data will be collected and analyzed for the purpose of answering questions in the following categories:

- Is the view of how this system will be used accurate?
- Is the system doing what the users want it to do?
- Is the design of the system a correct/effective implementation of the requirements?
- Are the individual hardware/software components working as predicted?

This data will then be used, in conjunction with user

feedback, to support simulation and redesign efforts aimed at improving the implementation for Phase II of the project.

Conclusions

This Phase I conceptual system is designed for .75 TB of 8 mm tape storage and 62 GB of disk in total. This system will have approximately 480 MIPS of batch and more than 600 MIPS in front end/file server computing. The tape storage system is intended only as a temporary, stop-gap measure until a more permanent solution is found for Phase II. Ideally, when the batch processors are finished computing they should have nothing more to do with the data, leaving that subtask to a system designed to do that job. Presumably such a system would provide more direct access to the file server and in addition be scalable to a 6 TB automated system, which this tape subsystem is not. Also the requirements call for support of about 250 users, 100 active at any one time. Since Phase I will not have enough front end machines to serve one per costumer, this ideal will probably have to wait until Phase II for its full realization. Current planning calls for full Phase I implementation around the middle of March 1991.

References

[1] G. Chartrand, L. Cormell, R. Hahn, D. Jacobson, H. Johnstad, P. Leibold, M. Marquez, B. Ramsey, L. Roberts, B. Scipioni, N. Shivapuja, G. Yost, "Physics and Detector Simulation Facility Specifications," SSCL-275, Attachment A, July 1990.

[2] M.G. D. Gilchriese, editor, "Report of the Task Force on Computing for the Superconducting Super Collider," SSC-N-579, December, 1988.

[3] L. Price, editor, "Report of the SSC Computer Planning Committee," SSC-N-691, January, 1990.

[4] L. Cormell, editor, "Physics and Detector Simulation Requirements", SSCL-259, March 1990.

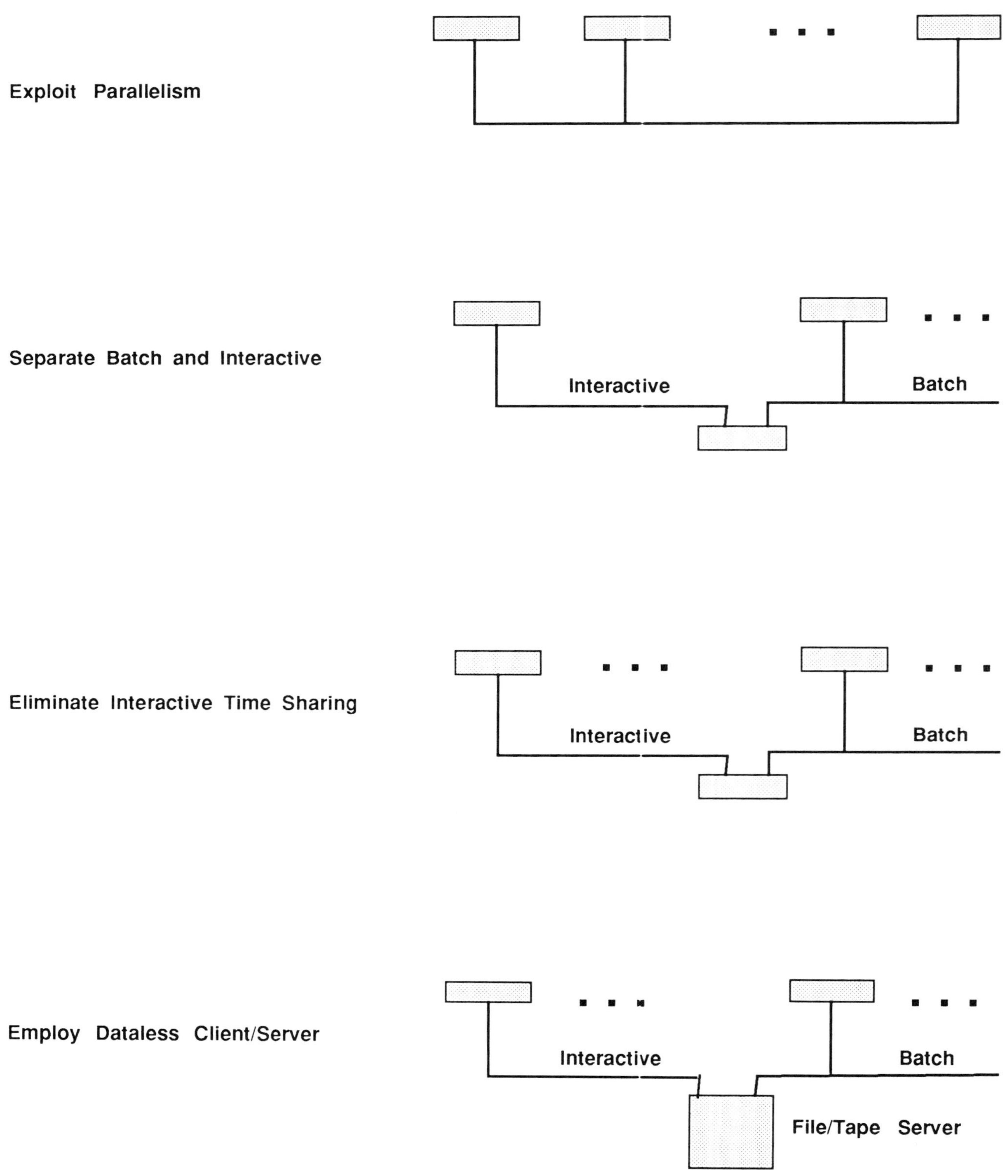

Figure 1. Architecture design strategy

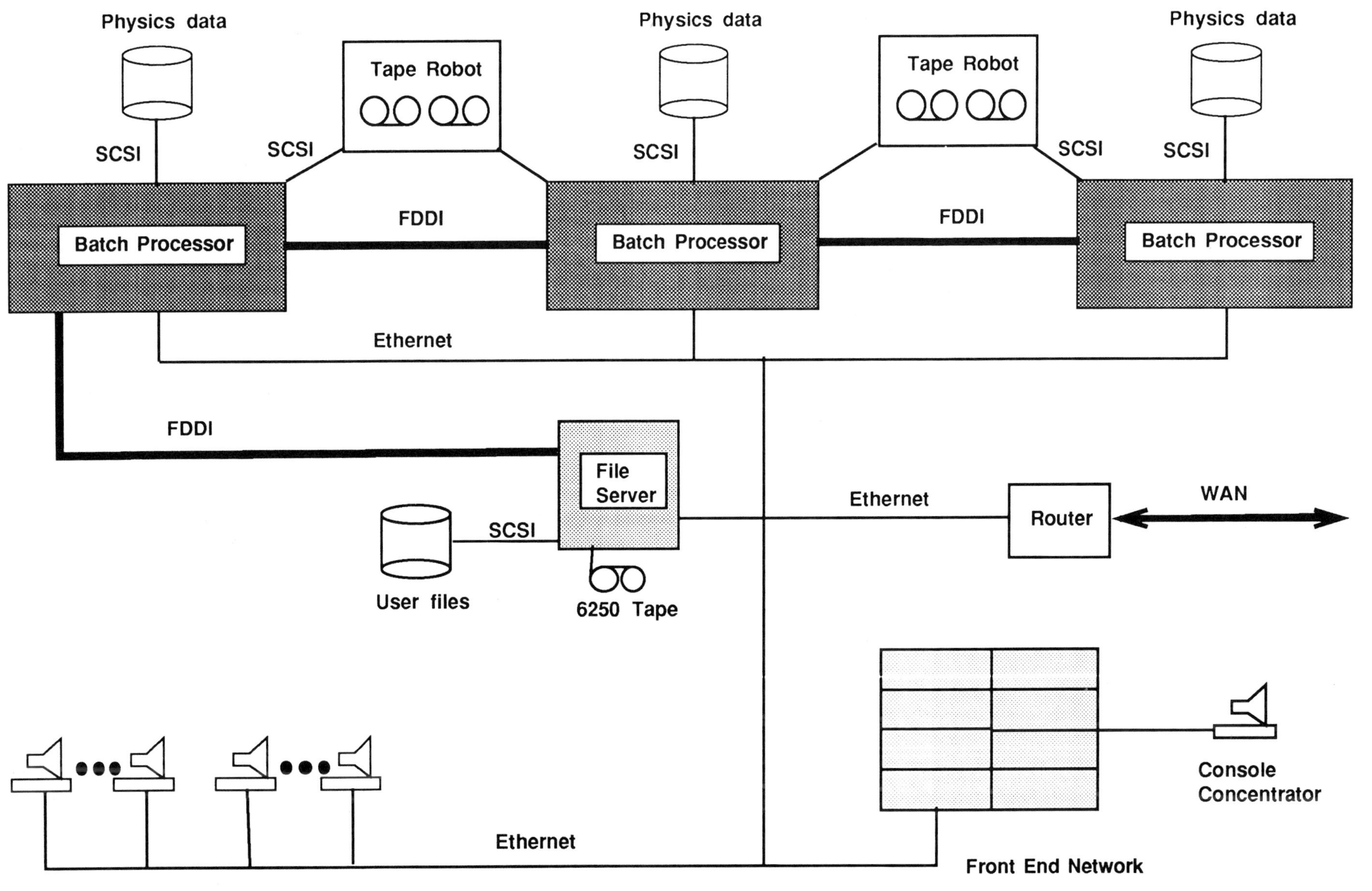

Figure 2. Functional Diagram of the Simulation Facility

MONTE CARLO AND DETECTOR SIMULATION IN OOP*

W. B. Atwood, R. Blankenbecler, and P. Kunz
Stanford Linear Accelerator Center
Stanford University, Stanford, CA 94309, USA

T. Burnett‡ and K. M. Storr
CERN, ECP Division, CH–1211 Geneva 23, Switzerland

ABSTRACT

Object-Oriented Programming techniques are explored with an eye towards applications in High Energy Physics codes. Two prototype examples are given: McOOP (a particle Monte Carlo generator) and GISMO (a detector simulation/analysis package).

1. Introduction

High Energy Physics (HEP) as a field came into being shortly after World War II. Accelerators replaced cosmic rays and computers were soon required both to gather data as well as assist in the analysis. The FORTRAN language was created to facilitate the translation of algorithms and formulas into machine executable code (FORTRAN stands for FORmula TRANslation language!) and was one of the first "higher-level" languages requiring a compilation step to produce machine code. HEP quickly adopted FORTRAN and today FORTRAN dominates our field. We have come to rely on many FORTRAN programs and packages, from histogrammers to memory managers, in achieving our research goals. Furthermore, entire systems of code such as GEANT [1] for the simulation of particle detectors and the LUND Monte Carlo [2] for generating events have become accepted starting points in the HEP code world.

FORTRAN code is referred to in computer science circles as a procedural language. The label is obvious and descriptive. Our programs for the most part are repetitively executed algorithms architected along client server lines with a hierarchy of drivers, subdrivers, ... and so on. Our style has changed very little, aside from trying to superimpose a few modern concepts such as memory management on top of FORTRAN.

Meanwhile the rest of the world continued to evolve, rather than refine and embellish. The UNIX operating system and its underlying language C resulted from efforts in the academic and engineering environments to achieve greater flexibility and open access to their minicomputers. The complexities of the graphical user interface employed on most small computers required further evolution. Object-Oriented Programming (OOP) techniques proved themselves equal to this challenge and are now used extensively. It's worthwhile to note that OOP

---

* Work supported by Department of Energy contract DE–AC03–76SF00515.

‡ Permanent address: University of Washington, Seattle, WA 98195, USA.

is driven by industrial needs, not hypothetical requirements from computer scientists.

The applicability of OOP to the computing problems in HEP is the subject of this paper [3]. Two separate projects have begun investigations into what typical HEP codes might "look" and "feel" like in OOP. The first is a Monte Carlo generator called McOOP, and the second is a detector simulation/analysis program called GISMO. As we shall see, OOP provides a very natural dialect for these problems. Before describing these two projects, we digress with a brief description of OOP especially designed for people having a FORTRAN background.

2. OOP for FORTRAN People

Object-oriented programming is a technique more than a language [4] and usually is based upon an existing language. Our experience has been in Objective-C [5] and as the name implies, it has the C language at its foundation. (There is also an Objective FORTRAN on the market, and others.) The program unit in OOP is something called a *class.* A class is defined by two parts: (1) a *header* file listing and typing the *instance variables* and defining the *method* names and their arguments, and (2) an *implementation* file fleshing out the various *methods.* As the program is run, instances of a class, called *objects*, are created in memory by the operating system using *factory methods.* This is when memory is allocated to the instance variables in contrast to FORTRAN where the memory is usually allocated at the compile-link stage. As the name *factory method* implies, multiple instances of a class may be present in memory simultaneously, each with its own set of instance variables, sharing the same form and methods but acting independently. Each object also contains a pointer to its associated class methods to allow the system to determine which routines to call for a particular method.

In contrast, the data in a FORTRAN program might be in COMMON blocks or in data structures managed by a memory management package (e.g., ZEBRA, JAZELLE, BOS, etc.). This data is manipulated and operated on by subroutines and functions. Now imagine breaking the data up into pieces along with the routines which operate on those pieces. These "chunks" of data/program might be termed objects. One of the basic ideas in OOP is to closely connect the code to the data that it deals with. For example, histograms and their associated routines could be recast as a class. The bins, bin limits, title, axis definitions, etc., would become the instance variables, while the operations of create, delete, clear, accumulate, and display would be the methods of this histogram class. Multiple instances of this class, i.e., several histograms, could be present simultaneously, each with its own bin size, title, accumulation variable, etc.

A method executes other methods in its own or other classes by sending *messages*, which are analogous to subroutine or function calls in FORTRAN. A message must contain an identifier of the class (i.e., a pointer to the target object), the name of the method, then possibily followed by arguments. In Objective-C, for example, the syntax is similar to Smalltalk:

[histogramaccum : x]

where *histogram* is a pointer to the target object, *accum:* is the accumulation method of the histogram class and x is the (single) value to be binned. The ":" is syntactical and separates methods and arguments from each other; messages are contained within brackets, "[.. ..]".

An important feature of OOP is a property called "inheritance." This is jargon for the fact that in OOP, the data, as well as the methods, are layered (i.e., form a hierarchical tree). This layering is much more extensive than in familiar procedural codes and does not refer to layering in the client server sense. In OOP, classes are *subclassed* to other classes. When this is done the new class "inherits" all the instance variables *as well as* the methods of ALL the classes lying above it in the hierarchy. This could be illustrated in the previous histogram example by subclassing our histogram class to a root class which allocates and deallocates memory. The "create" and "delete" methods of the histogram class would then be inherited from the root class. There would be no need for explicit code inside the histogram class to enable this functionality. On the other hand, if an action method needs to be modified for the subclass, it may be *overridden* by simply including new code in the subclass definition.

This property of inheritance is perhaps one of the most difficult aspects of OOP (for us old FORTRAN types) to use effectively. It requires (of us) a new level of abstraction in the creation of our programs not directly encountered in procedural code. It is believed by some "OOPers" that abstractions rarely evolve from the top down, but instead emerge from the bottom up. This is to say that the programmer is usually not aware of a new abstract class until he perceives common properties of several existing classes. He then creates a new abstract class with these common instance variables and methods and "bumps them upstairs." In spite of the difficulty (novelty?) in developing abstract classes, they usually prove to be quite valuable and efficient since they serve to establish the framework of the program as well as provide many useful reusable utilities which may be inherited by new classes.

Another OOP buzz word is "polymorphism." What this refers to is that the same function in different classes can, and usually should, have the same method name. For example, to deallocate an object and free up some memory, we might have a "free" method. The principle of polymorphism is that all objects can share the same name of the deallocation method, say, "free," but each class may implement it in its own way. This should be a familiar concept to us... it replaces "naming conventions" in procedural languages.

We end this section with a brief discussion on memory management and how data organization is handled in Objective-C. One important feature of OOP is the fact that memory management and bookkeeping are *not* the responsibility of the programmer but of the compiler and the run-time system. (The programmer must free objects appropriately, however, in order to keep memory usage under control.) First of all, data is bound to objects in OOP. Since many instances of various classes may be present simultaneously, a bookkeeping aid for objects is needed. This is provided in Objective-C by a class called List. List objects allow looping over the objects they contain, globally messaging each of them in turn, and providing edit functions to add, delete, and insert new objects into the List. An example usage of a List object might be to manage a group of histogram objects. This would allow global clearing of these histograms with a single message. Another more primitive data structure management tool is the Storage class. This object manages a group of identical data structures, similar to the concept in JAZELLE of *families* [6].

Examples of OOP programs are familiar to most of us although we may not have recognized them as such. These include popular drawing programs on personal computers with the windowed desktop screen layout, etc. Once aware of OOP it is easy to imagine that most of the things created on the screen are in fact objects, and indeed, multiple objects!

We shall see that OOP provides a very flexible coding framework, and now briefly describe our present formulation of two different applications. Keep in mind that this is prototype code; these examples can, however, provide the basis for a full implementation.

3. McOOP

One important characterization of numerical simulation problems is whether or not the "active" elements, or degrees of freedom, are fixed in number and characteristics, or whether they vary, perhaps even randomly, during the simulation process. We shall argue that OOP can offer many unique advantages in this latter situation [7].

The methodology described here, valid for general simulation problems, is one of our general points; the decay Monte Carlo application is one important example in particle physics. There are many other possible applications. The OOP characteristics discussed above and their uses here are: encapsulation, which will allow a simple treatment of the physics of each particle type, messaging and polymorphism, which will clarify the operation and writing of the code, and inheritance, which will play a very important role in shortening, simplifying, and insuring consistency among the physics objects.

A particular elementary particle is described by a set of simple parameters such as mass, charge, other quantum numbers, space-time position, momentum-energy, helicity, etc. These parameters are required for all particles; thus we are lead to the concept of a minimal or generic "particle" data set, or class. Unstable particles require more parameters; lifetimes and branching ratios for particular decay modes must be given. In addition to these variables, we also must add methods to allow a particle to perform an action, such as decay, in its own unique way.

We would like the code, in generating a new particle in the chain, to produce a generic particle, add any needed general properties, and finally, add the particular decay parameters specific to the produced particle type. To minimize the coding task, we will insist that the program use well-tested FORTRAN routines (e.g., phase space generators) whenever possible.

The highest class, a subclass of the abstract root class *Object*, will be denoted as *Particle*. In this class, all general and universal attributes shared by all particle types will be introduced as instance variables. Action methods of universal applicability and general utility will also be defined; this will therefore be the "largest" class. Many utility methods that are not used in the main code but are useful for examination, tweaking parameters, debugging, etc., are also included here for general availability. The general structure of the hierarchy is shown in Figure 1.

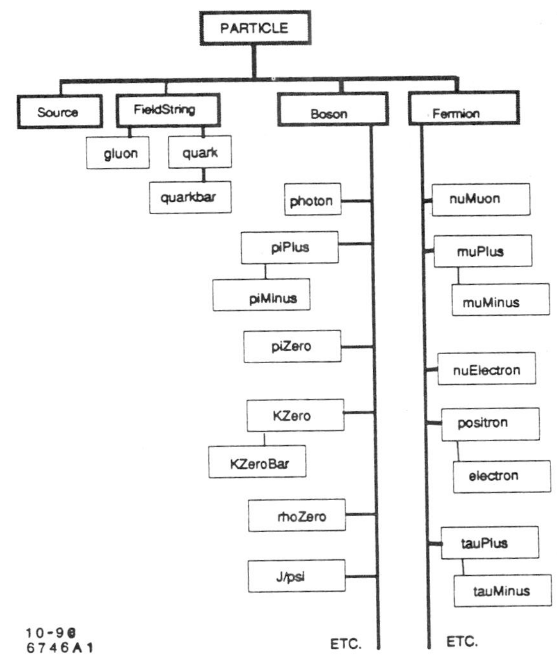

Figure 1. The decay Monte Carlo, McOOP, class hierarchy.

The top "generic" class *Particle* will have almost all the instance variables and action methods. This *class* is defined by giving a finite list of physical parameters such as charge, mass, energy, space-time position, etc., and a list of methods that act on these parameters. Action methods will be placed into the hierarchy at a physically appropriate level, although there is clearly some freedom here. Polymorphism will be extensively used.

Let us now move down the hierarchy and first describe the classes that contain the physical particles. Two natural subclasses of the Particle class are introduced, *Boson* and *Fermion*.

Next, most of the physical particle types, the photon, electron, pion, proton, etc., can now be introduced as a subclass of one of these two classes. Since ordinary particles decay sequentially and independently (except for mixing which can be separately handled), the decay methods are straightforward to implement.

```
#import <objc/Object.h>
#import <objc/List.h>

@interface Particle:Object
{
   char name[35];
   float charge;
   float hypercharge;
   float mass;
   float E;     /* energy */
   float px;
   ...........;
   float ptot;
   float lifeSpan;
   float lifetime;
   BOOL has_decayed;
   BOOL  is_Stable;
   id parent;
   id childList; /* progeny list */
   ...........;
 }
+ create:sender;
- (char *) name;
- (float) E;
- setE:(float) En;
- ................
- decay;
@end
```

```
#import "Particle.h"

@implementation Particle
+ create:sender
 {
  self = [ super new ];
  parent = sender;
  childList = [ List new];
  lifeSpan = rand();
  return self;
 }
- (float) parameter // many
 {                  // utility
  return parameter; // methods
 }                  // such as
- setparameter:(float) qx
 {
  px = qx;          // these.
  return self;
 }
 // ......etc................

- decay        // for stable
 {             // particles
  return self; // only.
 }
prints out decay chain.
```

```
@implementation piPlus
+ create:sender
 {
 self = [super create:sender];
 strcpy(name,"piPlus");
 charge = 1.0;
 hypercharge = 0.0;
 mass = 0.139;
 br_muon = 0.75;
 is_stable  = NO;
 has_decayed = NO;
 return self;
 } // end of create:

- decay
{ float rx;
if(has_decayed || is_stable)
return self;
rx = ranflat( 0 < rx < 1 );
if (rx < br_muon ) {//mode->muPlus+nuMuon
[childList addObject:[muPlus create:self]];
[childList addObject:[nuMuon create:self]];
else  {//mode -> positron+nuElectron.
[childList addObject:[positron   create:self]];
[childList addObject:[nuElectron create:self]];
}
   has_decayed = YES;
[childList makeObjectsPerform:@selector(decay)];
// sends decay message to each child on childList.
return self;
} // end of decay.
@end
```

10-90 6746A2

Figure 2. The (.h) and (.m) files of the Particle class, listed on the left and right, respectively, are shown at the top. The (.m) file for the class piPlus is shown at the bottom. In the decay method, one creates an instance of each particle class in the chosen mode, sets the parent id, and forms the childList by invoking a single compound message.

Using charge conjugation invariance to further simplify the code, the negatively charged member is chosen to be a subclass of its positively charged partner; thus physical parameters will occur in only one spot in the code—with the relevant positively charged member of the pair. These pairs possess the same number and values of branching ratios but have charge conjugate decay modes.

The class *FieldString* is introduced to handle the special properties of colored fields. For example, the decay properties of quarks and gluons is a collective phenomena in most models of hadronization; the decay of a rapidly separating quark-antiquark pair and their accompanying gluons is controlled in many models by the properties of the colored string(s) that are formed. Thus in the general case, the decay must be treated coherently. This correlated hadronization process is handled in the FieldString class, again introduced as a subclass of Particle as shown in Figure 1.

The class *Source* is introduced to describe the physical origin of the fields. This is specialized to the originating collision, i.e., proton-proton or electron-positron, etc. For the electron collider case, it includes the initial beam characteristics, momenta, widths and polarization, and randomly chooses the final state, i.e., a particular type of lepton or quark pair. It then messages the relevant class to produce instances of the produced particles, assigns them their correct kinematics, and finally, orders them to decay. For $p-p$ or $p-\overline{p}$ collisions, multiple instances of quark/gluon, etc., must be created according to the hadronic interaction model at hand.

An edited version of the class code is shown in Figure 2 for Particle and π^+. Note in particular the *create:* and the *decay* methods. Note also that the decay chain is stored NOT as a linked list, but rather as a "list linked by *Lists*."

This completes the discussion of the particle hierarchy illustrated in Figure 1. It is something of a compromise between having many generations for reasons of coding simplicity, and few generations for reasons of efficiency.

4. GISMO

GISMO (Graphical Interface for Simulation and Monte carlo with Objects) is a recent undertaking centered at CERN [8]. The objective of this project is to explore the applicability of OOP to particle detector modelling and production data analysis in HEP. Specifically we hope to learn if the modelling/analysis task is more easily broken up into pieces than is presently the case. Also, with GISMO, a goal is to "re-use" modelling code to perform functions in the analysis chain. Both of these goals are supposed to be features of OOP and could make an improvement on the way HEP collaborations write their software.

GISMO will have a Graphical User Interface (GUI) allowing for detector modification in a drawing program environment. The same GUI will serve as the event display.

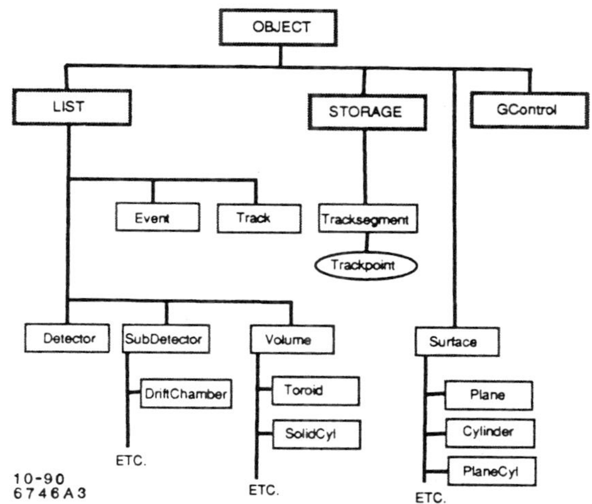

Figure 3. GISMO's class hierarchy.

The classes for GISMO are shown in Figure 3. Detector geometry is organized with the List class mentioned in Section 2. Detector is a "list" of SubDetectors, SubDetectors are a list of Volumes, and Volumes are a list of Surfaces which enclose them. Each of these classes is intended to be a superclass of specific types of detectors, volumes, and surfaces as indicated in the figure.

Inheritance plays a crucial role in this hierarchy. For example, Volume has instance variables for its title, a material-type index, and the maximum allowed step size for tracking. Variables to describe particular types of volumes are found in the classes subclassed to it. The Toroid class, for instance, has variables for its center, principle axis, length, inner and outer radii. No need to include title, material type, etc., as these attributes are inherited from Volume.

Recall that inheritance applies to not only the data, but to the methods as well. Volume is subclassed to List, hence anything a List can do (any messages a List will accept), Volumes can also do. This subclassing was done since most Volumes can be conveniently described by a list of surfaces (Volume is a List of Surfaces!). Volumes can be messaged to tell if a space point lies inside, how far along a helical trajectory to the boundary, as well as to be created and deleted. Most of these methods can be generalized (with the exception of the create method) from all Volumes and hence appear in the Volume superclass. For example, to determine if a space point lies inside a Volume, the Volume messages each of its Surfaces to tell if a radial vector passing through the space point intersects it. If the number of intersected surfaces is odd, the point lies inside. This is quite general and usually won't have to be superceded by code in each specific type of volume subclasses to Volume.

In GISMO, subdetector definitions are also treated in a hierarchy. At the top there is the List class to manage the Volumes in the SubDetector. Then comes SubDetector itself. It provides methods common to all subdetectors such as track propagation and management of geometry related questions such as arclength to swim, "inside," etc. SubDetector is a superclass for real detectors such as DriftChamber. In DriftChamber, for example, are the details of a typical axial/stereo wire chamber. This in turn could serve as a model from which to build drift chamber variants, each providing specialized instance variables and methods peculiar to themselves. The original code, List—SubDetector—DriftChamber, need not be touched (i.e., rewritten) but simply appended to and superceded by the subclassing process using inheritance.

GISMO simulates events as follows. Detector ingests a Monte Carlo event and creates an Event object which is a List of the Tracks to be processed. The Track class is similar to Particle in McOOP but also includes instance variables describing its spatial location, its present momentum, and so on. (In the next version of GISMO it would be interesting to subclass Track to the McOOP Particle class!) Track is subclassed to List because if it decays or interacts it will be a list of its daughter Tracks. The Track class has the "stepby" method (i.e., "swim") and takes as arguments a distance to step by and the material (index) in which to step (necessary to compute multiple scattering, interactions and energy loss). The physics of

```
/* ------------ Propagating the MCTrack through this subdetector ----------- */

- propagate: (Particle *) MCTrack
{

   id  trackseg, volume;
   float arclength;
   unsigned material, endptcode;
   BOOL alive;

   track = MCTrack;
     volume = [self locateVol];

   if(volume != nil) {
     trackseg = [Tracksegment new:  track];
     [tracksegList addObject: trackseg];
     [trackseg addTrackPoint: ENTRY ];

      material = [volume getMaterial];
      alive = YES; endptcode = ENTRY;
      while(alive & (endptcode!=LEAVE) ) {
         [self chooseStep: volume :&arclength : &endptcode];
         [track stepby: arclength in: material];
         alive = [track checkStatus];
         [trackseg addTrackPoint: (alive) ? endptcode: ENDPOINT];
        }
     }

    return self;
}
```

10-90
6746A4

Figure 4. The *propagate* method from the SubDetector class.

particles propagating in matter is part of the Track class definition.

Detector messages the first SubDetector with the id of the first Track in the event with the message to *propagate*. The Subdetector propagates this Track until it has decayed, interacted, or left the subdetector as illustrated by the code sample (see Figure 4). Detector proceeds to the next SubDetector and so on until it reaches the end of its list. Detector then processes the next Track until none are left "alive."

SubDetector creates a Tracksegment Object to manage the set of track points that are computed along a particles trajectory. Tracksegment is subclassed to the Storage class as track points are a fixed format data structure. Of course the number of track points in a Tracksegment is arbitrary. The Tracksegment class can display these track points for the event display and provides an implicit "loop" over trackpoints for modelling the subdetector's response. Tracksegments are managed in SubDetector by a List object.

An arclength is computed by taking the minimum of the maximum allowed step for the Volume, the distance to the wall, or some other criterion provide by a specific subdetector type.

Track is then messaged to "stepby" in the material of the Volume. At the end of each step the Tracksegment is messaged to add a track point to itself. The process is continued until the particle has interacted, decayed, or left the SubDetector. Once all the Tracks have been propagated through the Detector each subdetector will be messaged *simulateResponse* to compute and produce raw data facsimiles.

The analysis phase of GISMO has yet to be designated in detail. The idea will be to use the same object as in the simulation phase, adding methods to them to compare hypothetical tracks and interactions generated by a pattern recognition to those recorded either during simulation or actually measured in a real detector.

5. Conclusion

The OOP programmer does *no* explicit or detailed memory management *nor* other bookkeeping chores; hence, the writing, modification, and extension of the code is considerably simplified. Inheritance can be used to simplify the class definitions as well as the instance variables and action methods of each class; thus the work required to add new classes, parameters, or new methods is minimal.

The software industry is moving rapidly to OOP since it has been proven to improve programmer productivity, and promises even more for the future by providing truly reusable software. The High Energy Physics community clearly needs to follow this trend.

REFERENCES

[1] R. Brun et al., "GEANT: Simulation Program for Particle Physics for Particle Physics Experiments," CERN–DD/78/2 (July 1978).

[2] T. Sjostrand and M. Bengtsson, Comp. Phys. Com. **43**, 367 (1987).

[3] For a data analysis application, see: W. B. Atwood, R. Blankenbecler, P. F. Kunz, B. Mours, A. Weir, and G. Word, "The Reason Project," SLAC–PUB–5242 (April 1990).

[4] See, for example, P. F. Kunz, "Object-Oriented Programming," SLAC–PUB–5241 (April 1990); Invited paper given at 8th Conference on Computing in High Energy Physics, Santa Fe, NM, April 9–13, 1990.

[5] See B. J. Cox, *Object–Oriented Programming, An Evolutionary Approach.* Addison–Wesley, 1986. Objective-C is available from Stepstone Corporation, Sandy Hook, CT.

[6] A. S. Johnson et al., JAZELLE, *Proceedings of the 8th Computing in High Energy Physics Conference, AIP Conference 209,* 1989, p. 285; also see A. S. Johnson and D. J. Sherden, SLAC–PUB–263.

[7] R. Blankenbecler, "Object-Oriented Programming for Simulation Problems in Physics," SLAC–PUB–5251 (May 1990).

[8] W. B. Atwood, T. Burnett, R. Cailliau, D. Myers, and K. M. Storr, GISMO *(Graphical Interface for Simulation and Monte carlo with Objects), a CERN ECP Division Project.*

SOFTWARE METHODOLOGIES FOR THE SSC

Stewart C. Loken
Lawrence Berkeley Laboratory*
University of California
Berkeley, CA 94720

Abstract

We review the issues related to software development for the Superconducting Super Collider. These include the methods for software design as well as the use of specific software technologies and commercial products.

Introduction

This report describes some of the considerations that will determine how we develop software for the SSC. We will begin with a review of the general computing problem for SSC experiments and recent experiences in software engineering for the present generation of experiments. This leads to a discussion of the software technologies that will be critical for the SSC experiments. We will describe the emerging software standards and commercial products that may be useful in addressing the SSC needs. We conclude with some comments on how collaborations and the SSC Lab should approach the software development issue.

Parameters for SSC Software

In this section we review the characteristics of the SSC computing problem. The critical areas are the following:

- Storage and data management
- Computing environment
- Software development

What is missing from this list is a discussion of the computing power required to analyse the experiments. The recent developments in RISC computers and their use in loosely coupled systems make it clear that the CPU needs of experiments can be satisfied. It will, however, require some care to assemble systems that work efficiently.

Storage and Data Management

Each experiment will generate 10 – 1000 Terabytes per year of raw data. In addition, the typical event processing will double the amount of data. Most of these data will need to be accessible to physicists at remote institutions as well as at the SSC site.

The data management requirements are also severe. Some analysis will need to access very few variables for each event but need many events. On the other hand, to search for rare processes, it may be necessary to display all raw data from the detector in a 3-dimensional image.

The data must be organized to permit efficient access to both types of information. This presents a significant challenge to database technology.

Computing Environment

SSC physicists will need good interactive computing and efficient access to large databases. The environment should include statistical packages and powerful graphics. It will also be useful to have intuitive interfaces to analysis packages and data management tools. These should permit building complex applications from simple modules.

An important goal for the SSC software systems will be to make all programs appear as if they reside on the physicist's local workstation. The physicist should not care where the application runs or where the data are stored.

Software Development

Present experiments such as CDF now have approximately one million lines of Fortran code. It is typically developed by programmers who are not computing professionals and who are located at many different institutions in many countries.

The development of such large software systems would constitute a challenge to any group even using modern software engineering tools. In most experiments, however, the development process does not use formal methods or modern tools. As a result the software that exists is often unreliable and difficult to maintain.

*Supported by the U.S. Department of Energy under Contract No. DE-AC03-76SF00098.

the LEP experiments. These systems do satisfy, at least, some of the requirements of the machines and of experiments. There are, however, features that are needed and are not provided by the commercial systems and many groups have developed packages to provide the needed functions. Many of these have been built of the ZEBRA/RZ packages.

The optimal use of database systems requires the use of data modeling techniques to describe the objects in the database and the relationships among the objects. The Entity-relationship Model has been very successful in providing such a description. Some experiments have used this model as part of SA/SD or other software engineering. The ALEPH group has integrated data-modeling into their software design and has developed the ALEPH DAta MOdel (ADAMO). The ZEUS experiment has also adopted ADAMO.

The use of database systems and data modeling will continue to increase as the data management challenge increases. Groups need to experiment with these techniques and get experience with different products. To avoid becoming dependent on specific products, it is useful to adopt a high-level standard such as the Standardized Query Language (SQL).

Human interfaces

The development of intuitive user-interfaces is critical to improving the quality of scientific software. The popularity of graphical interfaces in the personal computer market demonstrates the need to follow this example in the development of programs for High Energy Physics. The Reason Project[10] has developed an interactive data-analysis package that utilizes the interface-building software that is supplied in the NeXT computer. The program does not yet have full functionality but it demonstrates the great potential for this approach. The ease with which the developers have developed the program indicates the benefits of using high-level software tools to develop scientific software.

Another new software product that is likely to have a significant impact on scientific computing is the Application Visualization System (AVS) from Stardent Computers. Using AVS, the physicist can develop sophisticated graphics applications without writing code. The D0 experiment has used AVS for 3-D event displays[11]. The approach shows great promise and will be used to develop more general tools.

The Use of Standards

The lifetime of experiment is now significantly longer than the career of a graduate student or post-doc. We need to recognize this in planning experiments. We cannot depend on students or post-docs to maintain code for the life of the experiment. We must design software that is reliable and can be maintained by others. The use of software engineering tools for design, documentation and testing will be critical.

The present generation of experiments already spans more than one cycle in computing technology. This introduces new challenges for the software developer. For the SSC and LHC era, the problems will be even greater. Experiments must plan from the beginning to evolve as the technology changes.

To ensure that systems will survive changes in technology, new software should be developed to utilize commercial products to the greatest extent possible. All software, whether developed or purchased, should adhere to standards. The industry will then assist in moving critical software to new technologies.

We are already seeing standards having a significant impact of computing in High Energy Physics. The use of UNIX is widespread, both for high-end workstations and as a source of cheap computing cycles. The Internet protocols (TCP/IP) which were virtually unknown in the High Energy Physics community a few years ago are now an important part of the network traffic. X-windows and new graphics standards (GKS and PHIGS) are gaining acceptance. This evolution will surely continue. The use of standards will ensure that investments in software will maintain their value as technology changes.

Getting it all together

The development of software systems for the SSC experiments will offer many new challenges and will require a change in the way that physicists work. The effort should be a collaboration of three groups, the experiments, the SSC lab and the computing indistry. Each of these can make a unique contribution.

The Role of Experiments

The experiments must plan and design computing systems with as much care as they put into any other aspect of the detector system. They should set priorities for funding all aspects of the detector including computing systems.

While the experiments will define the requirements for computing systems and software, they should work with laboratory management to define the scope of the SSC Lab central facilities. There should be a continuing interaction between the experiments and the SSC computing management to ensure that the laboratory support satisfies the needs of the experiments.

The Role of the Laboratory

The SSC Lab should provide centralized services for the experimental program. These are likely to include the following:

- Data storage and handling systems

Methodologies

The development of software systems for high energy physics experiments has been traditionally unstructured. Groups have defined some general guidelines regarding the use of common blocks for event data and other structures and some standards for FORTRAN code. Beyond this, individuals have been free to add software based on their own ideas of what was needed.

Recently, some groups has tried to introduce more formal methods into the process. These have been based on the "waterfall" model of software development. This assumes that the requirements can be clearly specified and that the design and code can procede from those.

Structured Methods

In the past, discussions on software engineering seemed to focus on one methodology, Structured Analysis/Structured Design (SA/SD). It is becoming clear, however, that this is only one possible part of the solution to the problem of engineering software systems. The full solution will require coding rules, code management and distribution, and many other pieces.

It is interesting, however, to review the success that projects have had using SA/SD. It has been applied to a number of experiments including ALEPH[1], D0[2] and ZEUS[3]. The results have been mixed but, in all cases, the groups agree that the methodology did improve the software design. In all cases, however, the lack of tools with which to maintain the analysis and design was a major problem. It was not possible to update the documentation easily and, as a result, it became obsolete. The SA/SD methodology was not followed through the coding phase.

An important aspect of the analysis/design efforts was the modeling of data using the Entity Relationship Model[4] or the ALEPH DAta MOdel (ADAMO)[5]. These tools provide a way to represent the complex relationships between data objects. Data modeling is an increasingly important part of the general problem of Database Management which is discussed below.

Problems with Structured Methods

As indicated, the introduction of SA/SD had some positive effects but was not a complete success. The lack of tools to support the methods was an important factor. Another important factor, however, was the fact that it is impossible to complete a full description of requirements early in the development cycle. The requirements will change as the experiment is designed. In addition, some of the software techniques must be tested before they can be introduced into the software.

These considerations lead to the idea of a prototyping step. This is similar to the method for designing hardware. Critical elements of the software are identified and tested in simple programs. The technologies can then be incorporated into larger system. Some of the key technologies that will be needed for SSC experiments are discussed in the following sections.

Software architectures

It is clear that systems are becoming larger and more complicated. We need to improve the level of management on computing projects and we need to become more aware of the need to develop an integrated system. It will be difficult to enforce engineering rules across the project and for this reason, we need to look at system architectures that will isolate people who do not follow the rules.

Architecture is an over-worked phrase these days. In our context we mean a software structure that allows us to build robust programs. It has been clear for some time that we need to design systems that have small modules that work, and can be tested, independently. This approach has been demonstrated to improve programmer productivity and reduce software maintenance costs.

One technology for building modular systems that interoperate is called the software bus[6]. There are now a number of commercial implementations of this technology and there is increasing interest in defining a standard for a software bus. A pilot project using one bus standard was described by Grieman[7]. While this approach requires more effort in the implementation phase, the result, so far, has been modules that are easier to maintain and to reuse in other applications.

A formal approach to the development of software modules is Object-oriented Programming. While this approach is not yet wide-spread in High Energy Physics, some pilot projects show considerable promise. The paper by Kunz[8] provides an introduction to Object-oriented Programming techniques. These techniques have been used very successfully in the Reason Project described below.

Commercial products

Many of the software products needed by future experiments can, and should, be supplied by industry. Some obvious examples are databases, human interfaces and graphics. The use of industry standards will ensure that software can evolve as future systems are introduced.

Databases

It has become increasingly clear that database management is an important component of High Energy Physics software. The problems will be even more severe for SSC and LHC experiments with more complex detectors and more events to deal with.

The use of commercial Relational Database Management Systems and of home-grown packages has been reviewed by Rimmer[9]. Commercial systems have been used successfully by the LEP builders and by some of

- Networking
- Data acquisition and control systems
- Widely applicable software

The laboratory should establish a mechanism by which the experiments can guide the choice of centrally supported systems and can review the quality of the support.

The laboratory should also premote appropriate software standards. It should monitor developments in the computing industry and should act as a focus for industry collaboration.

The Role of Industry

While much of the needed software and hardware should come from industry, it will be necessary in many cases to modify or extend existing products. Physicists and industry must work together to understand the requirements of the experiments and to evaluate existing commercial solutions. As new solutions are developed they must be tested in the SSC applications.

Conclusions

We are beginning to see a major change in the way that physicists do their computing. The availability of high-level interfaces and other commercial products will have a very significant impact of future projects. As with any new tool or technique, we need to get experience with the new software technologies.

Future software systems will be developed by collaborations of experimenters, the SSC Lab and industry. We should begin now to form these teams to address some of the critical software issues.

References

[1] G. Kellner, " Develpoment of Software for ALEPH using Structured Techniques," in *Computing in High Energy Physics: Proceedings of the International Conference on Computing in High Energy Physics, Asilomar*, 1987, p 229.

[2] J.J. Linnemann *et al.*, "The Use of SA/SD Methods in D0 Software Development, in *Computing in High Energy Physics: Proceedings of the International Conference on Computing in High Energy Physics, Asilomar*, 1987, p 245.

[3] R. Loveless *et al.*, "ZEUS Hardware Control System", in *Proceedings of the International Conference on Computing in High Energy Physics, Oxford, England*, 1989, p 313.

[4] P. P. Chen, ACM Transactions on Database Systems **1** 9, (1976).

[5] S. M. Fisher and P. Palazzi, " Using a data model from software design to data analysis: what have we learned?" in *Proceedings of the International Conference on Computing in High Energy Physics, Oxford, England*, 1989, p 169.

[6] D. E. Hall *et al.* " The Software Bus: A Vision for Scientific Software Development", in *Proceedings of the International Conference on Computing in High Energy Physics, Oxford, England*, 1989, p 313.

[7] W. Grieman *et al.* "Experience Using a Software Bus to Build Reuseable Scientific Software", in *Computing for High Luminosity and High Intensity Facilities: Proceedings of the Conference on Computing in High Energy Physics, Santa Fe*, 1990, p 276.

[8] P. F. Kunz, "An Introduction to Object- oriented Programming Techniques for High Energy Physics", in *Computing for High Luminosity and High Intensity Facilities: Proceedings of the Conference on Computing in High Energy Physics, Santa Fe*, 1990, p 264.

[9] E. M. Rimmer, "Databases for Large Detector Systems — Experiences at LEP and Future Needs" in *Computing for High Luminosity and High Intensity Facilities: Proceedings of the Conference on Computing in High Energy Physics, Santa Fe*, 1990, p 341.

[10] B. Atwood *et al.*, "The Reason Project", in *Computing for High Luminosity and High Intensity Facilities: Proceedings of the Conference on Computing in High Energy Physics, Santa Fe*, 1990, p 320.

[11] T. G. Trippe, Private communication.

CALOR89 CALORIMETER SIMULATIONS, BENCHMARKING, AND DESIGN CALCULATIONS

Thomas Handler
University of Tennessee

Job Panakkal and Jim Proudfoot
Argonne National Laboratory

Lucien Cremaldi, Brent Moore, and Jim Reidy
University of Mississippi

Tut Alsmiller, Peter Fu, and Tony Gabriel
Oak Ridge National Laboratory

Presented by Thomas Handler

Abstract

Results on CALOR89 benchmarking and design calculations utilizing the CALOR89 programs are presented. The benchmarking is done with respect to the ZEUS and D0 calorimeters. The design calculations were done for a variety of absorbers (depleted uranium, lead, and iron) of various thicknesses for a given scintillator thickness and for a fixed absorber thickness using various thicknesses for the scintillator. These studies indicate that a compensating calorimeter can be built using lead as the absorber, whereas a pure iron calorimeter would be non-compensating. A depleted uranium calorimeter would possibly be unsuitable if used in a large configuration and a high luminosity machine because of the delayed energy release from captured gammas.

Introduction

Since the calorimeters that are going to be used at the Superconducting Super Collider (SSC) are going to be large and expensive it would be prohibitive to build full scale prototypes. Therefore, the only avenue that is open is to build small scale prototypes that can be tested with the results being compared to the output of simulation programs. These simulation programs can then be used to design the calorimeters, once their accuracy has been established by benchmarking against the existing data. We describe here the CALOR89 simulation code system, the results of benchmarking it with respect to experimental data and results of design calculations involving various absorber materials and thicknesses and various scintillator thicknesses.

CALOR89

The CALOR89 code system [Fig. 1] consists of four primary programs: HETC88[1], SPECT89, EGS4[2], and MORSE[3] or MICAP[4], plus their ancillary programs and a final analysis program. HETC88 is used to generate and transport the hadronic particles through the calorimeter. HETC88 does the particle transport and generation in a three fold manner: 1) For energies less than 3 GeV particles are generated by means of an intermediate-energy Intranuclear-Cascade and Evaporation Model[5]; 2) From 3 GeV to approximately 10 GeV, particle generation is done by means of a scaling model. 3) From 10 GeV upwards particle generation is done by means of FLUKA[6] which uses a multi-chain fragmentation model. Inelastic nucleon hydrogen and charged-

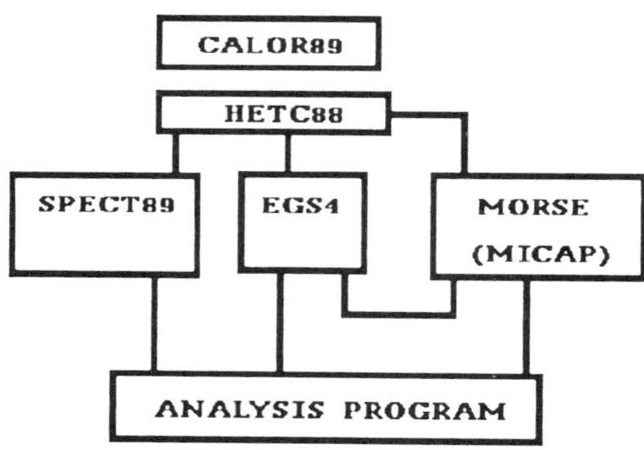

Figure 1: CALOR89 Code System

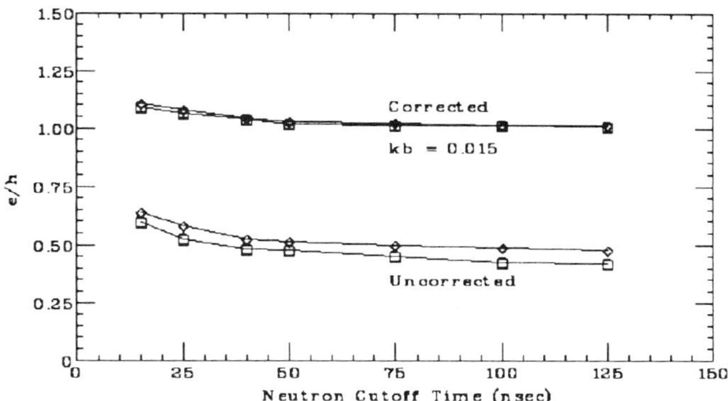

Figure 2: Scintillator Saturation Effects. Lower curves represent uncorrected scintillator signals, while upper curves have been corrected for saturation effects.

pion hydrogen collisions are done via the isobar model or a fragmentation model[1]. The boundary between the use of the Scaling Model and FLUKA is determined by a parameter, ESKALE, that is at the user's descretion. SPECT89 does the energy deposition of the hadrons in the calorimeter. The ancillary program, LIGHT, allows the user to take into account the non-linearity of the light pulse in the scintillator due to saturation effects within the active medium. This is done by use of Birks' law[7]. In the simulation of calorimeters and the comparison with the experimental test data it is imperative that saturation effects be taken into account. In Figure 2 we show the effects of saturation. The simulation is for a slab calorimeter made from a 4mm lead sheet followed by a 1mm sheet of scintillator repeated to a depth of 150cm. As is seen in the figure, not taking into account saturation effects, an overestimate is made of the hadronic signal.

EGS4 is used for the transport and energy deposition of electromagnetic particles in the calorimeter. The source data for EGS4 consists of direct photon production from hadron-nuclear collisions, photons from neutral pion decay, and electrons and positrons from muon decay. These are read from the HETC88 output tape. MORSE or MICAP is used to transport neutrons that are produced with energies less than 20 MeV and to generate the gamma rays from inelastic, fission and capture reactions. These subsequent gammas are then transported by EGS4. Both MORSE and MICAP have time dependence built into the code.

The programs SPECT89, EGS4, and MORSE (MICAP) do **NOT** explicitly incorporate many of the experimental details that are always there. These details include non-uniformity of light collection, electronic noise, pedastel cuts, material noise (natural fission noise), etc. These can be incorporated into the analysis program which combines the output of SPECT89, EGS4, and MORSE (MICAP).

Comparisons of experimental data with CALOR89 calculations should be done with caution. Unless the exact geometries are used, so that leakage out of the sides or the back of the calorimeter is the same, and the exact same materials in the same relative order and the same thicknesses are used, and the value for Birks' constant appropriate for the scintillator is used, the agreement between the simulation results and the experimental data will not be exact.

Benchmarking Results

The results obtained from the CALOR89 code system have been checked against experimental data from the ZEUS prototype lead-scintillator and depleted uranium-scintillator calorimeters, and CDHS iron-scintillator calorimeter and the D0 liquid argon-uranium calorimeter [8–11]. The results of the comparisons are presented in Table 1.

As can be seen, the overall agreement between the CALOR89 predictions and the experimental data is quite good.

Design Calculations

We have carried out a series of calculations to determine the sensitivity of the compensation and resolution characteristics of a prototypic calorimeter composed of various absorber materials (depleted uranium, lead, and iron) of various thicknesses (0.5, 1.0, 2.0, and 4.0 radiation length) in conjunction with various scintillator thicknesses (0.1, 0.25, 0.5, and 1.0 cm). The incoming particle used was a 10 GeV kinetic energy negative pion and a 10 GeV electron. Birks' constant for the scintillator saturation level was set at 0.0131. The calorimeter was 2m by 2m by 8 interaction lengths of absorber material. The complete matrix of absorber thicknesses and scintillator thicknesses was not explored. In one series of calculations the scintillator thickness was fixed at

Table 1. CALOR89 Benchmarking Results—e/h

Configuration	DU-SCSN	Pb-SCSN	Fe-SCSN	Liq. Arg.-DU
Thicknesses	0.33/0.26 cm	1.00/0.25 cm	2.50/0.25 cm	0.57/0.3 cm
CALOR89	1.01 + / − 0.03	0.99 + / − 0.03	1.17 + / − 0.03	1.06
Experiment	1.01 + / − 0.04	1.05 + / − 0.02	0.19 + / − 0.04	1.03 + / − 0.015

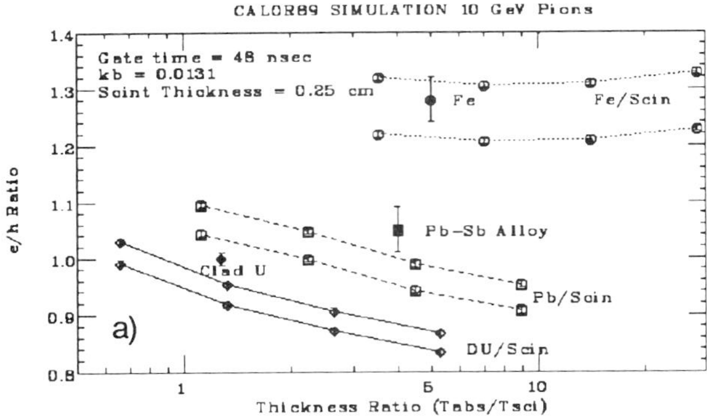

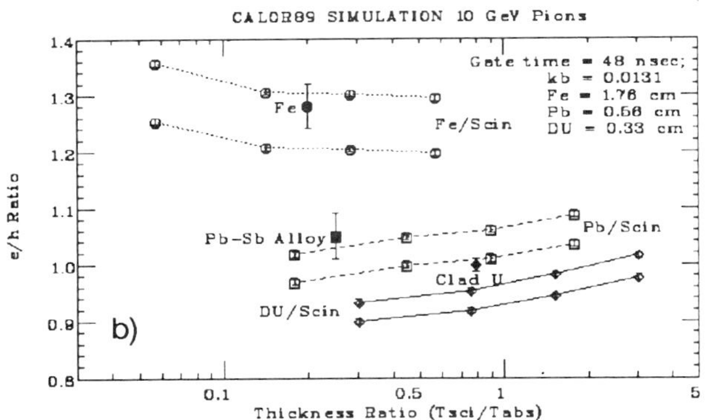

Figure 3: e/h is shown as a function of absorber thickness divided by a fixed scintillator thickness in 3a, while in 3b it is shown as a function of scintillator thickness divided by a fixed thickness of absorber.

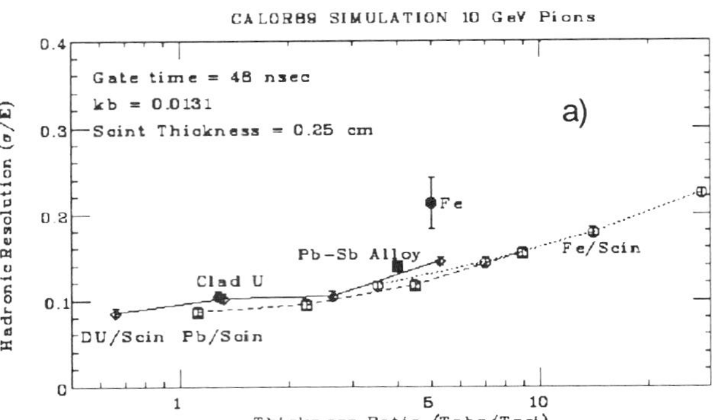

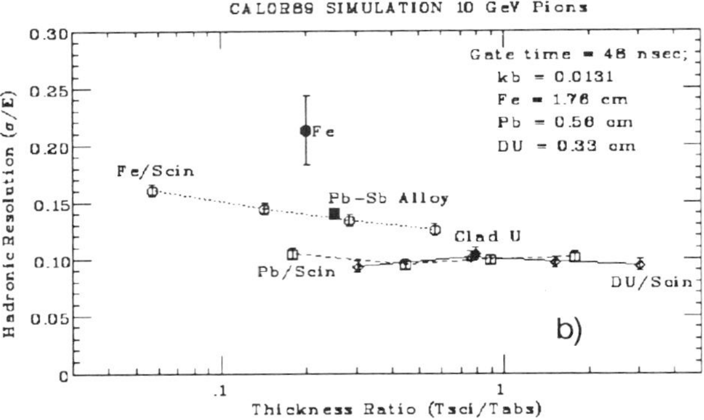

Figure 4: Hadronic resolution as a function of absorber divided by fixed scintillator in 4a and as a function of scintillator divided by fixed absorber in 4b.

0.25 cm and the various absorber thicknesses were used. In another series of calculations the absorber thickness was fixed at 1 radiation length and the scintillator thickness was varied. Various gate times were used in the low energy neutron and subsequent gamma analysis. The actual gate time has an uncertainty of approximately 5–10 nsec due to the fact that no timing is done in HETC88 or EGS4.

We present in Figure 3 the compensation results for the three absorbers as function of absorber thickness divided by scintillator thickness [Fig. 3a] when the scintillator was held fixed at 0.25 cm and as a function of scintillator thickness divided by absorber thickness [Fig. 3b] when the absorber was held fixed at 0.25 radiation lengths. The two curves for each of the absorber materials represent calculations done with ESKALE set at 5 GeV, the lower curve, and set at 15 GeV, the upper curve. The gate width used in the calculations was 48 nsec. For the iron and lead cases this gate width is sufficient to collect the large majority of the signal. However for the uranium case this is not true due to the fact that uranium through fission processes produces additional neutrons and that uranium has a very large capture cross section at energies less than 1 MeV. In a large calorimeter system and in a high luminosity environment, the produced gammas may yield a significant contribution to background noise and thus produce pile-up problems. The experimental data points, even though they are from calorimeters not simulated by these calculations, are plotted to show that the calculations agree with the general trends. In addition the iron experimental data point represents a "weighted" value. The unweighted value is $1.36 + / - 0.04$.

In Figure 3a we see that the iron calculations show that there seems to be no strong dependence of the compensation on absorber thickness. There is a stronger dependence for the lead and uranium cases. This is due, in part, to the strong Z dependence of the electromagnetic cross sections. In Figure 3b, the curves have to approach each other as the scintillator thickness becomes sufficiently large to contain both the electromagnetic and hadronic cascade.

In Figure 4a we present the hadronic resolution as a function of absorber thickness divided by the scintillator thickness which was held fixed at 0.25 cm and in Figure 4b for scintillator thickness divided by absorber thickness which was held at 0.25 radiation lengths. Only the calculation done at an ESKALE 5 GeV is presented as the values for ESKALE at 15 GeV lie basically on top of these values. The comparison with the uranium and lead resolution data appears quite good. We feel that

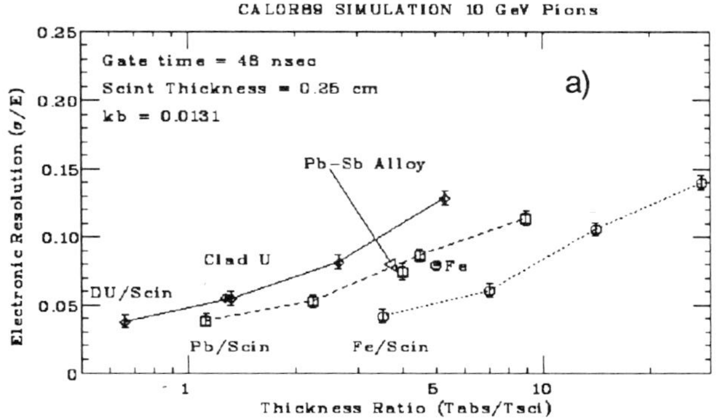

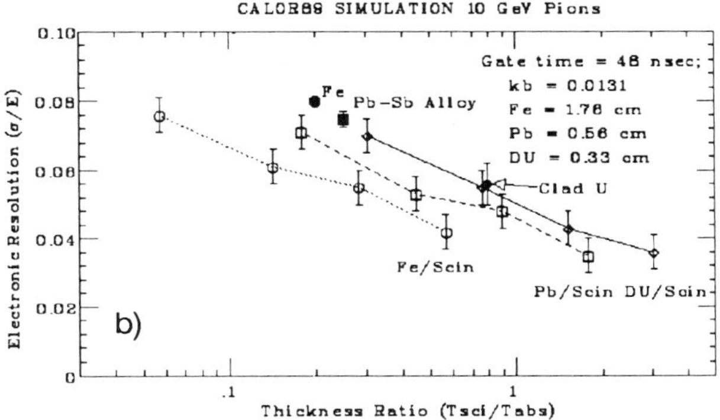

Figure 5: Electromagnetic resolution as a function of absorber divided by fixed scintillator in 5a and as a function of scintillator divided by fixed absorber in 5b.

the disagreement between the iron calculation and the iron data point is probably due to the strong data cuts applied and to the lateral size of the calorimeter.

In Figure 5 we present similar curves for the electromagnetic resolution. The agreement between the calculations and the experimental data are again good except for the iron data point. As in the case of the hadronic data point we feel that their experimental cuts are possibly the cause.

It has been suggested that by placing hydrogenous material on either side of a plastic scintillator in conjunction with an iron absorber, that such a calorimeter could be made compensating. Calculations utilizing CALOR89 indicate at 10 GeV at most a 5% improvement in the compensation characteristics. The calculations that were carried out were for an iron absorber of thickness 2.54 cm followed by 0.4 cm hydrogenous material (dead scintillator) followed by 0.2 cm of active scintillator followed by 0.4 cm of hydrogenous material. This was repeated for 8 interaction lengths of iron.

Summary

The CALOR89 code system has been used to compare with various calorimeter prototypes. In general, the agreement obtained has been satisfactory, for discrepancies can be traced to experimental cuts, biases, leakage, etc. This code system has been used to generate preliminary design data for a variety of absorber, scintillator configurations. These results give general trends on the anticipated changes in the energy resolution and compensation characteristics.

References

[1] R.G. Alsmiller Jr., F.S. Alsmiller, and O.W. Hermann, "The High Energy Transport Code—HETC88 and Comparison with Experimental Data." Submitted to Nuclear Instruments and Methods.

[2] W.R. Nelson, et al., SLAC–265, Stanford Linear Accelerator Center, (1985).

[3] M.B. Emmett, ORNL–4972, Oak Ridge National Laboratory, (1975). N.M. Greene et al., ORNL/TM–3706, Oak Ridge National Laboratory, (1973).

[4] J.O. Johnson and T.A. Gabriel, ORNL/TM–10196, Oak Ridge National Laboratory, (1987).

[5] H.W. Bertini, Phys. Rev. **188**, 1711, (1969).

[6] J. Ranft and S. Ritter, Z. Phys. C: Part. Fields **10**, 249, (1981). P.A. Aarnio, et al., "CERN TIS Divisional Report," TIS–RP/106-Rev, (1984). A. Capella and J. Tran Thanh Van, Phys. Letts., **93B**, 2, (1980).

[7] J.B. Birks, Proc. Phys. Soc., **A64**, 874, (1951).

[8] H. Abrahamowicz et al., Nucl. Inst. Meth., **180**, 429, (1980).

[9] E. Bernadi et al., Nucl. Inst. Meth., **A262**, 229, (1987).

[10] E. Ros and T. Tsurugai, ZEUS Note 88–086 (1988).

[11] M. Abolins, et al., Nucl. Inst. Meth., **A280**, 36, (1987).

DEVELOPMENT OF A ROBUST CAD-TO-GEANT INTERFACE*

A.M. Yacout, J.D. Simpkins, and W.L. Dunn
Quantum Research Services, Inc.
P.O. Box 52391
Durham, NC 27717-2391

S. Youssef and W.J. Womersley
Florida State University
Tallahassee, FL 32306

Abstract

We report on preliminary efforts to construct interface software to convert normal computer-aided-design (CAD) files into input files that are directly readable by the GEANT high energy physics detector simulation code. The advent of CAD and powerful graphics packages has provided a welcome assist to the detector simulation community; however, the benefit is far from fully realized, since detector geometry is subject to wasteful double entry – once into the CAD system, in its format, and then into GEANT, in its format. Our efforts involve construction of a user interface code, written in C, to read a CAD output file, perform conversions to GEANT representations, and prepare a GEANT input file for the geometry and material specification of detectors or detector subsystems. The user interface consists of a CAD-to-GEANT interpreter, a GEANT input generator, and a graphical user interface.

Introduction

Large, special-purpose computer codes, such as GEANT and CALOR89, perform detector simulations for high energy physics experiments. The GEANT code [1] has been in use, in some version, for well over a decade. Despite GEANT's wide use, generally good documentation, consistent support, and thorough benchmarking, the input to its geometry package is nontrivial to learn and tedious to use for the specification of large, complex detectors, such as those that will soon be developed for the Superconducting Super Collider (SSC). The work reported here addresses the well-recognized need for a user interface that simplifies the task of converting computer-aided design (CAD) system output to GEANT-code input.

The following considerations dictate that a robust (*i.e.*, faithful, fast, user-friendly) interface should be developed for GEANT, as soon as possible:

- GEANT is an essential tool for assisting in the design of new detectors; its use is driven by the desire to "get the best physics" out of the detectors while controlling detector costs, which can run into the hundreds of millions of dollars per detector.
- Detector geometry input to GEANT is time-consuming and labor intensive; it can take a seasoned GEANT user months to create a complete debugged GEANT input file for a new detector.
- Highly trained and talented high energy physicists would be better utilized doing physics than spending person-years, collectively, in tedious GEANT code input and data checking.
- The staggering simulation requirements for SSC detector design are imminent, and the critical design simulations will be performed over the next several years.

The need for a CAD-to-GEANT user interface has been discussed for several years. Indeed, R. Brun – one of the originators of the GEANT code – stated, at the 1990 SSC simulation workshop, that "GEANT geometry packages are hard to use by the novice. It is desirable to have geometric input using mouses much in the spirit of CAD/CAM packages. As the design is changed by engineers using CAD/CAM, it is desirable to have an expedient way of transferring that information to GEANT and vice versa. An interface to CAD/CAM geometric packages was felt to be necessary ."

* Work supported under DOE grant DE-FG05-90ER80972.

Scope

Quantum Research Services is working with Drs. Womersley and Youssef, both of Florida State University, to develop a powerful CAD-to-GEANT user interface. Our initial efforts have led to a Small Business Innovation Research grant. Long-term plans are to develop a professional, generic software package that will incorporate several advanced features (identification of data inconsistencies, incorporation of rules of good practice, inclusion of large but easily accessed libraries, estimation of detector costs), and operate within a variety of operating systems.

Within the Phase I effort, we are attempting to do the following:

- Use AutoCAD (as a representative CAD system) to specify several hypothetical detector configurations, based on an initial set of six geometrical primitives.
- Construct a user interface code to read the CAD DXF (Drawing Exchange File) output files, identify the volumes embedded in the CAD files, perform conversions to GEANT representations, prompt the user for material composition specifications, and prepare a GEANT input file containing geometry and material specification data. Initially, the interface code will be programmed to consider five of GEANT's allowable shapes and several of GEANT's materials; ultimately, the code will handle the full complement of GEANT volumes.
- Demonstrate the system by converting various hypothetical CAD detector geometries to GEANT inputs and having GEANT read the inputs. Some of the conversions will represent one-to-one conversions of allowable shapes; others will require that the code recognize incomplete or illegal CAD specifications and complete the conversions with user input.

System Architecture

The CAD-to-GEANT interface code will incorporate the following major components:

- Graphical user interface
- CAD-to-GEANT shape interpreter
- GEANT input generator.

The code is being written in C and employs a mouse-driven, window-based display that includes menus, icons, and dialog boxes. Each of these components is described below.

Graphical User Interface

The graphical user interface (GUI) is the communication link between the user and the code. It is used to indicate the status of processes, to input required information (*e.g.*, materials, names of entities, missing dimensions, etc.), to solicit user input in order to resolve inconsistencies that the code fails to resolve on its own (*e.g.*, shapes that are not represented in GEANT and cannot automatically be transformed by the shape interpreter), and to alert the user to improper inputs. The GUI also graphically displays the shapes, geometrical layout, and material overlays, as the transformation progresses from the CAD file to the final GEANT input file. Thus, for instance, it will be possible to graphically compare CAD and GEANT detector specifications before the GEANT input file is created. Most of the work on the GUI is now complete.

CAD-to-GEANT Shape Interpreter

The CAD-to-GEANT shape interpreter reads the user-supplied CAD file, which contains the CAD description of the detector geometry, and provides a graphic display; it then performs rule-based volume identification, from the 2-D (utilizing one or more views, as necessary) or 3-D CAD descriptions. The interpreter will eventually be able to read either DXF or IGES format. The well supported and documented DXF format is a neutral file format that can supply the interpreter with all the information about the drawing entities and attributes that define detector geometries and material compositions. IGES, though less widely used, is also a standard CAD format that will eventually be supported by the user interface.

The identification of volumes from DXF output is a multi-step process, which is beset by several difficulties. The process involves the following:

- entity (*e.g.*, line, polyline, arc) identification
- entity-view association – determining which entities are associated with which views
- shape construction – building up a view of a shape from various entities
- establishing view requirements – determining which sets of views are necessary and sufficient to completely specify a 3-D volume
- volume construction – identifying the volume from the collection of shapes or views
- dimensional consistency checking
- line type interpretation – distinguishing between solid and dashed (visible and hidden) lines.

The difficulties associated with this process are many; they arise largely from the following:

- the unrestricted nature of user specification – *e.g.*, a user may specify a rectangle as four connected lines, more than four lines, a 4-segment polyline, a combination of lines and polylines, or a combination of polylines.

- the fact that gaps must be resolved – *e.g.*, did the user intend not to close a polyline or was the specification simply not sufficiently precise?

- the diversity of representation modes – *e.g.*, one, two, or more planar views, 3-D perspective views, or mesh representations

- the variability in the number of entities and views required.

After the geometric primitive shapes in the CAD file are identified, they will be compared with the programmed GEANT volumes. If the primitive does not match one of the GEANT volumes, the interpreter will attempt to convert the primitive by subdividing it into volumes that are recognized by GEANT. Later, a more sophisticated conversion process will be constructed, which attempts to combine the primitive with other contiguous primitives. If the CAD primitive cannot be divided into GEANT shapes, the interpreter will alert the user, display the primitive in question, query the user for input, and – with user help – complete the conversion. This conversion process is challenging, since some CAD shape representations are very different from those in GEANT. We will attempt to construct a robust conversion protocol, *i.e.*, one that generally succeeds but never fails while appearing to succeed. Once all shapes in the CAD file have been identified and converted to GEANT shapes, control will transfer to the GEANT input generator.

Abstract data type specifications have been developed for manipulation of CAD entities, entity lists, and entity components. Implementation of these abstract data types is now underway.

GEANT Input Generator

The GEANT input generator will check to see if the CAD attributes specify the material types; if not, it will request that the user specify the materials to be assigned to the various zones (a zone may contain one or more volumes). The user will mark the zone to which he wishes to assign a material, then either select a material from a menu of standard materials or enter a detailed composition description by specifying elements and weight fractions. In the latter case, the user may specify that the new composition be added to the code's library, so that it will henceforth be selectable from the menu of standard materials.

The geometry representation in GEANT utilizes a geometrical tree structure that links all volumes in the structure. An entity is characterized by a name, a shape identifier that defines a "mother volume" (which will be the root of the subtree associated with this entity), a local reference system, physical properties, and a list of entities that are directly accessible from the root. The list of accessible entities can be defined by specifying a position in the mother volume of already defined entities, a translation and rotation matrix, and a different number for each occurrence of the same entity. Alternatively, these entities can be defined by dividing, partially or totally, the mother volume along any of the axes to generate new entities whose dimensions are computed according to the number of divisions and the step size for the slices.

During the processing of information, the input generator will check for inconsistencies and errors such as volume overlap between an entity and its mother volume or missing volume properties or dimensions. The user will be alerted to these errors and will be able to interactively provide input to correct them. The final processed geometry will be displayed to the user for verification before the GEANT input file is created.

Verification

The developed CAD-to-GEANT interface will be tested by creating various notional or hypothetical detector designs in AutoCAD and interactively using the interface to create files suitable for input to GEANT. It will also be demonstrated that the code is able to add material compositions, input by the user, to its internal library upon command. The designs will range in complexity from simple to relatively detailed, so that the following series of tests can be conducted:

- An automatic, one-to-one transformation from CAD geometric primitives to GEANT shapes.

- A conversion that requires the interface to recognize an incomplete CAD specification, prompt the user for the required information, and complete the conversion after receiving the necessary data from the user.

- A conversion that requires a CAD shape to be adjusted, automatically, for representation as a GEANT shape.

- Demonstration that the interface output files from the first three tests can be read by GEANT.

Summary

We are developing a robust CAD-to-GEANT interface, which will reduce the time necessary to generate GEANT input files and will help minimize wasteful double entry of detector geometry descriptions. When completed, the interface will work in both directions, allowing the user to read CAD or GEANT data sets. The interface should provide a major assist in reducing the time required to refine detector descriptions, perform detector design calculations, and change the granularity of simulations.

In its fianl form, we hope to provide a well maintained software package that will operate on various systems and will interface with the major CAD packages being used in high energy physics detector simulation. We intend to build into the code special features, such as checks against engineering and physics rules of good practice, cost-estimation options, and data consistency checks.

References

[1] R. Brun, et al., GEANT3 Users' Guide, CERN DD/EE/84-1 (May 1986).

INTEGRATED 3D MODELING, ENGINEERING ANALYSIS AND PHYSICS SIMULATION

K. Killian
Martin Marietta Astronautics
Denver, Colorado 80201
For The EMPACT Collaboration

Abstract

The status of the EMPACT integrated modeling, analysis and physics simulation tool kit is presented. Capabilities of the modeler, and the associated analysis codes are discussed, as is the flow and control of data among the elements. Particular emphasis is given to the methods developed to perform fast, accurate simulations based on parametrized longitudinal and transverse showers.

Introduction

Integrated computer aided engineering and design (CAE/CAD) is having a significant impact on the way design, modeling and analysis is performed, from system concept exploration and definition through final design and integration. Our experience with integrated CAE/CAD in high technology projects has helped us to construct an integrated computer-based design, modeling and analysis resource aimed specifically at SSC detector system development [1]. This resource emphasizes efficient design through the availability of relevant engineering data, minimizes mechanical contact with data, as well as modeling and analysis of components, sub-systems or systems with fidelity appropriate to the task.

We begin with the requirements for design, modeling and analysis during SSC detector development, and describe the resource architecture. We follow this with a description of the simulation analysis tool that we ar integraing into this environment. We present some results that are indicative of its performance and close with the status of the tool development.

Design, and Analysis Requirements

The development of a complex device or system proceeds through a number of distinct, well-defined phases. A constant feature of each project phase is the design, modeling and analysis cycle. This cycle is repeated continuously as the project proceeds, albeit with requirements and goals that are functions of the project phase. Concept exploration, for example, requires moderate fidelity analysis with a fast turnaround in order to explore a broad range of design possibilities. The final design, however, demands thorough, high fidelity modeling and detailed analysis of a single design on a longer time scale.

Project success depends strongly on the availability of the design, modeling and analysis tools appropriate to the task. It is important to stress the words available and appropriate. Just as simple modeling and analysis packages may be inadequate to an engineer performing a detailed part design, technically sophisticated analysis packages are of limited use in concept exploration. Simply put one needs the right tool for the job.

The Classical Approach: "Islands of Automation"

Individual design, modeling or analysis packages can be viewed as "Islands of Automation". Although, analyses is done by computer, data transfer, manipulation and translation between programs is often done by hand. This has several effects. The fidelity of modeling and analysis is limited by the time required for modeling, and manual setup, and not by the analysis or interpretation of the data. Different analyses performed are quite independent of each other, even using different revisions of the design. Since turnaround times between iterations are long compared to other project time scales, such as drawing preparation or management milestones, the analysis data generally lags the design decisions it was intended to support, and may even lag the purchase of hardware.

This is especially true for most forms of radiation analysis or detector simulation. Traditionally, these analysis have been performed by either greatly simplified calculations (on simplified models) or by many man-months of hand modelling the detector using a program like GEANT. The many man months of physics simulation added to the time needed to get released drawings results in a time lag of months to years between the development of a structurally sound design, and its evaluation for physics potential. By the time problems are found, it may be to late to change the design.

One situation where timely availability of simulation tools is crucial to the design process is in the interfacing of cryostats in a liquid argon calorimeter. This is illustrated by the DØ detector at Fermilab. The cryostat design is such that, for jets directed toward the region $0.8 < \eta < 1.4$, a large fraction of jet energy is deposited in the dead cryostat walls and stiffeners between the central and end calorimeters. The resulting degradation in the calorimeter resolution was discovered to late to effect the cryostat design. Fortunately, it was possible to solve the problem with additional detectors, because the alternatives would have been a costly redesign or a significant compromise in detector performance.

Integrated CAE/CAD

A design for the integrated CAE/CAD resource is shown in Figure 1. This resource provides designers and engineers with a linked design, modeling and analysis tools that can be tailored to the task at hand. The system is built around Structural Dynamics Research Corporation (SDRC) I-DEAS which provides a sophisticated 3D modeler and a flexible finite element interface. One provides interfaces that enhance the value of the data by emphasizing "value-added" contact with data. Analysts intervene in the flow of data between packages to contribute engineering or design judgment, not to participate in routine data transfer, translation or manipulation as in the classical approach.

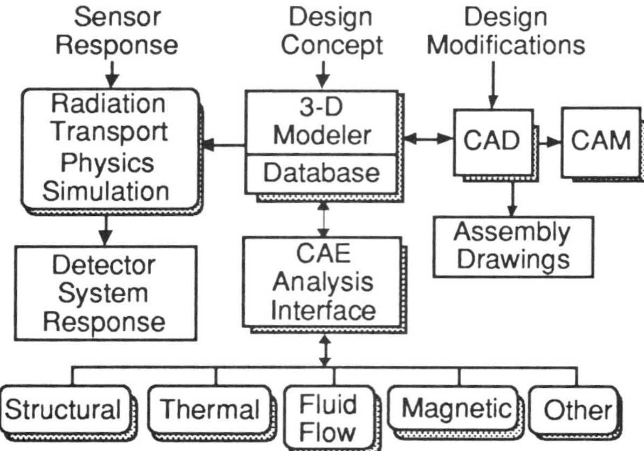

Figure 1. The architecture of the integrated CAE/CAD resource for detector system development.

The 3D interactive graphics are used extensively to input design, define modeling and analysis parameters and conditions and to evaluate results. The system is optimized to provide fast turnaround and to allow parallel analyses. The fidelity of modeling and analysis is determined by the task requirements and not by the tedium of setting up new grids or meshes, or by errors due to manual data manipulation are eliminated.

One of the more difficult interface tasks was to integrate physics simulation tools into the CAE/CAD architecture. This has been done by connecting the 3D modeler to a ray-tracing package which allows rays to be tracked through the detector from an arbitrary origin. Rays at arbitrary intervals in pseudorapidity and azimuth can be generated allowing detector structures to be probed in any detail desired. A post processor is used to evaluate this information according to several options. Calorimeters can be evaluated for their resolution using a one dimensional electron or pion shower approximation. The active depth of the calorimeter, the fraction of dead material, the entrance window thickness, and various other quantities can be evaluated in radiation lengths, areal density and other formats. Charged particle energy loss can be approximated through different structures. The post processor output has been used by other programs to prepare detailed plots of multiple scattering, or magnetic bending.

Despite the apparent simplicity of the data available, the advantage of the system lies in the fact that it provides information on the existing engineering models using a consistent set of assumptions. Since, the modeling overhead needed for raytracing on the 3D model is a matter of a few hours, simulation data is available to support design decisions at the trade study level. This ability has already proved useful in the development of a hermetic liquid argon calorimeter.

Calorimeter Design

We have used the tool to help us design cryostats for a hermetic liquid argon calorimeter [2]. In practice, the design cycle consisted of developing a shell model for structural analysis. Once the structural optimization was complete a 3D model was created for further thermal, structural analysis and for ray tracing. The ray tracing helped sort and eliminate a number of conceptually reasonable endcap cryostat designs on the basis of of their impact on the physics.

These design studies formed the basis of the LAC design presented by EMPACT in the EOI. Figure 2 is a CAD picture of the sampling rays overlaid on a hidden line cut through the calorimeter. This design met both the structural, and assembly requirements and presented a minimum impact on the physics measurements.

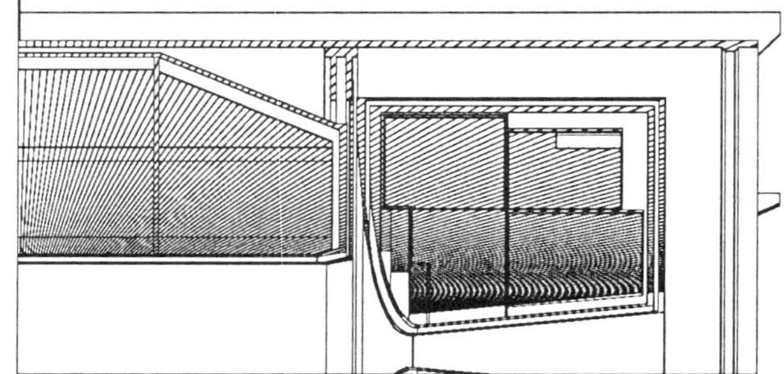

Figure 2. A Profile of the LAC With Rays Traced Radially, Starting From the Collision Point, Over the Interval $0 < \eta < 3.25$, at Intervals of $\Delta\eta = 0.02$

A preliminary indication of hermeticity given by the tool is shown in Figure 3. This is a plot of the depth of the dead material in front of the last active calorimeter layer (i.e. vessels, supports, etc.) expressed as a fraction of the active depth material. A comparison of Figures 2 and 3 shows the expected correlation between the physical geometry and the calorimeter design.

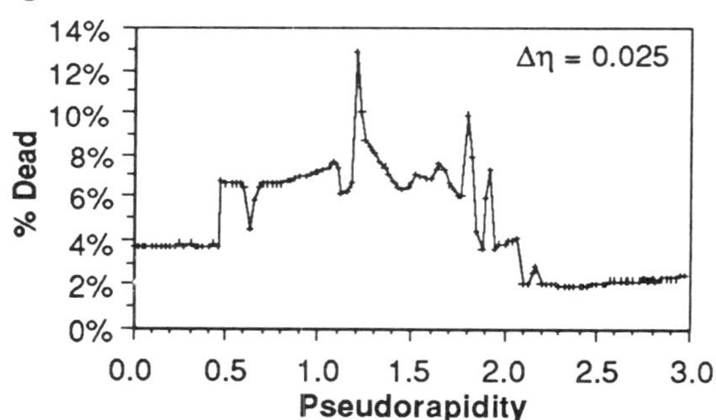

Figure 3. Percentage Dead Material, i.e. $\lambda_{dead}/(\lambda_{live} + \lambda_{dead})$, Between the Collision Point and the Last Active Calorimeter Layer

Calorimeter Resolution

For our design efforts we used shower resolution as a more direct indication of the calorimeter performance since it is a limiting factor in the reconstruction of the underlying physics event. The shower resolution approximation is calculated using longitudinal energy deposition approximations to estimate the energy deposited in each layer. The base resolution of the calorimeter modules are input to the code, then the lost energy is factored into the resolution according to models proposed by Womersley, and Furono [2]. Assuming a base resolution similar to the DØ values, we get electromagnetic and hadronic energy resolutions shown in Figure 4.

Muon Calculations

Another area where the data in the engineering models has proven useful to the physics analysis is in the prediction of the

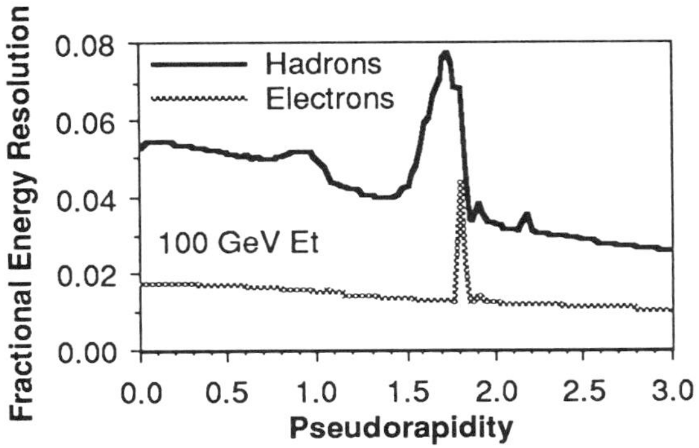

Figure 4. EMPACT LAC EM and Hadronic Resolution

muon system performance. The EMPACTbasline detector has air core toroids. These toroids are supported by aluminum webs at 5° increments. To determine whether multiple scattering in the web would significantly degrade the achievable resolution, we calculated ratio of the RMS multiple scattering for a muon passing through the web to the magnetic bend angle. Figure 5 shows that the multiple scattering is at most 10% of the magnetic bend. Figure 6 shows the interesting result that multiple scattering in the web is only four times larger than through the toroid coils. It is important to restate that these calculations include details like the conductor thickness, overlap of barrel and endcap toroids, and even the pumping holes in the webs.

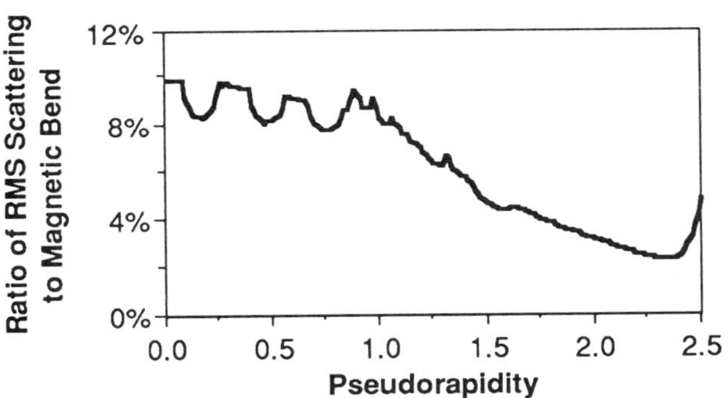

Figure 5. Ratio of RMS Multiple Scattering Angle in the Air Core Toroid to the Magnetic Bend Angle

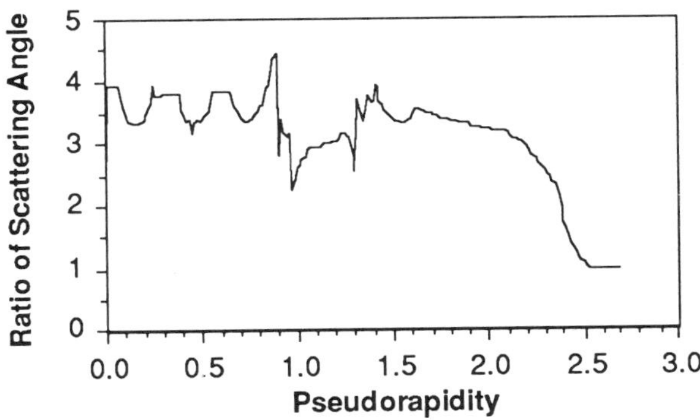

Figure 6. Ratio of RMS Multiple Scattering Angle in the Air Core Toroid to the Scattering in the Support Web

Future Developments

We are working on a stepwise development of this tool to expand its range of applicability. These improvements are driven by the need to integrate models for more detector elements into the code and to refine the existing models to keep the level of detail current with the design decisions being made. As an example, we are upgrading the post processors ability to do muon system calculations.

There is a major effort to develop a new subroutine level interface to the geometry data to allow Monte Carlo calculations to be made with the CAE/CAD geometry. We are also made an effort to address the speed issues involved with the geometry tracking through several techniques. We are taking advantage of the substructure in the 3D model, and by carefully working the data base access and structure issues. Lastly, we have implemented and tested the geometry routines on a parallel processor computer to ease future expansion to the proposed SSC computer "ranch".

This new geometry interface will allow us to use other shower algorithms to investigate more complex 3D designs and more direct estimates of the effects on resolution of energy loss in dead material than the present models. The new geometry routines should give us a powerful technique for specifying the tower segmentation relatively independent of the physical geometry. This segmentation ability, will allow us to look at digitizations and particle isolation studies to be made with various segmentation schemes.

Finally, we are looking at developing an interface with event generators, such as ISAJET, to allow simulation of missing E_t resolution. This will ultimately lead to the inclusion of detailed physics calculations into the CAE/CAD system and thereby completely integrating physics simulation into the design process.

References

[1] An Integrated 3D Design, Modeling and Analysis Resource for SSC Detector Systems, IISSC (1989).

[2] T.S.Adams et al., "An Engineering Design Study of a Hermetic Liquid Argon Calorimeter for the SSC", Submitted to NIMS, (1990).

PROGRESSIVE TRACK FINDING COUPLED WITH TRACK FITTING BY THE KALMAN FILTERING METHOD

Pierre BILLOIR
LPNHE, Universités Paris VI et VII, Paris, France

Sijin QIAN
CERN, Geneva, Switzerland/ INFN, Frascati, Italy

Abstract

A progressive track finding algorithm based on the Kalman filtering method has been proposed and tested for the track reconstruction in high energy physics experiment. The algorithm starts from small track segments with fitted track parameters and weight matrices or from fitted tracks of a neighboring detector, then extends the candidate tracks by adding measured points one by one. The track parameters and weight matrices of the candidate tracks are updated at the same time of each point addition. Therefore at the end of the track finding, the track fitting is accomplished simultaneously. The main procedures and its advantages of the algorithm are presented. Its performance is briefed.

1. Introduction

The charged track reconstruction in SSC experiment will face the challenge of the large track multiplicity. For the problems like how to reduce the combinatorial explosion during the pattern recognition, how to treat multiple scattering properly during the track fitting, how to avoid large matrix inversion and to speed up the reconstruction etc, the Kalman filtering (KF) method may provide an efficient way to cope with them all together.

The KF method applied in high energy physics experiment was originally proposed for track and vertex fitting [1]. However, since the formalism allows a progressive track following procedure, this feature has been used successfully to merge the track finding and fitting into a unique algorithm [2,3]. Unlike the traditional pattern recognition, this algorithm always carries and updates the information (i.e. track parameters P and weight matrix W which is the inverse of the error matrix) of candidate tracks along the track finding process, thus when finishing the procedure of the pattern recognition, the fitted track parameters and the associated weight matrix are obtained without further computation.

In this short paper, section 2 lists main procedures and advantages of the KF algorithm. More detailed mathematical formalism and the implementation for real detectors can be found in [2,3]. Its performance is briefed in section 3. See [2,3,4] for more comprehensive report.

2. Main procedures and major features of the KF algorithm

Figs. 1 and 2 depict the principle of the algorithm which includes the following main steps.

(1) Starting candidate tracks. There are several options to do it, e.g. by geometrically fitting first 3 points or a small set of points; or by following the fitted tracks of a neighboring detector, etc.

(2) Extrapolating the candidate tracks. This step provides the prediction point on the next measurement layer. If on the nth layer the fitted track parameter P_n^F and the associated weight matrix W_n are already obtained, this step extrapolates them to the $(n+1)^{th}$ layer and denotes the prediction as P_n^{Fe} and W_n^e.

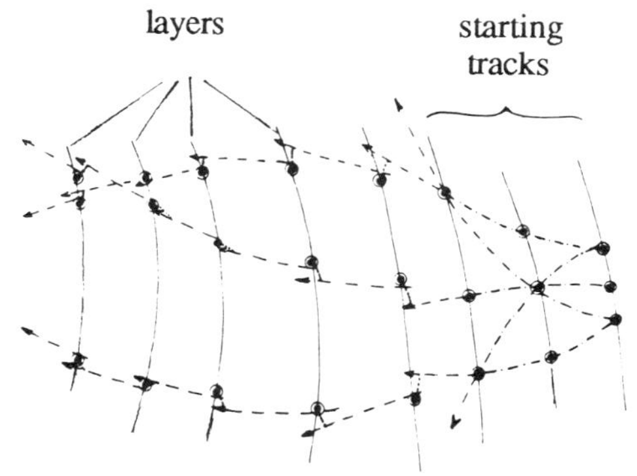

Fig. 1. Principle of the progressive track finding

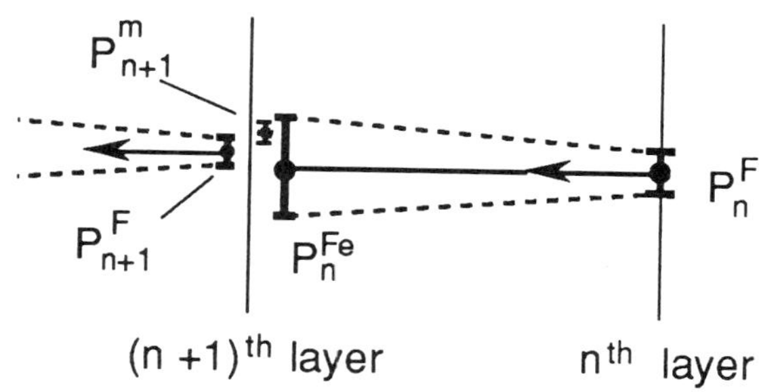

Fig. 2. Track extrapolation and updating

(3) Selecting measured points and updating the candidate tracks. Within a window around the prediction P_n^{Fe} on the (n+1)th layer, the closest measured point P_{n+1}^m (with the measured error matrix U) is selected, the updated (i.e. fitted) parameters P_{n+1}^F is obtained by solving a linear system of equation

$$(W_n^e + U)\,(P_{n+1}^F - P_n^{Fe}) = U(P_{n+1}^m - P_n^{Fe}) \ ;$$

the updated weight matrix $W_{n+1} = W_n^e + U$;

the increase of χ^2 is

$$(P_{n+1}^F - P_n^{Fe})^T W_n^e (P_{n+1}^F - P_n^{Fe}) + (P_{n+1}^F - P_{n+1}^m)^T U (P_{n+1}^F - P_{n+1}^m).$$

The steps 2 and 3 are repeated until the last layer is reached.

Some major advantages of the algorithm can be seen as follows.

(a) During track extension, the track parameters and weight matrix are updated when adding a new point on the track, this updated information gives an increasing precision on the prediction to the next measurement layer. This precision reduces ambiguity on the selection of the next point, therefore the combinatorial explosion at high track multiplicity can be minimized.

(b) The algorithm ensures the track parameters do not vary largely when adding a point, so that the coordinates of the measured point can be considered as linear functions of the track parameters in the neighborhood of the fitted values. Thus the minimization of χ^2 leads to solve a system of few linear equations, i.e. neither iteration nor large matrix inversion is needed. Therefore it is relatively fast comparing with other track fitting (especially the "global") method.

(c) The multiple scattering (m/s) is taken into account by updating the covariance matrix (i.e. the inversion of the weight matrix) wherever the m/s occurs during the track extension process. Thus the estimated track parameters follow the real path of the particle closely.

(d) The algorithm fulfils two tasks (i.e. pattern recognition and track fitting) simultaneously, and it is quite flexible to play roles in various situation [3] without introducing new tools.

The items (a), (b) and (d) are related with the speed of, whereas the (c) is related with the quality of, therefore hopefully the Kalman filtering method can provide an efficient way for, the track reconstruction.

3. Performance

The algorithm has been, so far, implemented in two experiments: DELPHI at LEP and ZEUS at HERA. The detector characteristic and brief results in Monte Carlo tests are listed in Table 1. More detailed report are in [2,3,4]. The Figs. 3 and 4 are graphic displays of two reconstructed events.

Table 1. KF algorithm applied in two detectors

	DELPHI TPC	ZEUS VXD
Total number of layers	16	12
Sensitive range in r	36.5 ~ 106.225 cm	10.65 ~ 13.95 cm
σ in r-ϕ	≈ 200 μm	35 ~ 55μm
σ in z	≈ 700 μm	no
Left-Right ambiguity	no	yes
Detector inefficiency	≈ 10%	4 ~ 6%
Background	≤ 50%	>30%
Track finding efficiency.	~ 98% (primary track) * ~ 80% (secondary track) *	> 95.7% (over all)
Speed	10 ~ 15 ms/event on VAX 6210	~100 ms/event (not optimize yet) on VAX 8800

*) The results are for early MC test, the lost secondaries were mainly low energy electrons. The efficiency on real events recorded in the past year was not yet evaluated independently of the hardware efficiency; it is about 95% or more.

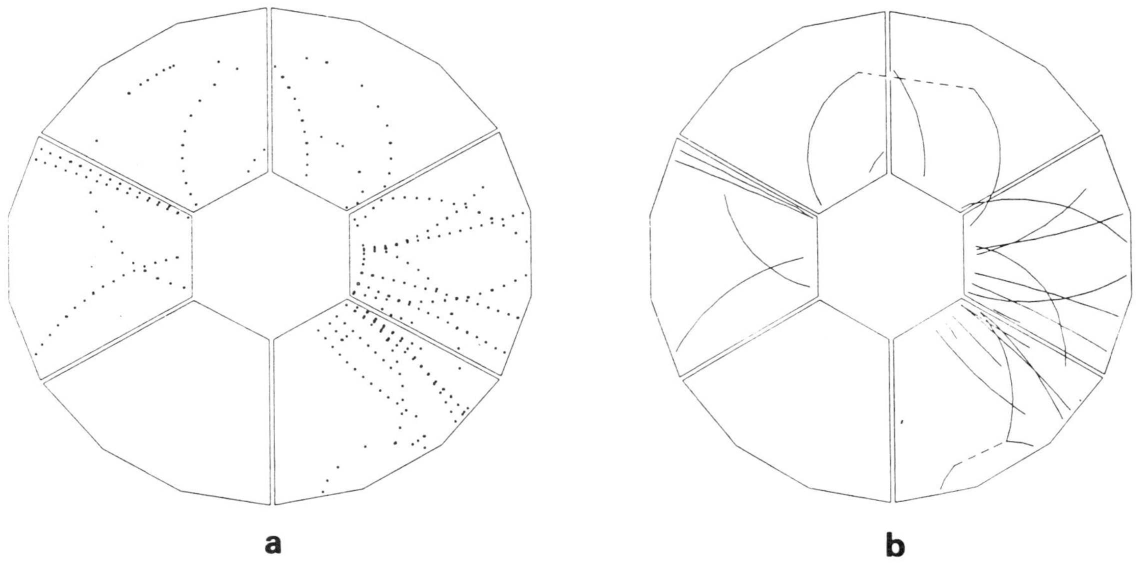

Fig.3. Reconstruction of tracks for a simulated event in DELPHI TPC: (a) space points; (b) tracks found.

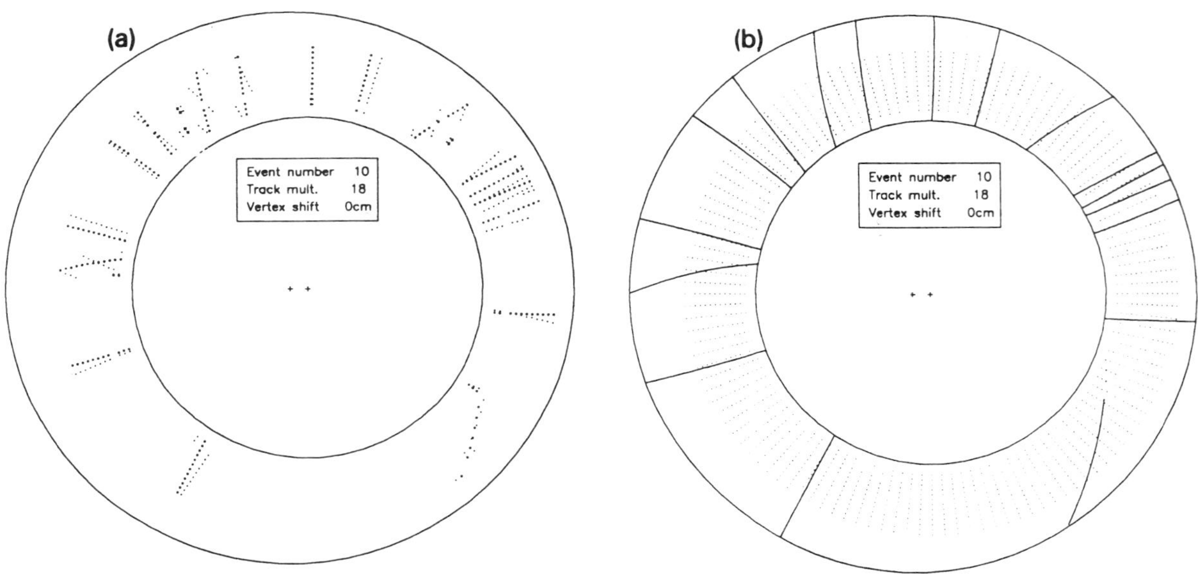

Fig. 4. Graphic display for a reconstructed event in ZEUS VXD. (a) input points: dark dots = real points; light dots = image points; crosses = backgrounds; plusses = beam-pipe centre and beam positions. (b) reconstructed event: solid lines = reconstructed tracks; dotted lines = cell walls.

References

[1] P. Billoir, Nucl. Inst. and Meth. 225 (1984) 352;
P. Billoir, R. Frühwirth and M. Regler, Nucl. Inst. and Meth. A241 (1985) 115;
R. Frühwirth, Nucl. Inst. and Meth. A262 (1987) 444.

[2] P. Billoir, Comput. Phys. Commun. 57 (1989) 390.

[3] P. Billoir and S. Qian, Nucl. Inst. and Meth. A294 (1990) 219.

[4] P. Billoir and S. Qian, Nucl. Inst. and Meth. A295 (1990) 492.

THE USE OF MASSIVELY PARALLEL COMPUTERS FOR THE DESIGN AND ANALYSIS OF SSC DETECTORS

R. Brower, D. Deutsch, J. Ross, E. Smith, R.J. Wilson

Departments of Physics, Computer, System and Electrical
Engineering and Computer Science
and
Information Technology
Boston University,
590 Commonwealth Avenue,
Boston, MA 02215

Abstract

We report on work-in-progress of an investigation into the use of massively parallel computers, of the Single Instruction Multiple Data (SIMD) kind, for the design, analysis and triggering of Superconducting Super Collider detectors. We have studied two dissimilar computer intensive tasks commonly encountered by experimental high energy physicists.

Introduction

We have examined the applicability of massively parallel SIMD (Single Instruction stream Multiple Data stream) processing to two problems representative of important analysis and design tools for SSC experiments. We have used a commercially available 32,000 processor "Connection Machine" (model CM-2)[1] for our studies. We chose one problem whose basic algorithm appeared to be ill-suited for this computer architecture (shower simulation) and another which appears more naturally parallel (pattern recognition). In this way we hope to develop an understanding of the wider utility of this powerful new approach to computation.

Electromagnetic Shower Simulation

The EGS4 (Electron Gamma Shower, version 4)[2] code system is a general purpose software package for the Monte Carlo simulation of the transport and interaction of electrons, photons, and positrons in specified materials for energies from a few keV up to several TeV. It is widely used by high-energy experimentalists as a tool in the design of calorimeters and other devices.

When a photon with energy much greater than the mass of an electron enters a non-vacuous region it can produce an electron-positron pair. These charged particles may then interact with the medium by radiating photons. These photons can, in turn, undergo further pair-production. This process continues with the particle multiplicity growing, whilst at the same time the mean energy per particle decreases. As the shower develops, however, the charged particle component loses energy to the medium through ionization of the atoms in the medium. Eventually all of the incident particle's energy is lost to the medium and the shower dissipates. The simulation of this process can be very time consuming, a rule of thumb is that it takes approximately 1 cpu second per GeV of incident energy using the EGS4 code on an IBM 3081.

An essential feature, as far as the coding algorithm is concerned, is that as each particle is generated it is pushed onto a stack which grows continuously until a particular branch reaches a user defined cut-off energy. The number of particles being transported can be in the thousands or even tens of thousands. If we could operate on them in parallel, rather than in series, we could speed up the simulation by very large factors.

The software environment for writing programs for the CM resides entirely in the front-end host computer, in our case a VAX 6320. In effect the CM is "just" a very powerful attached floating point processor array (up to 2092 units) with its own large memory (up to .5 Gigabytes). To integrate parallelized EGS routines into the existing EGS4 user environment, we have used the CM Fortran which is a version of the proposed Fortran 8x standard.

To map the serial EGS4 code onto the Connection Machine, we have built a simple interface which treats the CM as vector machine of fixed length equal to the number of (virtual*) processors. All communication and scheduling is handled "invisibly" by the interface which consists of less than 800 lines of code and comments. The key observation on the serial EGS code, which makes this simple interface possible, is the independence of a large number (stochastically determined) of processes (which correspond to subroutine calls in the Fortran context).

---

* The CM may be configured to behave as though it consists of many more "virtual" processors than there are physical ones. This is transparent to the programmer and is limited only by the memory available.

We exploit maximal parallelism by simulating concurrently a large number of injected parent particles. After each subroutine call the ensemble of produced particles are sorted and stored in vectors according to which subroutine should be applied next to each child. By this delayed application of the subroutine and a simple monitoring of the stored vectors each subroutine is called only when a vector of length equal to the number of virtual processors has been accumulated. In a steady state mode, new parent particles are injected frequently enough so that at least one full vector is available for at least one subroutine call.

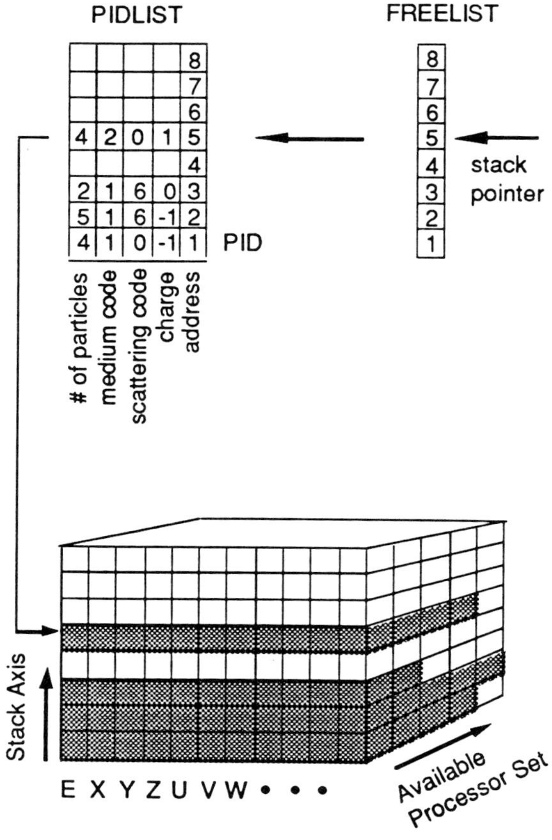

Figure 1. The new parallel memory scheme (filled stripes are shaded).

The new parallel scheme we have implemented for memory allocation/deallocation is represented graphically in Figure 1. Each particle is assigned an identification array, PID, which is an entry into a 2-d CM Fortran array called the PIDLIST. The elements of a PID are its address, charge, scattering code, medium code, and the number of current particles on the stack with this scattering code. The address of the PID is the stack index which not only points to the PID location on the PIDLIST, but also points to the location of a particle "stripe" --- the parallel analogue to an element on the stack. A stripe is the parallel version of the particle parameter set (E, X, Y, Z, U, V, W, . . .). Each variable is now a 1-d CM Fortran array distributed across the entire processor set of the Connection Machine, each processor now representing a particle. We have written routines to allocate and deallocate stripes corresponding to "push" and "pop" of a serial stack. The address of the next available stripe in memory is found on a new serial stack called FREELIST. This stack contains only the addresses of locations in parallel memory and not the particle parameter set as in the serial case.We have completed 50% of the translation of the cross-section subroutines into parallel form. The remaining routines are "wrapped" in parallel code so that they appear to be executing in parallel but the operation actually entails serial calls on the front-end computer. After each conversion we require that the output from the parallel code be identical to that from a standard serial simulation.

We have not yet performed any detailed timing studies with this code. However an earlier prototype code, which contained only the basic shower algorithm and look-up tables, gave a factor of 57 speed-up for a 32K CM over purely serial running on a SUN4.[3] No optimization was performed on this code and it is expected the full utilization of the CM would increase this ratio by a factor of 5 to 10.

Pattern Recognition

To study the application of the CM to a pattern recognition problem we have developed a parallel implementation of the Hough Transform (HT) inspired by optical image processing techniques. Such techniques may have applicability for track finding in both online (trigger/DAQ) and offline processing of SSC experiments.

The Hough Transform method is a global approach to edge finding in an image. The transform involves a many-to-one mapping of image points into a "feature" space such that image points which lie on a common curve all map to the same region of the feature space. A tally is kept of the number of image points that map to each point in feature space. We store the results of this tally in an "accumulator array". The procedure can be visualized by imagining an observer at some origin (e.g. the interaction point at the SSC), looking out in a particular direction. Straight lines oriented directly away from the observer would appear as a point (with an intensity proportional to the number of points on the line) whereas lines with a different orientation or origin, or random hits, would appear "smeared" out and be of lower intensity. To find tracks at different angles we apply sequential transforms, the number of such applications depends upon the resolution required. An efficient procedure for images with sparse data (as is the case here), is to use a coarse resolution initially (say, 10° sectors) followed by finer scans in those regions found to have tracks by the coarse scan.

We have implemented a Generalized Hough Transform (GHT) which is optimized for straight line segments (SLS-GHT) but which can used to find any arbitrary curve by means of a template table. For the GHT we transform points in the image onto overlapping *curves* in the feature space (as opposed to *points* in the normal HT case). This approach allows for an easy extension to curved track finding. To ensure that tests of our codes are realistic, we have used data generated by simulations of the Transition Radiation Detector (TRD) particle identification and tracking device being developed for the EMPACT collaboration (see the paper by J.S. Whitaker in these proceedings).

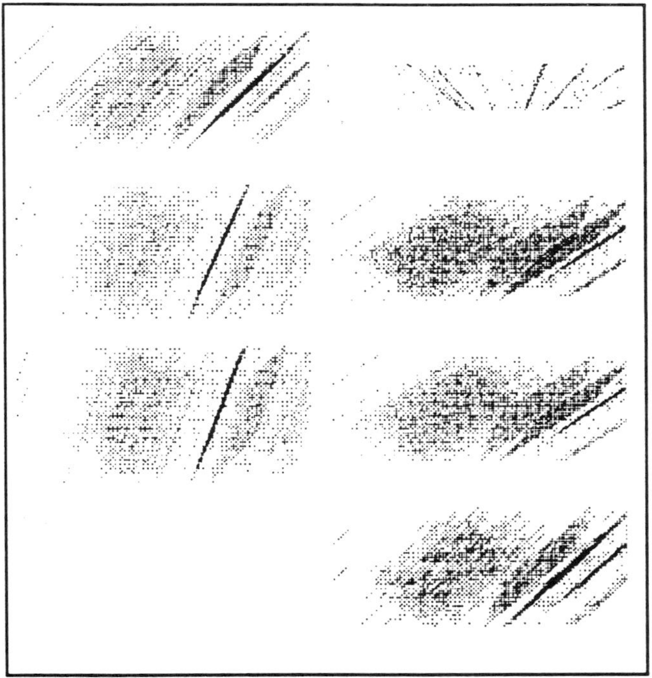

Figure 3. Original image (top right) and successive, transforms for identifying different track orientations (before threshold application).

In figure 3. we see an example of the accumulator array produced by application of the transform for six different track orientations, for one octant. Only those elements of the array above some threshold (determined by the amount of background) will subsequently be subject to an inverse transform to produce a list of image points associated with each line. In this example, of 43 generated tracks in all octants, SLS-GHT detected 39, 29 of which involved at least 10 edge-points. Rejection of the spurious lines was controlled by selection of the accumulator threshold, peak spreading and peak sharpness parameters.

The execution time for the transform accumulation is logarithmic in the total number of pixels to be drawn (i.e. sum of line segment lengths), independent of how many pixels are on each line and how many lines are drawn. As currently implemented, SLS-GHT exhibits time complexity linear in angular resolution due to it reliance on two-dimensional transform accumulators. With more memory, the transform can be made logarithmic in all parameters. Other functions contribute only I/O overhead, or constant factors on the logarithmic transform and the angle-iteration. For the SLS-GHT on the CM we find that it takes 511μsec per image point. On a per-image basis (179 transforms per image), elapsed time to execute the transform ranges from 40 to 70 seconds (exclusive of disk I/O). The CM is active for roughly 50% of the elapsed time (which includes CM operations that are not strictly part of a Hough transform and therefore excluded from the per-point comparison above).

Although the edge-point density of the octant-images is only about 1/2% we achieve excellent processor utilization while computing pixel coordinates. As expected for this architecture, we find that the limiting feature is the inter-processor communication time. Time complexity of random communication on a hypercube is good, $O(\log_2 N)$ for an N node hypercube but communications time still dominates over the computation time. As currently implemented this pixel oriented method shows no significant advantage over more traditional algorithms.

We are currently pursuing an alternative approach, inspired by our work on the EGS4 code. Here we treat the set of image points (hits cells) as a long vector so that each processor contains only hit detector cell coordinates. This is in contrast to a one-to-one geometrical correspondence of processors to all of the cells in the detector, as in the previous pixel oriented approach. The transform on this vector is stored in an array laid out on the CM in such a way as to minimize the costly communications steps.

Further Details

A report providing further details on the topics discussed in this paper can be obtained from the authors upon request.

References

[1] *Connection Machine Model CM-2 Technical Summary*, Thinking Machines Corporation Technical Report HA87-4

[2] W.R. Nelson, H. Hirayama and D.W.O. Rogers, *The EGS4 Code System*, Stanford Linear Accelerator Center Report SLAC-265, 1985

[3] R.J. Wilson *et al* "Massively Parallel Computing in High Energy Physics", Proposal to the SSC Laboratory, January 1990.

OPTICAL COMPUTER UTILIZATION AT THE SUPERCONDUCTING SUPER COLLIDER

Melinda B. Johnson, James K. Woosley, and Alphonsus J. Fennelly
Technology Applications Department, MS 50
Teledyne Brown Engineering, P.O. Box 070007, Huntsville AL 35807-7007

ABSTRACT

Optical computer systems offer the possibility of extremely high-speed, high efficiency processing for the SSC. The state of the art in optical computer system is described, with emphasis on the problems of timing, digitization, data readout, and storage. Particular emphasis is placed on the potential of utilizing detector optical signal readouts as a real-time trigger in a signal-rich environment (two to ten events per 16ns bunch crossing). A comparison of projected optical computer technology growth during the next decade and the capabilities required of SSC detectors and off-line processors is performed.

Introduction

One of the most stringent requirements for the Superconducting Super Collider (SSC) operation is high-speed data processing. At the initial operating luminosity of 10^{33}, approximately one interaction per bunch crossing (16ns) is expected. At the full operational luminosity of 10^{34}, the mean interaction rate is more than one interaction per 2 ns. The information provided by the detector, per particle, must be sufficient for pattern recognition and event reconstruction; the information content of the raw data is approximately 40 bits - the bandwidth potentially much larger than the the information content. Even with this minimum the detector is at least 2.5 Ghz - and the raw data output potentially much larger. The use of a layered trigger to reduce the output by four to five orders of magnitude simplifies postprocessing, but at the cost of requiring real-time processing and pattern recognition at large rates, with high accuracy. Post processing will still require a supercomputer with pattern recognition capability.

The use of the nascent technologies of optical computing [1] and optical signal processing may offer a technically feasible method of achieving the processing requirements for both preprocessing and post processing. The former case further offers the possibility of developing optical trigger hardware, analogous to the electronic triggers currently in use. There are potential advantages of an optical trigger in terms of processing speed, radiation hardness, and computational hardware.

The present paper addresses the basic technology of optical computation as it applies to high energy physics. The second section discusses the basic technology of optical processing - sources, connectors, switches, sensors, and integration. Current and projected capabilities for optical computers will then be given. The third section will describe the conceptual application of optical computing to on line trigger processing; technology limits will be identified.

Optical Processing Techniques

Current state of the art in optical computing has approximately 1 gigaflop speed in numerical processing and can have either parallel or neural architectures. Optical processing technology development is pursuing both analogue and digital techniques. This paper will focus on digital optical processing, used primarily for computational applications. The basic components of a digital optical system are sources (e.g. light emitting diodes, semiconducting diode lasers, and light from scintillating fibers), connectors (e.g. optical fibers and optical waveguides), switches (Fabry-Perot etalon devices, Mach Zenhder modulators and holograms), and sensors (e.g. photodiodes, photocathodes/ photomultipliers, and crystal detectors). Fabry-Perot etalon are dispersive, optically bistable, elements biased with a laser diode that convert few photon signals into high intensity signals. Critical performance parameters are listed in figure 1.

Several methods exist to create Boolean logic functions in optical systems. Electronically connected quantum well PIN diode modulators and detectors have been used to create optical logic gates that perform all of the Boolean functions, and use optical signals as a clock.[2] Of most interest to the Superconducting Supercollider is the availability of a parallel digital adder and subtractor for optical systems. Using a hybrid architecture (an electronic control unit and optical execution unit) a massively parallel adder and subtractor has been created that is capable of 10^9 arithmetic

Figure 1. Typical performance parameters of optical processor components

SOURCES: (GaAlAs/GaAs)	LD	LED
WAVELENGTH (μm)	.85	.85
PULSE DURATION (ns)	.002 - 1	.002 - 1
90% PULSE RISE TIME (ns)	10	10
PULSE POWER (mW)	200	200
PULSE ENERGY (fJ)	200	200
PULSE ENERGY (photons)	$8.6x10^{8}$	$8.6x10^{8}$
POWER EFFICIENCY	20 - 25%	1 - 2 %

CONNECTORS	
BANDWIDTH	0.3 μm
TRANSMISSION LOSSES	3-10 dB/km

SWITCHES	
THRESHOLD ENERGY	20 pJ
THRESHOLD ENERGY	$8.6x10^{7}$ PHOTONS
RECOVERY TIME	200 ps

SENSORS	
SENSITIVITY	5-20 pJ
RESPONSE TIME	10-100 ns

operations a second on 32 bit integer words. In the hybrid structure the clock pulse was supplied by the electronic components.[3]

Write Once Read Many (WORM) devices and magnetooptic read/write disk drives are commercially available for high density storage. These devices can display parallel access capability.[4] Although other types of optical memories, such as 3-D optical memory systems, are being explored, they are not expected to be available until after the year 2000. [5]

Optical computing offers many advantages for neural networks. An optical neural network could be capable of 3-D, high density interconnects, and rapid manipulation and addition of signals.[6]

Optical Computer Applications To The SSC

Optical computers evidently have several potential applications to SSC operations. With parallel architectures which already yield gigaflop processing performance in

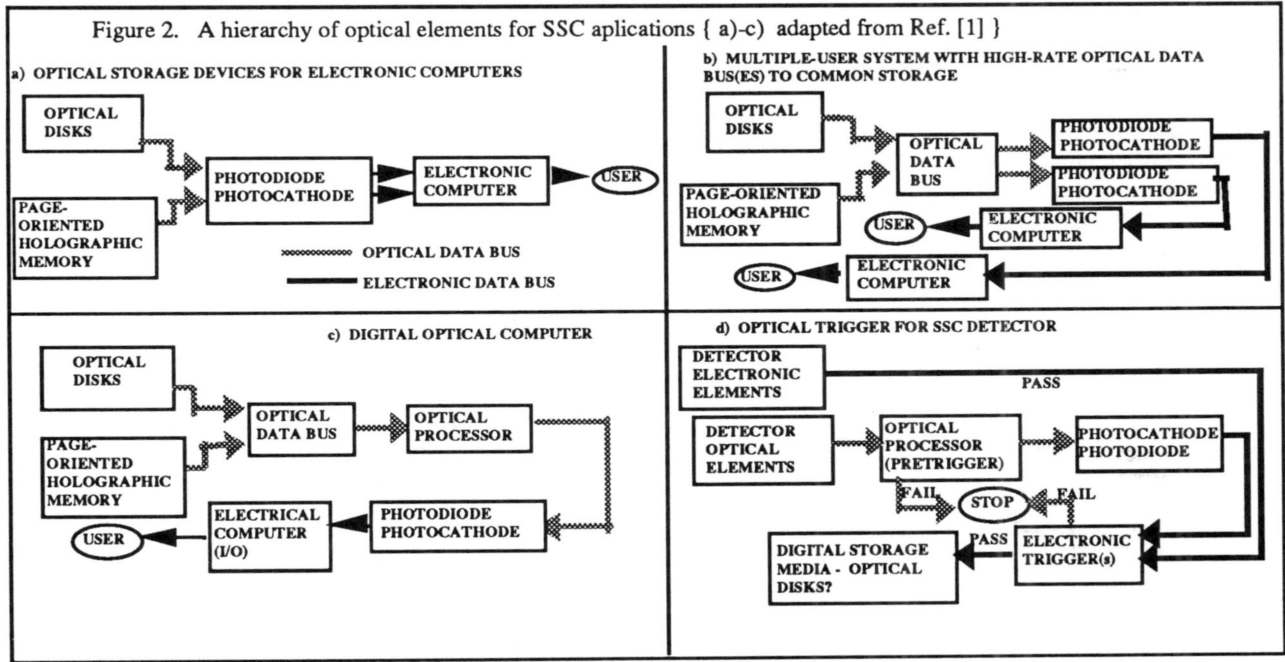

Figure 2. A hierarchy of optical elements for SSC aplications { a)-c) adapted from Ref. [1] }

addition and subtraction, and with the ease of developing optical neural network architectures, the utility of optical computers in postprocessing for pattern recognitions, and event reconstruction are evident. The potential application of on-going work involving neural network triggers [7] is obvious.

Another advantage of optical computer utilization comes from the potential radiation hardness of optical processor components to real-time computing - triggering. A typical first-line trigger uses calorimeter cell energy thresholds for.single-electron or single-jet triggering. Since most calorimeter approaches typically involve scintillation and total absorption of a quanta yields a high photon output, sufficient energy exists in the direct pulse to consider implementation of a neural optical trigger in real-time. The main limitation to such a neural trigger would be the pulse duration. The conceptual advantages in both speed and accuracy of such a pretrigger are obvious. One such device is described elsewhere in these proceedings [8]. As shown in figure 2, however, such an optical trigger would only be the first element in real time processing, due to the need for electronics in other elements of the detector.

Optically Bistable Signal Amplification and Neural (Retinal) Networking of Scintillating Fiber Bundles

Optical bistability can be exploited to significantly enhance the signal-to-noise ratio from particle interactions with scintillating fibers. A laser diode provides a bias intensity just below the etalon's bistable point by transmitting a steady bias down the scintillating fiber to the etalon, which is etched into the fiber. Only a slight increase of intensity in the scintillating fiber is needed to drastically amplify the transmitted intensity to the photodetector. Limiting all photodetectors to the same dynamic range will cause all fiber signals to start with the same intensity. Further signal processing can be done with the signals themselves by connecting a 3-dB coupler to the network fibers, which allows the signal amplitude received by a photodetector memory to indicate it's original position in the fiber bundle or fiber/photodetector array. Mamalian retina models [10] and neural network computers [11- 13] utilize a similar approach in their operations.

A fiber and complementary photodector network would create many advantages. For example higher-layer connectionscan be created which allow constant signal/ data comparison to set triggers, fire triggers and record events.

CONCLUSION

Optical processing is a promising technology that offers the SSC many benefits: real-time preprocessing, massively parallel processing, easy implementation of neural networks, and increased radiation hardness when compared to electronic components. We have illustrated this with two simple examples: (1) signal enhancement and conditioning and (2) direct array/network processing using the signals themselves in computations leading to higher layers of networks which accomplish detector decision making about setting/firing triggers, and preserve potentially unused data. These conceptually simple techniques are not very difficult to implement and will yield great dividends through reduction of data processing burdens as well as enhanced recognition of important events in SSC data.

References

[1] P. B. Berra, A. Ghafoor, M.Guizani, S. J. Marcinkowsi, and P.A. Mitkas, "Optics and Supercomputing," Proceedings of The IEEE, 1797-1815, (1989).

[2] A.L.Lentine, D.A.B. Miller, J.E. Henry, J.E> Cunningham, L.M. F. Chirovsky, and L.A. D'Asaro, "Optical Logic Using Electrically Connected Quantum Well PIN Diode Modulators and Detectors," Applied Optics, 2153-2163, (990).

[3] S.Fukushima, T. Kurokawa, and H. Suzuki, "Optical Implementation of Parallel Digital Adder and Subtractor," Applied Optics, 2099-2106, (1990)

[4] D.Psaltis, M. A. Neifeld, A. Yamamura, and S. Kobayashi, "Optical Memory Disks in Optical Information Processing," Applied Optics, 2038-2057, (1990).

[5] S. Hunter, F. Kiamilev, S. Esener, D. A. Parthenopoulos, and P.M. Rentzepis, "Potentials of Two-Photon Based 3-D Optical Memories for High Performance Computing," Applied Optics, 2058-2073, (1990).

[6] C.Wang, and B. K. Jenkins, "Subtracting Incoherent Optical Neuron Model: Analysis, Experiment, and Applications," Applied Optics, 2171-2186, (1990).

[7] C. Peterson, "Track Finding with Neural Networks," Nuclear Instruments and Methods in Physics Research , 537-545, (1989).

[8] M. Tourey. These proceedings

[9] H. M. Gibbs, Optical Bistability: Controlling Light with Light, New York, Academic Press, 1985.

[10] C. Mead, Analog VLSI and Neural Systems, Reading MA: Addison-Wesley, 1989, especially ch. 14 and 15.

[11] R. Hecht-Nielsen, Neurocomputing, Reading MA: Addison-Wesley, 1990.

[12] I. Aleksander, Neural Computing Architectures, Cambridge MA: MIT Press, 1989.

[13] T. Khoma, Foundations of Neural Networks, Reading MA: Addison-Wesley, 1990.

Particle Identification and Muon Systems

MUON SYSTEMS WITH A SOLENOIDAL DETECTOR

K. Heller
School of Physics and Astronomy
University of Minnesota
Minneapolis, Minnesota 55455

Abstract
The design considerations for a muon system for an SSC detector are outlined. R&D plans to develop a system meeting these criteria are given.

Purpose

The detection of muons will be an important part of any general purpose SSC detector for two very basic reasons: High transverse momentum (p_t) muons signal interesting physics appropriate to the SSC energy scale, and muons are possible to identify and measure in the high rate, high multiplicity environment of the SSC. The recent history of particle physics clearly shows the importance of such muons in discovering new particles. We have only to remember $\psi/J \to \mu\mu$, $\Upsilon \to \mu\mu$, $Z \to \mu\mu$, and $W \to \mu\nu$. This same history holds examples of detectors with inadequate muon resolution which missed discoveries and detectors with resolution much too good for the energy regions investigated. Unneeded resolution when scaled to the SSC could represent the waste of resources needed to bolster other systems. Clearly a careful design is needed to maximize discovery potential while minimizing cost and complexity of the muon system.

At the SSC, high p_t muons will signal the creation of new heavy particles which decay into at least one Z (i.e. $H \to ZZ$ or $\rho_T \to ZW$) or directly into leptons (i.e. $Z' \to \mu\mu$). The easily detected leptons are either electrons or muons so that 3/4 of all clean signature Higgs decays, $H \to ZZ$, $Z \to 2$ leptons, have muons in the final state. For discovering the Higgs, a single detector which has a good muon system as well as a good electron system is equivalent to building four detectors which have only one of these systems. These muons can be identified by requiring them to pass through material sufficient to shield the muon detector elements from the vast majority of the particles originating from an SSC collision. Hiding behind this material, the muon detectors will have the highest rate capability in the detector and, thus, the ability to explore the highest luminosity of the SSC. The system described briefly in this report is the product of the continuing work of the many people working together on the R&D for a muon system for a solenoidal detector.[1]

General Design

The muon system must provide information which will allow the electronic trigger to reduce an interaction rate of about 100 MHz to a data taking rate of a few Hz. Single muons with p_t above 50 GeV/c occur at a low enough rate to be collected but lower p_t muon signals must be combined with other information in the detector to form a final trigger.[2] Off-line, the detector must be able to measure the momenta of muons to a precision necessary to distinguish high mass states from the uninteresting low energy background. For all possible processes we have been able to imagine, a resolution of $\Delta p_t / p_t = 0.20\, p_t$ (in TeV/c) will be adequate.[3] This means that a muon must pass through approximately 14 interaction lengths of material for the trigger's particle identification while having its momentum measured with a minimum of multiple scattering for off-line resolution. A detector based on a solenoid, such as shown schematically in figure 1, accomplishes this in a natural way by using the tracking inside the solenoid to provide the high resolution momentum measurement and external muon detectors outside of the calorimetry to provide the other trigger information.

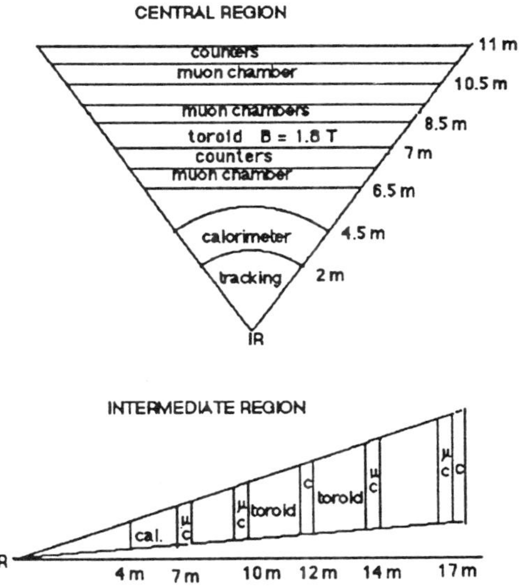

Figure 1. A general schematic showing the layout of a muon system surrounding a solenoidal detector.

A realistic design can be found in the SDC EOI.[3] An external muon system which makes a second measurement of the momentum of the muon will be able to avoid identifying a high momentum hadron in the inner tracker with an external low momentum muon from punchthrough. This same external muon momentum measurement can provide the low level trigger p_t tag.

The external muon detector separates into two systems which have, on the average, different backgrounds and occupancy. The central muon system ($|\eta| \leq 1.5$) covers the largest area, the barrel region, and has low occupancy while the intermediate muon system ($1.5 \leq |\eta|$

≤ 2.5) covers the ends of the detector and has a much higher occupancy and potential neutron background. For benchmark processes relevant at the SSC, the rapidity coverage ($|\eta| \leq 2.5$) appears adequate as illustrated by the examples in Table 1.

ACCEPTANCES FOR BENCHMARK PROCESSES

Process	Detector Region	Acceptance
$Z \to \mu\mu$	C - C	23%
	C - I	18%
	I - I	6%
	Total	47%
$Z'(200\ GeV) \to \mu\mu$	C - C	22%
	C - I	20%
	I - I	6%
	Total	48%
$Z'(500\ GeV) \to \mu\mu$	C - C	30%
	C - I	25%
	I - I	6%
	Total	61%
ZZ pairs $\to 2(\mu\mu)$	C - C, C - C	9%
	C - C, C - I	15%
	C - I, C - I	7%
	Total	31%
$H(400\ GeV) \to ZZ \to 2(\mu\mu)$	C - C, C - C	18%
	C -C, C - I	17%
	C- I, C- I	11%
	Total	46%
$H(800\ GeV) \to ZZ \to 2(\mu\mu)$	C - C, C - C	30%
	C - C, C - I	20%
	C- I, C- I	12%
	Total	62%

Table 1. Geometrical acceptance of central and intermediate detector regions for benchmark physics processes. At least 1/2 of the muons in each process required to have Pt > 20 GeV/c. C = $|\eta| < 1.5$, I = $1.5 < |\eta| < 2.5$.

Both subsystems must give a fast trigger by identifying muons and providing a p_t estimate. They also must give enough tracking information to identify the interesting muon track in the outer layers of the inner tracker to sharpen the muon momentum determination for inclusion in a higher level trigger. Off line the external muon system must measure the muon momentum well enough ($\Delta p_t / p_t = 0.25$) to reject punchthroughs and extend the charge determination capabilities of the inner tracker to higher momenta.

Complications

Because of its large size, the occupancy of the individual detectors which comprise the external muon system is relatively low. Conventional detectors such as scintillator or Cherenkov counters and drift modules will be effective. It is the details of the design of these elements which will determine the success of the muon system. The absorber, usually a calorimeter augmented with iron, used to identify muons gives rise to electromagnetic showers, both hard and soft, which accompany each muon and could confuse a trigger. Simulations show that about 30% of all muons emerging from the absorber material are accompanied by at least one charged particle.[4] Low energy photons which generate additional charged particles in the detector material are also present. Unfortunately, shower simulations are not necessarily reliable with respect to their low energy debris and must be checked by beam tests. The geometry and materials of the drift modules must be chosen to insure that these low energy particles do not defeat a simple trigger. In addition, a large number of low energy neutrons from hadron interactions in the detector material and beamline elements may make the use of scintillators impossible and limit the use of organic components of drift gas.[5] The material present in the external muon system also limits the momentum resolution of the trigger through multiple scattering. The overriding consideration in the muon system design is its large size and thus, its large cost. The cost per detector element must be kept low. Since the cost of the system is dominated by per channel electronics, low cost implies the longest possible wires and the widest possible cells. The difficulty of constructing such a system resides in understanding the technical details of the muon events, the detectors used, and the environment in which the detectors will operate.

The external muon system must be able to tag the interaction giving an interesting muon which implies a time resolution of better than the 16 nsec time between crossings. The first level trigger, muon identification and p_t tagging, must take no longer than about 1 μsec to be compatible with a reasonable electronic pipeline for the detector. The background environment of the external muon detectors is a combination of muons from hadron decay in the internal tracker and calorimeter, cosmic ray muons, muon shower products, and low energy neutrons. The total rate of muons from beam-beam interactions as a function transverse momentum in the rapidity region $|\eta| \leq 3$, calculated using both existing data and simulations, is shown in figure 2 as a function of transverse momentum after 14 interaction lengths.[2]

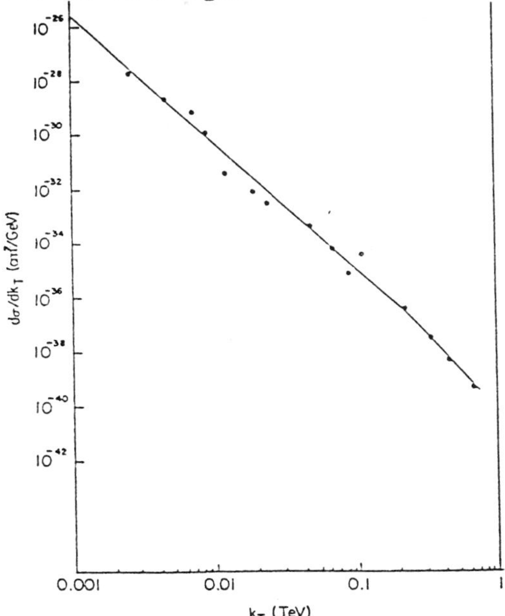

Figure 2. Muon rates from the interaction as a function of transverse momentum after 14 interaction lengths of material in the rapidity region $|\eta| \leq 3$.

After passing through 14 interaction lengths of material, hadronic punchthrough is no longer important at p_t's of interest, although they dominate at low p_t. In fact, for p_t > 20 GeV/c the number of muons is dominated by prompt decays of heavy quarks as shown in figure 3.[2]

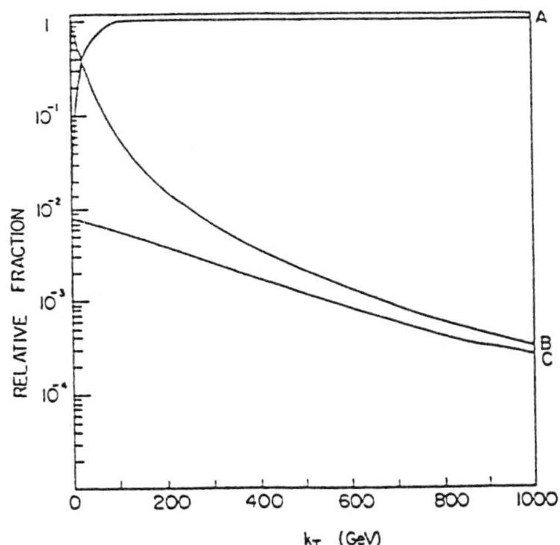

Figure 3. The relative fraction of particles exiting a 14 λ absorber with $|\eta| < 3.0$ as a function of k_T (p_t) from (A) prompt decays; (B) π/K decays; (C) hadron punchthrough.

Trigger Design

As shown in figure 1, a possible external muon system consists of 2 banks of scintillation counters and 4 sets of vector drift modules sandwiched around a 1.5 m thick iron toroid (1.8 T). The toroid bends muons toward the axis of the solenoid in a direction orthogonal to the deflection caused by the solenoid magnetic field. The purpose of the scintillator is to provide the fast trigger for muon identification, a beam interaction tag and an indication that the muon came from the interaction region with a minimum p_t of between 5 and 10 GeV/c. An example of a segmented scintillator hodoscope is shown in figure 4.

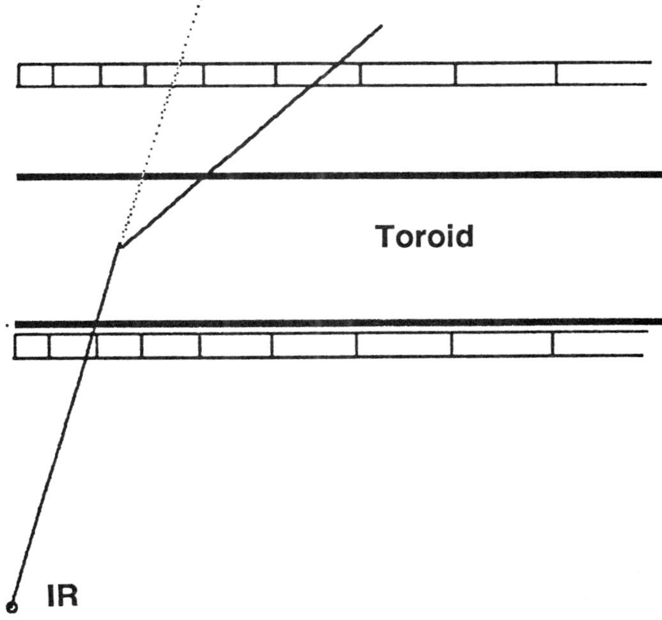

Figure 4. Fast trigger using a scintillator hodoscope. Not to scale.

An electronic matrix accepts those events for which a two counter coincidence points to the interaction region. A reasonable design,[6] using a counter segmentation no finer than 10 cm. along the solenoid axis each with a length of 4 m, gives a fast trigger resolution shown in figure 5.

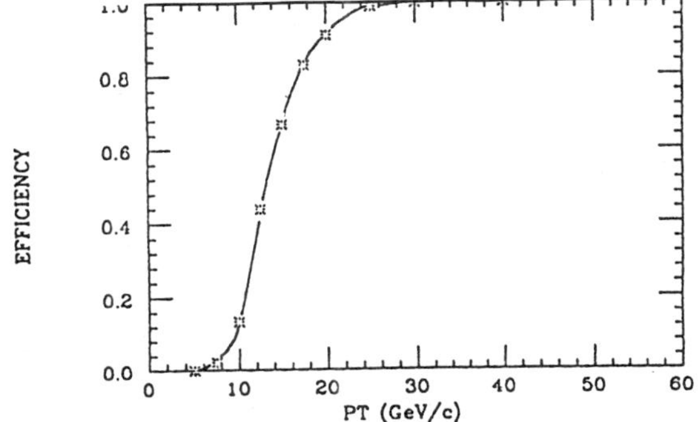

Figure 5. Trigger efficiency for the central region for a fast muon trigger using the muon scintillation counters.

For a luminosity of $10^{34}/(cm^2 s)$ the rate of muons at the outer bank of counters in the central region is about 1 MHz from interactions and 0.1 MHz from cosmic rays. The counter coincidence pointing to the interaction region essentially eliminates the cosmic rays and reduces the muon rate to about 10 KHz. Using the fast trigger as both a time and position marker, the drift modules are then used to give a level one momentum tag to the muon or muons passed by the counters.

Most of the available waiting time for the level 1 trigger is expended in drift time to maximize cell width, and thus minimize channel count. This means that the drift modules can have cell sizes no larger than about 8 cm (4 cm drift) to acquire the ionization signal in time to use the information in the level 1 trigger. At this level, the trigger does not compare hits in different modules to minimize signal transit times in cables. The transverse momentum tag is determined by a local vector within the outer drift module. An example of such a trigger is used in CDF.[7] The difference in the time of arrival of signals on projective wires gives the transverse momentum of the track. This scheme is illustrated in figure 6.

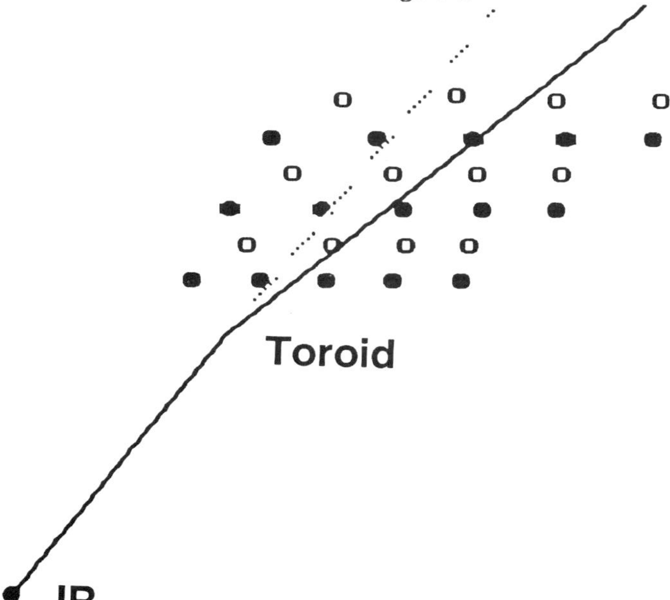

Figure 6. Local vector trigger using projective wires (not to scale). Only the signal wires are shown for a single drift module. Alternate layers of wires are offset (shown as solid and open dots) so that a fast signal appears in alternate layers. A track with no appreciable bending in the toroid passes an equal distance from a projective set of wires.

The position of the drift modules is primarily determined by the configuration of the entire detector and the desired resolution of the preliminary muon momentum tag from the level 2 trigger. It is at level 2 that the information from all four modules are used to determine the angular deflection of the track by the iron toroid as shown in figure 7.

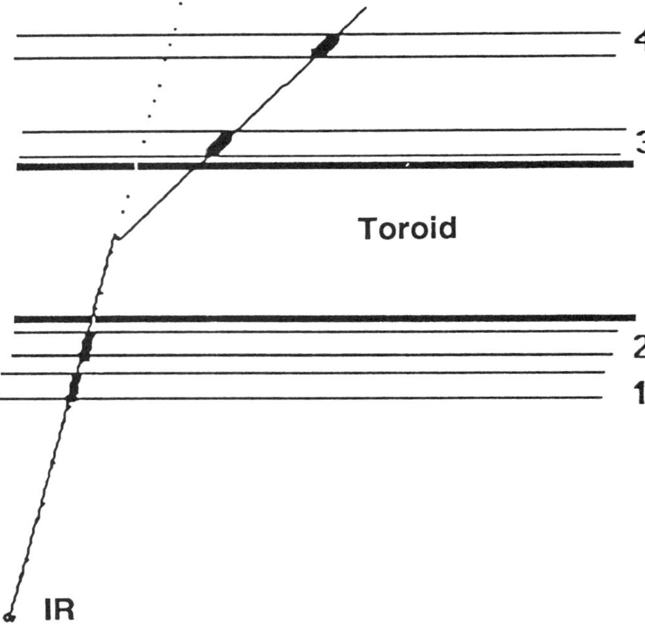

Figure 7. Difference of slopes determined from drift modules provides the level 2 trigger. Not to scale.

Four stations of drift modules separated by about 1 meter with an alignment accuracy of 250 μm gives a momentum resolution at this level of the trigger as shown in figure 8.[8]

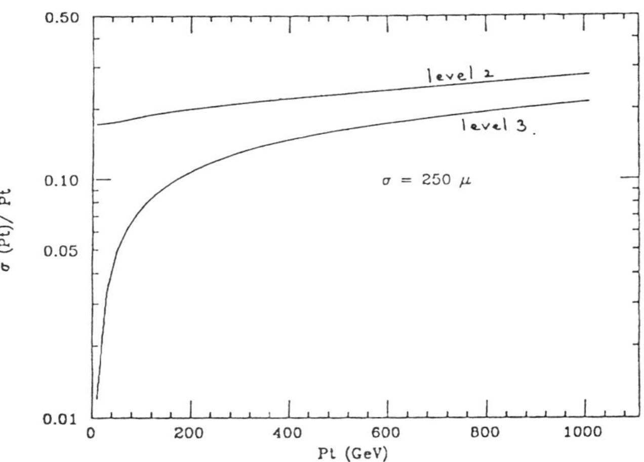

Figure 8. Momentum resolution of the central muon system as a stand alone trigger system (level 2) and using the beam position, the last superlayer of the central tracking chamber and the central muon system (level 3).

The resolution of the above configuration is about 20%. It is a slowly varying function of momentum with the minimum resolution set by the multiple scattering of the iron toroids. An angle - angle determination of the muon momentum depends only on the relative alignment precision of each set of two chambers on the same side of the toroid and not on the more difficult alignment across a toroid. An example of a level 2 trigger is a single muon with a momentum of greater than 70 GeV/c. If one requires that the trigger be 100% efficient at a p_t of 70 GeV/c, its efficiency as a function of p_t is shown in figure 9[8].

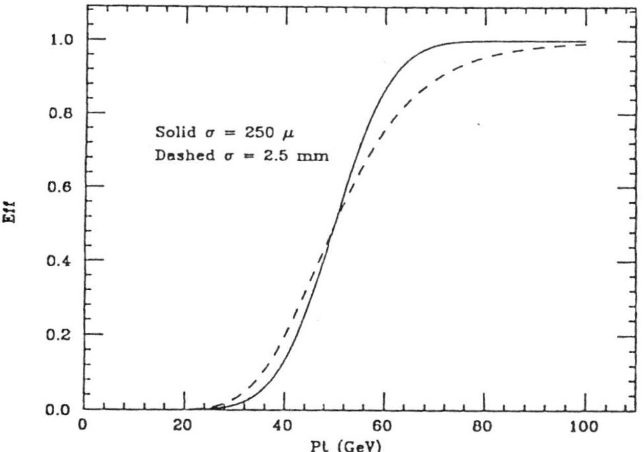

Figure 9. Trigger efficiency for the central region for a possible level-2 muon trigger which uses prompt hit information from the central muon tracking system to demand the presence of a muon with $p_t > 70$ GeV/c. The curves show the trigger efficiencies for a detection system with $\sigma = 250$ micron and $\sigma = 2.5$ mm spatial resolution.

Combining the above trigger efficiency with the muon p_t spectrum from figure 3 gives about 300 unwanted low p_t triggers for each muon above the desired 70 GeV/c threshold. Since the rate of 70 GeV/c muons is much less than 1 Hz even at $10^{34}/(\mathrm{cm}^2\mathrm{s})$, this stand-alone trigger rate would be less than 300 which is adequate to pass along to level 3 for refinement. The track from the muon drift modules is then followed into the inner tracker where an impact parameter measurement can be made using the relevant outer layers. Since multiple scattering is small in the central tracker, the level 3 momentum calculation can be sharp enough to keep this trigger rate at the Hz level. Figure 8 compares the resolution of a possible level 2 trigger using only the external muon system with a level 3 trigger using the outer layers of the inner tracker and the beam constraint.

Reconstruction Resolution

The off line muon resolution for the detector described above is determined primarily by the resolution of the inner tracker although the external muon system does contribute significantly for muons above about 500 GeV/c. To support the inner tracker in the intermediate region where the direction of the muons is closer to being along the solenoid field, we have considered an external intermediate region muon system using superconducting air core toroids as an alternative to iron toroids. Figure 10 compares the inner tracker resolution with that of an intermediate muon system based on air core toroids.[9] The resolution limit of an iron toroid solution is also shown.

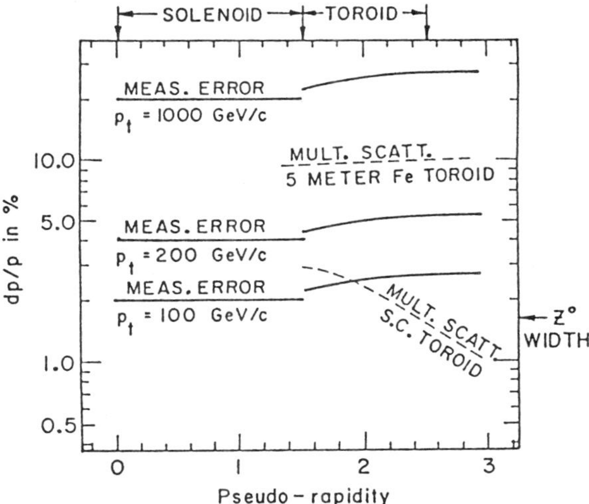

Figure 10. Comparison of the muon momentum resolution for the central tracker with that of the intermediate region superconducting air-core toroid system. The solid lines show the resolutions $dp_t = 0.20\ p_t^2\ (\mathrm{TeV/c})^{-1}$ for the inner tracker and $0.25\ (\mathrm{TeV/c})^{-1} \cos\theta$ for the air core toroids. The dashed lines show multiple coulomb scattering contributions to the resolution of an air core toroid and a 5 m long iron toroid. Also shown is the resolution required to reconstruct the $Z^0 \rightarrow \mu\mu$ mass with an uncertainty equal to the natural width of the Z^0 (2.5 GeV).

None of the benchmark processes we have studied require the extra measurement precision in this rapidity region gained by using air core instead of iron toroids. The possible use of air core toroids depends on whether or not they are significantly more expensive that iron toroids. It is true that air core toroids could save some alignment costs. Multiple scattering precludes using straight tracks for chamber alignment with iron toroids, but this option is possible with air core toroids. At the present time, no magnet of the type needed for an intermediate region air core toroid has ever been constructed. Preliminary engineering cost estimates vary wildly.[10]

R & D Plans

The muon system outlined above is large, ambitious, and expensive but certainly within our technical and financial capability. Specific drift module and fast counter designs must be chosen and prototyped. These designs will be optimized to reduce the effects of the soft electrons and photons accompanying a high energy muon emerging from material. For example, we must decide whether to use an open or closed cell design, whether it is useful to build the modules out of low Z material, and how to space the wires inside the module. Equipment for tests to gather the necessary data on which to base these decisions has been built and is now being installed in the Fermilab muon beam.[11] Data will be taken when the fixed target program resumes in 1991. Simulations and beam tests will also determine the extent of the problem of soft neutron albedo. An extremely high neutron flux will preclude the use of hydrogen based detectors such as scintillator, especially in the intermediate region. If so, gas Cherenkov counters could be used to give a fast trigger based on the deflection angle of the muon by the toroid magnetic field.[12] Equipment to make the neutron measurements was set up and tested during the first half of the current Fermilab fixed target run and will take data when that run resumes.[5] The neutron test apparatus is shown in figure 11 and a preliminary neutron energy spectrum shows the expected enhancement at 1 MeV.

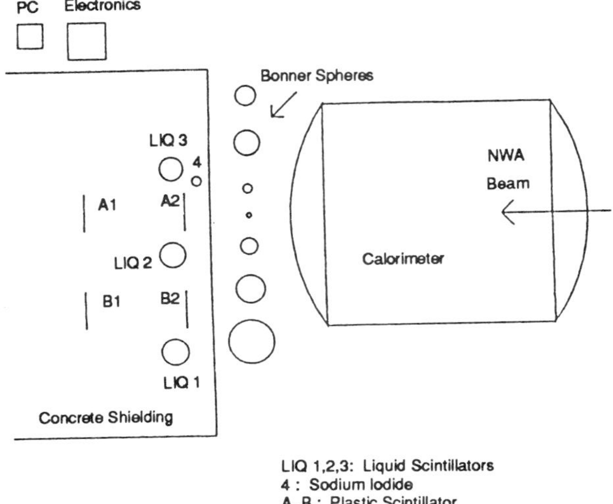

Figure 11. Low energy neutron test beam apparatus.

The configuration of the intermediate muon system is being investigated using simulations to see if one can shield against the high rate of neutrons from particles scattering from the inner edge of the forward calorimeter.

Above all we must now build prototypes to discover any hidden problems in the difficult combination of detector design criteria. The muon detectors must be robust enough for the assembly process and removal for access to the other detector systems. They must be possible to precisely align and keep aligned for long periods of time. Finally, the muon detectors must also be inexpensive enough to build in large quantities. Since the environment of the central region modules will be much less severe than the intermediate region, it is likely that different designs are more cost effective for each region.

Full size prototyping of several different module designs emphasizing different concerns is required to assess their relative advantages. These prototypes must include a prototype of the electronics used for the trigger. The effect of extra hits in the trigger drives the consideration of open cell modules (multiple projective wires) versus closed cell (one signal wire) and metal versus plastic cell walls. The speed of forming a trigger determines the maximum cell size and leads to the consideration of fast gas mixtures and an internal electrode structure to ensure an approximately constant drift speed necessary for trigger simplicity. The effectiveness of the momentum trigger is determined by the alignment which in turn will determine the maximum wire length. The question is, as always, one of cost/benefit.

The prototyping effort is separated into several subfunctions. Seven meter long extruded aluminum modules shown in figure 12 have been built and are now operating in the Soudan experiment.[13]

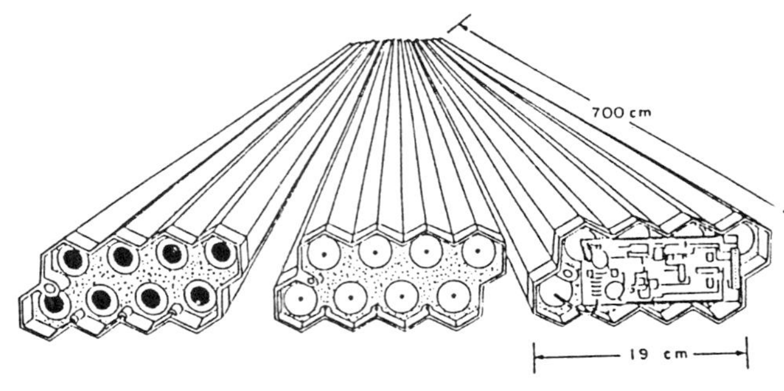

Figure 12. Extruded aluminum proportional tube manifolds of the 1800 m^2 active shield of Soudan II.

Because of wire sag, these modules do not have the alignment precision needed for the SSC but can be used to study the problem of inserting precision wire holders at intervals inside a tube. A preliminary design has a cage insert which contains the precision aligned wire support and auxiliary electrodes.[14] The insert, shown in figure 13, then slides into the extrusion which is then sealed at the ends.

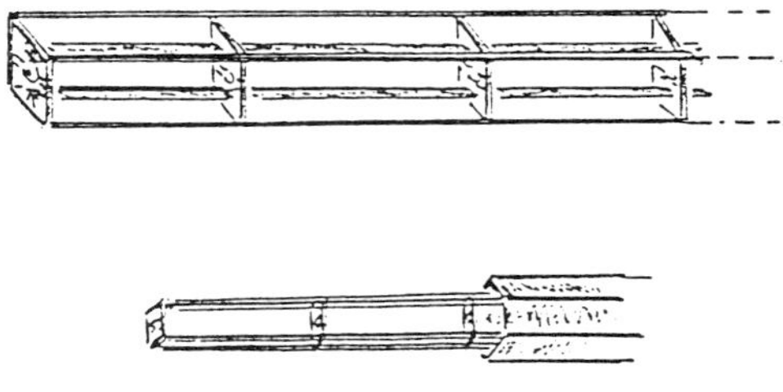

Figure 13. Rail-and-frame insert. Electrode strips and wire supports are mounted to the frame. The insert is then slid into the extrusion.

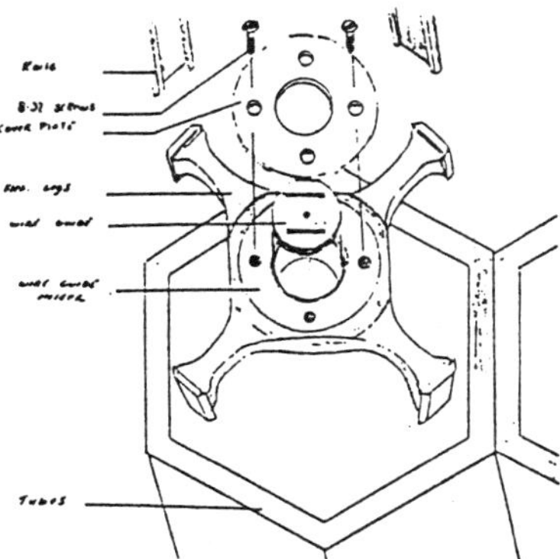

Figure 14. A design for a precision adjustable wire and electrode support.

The geometry shown is hexagonal but could clearly be almost any shape. Of course long prototypes of open structures which are sealed together, such as those used in CDF[7] and D0[15], will also be tested. Plastic chambers modeled on those used for Amy[16] will also be prototyped to determine if alignment can be maintained at a reasonable cost in a light weight but rigid chamber. A rigid plastic chamber would minimize the structural supports necessary to maintain the chamber alignment.

Keeping a linear time to distance relationship over as large a cell as possible is necessary for an inexpensive but sharp p_t trigger. Gas prototypes will be developed for the specific chamber geometries under consideration. Auxiliary electrode structures may be quite useful if they can be inexpensive and precise enough. We are currently considering designs using geometrical field shaping[17,18], for example as shown in figure 15, and potential difference field shaping using resistive plastic a la Amy[16] or Soudan.[19]

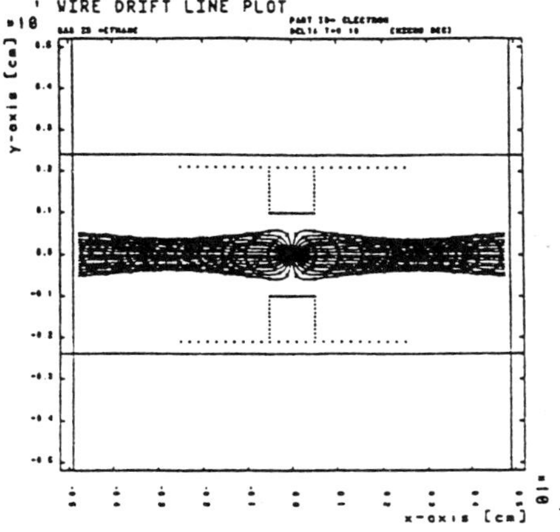

Figure 15. Drift lines of a 10 cm x 5 cm box with two focusing electrodes separated by 2 cm. The sense wire is a +6 kV, the field shaping electrodes at +4.5 kV, and the wall is at ground.

Summary

The purpose of these proposed prototypes is to address the very real but mundane problems of structural strength, precision alignment, triggering limitations and, above all, cost. Because the muon system is exterior to the other detection systems of a solenoid detector, it is affected by their design. It also affects the design of other systems because the muon detectors can prevent access to critical inner detector elements. The muon system is geometrically the largest system in the detector so that its total cost is high even if the cost per channel is low. Detailed cost / benefit analysis of the entire system must be undertaken. For example, a simpler drift tube design may be less expensive itself but may require a more expensive electronics for triggering or data acquisition, more extensive support structures to insure alignment, more material for shielding, a larger external magnetic field for background rejection, safety systems for potentially explosive gas, or redesign of the calorimetry for access. Replacement and repair might be extremely difficult. Any of these possibilities might make the actual cost of the entire detector greater than for a seemingly more expensive drift module design. Once the module geometry is chosen, the layout of the muon system must take into account the important reality of plumbing and wires which can create dead regions in the detector. Finally in such a large system, we must continue to be open to clever ideas to reduce the number of detectors and the cost per detector.

REFERENCES

[1] T. Fields et. al., "FY 1990 Summary Report and FY 1991 Proposal for the Development of a Muon Subsystem for a Solenoidal SSC Detector", UMN-EX-912/90

[2] D. Green and D. Hedin, "Muon Rates at the SSC", FERMILAB- Pub-90/18 , to appear in NIM,
Y. Sakai, "Muon Trigger Rates and Momentum Resolution Calculations", Proceedings of the International Workshop on Solenoidal Detectors for the SSC, p 321-434, April 23-25, 1990, KEK, Tsukuba, Japan, F. Abe, K. Hasegawa, eds.

[3] Expression of Interest by the Solenoidal Detector Collaboration to construct and operate a detector at the Superconducting Super Collider, p 50-53, May 24, 1990, G. H. Trilling, Spokesperson

[4] T. Yoshida, "A Proposal to Measure EM- Shower Associated with High Energy Muons in Iron Absorber", Proceedings of the International Workshop on Solenoidal Detectors for the SSC, p 401-403, April 23-25, 1990, KEK, Tsukuba, Japan, F. Abe, K. Hasegawa, eds,
D. Carlsmith, "Central Muon Trigger Resolution for the Solenoidal Detector", SDC-UWI-1, Jan. 9, 1990.

[5] D. Green and K. Johns, " Neutron Background Tests at Fermilab", Proceedings of the International Workshop on Solenoidal Detectors for the SSC, p 404-408, April 23-25, 1990, KEK, Tsukuba, Japan, F. Abe, K. Hasegawa, eds.

[6] R. Thun, "Iron-Scintillator Configurations for Level-1 Muon Triggers", University of Michigan, SDC-MU-UMI-1, Jan. 11, 1990.

[7] G. Ascoli et al., "CDF Central Muon Detector", NIM A268, 33, 1980

[8] J. Wiss and S. Errede, " Muon Detector Momentum Resolution", University of Illinois, SDC-MU-UIL-1, Nov. 15, 1989,
J. Wiss, "Barrel Muon Detector Design Considerations", University of Illinois, SDC-MU-UIL-3, Dec. 11, 1989,
Y. Sakai, "Muon Trigger Rates and Momentum Resolution Calculations", Proceedings of the International Workshop on Solenoidal Detectors for the SSC, p 321-434, April 23-25, 1990, KEK, Tsukuba, Japan, F. Abe, K. Hasegawa, eds.

[9] T. Fields, "Muon Momentum Accuracy - Matching the Forward Toroid to the Central Muon System", Argonne National Laboratory, SDC-MU-ANL-1, Jan. 31, 1990.

[10] "A Superconducting Toroidal Magnet for the SSC Detector", Advanced Cryo Magnetics Inc., May 1990, SDC-90-00062.
T. Fields, "Conceptual Design for a Superconducting Toroid", May 1990, SDC-90-00061
T. Fields, "Conceptual Design for a Superconducting Toroid", Proceedings of the International Workshop on Solenoidal Detectors for the SSC, p 435-437, April 23-25, 1990, KEK, Tsukuba, Japan, F. Abe, K. Hasegawa, eds.

[11] H. Lubatti, et al., "SSC Muon Detector SubSystem Beam Tests", SDC-MU-BTG-1, Jan. 31, 1990

[12] U. Nauenberg, "The Use of Gas Cerenkov Counters as part of a Trigger System for Muons in SSC Detectors", University of Colorado preprint, September 1990.

[13] W. Oliver et. al., NIM, A276, 371 (1989)

[14] T. Kafka et al., "Proposal to Develop Extrusion Cell Drift Tubes for the SSC Muon System", Tufts University preprint, August, 1990

[15] C. Brown et al., "D0 Muon System", NIM A279, 331, 1989

[16] A. Sill et al, "Graded Potential Drift Chambers", Proceedings of the 1990 meeting of the Division of Particles and Fields, Rice University, Houston, Texas, Jan. 1990

[17] T. Zhao, ``Tracking Muons at the SSC", University of Washington, SDC-MU-UWA-1, UW-PHY-125, May, 1990.

[18] Y. Teramoto, "Muon Chamber Design", Osaka City University preprint, October, 1990

[19] J. Thron, "The Soudan 2 Proton Decay Experiment", Proceedings of the 1989 Wire Chamber Conference, Vienna, Austria

GASES FOR AN SSC MUON DETECTOR[1]

L. G. Christophorou, P. G. Datskos, and J. G. Carter
Atomic, Molecular, and High Voltage Physics Group, Health and Safety Research Division, Oak Ridge National Laboratory, Oak Ridge, Tennessee 37831 USA, and Department of Physics and Astronomy, The University of Tennessee, Knoxville, Tennessee 37996 USA

Abstract

Recent measurements of electron drift velocities as a function of the density–reduced electric field E/N are reported for a number of unitary gases and the mixtures CO_2/CH_4 and $NH_3/CF_4/Ar$. Calculated values of the mean electron energy as a function of E/N are also reported for unitary gases and mixtures of CO_2/CH_4.

Introduction

Precision muon detectors for SSC put rather stringent constraints on the gaseous medium to be used in such large devices.[1,2] A muon detector gas is required to have good chemical properties (to be nonflammable, nonexplosive, nontoxic, and compatible with other materials). It also must have: low electron diffusion coefficient — transverse D_T and longitudinal D_L — ("cool" gas), high electron drift velocity w ("fast" gas), high gain ("efficient" gas), small Lorentz angle, convenient operating density–reduced electric field E/N, and high corona inception and breakdown voltages. It, also, must be non–electron attaching (especially below ~ 5 eV) and insensitive to changes in the electric field, magnetic field, temperature and composition/impurity content.

[1]Research sponsored by the U.S. Department of Energy, under Contract No. DE–AC05–84OR21400 with Martin Marietta Energy Systems, Inc., and the University of Tennessee, Knoxville, Tennessee, under Contract No. DE–AS05–76ER03956.

We have been exploring the development of multicomponent gas mixtures with optimal electron transport, gain, electrical and chemical properties focusing in particular on binary and ternary gas mixtures of Ar, N_2, CO, CO_2, CH_4, CF_4, $Si(CH_3)_4$, and NH_3. For the first six molecules, sets of electron scattering cross sections are available which can be used to compute electron transport parameters for their mixtures. In this paper, we report some of our results on w(E/N) and the mean electron energy $<\epsilon>$ (E/N).

Results and Discussion

Unitary Gases

In Fig. 1 are shown measurements of w(E/N) for a number of unitary gases which clearly show the superior properties[3] of CF_4. In Fig. 2 are presented the calculated $<\epsilon>$ (E/N) for Ar, CH_4, CF_4, N_2, and CO_2 and the characteristic energies ($\equiv 3/2\ (eD_T/\mu)$; μ = electron mobility) for C_2H_2 and NH_3; the former we calculated using available sets of electron scattering cross sections, measured electron transport coefficients and a Boltzmann–code analysis[4,5], while the latter we obtained from data on D_T/μ (E/N) in Refs. 6 and 7. Besides $<\epsilon>$ (E/N), we calculated

$w(E/N)$, D_T/μ (E/N) and D_L/μ (E/N) for a number of gases/mixtures. The data in Fig. 1 help identify "fast" gas components for multicomponent mixtures; those in Fig. 2 help identify "cool" gases, i.e., gases with strong electron thermalizing capabilities.

Binary Gas Mixtures

Examples of our measurements of $w(E/N)$ and calculations of $<\epsilon>$ (E/N) for binary gas mixtures are shown in Figs. 3 and 4 for CO_2/CH_4. The superior electron ther– malizing properties of CO_2 are clearly evident. From Figs. 3 and 4 it seems that mixtures containing ~ 20% of CO_2 in CH_4 have good $<\epsilon>(E/N)$ and $w(E/N)$ characteristics.

Ternary Gas Mixtures

In an effort to capitalize on both the large electron thermalizing capabilities of polar gases[6,7] and their possible beneficial effect on the detector conditioning properties[1,2], we have added various amounts ($\lesssim$ 10%) of NH_3 to the fast gas mixture 80% Ar + 20% CF_4. The measured $w(E/N)$ for these are shown in Fig. 5. We have not been able to compute $<\epsilon>(E/N)$ for the ternary mixtures due to lack of electron scattering cross sections for NH_3. Addition of NH_3 to CF_4/Ar causes electron "cooling" (and possibly improves the detector aging performance in a manner similar to that reported for water[2]); but may result in the formation of negative ions.

Overall Properties

The results presented in this paper, and similar unpublished findings on other systems, show that it is possible to develop multicomponent gas mixtures that are both fast ($w \sim 10^7$ cm s^{-1}) and cool ($<\epsilon> < 1$ eV) at convenient operating E/N values. While a number of mixtures appear promising based on their w and $<\epsilon>$ values, a number of the other desirable properties outlined in the introduction need to be investigated before an assessment can be made of their usability for muon detectors.

References

1. C. M. Ma, U. Becker, D. A. Ma, and A. H. Walenta, MIT Technical Report No. 129, October 22, 1982.

2. P. T. Ferbel, Experimental Techniques in High Energy Physics, Addison–Wesley Publishing Company, Inc., Menlo Park, California, 1987.

3. L. G. Christophorou, D. L. McCorkle, D. V. Maxey, and J. G. Carter, *Nucl. Instr. Meth.* 163, 141 (1979).

4. D. K. Gibson, *Austr. J. Phys.* 23, 683 (1970).

5. S. R. Hunter, J. G. Carter, and L. G. Christophorou, *Phys. Rev. A* **38**, 5539 (1988).

6. L. G. Christophorou, Atomic and Molecular Radiation Physics, Wiley, New York, 1971.

7. L. G. Christophorou (Ed.), Electron–Molecule Interactions and Their Ap– plications, Vols. 1 and 2, Academic, New York, 1984.

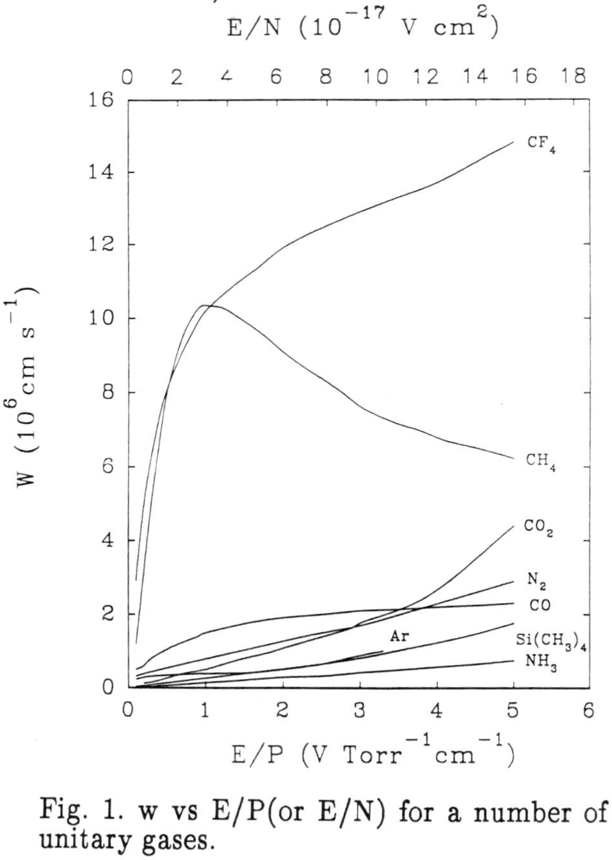

Fig. 1. w vs E/P(or E/N) for a number of unitary gases.

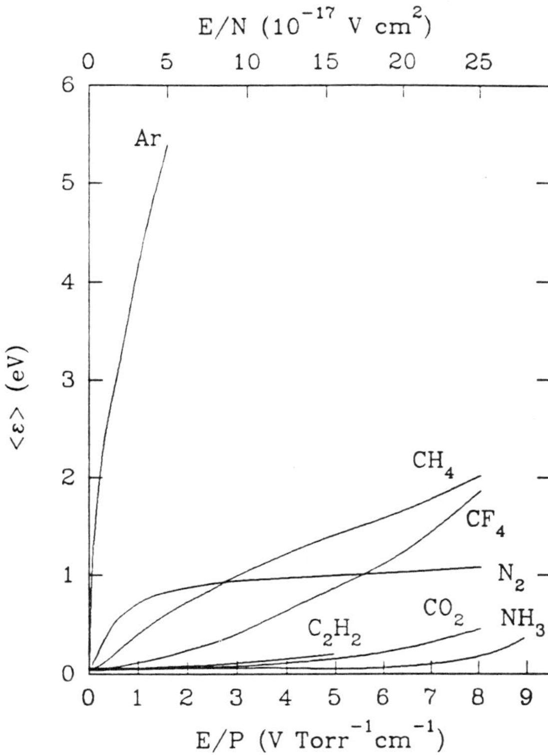

Fig. 2. <ε> vs E/P (or E/N) for a number of unitary gases.

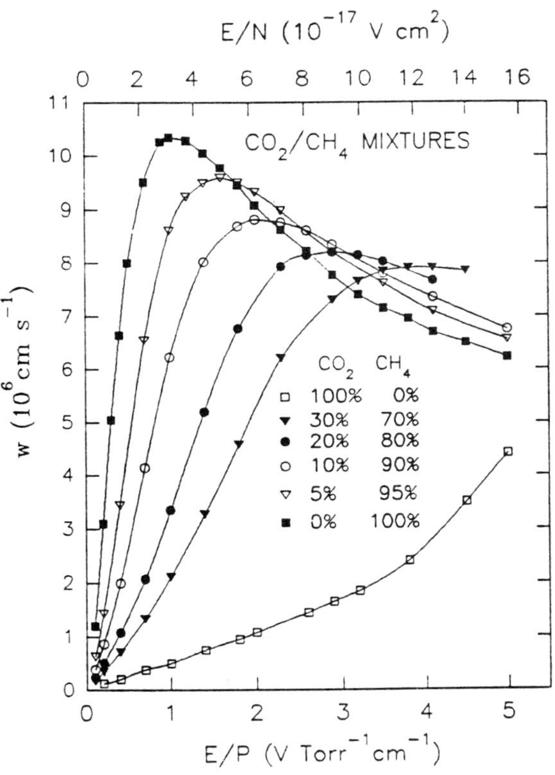

Fig. 3. w vs E/P (or (E/N) for CO_2/CH_4 mixtures.

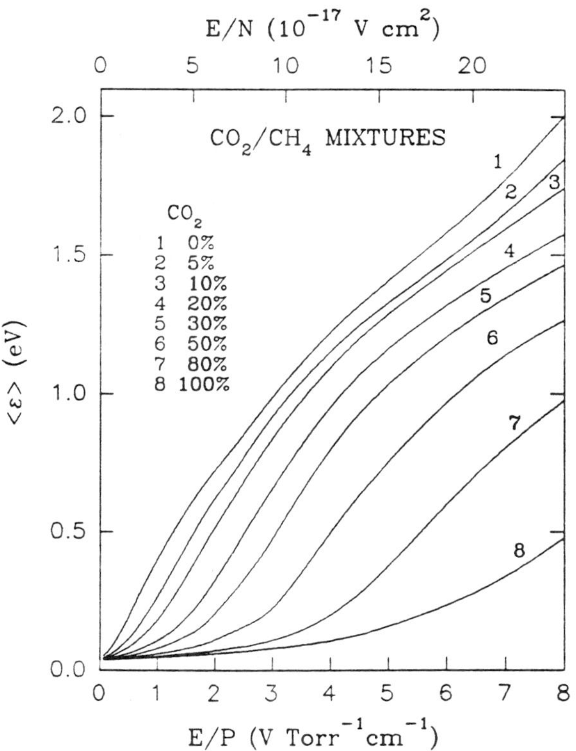

Fig. 4. <ε> vs E/P (or E/N) for CO_2/CH_4 mixtures.

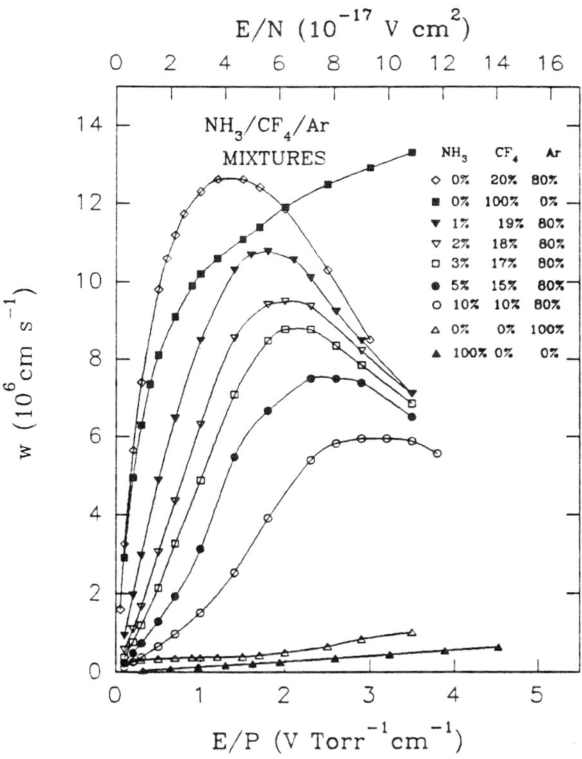

Fig. 5. w vs E/P (or (E/N) for $NH_3/CF_4/Ar$ mixtures.

STREAMER CHAMBERS FOR MUON TRACKING AT THE SSC

Kwong Lau, Jörg Pyrlik, and Roy Weinstein
Department of Physics and Institute for Beam Particle Dynamics
University of Houston
Science Research Building I
Houston, Texas 77204-5506

Abstract

The use of streamer tubes/chambers for muon tracking at the Superconducting Super Collider (SSC) is considered. The requirements on radiation hardness, spatial resolution, alignment precision, etc., are discussed. The highly automated Streamer Chamber Assembly and Research Facility (SCARF) at the University of Houston is described.

Introduction

Muon identification is a very important aspect of SSC physics [1]. Some noteworthy examples are:

1. Top discovery via high p_t muons plus jets.
2. Four muon final state of Higgs decay via two Z's.
3. Detection of supersymmetric particles, e.g., smuons via their decay to muons and photinos.

SSC environment presents new challenges for muon detection. The radiation level is high [2]. The high collision rate of the accelerator (16 ns between crossings) requires pipelined data readout and bucket identification. The need for large area of precision tracking system for muon detection is apparent. Wire chambers operating in the streamer mode is a strong candidate because of their large signal and ease of operation [3]. The performance of plastic streamer chambers is discussed in terms of SSC requirements. The highly automated Streamer Chamber Assembly and Research Facility (SCARF) developed by high energy physics groups from University of Houston and Northeastern University is capable of producing 100 eight-cell chambers per day. Ongoing R&D programs on streamer tubes at SCARF are also described.

Streamer Chambers for Muon Tracking

The radiation level at the SSC has been studied [2]. The low energy (~ 1 MeV) neutron fluence at muon chambers is $10^{10\text{-}12}$ neutrons per cm^2. The annual dosage corresponding to a typical gas gain of 10^5 for the streamer mode is estimated to be as high as 0.1 C/cm. Even though this level of radiation exposure does not seem to cause any measurable deterioration of the chambers [4] it does present challenge for using PVC streamer tubes in the forward region ($\eta > 6$). Most plastics exhibit loss of strength at a radiation level of $10^{16\text{-}17}$ neutrons per cm^2 [5].

The spatial resolution of wire chambers operating in the streamer mode has been measured [6]. The achieved resolution ranges from 100 to 150 μm, limited in both cases by systematic effects. The timing resolution obtainable from two layers of chambers staggered by half a cell width is about 4 ns, which represents a four standard deviations resolution in bunch crossing. The dead-time associated with each streamer is typically a few μsec over hundred μm. A test conducted at CERN with muon beam observed no loss of efficiency up to a rate of 200 Hz per cm^2 [7]. This is adequate for SSC applications where the muon rate is expected to be less than 20 Hz per cm^2.

Locating the wires to better than 100 μm requires careful considerations. One scheme is to attach the chambers to a rigid support and use muons from interactions to perform software surveys on an regular basis. This will monitor and correct the variations of the wire positions due to temperature changes.

SCARF

A standard Iarocci PVC streamer chamber consists of an extruded eight-channel open PVC profile coated with resistive graphite, enclosed in a gas tight PVC envelope [3]. The tube dimensions are 9 mm x 9 mm. The length can be up to eight meters long. The silver-plated Cu-Be 100 μm diameter anode wires are soldered to printed circuit boards at both ends of the chamber and supported every 50 cm by plastic spacers. The wired profile is inserted into a PVC envelope and sealed by plastic plugs at both ends for gas tightness. A cross section of eight-cell chamber is shown in Fig. 1. In order to produce the large number of streamer tubes for the LVD experiment (160,000 tubes in 20,000 eight-tube modules), the UH and NU groups have developed a highly automated plant to assemble streamer tubes at Houston. One of the most critical steps in the production is coating the walls of the eight-channel comb-shaped profile with a thin layer of resistive graphite. The performance of the chamber depends critically on the thickness and uniformity of this thin layer of graphite on the extruded PVC profile. The surface resistance should be in the range of 0.1 to 5 MΩ per unit area. The lower limit is determined by the cutoff in the transparency to the signal whereas the upper limit is to avoid too many unpainted areas which are known to cause continuous discharges.

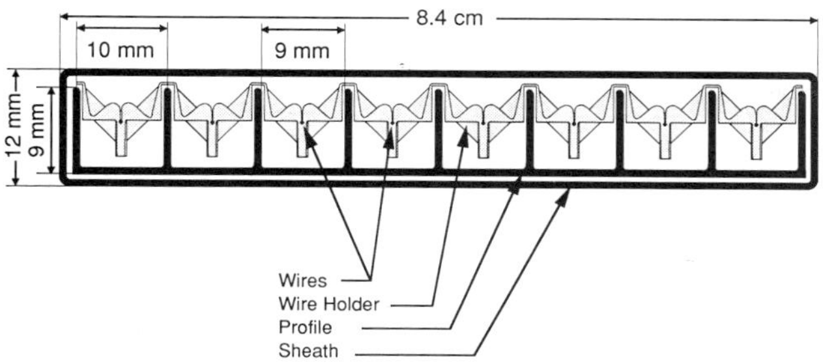

Fig. 1. Cross section of an eight-cell Iarocci streamer chamber.

We adopted the painting technique developed at Frascati which applies a graphite suspension (DAG) diluted with methyl-isobutyl-ketone (MIBK) to the profiles using a cellulose sponge. The profiles were driven through a stationary sponge wet with the paint using a machine developed by Bindi [8]. Slots are cut in the sponge so that both the bottom and the sides of the eight-channel profile can be painted simultaneously. Normally it takes one to two passes to achieve an acceptable surface resistance. As part of the quality control procedures, we measure the surface resistance of the painted profiles after they have been dried in air for more than ten hours. Resistance measurements are made at ten equally spaced positions on the bottom of the profile. The resistance spectrum of a random batch of unselected chambers produced at SCARF is compared with batch of selected chambers painted at Frascati in Fig. 2. The significantly narrower

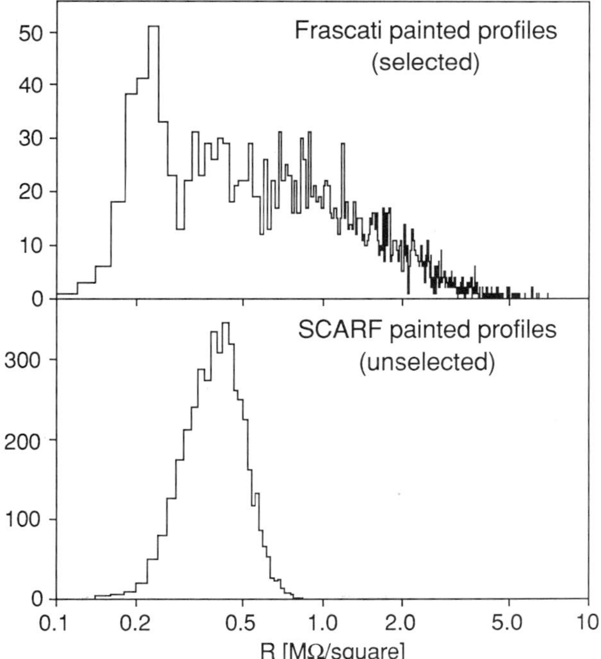

Fig. 2. Resistance spectra of SCARF painted profiles (bottom) and Frascati painted profiles (top).

distribution for the SCARF chambers is an indication of the improvement in the control of the painting process. The process of installing wires in the chambers is done by a machine developed at CERN which requires only one operator. We also improved on a manual staking machine developed by the Maryland group to secure wires onto plastic spacers by heat staking. The chambers are heat sealed by a machine developed by Bindi. The results are also very satisfactory. The end-sealing process is now fully automated.

We also developed a very thorough testing procedure to ensure the quality of the chambers. Primary leak test for each chamber is done by immersing the chamber in water with an overpressure of 15 inches of water (40 mbar). Only chambers releasing less than one bubble over the observation period of one minute are accepted and passed onto the next stage of the test. Fourteen chambers are connected in series for gas flow. The chambers are first evacuated to about ten Torr, then flushed with a gas mixture of 75% isobutane and 25% Argon. The mixture is carefully controlled by mass flow controllers [9] which were calibrated by a bubblemeter to about 1%.

After the chambers are flushed for ten volume changes, a 24-hour burn-in cycle for each chamber is performed with a computer-controlled high voltage power supply in batches of 40 chambers [10]. The voltage is gradually increased from an initial voltage of 4.4 to 5.3 kV in about 12 hours with a typical ramp-up rate of 100 volts per hour. The current drawn by each chamber is monitored and limited to less than 2 μA. Chambers passing the burn-in cycle are then subject to the operating voltage of 4.9 kV for 12 hours. Only chambers that draw less than an average current of 1.0 μA at the end of this burn-in cycle are considered acceptable. The present setup at SCARF can test up to 40 chambers per day, limited by the number of available computer-controlled high voltage channels. The test capability is easily expandable to 80-100 chambers per day.

Long-term failure rate of these chambers is a very important parameter because replacing failed chambers at the SSC will be difficult. We have therefore developed a long-term test facility at SCARF using a CAMAC system which monitors the current drawn by each chamber to a precision of 10 nA [11]. Out latest results indicate a long-term failure rate of less than 1% per year.

We have thus far completed 40,000 tubes (5,000 eight-tube chambers) for the LVD experiment. We will produce about 1,000 four-meter chambers for the SMC experiment at CERN.

R&D Program at SCARF

Nonfammable Gases

Streamer tubes constructed with a 9 mm x 9 mm cross section and 100 μm diameter wires require a heavy quenching gas of 75% isobutane and 25% Argon which unfortunately is flammable. Extensive searches have been made for a less or nonflammable gas mixture [12]. It is also important that this gas mixture will provide comparable or better pulse height and lifetime characteristics. Moromisato et al. have studied a few mixtures of Ar/Ne-CO_2-isobutane and found a suitable nonflammable mixture for the LVD streamer tubes [13]. A high statistics study of the plateau of SCARF production line chambers using this gas mixture was carried out at SCARF [14]. The characteristics are also compared with those of the standard gas mixture. The results of Moromisato et al. were confirmed and the new gas was found to give a usable plateau for a statistically significant sample of chambers produced at SCARF.

Nonhalogen Plastics

In the event of a fire, PVC releases 580 grams of HC1 per kilogram of PVC and presents a hazard to personnel and damage to the electronics. Alternative materials for streamer chambers have been suggested. Among them, NORYL (Diethyl Ether Oxide), LURANYL, and ABS (Acylnitril Butadien Styro) have been most extensively studied and used in some experiments. In the future, we plan to explore US sources of chamber parts based on these non-halogen materials and test our production techniques on them.

Failure Studies

About 10% of streamer chambers produced at SCARF failed the burn-in cycle as a result of drawing excessive current (> 1 μA) for an extended period of time. We believe that the continual discharges are caused by charges from ionizing particles, accumulated in unpainted areas, eventually resulting in a continuous corona. The discharges can sometimes be cured by switching off the high voltage for a short period of time, presumably allowing the accumulated charge to bleed off. The 24-hour short-term test results of chambers produced at SCARF are carefully recorded and analyzed. The loss of chambers due to mechanical failures, such as broken wires, is about 1%. We found a rather weak correlation between the average resistance and the failure rate, even though a stronger correlation is found if one considers the number of spots exceeding a large resistance (e.g. > 3 MΩ). However, we noticed a rather dramatic correlation between failure rate and some visual characteristics of the plastics of the profiles. It is observed that profiles with smooth surfaces, though difficult to paint evenly, result in a lower failure rate. On the other hand, plastics with a rough finish are easier to paint, but suffer larger burn-in failures. We attribute this to the uncovered $CaCO_3$ grains in the plastics which are common fillers in the PVC manufacturing process. This was pointed out by a recent microscope study of the PVC of the streamer chambers [15]. Our findings also corroborate the experience of the OPAL collaboration which applies a thin layer of glycerol based oil BREOX B-35 onto the painted chambers and found significant decrease in the failure rate [16].

Aging

Streamer tubes have been used in many experiments and exposures over 0.1 C/cm without loss of efficiency have been reported under controlled environments [4]. Since it is more economical to use plastic PVC streamer chambers, it is important to study the radiation effects on the PVC at SSC radiation level. It is also important to study the dependence of aging on the gas purity and hydrogen content, and to extrapolate the results to large systems (of typically 5,000 m^2).

Small-Cell Chamber and Fast Gas

Small-cell chambers operating with a fast gas have clear advantages at the SSC. We have developed a 4mm streamer chamber and tests are underway. The Ar-CO_2 mixture was found to have desirable speed properties. The possibility of producing small size tubes (e.g. 5 mm x 5 mm) with a similar technique now mastered at SCARF will be explored. Studies with small size streamer tubes operating in the proportional or streamer mode with fast gas will be studied.

This research is supported in part by a grant from the Department of Energy (DOE) and the Texas Advanced Research and Technology Program (TARTP).

References

[1] K. Heller, these proceedings, and previous Snomass proceedings.

[2] D.E. Groom, "Radiation Levels in the SSC Interaction Regions", **SSC-SR-1033**, 1988.

[3] For a recent review of streamer tubes and their applications, see E. Iarocci, NIM **217**, (1983), and references therein.

[4] J.P. DeWulf, et al., CERN-DP/86-26, February 1986; see also J.H. Moromisato, et al., in Ref. 13.

[5] J.N. Anno, "Notes on Radiation Effects on Materials", Spring-Verlag, 1984.

[6] K. Fujii, et al., "Study of Limited Streamer Drift Tube Performance," NIM **225**, 23-30, (1984); F. Graspanini, et al., "Drift Time Measurements in Limited Streamer Tubes," NIM **A267**, 89-92, (1988).

[7] K. Lau, et al., to be published.

[8] Costruzioni Meccaniche M. Bindi, San Giustino, Perugia, Italy.

[9] Model 825, Datametrics.

[10] 40-channel HV system, Model SY 127, Constrzioni Apparecchiature Elettroniche Nucleari (CAEN), Viareggio, Italy.

[11] R. Hammarstrom, "Anode Current Monitor Design," CERN-OPAL/EP-0228, Feb. 1986.

[12] P. Rapp, NIM A244 (1985), 430; T. Uebayashi, et al., NIM A265 (1988), 457; J.H. Moromisato, et al., NIM A274 (1989), 177; A. Antonelli, et al., (to be published).

[13] J.H. Moromisato, et al., NIM **A274**, (1989).

[14] K. Lau and D. Hungerford, et al., "Operations of Limited Streamer Mode Production-Line Chambers with Non-Flammable Gases," NIM **A286** 155-159, (1990).

[15] M. Ellila, NIM A 270 (1988), 393.

[16] F. Fabbri, et al., CERN/EP, 87-134, July 1987.

MANUFACTURING HIGH-GRADE ULTRA-HEAVY ELECTRICAL STEEL PLATE AND INVESTIGATION RESULTS ON HOMOGENEITY OF MAGNETIC PROPERTIES OF THE MATERIAL

Yukio Tomita, Yukio Tuda, Ryota Yamaba
Nagoya R&D Lab., Nippon Steel Corp., Tokai, Aichi, 476, Japan.
Tsuguyuki Kawai
Plate Technical Division, Nippon Steel Corp., Otemachi, Tokyo, 100, Japan.

Abstract

In an attempt to develop ultra-heavy electrical steel plates for particle accelerator magnet having homogeneity of magnetic and other properties, electrical steel plate of 550 mm in thickness and 56.3 tons in weight was manufactured from a 100-ton ingot of high-purity steel. Investigation was carried out into homogeneity of properties at many positions.

The high-grade ultra-heavy electrical steel plate has superb magnetic properties and good homogeneity of magnetic and other properties, and shall be the best iron material for the detector.

Introduction

With the recent rapid progress in high technology, the scope of application of high grade electrical steel plate with excellent magnetic properties is expanding. In huge particle accelerators represented by the SSC (superconducting super collider), electrical steel plate is used as material for magnetic poles and return yokes. Electrical steel plate used for these parts is required to show high magnetic properties homogeneously over the whole range from low to high magnetizing force. In developing high grade electrical steel plate, it is necessary to obtain high magnetic flux density in low to high magnetizing force. And particularly a sharp rise in magnetic flux density in the low magnetizing force and homogeneity of magnetic properties were intended in our development.

Studies on chemical compositions and manufacturing process conditions

Study on chemical compositions

The effect of chemical composition in the low magnetizing force was studied. For the experiment, 50 kg ingots were made by vacuum induction melting. The C content of steels were varied from 0.003% to 0.06%. The other elements, Mn, Al and N contents were varied from 0.001 to 0.15%, 0.001 to 0.04% and 0.001 to 0.009%. The ring-shaped test specimens were used for the measurement of magnetic properties.

Figure 1 shows the effect of C content on magnetic flux density at magnetizing of 80 A/m (1 Oe). Pearlite volume fraction is also shown in this figure. The magnetic flux density at low magnetizing force of 80 A/m shows a steep upward curve at C content of 0.01%, especially 0.005% or lower. The magnetic flux density in low magnetizing force of 80 A/m shows high value at Mn, Al and N contents of 0.05, 0.02 and 0.005% or lower. This study on chemical composition revealed the necessity of minimizing the contents of C, Mn, Al and N for the improvement of magnetic properties in the low magnetizing force at about 80 A/m. And especially the influence of C content is predominant. To reduce C content less than 0.005% is essential. Similar studies were conducted regarding the other elements. The results shows that the chemical compositions range shown in Table 1 is preferable.

Effect of grain size

Vacuum melted 50 kg ingots of pure iron type within the chemical composition range shown in Table 1 were hot rolled. Figure 2 shows the effect of grain size on magnetic flux density at the low magnetizing force of 80 A/m. The grain size is shown in terms of the ferrite grain size number. The lowest value of magnetic flux density increase with the decreasing grain size number., namely the increasing grain size. This experimental result shows the necessity of grain coarsening for the improvement of magnetic properties in the low magnetizing force.

Effect of manufacturing process conditions

Voids in the steel deteriorate the magnetic properties. And hydrogen in the steel increases the

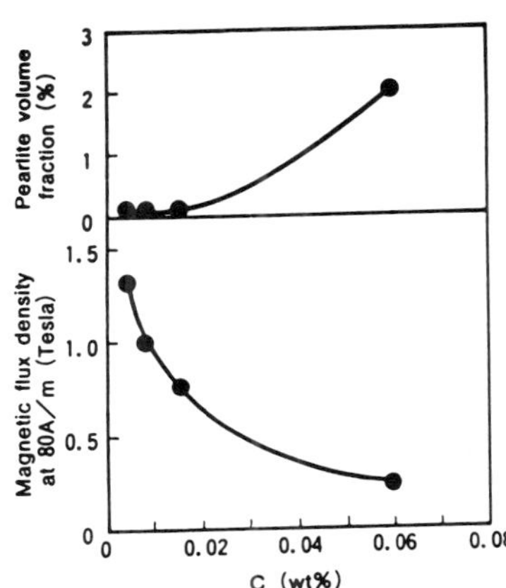

Figure 1 Relationship among C content, magnetic flux density and pearlite volume fraction

Table 1 Chemical composition range for high grade electrical steel plate

(wt%)

C	Si	Mn	P	S	Cu	Cr	Mo	Al	N
≤0.005	≤0.02	≤0.15	≤0.015	≤0.003	≤0.03	≤0.03	≤0.01	≤0.030	≤0.0050

voids and prevents the voids to be compressed by hot rolling. Voids should be minimize by decreasing the content of hydrogen and by compressing voids. High reduction rolling by heavy duty rolling mill is necessary to compress voids in the heavy gauge steel plate.

Investigation results on homogeneity of magnetic properties of high-grade ultra-heavy electrical steel plate - plate thickness: 550 mm -

Homogeneity of magnetic and other properties was investigated at different positions throughout ultra-heavy electrical steel plate (at positions corresponding to those of steel ingot) manufactured for particle accelerator magnet with high magnetic properties. The electrical steel plate was manufactured from 100-ton high-purity steel ingot. The ingot was subjected to heavy reduction rolling to a plate thickness of 550 mm. The heavy plate was then annealed. Since the plate is to be ultimately cut into fan-shape product, test samples were taken from the peripheral area of the plate product to investigate differences in magnetic or other properties at corresponding positions in ingot, positions through thickness, and positions in various directions to the rolling directions.

Manufacturing process for electrical steel plates

The following considerations are necessary for the manufacture of steel plates having high magnetic properties.

(1) The contents of the elements (C, Si, Mn, Cu, Cr, Mo, N) which increase the demagnetizing factor should be minimized.
(2) The grain size should be maximized.
(3) Nonmetallic inclusions and segregation should be minimized by decreasing the contents of P, S and O.
(4) Voids should be minimized by decreasing the content of Hydrogen and being compressed.

In order to meet the above requirements and develop ultra-heavy electrical steel plate having excellent magnetic properties, high-cleanliness steel production techniques (P<0.005%, S<0.002%, O<0.0015%), heavy-reduction rolling techniques (max. torque 700 ton-m, max. load 6400 tons, roll lift 1500 mm) and special heat treatment techniques were fully utilized by combinations.

Steel examined, experimental method and test results

Table 2 shows the chemical composition of the steel examined. The contents of elements like C, Mn, P and S

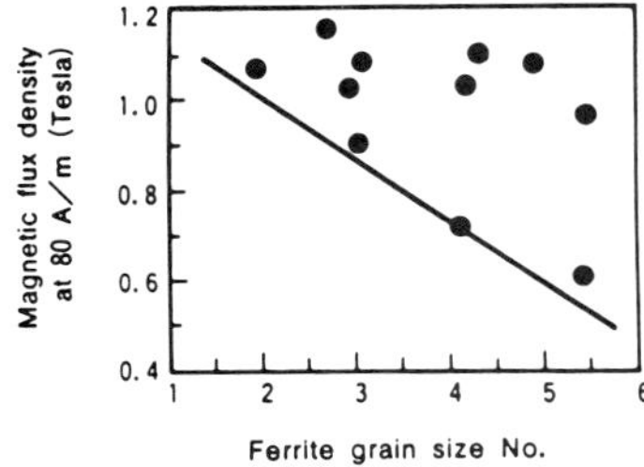

Figure 2 Effect of grain size on magnetic properties

reduced as much as possible. Ring-shaped test specimens for magnetism measurement were taken at the center and edge of the top and bottom of the ingot as shown in Figure 3. Furthermore, ring-shaped specimens were taken in the direction parallel and perpendicular to the rolling direction as shown in Figure 4. The size of these ring-shaped specimens was 20 mm in thickness, 120 mm in outside diameter and 100 mm in inside diameter. The magnetic properties were measured before and after annealing the specimens at 750°C for two hours.

Figure 5 shows the relationship between magnetizing force and magnetic flux density. Figure 6 shows the relationship between magnetic flux density and permeability. The annealed ring-shaped specimens which show the same magnetic properties as the annealed steel plate (specimens are annealed to relieve the stresses induced during the preparation of the specimens) show high and homogeneous magnetic flux density over the whole range from low to high magnetizing force. Figure 7 shows an example of the micro-structure. The grains are very coarse, as shown in this photo.

Variable shape of heavy plate cut with gas

Figure 8 shows the variable shape of heavy plate cut with gas. The tolerance of gas-cut plate are as followed: width and length are -0,+15mm(0.6in.), bevel angle is +1.5°(in the case thickness is less than 700mm).

Summary

In an attempt to develop ultra-heavy electrical steel plates for particle accelerator magnet having homogeneity of magnetic and other properties, electrical steel plate of 550 mm in thickness and 56.3 tons in weight was manufactured from a 100-ton ingot of high-purity steel. Investigation was carried out into homogeneity of properties of the plate mainly magnetic

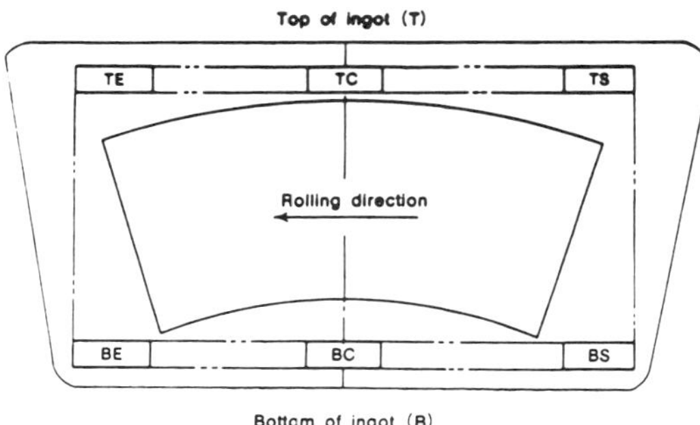

Figure 3 Locations where samples were taken

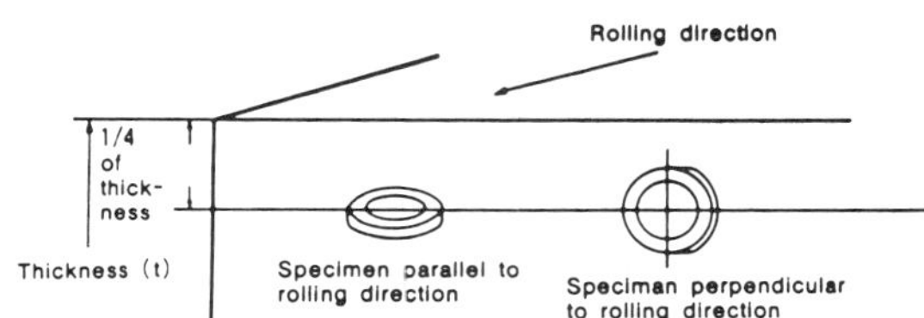

Figure 4 Positions and directions where ring-shaped specimens for magnetism measurement were taken (an example of 1/4 thickness)

Table 2 Chemical composition by ladle analysis

(wt%)

C	Si	Mn	P	S	Cu	Ni	Cr	Mo	Al	N
0.002	0.004	0.06	0.004	0.002	0.01	0.027	0.016	0.001	0.018	0.040

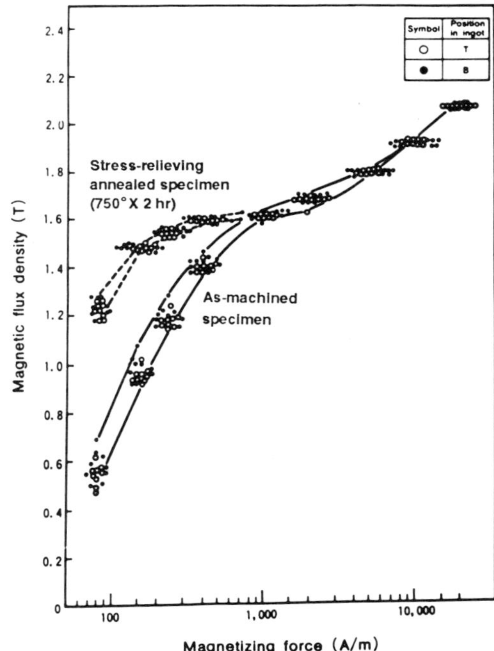

Figure 5 Relationship between magnetizing force and magnetic flux density

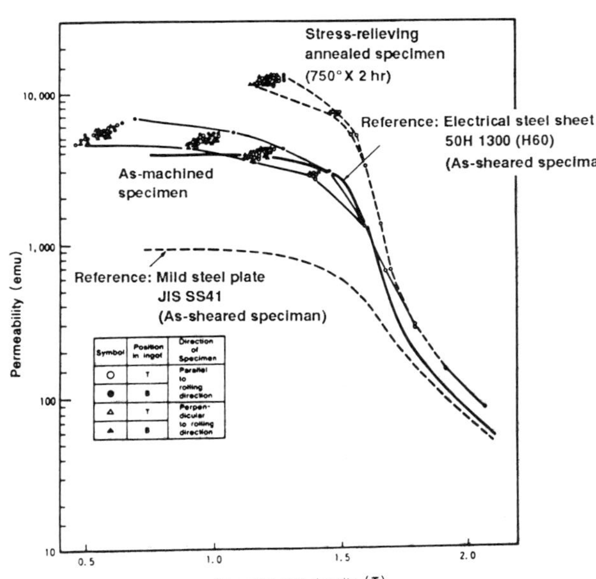

Figure 6 Relationship between magnetic flux density and permeability

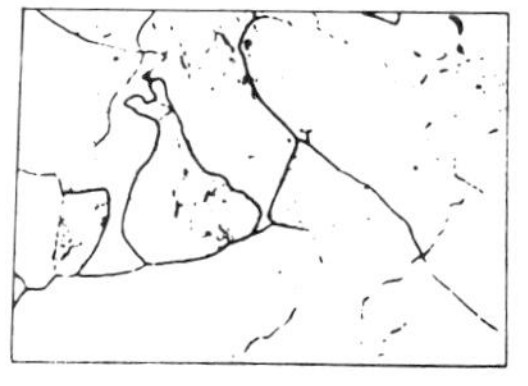

Figure 7 Microstructure

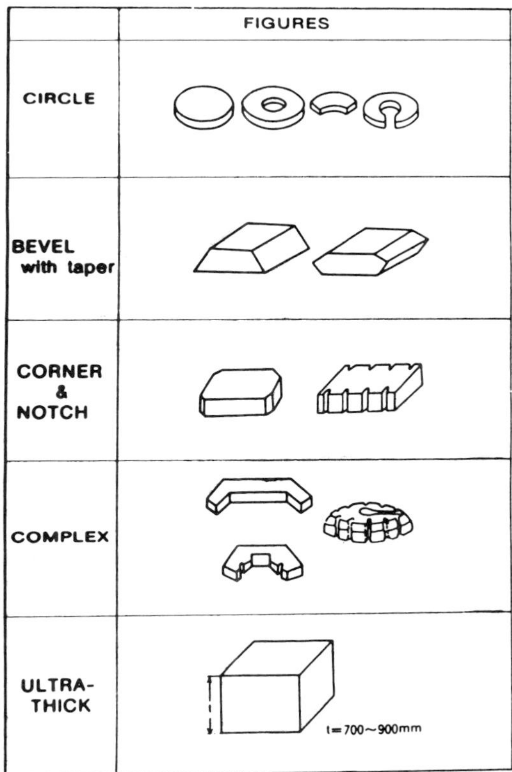

Figure 8 Variable shape of heavy plate cut with gas

properties at various positions and various direction.

Value obtained show superb magnetic properties at every sampling position of the electrical steel plate, i.e., at corresponding positions in the ingot, positions through thickness and positions parallel and perpendicular to the rolling direction, extremely good homogeneity of magnetic properties was observed.

It has been proved from the foregoing investigation results that the ultra-heavy electrical steel plate of 550 mm thickness manufactured has superb magnetic properties and homogeneity of magnetic properties.

And by these properties these electrical steel plates were used at the octagonal housing of the "TOPAZ" and at the hexagonal housing of the "AMY", two of the elementary particle detectors at the National Laboratory for High Energy Physics in Tsukuba Science City.

RESEARCH AND DEVELOPMENT TOWARDS A 50 PICOSECOND TIME-OF-FLIGHT

M.S. Alam F.A. AsSallami H. Severini J. Su P. N. Wang
M.M. Zoeller
State University of New York at Albany, Albany, New York 12222
D. Kaplan
Oklahoma University, Norman, Oklahoma 73019

October 10, 1990

Abstract

We have achieved a timing resolution of 85 picoseconds with a scintillation counter made from a piece of BICRON BC408 scintillator 2.8 meters long, 10 cm wide and 5 cm thick and viewed directly with two Hamamatsu R2490 photomultipliers glued directly to each edge. A timing resolution of 66 picoseconds is achieved if the particles traverse a 10 cm thickness.

Particle Identification is a very important issue for detectors being presently designed for the detailed study of B meson decays at both e^+e^- and $p\bar{p}$ colliders[1,2]. In both cases, $\pi - K$ and $\pi - p$ separation is important for momentum ranges up to around 3 GeV/c. The momentum region between 0.8 and 1.6 GeV/c is a region of confusion for hadron identification using specific ionization or dE/dx measurements. For a 1 meter radius time-of-flight, a scintillation counter system with 50 picosecond timing resolution would achieve 2σ $\pi - K$ separation at 2.0 GeV/c and 3σ separation $\pi - p$ separation at 3.0 GeV/c. Thus a combination of dE/dx and TOF could provide hadron identification over the desired momentum range.

The CLEO II barrel time-of-flight counter[3] is made from a slab of BICRON BC408 scintillator $2.8m \times 10cm \times 5cm$. Light incident on each edge is piped out by 1.2 m long adiabatic light guides to magnetic field free region outside the detector where it is detected by Amperex XP2020 photomultipliers. The timing resolution from each counter as measured in the lab at Harvard University varied between 110 to 120 picoseconds. In our design, we decided to glue two Hamamatsu R2490 photomultipliers directly on both edges of the scintillator. These photomultipliers have been used on the CLEO II endcap counters[4] and have been found to operate inside the 1.5 Tesla magnetic field with gain degradation factors of 100 to 200. The gain degradation can be restored by using fast preamplifiers with high gain and wide bandwidths without loss of timing resolution. In this paper, we present results from tests in zero magnetic field. Since we are doubling the number of time measurements, we expect the resolution to improve by a factor of $1/\sqrt{2}$. Another factor of 2 in light may be gained if particles go through the 10 cm thickness side.

Here we describe the test setup and the first set of results. Based on these results, we also describe future improved designs.

The TOF Test Setup

The test setup is shown in Figure 1. A cosmic ray trigger is formed using a start counter (START) at the top, two reference counters (REF1 and REF2) and two small counters B9 and B10 placed at equal distances above and below the test counter. The hit position along the length of the test counter is limited to a band of 5 cm or 50 cm around its center depending on whether the small counters are or are not part of the trigger, respectively. The TDC is started by an input from the start counter, while its stop inputs are triggered by signals from the reference and test counter PMT's. The total charge or pulse areas are recorded by an ADC. Thus for each PMT, a time and charge pair (t_i, Q_i) is recorded, where $i = 1$ *to* 8 as shown in Figure 1.

Data Analysis

The time measured at the TDC stop input for any of the PMT's consists of contributions from the start counter, from variations in time of emission of the triggering photon and its distance of travel to the photocathode, from the transit time jitter in the PMT and finally from the time walk correction due to varying PMT output pulse area. For a thick counter, there are contributions from the variations in the angle of incidence and the travel time of the particle in the scintillator. The time walk correction is taken out by fitting to the equation shown below:

$$t_i' = t_i - k/\sqrt{Q_i}$$

where t_i and t_i' are the measured and corrected times, Q_i is the pulse area and k is a calibration constant. It can be shown that the contribution from the start counter and that from different hit positions along the test counter length are cancelled out by using the proper choice of time differences such as $(t_5' - t_7')/2$ and $(t_6' - t_8')/2$ or $\{(t_5' + t_6')/4 - (t_7' + t_8')/4\}$. Let σ_{57}, σ_{78} and σ_{5678} denote the rms widths of the above distributions, respectively. Then it can be shown that these are the same as the rms widths for the mean times $(t_5'' + t_7'')/2$, $(t_6'' + t_8'')/2$ and $(t_5'' + t_6'' + t_7'' + t_8'')/4$ for tracks passing through the center of the test counter. The rms widths of the first two distributions are a measure of the timing resolution for a counter with two photomultipliers, while that of the third distribution corresponds to a counter with four photomultipliers. Further, one would expect $\sigma_{57} = \sigma_{68} = \sqrt{2}\ \sigma_{5678}$, since we have increased the number of measurements by a factor of 2.

Preliminary Results

We show the different configurations considered in Figure 2, referred to as Configurations A, B, C and D. Only the first two have been measured, the last two are projected designs to be measured later. Each of the first two configurations have been measured with and without the small counters. The results for these configurations are presented in Table 1. The rms widths of the time difference distributions are quoted.ed. The statistical errors are negligible. The errors quoted are only the systematic one. In configurations A and B, particles traverse a 5 cm and 10 cm thickness respectively.

Table 1. Measurements of Timing Resolution

Configuration	σ_{57}	σ_{68}	σ_{5678}
	ps	ps	ps
Config. A1	120	112	83
Config. A2	122	117	86
Tim. Resol.	$\sigma_{2PMT} =$	118 ± 6	$\sigma_{4PMT} = 85 \pm 2$
Config. B1	92	88	64
Config. B2	95	92	67
Tim. Resol.	$\sigma_{2PMT} =$	92 ± 4	$\sigma_{4PMT} = 66 \pm 2$

Summary and Conclusions

The timing resolutions for the broader beam are only slightly worse than the narrow beam case. Further, σ_{2PMT}, the timing resolution using only two PMT's is about a factor $\sqrt{2}$ larger than σ_{4PMT}, the timing resolution using four PMT's.

In conclusion, we have built and tested a large scintillation counter suitable for hadron identification in the next generation of particle detectors being designed for B meson factories at e^+e^- and $p\bar{p}$ colliders. Timing resolutions of 85 ± 2 and 66 ± 2 picoseconds have already been achieved using two PMT's and four PMT's, respectively. In Figure 2, Configurations C and D, we show two designs where we increase the number of measurements by a factor of 2 over the prototypes tested and hope to achieve an improvement of $1/\sqrt{2}$ in the timing resolutions. Timing resolutions of 46 to 60 picoseconds are expected for these configurations.

References

1 CLEO Collaboration, in *Prospects for a B Factory,*, Cornell University Preprint CLNS 89/962, 1989.

2 H. Castro *et al.*, in *Expression of Interest for a Bottom Collider Detector at the SSC*, Submitted to the SSCL, May 25, 1990

3 R.T. Giles *et al.*, in *Tests of Conventional and Micorchannel Plate Photomultipliers as Components of Scintillating Time-of-Flight System*, Nucl. Inst. Methods Vol. **A252** 41 – 52,(1988).

4 M.S.Alam, *et al.*, in *Final Design of the CLEO II Endcap Time-of-Flight Sytem*, CLEO Internal Report CBX 88-125, 1988

B9
#0
(START)
#1
REF1
#2
#5
#6
#7
TEST
#8
#3
REF2
#4
B10
DISC2 825
DISC3 321A
DISC1 825
TDC 2228A START
Logical Gate 365AL
ADC 2249A GATE
Le Croy 3500
DEC 3500
Le Croy 3921
8″ Floppy Drive

#0 Hamamatsu R2083

#1,2,3,4 Amperex XP2020

#5,6,7,8 Hamamatsu R2490

Figure 1. TOF Test Setup

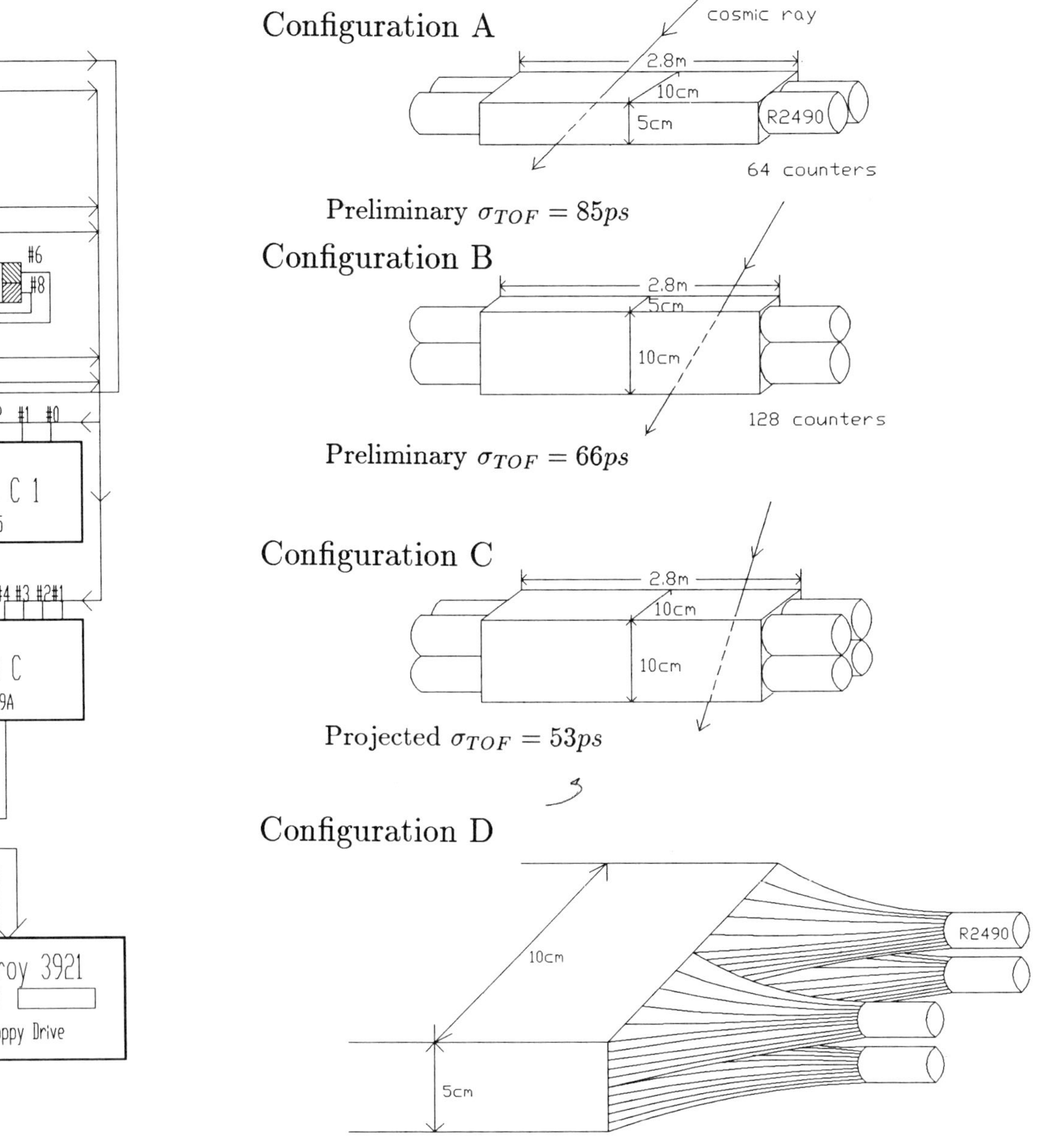

Figure 2. Different Experimental Configurations

THE RPC AS A MUON DETECTOR: THE E771 EXPERIENCE

Francesco Grancagnolo
Istituto Nazionale di Fisica Nucleare, Sezione di Lecce - Italy

Abstract

The parameters of the Resistive Plate Counters (RPC) have been studied at FNAL during the setup phase of the experiment E771. Results on efficiency and time resolution are reported as a function of the incident rate of particle.

Introduction

The RPC [1] is a dc operated gas device working in limited streamer mode. The high field necessary for its operation (40 KV/cm) is generated by a thin layer of graphite (surface resistivity 10^5 Ohm/square) which coats the outside surfaces of two parallel plates made out of a phenolic polymer (2 mm thick bakelite) of high volume resistivity (10^{10} – 10^{12} Ohm×cm).

When an ionizing particle crosses the thin (2 mm) gas gap between the two bakelite plates, the streamer, initiated by the liberated electrons, is quenched by the prompt switching off of the local field around the discharge point due to the high resistivity of the electrodes and by the action of the gas. The duration of the discharge is of the order of 20 nsec whereas the relaxation time of the bakelite electrodes is of the order of the product of the volume resistivity times the dielectric constant, i.e., 10 msec, thus ensuring, during the discharge, that the electrodes behave like insulators and the discharge is confined to a restricted area of the order of 10 mm^2 for a time of the order of 10 msec.

The high resistivity of the electrodes allows also to read out the generated pulse on external pick up electrodes. The spatial resolution of the device is dominated by the strip or pad size when read out in a digital mode.

The most interesting peculiarity of the RPC is due to the reduced size of the gas gap (2 mm) which allows for a very intense and uniform electric field. Under these conditions, the formation of the streamer occurs in a very short time and with minimal fluctuations. The time resolution one gets is of the order of 1 nsec or less and the time delay from the passage of a particle to the signal formation is of the order of 10 nsec making the RPC a suitable device for triggering purposes.

A further advantage of the RPC with respect to other devices is its economicity. The RPC are currently produced at a rate of about 30 m^2/week in standard moduli of 1×2 m^2 at a price of roughly 200 \$/m^2. Also, the cost of the front end electronics is greatly reduced because of the robustness of the read out signals, which are typically of the order of 100 pCoul per pulse, the duration of which is around 15–20 nsec.

The typical gas mixture used is 55% argon, 41% normal butane (for the absorption of the ultraviolet photons), 4% freon 13B1 (for the capture of peripheral electrons around the streamer in order to reduce the transversal size of the avalanche).

Figure 1 shows a sketch of the RPC stratigraphy.

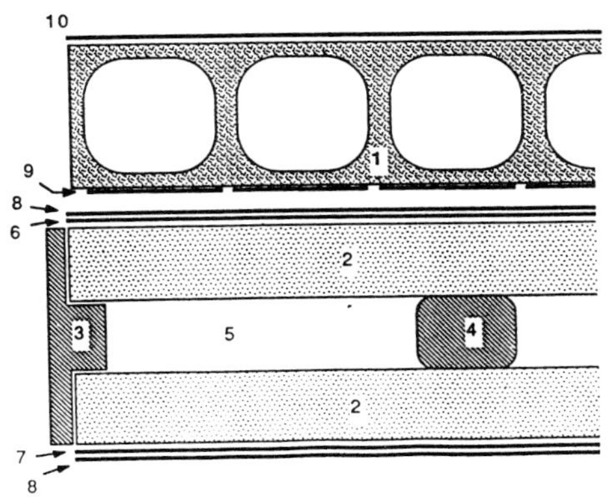

1) Polycarbonate Structure (6mm)
2) Phenolic polymer (bakelite, 2mm)
3) Frame (PVC)
4) Spacer (PVC)
5) Gas gap (2mm)
6) Graphite layer, high voltage
7) Graphite layer, ground
8) Insulator (polyetilene +mylar, 0.5mm)
9) Pads (copper)
10) Ground (copper)

Figure 1. Sketch of RPC stratigraphy.

Cosmic ray test results

Figure 2 shows the efficiency plateau for a 12×12 cm^2 pad RPC measured with a scintillator telescope and cosmic rays. Full efficiency is not reached because of misalignment and of the dead area around the spacers. To be noticed are the steep gain in efficiency, despite the relatively high threshold, and the stability of operation (length of the plateau well over 1500 V).

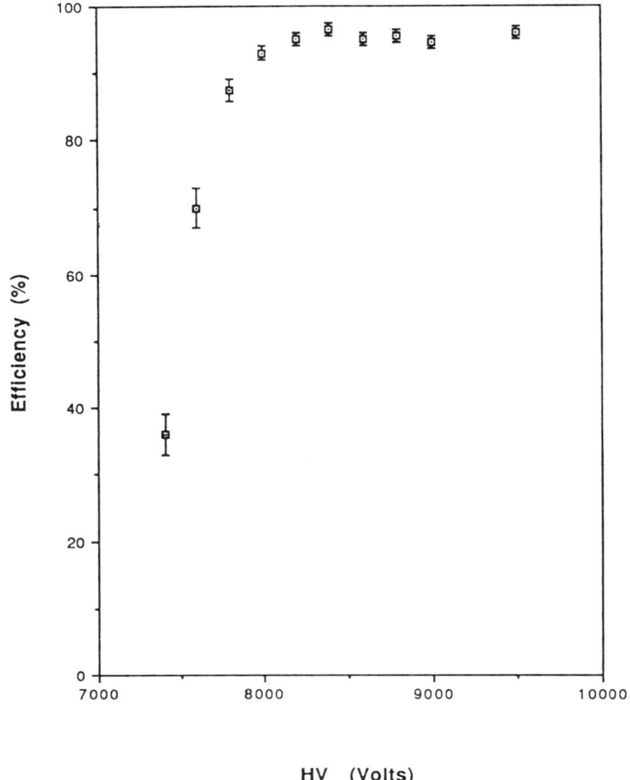

Figure 2. Efficiency plateau. Threshold = 90 mV.

Figures 3 and 4 represent, respectively, the distribution of the charge collected on a 12×12 cm^2 pad and the corresponding time distribution. The very little noise at low values of charge is associated with the long tail in the time distribution and can therefore be easily removed. The width of the charge distribution is typical of a streamer process (FWHM of about 40%). The time distribution is convoluted with the jitter of the scintillator telescope. By unfolding such a contribution one gets time resolutions whose behaviour with the high voltage is shown in Figure 5.

The behaviour of the RPC versus the gas composition, in particular versus the ratio of butane over argon and with different amount of freon, has been carefully studied as far as efficiency and resolution are concerned [2]. Detailed results will be reported in a forthcoming paper.

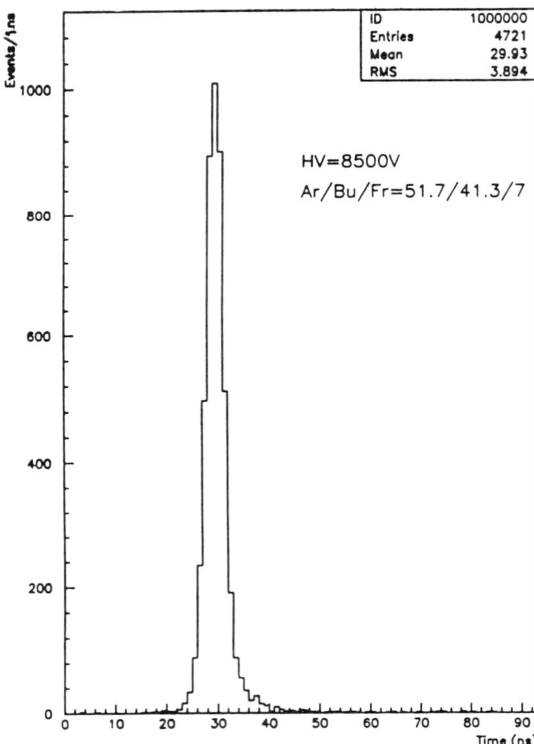

Figure 3. Charge distribution. 12×12 cm^2 pad.

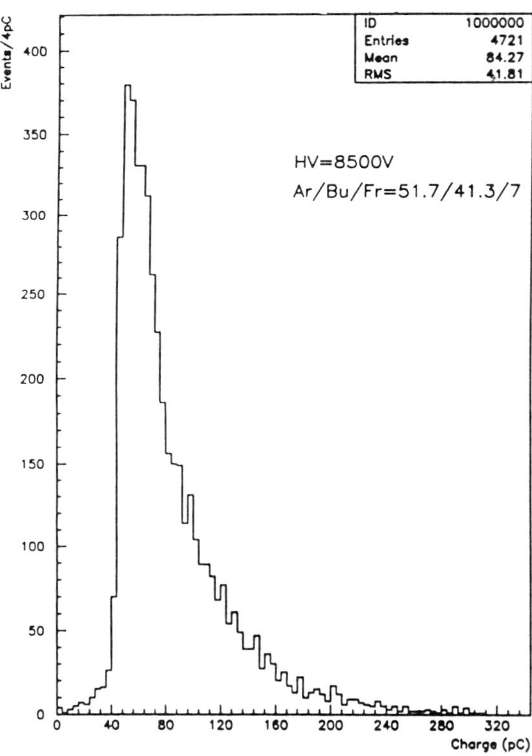

Figure 4. Time distribution. 12×12 cm^2 pad.

Beam test results

The RPC are used as the active element in the muon detector of the experiment E771 at Fermilab [3]. During the setup phase of the experiment, last summer, the entire system of RPC (29 for a total of about 54 m^2) has been exposed to a proton beam of variable intensity. Results on efficiency and time resolution versus rate had already been reported [4] for a test beam at CERN. Figure 6 summarizes such results. In the hottest region, around the proton beam, and in the first of the three planes of RPC, even at the highest interaction rate (10^7 interaction/sec) at which the experiment E771 will be exposed, the rate of particles per unit area will not be larger than 50 Hz/cm^2. At this rate and with the CERN setup (30 m long 50 Ohm cable before the discrimination of the signal at a threshold of 100 mV), the degradation of the efficiency is of the order of 2%.

Analogous results have been obtained at FNAL. Some recovery of the efficiency at high rate can be attempted by lowering the discriminator threshold and by acting simultaneously on the gas mixture and on the high voltage.

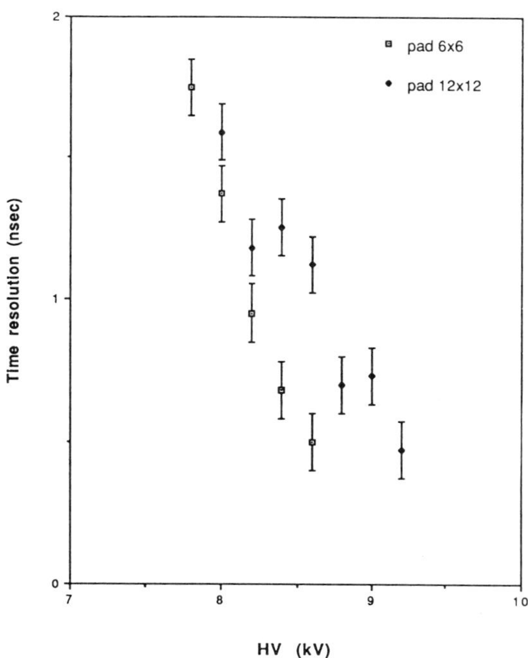

Figure 5. Time resolution vs. high voltage.

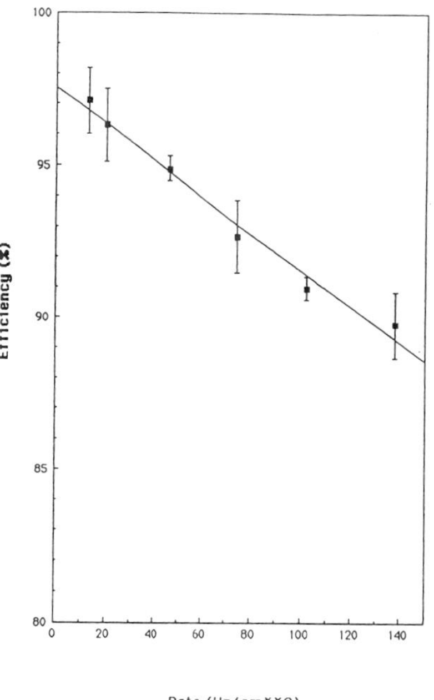

Figure 6. Degradation of efficiency with rate.

References

[1] R. Santonico and R.Cardarelli, NIM **187**(1981) 377
R. Cardarelli et al., NIM **A263** (1988) 20.

[2] F. Grancagnolo, "The RPC System of the Experiment E771, Cosmic Ray Test Results." Presented at the 4th Topical Seminar on Experimental Apparatus for High Energy Particle Physics and Astrophysics, San Miniato, Tuscany, Italy, 28 May–1 June, 1990.
E.Gorini, "The RPC Muon System of E771," presented at the 2nd International Conference on Advanced Technology and Particle Physics, Como, Italy, 11–15, June, 1990.

[3] B. Cox et al., "Experiment E771 : An Experiment to Study Beauty Production and Other Heavy Quark Physics Associated with Dimuon Production in 800 (925) GeV/c pp Interactions," Fermilab, P771, May 1986.
F. Grancagnolo, "Studies of B-Physics with FNAL E771", in *Proceedings of the XXV Rencontre de Moriond, "High Energy Hadronic Interactions"*, Les Arcs, Savoie, France, March 11–17, 1990 and INFN-AE/90-07, 10 luglio 1990.

[4] M.Bertino et al., NIM **A283** (1989) 654.

A FAST RING-IMAGING CERENKOV COUNTER FOR THE BOTTOM COLLIDER DETECTOR (BCD)

Jorge G. Morfín
Fermi National Accelerator Laboratory
Batavia, IL 60510

Abstract

The general requirements of a RICH counter for BCD at the SSC are summarized and a brief description of the recent work [1] of **B. Hoeneisen, D.F. Anderson,** and **S. Kwan** which has led to the development of a solid CsI-TMAE photocathode is briefly described.

1. Introduction

To **study,** as opposed to simply detect, CP-violation within the B system calls for the measurement of the CP-violating amplitude in as many decay channels as possible. This, in turn, requires **hadron identification** to distinguish pions from kaons from protons.

Earlier in this session we heard about the development of a 50 ps time-of-flight system, also for BCD, that would enable the distinction of pions from kaons out to about 2 GeV/c. However, even in the central region (30° <--> 90°) at the SSC, about 60% of the 2-body decay products have momentum above this value. The only way to efficiently identify high energy hadrons in both the central and forward regions is through the use of Ring Imaging Cerenkov techniques. The viability of the Ring-Imaging Cerenkov (RICH) technique, which was proposed by Ypsilantis and Sequinot[2], has been established by a number of groups[3,4] who have successfully built and tested devices for use in high-energy physics experiments. The basic principle of the RICH technique is based on the following expression:

$$\sin^2\theta_c = \frac{1}{\beta^2}\left[\frac{1}{\gamma_t^2} - \frac{1}{\gamma^2}\right] \quad \text{with} \quad \gamma_t^2 = \frac{\eta^2}{\eta^2 - 1}$$

Thus, a passing particle emits a cone of photons yielding a "ring" of hits on a detector with radius proportional to mass for a given momentum.

In general, when designing a RICH counter, the following factors have to be considered:

1 Length of radiator and radiator material
2 Desired momentum threshold
3 Desired Photoionizing medium
4 Speed of data acquisition
5 Expected multiplicities
6 Desired maximum momentum of m_1/m_2 discrimination.

The highest momentum covered by the time-of-flight system dictates the most efficient minimum value of the threshold (γ_t) of the RICH radiator for overall hadron identification. In the case of the radiator for the central detector, a value of γ_t = 5 would be ideal for π/K/P separation. Unfortunately, there are currently no radiator media with this value since normal gases are above and normal liquids fall below this value as shown in the following table.

Photocathode	Radiator	γ_t	$\sigma_{\theta c}$(μr)	($\sigma_{\theta c}/\theta_c$)
GAS				
TMAE	He	123	18	2.24 x10^{-3}
@ 1 bar	Ne	90	26	2.36
40°c	Ar	42	151	6.27
	Kr	33	274	9.00
	Xe	24	941	22.2
	N_2	41	153	6.18
	CF_4	32	60	1.94

Photocathode	Radiator	γ_t	$\sigma_{\theta c}(\mu r)$	$\sigma_{\theta c}/\theta_c$
	C_5F_{12}	17	66	2.84 x10^{-3}
	C2 F6	25	115	2.91
TEA	He	118	2.4	2.85
	He 25 bar	2.4	121	2.85
@ 1 bar	Ne	85	37	3.16
20°c	CF_4	31	61	1.88
	C_2F_6	24	116	2.79
	C_5F_{12}	17	161	2.71
Liquids				
	CF_4	1.71		
	C_5F_{12}	1.66		
	C_4F_{10}	1.63		
	C_6F_{14}	1.62	4.1 mr	
Solid				
	NaF	1.5		
	CaF_2	1.4		

A possibility is the use of aerogels which have already been used for this purpose at CERN. We intend to study this possibility in detail.

The central problem to be faced with a RICH counter for use at the SSC is obtaining stable operation in the anticipated high-rate high-multiplicity environment. The RICH detector must be sensitive to the single-photoelectron pulses from Cerenkov light, but insensitive to the ionization trails of through-going charged particles. Furthermore, it must be possible to extract signals using a narrow timing gate so as to minimize confusion from out-of-time events. In most devices built thus far, an intrinsic jitter in photoelectron detection time arises because the absorption length of the photosensitive gas is long enough that the photoelectrons are produced at varying distances from the amplification region and therefore arrive at varying times. For example, the absorption length of TMAE at 30 degrees is about 15 mm, resulting in a 200-300 ns variation in pulse arrival time, depending on the carrier gas. Since this is several SSC bunch-crossing times, a faster solution must be developed. The recent development of a solid photocathode detector, described in the following, offers an innovative solution to this problem.

2. Experimental Setup

The experimental detector of ultraviolet photons is shown schematically in Figure 1.

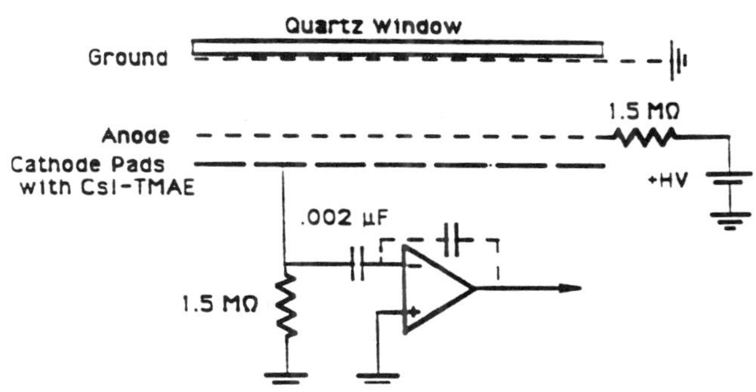

The photocathode metal pads (on a G10 printed circuit board) are sensitized by an evaporated CsI layer with adsorbed TMAE. The anode is either a mesh (of stainless steel wires 12.7 μm in diameter with 254 μm spacing, 90% transparent), or a wire plane (gold plated tungsten of 20μm diameter and 1 mm pitch). The gas in the chamber is typically at a pressure of 5 to 60 Torr.

The ultraviolet light source used is a hydrogen arc discharge lamp operating in a relaxation mode. The light source has a pickup which permits triggering on the light pulse. This light pulse has a full width at half maximum less than 3 ns. An ultraviolet photon traverses the quartz window, the ground wire mesh and the anode wire mesh, and releases a single electron on the surface of the solid photo-cathode. This electron drifts towards the anode which is connected to a positive voltage source. The applied electric field is sufficiently high that avalanche multiplication occurs.

3. Gases

Several gases transparent (or semi-transparent) down to the quartz cutoff wavelength have been tried. Sufficient gain (> 10^5) to observe single photo-electrons above the amplifier noise is obtained with the following gases (or mixtures with these gases): methane, ethane, pentane, hexane, ethanol, methanol, isopropyl alcohol, acetone and TEA (isopropyl alcohol shows significant ultraviolet absorption). Insufficient gain is obtained with the following pure gases: helium, argon, nitrogen and isobutane. Figure 2a shows the maximum gain as a function of pressure for methane with several cathode-anode spacings. There seems to be a yet-to-be-understood enhancement in the gain for a cathode-anode spacing of 1.59 mm. Figure 2b displays the gain for several gases for a constant 1.59 mm cathode-anode spacing. Note also that the highest gains ($\approx 10^7$) correspond to the relatively slow gases.

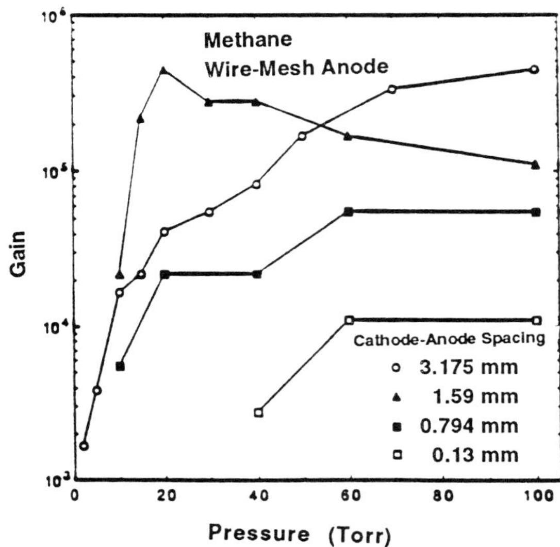

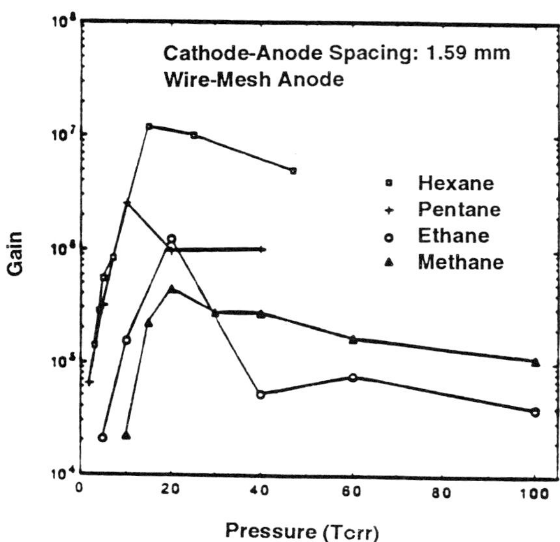

4. Signal Delay and Shape

The current pulse on a pad has a fast component due to the drift of the electrons and a slow component due to the positive ions. The fast component, observed directly on an oscilloscope with 50 Ω input impedance, is shown in Figure 3 for isopropyl alcohol at 7 Torr, an anode voltage HV = 410 V, and a cathode-anode spacing of 0.794 mm. This "fast" component has a **delay with respect to the light pulse of 4.5 ns, and a full width at half maximum of 2.5 ns** thus making it ideal for the expected timing of SSC events.

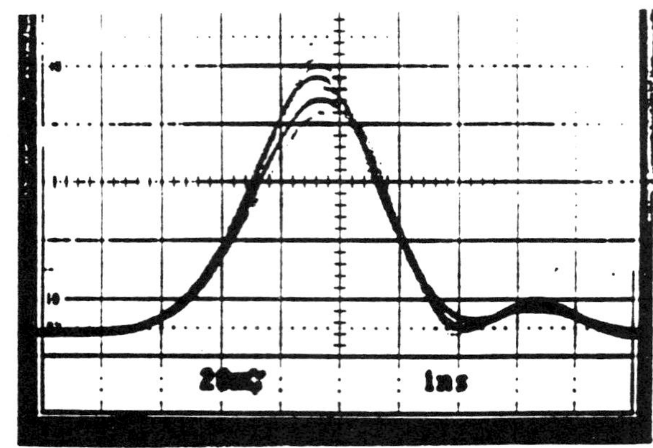

5. Photocathode Quantum Efficiency

Since the paper of reference 1 was submitted for publication, further research has revealed a method by which the quantum efficiency of the solid CsI-TMAE photocathode can be held at a **constant 45%.** A second publication describing this breakthrough is currently in preparation and will be submitted to NIM-A shortly.

SUMMARY

1. **A solid CsI - TMAE photocathode coupled to a low-pressure wire chamber has been developed by B. Hoeneisen, D.F. Anderson and S. Kwan which is very promising for SSC applications.**
2. **The wire chamber can achieve gains as high as 10^7, has 100% detection efficiency for single photoelectrons, and can produce a signal, ≈ 5 ns after the photon's arrival, which is only a few ns wide.**
3. **Single-step amplification simplifies construction of the photodetector and reduces the total thickness of the detector to ≈ 1 cm.**

[1] B. Hoeneisen, D.F. Anderson and S. Kwan, *A CsI-TMAE Photocathode with Low-Pressure Readout for RICH*, FERMILAB-PUB-90/182.

[2] J. Sequinot and T. Ypsilantis, Nucl. Instr. and Meth. 142, 377 (1977).

[3] M. Adams et al. Nucl. Instr. and Meth. 217, 237 (1983). R.J. Apsimon et al., IEEE NS 32, 674 (1985).

[4] D. Leith, Nucl. Instr. and Meth. A265, 120 (1988).

ADVANCED FIELD SHAPING DRIFT CHAMBERS FOR SSC FORWARD TRACKING

A. F. Sill
Department of Physics and Astronomy
University of Rochester
Rochester, NY 14627-0011

Abstract

Drift cell design and construction methods are described which can be used to form high-precision gas pixel drift chambers that have a linear voltage-to-distance relationship over large portions each of the drift cells. Use of resistive sheet film plastics, and alteration to form nonuniform patterns of resistance (and consequently uniform electric fields) is discussed. Examples are given with respect to triggering and track reconstruction in the intermediate angle region at the supercollider and other high rate experiments.

Introduction

Intermediate angle tracking, indeed tracking of any kind within the interior of a typical SSC detector is extremely challenging due to many considerations of rate, occupancy and instrumentation cost. Any device that has a large active area per instrumented channel will have problems due to the high rates; on the other hand, the present ideas for densely instrumented devices would appear to have associated with them very high costs.

In central rapidities, η less than 1.5 or so, possible solutions exist that involve tradeoffs between the densely packed cells of silicon strip and drift devices at close distances and the more cost-effective larger cells provided by straw tubes at larger values of distance from the interaction point. In the forward (or so-called "intermediate" angle) region, with rapidity between 1.5 and 2.5 (say), the solutions would appear to be a bit less obvious. Radial wire drift chambers[1] which incorporate densely instrumented cathodes or "crossing taggers" have been mentioned as a possible solution.

We would like to suggest an option based on two-dimensional gas-filled pixel devices with active field shaping, bonded on mechanical components that are stress-skin panels formed from a light core with a cladding of resistive or conductively-printed or etched metal-coated sheet plastic. Such devices extend from previous work[2,3] done to produce relatively conventional drift cells from similar materials and with similar methods.

The detectors we propose can be made entirely from commercially available materials to allow a construction that is light, strong, and economical to assemble. The goal is to combine the relatively cheap drift of gas-based technology with the dense channel count for readout provided by recent developments in silicon and straw-tube-based electronics. The result would be pixel gas drift devices with cells of nominal size on the order of millimeters to centimeters, depending on rates. It would be easy to arrange these in layers to provide projective towers of cells, giving a natural method to produce online triggering and track recognition in a high rate environment.

Basic Patterns

The simplest possible arrangement for the unit cell of such a device is shown for three possible cases in Figure 1. These show conductive lines which are assumed to be drawn on a resistive substrate such as those discussed in reference 3. The most basic would be just a central spot at ground potential, through which a sense wire is extended (held by an insulative feed-through), and surrounded by a conductive trace which is placed at some side field shaping potential. The shape of the cell can be either square or hexagonal in basic outline, though these borders need not be precisely rectilinear, and in fact would be altered to grow in outline as a function of distance from the beam line in order to maintain projective towers if that were desired.

The anode can be either a fairly conventional sense wire or a sharp point, as discussed below. The cell is closed on top with another similar resistive layer separated by a few millimeters and so form a gas drift pixel, here viewed from the top directly above the central anode.

Sense Wire Anodes and Tips

Until recently it was not possible to obtain a large quantity of sharply etched metal tips; however, recent development of the scanning tunneling microscope (STM) into a low-cost commercially available device has introduced a consequent need for suppliers of the sharp etched-wire metal tips that are used in such devices. Suppliers[5] for that industry now can provide quantities of several hundred tips, all generally sharp to better than 0.5 microns, out of materials such as tungsten and platinum-iridium as well as others), and of length up to a couple of centimeters. Wire diameters can be specified, and the rapidness of the taper of the wire to the tip can also be set by the user. Typical cost is on the order of a few US dollars per tip.

A side view of the gas pixels of Figure 1, along with the equipotentials these would produce, is shown in Figure 2. The case of a sense wire anode of radius 20 μm is shown, and also in the case of such a very sharp tip.

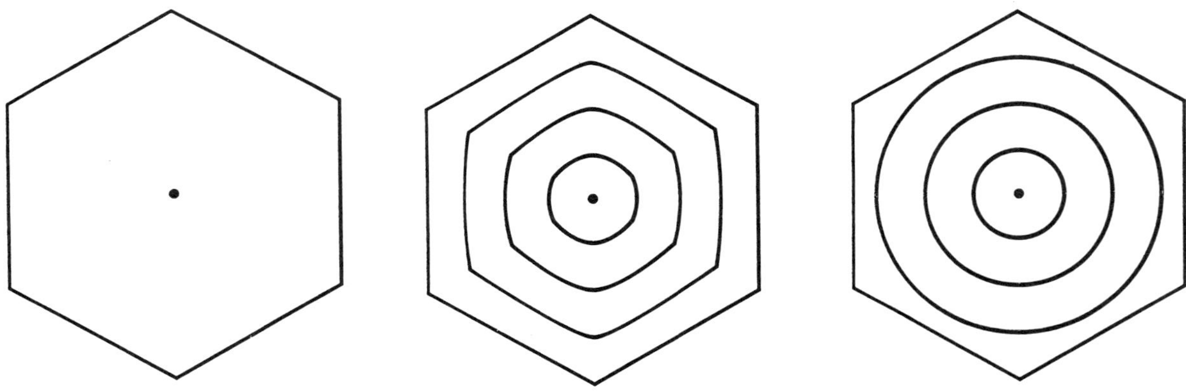

FIG. 1. Several possibilities for the basic layout of conductive patterns for the graded potential gas pixel drift chamber cells discussed in the text.

Resistive Field Gradient Alteration

Devices that are built for triggering should have an intrinsically linear time-to-distance relation for drift time as a function of distance from the anode, or at the least one that can be linearized with ease early in the electronic processing of the anode pulses. Because of simple geometrical effects this is not true for the simplest possible cell, and it would be helpful to consider the possibility of altering the resistance as a function of this distance (i.e. the radius from the sense wire). Of course the simplest method would be to produce conductive patterns of traces on an insulator, then tie them together electrically with a thin resistive strip running from the outside to the inside of the cell, passing through and connected to the intermediate traces as well.

More sophisticated methods of linearizing the voltage-vs.-radius relation would be to print or silkscreen a half-tone or random-dot pattern of very fine conductive dots onto the resistive surface, such that the resistance between two hypothetical traces concentric to the central dot and separated by a constant dr would stay the same regardless of their overall radius.

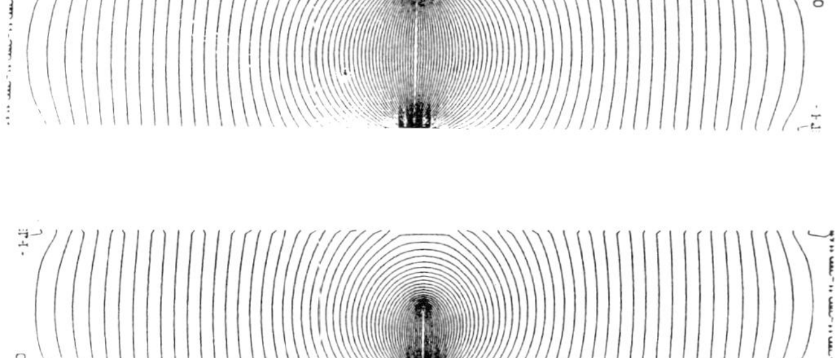

FIG. 2. Equipotentials for the gas pixel drift tiles shown in the previous figure. The plots show lines of constant voltage for the sense-wire and sharp-point versions of these cells as viewed from their sides.

This halftone screening method is shown in the first of the plots in Figure 3. The next two show an exaggerated view of a method which involves the printing or etching of conductive traces which grow in radial width (there would be many more in actual practice than shown here) as the radius becomes smaller.

The resistive plastic would have to be chosen to allow for fast dissipation of excess accumulated charge from within the chamber at high rates. As discussed in reference [3], this would imply a surface resistivity as low as possible without running into problems due to power deposited into the gas by resistive heating within the drift cell. Unlike the situation in long narrow cells, pixels have the advantage that the "squares" of resistance are mostly in parallel and so surface resistivities approaching the antistatic range can potentially be reached. The performance of such chambers in terms of ion clearing with respect to rate dependence will have to tested. We have begun a program to construct such prototypes to check this and other effects.

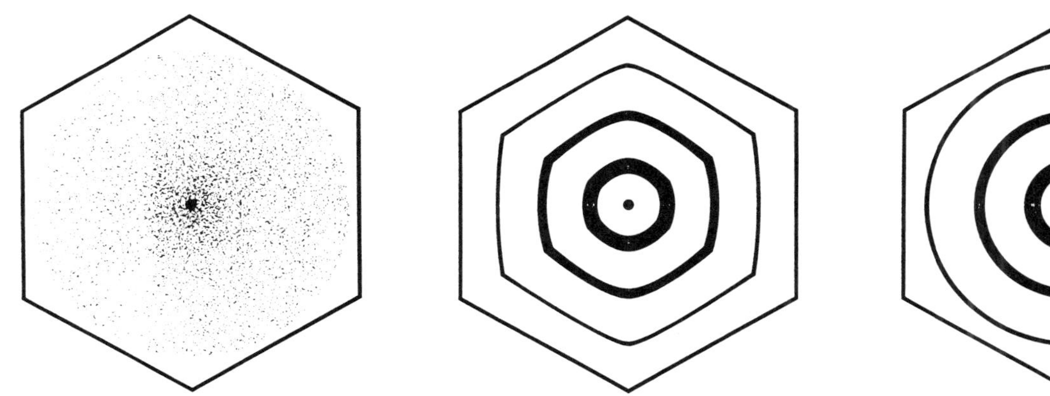

FIG. 3. Methods of providing a uniform drift field by altering the pattern of surface resistance to be proportional to the radius from the sense anode within the graded potential drift tile cell.

Application to SSC Forward Tracking

The greatest application of such a device would be to the online recognition of stiff projective tracks in a simple way. The trigger could even measure and cut on momentum if the alignment between adjacent layers of cells was done precisely enough to minimise the effects of misalignment during assembly. This would in turn imply a very stiff and easily aligned mechanical support structure and method. Possibilities for mechanical support include hexcell, graphite-reinforced composites, and polyimide foams. These should be selected to provide the maximum possible stability and accuracy in placement of the sense anodes.

Acknowledgement

I am extremely grateful to Dr. Robert Walker of Rochester and KEK, and to Dr. P. N. Kirk of Louisiana State University for enlightening conversations about drift chambers and other topics throughout the work described in this paper and reference [3].

Note Added

After presentation of this talk at the conference, my attention was called to recent work which appears to be along related lines being done at CERN with respect to the LAA project. Please see reference [6].

References

1. D. Saxon, this proceedings.
2. A. Sill, Proceedings of the Meeting of the Division of Particles and Fields of the American Physical Society, Rice University, Houston, TX, Jan. 5-9, 1990.
3. A.F. Sill, "Advanced Field Shaping Drift Chambers for SSC Muon Tracking," this proceedings.
4. XC series resistive Kapton^{TM} polyimide film, DuPont Corporation, Wilmington DE 19880.
5. Many such STM and biology support companies now exist; for instance, Longreach Scientific Resources, RFD 1, Box 549, Orr's Island, Maine 04066.
6. D. Mattern, M.C.S. Williams, A. Zichichi, CERN-EF-90-4, Jun. 1990, submitted to Nucl. Instrum. Methods

Radiation Damage

10

RADIATION EFFECTS ON ORGANIC SCINTILLATORS: STUDIES OF COLOR CENTER ANNEALING

R. L. Clough and J. S. Wallace
Sandia National Laboratories
Albuquerque, New Mexico

Abstract

The interaction of ionizing radiation with organic, polymer-based scintillators results in the discoloration of the material, which reduces the scintillation efficiency. This phenomenon can be described in terms of two types of color centers: annealable and nonannealable. For many common materials, at doses of a few Mrad, the annealable color centers account for a large fraction of the observed radiation-induced absorption. We are conducting studies to understand the mechanisms and kinetics of the annealing process. By conducting experiments on polystyrene in which polymerization of the monomer, γ-irradiation, and post-irradiation annealing were all carried out in a sealed, degassed tube, we have been able to resolve the question of whether annealing can occur in the absence of oxygen. We find that annealing does indeed occur under inert conditions. The rate of inert annealing at room temperature is drastically slower than oxygen-mediated annealing, whereas the rates for the two processes is only slightly different at 80°C. We have measured the annealing rate of color centers in γ-irradiated polystyrene in air over the temperature range of 30 to 90°C, and find a large increase in annealing rate with increasing temperature. Arrhenius plots of the annealing data show strong non-linearity, which can be ascribed to two different annealing mechanisms of very different activation energy: presumably oxidative and nonoxidative. Ongoing work involves deconvolution of the two annealing processes, and construction of a model for long-term annealing kinetics in these materials.

Introduction

Scope of Sandia Program

We have initiated a program to address the radiation stability of optical polymers and luminescent dyes used in scintillator fabrication. Primary goals include: identification of radiation-degradation mechanisms in scintillators, development of radiation-resistant systems, implementation of effective annealing schemes for scintillator recovery, and demonstration of predictive aging techniques for evaluating material lifetimes under low-level radiation conditions. Work carried out to date includes: 1) A survey of the radiation-induced color center formation in 21 different optical polymers, 2) an evaluation of radiation effects on the major classes of scintillator dyes, both in terms of their decomposition and their discoloration in a polystyrene matrix, 3) evaluation and synthesis of new red-shifted scintillator dyes, and 4) a detailed investigation of temperature and oxygen effects in the annealing of color centers in irradiated polystyrene. This report covers the results and conclusions to date on the annealing studies in polystyrene (number 4 above). Results on the other topics all involve extensive tables and discussion, and will be published elsewhere.

Major Issues in Color Center Annealing

There have been conflicting reports circulating on whether oxygen plays a beneficial or detrimental role in the radiation-induced changes in polymer-based scintillators [1]. The answer to this question can be expected to be complex, and will likely be affected by the particular polymers, dyes, environmental conditions and timescales involved. This issue is fundamental to the choice of the gaseous environment which will surround the scintillator media in future detectors, and is also crucial in the design of rad-hard scintillator materials. Another important materials issue is the ability to model the kinetics of color center annealing, for the purpose of predicting long-term effects of radiation exposure. This implies understanding the interrelationship of time and temperature in the annealing phenomenon. To address these primary annealing issues, we have conducted post-irradiation annealing experiments over a range of temperatures in vacuum and under air.

Results and Discussion

Annealing in the Absence of Oxygen

The concentration of annealable color centers which gives rise to the radiation-induced absorption is unknown; they may be very small in number. If so, then a tiny amount of oxygen contamination could potentially cause significant annealing. Thus, an "inert atmosphere" study which had even minimal oxygen exposure, and which revealed annealing at a rate slower than that found in air, would be open to the question: would the results actually represent oxygen annealing at a partial pressure lower than that of air? Because of this concern, we carried out experiments in which styrene monomer was subjected to repeated freeze-thaw degassing, and the glass tube in which it was contained was sealed under vacuum using a torch. This allowed for efficient and rapid deoxygenation, compared to what would be possible using polystyrene polymer, due to the extremely slow rate for gaseous permeation in the solid matrix [2]. All of the following operations were carried out on the sealed-tube sample: polymerization of the styrene monomer, γ-irradiation, and postirradiation annealing. Each absorption spectrum run was obtained using a separate sample which was open to the air immediately prior to the measurement.

Figure 1 presents data for the time-dependent change in absorbance, at 400 nm, for γ-irradiated polystyrene samples (15 Mrad dose under vacuum) which were subsequently annealed near room temperature under inert and air atmospheres. The samples held under vacuum do exhibit a slow, steady decrease in absorbance. This amounts to less than 10% after 40 days, and occurs more than an order of magnitude slower than annealing in the presence of air. Figure 2 presents data for annealing at 80°C under vacuum and air atmospheres. Substantial annealing is seen under both atmospheres, with the vacuum annealing occurring only moderately slower. In all cases, the annealing asymptotically approaches a lower residual value, corresponding to color centers which are permanent, and/or which at least have much longer annealing times.

From these data, it is clear that color center annealing does occur in the absence of oxygen. The difference in oxygen-mediated annealing and non-oxygen-mediated annealing is further apparent upon visual examination of samples. In the case of oxygen annealing, a distinct bleaching front moves through the samples with time, corresponding to the permeation of oxygen through the material. Thus, in a partially annealed sample, distinct colored regions remain in the interior of the specimen, while the edge regions appear nearly colorless. In the case of inert-atmosphere annealing, the sample coloration appears homogeneous throughout the annealing process.

It is likely that the annealable color centers involve reactive species (such as radicals, radical ions, trapped electrons, or similar) which are immobilized in the glassy polymer matrix. (A strong ESR signal is in fact observed for polystyrene samples subjected to γ-irradiation, and we are using the ESR technique to study the annealing phenomenon. This will not be discussed further in this report). The annealing mechanism in air must correspond to reaction with molecular oxygen to yield colorless (or less colored) products. Rapid reaction of oxygen with all of the species listed above would be expected. The annealing mechanism under inert atmosphere must correspond to mobility of the reactive species themselves within the matrix, leading to recombination reactions with other reactive species (i.e., radical-radical recombination; radical cation-radical anion recombination). Both the diffusion of oxygen and the diffusion of reactive species (including reactive sites on the macromolecules) can be expected to increase with increasing temperature, as is seen with the data. Thus, as the polystyrene is heated toward its glass transition temperature (~105°C), the mobility of species within the matrix is greatly increased. Under these conditions of low matrix rigidity, the annealable color centers have very short lifetimes.

An implication of these data is that elevated temperature may be useful for eliminating annealable color centers. Since oxygen may have an undesirable degrading effect on polymers [3] (possibly leading to higher concentrations of permanent color centers) and also on dyes (leading to oxidative degradation) [4], it may be desirable to utilize inert annealing processes. Clearly, even at room temperature, color centers diminish with time. Further kinetic studies should lead to predictions of decay time of the color centers under a given set of operating conditions (temperature, atmosphere, etc.), and should allow estimation of steady-state concentrations of annealable color centers under particular long-term operating conditions (dose rate, temperature) as irradiation is in progress.

Temperature-Dependence Plots

Figure 3 presents time-dependent decay curves for the radiation-induced absorbance of polystyrene at 400 nm, obtained in air atmosphere at temperatures ranging from 30 to 90°C. Progressively more rapid decay is seen at higher experimental temperatures. Figure 4 presents an Arrhenius plot for the data of Fig. 3, showing the time for recovery to three different extents of annealing as a function of reciprocal temperature. The three curves follow very much

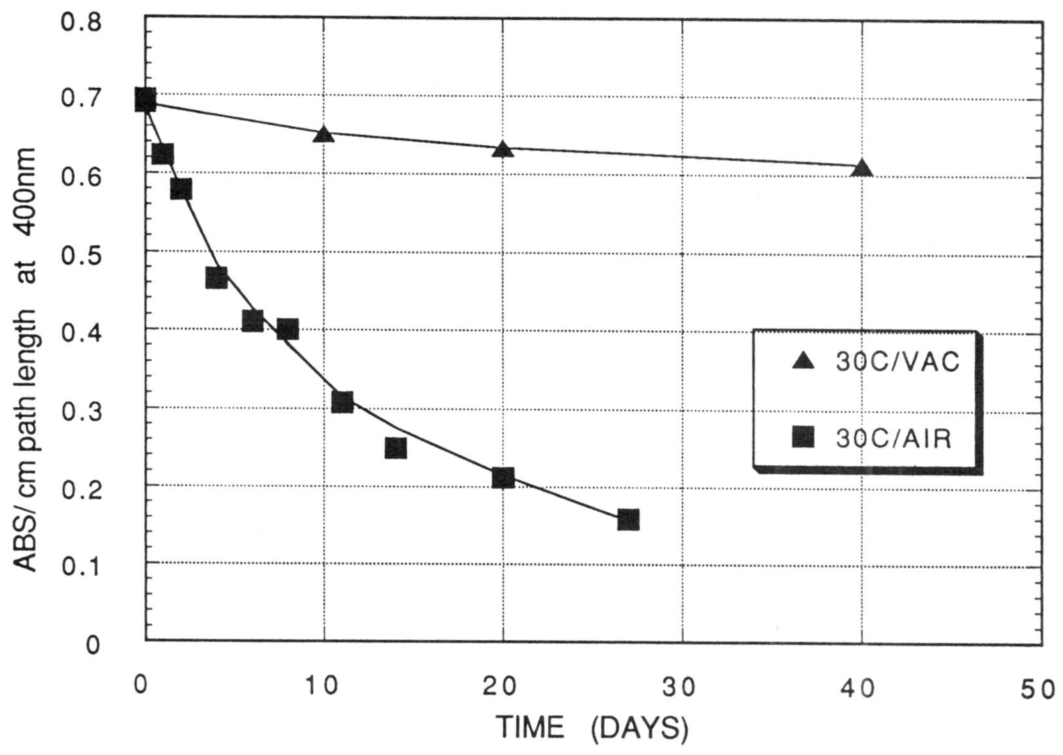

Figure 1. Air and vacuum annealing at 30° C of color centers in polystyrene after γ-irradiation in vacuum to 15 Mrad. Samples were polished disks 0.25 in (0.635 cm) thick and 0.612 in (1.55 cm) in diameter.

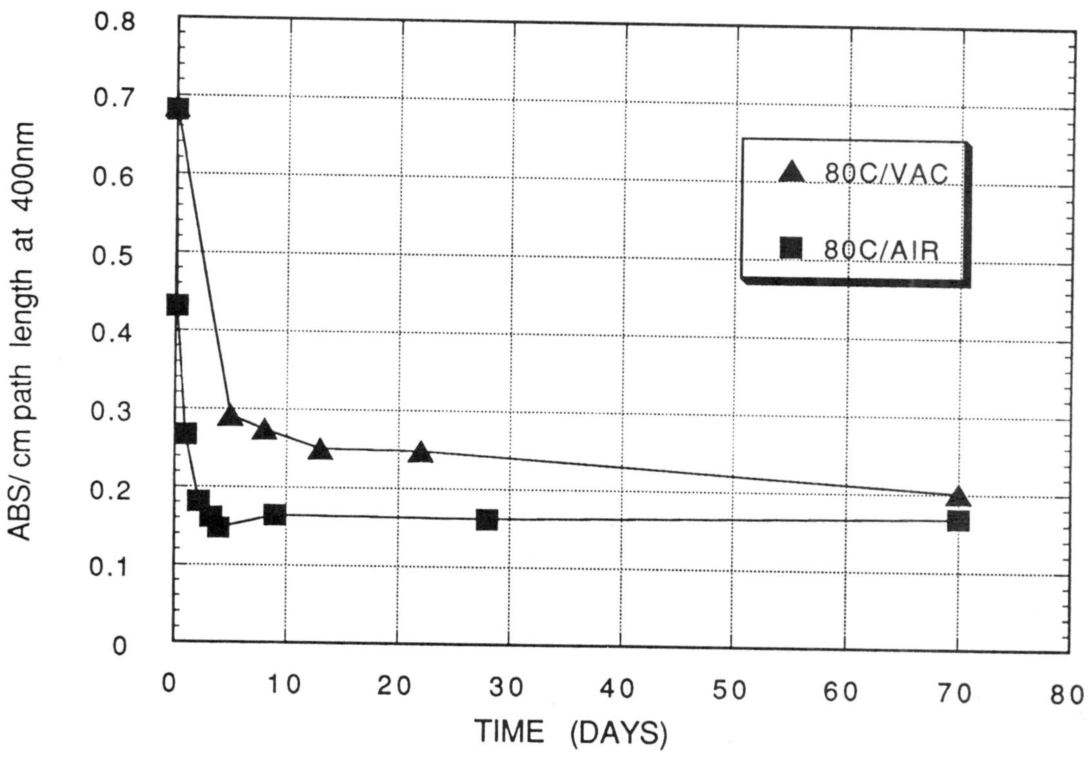

Figure 2. Air and vacuum annealing of radiation-induced color centers in polystyrene at 30° C. Irradiation and sample size as in Figure 1.

the same trend, and exhibit a strong non-linearity. Linearity in such a plot would be characteristic of a process having a single rate-determining step [5]. The curvature observed indicates the occurrence of two or more rate-limiting processes having substantially different activation energies. Such behavior would result from the competition of the two annealing mechanisms described earlier (oxygen-diffusion annealing, and inert-atmosphere recombination annealing), if these two process made proportionately different contributions at different temperatures. Comparison of the data in Figs. 1 and 2 would indicate that this is so. We are presently carrying out annealing experiments under both vacuum and air at a number of additional temperatures, from which we will construct Arrhenius plots for both annealing process. Deconvolution of the the two annealing processes will be carried out with the goal of utilizing the Arrhenius approach for modeling time-dependent annealing under both atmospheres. In addition, extensive experiments are currently being initiated to irradiate samples under various combinations of temperatures and low dose rates for construction of further long-term aging models utilizing the time-temperature-dose rate superposition approach which has been applied previously to the radiation-degradation of mechanical properties of polymers [5,6].

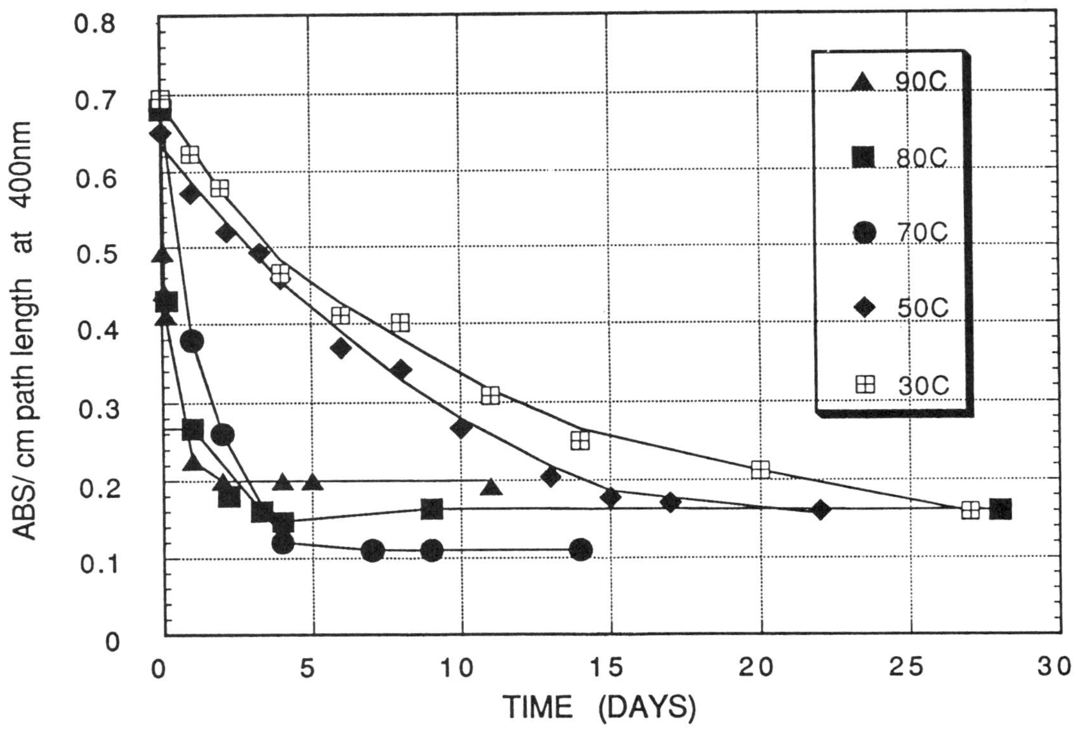

Figure 3. Temperature dependence of air annealed color centers. Irradiation and sample size as in Figure 1.

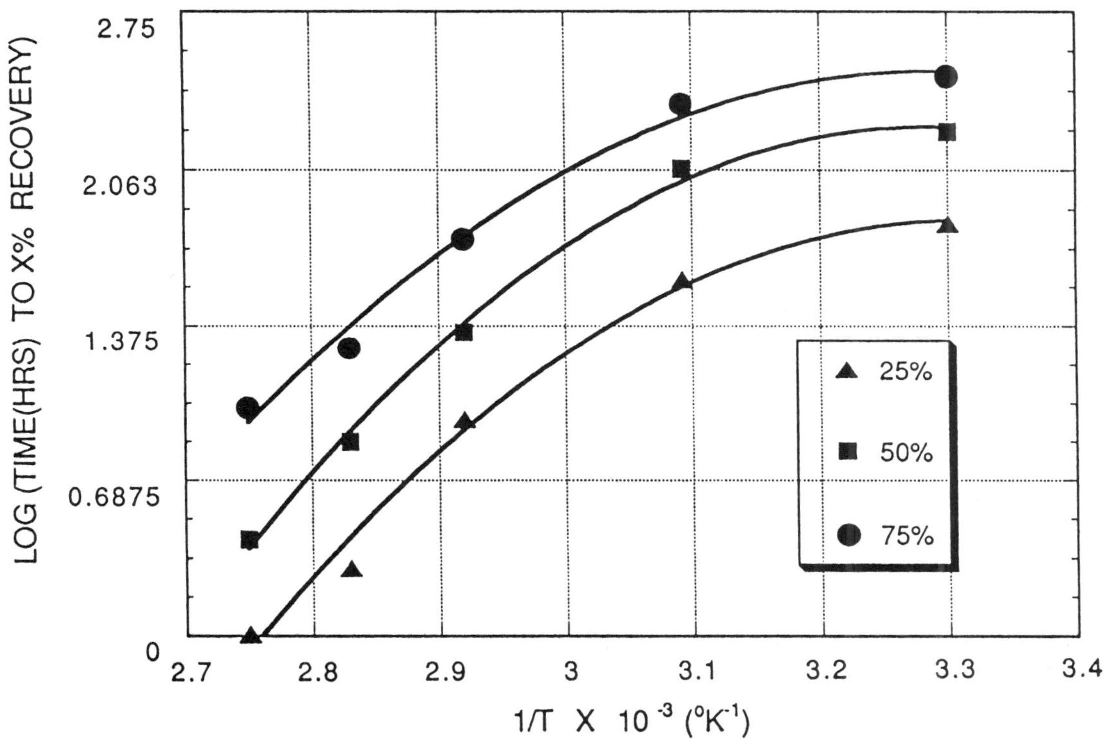

Figure 4. Arrhenius plot of the decrease in the initial post-irradiation concentration of color centers as a function of 1/T. Concentration decreases shown are 25%, 50%, and 75%.

References

[1] H. Schonbacher, "Review of Radiation Damage Studies on Scintillating Materials Used in High Energy Physics Experiments," in Radiation Effects at the SSC, M. G. D. Gilchriese, ed., SSC-SR-1035 (June, 1988), p. 215.

[2] K. T. Gillen and R. L. Clough, "Techniques for Monitoring Heterogeneous Oxidation of Polymers," in Handbook of Polymer Science and Technology, Vol. 2, N.P. Cheremisinoff, ed., Marcell Dekker, New York (1989) p. 167.

[3] R. L. Clough and K. T. Gillen, "γ-Radiation-Induced Oxidation and Mechanisms of its Inhibition," in Oxidation Inhibition in Organic Materials, Vol. II, J. Pospisil and P. P. Klemchuk, eds., CRC Press, Boca Raton (1990) p. 191.

[4] R. L. Clough, "γ-Radiation-Oxidation of Polycyclic Aromatic Hydrocarbons: Involvement of Singlet Oxygen," J. Amer. Chem. Soc., 102, 5242 (1980).

[5] K. T. Gillen and R. L. Clough, "Time-Temperature-Dose Rate Superposition: a Methodology for extrapolating Accelerated Radiation Aging Data to Low Dose Rate Conditions," Polym. Deg. and Stab., 24, 137 (1989).

[6] K. T. Gillen and R. L. Clough, "A Kinetic Model for Predicting Oxidative Degradation Rates in Combined Radiation-Thermal Environments," J. Polym. Sci., Polym. Chem. Ed. 23, 2683 (1985).

RESULTS OF A PRELIMINARY LOW DOSE RATE IRRADIATION OF SELECTED PLASTIC SCINTILLATING FIBERS

Carl Zorn, Brian Kross, Stan Majewski,
Joanna Radford, Randolph Wojcik
CEBAF *
Newport News, VA 23606

Abstract

A low dose rate irradiation of selected plastic scintillating fibers has been completed (700 rad/hr to 1 Mrad). A second run is planned in order to make a more thorough study. This first run has indicated that 3HF-doped fibers suffer significantly less damage than a set of standard blue-emitting scintillating fibers.

Introduction

Within the past few years, a concerted effort has been mounted to increase the radiation tolerance of organic scintillators for use in detectors at the proposed SSC and LHC high luminosity hadron colliders. In particular, there is a general interest in using plastic scintillaters in the form of optical fibers for both tracking and calorimetric applications. Such a fiber would have a diameter of about 0.5-1.0 mm and consist of a scintillating core (typically doped polystyrene) surrounded by a thin cladding of lower index of refraction material (usually acrylic) to provide the required waveguide character. It is accepted that standard blue-emitting fibers such as SCSF-38 and BCF-10 will suffer unacceptable levels of damage in the collider environment. As a result of this, newer, more radiation tolerant scintillators have been formulated. A detailed discussion of this work is reported elsewhere [1].

The standard mode of testing the radiation hardness of scintillators involves the use of an intense source of radiation in order to deliver a specific dose (say 1-100 Mrad) in a relatively short span of time (rates of 0.1 to 10 Mrad/hr). However this method produces a transient level of damage, that is, the level of damage is observed to decrease with time at a rate dependent upon such variables as thickness of scintillator, type of plastic and dopants, and especially upon atmosphere (that is, the recovery is most rapid if significant levels of oxygen are present). The level of damage reaches a plateau of residual damage which is chiefly dependent upon the absorbed dose. The recovery mechanism seems to be due to a chemical reaction(s) driven by atmospheric oxygen diffusing into the sample [2]. The problem with this method is that the residual level of damage after the initial recovery phase may not correspond to that found at the much lower dose rates for the actual SSC and LHC experiments. One body of work [3] suggest that for polystyrene-based scintillators, such a correspondence may be possible. However, for acrylic-based scintillators [4], a low dose rate irradiation produced considerably more damage than a high rate one. Since most scintillating fiber formulations consist of a polystyrene core surrounded with an acrylic cladding, it is of paramount importance to check the effects of a low dose irradiation as opposed to a high rate one to the same dose. In addition, since oxygen seems to play a strong role in both recovery and damage, it is also important to verify the role of atmospheric oxygen at realistic dose rates. In addition, it is important to test out if any of the fluors are especially sensitive at low dose rates (and if oxygen is reponsible for their sensitivity).

This paper reports the results of a first-order test of several fibers. Given the results so far, a second, more thorough, low dose rate experiment will be implemented.

Experimental Setup

The key parameters of interest are dose rate, presence of oxygen and total dose. To deal with oxygen, the sample set is split to two identical groups, one irradiated in an oxygen-free (argon) atmosphere, and other in an oxygenated atmosphere (air). One requires a total dose that is capable of damaging the scintillators. A dose of 1 Mrad seems to be a canonical minimum from high rate experiments. Choosing a dose rate is more difficult. This involves weighing the various factors such as (i) a reasonable time limit on the experiment, (ii) using a rate well below any used so far, and (iii) finding an experimental facility capable of delivering such a

*Operated by Southeastern Universities Research Association

rate. We have used the ^{60}Co irradiation facility at the Nuclear Reactor Facility of the University of Virginia (Charlottesville, Virginia). In the end, a rate of 700 rad/hr was used, thereby requiring two months for a 1 Mrad absorbed dose. The rate was determined with the use of the MicroShield simulation program [5]. An ionisation chamber experimentally verified this, although the small dose rate resulted in a low readout current in the chamber resulting in large (> 20%) uncertainties. The simulation program is estimated to produce a 10% uncertainty. (Most of this is due to the variation in activity levels of the chosen source rods which varied from 64-90 Ci in strength and were chosen randomly from the total supply of 70 rods.) In addition, samples of radiochromic dye film [6] were attached to the fibers as an independent check of the absorbed dose. Unfortunately, these films were found to suffer from both dose rate and oxygen sensitivity. The ones in air read 0.55 ± 0.01 Mrad while the ones in argon read 0.78 ± 0.01 Mrad. Since low dose rates and the presence of oxygen are known to desensitize the material [7], these results are consistent with the known property of such dye films and with the calculated dose. For the second experiment, we are currently investigating other possibilities, of which the best seems to be the use of luminescent crystals (TLDs). Such crystals are incapable of measuring doses of several Mrads, so they would be irradiated for a short period, and so be used to give a dose rate measurement. (In any case, we can still use the dye films for verifying the uniformity of the irradiation field.)

The chosen fibers were selected from an available set of samples and covers a range of different blue-emitting fibers and the more radiation-tolerant 3HF-doped green-emitting fibers. Unfortunately, the newer blue-emitting, radiation-tolerant fibers from Bicron Corporation (Bicron B and RH-1) were not available for this test. These will be included in the second experiment. However we did include several other new formulations, namely a PVT-based fiber from Nuclear Enterprise, and a 3HF-doped, PVT-based fiber from Bicron Corporation. It should be pointed out that these PVT-based fibers are an initial attempt at using such a base for fibers. Hence their inherent attenuation properties are not expected be superior to the polystyrene-based fibers. Each of the two sets included Kyowa 3HF (4 samples), Bicron 3HF (Type H) (4), Bicron 3HF (Type G) (4), Bicron 3HF (PVT base) (3), SCSF-38 (1), Optectron S101 (2), and a NE blue-emitting PVT-based fiber (2).

The effect of irradiation upon the fibers was measured in terms of the change in the attenuation property of the fiber. In order to obtain a reasonable measure of the attenuation length, the fibers were cut to a 1 meter length. Since cladding light is a significant contribution to the measured signal [8], it is best to apply a "cladding mode stripper" in the form of a strip of black paint at the PMT end of the fiber. With this, a more accurate and precise determination of the attenuation length is possible. However, since the effect of irradiation upon the black paint is unknown, it was decided to measure the fibers without the paint. Hence only the data beyond 40 cm from the PMT can reliably be used for an attenuation length determination. For the second experiment, the fiber length will be increased to 1.5 meters. In addition, the first 30 cm of the fiber (including the painted section) will be shielded by a lead wall (sufficient to reduce the absorbed dose by at least a factor of 100). In this way, we can also obtain the effect of irradiation upon the intrinsic scintillation yield of the fiber. In addition, if the paint has had catastrophic effects upon the fiber, that small portion can be cut off, and a new layer of paint applied without affecting the veracity of the scintillation yield and attenuation length results.

The irradiation setup was as follows. Each set of fibers was placed in an aluminum tube (3/4" diameter) and a gas circulation system hooked up to each of two tubes (one for dry air and the other for argon). The tubes were strapped together (in parallel) and placed in the pool 36 cm from a ^{60}Co line source made up of eight depleted ^{60}Co line sources placed inside an aluminum tube (each source 27 cm in length and 2.5 cm in diameter). The sources and samples were collinear with each other. The radiochromic dye film samples were placed at three equally spaced locations on each of the two fiber samples. The post-irradiation measurements of these films indicate that the irradiation field was uniform to within 2% (the computer simulation estimated an error of < 5%).

After two months, the fibers were removed from the pool, and their post-irradiation measurements made of their attenuation curves. Two sets of measurements were made, one within three days of removal and the other 2.5 weeks after removal in order to ascertain if any recovery phase was present (in high dose rate experiments, 2 weeks was found sufficient for complete recovery of such fibers).

Results

Figure 1 displays the overall results for the irradiated fibers. Post-irradiation attenuation lengths are compared to the pre-irradiation measurements (defined as 100%). The numbers are averaged over all samples of the fiber and over two sets of post-irradiation measurements separated by two weeks. No significant signs of the recovery phenomenon were observed. Despite the large error bars on some of the points, it is clear that the 3HF-doped, green-emitting fibers are clearly less affected by the irradiation than the blue-emitting fibers. No significant affect of oxygen vs no oxygen is seen in the data. Unfortunately, there is a paucity of published data of high dose rate irradiation of fibers to 1 Mrad. Figure 14 of reference [9] (not reproduced here) displays the results of the irradiation of an Optectron S101 fiber to 1 Mrad at a rate of 0.3 Mrad/min. It shows the atten-

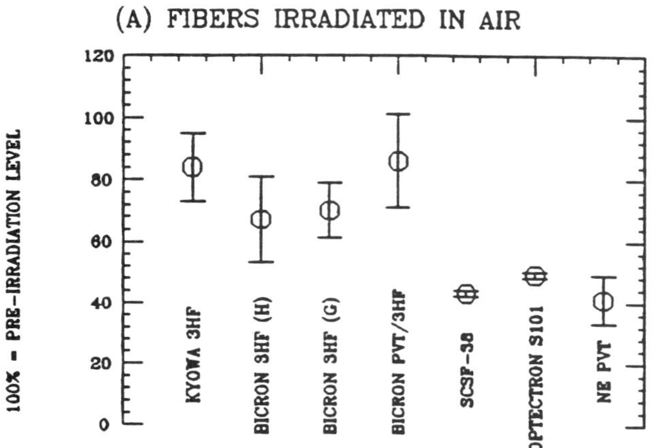

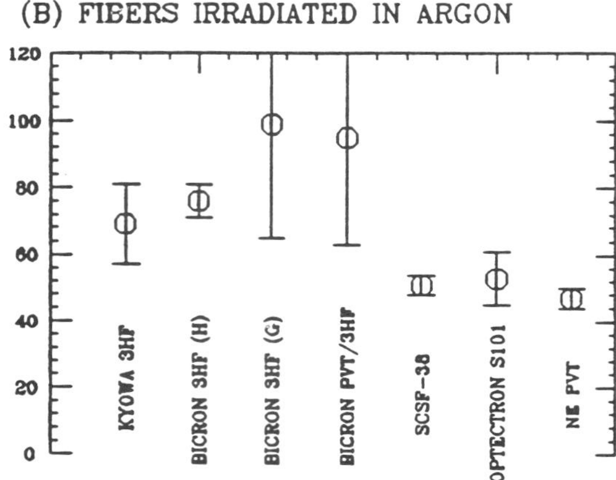

Figure 1: Low dose rate irradiation results for those fibers (a) irradiated in air and (b) in argon.

uation curve both before and soon after the the irradiation. No sign of damage is apparent from the viewpoint of attenuation changes. Considering the $\approx$ 50% loss seen in the low dose rate experiment, this may be a first indication that one should expect considerably more damage at realistic dose rates. On the other hand, there is unpublished data [10] on the irradiation of SCSF-38 fibers to 0.9 Mrad. In this case, the low and high dose rate results are consistent with each other. To study this matter further, the second planned experiment includes a high rate irradiation to the desired dose so that a comprehensive evaluation of high versus low dose rate irradiations (to the same integrated dose) can be accomplished.

Acknowledgements

We would like to thank the representatives of Bicron, Kuraray (formerly Kyowa), and Optectron for generously providing samples for this and the next irradiation test. We also thank Kurtis Johnson and Heather Whittaker of Florida State University for providing the radiochromic dye films as well as for measuring the absorbed dose. Thanks go to Geoffrey Stapleton of CEBAF for suggesting the use of the University of Virginia irradiation facility, and many thanks go to Paul Benneche, Bauvard Hosticka, Bob Muller, and Al Reynolds of the Nuclear Reactor Facility at the University of Virginia for their assistance with this experiment. One of us (J.R.) acknowledges the University of Virginia and the National Science Foundation for financial support.

References

[1] C. Zorn, "Designing a Radiation-Hard Plastic Scintillator for High Luminosity Hadron Colliders," published in the proceedings of the Workshop on Radiation Hardness of Plastic Scintillator, Tallahassee, Florida, March 19-21, 1990.

[2] K. Wick et al., *Nucl. Instr. Meth.* **A277** (1989) 251.

[3] U. Hölm and K. Wick, *IEEE Trans. Nucl. Sc.* **NS-36** (1989) 579.

[4] Y. Sirois and R. Wigmans, *Nucl. Instr. Meth.* **A240** (1985) 262.

[5] IBM-PC version of IBM mainframe program ISOSHLD available from Oak Ridge National Laboratory, Oak Ridge, Tennessee.

[6] Type FWT-60-00 available from Far West Technology, Goleta, California.

[7] Mr. Jim Humphreys, N.I.S.T., private communication.

[8] C.M. Hawkes et al., *Nucl. Instr. Meth.* **A292** (1990) 329.

[9] S. Majewski et al., *Nucl. Instr. Meth.* **A281** (1989) 500.

[10] K. Yasuoka, University of Tsukuba, "Effects of Gas Atmosphere on ^{60}Co-induced Radiation Damage of Scintillators," unpublished data presented at the Workshop on Radiation Hardness of Plastic Scintillator (1990).

RADIATION DAMAGE STUDIES OF STRAW TUBE AND SCINTILLATING FIBER ELEMENTS*

W.L. Dunn
Quantum Research Services, Inc.
P.O. Box 52391
Durham, NC 27717-2391

T.S. Elleman
Department of Nuclear Engineering
North Carolina State University
Raleigh, NC 27695-7909

A.T. Goshaw, S.H. Oh, W.J. Robertson
Department of Physics
Duke University
Durham, NC 27706

A. Grimes, I. Leedom, and S. Reucroft
Department of Physics
Northeastern University
Boston, MA 02115

Abstract

We report on the results of mixed-field irradiations of straw-tube, plastic scintillating fiber, and avalanche photodiode components. These irradiations are being carried out at the one-MW PULSTAR research reactor facility at North Carolina State University. A special sample holder was designed that allows relatively uniform irradiation of samples up to 5 ft long, without bending or coiling. A systematic irradiation program is underway that allows study of total fluence, fluence-rate, and neutron spectral effects. Samples have been exposed to neutron fluences as high as 2×10^{16} cm^{-2}.

Introduction

In support of the Hybrid Central Tracking Chamber (HCTC) concept, we undertook initial neutron radiation damage studies of straw tubes and a glue, of avalanche photodiodes (APD's), and of plastic scintillating fibers (PSF's). The HCTC is expected to occupy a cylindrical annular region extending in radius, r, from approximately 40 cm to almost 200 cm and over the pseudorapidity range $-1.6 \leq \eta \leq 1.6$. Citing the work of several investigators, Groom [1] estimates that the neutron albedo from an uranium scintillator calorimeter would vary, at a radius of 200 cm, between about 4×10^{11} and 5×10^{12} cm^{-2} yr^{-1}, over the pseudo rapidity range $\eta = 0$ to 1.6. Scaling by $1/r^2$, we estimate the maximum annual fluence (at $\eta = 1.6$, r = 40 cm) to be approximately 10^{14} cm^{-2} at design luminosity of 10^{33} cm^{-2} s^{-1}. Thus, assuming a ten-year active lifetime, the total annual neutron fluence within the HCTC volume is expected to vary over the approximate range 4×10^{12} to 10^{16} cm^{-2} (the latter figure accommodates a possible order of magnitude increase in luminosity).

The anticipated albedo neutron energy spectrum has been shown [1] to exhibit an approximately gaussian peak in its E dΦ/dE versus logE behavior (where E is neutron kinetic energy and Φ is energy-dependent flux), peaked near E = 1.3 MeV. This, combined with the fact that the silicon elastic scattering cross section exhibits a marked increase above E = 0.15 MeV, has led to considerable interest in "fast" neutron damage effects. However, we have chosen to study effects of both thermal and fast neutrons, for three primary reasons: (1) not all HCTC components will contain silicon (although electronic devices typically do); (2) early studies [2] suggest that the *number* flux of neutrons at thermal energies is comparable to that at the 1.3 MeV peak; and (3) thermal neutron interaction mechanisms (*e.g.*, absorption) are typically different in character from those for fast neutrons (primarily scatter).

---

* Work supported by the SSC Laboratory under the Subsystem R&D program through various DOE grants.

Facility Description

The PULSTAR reactor is a 1 MW swimming-pool type reactor operated at North Carolina State University for teaching, research, and service. The reactor operated 2,400 hours in calendar year 1989 and is available on a near full-time basis. The facility has four vertical exposure ports (VEP's), six horizontal beam tubes, and a bulk irradiation facility. The irradiations reported here utilized VEP Y, which has maximum full-power flux levels of 1.3×10^{13} cm^{-2}-sec^{-1}, thermal, and 2×10^{12} cm^{-2}-sec^{-1}, fast. Provision can be made for pneumatic transport and for specimen rotation (to enhance exposure uniformity). The reactor pool temperature is normally between 105°F and 109°F.

A water-tight outer irradiation capsule was designed and constructed that allows the simultaneous or sequential irradiation of PSF, straw tubes, and electronic devices. The capsule was machined from high purity aluminum and contains a cylindrical active irradiation volume, with a length of approximately five feet and a diameter of two inches. Both aluminum-tube and plastic-rod central supports were constructed. Circular plates surrounding the central supports serve to position them in the center of the capsule and to accommodate samples. A cadmium sleeve can be inserted in the space between the capsule wall and the guide plates, to remove thermal neutrons. Thermal and fast neutron fluxes were measured at 3-inch separations along the capsule tube length using indium and cadmium-covered indium foils, respectively (the "fast" flux is that above the cadmium threshold of 0.15 eV). The neutron dosimetry results are summarized in Fig. 1.

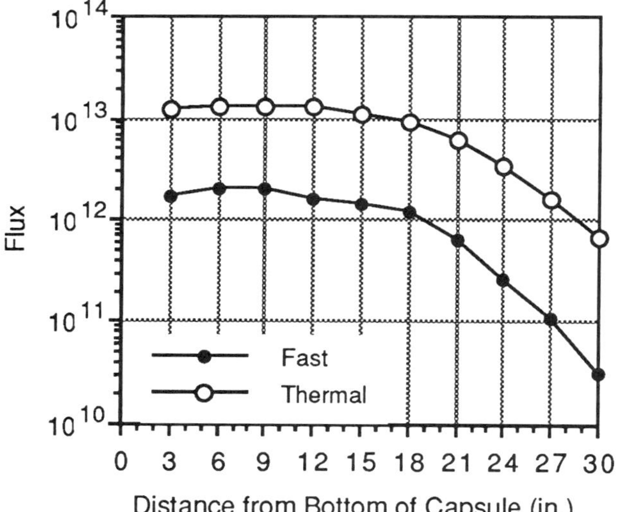

Figure 1. Measured full-power fast and thermal flux profiles at the capsule center in PULSTAR VEP Y; flux values are in units of cm^{-2} s^{-1}.

Straw Tube Component Tests

The work by Zhou, et al. [3] indicates that straw tubes can be expected to operate at neutron fluences up to at least 1.1×10^{13} cm^{-2}. Our initial tests examined the mechanical integrity of the straw tubes themselves, and of a glue used in fixing the straw tubes, before and after irradiations to neutron fluences as high as 2×10^{16} cm^{-2}. Four irradiations of 4-mm diameter straw tube samples were conducted. The straw tubes were fabricated by Stone Industrial of College Park, MD, from a spirally wrapped ribbon of 50 μm thick mylar film. The overlapping seams of adjacent spiral windings were bonded with a polyester adhesive.

A single 49-cm straw tube was irradiated for thirty minutes at full power. The straw received a neutron fluence profile that varied from about 1.1×10^{15} to 3.7×10^{15} cm^{-2}, fast, and from about 1.0×10^{16} to 2.3×10^{16} cm^{-2}, thermal. The gamma-ray dose was estimated to be between 0.1 and 0.5 Mrad. In order to test the integrity of the seams after the irradiation, the 49-cm long straw was pressurized to determine its burst strength. The straw was able to sustain a gauge pressure of 40 psi, which was the maximum to which it was subjected and represents many times the expected gas flow pressure in the tubes.

We have been assembling straw tube arrays by gluing adjacent tubes at 10 – 20 cm intervals with a cyanoacrylate adhesive (super glue). To test the ability of the adhesive to withstand the radiation environment anticipated at the SSC, pairs of 1 cm straw tube segments were glued together. Forty pairs were stressed to measure the load necessary to cause the glued joint to fail. The average failure load was 520±175 gm. In two separate ten-minute runs at reactor full power, thirty eight 1-cm straw tube section pairs were irradiated such that nineteen received neutron fluences of approximately 2×10^{14} cm^{-2}, fast, and 2×10^{15} cm^{-2}, thermal, and the other nineteen received fluences of approximately 10^{15} cm^{-2}, fast, and 8×10^{15} cm^{-2}, thermal. The straw sections received approximate gamma-ray doses between 30 and 150 krad. The average failure load of those irradiated to a thermal fluence 2×10^{15} cm^{-2} was measured to be 450±150 gm, while those irradiated to a thermal fluence of 8×10^{15} cm^{-2} failed at 430±100 gm.

Finally a collection of 25 straw tubes, which had been glued together into a trapezoidal array of 5 rows (7 straws in the bottom row, 3 straws in the top row), was irradiated for 30 minutes at full power. The array received nearly uniform neutron fluence distributions which varied between about 2.5-4.0×10^{15} cm^{-2}, fast, and 1.5-2.5×10^{16} cm^{-2}, thermal, and approximately 0.1-0.5 Mrad gamma-ray dose. In order to determine whether the exposure to which the straw tube array was subjected would result in a measurable dimensional change, the array was surveyed before and after irradiation with a traveling microscope that has a measurement precision of better than 20 microns. The width of the 40-cm long array of straws was measured at 5-cm intervals. Comparison of the measurements before and after irradiation show that to within the accuracy of the measurements no variation was observed. It is noted that the irradiated straw tube samples were mounted on $\frac{1}{8}$-in. diameter lucite rods to facilitate handling. After the radiation exposures, the lucite rods showed noticeable yellowing and had become brittle. In contrast to this, the mylar tubes showed no visible effects from the irradiations.

Scintillating Fiber Component Tests

Ten GE Electro Optics modified C30921S APD's were irradiated in the reactor at low power (1 watt) for various times from 10 to 240 sec. The following data were collected before and after irradiation: breakdown voltage (V_B), dark count rate at V_B+10 V, and dark current at V_B-10 V. Room temperature at the time of data collection was 72±1°F. An ultra-stable power supply and a high-precision voltmeter were used to record the variation inherent in dark count rate as the bias was varied by ±0.5 volts about the nominal V_B+10 V. This variation, and the statistical count-rate variances, were used to estimate the uncertainty in the ratio, R, of dark count rate after irradiation to that before.

The APD's were loaded inside the aluminum tube support at one of two selected levels using spacing rods; flux foils were inserted with the APD's. Since exposure times were rather short, the following irradiation procedure was used. The irradiation capsule was allowed to drop by gravity to the bottom of VEP Y; a preset timer was started as the capsule reached the core, as evidenced by the reactivity change displayed on the reactor instrumentation. When the preset time expired, the operator scrammed the reactor, immediately shutting off the neutron supply. The capsule was then withdrawn to the top of the pool and the samples removed.

The results of the ten exposures are summarized in Table 1. Estimates of gamma-ray dose received during the irradiations are 1-10 rads. We plot in Fig. 2 the dark count ratio as a function of fast fluence. The results indicate a damage threshold at approximately 2×10^6 cm^{-2}, as evidenced by a rapid increase in R above the no-damage value of unity, indicative of higher dark count rates due to radiation damage. The dark currents, on the other hand, were unaffected by the radiation exposures. In Test no. 7 the APD was covered entirely by cadmium, which filters out thermal neutrons. The results from tests 7 and 8, in which the same fast fluence was seen, indicate that the damage is due to fast neutrons, as expected. We have seen a modest recovery, of order 10%/mo, over two months of testing four of the damaged APD's. We have also begun PSF irradiations. Two of each of Bicron RH1, Type B, and Type H fibers were irradiated to fluences of about 2×10^{12} cm^{-2}, fast, and 10^{13} cm^{-2}, thermal. These fibers and one unirradiated fiber of each type were sent to Northeastern University for testing. One day after irradiation, no appreciable difference in light output or attenuation length was noticed between the irradiated and unirradiated fibers. Testing of five types of fibers at higher fluences is now underway.

Conclusions

It appears that straw tubes and glues are mechanically sufficiently radiation hard for use in central tracking. We have also shown that the GE APD's – operated in the geiger mode at room temperature – begin to exhibit noticeable radiation damage at fast fluences of approximately 2×10^6 cm^{-2}. If the APD's can be housed outside the calorimeter, coupled to the PSF's by optical fibers, adequate shielding could easily be achieved at modest cost. Damaged APD's appear to be capable of partial, slow recovery.

Table 1. Results of neutron irradiations of avalanche photodiodes

Test No.	Fluence (cm^{-2}) Fast	Thermal	Dark count Ratio, R	Dark current Ratio
1	3.5×10^5	7.3×10^7	1.13	0.86
2	7.0×10^5	1.5×10^7	0.96	1.00
3	1.6×10^6	3.2×10^7	1.30	0.90
4	1.6×10^6	3.2×10^7	3.68	1.00
5	3.3×10^6	6.4×10^7	3.63	1.10
6	5.4×10^6	1.0×10^8	3.16	0.94
7	7.8×10^6	–	7.93	1.00
8	7.8×10^6	1.5×10^8	5.79	0.87
9	1.1×10^7	8.5×10^7	11.04	0.86
10	2.0×10^7	1.6×10^8	13.11	0.83

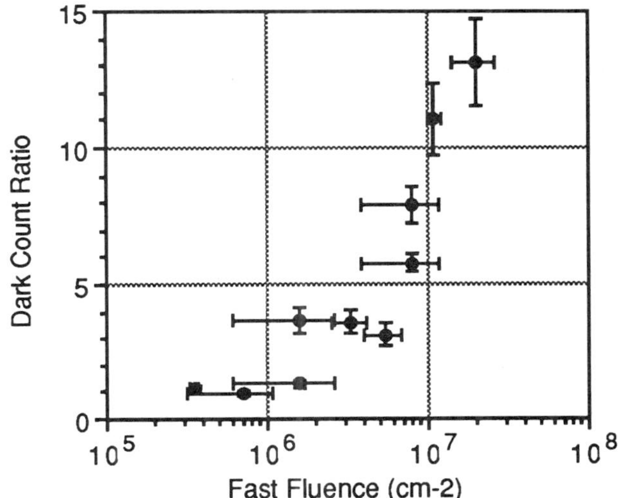

Figure 2. Variation with fast neutron fluence of the ratio of dark count rate after irradiation to that before irradiation.

References

[1] D.E. Groom (Ed.), Radiation Levels in the SSC Interaction Regions, SSC-SR-1033, June 10, 1988.

[2] D.E. Groom, Radiation Levels in SSC Detectors, SSC-201, January 1990.

[3] B. Zhou, D. Warner, J. Rohlf, D. Osborne, A. Marin, W.W. Lu, G. Hopkins, T. Coan, J. Beatty, and S. Ahlen, Performance of Small-Radius Thin-Wall Drift Tubes in an SSC Radiation Environment at the MIT Research Reactor, BUHEP-90-2, December 27, 1989.

RADIATION DAMAGE TESTS OF A SCINTILLATING FIBER CALORIMETER

K. F. Johnson
Florida State University, Tallahassee, FL 32306

D. W. Hertzog, S. A. Hughes, P. E. Reimer and R. L. Tayloe
University of Illinois at Urbana-Champaign, Urbana, IL 61801

Abstract

A scintillating fiber electromagnetic calorimeter module constructed from a new radiation hard fiber has been tested in a monoenergetic electron beam at alternating low and high currents. Energy resolution and light output measurements were done at low currents following each high-dose exposure.

The procedure was repeated until the detector resolution worsened to $14.3\%/\sqrt{E}$ from its initial value of $4.6\%/\sqrt{E}$. After four weeks recovery, the resolution was $7.8\%/\sqrt{E}$. After four weeks recovery, the resolution was $7.8\%/\sqrt{E}$.

For comparison, a previous test of a module based on a green-emitting fiber is presented.

Florida State University and the University of Illinois at Urbana-Champaign are pursuing investigations of radiation damage to assembled fiber calorimeter subunits. The intention of the project is to explore the limits of the rapidly developing radiation-hard scintillator technology and to check for synergistic effects which would normally not be revealed by testing individual components.

The construction of the fiber modules is described in detail in Ref. [1]. Briefly, the modules are built up of alternating layers of grooved lead-alloy plates and 1 mm diameter scintillating fibers, held together with adhesive. The dimensions of the plate and grooves are such that the completed device has a 50:40:10 volume proportion of scintillating fiber: load alloy: adhesive, respectively. The adhesive, polydimethylsiloxane, was especially selected for two qualities: it is highly immune to radiation, that is, it does not become chemically aggressive and attack the fiber cladding, and it is highly permeable to air. The intention is to allow air to diffuse through the adhesive along the channels, in which each fiber is embedded, to the damaged regions of the fiber, and thus enhance recovery [2].

The fibers used are designated RH-1 by their manufacturer, Bicron Corporation, and are an experimental fiber optimized for radiation hardness. Both the base material and fluor are proprietary to Bicron. RH-1 is the most radiation resistant blue fiber yet tested and retains, immediately after a 3 Megarad dose, almost ten times the attenuation length (50 cm) of the best green emitting fibers, Bicron G (6 cm). After recovery, however, the green fiber has a longer attenuation length (Figure 1).

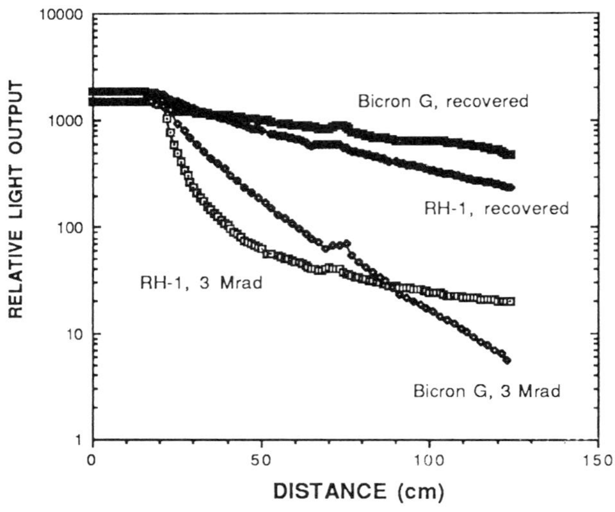

Fig. 1. Relative light yield of Bicron G and RH-1 fibers after 3 Mrad and after recovery.

The module is large enough (9 cm × 9 cm × 40 cm) to function as a totally absorbing Faraday cup for the 94 MeV electrons. Thus, the integral of the current to ground is a measure of the dose. To ensure that the module face was evenly irradiated, the beam was extracted and blown up with a lead foil diffuser before passing through a 6.1 cm diameter collimator to the module.

The energy resolution and pulse height were monitored by reducing the beam intensity to below 10^4 electrons/sec and digitizing the output when triggered by a set of three counters, one of which had a 1/4" hole in the center and was used as a veto. The trigger counters defined the impact point on the face of the module to 1/4". An RCA 8850 A photomultiplier tube viewed the calorimeter through a 6.5 cm × 10 cm long cylindrical Plexiglass lightguide.

The results are shown in Figure 2. For comparison, the results of a similar test of a calorimeter based on a green fiber are plotted as well. In this case, the fiber used was a PTP/3-HF polystyrene fiber manufactured by Kyowa Gas, Ltd., as described in Ref. [2]. The resolutions show a similar deterioration with increasing dose (Figure 2a). The dose is given in terms of microCoulombs. For the calorimeter module region within 0.5 radiation length of the shower maximum, full scale on the figure corresponds to 2.25 Megarad. Figure 2b shows the pulse height degradation as a function of dose where the unirradiated pulse height is unity. The initial dropoff of the RH-1 based calorimeter is much faster than the 3-HF based unit. The quantitative relationships are not yet disentangled and the reason for the rapid decay in light output is not yet clear. The main factors are believed to be the sharp initial increase in blue absorption upon irradiation and a recently observed "thermal degradation" effect. It was found that heating scintillator to be much more radiation soft [3]. Investigation is continuing.

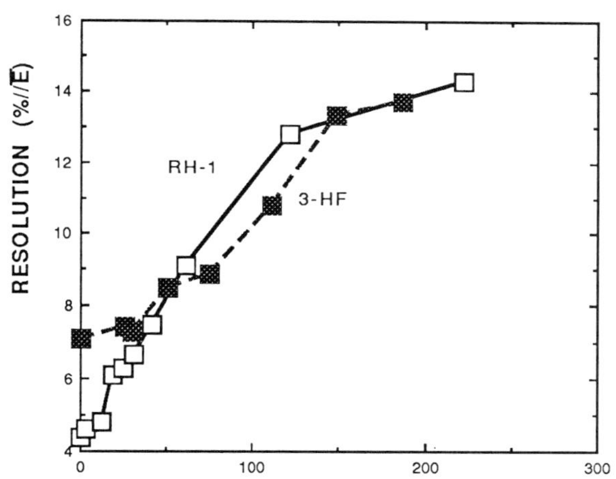

Fig. 2(a). Change in resolution as a function of dose for RH-1 and 3-HF based modules.

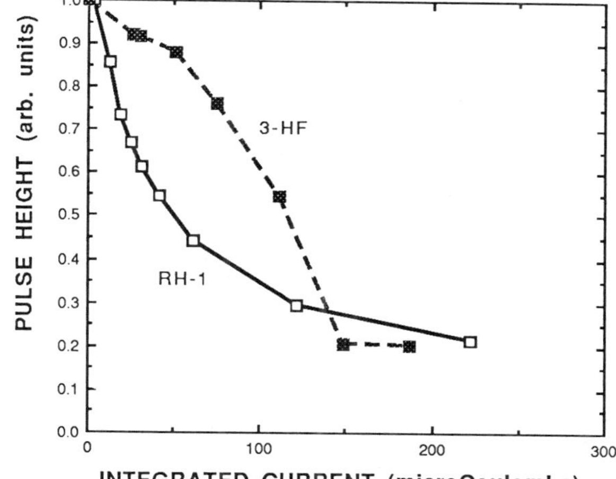

Fig. 2(b). Change in light output as a function of dose for RH-1 and 3-HF based modules.

The module recovered within about 4 weeks to a resolution of $7.8\%/\sqrt{E}$. This measurement was done with a different photomultiplier tube and ADC's than the original measurement and at 54 MeV, instead of 94 MeV, so we do not quote the recovery of light output.

REFERENCES

[1] D. W. Hertzog et al., A High-Resolution Lead/Scintillating Fiber Electromagnetic Calorimeter, ILL-(NPL)-90-001.

[2] K. F. Johnson et al., "High Dose Calorimetry with Scintillating Fibers," IEEE Trans. Nucl. Sci. **NS-37** (1990) 500.

[3] K. F. Johnson and H. L. Whitaker, "Thermal Treatment of Plastic Scintillator Affects Radiation Hardness," submitted to Nucl. Inst. and Methods (1990).

RADIATION DAMAGE TESTS OF NEW SCINTILLATING FIBERS AND PLATES

M. Bertoldi, M. Goforth, V. Hagopian, K. Hu, K.F. Johnson, P. Rulon 3, J. Thomaston, H. Wahl, H. Whitaker, J. Xu and C.A. Young

Department of Physics
Florida State University
Tallahassee, Florida, 32306
(Presented by V. Hagopian)

Abstract

The radiation hardness of various plastic scintillating fibers and plates have been investigated using a 3 MeV electron accelerator. The new fiber Bicron RH-1 is the most radiation resistant blue emitting fiber yet tested. Recovery is more complete in the green fibers, however, with the Bicrons 3-HF fibers being the most radiation resistant.

INTRODUCTION

Plastic scintillating fibers and plates are serious candidates for tracking and calorimetry at the SSC because of their triggerability and speed. Both applications critically depend on the ability of the active medium to receive large doses of radiation without degrading detector performance to an unaccep1table degree. We developed techniques for controlled irradiation at the FSU 3-MeV electron linac[1] and did comparative studies of the radiation resistance of scintillating fibers as well as composition and temperature effect studies. Much effort on the past half year has been expended on improving the precision and quality of our data. We can now irradiate 2.5 meter long fibers to any reasonable dose with extreme uniformity. Dosimetry is non-trivial for low energy electron beams, but we have refined our procedures so that we can measure the absolute dose given to a test object to an error of 10%.

Radiation damage manifests itself as a decrease in the attenuation length and a decrease in the local light yield. Our fiber investigations have led to partial remedies to both problems. The first fruitful result was the demonstration that the use of a large Stokes' shift fluor results in tremendous advantages, namely, a tenfold increase in radiation hardness and very long attenuation lengths. 3-HF, the first such fluor to be used, was originally investigated by FSU chemists. Its use as a fluor and the expected advantages to be accrued were first suggested by S. Majewski, who recognized that utilizing long wavelengths would by-pass color centers in a damaged plastic base, thus ameliorating the radiation induced decrease in attenuation length. (This is sometimes called the "better red than dead" strategy.) A partial remedy to the problem of a decrease in local light yield is to increase the concentration of the secondary fluor by a factor of 10 to 100. With most fluors this will result in very short attenuation lengths due to the re-absorption of emitted light. In the case of a large Stokes' shift fluor such as 3-HF, however, the re-absorption of emitted light is almost nil, and this technique suddenly becomes a viable option. First results showed only a 33% light yield loss after an exposure of 100 Mrad. Figure 1 shows the spectra of RH1 – a blue scintillant, Y7 wave shifter and 3HF wave shifter. Our measurements also show that the light yield of 1 mm fiber with 3HF (Bicron G) has 2.6 times the light output of SCSN81+Y7. This is an enormous improvement.

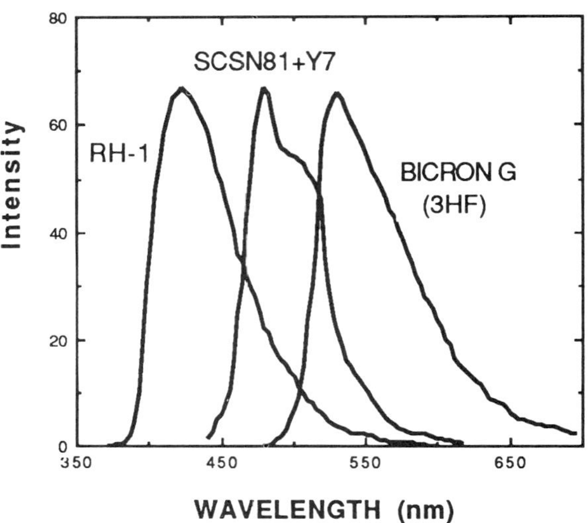

Fig.1. Light emission of fibers

A second area of investigation is the damage to the plastic base. Results from previous work proved that modifications to polystyrene base plastic could "harden" the base by an order of magnitude. BICRON Corporation, motivated by these results, has recently created an improved base material, designated RH-1, which we have tested.

At Florida State University fibers and plates are irradiated by the 3 MeV electron accelerator which can supply beams from 1 nanoamp to 1 microamp. The beam has a diameter of about 2.5 cm. Fibers are exposed on a rotating wheel of about 2.5 meter circumference. The dose is monitored on-line by Faraday cups and off-line with calibrated radiochromic dye films, which enable as to ascertain the dose to better than 10%. The first 20 cm of the fiber is shielded by lead and used as normalisation. In addition a 10 cm portion near the center of the fiber is also shielded by lead to separate the damage due to light attenuation from local light yield. Near the ends, the fiber is painted black to remove light that travels in the cladding. Fibers are measured with a phototube at one end. Scintillation was produced either by a UV light source or a 30KV X-ray source which moved on a 2 meter track. The light output was measured by a microammeter connected to the phototube anode. Figure 2 shows RH-1 (blue) and Bicron G (3HF doped-green) 1 mm fibers exposed to 3 Mrad radiation. The bump in the center is where the fiber was shielded by lead to separate the damage to light attenuation from local light yield [2]. The major difficulty in quoting a single attenuation length is due to fibers not exhibiting a pure single exponential form. A fluor emits a spectrum of wavelengths and the short wavelengths are preferentially absorbed by damaged scintillators. The value of the attenuation length depends therefore on the distance from the phototube. This can be seen clearly in figure 2. (See below for the results of transmission verus wavelength for plates).

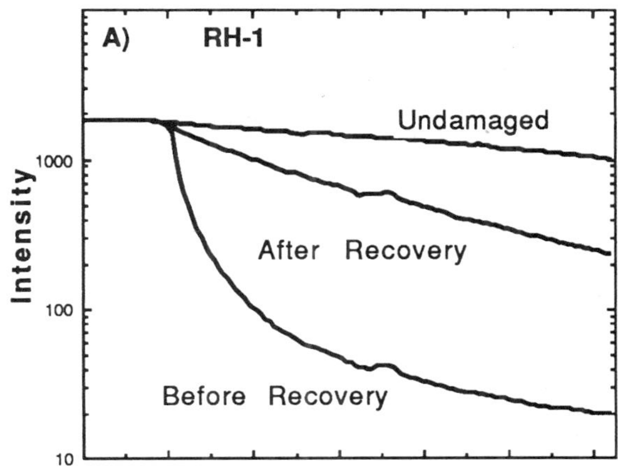

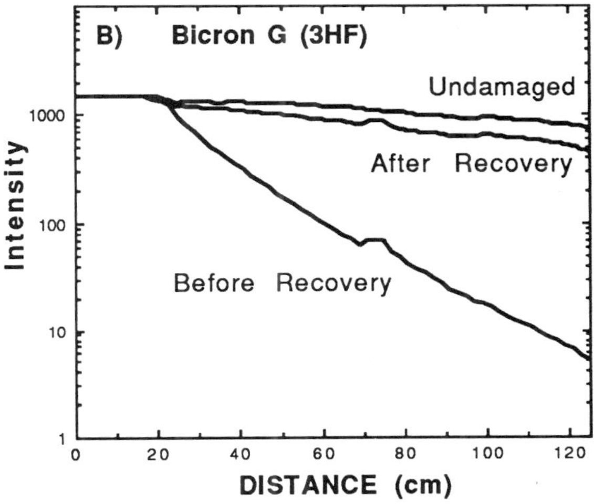

Fig.2. Light yield of fibers radiated to 3 Mrad.

Table 1.
Radiation damage of fibers

Fiber	Radiation Mrad	L_0 cm	L_1 cm	L_2 cm	R_0 %	R_1 %
Blue Emitting						
RH-1	10	235	<22	60	17	8
RH-1	3	235	<40	75	13	3
RH-1	1	235	<81	100	8	2
Green Emitting						
SCSN81+Y7	3	250	< 5	62	25	9
SCSN81+Y7	1	250	<33	95	7	2
BICRON G	3	190	<10	167	12	8

Notes: Attenuation lengths; L_0-undamaged L_1-irradiated no recovery; L_2-after recovery. Local light reduction:R_0-before recovery, R_1-after recovery. BICRON G has 3HF.

Table 1 shows the summary of the fiber radiations. The attenuation length quoted immediately after damage is the slope at the far end where essentially the longest wavelengths survive.

The relative light yields of 100 cm long 1mm diameter fibers are as follows: RH-1 = 1.0, BCF10 = 2.2, SCSN81+Y7 = 0.26, BICRON E(3HF) = 0.68, BICRON G(3HF) = 0.76, BICRON H(3HF) = 0.65, BICRON I(3HF) = 0.57.

In early July the first samples of 10 × 10 × 0.5 cm^3 RH-1 plate scintillator were received. These plated showed some crazing but were tested anyhow. For plate scintillators, the critical measurements are transmission as a function of wavelength and light yield. These plates were cut to 2.4 × 2.4 × 0.5 cm^3 pieces for transmission measurements. These were irradiated to 0.1, 0.3, 1, 3 and 10 Mrad. All were measured for transmission across the 0.5 cm as a function of wavelength over a period of about one month. All showed some damage but also showed recovery. As expected the recovery started at the surfaces and worked its way into the plate. In three weeks all but the 3 and 10 Mrad samples had completely recovered. Unfortunately as time went on, the crazing got worse, even to those samples not irradiated and recovery measurements became impossible after about a month. Visual inspection of the damage, which shows as yellow in the blue scintillator, continued to decrease with time. Figure 3 shows the transmission of $2 \times 2 \times 0.5$ cm^3 RH-1 for the 3 and 10 Mrad exposure samples.

To measure the light yield, the plates were cut and polished into $10 \times 2.4 \times 0.5$ cm^3 pieces. These were glued to a light guide on the 2.4×0.5 cm^2 side and the light output was measured, across the 10 cm length by using a UV lamp and a 30KV X-ray source. Two of these samples were exposed to 1 and 10 Mrad each, where the first 2 cm were not irradiated for normalization purposes. Both had damage but the 1 Mrad sample completely recovered in three weeks. Again deterioration due to crazing stopped these measurements too. Figure 4 shows the light yield as a function of distance of $2.4 \times 10 \times 0.5$ cm^3 RH-1 plates after 1 and 10 Mrad exposures. More recently newer samples of RH-1 plate were received. Even though these did not show crazing at the beginning, with time crazing appeared.

The plan is to continue radiation hardness studies as a function of temperature as well as newer samples as these become available.

We wish to thank Dr. C. Hurlbut of Bicron Corporation for providing us the samples. We would like to acknowledge the assistance of Dr. Stan Majewski in helping us get started and working with us until the end of 1989.

REFERENCES

[1] S. Majewski, et al., "Radiation Damage Studies in Plastic Scintillators With a 2.5 MeV Electron Beam", Nucl. Instr. and Meth. **A281**, 500 (1990).

[2] K.F. Johnson, et al., "High Dose Calorimetry With Scintillating Fibers", IEEE Trans. Nucl. Sci., **NS-37**, 500 (1990). C. Zorn, "Designing a Radiation-Hard Plastic Scintillator For High Luminosity Hadron Colliders" in Proceedings of the Workshop on Radiation Hardness of Plastic Scintillators, Florida State University, Tallahassee, Florida, March 19-21, 1990.

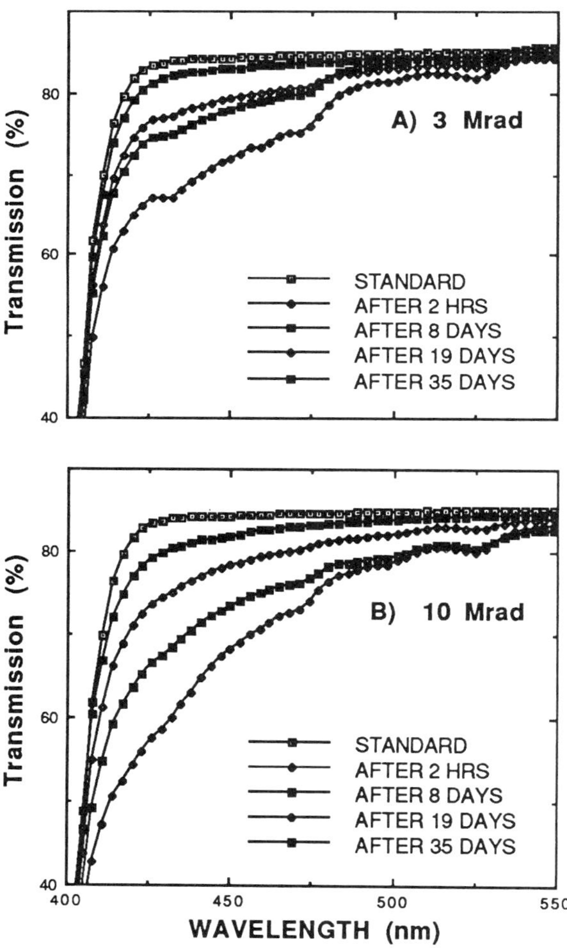

Fig.3. Transmission accross 5 mm RH-1 scintillator plate before and after radiation.

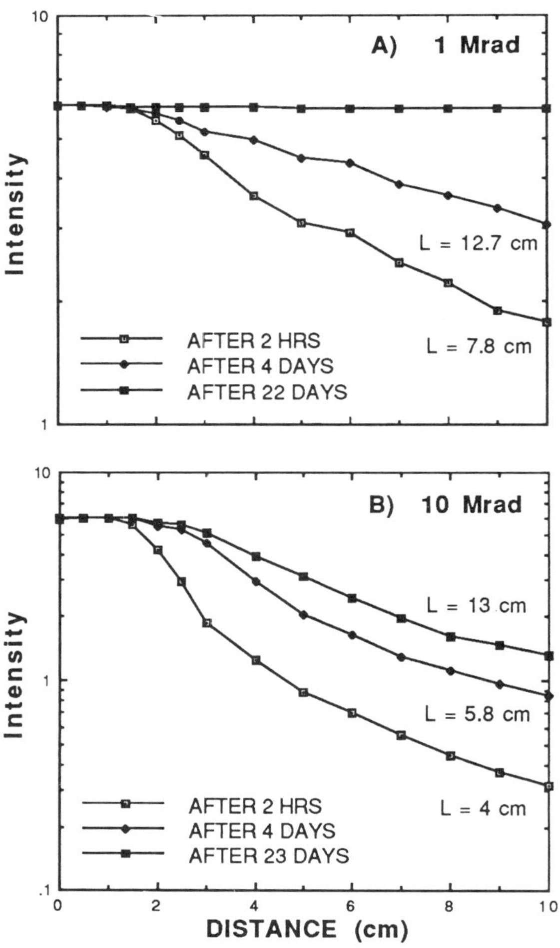

Fig.4. Light yield of 10x2x0.5 cm ^{3} RH-1 plate scintillator. L is attenuation length.

EFFECT OF RADIATION-INDUCED DISCOLORATION OF THE CLADDING ON LIQUID SCINTILLATING FIBERS*

P. Datte, G. Dudgeon, R. Huson, L. Peters, A. Sanzgiri, R. Webb, J. White and S. Zaman
Texas A&M University
College Station, TX 77843
and
V. Hagopian, K. Hu, K.F. Johnson, J. Thomason and H. Whitaker
Florida State University
Tallahassee, FL 32306

Abstract

In the extremely high radiation regions of SSC experiments, detectors employing scintillating fibers (such as for tracking or calorimetry) may require the use of liquid scintillating fibers in place of solid fibers. Although the liquid in such fibers may be circulated, the tube which contains the liquid and serves as the cladding is fixed and may become discolored from the radiation. Here is presented the results of a study of the effect of this discoloration on light transmission for the case where the tube is composed of a borosilicate glass (Pyrex). Indications are that the discoloration has little effect on the light transmission. It was also found that the glass contributes to the light production in such a way that light output as a function of distance could be tuned by pre-irradiation.

Introduction

The use of scintillating fibers for tracking and calorimetry has been widely investigated in recent years.[1] This technology is now being considered for use in the supercollider experiments and in other high-rate applications. One of the weaknesses is that currently available plastic fibers can withstand a maximum of about 10 Megarads per year before their performance is degraded to an unacceptable level. A possible solution to this problem is to switch to liquid fibers in regions where the radiation exceeds this level. Since liquid scintillators are inherently more radiation resistant and because they can be recirculated even if damaged, they should, in principle, be able to withstand even the worst environment.

However, there is still the problem of the radiation hardness of the material containing the liquid. This material acts as the fiber cladding for total internal reflection. It is possible that the radiation would degrade the optical properties of this cladding through discoloration or other damage to an extent that light transmission becomes unacceptably poor. (We are assuming here that materials such as Teflon that become brittle and fall apart are already excluded from consideration.) The purpose of this study was to determine what effect, if any, such damage would have for the particular case where the material is a borosilicate glass.

Materials and Measurement

For this study we used off-the-shelf pyrex tubing. The tubes were 120 cm in length with 3 mm outer diameters and 1.5 mm inner diameters. For fluid containment, one end was fitted with a plexiglas cap to allow light transmission and the other end was sealed with a heat shrinkable teflon tube which was clamped after the tube was filled. Both the plexiglas and the teflon tube were glued with epoxy to ensure leaktightness. The end with the clamped tube was intended to have a small reflection coefficient so as not to complicate the light versus distance curves.

The tubes were irradiated over a 30 cm section using a 3 MeV electron beam from a Van de Graaff accelerator at Florida State University (FSU). Sets of tubes were given nominal doses of 0, 10, 50 and 100 Megarads. The actual doses for the irradiated tubes were measured using films placed in several locations over the region and were found to be about 8.6, 44 and 84 Megarads respectively, near the center of the sections. These values fell by about 30% near the edges. The irradiated section

*Work supported in part by the DOE contracts DE-AS05-87ER40356 and DE-AS05-81ER40039 and through SSCintCal.

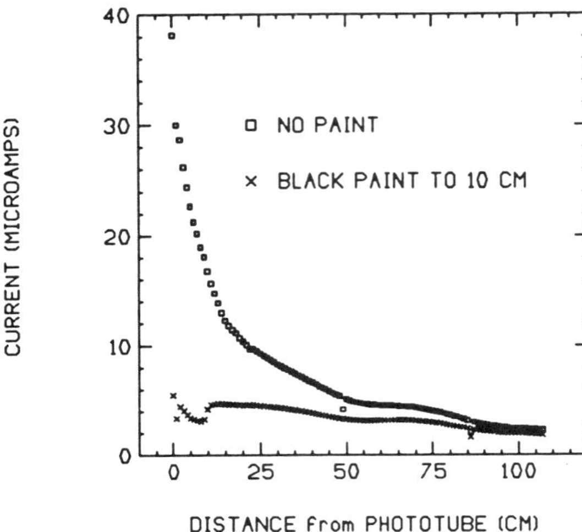

Figure 1: Light output as a function of distance for a non-irradiated tube both before and after addition of 10 cm of black paint near the phototube.

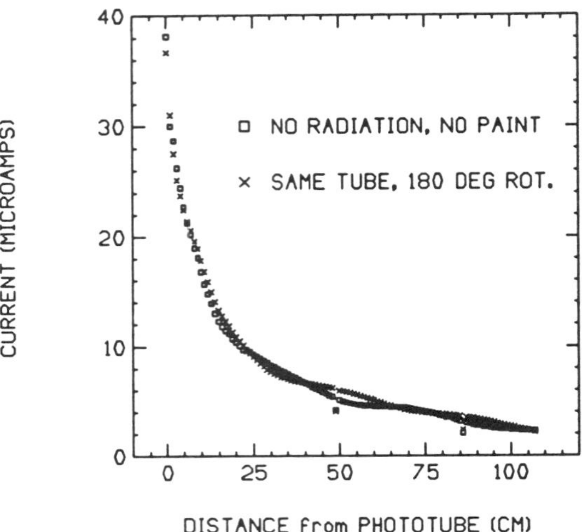

Figure 2: A non-irradiated tube measured in two orientations differing by a 180 degree rotation about the tube axis.

of the tubes became quite discolored ranging from lager to stout.

The tubes were filled with liquid scintillator (n=1.57, National Diagnostics) after irradiation so that only the damage to the tube, independent of any liquid, could be studied. The sample also contained a small amount of green wave shifter so the emitted light had a turquoise appearance. Since the glass has a refractive index of about 1.47, the total internal reflection angle was about 20 degrees.

In order to measure the light output as a function of distance, one end of the liquid filled tube was placed against a phototube and the liquid was excited using an x-ray beam. A PC was used to both drive the x-ray tube and to read out the current from the photomultiplier. This system was designed and built at FSU for use in studies of plastic fibers. It turned out to be ideal for this study as well.

Results

The light of interest is that trapped by total internal reflection at the liquid-glass interface. The light caught by the glass-air interface is sensitive to the external environment of the tube and must be eliminated for stable light output. The standard method it is to paint a small section of the tube near the phototube with a black paint. Light caught in the glass-air interface should then be absorbed when it bounces in the black paint region. Figure 1 shows the light verses distance curves for a tube (no irradiation) both before and after addition of $\approx$ 10 cm of black paint.

The sharp step on the lower curve at the 10 cm mark is caused by absorbtion of the x-rays by the paint. The sharp dips near the 50 and 85 cm marks are caused by the rubber bands holding the tube to a rail. The difference in the shape of the two curves dramatically illustrates the need to eliminate the component from the glass-air interface. To make sure this mode was being fully eliminated, a piece of plastic scintillator was glued to one end of a solid glass rod with dimensions identical to those of the tubes. The scintillator was then excited by the x-rays and the light output was measured both before and after application of five centimeters of black paint at the phototube end. The light output was reduced from a reading of 9.71 microamps before, to a value of 0.00 microamps after application of of the paint (a reduction of at least 10^3).

The curves were also found to contain undesirable long scale wiggles that appear to be caused by variations in the tube wall thickness. Figure 2 shows a tube measured in two orientations differing only by a 180 degree rotation about the axis of the tube. Because the rises and falls anticorrelate, it would appear that the wall gets thicker on one side while it gets thinner on the other.

The main qualitative results are illustrated in Figure 3. It shows the measurements for two tubes, one with no irradiation and one with a 10 Megarad dose, and neither containing black paint. Except for the dips caused by the rubber bands, the curve for the non-irradiated tube varies smoothly over the entire distance. The curve for the 10 Megarad tube drops abruptly near the 50 cm mark, followed by a flat region and then an abrupt rise at 80 cm to a level one might have expected had the tube not been irradiated! Two glaring facts that are apparent from this figure are:

- Since the light level drops in the irradiated region, the light output IS somehow affected by the damage.
- Since the light output returns about to the level it would have been if the tube had not been irradiated, it must propagate through the irradiated region with little effect.

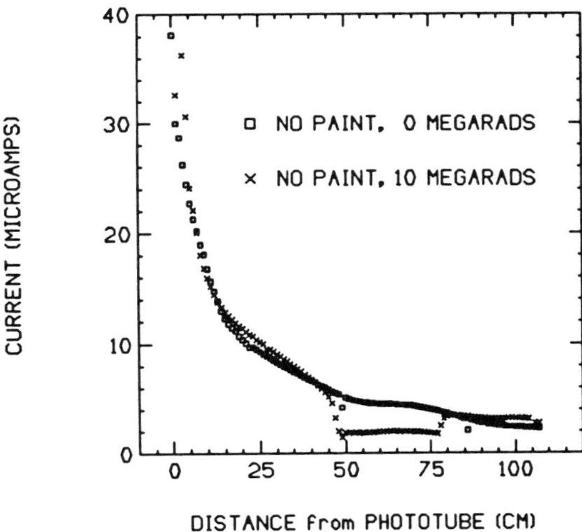

Figure 3: Comparison of two tubes with and without irradiation.

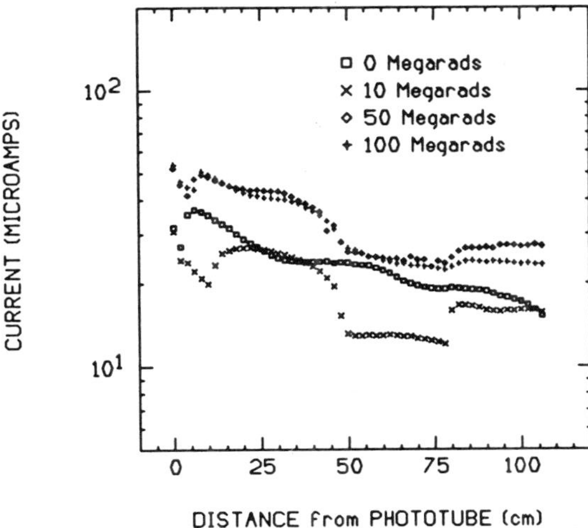

Figure 4: Measurements for tubes with nominal 0, 10, 50 and 100 Megarad doses.

The first effect is believed to be due to scintillation of the glass at short enough wavelengths to further excite the liquid. In the darkened region, this light does not reach the liquid. To verify this we will measure the scintillation spectrum of the glass directly in the near future. For a quick check, an empty tube was scanned and was seen to produce light.

Figure 4 shows the measurements for tubes with the four doses. For these plots, each tube had 10 cm of black paint near the phototube end. In all cases, the light dips where the tube was irradiated and returns to an "almost normal" level in the regions that were not irradiated.

The absolute level of light varies for the different curves for several reasons. The measurements were made over a two day time span. Over that time, the output of the x-ray tube varied and one phototube failed and had to be replaced. Also, the light coupling between the phototube and the fibers was not optimized since the shape of the curves was the primary concern for this study. The measurement setup will be improved in the future to allow correction for these variations.

The shapes of the curves are also not identical. Two primary reasons are the variation in glass thickness as shown in Figure 2 and scattering of the x-rays around the shielding. The non-irradiated sections were shielded by lead bricks, but, as evident in the curves for the higher doses, the boundaries are smeared considerably. Because of these effects, it is not possible to tell whether the higher doses are affecting the light propagation or whether any sagging of the curves is due to the other effects. All these problems will be corrected for the next set of measurements. Whichever the case, if there is an effect on propagation, it is not large even for the 100 Megarad dose.

Summary

The primary result is that radiation-induced discoloration of borosilicate glass has little effect on light propagation. It seems likely that this result will also apply to the cladding on solid fibers. In the near future, several other materials will be tested including plastics and other glasses (even colored ones). The problems with shielding, wall thickness variations and absolute light levels will be corrected to allow more quantitative studies.

It was also found that for the glass studied, the tube contributes to the actual light production. This effect should not be large for thin-walled tubing as one might use in an actual detector. It could also be eliminated by using uv-absorbing or colored glass. This component of the light output is affected by the discoloration which suggests that for special applications, it would be possible to tune the light verses distance curves by pre-irradiation of the tubes in a pattern (much like large NaI crystals are compensated by roughing of the surface).

Acknowledgements

We would like to thank the Texas Accelerator Center for use of its facilities to construct many of the parts for this study. We would also like to thank Walt Phillips of FSU for his help in operating the 3 MeV Van De Graaff and National Diagnostics for supplying us with a sample of high refractive index liquid scintillator.

[1] See, for example, the proceedings of this conference.

A REVIEW OF POLYMERIC SCINTILLATOR MATERIALS FOR USE AT THE SSC

Julie P. Harmon*
and
James K. Walker
Physics Department
University of Florida
Gainesville, Florida 32611

Abstract

We report a mechanistic study which defines a method of producing optically transparent polymers which retain transparency upon exposure to high energy radiation. The kinetics of recovery in PMMA and PS are characterized by boundary movement. The kinetics indicate that relaxation processes control recovery. Structures with low glass transition temperatures exhibit such rapid relaxations that annealable color centers do not persist. We altered the structures of PMMA and PS to yield optically radiation hard plastics. Low T_g polysiloxanes exhibit similar behavior.

Introduction

There is an increasing awareness of the importance of plastic scintillator for particle detection at the SSC. The fast nanosecond response time of plastic scintillator makes it highly suitable for tracking fiber and scintillating plate and fiber for calorimetry. A major problem up until recently has been the radiation induced loss in transmission in commercial scintillator base materials such as polystyrene and poly (vinyl toluene). These polymers exhibit severe discoloration that anneals with time (1). A recovery boundary moves inward from the sample surface, leaving colorless material behind. It is questionable how this recovery process will affect the stable operation of the detectors at the SSC.

The purpose of this paper is to report the successful advances in identifying optically radiation hard scintillator and wavelength shifter plastic base materials [2]. A kinetic study of the recovery process via boundary movement allows one to determine whether the recovery process is controlled by diffusion of say oxygen or by first order rate processes such as polymer relaxations or reactions. If Fickian diffusion controls recovery, one would expect that the the boundary would move inward at a rate proportional to the square root of time. When relaxations or first order reactions control recovery, the rate of boundary movement is proportional to time rather than the square root of time. It has been shown, for example, that when glassy polymers are subjected to mechanical swelling stresses, the swelling border moves inward at a rate proportional to time when relaxation rates control the process (3,4,5). Here, we present data that indicates that stress may play an important role in the recovery process. We study analogues of stress free, flexible PS and PMMA and report no annealable color centers in these polymers. Flexible siloxane polymers that have glass transitions below room temperature are shown to be optically radiation hard.

Experimental

PS 615 APR (Dow Chemical) and PMMA CP41 (Continental Polymers) were compression molded in a piston mold against two glass surfaces. The samples were 2.5 cm in diameter and 1 cm in height. Poly (n-hexyl methacrylate) was dissolved in toluene, poured into a quartz cuvette and the toluene was devolatilized. N-decyl styrene was polymerized in a cuvette with 0.1% Vazo 52 (Du Pont) for 8 hours at 90 °C. Polymethylphenylsiloxane (Dow Corning) was studied in a quartz cuvette. Samples were irradiated with a ^{60}Co gamma source at a dose rate of 1.5×10^5 rad per hour. Transmission spectra were recorded with a Hewlett Packard 8452 A spectrophotometer. Boundary motion was observed via a reticule with 0.1 mm divisions under fluorescent light.

Results and Discussion

The kinetics of recovery boundary movement in PMMA and PS were characterized for samples irradiated and recovered in an air environment. The boundary movement is linear with time. At room temperature the velocity of recovery for PMMA is 1.9×10^{-8} cm/sec and for PS is 28×10^{-8} cm/sec. Figure 1 depicts boundary velocities at 25,40,56 and 70 °C. The linear movement with time indicates that recovery may be controlled by a relaxation process rather than by a diffusion process. Scission and cross-linking reactions induced by radiation can highly stress glassy matrices. For example, when a chain undergoes scission, two new chain ends are created. The free volume required by each new end compresses neighboring molecules.

In some instances the stress may exceed the failure stress of the sample. Cracking was observed in a 1.5 X 1.5 X 5 cm piece of PVT scintillator exposed to 10 Mrad. Stress is relieved with time as the polymer relaxes. Relaxation proceeds from the surface of the sample inward. The border velocities for PS display Arrhenius Behavior. The activation energy for the relaxation process was calculated from our data on boundary velocity to be 9.9 kcal/mol. This is near to that reported for the secondary gamma relaxation in PS which is associated with phenyl group rotation assisted by cooperative back bone motion. This activation energy is reported to be 8 kcal/mole (6,7). With all of this in mind, stress and relaxations appear to play a fundamental role in radiation damage and recovery. We note, however, that oxygen too plays a role in recovery. In systems starved of oxygen, if diffusion is slower than relaxation, we expect the diffusion process to govern kinetics. When diffusion governs kinetics a sharp boundary is not noted [8]. Our samples were not oxygen starved, however, recovery in oxygen depleted PS has been reported to occur in the absence of a sharp boundary [9].

We were therefore motivated to modify the polymer matrix so that it responds or relaxes instantaneously. When a polymer is above a characteristic temperature, the glass transition temperature, T_g, backbone motions and relaxations are rapid. PMMA and PS have T_gs at 100 ° C. These structures were altered so that the T_gs were less than room temperature by attaching a plasticizing tail on each monomer unit. The structures are shown in figure 2. Figure 3 shows the transmission spectra of PMMA and "tailed" poly (n-hexyl methacrylate). The radiation induced optical damage is clearly minimal. Similar, excellent results were obtained for "tailed" PS.

The "tailed" polymers are high viscosity liquids. We changed these liquids into flexible, cross-linked elastomers by polymerizing in the presence of a cross-linking agent, DVB. These cross-linked rubbers do not show annealable radiation damage. By increasing the cross-link density, we can increase the polymer moduli. We have shown in the past that siloxane polymers with T_gs less that room temperature are optically radiation hard when cross-linked or in linear form.

Plastics which have the propensity to form annealable color centers can be modified to become radiation hard by depressing the glass transition temperatures. This does not, however, solve the problem of non-annealable color centers. Some plastics form stable chromophores upon irradiation. We have examined flexible polyurethanes which turn orange upon irradiation and do not recover.

In conclusion, we have expanded the number of optical polymers which can be used as scintillator. By synthesizing new polymers with glass transition temperatures less than room temperature we have tentatively identified radiation induced stress a key factor in light absorption by glassy polymers.

Acknowledgments

We acknowledge support from a Department of Energy Grant:DE - FGO5 - 86ER40272.

References

[1] V. D. Bezuglii and L. L. Nagnoraya, "Effect of Radiation on the Stability of Plastic Scintillators,"J. Nucl. Ener. A/B19, 490-494, (1965).

[2] The University of Florida has applied for a patent in this area.

[3] Julie P. Harmon, "Anisotropic Methanol Transport in PMMA After Mechanical Deformation," Polymer 29, 1221-1226, (1988).

[4] Julie P. Harmon, S. Lee and J.C.M. Li, "Methanol Transport in PMMA: The Effect of Mechanical Deformation," J. Polym. Sci. Part A, 25, 3215-3229, (1987).

[5] J. P. Harmon and J. C. M. Li, "Transport of Alcohols in Deformed PMMA,"Presented at Annuals New York Academy of Science Conference on Structure and Mobility in Molecular and Atomic Glasses, New York, 371, (1979).

[6] S. Reich and A. Eisenbiey, "Theoretical Approach to the Assignment of Molecular Mechanism for Cryogenic Loss Peaks in Polymers," J. Polym. Sci., A-2, 10, 1397-1399, (1972).

[7] V. Tanabe, "Phenyl Group Rotations in Polystyrene," J. Polym. Sci, Polymer Physics Ed. 23, 601-606, (1985).

[8] A. Peterlin, "Diffusion with Discontinuous Swelling," J. of Research NBS, 81A, Nos. 2 and 3, 243-250, (1977).

[9] R. Clough, Sandia National Laboratories, Albuquerque, NM. Private communication.

* To whom all inquiries should be directed.

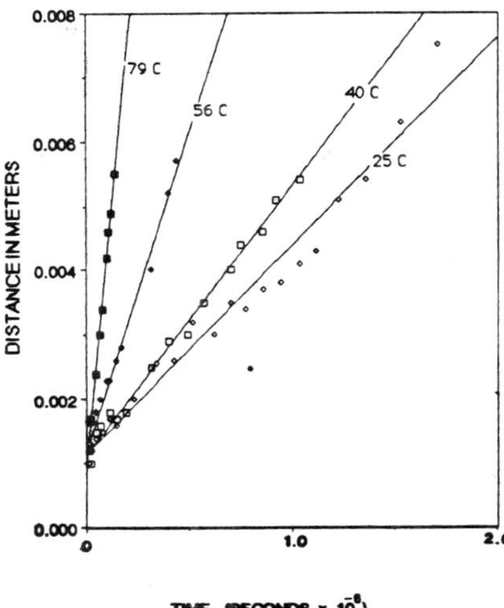

Figure 1. Boundary movement during PS recovery.

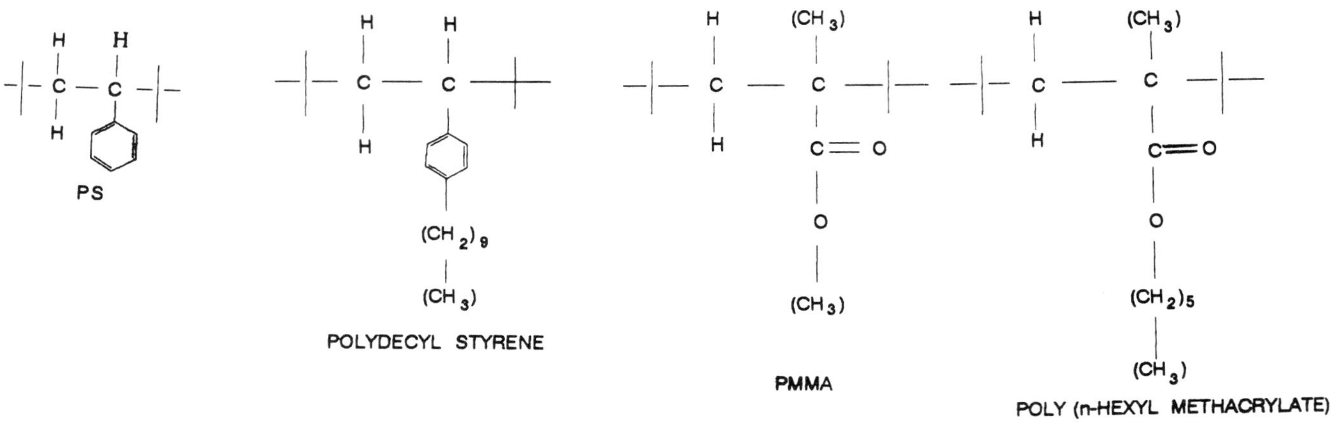

PS

POLYDECYL STYRENE

PMMA

POLY (n-HEXYL METHACRYLATE)

Figure 2. Structures of PS, PMMA and "tailed" versions of these polymers.

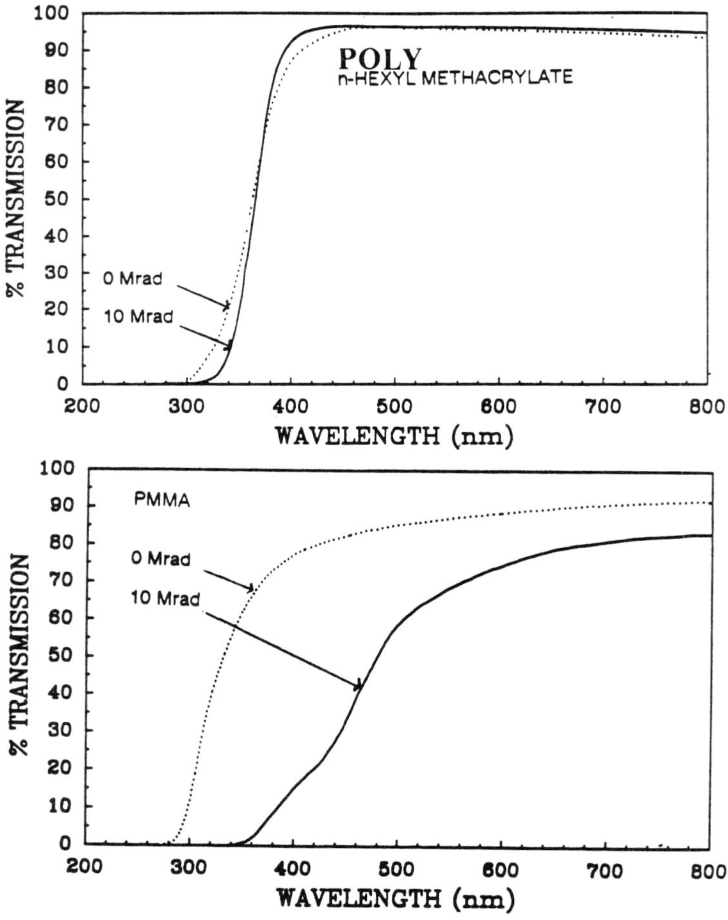

Figure 3. Poly (n-hexyl methacrylate) and PMMA before and after 10 Mrad.

NEUTRON FLUENCES IN IRON OR LEAD CALORIMETERS AT THE SSC

J. S. Russ
Carnegie Mellon University
Pittsburgh, PA 15213

A. Fassó and G. R. Stevenson
CERN
Geneva, Switzerland

M. Bruzzi, C. Furetta, and P.G. Rancoita
INFN
Milan, Italy

P. Giubellino and R. Steni
INFN
Turin, Italy

Abstract

A series of neutron flux measurements have been made for high energy hadrons interacting in a calorimeter-like dump structure, made of iron or lead. The neutron fluence as a function of energy and position is developed from integral neutron activation in materials having step-function energy response. From measurements in two materials and at two different energies, data to compare to Monte Carlo simulations of the SSC radiation field have been developed. Reasonable agreement is found, and a simple model is found to describe most of the features of the data.

Introduction

The neutron fluence within and backscattered from the detector calorimeters is a major design issue for 4π detectors at the SSC or LHC. The Monte Carlo simulation of the neutron generation process demands the correct matching of a high energy particle production model with a nuclear excitation and decay model to produce the myriad of low-energy particles, mostly neutrons and photons, that constitute the hadronic cascade. Prior to the inception of this series of experiments at CERN in 1988 there was very little data on the details of hadronic shower development. It is important to ensure that simulations provide accurate profiles of the neutron fluence within the calorimeter, in order to impose radiation tolerance standards on the components. It is equally important to define the energy and angle dependence of both this internal fluence and the albedo neutron flux that affects other components of the detector. The goal of this set of measurements has been to simulate a calorimeter structure and provide non-intrusive spatial and energy sampling of the neutron cascade initiated by high energy hadrons. The method chosen is that of neutron activation of inorganic elements. One chooses elements with sharp thresholds and overlapping cross sections in order to form a quasi-differential spectrum by subtraction.[1]

Iron Calorimeter Measurements

The iron calorimeter was a stack of twenty 5-cm. plates with 0.7-cm. gaps which allow the insertion of the activation foils. The total thickness was $6\lambda_{int}$. This dump was exposed to a 200 GeV/c positive beam (67% protons) in the North Area of the CERN SPS and to a 24 GeV proton beam from the PS. Similar activation detectors were used in the two exposures. The activation foils and their energy characteristics are listed in Table 1.

The major questions of interest are the energy profile of the neutrons in the cascade and the fluence, especially at the maximum of the cascade and at the front face (albedo). In this experiment the activation foils were disks on a crossed array. This permitted detection of and correction for centering errors of the beam. The increasing disk diameters roughly match the rate of flux decrease away from the beam axis. Statistical errors on all points were comparable and less than 8% in the worst case. Excellent agreement was observed between activation levels for disks at corresponding points in the array, so all disks at the same radii were combined to improve statistical precision.

In the SSC report on Radiation Levels in the Interaction Regions [2] there are approximate scaling laws developed for the albedo and maximum fluences as a function of bombarding energy, adjusted for material. One of the major interests in this work is to test these scaling estimates. They are expected to have different energy dependence. The neutron generation mechanism is chiefly a low-energy nuclear excitation process that leads to neutron evaporation. This will depend on the bombarding energy through the number of nuclei excited. Increasing incident energy implies that the cascade maximum moves deeper into the calorimeter. For albedo this moves the "source" of thealbedo neutrons farther away from the front face and imposes more attenuation material. This combination of solid angle reduction and increased absorption means that the

Table 1. Parameters of Activation Reactions

Reaction	Cross Section	Activating Hadron	Nominal Energy Window
$^{115}In\ (n,n')\ ^{115m}In$	120 mb	neutron	0.8-15 MeV
$^{32}S\ (n,p)\ ^{32}P$	300 mb	neutron	2.6-25 MeV
$^{27}Al\ (n,\alpha)\ ^{24}Na$	85 mb	neutron	6-25 MeV
$^{27}Al\ (h,x)\ ^{18}F$	8 mb	any hadron	> 35 MeV

increase in albedo neutrons is expected to be slower than the rise in the overall neutron fluence, e.g., that at maximum. The fluence at maximum will rise less rapidly than linear because of the loss of energy to electromagnetic processes, e.g., π^0 production and decay. Both of these effects are seen clearly in the data. Figure 1 shows the longitudinal profile of the hadronic cascade for the two bombarding energies as measured by different activation foils. The fluence is dominated by "1 MeV" neutrons, which activate the ^{115}In but not the other materials.

If one compares the ratio of the albedo fluence to the maximum fluence for 200 GeV/c and 24 GeV/c, the ratio is obviously larger for the lower energy, as the above arguments would imply. The quantitative results are in reasonable agreement with the scaling proposed in reference 2: albedo fluence scales as $E_{inc}^{0.5}$, but the maximum scales more nearly like $E_{inc}^{0.85}$ than $E_{inc}^{0.67}$ as seen from Table 2. The question of systematic error and the absolute normalization of these results must be addressed in making the energy comparison. The absolute normalization error of the activation measurement is quite small. The counting systems are calibrated against known standard sources before the run. The conversion from activation level to neutron fluence introduces systematic error. One assumes that the measured activation arises solely from neutrons in the cascade. The ^{115}In and ^{32}S reactions are neutron-specific, but ^{24}Na may arise from other processes. A systematic error of 50% to the 200 GeV/c data has been assigned. The error in the 24 GeV/c data is expected to be very similar. The neutron activation processes are low-energy phenomena and should have little variation with bombarding energy. Therefore, it is expected that the relative normalization of the 200 GeV/c and 24 GeV/c data will be better than the absolute normalization.

To compare to Monte Carlo simulations, use calculations presented by Brau and Gabriel in reference 2. They computed the neutron fluence versus neutron energy at various depths in an iron/silicon calorimeter for an incident energy of 20 GeV. This dump was designed to simulate a similar system, replacing Si by Al. Integrating the Monte Carlo spectrum over the various activation cross sections enables us to make a direct comparison between the data and the 20 GeV simulation. We present the 24 GeV/c data both as measured and scaled to 20 GeV. The 200 GeV/c data are scaled. This tests the absolute normalization both of the data and of the simulation. The results are listed in Table 2. There is a tendency for the Monte Carlo to underestimate the number of "high energy" (6-24 MeV) neutrons. This is of no consequence for radiation damage. It does have some

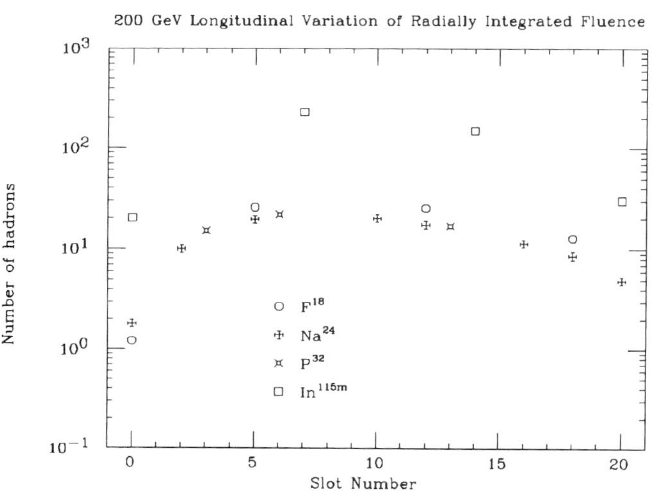

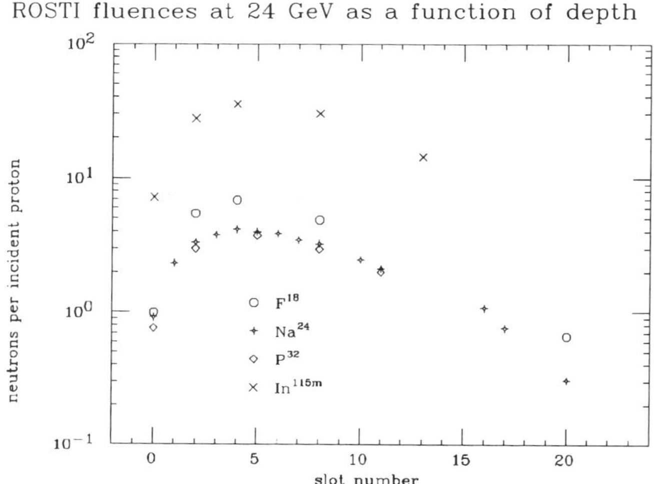

Figure 1. Neutron fluence (radially integrated) as a function of calorimeter depth for four different activation elements: ^{18}F(spallation reactions); ^{24}Na (>6 MeV); ^{32}P (>2.4 MeV) ; and ^{115m}In (> 0.8 MeV) for hadrons at 200 GeV and protons at 24 GeV

Table 2. Comparison of Predicted and Measured Neutron Fluences in Iron
Radially-integrated flux at cascade maximum

Data	^{115}In (n,n') ^{115m}In	^{32}S (n,p) ^{32}P	^{27}Al (n, α) ^{24}Na
200 GeV/c protons, scaled by $(E/20)^{0.85}$	51 neutrons/incident hadron	5.6 neutrons/hadron	5.1 neutrons/hadron
200 GeV/c protons, scaled by $(E/20)^{0.67}$	77 neutrons/hadron	8.6 neutrons/hadron	7.7 neutrons/hadron
24 GeV/c protons, scaled by $(E/20)^{0.85}$	50 neutrons/hadron	5.6 neutrons/hadron	6.0 neutrons/hadron
Brau-Gabriel spectrum folded with activation cross section	72 neutrons/hadron	10.6 neutrons/hadron	3.0 neutrons/hadron

import for silicon calorimetry and should be studied further in that application.

There are other simulations that should be noted. The Brau and Gabriel calculation uses the full details of the hadronic cascade development, as embedded in the Oak Ridge National Laboratory Code package CALOR.[3] This is an elaborate code. If nuclear evaporation processes excited chiefly by spallation reactions are the major source of neutrons, as these results indicate, then one can look at the high energy parts of the cascade, above 50 MeV, to simulate the deposition of energy via spallation. One can produce the evaporation neutrons by a simple isotropic decay model, using the observed attenuation length in the medium range to set the range of the slow neutrons. Figure 2 shows the results of a simple model applied to the "1 MeV" neutron radial profile at several depths in the stack, both at 200 GeV/c and 24 GeV/c. The match to the data is excellent, and the running time for such a code is far shorter than for CALOR.

In summary, the iron calorimeter data can be understood by a model in which the high energy interactions produce spallation-excited nuclei. The de-excitation process produces, on average, 10 neutrons per excitation, emitted isotropically. These neutrons propagate over a 10 cm mean free path in the material to generate the observed fluence. Due to the number and distribution of the spallation reactions in a cascade, the albedo becomes a smaller fraction of the total number of neutrons generated as the incident energy increases. Iron as an absorber has an advantage, in terms of neutron production, because the mean number of neutrons released per excitation is smaller than in heavier nuclei like lead.

Lead Calorimeter

In order to quantify the material dependence of neutron generation in the hadronic cascade, the group have built and exposed a lead dump of the same geometry as the iron dump described above. Iron and lead have almost the same interaction length for hadrons. The electromagnetic showers in lead are attenuated much faster, but they play little role in neutron generation.[4] Data were taken at 200 GeV/c in the same beam as for the iron dump and are now being analyzed.

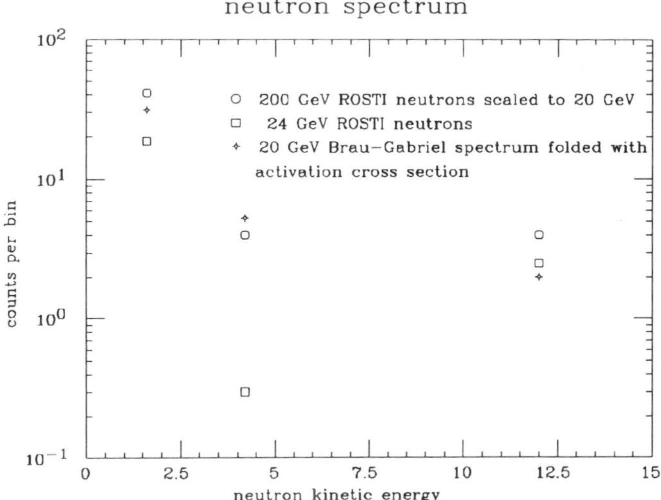

Figure 2. Comparison of the radial profiles for "1 MeV" neutrons, i.e., those detected by ^{115}In, at the front face (albedo) and at cascade maximum. Shown are the data points and also the absolute predictions of a simple evaporation model, using the ^{18}F spallation data as a source. The model is described in more detail in ref. 1.

One additional measurement made with the lead dump concerns the propect of attenuating the albedo neutron flux by inserting a CH_2 filter between the calorimeter and the detector volume inside. It has been estimated that a 10-cm absorber could reduce the albedo flux by an order of magnitude.[5] We have compared the albedo flux at the upstream face of the lead stack with the flux measured in another layer of ^{115}In located upstream of a 10-cm-thick CH_2 filter. Initial analysis indicates a flux reduction by a factor of 4. This is not inconsistent with the results of reference 5, which extended the neutron energy range significantly below the cutoff energy of ^{115}In. A more detailed comparison is underway.

In summary, the lead calorimeter exposure will provide a check of the material-dependence of neutron generation and a test of the simple ideas of evaporation-dominance of the neutron fluence developed from the study of the iron system. Results will be available within the next six months.

Acknowledgements

We thank the technical staff at the CERN SPS and PS and C. Leroy and the Montreal group for much assistance in the installation and execution of these runs. Thanks also go to the many others who helped in the measurements.

References

[1] J. S. Russ, G. R. Stevenson, A. Fasso, M. C. Nielsen, C. Furetta, P. G. Rancoita, and L. Vismara, *Low-energy neutron measurements in an iron calorimeter structure irradiated by 200 GeV hadrons,* CERN Report TIS-RP/89-02 (1989).

[2] D. E. Groom, ed., *Radiation Levels in the Interaction Regions,* SSC Report SSC-SR-1033 (1988).

[3] T. A. Gabriel, et al., *CALOR89 Monte Carlo Program Package for the Design and Analysis of Calorimeter Systems,* ORNL/TM-11185 (in press, 1990).

[4] H. Hirayama and W. R. Nelson, *Radiation Levels in Electromagnetic Calorimeter,* appendix 7 of ref. [2].

[5] T. A. Gabriel and R. A. Lillie, *Reduction of the Neutron Albedo in the Vertex Detector Region of a Generic SSC Detector,* SSC Central Design Group Internal Note SCC-N-545 (1988).

A PREDICTION OF THE NEUTRON AND CHARGED PARTICLE BACKGROUNDS IN THE L* DETECTOR*

D. M. Lee, W. W. Kinnison, and W. B. Wilson
Los Alamos National Laboratory
Mail Stop H846
Los Alamos, New Mexico 87545

Abstract

Monte Carlo calculations have been made of the neutron flux and activation in the forward and barrel calorimeters in the L* detector and of the neutron flux in the central detector volume. In addition estimates of the charged particle and neutron background rates in the vicinity of the muon chambers has been determined. The Los Alamos National Laboratory code system LAHET and CINDER,90 along with ISAJET and GEANT were used in these studies. The results indicate that neutron fluences as low as 2 x 10^{12} per SSC year can be achieved in the central volume.

Introduction

The design of detectors for the interaction regions of the SSC must take into account the potentially severe radiation problems that may exist due to the high energy and high luminosity. In the absence of a specific detector design, an SSC Central Design Group task force assessed the radiation levels in a 'generic' SSC calorimeter [1]. The levels of radiation from this study clearly show that any SSC detector design must address the problem of potential radiation damage to specific detector components.

The radiation levels in the L* detector have been addressed in considerable detail through detector simulation and modelling. The following is a report of this work.

Computational Program

The L* geometry was simulated in two codes. We used the GEANT code [2] with ISAJET [3] to produce minimum bias events from which we determined the angular and energy distributions of the hadrons entering the calorimeter. The distributions were determined for two ranges of pseudorapidity,-3.0 to 3.0 and 3.0 to 5.9 with the former used for the barrel portion and the later used for the forward calorimeter end caps. The hadron angular distribution was uniform in pseudorapidity. In general, the neutron flux in the central volume and barrel calorimeter is due to hadrons with momenta less than 5 GeV/c and in the forward calorimeter to momenta less than 20 GeV/c.

An exploded view of the L* detector is shown in figure 1. The electromagnetic calorimeter is 50 cm thick BaF_2 and the hadron calorimeter is a mixture of lead and iron and both cover the range of pseudorapidity of± 3.0. In the region at a pseudorapidity of ± 1 the hadron calorimeter is 193 cm thick while in the forward region with a pseudorapidity of 1.5 to 3 the thickness is 240 cm. The volume inside the electromagnetic calorimeter contains the central tracker and consists of silicon microstrip detectors, straw tubes, and scintillating fibers. The central tracker volume is 75 cm in radius and 1.5 meters long.

The Los Alamos National Laboratory code system LAHET [4] was used to transport the hadrons through the calorimeters. All secondaries produced were transported until they fell below some threshold (usually .1 to 1 MeV for charged particles or to 20 MeV for neutrons) by an enhanced version of the Oak Ridge National Laboratory code HETC [5]. All low energy neutrons (< 20 MeV) produced as a result of the intranuclear cascade were transported by the MCNP code [6] to thermal energies or until capture. Neutrons were allowed to survive for 1 ms at which time any survivors were killed. The LAHET code system allows for edits of particles in energy, angle, and time as well as over cells or surfaces defining the problem. Additional edits of residual masses were used in determining the activation.

Neutron Albedo in the Central Volume

The neutron fluence in the central volume is a very important parameter because the silicon tracker lifetime could be compromised if the fluence is to high. In analyzing the damaging effects of the return neutron albedo on the silicon microstrip detectors we have taken into account the neutron energy dependence of displacement damage in silicon. We have integrated the

* This work was performed under the auspices of the US Department of Energy, Contract W-7405, ENG-36.

neutron flux over two regions, above 100 keV and above 1 eV. The integral above 100 keV is the region most damaging to silicon and the region above 1 eV is important for other materials such as plastics.

The neutron flux in the central volume was calculated for the L* detector assuming the central tracker extended to the boudary of the electromagnetic calorimeter (radius=75 cm). However, it was anticipated that a soft neutron shield in front of the electromagnetic calorimeter might be beneficial in reducing the flux. Therefore, a composite shield of polyethylene and boron (0.13 radiation length, 5cm thick) was also investigated.

The time distribution of the neutron flux in the central volume for a single SSC bunch crossing is shown in figure 2. The maximum flux occurs 20-30 ns after the SSC bunch crossing. The distributions for all neutrons greater than 1 eV exhibits a long tail while the distributions for those neutrons greater than 100 keV falls more sharply with time. This is a result of the softening of the spectrum as the neutrons undergo repeated collisions in the shield or calorimeter. The tails of the distributions directly effect the equilibrium flux in the central volume since successive SSC crossings will add to the flux remaining from previous events.

The equilibrium fluence of neutrons in the central volume is tabulated in table 1.

Table 1.
Equilibrium Neutron Fluence
(neutrons/cm^2-SSC year)

Energy	poly shield	no shield
E > 1 eV	3.4×10^{12}	4.4×10^{13}
E > 100 keV	1.7×10^{12}	8.7×10^{12}

The polyethylene-boron shield is very effective in reducing the neutron fluence.

One possible option for the L* detector is to replace the BaF_2 calorimeter with a lead -iron calorimeter. We have studied this case and find that the neutron fluence in the central volume increases by factor of 1.4 over the similar BaF_2 case with a 5cm polyethylene shield.

An examination of a 50 cm central tracker radius was investigated with a 5 cm polyethylene shield in place. The neutron flux increased by 1.6 which does not scale with the square of the radius. This is because more neutrons escape when the radius is 50 cm than when it is 75 cm. If, however, the calorimeters extend only to eta of 2.5 than the flux in the central volume decreases by 1.5.

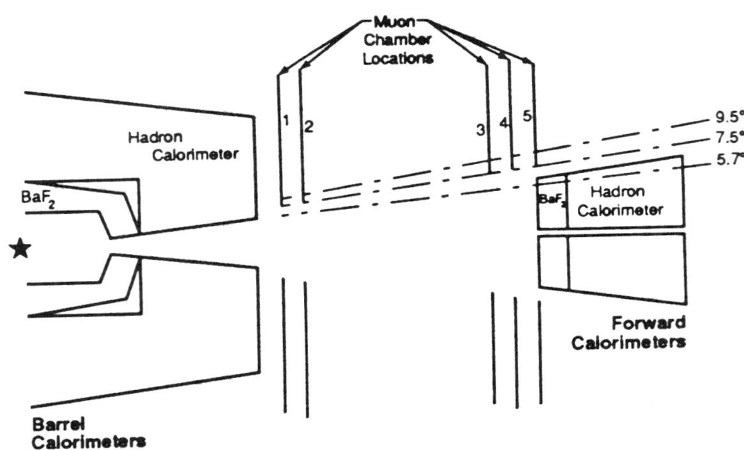

Figure 1. Cutaway view of the L* Detector

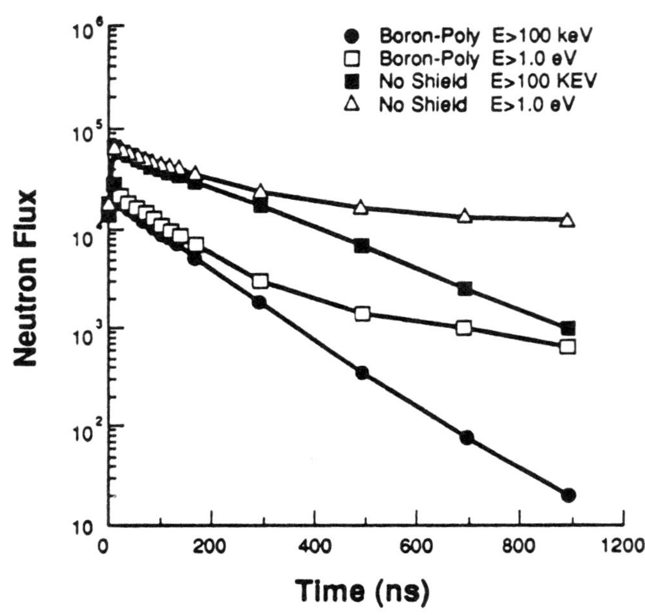

Figure 2. Neutron flux in the central volume

Neutron Fluence in the Calorimeters

The neutron fluence in the barrel and forward calorimeters are shown in figures 3 as a function of angle and depth for the energy region E > 1 eV. Maximum fluences occur in the BaF_2 region at levels of 6×10^{13} in the barrel calorimeter and 8×10^{14} in the forward calorimeter. The fluences for the energy region E > 100 keV are lower by a factor of 1.8.

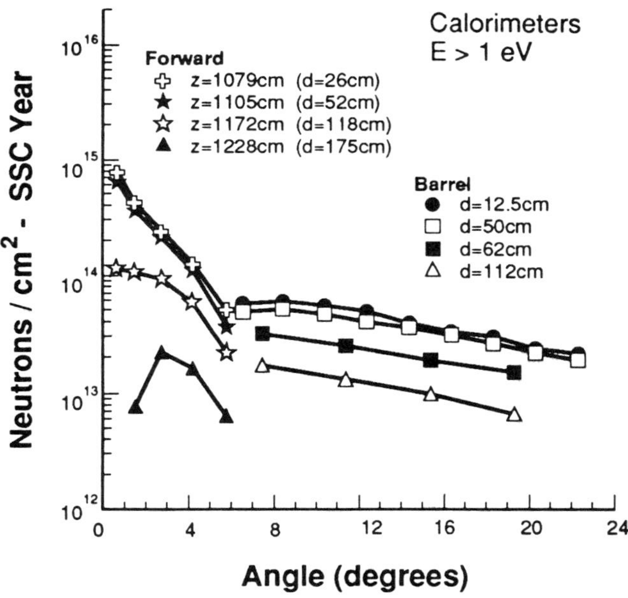

Figure 3. Neutron fluence in the calorimeters

Muon Detector Backgrounds

The backgrounds in the muon detectors will originate from a number of sources, backscatter from the forward calorimeter, punch through from the barrel calorimeter, and interactions in the exit hole portion of the barral calorimeters due to misalignments, as well as interactions in the beam pipe. The results of our study are shown in table 2. The values represent the contributions from the beam pipe and forward calorimeters in the angular region 7.5 degrees to 9.5 degrees. The 3 mm thick stainless steel beam pipe contributes greater than 80% of the background.

Table 2.
Background Rates in the Muon Chambers

(Particles/cm^2 - sec)

muon chamber	charged particle	neutron
1	8×10^3	1.6×10^4
2	1×10^4	1.5×10^4
3	6×10^3	1.5×10^4
4	5×10^3	1.2×10^4
5	4×10^3	7.2×10^3

A test for misalignments in the barrel calorimeter was simulated by indroducing a 1 cm step in the hadron calorimeter on the 5.7 degree edge, 40 cm from the back face. The results yielded rates per hadron that were 2 to 10 times lower than the above results.

Activation

In the forward calorimeter next to the beam pipe the flux levels are sufficiently high that activation of the calorimeter and beam pipe could present a problem. We have begun to study this problem. The residual masses as determined from the HETC code and the neutron flux determined from the MCNP code were input to the code CINDER,90 that predicts the residual nuclides and activation products. For this examination a small portion of the forward calorimeter was exactly specified as a 6 mm lead slab followed by an 18 mm stainless steel slab with a 3 mm gap of silene. In this region the neutron flux was 5.4×10^7. We assumed a 6 month beam on period followed by a 6 month beam off period and determined the activation at the end of 10 years. The activity in the lead and steel slabs was 1.6×10^5 disintegrations/cm^3-s and in the warm liquid 4.2×10^3. This activity reaches 80% of the 10 year value after the first 6 months. In the future we expect to determine the detectable count rate in the counting volume from the residual activity and from the photons produced during the neutron slowing down process.

References

[1] D. E. Groom, "Report of the Task Force on Radiation Levels in the SSC Interaction Region", SSC Central Design Report SSC-SR-1033 (June 1988)

[2] R. Brun, et. al.,"GEANT3", CERN Report CERN DD/EE/84-1

[3] F. E. Paige et. al., "ISAJET 6.25", CERN Pool W999

[4]R. E. Prael et. al., "User Guide to LCS: The LAHET Code System", Los Alamos National Laboratory Report LA-UR-89-3014 (Sept., 1988)

[5]Radiation Shielding Information Center, "HETC Monte Carlo High-Energy Nucleon-Meson Transport Code", Report CCC-178, Oak Ridge National Laboratory (August 1981)

[6]Group X-6, "MCNP-A General Monte Carlo Code for Neutron and Photon Transport", Los Alamos National Laboratory Report LA-7396-M Revised (April 1981)

CHARACTERIZING COMPLEX RADIATION ENVIRONMENTS USING MORE (MONITOR OF RADIATION EFFECTS)

P.J. McNulty and D.R. Roth
Department of Physics and Astronomy, Clemson University, Clemson SC 29634-1911

E.G. Stassinopoulos
NASA Goddard Spaceflight Center, Greenbelt MD

W.J. Stapor
Naval Research Laboratory
Beltsville, MD 20375-5000

Abstract

The MORE instrument can be used for testing microelectronic technologies for use in complex radiation environments such as those inside SSC detector assemblies or operational satellites.

1. Introduction

Particle detector assemblies proposed for the SSC resemble electronic systems flown on operational satellites in that they must be designed to operate for long periods in complex radiation environments with few if any opportunities for repairs. There are long lead times between the design of a system and its implementation in orbit or on the beam line. The radiation threats to microelectronics in both cases include the all-or-nothing single-event effects as well as the gradual degradation due to cumulative exposure. Like satellite system designers, SSC experimentalists will be forced into difficult choices between the guaranteed survivability obtained by building systems around rad-hard components and the greater performance at lower cost associated with commercial devices. The rad-hard approach will probably be followed in the earliest versions of the detector components, but the desire to improve will be almost immediate. Use of any commercial parts requires that one be able to predict with some accuracy both the single-event effects and the gradual degradation due to accumulated absorbed dose. This requires radiation testing of critical components combined with detailed knowledge of the particles making up the environment. Detailed measurement of the environment is not feasible. Nor can standard test procedures be corrected easily for dose-rate effects or synergistic effects. Trials of critical components in the identical or similar radiation environments would appear desirable before committing to specific technologies.

We describe in this paper an instrument designed to allow experimenters to characterize complex radiation environments in terms of their effects on microelectronic components of interest to the experimenter while simultaneously measuring the absorbed dose. This instrument, the Monitor of Radiation Effects (MORE) is designed to measure the rate of single event upsets (SEU), monitor how these rates would vary with device sensitivity, and measure circuit degradation due to chronic or acute exposures. It is designed to operate on operational satellites or within the limited empty spaces inside functioning SSC detectors.

2. Instrument Design

The proposed instrument has three separate components designed to carry out different types of measurements: The first component measures the SEU rate on an array of SRAMS. Soft errors or single event upsets are the most common form of single event phenomena in space, and they are the form most likely to be observed in microelectronics used inside detector assemblies. The name single event upsets comes from the fact that the upsets are typically the result of individual interactions. They can occur in logic as well as memory circuits, but they are most easily studied in memories. Often the electrical (and hence logic) states of more than one circuit element are changed as a result of one event. This is taken into account because the instrument provides the memory location, type of error, and time of occurrence for each event. Large memory arrays can be used in order to obtain reasonable statistics in the shortest possible time. Figure 1 is a schematic diagram of the SEU tester component of MORE. It requires only 6 chips including one of the memory devices being tested. An essential feature of this component is that the devices under test are interchangeable - the SRAMs can be of any technology from 4K to 256K (or higher when available). Changing SRAMs only requires switching socket adapters and the ROM chip.

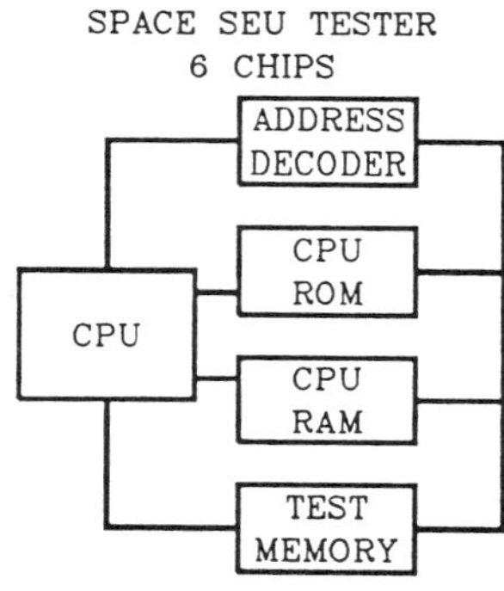

Fig. 1 Schematic of SEU tester of MORE.

Soft errors are believed to be a threshold phenomena with threshold being determined by the size of the pulse generated across the drain-substrate junction for most NMOS and CMOS devices. The cross section for SEU events is plotted in Fig. 2 versus the LET (or stopping power) of the incident particle for the IDT 6116V, a device used for testing the prototype of MORE. While the data taken from Ref. 1 appears to depend on the ion species, it does exhibit a threshold behavior in that there are no upsets at low LET values and then a sharp rise to a cross section which roughly equals the total cross sectional area of the array of SEU-sensitive junctions. The gradual increase beyond threshold is due to multiple upset events. The device has also been shown to upset when exposed to protons incident at energies as low as 30 MeV. The proton-induced upsets in SRAMs are due to spallation reactions at or near the SEU-sensitive junctions.

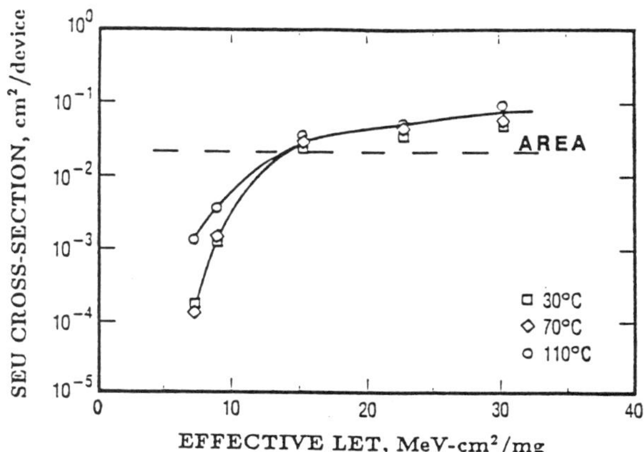

Fig. 2 Cross section for SEU events versus LET for the IDT 6116V. Data is from Ref. 1.

The second component of MORE conducts a pulse-height analysis of the pulses generated at the microjunctions of a parallel array of circuit elements as a result of radiation-induced events. The junctions used are identical or at least similar to the SEU-sensitive junctions of the SRAMs monitored by the first component. The SRAM used in the prototype was an NMOS device, the IDT 6116V, and the pulse height spectra were measured off the power pins of an identical device. In both cases, the events of interest are due to interactions at junctions from similar arrays. For CMOS devices, the pulse-height spectra must be measured from a parallel array of drain-substrate junctions. These test structure arrays are often availability from the vendors since they are used as test structures in yield studies. The experimental configuration used for the pulse-height measurements is shown schematically in Fig. 3. The pulse-height spectrum obtained with the prototype exposed to 4.8 MeV alphas is shown in Fig. 4. This spectrum was measured on an array of 32,678 junctions, each having dimensions of about 11 μm on a side. The area of the junctions drawn in Fig. 2 as a horizontal line was obtained from the ratio of the number of events under the peak in Fig. 4 to the fluence. This agreement between the area and the measured SEU cross section provides some confidence for using this arrangement for characterizing more complex radiation environments.

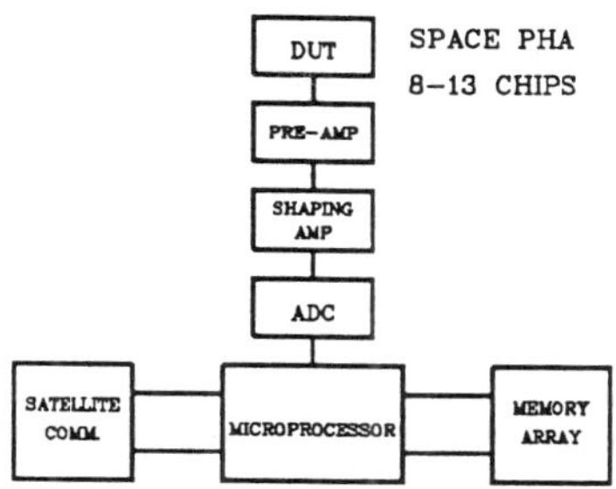

Fig. 3 Experimental configuration used for pulse-height measurements in MORE.

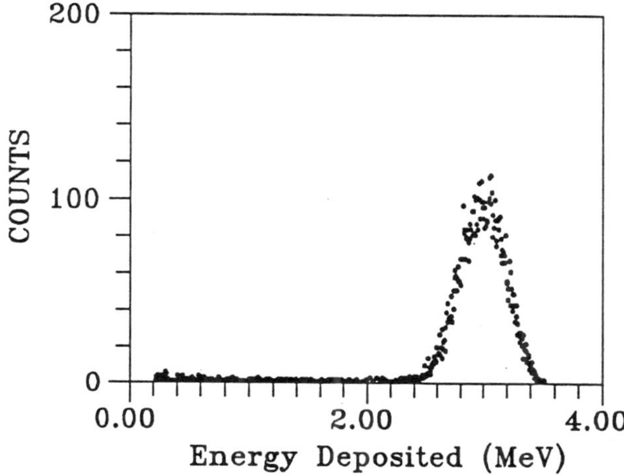

Fig. 4 Differential pulse-height spectrum obtained with more prototype exposed to 4.8 MeV alphas.

The pulse-height spectra obtained when the prototype was exposed to 148 MeV protons is shown in Fig. 5 where the data is plotted in integral form as the cross section for depositing at least a given energy E versus E. This spectrum is roughly typical of the form spectra will take in regions around the beam line within a detector assembly except that, cross sections cannot be determined in a complex radiation environment. Instead, the plots will be the number of events in which at least E is deposited versus E. Cross section cannot be calculated because the identity of the incident particles are not known. However, even without knowing the particles making up the incident radiation, the threshold for the device can be determined. This is illustrated in Fig. 5 where the SEU cross section measured for the IDT 6116V is represented as a horizontal line. The energy corresponding to threshold is marked on the abscissa where the line intersects the experimental curve. In a complex environment, the numbers of SEUs (scaled to the same junction area) should correspond to the largest events on the spectrum. Again, comparison of the measured SEU rate and the spectrum yields the value of threshold.

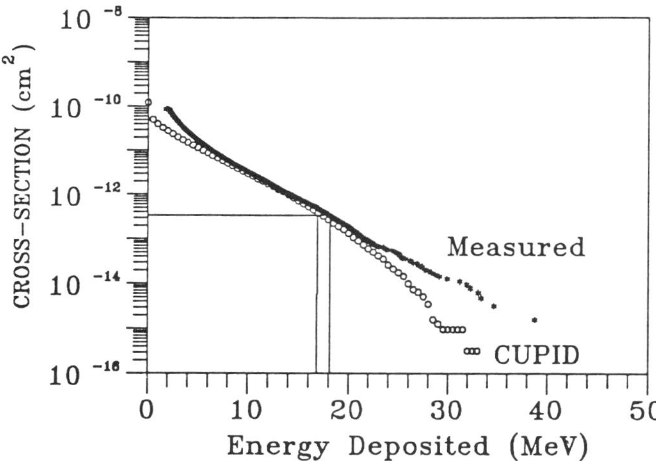

Fig. 5 Integral pulse-height spectrum obtained from exposure to 148 MeV protons.

The theoretical curve represents the simulations from CUPID (2) for the sensitive volume associated with the SEU-sensitive junction of the IDT 6116V with the dimensions of the sensitive volume obtained from the position of the peak and the number of events in Fig. 4 (3). The agreement between theory and experiment, besides testing the simulation model, provides some confidence in the use of pulse-height arrays to determine the SEU parameters of a technology.

The third component of MORE measures the total dose absorbed in SiO_2 while monitoring the effects on a test structure from a device made with the technology to be tested. The total dose is obtained by measuring the turn-on voltage on calibrated PMOS transistors following the procedures outlined by Holmes Siedle et al (4). The same procedures are used to measure the turn-on voltage on the test structures. A schematic cross section of a MOS transistor is shown in Fig. 6. The gate bias required to turn on or off the test structure is monitored. Changes in the turn-on voltage is the dominant failure mode for MOS devices exposed to ionizing radiation at low and moderate dose rates. It is principally the result of the buildup of positive charge in the thin oxide under the gate and negative interface states at the SiO_2/Si interface.

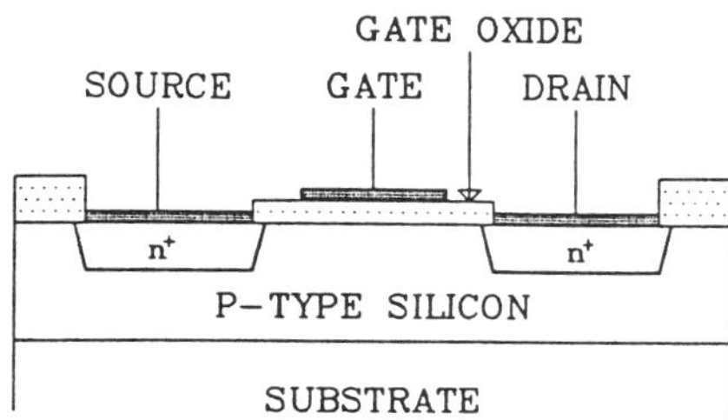

Fig. 6 Schematic cross section of a MOS transistor.

3. Specifications

The SEU, PHA, and Total Dose components of MORE require 2 W, 2.5 W, and 1.5 W of power, respectively. The total weight of the package is 6 lbs and fits inside a box having dimensions of 8″ x 6″ x 4″.

4. Summary and Conclusions

The radiation problems within the the detector assemblies proposed for the SSC will be similar to those experienced on operational satellites. Because of the absence of heavily ionizing cosmic rays around accelerators, the environment should be less severe than in deep space. However, both the single event upsets and gradual device degradation due to accumulated dose is to be expected. The use of rad-hard components developed for military systems flown in space should be sufficient for SSC applications for any locations except in or very close to the beam line. Use of commercial devices is risky even when the candidate parts are tested if the SEU rate to be expected must be accurately known. The MORE instrument allows trials of a technology of devices in complex radiation environments. Problems of synergistic effects and dose-rate effects are automatically included in the test data using this approach.

5. Acknowledgements

This work was funded in part by the Defense Nuclear Agency.

6. References

1. R. Koga, W.A. Kolasinski, J.V. Osborne, J.H. Elder, and R. Chitty, IEEE Trans. Nucl. Sci NS-35, 1638 (1988).
2. P.J. McNulty, G.E. Farrel, and W.P. Tucker, IEEE Trans Nucl. Sci. NS-28, 40007 (1981).
3. P.J. McNulty, D.R. Roth, W.J. Beauvais, W.G. Kader, and D.C. Dinge, Nuclear Tracks and Radiation Meas., to be published.
4. A. Holmes-Siedle and L. Adams, Radiat. Phys. Chem. 28, 235 (1986).

RADIATION EFFECTS ON JFETS, MOSFETS, AND BIPOLAR TRANSISTORS, AS RELATED TO SSC CIRCUIT DESIGN*

E. J. Kennedy, B. Gray, and A. Wu
Dept. of Electrical and Computer Engineering
The University of Tennessee
Knoxville, TN 37996

G. T. Alley and C. L. Britton, Jr.
Oak Ridge National Laboratory†
P. O. Box 2008, Oak Ridge, TN 37831-6006

P. L. Skubic
University of Oklahoma
Department of Physics & Astronomy
Norman, OK 73019

Abstract

Results of radiation effects on selected junction field-effect transistors, MOS transistors and bipolar junction transistors are presented. Evaluations include both dc parameters, as well as capacitance and noise evaluations. Detailed noise data were obtained both before and after 5-Mrad (gamma) total-dose exposure. SPICE radiation models are compared for a typical bipolar transistor charge-sensitive preamplifier.

Introduction

Superconducting Super Collider (SSC) circuits demand high frequency performance from the electronics, yet the need for thousands of circuits absolutely requires low power operation. Additionally, an absolute limitation to performance is the radiation due primarily to gamma and neutron bombardments that can exceed a total dose of 1 Mrad (gamma) and 10^{13} neutrons/cm^2 in a typical year of operation, dependant upon the distance from the beam [1]. The reported research was conducted at the Oak Ridge National Laboratory (ORNL) and the University of Tennessee, Knoxville, and was primarily concerned with gamma radiation effects on semiconductor devices and circuits. The source was a Gammacell 200 ^{60}Co irradiator, operating at a dose rate of 40 krads/hr. The circuits developed were primarily preamplifiers that could be used in large volume calorimeters having detector capacitances in the range of 20 to 500 pF.

Gamma Radiation Effects on JFETs

The JFET generally suffers the least damage in a radiation environment when compared with bipolar and MOS transistors. When using a JFET as the front-end device in a charge-sensitive (CS) preamplifier, the parameters of most interest are series noise, transconductance (g_m), capacitance, and output resistance [usually expressed as the ratio of the Early voltage (V_A) to the dc drain current (I_D)]. Generally five samples of each device were characterized both before and after a 5-Mrad ^{60}Co irradiation. The devices were irradiated in a passive mode (all leads shorted together), and postirradiation testing was accomplished within typically 2-3 days after irradiation. The n-channel JFET units evaluated were the NJ3600, NJ1800, NJ903, 2N6451 and 2N4858 from InterFet, Inc.; the 2N5432 and IMF5912 from Intersil; the 2SK152 from Sony; and the 2N4416A from Siliconix. Since space is limited, only data for the largest area (NJ3600) device is presented in this paper. All devices were characterized at three drain currents of 1 mA, 100 μA and 10 μA, and with a drain-source voltage of 2.5 V.

The equivalent noise voltage (ENV, or 'series' noise) of the NJ3600 is shown in Fig. 1 for I_D of 1 mA and 100 μA, both before and after 5-Mrad gamma radiation, expressed as an equivalent noise resistance. The lower straight line on the graphs indicates the theoretical midband noise resistance anticipated [$R_n = 4kT\ (0.66/g_m)$]. The "error bars" indicate the range of measurements (min., max.) for the 5 units of each device tested. As is apparent in Fig. 1 the region of midband noise is not achieved until >100kHz after radiation. The low-frequency 'noise corner', f_c, at which the series noise resistance is twice that of the midband noise resistance is shifted from ~ 15 Hz (before radiation) to ~ 70 kHz after radiation in the NJ3600, versus ~2 kHz (prerad) to ~ 500kHz (postrad) for the smallest area device tested, the 2N4416A.

---

*This research was funded by the DOE SSC Generic Detector Program.

†The Oak Ridge National Laboratory is operated for the U.S. Department of Energy by Martin Marietta Energy Systems under contract No. DE-AC05-84OR21400.

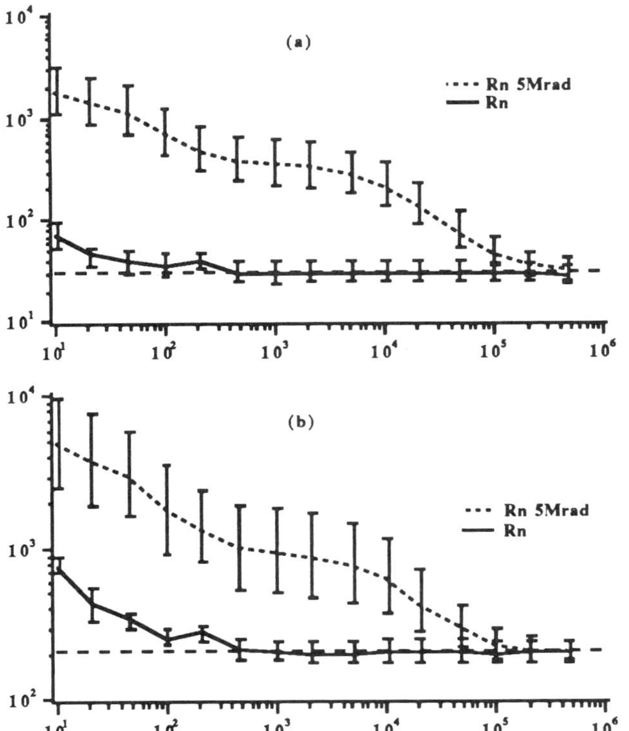

Figure 1. Series noise resistance versus frequency for the InterFet NJ3600L at (a) I_D = 1 mA and (b) I_D = 100 μA [V_{DS} = 2.5 V].

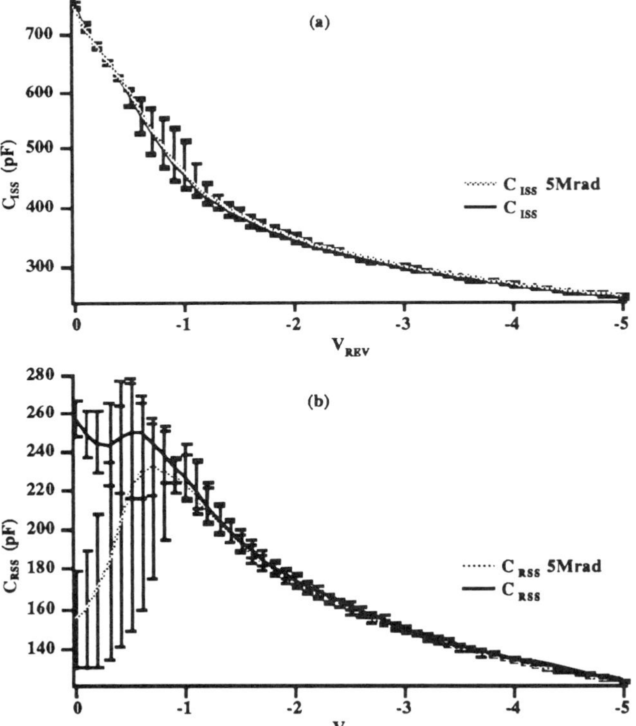

Figure 2. Capacitance variations for the NJ3600L JFET, both before and after a 5-Mrad dose. (a) Input capacitance C_{iss} and (b) gate-to-drain capacitance C_{rss}.

The measured input and output capacitance (C_{iss} and C_{rss}) for the JFETs generally indicated very slight changes in the postrad data relative to the initial values. Fig. 2 illustrates data comparisons for the NJ3600. The initial decrease of the C_{rss} values for $V_{reverse}$ < 1 V after radiation is puzzling. The largest postrad change in capacitance occurred with the 2N4416A devices, which indicated a 0.2 pF increase in C_{rss}.

Several other parameters of the JFETs tested indicated very little change with radiation, with pinch-off voltages changing < 1%, I_{DSS} values typically unchanged and gm values indicating perhaps a slight (<1.8%) decrease after radiation. The Early voltage values, however, increased from +18% for the NJ3600 to +160% for the 2N4416A devices. As anticipated, the gate leakage currents increased because of radiation from < 10^{-13} A to 0.4 nA for the 2N4416A, and for the largest area device (NJ3600) the change was from 0.3 nA to 6 nA after 5 Mrads.

Gamma Radiation Effects on Bipolar Transistors

The most significant effect of ^{60}Co gamma radiation on the high-frequency bipolar transistors tested was a decrease in the dc current gain, Beta (ß) or h_{FE}. The processes evaluated were those of the leading bipolar analog VHF semi-custom integrated circuit silicon foundries, namely AT&T, Harris Semiconductor and VTC, Inc. The AT&T ALA200 array is typically a 4.5 GHz npn and 3.8 GHz pnp f_T family. The Harris analog FASTRACK VHF process utilizes 1.2 GHz npns and 1 GHz pnps, while the VTC VJ900 process provides a typical maximum f_T of 6 GHz (npn) and 1.5 GHz (pnp).

Figure 3 compares the pre and postrad current gains (Beta) for the smallest area npn devices (an emitter size of 1 x 1, approximately 10 um x 10 um), while Fig. 4 compares Beta for the small area (1 x 1) pnps, although the VTC unit tested was only available in a 1 x 9 area. The lower f_T pnp devices suffer the most degradation due to the ionizing gamma radiation. There were only slight changes (<5%) in the C_{bc} and C_{be} depletion layer capacitances produced by the irradiations. The Early voltage (V_A) for npn devices typically decreased due to irradiation, from 3% to 10%, while V_A increased by 15% for the Harris pnps, but decreased by 5% for the AT&T pnps and by 25% for the VTC pnps.

An accurate SPICE model for each bipolar process was obtained, both before and after radiation. A comparison was made using each process to simulate a basic low-power 20 mW folded-cascode CS preamp [2] for use with a 2 cm x 2 cm silicon calorimeter detector (Cdet~ 112 pF). The results for each preamplifier are compared in Table 1, both before and after radiation.

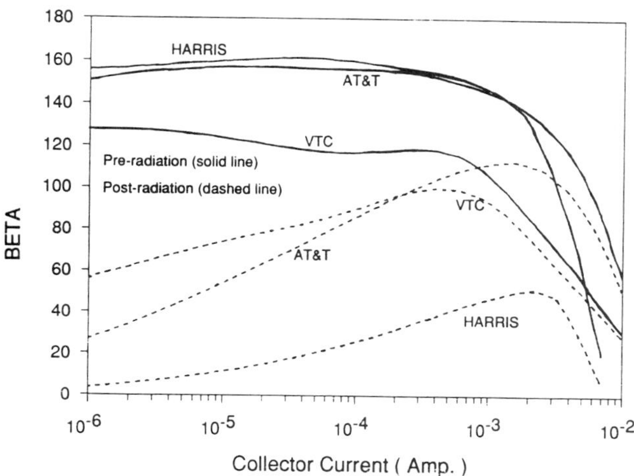

Figure 3. Beta vs. I_C for npn high-frequency transistors. V_{CE}= 2.5V.

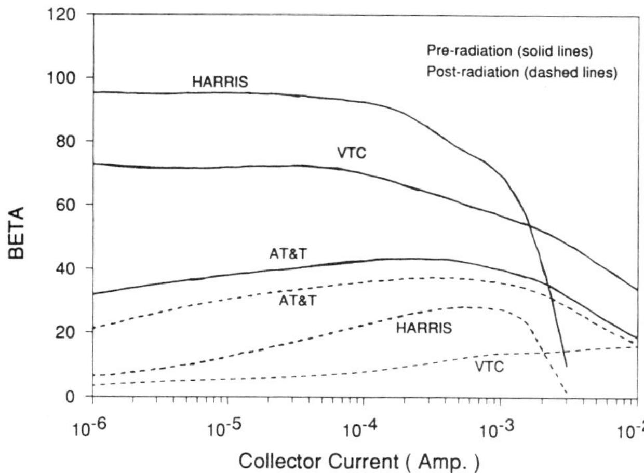

Figure 4. Beta vs. I_C for pnp high-frequency transistors. V_{CE}= 2.5V.

Table 1. SPICE simulation comparisons for a CS folded-cascode bipolar transistor preamplifier with a calorimeter detector capacitance of 112 pF. Shaping time-constant (CR-RC) of 50 nsec.

Process	Risetime (ns)	P_{diss} (mwatt)	Noise ENC_e (rms electrons)
	Preradiation		
AT&T	4.9	19.1	4098
Harris	5.5	20.0	5001
VTC	4.9	19.3	4026
	Postradiation (after 5 Mrad ^{60}Co)		
AT&T	7.0	18.3	4553
Harris	8.0	17.7	7821
VTC	7.0	18.5	5075

Gamma Radiation Effects on MOSFETS

Although a special radiation-hard CMOS process will be required for SSC circuits, we wanted to see how a typical MOSIS CMOS process would survive a gamma dose. Special NMOS and PMOS dual arrays were fabricated under the standard 2 um p-well Orbit Semiconductor process, using the MOSIS fabrication program. The dual transistors were biased in a normal conductive state (V_{DS} = 2.5 V|, I_D=1 mA, 100 uA and 10 uA) during the irradiation. The threshold voltage V_T was monitored, and devices were removed when the observed threshold voltage for the PMOS units had shifted to approximately 2 x V_T(prerad), which occurred at a total dose of ~1.3 Mrad. The PMOS devices had an average shift in V_T from -0.7 V to -1.5 V, while NMOS devices had V_T shifts from 0.9 V to 0.6 V, although undoubtedly the worst-case negative shift from the initial 0.9 V value was much larger. The large increase (approximately 5x) in the low-frequency noise due to radiation is shown in Fig. 5 for an NMOS device with W/L=100 μm/10 μm. There was a large increase in the offset voltage (V_{os}) of the dual units, with prerad values of ≤ 1 mV for both n-and p-channel units observed for 1000 um/10 um devices, while in a postrad measurement V_{os} values of ≥ 100 mV were obtained. The shift in g_m (at 1 mA) was significantly higher than that observed in JFETs, approximately -12% for PMOS and -19% for NMOS units after radiation. The Early voltages, similar to that for JFETs, increased after radiation from ~20 V to ~60 V for PMOS units, and from ~30 V to ~150 V for NMOS units.

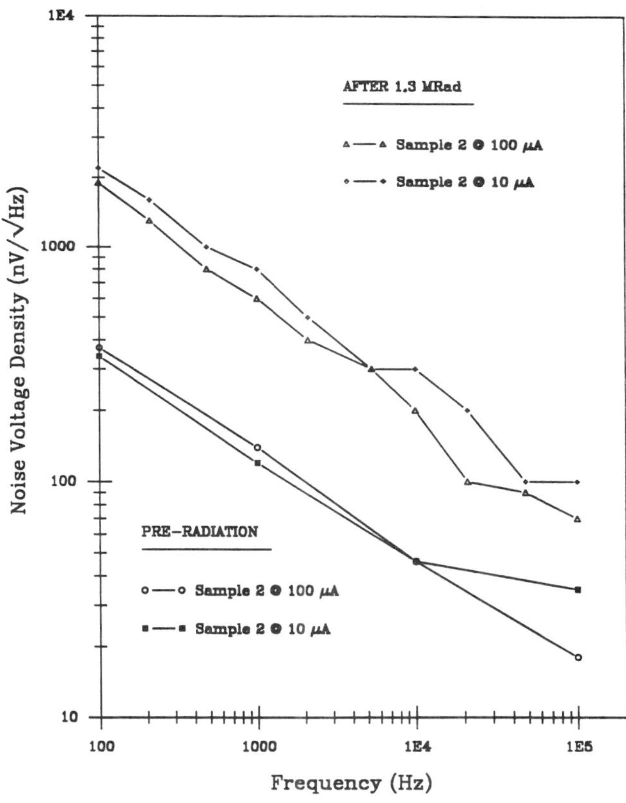

Figure 5. Noise spectra for a MOSIS NMOS transistor with W/L = 100μm/10μm, both before and after a 1.3 megarad (Co60) total dose.

Conclusions

Radiation using a ^{60}Co gamma source has shown a significant increase in the low-frequency noise for both JFETs and MOSFETs. JFETs are decidedly superior to MOSFETs where constancy of g_m and threshold voltage are concerned. Although the Beta degradation is severe, particularly for pnp transistors, the high-frequency bipolar transistor offers very low series noise when compared with an FET and shows promise as a useful technology for the short peaking times that will be required for the SSC.

Acknowledgements

The authors acknowledge the various semiconductor companies, namely InterFet, AT&T, Harris, and VTC, Inc. for providing sample transistors and arrays for testing. We also thank Ms. M. Kemper for her assistance in measurements and testing.

References

[1] D. E. Groom (ed.), "Radiation Levels in the SSC Interaction Regions," SSC-SR-1033, June 10, 1988.

[2] C. L. Britton, Jr.,et al.,"Bipolar Monolithic Preamplifiers for SSC Silicon Calorimetry," paper to be presented at this Conference.

RADIATION DAMAGE STUDIES ON HYBRID PREAMPLIFIERS

H. W. Kraner, J. A. Kierstead, P. W. Levy, S. Rescia, V. Radeka and A. Stevens*

Brookhaven National Laboratory, Upton, NY 11973
*Argonne National Laboratory, Argonne, IL 60439

Abstract

Low noise hybrid preamplifiers have been irradiated with 60Co gamma rays and fast neutrons at both room temperature and 120K to study radiation effects on the noise characteristics. Increase in low frequency parallel noise was observed in preamps powered during irradiation by 60Co at 1.45 Mrad at room temperature and at 12 Mrad at 120K and for neutron fluences at 10E13n/cm^2. Annealing and relaxation effects will be discussed.

Introduction

Low noise hybrid preamplifiers developed at BNL have proved to be excellent vehicles with which to study radiation effects on the noise characteristics of junction FET's. These studies also provide direct empiricle experience in a representative radiation environment. Figure 1 shows a variety of such hybrid preamplifiers, which have been applied to the readout technology of many detector systems ranging from silicon tracking detectors to liquid argon calorimeters.

A set of 19 hybrid preamplifies, IO 535-01 (typically the preamp in the lower right of Figure 1) were exposed to 1.45 Mrad 60Co over a period of six weeks at a rate of 5 × 10E3 rad/hr at room temperature. Half were powered, biased, half were not during the exposure. Four IO 475-1 preamplifiers, center, Figure 1, metal encased, but similar to the IO 535-01, were irraddiated to 12 Mrad at 120K in a vacuum cryostat, also over a period of up to five weeks; measurements were made at stages during the irradiation. Two were powered, two were not. Two IO 475-1 preamps plus a special preamp having a high resistivity channel. FET were exposed to 10E13 fast neutrons/cms during an eight hour shift at the U. of Lowell, MA, Van de Graaff accelerator.

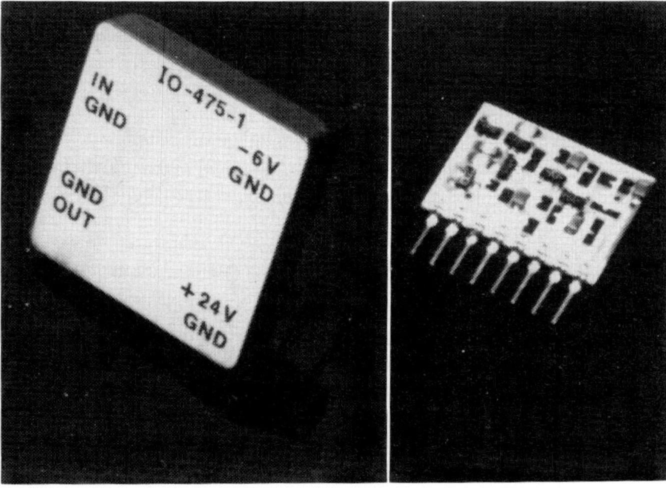

Figure 1. Hybrid preamplifiers designed at BNL for high energy physics applications. The metal encased IO475 is on the left, the IO535 on the right. The IO535 is ~ 1" in length.

Noise measurements were made by measuring the broadening of a calibrated injected charge (pulser peak) observed through a pulse amplifier (Ortec 450) and multichannel analyser. 100pf was added to the input to emphasize series noise. The noise is quoted in sigma, equivalent noise charge, electrons, as observed as the full width at half maximum of the pulser peak, =2.35sigma. The FWHM was measured over a range of shaping time constants, 0.25 μsec to 10 μsec to emphasize series noise at higher frequency band passes and parallel noise dominant at lower frequencies.

The preamplifiers are conventional low noise single stage charge sensitive loops with a folded cascode input stage and emitter follower line driver. The schematic for IO 535 is included as Figure 2; IO 475 is basically similar. They use a 2N6451 JFET (Interfet) as the input element and have a 2N4416 (Motorola) connected as a protection diode across the input.

Several preamps from the room temperature irradiation were given a 65C anneal for 1000 min and a 120C anneal for 1000 min at a later stage.

Results

The principle results from a 1.45 Mrad dose at room temperature are shown in Figure 3. An increase in low frequency noise of about a factor of two is observed in the powered devices compared with about a 20% increase over the full frequency range for unpowered devices.

Several devices were removed from the group for other purposes and the remainder were checked from time to time after irradiation up to 6 months (4580 hr) after the 1.45 Mrad exposure. Figure 4 shows the noise measurements during this period with the simple conclusion that the unpowered preamps did not anneal whereas the powered preamps recovered by perhaps 50% during this period.

The 2N6451 JFET is guard-ringed [1] and ought to be considerably more radiation tolerant than these results indicate; therefore the protection diode a 2N4416 JFET was suspected of being the low frequency parallel noise source in the powered preamps. The protection diodes were removed from two powered and unpowered preamplifiers and the powered preamp noise returned almost completely to its pre irradiation value; the unpowered preamps remained at their post irradiation values, slightly higher than the initial noise; this is the maximum effect of 1.45 Mrad that is attributable to the necessary components of the preamplifier.

Annealing beyond room temperature of four preamplifiers, two irradiated with power (3 and 10) and two without power (13 and 20) , was carried out at elevated temperatures, 65 and 120C. The data from the 120C anneal are shown in Figure 5. Both sets showed little change from their final room temperature anneal value (after 4580 hr at 22C) from 1000 min at 65C. However, a much greater improvement was evident after 1000 min at 120C, which is probably a prudent upper limit for anneals of this package, considering the solder and other fabrication materials. Especially pertinent to the powered preamps, one remembers that anneals are really directed towards the parallel noise from the unguarded 2N4416 across the JFET input.

An additional 10E5 rad was given in order to test the anneal mechanism. Pre and post anneal data are shown in figure 5, indicating again that the 120C anneal largely restores the pre irradiation noise response in the powered preamps of the powered devices it is recognized that only 10E5 rad places their noise spectrum well back to the post 1.45 Mrad, pre-anneal value. This effect is not unusual and is generally explained by suggesting that the anneal simply supplies electrons (from the silicon side of the interface) to the hole traps, neutralizing them and reducing the detrimental effect of a charged interface without effectively changing the structure which creates the traps. With a very little more ionization, the traps refill and a greatly worsened noise response immediately results. Some structural anneal may take place, but it is certainly incomplete at 120C, probably the limiting temperature for this package. This effect also clearly demonstrates the greater abundance of interface states initially generated in the powered device.

The four IO 475 preamplifiers irradiated at 120K up to 12 Mrad were measured at several stages during the irradiation and underwent several thermal cycles to room temperature over the five week period. Results of the two groups are shown in Figure 6 which reveals that if power is applied directly prior to measurement (after removing it to move the cryostat from the irradiation cell), little change is noted in the noise. However, after 2 hours of power on, bias applied, the low frequency noise in both groups is considerably increased. Power can be removed and reapplied while the devices remain cold and the cycle repeats; thus, the longer term result would be that observable during experimental conditions. After removal of the protection diode, the output voltage of the preamp which is the gate leakage current times the feedback resistance (20M) was found to decrease over the two hour period from 12.9V to 11.2V which implies an increase of gate leakage of almost 100na. It is this current that is presumably the origin of the increased parallel noise.

Three hybrid preamplifiers, type IO475-1, were irradiated immersed in liquid nitrogen with fast neutrons, receiving a fluence of 9.3 × 10E12 n/cm^2 at a van de Graaff accelerator from the 7Li(p,n) reaction [2]. Because the expected damage is displacement damage with little expected field dependence, no bias was applied during the irradiation, which lasted seven hours. Two preamplifiers (No.s 1 and 9) were standard types with a 2N6451 as the input JFET and has a 2N4416 protection diode across the input; the third preamp used a special type 132H JFET as the input element, which is a fet built on higher resistivity material and will contain a larger channel volume (for comparable pinch-off) than the 2N6451. After irradiation, the devices were studied cold but better noise measurements were not made until 70 days after the irradiation with several room temperature warming cycles in the interim. Figure 7 compares the noise power spectrum following irradiation with an average pre-irradiation spectrum of cold, IO475 preamplifiers. Also shown on Figure 7 is the post irradiation spectrum of the 132H preamplifier, however it should not be compared with the pre irradiation values of standard IO475's. With the increased channel length and volume, it will have generally higher noise values.

Discussion and Conclusions

1. For irradiations of warm preamplifiers up to 1.45 Mrad, the powered preamps increased in parallel noise by a factor of two but only slightly in series noise; the unpowered device increased also slightly uniformly as a function of frequency. By removing the protection transistor, an unguarded 2N4416, the parallel noise in the powered preamps was eliminated; no change was noted in the devices irradiated without power. This element can be replaced by a guarded device and little radiation effect would then be anticipated to well beyond the 1.45 Mrad used here.

The parallel noise produced by the unguarded 2N4416 is presumably due to increased gate leakage as has been observed by Allen et al [1]. One possibility is that the drain metalization overlying the oxide passivation over the channel, biased positively, forces holes from ionization in the oxide to the interface thus forming interface states as discussed by Winocur et al [3]. Without the guard band of p+ dopant, the area between the source and drain are inverted and an extraneous leakage path at the interface is formed. Increased gate leakage from the input FET, the 2N6451, guarded and easily measured, was not observed.

2. Very little effect of series noise was observed in any of these irradiations, warm or cold, indicating basic hardness above 10 Mrad for pulse measurement in the frequency domain beyond 10 MHz. This observation essentially qualifies these preamplifiers for use well above 10 Mrads, whether or not the protection diode is replaced. The low frequency characteristic noise from irradiated 2N4416s has previously been observed by Stephen [4].

3. No room temperature anneal was observed for the preamplifiers irradiated without bias; about a 40% anneal was observed in the parallel noise for those preamplifiers irradiated with power or bias on. Clearly much of the ionization in the oxide recombines without a bias field to direct it to the interface, which gives a very different interface from the powered devices.

4. A definite annealing stage for the parallel noise in the preamps irradiated warm was observed at 120C, with little noise improvement noticed for room temperature or 65C anneal. The annealing of trapped interface charged has been addressed recently by Lelis et al [5]. They describe the neutralization of interface hole traps by a tunneling mechanism from the silicon side, which does require 120C for both "hard and soft" oxides; it is thus probable that structural anneal, elimination of the traps themselves, does not take place. This agrees with the result that an additional 10E5 rad reestablishes the parallel noise to near it's pre-anneal value. The JFET's used were not particularly radiation "hardened".

5. An increase in low frequency parallel noise was observed only after 12 Mrad in the preamplifiers irradiated at 120K. A time dependent parallel noise together with increased gate leakage of the input FET with a characteristic time of 20 min was observed after returning bias onto the FETs that were irradiated cold. This effect was greater for the preamps powered during irradiation, but was also evident for the unpowered preamps. Because the transport of holes in the oxide is severely temperature dependent and restricted at lower temperatures [6], the nature of the hole traps either in the bulk or at the interface can not be compared to room temperature characteristics. Holes will not reach the interface to any similar extent forming fewer interface states and recombination must be a much larger factor (permitting 12 Mrad). The charging effect under bias and relaxation without may be a more likely indication of bulk trapping in the oxide with some inversion effect a the silicon side of the interface. In any event, the result from low temperature irradiation may be expected to be quit different from room temperature irradiation. The mechanism must also be active in the guarded structure, the 2N 6451, as well.

6. A general increase in the noise of about 25% was observed in IO 475 preamplifiers measured and irradiated near 77K by a fluence of about 10E13 fast n/cm^2. The 7Li(p,n) reaction produces neutrons evenly distributed between 10 Kev and 2.2 Mev at this machine energy. This spectrum is representative of neutron fluxes in calorimeters at high luminosity machines. These results essentially qualify hybrid preamplifiers to 10E13 fast n/cm^2 and it is expected that higher tolerances will be observed, especially in the higher frequency ranges for fast electronics.

References

[1] D. J. Allen, F. N. Coppage, G. L. Hash, D. K. Holck and T. F. Wrobel, "Gamma-induced leakage in junction field-effect transistors," IEEE Trans. Nucl. Sci. NS-31, 1487–1491, (1984).

[2] The irradiations were performed at the University of Lowell, MA, through the hospitality of Professor G. H. R. Kegel.

[3] P. S. Winokur, J. M. McGarrity and H. E. Boesch, Jr., "Dependence of interface-state buildup on hole generation and transport in irradiated MOS capacitors," IEEE Trans. Nucl. Sci. NS-23, 1580–1585, (1976).

[4] J. M. Stephen, "Low noise junction field effect transistors exposed to intense ionizing radiation," IEEE Trans. Nucl. Sci. NS-33, 1465–1470, (1986).

[5] A. J. Lelis, T. R. Oldham, H. E. Boesch, Jr. and F. B. McLean, "The nature of the trapped Hole annealing process," IEEE Trans. Nucl. Sci., NS-36, 1808–1815, (1989).

[6] T. P. Ma and P. V. Dressendorfer, *Ionizing Radiation Effects in MOS Devices and Circuits*. New York: John Wiley and Sons, 1989.

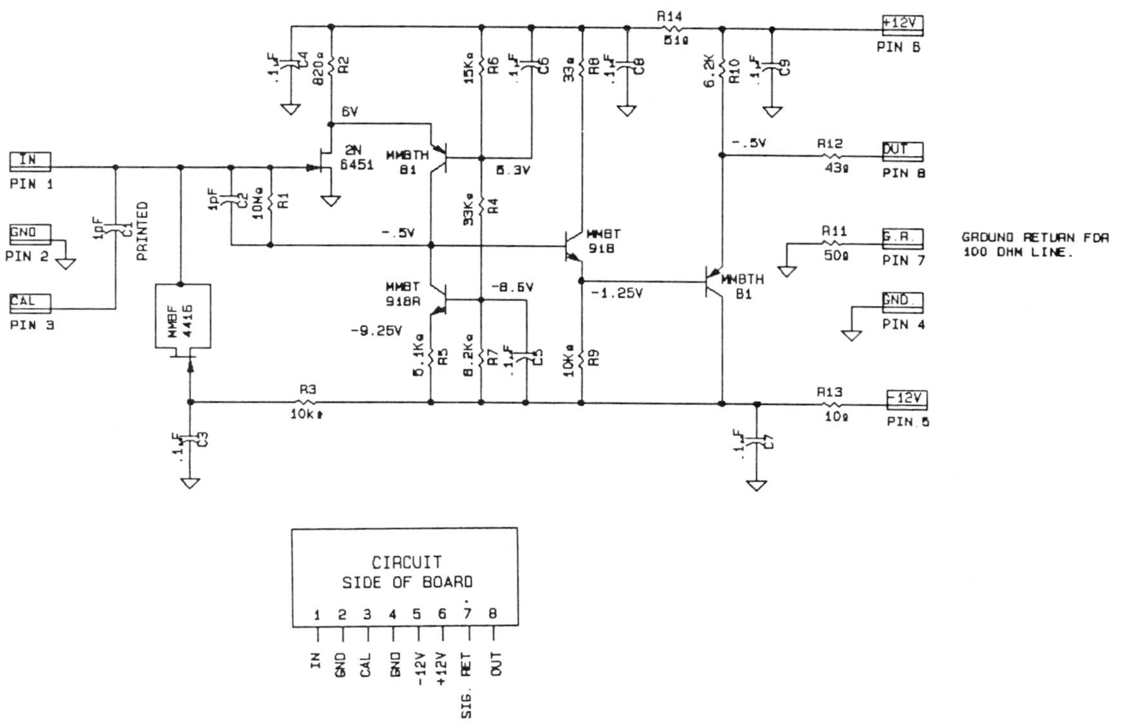

Figure 2. Schematic of the IO535 preamplifier.

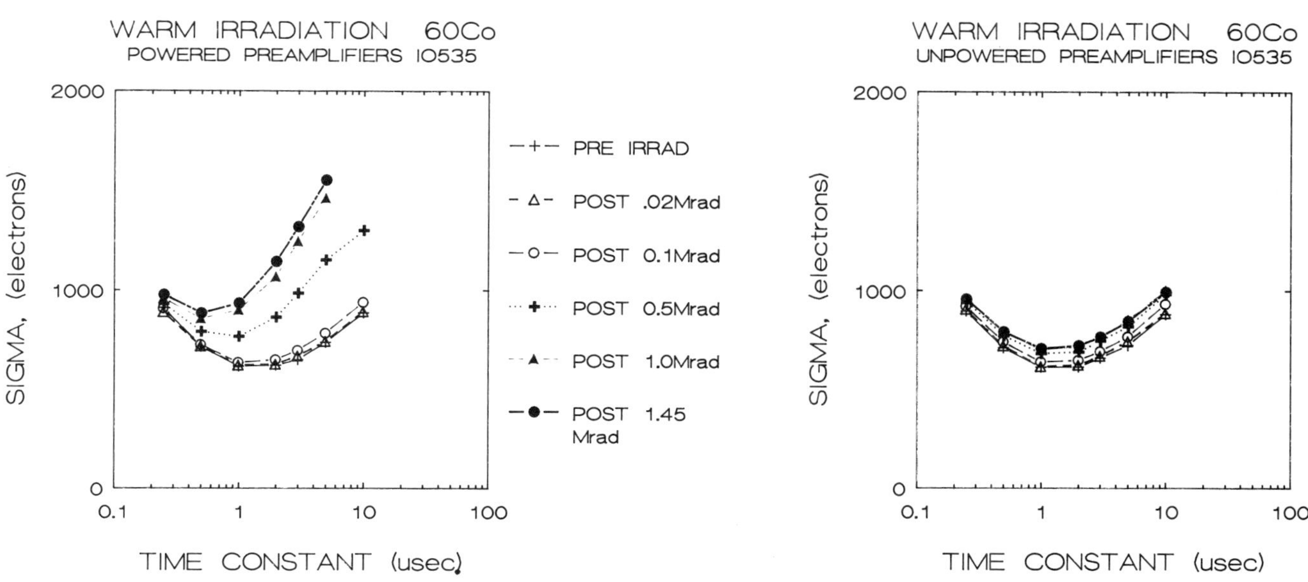

Figure 3. Results of room temperature irradiation with 60Co of IO 535 preamplifiers; nine were powered, ten were not.

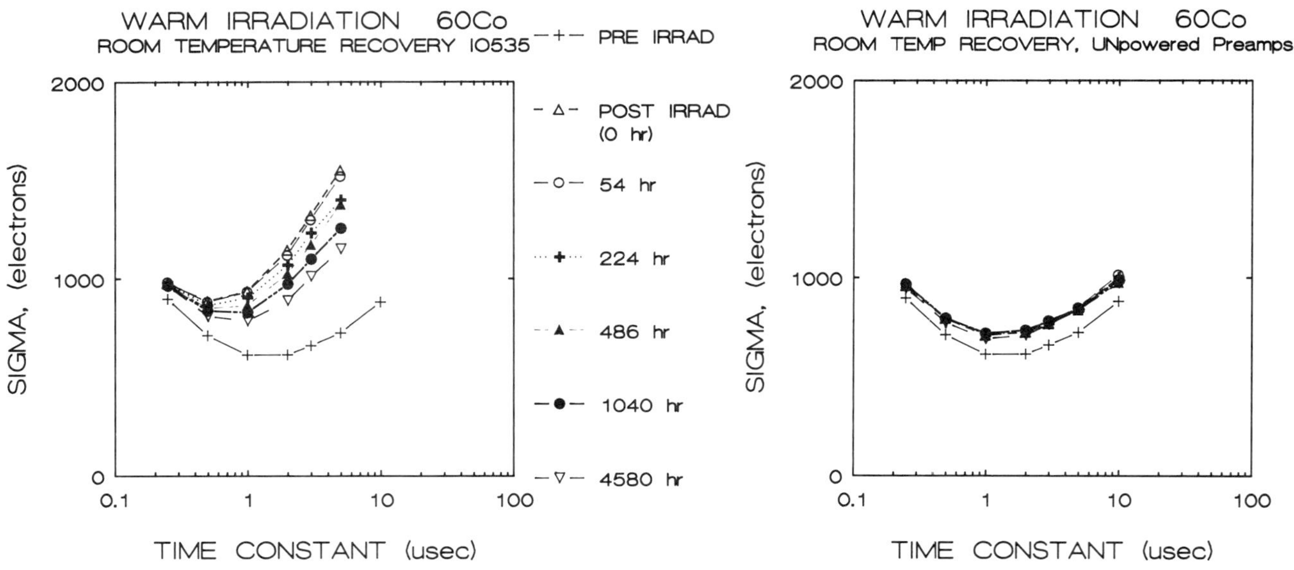

Figure 4. Results of room temperature anneal of IO 535 preamplifiers irradiated to 1.45 Mrad.

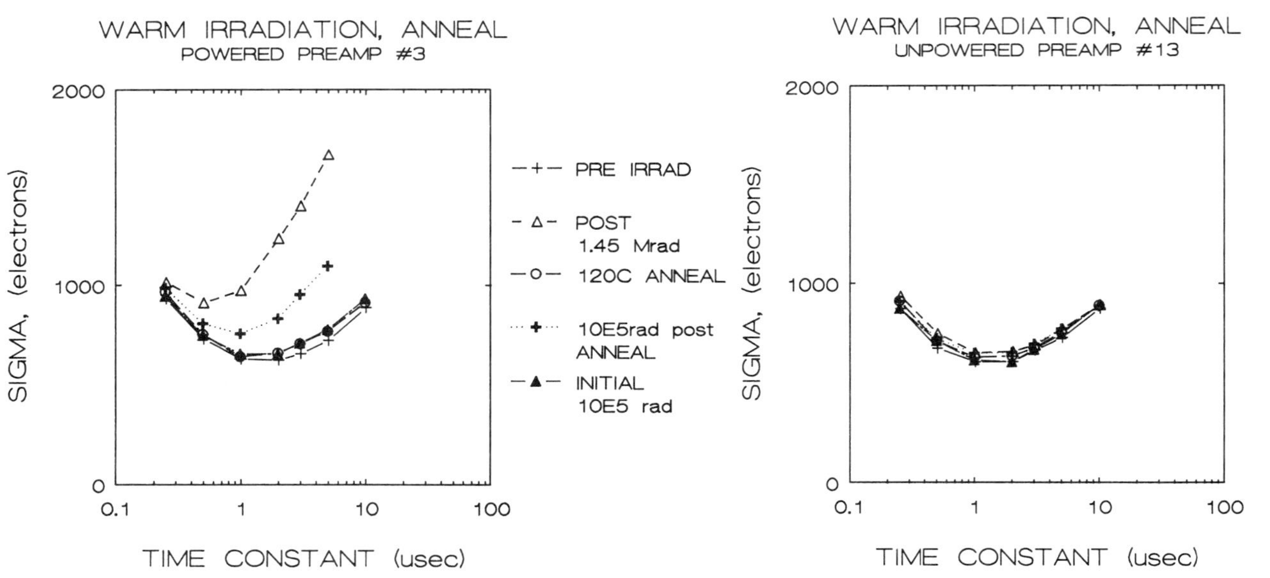

Figure 5. Elevated temperature anneal of IO 535 preamps irradiated at room temperature.

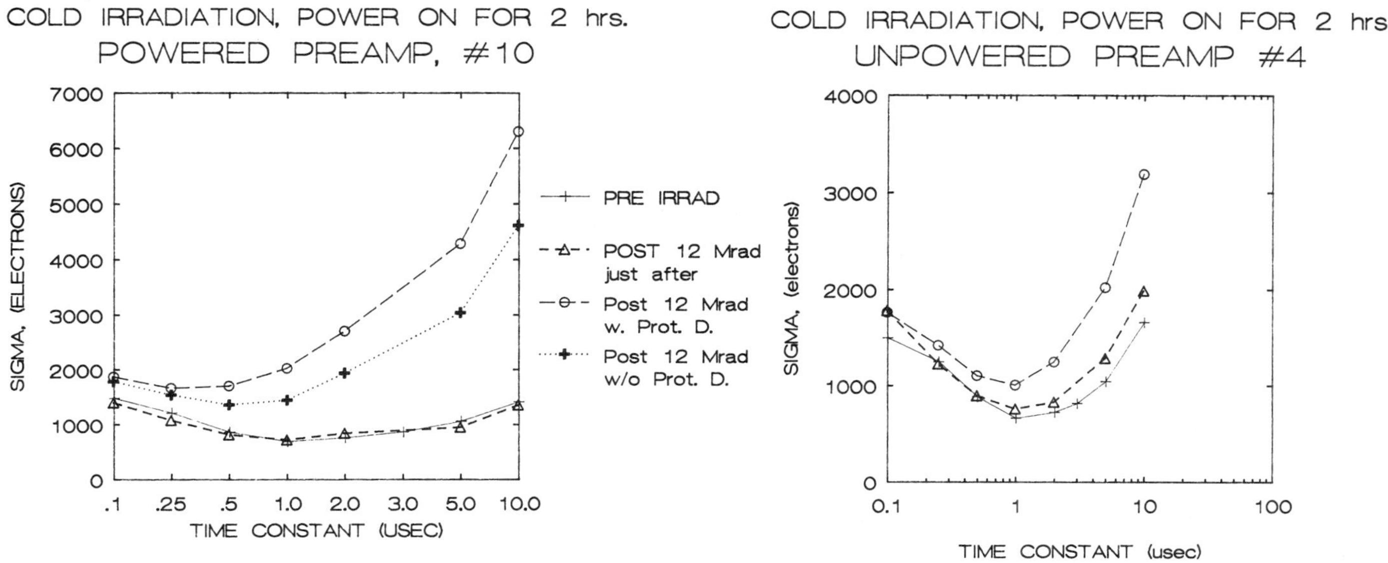

Figure 6. IO 475 preamplifiers irradiated with 60Co to 12 Mrad showing effect of measurements just after applying power and 2 hours later (open circles) and also result after removal of protection diode (crosses).

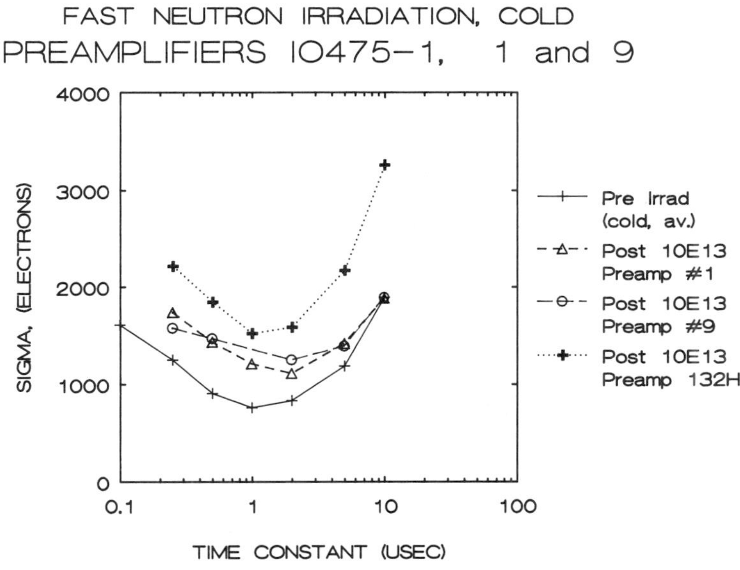

Figure 7. IO 475 preamplifiers irradiated with $9.3 \times 10E13$ fast n/cm^2.

MONOLITHIC JFET PREAMPLIFIER FOR IONIZATION CHAMBER CALORIMETER

V. Radeka, S. Rescia

Brookhaven National Laboratory, Upton, New York 11973

P. F. Manfredi, V. Speziali

Dipartimento di Elettronica, Universita di Pavia, Via Abbiategrasso, 209
27100 Pavia, Italy

Abstract

A monolithic charge sensitive preamplifier using exclusively n-channel diffused JFETs has been designed and is now being fabricated by INTERFET Corp. by means of a dielectrically isolated process which allows preserving as much as possible the technology upon which discrete JFETs are based. A first prototype built by means of junction isolated process has been delivered . The caracteristics of monolithically integrated JFETs compare favorably with descrete devices. First results of tests of a preamplifier which uses these devices are reported.

Characteristics and Noise Behavior of N-Channel Diffused JFETs

N-channel JFETs obtained by a p^+ gate diffusion into an n-doped epitaxial layer which constitutes the channel are widely used in high energy physics (HEP) and still provide the best noise performance as input devices in charge sensitive preamplifiers at processing times down to approximately 50–100 nsec. JFETs, unlike bipolar transistors, lend themselves to cryogenic operation. JFETs are superior to MOSFETs in ionizing radiation resistance and less sensitive than bipolar transistors to neutron damage, which makes them suitable for operation in the strong radiation field around the interaction region of hadron colliders. They are also less prone to damage due to electrostatic discharge (ESD) into the gate, a feature which dramatically increases the reliability of front-end circuits in applications where carrier drift is achieved by means of an applied high voltage. The amplification factor $g_m r_{ds}$ defined as the product of the transconductance and the differential source to drain resistance is quite large, from several hundreds to about one thousand, making it possible to achieve high dc open loop gains in charge sensitive preamplifier. Diffused JFETs have a predictable noise and gain behavior, with minimal variations from different production batches, a proof of how well established the production technology is. Among the class of field effect devices they have the smallest amount of $1/f$ noise in the channel current. The spectral power density of the thermal noise is remarkably close to the theoretical dependence on the transconductance g_m [1].

Silicon Processing

The monolithic integrated preamplifier is going to be fabricated by INTERFET Corp., by means of an innovative process meant to solve a difficult problem of gate to gate isolation [2].

* This research was supported in part by the U. S. Department of Energy: Contract No. DE-AC02-76CH00016.

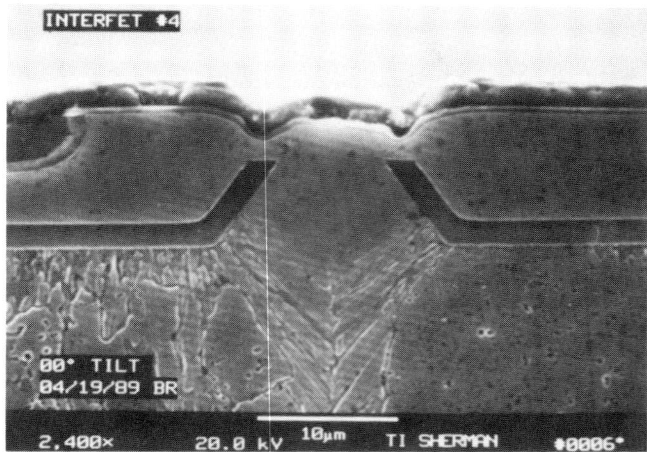

Figure 1: Cross section of a Dielectrically Isolated Wafer The SEM picture (at ×2400 magnification) shows the region between two adjacent tubs. The n-epitaxial layer grows as polysilicon over the silicon oxide separating adjacent tubs.

The new technique will still allow the use of an epitaxially formed channel and a diffused gate, as in standard JFET processing. The epitaxially grown lattice is free of defects created by ion implantation which only in part can be annealed. Also, each doping polarity reversal requires greater and greater concentrations of compensating dopant element. Two or three such reversals are required for the ion implantation method while only one is necessary if an epitaxially grown channel is used.

A. Dielectric Isolation Method

A standard slice of the needed resistivity is first oxidized and then patterned through a photoresist process. The slice is then etched by means of an orientation dependent etching process to form "V" grooves which will achieve electrical insulation. Another layer of oxide is then applied followed by a polysilicon grow to fill the grooves and to build up a thickness comparable to the original wafer. The slice is then turned over and most of

Table 1: Comparison of Discrete and Monolithic Diffused JFETs

	Short Channel		Long Channel	
W	11400 μm		400 μm	
L	5 μm		7 μm	
	Monolithic	Discrete	Monolithic	Discrete
I_{DSS} ($V_{DS} = 5$ V)	54 mA	37 mA	1.7 mA	1.1 mA
V_T	−1.3 V	−1 V	−1.25 V	−1 V
g_m ($I_D = 2$ mA, $V_{DS} = 3$ V)	20.1 mS	21 mS	2.1 mS	1.6 mS
I_G	1 pA	1 pA	0.4 pA	0.2 pA
R_{ds} ($I_D = 2$ mA, $V_{DS} = 3$ V)	12 kΩ	60 kΩ	57 kΩ	65 kΩ
μ ($I_D = 2$ mA, $V_{DS} = 3$,V)	240	1260	120	105

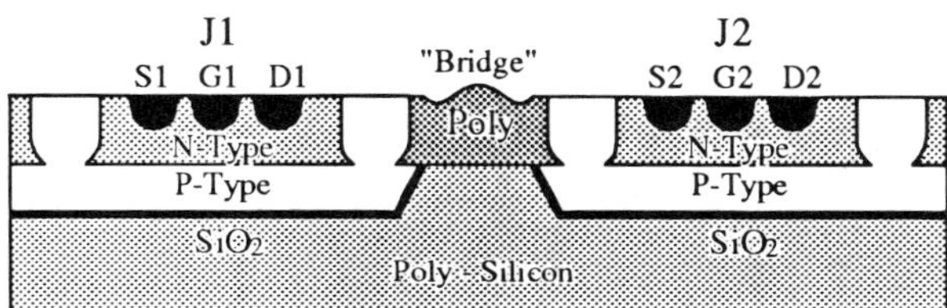

Figure 2: Dielectric Isolation Process.
Devices built in adjacent "tubs" are connected by back-to-back diodes in series with a high value resistor due to the polysilicon bridge shown in Fig. 1. If true isolation is needed, anisotropic etching can be employed to leave free standing "mesa" of silicon on a silicon dioxide floor. In this case only a single large tub is necessary.

Figure 3: Double-epi, V-groove Process.
Two successive epitaxial layers of p and n type silicon are grown on top of an n-type substrate. Anisotropic etching is used to reach into the substrate, thereby separating the backside gates. Adjacent devices are connected by back-to-back diodes through the substrate.

the original silicon is lapped away until the tips of the V-grooves are exposed. This results in single crystal silicon islands ("tubs") isolated by silicon oxide and supported by a thick polysilicon substrate.

A single layer of n-type epitaxial silicon is grown over the dielectric isolation substrate and provides a uniform channel of controlled thickness for the JFET.

Figure 1 shows a cross section of a dielectrically isolated wafer in the region between two adjacent tubs. The epitaxial layer grows as doped polysilicon over the silicon oxide separating the tubs, thus connecting adjacent tubs. At this point several process options are possible.

i. *Planar Process*

If a standard planar process is used with multiple tubs, adjacent JFETs will be connected by back-to-back diodes and a high value resistor due to the polysilicon bridge like in Fig. 2. This should not cause isolation problems except perhaps increased noise in the input device.

ii. *"Mesa" Isolation*

If true isolation is needed, V-grooves can be etched to leave a free standing "mesa" of silicon on a silicon dioxide base. Metallization across this groove to connect devices on adjacent "mesas" degrades the yield, so it is better to use whenever it is possible the simpler planar process.

If grooves or mesas are necessary or if metallization over the grooves turns out to be a high yield step then there is no need to produce multiple dielectric isolation tubs. Multiple silicon mesas over a flat, large single tub can be used.

B. Double Epitaxy and V-Groove Isolation Method

This method utilizes an n-type silicon substrate on top of which a p-type layer for the back gate and an n-type layer for the channel are epitaxially grown. Next

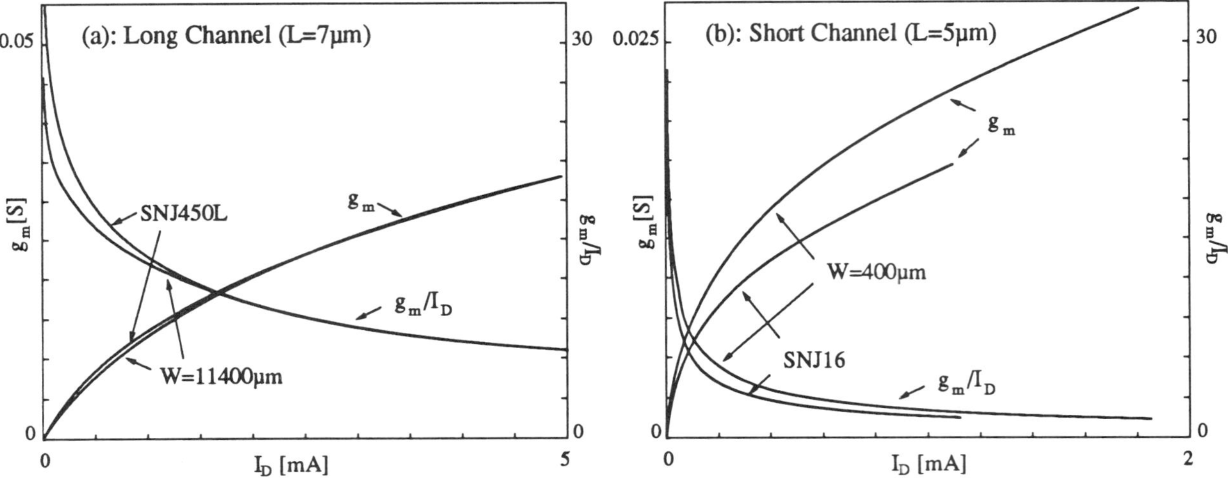

Figure 4: Comparison of Discrete and Monolithic JFETs. (a) Short channel devices, $L = 5\,\mu$m, $W = 11400\,\mu$m. (b) Long channel devices, $L = 7\,\mu$m, $W = 400\,\mu$m.

V-shaped grooves are etched deep enough to reach into the n-substrate thereby separating the backside gates, Fig. 3.

A true isolation is not achieved with this method since adjacent devices are connected by back-to-back diodes through the common substrate. Moreover, parasitic n–p–n transistors and p–n–p–n SCRs may cause latch-up problems under adverse conditions.

First Experimental Results

Up to the present only a preliminary silicon lot using the double epitaxial method has been manufactured, packaged and delivered. This first lot was manufactured on available material, which has a higher p-epi resistivity (about $1\,\Omega$cm, instead of the design value of about $0.01\,\Omega$cm). This increases the gate spreading resistance in series with the back gate, thus degrading the noise performance, a problem which will be corrected in the future production runs.

Figure 4(a) compares the g_m and g_m/I_D characteristics for a monolithic and discrete JFET of $W = 11400\,\mu$m and drain to source spacing of $16\,\mu$m (corresponding to an effective gate length of about $5\,\mu$m). At higher currents ($I_D > 1.5$ mA) they are almost identical.

Figure 4(b) shows the same traces for long channel cycle device ($25\,\mu$m) corresponding to a gate length $L \simeq 7\,\mu$m) and for a gate width $W \simeq 400\,\mu$m. Also in this case they are very similar, with a slight superiority of the monolithic one.

Table 1 compares important dc parameters of monolithic and discrete devices. The only parameter which is somehow degraded is the output impedance, especially for short channel devices. This is a consequence of the low back gate doping and will be corrected in future runs.

Another parameter of key importance is the local matching of pinch-off voltage, V_T, over an area equivalent to the one of the preamplifier.

The median value of V_T is about -1.3 V, within specification but higher than the nominal value of -1 V, with an outstanding narrow spread of only ± 25 mV. The yield of this first lot, which uses V-groove trenches, is very high, so that tighter specifications can be set in the future.

Preamplifier Circuit and First Experimental Results

The charge sensitive preamplifier schematic is shown in Fig. 5 along with the channel widths W and is discussed in depth in Ref. [3].

The preamplifier consists of a cascode stage (J1 – J2) driving an active load made of J3 and J4. J7 serves the double purpose of bootstrapping the gate of J4 in order to allow the operation of J3 in the saturation region (V_{DS} for J3 is the sum of the V_{GS} of J4 and J7) and of buffering the high impedance point on the drain of J2. J9 is a large transistor and acts as an output driver. The signal feedback path includes J9 in order to improve the circuit non-linearity. The dc feedback is resistive through a voltage shifter made of eight series connected diodes to adjust properly the voltage level. A matched resistors voltage shifter will be explored in the future. In this first prototype both the feedback resistor and capacitor will be external components even though integrated high value resistors and capacitors have been built on a test portion of the chip. On the preliminary version, on double epi material the diode voltage shifter is not working properly, which prevents the operation of the preamplifier. An amplifier was nevertheless wired using test devices and an external diode voltage shifter.

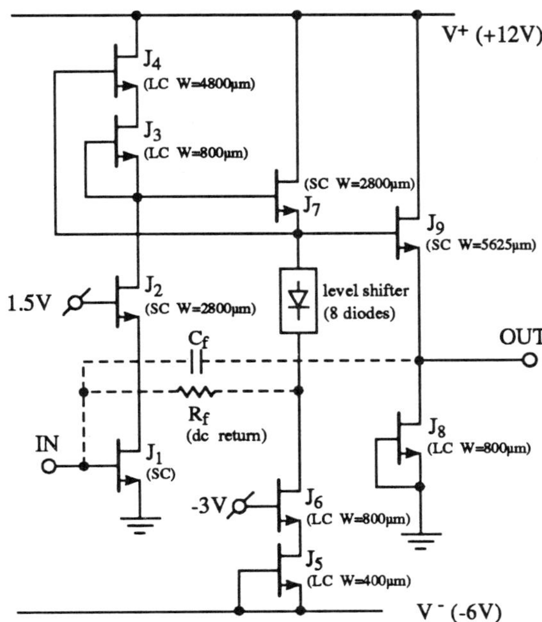

Figure 5: Circuit Schematic of the JFET Charge Sensitive Preamplifier. SC: devices with short channel, $L = 5\,\mu$m; LC: devices with long channel, $L = 7\,\mu$m. The input device J1 has a channel width $W = 11400\,\mu$m. The feedback resistor and capacitor (R_f and C_f) are shown connected with dotted lines since they are outside the monolithic chip.

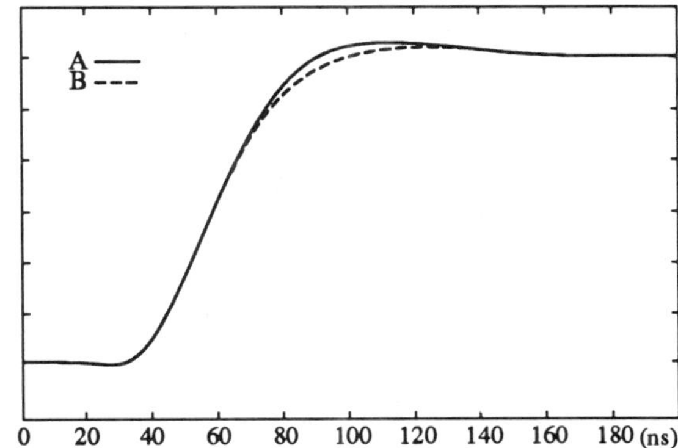

Figure 6: δ response of the charge sensitive preamplifier. Trace A: 25 mV/div; Trace B: 250 mV/div.

Figure 6 shows the impulse response, normalized to the same displayed amplitude, for a 150 mV and a 1.5 V output pulse. A different rise time (40 nsec and 44 nsec, respectively) is evident, as predicted by the computer simulations and breadboard tests [3]. The noise performance has been evaluated and is consistent with the additional gate spreading resistance in series with the back gate due to the low *p*-epi doping.

Future Improvements

Test structures for the integration of high value feedback resistors ($R_F > 1\,\mathrm{M\Omega}$) according to a scheme pioneered for the integration of front-end electronics on high resistivity silicon [4] and of the feedback capacitor have been included and will be evaluated.

As soon as a working version of the preamplifier is ready, it will undergo low temperature testing and evaluation of its resistance to neutron and ionizing radiation as well to electrostatic discharge. Test structures have been included to study the technology limits to achieve shorter channel lengths, which should improve the noise and speed performance of the circuit.

A systematic study of the optimum doping levels in the channel region is necessary for the development of a JFET with optimum noise performance near liquid argon temperature (90 K). In all alternative development, low thermal conductivity packages can improve the noise performance of the integrated circuit by raising its temperature by self heating.

A version of the circuit for transformer coupled systems will also be designed.

Conclusions

A preliminary run of a process capable of monolithic integration of JFETs has been evaluated. Performance, tolerances and matching of the devices are remarkably good, and the excess noise can be reduced in future runs. A preamplifier has been wired using monolithic JFETs and is working according to predictions. Full characterization of the different process options is being carried out. The preamplifier is intended for applications with ionization chamber calorimeters in the SSC environment at processing times of about 50 nsec.

References

1. V. Radeka, "Semiconductor Position Sensitive Detectors," *Nucl. Instrum. Methods* 226, 209-218, 1984.

2. L. A. Rehn, D. E. Roberts, "Monolithic JFET Preamplifier for Ionization Chamber Calorimeters," Proc. Intl. Industrial Symp. on the Super Collider, Miami Beach, FL, March 14-16, 1990 (to be published).

3. M. Demicheli et al., "Design of Monolithic Preamplifiers Employing Diffused *N*-JFETs for Ionization Chamber Calorimeters," to be published in *Nucl. Instrum. Methods* .

4. V. Radeka et al., "Design of a Charge Sensitive Preamplifier on High Resistivity Silicon," *IEEE Trans. Nucl. Sci.* NS-35 (1), 155-159, 1988.

RADIATION DAMAGE TESTING OF TRANSISTORS FOR SSC FRONT-END ELECTRONICS

J. Dawson, T. Ekenberg, A. Stevens
Argonne National Laboratory
High Energy Physics Division
Argonne, IL 60439

H. Kraner, V. Radeka, S. Rescia
Brookhaven National Laboratory
Instrumentation Division
Upton, NY 11973

S. Kerns
Vanderbilt University
Department of Electrical Engineering
Nashville, TN 37212

October 15, 1990

Abstract

Over the ten year expected lifetime of a typical SSC detector operating at the design luminosity of 10^{33} cm^{-2}s^{-1}, the front-end electronics at large pseudo-rapidities may receive total doses as high as 20 MRad(Si) of ionizing radiation and 10^{16} neutrons/cm^2. Discrete JFETs and monolithic MOS and bipolar transistors have been irradiated at 10 MRad(Si) and 10^{14} neutrons/cm^2, and the effect on transfer characteristics and noise performance have been measured. All transistors were still functional after irradiation but suffered increased noise and the MOS transistors showed significant threshold shifts and increased leakage currents.

Introduction

At a luminosity of 10^{33} cm^{-2}s^{-1} the radiation doses inside a SSC detector may reach 20 MRad(Si) of ionizing radiation and 10^{16} neutrons/cm^2 over the ten year lifetime of the detector. The dose is strongly dependent on the geometry of the detector and the exact location of the electronics [1]. Any upgrades of the machine from the design luminosity will further increase the radiation levels inside the detector.

Over the last 30 years, the effects of radiation damage have been researched by the military and space-scientists, as well as in the high-energy physics community, and a large body of knowledge has been acquired [2,3]. This can be taken as the point of departure for the study of the specific SSC environment. Several industrial vendors have also developed the capability of manufacturing radiation-hardened devices, so the technological know-how for fabrication of SSC front-end electronics already exists.

The specific requirements of the SSC radiation environment, *i.e.* low dose rate but high total dose, are different from the military and space-science applications. This indicates that extensive study of avaliable fabrication processes must be made in order to evaluate their applicability for the SSC. General properties of radiation damage, such as the effect of low-temperature operation and short and long term annealing must also be investigated.

Argonne National Laboratory has several facilities that are well suited for radiation damage research. These facilities include a fast neutron generator, several cobalt-60 sources that can deliver up to 2 MRad(Si) per hour, and a 22-MeV electron linac. The facilities available for radiation damage research and testing are extensively described in [4].

At the Argonne radiation facilities the dose rates can be quite high, allowing the life-time dose at the SSC detector, *i.e.* the full ten years of operation, to be received during 4–8 hours of irradiation. This makes it easy to perform experiments which demands high total dose, but introduces dose rate effects that are difficult to predict. We do not make any attempt to take dose rate effects into account in this paper.

Measurements

Test samples of JFETs, bipolar, and MOS transistors were obtained from several manufacturers. The JFETs were commerciallly available discrete components, while the bipolar and MOS transistors were from custom rad-hard monolithic CMOS and BiCMOS processes. Both n- and p-channel devices were tested.

JFETs were irradiated with ionizing radiation at a total dose of 1 MRad(Si) and 12 MRad(Si) and neutrons at doses between 10^{13} and 10^{15} neutrons/cm^2. The transistors were biased at $I_D = 1$ mA and $V_{GS} = 5$ V, and irradiations were done both at room temperature and at liquid nitrogen temperatues (120 K). DC characteristics, as well as small signal parameters and noise performance, were measured approximately one week after exposure to allow for all annealing effects to take place.

The monolithic MOS and bipolar transistors were biased at I_D (or I_C) = 100 μA and V_{GS} (or V_{CE}) = 5 V. Total doses received by the integrated devices were $\sim 10^{14}$ neutrons/cm^2 and 1 MRad(Si) to 10 MRad(Si). All irradiation and mesurements were done at room temperature.

Noise performance and forward β for the bipolar transistors, and g_m, noise characteristics, I-V curves (I_D vs. V_{GS} and V_{DS}), and V_{th} was measured for the MOS transistors.

Results

The results of radiation damage testing for JFETs have been extensively covered in [4]. The main findings included an increase in equivalent noise charge of 25 percent at 100 ns shaping time, and an increase in gate leakage current of one order of magnitude after 10^{14} neutrons/cm^2. The equivalent noise charge increased by less than 20 percent for 100 ns shaping time after 12 MRad(Si) of γ-rays.

Bipolar monolithic transistors suffered a reduction in forward β of between 15 and 25 percent, but still remained above 75 for both npn and pnp transistors for a dose of 10^{14} neutrons/cm^2. The $1/f$ noise increased by a factor of 4, but will have a very small effect at the frequency range of interest. 10 MRad(Si) of ionizing radiation caused the β to be reduced by 30–40 percent, bringing the pnp β down to approximately 50, while the npn β stayed above 75 (see Fig. 1). Ionizing radiation had a negligble effect on the equivalent input noise of the bipolar devices. Both the pnp and the npn transistors were still fully functional after irradiation and the decrease of β can be taken into account in the design process.

Radiation damage in MOS transistors is caused mainly by ionizing radiation that shifts the threshold voltages more negative, thus making the n-channel $|V_{th}|$ smaller and the p-channel $|V_{th}|$ larger. This is due to the build up of charge Q_{ss} in the gate oxide, which influences the threshold voltage according to

$$V_{th} = \Phi_{ms} + 2\Phi_f + \frac{Q_b}{C_{ox}} - \frac{Q_{ss}}{C_{ox}}.$$

The shift in V_{th} is up to 700 mV for 10 MRad(Si) of γ-rays for both n-channel and p-channel devices. An example of the threshold shift can be seen in Fig. 2. The same dose also causes the g_m to decrease by approximately a factor of 2 and increases the noise of both the n- and p-channel devices by a factor of 5.

A neutron dose of 4×10^{14} neutrons/cm^2 causes a threshold shift of $\sim$ 200 mV. The same dose leaves g_m and the noise of the n-channel transistors unchanged, while the noise of the p-channel transistors increased by a factor of two.

Conclusions

Radation damage testing has been done on discrete JFETs and monolithic bipolar and MOS transitors to evaluate their suitability for SSC front-end electronics. The transistors were exposed to total doses of up to 12 MRad(Si) of ionizing radiation and 10^{14} neutrons/cm^2. All transistors were found functional after irradiation.

JFETs are intrinsically the most radiation-hard devices and the samples tested were from a non rad-hard commercial process. They suffered little or no degradation in DC characteristics and small-signal behavior, but noise increased between 20 and 25 percent.

The bipolar and MOS transistors were acquired from vendors with rad-hard CMOS and BiCMOS processes. Forward β of the bipolar devices decreased but remained over 75 for both neutron radiation and ionizing radiation, and noise performance was not affected. The MOS transistors suffered shifts in V_{th} of up to 700 mV, increased noise and decreased g_m.

Acknowledgements

Thanks to G. Holmblad and A. Smith for the use of sources and help with testing and dosimetry.

References

[1] D. E. Groom, ed., "Radiation Levels in the SSC Interaction Regions," *SSC-SR-1033*, June 10, 1988.

[2] G. C. Messenger, M. S. Ash, *The Effects of Radiation on Electronics Systems*, Van Nostrand Reinhold Co., New York, 1986.

[3] H. W. Kraner, Z. Li, K. U. Posnecker, "Fast Neutron Damage in Silicon Detectors," *Nucl. Instr. and Meth.*, A279(1989), pp. 266-271.

[4] A. Stevens, J. Dawson, et al. "Rad-Hard Electronics Development Program for SSC Liquid-Argon Calorimeters," to be published in *The 1990 International Industrial Symposium on The Supercollider 2*, Plenum Press, New York.

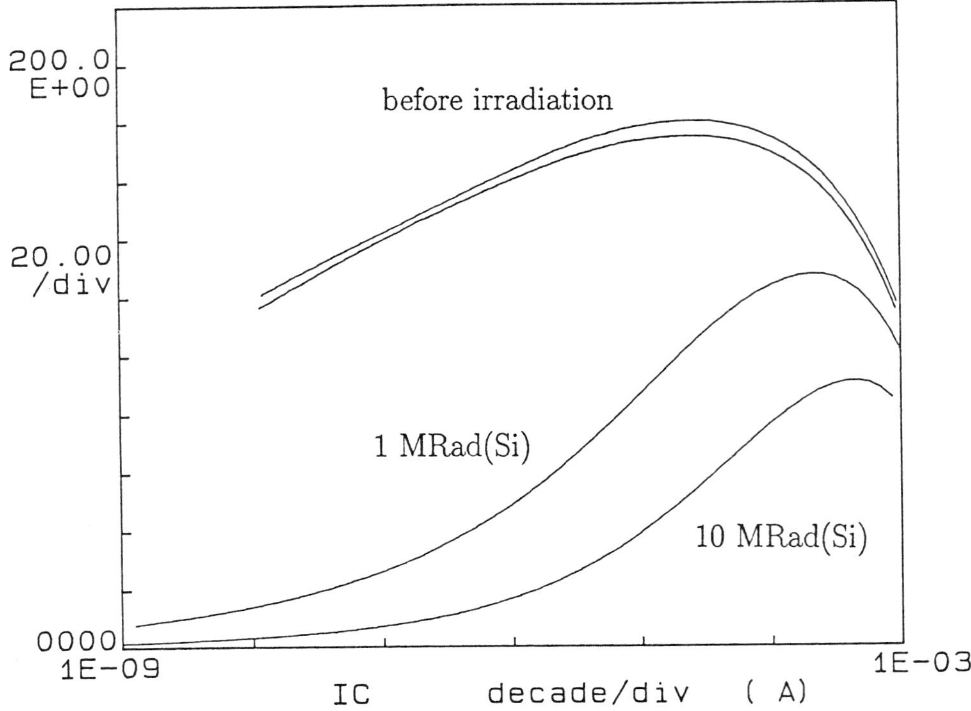

Figure 1: Forward β of npn bipolar transistor from Hughes' CBIT process (a 1.4 μm feature size BiCMOS process) before **irradiation and after** 1 **MRad**(Si) and 10 **MRad**(Si) **of ionizing radiation.**

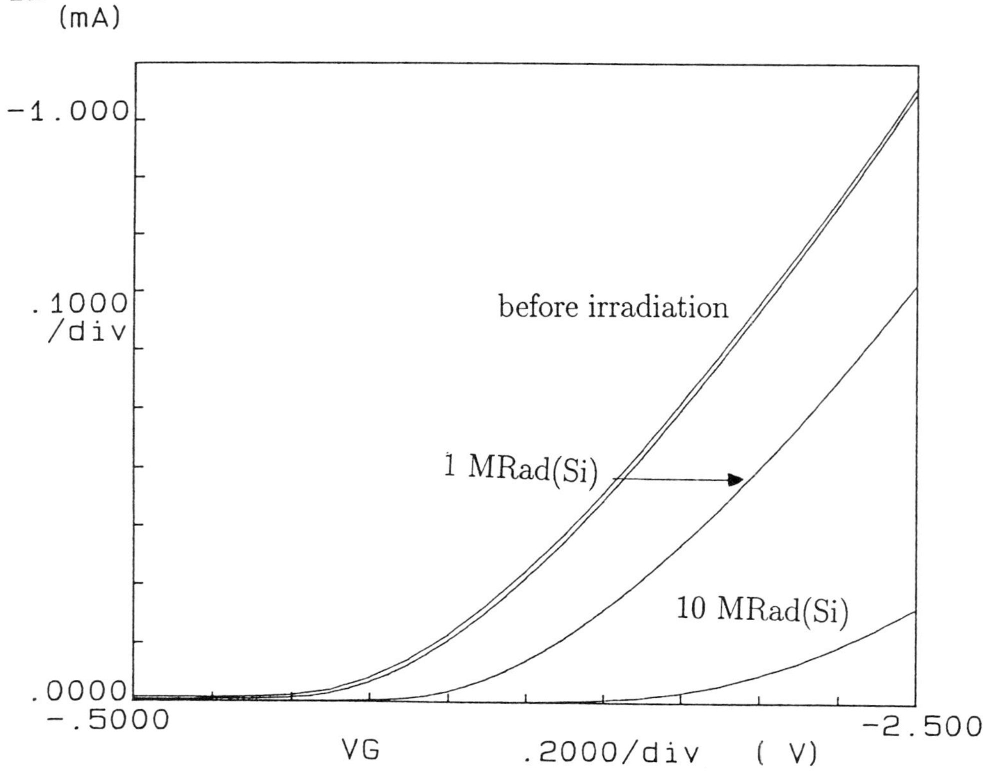

Figure 2: V_{th} of an p-channel MOS transistor from Hughes' CBIT process before irradiation and after 1 MRad(Si) and 10 MRad(Si) of ionizing radiation.

RADIATION STUDIES OF LECROY HIGH ENERGY CIRCUITS

Alvin S. Kanofsky
Department of Physics, Lehigh University
Bethlehem, PA 18015

and

Bert Yost and Werner Farr
LeCroy Corporation
Chestnut Ridge, N.Y. 10977

Introduction

We have studied the effects of Radiation on LeCroy circuits in common use in high energy experiments. The LeCroy circuits studied include the TRA402–4 Channel Transresistance Preamplifier, the MVL407 Voltage Comparator, the TRA1000 Charge/Current Pulse Preamplifier, the MLL400 1024 Bit Static Shift Register, and the MVV 320/640 Cell Analog Shift Register. These circuits were exposed to both continuous and pulsed electron beam irradiation, and the performance of the circuits with dosage were monitored.

MVV200

The MVV200 is a high accuracy analog shift register optimized for use where analog signals must be accepted (written) at a high clock rate but utilized (read) at a lower rate. The geometry of the CCD has been designed to optimize the dynamic range response and device speed (both input slewing and clock speed). The MVV200 consists of two 320-cell analog shift registers, each with its own voltage input, transport clock and output port, allowing the device to be used as two 320-cell or one 640-cell analog shift register. Although the device utilizes charge in its shift register, the input is a voltage sampling device and the output is a voltage source. The MVV200 can be used in applications requiring time base expansion in which analog channels require high speed sampling and temporary storage, allowing later conversion to digital by a lower speed low cost ADC. In particular, applications include Time Projection Chambers (TPC's), Jet Chambers, and other imaging detectors.

MLL400

The MLL400 is a digital shift register organized as 4 channels of 256 bits each. It is ideal in applications where data must be sampled at rates to 250 MHz but can be processed at lower rates. When the clocking is stopped, data can be held indefinitely because the device is static. Readout can then occur at rates up to 80 MHz. Applications include time expansion for Time-to-Digital Conversion and as temporary storage of Flash ADC or other data.

The MLL400 has 4 identical channels. Each has an active input and a dummy input. The dummy has no function within the device but may be used to symmetrically load the differential output stage of the device driving the MLL400. In applications where the MLL400 must be in close proximity to sensitive high gain circuitry, such differential drive can help eliminate feedback from the clock. The four inputs feed the four 256-bit shift registers.

The shift register is clocked by an externally supplied four-phase clocking scheme. Each phase has a frequency of one fourth of the effective clock rate. This makes generation of CMOS level clock signals feasible. All four phases are required to operate each of the 4 channels.

MVL407: Quad Ultrafast Voltage Comparator

The MVL407 is a quad voltage comparator designed for applications in which ultra high speed and accurate timing are most important. The device is manufactured using a state-of-the art high speed bipolar process which results in an extremely short (2.5 nsec) propagation delay with operation at speeds in excess of 400 MHz.

Each channel provides differential inputs and complementary outputs compatible with the ECL logic family. The outputs can drive 50 ohm loads or 100 ohm twisted pair directly.

The MVL407 incorporates a unique hysteresis feature for exceptionally clean operation. When the comparator changes state, an internal differential input offset of about 3.5 mV is generated. The positive feedback drives the device quickly through its switching region, greatly reducing the possibility of oscillation or output chatter with small or slowly changing inputs.

The propagation delay is typically 2.5 nsec and changes by only 100 psec for 5 to 100 mV range of overdrive. This very low delay variation makes the MVL407 extremely useful in critical timing applications.

Monolith Model TRA402–4:
Channel Transresistance Preamplifier

The TRA402 is a monolithic four channel, fast, low noise preamplifier. Its principal application is amplification of wire chamber signals for time-resolved measurements. Its high density makes it a practical solution, especially if space at the detector is at a premium, yet chamber mounting is a necessity.

The amplifier provides a gain of 25 mV/microamp with a risetime of 3 nsec. The low input impedance of < 100 ohms and low chamber noise of < 100 nA RMS is ideally suited to chamber applications, providing little integration of the current pulses. The special input geometry of the TRA402 makes it useful even with high capacitance detectors such as liquid argon calorimeters and wire chamber strips and pads. Because the risetime of the wire input is maintained by the TRA402, the user has greater freedom in selecting the RC coupling to the subsequent circuitry, allowing the risetime/noise trade-off to be optimized.

The TRA402 provides both inverting and non-inverting inputs for all four channels. This allows use of the amplifier for both positive and negative inputs provided by the cathodes and anodes of a wire chamber respectively.

Model TRA 1000: Charge/Current Pulse Preamplifier

The LeCroy Model TRA1000 monolithic preamplifier is a versatile, economical, low-noise device which can be used either as a current-to-voltage preamplifier or as a charge-to-voltage preamplifier. The device has been designed for use with negative input signals; however, it may also be configured for operation with positive inputs. The various options are selected through the use of external components.

The TRA 1000 has been designed for direct connection to a variety of detectors. Its low noise and low input impedance make it ideal for use with proportional wire chambers even when resistive wire is used for position measurements. The device also finds application with photomultipliers when the economy of low-gain tubes is a factor or dynamic range considerations are important. When used in conjunction with both the last dynode and the anode signals, exceptional dynamic range can be achieved by selecting different gains for the two devices. In this way, high and low sensitivity channels are configured.

The linearity of the preamplifier is excellent for a wide range of input risetimes and selected gains. The linear range of the output is 0 to 1 V, even for 50 ohm loads. Linearity has been measured in conjunction with a LeCroy 2280 Series 12 Bit ADC. The result is typically < 0.5% of reading with proper compensation.

The equivalent input noise of the preamplifier is as low as 2 pA in some configurations. Noise measurements are quoted in pC, referred to the preamplifier input. The LeCroy 2280- Series current-integrating ADC's employing a 500 nsec wide gate, were used to characterize the current-to-voltage configurations. For the charge-to-voltage configurations a peaking sensing 2280 series ADC was used.

The low price, compact packaging, and low power dissipation of the Model TRA1000 make it an ideal choice for use in large-scale systems applications. It is particularly suited for use as a current-sensitive preamplifier for MWPC analog position measurements. The Model TRA1000 is a low-cost, high performance answer to a wide range of charge-sensitive and current sensitive preamplifier needs.

IRRADIATION TECHNIQUES

The 3 MeV High Voltage KN3000 Van de Graaff at Lehigh University was used for the irradiations. This machine provided beams of 2.5 MeV electrons which were directly used. We looked for slowly changing time effects using a continuous beam as well as transient effects with a pulsed beam arrangement. Figure 1 shows the experimental arrangement.

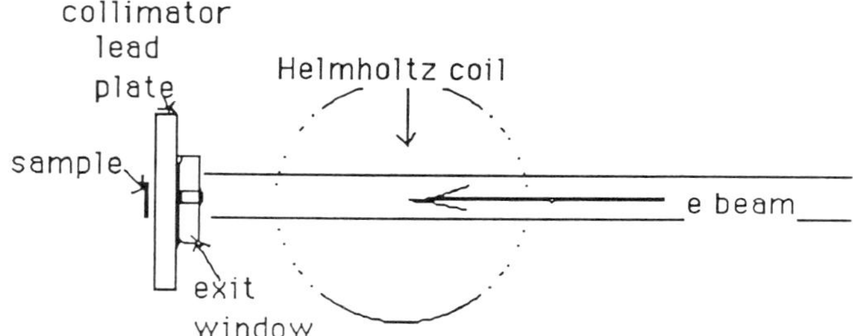

Figure 1. Electron Beam Irradiation

The electron beam dosage was measured by determining the beam profile using the luminescence from a glass plate at the exit window, placing the collimator hole at the beam location with the sample over the collimator hole, and measuring the total charge collected from the beam on a Faraday cup behind the sample.

MVV200: Results

Two MVV200 chips were irradiated in the Van de Graaff beam. One received a dose of 10 Mrads in about 10 seconds. The other received a dose of 100 Mrads in 100 seconds. The chip pins were all shorted together while irradiation occurred. The chips were then tested about 28 hours later in a 2262 waveform recorder equivalent circuit. Neither chip functioned at any level–i.e. there was total catastrophic failure. We are now testing chips exposed at lower levels of radiation to determine where the threshold occurs for failure with radiation.

MLL400: Shift Register Results

Three MLL400 shift register chips were irradiated at levels of 10 Mrads, 100 Mrads, and 1000 Mrads. The dose rate was about 1 Mrad/sec. The chips were then tested 28 hours later with matched pair chips in an 1879 unit. None of the chips functioned–failure was catastrophic in all cases. In the case of the MLL400 chip, which uses silicon on saphire technology, we would expect it to be harder to irradiate. Also, we did not see any annealing effects with this coarse scan of radiation levels. We are now radiating chips at lower levels to determine where the failure threshold occurs.

With both the MVV200 and MLL400 results, it is important to realize that the doses and dose rates used are much greater than would be expected. The SSC doses of one Mrad or less over one year are expected except in the very forward directions close to the beams.

MVL407: Comparator Results

With the initial application of beam, the comparator output jumped from 18 mV to 21 mV and remained there for 126 seconds, or equivalently, 126 Mrads. Then the output dropped slowly over the next 80 seconds or 80 Mrads to zero. The shutter was then closed and reopened in 20 seconds. The output then reappeared at the 3 mV level and remained at this level roughly for the next 320 seconds at which point it dropped to zero. After another 100 Mrads of irradiation, during which the signal remained at the 1 or 2 mV level, the output ceased and no longer returned. During the entire irradiation, the position of the signal did not visibly shift except at the very end, when the output was less than 1 mV. These results are shown in figure 2.

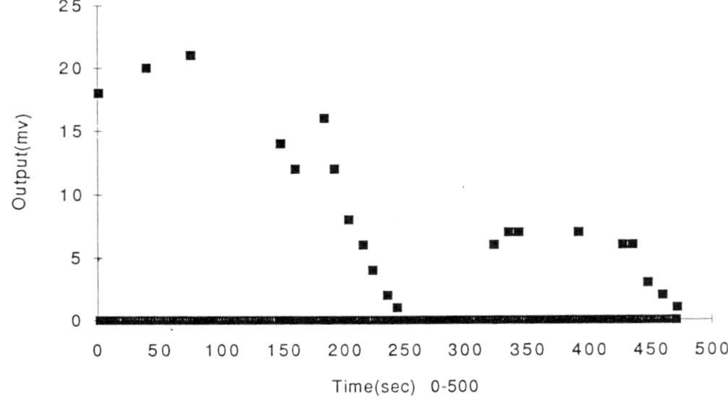

Figure 2. MVL407—Output vs. Time (sec) Mrad/sec.

TRA1000: Results

We bench tested the gain for the TRA1000 amplifier to check the linearity and waveforms under normal conditions. Figure 3 shows the test set-up. Figure 4 shows typical waveforms, as obtained on a LeCroy 9450/20 digital oscilloscope.

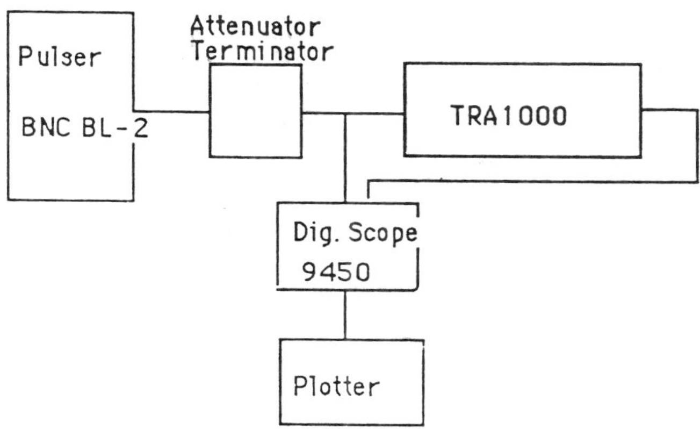

Figure 3. Circuit Test Set-Up

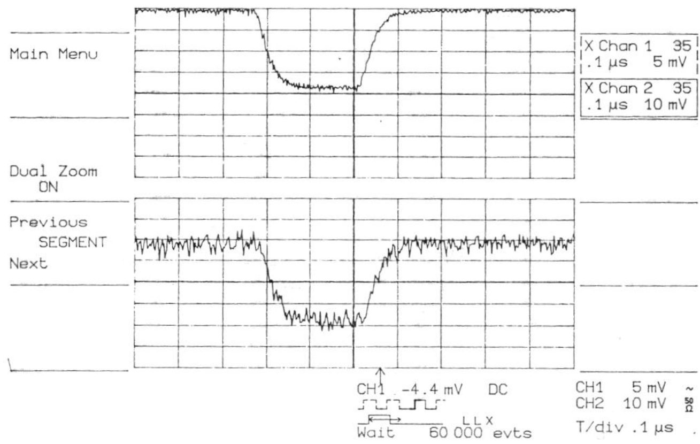

Figure 4. LeCroy Test Board

We monitored the signal input and output of the TRA1000 while the device was being irradiated. Immediately upon irradiation, the device failed, within 2/3 of a second. This occurred at a dose rate of about 1 Mrad/sec. The device then returned to operation with an output signal equal to 1/10 the original signal—40 mV. With additional irradiation, the output was reduced to 20 mV. We then let the device anneal somewhat and after twelve minutes we exposed it to beam again. The initial output pulse was 40 mV and after a minute of exposure it had increased to 100 mV. After three minutes of exposure it had increased to 400 mV, but the output pulse width had increased to about three times what it had been originally. After four minutes, the output pulse peaked at 650 mV and slowly decreased to 100 mV after 13 minutes, and then rapidly dropped to zero in another 3 minutes. After about half an hour, the pulse recovered to 5 mV, and now with irradiation, the signal increased further to 120 mV after about

1 minute of irradiation. It then dropped rapidly to zero in the next half minute. A plot of the output as a function of radiation for the main anneal is shown in figure 5. These results are consistent with earlier results we had reported at the Miami conference (Miami Industrial SSC conf.—to be published), which indicate a radiation assisted annealing process is occurring.

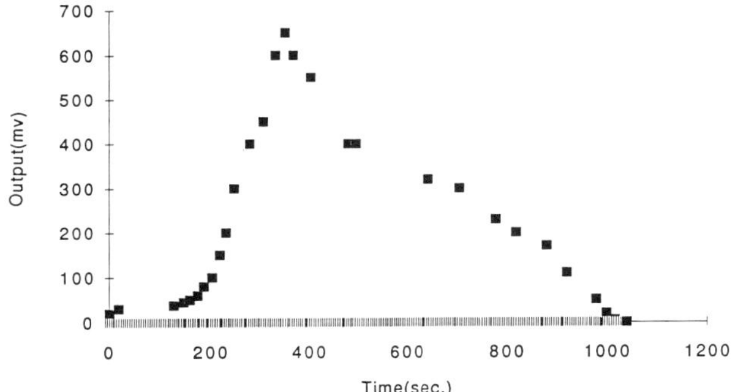

Figure 5. TRA1000—Output (mV) vs. Time (sec) Mrad/sec.

TRA402: Results

The TRA402 amplifier was set up identical to the TRA1000 amplifier using the LeCroy test board. The gain was checked to be nominally the same as the specifications, using the same method as that of the TRA1000.

The TRA402 was monitored in the same way as the TRA1000 while it was being irradiated. The output remained essentially constant for 120 minutes at a dose rate of about 1 Mrads/sec., at which point the output signal dropped rapidly to zero within 8 seconds. We were not able to recover any signal after waiting without irradiating, or with irradiation. This result is consistent with what was observed before with this circuit and reported at the Miami conference. Figure 6 shows the output of the TRA402 as a function of time.

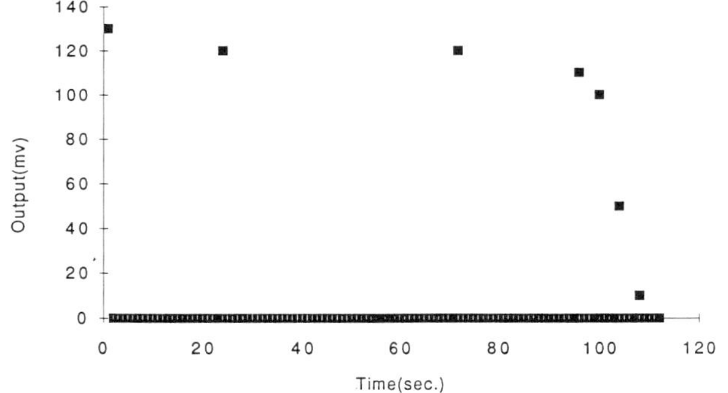

Figure 6. TRA402—Output vs. Time (sec) Mrad/sec.

Thus, there was no indication of a radiation hardening process occurring, as with the TRA1000 amplifier, but rather a radiation tolerance up to about 100 Mrads, and then a uniform deterioration for the next 8 Mrads.

Conclusions

The two amplifier chips showed markedly different behavior. With the first Mrad of radiation, we notice a degradation of the TRA1000. Beyond this, the TRA1000 showed erratic behavior initially with radiation up to the first 200 Mrads of irradiation and then stabilized for a period of an additional 500 Mrads, before ceasing to function. This indicates the possibility of a radiation hardening process occurring.

The TRA402 maintained its performance for about 100 Mrads and then deteriorated uniformly for the next 8 Mrads, indicating no radiation hardening, but rather a tolerance to radiation. Thus it seems the performance under radiation was different for the two amplifier circuits, and that initial radiation hardening in an electron beam may be a way to obtain radiation tolerances in circuits that initially are susceptible to radiation.

The MVL407 comparator showed a fairly constant output for the first 200 Mrads of irradiation, then dropped to zero in the next 50 Mrads, with a recovery at a lower output level for the next 100 Mrads. Thus, it might be possible to use the device beyond the initial limits of where it first drops to zero.

The MVV200 chip was catastrophically damaged at levels of 10 and 100 Mrad, and thus we only know that the radiation threshold for failure is below these values. Further tests are in progress.

Similarly, the MLL400 chip catastrophically failed for levels of 10, 100, and 1000 Mrads, and thus we will need to run additional tests at lower exposures to determine the radiation threshold for failure.

As well, we are investigating all of the observed phenomena further, both at the microscopic and gross circuit level, as well as obtaining more measurements on these and additional circuits.

Acknowledgements

This work was partially supported by the D.O.E., Bell Laboratories, and the Navy. We also thank Mr. Walter LeCroy for his support and encouragement.

11 Detector Engineering

AN ENGINEERING DESIGN NETWORK FOR SSC DETECTOR DEVELOPMENT

N. J. DiGiacomo
Martin Marietta Astronautics
Denver, CO 80201

For The EMPACT Collaboration

Abstract

An engineering design network that distributes state-of-the-art modeling, analysis simulation and configuration management capabilities to universities, laboratories and industries involved in SSC detector development is discussed.

Introduction

The detector systems that are being proposed to exploit the capabilities of the SSC are of a scale and scope that will make them among the most complex devices ever built. To successfully design and build these systems over the next decade, we must make use of integrated state of the art computer aided engineering and design (CAE/CAD) tools that have been developed and employed in industry. The challenge, represented in Figure 1, is to make these tools and associated engineering resources available to the spectrum of institutions - large and small universities, industries and national labs - involved in SSC detector development in such a way that each may contribute and participate in the most effective manner. We believe that powerful workstations running sophisticated modeling, analysis and simulation software, linked by high speed data networks and governed by modern configuration management methods offer the ideal means of arriving at the optimum detector configuration for physics at the SSC.

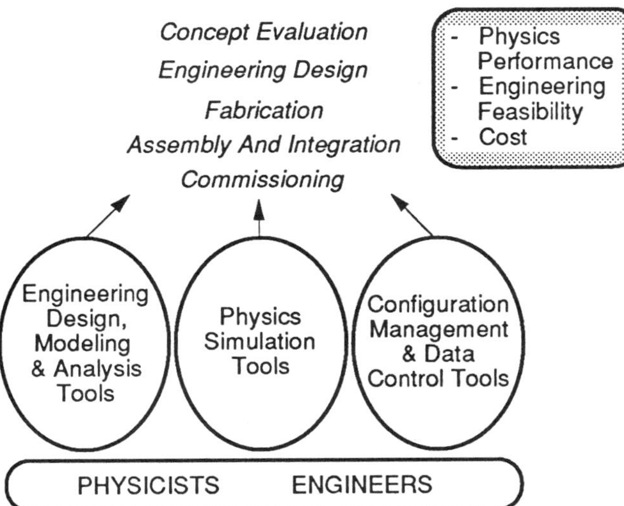

Figure 1. An integrated, computer-based engineering and design resource would utilize high speed data networks to connect workstations that run engineering, design, simulation and configuration management tools. Physicists and engineers at universities, national laboratories and industrial locations will use the resource to evaluate detector concepts for physics performance, engineering feasibility and cost.

Integrated Computer Aided Engineering and Design

Integrated CAE/CAD provides designers and engineers in all phases of the project with linked design, modeling and analysis tools that can be tailored to the task at hand. The emphasis is on value-added contact with data. One only intervenes in the flow of data between packages to contribute engineering or design judgment. The user does not participate in routine data transfer, translation and manipulation. Interactive graphics are used extensively to input design, define modeling and analysis parameters and conditions and to evaluate results. The system is optimized to provide fast turnaround and to allow parallel analyses. The fidelity of modeling and analysis is determined by the task requirements and not by the tedium of setting up new grids or meshes, and errors due to manual data manipulation are eliminated.

Figure 2 is a schematic representation of the Martin Marietta EMPACT integrated design, modeling, analysis and simulation tool kit for detector system development.

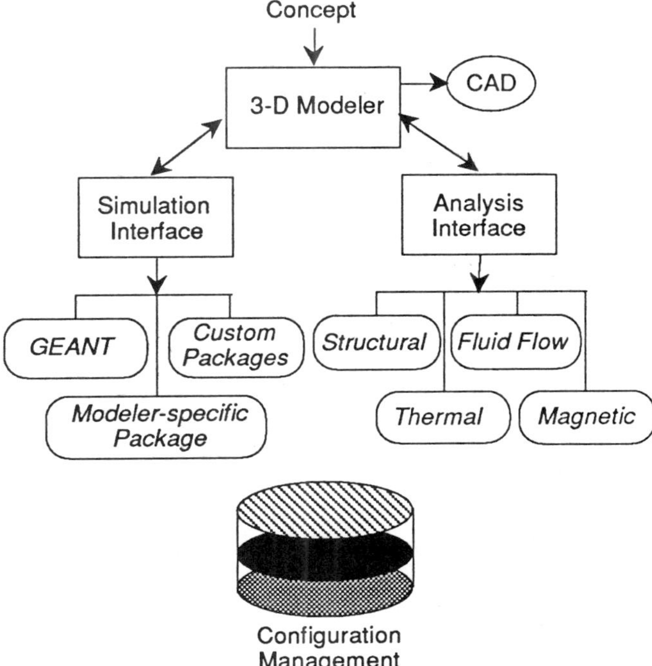

Figure 2. Architecture of the Martin Marietta EMPACT integrated design, modeling, analysis and simulation resource for detector system development.

The resource has five basic functional areas: 3D modeler and database, CAD/CAM, CAE interface, analysis codes, and physics simulation. We discuss below the capabilities of each functional area, and the flow of data and information between them. The 3D modeler and database has:

- Easy entry of designs at all levels (piece-part through complete system), with instancing
- 3D interference and fit checking at any stage of motion
- Material and mass properties
- A user-friendly operating environment
- Structured links to major CAD packages and to an analysis interface

The CAD/CAM packages allow generation of detailed dimensioned drawings and CAM sequences from the 3D design database. The 3D modeler is capable of two way communication with a number of CAD packages. The CAD package also has electrical design and cable management capability, as depicted in Figure 3.

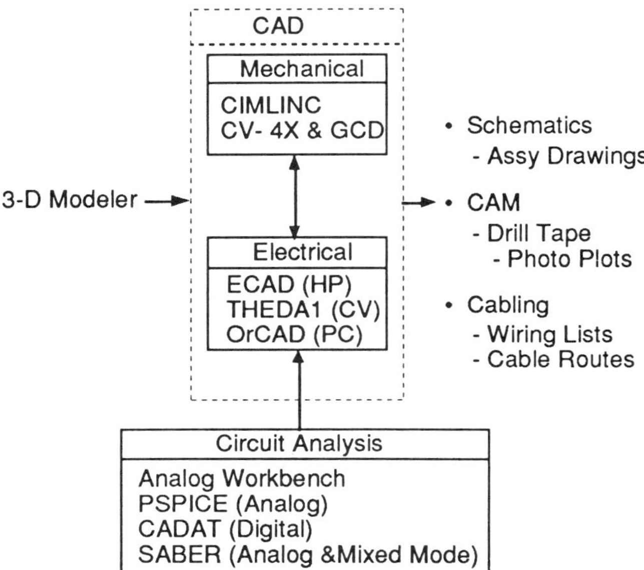

Figure 3. Electrical CAD Design and Analysis Architecture

The CAE interface allows the engineer to mesh or grid designs for analysis, to view the results and to make 3D drawings and representations of designs both before and after analysis. This package is central to the efficient operation of the resource, since it is here that one can minimize tedium and maximize "value-added" interaction with the data. The CAE interface allows the use of a common grid or mesh by different analysis tasks. The analysis codes are capable of performing three-dimensional analyses. Codes are currently being utilized for the following areas, with additional tools being developed as required:

- Mechanical: Linear and non-linear stress, strain, loading, contacts and assembly
- Thermal and fluid flow: Forces, radiation, convection, conduction, mass transfer and distortions
- Magnetic: Field shape, quality and resistant forces

We currently utilize a variety of codes for each analysis function, given variations in algorithms, methods and capabilities (e.g. the engineering team uses MSC/NASTRAN, ABAQUS and ANSYS for structural analysis). As shown in Figure 3, a wide variety of circuit analysis codes and electrical CAD packages are also in use.

The detector physics performance section of the resource performs radiation transport, physics simulation, calculates coverage, shadowing and hermeticity. It uses as input the 3D modeler database, individual sensor responses and physics event simulators. Interface with other simulation packages such as GEANT or custom-application codes is also in principle possible via geometry translation from the 3D modeler. The result is the overall detector system response and performance.

Configuration Management

The evolution of a configuration for a major detector requires careful management of the designs, models, engineering analyses and physics simulations. Without an orderly means of tracking and controlling the flow of information, communication between groups developing various sub-systems or components is too often "after the fact". This is particularly important in an international project such as an SSC detector, where communication difficulties due to distances between groups are compounded by the use of a variety of engineering and measurement standards.

Computer-aided engineering and design makes configuration control easier from the logistical perspective, but the discipline must still come from the physicists and engineers involved in the project. Figure 4 depicts the process flow and associated control of a conceptual design model. A similar flow is required for for the models created to perform physics simulation analyses.

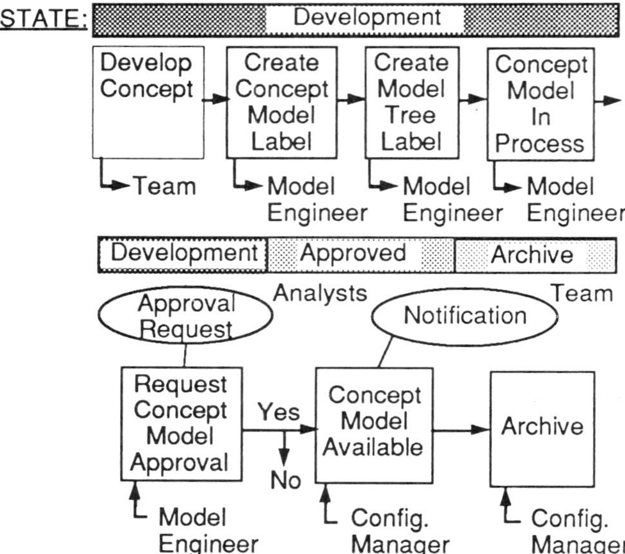

Figure 4. Configuration Process Flow of Conceptual Model

The Engineering Design Network

In designing our engineering network, we have sought to develop a system that incorporates the organizational advantages of a centralized system, yet allows us to make efficient use of human resources at remote sites. The proposed system, sketched in Figure 6, consists of an Engineering Design Center - the central repository of engineering design resources as well as the design databases, and a collection of remote workstations connected to the Engineering Design Center via a high-speed network.

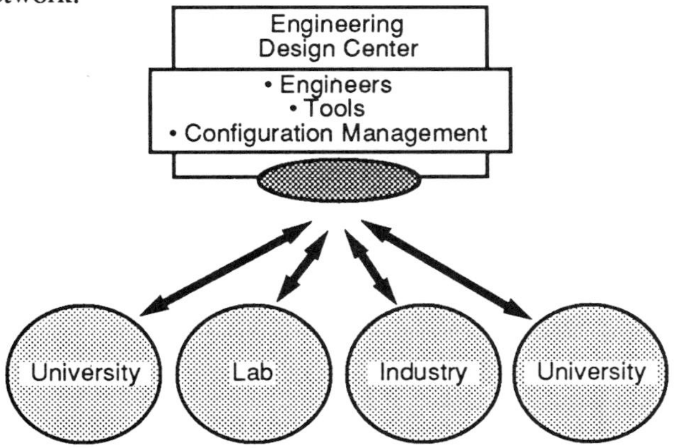

Figure 5. The Architecture of the Engineering Design Network.

The Engineering Design Center

The Engineering Design Center is responsible for the overall detailed engineering of the detector system, as well as the coordination and controlled evolution of the data that define the detector system configuration. The Engineering Design Center should have network capacity to permit several simultaneous transfers for collection or distribution of engineering or simulation models to the collaborators. A high-speed connections to ESnet and support of both sets of prototcols at the Engineering Design Center would give the collaborators a choice when tuning their individual connections to the network.

The present EMPACT Engineering Design Center at Martin Marietta Denver is shown in Figure 6. It consists of an Ethernet network connecting a VAXstation 3500 with a number of 3D workstations (e.g. Tektronix 4236 and Silicon Graphics Personal Iris). A VAXstation 3540 is being used for a beta test of the latest level of I-DEAS software. A VAX cluster with a disk farm is used to store drawings. Apple Macintoshs connected to the network provide easy-to-use, low-cost workstations for simple modeling, graphics, coding, and desktop publishing. Finally, a network connection to the INTERNET is made thru a microVAX, SUN and a cisco router.

The engineering design package I-DEAS is resident on the VAX cluster with some individual I-DEAS modules and drawings on the workstations. Some workstations are stand-alone with local disk storage. The VAX cluster is used for long term storage and backup. The Tektronix workstation does not have local storage and only functions as a graphics terminal. The VAX cluster, running both DECnet and WINS TCP/IP from Wollongong performs the function of central data base, which is backed up regularly, for all the workstations.

Typically, each engineer/designer at the Engineering Design Center requires about 300MB of disk storage for drawings. The operating system and the full complement of the integrated I-DEAS package takes an additional 300MB. In general, one should select computer systems that offer growth capability in memory, disk space and card slots, since the

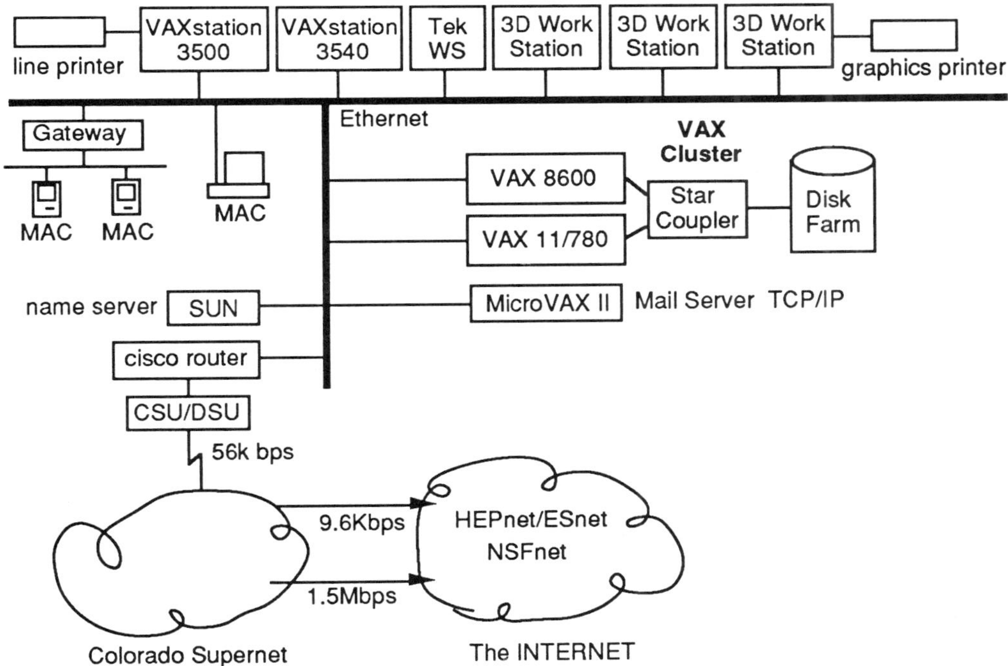

Figure 6. Martin Marietta Astronautics EMPACT Design Network

space requirements of the design tools, network products and operating systems will increase with time.

Collaborating Institutions

Collaborating universities will, in general, have a 3D workstation that can run the software in use at the Engineering Design Center, as well as most physics simulation packages. In addition, local storage and a tape backup unit will be required. The software actually resident (and licensed) to each university collaborator system will depend on the work being done at that site.

Once detailed design starts, it is expected that each collaborator would send or retrieve a package of models or CAD drawings once a day. The engineering drawings will grow as the project progresses and more and more detail is added; five to ten mega byte files will be common. This amounts to 20 10MB files per day. If a 10MB file takes one-half hour to transfer at 56K, then it would take 10 hours, more than one working day, to transfer the files. Given these assumptions, multiple 56k circuits or fractional 1.5Mbps circuits to the Engineering Design Center would give the collaborators adequate access to the central data base.

Each collaborator needs a network connection such that they can reliably transfer their computerized models and designs in a reasonable amount of time. This could range from a few minutes to an hour. The transfer times will depend on the bandwidth of the slowest segment of the network, the network's utilization and the performance of any gateways. The collaborators will consider the existing utilization of shared networks and gateways when estimating the time to transfer a drawing.

The High Speed Backbone

High speed wide area networking is a prerequisite for integrating the spectrum of institutions collaborating on the SSC detector design. Obtaining the benefits of an integrated design, modeling, analysis, and simulation resource demands that the network provide collaborators turnaround appropriate to the task. The timeliness of an analysis, for example, would be lost if it took many days to communicate the details to other collaborators.

The Department of Energy's Energy Science Network, ESnet (see Figure 7), is one solution for networking SSC detector design collaborators. Its design had the benefit of the experience with the High Energy Physics network, HEPnet. It was enhanced in 1989 to accommodate the network protocols of the two operating systems used by scientists and engineers today, UNIX's TCP/IP and DEC's DECnet. Also its speed was increased nationwide to 1.5meg bits/second, or T1. As seen in the network diagram of ESnet in Figure 7, many of the major physics research centers are on this backbone.

While many other networks have gateways to ESnet, the use of these gateways is not recommended for large file transfers. Only one major network, the National Science Foundation's NSFNET (see Figure 8), has the speed, 1.5meg bits/sec, and connectivity to permit the large file transfers required by the SSC detector collaborators. NSFNET is a major component of a larger collection of networks called INTERNET. Their wide spread use by universities may permit existing network connections to be used by some collaborators. However, NSFNET only supports TCP/IP protocols, so

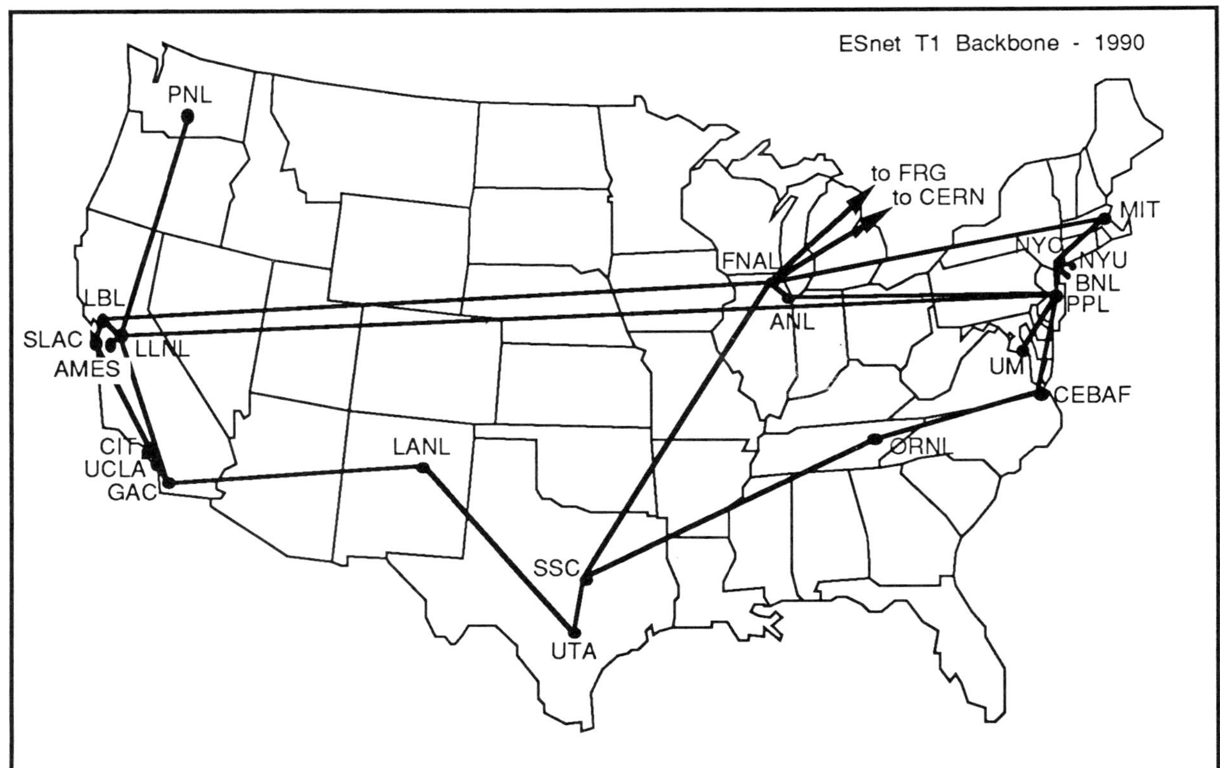

Figure 7. Schematic map of the ESnet connections in the United States.

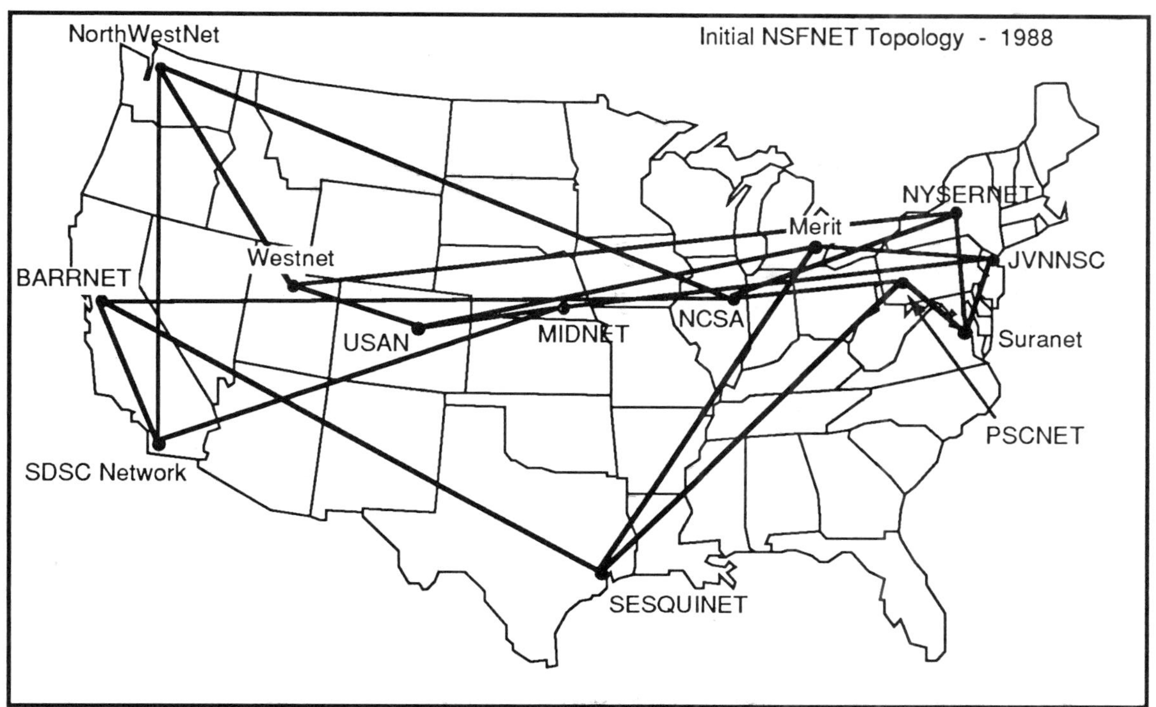

Figure 8. Schematic map of the NSFnet connections in the United States.

collaborators wishing to use NSFNET will use UNIX's native TCP/IP or add TCP/IP to their computer systems.

Conclusions

The utility of integrated design, modeling, analysis and simulation has already been clearly demonstrated in the EMPACT Expression of Interest to the SSC Laboratory. This work involved the construction of many three-dimensional models of detector sub-systems (e.g. muon chambers, toroid magnets and calorimeters), with the accompanying engineering analyses, particle tracking, physics simulation and system assembly studies. The capability to perform rapid and accurate turnarounds on various design options will be invaluable as sub-systems proceed toward detailed design and are ultimately ntegrated into complete detector systems.

The development of a useful network that connects all collaborators within EMPACT and allows access to the engineering design center is a challenge that has only begun to be addressed. To complete the network, a significant amount of hardware and software is required to outfit each participating institution. In addition, individuals must be trained in the use of the software, be they physicists, engineers or system management profesionals. Perhaps the greatest challenge involves connecting all institutions over a network that supports both TCPIP and DECNET protocols and allows data transfer at rates that permit mega-byte files to be sent and received as a matter of course.

MECHANICAL ENGINEERING AND DESIGN OF SILICON-BASED PARTICLE TRACKING DEVICES

W. O. Miller, T. C. Thompson, M. T. Gamble,
R. S. Reid, K. A. Woloshun, G. D. Dransfield, and H. J. Ziock
Los Alamos National Laboratory
Los Alamos, NM 87545

ABSTRACT

The Mechanical Engineering and Electronics Division of the Los Alamos National Laboratory has been investigating silicon-based particle tracking device technology as part of the Superconducting Super Collider-sponsored silicon subsystem collaboration. Structural, thermal, and materials issues have been addressed. This paper discusses detector structural integrity and stability, including detailed finite element models of the silicon chip support and predictive methods used in designing with advanced composite materials. Electronic thermal loading and efficient dissipation of such energy using heat pipe technology has been investigated. The use of materials whose coefficients of thermal expansion are engineered to match silicon or to be near zero, as appropriate, have been explored. Material analysis and test results from radiation, chemical, and static loading are compared with analytical predictions and discussed.

Introduction

Los Alamos National Laboratory (LANL) has been investigating silicon-based particle tracking device technology on behalf of the Superconducting Super Collider (SSC)-sponsored silicon subsystem collaboration.[1] The overall program objective is to prove the mechanical and cooling feasibility and demonstrate a viable design, by means of analysis and prototype testing, for a silicon tracking device. Particular attention has been given to the highly stable mechanical structure needed to support and cool all detector elements and related electronics. The current design exploits the inherent physics advantages of silicon microstrip detectors by accomodating silicon wafer placement of extreme accuracy and stability.

The silicon tracking system research and development effort addresses the numerous engineering issues facing successful deployment of a precision silicon tracking device. The stringent design requirements are found in Table 1. At this interim stage of the mechanical design study, the central issues that have been addressed are the (1) overall system design approach, (2) radiation length calculations, (3) material candidates selection, (4) material specimen testing initiation, (5) critical systems structural analysis, (6) alignment concepts for verifying wafer position, (7) cabling concepts for data transmission and electronic circuit power, and (8) cooling system design for the individual wafer electronic circuits. The significant progress in these areas has made possible extensive prototype testing during fiscal year (FY) 1991 that will culminate in verification of the feasibility of a silicon tracking device and a viable design for such a system.

Our current design comprises a metal matrix truss space frame to provide a very sparse support structure for the complex array of silicon super layers while effectively satisfying structural integrity requirements in a 10 Mrad environment and not exceeding the stringent radiation length constraints. The silicon superlayers cradled inside the space frame are displayed in Fig. 1. The framework draws upon material technologies being developed for the National Aerospace Plane and Strategic Defense Initiative programs. The silicon superlayers are formed into silicon sandwich-like constructions, which are combined in cylindrical silicon shell configurations for the central region and flat panel arrays for the forward region. Coolant rings that support the silicon also serve as an effective means of incorporating a heat pipe network. Use of heat pipe technology is another innovative use of advanced technology for achieving the stringent stability requirements demanded of the silicon tracking system. Meeting these requirements is expected to produce major advances in areas including the (1) application of metal

Table 1. Baseline design requirements.

Detector characteristics	Silicon, longstrip technology 50-μm pitch, double-sided Electronics surface area, 3.52 × 0.6 cm^2
Detector assembly	Maximum length, 36 cm (six wafers edge bonded) Mounting precision R, ϕ, Z — 25μm's Angular, 1 milliradian Positional stability R, ϕ, Z — 5μm's Angular, 0.2 milliradian Detector cant angle, 11° Operational temperature, 0°C
Detector heat load	Heat flux, 1.8 W/cm^2 40 kW Total
Radiation exposure	10 Mrads over 10 year service life No measureable stability effects >5 μm's
Material radiation length budget	Structure and related service connections 2.5% (except in limited regions, e.g., transition from central to forward)

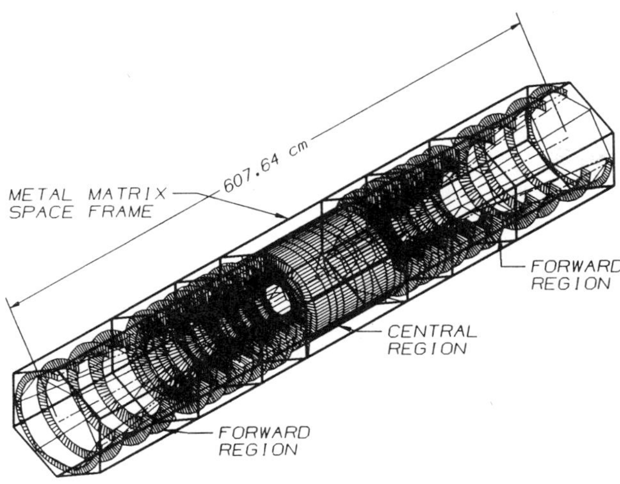

Figure 1. Metal matrix space frame surrounding silicon superlayers.

matrix composites, (2) successful development of silicon sandwich, shell structures, and silicon planar array, (3) implementation of heat pipe cooling technology for isothermal stabilization of the structure and silicon, and (4) innovative use of fiber optics for rapid data transmission.

Design Features

The Silicon Tracking Subsystem overall length is 6.2 m and the outer diameter including the shell enclosure is approximately 1.4 m. The silicon wafers are edge-bonded into long strips. Strip length will be optimized to yield a maximum number of similar individual wafers. The silicon layer is bridged at discrete points between the end supports to preclude excessive sag. Both ends of the silicon layers are bonded to rings. The rings, in addition to providing structural support, provide a passage for an internal heat pipe artery, as shown in Fig. 2. The silicon modules are mounted onto the support structure of the silicon detector central region. These modules are supported on kinematic supports to avoid introducing distortions. Consideration is being given to using a shell support structure material such as graphite/epoxy or a metal matrix composite that can be engineered to exhibit a zero coefficient of thermal expansion (CTE).

Cooling rings, which support pie-shaped silicon layers, must track the contraction of silicon or will introduce the possibility of distortion. Our calculations show that in the central region the cooling ring CTE must also match the silicon CTE to avoid wafer distortion. Tailoring the expansivity of a graphite metal matrix, or a graphite/epoxy composite to match that of silicon are options for achieving the desired CTE equality. Prototypes of the envisioned ring construction techniques are being produced in this study phase.

The space frame provides structural support for the silicon wafers and related cooling rings. The design criteria applied to the development of this structure are based largely upon dimensional stability, high stiffness to weight ratio, minimal intrusion on the paths of energetic particles emanating from the interaction region, and structural

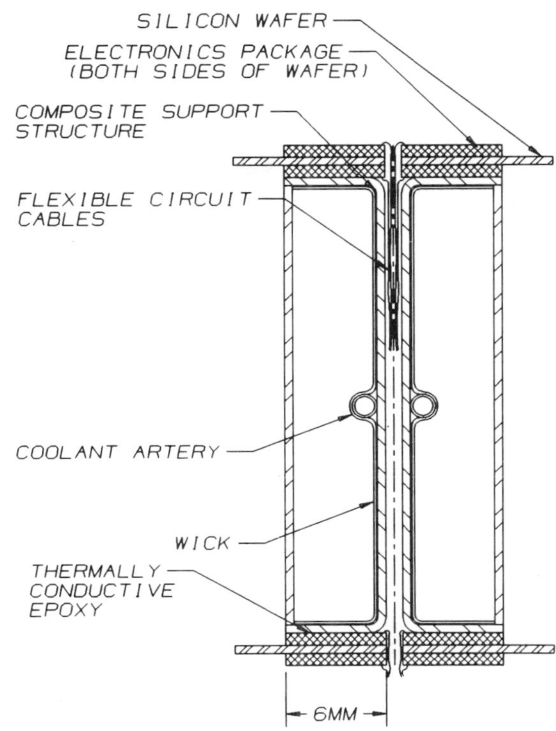

Figure 2. Silicon support ring with integrated heat pipe technology (cross-sectional view).

integrity. Consideration of each of these is important for assuring the silicon tracking system is capable of achieving its operational objectives. However, the interrelation of these with particular materials and constructions is complicated.

Radiation path length is a function of radius, r, longitudinal angle, θ, and azimuthal circumferential angle, ϕ. The angle ϕ determines when discrete tubular members of the frame will be intercepted, while θ is an indicator of deviation to the surface normal, effectively increasing the material thickness in the forward region. For most angles the radiation length restriction is met with exceptions at rather shallow angles, $\theta \sim 10°$ and for ϕ angles corresponding to tubular space frame members. For a structure comprised entirely of Be the radiation length would be less than 1.5%. Alternative construction materials, e.g., graphite/epoxy and Mg-MMC composites would increase this value to 2% nominally.

Thermal Design

Early studies demonstrated that air convective cooling would be insufficient to maintain the system at the design operating temperature, in the presence of the anticipated heat flux associated with the electronics, without introducing undesirable thermal gradients. Therefore, an evaporative heat pipe cooling system concept using hydrocarbon fluid candidates as means to achieving the 0° C operational temperature was initiated.

The heat pipe concept underwent a conceptual definition and several possible working fluid candidates were selected. Wick materials that are consistent with the materials chosen for the structures are under experimental investigation. During the conceptual design phase, the analysis tasks included wick sizing, facets of wafer cooling, heat pipe condenser design, and system operation.

Working fluid, such as butane, is wicked to an area directly beneath the electronics package, as shown in Fig. 2. Heat flows through this thin-walled area and evaporates the cooling medium. This technique greatly simplifies the process of cooling the enormous number of individual wafers. The 1.8 W/cm^2 heat flux, associated with the wafer electronic package, can be handled effectively with a minimum temperature gradient by minimizing the conductive path. Distortions in the silicon shell substructure are avoided by minimizing the thermal gradient and through an appropriate material choice for the cooling ring.

As long as the vapor space around the various detector elements is at a constant pressure, and the pressure loss in individual vapor return lines for all silicon modules are equal, the structure will thermally equilibrate. Our design goal for structure temperature uniformity is 0.25°C.

The wick design requires the integration of an artery, an open flow passage that runs circumferentially around the cooling ring. The artery provides a low-impedance flow path through, and hydrologically connected to, the wick structure. For a sprayed-on graphite wick, the artery may be a porous tube placed on the cooling ring prior to spraying. The pores in this tube must be small enough so that the graphite does not penetrate, but it adheres to that surface in such a way as to form a hydrologically connected capillary path. This concept is being experimentally investigated.

Materials Studies

In most design applications many factors are considered in material selection; however, in the instance of the SSC silicon tracking subsystem, material selection is driven by three primary factors. These are (1) 5-micron material stability, (2) 2.0% radiation length, for the structure, and (3) radiation resistance in the presence of hydrocarbon fluids.

Our preliminary studies have shown that to achieve these goals application of state-of-the-art material composites and high stiffness to weight ratio construction techniques will be required. An extensive discussion of candidate materials and properties of interest has been compiled.[1]

Radiation Damage

A maximum radiation dosage of ten megarads is anticipated over the ten year life of the detector. One of the primary issues in the study of the feasibility of the silicon tracking system is that of radiation damage. Not only is there a question of the survivability of the detectors and their associated front-end electronics, but one must also verify that the support structure is capable of maintaining the highly precise (~5 micron) positional stability of the detectors in the radiation environment. Ultra-lightweight graphite/epoxy sandwich panels have been exposed to radiation levels of 3×10^{13} neutrons/cm^2 and more recently to 10^{15} neutrons/cm^2. The early test at 10^{13} neutrons/cm^2 did not disclose any serious effects. Evaluations after exposure to 10^{15} neutrons/cm^2 are pending.

The mechanical support structure for the silicon tracker will be subjected to radiation damage from charged particles and photons emanating directly from the interaction point and neutrons resulting from the interactions of the primary particles and their interaction products in the calorimeter. Over the ten-year lifetime of the experiment, at a radial distance of 10 cm, the total fluences are expected to consist of approximately 10^{14} charged particles/cm^2, an equivalent fluence of photons, and on the order of 10^{13} neutrons/cm^2 for a luminosity of 10^{33} particles/cm^2. Although those radiation levels are low compared to those commonly encountered and tested for in fission power reactor facilities, the problem here is of a different scope and involves unique materials and structures. We are interested in effects such as creep, warping, and swelling, which lead to the slightest dimensional changes of items it is supporting. Our ultimate goal is to achieve a design capable of withstanding the radiation environment associated with a luminosity of 10^{34} particles/cm^2.

Fiscal Year 1991 Tasks

The mechanical engineering program for FY1991 focuses on prototyping with the intent to prove the feasibility and viability of material selections, cooling schemes, and structural designs developed during FY1990. Accomplishing these goals requires the fabrication of a full-scale subsection of the barrel region of the detector. This 54-cm diameter replica is composed of approximately 1,200 silicon wafers arranged as two superlayers connected by a polymeric matrix composite material cooling rings with associated heat pipe cooling. The two silicon superlayers will be supported by a low CTE graphite/epoxy sandwich shell structure and a zero CTE metal matrix space frame. This combination will provide an opportunity to investigate all mechanical and structural effects anticipated in a full-scale silicon tracking system. It will be fully equipped with electronics-equivalent resistive heat loading to simulate thermal conditions of the operating silicon detector system. An interferometric measurement system will detect minute motions of the silicon detectors. The performance of the metal matrix composite space frame and fully operational heat pipe cooling system can be inferred from such measurements.

Critical issues that will be resolved by these tests include (1) cooling ring construction and wick development and their optimization, (2) materials characteristics and stability, (3) selectively enhanced polymeric and metal matrix composite engineering properties, such as CTE and elastic modulus, (4) silicon module stability under electrical load simulation, and (5) material and component radiation and chemical exposure compatibility. Design and analysis activities will support these tests.

Reference

1. W. O. Miller, T. C. Thompson, M. T. Gamble, et al., "Superconducting Super Collider Silicon Tracking Subsystem Research and Development Report," October 1990.

MECHANICAL DESIGN OF THE L* CENTRAL HADRON CALORIMETER

M. J. Rennich,[1] W. M. Bugg,[2] F. Plasil,[1] S. C. Berridge[2]
1. Oak Ridge National Laboratory,* Oak Ridge, Tennessee 37831
2. University of Tennessee, Knoxville, Tennessee 37996

Abstract

The L* collaboration proposes to use a silicon calorimeter to provide precise and fast hadron measurements. The primary design challenge of this detector is the efficient and reliable removal of heat generated by the readout electronics located inside the detector. The preliminary design and analysis for the proposed heat transfer system demonstrates that such a system can be built using commercially available components and manufacturing processes.

Introduction

The L* detector proposed for the SSC provides extremely precise measurements of electrons and muons coupled with precise and fast hadron calorimetry measurements. Two active detector media have been proposed for the hadron calorimeter: liquid scintillator and silicon. The silicon calorimeter is shown in Figure 1. It offers advantages of absolute gain, linearity of response, insensitivity to magnetic fields, easy monitoring of gain stability with radioactive sources, convenient transverse segmentation, low voltage operation and ease of construction.

The proposed Silicon calorimeter is composed of 15 doughnut-shaped barrel modules having 16 mechanical towers each, and 2 cylindrical end caps with 8 mechanical towers each. The mechanical towers are segmented into layers ($2X_0$ to 5λ and $4X_0$ to 10λ) and $\eta - \phi$ partitions of 0.04×0.04. The absorber materials will be lead and brass plates.

Thermal Control of a Silicon Hadron Calorimeter

The silicon hadron calorimeter, like many other detectors proposed for the SSC, presents a thermal control challenge because of the need to locate very fast, high-power electronics near the detector medium. In the case of the L* detector the hadron calorimeter is expected to generate up to 500 KW of power internally, primarily from the preamps which will be attached directly to the silicon wafers inside the detector. Additional thermal requirements specify an average operating temperature of -20°C and an operating range of ±2°C The low temperature is expected to substantially reduce silicon radiation damage and the limited operating range will permit uniform measurements throughtout the detector.

A review of possible methods of removing the heat from the inside of the detector was undertaken as part of the early mechanical design process since this issue is one of the principal design concerns. Initially, the three basic mediums for removing heat were considered: forced gas convection, phase change refrigerants and liquid coolants. Gases could not be used since the detector is designed to be compact and therefore lacks the necessary openings for gas transport. Refrigerants were eliminated at the suggestion of several vendors due to problems with uniform circulation of lubricants within the piping and heat exchanger system. Thus, liquid coolants were selected as the best alternative provided the coolant could be contained in closed loop heat exchangers rather than circulated throughout the interior of the detector.

Thermal Control Analysis

An analysis of the local heat transfer within the layers of the calorimeter indicated that liquid coolants can reliably cool the silicon sense layers if the generated heat is conducted to the edge of layers where conventional panel heat exchanges can be located without seriously reducing the physics performance of the detector. Figure 2 shows the proposed layer assembly. A thermal analysis was performed to confirm that this could successfully function with air convectively transferring the heat to a brass absorber plate which would then conduct the heat to the heat exchanger. Figure 3 shows the calculated temperature profile of a worst-case layer assembly. The analysis also indicated that the brass had to have a zinc content of 5% or less to maintain the total temperature differential at less than 3°C. Further, the analysis

*Operated by Martin Marietta Energy Systems under contract DE-AC05-84OR21400 with the U.S. Department of Energy.

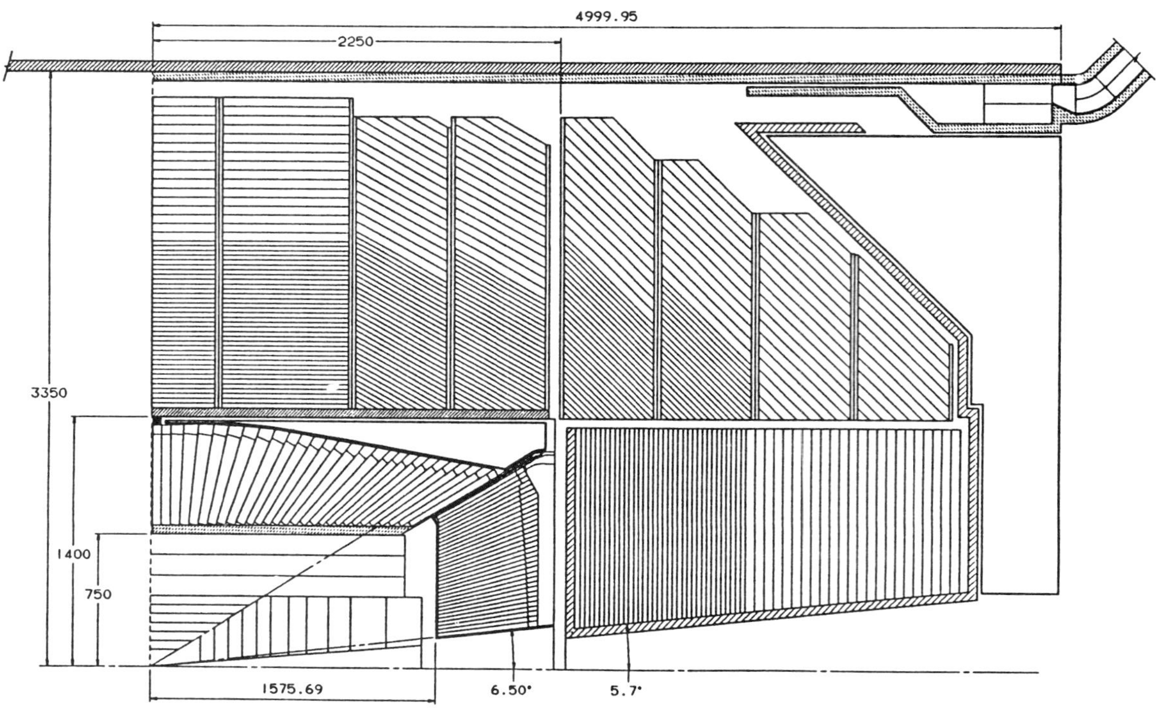

Figure 1: L* hadron calorimeter

indicated that the thermal connection between the brass and the heat exchanger requires a special thermal bond. Such bonding materials are common in the electronics industry and radiation damage studies are underway to verify candidates for use in the L* detector.

There are several commercially available liquid coolants which can be used to control the heat inside a silicon hadron calorimeter. Figure 4 provides a partial tabulation of liquid coolants representing several families including salt brines, gylcols, alcohols, solvents, silicones and synthetic hydrocarbon based systems. While all those listed will function at the specified -20° to -40°C most have individual characteristics which limit their use inside an SSC detector, most notably the corrosive salt brines, hazardous solvents andflammable alcohols. Of the remaining candidates; cost is the primary concern, with the more expensive candidates offering slightly better performance.

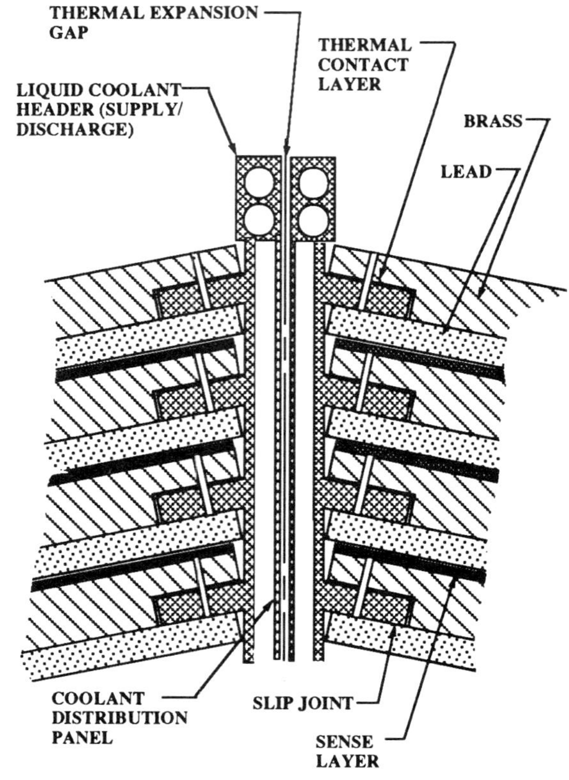

Figure 2: Layer assembly

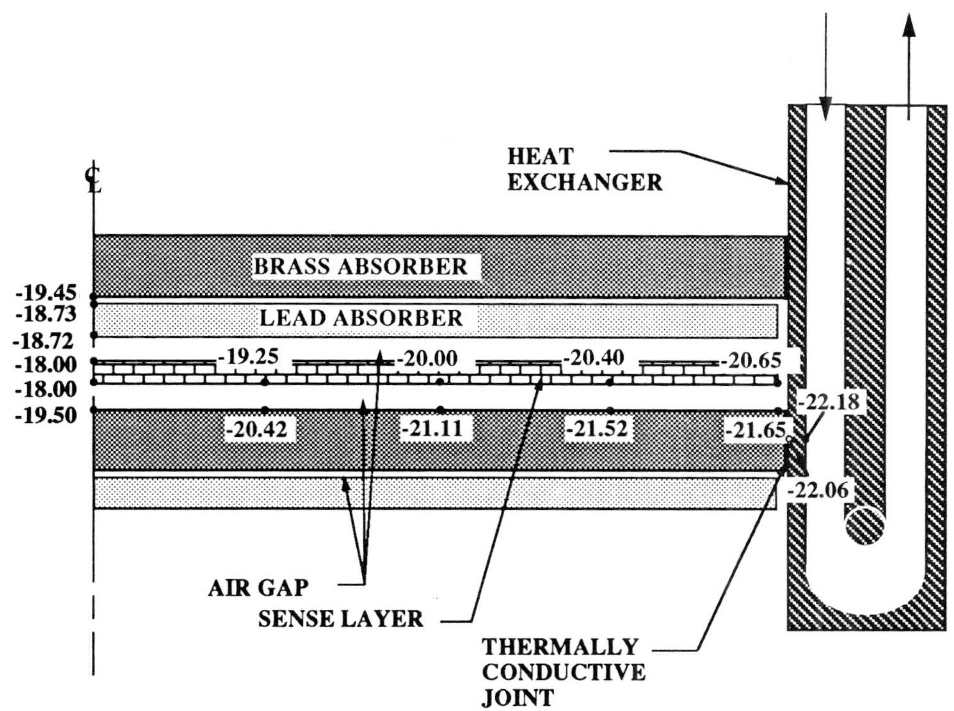

Figure 3: Temperature distribution within a typical layer

COOLANT	*SPEC. HEAT*	*VISCOSITY*	*FLAMM*	*TOXICITY*	*CORR*	*SOLVENT*	*COST*
	(CAL/GM_CAL)	(CENTIPOISE)					
SILICON	0.37	65	LOW	LOW	LOW	LOW	HIGH
ETH GLYCOL (60%)	0.7	100+	MEDIUM	HIGH	LOW	LOW	LOW
MINERAL OIL	0.45	LOW-HIGH	NOTE: NOT RECOM FOR USE WITH COPPER BASED ALLOYS				
SYN HYDROCARBON	0.55	100+	LOW	LOW	LOW	LOW	HIGH
SALT BRINE	0.64	15+	NO	LOW	HIGH	NO	LOW
SOLVENTS	1.0+	<1	LOW	MILD(100+)	LOW	HIGH	LOW
FLORINATED LIQ	0.22		LOW	LOW	LOW	LOW	HIGH
METHANOL	0.82	6.5	HIGH	MILD	NO		LOW

Figure 4: Liquid coolants for use in the L* calorimeter

Conclusion

The thermal control of all SSC detectors will be a primary design consideration. A preliminary thermal analysis of a silicon hadron calorimeter has shown that heat can be efficiently removed from the inside of the detector using commercially available liquid coolants and heat transfer systems. It has also been shown that heat can be uniformly removed from the detector layers with a conventional, closed circuit, panel heat exchanger system, thus increasing the reliability and decreasing the complexity of the overall detector design.

SCINTILLATING PLATE CALORIMETER MECHANICAL DESIGN*

A. Buehring, N. Hill, T. Kirk, J. Nasiatka, E. Petereit, L. Price, J. Proudfoot, H. Spinka, and D. Underwood
High Energy Physics Division
Argonne National Laboratory
9700 South Cass Avenue
Argonne, Illinois 60439

M. Burke, D. Hackworth, T. Hordubay, D. Marshik, D. Scherbarth, and R. Swensrud
Science and Technology Center
Westinghouse Electric Corporation
Pittsburgh, Pennsylvania 15235

Presented by H. Spinka

Abstract

Progress on designs for compensating scintillator plate calorimeters will be presented. One design includes a lead composite absorber, fiber readout, and radiation hardened scintillator plates, and the second design has depleted uranium absorbers, wavelength shifter plate readout, and scintillator plates. The lead absorber is cast with slots to accept the scintillator in the first design, while the depleted uranium is in the form of plates in the second design.

Two mechanical designs for compensating scintillator plate calorimeters for the SSC have been developed based on different absorber materials and readout methods. The first uses lead cast into steel-reinforced modules with slots for the scintillator tiles. Wavelength shifter fiber is embedded in the tiles, and clear fiber brings the collected light to photodetectors outside the calorimeter. The second design has depleted uranium (DU) plates alternating with scintillator tiles. A wavelength shifter (WLS) plate is used to transmit the light to the photodetectors.

For both designs, the calorimeter is separated into one electromagnetic (EMC, $25\,X_o$) and two hadronic (HACI, HACII) sections. The HAC sections are approximately projective, whereas the EMC section is rotated 8° from purely radial. The total absorber thickness is a minimum of 9λ in the barrel calorimeter and 11λ in the endcap calorimeter. The scintillator plates are approximately $11 \times 11\,\mathrm{cm}^2$ at the inside diameter of the barrel (450 cm), and correspond to $\Delta\eta = 0.05$ and $\Delta\phi = 0.05$. A solenoidal magnet is assumed to be present inside the calorimeter. Magnetic flux return and phototube shielding in the barrel calorimeter are accomplished in part by a 30 cm thick iron cylinder located outside HACII. Additional iron will probably be necessary for the full flux return.

The scintillator plate thickness was assumed to be 2.5 mm in both designs. The lead thickness was 5.0 mm in the EMC and 10.0 mm in the HAC sections, corresponding to approximately compensating ($e/h = 1.00 \pm 0.05$) values. Similarly, the DU plate thickness was taken to be 3.3 mm for compensation.

Pure lead or lead alloys will not work as the absorber for the HAC section of the barrel calorimeter. Analytical and finite element analysis (FEA) calculations indicate that stresses would be beyond yield strengths for some locations with either lead plates or cast lead modules. Lead composites, reinforced with steel or some other material, will be required. It is not yet clear whether composites will be necessary in the EMC section.

Casting lead in a steel frame with steel reinforcing and slots for the scintillator tiles (see Figure 1) appears most economical for the HAC section of the barrel calorimeter. The slots can be cast around teflon forms. Such a casting technique has already been demonstrated

*Work supported in part by the U.S. Department of Energy, Division of High Energy Physics, Contract W-31-109-ENG-38.

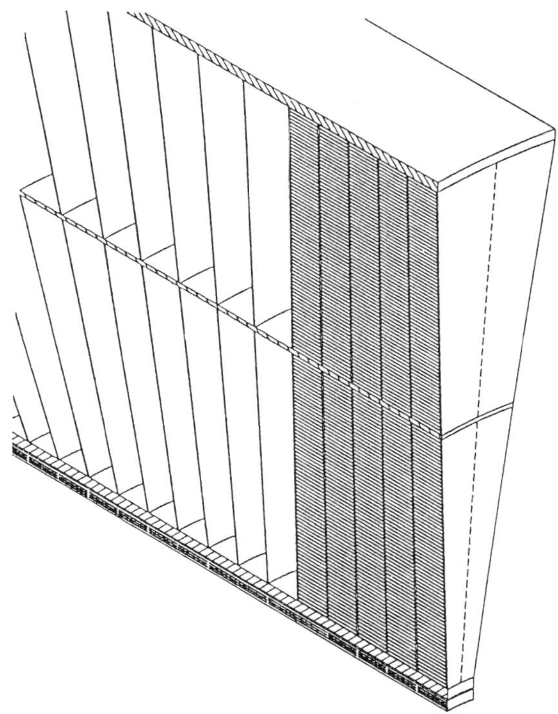

Figure 1. Conceptual design of a cast lead module. Steel plates at the inner, intermediate, and outer diameter of the HAC sections are shown as well as thin steel reinforcing between towers. The lead is cast with slots in the steel framework. A similar module design has been developed for DU absorber plates.

for 6 mm diameter teflon rods, which were easily removable able after cooling. It has not yet been established that the slot thicknesses or the slot spacing can be made sufficiently uniform so they do not limit the calorimeter energy resolution.

A calorimeter design with DU absorber and WLS plate readout, similar to ZEUS, seems to work acceptably based on analytical and FEA calculations. However, the DU processing costs and the poor fiducial volume resulting from the WLS plates are primary problems of this design compared to the design with fiber readout and lead absorber.

A variety of schemes to take advantage of the small diameter and good flexibility of the fibers exist. The lead absorber design shows reduced cracks between modules with the use of fiber readout. The photodetector location can be more accessible with fibers than with WLS plate readout. A scheme for reading multiple scintillator tiles in series has also been developed. A method of routing the fibers projectively to the outside of the calorimeter has been studied. However, it has not yet been established whether the routing method will permit a sufficiently uniform response in the scintillator tiles.

Having the EMC barrel calorimeter modules separate from the HAC modules, will probably reduce the size of the cracks in the EMC section, and also minimize material inside the first EMC scintillator layer. The EMC modules would not necessarily be removable. Fiber readout to the module ends would be employed.

The HAC section of the barrel calorimeter modules are similar for the two conceptual designs studied. Both designs have evolved toward a module that is two towers wide in ϕ and many towers long in η. Both contain intermediate and inner steel reinforcing plates, and both utilize the outer 30 cm steel for structural strength as well as magnetic flux return and magnetic shielding for the photodetectors.

The barrel calorimeter will need to be in two or more sections longitudinally for the modules to be easily transportable. For three longitudinal sections, a scheme exists to reduce the effects of cracks and structural material at the ends of the modules. This involves HAC towers that are quasiprojective in η.

Two different methods for supporting the barrel calorimeter have been considered. Pure radial supports to external muon iron take up too much space and will be difficult to access and adjust. An alternate method, cradle or saddle supports from the bottom, will restrict the muon coverage in some regions asymmetrically, but appears to be the preferred solution.

Work has begun on a conceptual design for the endcap calorimeters. The endcap towers are assumed to be projective. Difficulties with the design at the barrel/endcap interface are being studied, as well as the optimum method to route cables from the central tracking detectors and the solenoid cryogenic system and current leads. Additional work is needed to develop a satisfactory endcap design.

Estimates of the materials for the cast lead design include a total lead weight of 5100 tons, a total iron weight of 1800 tons, a scintillator area of 4.7×10^8 cm^3, a WLS fiber length of $\approx$ 700 km and a clear fiber length of $\approx$ 8400 km. For the DU absorber design, the total DU weight is 5200 tons, the total iron weight is 1800 tons, the scintillator area is 7.6×10^8 cm^3, and the WLS plate area is $\approx 100 \times 10^6$ cm^2. The sizes of the two calorimeters are very nearly the same. However, using an approximate cost of \$6 per pound for processing the DU, and of \$1.75 per pound for the lead material and processing, the cost difference for the absorber in the two designs is about \$45M. The scintillator area is also much larger in the DU design. Work is continuing on the lead design, partly because of the much higher DU design costs.

These mechanical design studies are supported in part from the Compensating Scintillator Plate SSC Subsystem funds. Additional members of the collaboration include physicists and engineers from Ames Laboratory at Iowa State University, BICRON Corporation, Florida State University, Louisiana State University, University of Mississippi, Oak Ridge National Laboratory, University of Tennessee, Virginia Polytechnic University, and the University of Wisconsin. We are indebted to many of these collaborators and to other members of the SDC collaboration for helpful suggestions.

ENGINEERING DESIGN STUDY FOR A WARM LIQUID CALORIMETER*

Teresa Ballinger and Artie Debeling
Mechanical Engineering
EG&G Energy Measurements, Inc.
5667 Gibraltar Drive
Pleasanton, CA 94566

Abstract

A systematic engineering study was conducted to quantitatively evaluate the feasibility of the Warm Liquid Calorimeter (WLC) concept for the Superconducting Super Collider (SSC). The baseline constraints were that each detector module could not exceed a weight of 20 tons, and that the main support structure plus detector modules could be assembled after the magnetic coil was emplaced. A systematic approach was developed that required a feedback loop that optimized the efficiency of the finite element structural analysis, assembly procedures, and hermeticity considerations. This approach is presented plus the results of the study.

Introduction

This paper presents the systematic approach used to assist in the generation of the report "Engineering Design Study of Warm Liquid Calorimeter Concept for a Detector Experiment at the Superconducting Super Collider".[1] The final WLC conceptual design and baseline concepts are also presented. The initial challenge project engineers face at the inception of any project is arranging information in order of importance and identifying coupling parameters. The degree of difficulty encountered in producing a successful design is usually in direct proportion to the complexity of these two tasks. In the case of designing a forefront scientific instrument, the tasks are amplified due to the transitory nature of both the scientific and engineering data. Therefore, some type of methodical approach is essential to assure that project objectives are updated and remain focused.

WLC Baseline Concepts

In the early phase of the WLC study, it was determined that baseline concepts were needed if scheduling milestones were to be achieved. These baseline concepts would be rigidly held to unless judged to be unacceptable as the study progressed. It was realized that by defining baseline concepts, many design concepts and assembly schemes that may have eventually proven to be advantageous to the final WLC design, were not considered. The following lists the five initial baseline concepts:

(1) individual hadronic modules possess an upper limit of 20 tons,
(2) hadronic bays will be tested outside the main barrel before final emplacement,
(3) construction material is aluminum or stainless steel with aluminum preferred,
(4) main support structure is a barrel which is supported only at its ends,
(5) the main barrel and all internal components can be assembled after the magnetic coil is emplaced.

Systematic Approach

The project team realized that the hadronic modules and support washer designs were heavily interconnected, and that the development of the hadronic modules would require that many designs be considered. Therefore, some type of systematic approach was necessary to assure that design changes to either the support washers or hadronic modules did not adversely affect the project. The systematic approach developed incorporates feedback loops driven by design reviews. These review processes guarantee the continuous input of new ideas from both the physics and engineering communities.

The systematic approach applied to the design process of both these components is outlined in the following paragraph. [This approach was not used in the main barrel or end cap design because of weak interconnecting parameters existing between components.]

*Prepared for the Department of Energy under Contract No. DE-AC08-88NV10617.

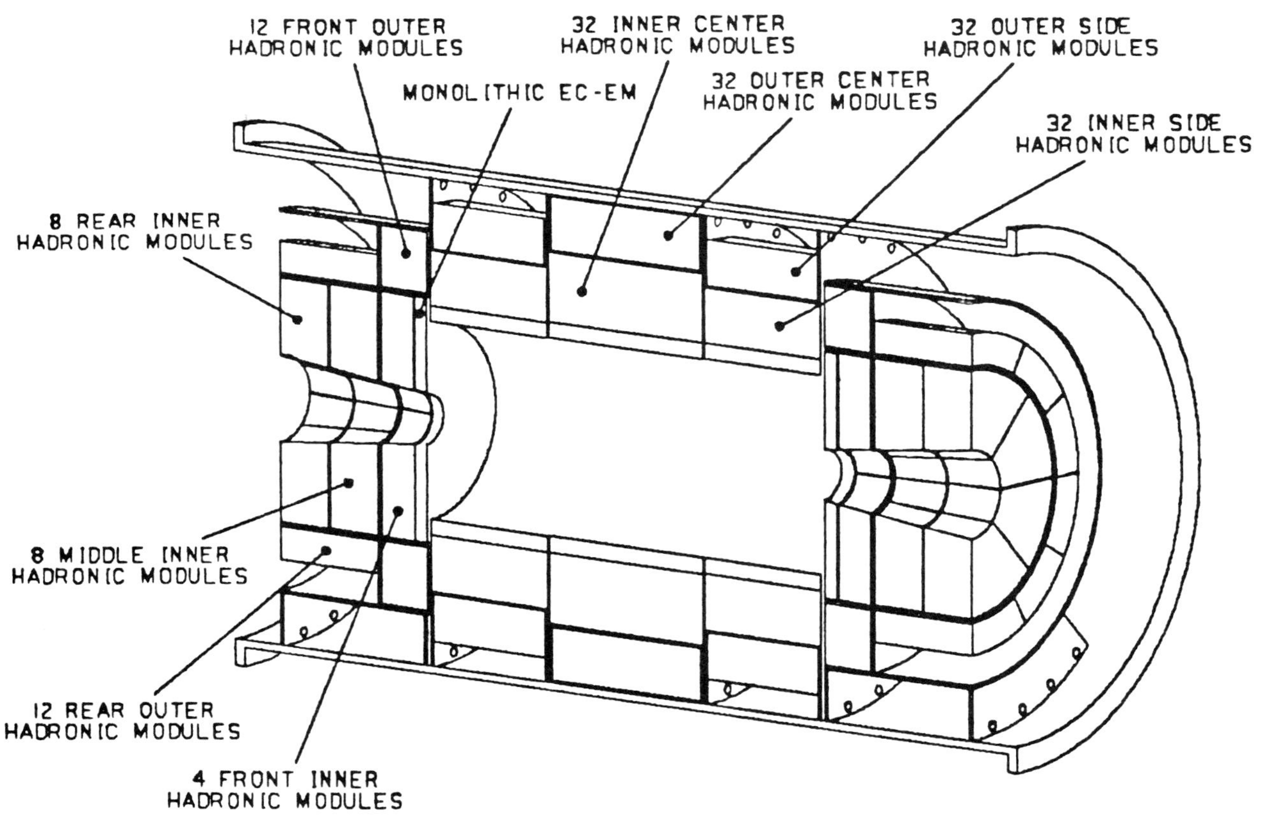

Figure 1. Warm Liquid Calorimeter Conceptual Design.

After the calorimeter baseline concepts were established, initial hadronic and washer designs were produced that satisfied both structural and assembly concerns. These designs were then subjected to a design review that allowed both physics and engineering personnel to focus on conflicting parameters and agree on solutions. The updated designs were then reevaluated to assure engineering feasibility and resubmitted to the review process. This process continued throughout the WLC study. Once final component designs were achieved, all components were assembled and depicted by a mathematical model for a final hermeticity evaluation. Unlike the hermeticity analysis that was done by inspection for each of the component designs, the final hermeticity evaluation was performed with the aid of a calorimeter resolution software package.[2]

WLC Final Conceptual Design

Figure 1 illustrates the final WLC conceptual design. The design can be separated into three major components: Main Barrel, Central Calorimeter, and End Caps. The main barrel is 16 M long, has an ID of 8 M and a shell thickness of 127 mm. It must support 4431 tons of hardware and for the analysis it was assumed to be constructed out of aluminum.

The central calorimeter consists of a center bay and two side bays. Each bay consists of 32 inner and 32 outer hadronic modules separated by two 63.5 mm thick aluminum support washers. The difference between the center and side bays is that the center bay is larger and weighs 1280.0 tons compared to 742.4 tons for each side bay. Figure 2 provides a detailed view of the three bays making up the central calorimeter. Figure 3 illustrates the pinwheel geometry of the hadronic modules about the azimuthal axis, this arrangement is the same for all bays.

The WLC design required two end cap assemblies each consisting of 44 hadronic modules and a monolithic EC-EM which are illustrated in Figure 1. The hadronic module geometry is symmetrical about the azimuthal axis although the hadronic modules were sized differently to reduce exiting paths. The aluminum structure designed to support the EM and hadronic modules consists of an inner and outer shell that are held in place by a washer. The inner shell and washer is 50.9 mm thick, while the outer shell thickness is 76.2 mm. The end cap has an outside diameter of 6.2 M and is 3.0 M long. Support beams, not shown in Figure 1, will connect the inner and outer shells on the outer end of the end cap. Each end cap weighs 720 tons.

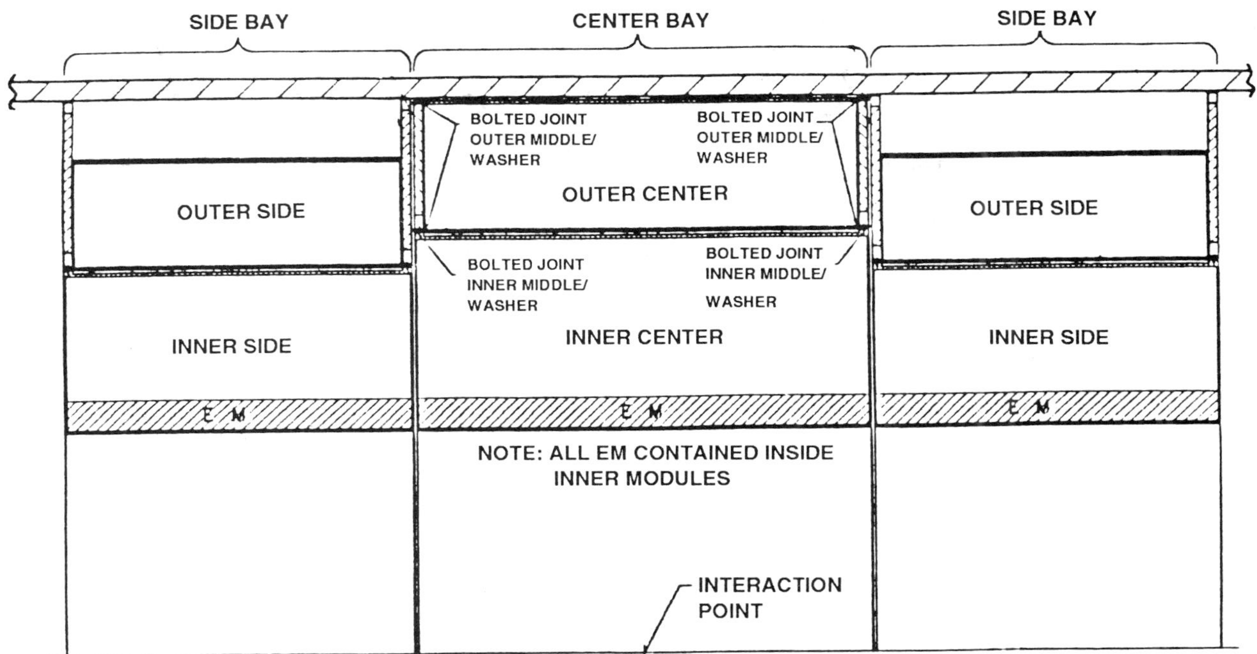

Figure 2. Central Calorimeter.

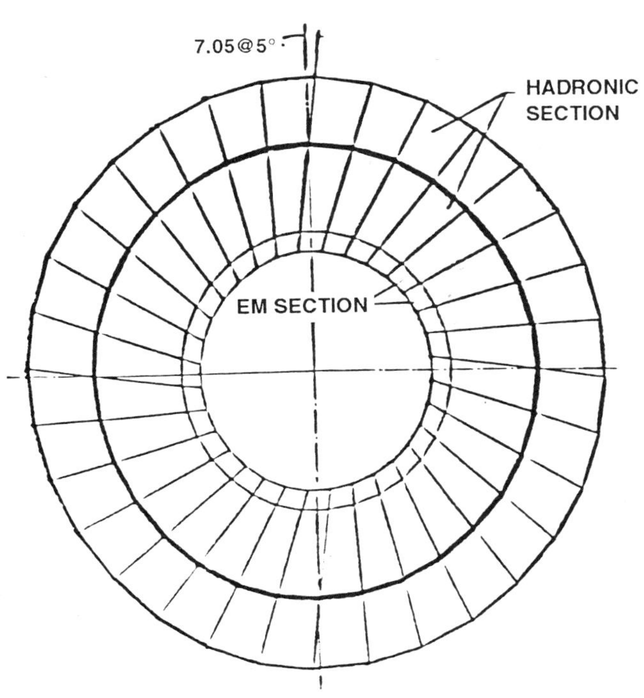

Figure 3. Hadronic Pinwheel Geometry.

Hermeticity Study

The purpose of this investigation was to obtain the impact of different structural material and geometry on the detector's resolution. Both the hadronic and EM modules require an ionization liquid, for this study it was assumed to be tetramethylpentane [TMP]. The calorimeter resolution code evaluates resolution and energy deposition layer by layer. Once the resolution-versus-rapidity (function of calorimeter position) plots were complete for eight azimuthal angles, the detectors hermeticity could be evaluated. Eight azimuthal angles were chosen due to the "pin-wheel" geometry of the detector. The position of the peaks can then be transferred back to the detector in order to improve upon the hermeticity.

The results of this study showed two peaks in resolution which coincided with a rapidity of 0.85 and 1.62. These peaks were due to a decrease in electromagnetic (EM) material. It occurred at the interface between the inner center hadronic modules and the outer side hadronic modules and again at the interface between the inner side hadronic modules and the end cap (see Figure 1).

References

[1]T. Ballinger, A. Debeling, S. Gonzales, T. Thurston and W. Wade, "Engineering Design Study of a Warm Liquid Calorimeter Concept for a Detector Experiment at the Superconducting Super Collider", AVO-395, (1989).

[2]W. J. Womersley, RAY Computer Program, July 17, 1989.

PROTOTYPE DOCUMENT TRACKING AND STORAGE SYSTEM FOR THE SUPERCONDUCTING SUPER COLLIDER LABORATORY (SSCL)*

Norman E. Wells
Physics Research Division
Superconducting Super Collider Laboratory†
2550 Beckleymeade Ave, MS-2000
Dallas, TX 75237

Abstract

The design and development of the detectors for the SSCL will be facilitated by a central system to track and store the design documents and drawings. Collaborators and SSCL personnel need a single system that they can access from a terminal on their desk and be able to locate and view the latest version of a document. The SSCL is developing a prototype system to track documents and drawings, and to make them accessible via the local and wide-area networks. The prototype will be used to refine the requirements and to develop the procedures for a larger system that will be acquired later. This paper discusses the system requirements, architecture, and status.

INTRODUCTION

The SSCL and the collaborations have a requirement for a system to track and store the documents and drawings being used to design and develop the detectors. The system should also be usable for tracking and storing physics papers. With the geographical separation of detector developers, it is important that a central system be provided so the latest version of a document or drawing is available for a collaborator working on that component or on the interface to another component. The goal is to have a system that can be accessed via the SSCL local area net or the E-Mail network so the information is available to physicists and engineers from their desks. To be useful, the system not only has to handle information on a detector, but on the detector's interfaces to the detector hall and its utilities, and to the accelerator components.

Current technology does not support the error-free exchange of information between the many computer-aided design (CAD) and word processing systems in use at the institutions involved. In a limited survey of the processing systems in use by the Solenoidal Detector Collaboration (SDC), 20 different CAD and 15 separate word processing systems were in use by the collaborators. The challenge is to develop a system that provides interoperability and does not require the replacement of hardware and software currently in use.

*This work was supported in part by Tom Kirk and the Argonne National Laboratory for the U.S. Department of Energy under Contract No. W-31-109-ENG-48.

†Operated by the Universities Research Association, Inc, for the U.S. Department of Energy under Contract No. DE-AC02-89ER40486.

Operating procedures within the SSCL and in some of the collaborations are still being developed. With the human interface and the procedures for use of the system being such important considerations, a prototype system is being developed in-house to ensure that a large and expensive system is not acquired until we know exactly how it will be used. The prototype also provides experience that will help specify the complete system that will be developed or acquired later. The plan is to get a limited-capability system on line as soon as possible to work out the human/procedural issues and refine the technical requirements for the larger system to come later.

REQUIREMENTS

System requirements include the following:

- A database management system (DBMS) for tracking the documents and drawings.
- Processing and storage devices for hosting the DBMS and the documents/drawings.
- Local and wide-area networks for communicating with the system.
- Document entering and retrieval via menu selection and not by use of a database language.
- Access controls to keep unauthorized personnel out of the system. A public area that will contain documents and drawings available to everyone. A private area where collaborations and collaborators can store documents and have the access restricted.

- A system to control changes to documents so unauthorized personnel do not modify them without permission.
- The capability to scan and store A and B size documents and drawings.
- An inexpensive path for upgrades and enhancements.
- The capability to add a configuration management capability to the initial system.

ARCHITECTURE

The system architecture is shown in Figure 1. Users log into the system from their terminal at their desk just like they would do for accessing the SSCVX1 VAX except that the log in address is different. The current E Mail system and procedures will continue to be used. (The address and details will be distributed when the system is on line.)

Two Sun SparcStation 1+ workstations with 2.5 GB of storage each are on the network. Each has tape capability for backups and for mailing large files. The Sybase database management system is on one of the workstations and contains the index of all of the documents and drawings in the system. This database (similar to a card catalog in a library) can be searched for subjects, authors, keywords, etc. When the desired document or drawing is found, it can be accessed electronically.

There are many problems to be solved so a user can view or transfer a document or drawing at his/her workstation without having to replace the existing hardware and software. Current technology does not provide error-free translations between the many word processors and CAD systems in use by the physicists and engineers developing the collider and detectors. The prototype will initially use ASCII for text documents and IGES (Initial Graphics Exchange Specification) for drawings as the standards for interoperability. Other standards will be evaluated to determine the best way to provide a format to meet the needs of most users.

The database and document storage will provide the capability for universal access to all data in the public area and for restricted access so collaborators can keep documents private until they are ready to release them. Users will have access to the public area and to the appropriate restricted areas of the database.

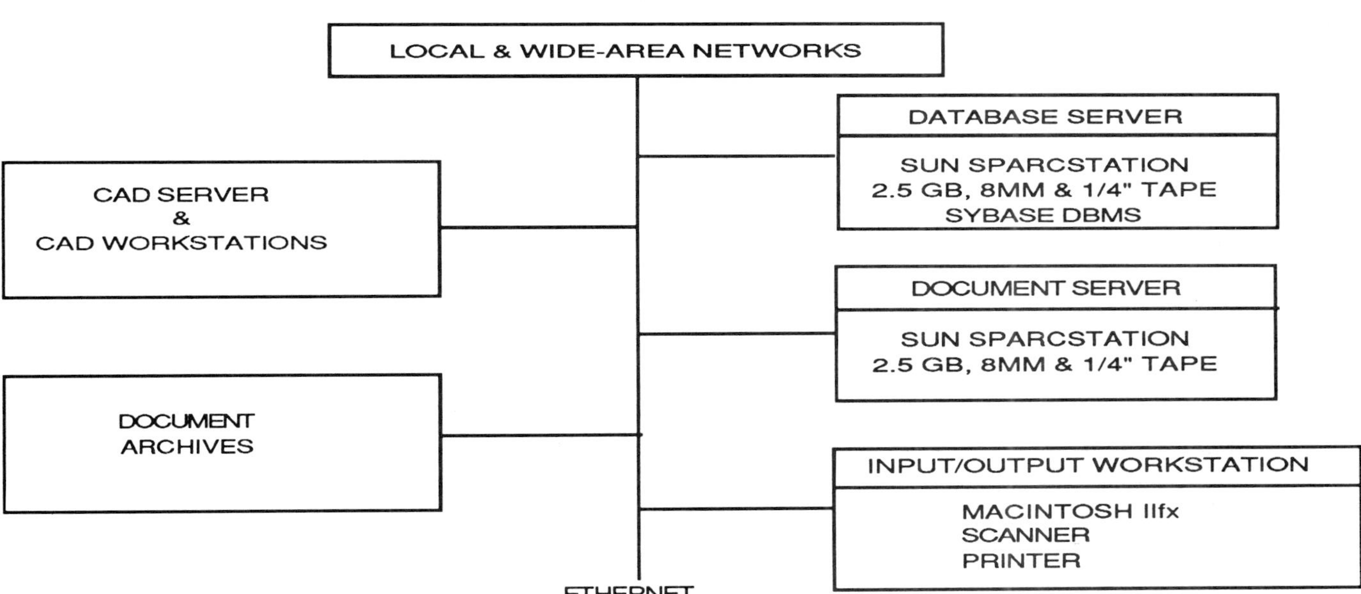

FIGURE 1: SYSTEM ARCHITECTURE

Current plans are for the users to enter the data into the system and to E Mail the documents into the storage system. Data entry will be done by filling in the blanks on a screen. Knowledge of the database language will not be necessary. Electronically transmitting the document files to the SSCL will follow the procedures currently in use by the E Mail system.

Users will have to translate text documents into ASCII and drawings into IGES on their workstations prior to sending them to the prototype for storage. Future upgrades may do away with this requirement.

An input-output workstation in the form of a Macintosh IIfx with an A/B size scanner and laser printer is also part of the system. The scanner will provide the capability to scan documents that are not received or generated electronically. Hand-written documents can be entered in to the system in this manner.

Intergraph workstations are connected to the system via the local area net and provide the capability to generate and translate CAD drawings.

The document and drawing archive system of the SSCL will be accessible via the local area net and will be able to store inactive documents.

STATUS

The system hardware has been installed and is being used to develop the application software with the Sybase database management system. An initial capability is expected by December 1990. The prototype will be developed incrementally by starting with a limited group of users and a subset of the final capabilities to work out the problems. Enhancements and increased capabilities will be added to meet the full requirements of the system.

Procedures will be developed both for system operation and for information transfer among the detector and collider developers. These procedures developed for the prototype will also be used to help determine the requirements for the larger system that will be needed to provide such features as configuration management.

A major issue to be resolved is how to achieve interoperability among the various word processing and drawing systems in use. The SSCL has demonstrated that a text and graphics file could be generated in Microsoft Word and transmitted via the E Mail system to another location where it was successfully read. However, since many word processors are in use, additional work is necessary to devise a system that will provide interoperability among them. ASCII files will be used initially for text transfer.

Two–dimensional CAD drawings have been transferred between systems with few problems. Three-dimensional CAD transfers are a known problem and will require more work.

SUMMARY

A prototype document/drawing tracking and storage system is being developed at the SSCL to provide a single source for accessing the latest collider and detector design data. The hardware is in place with the application software in development. An initial capability is expected by December 1990.

The system will provide a capability quickly and will be used to develop the procedures for transferring information electronically among the detector and collider developers. The prototype will also help develop the technical capabilities for transferring data among the various word processing and CAD systems in use by the SSCL and the collaborations.

STABLE BASE SUPPORT FOR CENTRAL TRACKING CHAMBER COMPONENTS

David Vandergriff
Oak Ridge National Laboratory
Post Office Box 2003
Oak Ridge, Tennessee 37831-7520

ABSTRACT

The Central Tracking Chamber (CTC) for a Solenoid Detector requires an extremely stable platform onto which detector components are mounted. One of the most efficient ways to provide a stable base is by using a lightweight cylindrical tube which satisfies all the physics criteria. This presentation will focus on the design parameters used in the development of such a tube for the SSC Solenoid Detector.

Introduction

The Central Tracking Chamber for the SSC requires a stable structure to mount the various straws and fibers used for particle tracking. The use of a carbon composite cylinder will provide this stable base. The goal is to prove the feasibility of this design.

Basic Requirements

The stable base cylinder must provide a lightweight, low radiation length structure, radiation hardened, and with superior thermal stability. The alignment of each superlayer is simplified when each element can be registered from a common surface. This registration requires maximum stability under load. In addition to the requirements of the physics involved, the base structure must be simple to fabricate and assemble. This article will attempt to illustrate the overall structural integrity of the stable base cylinder design.

Configuration

When all the requirements were taken into consideration, it appeared that a cylindrical carbon-based tube best met the criteria. The Solenoid Detector is itself cylindrical, which makes it structurally efficient and capable of carrying loads in the expected range. Additionally, the properties of a structural tube are widely understood.

Considerations

Any cylinder used for the stable base had to show minimum bending throughout its length, and resist compression. In addition, it could not sag or show local distortions. The ends of the tube must have sufficient hoop strength to sustain the end plates, and finally, the carbon fibers used to make up the tube had to be stable.

Initial Assessment

The cylinder size chosen for assessment was based on the CTC description in the 24 May 1990 Expression of Interest (EOI). This is an 80 cm radius cylinder, 6 meters long. This cylinder is representative of the inner superlayer in the outer tracking volume of the CTC. It is also the smallest cross section cylinder to be used and therefore the most conservative in regards to the results of the bending analysis. The initial thickness chosen for the cylinder was 2.54 cm, based on the results of the standard beam deflection calculation shown below.

$$\text{Deflection} = \frac{5WL^4}{384EI}$$

where:

$W = .59$ KG/CM
$L = 600$ CM
$E = .703 \times 10^6$ KG/CM^2
$I = 2297116$ CM^4

This analysis assumes that the tube is simply supported, that the tube retains its circular cross section throughout its entire length, and that the 2.54 cm thickness is solid and has the modulus of elasticity of aluminum. Based on these assumptions, the deflection was found to be less than 50 microns. Shear deflection of the cylinder was not evaluated in this simple analysis, but was considered in subsequent models.

Stable Base Analyses

A series of finite element analyses (FEA) were done on the cylinder using increasingly accurate and refined values for material properties and layup across the 2.54 cm thickness.

The first FEA run duplicated the conditions described above and the results were within 20%. This verified both the FEA model geometry as well as the hand calculations. Subsequent calculations lead to the final analysis presented here, based on the superlayer cross section described above. For analysis purposes, the material tensile modulus of the two carbon face sheets was set at .7 x 10^6 KG/CM^2. This value was determined using the GENRAD composite material analyzer code. This program allows the user to input varying composite materials and layup orientations.

For this analysis, standard strength IM6 composite material was chosen, having a wind pattern of [0/45/45/0]s, a symmetric layup giving the properties of a quasi-isotropic material, i.e., the same material properties in both the axial and transverse directions.

The carbon honeycomb core was based on the properties of the Hexcel Corporation HFT-G-3/16-6.0 Graphite Reinforced Honeycomb core, however, any similar core material may be chosen because the analysis code, MSC/NASTRAN, ignores the bending properties of core material and only requires input of the shear modulus. The number included in this analysis is sufficiently low that it does not affect the results of the bending analysis and is representative of any number of core materials, including foam. Core material properties only affect the shear deflection.

Verification

The FEA performed on the stable base cylinder calculated the bending and sag properties only. The FEA was verified using NASTRAN and the MSC/CAASE program. Each method proved repeatable, and the results of each method agreed, providing a high degree of confidence in the results obtained.

Similar iterative analyses were performed for each of the other conditions: local distortion, fiber stability, hoop strength, and thermal stability.

Results

Based on the accumulated results of these analyses, it can be seen that the basic cylinder is sufficiently stable to be used with confidence as a support structure for the straw tubes and scintillating fibers. The minimal deflection provides sufficient base while not sacrificing the need for low radiation length supports. Additional analysis is needed to optimize the material sizing and further reduce the radiation lengths. Radiation lengths for the layup described here have been calculated.

Fabrication and Assembly

Fabrication and assembly of the cylinder will be accomplished in a series of operations. Each operation will require careful planning and study to determine the best method of proceeding. Some general steps in the process are given in the vu-graphs, with each of these steps containing sub-steps best left to the expertise of the fabricator. Production of a prototype cylinder would be helpful in analyzing these concerns as well as establishing fabrication techniques needed to produce the cylinders in larger quantities with the accuracy required by the SSC.

MECHANICAL ENGINEERING AND DESIGN OF SILICON TRACKING DEVICES*

W. O. Miller, T. C. Thompson, M. T. Gamble†,
R. S. Reid, K. A. Woloshun, G. D. Dransfield, and H. J. Ziock
Los Alamos National Laboratory
Los Alamos, New Mexico 87545, USA

Abstract

The Los Alamos National Laboratory, Mechanical Engineering and Electronics Division has been investigating silicon-based particle tracking device technology as part of the Superconducting Super Collider-sponsored silicon subsystem collaboration. Structural, thermal, and materials issues have been addressed. This paper explores detector structural integrity and stability, including detailed finite element models of the silicon chip support and predictive methods used in designing with advanced composite materials. Electronic thermal loading and efficient dissipation of such energy using heat pipe technology are discussed. The use of materials whose coefficients of thermal expansion are engineered to be near zero, or to match silicon, as appropriate, are explored. Material analysis and test results from radiation, chemical, and static loading are compared with analytical predictions and discussed.

Introduction

The Los Alamos National Laboratory, Mechanical Engineering and Electronics Division has been investigating silicon-based elementary particle tracking device technology on behalf of the Superconducting Super Collider (SSC)-sponsored Silicon Tracking Subsystem Collaboration. Analysis and prototype testing are being used as a means to develop a mechanical design for a silicon tracking device. Emphasis has been placed upon designing a highly stable mechanical structure to support and cool all silicon microstrip detector elements and related electronics. The current design exploits the inherent physics advantages of silicon microstrip detectors by providing extremely accurate silicon wafer placement and ensuring excellent mechanical stability.[1,2]

The mechanical design study task consists of conceptual design and preliminary design. During conceptual design, various arrangements of the mechanical support structure were studied to minimize the mass in the particle path in an effort to reduce particle interactions. Exclusive of the detectors themselves, a restriction on material encountered was set at a maximum of 2.5% of a radiation length with most of the detector being well below this limit. This requirement restricts the materials choices, basic structural arrangement, cooling system, electrical cabling, and connections.

Strain-free mounting of the silicon, aligning substructures, manufacturability, assembly, and material stability were considered in more detail in the preliminary design phase. The structural loads used in our conceptual design calculations were updated and analyses revised as these parameters were better defined.

* Work supported by Department of Energy Contract No. KS-3501, Superconducting Super Collider Detector Research.

† Currently at the Massachussetts Institute of Technology, 77 Massachussetts Avenue, 9-428, Cambridge, MA 02139.

Design Considerations

The silicon tracking system research and development effort addresses numerous mechanical engineering issues facing successful deployment of a precision silicon tracking device. Design requirements are found in Table 1 and are comparable to the stringent requirements that are currently placed upon much smaller silicon-based vertex detectors.[3,4] At this interim stage of the mechanical design study, the central issues that have been addressed are the (1) conceptual system design approach, (2) radiation length calculations, (3) material candidates selection, (4) material specimen testing initiation, (5) critical systems structural analysis, (6) alignment concepts for verifying wafer position, (7) cabling concepts for data transmission and electronic circuit power, and (8) cooling system design for the individual wafer electronic circuits. Progress in these areas makes possible extensive prototype testing during fiscal year (FY) 1991 to verify the mechanical design of a silicon tracking device for the SSC.

The current design comprises a metal matrix truss space frame to provide a sparse support structure for the complex array of silicon superlayers. The design effectively satisfies structural integrity requirements in a 10 Mrad environment without exceeding the stringent radiation length constraints. The silicon superlayers cradled inside the space frame are displayed in Fig. 1. The framework draws upon graphite/epoxy and metal matrix composite materials technologies being developed for the National Aerospace Plane (NASP) and Strategic Defense Initiative (SDI) programs. The silicon superlayers are formed into silicon sandwich-like constructions which are combined in cylindrical silicon shell configurations for the central region and flat panel arrays for the forward region. Structural rings which support the bonded silicon detectors also serve as an effective means of incorporating a heat pipe cooling network. Heat pipe technology is required for achieving the thermal stability requirements demanded for silicon tracking systems.

TABLE 1. Baseline Design Requirements

Detector characteristics	Silicon, longstrip technology 50 μm pitch, double-sided Electronics surface area, 3.52×0.6 cm^2
Detector assembly	Maximum length, 36 cm (six wafers edge bonded) Mounting precision R, ϕ, Z - 25μm's Angular, 1 milliradian Positional stability R, ϕ, Z - 5μm's Angular, 0.2 milliradian Detector cant angle, 7.4° Operational temperature, 0°C
Detector heat load	Heat flux, 1.8 W/cm^2 40 kW Total
Radiation exposure	10 Mrads over 10 year service life No stability effects >5 μm's
Material radiation length budget	Structure and service connections 2.5% (except in limited regions e.g., transition from central to forward)

Meeting the design requirements requires exploiting advances in the (1) application of metal matrix composites, (2) development of "silicon sandwich" shell structures, and silicon planar arrays, (3) implementation of heat pipe cooling technology for isothermal stabilization of the structure and silicon, and (4) innovative use of fiber optics for rapid data transmission.

A full scale prototype will be developed in FY1991 to ensure that all materials, structural, and heat pipe cooling issues are resolved. This prototype will demonstrate the feasibility of a silicon tracking device and solidify a design based upon proven concepts. The detector structural, thermal, and materials issues will be examined in detail following a brief description of physics considerations germane to the detector's engineering.

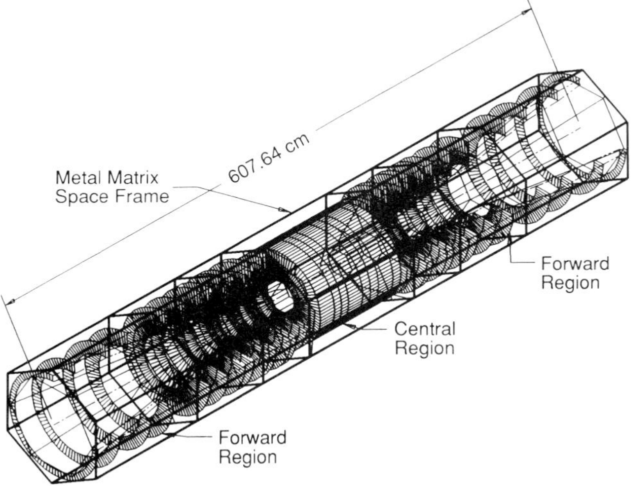

Fig. 1. Metal matrix space frame surrounding silicon. superlayers.

Physics Considerations

A dominant factor in the design is to minimize the space frame structural mass and maximize radiation length. This study is based on a design limit for the structure and associated cooling components of 2% of a radiation length. This is a very stringent requirement, necessitating a trade-off between geometry and material properties. Cabling is limited to an additional 1% of a radiation length.

One radiation length is defined as the average distance, L_R, over which a high energy electron ($E_{e^\pm} \geq$ 1GeV) loses all but $1/e$, 37%, of its energy to bremsstrahlung.[5] Similarly, for photons of infinite energy, the total cross section for producing electron-positron pairs, σ_{pp}, is

$$\sigma_{pp} = 7A/(9L_R N_A) \tag{1}$$

where N_A is Avogadro's number, and A is the atomic weight. The inverse L_R ($1/L_R$) dependence of the cross section, coupled with the approximately inverse square dependence of L_R on the atomic number, Z, and direct proportionality to A (A/Z^2), clearly demonstrates the need for a sparse mechanical structure of very low Z composition. A single 300 μm thick silicon detector represents $0.0032L_R$. Ten such layers comprise 3.2% of a radiation length. This ensures the average number of radiation lengths in the support structure is less than or equal to the number of radiation lengths in the detection medium. The term average is required because certain particle trajectories that intercept the tubular space frame, electronics, or cooling rings, which perform essential services, will encounter somewhat more than the nominal quantity of material, which is approximately 1% of a radiation length. Every attempt is being made to minimize the fractional volume where this occurs. These very stringent requirements have stimulated a great deal of interest in highly advanced composite materials such as metal matrix tubing and sparse core sandwich constructions. However, for compounds such as polymer-based epoxies, each constituent's radiation length must be weighted by its weight fraction present in the compound. The compound's resultant radiation length is then expressed as a summation

$$\frac{1}{L_R} = \sum_i \frac{W_i}{L_i} \tag{2}$$

where W_i is the weight fraction of each constituent and L_i is the radiation length of each.[6] For an advanced composite material such as a fiber reinforced metal matrix, the resultant composite radiation length is also similarly weighted. The most useful figure of merit for a material to be used in the central tracking region is its effective radiation length, L_{Re}, defined as the radiation length divided by the density of the material. L_{Re} not only quantifies the Z of each atom of material but also accounts for the number of such atoms in a unit volume. Hence, L_{Re} has units of length and can be used directly in the energy equation $E(x) = E_0 e^{-x/L_{Re}}$ to obtain estimates of the energy lost by an electron traversing a thickness x of a particular medium. Calculations of effective radiation lengths for selected candidate materials have been performed and appear in Table 2. Among

TABLE 2. Selected Material Properties

Material	Elastic Modulus E (GPa)	Density ρ (gm/cm^3)	Specific Stiffness E/ρ (10^8cm)	CTE α (ppm/K)	Thermal Conduct. k (W/m-K)	Specific Heat C_p (J/gm-K)	Radiation Leng. (eff.) L_{Re} (cm)
			Material Candidates				
Beryllium	290.0	1.84	16.08	11.60	146.0	1.88	35.43
Boron Carbide	448.2	2.52	18.15	4.73	43.0	0.90	19.90
Carbon/Carbon*	152.0	1.90	8.16	0.10	246.0	1.00	18.80
Graphite/Epoxy†	311.7	1.68	20.00	−1.13	46.2	0.92	25.00
Mg-MMC**	190.3	2.03	9.56	−0.35	412.0	1.00	16.80
Al-MMC**	196.5	2.39	8.39	0	409.0	1.02	13.22
			Reference Materials				
Silicon	**131.0**	**2.33**	**5.74**	**2.60**	**129.0**	**0.70**	**9.37**
Aluminum	68.9	2.70	2.59	23.58	168.0	0.90	8.89
Boron	344.7	2.34	15.03	8.30	1.0	1.29	22.52
Fused Silica	73.0	2.20	3.39	0.50	1.0	0.92	12.30
Copper	117.2	8.91	1.34	16.90	400.0	0.39	1.44
SiC	325.0	3.00	11.05	2.30	200.0	0.89	8.52

*Unidirectional properties.
**Graphite Fiber (P120) MMC-Metal Matrix Composite, quasi-isotropic, ~60% fiber.
†Graphite Fiber (P75)/Epoxy matrix (1939-3), unidirectional, ~60% fiber.

these are silicon, graphite/epoxy composite, magnesium metal matrix composite, and beryllium, with L_{Re} values of 9.37 cm, 25.0 cm, 16.8 cm, 35.43 cm, respectively. Clearly, based solely on radiation length considerations, a support structure composed of beryllium would be superior. The calculated detector L_{Re} is based solely on the structural arrangement and as such does not include details such as electronic packaging and cabling. As a point of reference, 0.71 cm of beryllium adds 2% to the radiation length.

Radiation path length is a function of radius, r, longitudinal angle, θ, and azimuthal circumferential angle, ϕ. The angle ϕ determines when discrete tubular members of the frame will be intercepted, while θ denotes deviation to the surface normal, effectively increasing the material thickness in the forward region. For most angles, the radiation length restriction is met with exceptions at rather shallow angles, $\theta \sim 10°$, and for ϕ-angles corresponding to tubular space frame members. For a structure comprised entirely of beryllium, the radiation length would be less than 1.5%. Alternative construction materials, e.g., graphite/epoxy and Mg-MMC composites would increase this to a nominal value of 2%. Figure 2 displays effective radiation length plotted as a function of incident angle.

Structural Considerations

The Silicon Tracking Subsystem (STS) for the subsystem research and development program overall length is a hierarchy of structural units 6.2 m in overall length and 1.4 m in diameter. The myriad silicon sandwich detector units within are bonded to cooling rings that are in turn supported by an ultrastable space frame.

Silicon Assembly

The silicon wafers are edge-bonded into long strips. The maximum active strip length is now 18 cm, and to achieve these strip lengths the individual wafers are wired bonded together, as well as edge bonded structurally together. Two such assemblies are bonded together to form one 36 cm- long module. The use of silicon in this manner may be credited to Michal Turala of the Institute of Nuclear Physics, Krakow, Poland.[7] We have investigated combining these long module assemblies into a sandwich arrangement to maximize the wafer stiffness. The sandwich arrangement is essential, e.g., a 30.5 cm-long wafer will sag under its own weight, 0.235 cm. Deflections calculated using a finite element model, shown in Fig. 3, compare reasonably well with these measured results. The linear elastic solution overpredicts the deflection by 12%. It is important to note that the analytical model did not simulate the potential effect of the three, 50 micron adhesive joints. From all outward appearances these bondlines do not contribute to any additional sag.

A three-point bend test of three 8 cm-long edge bonded wafers was performed to explore the adhesive joint compliance. Again, our linear elastic model over predicted the measured 0.198 cm deflection, this time by 13%. These finite element results can be seen in Fig. 4.

The correlation between predicted and measured wafer bending could be improved by constructing a non-linear large deflection model which accounts for in-plane stiffening effects. However, the silicon sandwich design proposed for the silicon superlayers will not be a large deflection problem, but entirely linear elastic, otherwise damage to the wafers is very likely. Verifying that the adhesive bond joints could be ignored in the linear elastic model was the main intent .

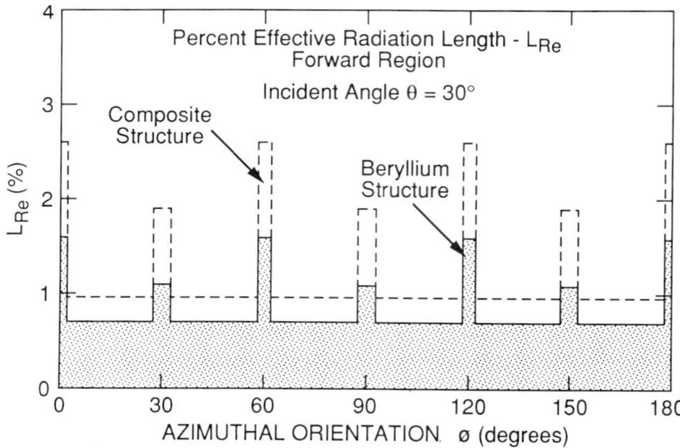

Fig. 2. Effective radiation length expressed as a function of azimuthal orientation angle, ϕ.

The strength of the adhesive joint was tested. Tensile strength was measured to be nominally 5000 psi with Epotek H74, which is an electrically nonconductive material. For our initial tests at fabricating silicon sandwich structures we chose Epotek H72 adhesive, a thermally conductive material rated at 10,000 psi. A support structure material CTE near that of silicon minimizes residual stresses after curing and cooling to the operational temperature of 0°C.

Space Frame

The central and forward detector regions must be supported by a frame network that possesses high dimensional stability throughout the detector life. Problems which must be considered are (1) thermal distortion, (2) physical property changes due to radiation effects and/or chemical attack from the heat pipe working fluid, (3) creep, and (4) radiation length restrictions.

Physical material data for proposed composite candidates are meager especially at the precision levels needed for our application. Consequently, structural analysis activities included provisions for predicting materials property data as well as the behavior of high stiffness to weight sandwich structures made from these materials, and finally, the properties of complex constructions made from sandwich assemblies. These predictive techniques will be validated by conducting material and structural tests. Construction materials chosen to have a long radiation length have been reviewed and available engineering property data assimilated. Materials of current interest are (1) aluminum or magnesium metal matrix composite, (2) graphite/epoxy composite, and (3) beryllium. Each material offers advantages, such as ease of fabricability, a desirable coefficient of thermal expansion, or a large effective radiation length. As the design study progressed, each candidate material's performance was evaluated to achieve the best combination of overall attributes.

A parametric study of space frame tube modules and size was undertaken. Maximum deflection and first bending mode, the fundamental vibrational mode, expressed as a function of elastic modulus, tube size, and wall thickness, and number of support points were determined. A stiff structure is characterized by low deflection and high fundamental frequency. The natural frequency and displacement results are displayed in Fig. 5. For example, if the space frame were constructed solely of 33.7 mm (1 3/8 in) diameter tubing with a wall thickness of 1.1 mm (0.045 in) an elastic modulus of 515 GPa, and a four point support scheme, the maximum deflection would be 0.22 mm and its fundamental natural mode of vibration would be a bending mode occurring at approximately 35.7 Hz. The gravity deformed configuration is shown in Fig. 6. Engineering design will seek to maintain low deflections and high fundamental frequencies achieved by this baseline case while minimizing material mass and maximizing the effective radiation length of this structure.

Cooling Rings

Both ends of the silicon layers are bonded to rings. The rings, in addition to providing structural support, provide a passage for an internal heat pipe artery for cooling, as shown in Fig. 7. The modules of 2 rings with silicon between them are mounted onto the shell support structure of the silicon detector central region. These modules are supported on kinematic supports to avoid introducing distortions. Consideration is being given to using a shell support structure material such as graphite/epoxy or a metal matrix composite that can be engineered to exhibit a zero CTE.[8]

In the forward region, the cooling rings, which support wedge-shaped silicon layers, must track the contraction of the silicon or they may introduce distortion. The central region cooling ring CTE must also match that of silicon to avoid wafer distortion. Tailoring the expansion of a graphite metal matrix or a graphite/epoxy composite to match that of silicon are options for achieving the desired CTE equality. Construction techniques for prototype rings were developed in the preliminary design phase. A full-scale prototype comprising cooling rings, thermal sources, and silicon chips is under development and shown as Fig. 8.

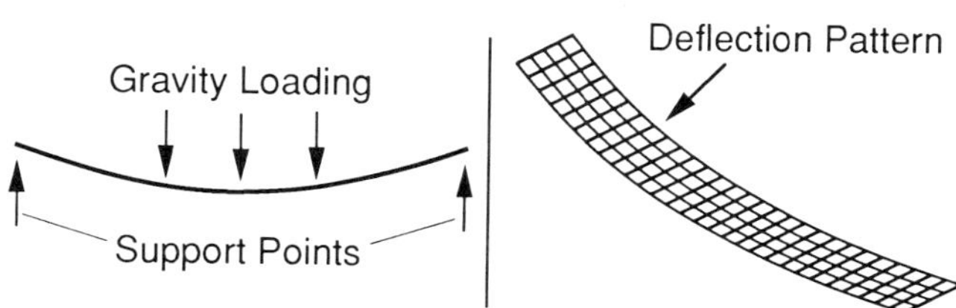

Fig. 3. Finite element and experimental maximum deflection results for a 30 cm-long, simply supported, bonded wafer (0.264 cm vs 0.235 cm measured).

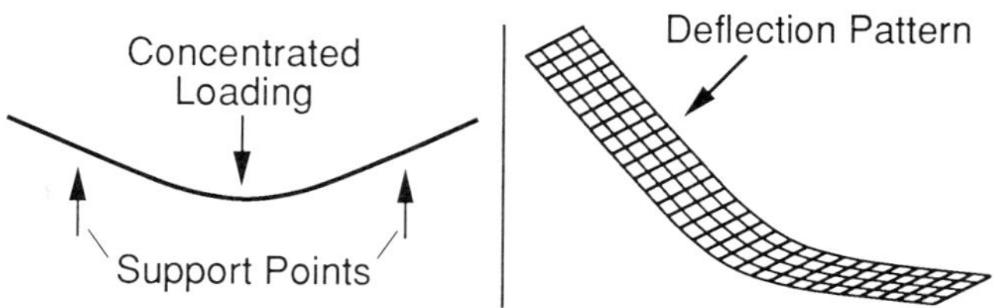

Fig. 4. Finite element and experimental maximum deflection results for a 24 cm-long, three-point bend tested, bonded wafer (0.223 cm vs 0.198 cm measured).

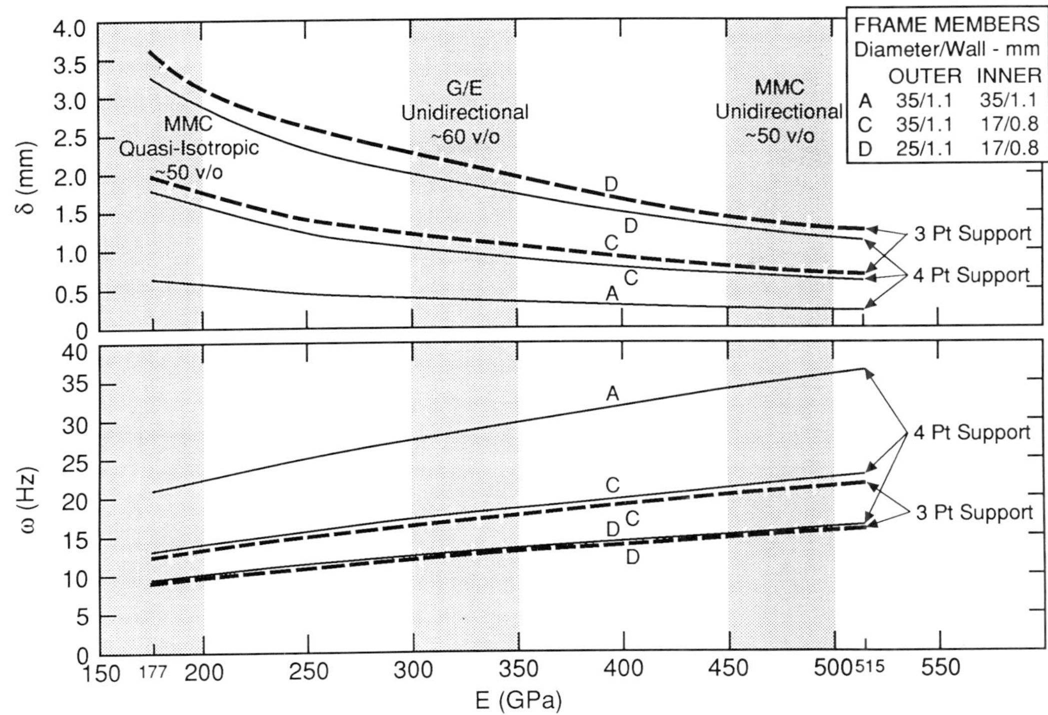

Fig. 5. Space frame gravity induced displacements, δ, and fundamental vibration bending frequency, f.

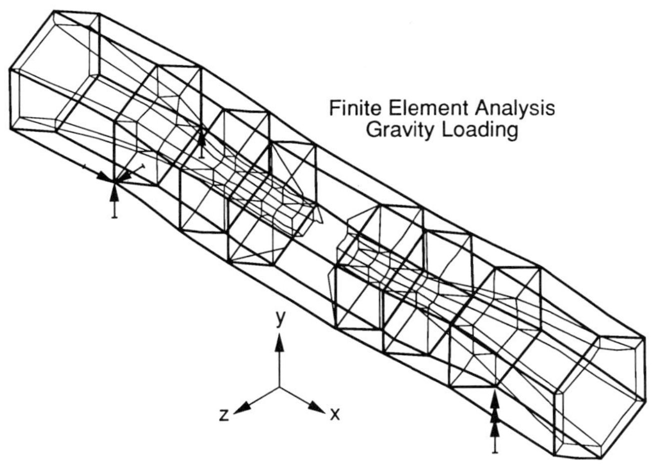

Fig. 6. Metal matrix space frame (deformed configuration).

Thermal Design

Early studies demonstrated that air convective cooling could not maintain the system at the design operating temperature, in the presence of the anticipated heat load, without introducing undesirable thermal gradients. Therefore, an evaporative heat pipe cooling system study employing a sprayed-on graphite wick structure was initiated.

Heat Pipe Cooling Concept

The heat pipe concept underwent a conceptual definition and several possible working fluid candidates were selected. Wick materials that are consistent with the materials chosen for the structures are under experimental investigation. During the conceptual design phase, the analysis tasks included wick sizing, facets of wafer cooting, heat pipe condenser design, and system operation.

The SSC detector array stability requirements dictate precise thermal control of the entire array. The design goal is to control the overall temperature excursions within the system to ± 0.10° C with a maximum thermal gradient of 0.25° C within the detector array because nonisothermalities will detract from the mechanical stability. These temperature controls maintain detector stability within 5 microns. Evaporative cooling is the only option for realizing these design objectives. Heat pipe cooling is a liquid evaporation cooling process. The liquid is distributed over the surfaces to be cooled by a combination of hydrostatic and capillary forces and the evaporated liquid vapor is removed by the pressure gradient generated by the condensing process that closes the fluid loop. This fluid cycle is fully contained in a hermetic enclosure at a positive pressure with respect to the outside atmosphere. Noncondensible gases are not used. The entire detector array is in contact with the hermetic enclosure. A manifold filled with flowing gaseous nitrogen surrounds the seals of the hermetic enclosure to prevent leaks between the enclosure and the inside of the detector building.

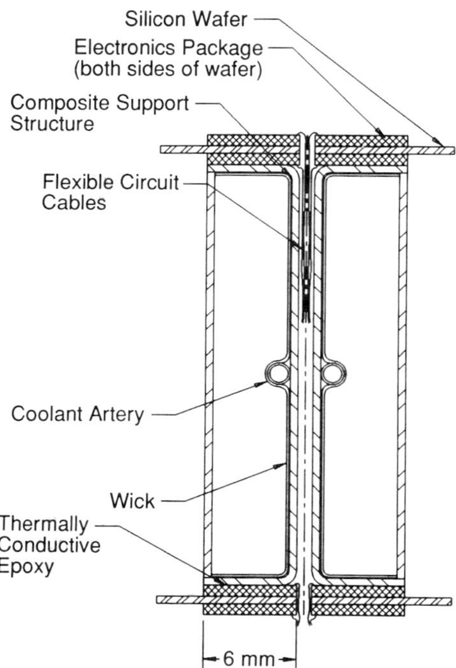

Fig. 7. Silicon support ring with integrated heat pipe cooling technology (cross-sectional view).

Distortions in the silicon shell substructure are avoided by minimizing the thermal gradient and through an appropriate material choice for the cooling ring. Working fluid is wicked to an area directly beneath the electronics package, as shown in Fig. 7. Heat flows through this thin walled area and evaporates the cooling liquid. This technique greatly simplifies the process of cooling the enormous number of individual wafers. A 1.8 W/cm^2 heat flux, associated with the wafer electronics package, can be effectively handled with a minimum temperature gradient. A comparison of working fluid candidates may be found in Table 3.

The long strip wafer technology imposes a significant design heat load; a maximum value of 40 kW was chosen as this design value. The heat load is highly localized. Helium cooling was considered as a simple alternative, but it cannot equilibrate the system adequately to avoid excessive thermal gradients and hence, physical distortions. A design goal is to optimize placement of heat pipes and thereafter minimize pressure drop in the vapor return path to the condenser. As long as the vapor space around the various detector elements is at a constant pressure, and the pressure loss in individual vapor return lines for all silicon modules are equal, the structure will thermally equilibrate.

Wick Design

The wick, composed of graphite particles sprayed onto a suitable substrate, requires the integration of an artery, an open flow passage that runs circumferentially around the support structure. The artery provides a low-impedance flow path through, and is hydrologically connected to, the wick structure.The artery may be a porous tube placed on the cooling ring prior to spraying.

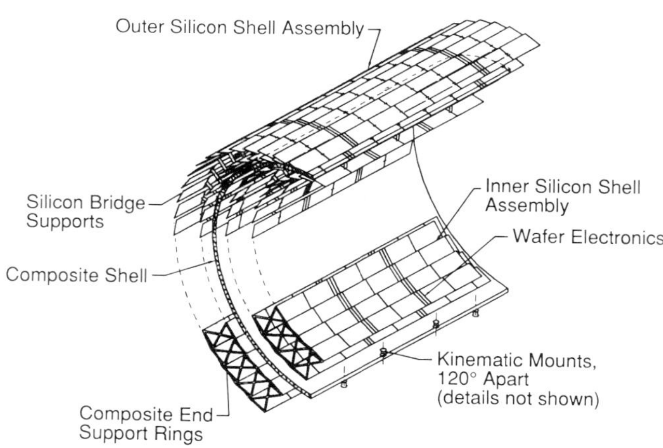

Fig. 8. Silicon tracking system full-scale, fully operational and instrumented prototype (cutaway view).

The tube pores must be small enough so that the graphite does not penetrate, but adheres to the surface forming a hydrologically connected capillary path. Another possible approach is to form the wick over a tube or wire that is subsequently dissolved to leave an open flow passage. Paper wick arteries can be incorporated into the pliant wick structure. The artery must possess sufficient strength to withstand the forces due to the hydrostatic pressure.

The wick thickness required to distribute an adequate supply of liquid over the heated surface is a function of wick pore size, wick permeability, capillary pumping distance, and power density. Maximum available capillary pumping pressure, is given by

$$P_{cap} = \frac{2\sigma}{r_p}, \tag{3}$$

where σ is the fluid surface tension and r_p is the capillary pore size assuming perfect wetting. The maximum liquid pressure in the system is the local vapor pressure. If the liquid pressure exceeds the vapor pressure at any point the liquid meniscus at the capillary pores is curved into the vapor space. It is theoretically possible to hold the liquid in the wick with a convex meniscus but, in practice the convex menisci coalesce to form a continuous film. In the presence of gravity, this film will flow off of or bleed from the wick surface.

In the interest of developing the thinnest, adequate wick it is important to study the effect on wick thickness of variations in system parameters. Minimum wick thickness as a function of pore size is shown in Fig. 9 for three different vertical heights, with the hydrocarbon butane as the cooling medium, and assuming typical system values for power density, permeability, and capillary pumping distance. Although it is not obvious from the graph, there is a maximum pore size above which there is no physical solution to the governing equation. Those pore sizes are 4.2, 8.5, and 16 microns for 1.09, 0.54, and 0.29 m wick heights, respectively. The figure demonstrates that there is a minimum

TABLE 3. Comparison of Candidate Working Fluids

Fluid	BP†	MW††	ρ (g/cc)	g/s-kW	V/kW*	V/kW**
Butane	–0.5	58.1	0.601	2.60	3.96	923
1,2 Butadiene	10.9	54.1	0.676	1.98	2.92	996
2,2 Dimethylpropane	9.5	72.2	0.613	3.05	4.98	948
cis-2-Butene	3.7	56.1	0.621	2.09	3.37	868
Ethylamine	16.6	45.1	0.683	1.57	2.30	811
Methylamine	–6.3	31.1	0.699	1.15	1.64	861
Dimethylamine	7.4	45.1	0.680	1.61	2.37	831

† Normal boiling point, °C.
†† Molecular weight.
* Liquid at normal boiling point. V is volume flow rate in cc/s.
** Vapor at 10°C.

wick thickness, and that thickness is a strong function of vertical height There is a relatively broad band of pore sizes for which there is little change in minimum wick thickness. Other factors affecting wick thickness have the following ramifications. Doubling the power density will double the minimum wick thickness. Moving the arteries closer to the heated surface could significantly reduce the minimum wick thickness. If two arteries are placed in each cooling ring, each as close as possible to the electronics on the inner and outer circumference, the effective length can be reduced to 0.3 cm in the limit, reducing the required wick thickness by a factor of 7. The total amount of wick material would also be substantially reduced by not wicking the adiabatic portion of the cooling ring. Placing the arteries directly on the center of the heated part of the cooling ring would gain another factor of 2. This would probably be impractical due to both the complex geometry and the possibility of generating vapor directly in the artery.

It should be noted that the wick thicknesses shown in the figure are required only at the top of the cooling ring. At all other lower points, a thinner wick would be adequate resulting from gravitational assistance. The thinner the wick, the steeper the path of least resistance. But at some distance from the highest point in the cooling ring, it is possible to significantly reduce the wick thickness. This advantage will be examined in detail as wick data become available.

It is desirable to keep flow induced pressure drop in the artery to a minimum. The total thermal load on a 108 cm diameter ring with a power distribution of 1.1 W/cm is about 740 W assuming 6 mm-wide double-sided electronics on 2 sides of the cooling ring. The mass flow rate of butane into this cooling ring would be 1.76 g/s, or 2.9 cc/s liquid (from Table 3). Locally, the flow out of the artery is about 2.4 mg/s per cm of length. Since the wick pores are sized to prevent wick bleeding with no flow, it is permissible to accept a flow induced pressure drop equal to the hydrostatic pressure at the bottom of the cooling ring. For a single artery system the artery diameter must be about 1 mm. Because artery pressure drop is inversely proportional to the fourth power of the diameter, a double artery system, which divides the flow in half, only reduces the required artery diameter to 0.9 mm.

The preceding discussion focused on the most severe cooling ring requirements, those of the largest diameter cooling ring. It may ultimately prove worthwhile to design a uniquely sized wick and artery for each cooling ring diameter. It may also be desirable or necessary to make multiple connections to the larger cooling rings and a single connection to the others.The mass savings would have to be traded off with the necessity of making more liquid line connections. These decisions should be deferred until the wicks have been further characterized.

Materials Studies

Most design applications require many factors be considered in material selection. However, in the instance of the SSC silicon tracking subsystem, material selection is governed by three primary factors. These are (1) 5 micron material stability, (2) 2.0% effective radiation length for structure, and (3) radiation resistance in the presence of hydrocarbon fluids. Our preliminary studies have shown that to achieve these goals application of state-of-the-art material composites and high stiffness to weight ratio construction techniques will be required.

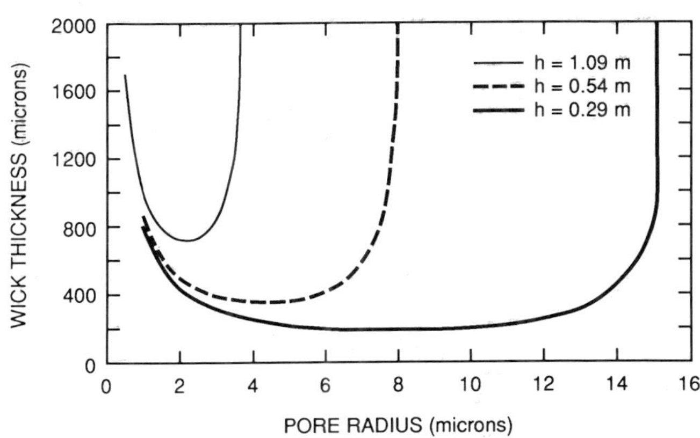

Fig. 9. Minimum wick thickness as a function of pore radius for 3 vertical artery lengths, assuming a porosity of 0.5.

Materials Trade-offs

Table 2 summarizes pertinent data for candidate materials for the structures under investigation. If material specific stiffness were a singular important attribute, then boron carbide would be an ideal candidate. It is rad hard and does not show any evidence of creep under load. The material, however, is brittle and exceedingly hard; its hardness is surpassed only by that of diamond. Attainment of precise geometric shapes would be difficult and expensive. Another factor that detracts from its choice for this application is its modest effective radiation length. Beryllium's effective radiation length is nominally 1.78 times that of boron carbide. As previously discussed, effective radiation length is one of the primary factors in selecting a suitable material and beryllium clearly stands out as the best choice in this singular regard. However, beryllium has a large CTE and possibly creeps. Graphite/epoxy composites offer the next best alternative for achieving the radiation length design goal. Silicon carbide is used extensively in highly stable mirror applications. It can be reaction sintered to complex shapes, but limited material can be used in configuring a structural design. Silicon carbide has a very low effective radiation length, indicating it is not a low-Z material by comparison. However, silicon carbide has a coefficient of thermal expansion that closely matches that of silicon and this is a very desirable characteristic as a structure for mounting silicon.

To ensure axial dimensional stability of the 6 m-long space frame, a CTE approximately equal to zero is required. This can be accomplished with a unidirectional or quasi-isotropic construction composed of either a polymeric or a metal graphite-reinforced matrix. The quasi-isotropic construction would provide a nearly zero CTE in both the axial and transverse dimensions of the body. The effective elastic modulus of the structure would be much less than if a unidirectional construction possessing a CTE near zero were used. In fact, the tubular design of the space frame reduces the importance of the transverse CTE, because of the small lengths for expansion present in the tubes radial directions. Based upon this scenario, the magnesium metal matrix (Mg-MMC) construction composed of approximately 55% graphite fiber with a CTE nearly zero and an axial elastic modulus of about 490 GPa is the baseline choice for the space frame's structural members.

The information presented in Table 2 includes calculations of the specific stiffness of selected materials. Among these are monolithic metals as well as polymeric and metal fiber-reinforced matrix composites. Larger specific stiffness values are desired because they represent an efficient use of structural materials. It is clear from the table that based solely upon specific stiffness criteria, the graphite and epoxy composite material is preferable. However, it should be noted that the specific stiffness value of 20.0×10^8 cm shown for P75 graphite and epoxy is a unidirectional value. This implies that only the axial direction possesses such a high value. An isotropic material such as boron-carbide with a specific stiffness of about 18.15×10^8 cm has no preferential direction of excellence. However, the CTE of boron-carbide is approximately 4.73 ppm/°C and cannot be altered. Accounting for all considerations introduced thus far, polymeric and metal fiber reinforced matrix composites are the best overall.

Structural panels composed of individual graphite/epoxy plies with fiber orientations arranged to achieve quasi-isotropic properties possess noticeably lower stiffness properties. Table 4 depicts the quasi-isotropic engineering parameters for some graphite/epoxy combinations of interest. The silicon tracking subsystem can effectively use both unidirectional and quasi-isotropic constructions. The internal truss space frame is an example of the use of a unidirectional laminate. A zero CTE in the direction of the truss member is desired but requires a trade-off in fiber properties and fiber to epoxy volume fraction. The table illustrates candidates under consideration for shell and ring structures requiring a quasi-isotropic CTE or matching a particular CTE, e.g., IM6/3501-6 (2.9 ppm/°C) compares favorably with silicon (2.6 ppm/ °C).

Sandwich Panels

Low weight requirements for the central detector mandate sparse support of the silicon wafers. This support structure, the composite shell, shown in Fig. 8, will be composed of sparse core sandwich constructions. Even a very low effective density and high efficiency construction using a honeycomb core, sandwich geometry will not suffice. The need for high specific stiffness and high radiation length have precluded the use of conventional materials and construction techniques. An optimization study for a graphite/epoxy panel has been performed. The objective of the

Table 4. Graphite/Epoxy Composite Material Systems Quasi-Isotropic Laminate Engineered Parameters

System	Elastic Modulus E (GPa)	Density ρ (g/cm^3)	Specific Stiffness E/ρ (10^8 cm)	CTE α (ppm/°C)	Poisson Ratio ν	Shear Modulus G (GPa)
P75/1939-3	109.2	1.68	6.6	–0.43	0.321	41.3
P75/3501-6	107.8	1.68	6.5	–0.5	0.321	40.8
GY70/3501-6	98.9	1.66	6.1	–0.016	0.318	37.5
UHM3501-6	95.8	1.6	6.1	0.18	0.314	36.4
IM6/3501-6	78.3	1.56	5.1	2.9	0.296	30.2
AS4/3501-6	55.9	1.56	3.7	1.96	0.30	21.5
AS1/3501-6	51.7	1.56	3.4	2.32	0.298	19.9

study was to parametrically investigate the effects of material geometry trade-offs on the panels' physical properties.

The first optimization trade-off was to explore the effect of core thickness on the core shear modulus. Three cases using core angles of 45°, 60° and 75° were studied. Results indicate the 45° case has substantially higher core shear modulus for a given thickness as shown in Fig. 10. Another interesting phenomenon is the rapid decay of the core shear modulus with increasing core thickness. Low shear modulus results in a high degree of coupling between deflection and core shear modulus.

The second optimization trade-off study was to balance panel deflection resulting from bending and shear effects. A plot of deflection as a function of panel thickness reveals a deflection minimum for each of the three cases as displayed in Fig. 11. The optimization trades core shear stiffness for face bending stiffness. The two effects are independent and can be linearly superimposed. Two interesting phenomena were observed. The first is that with increasing panel thickness beyond the minima, shear effects dominate, and decreasing panel thickness below the minima bending effects dominate. The second interesting result clearly demonstrates the 45° case to be stiffer for a flat panel construction. This optimization process will be similarly conducted for a circular shell.

Two finite element models were constructed. The first consists of the entire panel modeled as a sandwich and the second a detailed section of the panel as seen in Fig. 12. The sandwich model assumes prior knowledge of the core material properties. Because the core is sparse, and the only known properties are those of the laminate used to form it, a second numerical model was needed to obtain the adjusted core properties. This detailed finite element model of the rib was used to recover the core engineering properties.

The full panel was then used to predict deflection resulting from a four-point bend test. Hardware was prepared for testing and used to verify the analytical model. This step was essential for further designs to ensure the behavior of the panel was understood. The calculated core shear modulus was 2.0 MPa using the theoretical model. Figure 13 illustrates the numerical results for the test panel. Using the calculated shear modulus the predicted deflection is 0.624 mm. The actual measured deflection (average of two outer measurements and one midpoint measurement) was 0.592 mm. This excellent agreement, within 5%, was achieved even though deflection is a function of rapidly varying shear modulus for the case analyzed.

Materials Testing

Fifteen material swatches were fabricated to undergo tensile speciment testing. Ten were used in chemical baths to simulate the two cooling fluid candidates, and five will be used for high dose radiation damage testing. Baseline tensile tests were conducted prior to soaking, and the same tests were performed after soaking. The results indicate there are no apparent incompatibilites between the graphite/epoxy composite and working fluids.

The mechanical support structure for the silicon tracker will be subjected to radiation damage from charged particles and photons emanating directly from the interaction point and neutrons resulting from the interactions of the primary particles and their interaction products in the calorimeter. Over the ten-year lifetime of the experiment, a maximum radiation dosage of ten megarads is anticipated. At a radial distance of 10 cm, the total fluences are expected to consist of approximately 10^{14} charged particles/cm^2, an equivalent fluence of photons, and on the order of 10^{13} neutrons/cm^2 for a luminosity of 10^{33} particles/cm^2. Ultralightweight graphite/epoxy sandwich panels have been exposed to radiation levels of 3×10^{13} neutrons/cm^2 and more recently to 10^{15} neutrons/cm^2. The early test at 10^{13} neutrons/cm^2 showed no serious effects and evaluations after exposure to 10^{15} neutrons/cm^2 are pending; ^{60}Co gamma ray exposures are also planned. Although these radiation levels are low compared to those commonly encountered and tested for in fission power reactor facilities, the problem here is of a different scope and involves unique materials and structures. Effects such as creep, warping, and swelling, which lead to slight dimensional changes of supported entities, are of interest. The ultimate goal is to acheive a design capable of withstanding the radiation environment associated with a luminosity of 10^{34} particles/cm^2.

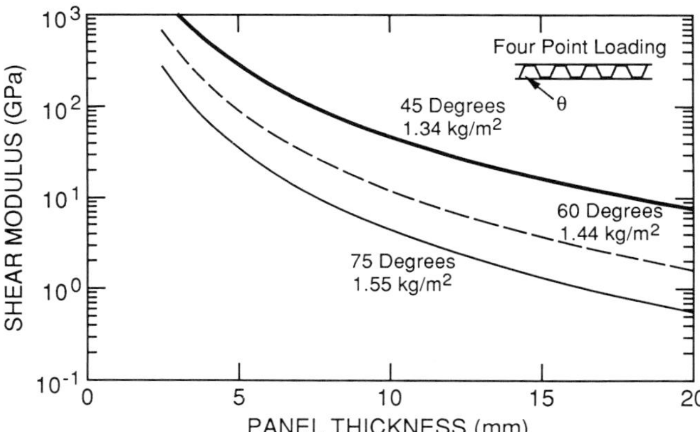

Fig. 10. Shear modulus optimization for three values of areal density.

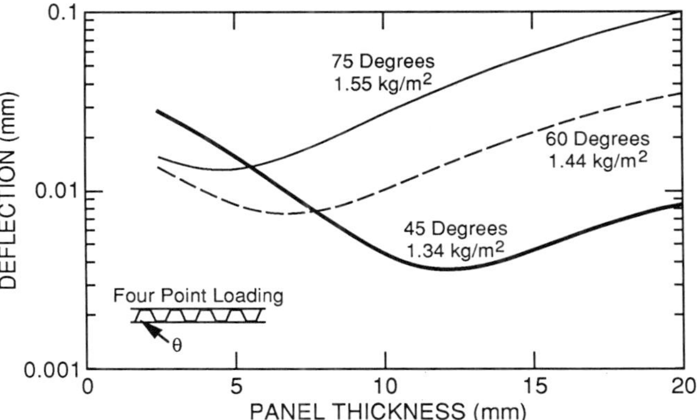

Fig. 11. Sandwich panel geometry optimization for three values of areal density.

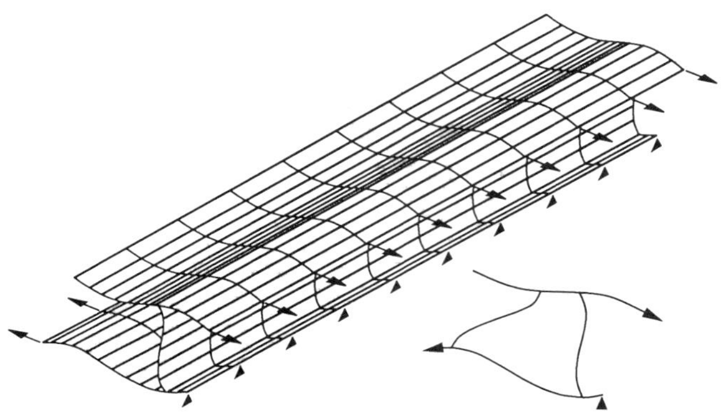

Fig. 12. Detailed finite element model used to predict core shear modulus (deformed configuration, pictorial and end views).

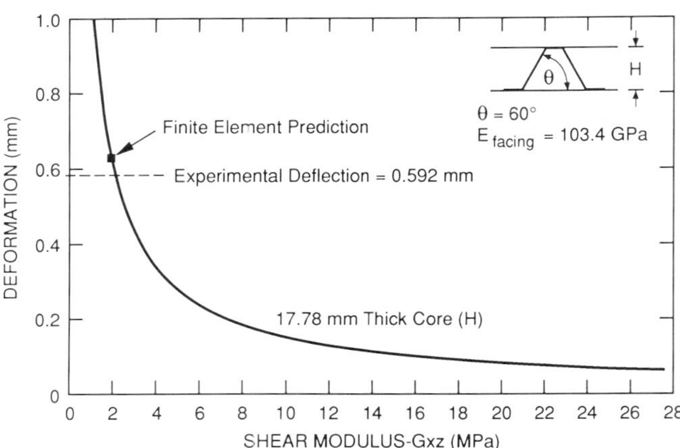

Fig. 13. Numerical and experimental results of four-point core shear modulus (pictorial and end views).

1991 Program

The mechanical engineering program for FY1991 focuses on producing prototypes to prove the viability of material selections, cooling schemes, and structural designs developed during FY1990. Meeting these goals requires the fabrication of a full-scale, full cylindrical subsection of the barrel region of the detector depicted in Figure 8. This 54 cm-diameter replica will be composed of approximately 1,200 silicon wafers arranged as two superlayers connected by polymeric matrix composite material cooling rings with associated heat pipe cooling. The two silicon superlayers will be supported by a low CTE graphite/epoxy sandwich shell structure and a zero CTE metal matrix space frame. This combination will provide an opportunity to investigate all mechanical and structural effects anticipated in a full-scale silicon tracking system. It will be fully equipped with electronics-equivalent resistive heat loading to simulate thermal conditions of the operating silicon detector system. An interferometric measurement system will detect minute motions of the silicon detectors and infer the performance of the metal matrix composite space frame and fully operational heat pipe cooling system.

Critical issues that will be resolved by these tests include (1) cooling ring construction and wick development and their optimization, (2) materials characteristics and stability, (3) selectively enhanced polymeric and metal matrix composite engineering properties, such as CTE and elastic modulus, (4) silicon module stability under electrical load simulation, and (5) material and component radiation and chemical exposure compatibility. Design and analysis activities will be performed as needed to support these tests.

References

1. W. O. Miller, M. T. Gamble, T. C. Thompson, et al., *Superconducting Super Collider Silicon Tracking Subsystem Research and Development Report*, LANL LA-MS Report, in press.
2. A. Seiden, "Silicon tracker for the SSC," *Proceedings of the International Workshop on Solenoidal Detectors for the SSC*, 1990, p. 330.
3. C. Adolphsen, G. Gratta, et al., "An alignment method for the Mark II silicon strip vertex detector using an x-ray beam," Nuclear Instruments and Methods in Physics Research, **A288**, 1990, p. 257.
4. A. Litke, C. Adolphsen, et al., "A silicon strip detector for the MARK II experiment at the SLAC linear collider," Nuclear Instruments and Methods in Physics Research, **A265**, 1988, p. 93.
5. Y. S. Tsai, "Pair production and bremsstrahlung of charged leptons," Review of Modern Physics, **46**:4, (1974), p. 828.
6. Particle Data Group, "Review of particle properties, particle detectors, absorbers, and ranges," Review of Modern Physics, **56**:2, (1984).
7. M. Turala, "Tracking with silicon detectors," *Proceedings of the XXIV International Conference on High Energy Physics*, 1989, p. 1240.
8. R. M. Jones, *Mechanics of Composite Materials*, Hemisphere Publishing Corp., 1975, p. 196.

Engineering Challenges for Scintillating Fiber Tracking at the SSC*

David Koltick
Physics Department, Purdue University
West Lafayette, Indiana 47907

Abstract

A scintillating fiber charged particle tracking detector for use at the SSC is presented. The design is highly segmented as well as capable of very high data rates. It's more than 10^6 channels are simple in design so that industrialization of its production can easily be accomplished. The scintillating fibers used range between 500µm and 1mm. Upon passage of a charged particle through a fiber light is emitted. For a faction of the photons the fiber acts like a wave guide, transmitting the photons to a transducer outside the active detection region. The tracker extends 8 meter in length by 3.2 meters in diameter.

For many reasons it is desirable to use scintillating fibers at the SSC to build a charged particle tracker. The engineering challenge necessary to reap the full physics capability of this technology will ultimately set the pace of its application.

Present charged particle tracking technology is dominated by drifting electrons through a gas in the presence of a high electric field. In this wire chamber technology electrons drift towards an anode typically ~25 µm in diameter. Because the wire is tensioned between two accurately machined endplates, the precision comes about from the 2-dimension machining accuracy of the endplates. Groups building vertex chambers have obtained accuracies of 1µm-2µm in the machining of these endplates. Vertex chambers have obtained absolute measurement errors of 20µm-30µm, limited not by the construction errors but by the electron drift dispersion in the gas.

While wire chamber technology is mature, being over 50 years old, it has many disadvantages; electron drift is inherently slow, limiting the size of the drift cell in high speed operation. The use of gases, often explosive or flammable, and the need to flow these gases over a large system requires complex manifold and endplate structures. The use of high voltage requires complex endplate design and electronic components to be placed within the gas volume. The very thin sense wires are very fragile so that wire breakage can be a problem. A single wire breakage can cause an entire section of a chamber to fail because of the common high voltage line used for that section. A major construction problem is that a large chamber will not work unless its component parts lie absolutely along a straight line to within very high tolerances (~100µm), a difficult task for a large SSC system ~4 meters long.

None of these problems apply to a scintillating fiber tracking system. A scintillating fiber system is a totally passive system. Once built and installed, it should need no maintenance. There are no gases, no high voltages nor delicate parts to break. If a single fiber should "fail" for some reason, it will not effect any other fibers since each acts independently. The major problem with the use of scintillating fibers is the 3-dimensional nature of the construction problem that they present. This problem comes about because polystyrene or a mechanically similar material will be used as the base material for the fiber. Because of this material's plastic properties the construction problem cannot be limited to 2-dimensions by the use of an endplate. The full 3-dimensional accuracy of the tracker must be addressed. Nonetheless for physics reasons similar measuring resolutions, as obtained with the best wire chambers are desirable and theoretically

* This research was conduced by the Fiber Tracking Group, a list of authors can be found in "Scintillating Fiber Tracking for the SSC", R. Ruchti, et. al., these proceedings

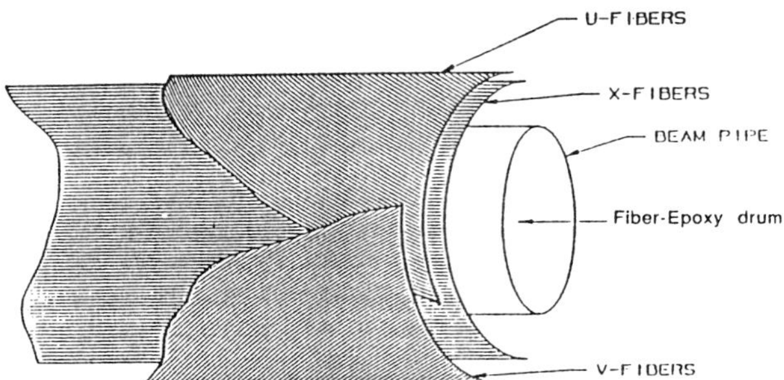

Figure 1: Schematic showing the ease of constructing the scintillating fiber tracker.

obtainable with scintillating fibers. To achieve such accuracy, fiber placement errors no greater than 25μm over a 3 dimensional surface are required, using fibers ranging in diameters between 500μm - 1mm. One saving grace of the fibers is that one does not have to position the fiber along a straight line. It is the position of the fiber at any point that must be known to 25μm. This extra freedom allows many more design solutions. For example, fibers can twist and bend in 2 or 3 dimensions.

For the central region the scintillating fibers will be formed onto cylinders as shown in Fig. 1. Three orientations of the fibers will be used. The main orientation will lie parallel to the axis of the drum while 2 others known as stereo orientations will lie at plus and minus 5 degrees with respect to the cylinders axis. It would be advantageous to build up the cylinders using modules, but unless each sublayer of the cylinder can be built individually, the use of stereo layering calls for unimodular construction of a superlayer as shown in Fig. 2.

At present we have taken the attitude that we must remove one degree of freedom at a time during the construction of the tracker. First the fiber's movement relative to one another will be removed by forming the fibers into ribbons. Fibers are now being manufactured to tolerances of 1% in their diameters. While this is excellent it will not allow the ribbons to be built up by simply placing the fibers one next to another. There must be a gap between fibers to allow for run out errors if high resolution measurements are to be made. A gap of 1.5% of the fibers nominal diameter will be adequate.

The ribbons will be place accurately on a fiber-epoxy composite drum. The drum needs to have a high precision radius. It is important that the layer to layer placement of the ribbons also be of high accuracy. Presently we are planning a sublayer to consist of 4 individual layers of fibers each staggered by 1/4 fiber diameter. The need to correlate the layers becomes clear by

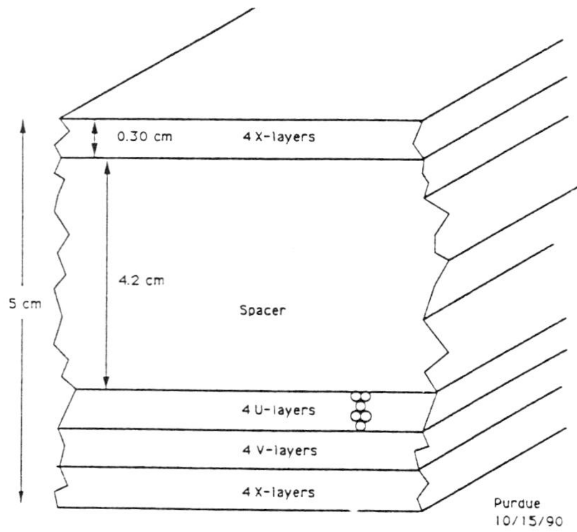

Figure 2: Cross sectional view of a central region superlayer.

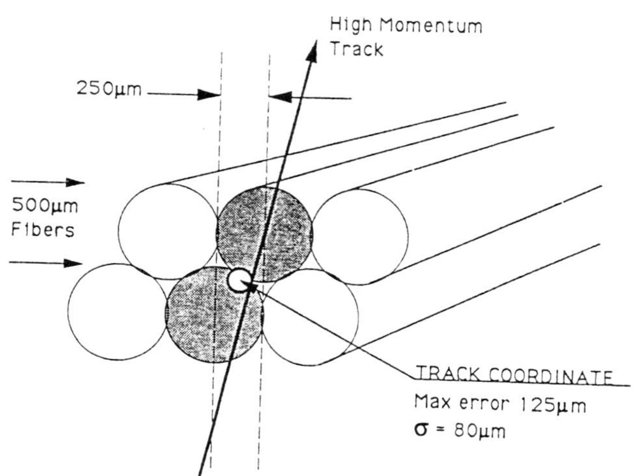

Figure 3: Schematically shown is the resolution obtainable using 500 μm fibers. By staggering the fiber layers, a passing track can be localized to a strip only 250 μm wide. By taking the coordinate at the center of the strip, the maximum error reduces to 125 μm. This yields a measuring resolution of 80 μm for two layers and 40 μm for four layers.

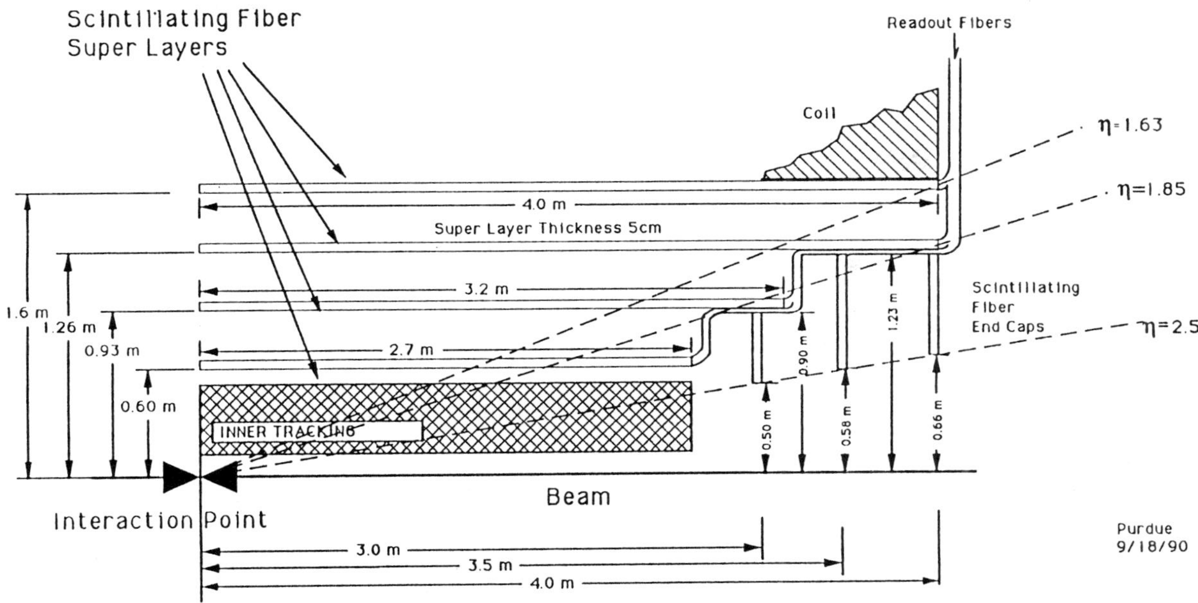

Figure 5: Cross sectional view of the scintillating fiber tracking system. The break in the tracker at the interaction point is done so that the fibers respond to only one beam crossing.

studying Fig. 3. The correlated layers have a measuring resolution which improves much faster than by the square root of the number of measuring layers, the expected rate of improvement for uncorrelated layers. By using 4 layers and their pattern of hits, a sublayer made of 500μm fibers is capable of achieving ~40μm resolution, whereas we would expect only ~125μm otherwise.

The structure of the superlayers in the forward direction takes full advantage of the flexibility of the fibers, Fig. 4. Here an area filling spiral design is used. The spiral has a left hand twist on two of the sublayers and a right hand twist on the next two, followed by two layers of fibers that lie on a constant radius. Again these spiral structures will have to be laid down with a precision of 25μm. However, the forward tracker can be constructed on 2-dimensional endplates so that the construction for this region should be conventional.

The overall tracker design is shown in Fig. 5. The scintillating fibers cannot be used as wave guides because the light has a very short attenuation length within these fibers. A clear wave guide fiber will be used which consists of the same base material as the scintillating fiber except the dopant (~1% by weight) will be left out. The attachment of the wave guide fiber to the scintillating fiber has been successfully accomplished with typical transmission efficiencies of 97%. But the mass production of fiber ribbons which make this transition is yet to be achieved. Ideas to accomplish this task include the production of boules containing alternating layers of clear and scintillating material to be pulled into fibers which would readily be turned into ribbons. Another idea would have a machine fuse individual clear and scintillating fiber as a basis for ribbon making. Still another approach would have wave guide and scintillating ribbons fused together using a laser.

The engineering challenges that confront the application of scintillating fiber to solve the physics problems of the SSC are clear. The job now is to find the best approach to accomplish the task.

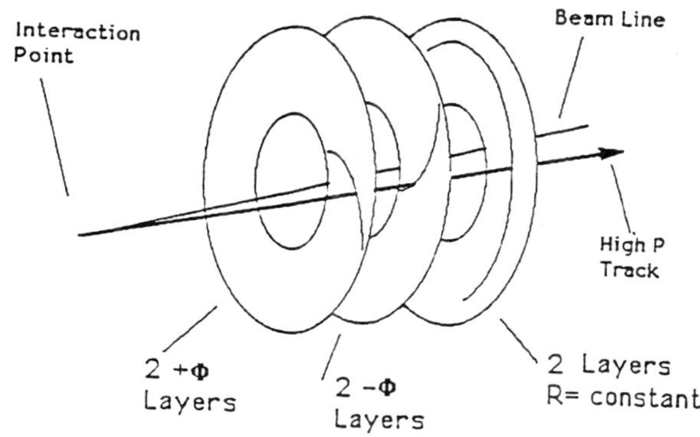

Figure 4: Super sublayer of the intermediate tracker.

ENGINEERING DESIGN OF THE EMPACT LIQUID ARGON CALORIMETER

L. Mason
Martin Marietta Astronautics
Denver, CO 80201
For The EMPACT Collaboration

Abstract

The Engineering design of the EMPACT liquid argon calorimeter is presented. The studies and analysis that led to the overall configuration as well as the detailed shape of the vessels and the location of the support structure are discussed, as are the engineering R&D issues the must be resolved to proceed to a detailed design.

Introduction

The goal of this study is to establish a liquid argon calorimeter (LAC) design baseline for the EMPACT [1] Detector. The concept is based on the generic SSC Hermetic Liquid Argon Calorimeter (HLAC) [2].

Requirements

Physics requirements for the EMPACT Baseline Liquid Argon Calorimeter are:

- Coverage: $0 < \eta < 3.0$
- Calorimetry depth as follows:
  - EM plus hadron depth:
    - 12 λ at $0.0 \le \eta \le \sim 1.5$
    - 12 λ at $\eta \sim 1.5$ progressing to 14 λ at $\eta = 3.0$
- EM Segmentation:
  - $\Delta\eta x \Delta\varphi = 0.025x0.025$ for $0 \le \eta \le 2.35$
- Hadron segmentation:
  - $\Delta\eta x \Delta\varphi = 0.05x0.05$ for $0 \le \eta \le 3.0$

Engineering requirements for the EMPACT Baseline Liquid Argon Calorimeter are:

- Dimensional constraint allowed by overall EMPACT Detector. See Figure 1.

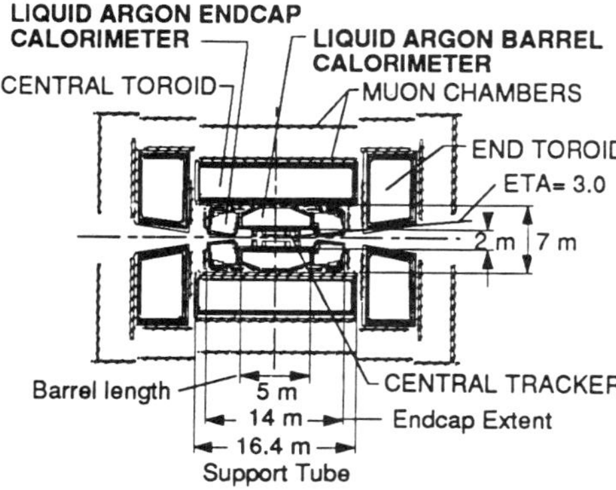

Figure 1. EMPACT LAC Envelope

- Calorimeter weight carried by the central toroid
- Access at both ends of the tracker is needed for servicing
  - Services area per tracker end is 0.5 square meter
- Design must comply with Section VIII of the ASME Boiler and Pressure Vessel Code
- Structural materials:
  - Vessels: Aluminum (5000 series)
  - Support Structure: Stainless steel (300 series), Aluminum (5000 series), Titanium (TBD series)
- Argon vessel internal design pressures are +55.0 psid and -14.7 psid
- Vacuum vessel design pressures are +12.0 psid and -14.7 psid

Design and Analysis

The integrated design, modeling, and analysis system we use is built around the Structural Dynamics Research Corporation I-DEAS packages Geomod, a sophisticated solid modeler and database, and Supertab, a flexible analysis interface to a variety of structural, thermal, fluid flow, magnetic analysis codes and most commercial computer aided design (CAD) systems. We have also written, and interfaced to Geomod, a ray tracing and tracking package that collects pertinent data along the path of the ray through the solid model, and a post processor that uses the geometric profiles and selected attributes and shower parametrics to evaluate the performance of the LAC.

Model Development

The design engineer develops a model of the calorimeter with a level of complexity appropriate for the required analysis and communicates that information via electronic data transfer and instructions to analysis engineers. An analysis is performed and the results are communicated back to the designer. The first models are simple, having only smeared material properties that allow an evaluation of the overall performance of various configurations of vessels and modules. With the preferred configuration chosen, more sophisticated models are developed to analyze structural and thermal performance. As the design matures, so do the model and analysis techniques.

Design Optimization Process

The designer develops a shell model representing the center line of vessel walls and then utilizes the IDEA-S Model Solution optimizer to determine wall thickness using Von Mises stress limits based on ASME allowables. Von Mises stresses are generally conservative in the analysis process. Non-compliances are noted and will require special consideration. Once new wall thicknesses are determined, a solid model is generated for further thermal and structural analysis, and hermeticity evaluation. This process is iterated to optimize the design. The present configuration is shown in Figure 2. The LAC vessel wall thicknesses and masses are shown in separated views in Figure 3.

Support Tube

A support tube carries the weight of the barrel and both endcap calorimeters to the ends of the central toroid and attaches to the central toroid vacuum vessel via closure plates. The closure plates, at opposite ends of the calorimeter, fill alternate spaces leaving access to the muon chambers from one end or the other of the calorimeter.

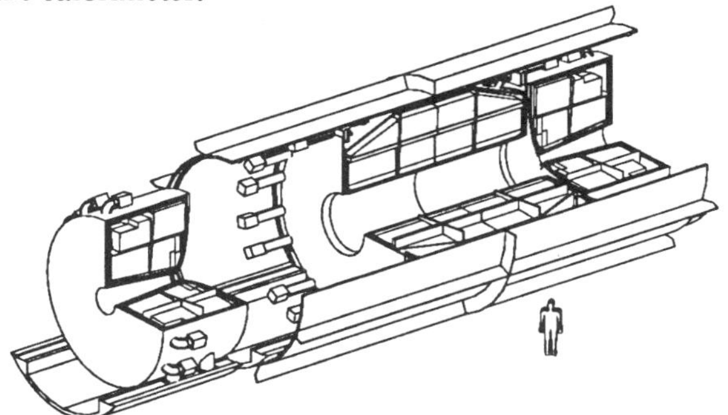

Figure 2. Cutaway View of LAC, Showing Barrel and Endcaps.

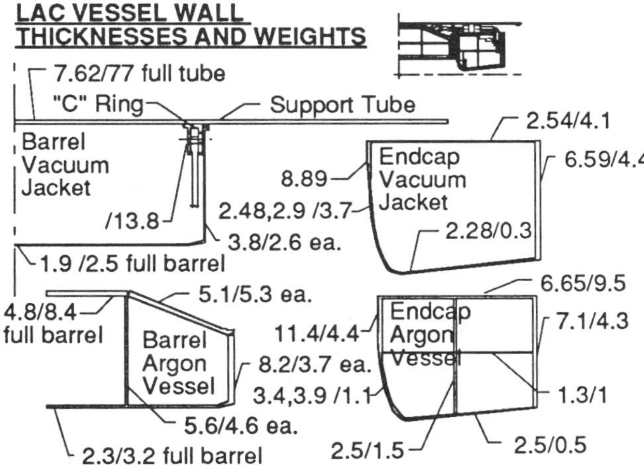

Dimensions in centimeters - Weight in metric tonnes (cm/MT)

Figure 3. Liquid Argon Vessel Thicknesses and Masses.

Barrel Calorimeter

The barrel calorimeter utilizes flat plate heads and two structural washers, forming three bays for modules. There are two layers of modules in each bay. Figure 4 shows the truncated end of the barrel calorimeter with one washer and one flat plate head.

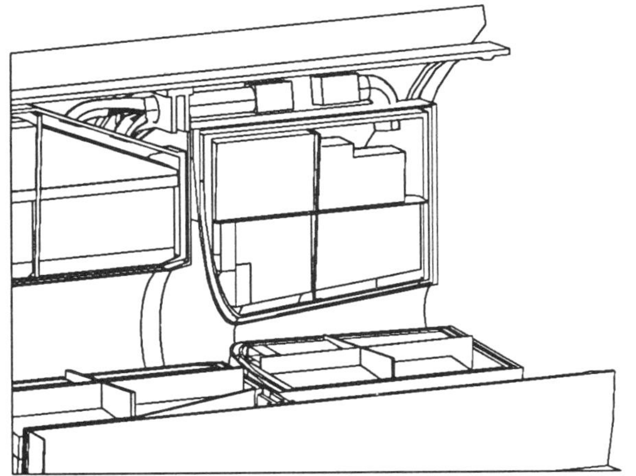

Figure 4. Close-up View of Barrel End and Endcap

The outer layer of the end modules is truncated as the full radial thickness is not required for the underlying physics processes. The outer shell of the argon containment vessel is truncated to match the modules. This reduces the required outer radius at the argon vessel ends to allow space for the supports and the electronic feed-throughs. The barrel calorimeter is supported by straps. The straps are attached to the support tube at stiffening rings located at each end of the argon vessel, with the strap passing under the heads. Alignment is provided by adjustable cam pins at the strap to stiffening ring interfaces.

Endcap Calorimeters

The heads of the endcap vessels, toward the interaction point, are a combination of flat and elliptical shapes. The elliptical portion of the argon and vacuum vessel heads is toward the beam line. The shaped heads significantly reduce the overall head thickness, permitting the active portion of the endcap calorimeter to fit closer to the barrel calorimeter. The elliptical head is designed for strength and structural stability but falls outside the ASME guide lines for analysis and will require special consideration in the request for ASME certification. The endcap calorimeters are supported on stanchions similar to the stanchions used at Fermi Laboratory D "Zero" detector. The endcap calorimeters slide on rails within the support tube to allow access to the central tracker.

Modules

The preliminary calorimeter module design has a strongback and end plates to support the stack. The module mates to the argon vessel heads and washers. The module shapes, weights, and quantities are shown in Figure 5.

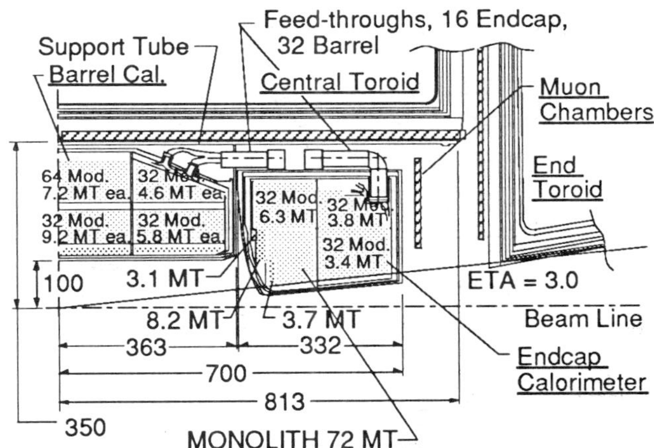

Figure 5. Sections Showing Module Quantities, Weights and Shapes

Cryogenics

It is likely that the electronics may emit 70 to 200 mW of heat per channel resulting in a total heat rejection requirement of 5 to 15 kW per argon vessel. A design concept for cooling the nearly 100,000 channels of electronics per barrel and endcap calorimeter will be resolved during the LAC R&D effort. Cool down and heat intercept cryogenics for the barrel and endcap argon vessels will be via LN_2 cryogenic tubing attached to the exterior surfaces of the Argon vessels. The tubing is plumbed in zones with a controlled flow between parallel paths. Each flow path will be oriented to prevent gas traps. The argon system is a closed system conditioned with liquid nitrogen, to require a minimal amount of circulation. Safety devices prevent over pressurizing the argon vessels and nitrogen systems.

Physics Performance Analysis

We consider a hermetic design to be one in which performance (e.g. single particle or jet resolution) approaches that of a pure calorimeter. To evaluate the performance, we use the ray tracing capability of our engineering analysis tools to sample the engineering model and the results of this analysis are shown in Figures 6 - 9. Figure 6 depicts rays traced through the baseline elliptical head LAC geometry at pseudorapidity (η) intervals of 0.025 out to η = 3.00.

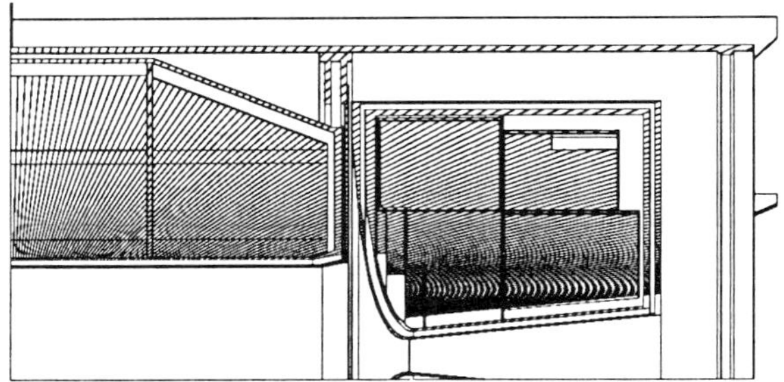

Figure 6. A Profile of the LAC With Rays.

Figure 7 shows the fraction of dead material (i.e. vessels, supports, etc.) between the interaction point and the last active calorimeter layer.

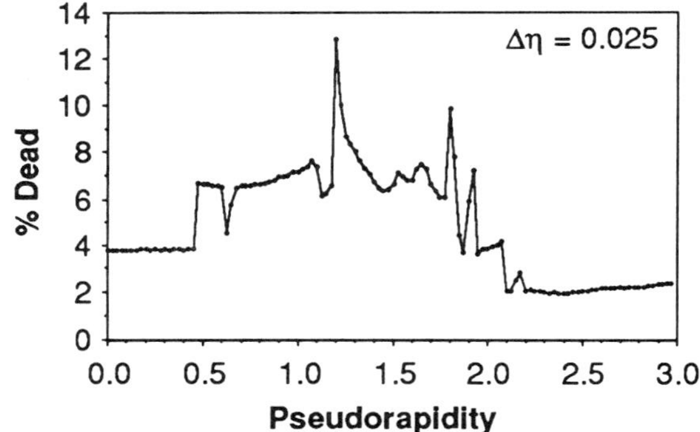

Figure 7. Percentage Dead Material, i.e. $\lambda_{dead}/(\lambda_{live} + \lambda_{dead})$, Between the Collision Point and the Last Active Calorimeter Layer

Figure 8 shows the active depth of the calorimeter, not including the depth contributed by vessels, structure, wiring, etc. A comparison of Figures 6 and 7 allows direct correlation of the geometry with the dead fraction. We emphasize, however, that the total amount of the dead material is not a particularly relevant measure of LAC performance, as it does not consider the showers longitudinal or transverse energy deposition profiles. We use the energy resolution as our performance measure since it is a limiting factor in the reconstruction of the underlying physics event. We estimate the LAC performance by using an analytic shower deposition model to calculate the energy deposited in a dead layer. This lost energy contributes to a change in the LAC resolution. The energy deposition is presently calculated by assuming an isolated electron or pion shower with no transverse extent, but simply a longitudinal profile [2] and is shown in Figure 9.

Additional design information can be found in References 2 and 3.

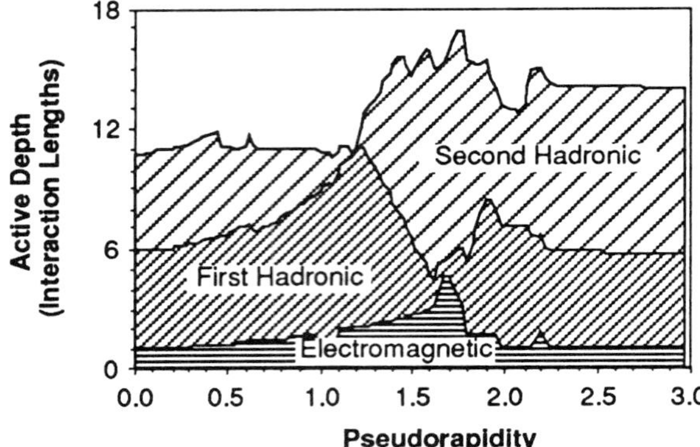

Figure 8. Active Interaction Lengths of Calorimeter

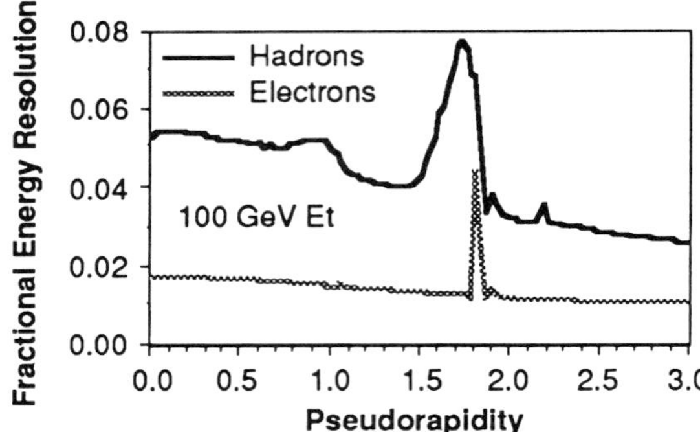

Figure 9. EM and Hadronic Resolution for the Baseline Calorimeter.

Conclusions

This study has developed a baseline liquid argon calorimeter that meets the specified physics and engineering requirements. Assembly and fabrication concepts for the vessels and supports are viable using state of the art methods and equipment. However, much work remains to refine this baseline concept into a detailed design. The major mechanical issues follow: Certainly the potential of heat generated by the electronics contained in the argon vessels is of major concern and will require R&D to resolve. Inherent to the electronic feed-throughs are potential fluid leak paths and electrical anomalies that could render the calorimeter inoperable if not properly researched, designed, and constructed. The vessel head shapes need further analysis and testing to verify the analytical approach to the design that satisfies ASME requirements. Considerable work needs to be accomplished on the support systems throughout the Calorimeter.

References

[1] M. Marx *et al.*, "EMPACT: Electrons Muons Partons with Air Core Toroids, An Expression of Interest for an Experiment to be performed at the Superconducting Super Collider", SSC-EOI0006, (May 25, 1990).

[2] T. Adams et al., "Evolution of a Reference Design for a Hermetic Liquid Argon Calorimeter for the SSC", January 1990.

[3] T. Adams et al., Status of EMPACT Liquid Argon Calorimeter Engineering Design", EMPACT Technical Note 218, June 6, 1990.

R & D AND MANUFACTURING ISSUES FOR SPACAL CALORIMETRY

J. Pohlen
Martin Marietta Astronautics
Denver, CO 80201

For The EMPACT Collaboration

Abstract

The development of a scintillating fiber-lead (SPACAL) calorimeter based on the "super-ring" approach is presented and discussed, with particular emphasis on R&D required to validate the EMPACT SPACAL conceptual design. These include module manufacturing and readout assembly, super ring construction and long term stability, and system assembly.

EMPACT Spaghetti Calorimeter Requirements

A baseline scintillating fiber/lead calorimeter design has been developed for EMPACT that consists of structurally independent 100 to 200 metric ton "super-rings" of glued fiber/ lead modules assembled into a barrel (1200 metric tons) and endcap (650 metric tons) configuration. The physics requirements, summarized in the EMPACT Expression of Interest (EOI) [1], include physics performance goals to maximize the calorimeter hermiticity and depth of active calorimetry for the pseudorapidity region $|\eta| < 3$. The active calorimeter depth goal is 11 interaction lengths at $|\eta| = 0$ and 14 at $|\eta| = 3$. Derived engineering requirements for the baseline design include:

- **Configuration**
  - The calorimeter must fit within the calorimeter region specified in the EMPACT EOI [1] baseline design i.e. inner radius 1.0M, outer radius 3.5 M, total length 14.0M.
  - The support structure for the calorimeter must not penetrate or pass loads through the center and end toroids.
- **Accessibility**
  Provide maintenance access to the central tracking system without major disassembly.
- **Tracking Services**
  Provide a total of one square meter of exit space for tracking signal cables and utilities.
- **Assembly**
  Simple and reliable overall structure assembly, module installing, maintenance, and signal cable routing.
- **Reliability**
  Lead/fiber matrix structural integrity and fiber radiation resistance such that system performance life is 10 years.

Configuration Baseline Features

An overall view of the SPACAL calorimeter installed in EMPACT is shown in Figure 1.

- **Independent Barrel And Endcap Calorimeters**
  The requirement for access to the tracking system drove selection of independent barrel and endcap calorimeters. The endcap calorimeters are fitted with roller assemblies to facilitate assembly and disassembly for tracker access.
- **Super-Ring Construction**
  The barrel and endcap calorimeters are fabricated in "super-ring" [2] subassemblies to facilitate handling and assembly. The barrel calorimeter super-rings are installed individually into the calorimeter support tube. The endcap calorimeters are pre-assembled prior to installation, with joints at the super-ring interfaces that will permit future replacement of individual super-rings.

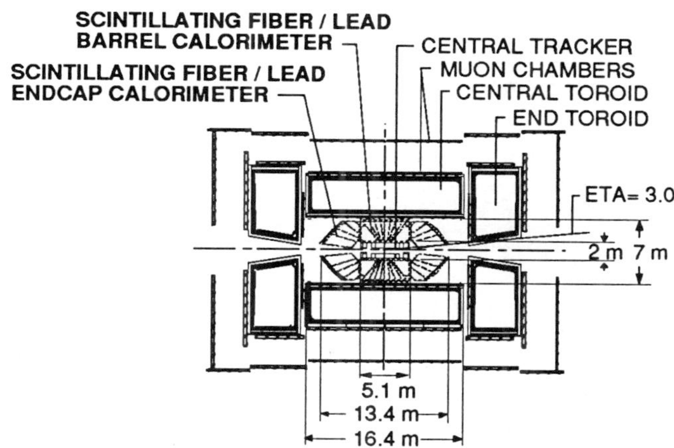

Figure 1. SPACAL Calorimeter in EMPACT Detector

- **Si Preconverter**
  A silicon-lead-steel preconverter array is provided on a thin steel tube inside the central barrel and on cones in front of the endcaps.

Overview of the Baseline Calorimeter Design

Figures 2 and 3 show the overall assembled system inside an aluminum calorimeter support tube, and an individual super-ring, respectively.

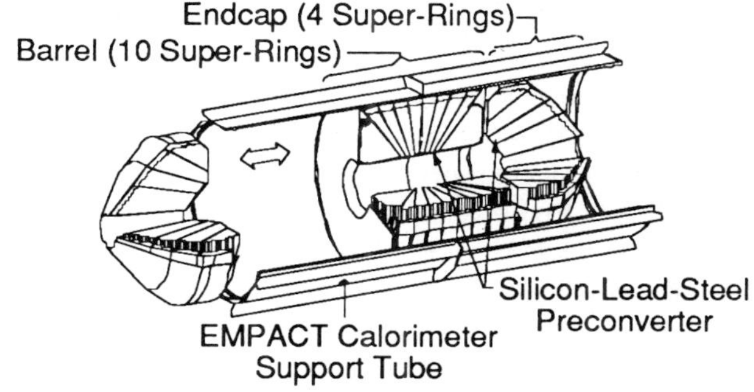

Figure 2. EMPACT Baseline SPACAL with Si Preconverter assembled in Calorimeter Support Tube

Also represented in Figure 2 is a 2.5 cm thick silicon-lead steel preconverter array mounted on thin steel support tube and cones inside the barrel and endcaps. Figure 4 shows the symmetrical baseline calorimeter geometry and super-ring sizing.

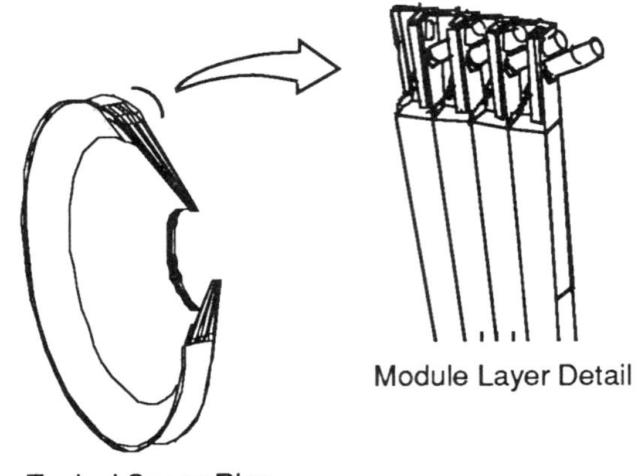

Figure 3. Individual Super-Ring Showing Assembled Modules (Towers) and Layers

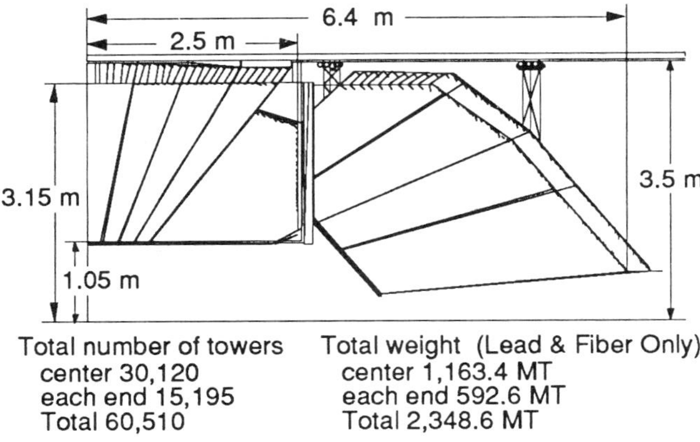

Figure 4. Baseline Geometry for SPACAL with Si Preconverter

The super-rings are attached to and supported by an aluminum tube that carries calorimeter loads to the toroid structure and contains the inner layer of muon chambers. Super-rings are assembled from individual modules built of 2mm fibers and 4mm lead sheets with attached light pipes, phototubes, and aluminum end fittings as shown in Figure 5. These individual modules, each with a mass of approximately 65 kg, are glued into multiple layers of up to 251 modules each to form a basic super-ring. Modules in the barrel section super-rings are assembled pointing approximately 7.5 cm from the beam axis and 4 degrees off the true radial projection line, to prevent particles from channeling in the fibers. Each subsequent layer changes the orientation from 4 degrees to -4 degrees and vice versa, as shown in Figure 6, which increases the mechanical strength and eliminates helicity from the calorimeter. [2]

At the barrel/endcap interface, light from the truncated modules in the barrel is brought to phototubes located in the obvious notch in the outer barrel radius via non-scintillating fiber optics. Signal cables from the pre-converter are brought out through a 5 cm gap between the barrel and endcaps.

Component Issues

Fiber Issues

Fiber characteristics requiring further R & D to confirm the EMPACT SPACAL design concept include fiber diameter,

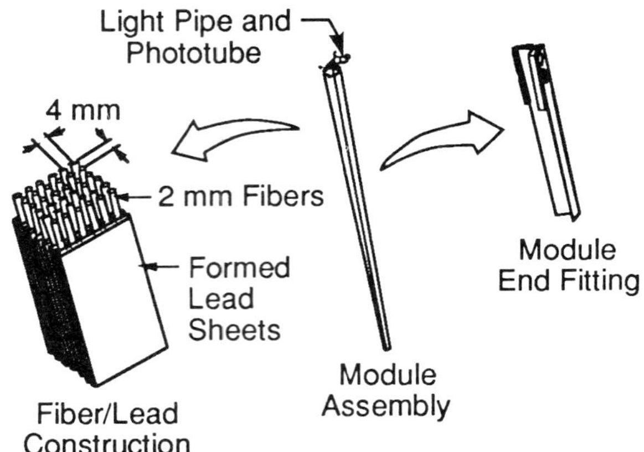

Figure 5. Individual Module (Tower) Construction Details

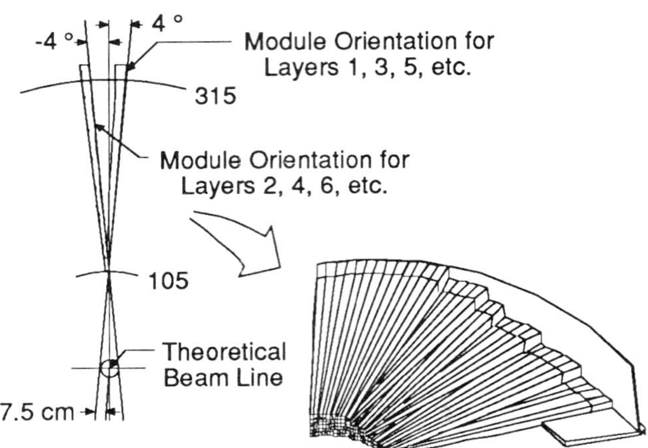

Figure 6. Non-Projective Module Assembly Orientation Showing Alternation Between Layers.

radiation degradation, mirrored front face requirements, protective coatings, and compatibility with other materials.

The 2mm fiber diameter used in the EOI baseline was selected based on recent prototype work by the SPACAL collaboration. Consideration of even larger fibers (3mm and up) could potentially offer further savings at the expense of physics performance and the additional space penalty required to gather the fibers with larger bend radii.

Radiation degradation is currently driving the requirement to disassemble at least portions of the end caps to replace damaged sections. Further investigation of the fibers ability to tolerate radiation could relieve this constraint and thereby simplify the construction.

The mirrored front face and external coatings on the fibers impose special handling and positioning requirements during fabrication. Protection requirements must be developed based on investigation of materials compatibilities and sensitivities.

Lead Issues

Lead fabrication R & D issues to be addressed include forming, fastening, strength, and materials compatibility. Forming and fastening techniques must be developed that are compatible with close tolerances, high production rates, and safety requirements. Material strength and compatibility issues include short term limits for fabrication and assembly, and long term compatibility and creep assessment.

Glue Issues

Glue development will involve a specification of the engineering properties requirements and evaluation of glue compounds. Issues include strength, cure time and temperature, lead filler mix, radiation degradation, bond line thickness sensitivity and control, and materials compatibility.

Photomultiplier Tubes

Photomultiplier (PM) tube development issues include size miniaturization and optimization of power, temperature, and cabling requirements. The current EOI baseline depicts a tilted installation, shown in Figure 3, that minimizes the impact of the relatively long photomultiplier on the calorimeter depth, at the expense of assembly complexity. Development of a shorter PM can provide the same economy of space without the added complexity.

Subsystem Issues

Tower Fabrication

Tower fabrication cost and tolerance issues are highly magnified just by the quantity of assemblies involved. A critical portion of this R & D task will be to evaluate tolerance accumulations and allowables with respect to system performance and fabrication costs, and allocate these tolerance limits to tower components. Manufacturing techniques must then be developed and verified to control the final tower shape within these limits. Optimization of the tower size and integration of components such as interface fittings, fiber termination, photomultipliers, and cabling must also be included as an integral part of this development task.

One of the major complications to the fabrication and shape control of the towers is introduced by the non-projective orientation of the towers within the super ring. The positioning of the towers off the true radial projection line generates a twist in the tower to conform to the conical shape of the super ring. This twist becomes more pronounced as the cone angle increases.

Super Ring Fabrication

Super ring segment definition and support concepts need to be finalized in conjunction with tower shape and tolerance development described above. Physical interfaces between super rings, the support tube, and the central tracker must be evaluated and baseline cable routing and envelopes established. Evaluation and prototype testing of bonding (glueing) techniques for super ring assemblies must also be accomplished to validate fabrication techniques and glue development.

The impact of cooling requirements will also have to be evaluated during the development of super ring fabrication and tooling techniques.

System Issues

Central Barrel

The initial concept of barrel section assembly, shown in Figure 7, utilizes the fabrication tooling to sequentially assemble the super rings in the support tube.[3] Issues that need to be addressed include tooling, assembly, and structural support requirements for both installation and transport. Cable routing, plumbing (if cooling is required), and checkout concepts must also be developed.

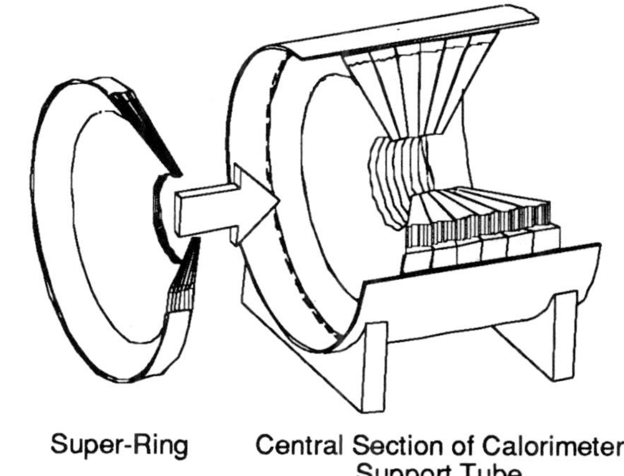

Figure 7. Assembly of Barrel Calorimeter Super Rings in Support Tube

End Caps

The end cap calorimeters are preassembled and installed in the support tube as shown in Figure 8. In addition to assembly, structural support, and cable routing issues, the end cap development must also consider such mobility issues as distribution of roller loads into the calorimeter, rail requirements and support tube ovalling, and removal and replacement of calorimeter sections exposed to high radiation levels.

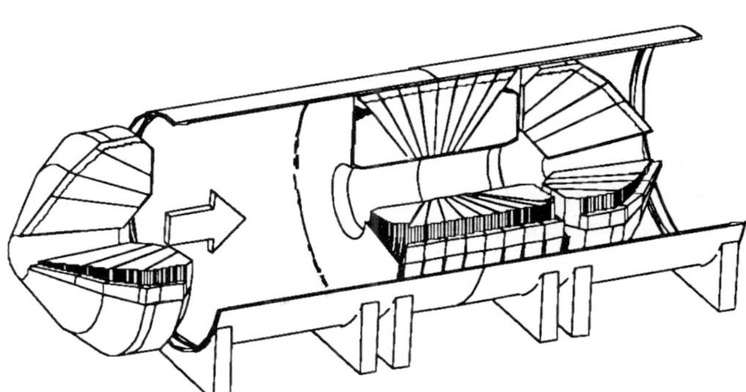

Figure 8. Installation of Endcap Calorimeter in Support Tube

Conclusion

While none of the R & D issues appear to be beyond current technology, the development of manufacturing techniques to control the end products and verification of these techniques is critical for the confirmation of the SPACAL concept.

References

[1] M. Marx et al., "EMPACT: Electrons Muons Partons with Air Core Toroids, An Expression of Interest for an Experiment to be Performed at the Superconducting Super Collider", SSC-EOI0006, (May 25, 1990).

[2] R. DeSalvo, Dream On A Supercollider Spaghetti Calorimeter, CERN/LAA/HC/89-004, (1989).

[3] T. Adams et al., "Status of EMPACT SPACAL Calorimeter Engineering Design", EMPACT Technical Note 219, June 1990.

ASSEMBLY AND INTEGRATION ISSUES FOR EMPACT

G. Guenterberg
Martin Marietta Astronautics
Denver CO 80201

For The EMPACT Collaboration

Abstract

Trade studies and analyses involved in determining assembly and integration scenarios for SSC detectors are presented. In particular, the scenario in which most major components of the EMPACT detector are assembled above ground and subsequently installed as complete sub-systems is discussed.

Introduction

This paper describes how the major elements of the EMPACT detector form an integrated concept with provisions for signal cables, mechanical supports, and service and assembly access. An assembly plan is presented which determines baseline concepts for the detector hall and surface facilities. Consideration is given to the use of these facilities for the data-taking phase of the experiment and during periods of routine maintenance, unscheduled major repairs, and detector upgrades. The baseline has been developed in sufficient detail to determine the critical requirements. Significant work has been accomplished, however, trade studies for future optimization are indicated.

Integration

The EMPACT detector design philosophy incorporates systems integration requirements and constraints from the start of conceptual development. Requirements driven by component designs, assembly, installation, and maintenance govern evolution of the total system. The 3-D model in Figure 1 shows the EMPACT assembly supported from an external structure that minimizes support structure-induced dead areas in the detector coverage.

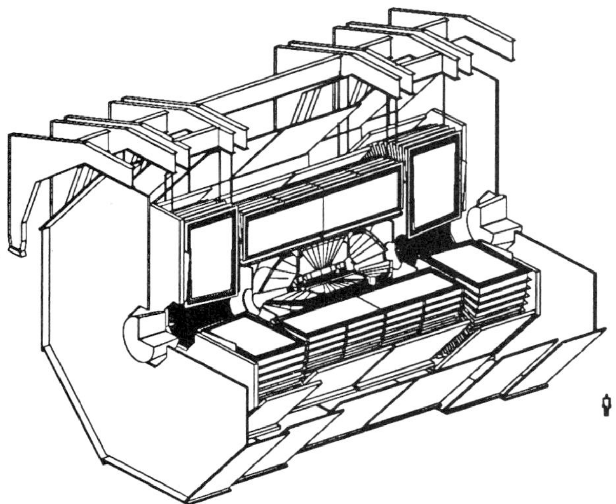

Figure 1. Cut-away view of the EMPACT detector showing parts of the overhead frames that suspended the detector.

The relatively low weight of the detector (5600 metric tons) permits suspension of the major assemblies with adjustable tension members from above rather than from massive internal columns required for support from below. The tension members pass between muon chamber panels and require only local truncation of panel corners or edges in upper detector quadrants. A double-walled tube supports the calorimeter and forms the inner vacuum vessel wall of the central toroid and the outer vacuum vessel wall of the barrel calorimeter. The inner barrel muon chambers are located between these two walls. The number of cavities around the circumference for muon chambers is driven by issues of structural strength versus thickness, spacing of stiffening ribs, and optimization of the number of muon channels.

The dimensions of the detector elements and the spaces between them provide room for support members, signal cables, high voltage cables, low voltage cables for electronics, muon chamber gas feeds, cryogen feeds, and vacuum lines. The structure, cabling, and utilities that interpose dead material in the sensor paths of the detector are included in physics models to determine detector performance degradation. Design iterations using results from these models have provided initial sizing and routing optimization in the detector. Services that are routed between the central and end toroids egress from the upper 135 degree azimuth sector. This preliminary concept minimizes interference with muon coverage. Further optimization of the interaction between the supports and the muon chambers will be necessary, especially to answer questions regarding the overlap of the barrel and plug muon chambers.

Assembly Plan

The point-of-departure detector assembly plan and the corresponding hall concept [1] incorporate above ground fabrication, assembly, and test of all major detector components followed by integration and system tests below ground in a minimum-size hall. Initial studies indicate this concept offers lower overall facility and installation costs through minimization of high-cost hall volume coupled with some increase in low-cost surface facility volume. Detector construction proceeds in surface buildings in parallel with hall construction thereby providing maximum schedule flexibility. In contrast, an alternative plan where most fabrication and assembly work is performed in the hall is relatively inefficient and the hall volume must be enlarged (up to 60%) to provide space for parallel construction tasks. Detailed trade studies in cooperation with the SSCL will evaluate alternative hall and assembly concepts.

Figure 2 is a cutaway view of the integrated EMPACT detector in the collision hall with one end toroid moved back as required during construction or maintenance. The calorimeters, and central toroid components are supported from stationary overhead frames, while the end toroid assembly and forward calorimeter assemblies are supported from movable frames. The integrated model of Figure 2 contains significant detail, and further detail will be added for the support frames, muon chamber support structures, and service interfaces for vacuum, cryogenics, and electrical equipment to support trade studies. These details will also be included in site design models for studying detector integration, assembly, and service requirements.

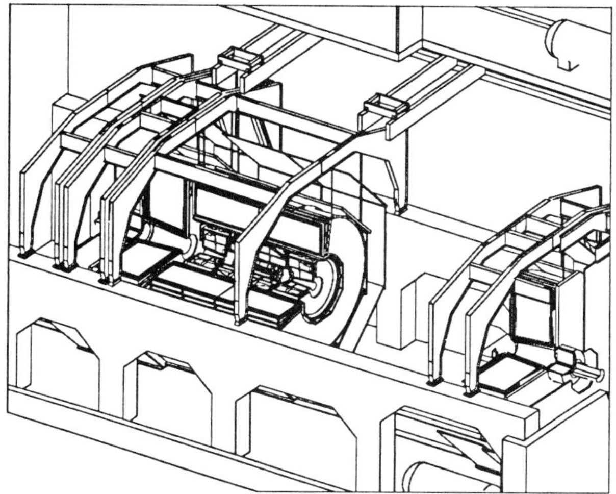

Figure 2. Cut-away view of the EMPACT detector in the collision hall. One end toroid with attached muon cap chambers and forward calorimeter is moved from the operational position as during assembly or maintenance. Overhead cranes and one bearing wall for the overhead support frames show.

Surface Activities

Most assembly operations use modules and sub-assemblies, fabricated at vendor sites or at collaborator institutions, which are then assembled in surface facilities near the collision hall. For example, the calorimeter modules, fabricated offsite, are installed in a surface building. The toroid modules and muon chambers are assembled in adjacent buildings. All completed assemblies are moved to one of two 9 x 27 meter vertical access shafts over the collision hall on a track-mounted transporter and lowered using a temporary, heavy-lift device. A temporary portable enclosure provides protection during transport and lowering operations.

Surface facilities include construction sheds for the central and end toroids, the muon chambers, and the calorimeters. The calorimeter shed contains clean rooms for installation of the calorimeter modules into the barrel, end, and forward calorimeter assemblies. Figure 3 shows a cutaway view of the collision hall with the EMPACT detector in place at the interaction point; it also shows the access shafts. Two large shafts (9 x 27 meters) are for lowering major components and two smaller shafts (7.2 meters inside diameter) for routine access of equipment, personnel, and services. Also shown are dimensioned plan and elevation views. The assembly approach shown here minimizes the size of the costly collision hall. Some surface buildings can be converted to control room, office, and laboratory space after the construction phase.

The baseline approach is to install the barrel calorimeter and central tracker into the double wall support tube before lowering the assembly into the hall. The total weight of this 16 meter long assembly is 1710 metric tons. The fully assembled endcap calorimeters are also lowered; each weighs 560 metric tons.

Hall Activities

Eight central toroid modules, delivered to the site as 45 degree octants weighing 94 metric tons each, are lowered into the hall where they are assembled into the complete central toroid using temporary fixtures. During these operations the support frames for the central toroid/calorimeter assembly and the end toroids are installed in the hall. Tie rods are attached to the frames to support the toroid/calorimeter barrel assembly before the temporary fixture is dismantled. The support tube, with or without the central calorimeter, is then installed in the central toroid using movable beams and the calorimeter end caps are installed. Two independent 55 metric ton bridge cranes are provided in the hall for routine lifting.

In conjunction with the assembly of the central toroid octants in the hall, the end toroid magnets are assembled in a surface building using the same 45 degree octant approach. They are approximately 18 meters in diameter and each weighs 528 metric tons.

While assembly of the calorimeters and toroids proceeds, the muon chambers are assembled in the surface facility. Upon completion of the muon chamber sets, each pair is joined in a truss structure and initially aligned prior to lowering into the hall. The largest muon chamber assemblies (27-meter diameter cap ends) determine the access shaft dimensions. They are moved to the hall prior to lowering the end toroids. After the toroids are in the beamline position the muon chambers are installed and aligned.

Service equipment such as vacuum pumps, cryogenic lines, electrical cabling, and instrumentation are the last items to be assembled and installed in the hall. Much of this equipment will be located on platforms and routed on cable trays that are part of the central, fixed support structure. When the detector is fully assembled, the beam line elements within the collision hall can be installed on fixed platforms [2].

A cross tunnel connecting the horizontal access shafts near the interaction region houses controls, trigger electronics, and power supplies requiring minimal cable lengths and access during operations. Shielding doors allow occupancy during beam operations.

Facilities/Services

The hall, access shafts, and surface buildings necessary for EMPACT construction meet space requirements for experiment operation, routine maintenance, major unscheduled repairs, and upgrades. Surface buildings will be converted to offices and maintenance labs.

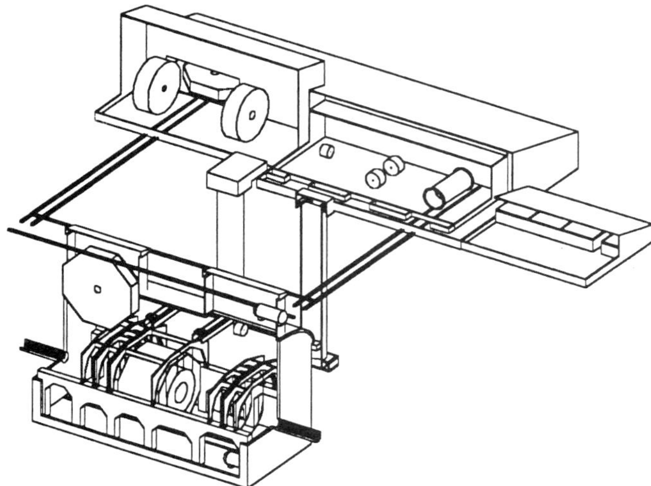

Perspective View

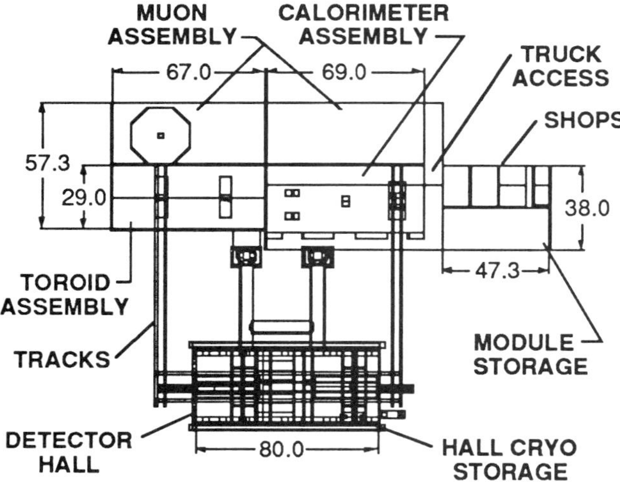

Plan View

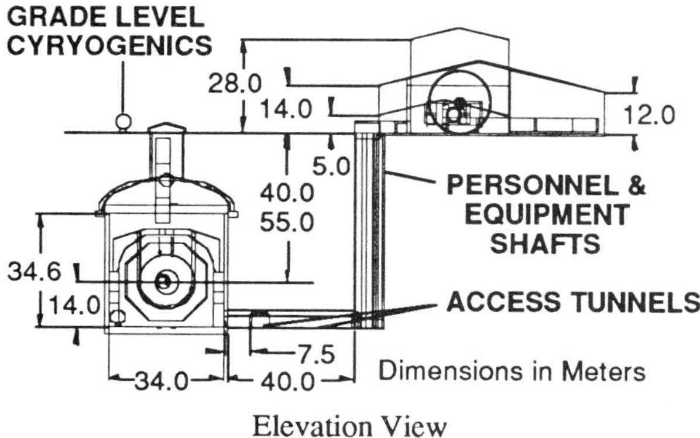

Elevation View

Figure 3. EMPACT baseline assembly scenario. The perspective cut-away view (top) shows the collision hall with two large access shafts (covers in place), rails for moving assemblies to the shafts, smaller utility shafts, and surface sheds (captions indicate use during assembly). A plan view (center) and an elevation view normal to the beam line (bottom) show overall dimensions.

Assembly Issues

A number of issues are being investigated to optimize the assembly scenario and reduce building requirements. The primary driver for the large shaft size is surface assembly of the cap muon chambers. We are now studying the option of fabricating smaller sub-assemblies on the surface, and installing them on support frames in the hall. This will reduce the shaft dimensions, and may simplify subsequent maintenance and disassembly operations of the cap chambers.

The option of installing the support tube into the assembled central toroid without the barrel calorimeter may simplify assembly and handling operations in the hall. This also provides the option of completing either the central toroid or central calorimeter first.

Site Geology Issues

Surveys of some hall sites on the west side of the SSC, tentatively assigned to large detectors, show a layer of relatively high-strength Austin Chalk to depths of 55 to 70 meters over a layer of unstable low-strength Eagle Ford Shale. Figure 4 shows that the shale strata intercepts the lower half of the EMPACT hall and illustrates that a simple adaptation of the EMPACT top-support concept can efficiently transfer detector loads into the chalk layer along terraces created during the planned hall excavation process. Initial estimates indicate a potential saving of $3M in foundation costs compared to floor-supported detector concepts that require elaborate footings into the shale to distribute loads and to provide necessary detector position stability.

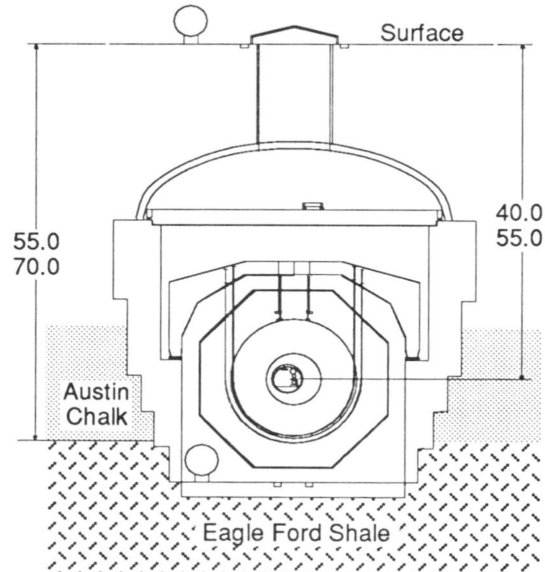

Figure 4. Site geology and load transfer concept.

References

[1] EMPACT Technical Note #216, L. Mason and J. Rutherfoord, "EMPACT Detector Assembly Scenario", 5/90.

[2] EMPACT Technical Note #215, J. Rutherfoord, "Accelerator Constraints in an SSC High Luminosity Collision Hall", 5/90.

MONOLITHIC HIGH T_c SUPERCONDUCTING PERMANENT MAGNETS

Roy Weinstein, In-Gann Chen, and Jia Liu
Institute for Beam Particle Dynamics and
Texas Center for Superconductivity
University of Houston, Houston, Texas 77204-5506

Abstract

Research on the Very Incomplete Meissner Effect has been pursued with the aim of learning whether monolithic high Tc superconductor materials can be developed to provide candidate magnets for external beam lines. Presently, materials exist to fabricate magnets of 1.5T, with materials for 3.0T probable within a year, and 6.0T possible in the future. Stability of the order of a few percent per year is readily attainable.

Introduction

An R and D program to develop practical magnet replicas based upon high T_c superconductors has been in progress for about two years. Usually superconductors (SC's) expel magnetic field (Meissner Effect), but type II SC's may trap the field. The field can be "expelled" into the small imperfections in the SC. When the external field is turned off, the collapse of such internal flux lines is prevented by Faraday's Law, and their movement is restrained by pinning forces. The effect is referred to as the Very Incomplete Meissner Effect (VIME) [1]. VIME was studied in low T_c SC's in the early 1970's [1]. We study VIME in high T_c materials. Magnets based upon VIME in high T_c materials have several promising advantages: (i) Replication: If there are many trapping centers per unit volume, the captured field, for example inside a cylinder, is an essentially perfect replica of even a complex activating field. (ii) Weight: The resulting magnet replicas are light. For example, an iron electromagnet for beam steering, weighing several tons, can be replaced by a replica composed of $YBa_2Cu_3O_7$ (1,2,3 material) weighing a few hundred pounds. (iii) Power: No power is needed for magnet replicas, once activated, except for replacement of liquid nitrogen (LN). (iv) Cryostats: Cryostat requirements are greatly simplified due to use of LN. (v) High Field: The only upper limit known at the present time to high T_c replica fields is about 300,000 G. Recent data indicates that magnet replicas of at least 60,000 Gauss are achievable. As we shall describe below, materials for 15,000 G magnets have already been developed. (vi) Low Inventory: For steering magnet applications, the same magnet replica can be activated as a quadrupole in one application, and a dipole or sextupole in the next. (vii) Low Capital Cost: Replica cost estimates typically run about 10-30% of the cost of the parent magnet. (viii) No Wire Needed: Magnet replicas require only bulk material and can be made even if the wire problem in 1,2,3 material proves insurmountable.

Progress in Increasing Trapped Field

Our work began with sintered samples in which 0.3 Gauss was trapped. Table 1 shows our progress to date. Prior to spring 1990, a steady improvement of materials resulted in an increase in trapped field by a factor of 20,000, from 0.3 to 6,000 Gauss [2]. At that time the best materials were melt-textured samples of $YBa_2Cu_3O_7$ (1,2,3 material), following the method of Salama, et.al. [3]. More recently, in an experiment done at the Harvard Cyclotron, we have increased the number of pinning centers in the SC by proton bombardment. Small tiles of sintered and melt-textured material were exposed to a beam of 160 MeV protons. The melt-textured samples which were used in this bombardment were capable, before exposure, of fabrication to produce a 3,000 Gauss magnet. Each typical small tile could itself capture about 300 Gauss. After exposure to about 10^{16} protons/cm^2 single tiles retained as much as 1,500 Gauss. With the bombarded materials it would be possible to fabricate a 15,000 Gauss magnet replica. If one defines the ratio of the retained magnetic field after bombardment to that before bombardment to be R, the bombardment results are reasonably well represented by [4]

$$R \approx 1 + (R_{max} - 1)(1 - e^{-F/F_0}), \quad (1)$$

where F is the fluence of protons in units of protons/cm^2, and $F_0 \approx (6 \pm 2) \times 10^{15}$ protons/cm^2.

A prototype mini-magnet was made from several bombarded tiles. The predicted and observed values of the trapped field in this mini-magnet were about 4,300 Gauss. Fig. 1 shows activation curves for this mini-magnet and for a single tile. What is shown is the trapped field, B_T, as a function of the activating field, B_0. In our early research [2] it was discovered that B_T and B_0 are related by

$$B_T = B_{T,max}(1 - e^{-B_0/B_{T,max}}) \quad (2)$$

Table 1. Progress of Magnetic Field Trapped within Y-123 HTS at 77K at IBPD/TCSUH

Date	Field Trapped on Disk (Gauss)	Field in Cylinder (Gauss)	Comment
12/88	--	0.3*	Poor Y-123 sintered ceramic
2/89	2	10*	Good Y-123 sintered ceramic
4/89	4.5	18*	Optimum thickness Y-123 sintered
5/89	10	50	Y-123/Ag composite, sintered
6/89	80	700	Y-123/Ag composite, partially melt grown, sub-mm size grain
8/89	380	4,000	Melt texture Y-123, sub-cm size grain, off ab plane
9/89	600	6,000	Melt texture Y-123, sub-cm size grain, ab plane
10/89	1,280	6,000	Same as 9/89, with limited fabrication
6/90	1,500	15,000	Proton irradiated, melt texture Y-123, cm size grain, ab plane
7/90	4,360	15,000	Same as 6/90, with limited fabrication

* Cylindrical sample tested. (Entries under cylinders, having no *, are calculations based on Vortex current and Persistent current model including non-linear magnet field effect.

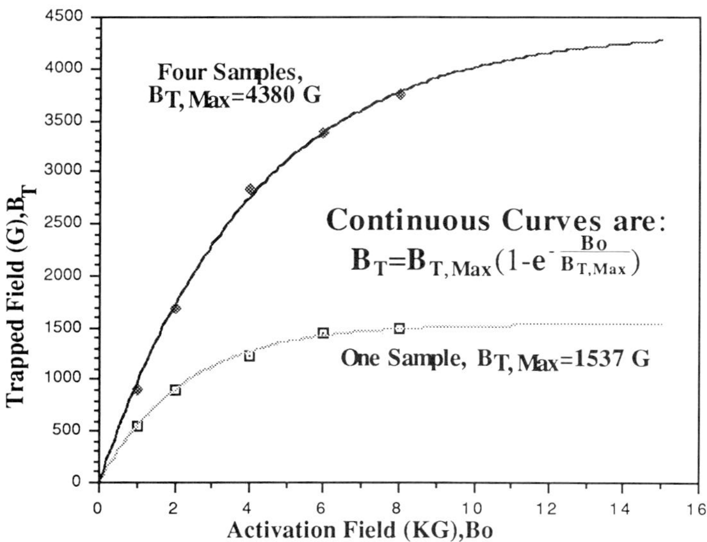

Fig. 1. Activation curve of one sample, and mini-magnet composed of four samples.

where $B_{T,max}$ is the maximum trappable field in the sample used. This phenomenological law, which constrains the design of the activating magnet, has held with great accuracy for all of the materials we have tried.

Stability

It is known that the field trapped in a magnet replica decreases with time according to the law

$$B_T(t) = B_T(0) - C \log t \tag{3}$$

We experimented [2] with a wide variety of 1,2,3 material including sintered, silver mixed sintered, and melt textured, and found that $C/B_T(O)$ was almost identical for all of these materials _if_ they were initially activated to $B_{T,max}$. If t, in Eq. 3, is in minutes, our observations are consistent with

$$C/B_{T,max} \sim 0.043 \pm 0.009 \tag{4}$$

Eqs. 3 and 4 indicate that a magnet replica can be stabilized to the order of a few percent per year by simply "aging" it, for example for several months, before it is used. Another method of stabilization was found [2]. Eq. 4 holds if $B_T(0) = B_{T,max}$. We also learned that if $B_T(0) < B_{T,max}$

$$C \propto (B_T(0)/B_{T,max})^2 \tag{5}$$

Figure 2 is a graph showing the time behavior of one sample with two initial activations, $B_{T,max}$, and 27% of $B_{T,max}$.

Models of Current Flow, and Measurement of J_c

We have measured the trapped fields surrounding sample tiles and modeled the current flow in the material in an attempt to theoretically calculate the fields. It is found that

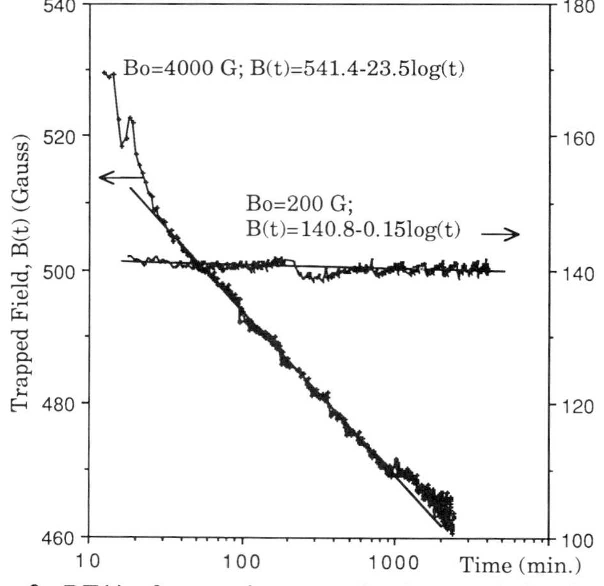

Fig. 2. BT(t), for maximum activation, and for 27% activation.

the observed field can be very well reproduced by a current model with two components [5]. First, we assume a surface current density, J_S, on the sample. This Amperean current successfully models the fields around, for example, SmCo permanent magnets. In melt-textured 1,2,3 material, however, a second current component is needed. We assume for this a uniform volume current, J_V, flowing in concentric circles in the sample, axially around the field direction (see Fig. 3a). A fit to the field is required, and best values of J_S and J_V are deduced. These fits can be seen, for one example, in Fig. 3. The presence of the J_V term makes the trapped field around a tile of 1,2,3 material more peaked than for a permanent ferromagnet of the same aspect ratio.

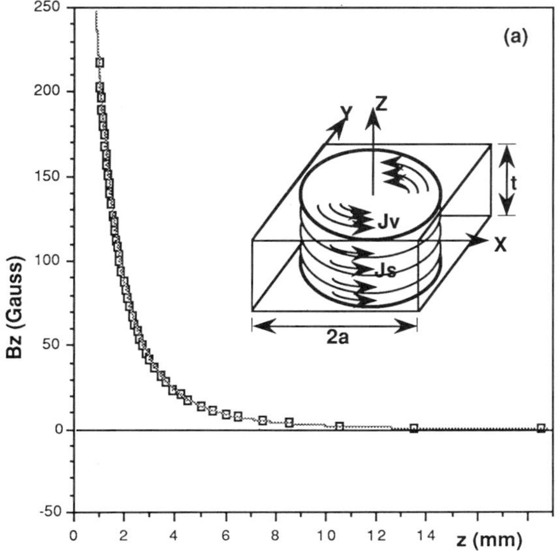

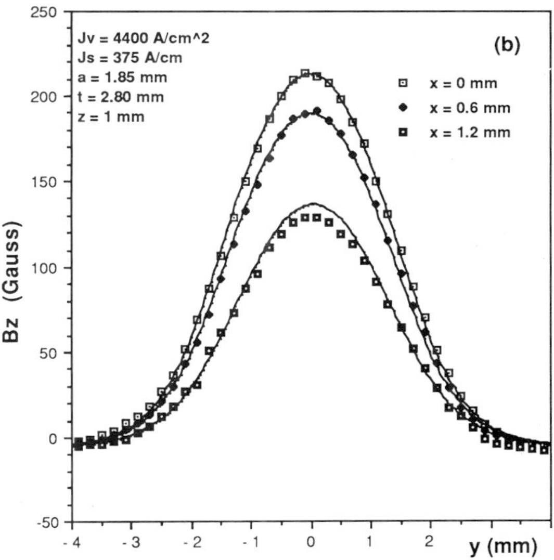

Fig. 3. Measured and calculated magnetic field of (a) varying z, and (b) varying x and y. Insert shows the schematic illustration of Js + Jv model.

Once J_S and J_V are known, the magnetic moment of the sample can readily be calculated [5], and is found to be $m = \pi a^2 t/10 (J_V a + J_S)$. We find the moment, calculated in this way, is the same as that found experimentally by the use of Vibrating Sample Magnetometer (VSM) or SQUID devices. In VSM measurements, the magnetic moment is interpreted by the Bean critical state model [6] to yield J_C. In the same way, magnetic profile measurements of B(x,y,z) around a magnetized sample yield the magnetic moment, and from this, J_C. As an example of the J_C's in recent samples, we find that a melt textured, irradiated sample, capable of making a tile which trapped 1,300 G, had $J_C \approx 26{,}000$ A/cm^2. The method described here, developed as an offshoot of the magnet replica work, appears to be more suitable to industrial production of large 1,2,3 monoliths than are VSM or SQUID methods.

Prototypes

Initially, because of budget, we have focused on prototyping very small devices. The mini-magnet characterized in Fig. 1 was discussed previously. We have also taken a small electric motor, excised its magnet, and replaced it with a mini-replica of about 1,000 G. We then operated the hybrid motor, drawing 3.5 Watts.

Conclusion

Magnet replicas, based upon VIME can produce high fields. Materials have been developed, using proton bombardment, which presently allow fabrication of replicas of 15,000 G. Material for thirty-thousand Gauss replicas appear probable within a year, and the collection of all present experimental evidence indicates that 60,000 G. will soon be achievable. Laws governing the activation, decay, and fabrication of magnetic replicas have been found. Decay of replica fields can be controlled by aging, or reduced by activating to less than the maximum trappable field.

References

[1] M. Rabinowitz, et.al., "Very Incomplete Meissner Effect", Lett. al Nuovo Cimento **7**, 1, (1973); Appl. Phys.Lett. **22**, 599, (1973); Appl. Phys. Lett. **30**, 607, (1977); IEEE Magn. **11**, 548, (1975).

[2] R. Weinstein, I.G. Chen, J. Liu, and D. Parks, "Persistent Magnetic Fields Trapped in High T_CSuperconductors" Appl. Phys. Lett. **56**, 1475, (1990).

[3] K. Salama, V. Selvamanickam, L. Gao, and K. Sun, "High J in Bulk Y1,2,3 Superconductor," Appl. Phys. Lett. **54**, 2,352, (1989).

[4] R. Weinstein, I.G. Chen, J.Liu, "Permanent Magnets of Y1,2,3 Material Bombarded with 160MeV Protons" to be submitted to Appl. Phys. Lett.

[5] I.G. Chen, J. Liu, and R. Weinstein, "Trapped Magnetic Field and Related Transport Critical Current Density in Y1,2,3 Superconductors" submitted to Physics C.

[6] C.P. Bean, "Magnetization of Hard Superconductor," Phys. Rev. Letters **8**, 250, (1962).

Conference Program

Monday, October 15 • Morning Session

8:30 A.M.	Plenary Session	Atlantic and Gulf Rooms
8:30 A.M.	Welcome	*Bob Bolen* *Mayor of Fort Worth*
8:45 A.M.	DOE's Welcome	*Gregory Haas* *Chief Scientist,* *SSC Project Office, U. S. DOE*
9:00 A.M.	The SSC Project	*Roy Schwitters* *SSC Laboratory*
9:45 A.M.	Research Program at the SSC	*Fred Gilman* *SSC Laboratory*
11:00 A.M.	SSC Detector R&D Program	*Hugh Williams* *Univ. of Pennsylvania*

Monday, October 15 • Afternoon Session

1:30 P.M. Parallel Sessions

Room	Session	Chairman
Room L3	Silicon Tracking Devices	*H. Sadrozinski* *Univ. of California–Santa Cruz*
Room L2	Wire Chambers	*W. Ford* *University of Colorado*
Room L3	Scintillating Fiber Tracking	*J. Elias* *Fermilab*
Ballroom	Liquid Ionization and Silicon Calorimetry	*P. Mockett* *Univ. of Washington*
Room L6	Scintillation Calorimetry	*H. Kasha* *SSC Laboratory*
Room L1	Front-End, DAQ, and Triggering Electronics	*J. Dorenbosch* *SSC Laboratory*
Room L7	Superconducting Magnets	*T. Fields* *Argonne National Laboratory*
Room L7	Computing and Simulations	*A. Gabriel* *Oak Ridge National Laboratory*

Room L6	Particle Identitication and Muon Systems	*A. Skuja* *University of Maryland*
Room L5	Radiation Damage	*D. Groom* *LBL*
Room L5	Detector Engineering	*R. Stefanski* *SSC Laboratory*

Tuesday, October 16 • Morning Session

8:30 A.M.	Parallel Sessions	
10:00 A.M.	Industrial Exhibits	Great Lakes and Pacific Rooms

Tuesday, October 16 • Afternoon Session

1:30 P.M.	Parallel Sessions	

Wednesday, October 17 • Morning Session

7:45 A.M.	SSC Coalition Panel	Atlantic and Gulf Rooms
	Panel Members:	*Hollye Doane* *National SSC Coalition*
		Henry Gandy *Americans for the SSC*
		Gordon Kane *University of Michigan*
		Joe Karpinski *U. S. Department of Energy*
		Roy Schwitters *SSC Laboratory*
8:30 A.M.	Parallel Sessions	
10:30 A.M.	Plenary Sessions	Atlantic and Gulf Rooms
10 :30 A.M.	Warm Liquid Calorimetry	*Morris Pripstein* *LBL*
11:00 A.M.	Scintillating Fiber Calorimetry	*Richard Wigmans* *CERN*
11:30 A.M.	Plastic Scintillating Calorimetry	*G. Wliiam Foster* *Fermilab*

Wednesday, October 17 • Afternoon Session

1:30 P.M.	Liquid Argon Calorimetry	*Howard Gordon* *BNL*
2:00 P.M.	Silicon Calorimetry	*Fier Giorgio Rancoita* *INFN-Universita de Milano*
2:30 P.M.	Precision Electromagnetic Shower Detectors	*Harvey Newman* *CalTech*
3:30 P.M.	Calorimetry at High Luminosity	*Veljko Radeka* *BNL*
4:00 P.M.	VLSI Electronics Developments	*Michael Levi* *LBL*

Thursday, October 18 • Morning Session

8:30 A.M.	Plenary Session	Atlantic and Gulf Rooms
8:30 A.M.	Wire Chamber Tracking	*Harold Ogren* *Indiana University*
9:00 A.M.	Silicon Tracking	*Abraham Seiden* *Univ. of California - Santa Cruz*
9:30 A.M.	Scintillating FiberTracking	*Randall Ruchti* *Univ. of Notre Dame*
10:00 A.M.	Precision Muon Tracking	*Ulrich Becker* *M.I.T.*
11:00 A.M.	Superconducting Toroids	*Michael Marx* *SUNY at Stony Brook*
11:30 A.M.	The Challenge of High Luminosity	*Murdock Gilchriese* *LBL*

Dr. Roy Schwitters, Director of the SSC Laboratory, pours a toast to Dr. Jerome Friedman at the Symposium, following announcement of Friedman's Nobel Prize.

Valerie Kelly
5-94

List of Plenary and Parallel Sessions

Plenary Talks (Ballroom)

	October 15		
9:00	R. Schwitters	SSCL	The SSC Project
9:45	F. Gilman	SSCL	The Research Program at the SSC
11:00	H. Williams	Pennsylvania	The SSC Detector R&D Program
	October 16		
1:00	M. Meyerson	TNRLC	Texas Supports the SSC
	October 17		
10:30	M. Pripstein	LBL	Recent Developments in Warm Liquid Calorimetry
11:00	R. Wigmans	CERN	Scintillating Fiber Calorimetry
11:30	G. W. Foster	Fermilab	Scintillating Tile Calorimetry
1:30	H. Gordon	BNL	Progress in Liquid Argon Calorimetry
2:00	P-G. Rancoita	Milan	Silicon Calorimetry
2:30	H. Newman	Cal Tech	Electromagnetic Shower Detectors
3:30	V. Radeka	BNL	Calorimetry at High Luminosity
4:00	M. Levi	LBL	VLSI Electronics Development
	October 18		
8:30	H. Ogren	Indiana	Wire Drift Chamber Tracking at the SSC
9:00	A. Seiden	UC Santa Cruz	Developing Silicon Tracking for the SSC
9:30	R. Ruchti	Notre Dame	Scintillating Fiber Tracking
10:00	U. Becker	MIT	Precision Muon Tracking
11:00	M. Marx	SUNY-SB	Superconducting Toroids
11:30	M.G.D. Gilchriese	LBL	The Challenge of High Luminosity

Silicon Tracking Devices (Room L3)
H. Sadrozinski, University of California Santa Cruz, Chairman

	October 15		
1:30	H. Spieler	LBL	Short Strip Detectors and Integrated Readout Electronics
2:00	P. Skubic	Oklahoma	Vertex Detector Technology for Bottom Physics at the SSC
2:20 VLSI	H. Attias	Yale	Beam Tests of Silicon Microstrip Detectors With Readout
2:40	V. Cook	Washington	Precision Vertex Chamber for Detector Without a Central Magnetic Field
3:30	P. Rehak	BNL	Silicon Drift Detectors for the SSC Environment
4:00	C. Lu	Princeton	Silicon Drift Chambers
4:15	D. Nygren	LBL	Pixel Vertex Detector Technology
4:45	J. Arens	LBL	Performance Measurements of Hybrid Pin Diode Arrays
5:00	O. Barkan	Hughes	Development of a Customized SSC Pixel Detector Readout for Vertex Tracking
5:15	V. Perez-Mendez	LBL	Position Sensitive Detection of Charged Particles by Amorphous Silicon Diode Layers

**This list does not reflect the correct order in which talks were given at the Symposium.*

October 16

8:30	Z. Li	BNL	Radiation Damage Studies on Silicon Detectors at BNL
8:45	D. Pellet	UC Davis	Study of Radiation Damage to PIN Photodiodes Using 63 MeV Protons
9:00	J. Kapustinsky	LANL	Operation of Silicon Microstrip Detectors in a High Radiation Environment
9:15	M. Laakso	Helsinki	Radiation Damage Measurements in Double-Sided Si Strip Detectors
9:30	C. Storment	Stanford-CIS	Indium Bump Bonding for Silicon Tracking Devices
9:45	H. Sadrozinski	UC Santa Cruz	Radiation-Hard Electronics and Silicon-Strip Detectors
11:00	H. Ziock	LANL	Cabling for an SSC Central Tracking Silicon Detector System
11:15	C. Pfeiffer	Hughes	A Signal Processing Architecture to Support Fast, Primary and Secondary Vertex Finding
11:30	W. Kinnison	LANL	Monte Carlo Simulations of Central Tracking Sub-System

Wire Chambers (Room L2)
W. Ford, University of Colorado, Chairman

October 15

1:30	V. Polychronakos	BNL	Particle Identification and Tracking with Transition Radiation Detectors
2:00	J. Palathingal	Puerto Rico	Optimization of the Transition Radiation Detector
2:15	R. Bock	CERN	Transition Radiation Detector as a Triggering Device
2:30	D. Saxon	Glasgow	Forward Tracking with Enhanced Electron Identification
2:45	R. Debbe	BNL	Study of MWPCs with Interpolating Pad Readout for High Multiplicity Charged Particle Detection
3:30	H. Van der Graaf	NIKHEF-H	Results of the Test of the Honeycomb Strip Chamber
3:45 for	P. McIntyre	Texas A&M	Knife-Edge Chamber: High Resolution Tracking SSC Detectors
4:00	A. Maki	KEK	The Knife-Edge Chamber R&D at KEK

October 16

8:30	W. Robertson	Duke	Study of the Electrostatic Stability and Wire Support in Long Straw Drift Tubes
8:45	S. Oh	Duke	Construction of a Prototype 3 Meter Long Straw Tube Superlayer
9:00	S. Oh	Duke	Performance of a Prototype 3 Meter Long Straw Tube Superlayer
9:15	C. Lu	Princeton	Investigations of a Single-electron's Behavior in a Proportional Drift Tube
9:30	W. Ford	Colorado	Design of Readout Electronics Interface for High Density Straw Tube Arrays
9:45	R. Henderson	TRIUMF	Etching of Anode Wire Deposits with CF4/Isobutane
10:00	J. Kadyk	LBL	Recent Work on Radiation-Hard Gases and Straw Tubes
1:30	S. Oh	Duke	Hybrid Tracking Systems
2:00	G. Hanson	Indiana	Design of a Tracking System for a Solenoidal Detector
2:15	W. Ford	Colorado	Pattern Recognition in Tracking Systems with Small Drift Cell Arrays

2:30	M. Corden	FSU-SCRI	Properties of a Straw Tube/Scintillating Fiber Hybrid Central Tracking Chamber from Computer Simulation
2:45	M. Mermikides	FSU-SCRI	Track Finding in a Hybrid Central Tracking Chamber
3:30	J. Armitage	Princeton	A Straw-Tube Tracking System
3:45	J. Bailey	Liverpool	Measurements of Electron Drift in Fast Gases
4:00	A. Sill	Rochester	Advanced Field Shaping Drift Chambers for SSC Muon Tracking
4:15	B. Yuldashev	Physical Technical Institute	A Drift Velocity Monitor System for Wire Chambers

Scintillating Fiber Tracking (Room L3)
J. Elias, Fermi National Laboratory, Chairman

October 16

1:30	C. Anway	UC Los Angeles	Simulations of SSC Events with Scintillating Fiber Tracking
1:50	D. Adams	Rice	Simulation of a Scintillating Fiber Tracker for the SSC
2:10	R. Lewis	Pennsylvania State	Simulation Studies for a Scintillating Fiber Tracker
2:30	R. Mountain	Notre Dame	Monte Carlo Simulations of Fiber Waveguides
3:20	I. Leedom	Northeastern	A Plastic Scintillating Fiber Beam Hodoscope
3:40	K. Arisaka	UC Los Angeles	High Rate Tracking Detectors Based on Scintillating Fiber Technology
4:00	J. Piekarz	Notre Dame	Measurements of Flourescence Efficiency of Scintillators Before and After Irradiation
4:20	C. Park	Florida	Fabrication of a Step-Index Scintillating Optical Fiber By a Coextrusion Process
4:40	S. Reucroft	Northeastern	Avalanche Photodiodes as Plastic Scintillating Fiber Readout Devices
5:00	M. Salomon	TRIUMF	Preliminary Results on Avalanche Photodiodes for Scintillating Fiber Readout
5:20	S. Suzuki	Hamamatsu	New Position Sensitive Photomultiplier Tubes

October 17

8:30	M. Atac	FERMILAB	Status of Solid State Photomultipliers for Scintillating Fiber Tracking
9:00	E. Fenyves	UT-Dallas	Cryogenic Preamplifiers for SSPM/VLPC Devices for Scintillating Fiber Tracking
9:20	S. Margulies	UI-Chicago	Cryogenics Issues for SSPM Operation
9:40	J. Woosley	Teledyne	Optical Subsystems for Scintillating Fiber Readout

Liquid Ionization and Silicon Calorimetry (Ballroom)
P. Mockett, University of Washington, Chairman

October 15

1:30	J. Brau	Oregon	Recent Developments in Silicon Calorimetry
2:15	S. Berridge	Tennessee	The Design and Construction of a 3.5-m Squared EM Silicon Calorimeter
2:30	W. Bugg	Tennessee	Radiation Effects in Silicon Calorimetry
2:45	J. Russ	Carnegie-Mellon	Radiation Damage Effects in Silicon Electromagnetic Calorimetry
3:30	J. Korde	ORNL	Low Cost Large Area Silicon Detectors for Calorimetry
3:45	S. Berridge	Tennessee	The Design and Construction of an SSC Silicon Electromagnetic Calorimeter Prototype
4:00	G. Lindstroem	Hamburg	Application of Silicon Detectors for Calorimetry and Monte Carlo Studies

4:15	S. Schnetzer	Rutgers	Development of CVD Diamond Radiation Detectors
4:30	C. Beetz	Fairfield	Diamond Film Particle Detectors
4:45	R-Y. Zhu	Cal Tech	A Precision Barium Fluoride Crystal Calorimeter for the SSC

<u>**October 16**</u>

8:30	W. Cleland	Pittsburgh	Pileup and Thermal Noise in a Liquid Ionization Calorimeter
9:15	N. Giokaris	Rockefeller	High-Pressure Ionization Chambers for Calorimetry in High-Energy Physics
9:30	D. McCorkle	Tennessee	Cryogenic Liquid Mixtures for SSC Calorimetry
9:45	T. Adams	Martin-Marietta	Test and Modellingt of Heat Tranport in Liquid Argon
10:00	P. Lebedev	Novosibirsk	Time and Coordinate Resolution of LKr and LXe Ionization Chambers
10:15	J. Makowiecki	BNL	Liquid Argon/Lead Calorimeter Test
11:00	P. Mockett	Washington	Tests of a Novel Segmentation Scheme for Charge Sampling Calorimeters
11:15	G. Abrams	LBL	Liquid Argon Calorimeter Readout Via the Electrostatic Transformer
11:30	D. Rahm	BNL	A Novel Structure for a Liquid Argon Electromagnetic Calorimeter
1:30	R. Holroyd	BNL	Free Ion Yields for Several Silicon, Germanium and Tin Containing Liquids
1:45	H. Faidas	ORNL	Warm Liquids for SSC Calorimetry: Electron Transport and Electrical Properties
2:00	R. Hollebeek	Pennsylvania	Tests of Materials Compatability and Purification Procedures in TMS
2:15	H. Yuta	Tohoku	Material Compatibility in a Warm Liquid Calorimeter
2:30	H. Yuta	Tohoku	Radiation Damage in TMP and TMGe
2:45	G. Brandenburg	Harvard	Warm Liquid Calorimeter Prototype
3;30	K. Masuda	Saitama	Lifetime Measurement for Warm Liquid Ionization Chamber
3:45	K. Furuno	Oregon	Neutron Flux Suppression with Polyethylene Moderators in the EMPACT Silicon EM Calorimeter
4:00	P. Mockett	Washington	Electromagnetic Shower sampling Efficiency for Iron Clad Lead Plate Calorimeters
4:15	D. Groom	LBL	Calorimeter Response to an Incident Jet
4:30	T. Ferbel	Rochester	Monte Carlo Studies in Calorimetry
4:45	Y. Kamyshkov	ITEP	Results of Beam Tests of a Prototype Sampling Hadron Calorimeter with TMS
5:00	Y. Kamyshkov	ITEP	Planar Proportional Chambers for Fast Calorimeter Response

<u>**Scintillation Calorimetry**</u> **(Room L6)**

H. Kasha, Superconducting Super Collider Laboratory, Chairman

<u>**October 16**</u>

8:30	H. Paar	UC San Diego	Test Beam Results of a 20-Ton Lead-Fiber Calorimeter
9:00	A. Para	Fermilab	Performance of a Projective Scintillating Fiber Calorimeter
9:30	J. Woosley	Teledyne	Signal and Timing Requirements for a Single-Tier Scintillating Fiber Calorimeter
9:45	D. Winn	Fairfield	Properties of Photodetectors for scintillating Calorimeters at the SSC
10:00	P. Cushman	Yale	Pre-Radiators for Electron Identification

10:15	R. Rusack	Rockefeller	Results from the Fermilab Pre-Radiator Beam Tests
11:00	Y. Kamyshkov	ITEP	Liquid Scintillator Used as an Active Medium in a Sampling Calorimeter
11:15	W. Worstell	Boston	Cast Lead-Eutectic Solid and Liquid Scintillating Fiber Shower Calorimeters
11:30	C. Bromberg	Michigan State	A Dichromatic Fiber Calorimeter
1:30	M. Chen	MIT	Precision EM Detector Using Scintillating Liquid Xenon
2:00	J. Proudfoot	ANL	Simulation Studies for Design Optimization of a Scintillator Plate Calorimeter
2:30	D. Reeder	Wisconsin	Evaluation of Absorber Materials for a Scintillating Plate Calorimeter
2:45	R. McNeil	Louisiana State	Scintillating Plate Calorimeter Optical Design
3:30	E. Rosenberg	Iowa State	Design of Readout Electronics for a Scintillating Plate Calorimeter
3:45	V. Barnes	Purdue	Scintillator Calorimeter
4:00	P. Altiec	Texas A&M	A Liquid Scintillator Alternative to Scintillating Fiber Calorimetry for the SSC
4:15	J. White	Texas A&M	A Liquid Scintillator Spaghetti Calorimeter EM Module for Forward Calorimetry at the SSC
4:30	A. Sanzgiri	Texas A&M	Tests of a Prototype Liquid Scintillator Based Scintillating Fiber Electromagnetic Calorimeter
4:45	W. Frisken	York	The ZEUS Calorimeters: Pro-duction and Test Beam Calibration

Front-End, DAQ, and Triggering Electronics (Room L1)
J. Dorenbosch, Superconducting Super Collider Laboratory, Chairman

October 15

1:30	C. Britton	ORNL	Bipolar Monolithic Preamplifiers for SSC Silicon Calorimetry
1:48	A. Wintenberg	ORNL	Considerations on the Design of Front-End Electronics for SSC Silicon Calorimetry
2:06	V. Radeka	BNL	Development of Low Noise Junction Field Effect Transistors (JFET) and Monolithic JFET Preamplifiers
2:24	E. Stebbins	Hypres, Inc.	Applications of Superconducting Electronics for the SSC
2:42	V. Budinhartono	Pennsylvania	CMOS Analog Building Blocks for SSC Applications
3:30	F. Newcomer	Pennsylvania	The Low Power Timing Discriminator for SSC Applications
3:50	F. Newcomer	Pennsylvania	A Prototype Monolithic Preamp and Shaper for Straw Tracking
4:10	J. Paulos	North Carolina State	Front-End Electronics for Long Straw Tube Systems
4:30	L. Callewaert	Pennsylvania	A Time to Voltage Converter and Analog Memory Unit for Straw Tracking Detectors
4:50	P. Sinervo	Pennsylvania	A Model for the Architecture of a Data Driven Front-End Circuit
5:10	M. Levi	LBL	A 16-Channel Analog Pipeline Chip

October 16

8:30	M. Lowry	LLNL	Electro-Optic Transient Imaging Instrumentation Development at LLNL: Implications for SSC Instrumentation Development
8:50	M. Lowry	LLNL	An Electro-Optical Imaging Approach to Prompt Signal Processing Problem of Mega-Channel SSC Detector Arrays
9:10	B. Leskovar	LBL	Optical Fiber Networks for the SSC Detector
9:30	B. Cook	LANL	SSC High Speed Communication Links and Interconnects
9:50	J. Ogle	LANL	Applications of Electyro-Optic Modulators for Wide Bandwidth Recording of Transient Signals
10:10	Y. Ofek	IBM	High Speed Network for Real-Time Data Acquisition and Control with Lossless Balanced Self-Routing
11:00	M. Atiya	BNL	A Study of Electro-Optical Modulators for Transfer of Detector Signals by Optical Fibers
11:15	M. Johnson	Teleldyne	Use of Non-Volatile Memories for SSC Detector Readout
11:30	E. Barsotti	Fermilab	Switch-Based DAQ System Prototype Project: A Progress Report and the Application of Switches from Industry to High-Energy Physics Event Building
1:30	S. Li	Florida	An Ultra High-Gain Low-Noise AlGaAs/GaAs SAM Avalanche Photodiode for Use in SSC Light Detection from Scintillating Plastic Fibers
1:48	W. Nunnally	UT-Arlington	Optically Controlled Switches
2:06	J. Elias	Fermilab	First-Level Triggering with Scintillating Fibers
2:24	K. Kondo	Tsukuba	Analog Triggering with Scintillating Fibers
2:42	K. Lackner	LANL	Data Processing at the SSC With Structured Neural Nets
3:30	G. W. Foster	Fermilab	Pipelined Fast Trigger with Scintillating Fibers
3:52	S. Clearwater	Pittsburgh	Studies of Trigger Requirements Using a Rule Learning Program
4:15	M. Campbell	Michigan	Neural Net Based B-Trigger
4:45	C. Peterson	Lund	Using Neural Networks to Identify the Origin of Jets
5:07	N. Nikityuk	Dubna	Some Questions of Using the Algebraic Coding Theory for the Construction of Special-Purpose Processors in High Energy Physics Spectrometers

October 17

8:30	H. Crawley	Iowa State	Performance of Flash ADCs in the 100 MHz Range
8:50	D. Bryman	TRIUMF	GaAs CCD 500 MHz Digitizers
9:10	E. Heijne	CERN	HARP: Hierarchial Analog Readout Processor with Analog Pipelining in CMOS
9:40	H. Lubatti	Washington	Acoustic Charge Transport (ACT) Delay Lines for SSC Applications

Superconducting Magnets (Room L7)

T. Fields, Argonne National Laboratory, Chairman

October 16

1:30	T. Brown	Grumman	Toroid Design for the EMPACT Detector
2:00	J. Luton	ORNL	Feasibility of Low-Mass Conductors for Toroids
2:15	J. Purcell	Cryo-Mags	Conceptual Design of a Toroid for the SDC Detector
2:35	S. Mori	KEK	SDC Toroid Design Work in Japan
3:30	G. Susinno	CERN	Toroid Designs for LHC Detectors
4:00	C. Henning	LLNL	SC Toroids for Fusion and Detectors

4:30	*Roundtable:*		Discussion of Toroid Designs
	Z. Stekly	Intermag	
	P. Marston	MIT	
	C. Henning	LLNL	
	R. Coombes	SSCL	
	October 17		
8:30	P. Bonanos	Princeton	Design of Radiation Thin Air-Core Toroids
8:45	J. Pusteri	Grumman	Overview of a Muon Spectrometer Based on SC Toroids
9:00	S. Gottesman	Grumman	Monte Carlo Study of a SC Air Core Toroid Spectrometer
9:15	T. Fields	ANL	Matching Forward Toroids to a Central Solenoid
9:30	P. Marston	MIT	Magnet Design and Technology for Large Scale Detectors
9:45	R. Fast	FERMILAB	Comparison of Solenoids for SSC Detector

Computing and Simulations (Room L7)
A. Gabriel, Oak Ridge National Laboratory, Chairman

	October 15		
1:30	A. Gabriel	ORNL	Introductory Comments
1:40	P. Leibold/ B. Scipioni	SSCL	Computing Facility at SSC for Detectors
2:10	L. Price	ANL	Detector Simulations
2:40	P. Keener	Pennsylvania	Benchmarks for an i-860 Parallel Computing Ranch
3:30	R. Wilson	Boston	Massively Parallel Computing in High Energy Physics
3:50	T. Handler	Tennessee	CALOR89 Benchmarking Results
4:30	S. Youssef	FSU-SCRI	Detecting Errors in GEANT Geometries
4:50	A. Yacout	Quantum	Development of a Robust CAD-to-GEANT Interface
5:10	A. Palounek	LBL	Electron ID at the SDC Detector
5:30	A. Para	Fermilab	A Fast Package for SSC Detector Simulation
5:50	Concluding Discussion		
	October 16		
8:30	S. Loken	LBL	Software Methodologies for the SSC
9:00	W. Atwood	SLAC	Simulations Using Object Oriented Programming
9:30	S. Linn	FSU-SCRI	A Detector Simulation Package for SSC Detectors
9:50	L. Cremaldi	Mississippi	Systematic Errors in Calorimetry Simulations
10:10	K. Killian	Martin-Marietta	Integrated 3D Modelling, Engineering Analysis and Physics Simulation
11:00	S. Qian	CERN	Simultaneous Pattern Recognition and Track Fitting By Kalman Filtering
11:20	M. Johnson	Teledyne	Optical Computering Utilization at the SSC
11:40	C. Glover	ORNL	Pattern Recognition Utilizing Neural Network Algorithms

Particle Identification and Muon Systems (Room L6)
A. Skuja, University of Maryland, Chairman

	October 15		
1:30	K. Heller	Minnesota	Muon Systems
2:00	K. Lau	Houston	Streamer Chambers for Muon Tracking at the SSC
2:15	Y. Tsuda	Nippon Steel	Manufacturing High-Grade Ultra-Heavy Electrical Steel Plate and Investigation Results
2:30	L. Christophorou	ORNL	Gases for an SSC Muon Detector

2:45	M. Alam	SUNY-A	Design of a 50ps Time-of-Flight System
3:30	F. Grancagnolo	INFN-Lecce	The Resistive Plate Counters (RPC) as a Muon Detector
3:45	A. Pemmaraju	Stevens	Megagauss Physics and the SSC
4:00	J. Morfin	Fermilab	A Fast RICH Counter for BCD
4:15	R. Lander	UC Davis	Muon Systems
4:30	A. Sill	Rochester	Advanced Field Shaping Drift Chambers for SSC Muon Tracking

Radiation Damage (Room L5)
D. Groom, Lawrence Berkeley Laboratory, Chairman

October 15

1:30	R. Clough	Sandia	Radiation Degradation of Organic Scintillators
2:00	A. Bross	Fermilab	Radiation Damage to Organic Scintillator
2:15	C. Zorn	CEBAF	Results of High and Low Dose Rate Irradiations of Prototype Radiation-Resistant Plastic Scintillating Fibers
2:30	W. Dunn	Quantum	Radiation Damage Studies of Straw Tube and Scintillating Fiber Elements
2:45	K. Johnson	Florida State	Radiation Hardness Tests of a Pb/SciFi Calorimeter
3:30	V. Hagopian	Florida State	Radiation Damage Tests of New Scintillating Fibers
3:45	M. Contreras	Chicago	Radiation Damage to SCSN23, SCSN81 and SCSN81+Y7 Scintillators
4:00	J. White	Texas A&M	Effects of Radiation-Induced Discoloration of the Cladding on Liquid Scintillating Fibers
4:15	A. Sanzgiri	Texas A&M	A Study of the Radiation Hardness of Materials for Use in a Liquid Scintillator Based Calorimeter
4:30	J. Walker	Florida	Scintillator Development at University of Florida
4:50	J. Harmon	Florida	Polymeric Materials for Scintillator at the SSC

October 16

8:30	J. Russ	Carnegie-Mellon	Neutron Fluences in Iron or Lead Calorimeters at the SSC
8:45	F. Alsmiller	ORNL	HECC: A Phenomenonological Hadron-Nucleus High Energy Cascade Collision Model
9:00	D. Lee	LANL	A Prediction of the Neutron and Charged Particle Backgrounds in the L* Detector
9:15	P. McNulty	Clemson	Characterizing Complex Radiation Environments Using MORE (Monitor of Radiation Effects)
9:30	H. Sadrozinski	UC Santa Cruz	Radiation Hard Electronics
9:45	E. Kennedy	ORNL	Radiation Effects on JFETs, MOSFETs and Bipolar Transistors as Related to SSC
10:30	S. Rescia	BNL	Radiation Damage Studies on Hybrid Preamplifiers
10:45	A. Stevens	ANL	Radiation Damage Testing of Transistors for SSC Front-End Electronics
11:00	A. Kanofsky	Lehigh	Radiation Studies of LeCroy High Energy Circuits

Detector Engineering (Room L5)
R. Stefanski, Superconducting Super Collider Laboratory, Chairman

October 16

1:30	M. Harris	SSCL	Detector Engineering
2:00	M. Rennich	Martin Marietta	Mechanical Design of the L* Central Hadron Calorimeter
2:15	H. Spinka	ANL	Scintillating Plate Calorimeter Mechanical Design
2:30	R. Swensrud	Westinghouse	SDC Cast Lead Calorimeter for the SSC

2:45	T. Ballinger	EG&G	Engineering Design Study for a Warm Liquid Calorimeter
3:30	N. Wells	SSCL	System Integration Experience
3:45	D. Vandergriff	ORNL	Stable Base Support for Central Tracking Chamber Components
4:00	R. Swensrud	Westinghouse	SDC Central Wire Tracking for the SSC
4:15	W. Stoeffl	LLNL	Alignment Calibration for Central Trackers
4:30	W. Miller	LANL	Mechanical Engineering and Design of Silicon Tracking Systems
4:45	D. Koltick	Purdue	Engineering Issues for Scintillating Fiber Tracking
5:00	D. Slack	LLNL	A Cryogenic System for the SSC L* Detector
5:15	L. Sulack	Boston	Fiber Calorimetry
	October 17		
8:30	N. DiGiacomo	Martin-Marietta	An Engineering Design Network for SSC Detector Development
9:00	L. Mason	Martin-Marietta	Engineering Design of the EMPACT Liquid Argon Calorimeter
9:15	J. Pohlen	Martin-Marietta	R&D and Manufacturing Issues for SPACAL Calorimetry
9:30	G. Guenterberg	Martin-Marietta	Assembly and Integration Issues for EMPACT
9:45	R. Weinstein	Houston	Monolithic High T_C Superconducting Permanent Magnets

List of Participants

ELDEN ABLES	LAWRENCE LIVERMORE NATIONAL LABORATORY
MARIS ABOLINS	MICHIGAN STATE UNIVERSITY
DARIN ACOSTA	UNIVERSITY OF CALIFORNIA - SAN DIEGO
DAVID ADAMS	RICE UNIVERSITY
TOM ADAMS	MARTIN MARIETTA
MOHAMMAD ALAM	SUNY - ALBANY
KAMBIZ ALAVI	UNIVERSITY OF TEXAS AT ARLINGTON
THEODOROS ALEXOPOULOS	UNIVERSITY OF WISCONSIN
GARY ALLEY	OAK RIDGE NATIONAL LABORATORY
FARINAZ AMIRANI	MITSUI PLASTICS, INC.
CHARLES ANDERSON	AIR PRODUCTS AND CHEMICALS, INC.
KATHY ANDERSON	SSC LABORATORY
CAROL ANWAY	UNIVERSITY OF CALIFORNIA - LOS ANGELES
STANISLAW AUGUSTYNOWICZ	SSC LABORATORY
JOHN ARENS	UNIVERSITY OF CALIFORNIA - BERKELEY
JOHN ARMITAGE	PRINCETON UNIVERSITY
KATSUSHI ARISAKA	UNIVERSITY OF CALIFORNIA - LOS ANGELES
MUZAFFER ATAC	FERMI NATIONAL ACCELERATOR LABORATORY
WILLIAM ATWOOD	STANFORD UNIVERSITY
NEIL BAGGETT	SSC LABORATORY
JOHN BAILEY	UNIVERSITY OF LIVERPOOL
SAMUEL BAKER	SSC LABORATORY
JON BAKKEN	MARTIN MARIETTA
ALBERTO BALDINI	I.N.F.N. - PISA
JOSEPH BALLAM	STANFORD LINEAR ACCELERATOR CENTER/SSC
CLINTON BALLINGER	LAWRENCE LIVERMORE NATIONAL LABORATORY
TERESA BALLINGER	EG&G/ENERGY MEASUREMENTS
PETER BONANOS	PRINCETON UNIVERSITY
RON BARBER	LOS ALAMOS NATIONAL LABORATORY
PHILIP BARINGER	UNIVERSITY OF KANSAS
OZDAL BARKAN	HUGHES AIRCRAFT COMPANY
VIRGIL BARNES	PURDUE UNIVERSITY
JOSEPH BARFOOT	ORION INTERNATIONAL TECHNOLOGIES
ED BARSOTTI	FERMI NATIONAL ACCELERATOR LABORATORY
MARTIN BAUER	OAK RIDGE NATIONAL LABORATORY
BARRY BAUMBAUGH	UNIVERSITY OF NOTRE DAME
ULRICH BECKER	MASSACHUSETTS INSTITUTE OF TECHNOLOGY
STEVE BERRIDGE	UNIVERSITY OF TENNESSEE
DAVID BERLEY	NATIONAL SCIENCE FOUNDATION
RAMESH BHARAT	ROCKWELL INTERNATIONAL CORPORATION
RICHARD BIONTA	LAWRENCE LIVERMORE NATIONAL LABORATORY
NIPREN BISWAS	UNIVERSITY OF NOTRE DAME
R.J. BLANKEN	IBM CORPORATION
CRAIG BLOCKER	BRANDEIS UNIVERSITY
RUDOLPH BOCK	CERN
PETER BONANOS	PRINCETON UNIVERSITY
BILLY BONNER	RICE UNIVERSITY
MICHI BOTLO	SSC LABORATORY
MAURICE BOURQUIN	UNIVERSITY OF GENEVA

JAMES BRAU	UNIVERSITY OF OREGON
CHARLES BRITTON	OAK RIDGE NATIONAL LABORATORY
CARL BROMBERG	MICHIGAN STATE UNIVERSITY
JOHN R. BROOKS	LECROY CORPORATION
THOMAS BROWN	GRUMMAN SPACE SYSTEM DIVISION
DOUGLAS BRYMAN	TRIUMF
WILLIAM BUGG	UNIVERSITY OF TENNESSEE
JOSEPH BURGER	MASSACHUSETTS INSTITUTE OF TECHNOLOGY
JEFFREY BULL	SSC LABORATORY
THOMAS BUSH	SSC LABORATORY
MARTIN BUTALLA	SSC LABORATORY
JULIAN BUDAGOV	JINR - DUBNA
KATERI CALLAHAN	INDUSTRY ASSOCIATION FOR THE SSC (IASSC)
MYRON CAMPBELL	UNIVERSITY OF MICHIGAN
CLAUDIO CAPRIO	IBM CORPORATION
DAVID CASSEL	CORNELL UNIVERSITY
ROY CHANEY	THE UNIVERSITY OF TEXAS AT DALLAS
YUANHANN CHANG	MASSACHUSETTS INSTITUTE OF TECHNOLOGY
GREG CHARTRAND	SSC LABORATORY
MIN CHEN	MASSACHUSETTS INSTITUTE OF TECHNOLOGY
LOUCAS CHRISTOPHOROU	OAK RIDGE NATIONAL LABORATORY
SELCUK CHIANGIR	FERMI NATIONAL ACCELERATOR LABORATORY
ALESSANDRA CIOCIO	LAWRENCE BERKELEY LABORATORY
SUSAN CLAUDE	SHEDAHL, INC.
SCOTT CLEARWATER	UNIVERSITY OF PITTSBURGH
WILFRED CLELAND	UNIVERSITY OF PITTSBURGH
MARK CLEMEN	UNIVERSITY OF PITTSBURGH
DAVID CLINE	UNIVERSITY OF CALIFORNIA - LOS ANGELES
ROGER CLOUGH	SANDIA NATIONAL LABORATORIES
TONY COCHRANE	DELFT ELECTRONICS CORPORATION
HANS COHN	OAK RIDGE NATIONAL LABORATORY
TIM COLLINS	HUGHES AIRCRAFT COMPANY
BRADLEY COOKE	LOS ALAMOS NATIONAL LABORATORY
VICTOR COOK	UNIVERSITY OF WASHINGTON
MARTYN CORDEN	FLORIDA STATE UNIVERSITY
LAIRD CORMELL	SSC LABORATORY
LARRY COULSON	SSC LABORATORY
DAVID COUPAL	SSC LABORATORY
BILL COURTNEY	IBM CORPORATION
BILL COVINGTON	SAM HOUSTON STATE UNIVERSITY
J. JOSEPH COYNE	NATIONAL INSTITUTE OF STANDARDS AND TECHNOLOGY
PRISCILLA CUSHMAN	YALE UNIVERSITY
JOHN DAINTON	LIVERPOOL UNIVERSITY
FREDERICA DAREMA	IBM CORPORATION
HENRI DAUTET	GE ELECTRO OPTICS
CHRISTOPHER DAY	LAWRENCE BERKELEY LABORATORY
PAWEL deBARABARO	UNIVERSITY OF ROCHESTER
ARTIE DIEBLING	EG&G
SERGEI DENISOV	IHEP - PROTOVINO
ALBERT DEROECK	DESY
NICK DIGIACOMO	MARTIN MARIETTA
HOLLYE DOANE	SSC NATIONAL COALITION

THOMAS DOMBECK	SSC LABORATORY
JHEROEN DORENBOSCH	SSC LABORATORY
WILLIAM DUNN	QUANTUM RESEARCH SERVICES, INC.
JIM DUNLEA	UNIVERSITY OF ROCHESTER
PEDRO DUTRA	BRAZILIAN INDUSTRIES
FRIEDRICH DYDAK	CERN
MICHAEL EDWARDS	RUTHERFORD APPLETON LABORATORY
WILLIAM EDWARDS	LAWRENCE BERKELEY LABORATORY
CHARLES EINOLF	WESTINGHOUSE ELECTRIC CORPORATION
TOR EKENBERG	ARGONNE NATIONAL LABORATORY
ANTHONY ELAM	IBM FEDERAL SYSTEMS
JOHN ELIAS	FERMI NATIONAL ACCELERATOR LABORATORY
TOM ELIOFF	SSC LABORATORY
ROGER ELLEFSON	IBM CORPORATION
TIMO ERKOLAHTI	OUTOKUMPU COPPER (USA), INC.
ALBERT ERWIN	UNIVERSITY OF WISCONSIN
DAVID ETHERTON	SSC LABORATORY
W.M. EVANS	RUTHERFORD APPLETON LABORATORY
CHRISTIAN FABJAN	CERN
ORRIN FACKLER	LAWRENCE LIVERMORE NATIONAL LABORATORY
HOMER FAIDAS	UNIVERSITY OF TENNESSEE
ALI FAZELY	LOUISIANA STATE UNIVERSITY
HOWARD FENKER	SSC LABORATORY
ERVIN FENYVES	THE UNIVERSITY OF TEXAS AT DALLAS
THOMAS FERBEL	UNIVERSITY OF ROCHESTER
PAULO FERREIRA	ANDRADE GUTIERREZ COMPANY
THOMAS FIELDS	ARGONNE NATIONAL LABORATORY
MICHELE FISHER	IBM CORPORATION
WILLIAM FORD	UNIVERSITY OF COLORADO
GEORGE FOSTER	FERMI NATIONAL ACCELERATOR LABORATORY
SOREN FREDRIKSEN	SSC LABORATORY
ALAN FRY	SSC LABORATORY
S. O. FLYCKT	PHILLIPS COMPONENTS
JEROME FRIEDMAN	MASSACHUSETTS INSTITUTE OF TECHNOLOGY
BILL FRISKEN	YORK UNIVERSITY
KOICHIRO FURUNO	UNIVERSITY OF OREGON
TONY GABRIEL	OAK RIDGE NATIONAL LABORATORY
MIKE GAMBLE	LOS ALAMOS NATIONAL LABORATORY
YURI GALAKTIONOV	INSTITUTE OF THEORETICAL & EXPERIMENTAL PHYSICS - MOSCOW
EDUARDO GAYTON	STATE OF TEXAS
STEPHEN GEER	HARVARD UNIVERSITY
VAHE GHAZIKHANIAN	UNIVERSITY OF CALIFORNIA - LOS ANGELES
PHILIPPE GHEZ	LAB. D'ANNECY-LE-VIEUX DE PHY. DES PARTICULES
PAUL GILBERT	PARSONS BRINCKERHOFF QUADE & DOUGLAS, INC.
MURDOCK GILCHRIESE	LAWRENCE BERKELEY LABORATORY
NIKOS GIOKARIS	THE ROCKEFELLER UNIVERSITY
CHARLES GLOVER	OAK RIDGE NATIONAL LABORATORY
STEPHEN GOTTESMAN	GRUMMAN CORPORATION
HOWARD GORDON	BROOKHAVEN NATIONAL LABORATORY
ALFRED GOSHAW	DUKE UNIVERSITY
FRANCESCO GRANCOGNOLO	I.N.F.N. - LECCE
DAVID GREEN	SAES GETTERS/USA, INC.

DON GROOM	LAWRENCE BERKELEY LABORATORY
CHARLES GRUHN	LAWRENCE BERKELEY LABORATORY
JOHN GRUVER	INDUSTRY ASSOCIATION FOR THE SSC (IASSC)
GLENN GUENTERBERG	MARTIN MARIETTA
SAM GUROL	GENERAL DYNAMICS
AHMET GURSOY	PARSONS BRINCKERHOFF QUADE & DOUGLAS, INC.
R. GUSTAFERSON	UNIVERSITY OF MICHIGAN
LASZLO GUTAY	PURDUE UNIVERSITY
DONALD HACKWORTH	WESTINGHOUSE SCIENCE & TECHNOLOGY CTR.
VASKEN HAGOPIAN	FLORIDA STATE UNIVERSITY
HILTON HAMMACK	THE UNIVERSITY OF TEXAS AT DALLAS
THOMAS HANDLER	UNIVERSITY OF TENNESSEE
GAIL HANSON	INDIANA UNIVERSITY
JULIE HARMON	UNIVERSITY OF FLORIDA
DAVID HARRIS	C. ITOH & CO.
EVERETTE HARVEY	LAWRENCE BERKELEY LABORATORY
ERIC HAZEN	BOSTON UNIVERSITY
CHRISTOPHER HEARTY	LAWRENCE BERKELEY LABORATORY
ROBERT HELART	BALL AEROSPACE
KENNETH HELLER	UNIVERSITY OF MINNESOTA
ROBERT HENDERSON	TRIUMF
ERIK HEIJNE	CERN
CARL HENNING	LAWRENCE LIVERMORE NATIONAL LABORTORY
GREGOR HERTEN	MASSACHUSETTS INSTITUTE OF TECHNOLOGY
CLEMENS HEUSCH	UNIVERSITY OF CALIFORNIA - SANTA CRUZ
STEVE HICKS	IBM CORPORATION
JOHN HOFTIEZER	LECROY CORPORATION
CARLOS HOJVAT	FERMI NATIONAL ACCELERATOR LABORATORY
RICHARD HOLROYD	BROOKHAVEN NATIONAL LABORATORY
BERNARD HILDEBRAND	SOUTHEASTERN UNIVERSITIES RESEARCH ASSOCIATION
JAIME G. HONTORIA	FSD TRADING CORPORATION
MICHAEL HOSKINS	ELECTRONIC DECISION INC.
CHARLES HURLBUT	BICRON CORPORATION
JOEY HUSTON	MICHIGAN STATE UNIVERSITY
KENRO IKEDA	NIPPON STEEL CORPORATION
ANDREW JARABAK	WESTINGHOUSE ELECTRIC CORPORATION
JOHN JENNINGS	UNIVERSITY OF WISCONSIN
KURTIS JOHNSON	FLORIDA STATE UNIVERSITY
NORMAN JOHNSON	THE UNIVERSITY OF TEXAS AT DALLAS
HARALD JOHNSTAD	SSC LABORATORY
LAWRENCE JONES	UNIVERSITY OF MICHIGAN
DENNIS JUDD	PRAIRIE VIEW A&M UNIVERSITY
JOHN KADYK	LAWRENCE BERKELEY LABORATORY
AARON KAMIYA	HAMAMATSU PHOTONICS K.K
YURI KAMYSHKOV	INSTIT. OF THEORETICAL AND EXPERI. PHYSICS - MOSCOW
GORDON KANE	UNIVERSITY OF MICHIGAN
ALVIN KANOFSKY	LEHIGH UNIVERSITY
JON KAPUSTINKSY	LOS ALAMOS NATIONAL LABORATORY
PAUL KARCIN	YALE UNIVERSITY
HENRY KASHA	SSC LABORATORY
RAPHAEL KASPER	SSC LABORATORY
RICHARD KASS	OHIO STATE UNIVERSITY

PAUL KEENER	UNIVERSITY OF PENNSYLVANIA
WALTER KEHOE	MASSACHUSETTS INSTITUTE OF TECHNOLOGY
ED KENNEDY	OAK RIDGE NATIONAL LABORATORY
V. PAUL KENNEY	UNIVERSITY OF NOTRE DAME
ROBERT KEPHART	FERMI NATIONAL ACCELERATOR LABORATORY
KENDRICK KILLIAN	MARTIN MARIETTA
W. WAYNE KINNISON	LOS ALAMOS NATIONAL LABORATORY
DAVID KOLTICK	PURDUE UNIVERSITY
EVGENI KLIMENKO	KURCHATOV INSTITUTE OF ATOMIC ENERGY
TAKAHIKO KONDO	KEK
RAJ KORDE	INTERNATIONAL RADIATION DETECTORS
DENNIS KORTE	HUGHES AIRCRAFT COMPANY
VALERY KUBAROVSKY	IHEP, PROTVINO
GORDON KRAMER	HUGHES AIRCRAFT COMPANY
HOBART KRANER	BROOKHAVEN NATIONAL LABORATORY
NOWHAN KWAK	UNIVERSITY OF KANSAS
THADDEUS KYCIA	BROOKHAVEN NATIONAL LABORATORY
MIKKO LAAKSO	UNIVERSITY OF HELSINKI - FINLAND
KLAUS LACKNER	LOS ALAMOS NATIONAL LABORATORY
STEFANO LAMI	I.N.F.N. - PISA
KEJIAN LAN	UNIVERSITY OF HOUSTON
ANDREW LANKFORD	UNIVERSITY OF CALIFORNIA - IRVINE
KWONG LAU	UNIVERSITY OF HOUSTON
MAURY LAWSON	IBM CORPORATION
ALEXI LEBEDEV	INSTITUTE FOR HIGH ENERGY PHYSICS, SERPUKHOV
DAVID LEE	LOS ALAMOS NATIONAL LABORATORY
IAN LEEDOM	NORTHEASTERN UNIVERSITY
PHIL LEIBOLD	SSC LABORATORY
LAWRENCE LEIPUNER	BROOKHAVEN NATIONAL LABORATORY
ROB LEITCH	MARTIN MARIETTA
BRANKO LESKOVAR	LAWRENCE BERKELEY LABORATORY
MICHAEL LEVI	LAWRENCE BERKELEY LABORATORY
RAYMOND LEWIS	PENN STATE UNIVERSITY
SHENG LI	UNIVERSITY OF FLORIDA
ZHENG LI	BROOKHAVEN NATIONAL LABORATORY
QITANG LIN	CERN
GUNNER LINDSTROEM	UNIVERSITY OF HAMBURG
JOE LISCOUSKI	DIGITAL EQUIPMENT CORPORATION
CHRISTOPHR LIRAKIS	NORTHEASTERN UNIVERSITY
JIN-TU LIU	IBM CORPORATION
WAYMOND LOWERY	THE UNIVERSITY OF TEXAS AT DALLAS
MARK LOWRY	LAWRENCE LIVERMORE NATIONAL LABORATORY
CHANGGUO LU	PRINCETON UNIVERSITY
HENRY LUBATTI	UNIVERSITY OF WASHINGTON
VERA LUTH	STANFORD LINEAR ACCELERATOR CENTER
JAMES LUTON	OAK RIDGE NATIONAL LABORATORY
HY LYON	SSC II
HONG MA	BROOKHAVEN NATIONAL LABORATORY
STAN MAJEWSKI	CEBAF
AKIHIRO MAKI	KEK
DONALD MAKOWIECKI	BROOKHAVEN NATIONAL LABORATORY
PHILLIPE MANGEOT	CEN SACLAY

SEYMOUR MARGULIES	UNIVERSITY OF ILLINOIS AT CHICAGO
NIKOLAI MARTOVETSKY	KURCHATOV INSTITUTE OF ATOMIC ENERGY
PETER MARTSON	MASSACHUSETTS INSTITUTE OF TECHNOLOGY
MICHAEL MARX	SUNY - STONY BROOK
LYLE MASON	MARTIN MARIETTA
MARK MENTZER	BURLE INDUSTRIES, INC.
MICHAEL MERMIKIDES	FLORIDA STATE UNIVERSITY
GEOFFREY MILLS	LOS ALAMOS NATIONAL LABORATORY
WILLIAM MILLER	LOS ALAMOS NATIONAL LABORATORY
DAVID MILLER	PURDUE UNIVERSITY
ROBERT MITCHELL	UNIVERSITY OF TEXAS AT ARLINGTON
GENA MITSELMAHKER	JINR - DUBNA
LUKE MO	VIRGINIA POLYTECHNIC INSTITUTE & STATE UNIVERSITY
PAUL MOCKETT	UNIVERSITY OF WASHINGTON
GERARDO MORENO	UNIVERSIDAD DE GUANAJUATO
JORGE MORFIN	FERMI NATIONAL ACCELERATOR LABORATORY
KATE MORGAN	SSC LABORATORY
SHIGEKI MORI	UNIVERSITY OF TSUKUBA
FRED MORSE	LOS ALAMOS NATIONAL LABORATORY
RAYMOND MOUNTAIN	UNIVERSITY OF NOTRE DAME
DUNCAN MACFARLANE	THE UNIVERSITY OF TEXAS AT DALLAS
CHUCK McCONAGHY	LAWRENCE LIVERMORE NATIONAL LABORATORY
KIRK McDONALD	PRINCETON UNIVERSITY
KENNETH McFARLANE	SSC LABORATORY
PETER McINTYRE	TEXAS A&M UNIVERSITY
ROBERT McINTYRE	GE ELECTRO OPTICS
ROGER McNEIL	LOUISIANA STATE UNIVERSITY
PETER McNULTY	CLEMSON UNIVERSITY
THOMAS MOORE	LAWRENCE LIVERMORE NATIONAL LABORATORY
RICHARD MORSE	MASSACHUSETTS INSTITUTE OF TECHNOLOGY
RAYMOND MULLER	HAMAMATSU CORPORATION
DAVID NAHMIAS	AIR PRODUCTS AND CHEMICALS, INC.
URIEL NAUENBERG	UNIVERSITY OF COLORADO
KENNETH NELSON	UNIVERSITY OF VIRGINIA
HARVEY NEWMAN	CALTECH
BOGDAN NICZYPORUK	CEBAF
MARYANN NOVAK	PARSON BRINCKERHOFF QUADE AND DOUGLAS, INC.
WILLIAM NUNNALLY	UNIVERSITY OF TEXAS AT ARLINGTON
F. GERALD OAKHAM	CARLETON UNIVERSITY
YORAM OFEK	IBM CORPORATION
HAROLD OGREN	INDIANA UNIVERSITY
TOKIO OHSKA	KEK
SEOG OH	DUKE UNIVERSITY
YAS ONEL	UNIVERSITY OF IOWA
LUIZ OLIVEIRA	CENTRO BRASILEIRO DE PESQUISAS FISICAS
KENNETH OLSEN	MARTIN MARIETTA
BEALE ONG	UNIVERSITY OF CALIFORNIA - SAN DIEGO
JOSEPH ORGERON	THE UNIVERSITY OF TEXAS AT DALLAS
HANS PAAR	UNIVERSITY OF CALIFORNIA - SAN DIEGO
JOSE PALATHINGAL	UNIVERSITY OF PUERTO RICO
DAVID PELLETT	UNIVERSITY CALIFORNIA - DAVIS
ADAM PARA	FERMI NATIONAL ACCELERATOR LABORATORY

JOSEPH PARADISO	C.S. DRAPER LABORATORY
CHANG WANG PARK	UNIVERSITY OF FLORIDA
SHERWOOD PARKER	UNIVERSITY OF HAWAII
MARK PATTERSON	EBCO INDUSTRIES
JOHN PAULOS	NORTH CAROLINA STATE UNIVERSITY
STEVE PAXTON	HEWLETT PACKARD
AMMIRAJU PEMMARAJU	STEVENS INSTITUTE OF TECHNOLOGY
ROBERTO PENCO	ANSALDO ABB COMPONENTI
PETER PENFOLD	OXFORD INSTRUMENTS N.A., INC.
V. PEREZ-MENDEZ	LAWRENCE BERKELEY LABORATORY
CARSTEN PETERSON	UNIVERSITY OF LUND - SWEDEN
VINCENT PETERSON	UNIVERSITY OF HAWAII
CARL PFEIFFER	HUGHES AIRCRAFT COMPANY
JADWIGA PIEKARZ	UNIVERSITY OF NOTRE DAME
JAMES PEOPLES	CVI, INC.
MICHAEL PETROFF	ROCKWELL INTERNATIONAL CORPORATION
FRANZ PLASIL	OAK RIDGE NATIONAL LABORATORY
IRWIN PLESS	MASSACHUSETTS INSTITUTE OF TECHNOLOGY
JOE POHLEN	MARTIN MARIETTA
VENETIOS POLYCHRONAKOS	BROOKHAVEN NATIONAL LABORATORY
LAWRENCE PRICE	ARGONNE NATIONAL LABORATORY
MORRIS PRIPSTEIN	LAWRENCE BERKELEY LABORATORY
JAMES PROUDFOOT	ARGONNE NATIONAL LABORATORY
GEORGE PRYTULA	GRUMMAN CORPORATION
JOHN PURCELL	KMS FUSION, INC.
SIJIN QIAN	I.N.F.N. - FRASCATI/CERN
VELJKO RADEKA	BROOKHAVEN NATIONAL LABORATORY
PIER-GIORGIO RANCOITA	UNIVERSITY OF MILAN
PAVEL REHAK	BROOKHAVEN NATIONAL LABORATORY
LARRY REHN	INTERFET CORPORATION
ROBERT REID	LOS ALAMOS NATIONAL LABORATORY
JAMES REIDY	UNIVERSITY OF MISSISSIPPI
MARK RENNICH	MARTIN MARIETTA ENERGY SYSTEMS
STEPHEN REUCROFT	NORTHEASTERN UNIVERSITY
ERNST RINGLE	NOELL, INC.
DAN ROBERTS	INTERFET CORPORATION
LEE ROBERTS	SSC LABORATORY
S. PETER ROSEN	UNIVERSITY OF TEXAS AT ARLINGTON
ELI ROSENBERG	IOWA STATE UNIVERSITY
JOHN ROSS	OXFORD INSTRUMENTS
CHRISTOPHER ROWAN	OPTOVAC
RANDAL RUCHTI	UNIVERSITY OF NOTRE DAME
JAMES RUSS	CARNEGIE MELLON UNIVERSITY
HANS RYKACZEWDKI	ETH ZURICH
HARMUT SADROZINSKI	UNIVERSITY OF CALIFORNIA - SANTA CRUZ
MARTIN SALOMON	TRIUMF
GARY SANDERS	LOS ALAMOS NATIONAL LABORATORY
JACK SANDWEISS	YALE UNIVERSITY
ALBERTO SANTORO	CENTRO BRASILEIRO DE PESQUISAS FISICAS
OSAMU SASAKI	KEK
DAVID SAXON	GLASGOW UNIVERSITY
LEONARD SCHIEBER	KOLLMORGEN CORPORATION

RAFE SCHINDLER	STANFORD LINEAR ACCELERATOR CENTER
ROY SCHWITTERS	SSC LABORATORY
BRIAN SCIPIONI	SSC LABORATORY
ABRAHAM SEIDEN	UNIVERSITY OF CALIFORNIA - SANTA CRUZ
JAMES SHANK	BOSTON UNIVERSITY
STEVE SHAPIRO	STANFORD LINEAR ACCELERATOR CENTER
JAMES SHEERIN	KMS FUSION, INC.
DAVID SHARP	LOS ALAMOS NATIONAL LABORATORY
PETER SHARP	RUTHERFORD APPLETON LABORATORY
MARLEIGH SHEAFF	UNIVERSITY OF WISCONSIN
EDWARD SHIBATA	PURDUE UNIVERSITY
TOHRU SHIMIZU	KURARAY INTERNATIONAL CORPORATION
YUJI SHINODA	HAMAMATSU CORPORATION
IAN SHIPSEY	PURDUE UNIVERSITY
FRANK SHIVELY	AMPARO CORPORATION
JACK SHULTZ	TEMPEL STEEL
JAMES SIEGRIST	SSC LABORATORY
ALAN F. SILL	UNIVERSITY OF ROCHESTER
PEKKA SINERVO	UNIVERSITY OF TORONTO
MICHAEL SIVERTZ	UNIVERSITY OF CALIFORNIA - SAN DIEGO
PATRICK SKUBIC	UNIVERSITY OF OKLAHOMA
ANDRIS SKUJA	UNIVERSITY OF MARYLAND
DONALD SLANINA	IBM CORPORATION
A. JEAN SLAUGHTER	YALE UNIVERSITY
A.J. SMITH	PRINCETON UNIVERSITY
BRIAN SMITH	LOS ALAMOS NATIONAL LABORATORY
GERALD SMITH	DEPARTMENT OF ENERGY
WESLEY SMITH	UNIVERSITY OF WISCONSIN
PETER SOLIZ	ORION INTERNATIONAL TECHNOLOGIES
ROBERT SPARROW	OPTOVAC
HELMUTH SPIELER	LAWRENCE BERKELEY LABORATORY
HAROLD SPINKA	ARGONNE NATIONAL LABORATORY
NED STEBBINS	HYPRES, INC.
ELIZABETH STEFANSKI	SSC LABORATORY
RAY STEFANSKI	SSC LABORATORY
JOHN STEKLY	INTERMAGNETICS GENERAL
ERIC STERN	UNIVERSITY OF PITTSBURGH
FRANCOIS STOCKER	SSC LABORATORY
WOLFGANG STOEFFL	LAWRENCE BERKELEY LABORATORY
JACK STORY	SSC LABORATORY
NORMAN STRANBERG	GENERAL DYNAMICS
MARK STROVINK	LAWRENCE BERKELEY LABORATORY
LAWRENCE SULAK	BOSTON UNIVERSITY
JAMES SULLIVAN	MASSACHUSETTS INSTITUTE OF TECHNOLOGY
RICHARD SUMNER	LECROY CORPORATION
SEIJI SUZUKI	HAMAMATSU PHOTONICS K.K.
ROGER SWENSRUD	WESTINGHOUSE SCIENCE & TECHNOLOGY CENTER
VITO SYLVESTER	GTE PRODUCTS CORPORATION
ROBERT THEWS	UNIVERSITY OF OREGON
TIM THOMPSON	LOS ALAMOS NATIONAL LABORATORY
TIM THURSTON	SSC LABORATORY
HAROLD TICHO	UNIVERSITY OF CALIFORNIA - SAN DIEGO

JOHN TIMBROOH	MARTIN MARIETTA
SAMUEL TING	MASSACHUSETTS INSTITUTE OF TECHNOLOGY
JOHN TOLL	UNIVERSITIES RESEARCH ASSOCIATION, INC.
YUKIO TOMITA	NIPPON STEEL CORPORATION
T. TORTSCHANOFF	CERN
MARC TURCOTTE	SSC LABORATORY
BERNARD TRAVERSAT	SSC LABORATORY
GEORGE TRILLING	LAWRENCE BERKELEY LABORATORY
EDUARDO TSYGANOV	JINR - DUBNA
GEORGE URICH	EVERSON ELECTRIC COMPANY
SUBHASH VAIDYA	GARDNER CRYOGENICS
RICHARD VAN BERG	UNIVERSITY OF PENNSYLVANIA
HARDY LANGE VAN RAVENSWAY	ATOMIC ENERGY OF CANADA LIMITED
DAVID VANDERGRIFF	OAK RIDGE NATIONAL LABORATORY
GUY VANSTRAELEN	IMEC
LUIS VILLASENOR	UNIVERSIDAD DE GUANAJUATO
A. VODOPIANOV	JINR - DUBNA
HELMUT VOGEL	CARNEGIE MELLON UNIVERSITY
ANTON VOLODKOV	JINR - DUBNA
ALBRECHT WAGNER	CERN
DAVID WAGONER	PRAIRIE VIEW A&M UNIVERSITY
JAMES WALKER	UNIVERSITY OF FLORIDA, GAINESVILLE
WOLFGANG WALLRAFF	RWTH AACHEN
GULLY WALTER	TEMPEL STEEL
ROBERT WARREN	HUGHES AIRCRAFT COMPANY
GENE WECKLER	EG&G SOLID STATE PRODUCTS GROUP
TORRE WENAUS	CERN
ROY WEST	UNIVERSITY OF TEXAS AT ARLINGTON
J. SCOTT WHITAKER	BOSTON UNIVERSITY
JAMES WHITE	TEXAS A & M
HOWARD WIEMAN	LAWRENCE BERKELEY LABORATORY
HUGH WILLIAMS	UNIVERSITY OF PENNSYLVANIA
KENNETH WILSON	CVI INCORPORATED
ROBERT WILSON	BOSTON UNIVERSITY
PHILIP WINKLER	AIR PRODUCTS AND CHEMICALS, INC.
DAVID WINN	FAIRFIELD UNIVERSITY
ALAN WINTENBERG	OAK RIDGE NATIONAL LABORATORY
ZACHARY WOLF	SSC LABORATORY
W. JOHN WOMERSLEY	FLORIDA STATE UNIVERSITY
CRAIG WOODY	BROOKHAVEN NATIONAL LABORATORY
JAMES WOOSLEY	TELEDYNE BROWN ENGINEERING
STUART WORLEY	HUGHES AIRCRAFT COMPANY
GEORGE YOST	SSC LABORATORY
SAUL YOUSSEF	FLORIDA STATE UNIVERSITY
BO YU	BROOKHAVEN NATIONAL LABORATORY
BING ZHOU	BOSTON UNIVERSITY
TIANCHI ZHAO	UNIVERSITY OF WASHINGTON
REN-YUAN ZHU	CALTECH
HANS-JOACHIM ZIOCK	LOS ALAMOS NATIONAL LABORATORY
KLAUS ZIOCK	UNIVERSITY OF VIRGINIA
CARL ZORN	CEBAF

Organizing Committees

INTERNATIONAL ADVISORY COMMITTEE

Rudolf Bock • CERN

Myron Campbell • University of Michigan

Boris Dolgoshein • Engineering Physical Institute - Moscow

Murdock Gilchriese • Lawrence Berkeley Laboratory

Howard Gordon • Brookhaven National Laboratory

Daniel Green • Fermi National Accelerator Laboratory

Vera Luth • SLAC

Takashi Ohsugi • Hiroshima University

Fabio Sauli • CERN

David Saxon • Rutherford Appleton Laboratory

James Siegrist • SSC Laboratory

Mark Strovink • Lawrence Berkeley Laboratory

J. Scott Whitaker • Boston University

LOCAL ORGANIZING COMMITTEE

Laird Cormell • SSC Laboratory

Tom Dombeck • SSC Laboratory

Phyllis Hale • SSC Laboratory

Valerie Kelly • SSC Laboratory

George Yost • SSC Laboratory

Exhibitors

AT&T Network Systems

Best-Southwest Partnership

Canberra Nuclear Products Group

Digital Equipment Corporation

DSP Technology, Inc.

Hamamatsu Corporation

Hewlett-Packard Company

IBM Corporation

Keithley Instruments

Kuraray International Corporation

LeCroy Corporation

Martin Marietta

Micron Semiconductor, Inc.

MKS Instruments

Optovac

Philips Components

Phillips Scientific

Scientific International, Inc.

Silicon Graphics, Inc.

Southland Cryogenics, Inc.

Sun Microsystems, Inc.

Superconducting Super Collider Laboratory

Texas National Research Laboratory Commission

The Chambers of Ellis County

Author Index

Overlooking the Trinity River, Fort Worth, Texas
(Photo courtesy of Fort Worth Convention and Visitors Bureau)